Phase Transition Dynamics

Phase transition dynamics is of central importance in current condensed matter physics. Akira Onuki provides a systematic treatment of a wide variety of topics including critical dynamics, phase ordering, defect dynamics, nucleation, and pattern formation by constructing time-dependent Ginzburg–Landau models for various systems in physics, metallurgy, and polymer science.

The book begins with a summary of advanced statistical–mechanical theories including the renormalization group theory applied to spin and fluid systems. Fundamental dynamical theories are then reviewed before the kinetics of phase ordering, spinodal decomposition, and nucleation are covered in depth in the main part of the book. The phase transition dynamics of real systems are discussed, treating interdisciplinary problems in a unified manner. New topics include supercritical fluid dynamics, boiling near the critical point, stress–diffusion coupling in polymers, patterns and heterogeneities in gels, and mesoscopic dynamics at structural phase transitions in solids. In the final chapter, theoretical and experimental approaches to shear flow problems in fluids are reviewed.

Phase Transition Dynamics provides a comprehensive treatment of the study of phase transitions. Building on the statics of phase transitions, covered in many introductory textbooks, it will be essential reading for researchers and advanced graduate students in physics, chemistry, metallurgy and polymer science.

AKIRA ONUKI obtained his PhD from the University of Tokyo. Since 1983 he has held a position at Kyoto University, taking up his current professorship in 1991. He has made important contributions to the study of phase transition dynamics in both fluid and solid systems.

T0207023

Phase Transition Dynamics

AKIRA ONUKI
Kyoto University

CAMBRIDGE
UNIVERSITY PRESS

CAMBRIDGE UNIVERSITY PRESS
Cambridge, New York, Melbourne, Madrid, Cape Town, Singapore, São Paulo

Cambridge University Press
The Edinburgh Building, Cambridge CB2 8RU, UK

Published in the United States of America by Cambridge University Press, New York

www.cambridge.org
Information on this title: www.cambridge.org/9780521572934

First published 2002
Reprinted 2004
This digitally printed version 2007

A catalogue record for this publication is available from the British Library

Library of Congress Cataloguing in Publication data

Onuki, Akira.
Phase transition dynamics / Akira Onuki
p. cm.
Includes bibliographical references and index.
ISBN 0 521 57293 2
1. Phase transformations (Statistical physics). 2. Condensed matter.
I. Title

QC175.16.P5 O58 2002
530.4′14–dc21 2001037340

ISBN 978-0-521-57293-4 hardback
ISBN 978-0-521-03905-5 paperback

Contents

Preface

This book aims to elucidate the current status of research in phase transition dynamics. Because the topics treated are very wide, a unified phenomenological time-dependent Ginzburg–Landau approach is used, and applied to dynamics near the critical point. Into the simple Ginzburg–Landau theory for a certain order parameter, we introduce a new property or situation such as elasticity in solids, viscoelasticity in polymers, shear flow in fluids, or heat flow in ^{4}He near the superfluid transition. By doing so, we encounter a rich class of problems on mesoscopic spatial scales. A merit of this approach is that we can understand such diverse problems in depth using universal concepts.

The first four chapters (Part one) deal with static situations, mainly of critical phenomena, and introduce some new results that would stand by themselves. However, the main purpose of Part one is to present the definitions of many fundamental quantities and introduce various phase transitions. So it should be read before Parts two and three which deal with dynamic situations. Chapter 5 is also introductory, reviewing fundamental dynamic theories, the scheme of Langevin equations and the linear response theory. Chapter 6 treats critical dynamics in (i) classical fluids near the gas–liquid and consolute critical points and (ii) ^{4}He near the superfluid transition. Chapter 7 focuses on rather special problems in complex fluids: (i) effects of viscoelasticity on composition fluctuations in polymer systems; and (ii) volume phase transitions and heterogeneity effects in gels. Chapters 8 and 9 (in Part three) constitute the main part of this book, and consider the kinetics of phase ordering, spinodal decomposition, and nucleation. Motions of interfaces and vortices are examined in the Ginzburg–Landau models. Chapter 10 focuses on dynamics in solids, including phase separation, order–disorder and martensitic transitions, shape instability in hydrogen–metal systems, and surface instability in metal films. These problems have hitherto been very inadequately studied and most papers are difficult to understand for those outside the field, so it was important to write this chapter in a coherent fashion, though it has turned out to be a most difficult task. I believe that many interesting dynamical problems remain virtually unexplored in solids, because such phenomena have been examined either too microscopically in solid-state physics without giving due respect to long-range elastic effects or with technologically-oriented objectives in engineering. Chapter 11 is on shear flow problems in fluids, a topic on which a great number of theoretical and experimental papers appeared in the 1980s and 1990s. This book thus covers a wide range of phase transition dynamics. Of course, many important problems had to be omitted.

I have benefited from discussions with many people working in the fields of low-temperature physics, statistical physics, polymer science, and metallurgy. Particularly useful suggestions were given by H. Meyer, Y. Oono, K. Kawasaki, T. Ohta, M. Doi, T. Hashimoto, H. Tanaka, M. Shibayama, T. Miyazaki, T. Koyama, and Y. Yamada. Thanks are due to R. Yamamoto, K. Kanemitsu, and A. Furukawa for drawing some of the figures. It is with deep sadness that I record the deaths of T. Tanaka and K. Hamano. It is a great pleasure to be able to acknowledge their memorable contributions to Chapters 7 and 11, respectively. Finally, I apologize to my students, colleagues, and family, for any difficulty they may have experienced because I have been so busy with this extremely time-consuming undertaking.

Akira Onuki
Kyoto, Japan

Part one

Statics

1

Spin systems and fluids

To study equilibrium statistical physics, we will start with Ising spin systems (hereafter referred to as Ising systems), because they serve as important reference systems in understanding various phase transitions [1]–[7].[1] We will then proceed to one- and two-component fluids with short-range interaction, which are believed to be isomorphic to Ising systems with respect to static critical behavior. We will treat equilibrium averages of physical quantities such as the spin, number, and energy density and then show that thermodynamic derivatives can be expressed in terms of fluctuation variances of some density variables. Simple examples are the magnetic susceptibility in Ising systems and the isothermal compressibility in one-component fluids expressed in terms of the correlation function of the spin and density, respectively. More complex examples are the constant-volume specific heat and the adiabatic compressibility in one- and two-component fluids. For our purposes, as far as the thermodynamics is concerned, we need equal-time correlations only in the long-wavelength limit. These relations have not been adequately discussed in textbooks, and must be developed here to help us to correctly interpret various experiments of thermodynamic derivatives. They will also be used in dynamic theories in this book. We briefly summarize equilibrium thermodynamics in the light of these equilibrium relations for Ising spin systems in Section 1.1, for one-component fluids in Section 1.2, and for binary fluid mixtures in Section 1.3.

1.1 Spin models

1.1.1 Ising hamiltonian

Let each lattice point of a crystal lattice have two microscopic states. It is convenient to introduce a spin variable s_i, which assumes the values 1 or -1 at lattice point i. The microscopic energy of this system, called the Ising spin hamiltonian, is composed of the exchange interaction energy and the magnetic field energy,

$$\mathcal{H}\{s\} = \mathcal{H}_{\rm ex} + \mathcal{H}_{\rm mag}, \tag{1.1.1}$$

where

$$\mathcal{H}_{\rm ex} = -\sum_{<i,j>} J s_i s_j, \tag{1.1.2}$$

[1] References are to be found at the end of each chapter.

$$\mathcal{H}_{\text{mag}} = -H \sum_i s_i. \tag{1.1.3}$$

The interaction between different spins is short-ranged and the summation in $\mathcal{H}_{\text{ex}}$ is taken over the nearest neighbor pairs i, j of the lattice points. The interaction energy between spins is then $-J$ if paired spins have the same sign, while it is J for different signs. In the case $J > 0$ the interaction is ferromagnetic, where all the spins align in one direction at zero temperature. The magnetic field H is scaled appropriately such that it has the dimension of energy. At zero magnetic field the system undergoes a second-order phase transition at a critical temperature T_c. The hamiltonian $\mathcal{H}$ mimics ferromagnetic systems with uniaxial anisotropy.

In the case $J < 0$, the interaction is antiferromagnetic, where the neighboring paired spins tend to be antiparallel at low temperatures. Let us consider a cubic lattice, which may be divided into two sublattices, A and B, such that each lattice point and its nearest neighbors belong to different sublattices. Here, we define the staggered spin variables S_i by

$$S_i = s_i \quad (i \in A), \qquad S_i = -s_i \quad (i \in B). \tag{1.1.4}$$

Then, $\mathcal{H}_{\text{ex}}$ in terms of $\{S_i\}$ has the positive coupling $|J|$ and is isomorphic to the ferromagnetic exchange hamiltonian.

The Ising model may also describe a phase transition of binary alloys consisting of atoms 1 and 2, such as Cu–Zn alloys. If each lattice point i is occupied by a single atom of either of the two species, the occupation numbers n_{1i} and n_{2i} satisfy $n_{1i} + n_{2i} = 1$. Vacancies and interstitials are assumed to be nonexistent. If the nearest neighbor pairs have an interaction energy ϵ_{KL} $(K, L = 1, 2)$, the hamiltonian is written as

$$\mathcal{H}\{n\} = \sum_{<i,j>} \sum_{K,L} \epsilon_{KL} n_{Ki} n_{Lj} - \sum_i \sum_K \mu_K n_{Ki}, \tag{1.1.5}$$

where μ_1 and μ_2 are the chemical potentials of the two components. From (1.1.4) we may introduce a spin variable,

$$s_i = 2n_{1i} - 1 = 1 - 2n_{2i}, \tag{1.1.6}$$

to obtain the Ising model (1.1.1) with

$$J = \frac{1}{4}(-\epsilon_{11} - \epsilon_{22} + 2\epsilon_{12}), \qquad H = \frac{1}{2}(\mu_1 - \mu_2) - \frac{z}{4}(\epsilon_{11} - \epsilon_{22}), \tag{1.1.7}$$

where z is the number of nearest neighbors with respect to each lattice point and is called the coordination number.

1.1.2 Vector spin models

Many variations of spin models defined on lattices have been studied in the literature [8]. If the spin $\boldsymbol{s}_i = (s_{1i}, \ldots, s_{ni})$ on each lattice point is an n-component vector, its simplest hamiltonian reads

$$\mathcal{H}\{s\} = -\sum_{<i,j>} J s_i \cdot s_j - H \sum_i s_{1i}. \quad (1.1.8)$$

The first term, the exchange interaction, is assumed to be invariant with respect to rotation in the spin space. The magnetic field H favors ordering of the first spin components s_{1i}. The model with $n = 2$ is called the xy model, and the model with $n = 3$ the Heisenberg model. It is known that the static critical behavior of the three-dimensional xy model is isomorphic to that of ^{4}He and ^{3}He–^{4}He mixtures near the superfluid transition, as will be discussed later. However, there are many cases in which there is some anisotropy in the spin space and, if one direction is energetically favored, the model reduces to the Ising model asymptotically close to the critical point. Such anisotropy becomes increasingly important near the critical point (or *relevant* in the terminology of renormalization group theory). As another relevant perturbation, we may introduce a long-range interaction such as a dipolar interaction.

1.1.3 Thermodynamics of Ising models

Each microscopic state of the Ising system is determined if all the values of spins $\{s\}$ are given. In thermal equilibrium, the probability of each microscopic state being realized is given by the Boltzmann weight,

$$P_{eq}(\{s\}) = Z^{-1} \exp(-\beta\mathcal{H}\{s\}), \quad (1.1.9)$$

where

$$\beta = 1/T. \quad (1.1.10)$$

In this book the absolute temperature multiplied by the Boltzmann constant $k_B = 1.381 \times 10^{-16}$ erg/K is simply written as T and is called the temperature [1], so T has the dimension of energy. The normalization factor Z in (1.1.9) is called the partition function,

$$Z = \sum_{\{s\}} \exp(-\beta\mathcal{H}\{s\}), \quad (1.1.11)$$

where the summation is taken over all the microscopic states. The differential form for the logarithm ln Z becomes

$$d(\ln Z) = -\langle\mathcal{H}\rangle d\beta + \beta\langle\mathcal{M}\rangle dH = -\langle\mathcal{H}_{ex}\rangle d\beta + \langle\mathcal{M}\rangle dh, \quad (1.1.12)$$

where the increments are infinitesimal,

$$h = \beta H = H/T, \quad (1.1.13)$$

and $\mathcal{M}$ is the sum of the total spins,[2]

$$\mathcal{M} = \sum_i s_i. \quad (1.1.14)$$

[2] In this book the quantities, $\mathcal{H}$, $\mathcal{M}$, $\mathcal{N}$, ... in script, are fluctuating variables (dependent on the microscopic degrees of freedom) and not thermodynamic ones.

Hereafter $\langle \cdots \rangle$ is the average over the Boltzmann distribution (1.1.9). The usual choice of the thermodynamic potential is the free energy,

$$F = -T \ln Z, \tag{1.1.15}$$

and the independent intensive variables are T and H with

$$dF = -SdT - \langle \mathcal{M} \rangle dH, \tag{1.1.16}$$

where $S = (\langle \mathcal{H} \rangle - F)/T$ is the entropy of the system.

We also consider the small change of the microscopic canonical distribution in (1.1.9) for small changes, $\beta \to \beta + \delta\beta$ and $h \to h + \delta h$. Explicitly writing its dependences on β and h, we obtain

$$P_{\text{eq}}(\{s\}; \beta + \delta\beta, h + \delta h) = P_{\text{eq}}(\{s\}; \beta, h) \exp\big[-\delta\mathcal{H}_{\text{ex}}\delta\beta + \delta\mathcal{M}\delta h + \cdots\big], \tag{1.1.17}$$

where $\delta\mathcal{H}_{\text{ex}} = \mathcal{H}_{\text{ex}} - \langle \mathcal{H}_{\text{ex}} \rangle$ and $\delta\mathcal{M} = \mathcal{M} - \langle \mathcal{M} \rangle$. To linear order in $\delta\beta$ and δh, the change of the distribution is of the form,

$$\delta P_{\text{eq}}(\{s\}) = P_{\text{eq}}(\{s\})\big[-\delta\mathcal{H}_{\text{ex}}\delta\beta + \delta\mathcal{M}\delta h + \cdots\big]. \tag{1.1.18}$$

Therefore, the average of any physical variable $\mathcal{A} = \mathcal{A}\{s\}$ dependent on the spin configurations is altered with respect to the change (1.1.18) as

$$\delta\langle \mathcal{A} \rangle = -\langle \mathcal{A}\delta\mathcal{H}_{\text{ex}} \rangle \delta\beta + \langle \mathcal{A}\delta\mathcal{M} \rangle \delta h + \cdots. \tag{1.1.19}$$

We set $\mathcal{A} = \mathcal{M}$ and $\mathcal{H}_{\text{ex}}$ to obtain

$$V\chi = \frac{\partial^2 \ln Z}{\partial h^2} = \frac{\partial \langle \mathcal{M} \rangle}{\partial h} = \langle (\delta\mathcal{M})^2 \rangle, \tag{1.1.20}$$

$$\frac{\partial^2 \ln Z}{\partial \beta^2} = -\frac{\partial \langle \mathcal{H}_{\text{ex}} \rangle}{\partial \beta} = \langle (\delta\mathcal{H}_{\text{ex}})^2 \rangle, \tag{1.1.21}$$

$$\frac{\partial^2 \ln Z}{\partial h \partial \beta} = \frac{\partial \langle M \rangle}{\partial \beta} = -\frac{\partial \langle \mathcal{H}_{\text{ex}} \rangle}{\partial h} = -\langle \delta\mathcal{M}\delta\mathcal{H}_{\text{ex}} \rangle, \tag{1.1.22}$$

where V is the volume of the system, χ is the isothermal magnetic susceptibility per unit volume, h and β are treated as independent variables, and use has been made of (1.1.12). Another frequently discussed quantity is the specific heat C_H at constant magnetic field defined by[3]

$$C_H = \frac{T}{V}\left(\frac{\partial S}{\partial T}\right)_H = \frac{1}{V}\left(\frac{\partial \langle \mathcal{H} \rangle}{\partial T}\right)_H. \tag{1.1.23}$$

Here we use $-(\partial\langle \mathcal{H} \rangle/\partial\beta)_H = (\partial^2 \ln Z/\partial\beta^2)_H$ to obtain

$$C_H = \langle (\delta\mathcal{H})^2 \rangle / T^2 V. \tag{1.1.24}$$

[3] In this book all the specific heats in spin systems and fluids have the dimension of a number density.

Namely, C_H is proportional to the variance of the total energy. We also introduce the specific heat C_M at constant magnetization $\langle \mathcal{M} \rangle$ by

$$VC_M = T\left(\frac{\partial S}{\partial T}\right)_M = VC_H - T\left(\frac{\partial \langle \mathcal{M} \rangle}{\partial T}\right)_H^2 \Big/ \left(\frac{\partial \langle \mathcal{M} \rangle}{\partial H}\right)_T. \tag{1.1.25}$$

From $(\partial \langle \mathcal{M} \rangle / \partial \beta)_H = -\langle \delta \mathcal{H} \delta \mathcal{M} \rangle$ we obtain

$$C_M = \left[\langle (\delta \mathcal{H})^2 \rangle - \langle \delta \mathcal{H} \delta \mathcal{M} \rangle^2 / \langle (\delta \mathcal{M})^2 \rangle\right] / VT^2, \tag{1.1.26}$$

where $\delta \mathcal{H}$ may be replaced by $\delta \mathcal{H}_{\text{ex}}$ because $\delta \mathcal{H} - \delta \mathcal{H}_{\text{ex}} = -H \delta \mathcal{M}$ is linearly proportional to $\mathcal{M}$. It holds the inequality $C_H \geq C_M$. These two specific heats coincide in the disordered phase at $H = 0$ where $\langle \delta \mathcal{H} \delta \mathcal{M} \rangle = 0$. We shall see that C_M in spin systems corresponds to the specific heat C_V at constant volume in one-component fluids.

Positivity of C_M

Combinations of the variances of the form,

$$C_{AB} = \langle (\delta \mathcal{A})^2 \rangle - \langle \delta \mathcal{A} \delta \mathcal{B} \rangle^2 / \langle (\delta \mathcal{B})^2 \rangle \geq 0, \tag{1.1.27}$$

will frequently appear in expressions for thermodynamic derivatives. Obviously C_{AB} is the minimum value of $\langle (\delta \mathcal{A} - x \delta \mathcal{B})^2 \rangle = \langle (\delta \mathcal{A})^2 \rangle - 2x \langle \delta \mathcal{A} \delta \mathcal{B} \rangle + x^2 \langle (\delta \mathcal{B})^2 \rangle \geq 0$ as a function of x, so it is positive-definite unless the ratio $\delta \mathcal{A} / \delta \mathcal{B}$ is a constant. Thus we have $C_M > 0$.

1.1.4 Spin density and energy density variables

We may define the spin density variable $\hat{s}(\boldsymbol{r})$ by[4]

$$\hat{\psi}(\boldsymbol{r}) = \sum_i s_i \delta(\boldsymbol{r} - \boldsymbol{r}_i), \tag{1.1.28}$$

where $\boldsymbol{r}_i$ is the position vector of the lattice site i. Then $\mathcal{M} = \int d\boldsymbol{r} \hat{\psi}(\boldsymbol{r})$ is the total spin sum in (1.1.14). Through to Chapter 5 the equilibrium equal-time correlation functions will be considered and the time variable will be suppressed. For the deviation $\delta \hat{\psi} = \hat{\psi} - \langle \hat{\psi} \rangle$ of the spin density, the pair correlation is defined by

$$g(\boldsymbol{r} - \boldsymbol{r}') = \langle \delta \hat{\psi}(\boldsymbol{r}) \delta \hat{\psi}(\boldsymbol{r}') \rangle, \tag{1.1.29}$$

which is expected to decay to zero for a distance $|\boldsymbol{r} - \boldsymbol{r}'|$ much longer than a correlation length in the thermodynamic limit ($V \to \infty$). The Fourier transformation of $g(\boldsymbol{r})$ is called the structure factor,

$$I(\boldsymbol{k}) = \int d\boldsymbol{r} g(\boldsymbol{r}) \exp(i\boldsymbol{k} \cdot \boldsymbol{r}), \tag{1.1.30}$$

[4] Hereafter, the quantities with a circumflex such as $\hat{\psi}, \hat{m}, \hat{n}, \ldots$ are fluctuating quantities together with those in script such as $\mathcal{H}, \mathcal{A}, \mathcal{B}, \ldots$. However, the circumflex will be omitted from Chapter 3 onward, to avoid cumbersome notation.

which is expected to be isotropic (or independent of the direction of $\boldsymbol{k}$) at long wavelengths ($ka \ll 1$, a being the lattice constant). The susceptibility (1.1.20) is expressed as

$$\chi = \int d\boldsymbol{r} g(\boldsymbol{r}) = \lim_{k \to 0} I(\boldsymbol{k}). \tag{1.1.31}$$

However, in the thermodynamic limit, χ is long-range and the space integral in (1.1.31) is divergent at the critical point. We may also introduce the exchange energy density $\hat{e}(\boldsymbol{r})$ by

$$\hat{e}(\boldsymbol{r}) = -\sum_{<i,j>} J s_i s_j \delta(\boldsymbol{r} - \boldsymbol{r}_i). \tag{1.1.32}$$

Then, $\int d\boldsymbol{r}\hat{e}(\boldsymbol{r}) = \mathcal{H}_{\text{ex}}$, and the (total) energy density is

$$\hat{e}_{\text{T}}(\boldsymbol{r}) = \hat{e}(\boldsymbol{r}) - H\hat{\psi}(\boldsymbol{r}), \tag{1.1.33}$$

including the magnetic field energy. From (1.1.24) C_H is expressed in terms of the deviation $\delta\hat{e}_{\text{T}} = \hat{e}_{\text{T}} - \langle e_{\text{T}} \rangle$ as

$$C_H = T^{-2} \int d\boldsymbol{r} \langle \delta\hat{e}_{\text{T}}(\boldsymbol{r} + \boldsymbol{r}_0)\delta\hat{e}_{\text{T}}(\boldsymbol{r}_0) \rangle, \tag{1.1.34}$$

which is independent of $\boldsymbol{r}_0$ in the thermodynamic limit.

Hereafter, we will use the following abbreviated notation (also for fluid systems),

$$\langle \hat{a} : \hat{b} \rangle = \int d\boldsymbol{r} \langle \delta\hat{a}(\boldsymbol{r})\delta\hat{b}(\boldsymbol{r}') \rangle, \tag{1.1.35}$$

defined for arbitrary density variables $\hat{a}(\boldsymbol{r})$ and $\hat{b}(\boldsymbol{r})$, which are determined by the microscopic degrees of freedom at the space position $\boldsymbol{r}$. The space correlation $\langle \delta\hat{a}(\boldsymbol{r})\delta\hat{b}(\boldsymbol{r}') \rangle$ is taken as its thermodynamic limit, and it is assumed to decay sufficiently rapidly for large $|\boldsymbol{r} - \boldsymbol{r}'|$ ensuring the existence of the long-wavelength limit (1.1.35). Furthermore, for any thermodynamic function $a = a(\psi, e)$, we may introduce a fluctuating variable by

$$\hat{a}(\boldsymbol{r}) = a + \left(\frac{\partial a}{\partial \psi}\right)_e \delta\hat{\psi}(\boldsymbol{r}) + \left(\frac{\partial a}{\partial e}\right)_\psi \delta\hat{e}(\boldsymbol{r}), \tag{1.1.36}$$

where a is treated as a function of the thermodynamic averages $\psi = \langle \hat{\psi} \rangle$ and $e = \langle \hat{e} \rangle$. From (1.1.19) its incremental change for small variations, $\delta\beta = -\delta T/T^2$ and δh, is written as

$$\delta\langle \hat{a} \rangle = \langle \hat{a} : \hat{e} \rangle \frac{\delta T}{T^2} + \langle \hat{a} : \hat{\psi} \rangle \delta h + \cdots. \tag{1.1.37}$$

From the definition, the above quantity is equal to $\delta a = (\partial a/\partial T)_h \delta T + (\partial a/\partial h)_{\text{T}} \delta h$. Thus,

$$T^2 \left(\frac{\partial a}{\partial T}\right)_h = \langle \hat{a} : \hat{e} \rangle, \quad \left(\frac{\partial a}{\partial h}\right)_T = \langle \hat{a} : \hat{\psi} \rangle. \tag{1.1.38}$$

The variances among $\hat{\psi}$ and $\hat{e}$ are expressed as

$$\chi = \left(\frac{\partial \psi}{\partial h}\right)_T = \langle \hat{\psi} : \hat{\psi} \rangle, \quad T^2\left(\frac{\partial e}{\partial T}\right)_h = \langle \hat{e} : \hat{e} \rangle,$$

$$T^2\left(\frac{\partial \psi}{\partial T}\right)_h = \left(\frac{\partial e}{\partial h}\right)_T = \langle \hat{\psi} : \hat{e} \rangle. \tag{1.1.39}$$

The specific heats are rewritten as

$$C_H = \frac{1}{T^2}\langle \hat{e}_{\mathrm{T}} : \hat{e}_{\mathrm{T}} \rangle, \quad C_M = \frac{1}{T^2}[\langle \hat{e} : \hat{e} \rangle - \langle \hat{e} : \hat{\psi} \rangle^2 / \langle \hat{\psi} : \hat{\psi} \rangle]. \tag{1.1.40}$$

1.1.5 Hydrodynamic fluctuations of temperature and magnetic field

In the book by Landau and Lifshitz (Ref. [1], Chap. 12), long-wavelength (or hydrodynamic) fluctuations of the temperature and pressure are introduced for one-component fluids. For spin systems we may also consider fluctuations of the temperature and magnetic field around an equilibrium reference state. As special cases of (1.1.36) we define

$$\delta \hat{T}(\boldsymbol{r}) = \left(\frac{\partial T}{\partial \psi}\right)_e \delta \hat{\psi}(\boldsymbol{r}) + \left(\frac{\partial T}{\partial e}\right)_\psi \delta \hat{e}(\boldsymbol{r}), \tag{1.1.41}$$

$$\delta \hat{h}(\boldsymbol{r}) = \left(\frac{\partial h}{\partial \psi}\right)_e \delta \hat{\psi}(\boldsymbol{r}) + \left(\frac{\partial h}{\partial e}\right)_\psi \delta \hat{e}(\boldsymbol{r}). \tag{1.1.42}$$

We may regard $\delta \hat{T}$ and $\delta \hat{H} = T\delta \hat{h} + h\delta \hat{T}$ as local fluctuations superimposed on the homogeneous temperature T and magnetic field $H = Th$, respectively. Therefore, (1.1.38) yields

$$\langle \hat{h} : \hat{\psi} \rangle = \frac{1}{T^2}\langle \hat{T} : \hat{e} \rangle = 1, \quad \langle \hat{h} : \hat{e} \rangle = \langle \hat{T} : \hat{\psi} \rangle = 0. \tag{1.1.43}$$

More generally, the density variable $\hat{a}$ in the form of (1.1.36) satisfies

$$\langle \hat{a} : \hat{T} \rangle = T^2\left(\frac{\partial a}{\partial e}\right)_\psi, \quad \langle \hat{a} : \hat{h} \rangle = \left(\frac{\partial a}{\partial \psi}\right)_e. \tag{1.1.44}$$

In particular, the temperature variance reads[5]

$$\langle \hat{T} : \hat{T} \rangle = T^2 / C_M. \tag{1.1.45}$$

The variances among $\delta \hat{h}$ and $\delta \hat{T}/T$ constitute the inverse matrix of those among $\delta \hat{\psi}$ and $\delta \hat{e}/T$. To write them down, it is convenient to define the determinant,

$$\mathcal{D} = \frac{1}{T^2}[\langle \hat{\psi} : \hat{\psi} \rangle \langle \hat{e} : \hat{e} \rangle - \langle \hat{\psi} : \hat{e} \rangle^2] = \chi C_M. \tag{1.1.46}$$

[5] In the counterpart of this relation, C_M will be replaced by C_V in (1.2.64) for one-component fluids and by C_{VX} in (1.3.44) for binary fluid mixtures.

The elements of the inverse matrix are written as[6]

$$V_{\tau\tau} \equiv \frac{1}{T^2}\langle\hat{T}:\hat{T}\rangle = \frac{1}{C_M}, \quad V_{hh} \equiv \langle\hat{h}:\hat{h}\rangle = \langle\hat{e}:\hat{e}\rangle/T^2\mathcal{D},$$

$$V_{h\tau} \equiv \frac{1}{T}\langle\hat{T}:\hat{h}\rangle = -\langle\hat{\psi}:\hat{e}\rangle/T\mathcal{D}. \tag{1.1.47}$$

In the disordered phase with $T > T_c$ and $H = 0$, we have no cross correlation $\langle\hat{\psi}:\hat{e}\rangle = 0$, so that $V_{\tau\tau} = 1/C_H$, $V_{hh} = 1/\chi$, and $V_{h\tau} = 0$. For other values of T and H, there is a nonvanishing cross correlation ($V_{h\tau} \neq 0$). The following dimensionless ratio represents the degree of mixing of the two variables,

$$\begin{aligned} R_v &= \langle\hat{\psi}:\hat{e}\rangle^2/[\langle\hat{\psi}:\hat{\psi}\rangle\langle\hat{e}:\hat{e}\rangle] \\ &= T^2\left(\frac{\partial\psi}{\partial T}\right)_h^2 \bigg/ \left(\frac{\partial\psi}{\partial h}\right)_T\left(\frac{\partial e}{\partial T}\right)_h, \end{aligned} \tag{1.1.48}$$

where $0 \le R_v \le 1$ and use has been made of (1.1.39) in the second line. From (1.1.40) we have

$$C_M = C_H(1 - R_v), \tag{1.1.49}$$

for $h = 0$ (or for sufficiently small h, as in the critical region). In Chapter 4 we shall see that $R_v \cong 1/2$ as $T \to T_c$ on the coexistence curve ($T < T_c$ and $h = 0$) in 3D Ising systems.

In the long-wavelength limit, the probability distribution of the gross variables, $\hat{\psi}(\boldsymbol{r})$ and $\hat{m}(\boldsymbol{r})$, tends to be gaussian with the form $\exp(-\beta\mathcal{H}_{\text{hyd}})$, where the fluctuations with wavelengths shorter than the correlation length have been coarse-grained. From (1.1.39), (1.1.43), and (1.1.46) the *hydrodynamic hamiltonian* $\mathcal{H}_{\text{hyd}}$ in terms of $\delta\hat{\psi}$ and $\delta\hat{T}$ is expressed as

$$\mathcal{H}_{\text{hyd}} = T\int d\boldsymbol{r}\left\{\frac{1}{2\chi}[\delta\hat{\psi}(\boldsymbol{r})]^2 + \frac{1}{2T^2}C_M[\delta\hat{T}(\boldsymbol{r})]^2\right\}. \tag{1.1.50}$$

Another expression for $\mathcal{H}_{\text{hyd}}$ can also be constructed in terms of $\delta\hat{e}$ and $\delta\hat{h}$.

1.2 One-component fluids

1.2.1 Canonical ensemble

Nearly-spherical molecules, such as rare-gas atoms, may be assumed to interact via a pairwise potential $v(r)$ dependent only on the distance r between the two particles [4]–[6]. It consists of a short-range hard-core-like repulsion ($r \lesssim \sigma$) and a long-range attraction ($r \gtrsim \sigma$). These two behaviors may be incorporated in the Lenard-Jones potential,

$$v(r) = 4\epsilon\left[\left(\frac{\sigma}{r}\right)^{12} - \left(\frac{\sigma}{r}\right)^6\right]. \tag{1.2.1}$$

[6] These relations will be used in (2.2.29)–(2.2.36) for one-component fluids and in (2.3.33)–(2.3.38) for binary fluid mixtures after setting up mapping relations between spin and fluid systems.

This pairwise potential is characterized by the core radius σ and the minimum $-\epsilon$ attained at $r = 2^{1/6}\sigma$. In classical mechanics, the hamiltonian for N identical particles with mass m_0 is written as

$$\mathcal{H} = \frac{1}{2m_0}\sum_i |\boldsymbol{p}_i|^2 + \sum_{<i,j>} v(r_{ij}), \tag{1.2.2}$$

where $\boldsymbol{p}_i$ is the momentum vector of the ith particle, r_{ij} is the distance between the particle pair i, j, and $<i, j>$ denotes summation over particle pairs. The particles are confined in a container with a fixed volume V and the wall potential is not written explicitly in (1.2.2).

In the canonical ensemble T, V, and N are fixed, and the statistical distribution is proportional to the Boltzmann weight as [1]–[3]

$$P_{\text{ca}}(\Gamma) = \frac{1}{Z_N}\exp[-\beta\mathcal{H}], \tag{1.2.3}$$

in the $2dN$-dimensional phase space $\Gamma = (\boldsymbol{p}_1 \cdots \boldsymbol{p}_N, \boldsymbol{r}_1 \cdots \boldsymbol{r}_N)$ (sometimes called the Γ-space). The spatial dimensionality is written as d and may be general. The partition function Z_N of N particles for the canonical ensemble is then given by the multiple integrations,

$$\begin{aligned} Z_N &= \frac{1}{N!(2\pi\hbar)^{dN}}\int d\boldsymbol{p}_1 \cdots \int d\boldsymbol{p}_N \int d\boldsymbol{r}_1 \cdots \int d\boldsymbol{r}_N \exp(-\beta\mathcal{H}) \\ &= \frac{1}{N!\lambda_{\text{th}}^{dN}}\int d\boldsymbol{r}_1 \cdots \int d\boldsymbol{r}_N \exp(-\beta\mathcal{U}), \end{aligned} \tag{1.2.4}$$

where $\hbar = 1.054\,57 \times 10^{-27}$ erg s is the Planck constant. In the second line the momentum integrations over the maxwellian distribution have been performed, where

$$\lambda_{\text{th}} = \hbar(2\pi/m_0 T)^{1/2} \tag{1.2.5}$$

is called the thermal de Broglie wavelength, and

$$\mathcal{U} = \sum_{<i,j>} v(r_{ij}) \tag{1.2.6}$$

is the potential part of the hamiltonian.

The Helmholtz free energy is given by $F = -T \ln Z_N$. The factor $1/N!(2\pi\hbar)^{dN}$ in (1.2.4) naturally arises in the classical limit ($\hbar \to 0$) of the quantum mechanical partition function [2]. Physically, the factor $1/N!$ represents the indistinguishability between particles, which assures the extensive property of the entropy. That is, a set of classical microscopic states obtainable only by the particle exchange, $i \to j$ and $j \to i$, corresponds to a single quantum microscopic state.[7] The factor $1/(2\pi\hbar)^{dN}$ is ascribed to the uncertainty principle ($\Delta p \Delta x \sim 2\pi\hbar$).

[7] The concept of indistinguishability is intrinsically of quantum mechanical origin as well as the uncertainty principle. It is not necessarily required in the realm of classical statistical mechanics. Observable quantities such as the pressure are not affected by the factor $1/N!$.

1.2.2 Grand canonical ensemble

A fluid region can be in contact with a mass reservoir characterized by a chemical potential μ as well as with a heat reservoir at a temperature T. As an example of such a system, we may choose an arbitrary macroscopic subsystem with a volume much smaller than the volume of the total system. In this case we should consider the grand canonical distribution, in which T, μ, and V are fixed and the energy and the particle number are fluctuating quantities. To make this explicit, the particle number will be written as $\mathcal{N}$ and, to avoid too many symbols, the average $\langle\mathcal{N}\rangle$ will be denoted by N which is now a function of T and μ. The statistical probability of each microscopic state with $\mathcal{N}$ particles being realized is given by [1]–[3]

$$P_{\text{gra}}(\Gamma) = \frac{1}{\Xi}\exp[-\beta\mathcal{H} + \beta\mu\mathcal{N}]. \tag{1.2.7}$$

The equilibrium average is written as $\langle\cdots\rangle = \int d\Gamma(\cdots)P_{\text{gra}}(\Gamma)$, where

$$\int d\Gamma = \sum_{\mathcal{N}} \frac{1}{\mathcal{N}!(2\pi\hbar)^{d\mathcal{N}}} \int d\boldsymbol{p}_1 \cdots \int d\boldsymbol{p}_{\mathcal{N}} \int d\boldsymbol{r}_1 \cdots \int d\boldsymbol{r}_{\mathcal{N}} \tag{1.2.8}$$

represents the integration of the configurations in the Γ-space. The normalization factor or the grand partition function Ξ is expressed as

$$\Xi = \sum_{\mathcal{N}} Z_{\mathcal{N}} \exp(\mathcal{N}\beta\mu). \tag{1.2.9}$$

In this summation the contribution around $\mathcal{N} \cong N = \langle\mathcal{N}\rangle$ is dominant for large N, and the logarithm $\Omega \equiv \ln \Xi$ satisfies

$$\Omega = \ln Z_N + N\beta\mu = pV/T, \tag{1.2.10}$$

in the thermodynamic limit $N \to \infty$. Use has been made of the fact that $G = N\mu$ is the Gibbs free energy.

We may choose Ω as a thermodynamic potential dependent on β and

$$\nu = \beta\mu = \mu/T. \tag{1.2.11}$$

Then, analogous to (1.1.12) for Ising systems, the differential form for Ω is written as [9, 10]

$$d\Omega = -\langle\mathcal{H}\rangle d\beta + \langle\mathcal{N}\rangle d\nu, \tag{1.2.12}$$

where

$$\langle\mathcal{H}\rangle = \frac{3}{2}\langle\mathcal{N}\rangle T + \langle\mathcal{U}\rangle \tag{1.2.13}$$

is the energy consisting of the average kinetic energy and the average potential energy. Notice that (1.2.12) may be transformed into the well-known Gibbs–Duhem relation,

$$d\mu = \frac{1}{n}dp - s\,dT, \tag{1.2.14}$$

where $n = \langle \mathcal{N} \rangle / V$ is the average number density and $s = (\langle \mathcal{H} \rangle - F)/NT$ is the entropy per particle.

We then find the counterparts of (1.1.20)–(1.1.22) among the thermodynamic derivatives and the fluctuation variances of $\delta\mathcal{N} = \mathcal{N} - \langle\mathcal{N}\rangle$ and $\delta\mathcal{H} = \mathcal{H} - \langle\mathcal{H}\rangle$ as

$$\frac{\partial^2 \Omega}{\partial \nu^2} = \frac{\partial \langle \mathcal{N} \rangle}{\partial \nu} = \langle (\delta\mathcal{N})^2 \rangle, \tag{1.2.15}$$

$$\frac{\partial^2 \Omega}{\partial \beta^2} = -\frac{\partial \langle \mathcal{H} \rangle}{\partial \beta} = \langle (\delta\mathcal{H})^2 \rangle, \tag{1.2.16}$$

$$-\frac{\partial^2 \Omega}{\partial \nu \partial \beta} = -\frac{\partial \langle \mathcal{N} \rangle}{\partial \beta} = \frac{\partial \langle \mathcal{H} \rangle}{\partial \nu} = \langle \delta\mathcal{N} \delta\mathcal{H} \rangle, \tag{1.2.17}$$

where all the quantities are regarded as functions of β, and $\nu = \beta\mu$ and the volume V is fixed.

The isothermal compressibility is expressed as

$$K_T = \frac{1}{n}\left(\frac{\partial n}{\partial p}\right)_{VT} = \frac{\beta}{n^2}\left(\frac{\partial}{\partial \nu}\frac{\langle \mathcal{N} \rangle}{V}\right)_\beta, \tag{1.2.18}$$

where $n = \langle \mathcal{N} \rangle / V$ is the average number density and use has been made of (1.2.14). The fluctuation variance of $\delta\mathcal{N} = \mathcal{N} - \langle\mathcal{N}\rangle$ is expressed in terms of K_T as

$$\langle (\delta\mathcal{N})^2 \rangle = V n^2 T K_T \quad \text{(grand canonical)}. \tag{1.2.19}$$

As for C_M in (1.1.26), the constant-volume specific heat $C_V = (\partial\langle\mathcal{H}\rangle/\partial T)_{VN}/V$ per unit volume can be calculated in terms of the fluctuation variances as

$$C_V = \left[\langle (\delta\mathcal{H})^2 \rangle - \langle \delta\mathcal{H}\delta\mathcal{N} \rangle^2 / \langle (\delta\mathcal{N})^2 \rangle\right] / V T^2 \quad \text{(grand canonical)}, \tag{1.2.20}$$

where use has been made of

$$(\partial\langle\mathcal{H}\rangle/\partial T)_N = (\partial\langle\mathcal{H}\rangle/\partial T)_\nu + (\partial\langle\mathcal{H}\rangle/\partial N)_{\mathrm{T}}(\partial N/\partial T)_\nu.$$

Field variables and density variables

Following Griffiths and Wheeler [10] and Fisher [11], we refer to T (or β) and h in spin systems and T (or β), p, ν, ... in fluids as *fields*, which have identical values in two coexisting phases. We refer to the spin and energy densities in spin systems and the densities of number, energy, entropy, ...in fluids as *densities*. In spin systems, the average spin is discontinuous between the two coexisting phases, but the average energy is continuous. In fluids, the density variables usually have different average values in the two coexisting phases, but can be continuous in accidental cases such as the azeotropic case (see Section 2.3). In this book the density variables (even the entropy and concentration) have microscopic expressions in terms of the spins or the particle positions and momenta. Their equilibrium averages become the usual thermodynamic variables, and their equilibrium fluctuation variances can be related to some thermodynamic derivatives in the long-wavelength limit.

Shift of the origin of the one-particle energy

It would also be appropriate to remark on the arbitrariness of the origin of the energy supported by each particle. That is, let us shift the hamiltonian as

$$\mathcal{H} \to \mathcal{H} + \epsilon_0 \mathcal{N} \tag{1.2.21}$$

and the chemical potential from μ to $\mu + \epsilon_0$. Then, ϵ_0 vanishes in the grand canonical distribution and hence measurable quantities such as the pressure p should remain invariant or independent of ϵ_0 as long as they do not involve the origin of the one-particle energy. We can see that the terms involving ϵ_0 cancel in the variance combination (1.2.20), so C_V is clearly independent of ϵ_0.

Lattice gas model

In the lattice gas model [12], particles are distributed on fixed lattice points in evaluating the potential energy contribution to Ξ. The lattice constant a is taken to be the hard-core size of the pair potential, so each lattice point is supposed to be either vacant ($n_i = 0$) or occupied ($n_i = 1$) by a single particle. Then Ξ is approximated as

$$\Xi = \sum_{\{n\}} \exp(-\beta \mathcal{H}\{n\}), \tag{1.2.22}$$

with

$$\mathcal{H}\{n\} = -\sum_{<i,j>} \epsilon n_i n_j - (\mu - dT \ln \lambda_{\text{th}}) \sum_i n_i, \tag{1.2.23}$$

where the summation in the first term is taken over the nearest neighbor pairs and ϵ represents the magnitude of the attractive part of the pair potential. Obviously, if we set $s_i = 2n_i - 1$, the above hamiltonian becomes isomorphic to the spin hamiltonian (1.1.1) under $J = \epsilon/4$ and

$$H = \frac{1}{2}\mu - \frac{d}{2} T \ln \lambda_{\text{th}} + \frac{1}{4} z\epsilon = \frac{1}{2}\mu + \frac{d}{4} T \ln T + \text{const.}, \tag{1.2.24}$$

z being the coordination number. The pressure p in the lattice gas model is related to the free energy F_{Ising} of the corresponding Ising spin system by

$$p = -V^{-1} F_{\text{Ising}} + a^{-d}\left(H + \frac{1}{8} z\epsilon\right). \tag{1.2.25}$$

1.2.3 Thermodynamic derivatives and fluctuation variances

Analogously to the spin case (1.1.18), the grand canonical distribution function $P_{\text{gra}}(\Gamma)$ in (1.2.7) is changed against small changes, $\beta \to \beta + \delta\beta$ and $\nu \to \nu + \delta\nu$, as [9]

$$\delta P_{\text{gra}} = [-\delta \mathcal{H} \delta\beta + \delta \mathcal{N} \delta\nu] P_{\text{gra}}, \tag{1.2.26}$$

where only the linear deviations are written. Because the choice of β and ν as independent field variables is not usual, we may switch to the usual choice, T and p. Here $\delta T = -T^2\delta\beta$ and

$$\delta p = nT(\delta\nu - \bar{H}\delta\beta), \tag{1.2.27}$$

where

$$\bar{H} = \mu + Ts \tag{1.2.28}$$

is the enthalpy per particle and should not be confused with the magnetic field in the spin system, and s is the entropy per particle. Then (1.2.26) is rewritten as

$$\delta P_{\rm gra} = \left[n\delta\mathcal{S}\frac{\delta T}{T} + \delta\mathcal{N}\frac{\delta p}{nT}\right] P_{\rm gra}, \tag{1.2.29}$$

where

$$\delta\mathcal{S} = \frac{1}{nT}[\delta\mathcal{H} - \bar{H}\delta\mathcal{N}] \tag{1.2.30}$$

is the space integral of the entropy density variable to be introduced in (1.2.46) below. Thus, the thermodynamic average of any fluctuating quantity $\mathcal{A}$ changes as

$$\begin{aligned} \delta\langle\mathcal{A}\rangle &= -\langle\mathcal{A}\delta\mathcal{H}\rangle\delta\beta + \langle\mathcal{A}\delta\mathcal{N}\rangle\delta\nu + \cdots, \\ &= \langle\mathcal{A}\delta\mathcal{S}\rangle n\frac{\delta T}{T} + \langle\mathcal{A}\delta\mathcal{N}\rangle\frac{\delta p}{nT} + \cdots. \end{aligned} \tag{1.2.31}$$

Note that $\delta\mathcal{S}$ is invariant with respect to the energy shift in (1.2.21) because the enthalpy $\bar{H}$ is also shifted by ϵ_0.

The familiar constant-pressure specific heat $C_p = nT(\partial s/\partial T)_p$ per unit volume is obtained from $VC_p = nT\lim_{\delta T\to 0}\langle\delta\mathcal{S}\rangle/\delta T$ with $\delta p = 0$. From the second line of (1.2.31) C_p becomes

$$C_p = n^2\langle(\delta\mathcal{S})^2\rangle/V = \langle(\delta\mathcal{H} - \bar{H}\delta\mathcal{N})^2\rangle/VT^2 \quad \text{(grand canonical).} \tag{1.2.32}$$

In terms of $\delta\mathcal{S}$, the constant-volume specific heat C_V is also expressed as

$$C_V = n^2\left[\langle(\delta\mathcal{S})^2\rangle - \langle\delta\mathcal{S}\delta\mathcal{N}\rangle^2/\langle(\delta\mathcal{N})^2\rangle\right]/V \quad \text{(grand canonical),} \tag{1.2.33}$$

which is equivalent to (1.2.20). It leads to the inequality $C_p \geq C_V$. Use of the thermodynamic identity $C_p/C_V = K_T/K_s$ yields the adiabatic compressibility $K_s = (\partial n/\partial p)_s/n$ in the form

$$K_s = \left[\langle(\delta\mathcal{N})^2\rangle - \langle\delta\mathcal{S}\delta\mathcal{N}\rangle^2/\langle(\delta\mathcal{S})^2\rangle\right]/Vn^2T \quad \text{(grand canonical).} \tag{1.2.34}$$

The sound velocity c is given by $c = (\rho K_s)^{-1/2}$, $\rho = m_0 n$ being the mass density.

1.2.4 Gaussian distribution in the long-wavelength limit

We next consider the equilibrium statistical distribution function for the macroscopic gross variables, $\mathcal{H}$ and $\mathcal{N}$, for one-component fluids, which we write as $P(\mathcal{H}, \mathcal{N})$. The entropy $S(E, N)$ as a function of E and N is the logarithm of the number of microscopic configurations at $\mathcal{H} = E$ and $\mathcal{N} = N$. It may be written as

$$\exp[S(E, N)] = \int d\Gamma \delta(\mathcal{H} - E)\delta(\mathcal{N} - N), \tag{1.2.35}$$

where $d\Gamma$ is the configuration integral (1.2.8). This grouping of the microscopic states gives

$$P(\mathcal{H}, \mathcal{N}) = \frac{1}{\Xi} \exp[S(\mathcal{H}, \mathcal{N}) - \beta\mathcal{H} + \nu\mathcal{N}], \tag{1.2.36}$$

with the grand canonical partition function,

$$\Xi = \int d\mathcal{H} \int d\mathcal{N} \exp[S(\mathcal{H}, \mathcal{N}) - \beta\mathcal{H} + \nu\mathcal{N}]. \tag{1.2.37}$$

Each thermodynamic state is characterized by β and ν or by $E = \langle\mathcal{H}\rangle$ and $N = \langle\mathcal{N}\rangle$. We then expand $S(\mathcal{H}, \mathcal{N})$ with respect to the deviations $\delta\mathcal{H} = \mathcal{H} - E$ and $\delta\mathcal{N} = \mathcal{N} - N$ as

$$S(\mathcal{H}, \mathcal{N}) = S(E, N) + \beta\delta\mathcal{H} - \nu\delta\mathcal{N} + (\Delta S)_2 + \cdots, \tag{1.2.38}$$

where $(\delta S)_2$ is the bilinear part,

$$(\Delta S)_2 = \frac{1}{2}\left(\frac{\partial^2 S}{\partial E^2}\right)(\delta\mathcal{H})^2 + \left(\frac{\partial^2 S}{\partial E \partial N}\right)\delta\mathcal{H}\delta\mathcal{N} + \frac{1}{2}\left(\frac{\partial^2 S}{\partial N^2}\right)(\delta\mathcal{N})^2. \tag{1.2.39}$$

In the probability distribution (1.2.36) the linear terms cancel if (1.2.38) is substituted, so the distribution becomes the following well-known gaussian form [1, 3, 7]:

$$P(\mathcal{H}, \mathcal{N}) \propto \exp[(\Delta S)_2]. \tag{1.2.40}$$

From this distribution we can re-derive (1.2.15)–(1.2.17) by using the relations,

$$\begin{aligned} \alpha_{ee} &\equiv V\frac{\partial^2 S}{\partial E^2} = \frac{\partial \beta}{\partial e}, \quad \alpha_{nn} \equiv V\frac{\partial^2 S}{\partial N^2} = -\frac{\partial \nu}{\partial n}, \\ \alpha_{en} &\equiv V\frac{\partial^2 S}{\partial N \partial E} = \frac{\partial \beta}{\partial n} = -\frac{\partial \nu}{\partial e}, \end{aligned} \tag{1.2.41}$$

where β and ν are regarded as functions of $n = N/V$ and $e = E/V$. The three coefficients in (1.2.41) divided by $-V$ constitute the inverse of the matrix whose elements are the variances among $\mathcal{H}$ and $\mathcal{N}$.

Weakly inhomogeneous cases

The above result may be generalized for weakly inhomogeneous cases as follows. Let us consider a small fluid element whose linear dimension is much longer than the correlation length. Because the thermodynamics in the element is described by the grand canonical ensemble, the long-wavelength, number and energy density fluctuations, $\delta\hat{n}(\boldsymbol{r})$ and $\delta\hat{e}(\boldsymbol{r})$, obey a gaussian distribution of the form (1.2.40) with

$$(\Delta S)_2 = \int d\boldsymbol{r}\left[\frac{1}{2}\alpha_{ee}(\delta\hat{e}(\boldsymbol{r}))^2 + \alpha_{en}\delta\hat{e}(\boldsymbol{r})\delta\hat{n}(\boldsymbol{r}) + \frac{1}{2}\alpha_{nn}(\delta\hat{n}(\boldsymbol{r}))^2\right]. \tag{1.2.42}$$

Thermodynamic stability

It has been taken for granted that the probability distribution (1.2.36) is maximum for the equilibrium values, which results in the positive-definiteness of the matrix composed of the coefficients in (1.2.41). In thermodynamics [2, 13] this positive-definiteness (implying the positivity of C_V, K_T, etc.) follows from the thermodynamic stability of equilibrium states. In this book, because we start with statistical–mechanical principles, their positivity is an obvious consequence evident from their variance expressions.

1.2.5 Fluctuating space-dependent variables

The number density variable $\hat{n}(\boldsymbol{r})$ and the energy density variable $\hat{e}(\boldsymbol{r})$ have microscopic expressions,

$$\hat{n}(\boldsymbol{r}) = \sum_i \delta(\boldsymbol{r} - \boldsymbol{r}_i), \tag{1.2.43}$$

$$\hat{e}(\boldsymbol{r}) = \sum_i \frac{1}{2m_0}|\boldsymbol{p}_i|^2\delta(\boldsymbol{r} - \boldsymbol{r}_i) + \frac{1}{2}\sum_{i\neq j} v(r_{ij})\delta(\boldsymbol{r} - \boldsymbol{r}_i), \tag{1.2.44}$$

in terms of the particle positions and momenta. As in (1.1.36) we may introduce a fluctuating variable by

$$\hat{a}(\boldsymbol{r}) = a + \left(\frac{\partial a}{\partial n}\right)_e \delta\hat{n}(\boldsymbol{r}) + \left(\frac{\partial a}{\partial e}\right)_n \delta\hat{e}(\boldsymbol{r}), \tag{1.2.45}$$

for any thermodynamic variable a given as a function of the averages $n = \langle\hat{n}\rangle$ and $e = \langle\hat{e}\rangle$. The nonlinear terms such as $(\partial^2 a/\partial n^2)(\delta\hat{n})^2$ are not included in the definition. From $ds = (de - \bar{H}dn)/nT$ the space-dependent entropy variable is introduced by

$$\hat{s}(\boldsymbol{r}) = s + \frac{1}{nT}[\delta\hat{e}(\boldsymbol{r}) - \bar{H}\delta\hat{n}(\boldsymbol{r})], \tag{1.2.46}$$

where $\bar{H} = \mu + Ts = (e + p)/n$ is the enthalpy per particle. The space integral of $\delta\hat{s}(\boldsymbol{r}) = \hat{s}(\boldsymbol{r}) - s$ is equal to $\delta\mathcal{S}$ in (1.2.30). In terms of these density variables, the incremental change of the grand canonical distribution in (1.2.26) and (1.2.29) is expressed as

$$\begin{aligned} \delta P_{\text{gra}} &= P_{\text{gra}}\int d\boldsymbol{r}[-\delta\hat{e}(\boldsymbol{r})\delta\beta + \delta\hat{n}(\boldsymbol{r})\delta\nu] \\ &= P_{\text{gra}}\int d\boldsymbol{r}\left[n\delta\hat{s}(\boldsymbol{r})\frac{\delta T}{T} + \delta\hat{n}(\boldsymbol{r})\frac{\delta p}{nT}\right], \end{aligned} \tag{1.2.47}$$

where δp is the pressure deviation defined in (1.2.27). With these two expressions we may express any thermodynamic derivatives in terms of fluctuation variances of $\hat{n}$, $\hat{e}$, and $\hat{s}$ in the long-wavelength limit. Using the notation $\langle\ :\ \rangle$, as in (1.1.35), we have

$$K_T = (n^2T)^{-1}\langle \hat{n} : \hat{n}\rangle, \quad C_p = n^2\langle \hat{s} : \hat{s}\rangle, \quad \alpha_p = -T^{-1}\langle \hat{s} : \hat{n}\rangle, \tag{1.2.48}$$

where $\alpha_p = -(\partial n/\partial T)_p/n$ is the thermal expansion coefficient. From (1.2.20) and (1.2.33) the constant-volume specific heat is expressed as

$$\begin{aligned} C_V &= T^{-2}\big[\langle \hat{e} : \hat{e}\rangle - \langle \hat{e} : \hat{n}\rangle^2/\langle \hat{n} : \hat{n}\rangle\big] \\ &= n^2\big[\langle \hat{s} : \hat{s}\rangle - \langle \hat{s} : \hat{n}\rangle^2/\langle \hat{n} : \hat{n}\rangle\big]. \end{aligned} \tag{1.2.49}$$

The first line was obtained by Schofield [see Ref. 18]. From (1.2.34) the adiabatic compressibility is expressed as

$$K_s = (\rho c^2)^{-1} = \big[\langle \hat{n} : \hat{n}\rangle - \langle \hat{n} : \hat{s}\rangle^2/\langle \hat{s} : \hat{s}\rangle\big]/n^2T. \tag{1.2.50}$$

These expressions are in terms of the long-wavelength limit of the correlation functions. Hence, to their merit, they tend to unique thermodynamic limits, whether the ensemble is canonical or grand canonical, as $N, V \to \infty$ with a fixed density $n = N/V$.

More generally, for any density variable $\hat{a}$ in the form of (1.2.45), we obtain

$$\langle \hat{a} : \hat{e}\rangle = T^2\left(\frac{\partial a}{\partial T}\right)_v, \quad \langle \hat{a} : \hat{n}\rangle = nT\left(\frac{\partial a}{\partial p}\right)_T, \quad \langle \hat{a} : \hat{s}\rangle = \frac{1}{n}T\left(\frac{\partial a}{\partial T}\right)_p. \tag{1.2.51}$$

It then follows that

$$\left(\frac{\partial p}{\partial T}\right)_a = -\left(\frac{\partial a}{\partial T}\right)_p\Big/\left(\frac{\partial a}{\partial p}\right)_T = -n^2\langle \hat{a} : \hat{s}\rangle/\langle \hat{a} : \hat{n}\rangle. \tag{1.2.52}$$

Finally, we give some thermodynamic identities,

$$\rho c^2 C_V = T\left(\frac{\partial p}{\partial T}\right)_s\left(\frac{\partial p}{\partial T}\right)_n = T\left(\frac{\partial p}{\partial T}\right)_s^2(1 - C_V/C_p), \tag{1.2.53}$$

$$C_V/C_p = K_s/K_T = 1 - \left(\frac{\partial p}{\partial T}\right)_n\Big/\left(\frac{\partial p}{\partial T}\right)_s. \tag{1.2.54}$$

These are usually proved with the Maxwell relations but can also be derived from the variance relations (1.2.48)–(1.2.54).

1.2.6 Density correlation

In the literature [4]–[6] special attention has been paid to the radial distribution function $g(r)$ defined by

$$\begin{aligned} n^2 g(|\mathbf{r} - \mathbf{r}'|) &= \sum_{i\neq j}\langle \delta(\mathbf{r} - \mathbf{r}_i)\delta(\mathbf{r}' - \mathbf{r}_j)\rangle \\ &= \langle \hat{n}(\mathbf{r})\hat{n}(\mathbf{r}')\rangle - n\delta(\mathbf{r} - \mathbf{r}'), \end{aligned} \tag{1.2.55}$$

where the self-part ($i = j$) has been subtracted and $g(r) \to 1$ at long distance in the thermodynamic limit.[8] The structure factor is expressed as

$$I(k) = \int d\boldsymbol{r} e^{i\boldsymbol{k}\cdot\boldsymbol{r}} \langle \delta\hat{n}(\boldsymbol{r})\delta\hat{n}(\boldsymbol{0})\rangle = n + n^2 \int d\boldsymbol{r} e^{i\boldsymbol{k}\cdot\boldsymbol{r}}[g(r) - 1]. \tag{1.2.56}$$

An example of $I(k)$ can be found in Fig. 2.3. The isothermal compressibility (1.2.18) is expressed as

$$K_T = (n^2 T)^{-1} \lim_{k\to 0} I(k) = (nT)^{-1} + T^{-1} \int d\boldsymbol{r}[g(r) - 1]. \tag{1.2.57}$$

The physical meaning of $g(r)$ is as follows. We place a particle at the origin of the reference frame and consider a volume element $d\boldsymbol{r}$ at a position $\boldsymbol{r}$; then, $ng(r)d\boldsymbol{r}$ is the average particle number in the volume element. In liquid theories another important quantity is the direct correlation function $C(r)$ defined by

$$g(r) - 1 = C(r) + \int d\boldsymbol{r}' C(|\boldsymbol{r} - \boldsymbol{r}'|)n[g(|\boldsymbol{r}'|) - 1]. \tag{1.2.58}$$

Its Fourier transformation C_k satisfies

$$I(k) = n/(1 - nC_k). \tag{1.2.59}$$

Let us assume naively that $C(r)$ decays more rapidly than the pair correlation function $g(r)$ at long distances and C_k can be expanded as $C_k = C_0 - C_1 k^2 + \cdots$ at small k with $C_1 > 0$ [14]. Then, (1.2.59) yields a well-known expression called the Ornstein–Zernike form,

$$I(k) \cong n/(1 - nC_0 + nC_1 k^2), \tag{1.2.60}$$

at small k. Notice that $C_0 = \lim_{k\to 0} C_k$ approaches to n^{-1} as the critical point (or the *spinodal line* more generally) is approached. The direct correlation functions for binary mixtures will be discussed at the end of Section 1.3.

1.2.7 Hydrodynamic temperature and pressure fluctuations

As in the book by Landau and Lifshitz [1], we introduce the temperature fluctuation $\delta\hat{T}$ as a space-dependent variable by

$$\begin{aligned} \delta\hat{T}(\boldsymbol{r}) &= \left(\frac{\partial T}{\partial e}\right)_n \delta\hat{e}(\boldsymbol{r}) + \left(\frac{\partial T}{\partial n}\right)_e \delta\hat{n}(\boldsymbol{r}) \\ &= \frac{nT}{C_V}\left[\delta\hat{s}(\boldsymbol{r}) + \frac{1}{n^2}\left(\frac{\partial p}{\partial T}\right)_n \delta\hat{n}(\boldsymbol{r})\right], \end{aligned} \tag{1.2.61}$$

where the energy density $\hat{e}(\boldsymbol{r})$, the number density $\hat{n}(\boldsymbol{r})$, and the entropy density $\hat{s}(\boldsymbol{r})$ are defined by (1.2.45)–(1.2.47), and use has been made of $(\partial s/\partial n^{-1})_T = (\partial p/\partial T)_n$. We assume that these density variables consist only of the Fourier components with wavelengths

[8] In a finite system, the space integral of (1.2.55) in the volume V would become $N(N-1)/V$, in apparent contradiction to (1.2.57).

much longer than any correlation lengths ($q \ll \xi^{-1}$, near the critical point, ξ being the correlation length). Then $\hat{a}$ in the form of (1.2.45) satisfies

$$\langle \hat{a} : \hat{T} \rangle = \frac{T}{n}\left(\frac{\partial a}{\partial s}\right)_n = \frac{T^2}{C_V}\left(\frac{\partial a}{\partial T}\right)_n. \tag{1.2.62}$$

This relation gives [1]

$$\langle \hat{n} : \hat{T} \rangle = 0, \quad \langle \hat{s} : \hat{T} \rangle = T/n, \tag{1.2.63}$$

$$\langle \hat{T} : \hat{T} \rangle = T^2/C_V, \tag{1.2.64}$$

The long-wavelength fluctuations obey a gaussian distribution $\propto \exp[-\beta\mathcal{H}_{\text{hyd}}]$. The hydrodynamic hamiltonian is written as

$$\mathcal{H}_{\text{hyd}} = \int d\boldsymbol{r}\left\{\frac{C_V}{2T}[\delta\hat{T}(\boldsymbol{r})]^2 + \frac{1}{2n^2K_T}[\delta\hat{n}(\boldsymbol{r})]^2\right\}, \tag{1.2.65}$$

which is analogous to (1.1.50) for Ising systems.

We may also introduce a hydrodynamic pressure variable $\delta\hat{p}(\boldsymbol{r})$ by

$$\begin{aligned}\delta\hat{p}(\boldsymbol{r}) &= \left(\frac{\partial p}{\partial e}\right)_n \delta\hat{e}(\boldsymbol{r}) + \left(\frac{\partial p}{\partial n}\right)_e \delta\hat{n}(\boldsymbol{r}) \\ &= \rho c^2\left[\frac{1}{n}\delta\hat{n}(\boldsymbol{r}) + n\left(\frac{\partial T}{\partial p}\right)_s \delta\hat{s}(\boldsymbol{r})\right],\end{aligned} \tag{1.2.66}$$

where ρ is the mass density and use has been made of $(\partial n^{-1}/\partial s)_p = (\partial T/\partial p)_s$. For $\hat{a}(\boldsymbol{r})$ in the form of (1.2.45) we obtain

$$\langle \hat{a} : \hat{p} \rangle = Tn\left(\frac{\partial a}{\partial n}\right)_s = T\rho c^2\left(\frac{\partial a}{\partial p}\right)_s. \tag{1.2.67}$$

Substituting $\hat{a} = \hat{p}$ and $\hat{T}$ yields

$$\langle \hat{p} : \hat{p} \rangle = \rho c^2 T, \tag{1.2.68}$$

$$\langle \hat{p} : \hat{T} \rangle = T\rho c^2\left(\frac{\partial T}{\partial p}\right)_s = \frac{T^2}{C_V}\left(\frac{\partial p}{\partial T}\right)_n. \tag{1.2.69}$$

By setting $\hat{a} = \hat{s}$ and $\hat{n}$ we also notice

$$\langle \hat{s} : \hat{p} \rangle = 0, \quad \langle \hat{n} : \hat{p} \rangle = nT. \tag{1.2.70}$$

The $\mathcal{H}_{\text{hyd}}$ may be rewritten in another orthogonal form,

$$\mathcal{H}_{\text{hyd}} = \int d\boldsymbol{r}\left\{\frac{1}{2\rho c^2}[\delta\hat{p}(\boldsymbol{r})]^2 + \frac{n^2T}{2C_p}[\delta\hat{s}(\boldsymbol{r})]^2\right\}. \tag{1.2.71}$$

It goes without saying that $(\Delta S)_2$ in (1.2.42) coincides with $-\beta\mathcal{H}_{\text{hyd}}$.

1.2.8 Projection onto gross variables in the hydrodynamic regime

The pressure fluctuation variable $\delta\hat{p}(\boldsymbol{r})$ in (1.2.66) may be interpreted as the *projection* of the microscopic stress tensor $\hat{\Pi}_{\alpha\beta}(\boldsymbol{r})$ $(\alpha, \beta = x, y, z)$ onto the gross variables $\delta\hat{e}$ (or $\delta\hat{s}$) and $\delta\hat{n}$.[9] In the hydrodynamic regime, for any fluctuating variable $\hat{a}(\boldsymbol{r})$ dependent on space, the projection operator $\mathcal{P}$ is defined as

$$\mathcal{P}\hat{a}(\boldsymbol{r}) = \langle\hat{a}\rangle + A_{en}\delta\hat{e}(\boldsymbol{r}) + A_{ne}\delta\hat{n}(\boldsymbol{r}). \tag{1.2.72}$$

The two coefficients A_{en} and A_{ne} are determined such that the right-hand side and $\delta\hat{a}$ have the same correlations with $\delta\hat{e}$ and $\delta\hat{n}$. Then $\mathcal{P}^2 = \mathcal{P}$. If $\hat{a}$ is of the form (1.2.45), we have $\mathcal{P}\hat{a} = \hat{a}$. We neglect nonlocality in (1.2.72) assuming that $\delta\hat{e}$ and $\delta\hat{n}$ consist of the Fourier components with an upper cut-off wave number Λ much smaller than the inverse thermal correlation length. The calculation of the coefficients is simplified if the above relation is rewritten in terms of $\delta\hat{p}$ and $\delta\hat{s}$ as

$$\mathcal{P}\delta\hat{a}(\boldsymbol{r}) = A_{ps}\delta\hat{p}(\boldsymbol{r}) + A_{sp}\delta\hat{s}(\boldsymbol{r}). \tag{1.2.73}$$

Using $\langle\hat{s}:\hat{p}\rangle = 0$, we find

$$A_{ps} = \langle\hat{a}:\hat{p}\rangle/\langle\hat{p}:\hat{p}\rangle, \quad A_{sp} = \langle\hat{a}:\hat{s}\rangle/\langle\hat{s}:\hat{s}\rangle. \tag{1.2.74}$$

From (1A.11) and (1A.12) in Appendix 1A, we may derive the following variance relations,

$$\langle\hat{n}:\hat{\Pi}_{\alpha\beta}\rangle = nT\delta_{\alpha\beta}, \quad \langle\hat{e}:\hat{\Pi}_{\alpha\beta}\rangle = (e+p)T\delta_{\alpha\beta}. \tag{1.2.75}$$

Then, from the definitions of $\hat{s}$ in (1.2.46) and $\hat{p}$ in (1.2.66) we obtain

$$\langle\hat{s}:\hat{\Pi}_{\alpha\beta}\rangle = 0, \quad \langle\hat{p}:\hat{\Pi}_{\alpha\beta}\rangle = \rho c^2 T\delta_{\alpha\beta}. \tag{1.2.76}$$

Hence, we arrive at

$$\mathcal{P}\delta\hat{\Pi}_{\alpha\beta}(\boldsymbol{r}) = \delta_{\alpha\beta}\delta\hat{p}(\boldsymbol{r}). \tag{1.2.77}$$

This leads to the inequality

$$\rho c^2 \le K_\infty \equiv \left\langle \sum_\alpha \hat{\Pi}_{\alpha\alpha} : \sum_\beta \hat{\Pi}_{\beta\beta} \right\rangle / d^2 T. \tag{1.2.78}$$

See (1.2.84) below for K_∞ [18]. In fact, at the gas–liquid critical point the sound velocity c goes to zero but K_∞ remains finite. These are consistent with the inequality in (1.2.78).

1.2.9 Pressure, energy, and elastic moduli in terms of $g(r)$

In Appendix 5E we will give the space-dependent microscopic expression for the stress tensor $\hat{\Pi}_{\alpha\beta}(\boldsymbol{r})$. Its space integral has the following microscopic expression [5, 6],

$$\int d\boldsymbol{r}\hat{\Pi}_{\alpha\beta}(\boldsymbol{r}) = \sum_i \frac{p_{i\alpha}p_{i\beta}}{m_0} - \sum_{<i,j>} v'(r_{ij})\frac{1}{r_{ij}}x_{ij\alpha}x_{ij\beta}, \tag{1.2.79}$$

9 As will be discussed in Chapter 5, the projection operator method has been developed in the study of irreversible processes.

where $v'(r) = dv(r)/dr$, $x_{i\alpha}$ ($\alpha = x, y, z$) are the cartesian coordinates of the particle position $\boldsymbol{r}_i$, and $x_{ij\alpha} = x_{i\alpha} - x_{j\alpha}$. The pressure is then expressed in terms of the radial distribution function $g(r)$ in (1.2.55) as

$$p = nT - \frac{1}{2d} J_1, \tag{1.2.80}$$

with

$$J_1 = \int dr n^2 g(r) r v'(r), \tag{1.2.81}$$

where d in (1.2.80) is the spatial dimensionality. In addition, the internal energy density is expressed as

$$e = \langle \hat{e} \rangle = \frac{d}{2} nT + \frac{1}{2} \int dr n^2 g(r) v(r). \tag{1.2.82}$$

In an isotropic equilibrium state the variances among the stress tensor $\hat{\Pi}_{\alpha\beta}$ in the long-wavelength limit are written as

$$\frac{1}{T} \langle \hat{\Pi}_{\alpha\beta} : \hat{\Pi}_{\gamma\delta} \rangle = (\delta_{\alpha\gamma}\delta_{\beta\delta} + \delta_{\alpha\delta}\delta_{\beta\gamma}) G_\infty + \delta_{\alpha\beta}\delta_{\gamma\delta} \left(K_\infty - \frac{2}{d} G_\infty \right). \tag{1.2.83}$$

Here K_∞ and G_∞ are called the *elastic moduli* of fluids [6], [15]–[18]. Although elastic deformations are not well defined in fluids, they were interpreted as the infinite-frequency elastic moduli of fluids [17].[10] Interestingly, they can be expressed in terms of $g(r)$ as [17, 18]

$$K_\infty = \frac{1}{d^2 T} \left\langle \sum_\alpha \hat{\Pi}_{\alpha\alpha} : \sum_\beta \hat{\Pi}_{\beta\beta} \right\rangle = \left(1 + \frac{2}{d} \right) nT - \frac{d-1}{2d^2} J_1 + \frac{1}{2d^2} J_2, \tag{1.2.84}$$

$$G_\infty = \frac{1}{T} \langle \hat{\Pi}_{xy} : \hat{\Pi}_{xy} \rangle = nT + \frac{1}{2d(d+2)} \left[(d+1) J_1 + J_2 \right], \tag{1.2.85}$$

where J_1 is defined by (1.2.81) and

$$J_2 = \int dr n^2 g(r) r^2 v''(r), \tag{1.2.86}$$

with $v''(r) = d^2 v(r)/dr^2$. Elimination of J_1 and J_2 yields a general relation,

$$K_\infty - \left(1 + \frac{2}{d} \right) G_\infty = 2(p - nT). \tag{1.2.87}$$

It is not trivial that K_∞ and G_∞ can be expressed in terms of the radial distribution function, although they involve correlations among four particles. We will present a general theory for calculating correlation functions involving the stress tensor in Appendix 1A.

Schofield calculated more general wave number-dependent correlation functions among

[10] In highly supercooled fluids, a shear modulus becomes well defined and measurable. It is smaller than G_∞ but larger than nT. See Fig. 11.33 and its explanation in Section 11.4.

the stress components [18]. He considered the projection of the time derivative of the Fourier component $\hat{\Pi}_{\alpha\beta}(\boldsymbol{k})$ of the stress tensor,

$$\mathcal{P}\left[\frac{\partial}{\partial t}\hat{\Pi}_{\alpha\beta}(\boldsymbol{k})\right] = \sum_{\gamma\delta} C_{\alpha\beta\gamma\delta}(k)\epsilon_{\gamma\delta}(\boldsymbol{k}), \tag{1.2.88}$$

onto the Fourier component of the *strain* tensor, $\epsilon_{\alpha\beta}(\boldsymbol{k}) \equiv ik_\alpha J_\beta(\boldsymbol{k}) + ik_\beta J_\alpha(\boldsymbol{k})$, where $\boldsymbol{J}$ is the mass current. Then the coefficients $C_{\alpha\beta\gamma\delta}(k)$ become the correlation functions among $\hat{\Pi}_{\alpha\beta}(\boldsymbol{k})$, and their small-$k$ limits are linear combinations of K_∞ and G_∞ introduced above. Numerical analysis of these nonlocal elastic moduli was performed subsequently [19].

Generalization to the binary fluid mixture case

For binary fluid mixtures interacting with the pair potentials $v_{ij}(r)$, the expressions for p, K_∞, and G_∞ are still given by (1.2.80), (1.2.84) and (1.2.85), respectively, in terms of J_1 and J_2 if we re-define

$$\begin{aligned} J_1 &= \int d\boldsymbol{r} \sum_{i,j=1,2} n_i n_j g_{ij}(r) r v'_{ij}(r), \\ J_2 &= \int d\boldsymbol{r} \sum_{i,j=1,2} n_i n_j g_{ij}(r) r^2 v''_{ij}(r). \end{aligned} \tag{1.2.89}$$

Here $i, j = 1, 2$ represent the particle species, and $g_{ij}(r)$ are the radial distribution functions defined in (1.3.12) below. The expression for e is obtained if $n^2 g(r)v(r)$ is replaced by $\sum_{i,j=1,2} n_i n_j g_{ij}(r) v_{ij}(r)$ in (1.2.82).

1.3 Binary fluid mixtures

The thermodynamics of binary fluid mixtures composed of two species 1 and 2 interacting with short-range pair potentials will be considered. Although it is a straightforward generalization of that for one-component fluids, it becomes much more complicated and has rarely been discussed in detail [16]. We will show that its structure can be elucidated using variance relations among the density variables. Readers who do not work on fluid binary mixtures may skip this section now and return to it later when the information is needed in Chapters 2 and 6.

1.3.1 Grand canonical ensemble

As in the one-component fluid case, we choose $\Omega = pV/T = \ln \Xi$ as the thermodynamic potential, where Ξ is the grand canonical partition function. The independent field variables are β, $\nu_1 = \mu_1/T$, and $\nu_2 = \mu_2/T$, where μ_1 and μ_2 are the chemical potentials per particle. The incremental change of Ω is written as [20]

$$d\Omega = -\langle\mathcal{H}\rangle d\beta + \langle\mathcal{N}_1\rangle d\nu_1 + \langle\mathcal{N}_2\rangle d\nu_2, \tag{1.3.1}$$

where $\mathcal{N}_1$ and $\mathcal{N}_2$ are the particle numbers treated as fluctuating variables in the grand canonical ensemble. This relation is equivalent to the Gibbs–Duhem relation,

$$\frac{1}{n}dp = sdT + \frac{n_1}{n}d\mu_1 + \frac{n_2}{n}d\mu_2, \tag{1.3.2}$$

where $n_i = \langle \mathcal{N}_i \rangle / V$ $(i = 1, 2)$, and $n = n_1 + n_2$. The entropy s per particle satisfies $s = (e + p - n_1\mu_1 - n_2\mu_2)/nT$, where $e = \langle \mathcal{H} \rangle / V$. Sometimes μ_2 is treated as the potential; then, (1.3.2) is rewritten as

$$d\mu_2 = \frac{1}{n}dp - sdT - Xd\Delta, \tag{1.3.3}$$

where the independent field variables [10, 20] are p, T, and the chemical potential difference,

$$\Delta = \mu_1 - \mu_2. \tag{1.3.4}$$

The energy density variable $\hat{e}(\boldsymbol{r})$ and the number density variables $\hat{n}_i(\boldsymbol{r})$ have well-defined microscopic expressions, as in the one-component fluid case (1.2.43) and (1.2.44). Using the notation (1.1.35), the counterparts of (1.2.15)–(1.2.17) are of the forms [21]–[23]

$$\frac{\partial^2}{\partial \nu_i \partial \nu_j}\left(\frac{p}{T}\right) = \frac{\partial n_i}{\partial \nu_j} = \langle \hat{n}_i : \hat{n}_j \rangle, \tag{1.3.5}$$

$$\frac{\partial^2}{\partial \beta^2}\left(\frac{p}{T}\right) = -\frac{\partial e}{\partial \beta} = \langle \hat{e} : \hat{e} \rangle, \tag{1.3.6}$$

$$-\frac{\partial^2}{\partial \nu_i \partial \beta}\left(\frac{p}{T}\right) = -\frac{\partial n_i}{\partial \beta} = \frac{\partial e}{\partial \nu_i} = \langle \hat{n}_i : \hat{e} \rangle. \tag{1.3.7}$$

As an application of the above results, let us consider the specific heat $C_{VX} = (\partial e/\partial T)_{VNX}$ at constant volume V and concentration X. Since V is fixed,

$$C_{VX} = \left(\frac{\partial e}{\partial T}\right)_{n_1 n_2} = -1 \Big/ \left[T^2\left(\frac{\partial \beta}{\partial e}\right)_{n_1 n_2}\right]. \tag{1.3.8}$$

We should note that $-(\partial \beta/\partial e)_{n_1 n_2}$ is equal to the 33 element I^{33} of the inverse of the matrix $\{I_{ij}\}$ defined by

$$I_{ij} = \langle \hat{n}_i : \hat{n}_j \rangle, \quad I_{3i} = \langle \hat{n}_i : \hat{e} \rangle, \quad I_{33} = \langle \hat{e} : \hat{e} \rangle, \tag{1.3.9}$$

with $i, j = 1, 2$. Then we may express C_{VX} as

$$C_{VX} = \det I / T^2 [I_{11} I_{22} - I_{12}^2], \tag{1.3.10}$$

where

$$\det I = \det\{I_{ij}\} = \frac{\partial(n_1, n_2, e)}{\partial(\nu_1, \nu_2, -\beta)} \tag{1.3.11}$$

is the determinant of the 3×3 matrix $\{I_{ij}\}$. This expression is much more complicated than (1.2.49) for C_V in one-component fluids.

1.3.2 Fluctuating density variables

The radial distribution functions $g_{ij}(r)$ defined from the density correlation functions,

$$\langle \hat{n}_i(\boldsymbol{r})\hat{n}_j(\boldsymbol{r}')\rangle = n_i n_j g_{ij}(|\boldsymbol{r}-\boldsymbol{r}'|) + \delta_{ij} n_i \delta(\boldsymbol{r}-\boldsymbol{r}'), \tag{1.3.12}$$

have been studied in liquid theories [4]–[6]. Their numerically calculated profiles will be given in Fig. 11.26 for a supercooled state. The Fourier transformation yields the 2×2 matrix of the structure factors,

$$I_{ij}(k) = \delta_{ij} n_i + n_i n_j \int d\boldsymbol{r} e^{i\boldsymbol{k}\cdot\boldsymbol{r}}[g_{ij}(r) - \delta_{ij}]. \tag{1.3.13}$$

Their long-wavelength limits are $\langle \hat{n}_i : \hat{n}_j\rangle$ in (1.3.9):

$$I_{ij} \equiv \lim_{k\to 0} I_{ij}(k) = \langle \hat{n}_i : \hat{n}_j\rangle = (\partial n_i/\partial \nu_j)_{\mathrm{T}} = (\partial n_j/\partial \nu_i)_{\mathrm{T}}. \tag{1.3.14}$$

As in (1.2.45) for the one-component case, we may introduce a fluctuating variable $\hat{a}$ by

$$\hat{a}(\boldsymbol{r}) = a + \left(\frac{\partial a}{\partial n_1}\right)_{en_2} \delta\hat{n}_1(\boldsymbol{r}) + \left(\frac{\partial a}{\partial n_2}\right)_{en_1} \delta\hat{n}_2(\boldsymbol{r}) + \left(\frac{\partial a}{\partial e}\right)_{n_1 n_2} \delta\hat{e}(\boldsymbol{r}), \tag{1.3.15}$$

for any thermodynamic variable $a = a(n_1, n_2, e)$ given as a function of the averages $n_1 = \langle \hat{n}_1\rangle$, $n_2 = \langle \hat{n}_2\rangle$, and $e = \langle \hat{e}\rangle$. We may define fluctuating entropy and concentration variables as [23]

$$\hat{s}(\boldsymbol{r}) = s + \frac{1}{nT}\big[\delta\hat{e}(\boldsymbol{r}) - Ts\delta\hat{n}(\boldsymbol{r}) - \mu_1\delta\hat{n}_1(\boldsymbol{r}) - \mu_2\delta\hat{n}_2(\boldsymbol{r})\big], \tag{1.3.16}$$

$$\hat{X}(\boldsymbol{r}) = X + \frac{1}{n}\big[(1-X)\delta\hat{n}_1(\boldsymbol{r}) - X\delta\hat{n}_2(\boldsymbol{r})\big], \tag{1.3.17}$$

where

$$\hat{n}(\boldsymbol{r}) = \hat{n}_1(\boldsymbol{r}) + \hat{n}_2(\boldsymbol{r}) \tag{1.3.18}$$

is the (total) number density variable. The ratio $X = n_1/n$ is called the molar concentration, in terms of which the average number densities are expressed as

$$n_1 = nX, \quad n_2 = n(1-X). \tag{1.3.19}$$

For small variations of the field variables the microscopic grand canonical distribution P_{gra} changes as

$$\begin{aligned} \delta P_{\mathrm{gra}} &= P_{\mathrm{gra}} \int d\boldsymbol{r}\big[-\delta\hat{e}(\boldsymbol{r})\delta\beta + \delta\hat{n}_1(\boldsymbol{r})\delta\nu_1 + \delta\hat{n}_2(\boldsymbol{r})\delta\nu_2\big] \\ &= P_{\mathrm{gra}} \int d\boldsymbol{r}\left[n\delta\hat{s}(\boldsymbol{r})\frac{\delta T}{T} + \delta\hat{n}(\boldsymbol{r})\frac{\delta p}{nT} + n\delta\hat{X}(\boldsymbol{r})\frac{\delta\Delta}{T}\right], \end{aligned} \tag{1.3.20}$$

where $\delta p = ns\delta T + n_1\delta\mu_1 + n_2\delta\mu_2$ is the pressure deviation. The above relations are generalizations of (1.2.47), which is for one-component fluids. The second line of (1.3.20) implies that the conjugate fields of $\hat{s}(\boldsymbol{r})$, $\hat{n}(\boldsymbol{r})$, and $\hat{X}(\boldsymbol{r})$ are the deviations $n\delta T$, $n^{-1}\delta p$, and $n\delta\Delta$, respectively. As in (1.2.51), for $\hat{a}(\boldsymbol{r})$ in the form of (1.3.15), we find

$$\left(\frac{\partial a}{\partial T}\right)_{p\Delta} = \frac{n}{T}\langle \hat{a} : \hat{s}\rangle, \quad \left(\frac{\partial a}{\partial p}\right)_{T\Delta} = \frac{1}{nT}\langle \hat{a} : \hat{n}\rangle, \quad \left(\frac{\partial a}{\partial \Delta}\right)_{pT} = \frac{n}{T}\langle \hat{a} : \hat{X}\rangle. \tag{1.3.21}$$

In particular,

$$\left(\frac{\partial X}{\partial \Delta}\right)_{pT} = \frac{n}{T}\langle \hat{X} : \hat{X}\rangle \tag{1.3.22}$$

is called the concentration susceptibility, representing the strength of the concentration fluctuations.

In most experiments, however, X is fixed instead of Δ. The first two equations of (1.3.21) may then be changed to [21]–[23]

$$\left(\frac{\partial a}{\partial T}\right)_{pX} = \frac{n}{T}\big[\langle \hat{a} : \hat{s}\rangle - \langle \hat{a} : \hat{X}\rangle\langle \hat{s} : \hat{X}\rangle/\langle \hat{X} : \hat{X}\rangle\big],$$

$$\left(\frac{\partial a}{\partial p}\right)_{TX} = \frac{1}{nT}\big[\langle \hat{a} : \hat{n}\rangle - \langle \hat{a} : \hat{X}\rangle\langle \hat{n} : \hat{X}\rangle/\langle \hat{X} : \hat{X}\rangle\big]. \tag{1.3.23}$$

Then the specific heat $C_{pX} = nT(\partial s/\partial T)_{pX}$, the compressibility $K_{TX} = (\partial n/\partial p)_{TX}/n$, and the thermal expansion coefficient $\alpha_{pX} = -(\partial n/\partial T)_{pX}/n$ at fixed concentration X are written as

$$C_{pX} = n^2\big[\langle \hat{s} : \hat{s}\rangle - \langle \hat{s} : \hat{X}\rangle^2/\langle \hat{X} : \hat{X}\rangle\big], \tag{1.3.24}$$

$$K_{TX} = \frac{1}{n^2 T}\big[\langle \hat{n} : \hat{n}\rangle - \langle \hat{n} : \hat{X}\rangle^2/\langle \hat{X} : \hat{X}\rangle\big], \tag{1.3.25}$$

$$\alpha_{pX} = -\frac{1}{T}\big[\langle \hat{s} : \hat{n}\rangle - \langle \hat{s} : \hat{X}\rangle\langle \hat{n} : \hat{X}\rangle/\langle \hat{X} : \hat{X}\rangle\big]. \tag{1.3.26}$$

From (1.3.17) and (1.3.18) $\hat{n}$ and $\hat{X}$ are expressed in terms of $\hat{n}_1$ and $\hat{n}_2$. It leads to the identity,

$$\langle \hat{n} : \hat{n}\rangle\langle \hat{X} : \hat{X}\rangle - \langle \hat{n} : \hat{X}\rangle^2 = n^{-2}\big[I_{11}I_{22} - I_{12}^2\big], \tag{1.3.27}$$

where $I_{ij} = \langle \hat{n}_i : \hat{n}_j\rangle$. Therefore, K_{TX} may also be rewritten as

$$K_{TX} = \big[I_{11}I_{22} - I_{12}^2\big]/n^4 T\langle \hat{X} : \hat{X}\rangle. \tag{1.3.28}$$

Expressions equivalent to (1.3.25) and (1.3.28) were derived by Kirkwood and Buff [21]. The positivity of C_{pX} and K_{TX} is evident from (1.1.27).

1.3.3 Molar and mass concentrations

So far we have used the molar concentration. However, in many experimental papers, use has often been made of the mass concentration,

$$x = \frac{m_{01}n_1}{m_{01}n_1 + m_{02}n_2} = \frac{m_{01}X}{m_{01}X + m_{02}(1-X)}, \tag{1.3.29}$$

where m_{01} and m_{02} are the molecular masses. The corresponding field variable is the chemical potential difference,

$$\bar{\Delta} = \frac{1}{m_{01}}\mu_1 - \frac{1}{m_{02}}\mu_2, \tag{1.3.30}$$

per unit mass. The mass densities of the two components are $\rho_1 = \rho x$ and $\rho_2 = \rho(1-x)$, respectively, where $\rho = \rho_1 + \rho_2 = m_{01}n_1 + m_{02}n_2$ is the (total) mass density. Because x depends only on X as $dx/dX = m_{01}m_{02}(n/\rho)^2$, there arises no essential difference between these two choices. That is, expressions in one of these two choices are transformed into those in the other choice with multiplication of some factors. For example, the square of the sound velocity and the concentration susceptibilities in the two choices are expressed as

$$c^2 = \left(\frac{\partial p}{\partial \rho}\right)_{sx} = \frac{n}{\rho}\left(\frac{\partial p}{\partial n}\right)_{sX}, \quad \left(\frac{\partial x}{\partial \bar{\Delta}}\right)_{pT} = \left(\frac{n}{\rho}\right)^3\left(\frac{m_{01}}{m_{02}}\right)^2\left(\frac{\partial X}{\partial \Delta}\right)_{pT}. \tag{1.3.31}$$

The second relation is because $n_1 d\mu_1 + n_2 d\mu_2 = 0$ from (1.3.2) and $d\bar{\Delta} = (\rho/m_{01}m_{02}n)d\Delta$ from (1.3.30) for $dT = dp = 0$.

1.3.4 Hydrodynamic fluctuations of the field variables

We next introduce the fluctuating temperature and pressure variables $\delta\hat{T}(\boldsymbol{r})$ and $\delta\hat{p}(\boldsymbol{r})$ and examine their statistical properties. To this end we need some matrix analysis. We define

$$\hat{m}_1(\boldsymbol{r}) = \hat{s}(\boldsymbol{r}), \quad \hat{m}_2(\boldsymbol{r}) = \hat{n}(\boldsymbol{r}), \quad \hat{m}_3(\boldsymbol{r}) = \hat{X}(\boldsymbol{r}) \tag{1.3.32}$$

and write their fluctuation variances as $A_{ij} = \langle \hat{m}_i : \hat{m}_j \rangle$. Then, from (1.3.21) we have

$$A_{1i} = \frac{T}{n}\frac{\partial m_i}{\partial T}, \quad A_{2i} = nT\frac{\partial m_i}{\partial p}, \quad A_{3i} = \frac{T}{n}\frac{\partial m_i}{\partial \Delta}, \tag{1.3.33}$$

where $m_1 = s, m_2 = n$, and $m_3 = X$ are the thermodynamic quantities regarded as functions of T, p, and Δ. The inverse matrix of A_{ij} is written as A^{ij}. It may be expressed as

$$A^{1i} = \frac{n}{T}\frac{\partial T}{\partial m_i}, \quad A^{2i} = \frac{1}{nT}\frac{\partial p}{\partial m_i}, \quad A^{3i} = \frac{n}{T}\frac{\partial \Delta}{\partial m_i}, \tag{1.3.34}$$

where T, p, and Δ are regarded as functions of s, n, and X. In particular,

$$A^{11} = [A_{22}A_{33} - A_{23}^2]/\det A = n^2/C_{VX}, \tag{1.3.35}$$

$$A^{22} = [A_{11}A_{33} - A_{13}^2]/\det A = \rho c^2/n^2 T, \tag{1.3.36}$$

where $\det A$ is the determinant of the matrix $\{A_{ij}\}$. The first relation (1.3.35) may be transformed into (1.3.10) if use is made of (1.3.27) and the relations between the two determinants,

$$\det A = \frac{T^3}{n}\frac{\partial(s, n, X)}{\partial(T, p, \Delta)} = \frac{1}{n^4 T^2}\det I, \tag{1.3.37}$$

which follows from the definitions (1.3.16)–(1.3.18). The second relation (1.3.36) is rewritten as

$$\rho c^2 = n^2 T\big[\langle \hat{s} : \hat{s}\rangle\langle \hat{X} : \hat{X}\rangle - \langle \hat{s} : \hat{X}\rangle^2\big]/\det A, \tag{1.3.38}$$

which gives

$$\det A = T^2 C_{pX}\left(\frac{\partial X}{\partial \Delta}\right)_{pT}/n\rho c^2 = T^2 C_{VX} K_{TX}\left(\frac{\partial X}{\partial \Delta}\right)_{pT}/n, \tag{1.3.39}$$

if use is made of (1.3.22) and (1.3.24). It also follows the thermodynamic identity for the specific heat ratio,

$$\gamma_X \equiv C_{pX}/C_{VX} = \rho c^2 K_{TX}. \tag{1.3.40}$$

The fluctuating temperature variable is defined by

$$\delta\hat{T}(\boldsymbol{r}) = \frac{T}{n}\sum_{i=1}^{3} A^{1i}\delta\hat{m}_i(\boldsymbol{r}) = \sum_{i=1}^{3}\frac{\partial T}{\partial m_i}\delta\hat{m}_i(\boldsymbol{r}). \tag{1.3.41}$$

For $\hat{a}$ in the form of (1.3.15) we obtain

$$\langle \hat{a} : \hat{T}\rangle = \frac{T}{n}\left(\frac{\partial a}{\partial s}\right)_{n_1 n_2} = \frac{T^2}{C_{VX}}\left(\frac{\partial a}{\partial T}\right)_{n_1 n_2}, \tag{1.3.42}$$

where C_{VX} is the specific heat at constant n and X per unit volume discussed in (1.3.8)–(1.3.11). Substituting $\hat{a} = \hat{n}_i$ $(i = 1, 2)$, $\hat{s}$, and $\hat{T}$, we obtain

$$\langle \hat{n}_i : \hat{T}\rangle = 0, \quad \langle \hat{s} : \hat{T}\rangle = T/n, \tag{1.3.43}$$

$$\langle \hat{T} : \hat{T}\rangle = T^2/C_{VX}. \tag{1.3.44}$$

Thus $\delta\hat{T}$ is orthogonal to the number densities.

The fluctuating pressure variable is defined by

$$\delta\hat{p}(\boldsymbol{r}) = nT\sum_{i=1}^{3} A^{2i}\delta\hat{m}_i(\boldsymbol{r}) = \sum_{i=1}^{3}\frac{\partial p}{\partial m_i}\delta\hat{m}_i(\boldsymbol{r}). \tag{1.3.45}$$

For any fluctuation variable $\hat{a}(\boldsymbol{r})$ we obtain

$$\langle \hat{a} : \hat{p}\rangle = nT\left(\frac{\partial a}{\partial n}\right)_{sX} = T\rho c^2\left(\frac{\partial a}{\partial p}\right)_{sX}. \tag{1.3.46}$$

On the other hand, by setting $\hat{a} = \hat{s}$, $\hat{X}$, and $\hat{n}$ we have

$$\langle \hat{s} : \hat{p}\rangle = \langle \hat{X} : \hat{p}\rangle = 0, \quad \langle \hat{n} : \hat{p}\rangle = nT. \tag{1.3.47}$$

On the other hand, for $\hat{a} = \hat{T}$ and $\hat{p}$, we derive

$$\langle \hat{T} : \hat{p} \rangle = T\rho c^2 \left(\frac{\partial T}{\partial p} \right)_{sX} = \frac{T^2}{C_{VX}} \left(\frac{\partial p}{\partial T} \right)_{nX}, \tag{1.3.48}$$

$$\langle \hat{p} : \hat{p} \rangle = T\rho c^2. \tag{1.3.49}$$

These relations are straightforward generalizations of those in the one-component case.

We may also introduce the fluctuation of the chemical potential difference $\hat{\Delta}$ as

$$\delta \hat{\Delta}(\boldsymbol{r}) = \frac{T}{n} \sum_{i=1}^{3} A^{3i} \delta \hat{m}_i(\boldsymbol{r}) = \sum_{i=1}^{3} \frac{\partial \Delta}{\partial m_i} \delta \hat{m}_i(\boldsymbol{r}). \tag{1.3.50}$$

For $\hat{a}(\boldsymbol{r})$ in the form of (1.3.15) we obtain

$$\langle \hat{a} : \hat{\Delta} \rangle = \frac{T}{n} \left(\frac{\partial a}{\partial X} \right)_{sn}. \tag{1.3.51}$$

Substitutions $\hat{a} = \hat{s}, \hat{n}$, and $\hat{X}$ give

$$\langle \hat{s} : \hat{\Delta} \rangle = \langle \hat{n} : \hat{\Delta} \rangle = 0, \quad \langle \hat{X} : \hat{\Delta} \rangle = T/n. \tag{1.3.52}$$

Projection of $\hat{\Pi}_{\alpha\beta}$ onto the hydrodynamic variables

By generalizing the calculation in Appendix 1A to the binary fluid mixture case, we may readily show that the inner products of the microscopic tensor $\Pi_{\alpha\beta}$ with the hydrodynamic variables are expressed as

$$\langle \hat{n}_i : \hat{\Pi}_{\alpha\beta} \rangle = n_i T \delta_{\alpha\beta}, \quad \langle \hat{e} : \hat{\Pi}_{\alpha\beta} \rangle = (e + p) T \delta_{\alpha\beta}. \tag{1.3.53}$$

Then, from the definitions of $\hat{s}$ and $\hat{X}$ in (1.3.16) and (1.3.17), respectively, we obtain $\langle \hat{s} : \hat{\Pi}_{\alpha\beta} \rangle = 0$ and $\langle \hat{X} : \hat{\Pi}_{\alpha\beta} \rangle = 0$, so that $\mathcal{P}\delta\hat{\Pi}_{\alpha\beta} = \delta_{\alpha\beta}\delta\hat{p}$ as in (1.2.77) for one-component fluids.

1.3.5 The direct correlation functions and the hydrodynamic hamiltonian

We introduce the direct correlation functions $C_{ij}(r)$ for binary fluid mixtures [5]. The Fourier transformations of $C_{ij}(r)$ are related to the structure factors $I_{ij}(k)$ in (1.3.13) by

$$\frac{1}{n_i} I_{ij}(k) - \sum_{\ell} C_{i\ell}(k) I_{\ell j}(k) = \delta_{ij}. \tag{1.3.54}$$

The physical meaning of $C_{ij}(r)$ becomes apparent if the radial distribution functions $g_{ij}(r)$ are expressed as

$$g_{ij}(r) - 1 = C_{ij}(r) + \sum_{\ell} \int d\boldsymbol{r}' C_{i\ell}(|\boldsymbol{r} - \boldsymbol{r}'|) n_\ell g_{\ell j}(|\boldsymbol{r}'|) - 1. \tag{1.3.55}$$

The first term represents the *direct* correlations, while the second term arises from superposition of the *indirect* correlations. In one-component fluids the direct correlation function $C(r)$ has been introduced in (1.2.58) and its Fourier transformation in (1.2.59).

Next, using (1.3.43) and (1.3.44), we may generalize (1.2.65) to obtain the hydrodynamic hamiltonian for binary fluid mixtures,

$$\mathcal{H}_{\text{hyd}} = \int d\boldsymbol{r} \left\{ \frac{1}{2T} C_{VX} [\delta \hat{T}(\boldsymbol{r})]^2 + \frac{T}{2} \sum_{ij} I^{ij} \delta \hat{n}_i(\boldsymbol{r}) \delta \hat{n}_j(\boldsymbol{r}) \right\}. \tag{1.3.56}$$

Here $\{I^{ij}\}$ is the inverse of the matrix $I_{ij} = \langle \hat{n}_i : \hat{n}_j \rangle$ in (1.3.14), so

$$I^{ij} = (\partial \nu_j / \partial n_i)_{\text{T}} = (\partial \nu_i / \partial n_j)_{\text{T}}, \tag{1.3.57}$$

where $\nu_i = \mu_i / T$ are regarded as functions of n_1, n_2, and T. In the long-wavelength limit we have

$$I^{ij} = \frac{1}{n_i} \delta_{ij} - C_{ij}(0), \tag{1.3.58}$$

where $C_{ij}(0) = \lim_{k \to 0} C_{ij}(k)$. Using (1.3.53) and the Gibbs–Duhem relation (1.3.2) we also notice

$$\sum_j I^{ij} n_j = 1 - \sum_j C_{ij}(0) n_j = \frac{1}{T} \left(\frac{\partial p}{\partial n_i} \right)_T = n \left(\frac{\partial \nu_i}{\partial n} \right)_{TX}, \tag{1.3.59}$$

where p is regarded as a function of n_1, n_2, and T. Furthermore, multiplying (1.3.58) by n_i and summing over i, we may relate the compressibility K_{TX} to $C_{ij}(0)$ as

$$\frac{1}{K_{TX}} = n \left(\frac{\partial p}{\partial n} \right)_{TX} = nT - T \sum_{ij} C_{ij}(0) n_i n_j. \tag{1.3.60}$$

This is a generalization of the well-known compressibility relation for one-component fluids, (1.2.57).

Appendix 1A Correlations with the stress tensor

There is a general method of calculating the correlation function between the stress tensor $\hat{\Pi}_{\alpha\beta}(\boldsymbol{r}, \Gamma)$ and any local variable $\mathcal{A}(\boldsymbol{r}, \Gamma)$, where we explicitly write the dependence on the phase space point $\Gamma = (\boldsymbol{r}_1, \boldsymbol{p}_1, \ldots, \boldsymbol{r}_N, \boldsymbol{p}_N)$. For simplicity we consider the correlations in the long-wavelength limit in one-component fluids. It is straightforward to generalize the following results to binary fluid mixtures. We only need to replace $n^2 g(r)$ by $\sum_{ij} n_i n_j g_{ij}(r)$ in the following expressions.

Let us slightly perturb the hamiltonian as

$$\mathcal{H}'(\Gamma) = \mathcal{H}(\Gamma) - \int d\boldsymbol{r} \sum_{\alpha\beta} \hat{\Pi}_{\alpha\beta}(\boldsymbol{r}, \Gamma) \mathcal{D}_{\alpha\beta}(\boldsymbol{r}), \tag{1A.1}$$

where $\mathcal{D}_{\alpha\beta} = \partial u_\alpha / \partial x_\beta$ is the gradient tensor of a small, slowly varying displacement vector $\boldsymbol{u}(\boldsymbol{r})$. We then slightly shift the momenta and positions, $\boldsymbol{r}_i = (x_{i1}, \ldots, x_{id})$ and $\boldsymbol{p}_i = (p_{i1}, \ldots, p_{id})$ $(i = 1, \ldots, N)$, as

$$p'_{i\alpha} = p_{i\alpha} - \sum_\beta \mathcal{D}_{\alpha\beta}(\boldsymbol{r}_i) p_{i\beta}, \quad x'_{i\alpha} = x_{i\alpha} + u_\alpha(\boldsymbol{r}_i). \tag{1A.2}$$

It is important that the perturbed hamiltonian $\mathcal{H}'(\Gamma)$ becomes of the same form as the unperturbed hamiltonian,

$$\mathcal{H}'(\Gamma) = \mathcal{H}(\Gamma') + O(\boldsymbol{u}^2) \tag{1A.3}$$

in terms of the displaced phase space point $\Gamma' = (\boldsymbol{r}'_1, \boldsymbol{p}'_1, \ldots, \boldsymbol{r}'_N, \boldsymbol{p}'_N)$ to first order in $\boldsymbol{u}$. We assume that $\boldsymbol{u}$ vanishes at the boundary of the system, so that the displaced positions $\boldsymbol{r}'_i$ are also within the same fluid container. From (1A.1) the average over the equilibrium distribution for the perturbed hamiltonian is written as

$$\begin{aligned} \langle \mathcal{A}(\boldsymbol{r}, \Gamma) \rangle' &= \frac{1}{Z'_N} \int d\Gamma \mathcal{A}(\boldsymbol{r}, \Gamma) \exp\left[-\frac{1}{T} \mathcal{H}'(\Gamma) \right] \\ &= \langle \mathcal{A}(\boldsymbol{r}, \Gamma) \rangle + \frac{1}{T} \sum_{\alpha\beta} \langle \mathcal{A} : \hat{\Pi}_{\alpha\beta} \rangle \mathcal{D}_{\alpha\beta}(\boldsymbol{r}), \end{aligned} \tag{1A.4}$$

where $Z'_N = \int d\Gamma \exp[-\mathcal{H}'(\Gamma)/T]$, and $\langle \cdots \rangle$ is the equilibrium average for the unperturbed hamiltonian $\mathcal{H}(\Gamma)$. In the second line use has been made of the fact that $\mathcal{D}_{\alpha\beta}(\boldsymbol{r})$ change slowly compared with any correlation lengths. Also, from (1A.3) we obtain

$$\langle \mathcal{A}(\boldsymbol{r}, \Gamma) \rangle' = \frac{1}{Z_N} \int d\Gamma' \mathcal{A}(\boldsymbol{r}, \Gamma) \exp\left[-\frac{1}{T} \mathcal{H}(\Gamma') \right], \tag{1A.5}$$

where $Z_N = \int d\Gamma' \exp[-\mathcal{H}(\Gamma')/T] = \int d\Gamma \exp[-\mathcal{H}(\Gamma)/T]$ is the partition function for the unperturbed hamiltonian. Here,

$$\mathcal{A}(\boldsymbol{r}, \Gamma) = \mathcal{A}(\boldsymbol{r}, \Gamma') + \sum_{\alpha\beta} \mathcal{D}_{\alpha\beta} \mathcal{A}_{\alpha\beta}(\boldsymbol{r}, \Gamma') + \cdots, \tag{1A.6}$$

with

$$\mathcal{A}_{\alpha\beta}(\boldsymbol{r}, \Gamma) = \sum_i \left(p_{i\beta} \frac{\partial}{\partial p_{i\alpha}} - x_{i\beta} \frac{\partial}{\partial x_{i\alpha}} \right) \mathcal{A}(\boldsymbol{r}, \Gamma). \tag{1A.7}$$

Comparing (1A.4) and (1A.5) we find the desired result,

$$\langle \mathcal{A} : \hat{\Pi}_{\alpha\beta} \rangle = \int d\boldsymbol{r}_1 \langle \delta \mathcal{A}(\boldsymbol{r}) \delta \hat{\Pi}_{\alpha\beta}(\boldsymbol{r}_1) \rangle = T \langle \mathcal{A}_{\alpha\beta}(\boldsymbol{r}, \Gamma) \rangle, \tag{1A.8}$$

where the integrand is assumed to decay sufficiently rapidly for large $|\boldsymbol{r} - \boldsymbol{r}_1|$ in the thermodynamic limit $V \to \infty$. An equivalent formula can be found in Ref. [24].

For example, we consider the fluctuations of the one-body distribution,

$$\hat{f}(\boldsymbol{r}, \boldsymbol{p}) = \sum_i \delta(\boldsymbol{r} - \boldsymbol{r}_i) \delta(\boldsymbol{p} - \boldsymbol{p}_i), \tag{1A.9}$$

and the pair distribution,

$$\hat{g}(\boldsymbol{r},\boldsymbol{p},\boldsymbol{r}',\boldsymbol{p}') = \sum_{i\neq j} \delta(\boldsymbol{r}-\boldsymbol{r}_i)\delta(\boldsymbol{p}-\boldsymbol{p}_i)\delta(\boldsymbol{r}'-\boldsymbol{r}_j)\delta(\boldsymbol{p}'-\boldsymbol{p}_j), \tag{1A.10}$$

in the $(\boldsymbol{r},\boldsymbol{p})$ space. Use of (1A.8) yields[11]

$$\langle \hat{f}(\boldsymbol{r},\boldsymbol{p}) : \hat{\Pi}_{\alpha\beta}\rangle = \frac{1}{m_0} p_\alpha p_\beta f_0(p), \tag{1A.11}$$

$$\langle \hat{g}(\boldsymbol{r},\boldsymbol{p},\boldsymbol{r}',\boldsymbol{p}') : \hat{\Pi}_{\alpha\beta}\rangle = \left[\frac{p_\alpha p_\beta + p'_\alpha p'_\beta}{m_0} + T(x_\alpha - x'_\alpha)\frac{\partial}{\partial x_\beta}\right] f_0(p) f_0(p') g(|\boldsymbol{r}-\boldsymbol{r}'|), \tag{1A.12}$$

where $f_0(p) = n(2\pi m_0 T)^{-d/2}\exp(-p^2/2m_0T)$ is the Maxwell distribution. Note that (1A.11) is independent of $\boldsymbol{r}$, while (1A.12) is independent of $\frac{1}{2}(\boldsymbol{r}+\boldsymbol{r}')$ in the thermodynamic limit. Now using the microscopic expression for $\hat{\Pi}_{\alpha\beta}(\boldsymbol{r})$ we may derive the expressions for K_∞ and G_∞ as in (1.2.84) and (1.2.85), respectively. Furthermore, integrations of (1A.11) and (1A.12) over the momenta lead to (1.2.75) with the aid of (1.2.80)–(1.2.82).

References

[1] L. D. Landau and E. M. Lifshitz, *Statistical Physics* (Pergamon, New York, 1964).

[2] R. Kubo, *Statistical Mechanics* (North-Holland, New York, 1965).

[3] L. E. Reichel, *Modern Course in Statistical Physics* (University of Texas Press, 1980).

[4] P. A. Egelstaff, *An Introduction to the Liquid State* (Academic, New York, 1967).

[5] J. P. Hansen and I. R. McDonald, *Theory of Simple Liquids* (Academic, New York, 1986).

[6] J. P. Boon and S. Yip, *Molecular Hydrodynamics* (Dover, 1980).

[7] P. M. Chaikin and T. C. Lubensky, *Principles of Condensed Matter Physics* (Cambridge University Press, 1995).

[8] A. Aharony, in *Phase Transitions in Critical Phenomena*, eds. C. Domb and J. L. Lebowitz (Academic, New York, 1976), Vol. 6, p. 358.

[9] L. P. Kadanoff and P. C. Martin, *Ann. Phys.* (N.Y.) **24**, 419 (1963).

[10] R. B. Griffiths and J. C. Wheeler, *Phys. Rev. A* **2**, 1047 (1970).

[11] M. E. Fisher, in *Critical Phenomena*, ed. M. S. Green, Proceedings of Enrico Fermi Summer School, Varenna, 1970 (Academic, New York, 1971), p. 1.

[12] C. N. Yang and T. D. Lee, *Phys. Rev.* **87**, 404 (1952); T. D. Lee and C. N. Yang, *ibid.* **87**, 410 (1952).

[13] H. B. Callen, *Thermodynamics* (John Wiley & Sons, New York, 1960).

[14] M. E. Fisher, *J. Math. Phys.* **5**, 944 (1964).

[15] H. S. Green, *The Molecular Theory of Fluids* (North-Holland, Amsterdam, 1952).

[16] J. S. Rowlinson, *Liquids and Liquid Mixtures* (Butterworths, London, 1959).

[11] We will use (1A.12) in the mode coupling calculations in (6.2.32).

[17] R. Zwanzig and R. D. Mountain, *J. Chem. Phys.* **43**, 4464 (1965).

[18] P. Schofield, *Proc. Phys. Soc.* **88**, 149 (1966).

[19] A. Z. Akcasu and F. Daniels, *Phys. Rev. A*, **2**, 962 (1970).

[20] S. S. Leung and R. B. Griffiths, *Phys. Rev. A* **8**, 2670 (1973).

[21] J. G. Kirkwood and F. P. Buff, *J. Chem. Phys.* **19**, 774 (1951).

[22] W. F. Saam, *Phys. Rev. A* **2**, 1461 (1970).

[23] A. Onuki, *J. Low Temp. Phys.* **61**, 101 (1985).

[24] K. Kawasaki, *Phys. Rev.* **150**, 291 (1966).

2

Critical phenomena and scaling

General aspects of static critical behavior [1]–[5] will be summarized using fractal concepts in Section 2.1. The mapping relations between the critical behavior of one- and two-component fluids and that of Ising systems will be discussed in Sections 2.2 and 2.3. They are useful in understanding a variety of thermodynamic experiments in fluids and will be the basis of the dynamical theories developed in Chapter 6. As another kind of critical behavior of xy symmetry, ^{4}He near the superfluid transition will be treated in our scheme in Section 2.4. Gravity effects on the critical behavior in one-component fluids and ^{4}He will also be discussed.

2.1 General aspects

First we provide the reader with snapshots of critical fluctuations whose characteristic features are strikingly similar in both Ising spin systems and fluids. Figure 2.1 shows a 128×128 spin configuration generated by a Monte Carlo simulation of a 2D Ising spin system in a disordered phase very close to the critical point. Figure 2.2 displays particle positions realized in a molecular dynamics simulation of a 2D one-component fluid in a one-phase state close to the gas–liquid critical point. In the latter simulation, the pair potential $v(r)$ is of the Lenard-Jones form (1.2.1) cut off at $r/\sigma = 2.5$ and characterized by ϵ and σ. The temperature and average number density are $T = 0.48\epsilon$ and $n = 0.325\sigma^{-2}$, respectively. In Fig. 2.3 we plot the structure factors, $\langle|\hat{n}_{\mathbf{k}}|^2\rangle/n$ and $\langle|\hat{e}_{\mathbf{k}}|^2\rangle/n\epsilon^2$, for the number and energy density fluctuations, respectively, in the same Lenard-Jones model fluid with the same parameter values. We can see strong critical enhancement at small wave numbers, which indicates large compressibility because the small-k limit of $\langle|\hat{n}_{\mathbf{k}}|^2\rangle/n$ is equal to nTK_T from (1.2.57). We also show, in Fig. 2.4, weak critical enhancement of the specific heat C_V with the critical exponent α in oxygen measured in an early period of the research in this field [6].[1]

[1] In this experiment, stirring was used successfully to suppress the gravity effect. Effects of stirring on the critical behavior will be discussed in Chapter 11. See also the last part of Section 4.3 for further discussions of C_V.

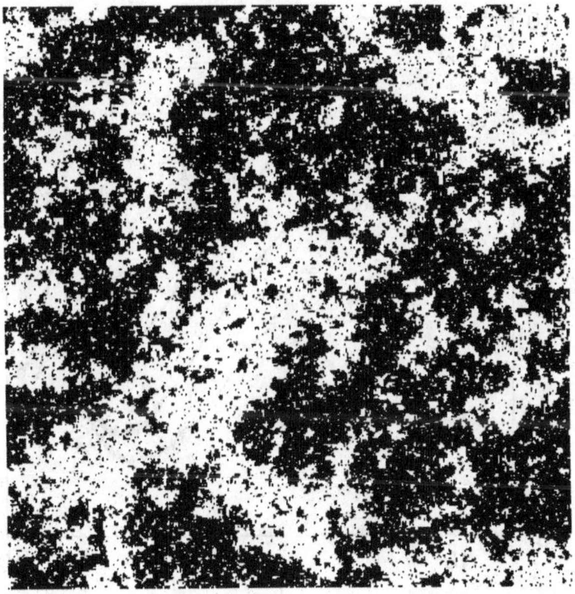

Fig. 2.1. Spin configuration in a 2D Ising system close to the critical point obtained by Monte Carlo simulation (courtesy of Mr K. Kanemitsu). The correlation length is of the order of the system dimension.

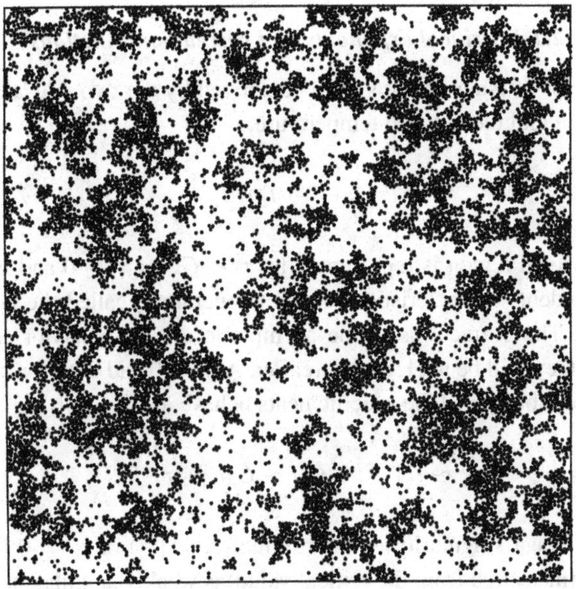

Fig. 2.2. Snapshot of particle positions in a 2D Lenard-Jones fluid close to the gas–liquid critical point obtained by molecular dynamics simulation (courtesy of Dr R. Yamamoto). A quarter of the total system ($L = 3926\sigma$ and $N = 5 \times 10^4$) is shown.

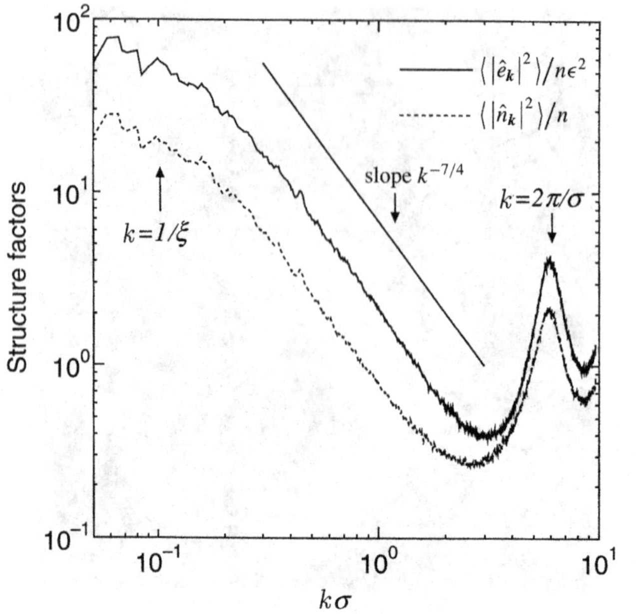

Fig. 2.3. The structure factors of the density and energy fluctuations vs $k\sigma$ for a 2D Lenard-Jones fluid close to the gas–liquid critical point. The long-wavelength parts ($k\sigma \lesssim 2$) represent the critical fluctuations. A line with a slope of $-7/4 = -(2-\eta)$ is included as a guide.

2.1.1 Critical exponents and correlation functions

The critical behavior of Ising systems is characterized by the two relevant field variables, the magnetic field h and the reduced temperature,

$$\tau = (T - T_c)/T_c. \tag{2.1.1}$$

The asymptotic critical region is represented by $\tau < Gi$ for $\tau > 0$ and $h = 0$, where Gi is a (system-dependent) characteristic reduced temperature, called the Ginzburg number (see Section 4.1). The corrections to the asymptotic critical behavior can be discussed generally [7], but they will be neglected hereafter. At $h = 0$ and both for $\tau > 0$ and $\tau < 0$, the magnetic susceptibility and the specific heats behave as

$$\chi \sim |\tau|^{-\gamma}, \quad C_H \sim C_M \sim |\tau|^{-\alpha}. \tag{2.1.2}$$

In 2D, the specific-heat singularity is logarithmic ($\propto \ln|\tau|$) or $\alpha = 0$ [8]. The average energy density m (measured from the critical value) is weakly singular at $h = 0$ as

$$m = \langle\hat{m}\rangle \sim |\tau|^{1-\alpha}, \tag{2.1.3}$$

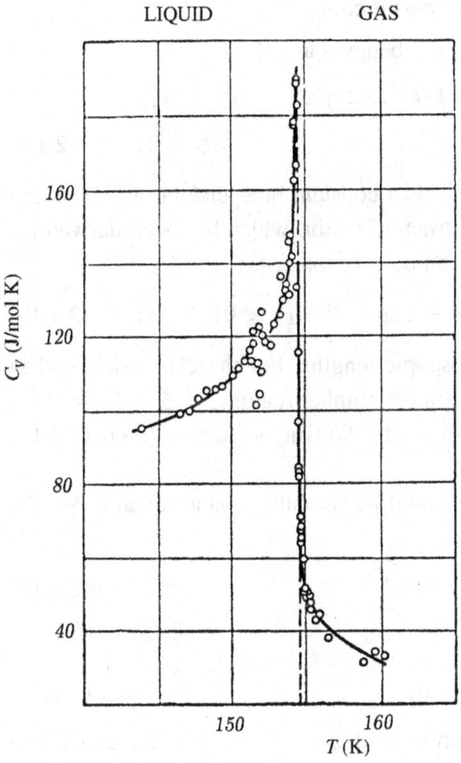

Fig. 2.4. Temperature dependence of the constant-volume specific heat at the critical density in oxygen [6].

which is consistent with the specific-heat behavior. On the coexistence line, where $h = 0$ and $\tau < 0$, the average spin $\psi = \langle \hat{\psi} \rangle$ is nonvanishing as

$$\psi \cong \pm B_0 |\tau|^\beta, \tag{2.1.4}$$

where B_0 is called the critical amplitude. The exponent β should not be confused with the inverse temperature. When $\tau = 0$ and $h \neq 0$, ψ has the same sign as h and

$$|h| \sim |\psi|^\delta. \tag{2.1.5}$$

Between the critical exponents, γ, α, β, and δ, the following relations are well known:

$$\alpha + 2\beta + \gamma = 2, \tag{2.1.6}$$

$$\delta = 1 + \gamma/\beta. \tag{2.1.7}$$

In Ising spin systems the critical exponents are

$$\begin{aligned} \gamma &\cong 1.24, \quad \alpha \cong 0.10, \quad \beta \cong 0.33, \quad \delta \cong 4.75, \qquad \text{(3D)}, \\ \gamma &= 7/4, \quad \alpha = 0, \quad \beta = 1/8, \quad \delta = 15 \qquad \text{(2D)}. \end{aligned} \tag{2.1.8}$$

Order parameter correlation

The structure factor $I(k) = \langle|\hat{\psi}_{\boldsymbol{k}}|^2\rangle$ asymptotically behaves as [9]

$$\begin{aligned} I(k) &\cong \chi/(1+k^2\xi^2) = \chi(1-k^2\xi^2+\cdots) && (k\xi \lesssim 1), \\ &\cong C_\infty/k^{2-\eta} && (k\xi \gg 1), \end{aligned} \tag{2.1.9}$$

where ξ is called the correlation length and C_∞ is a constant independent of ξ. The first line gives the Ornstein–Zernike form for the structure factor, which has been derived for fluids in (1.2.60). At $h = 0$ and for small τ, ξ can be very long as

$$\xi \cong \xi_{+0}\tau^{-\nu} \quad (\tau > 0), \quad \xi \cong \xi_{-0}|\tau|^{-\nu} \quad (\tau < 0), \tag{2.1.10}$$

where the coefficients ξ_{+0} and ξ_{-0} are microscopic lengths. For the 2D model fluid in Fig. 2.3 the power law $k^{-7/4}$ can be seen in the wave number region $\xi^{-1} \lesssim k \lesssim 2\sigma^{-1}$, as shown in the figure. This behavior is consistent with the 2D Ising value $\eta = 1/4$ in (2.1.17) below.

Since the two limiting behaviors in (2.1.9) should be smoothly connected at $k \sim \xi^{-1}$, the following scaling relation holds,

$$\gamma = (2-\eta)\nu. \tag{2.1.11}$$

The following relation is also well known:

$$d\nu = 2-\alpha, \tag{2.1.12}$$

where d is the space dimensionality. This relation holds for $d \le 4$. With the above two relations, β and δ may also be expressed in terms of ν and η as

$$\beta = \frac{1}{2}(d-2+\eta)\nu, \tag{2.1.13}$$

$$\delta = (d+2-\eta)/(d-2+\eta). \tag{2.1.14}$$

The structure factor may be written in the scaling form,

$$I(k) = \chi I^*(k\xi), \tag{2.1.15}$$

where the scaling function $I^*(x)$ behaves as $I^*(0) = 1$ and $I^*(x) \sim x^{-2+\eta}$ for $x \gg 1$ from (2.1.9). The corrections to the above scaling expression becomes negligible (or irrelevant) close to the critical point. The pair correlation function $g(r)$ is written as

$$g(r) = \frac{1}{r^{d-2+\eta}} G^*(r/\xi), \tag{2.1.16}$$

where $G^*(x)$ is a constant for $x \ll 1$ and decays exponentially as $\exp(-x)$ for $x \gg 1$. In Ising systems the critical exponents η and ν are given by

$$\begin{aligned} \eta &= 0.03\text{–}0.05, && \nu \cong 0.63\ (\cong 5/8) && (\text{3D}), \\ \eta &= 1/4, && \nu = 1 && (\text{2D}). \end{aligned} \tag{2.1.17}$$

In 3D, η is very small and is in many cases negligible.

Energy correlation

Similar scaling relations hold for the correlation function $g_e(r)$ of the energy density $\hat{m}(\mathbf{r})$ measured from the critical value and divided by T_c [10]:

$$g_e(r) = \langle \delta\hat{m}(\mathbf{r}+\mathbf{r}_0)\delta\hat{m}(\mathbf{r}_0)\rangle. \tag{2.1.18}$$

For Ising spin systems we introduced the exchange energy density $\hat{e}(\mathbf{r})$ in (1.1.32) and we here have $\hat{m}(\mathbf{r}) = [\hat{e}(\mathbf{r}) - e_c]/T_c$, where e_c is the critical value. Near the critical point, $g_e(r)$ is scaled as

$$g_e(r) = \frac{1}{r^{d-\alpha/\nu}} G_e^*(r/\xi). \tag{2.1.19}$$

Its Fourier transformation gives

$$I_e(k) = C_H I_e^*(k\xi), \tag{2.1.20}$$

where C_H is the specific heat written in the variance form in (1.1.24) or (1.1.34). The scaling function $I_e^*(x)$ tends to 1 for $x \ll 1$ and to const.$x^{-\alpha/\nu}$ for $x \gg 1$. It is well known that, as far as the static properties are concerned, we can set

$$\hat{m}(\mathbf{r}) \cong \text{const.}\hat{\psi}(\mathbf{r})^2, \tag{2.1.21}$$

where the coefficient is a constant independent of τ. Then $g_e(r)$ becomes the correlation function of $\hat{\psi}^2$. For Ising spin systems this means that the microscopic expression $\hat{e}(\mathbf{r})$ in (1.1.32) is coarse-grained in the form of const.$\hat{\psi}^2$ on spatial scales much longer than a.

2.1.2 Fractal dimensions

In Figs 2.1 and 2.2 we can see clusters of various sizes. If we consider the clusters with linear dimension λa in the intermediate range,

$$1 \ll \lambda \ll \xi/a, \tag{2.1.22}$$

they are self-similar with respect to appropriate scale changes. The system is assumed to extend to infinity. The geometrical characteristics of the clusters may be understood using the concept of *fractals* [11]–[13]. Following Suzuki, we introduce the Hausdorff fractal dimension D to characterize the critical clusters.

Let us consider the spin sum S_λ in a volume $V_\lambda = (\lambda a)^d$ with linear dimension λ,

$$S_\lambda = \int_{V_\lambda} d\mathbf{r}\hat{\psi}(\mathbf{r}) = \int_{V_\lambda} d\mathbf{r}\delta\hat{\psi}(\mathbf{r}) + V_\lambda\langle\hat{\psi}\rangle, \tag{2.1.23}$$

where λ satisfies (2.1.22), $\delta\psi = \hat{\psi} - \langle\hat{\psi}\rangle$ is the deviation, and the second term gives the average $\langle S_\lambda\rangle$. We may then consider the probability of finding the deviation $\delta S_\lambda = S_\lambda - \langle S_\lambda\rangle$ at S. The distribution function is written as $P(S)$ and is of the following scaling form,

$$P(S) = \lambda^{-D} P^*(S/\lambda^D), \tag{2.1.24}$$

where $P^*(x)$ is a scaling function independent of λ, and D is called the fractal dimension. This implies that δS_λ is typically of order λ^D. In the range (2.1.22) the variance of δS_λ is estimated as

$$\langle \delta S_\lambda^2 \rangle \sim \int_{V_\lambda} d\boldsymbol{r} \int_{V_\lambda} d\boldsymbol{r}' \frac{1}{|\boldsymbol{r}-\boldsymbol{r}'|^{d-2+\eta}} \sim \lambda^{2d}/\lambda^{d-2+\eta} \sim \lambda^{d+2-\eta}, \tag{2.1.25}$$

where use has been made of (2.1.16). Thus we can express D in terms of η as

$$D = \frac{1}{2}(d+2-\eta). \tag{2.1.26}$$

Therefore,

$$D \cong 2.5 \quad (3\text{D}), \qquad D = 15/8 \quad (2\text{D}). \tag{2.1.27}$$

for Ising models. From (2.1.13), (2.1.14), and (2.1.26) we also obtain

$$D = d - \beta/\nu = \beta\delta/\nu. \tag{2.1.28}$$

In 3D, $D \cong 2.5$ holds for any n-component spin system, because the exponent η is very small for any n, and the clusters are *ramified* objects [13]. In 2D Ising systems, $D = 15/8$ is close to the geometrical dimension 2 and the clusters are rather *compact* objects. This aspect is apparent in Figs 2.1 and 2.2.

We may also introduce the fractal dimension D_e for the energy density fluctuations $\hat{m}$ [12]. We consider its space integral E_λ in a region with linear dimension λa,

$$E_\lambda = \int_{V_\lambda} d\boldsymbol{r}\hat{m}(\boldsymbol{r}) \sim \int_{V_\lambda} d\boldsymbol{r}\hat{\psi}(\boldsymbol{r})^2. \tag{2.1.29}$$

From (2.1.19) we estimate

$$\langle \delta E_\lambda^2 \rangle \sim \lambda^{2d}/\lambda^{d-\alpha/\nu} \sim \lambda^{2/\nu}, \tag{2.1.30}$$

so that δE_λ is typically of order $\lambda^{1/\nu}$. This means

$$D_e = 1/\nu \quad \text{or} \quad D/D_e = 1 + \frac{1}{2}(\gamma - \alpha) > 1. \tag{2.1.31}$$

We next compare the averages, $\langle S_\lambda \rangle = V_\lambda \langle \hat{\psi} \rangle$ and $\langle E_\lambda \rangle = V_\lambda \langle \hat{m} \rangle$, and the typical fluctuation magnitudes on the coexistence curve very close to the critical point. Use of (2.1.3), (2.1.4), and the exponent relations yields

$$\sqrt{\langle \delta S_\lambda^2 \rangle}/\langle S_\lambda \rangle \sim \sqrt{\langle \delta E_\lambda^2 \rangle}/\langle E_\lambda \rangle \sim (\xi/\lambda a)^d. \tag{2.1.32}$$

Therefore, the averages are smaller than the typical magnitudes of the fluctuations for $\lambda a \ll \xi$. In the reverse case, $\lambda a \gg \xi$, the averages are much larger than the fluctuations. This means that domains appearing in phase separation are *compact* (not fractal) on spatial scales much longer than ξ. In some systems the crossover from mean field to asymptotic critical behavior occurs at a small value of the Ginzburg number Gi, where the condition $\langle S_\lambda \rangle \gg \sqrt{\langle \delta S_\lambda^2 \rangle}$ holds at $\lambda = \xi/a$ in a sizable temperature region $|T/T_c - 1| \gg Gi$ on the coexistence curve. This is the famous *Ginzburg criterion*, which assures mean field critical behavior, see Section 4.1.

Finite systems at the critical point

We have supposed infinite systems in the above arguments. However, finiteness of the system dimension L itself gives rise to some interesting effects. In particular, it is inevitable in simulations. If the bulk correlation length is much longer than L, the total spin sum S obeys a distribution of the finite-size scaling form,

$$P(S) = L^{-D}\tilde{P}(S/L^D), \tag{2.1.33}$$

which is analogous to (2.1.24) [14]–[17]. The scaling function $\tilde{P}(x)$ depends on the space dimensionality d and the boundary condition. In particular, in 2D at the bulk critical point under the periodic boundary condition, Ito and Suzuki [15] observed that S evolves in time between positive values of order $(L/a)^D$ and negative values of order $-(L/a)^D$, resulting in a doubly peaked distribution of S on the average. In 3D under the periodic boundary condition, $\tilde{P}(S)$ has a wing-like form peaked at $\pm(L/a)^D$ [14, 16].

Fisher cluster model

We mention here the cluster or droplet model due to Fisher [18]. He considered the statistical distribution of liquid clusters with ℓ molecules which are thermally activated in a gas phase close to the gas–liquid coexistence curve. His theory was subsequently confirmed in computer simulations on Ising spin systems. Such clusters with linear dimensions not exceeding ξ are fractal objects close to the critical point, as previously discussed. This model will be mentioned again in Section 9.1 in the context of nucleation.

2.1.3 Scaling ansatz

We now argue why the relations (2.1.6), (2.1.7), (2.1.13), and (2.1.14) between the critical exponents hold. Following Kadanoff [1, 19], we reduce the scale of the lengths by λ and consider a coarse-grained lattice whose lattice constant is λa. The coarse-grained spin configurations are assumed to correspond to those of the original spin system with larger scaling fields,

$$\tau' = \lambda^x \tau, \qquad h' = \lambda^y h, \tag{2.1.34}$$

where x and y are exponents. The probability distributions (= the canonical distributions) of the two sets of spin configurations should be nearly the same, so we require

$$h'\hat{\psi}' \sim h S_\lambda, \quad \tau'\hat{m}' \sim \tau E_\lambda, \tag{2.1.35}$$

where S_λ and E_λ are defined by (2.1.23) and (2.1.29), respectively. The $\delta\hat{\psi}'$ and $\delta\hat{m}'$ are the spin and energy variables in the coarse-grained lattice and their typical amplitudes should be independent of λ. From (2.1.25) and (2.1.30) we thus obtain

$$x = 1/\nu, \quad y = D. \tag{2.1.36}$$

This mapping relationship also means that the singular part of the free-energy density $f_{\rm sing}(h,\tau)$ divided by T satisfies

$$f_{\rm sing}(\tau,h)=\lambda^{-d}f_{\rm sing}(\lambda^{1/\nu}\tau,\lambda^{D}h). \qquad (2.1.37)$$

Since λ is arbitrary and may be set equal to $|\tau|^{-\nu}(\gg 1)$, we obtain

$$f_{\rm sing}(\tau,h)=|\tau|^{\nu d}f_{\rm sing}\left(\frac{\tau}{|\tau|},\frac{h}{|\tau|^{\nu D}}\right). \qquad (2.1.38)$$

Differentiations of $f_{\rm sing}(h,\tau)$ with respect to τ and h yield the exponent relations presented so far. For example,

$$C_H\sim(\partial^2 f_{\rm sing}/\partial\tau^2)_{h=0}\sim\tau^{d\nu-2}\quad(\tau>0,h=0), \qquad (2.1.39)$$

leading to (2.1.12). As $h/|\tau|^{\nu D}\to\infty$, $f_{\rm sing}$ should become independent of τ and

$$f_{\rm sing}\sim|h|^{d/D}, \qquad (2.1.40)$$

which leads to another expression for δ,

$$1/\delta=d/D-1\quad\text{or}\quad\delta=D/(d-D). \qquad (2.1.41)$$

This relation can also be derived from (2.1.14) and (2.1.26).

2.1.4 Two-scale-factor universality

The scaling arguments themselves do not give the concrete functional form of the singular free energy $f_{\rm sing}$. However, it is natural that the singular free energy $\xi^d f_{\rm sing}/T_{\rm c}$ divided by $T_{\rm c}$ in the volume ξ^d is a universal quantity. In fact, the renormalization group theory in Chapter 4 will confirm this expectation in the form,

$$f_{\rm sing}(\tau,h)=T_{\rm c}\xi^{-d}\mathcal{F}_{\rm sing}\left(x_0\frac{\tau}{|h|^{1/\nu D}}\right), \qquad (2.1.42)$$

where $\mathcal{F}_{\rm sing}(x)$ is a universal scaling function independent of material type and x_0 is a material-dependent constant. See Section 4.3 for its calculation. This form indicates that the singular part of the free-energy density is of order T/ξ^d with the coefficient being universal. This is a very natural and important consequence of the renormalization group theory. In particular, at $h=0$ and $\tau>0$, we have

$$f_{\rm sing}(\tau,0)=-T_{\rm c}A_{+\infty}\xi_{+0}^{-d}\tau^{2-\alpha}, \qquad (2.1.43)$$

where $A_{+\infty}$ is a universal number of order 0.1 and ξ_{+0} was introduced in (2.1.10). The specific heat C_H in (1.1.24) is then written as

$$C_H\cong -T_{\rm c}\frac{\partial^2 f_{\rm sing}}{\partial T^2}\cong(2-\alpha)(1-\alpha)A_{+\infty}\xi_{+0}^{-d}\tau^{-\alpha}. \qquad (2.1.44)$$

Therefore, we arrive at a universal number extensively discussed in the literature [5], [20]–[26],

$$R_\xi = \lim_{\tau\to+0} \xi(\alpha\tau^2 C_H)^{1/d} = \left[(2-\alpha)(1-\alpha)\alpha A_{+\infty}\right]^{1/d}, \tag{2.1.45}$$

at $h = 0$ and $\tau > 0$. It is known that $R_\xi \cong 0.25$ at $d = 3$ theoretically and experimentally. In fluids, we will define R_ξ using C_V for one-component fluids in (2.2.28) [23], C_{pX} for binary fluid mixtures in (2.3.64) [27]–[29], and C_p for ^{4}He near the superfluid transition in (2.4.4) [21, 30]. The above theory shows that ξ_{+0} can be obtained from specific-heat measurements only. Experimentalists can compare their data with the scaling form (2.1.42) if they have determined the scale factor for the magnetic field. Moreover, with data of ξ_{+0} from scattering experiments, they can check the validity of the theory when applied to a specific material. Discrepancy very close to criticality indicates that the material might not belong to the Ising universality class.

The two-scale-factor universality also implies that the typical magnitude of the temperature fluctuations are much smaller than the reduced temperature on spatial scales much longer than ξ. To show this we introduce the smoothed temperature fluctuation by $(\delta\hat{T})_\ell = \ell^{-d}\int_{\ell^d} d\boldsymbol{r}\delta\hat{T}(\boldsymbol{r})$ in a finite region with length ℓ longer than ξ. From (1.1.45) we then obtain

$$\langle(\delta\hat{T})_\ell^2\rangle = T^2\ell^{-d}C_M^{-1} = \alpha(T-T_c)^2 R_\xi^{-d}(\xi/\ell)^d, \tag{2.1.46}$$

at $h = 0$ above T_c. As long as $\ell \gg \xi$, we thus have $\langle(\delta\hat{T})_\ell^2\rangle \ll (T-T_c)^2$.

2.1.5 Parametric representation of equations of state

Ising systems

The linear parametric model [31] provides the equation of state and thermodynamic derivatives of Ising-like systems in remarkably compact forms. As illustrated in Fig. 2.5, it uses two parametric variables, r and θ, with $r \geq 0$ and $|\theta| \leq 1$; Parametric r represents the *distance* from the critical point (the origin) and θ the *angle* around it. The usual field variables h and τ are expressed as

$$h = a\theta(1-\theta^2)r^{\beta\delta}, \tag{2.1.47}$$

$$\tau = (1-b^2\theta^2)r, \tag{2.1.48}$$

with

$$b^2 = (\delta-3)/[(\delta-1)(1-2\beta)] \cong 1.4. \tag{2.1.49}$$

The average spin ψ is given by

$$\psi = c\theta r^\beta. \tag{2.1.50}$$

Here a and c are positive constants. The case $\theta = 0$ corresponds to $\tau > 0$ and $h = 0$, $\theta = \pm 1/b$ to $\tau = 0$, and $\theta = \pm 1$ to the coexistence curve ($h = 0$ and $\tau < 0$). We may

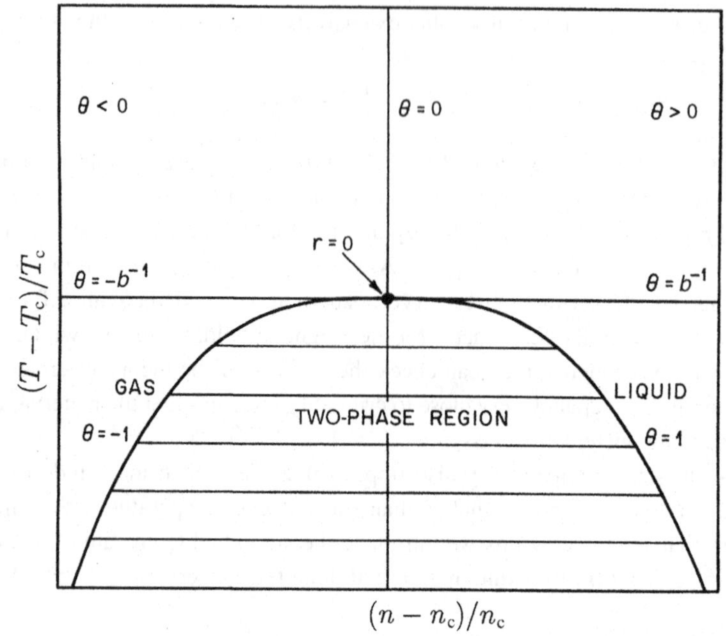

Fig. 2.5. Parametric representation of the equation of state near the critical point. The temperature–number-density plane is divided into several regions depending on the value of θ. The distance from the origin (the critical point) is denoted by r.

then calculate the free energy, entropy, and magnetic susceptibility. The scaling relations are satisfied in all these cases. Though the value of b is arbitrary within the model, the choice of b as in (2.1.49) yields simple expressions for the critical amplitude ratios, in close agreement with experimental values [31]. In particular, the specific heat C_M at constant magnetization does not depend on θ:

$$C_M = (ac)\left[\gamma(\gamma - 1)/2\alpha b^2\right] r^{-\alpha}. \tag{2.1.51}$$

The magnetic susceptibility $\chi = (\partial\psi/\partial h)_\tau$ also simplifies as

$$\chi = (c/a)\left[1 + (2\beta\delta - 3)\theta^2/(1 - 2\beta)\right]^{-1} r^{-\gamma}. \tag{2.1.52}$$

In these expressions the background parts are neglected.

The linear parametric model given by (2.1.47)–(2.1.50) may be verified to be a good approximation in the scheme of the renormalization group theory. Wallace [32] showed that it is exact up to order ϵ^2 for Ising systems, where $\epsilon = 4 - d$ is an expansion parameter to be explained in Chapter 4. The two-scale-factor universality (2.1.45) furthermore shows that the combination $(ac)\xi_{+0}^d$ of the coefficients is a universal number from (2.1.51), where ξ_{+0} is the microscopic length appearing in (2.1.10).

One-component fluids

Using this model the scaled equations of state of one-component fluids have been represented [33] by

$$\mu(n, T) - \mu(n_c, T) = (p_c/n_c)h, \tag{2.1.53}$$

$$(T - T_c)/T_c = \tau. \tag{2.1.54}$$

Here h and τ are the field variables given by (2.1.47) and (2.1.48) in the corresponding Ising system, $\mu(n, T)$ is the chemical potential per particle regarded as a function of n and T, and n_c and p_c are the critical values. The coefficient in front of h in (2.1.53) may be taken arbitrarily. In fluids, the number density n is assumed to correspond to the spin variable, so (2.1.50) yields

$$(n - n_c)/n_c = k\theta r^{\beta}, \tag{2.1.55}$$

where k is a positive constant. Then the constant-volume specific heat C_V in fluids corresponds to C_M in (2.1.51) for Ising systems. The critical isochore above T_c is given by $\theta = 0$. The coefficients a and k are dimensionless numbers of order unity. From (1.2.47) we require $h\psi = (\mu - \mu_c)(n - n_c)/T_c$ at $T = T_c$, which yields $c/k = p_c/T_c$.

The isothermal compressibility $K_T = (\partial n/\partial p)_T/n = (\partial n/\partial \mu)_T/n^2$ is proportional to the susceptibility χ in (2.1.52) as

$$n^2 K_T = (n_c^2/p_c)(k/a)\left[1 + (2\beta\delta - 3)\theta^2/(1 - 2\beta)\right]^{-1} r^{-\gamma}. \tag{2.1.56}$$

The Helmholtz free energy A per unit volume is obtained by integration of $dA = -SdT + \mu dp$. The entropy S per unit volume consists of a background term and a singular term ($\propto r^{1-\alpha}$), yielding

$$C_V = (n_c p_c/T_c^2)(T/n)(ak)\left[\gamma(\gamma - 1)/2\alpha b^2\right] r^{-\alpha} + C_B, \tag{2.1.57}$$

where C_B is the background specific heat. The first term is proportional to C_M in (2.1.51). Therefore, in the parametric model, C_V in one-component fluids corresponds to C_M in Ising systems.

2.2 Critical phenomena in one-component fluids

In Section 2.2 we have shown that the choice of $\Omega = Vp/T$ as the thermodynamic potential is most natural theoretically because it is the logarithm of the grand canonical partition function. Therefore, $\omega = \Omega/V = p/T$ corresponds to $-f/T$ of Ising systems, where f is the free-energy density in Ising systems. This correspondence is exact for the lattice gas model as can be seen in (1.2.25). We may assume that ω consists of a singular part and a regular part dependent on two relevant field variables, h and τ, as [5, 34, 35]

$$\omega = \omega_{\text{sing}}(h, \tau) + \omega_{\text{reg}}(h, \tau), \tag{2.2.1}$$

where $\omega_{\text{sing}}(h, \tau)$ coincides with $-f_{\text{sing}}(h, \tau)/T$ in (2.1.38) or (2.1.42). We neglect the corrections to the asymptotic scaling behavior [7].

2.2.1 Mapping relations

Now, how should we determine h and τ for fluids? Our postulate is that they are expressed as regular functions of $\delta T = T - T_c$ and $\delta\nu = \nu - \nu_c$ in one-component fluids, where β and ν should not be confused with the usual critical exponents. Near the critical point, we have linear relations [3], [34]–[37],

$$h = \alpha_1 \delta\nu + \alpha_2 \delta T/T_c, \tag{2.2.2}$$

$$\tau = \beta_1 \delta\nu + \beta_2 \delta T/T_c. \tag{2.2.3}$$

The coefficients in (2.2.2) and (2.2.3) are expressed as

$$\alpha_1 = \left(\frac{\partial h}{\partial \nu}\right)_T, \quad \beta_1 = \left(\frac{\partial \tau}{\partial \nu}\right)_T, \tag{2.2.4}$$

$$\alpha_2 = T_c\left(\frac{\partial h}{\partial T}\right)_\nu, \quad \beta_2 = T_c\left(\frac{\partial \tau}{\partial T}\right)_\nu. \tag{2.2.5}$$

Because $h = 0$ on the coexistence curve, we have $\alpha_2/\alpha_1 = -T_c(\partial\nu/\partial T)_{cx}$, where $(\partial\ /\partial\ )_{cx}$ is the derivative on the coexistence curve in the limit $T \to T_c$. We stress that $\delta T/T_c$ on the right-hand sides of (2.2.2) and (2.2.3) is the reduced temperature in fluids, whereas τ is the reduced temperature in the corresponding Ising system. With the postulates (2.2.2) and (2.2.3), we can now map the critical behavior of one-component fluids onto that of Ising spin systems. It follows that one-component fluids belong to the Ising universality class in static critical behavior. Note that two of the four coefficients in (2.2.2) and (2.2.3) may be taken arbitrarily by scale changes without loss of generality. In particular, in the original parametric model, the postulates (2.1.53) and (2.1.54) imply the special choice: $\alpha_1 = n_c T_c/p_c$, $\beta_1 = 0$, and $\beta_2 = 1$, while α_2 is determined from $\alpha_2/\alpha_1 = -T_c(\partial\nu/\partial T)_{cx}$. No mixing ($\beta_1 = 0$) is assumed. Similar mapping relationships with $\beta_1 = 0$ hold for the lattice gas model and the van der Waals fluid model. The latter will be discussed in section 3.4.

Next, we express the deviations $\delta\hat{e} = \hat{e} - e_c$ and $\delta\hat{n} = \hat{n} - n_c$ in fluids in terms of $\hat{\psi}$ and $\hat{m}$ in the corresponding Ising system by requiring [3, 37]

$$h\hat{\psi} + \tau\hat{m} = \delta\nu\delta\hat{n} + T_c^{-2}\delta T\delta\hat{e}, \tag{2.2.6}$$

where n_c and e_c are the critical values. This relation stems from (1.1.17) and (1.2.26) (or (1.2.47)), which describe the changes of the microscopic distribution P_Γ against variations of the field variables. The averages of $\hat{\psi}$ and $\hat{m}$ are taken to vanish at the critical point (by measuring them from the critical values). By substituting (2.2.2) and (2.2.3) into the above relation we obtain

$$\delta\hat{n} = \alpha_1\hat{\psi} + \beta_1\hat{m}, \tag{2.2.7}$$

$$T_c^{-1}\delta\hat{e} = \alpha_2\hat{\psi} + \beta_2\hat{m}. \tag{2.2.8}$$

To support (2.2.8), Fig. 2.3 demonstrates the linear relation $\hat{n}_k \propto \hat{e}_k$ with $\alpha_2/\alpha_1 \sim 1$ at long wavelengths for the 2D Lenard-Jones system. Similar numerical analysis was also made in Ref. [17]. A simple example of the relations (2.2.7) and (2.2.8) will be given in Section 3.4.

From (1.2.46) the entropy density variable may be written as

$$n\delta\hat{s} = \alpha_s\hat{\psi} + \beta_s\hat{m}, \tag{2.2.9}$$

with

$$\alpha_s = \alpha_2 - (H/T)\alpha_1, \quad \beta_s = \beta_2 - (H/T)\beta_1. \tag{2.2.10}$$

Using the pressure deviation $\delta p = p - p_c$ and eliminating δv, we may rewrite (2.2.2) and (2.2.3) as

$$T_c h = \alpha_1 n^{-1}\delta p + \alpha_s \delta T, \tag{2.2.11}$$

$$T_c \tau = \beta_1 n^{-1}\delta p + \beta_s \delta T. \tag{2.2.12}$$

Thus we have $\alpha_s = T_c(\partial h/\partial T)_p$ and $\beta_s = T_c(\partial \tau/\partial T)_p$. It leads to the relation

$$\frac{\alpha_s}{\alpha_1} = -n^{-1}\left(\frac{\partial p}{\partial T}\right)_{\text{cx}}. \tag{2.2.13}$$

Here we note that the energy variable $\hat{e}$ can be arbitrarily changed to $\hat{e} + \epsilon_0\hat{n}$ with respect to the shift (1.2.21). If we consider the following shifted energy variable,

$$\delta\hat{e} - (T_c\alpha_2/\alpha_1)\delta\hat{n} = T_c b_c \hat{m} \tag{2.2.14}$$

it becomes proportional to $\hat{m}$ from (2.2.7) and (2.2.8). The coefficient b_c is given by

$$b_c = \beta_2 - \beta_1\alpha_2/\alpha_1 = \beta_s - \beta_1\alpha_s/\alpha_1 = T_c\left(\frac{\partial \tau}{\partial T}\right)_h. \tag{2.2.15}$$

Therefore, by applying this energy shift, α_2 may be set equal to zero from the outset in (2.2.2) or (2.2.8).

Critical isochore

In many experimental situations, $h = 0$ nearly holds in the corresponding spin system. In such cases we can eliminate δv from (2.2.2) and (2.2.3) and can relate the two reduced temperatures as

$$\tau = b_c(T/T_c - 1), \tag{2.2.16}$$

where b_c is the constant defined by (2.2.15) and can be assumed to be positive. Because of this relation, b_c will frequently appear in the book. Note that it may be set equal to 1 (without loss of generality) by rescaling of m as $m \to b_c^{-1}m$. Let us consider the critical isochore case above T_c ($n = n_c$ and $T > T_c$). Here h remains extremely small. In fact, from (2.2.7) we have $dn = \alpha_1 d\psi + \beta_1 dm = 0$ with $\psi = \langle\hat{\psi}\rangle$ and $m = \langle\hat{m}\rangle$, so

$$h \sim \langle\hat{m} : \hat{m}\rangle\tau/\langle\hat{\psi} : \hat{\psi}\rangle \sim \tau^{\gamma-\alpha+1}. \tag{2.2.17}$$

The scaling variable $h/\tau^{\nu D}$ in (2.1.38) is of order $\tau^{1-\alpha-\beta} \ll 1$ and is very small.

Coexistence-curve diameter

The relation (2.2.7) indicates that the average number density on the coexistence curve ($h = 0$ and $\tau < 0$) behaves as

$$n - n_c = \pm\alpha_1 B_0|\tau|^{\beta} - \beta_1(1-\alpha)^{-1}A_0'|\tau|^{1-\alpha} + \cdots, \tag{2.2.18}$$

where the plus sign is for the liquid density $n = n_\ell$, the minus sign is for the gas density $n = n_g$, and $\tau = b_c(T/T_c - 1)$ as (2.2.16). Here the coefficient B_0 appears in (2.1.4), and A_0' is the critical amplitude in $C_H = \partial\langle\hat{m}\rangle/\partial\tau = A_0'|\tau|^{-\alpha}$ on the coexistence line. The cross coefficient β_1 is often referred to as the *mixing parameter* [35]–[39] and gives rise to the second term in (2.2.19). It causes singular asymptotic behavior of the coexistence-curve diameter,[2]

$$\frac{1}{2n_c}(n_\ell + n_g) - 1 = A_{1-\alpha}(1 - T/T_c)^{1-\alpha} + A_1(1 - T/T_c) + \cdots, \tag{2.2.19}$$

where $A_{1-\alpha} = -\beta_1 A_0' b^{1-\alpha}/(1-\alpha)n_c$. The coefficient $A_{1-\alpha}$ is relatively small in simple insulating fluids ($\sim$0.2 for Ne) and is considerably larger in liquid metals ($\sim$2 for Ru and Cs) [41]. However, in liquid metals and in ionic fluids [42] the effect of charges on critical phenomena (particularly on critical dynamics) is not yet well understood.

The Clausius–Clapeyron relation

If a gas phase and a liquid phase coexist, the entropy difference $\Delta s = s_g - s_\ell$ and the density difference $\Delta n = n_g - n_\ell$ are given by

$$\Delta s = n_c^{-1}\alpha_s\Delta\psi, \quad \Delta n = \alpha_1\Delta\psi, \tag{2.2.20}$$

from (2.2.7) and (2.2.9) with $\Delta\psi = 2B_0|\tau|^\beta \propto |T_c - T|^\beta$. The above relations are consistent with the Gibbs–Duhem relation, which relates Δs and the volume difference $\Delta n^{-1} = n_g^{-1} - n_\ell^{-1}$ via the Clausius–Clapeyron relation,

$$\Delta s = \left(\frac{\partial p}{\partial T}\right)_c \Delta n^{-1} \cong -\left(\frac{\partial p}{\partial T}\right)_c \frac{\Delta n}{n_c^2}. \tag{2.2.21}$$

2.2.2 Thermodynamic derivatives and the two-scale-factor universality

As far as the most singular critical divergence is concerned, we may set

$$\delta\hat{n} \cong \alpha_1\hat{\psi}, \quad \delta\hat{e} \cong \alpha_2\hat{\psi}, \quad \delta\hat{s} \cong n_c^{-1}\alpha_s\hat{\psi}, \tag{2.2.22}$$

from (2.2.7)–(2.2.9). The thermodynamic derivatives C_p, K_T, and α_p in (1.2.48) behave as

$$C_p \cong \alpha_s^2\chi, \quad n^2TK_T \cong \alpha_1^2\chi, \quad nT\alpha_p \cong -\alpha_1\alpha_s\chi, \tag{2.2.23}$$

where $\chi = \langle\hat{\psi} : \hat{\psi}\rangle$ is the magnetic susceptibility in the corresponding Ising spin system.

[2] If a fifth-order term is present in the Landau expansion of the free energy, it can also give rise to the first singular term in (2.2.19) [40]. See discussions in Section 3.4.

Next we examine the constant-volume specific heat C_V. From (2.2.7) and (2.2.8) we obtain

$$\langle \hat{e} : \hat{e} \rangle \langle \hat{n} : \hat{n} \rangle - \langle \hat{e} : \hat{n} \rangle^2 = T_c^2(\alpha_1\beta_2 - \alpha_2\beta_1)^2 \mathcal{D}, \tag{2.2.24}$$

where $\mathcal{D} = \chi C_M$ is the determinant (1.1.46). Therefore, from (1.2.49) and (2.2.15) we find a very simple result,

$$C_V \cong (\alpha_1\beta_2 - \alpha_2\beta_1)^2 \chi C_M / \langle \hat{n} : \hat{n} \rangle \cong b_c^2 C_M. \tag{2.2.25}$$

Using (1.2.54) and (2.2.13) we obtain

$$\rho c^2 C_V = C_p/K_T \cong T n^2 \left(\frac{\alpha_s}{\alpha_1} \right)^2 \cong T \left(\frac{\partial p}{\partial T} \right)^2_{\rm cx}, \tag{2.2.26}$$

from which the behavior of the sound velocity c is also known. On the critical isochore above T_c we have $C_M = C_H$ and

$$\begin{aligned} \tau^2 C_H &= (T/T_c - 1)^2 C_V \\ &= (p - p_c)^2 / T_c \rho c^2, \end{aligned} \tag{2.2.27}$$

where b_c is cancelled, $T - T_c \cong (\partial T/\partial p)_{\rm cx}(p - p_c)$, and use has been made of (2.2.16) and (2.2.26) in the second line. The right-hand sides of (2.2.27) consist of the quantities in fluids on the critical isochore above T_c, while the left-hand side contains those of the corresponding Ising system for $h = 0$ and $\tau > 0$. The two-scale-factor universality (2.1.45) in Ising systems is translated as [23]

$$R_\xi = \xi \left[\alpha (T/T_c - 1)^2 C_V \right]^{1/d}, \tag{2.2.28}$$

on the critical isochore above T_c in one-component fluids. We may use c^2 instead of C_V if use is made of the second line of (2.2.27).

2.2.3 Temperature and pressure fluctuations

In Section 1.1 we introduced the fluctuating variables $\delta\hat{T}$ and $\delta\hat{h}$ in (1.1.41) and (1.1.42), respectively, for Ising systems. In one-component fluids the temperature and pressure fluctuations in the long-wavelength limit are expressed as

$$\delta\hat{T} = \left(\frac{\partial T}{\partial h} \right)_\tau \delta\hat{h} + \left(\frac{\partial T}{\partial \tau} \right)_h \delta\hat{\tau}, \tag{2.2.29}$$

$$\delta\hat{p} = \left(\frac{\partial p}{\partial h} \right)_\tau \delta\hat{h} + \left(\frac{\partial p}{\partial \tau} \right)_h \delta\hat{\tau}, \tag{2.2.30}$$

where the coefficients constitute the inverse of the matrix composed of the coefficients in (2.2.11) and (2.2.12), and $\delta\hat{\tau} = \delta\hat{T}/T_c$ represents the reduced temperature fluctuation in the corresponding Ising spin system. Near the critical point the second terms ($\propto \delta\hat{\tau}$) in these relations dominate the first terms ($\propto \delta\hat{h}$). In fact, from (1.1.47) the variances of $\delta\hat{T}$ and $\delta\hat{p}$ can be expressed as

$$\frac{T^2}{C_V} = \left(\frac{\partial T}{\partial h}\right)_\tau^2 V_{hh} + 2\left(\frac{\partial T}{\partial h}\right)_\tau \left(\frac{\partial T}{\partial \tau}\right)_h V_{h\tau} + \left(\frac{\partial T}{\partial \tau}\right)_h^2 V_{\tau\tau}, \tag{2.2.31}$$

$$T\rho c^2 = \left(\frac{\partial p}{\partial h}\right)_\tau^2 V_{hh} + 2\left(\frac{\partial p}{\partial h}\right)_\tau \left(\frac{\partial p}{\partial \tau}\right)_h V_{h\tau} + \left(\frac{\partial p}{\partial \tau}\right)_h^2 V_{\tau\tau}. \tag{2.2.32}$$

where use has been made of (1.2.64) and (1.2.68). In the above relations the last terms are dominant and

$$\frac{T^2}{C_V} \cong \left(\frac{\partial T}{\partial \tau}\right)_h^2 \frac{1}{C_M}, \quad T\rho c^2 \cong \left(\frac{\partial p}{\partial \tau}\right)_h^2 \frac{1}{C_M}, \tag{2.2.33}$$

which are consistent with (2.2.25) and (2.2.26). The cross correlation $\langle \hat{T} : \hat{p}\rangle$ in (1.2.69) may also be calculated in the same manner.

Adiabatic T–p relation on the coexistence curve

As an application, we give the expansion of the adiabatic coefficient,

$$\left(\frac{\partial T}{\partial p}\right)_s = \frac{\langle \hat{T} : \hat{p}\rangle}{\langle \hat{p} : \hat{p}\rangle} = \left(\frac{\partial T}{\partial p}\right)_{\rm cx}\left[1 + A\frac{V_{h\tau}}{V_{\tau\tau}} + O\left(\frac{C_V}{C_p}\right)\right], \tag{2.2.34}$$

where

$$A = \left(\frac{\partial \tau}{\partial h}\right)_p - \left(\frac{\partial \tau}{\partial h}\right)_T = \left(\frac{\partial \tau}{\partial T}\right)_h \left(\frac{\partial T}{\partial h}\right)_p = \frac{b_{\rm c}}{\alpha_s} \tag{2.2.35}$$

from (2.2.11) and (2.2.15). We are interested in the leading correction of order $V_{h\tau}/V_{\tau\tau} = -\langle \hat{\psi} : \hat{m}\rangle / \langle \hat{\psi} : \hat{\psi}\rangle$, although it vanishes on the critical isochore above $T_{\rm c}$. On the coexistence curve ($T < T_{\rm c}$), we find a convenient form [37],

$$\left(\frac{\partial T}{\partial p}\right)_s = \left(\frac{\partial T}{\partial p}\right)_{\rm cx}\left[1 \pm a_{\rm c}\left(\frac{C_V}{C_p}\right)^{1/2} + \cdots\right], \tag{2.2.36}$$

where the plus (minus) sign is for the gas (liquid) phase, and the coefficient $a_{\rm c}$ is related to the universal number $R_{\rm v}$ in (1.1.48) as

$$a_{\rm c}^2 = \langle \hat{\psi} : \hat{m}\rangle^2 / C_M \chi = R_{\rm v}/(1 - R_{\rm v}). \tag{2.2.37}$$

In 3D, we have $a_{\rm c} \cong 1$ near the critical point.[3] We note that $(\partial T/\partial p)_n$ also satisfies (2.2.36) on the coexistence curve because its difference from $(\partial T/\partial p)_s$ is of order C_V/C_p from (1.2.53).

The above derivation of (2.2.36) might look complicated. A simpler one is to rewrite the identity $ds = (\partial s/\partial T)_p[dT - (\partial T/\partial p)_s dp]$ as

$$\left(\frac{\partial T}{\partial p}\right)_s \left(\frac{\partial p}{\partial T}\right)_{\rm cx} - 1 = -\frac{nT}{C_p}\left(\frac{\partial s}{\partial T}\right)_{\rm cx}. \tag{2.2.38}$$

[3] Let $\chi = \Gamma_0'|\tau|^{-\gamma}$, $C_H = A_0'|\tau|^{-\alpha}$, and $\langle\psi\rangle = B_0|\tau|^\beta$ below $T_{\rm c}$ on the coexistence curve; then, $R_{\rm v} = (\beta B_0)^2/A_0'\Gamma_0'$ from the second line of (1.1.48). This combination is about 0.5 [24, 26].

Similarly, exchanging $\{T, s\} \leftrightarrow \{p, n\}$, we also notice

$$\left(\frac{\partial p}{\partial T}\right)_n \left(\frac{\partial T}{\partial p}\right)_{\text{cx}} - 1 = -\frac{1}{nK_T}\left(\frac{\partial n}{\partial p}\right)_{\text{cx}}. \tag{2.2.39}$$

These relations hold both in liquid and gas phases on the coexistence curve at any temperature. Near the critical point we have $(\partial s/\partial T)_{\text{cx}} = -\beta(s - s_c)/(T_c - T)$ because $s - s_c \propto \pm(T_c - T)^\beta$ with the plus (minus) sign for the gas (liquid) phase. The origin of $\pm$ in (2.2.36) is then obvious. Comparison of (2.3.36) and (2.3.38) yields

$$a_c = \frac{\beta}{2}|\Delta s| n/\left[\sqrt{C_V C_p}(1 - T/T_c)\right] = \frac{\beta}{2}|\Delta n|/n\left[\sqrt{K_s K_p}(1 - p/p_c)\right], \tag{2.2.40}$$

where Δs and Δn are the entropy and number density differences between the two coexisting phases. The above relations hold in the limit $T \to T_c$, leading to (2.2.37) with the aid of the mapping relations (2.2.20), (2.2.23), and (2.2.25). Physically, (2.2.36) implies that a pressure change in two-phase coexistence gives rise to a temperature difference between the two phases. This effect will be important in studying the specific heat in two-phase coexistence in Appendix 4F, thermal equilibration in two-phase coexistence in Section 6.3, and nucleation and sound propagation in two phase states in Section 9.4.

2.2.4 *Gravity effects in one-component fluids*

In one-component fluids near the gas–liquid critical point, density stratification in gravity becomes quite large in equilibrium due to the diverging isothermal compressibility K_T [33, 43]. The average pressure decreases with increasing height z as

$$\frac{dp}{dz} = \frac{1}{nK_T}\frac{dn}{dz} = -\rho g, \tag{2.2.41}$$

where the local equilibrium relation between p and n at homogeneous T is assumed, and $\rho = m_0 n$ with m_0 being the molecular mass. As an example, see Fig. 2.6 for optically measured density profiles in N_2O [44]. This severe stratification prevents precise measurements of the critical behavior in one-component fluids. For example, in quiescent fluids (without stirring), C_V exhibits only a broad rounded peak at $T \cong T_c$ even if the average density in the container is at the critical value.

Hohenberg and Barmatz [33] studied the equilibrium gravity effects using the parametric model in Subsection 2.1.5 and assuming the local equilibrium relation,

$$\frac{d\mu}{dz} = -m_0 g \quad \text{or} \quad \mu(n, T) - \mu(n_c, T) = -m_0 g(z - z_0), \tag{2.2.42}$$

where $\mu(n, T)$ is the chemical potential per particle given in (2.1.53), and z_0 is a constant height at which $n = n_c$. The z axis is taken in the upward direction. They calculated the space average of C_V as a function of the experimental cell size L and examined two-phase coexistence for $T < T_c$. In particular, we consider an equilibrium fluid above T_c. In gravity the number-density deviation in the cell is of order $nK_T\rho g L$. If it is much smaller than

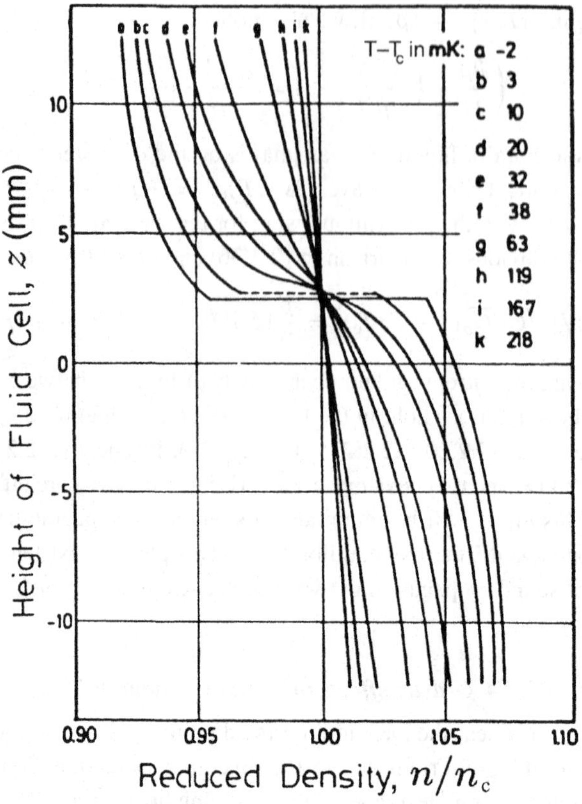

Fig. 2.6. Density profiles of N_2O near the critical point measured with a refractometer [44]. Here $\text{mK} = 10^{-3}$ K.

$n_c\tau^\beta$ with $\tau = T/T_c - 1$, the parameter θ in (2.1.55) is nearly constant as is τ. Thus, the fluid critical behavior is nearly homogeneous in the cell when

$$T/T_c - 1 > (\rho_c g L/p_c)^{1/(\beta+\gamma)}. \tag{2.2.43}$$

The right-hand side is of order 10^{-4} at $L = 1$ cm for Xe on earth. If $T/T_c - 1$ is smaller than the right-hand side, gravity-induced inhomogeneity becomes important.

Theoretically, however, it is necessary to clarify the condition of the validity of the local equilibrium assumption (2.2.42) [43]. To this end, let us calculate the height-dependent correlation length $\xi(z)$ for the case $T = T_c$ or $\tau = 0$ using (2.2.42). The parameter θ in (2.1.48) is equal to $1/b$ for $z < z_0$ and to $-1/b$ for $z > z_0$. Then, from (2.1.47) and (2.2.42), the distance from the criticality r becomes a function of z as

$$(p_c a/b)(1 - b^{-2})r^{\beta\delta} = \rho_c g|z - z_0|, \tag{2.2.44}$$

where $\rho_c = m_0 n_c$ is the critical value of the mass density. Since $a \sim 1$ and $b^2 \cong 1.4$, we have $r^{\beta\delta} \sim \rho_c g|z - z_0|/p_c$. As a result, the local correlation length behaves as

$$\xi \sim \xi_{+0} r^{-\nu} \sim \ell_g \left(\frac{|z - z_0|}{\ell_g} \right)^{-\nu/\beta\delta}, \tag{2.2.45}$$

where ξ_{+0} is the microscopic length in (2.1.10) and ℓ_g is a characteristic length in gravity defined by

$$\ell_g = \xi_{+0} (\xi_{+0} \rho_c g / p_c)^{-\nu/(\beta\delta+\nu)}, \tag{2.2.46}$$

with $\nu/(\beta\delta + \nu) \cong 0.28$. For Xe we have $\ell_g = 4 \times 10^{-5}$ cm on earth. The local equilibrium assumption is valid if the number density change on the length scale ξ is negligibly small compared with $n - n_c$. This condition is expressed as $\xi |dn/dz| / |n - n_c| \sim (\xi/\ell_g)^{(\beta\delta/\nu+1)} \ll 1$. Thus (2.2.44) is valid only in the region,

$$\xi \ll \ell_g \quad \text{or} \quad |z - z_0| \gg \ell_g. \tag{2.2.47}$$

Therefore, ℓ_g is the maximally attainable correlation length in gravity. In the transition region $|z - z_0| \lesssim \ell_g$, nonlocal effects are crucial, where the density profile need to be calculated in the Ginzburg–Landau scheme. We may also introduce a characteristic reduced temperature τ_g by $\ell_g = \xi_{+0} \tau_g^{-\nu}$ [43]. It is written as

$$\tau_g = (\xi_{+0} \rho_c g / p_c)^{1/(\beta\delta+\nu)}, \tag{2.2.48}$$

which is 1.8×10^{-6} for Xe on earth.

It is easy to extend the above arguments for the case $T \neq T_c$. The local equilibrium holds in the spatial region where $\xi \gg \ell_g$. If $|T - T_c| \gg T_c \tau_g$, the local equilibrium approximation is valid in the whole space region.

2.3 Critical phenomena in binary fluid mixtures

In binary fluid mixtures, there are liquid–liquid, gas–liquid, and gas–gas phase equilibria. There are no absolute differences between these three types of phase transitions [4, 45]. Figure 2.7 shows a simple geometrical representation of a gas–liquid transition in the space of p, T, and the fugacity $f_2 = \exp(\mu_2/T)$ of the second component. The geometrical representation of coexistence surfaces and critical lines is, in general, very complicated. If visualized in the space of three field variables, coexistence surfaces terminate at critical lines and, on an arbitrary plane cutting a critical line, critical phenomena are believed to be isomorphic to those of Ising systems [46, 47]. In binary fluid mixtures, however, experimentally measurable quantities are mostly those at fixed concentration, and complicated crossover effects take place. Seemingly exceptional cases have often been observed when the critical line and the coexistence surface bear special relationship with the coordinates of the field variables. Among various types of binary mixtures, we focus our attention on nearly *azeotropic* binary mixtures along the gas–liquid critical line and nearly *incompressible* binary mixtures along the consolute critical line. These constitute two important classes of extensively studied binary fluid mixtures. As representative examples, we show

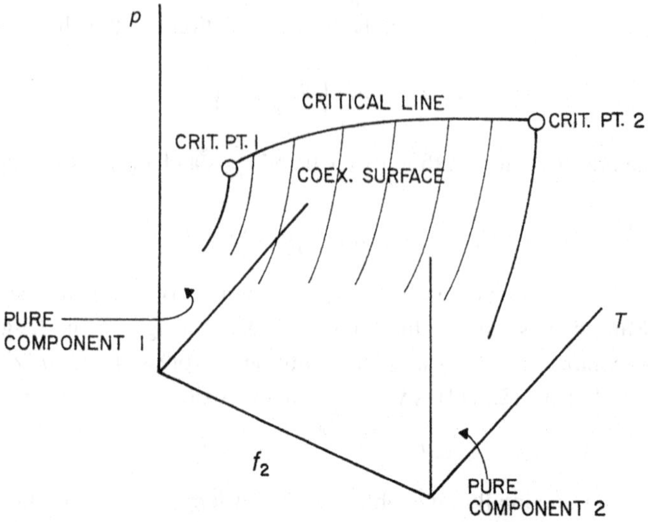

Fig. 2.7. The gas–liquid coexistence surface and critical line of a binary fluid mixture in the space of pressure p, temperature T, and fugacity f_2 [5].

isobaric T–X phase diagrams of ^{3}He–^{4}He [48] and 3-methylpentane + nitroethane [49] in Figs 2.8 and 2.9, respectively.

We will take a novel approach to these complicated effects by introducing a density variable $\hat{q}$ conjugate to the coordinate ζ along the critical line. We shall see that the asymptotic critical behavior of various thermodynamic quantities is determined by the fluctuations of $\hat{q}$.

2.3.1 Mapping relations

In addition to h and τ, another field variable ζ is needed. It is convenient to take ζ to be the coordinate along the critical line in the neighborhood of a critical point represented by $\zeta = 0$. As a generalization of (2.2.1), the thermodynamic potential $\omega = p/T$ is written as

$$\omega = \omega_{\text{sing}}(h, \tau) + \omega_{\text{reg}}(h, \tau) + \frac{1}{2} Q_0 \zeta^2, \tag{2.3.1}$$

where $\omega_{\text{sing}}(h, \tau)$ is the same as in the one-component case, $\omega_{\text{reg}}(h, \tau)$ is a regular function of h and τ, and Q_0 is a positive constant. We neglect the corrections to the asymptotic critical behavior. Because there are three density variables conjugate to three field variables, we may suppose the presence of a density variable $\hat{q}$ conjugate to ζ.[4] Then the equilibrium

[4] The Ginzburg–Landau–Wilson hamiltonian for the three variables $\hat{\psi}$, $\hat{m}$, and $\hat{q}$ will be set up in (4.2.6).

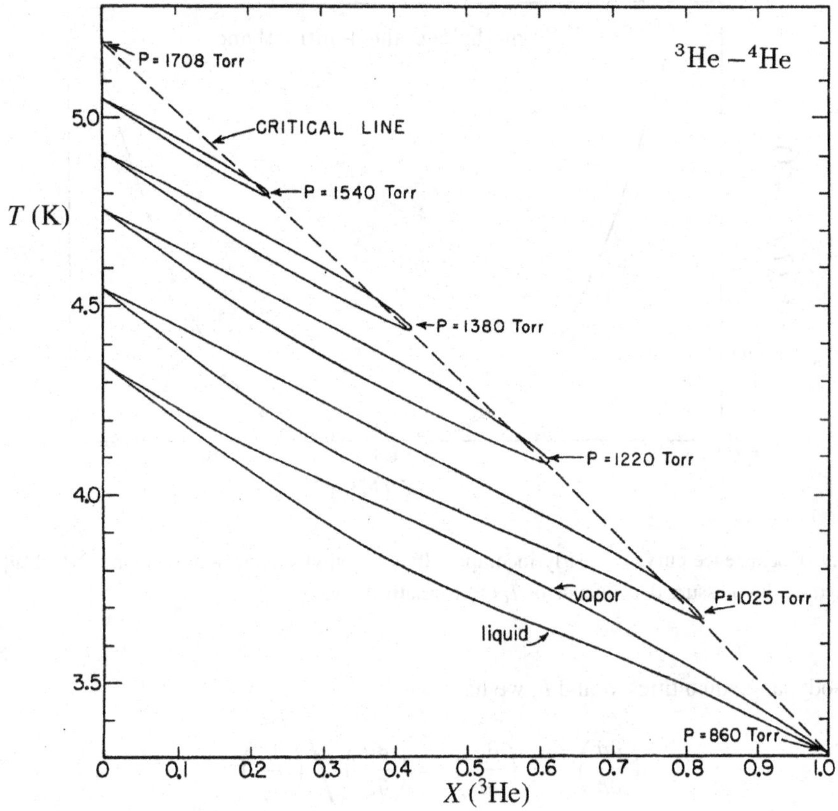

Fig. 2.8. Phase diagram for constant-pressure projections in nearly azeotropic ^{3}He–^{4}He [48], characterized by narrow lens-like coexistence regions. We can see that the gas–liquid critical line intersects the T–X loops at temperature minima given by $(\partial T/\partial X)_{\mathrm{cx},p} = 0$. A similar phase diagram can be drawn for constant-temperature projections, where the gas–liquid critical line intersects the p–X loops at the pressure maxima.

average and variance of $\hat{q}$ are

$$\langle \hat{q} \rangle = \left(\frac{\partial \omega}{\partial \zeta} \right)_{h\tau} = Q_0 \zeta, \quad \langle \hat{q} : \hat{q} \rangle = \left(\frac{\partial^2 \omega}{\partial \zeta^2} \right)_{h\tau} = Q_0. \tag{2.3.2}$$

We may define $\hat{q}$ such that it is statistically independent of $\hat{\psi}$ and $\hat{m}$ as

$$\langle \hat{q}; \hat{\psi} \rangle = \left(\frac{\partial^2 \omega}{\partial \zeta \partial h} \right)_{h\tau} = 0, \quad \langle \hat{q} : \hat{m} \rangle = \left(\frac{\partial^2 \omega}{\partial \zeta \partial \tau} \right)_{h\tau} = 0, \tag{2.3.3}$$

in the vicinity of the reference critical point. The average of $\hat{q}$ (as well as that of $\hat{m}$) has no discontinuity in two-phase coexistence. Derivatives with fixed h and τ are nearly equal to those along the critical line ($h = \tau = 0$). They will be written as $(\partial \cdots / \partial \cdots)_{\mathrm{c}}$. For any

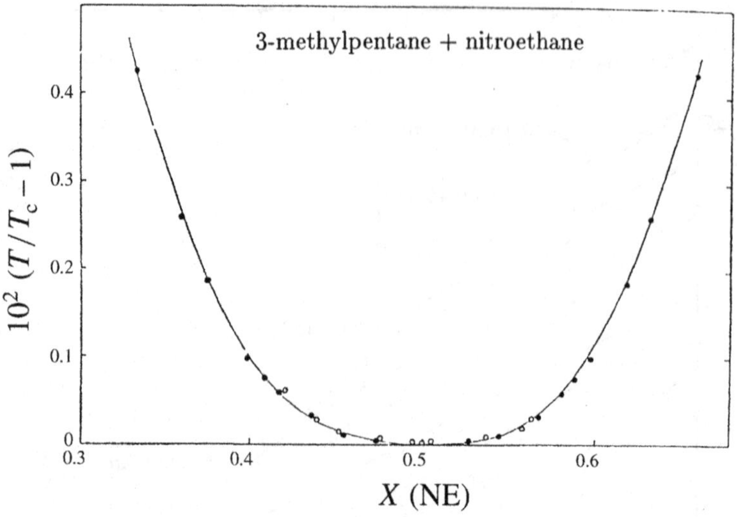

Fig. 2.9. Coexistence curve of nearly incompressible 3-methylpentane + nitroethane (NE) [49]. In this mixture the pressure dependence of $T_c(p)$ is relatively weak.

thermodynamic quantities, a and b, we have

$$\left(\frac{\partial a}{\partial b}\right)_{h\tau} \cong \left(\frac{\partial a}{\partial b}\right)_c = \left(\frac{\partial a}{\partial \zeta}\right)_c \bigg/ \left(\frac{\partial b}{\partial \zeta}\right)_c. \tag{2.3.4}$$

As in the one-component case, for the field deviations $\delta\nu_1 = \nu_1 - \nu_{1c}$, $\delta\nu_2 = \nu_2 - \nu_{2c}$, and $\delta T = T - T_c$ we assume the following mapping relations:

$$h = \alpha_1\delta\nu_1 + \alpha_2\delta\nu_2 + \alpha_3\delta T/T_c, \tag{2.3.5}$$

$$\tau = \beta_1\delta\nu_1 + \beta_2\delta\nu_2 + \beta_3\delta T/T_c, \tag{2.3.6}$$

$$\zeta = \gamma_1\delta\nu_1 + \gamma_2\delta\nu_2 + \gamma_3\delta T/T_c. \tag{2.3.7}$$

For the density deviations $\delta\hat{n}_1 = \hat{n}_1 - n_{1c}$, $\delta\hat{n}_2 = \hat{n}_2 - n_{2c}$, $\delta\hat{e} = \hat{e} - e_c$, we require

$$h\hat{\psi} + \tau\hat{m} + \zeta\hat{q} = \delta\nu_1\delta\hat{n}_1 + \delta\nu_2\delta\hat{n}_2 + T_c^{-2}\delta T\delta\hat{e} \tag{2.3.8}$$

to obtain

$$\delta\hat{n}_1 = \alpha_1\hat{\psi} + \beta_1\hat{m} + \gamma_1\hat{q}, \tag{2.3.9}$$

$$\delta\hat{n}_2 = \alpha_2\hat{\psi} + \beta_2\hat{m} + \gamma_2\hat{q}, \tag{2.3.10}$$

$$T_c^{-1}\delta\hat{e} = \alpha_3\hat{\psi} + \beta_3\hat{m} + \gamma_3\hat{q}. \tag{2.3.11}$$

From (1.3.16)–(1.3.18) the deviations of the entropy density, number density, and concentration may be defined by

$$n\delta\hat{s} = \alpha_s\hat{\psi} + \beta_s\hat{m} + \gamma_s\hat{q}, \tag{2.3.12}$$

$$\delta\hat{n} = \alpha_n\hat{\psi} + \beta_n\hat{m} + \gamma_n\hat{q}, \tag{2.3.13}$$

$$n\delta\hat{X} = \alpha_X\hat{\psi} + \beta_X\hat{m} + \gamma_X\hat{q}. \tag{2.3.14}$$

The critical values n_c, e_c, ... are those at the reference critical point $h = \tau = \zeta = 0$. The coefficients α_K $(K = s, X, n)$ are linear combinations of α_i $(i = 1, 2, 3)$ as

$$\begin{aligned} \alpha_s &= \alpha_3 - s(\alpha_1 + \alpha_2) - \nu_1\alpha_1 - \nu_2\alpha_2, \\ \alpha_n &= \alpha_1 + \alpha_2, \\ \alpha_X &= (1 - X)\alpha_1 - X\alpha_2. \end{aligned} \tag{2.3.15}$$

We may express h, τ, and ζ in terms of the deviations δp, δT, and $\delta\Delta$ as a generalization of the one-component fluid version (2.2.11) and (2.2.12). We require

$$h\hat{\psi} + \tau\hat{m} + \zeta\hat{q} = T_c^{-1}\left[(\delta T)n\delta\hat{s} + n^{-1}\delta p\delta\hat{n} + (\delta\Delta)n\delta\hat{X}\right], \tag{2.3.16}$$

which arises from (1.3.20) and is equivalent to (2.3.8). The deviations $\delta p = p - p_c$ and $\delta\Delta = \Delta - \Delta_c$ are measured from the critical values. Then we obtain

$$T_c h = \alpha_n n^{-1}\delta p + \alpha_s\delta T + \alpha_X\delta\Delta, \tag{2.3.17}$$

$$T_c\tau = \beta_n n^{-1}\delta p + \beta_s\delta T + \beta_X\delta\Delta, \tag{2.3.18}$$

$$T_c\zeta = \gamma_n n^{-1}\delta p + \gamma_s\delta T + \gamma_X\delta\Delta. \tag{2.3.19}$$

As in (2.2.13) for the one-component case, the ratios among the α_K are expressed in terms of derivatives on the coexistence surface,

$$\frac{\alpha_s}{\alpha_n} \cong -n_c^{-1}\left(\frac{\partial p}{\partial T}\right)_{\Delta,\mathrm{cx}}, \quad \frac{\alpha_X}{\alpha_n} \cong -n_c^{-1}\left(\frac{\partial p}{\partial \Delta}\right)_{T,\mathrm{cx}}. \tag{2.3.20}$$

Along the critical line, the average entropy, density, and concentration change as

$$n_c\left(\frac{\partial s}{\partial \zeta}\right)_c = Q_0\gamma_s, \quad \left(\frac{\partial n}{\partial \zeta}\right)_c = Q_0\gamma_n, \quad n_c\left(\frac{\partial X}{\partial \zeta}\right)_c = Q_0\gamma_X, \tag{2.3.21}$$

which can be known if the averages of (2.3.12)–(2.3.14) are taken along $h = \tau = 0$.

The previous literature has used the determinant of the variances, $\det I$ in (1.3.11) or $\det A$ in (1.3.39) (which are related by (1.3.37)), to examine the asymptotic thermodynamic properties [47, 5]. With the linear mapping relations it is obvious that

$$\det A = (D_0^2 Q_0)C_M\chi \propto C_M\chi. \tag{2.3.22}$$

The coefficient D_0 is expressed in terms of the determinants of the mapping matrices as

$$D_0 = \frac{T^3}{n}\frac{\partial(h, \tau, \zeta)}{\partial(T, \Delta, p)} = \frac{T}{n^2}\frac{\partial(h, \tau, \zeta)}{\partial(T, \nu_1, \nu_2)}, \tag{2.3.23}$$

where use has been made of $\partial(\Delta, p)/\partial(\nu_1, \nu_2) = nT^2$ at fixed T. The determinants, (1.1.46) for spin systems and (2.2.24) for one-component fluids, and (2.3.22) for binary fluids, all behave as const.$C_M \chi$. It is important that D_0 and Q_0 are insensitive to the relationship of the coexistence surface and the critical line with respect to the axes of the field variables. For example, they will be treated as nonvanishing constants at critical azeotropy.

Leung and Griffiths' theory

Leung and Griffiths [47] constructed a phenomenological model for ^{3}He–^{4}He mixtures, where the potential $\omega = p/T$ is expressed in terms of the three field variables h, τ, and ζ. Using a number of fitting parameters, it describes the global thermodynamics along the critical line which connects the two critical points of pure ^{3}He and ^{4}He as in Fig. 2.7. In particular, they set

$$\zeta = 1/[1 + A_0 \exp(\Delta/T)], \tag{2.3.24}$$

where A_0 is a constant. Then $0 \le \zeta \le 1$; $\zeta = 1$ for pure ^{3}He and $\zeta = 0$ for pure ^{4}He, because $\nu_1 \cong \ln X$ (or $\nu_2 \cong \ln(1 - X)$) in the dilute limit $X \to 0$ (or $X \to 1$). However, such a global parametrization is feasible only in nearly azeotropic binary fluids, as explained below [45]. Our local parametrization is much simpler but is valid only in a narrow region around a particular critical point in the three-dimensional space of the field variables.

2.3.2 Concentration fluctuations

With the above relationship it is straightforward to examine the critical behavior of various thermodynamic derivatives. For example, the variances among $\delta\hat{s}$, $\delta\hat{n}$, and $\delta\hat{X}$ diverge strongly as χ on approaching the critical line. A consolute critical point is characterized by $|\alpha_X/\alpha_n| \gtrsim 1$, and in its vicinity the concentration fluctuations are strongly enhanced as

$$\delta\hat{X} \cong \alpha_X \psi, \tag{2.3.25}$$

with the concentration susceptibility (1.3.22) of the form,

$$\langle \hat{X} : \hat{X} \rangle = \frac{T}{n} \left(\frac{\partial X}{\partial \Delta} \right)_{pT} \cong \alpha_X^2 \chi. \tag{2.3.26}$$

We have neglected the background part. However, in azeotropic cases where α_X is small, this approximation is allowable only very close to the critical line, as will be shown in (2.3.50).

If α_X is not small or if the mixture is non-azeotropic, (2.3.26) holds in a sizable temperature region and C_{pX}, K_{TX}, and α_{pX} in (1.3.24)–(1.3.26) behave as

$$C_{pX} \cong \bar{\beta}_s^2 C_M + C_B, \tag{2.3.27}$$

$$T K_{TX} \cong B_c^2 (\bar{\beta}_s^2 C_M) + A_c^2 C_B, \tag{2.3.28}$$

$$T \alpha_{pX} \cong B_c (\bar{\beta}_s^2 C_M) + A_c C_B, \tag{2.3.29}$$

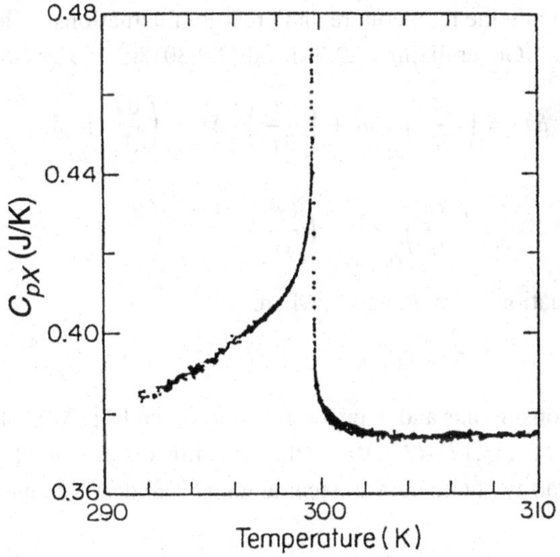

Fig. 2.10. C_{pX} in a nearly incompressible binary mixture of 3-methylpentane + nitroethane at the critical concentration [29].

where the first terms are weakly divergent, the second terms are nonsingular, and

$$\bar{\beta}_s = T_c\left(\frac{\partial \tau}{\partial T}\right)_{hp}, \quad C_B = T_c^2\left(\frac{\partial \zeta}{\partial T}\right)_{hp}^2 Q_0. \tag{2.3.30}$$

The two coefficients A_c and B_c will appear in many relations below and are defined by

$$A_c = \left(\frac{\partial T}{\partial p}\right)_{h\zeta}, \quad B_c = \left(\frac{\partial T}{\partial p}\right)_{h\tau} = \left(\frac{\partial T}{\partial p}\right)_c. \tag{2.3.31}$$

See Appendix 2A for the derivation of (2.3.27)–(2.3.29). In Fig. 2.10 we show an example of C_{pX} in a critical binary mixture of 3-methylpentane + nitroethane [29]. See (2.3.59) and Fig. 2.13 for data of other thermodynamic derivatives of this mixture.

Moreover, the thermodynamic identities in (1.3.39) yield

$$\begin{aligned} C_{pX}/\rho c^2 &= C_{VX}K_{TX} = \text{const.}\chi C_M \bigg/ \left(\frac{\partial X}{\partial \Delta}\right)_{pT} \\ &\cong \text{const.}C_M. \end{aligned} \tag{2.3.32}$$

The first line are the identities and the second line holds under (2.3.26).

2.3.3 Temperature and pressure fluctuations

From (1.3.38) and (1.3.39) with the aid of (2.3.22), C_{VX} and ρc^2 are known to tend to finite constant values on the critical line. However, their behavior can be more conveniently examined by introducing the temperature and pressure fluctuations in the long-wavelength limit as in Section 1.3. Generalizing (2.2.29) and (2.2.30) we express them as

$$\delta\hat{T} = \left(\frac{\partial T}{\partial h}\right)_{\tau\zeta}\delta\hat{h} + \left(\frac{\partial T}{\partial \tau}\right)_{h\zeta}\delta\hat{\tau} + \left(\frac{\partial T}{\partial \zeta}\right)_{h\tau}\delta\hat{\zeta}, \tag{2.3.33}$$

$$\delta\hat{p} = \left(\frac{\partial p}{\partial h}\right)_{\tau\zeta}\delta\hat{h} + \left(\frac{\partial p}{\partial \tau}\right)_{h\zeta}\delta\hat{\tau} + \left(\frac{\partial p}{\partial \zeta}\right)_{h\tau}\delta\hat{\zeta}, \tag{2.3.34}$$

in terms of the fluctuations of τ, h, and ζ, where

$$\delta\hat{\zeta} = Q_0^{-1}\hat{q} - \zeta, \quad \langle\hat{\zeta} : \hat{\zeta}\rangle = 1/Q_0 \tag{2.3.35}$$

The variable $\delta\hat{\zeta}$ is nonsingular and is uncorrelated to $\delta\hat{h}$ and $\delta\hat{\tau}$. We can see that these are the inverse relations of (2.3.17)–(2.3.19) (if the circumflex is put on all the field variables). Using (1.1.47) we readily obtain the variances among $\delta\hat{T}$ and $\delta\hat{p}$ given in (1.3.44), (1.3.48), and (1.3.49):

$$\langle\hat{T} : \hat{T}\rangle = \frac{T^2}{C_{VX}} \cong \left(\frac{\partial T}{\partial \zeta}\right)_{\rm c}^2 \frac{1}{Q_0} + \left(\frac{\partial T}{\partial \tau}\right)_{h\zeta}^2 \frac{1}{C_M}, \tag{2.3.36}$$

$$\langle\hat{p} : \hat{p}\rangle = T\rho c^2 \cong \left(\frac{\partial p}{\partial \zeta}\right)_{\rm c}^2 \frac{1}{Q_0} + \left(\frac{\partial p}{\partial \tau}\right)_{h\zeta}^2 \frac{1}{C_M}, \tag{2.3.37}$$

$$\begin{aligned}\langle\hat{T} : \hat{p}\rangle &= nT\left(\frac{\partial T}{\partial n}\right)_{sX} = n^{-1}T\left(\frac{\partial p}{\partial s}\right)_{nX} \\ &\cong \left(\frac{\partial T}{\partial \zeta}\right)_{\rm c}\left(\frac{\partial p}{\partial \zeta}\right)_{\rm c}\frac{1}{Q_0} + \left(\frac{\partial T}{\partial \tau}\right)_{h\zeta}\left(\frac{\partial p}{\partial \tau}\right)_{h\zeta}\frac{1}{C_M}.\end{aligned} \tag{2.3.38}$$

The first terms in these relations are the fluctuation contributions along the critical line unique to fluid mixtures and remain nonvanishing on the critical line, while the second terms are weakly singular and common to one- and two-component fluids. The leading terms we have not written are of order $V_{hh} \sim 1/\chi$ in one-phase states at $h = 0$ and are of order $V_{h\tau} \sim (\chi C_M)^{-1/2}$ on the coexistence curve from (1.1.47). The critical-point values of C_{VX} and ρc^2 are expressed as

$$(C_{VX})_{\rm c} = T_{\rm c}^2\left(\frac{\partial \zeta}{\partial T}\right)_{\rm c}^2 Q_0 = (1 - A_{\rm c}/B_{\rm c})^2 C_{\rm B}, \tag{2.3.39}$$

$$\rho_{\rm c}c_{\rm c}^2 = \left(\frac{\partial p}{\partial \zeta}\right)_{\rm c}^2 \frac{1}{T_{\rm c}Q_0} = T_{\rm c}(A_{\rm c} - B_{\rm c})^{-2}C_{\rm B}^{-1}, \tag{2.3.40}$$

where $C_{\rm B}$ is defined by (2.3.30) and $A_{\rm c}$ and $B_{\rm c}$ by (2.3.31). Here Q_0 (or $C_{\rm B}$) is eliminated in the product,

$$(C_{VX})_{\rm c}\rho_{\rm c}c_{\rm c}^2 = T_{\rm c}B_{\rm c}^{-2} = T_{\rm c}\left(\frac{\partial p}{\partial T}\right)_{\rm c}^2, \tag{2.3.41}$$

which is a well-known relation [46]. The differences, $1/C_{VX} - 1/(C_{VX})_c$ and $\rho c^2 - \rho_c c_c^2$, behave as C_V^{-1} or ρc^2 in one-component fluids. In particular, when (2.3.27) holds in non-azeotropic mixtures, we find a simple relation,

$$(\rho c^2 - \rho_c c_c^2)/\rho_c c_c^2 \cong C_B/(C_{pX} - C_B), \tag{2.3.42}$$

using (2.3.30) and (2.3.37). From the cross correlation (2.3.38) we also obtain

$$\left(\frac{\partial T}{\partial p}\right)_{sX} = \frac{\langle \hat{T} : \hat{p}\rangle}{\langle \hat{p} : \hat{p}\rangle} \cong A_c + (B_c - A_c)\frac{\rho_c c_c^2}{\rho c^2}, \tag{2.3.43}$$

$$\left(\frac{\partial p}{\partial T}\right)_{nX} = \frac{\langle \hat{p} : \hat{T}\rangle}{\langle \hat{T} : \hat{T}\rangle} \cong \frac{1}{A_c} + \left(\frac{1}{B_c} - \frac{1}{A_c}\right)\frac{C_{VX}}{(C_{VX})_c}. \tag{2.3.44}$$

Thus, we conclude

$$\left(\frac{\partial T}{\partial p}\right)_{sX} \to \left(\frac{\partial T}{\partial p}\right)_c, \quad \left(\frac{\partial p}{\partial T}\right)_{nX} \to \left(\frac{\partial p}{\partial T}\right)_c, \tag{2.3.45}$$

on approaching the critical line [46]. The first relation (2.3.43) characterizes the temperature variation against pressure changes in adiabatic conditions (see Chapter 6). The second relation (2.3.44) can be important in measurements at a fixed volume. In deriving the above relations use has been made of the fact that the coefficients A_c and B_c defined by (2.3.31) satisfy

$$\left(\frac{\partial \zeta}{\partial T}\right)_c \left(\frac{\partial T}{\partial \zeta}\right)_{hp} = \left(\frac{\partial p}{\partial \tau}\right)_{hT} \left(\frac{\partial \tau}{\partial p}\right)_{h\zeta} = 1 - \frac{A_c}{B_c},$$

$$\left(\frac{\partial \zeta}{\partial p}\right)_c \left(\frac{\partial p}{\partial \zeta}\right)_{hT} = \left(\frac{\partial T}{\partial \tau}\right)_{hp} \left(\frac{\partial \tau}{\partial T}\right)_{h\zeta} = 1 - \frac{B_c}{A_c}, \tag{2.3.46}$$

where the second line follows from the first line by exchange of T and p.

There are some exceptional cases in which one of the coefficients in the relations (2.3.36)–(2.3.38) vanishes. In some mixtures such as CH_4–C_2H_6, the critical pressure P_c as a function of the concentration has a maximum or minimum or $(\partial p/\partial T)_c = 0$ [46], where $c^2 \propto 1/C_M$ as in one-component fluids. The reverse case in which $(\partial T/\partial p)_c$ is small is encountered in many binary mixtures, which we will discuss in the vicinity of (2.3.55).

2.3.4 Azeotropy and the dilute limit

A gas–liquid critical line is characterized by $|\alpha_X/\alpha_n| \lesssim 1$ in terms of the coefficients α_n and α_X in (2.3.13) and (2.3.14), respectively. Here the concentration fluctuations are relatively small. As an extreme case, $\alpha_X = 0$ is realized at a critical point in a number of binary fluid mixtures such as CO_2–C_2H_4 [46, 50]. This leads to the *critical azeotropy*, where $(\partial X/\partial \Delta)_{pX}$ diverges only weakly with the critical exponent α (if $\beta_X \neq 0$) and hence C_{pX} and K_{TX} diverge strongly with the critical exponent γ along the critical line. However, in contrast to the one-component fluid case, the specific heat C_{VX} and the sound

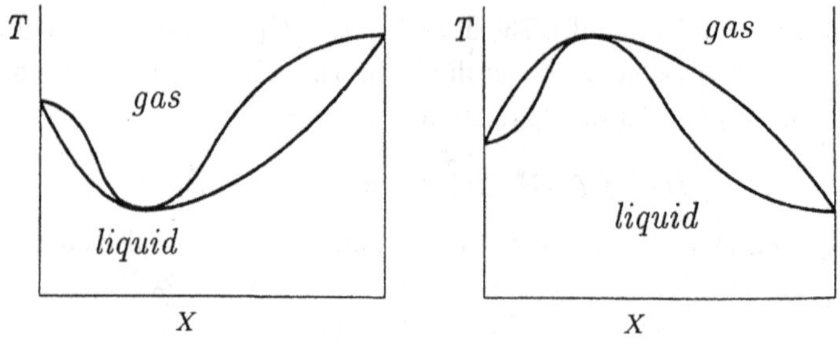

Fig. 2.11. Coexistence curves of gas and liquid phases in the temperature–concentration plane at fixed pressure. At the extremum points the azeotropic condition holds.

velocity c tend to constants at azeotropic criticality owing to the fluctuations of $\hat{\zeta}$ along the critical line.

Technologically, a line of azeotrope on the coexistence surface is of great importance, along which there is no composition difference between the two coexisting phases [51]. If it intersects the critical line, an azeotropic critical point is realized. On that line, because $X_{\rm g} = X_\ell$, the thermodynamic relation (1.3.3) yields

$$\left(\frac{1}{n_{\rm g}} - \frac{1}{n_\ell}\right)dp - (s_{\rm g} - s_\ell)dT = 0, \tag{2.3.47}$$

when a gas phase (g) and a liquid phase (ℓ) coexist. As shown in Fig. 2.11, if T (or p) is plotted vs X at fixed p (or T) in two-phase coexistence, the two curves in the gas and liquid phases (T vs $X_{\rm g}$ and T vs X_ℓ) touch and assume an extremum at an azeotropic point [51].

In general, if two components are alike, we expect small α_X. The degree of azeotropy is represented by [52]

$$\epsilon_{\rm az} = \frac{\alpha_X}{\alpha_1} = -\frac{1}{n}\left(\frac{\partial p}{\partial \Delta}\right)_{T,{\rm cx}} \tag{2.3.48}$$

where $(\partial \cdots / \partial \cdots)_{T,{\rm cx}}$ is the derivative on the coexistence surface at fixed T.[5] In two-phase coexistence, (1.3.3) gives

$$\epsilon_{\rm az} = n_{\rm c}(X_{\rm g} - X_\ell)/(n_{\rm g} - n_\ell) \tag{2.3.49}$$

in terms of the differences between the two phases. This parameter has been recognized to conveniently characterize the nature of critical lines for a number of binary fluids [45]. If $\epsilon_{\rm az}$ is small, the concentration susceptibility behaves as

$$T\left(\frac{\partial X}{\partial \Delta}\right)_{Tp} \cong A_X + n^{-1}\alpha_X^2 \chi, \tag{2.3.50}$$

[5] In Refs [52, 45] α_1 is taken to be 1, so α_2 is the degree of azeotropy.

where A_X is the background part. We may introduce a crossover reduced temperature τ_{s1} by

$$\tau_{s1}^{\gamma} \sim \epsilon_{az}^2 / A_X. \tag{2.3.51}$$

From (1.3.22), (1.3.24), and (1.3.25) we can see that C_{pX} and K_{TX} increase strongly with the exponent γ for $T/T_c - 1 \gtrsim \tau_{s1}$ and increase weakly with the exponent α (or nearly saturate) for $T/T_c - 1 \lesssim \tau_{s1}$.

^{3}He–^{4}He mixtures near the gas–liquid critical line

As a typical example, ^{3}He–^{4}He mixtures are nearly azeotropic at any X since $\epsilon_{az} \cong -\frac{1}{3}X(1-X)$ along the gas–liquid critical line ($0 < X < 1$), while $A_X \cong X(1-X)$ [52]. Thus, $\tau_{s1}^{\gamma} \sim \epsilon_{az}^2 / X(1-X) \sim 0.1X(1-X)$, which explains the observed behavior of K_{TX} [48] and C_{pX} [53]. In addition, in C_{VX} given by (2.3.36), the first constant term is smaller than the second weakly singular term except very close to the critical line. Comparison of these two terms gives another crossover reduced temperature τ_{s2} as $\tau_{s2}^{\alpha} \sim 0.2X(1-X)$. Thus τ_{s2} is extremely small ($< 10^{-12}$ for any X), so the saturation of C_{VX} cannot be observed in realistic conditions [54]. Such an extremely slow crossover of C_{VX} is expected in many binary mixtures near the gas–liquid critical line.

Dilute mixtures

In the dilute case $X \ll 1$, α_X and β_X in (2.3.14) are both of order X and the concentration fluctuations become much suppressed. To examine this case, we set $\zeta = \exp(\Delta/T) +$ const. along the critical line as in the Leung–Griffith parametrization (2.3.24), for which the derivatives with respect to ζ have well-defined limits as $X \to 0$. Then, because $\delta\hat{X}\delta\Delta/T \cong \hat{q}\zeta$ from (1.3.20), we obtain

$$\hat{q} \cong \frac{1}{X}\delta\hat{X}, \quad Q_0 \cong \frac{1}{nX}. \tag{2.3.52}$$

Note that $\langle\hat{X} : \hat{X}\rangle \to X/n$ as $X \to 0$. From (2.3.36), (2.3.37), and (2.3.50) we obtain the small-X behavior,

$$\begin{aligned} 1/C_{VX} &= a_1 X + a_2/C_M, \\ \rho c^2 &= b_1 X + b_2/C_M, \\ T\left(\frac{\partial X}{\partial \Delta}\right)_{pT} &= X + c_1 X^2 \chi, \end{aligned} \tag{2.3.53}$$

where a_1, a_2, b_1, b_2, and c_1 are constants independent of X. Therefore, $(C_{VX})_{cx} \propto X^{-1}$ and $c_c \propto X^{1/2}$ on the critical line. From the first line of (2.3.32) we can also find the behavior of C_{pX} and K_{TX} in dilute mixtures. With the above formulas, two crossover reduced temperatures τ_{s1} and τ_{s2} may be introduced by [36]

$$\tau_{s1}^{\gamma} \sim X, \quad \tau_{s2}^{\alpha} \sim X, \tag{2.3.54}$$

where $\tau_{s2} \ll \tau_{s1} \ll 1$. In most cases τ_{s2} is inaccessibly small.

2.3.5 Incompressible limit

In non-azeotropic binary mixtures, the compressibility $K_{TX}(\propto C_M$ asymptotically) is already much more suppressed than in one-component fluids (where $K_T \propto \chi$). Moreover, it is usual along a consolute critical line that the critical temperature $T_c(p)$ depends only weakly on p. The degree of compressibility, ϵ_{in} is then represented by [4, 36, 55]

$$\epsilon_{in} = n_c B_c = n_c \left(\frac{\partial T}{\partial p} \right)_c . \tag{2.3.55}$$

If $|\epsilon_{in}| \ll 1$, the singular parts of $n^2 T K_{TX}$ and $nT\alpha_{pX}$ are smaller than that of C_{pX} by ϵ_{in}^2 and ϵ_{in}, respectively, from (2.3.27)–(2.3.29). Therefore, even when C_{pX} grows as C_M, K_{TX} remains close to its small background value. On the other hand, (2.3.32) indicates $C_{VX} \propto C_M$ approximately. Using (2.3.46) their explicit expressions are known to be

$$K_{TX} \cong (1 - B_c/A_c)^{-2}/\rho_c c_c^2, \tag{2.3.56}$$

$$C_{VX} \cong (1 - B_c/A_c)^2 (C_{pX} - C_B) \tag{2.3.57}$$

except extremely close to the criticality. The asymptotic critical value $(C_{VX})_c$ grows as B_c^{-2} from (2.3.39) and cannot be reached in practice, whereas ρc^2 behaves as (2.3.42). Furthermore, we expect that B_c/A_c is of order ϵ_{in} and is small except for accidental cases. Then, $1 - B_c/A_c$ may be replaced by 1 in the above expressions to give

$$K_{TX} \cong 1/\rho_c c_c^2, \quad C_{VX} \cong C_{pX} - C_B. \tag{2.3.58}$$

Using (2.3.42) we can see the relation $\gamma_X = C_{pX}/C_{VX} \cong \rho c^2/\rho_c c_c^2$. Thus the specific-heat ratio remains close to 1 near the critical point.

Anisimov and coworkers [4, 36, 55] observed singular enhancement of C_{VX} in such *incompressible* mixtures. There, $\epsilon_{in} \cong 0.03$ for methanol + cyclohexane and of order 10^{-3} for iso-octane + nitroethane. Figure 2.12 shows their data of C_{pX} and C_{VX} in the latter mixture, which indicate (2.3.58) or $C_{pX} - C_{VX} = \text{const}$. In the context of studying adiabatically induced spinodal decomposition, Clerke and Sengers [56] examined the adiabatic coefficient $(\partial T/\partial p)_{sX} = T\alpha_{pX}/C_{pX}$ at the critical composition for 3-methylpentane + nitroethane and isobutyric acid + water. For the former mixture ϵ_{in} is expected to be of order 0.1 and

$$\begin{aligned} C_{pX} &\cong [\,1.8(T/T_c - 1)^{-0.11} + 9.5\,] \times 10^{22}\ \text{cm}^{-3}, \\ T_c\alpha_{pX} &\cong 0.009(T/T_c - 1)^{-0.11} + 0.38. \end{aligned} \tag{2.3.59}$$

See Fig. 2.13 for the curve of $(\partial T/\partial p)_{sX}$. Small ϵ_{in} is indicated by the small singular term of $T_c\alpha_{pX}$. These relations are also consistent with the data, $(\partial T/\partial p)_c = 3.67$ mK/bar or $B_c = 5.07 \times 10^{-25}$ cm^3 [56]. We then find $B_c/A_c \cong 0.09$ by neglecting the constant part of $\bar{\beta}_s^2 C_M$ as compared to C_B in (2.3.27). Similar arguments can also be made for near-critical binary mixtures of isobutyric acid + water, where $(\partial T/\partial p)_c = -55$ mK/bar is negative [56]. In incompressible binary fluids, $(\partial T/\partial p)_{sX}$ tends to the critical point value $(\partial T/\partial p)_c$ only very close to the critical line due to the slow crossover of α_{pX}.

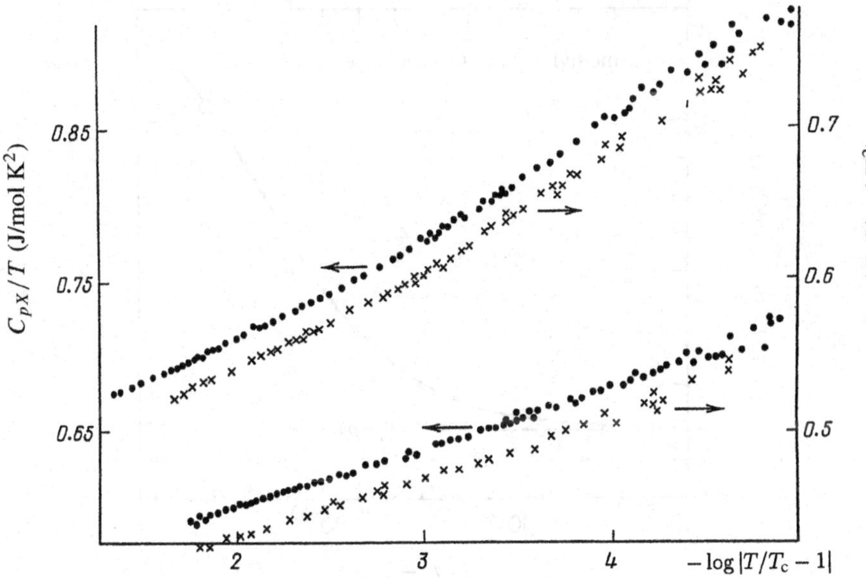

Fig. 2.12. Isochoric (constant-volume) and isobaric (constant-pressure) specific heats in a nearly *incompressible* mixture of iso-octane + nitroethane near the consolute critical point [4]. The upper curves are those below T_c, while the lower ones are those above T_c.

It is worth noting that a similar incompressible limit is attained in ^{4}He near the superfluid transition (see Section 2.4).

2.3.6 Two-scale-factor universality in the isobaric case

Many experiments of binary fluid mixtures have been performed under a constant pressure (isobaric condition). Near a consolute critical line, it is usual to perform experiments in the presence of a noncritical gas phase [4, 5]. In such cases, since the gas phase is highly compressible, the pressure of the total system is kept almost constant. The consolute critical line here meets a coexistence surface in the three-dimensional space of field variables, giving rise to a *critical end point* and a three-phase coexistence line.

In these isobaric cases two field variables are independent because

$$\zeta = \left(\frac{\partial \zeta}{\partial h}\right)_{\tau p} h + \left(\frac{\partial \zeta}{\partial \tau}\right)_{hp} \tau. \tag{2.3.60}$$

We may write the temperature deviation as

$$T - T_c(p) = \left(\frac{\partial T}{\partial h}\right)_{\tau p} h + \left(\frac{\partial T}{\partial \tau}\right)_{hp} \tau. \tag{2.3.61}$$

Furthermore, the condition $h \cong 0$ is realized at the critical composition because the

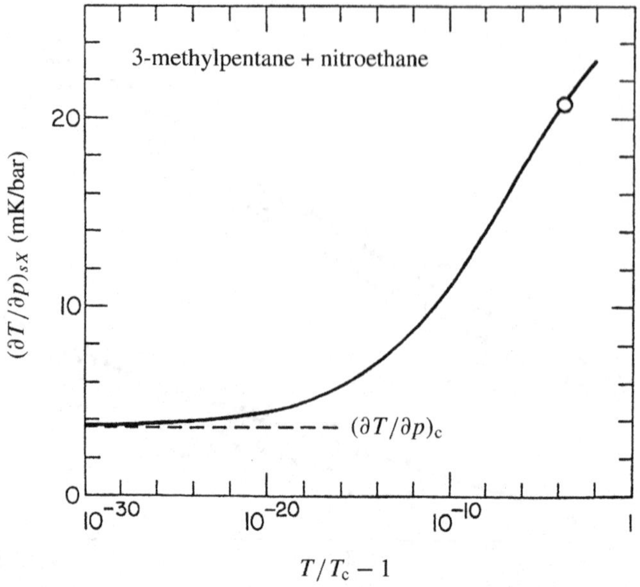

Fig. 2.13. The slope of $(\partial T/\partial p)_{sX}$ at the critical composition in 3-methylpentane + nitroethane as a function of $T/T_c - 1$ calculated from data of C_{pX} and α_{pX}. At the datum point at $T/T_c - 1 = 1.5 \times 10^{-4}$, a quench experiment was performed [56] (see Section 8.5).

estimation (2.2.17) is also applicable here, so that

$$\tau \cong \bar{\beta}_s(T/T_c - 1), \tag{2.3.62}$$

where $\bar{\beta}_s$ is defined in (2.3.30). This relation is analogous to that for one-component fluids, (2.2.16). Recall the expression for C_{pX} in (2.3.27) for non-azeotropic binary fluids. If the first term dominates the second term there, we have $C_{pX} \cong \bar{\beta}_s^2 C_M$, analogous to (2.2.25). Using (2.3.62) we find

$$(T/T_c - 1)^2 C_{pX} \cong \tau^2 C_M. \tag{2.3.63}$$

As a result, the relation of the two-scale-factor universality (2.1.45) becomes

$$R_\xi = \lim_{T\to T_c} \xi\left[\alpha(T/T_c - 1)^2 C_{pX}\right]^{1/d}, \tag{2.3.64}$$

at the critical composition in the isobaric case in binary fluids [27]–[29]. Indeed (2.3.59) and $\xi = 2.16\times 10^{-8}(T/T_c - 1)^{-0.63}$ cm give $R_\xi \cong 0.27$ for 3-methylpentane + nitroethane, in agreement with the theory.

2.4 ^{4}He near the superfluid transition

As pointed out by Anisimov [4], liquid ^{4}He near the superfluid transition is analogous to incompressible binary fluid mixtures, where the logarithmic specific-heat singularity is

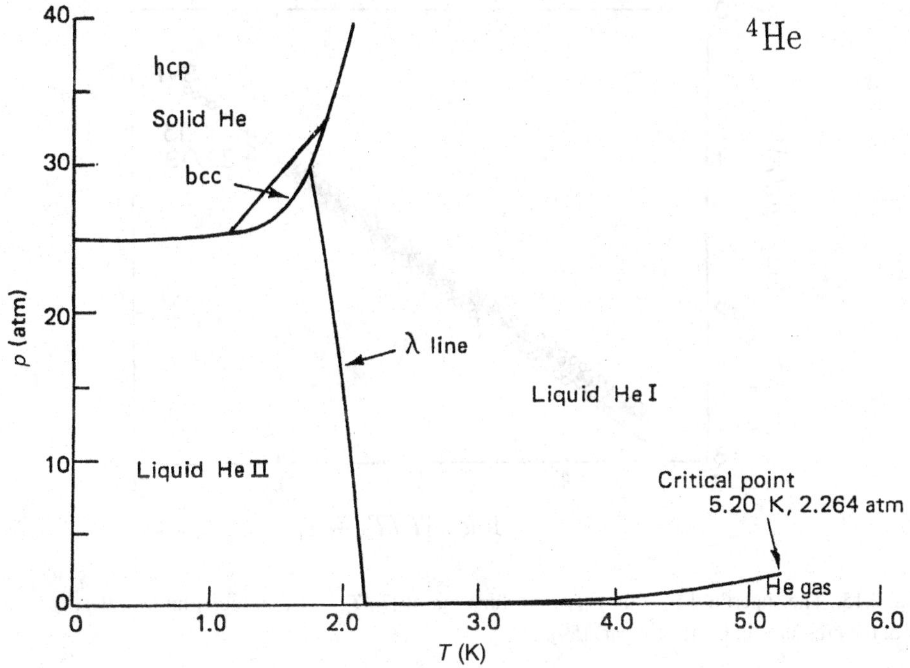

Fig. 2.14. The p–T phase diagram of ^{4}He.

marked but the compressibility is nearly nonsingular. The smallness parameter is again given by ϵ_{in} in (2.3.55) if the derivative is taken along the λ line for ^{4}He (see below). We will develop this idea to understand the static critical behavior of ^{4}He. To this end we will introduce a weakly singular variable $\hat{m}$ and a nonsingular variable $\hat{q}$ as we did in the case of binary fluid mixtures. Then the number density deviation $\delta\hat{n}$ nearly coincides with $\hat{q}$ with a small fraction of $\hat{m}$ superimposed.

2.4.1 *Singular and nonsingular density variables*

As shown in Fig. 2.14, when liquid ^{4}He (He I) is cooled at a fixed pressure p below 25 atm (25 bar), it undergoes a second-order phase transition at the critical point $T_\lambda(p)$ [57, 58], below which ^{4}He becomes a superfluid (He II). This transition has been called the λ transition because the curve of the specific heat vs $T - T_\lambda$ assumes a form of λ. The order below the transition is characterized by a nonvanishing complex order parameter $\psi = \psi_1 + i\psi_2$ originating from quantum Bose–Einstein condensation. Its square $|\psi|^2$ is proportional to the superfluid density ρ_s in the two-fluid hydrodynamic description of superfluidity,

$$|\psi|^2 \propto \rho_s = \rho_{s0}|\tau|^{2\beta}, \tag{2.4.1}$$

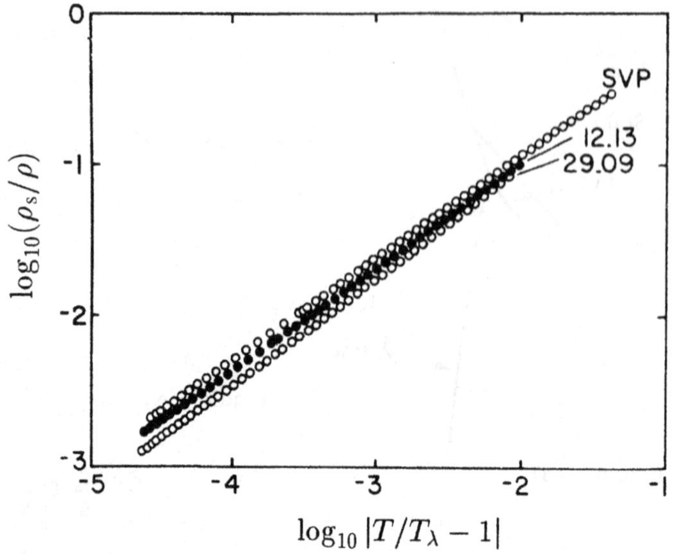

Fig. 2.15. The superfluid fraction ρ_s/ρ as a function of $|T/T_\lambda - 1|$, on logarithmic scales, at SVP or at the pressures indicated (in bar) [59].

where $\tau = T/T_\lambda - 1$, $2\beta \cong 2/3$, and the coefficient ρ_{s0} is of the same order as the mass density $\rho \sim 0.1$ g/cm^3. In Fig. 2.15 we show ρ_s obtained from second-sound measurements [59].[6] In this system, the specific heat C_p behaves nearly logarithmically as

$$\begin{aligned} C_p &\cong -A \ln \tau + B \qquad (T > T_\lambda), \\ &\cong -A' \ln |\tau| + B' \qquad (T < T_\lambda), \end{aligned} \tag{2.4.2}$$

with $A' \cong A > 0$ and $B' \sim -2B > 0$. At saturated vapor pressure (SVP) we have $C_p/n_\lambda \cong -0.64 \ln \tau - 0.9$ per particle, where $n_\lambda = 0.23 \times 10^{23}$ cm^{-3} [57]. See Fig. 2.16 for a recent precise measurement of the specific heat at SVP [60]. The critical exponent α for the specific heat is nearly equal to zero, giving rise to the logarithmic singularity, and C_p can also be expressed as

$$C_p = A\frac{1}{\alpha}\left(\tau^{-\alpha} - \tau_0^{-\alpha}\right) + C_{p0} \qquad (T > T_\lambda), \tag{2.4.3}$$

where τ_0 is an appropriate reduced temperature (theoretically equal to the Ginzburg number Gi for general n-component systems as will be found in Section 4.3), and $C_{p0} = B - A \ln \tau_0$

[6] The data can be excellently fitted to the form $\rho_s/\rho = k(p)|\tau|^{2/3}(1+a(P)|\tau|^{0.5})$, where $k(p)$ and $a(p)$ are pressure-dependent coefficients and the correction with the exponent 0.5 agrees with a prediction of a renormalization group theory [7].

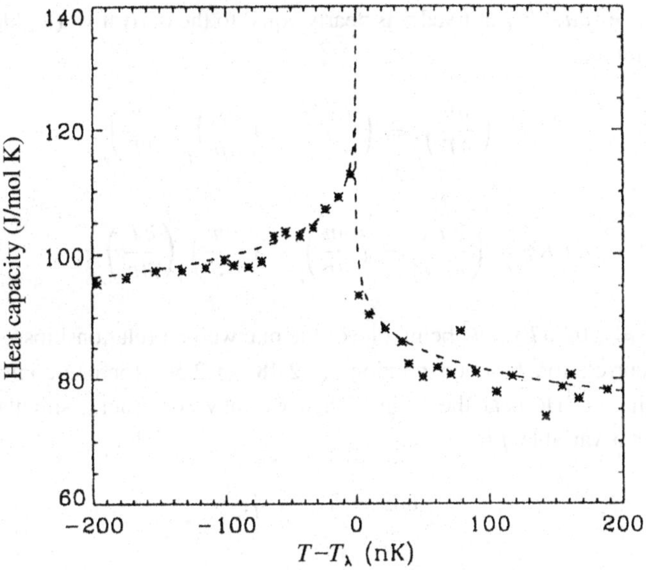

Fig. 2.16. High-resolution specific-heat capacity results near the superfluid transition taken in the *Space Shuttle* [60]. Note that the temperature is measured in units of nK, where nK = 10^{-9} K.

is the background. The two-scale-factor universality relation (2.1.45) for this case ($\alpha \cong 0$) may be expressed as[7]

$$R_\xi = \xi_{+0} A^{1/d}. \tag{2.4.4}$$

The correlation length $\xi \cong \xi_{+0}\tau^{-\nu}$ above T_λ behaves as (2.1.10) with $\nu \cong 2/3$. In ^{4}He, while ξ_{+0} cannot be directly measured, the theoretical estimate $R_\xi \cong 0.36$ and the data for A yield $\xi_{+0} \cong 1.4$ Å at SVP [22].

The corresponding spin system is the xy model in 3D given by (1.1.8) with $\langle s_{1i}\rangle = \psi_1$ and $\langle s_{2i}\rangle = \psi_2$. In ^{4}He, however, there is no physically realizable ordering field corresponding to a magnetic field ($h = 0$). Thus there remains only one relevant scaling field,

$$\tau = \frac{T}{T_\lambda(p)} - 1 \cong \left(\frac{T}{T_{\lambda 0}} - 1\right) - \frac{1}{T_{\lambda 0}}\left(\frac{\partial T}{\partial p}\right)_\lambda (p - p_0), \tag{2.4.5}$$

which is the expansion around a reference λ point, $p = p_0$ and $T = T_{\lambda 0} = T_\lambda(p_0)$. Hereafter $(\partial a/\partial b)_\lambda$ is the derivative along the λ line, $T = T_\lambda(p)$, for any a and b. Relationships between various thermodynamic derivatives can be understood in terms of the Pippard–Buckingham–Fairbank relations [61, 62], which arises from the observation

7 Another form of the two-scale-factor universality [21] in terms of the superfluid density was confirmed in experiments [30]. See Section 4.3 for more discussions.

that the derivative $(\partial a/\partial b)_\tau$ at fixed τ is nearly equal to the derivative $(\partial a/\partial b)_\lambda$ along the λ line. For example,

$$\frac{C_p}{T} = \left(\frac{\partial s}{\partial T}\right)_p \cong \left(\frac{\partial s}{\partial T}\right)_\lambda - \left(\frac{\partial s}{\partial p}\right)_T \left(\frac{\partial p}{\partial T}\right)_\lambda, \tag{2.4.6}$$

$$nK_T = \left(\frac{\partial n}{\partial p}\right)_T \cong \left(\frac{\partial n}{\partial p}\right)_\lambda - \left(\frac{\partial n}{\partial T}\right)_p \left(\frac{\partial T}{\partial p}\right)_\lambda, \tag{2.4.7}$$

with $(\partial s/\partial p)_T = (\partial n/\partial T)_p/n^2$ being one of the maxwellian relationships.

Recall the correlation function relations (1.2.48)–(1.2.50) for one-component fluids. They hold in liquid ^{4}He near the λ line. Then we may construct a singular variable $\hat{m}$ and a nonsingular variable $\hat{q}$ by

$$n\delta\hat{s} = \hat{m} - A_\lambda \hat{q}, \tag{2.4.8}$$

$$\delta\hat{n} = \hat{q} - \epsilon_{\text{in}}\hat{m}, \tag{2.4.9}$$

with[8]

$$A_\lambda = -n\left(\frac{\partial s}{\partial n}\right)_\lambda, \quad \epsilon_{\text{in}} = n\left(\frac{\partial T}{\partial p}\right)_\lambda. \tag{2.4.10}$$

Here $\delta\hat{s}$ is the deviation of the entropy in (1.2.46) and $\delta\hat{n}$ is the deviation of the number density. They are measured from the reference λ-point values. The equilibrium fluctuations of the two variables $\hat{m}$ and $\hat{q}$ are independent of each other,

$$\langle \hat{m} : \hat{m} \rangle = C, \quad \langle \hat{m} : \hat{q} \rangle = 0, \quad \langle \hat{q} : \hat{q} \rangle = Q_0, \tag{2.4.11}$$

where C is logarithmically dependent on $|\tau|$ and

$$Q_0 = (1 - A_\lambda \epsilon_{\text{in}})^{-1} nT \left(\frac{\partial n}{\partial p}\right)_\lambda \tag{2.4.12}$$

is nonsingular. In the corresponding xy model, $\hat{m}$ is the energy density (divided by T) and C is the specific heat. Near SVP we estimate

$$A_\lambda \sim 0.8, \quad \epsilon_{\text{in}} \sim -0.04, \quad Q_0/n_\lambda \sim 0.1. \tag{2.4.13}$$

Thus, in ^{4}He, (i) ϵ_{in} is small and is analogous to the incompressibility parameter (2.3.55) for binary fluids and (ii) the variance Q_0 of the nonsingular variable is relatively small, in contrast to the usual nearly incompressible binary mixtures.

8 We have $(\partial T/\partial p)_\zeta = 1/nA_\lambda$ from (2.4.30) below, which corresponds to A_c for binary fluid mixtures, given in (2.3.31).

2.4.2 Thermodynamic derivatives

Now, from (1.2.48) we may express thermodynamic derivatives in terms of the parameters defined above:

$$C_p = n^2\langle\hat{s}:\hat{s}\rangle = C + A_\lambda^2 Q_0, \tag{2.4.14}$$

$$n^2 T K_T = \langle\hat{n}:\hat{n}\rangle = \epsilon_{\rm in}^2 C + Q_0, \tag{2.4.15}$$

$$nT\alpha_p = -n\langle\hat{n}:\hat{s}\rangle = \epsilon_{\rm in} C + A_\lambda Q_0, \tag{2.4.16}$$

where $\alpha_p = -(\partial n/\partial T)_p/n$ is the thermal expansion coefficient. These expressions are analogous to (2.3.27)–(2.3.29) and satisfy the Pippard–Buckingham–Fairbank relations (2.4.6) and (2.4.7). In accord with experiments [57], the singular part of K_T is very small ($\propto \epsilon_{\rm in}^2$) compared with the background part Q_0/n^2T, and α_p changes its sign at $T = T_0 > T_\lambda$ above the λ line (which occurs at $C \cong A_\lambda Q_0/|\epsilon_{\rm in}|$). The latter fact leads to some interesting consequences on hydrodynamic convection in gravity.

To calculate C_V we use the relation,

$$\langle\hat{n}:\hat{n}\rangle\langle\hat{s}:\hat{s}\rangle - \langle\hat{n}:\hat{s}\rangle^2 = n^{-2}(1-\epsilon_{\rm in}A_\lambda)^2 C Q_0, \tag{2.4.17}$$

which readily follows from (2.4.8) and (2.4.9). From (1.2.49) C_V is then given by

$$C_V = (1 - A_\lambda\epsilon_{\rm in})^2 C\big/(1 + \epsilon_{\rm in}^2 C/Q_0). \tag{2.4.18}$$

If C tends to ∞ at the λ point (or if $\alpha > 0$), C_V in principle saturates to a λ-point value ($\cong Q_0/\epsilon_{\rm in}^2$). In realistic conditions, however, $\epsilon_{\rm in}^2 C \ll Q_0$ holds, so that

$$C_V \cong (1 - A_\lambda\epsilon_{\rm in})^2 C, \tag{2.4.19}$$

$$C_p - C_V \cong (\epsilon_{\rm in} C + A_\lambda Q_0)^2/Q_0, \tag{2.4.20}$$

except extremely close to the critical point. In the difference $C_p - C_V$ we cannot set $\epsilon_{\rm in} = 0$ in the numerator due to the small size of Q_0/C, while $C_{pX} - C_{pV} = $ const. in the case of a binary mixture, as shown in Fig. 2.12. From (1.2.50) the sound velocity c behaves as

$$\rho c^2 = Tn^2(1-\epsilon_{\rm in}A_\lambda)^{-2}\left(\frac{1}{Q_0} + A_\lambda^2\frac{1}{C}\right). \tag{2.4.21}$$

This is consistent with the thermodynamic identity $\rho c^2 K_T = C_p/C_V$ in (1.2.54). The singular part of $K_s = 1/\rho c^2$ is not small, whereas K_T is almost nonsingular.

The nonsingular variable $\hat{q}$ is analogous to $\hat{q}$ for classical binary fluid mixtures in Section 2.3. We may also introduce a field variable ζ representing the coordinate along the λ line by requiring

$$\tau\hat{m} + \zeta\hat{q} = (T/T_{\lambda 0} - 1)n\delta\hat{s} + \frac{1}{nT}(p - p_0)\delta\hat{n}, \tag{2.4.22}$$

which follows from (1.2.47). Substitution of (2.4.8) and (2.4.9) yields τ given by (2.4.5) and

$$\zeta = \frac{1}{nT}(p - p_0) - A_\lambda(T/T_{\lambda 0} - 1). \tag{2.4.23}$$

The equilibrium averages $\langle \hat{m} \rangle$ and $\langle \hat{q} \rangle$ then behave as

$$\langle \hat{m} \rangle = C\tau, \quad \langle \hat{q} \rangle = Q_0 \zeta. \tag{2.4.24}$$

2.4.3 Temperature and pressure fluctuations

As in binary fluid mixtures we introduce the fluctuations of the field variables. Noting that the ordering field for the order parameter identically vanishes ($h = \delta \hat{h} = 0$), we define

$$\delta \hat{\tau} = \frac{1}{C}\delta \hat{m}, \quad \delta \hat{\zeta} = Q_0^{-1}\hat{q} - \zeta, \tag{2.4.25}$$

which are superimposed on the homogeneous averages τ and ζ and satisfy

$$\langle \hat{\tau} : \hat{\tau} \rangle = C^{-1}, \quad \langle \hat{\zeta} : \hat{\zeta} \rangle = Q_0^{-1}, \quad \langle \hat{\tau} : \hat{\zeta} \rangle = 0. \tag{2.4.26}$$

In ^{4}He the temperature and pressure fluctuations in the long-wavelength limit are defined as (1.2.61) and (1.2.66). They are rewritten in terms of $\delta\hat{\tau}$ and $\delta\hat{\zeta}$ as

$$\delta \hat{T} = T(1 - \epsilon_{\rm in} A_\lambda)^{-1}\big[\delta\hat{\tau} + \epsilon_{\rm in}\delta\hat{\zeta}\big], \tag{2.4.27}$$

$$\delta \hat{p} = nT(1 - \epsilon_{\rm in} A_\lambda)^{-1}\big[A_\lambda\delta\hat{\tau} + \delta\hat{\zeta}\big]. \tag{2.4.28}$$

These relations are the counterparts of (2.3.33) and (2.3.34) for binary fluid mixtures, and they can yield the variance relations (2.4.18) and (2.4.21) as in (2.3.36)–(2.3.38). We may also express $\delta\hat{\tau}$ and $\delta\hat{\zeta}$ as $T\delta\hat{\tau} = \delta\hat{T} - (\partial T/\partial p)_\lambda \delta\hat{p}$ and $T\delta\hat{\zeta} = \delta\hat{p}/n - A_\lambda\delta\hat{T}$, which are of the same forms as (2.4.5) and (2.4.23) (with the circumflex).

With (2.4.26)–(2.4.28) it is easy to confirm the variance relations, (1.2.62)–(1.2.64) and (1.2.68)–(1.2.70), which we derived for classical one-component fluids in Section 1.2. In particular, the temperature variance is written as $\langle \hat{T} : \hat{T} \rangle = T^2/C_V$ also in ^{4}He. Theoretically however, close to the λ point, we need to show that the temperature fluctuations are much smaller than the (average) reduced temperature $|T - T_\lambda|$ over spatial scales much longer than ξ. As in (2.1.46) we define the coarse-grained average $(\delta\hat{T})_\ell = \ell^{-d}\int_{\ell^d} d\boldsymbol{r}\delta\hat{T}(\boldsymbol{r})$ where the integral is over a volume element with linear dimension $\ell(\gg \xi)$. Above T_λ, (2.4.4) and (2.4.20) give

$$\langle (\delta\hat{T})_\ell^2 \rangle/(T - T_\lambda)^2 = (\xi/R_\xi \ell)^d, \tag{2.4.29}$$

which is clearly less than 1 for $\ell > \xi/R_\xi$.

2.4.4 Gravity effects in ^{4}He

Height-dependent reduced temperature

In equilibrium on earth, the pressure depends on the height as $dp/dz = -\rho g$ with g being the gravitational acceleration. This gives the height-dependent transition temperature,

$$T_\lambda(p) = T_\lambda(p_0) - \left(\frac{\partial T}{\partial p}\right)_\lambda \rho_\lambda g z. \tag{2.4.30}$$

The z axis is in the upward direction with the origin taken appropriately. Then the local reduced temperature depends on z even in equilibrium as

$$\tau(z) \equiv T/T_\lambda(p) - 1 \cong (T/T_{\lambda 0} - 1) - Gz \tag{2.4.31}$$

where

$$G = \rho_\lambda g |(\partial T/\partial p)_\lambda|/T_\lambda \tag{2.4.32}$$

is 0.6×10^{-6}/cm at SVP on earth. Equilibrium states on earth become noticeably inhomogeneous in the following temperature region [58],

$$|\tau| \lesssim GL \sim 10^{-6} L, \tag{2.4.33}$$

where L is the vertical cell length (in units of cm). The presently attained precision of temperature measurements is exceedingly high for helium ($\sim 10^{-9}$ deg) [58]. Therefore, the pressure dependence of the critical temperature is the main cause preventing precise measurements of the critical phenomena in ^{4}He.

Gravity-induced two-phase coexistence

An interesting effect brought about by gravity is that, if $\tau = 0$ at a middle point ($z = z_0$) of the container, two-phase coexistence may be realized with a superfluid in the upper region ($z > z_0$) and a normal fluid in the lower region ($z < z_0$). Such coexistence is detectable because these two regions react to an applied heat flow very differently [63]. As shown in Fig. 2.17, a gradual change from a normal fluid to a superfluid occurs in an interface or in a transition region ($|z - z_0| \lesssim \ell_g$). Its thickness ℓ_g and the typical reduced temperature τ_g in the interface region are determined from the following scaling relations,

$$\ell_g = \xi_{+0}\tau_g^{-\nu}, \quad \tau_g = G\ell_g. \tag{2.4.34}$$

These equations are solved to give

$$\ell_g = \xi_{+0}(\xi_{+0}G)^{-\nu/(1+\nu)}, \tag{2.4.35}$$

$$\tau_g = (\xi_{+0}G)^{1/(1+\nu)}, \tag{2.4.36}$$

where $\nu/(1+\nu) \cong 0.4$. We notice that the local correlation length $\xi(z)$ attains a maximum of order ℓ_g in the transition region. On earth, we have $\tau_g \sim 10^{-9}$ and $\ell_g \sim 10^{-2}$ cm.

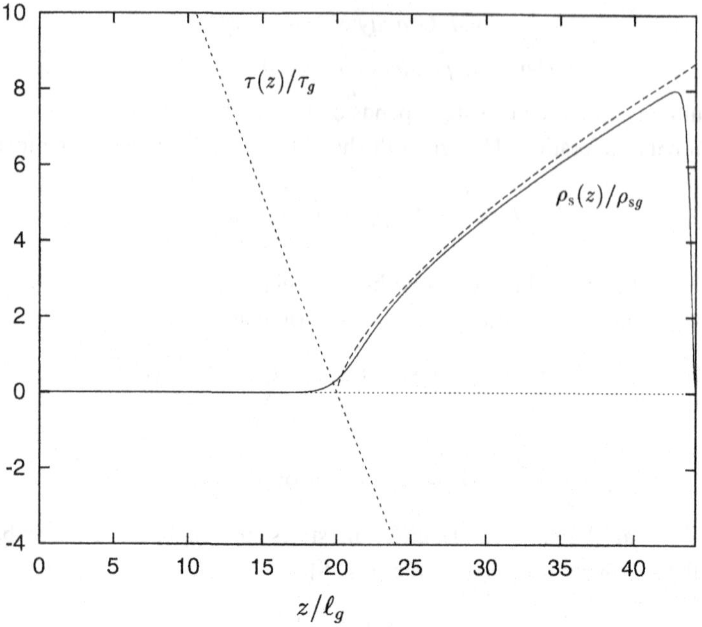

Fig. 2.17. Dimensionless gravity-induced superfluid density $\rho_s(z)/\rho_{sg}$ (solid line) in a thin film with thickness $L = 44.15\ell_g$ of ^{4}He, where $\rho_{sg} = \rho_{s0}\tau_g^{2/3}$ and the space is measured in units of ℓ_g. The height-dependent reduced temperature $\tau(z)$ in (2.4.31) is also plotted in units of τ_g (short-dash line). The system is a superfluid in $z_0 < z < L$ and a normal fluid in $0 < z < z_0$ where $z_0 = 20\ell_g$. At the boundaries $z = 0$ and L we impose the condition $\rho_s = 0$. We compare the calculated profile with the local equilibrium profile $\rho_{sg}[(z - z_0)/\ell_g]^{2/3}$ $(z > z_0)$ in (2.4.37) (long-dash line).

Outside the interface, the local equilibrium holds; namely, the thermodynamic relations such as (2.4.1) and (2.4.2) are valid if use is made of the local reduced temperature. For example, we have

$$\rho_s \cong \rho_{s0}[G(z - z_0)]^{2/3} \qquad (z - z_0 \gg \ell_g). \tag{2.4.37}$$

The profile in Fig. 2.17 has been calculated in the Ginzburg–Landau theory, as will be explained below (4.2.51).

Appendix 2A Calculation in non-azeotropic cases

As an illustration, we express C_{pX}, K_{TX}, and α_{pX} in terms of the variances using (2.3.12)–(2.3.14) when the concentration fluctuations are much enhanced as in (2.3.26). We notice that (1.3.24)–(1.3.26) remain unchanged with respect to replacements, $\delta\hat{s} \to \delta\hat{s} - (\alpha_s/n\alpha_X)\delta\hat{X}$ and $\delta\hat{n} \to \delta\hat{n} - (\alpha_n/\alpha_X)\delta\hat{X}$, which are linear combinations of $\hat{m}$ and $\hat{q}$.

The following expressions readily follow:

$$C_{pX} = \bar{\beta}_s^2 C_M + \bar{\gamma}_s^2 Q_0 + \cdots, \tag{2A.1}$$

$$n^2 T K_{TX} = \bar{\beta}_n^2 C_M + \bar{\gamma}_n^2 Q_0 + \cdots, \tag{2A.2}$$

$$-nT\alpha_{pX} = \bar{\beta}_s \bar{\beta}_n C_M + \bar{\gamma}_s \bar{\gamma}_n Q_0 + \cdots, \tag{2A.3}$$

where $\bar{\beta}_s = \beta_s - \alpha_s \beta_X/\alpha_X$, $\bar{\gamma}_s = \gamma_s - \alpha_s \gamma_X/\alpha_X$, $\bar{\beta}_n = \beta_n - \alpha_n \beta_X/\alpha_X$, and $\bar{\gamma}_n \equiv \gamma_n - \alpha_n \gamma_X/\alpha_X$. These coefficients can also be expressed as

$$\bar{\beta}_s = T\left(\frac{\partial \tau}{\partial T}\right)_{hp}, \quad \bar{\gamma}_s = T\left(\frac{\partial \zeta}{\partial T}\right)_{hp}, \quad \bar{\beta}_n = nT\left(\frac{\partial \tau}{\partial p}\right)_{hT}, \quad \bar{\gamma}_n = nT\left(\frac{\partial \zeta}{\partial p}\right)_{hT}. \tag{2A.4}$$

From (2.3.19)–(2.3.21) the first two relations follow under $h = \delta p = 0$, while the last two follow under $h = \delta T = 0$. We notice the relations,

$$\frac{\bar{\beta}_n}{\bar{\beta}_s} = -n\left(\frac{\partial T}{\partial p}\right)_{h\tau} = -nB_c, \quad \frac{\bar{\gamma}_n}{\bar{\gamma}_s} = -n\left(\frac{\partial T}{\partial p}\right)_{h\zeta} = -nA_c, \tag{2A.5}$$

where A_c and B_c are defined by (2.3.31). These relations lead to (2.3.27)–(2.3.30).

References

[1] M. S. Green (ed.) *Critical Phenomena*, Proceedings of Enrico Fermi Summer School, Varenna, 1970 (Academic, New York, 1971).

[2] H. E. Stanley, *Introduction to Phase Transition and Critical Phenomena* (Oxford University Press, 1973).

[3] A. Z. Patashinskii and V. L. Pokrovskii, *Fluctuation Theory of Phase Transitions* (Pergamon, Oxford, 1979).

[4] M. A. Anisimov, *Critical Phenomena in Liquids and Liquid Crystals* (Gordon and Breach, Philadelphia, 1991).

[5] J. V. Sengers and J. M. H. Levelt Sengers, in *Progress in Liquid Physics,* ed. C. A. Croxton (John Wiley & Sons, Chichester, England, 1978).

[6] A. V. Voronel, Yu. R. Chashkin, V. A. Popov, and V. G. Simkin, *Sov. Phys. JETP* **18**, 568 (1964); A. V. Voronel, *Physica* **73**, 195 (1974).

[7] F. J. Wegner, in *Phase Transitions and Critical Phenomena*, Vol. 6, eds. C. Domb and J. L. Lebowitz (Academic, London, 1976), p. 8.

[8] L. Onsager, *Phys. Rev.* **65**, 117 (1944).

[9] M. E. Fisher and A. Aharony, *Phys. Rev. Lett.* **31**, 1238 (1973).

[10] J. F. Nicoll, *Phys. Rev. B* **20**, 4527 (1979).

[11] B. B. Mandelbrot, *Fractals: Form, Chance and Dimension* (Freeman, San Francisco, CA, 1977).

[12] M. Suzuki, *Prog. Theor. Phys.* **69**, 65 (1983).

[13] F. Family, *J. Stat. Phys.* **36**, 881 (1984).

[14] K. Binder, *Z. Phys. B* **43**, 119 (1981).

[15] N. Ito and M. Suzuki, *Prog. Theor. Phys.* **77**, 1391 (1987).

[16] R. Hilfer and N. B. Wilding, *J. Phys. A: Math. Gen.* **28**, L281(1995).

[17] N. B. Wilding, *J. Phys. C: Condens. Matter* **9**, 585 (1997).

[18] M. E. Fisher, *Physics* **3**, 255 (1967); *Rep. Prog. Theor.* **30**, 615 (1967).

[19] L. P. Kadanoff, in *Phase Transitions and Critical Phenomena*, Vol. 5A, eds. C. Domb and J. L. Lebowitz (Academic Press, London, 1976), p. 1.

[20] D. Stauffer, D. Ferer, and M. Wortis, *Phys. Rev. Lett.* **29**, 345 (1972).

[21] D. Ferer, *Phys. Rev. Lett.* **33**, 21 (1974).

[22] P. C. Hohenberg, A. Aharony, B. I. Halperin, and E. D. Siggia, *Phys. Rev. B* **13**, 2986 (1976).

[23] J. V. Sengers and M. R. Moldover, *Phys. Lett.* **66A**, 44 (1978).

[24] A. Aharony and P. C. Hohenberg, *Phys. Rev. B* **13**, 2110 (1976).

[25] C. Bagnuls and C. Bervillier, *J. Physique Lett.* **45**, L95–100 (1984).

[26] A. J. Liu and M. E. Fisher, *Physica A* **156**, 35 (1989).

[27] D. Beysens, in *Phase Transitions* : Cargèse 1980, eds. M. Levy, J.-C. Le Guillou, and J. Zinn-Justin (Plenum, New York, 1981), p. 25.

[28] E. Bloemen, J. Thoen, and W. van Dael, *J. Chem. Phys.* **73**, 4628 (1980); J. Thoen, J. Hamelin, and T. K. Bose, *Phys. Rev. E* **53**, 6264 (1996).

[29] G. Sanchez, M. Meichle, and C. W. Garland, *Phys. Rev. A* **28**, 1647 (1983).

[30] A. Singsaas and G. Ahlers, *Phys. Rev. B* **30**, 5103 (1984).

[31] P. Schofield, J. D. Lister, and J. T. Ho, *Phys. Rev. Lett.* **23**, 1098 (1969).

[32] D. J. Wallace, in *Phase Transitions and Critical Phenomena*, Vol. 6, eds. C. Domb and J. L. Lebowitz (Academic, London, 1976), p. 294.

[33] P. C. Hohenberg and M. Barmatz, *Phys. Rev. A* **6**, 289 (1972).

[34] M. S. Green, M. J. Cooper, and J. M. H. Sengers, *Phys. Rev. Lett.* **26**, 492 (1971); M. Ley-Koo and M. S. Green, *Phys. Rev. A* **16**, 2483 (1977).

[35] M. R. Moldover, in *Phase Transitions* : Cargèse 1980, eds. M. Levy, J.-C. Le Guillou, and J. Zinn-Justin (Plenum, New York, 1981), p. 63; J. V. Sengers, *ibid.*, p. 95.

[36] M. A. Anisimov, E. E. Gorodetskii, V. D. Kulikov, and J. V. Sengers, *Phys. Rev. E* **51**, 1199 (1995); M. A. Anisimov, E. E. Gorodetskii, V. D. Kulikov, A. A. Povodyrev, and J. V. Sengers, *Physica A* **220**, 277 (1995).

[37] A. Onuki, *Phys. Rev. E* **55**, 403 (1997).

[38] N. D. Mermin, *Phys. Rev. Lett.* **26**, 169 (1971); J. J. Rehr and N. D. Mermin, *Phys. Rev. A* **8**, 472 (1973).

[39] N. B. Wilding and A. D. Bruce, *J. Phys. C*, **4**, 3087 (1992).

[40] J. F. Nicoll, *Phys. Rev. A* **24**, 2203 (1981).

[41] M. W. Pestak, R. E. Goldstein, M. H. W. Chan, J. R. de Bruyn, D. A. Balzarini, and N. W. Ashcroft, *Phys. Rev. B* **36**, 599 (1987); S. Jüngst, B. Knuth, and F. Hensel, *Phys. Rev. Lett.* **20**, 2160 (1985).

[42] H. Weingärtner and W. Schröer, in *Advances in Chemical Physics*, eds. I. Prigogine and S. A. Rice, Vol. 116 (John Wiley & Sons, Inc., New York, 2001), p. 1.

[43] M. R. Moldover, J. V. Sengers, R. W. Gammon, and R. J. Hocken, *Rev. Mod. Phys.* **51**, 79 (1979); J. H. Sikkenk, J. M. J. van Leeuwen, and J. V. Sengers, *Physica A* **139**, 1 (1986).

[44] J. Straub, Thesis, Technische Universität München (1966); Habilitation Technische Universität, München (1967).

[45] J. C. Rainwater, in *Supercritical Fluid Technology* eds. T. J. Bruno and J. F. Fly (CRC Press, Boca Raton, FL, 1991), p. 57.

[46] R. B. Griffiths and J. C. Wheeler, *Phys. Rev. A* **2**, 1047 (1970).

[47] S. S. Leung and R. B. Griffiths, *Phys. Rev. A* **8**, 2670 (1973).

[48] B. Wallace, Jr and H. Meyer, *Phys. Rev. A* **5**, 953 (1972).

[49] A. M. Wims, D. McIntyre, and F. Hynne, *J. Chem. Phys.* **50**, 616 (1969).

[50] G. D'Arrigo, L. Mistura, and P. Tartaglia, *Phys. Rev. A* **12**, 2587 (1975).

[51] L. D. Landau and E. M. Lifshitz, *Statistical Physics* (Pergamon, New York, 1964).

[52] A. Onuki, *J. Low Temp. Phys.* **61**, 101 (1985).

[53] B. A. Wallace and H. Meyer, *Phys. Rev. A* **5**, 953 (1972).

[54] G. R. Brown and H. Meyer, *Phys. Rev. A* **6**, 364 (1972)

[55] M. A. Anisimov, A. V. Voronel, and T. M. Ovodova, *Sov. Phys. JETP* **35**, 536 (1972) [*Zh. Eksp. Teor. Fiz.* **62**, 1015 (1972)].

[56] E. A. Clerke and J. V. Sengers, *Physica* **118A**, 360 (1983).

[57] G. Ahlers, in *The Physics of Liquid and Solid Helium*, Part I, eds. K. H. Bennemann and J. B. Ketterson (John Wiley & Sons, Inc., New York, 1976), p. 85; in *Quantum Liquids*, eds. J. Ruvalds and T. Regge (North-Holland, Amsterdam, 1978), p. 1; *Phys. Rev. A* **8**, 530 (1973).

[58] G. Ahlers, *Rev. Mod. Phys.* **52**, 489 (1980).

[59] D. S. Greywall and G. Ahlers, *Phys. Rev. A* **7**, 2145 (1973).

[60] J. A. Lipa, D. R. Swanson, J. A. Nissen, and T. C. P. Chui, *Physica B* **197**, 239 (1994).

[61] A. B. Pippard, *The Elements of Classical Thermodynamics* (Cambridge University Press, 1957), Chapter IX.

[62] M. J. Buckingham and W. M. Fairbank, in *Progress in Low Temperature Physics*, ed. C. J. Gorter (North-Holland, Amsterdam, 1961), Vol. III, p. 80.

[63] G. Ahlers, *Phys. Rev.* **171**, 275 (1968).

3

Mean field theories

In this chapter we will introduce the simplest theory of phase transitions, the Landau theory [1]–[4]. It assumes a free energy $\mathcal{H}(\psi)$, called the Landau free energy, which depends on the order parameter ψ as well as the temperature and the magnetic field. The thermodynamic free energy F is the minimum of $\mathcal{H}(\psi)$ as a function of ψ. This minimization procedure gives rise to the mean field critical behavior. Historically, a number of mean field theories have been presented to explain phase transitions in various systems. They reduce to the Landau theory near the critical point. Examples we will treat are the Bragg–Williams theory [5] for Ising spin systems and alloys undergoing order–disorder phase transitions, the van der Waals theory of the gas–liquid transition [6], the Flory–Huggins theory and the classical rubber theory for polymers and gels. We will also discuss tricritical phenomena in the scheme of the Landau theory. In Appendix 3A elastic theory for finite strain will be considered, which will be needed to understand the volume-phase transition in gels.

3.1 Landau theory

3.1.1 Order parameter and constrained free energy

It is desirable to sum up the spin configurations in (1.1.9) to exactly determine the thermodynamic limit. This attempt has not been successful for the 3D Ising model, while it was successful for 2D and is a simple exercise for 1D [3]. Another approach is a phenomenological one, known as the Landau theory, in which the key quantity is the order parameter ψ.[1] Assuming that the system is homogeneous on average, we define ψ as the space average of the spins,

$$\psi = \frac{1}{V}\sum_i s_i. \tag{3.1.1}$$

In Chapter 4 we will give a more appropriate definition of the order parameter taking into account spatially inhomogeneous fluctuations. For now, we introduce a *constrained* free energy $\mathcal{H}(\psi)$ obtained by partial summation with ψ held fixed. That is, we sum up only the spin configurations in which (3.1.1) is satisfied:

$$\exp[-\beta\mathcal{H}(\psi)] = \sum_{\{s\}} \delta\left(\psi - \frac{1}{V}\sum_i s_i\right) \exp[-\beta\mathcal{H}\{s\}]. \tag{3.1.2}$$

[1] We have denoted the fluctuating spin variable and energy variable by $\hat{\psi}$ and $\hat{m}$ with the circumflex in Chapter 2. Hereafter, we will write them as ψ and m to avoid cumbersome notation.

Then $\mathcal{H}(\psi)$ is dependent on the order parameter ψ as well as on the temperature field T and the magnetic field H. It will be called the Landau free energy. From the definition, $\exp[-\beta\mathcal{H}(\psi)]/Z$ is the equilibrium distribution of the order parameter fluctuations. For sufficiently large systems ψ may be treated as a continuous variable and then the true thermodynamic free energy F is given by

$$\exp[-\beta F] = \int d\psi \exp[-\beta\mathcal{H}(\psi)]. \tag{3.1.3}$$

For simplicity, let $\mathcal{H}(\psi)$ have a single minimum at $\psi = \psi^*$. Then it is expanded around the minimum as

$$\mathcal{H}(\psi) = \mathcal{H}(\psi^*) + V\frac{T}{2\chi}(\psi - \psi^*)^2 + \cdots, \tag{3.1.4}$$

where χ is the magnetic susceptibility. For large systems, the integration from the narrow region $|\psi - \psi^*| \lesssim (\chi/V)^{1/2}$ is dominant in (3.1.3), leading to

$$F \cong \mathcal{H}(\psi^*) - \frac{1}{2}T\ln(2\pi\chi/V) \cong \mathcal{H}(\psi^*). \tag{3.1.5}$$

The logarithmic correction is negligible in the limit $V \to \infty$. Thus, the thermodynamic Helmholtz free energy F is obtained by minimization of $\mathcal{H}(\psi)$ with respect to ψ. This is a very important step in the Landau theory.

3.1.2 Regular expansion of the Landau free energy

The assumption Landau made is that $\mathcal{H}(\psi)$ is an analytic function of the order parameter ψ near the critical point, $\tau = 0$ and $h = 0$, where $\tau = T/T_c - 1$ and $h = H/T \cong H/T_c$ are the two relevant field variables. Then the free-energy density $f(\psi) \equiv \mathcal{H}(\psi)/V$ may be expanded as

$$f(\psi) = f_{\text{reg}} + T_c\left[\frac{1}{2}r\psi^2 + \frac{1}{4}u_0\psi^4 - h\psi\right] + \cdots, \tag{3.1.6}$$

where the coefficient r is proportional to the reduced temperature τ as

$$r = a_0\tau. \tag{3.1.7}$$

The coefficient a_0 will be assumed to be positive[2] as well as the coefficient u_0. The first term is a regular function of τ expanded as

$$f_{\text{reg}} = f_c - T_c s_c\tau - \frac{1}{2}T_c C_0\tau^2 + \cdots. \tag{3.1.8}$$

The coefficients f_c and s_c are the critical values of the free-energy density and the entropy density, respectively, and C_0 is the background specific heat. From the definition (3.1.2), $f(\psi) + H\psi$ is an even function of ψ, so the cubic term does not appear in the Landau

[2] However, in some fluid mixtures such as lutidine + water, the coexistence curve at fixed p is *inverted* with the critical point located at the minimum in the temperature–concentration phase diagram. In such cases, a_0 is negative.

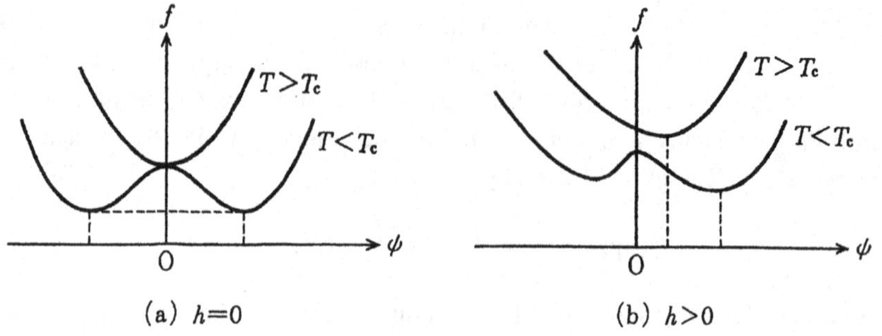

Fig. 3.1. The Landau free-energy density $f(\psi)$ near the critical point for typical cases.

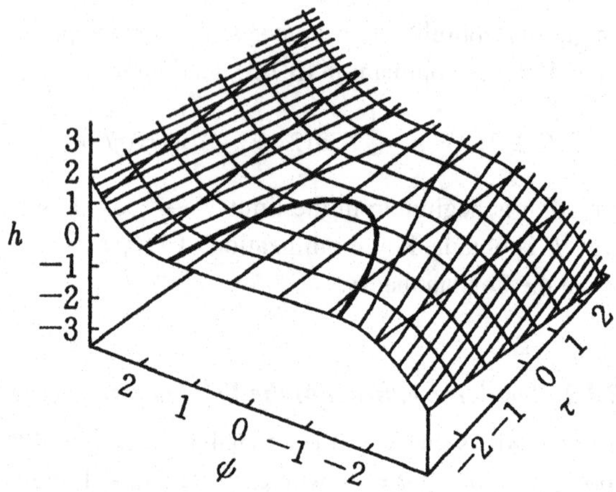

Fig. 3.2. The equation of state obtained from (3.1.9) with $u_0 = 1$. The (bold) parabolic curve on the surface represents the coexistence curve.

expansion (3.1.6). Figure 3.1 illustrates the Landau free-energy density $f(\psi)$ in typical cases.

We have shown that the equilibrium value ψ^* is given by minimization of $f(\psi)$ for each given T and H (or τ and h). Therefore, at $\psi = \psi^*$, we require

$$\beta f'(\psi) = r\psi + u_0\psi^3 - h = 0, \tag{3.1.9}$$

$$\beta f''(\psi) = r + 3u_0\psi^2 > 0, \tag{3.1.10}$$

where $f' = \partial f/\partial\psi$ and $f'' = \partial^2 f/\partial\psi^2$. Hereafter, the equilibrium value ψ^* will be written as ψ for simplicity. When $h = 0$, the equilibrium is attained at $\psi = 0$ for $r > 0$, while

$$\psi = \pm(|r|/u_0)^{1/2} \propto \pm(T_c - T)^{1/2} \tag{3.1.11}$$

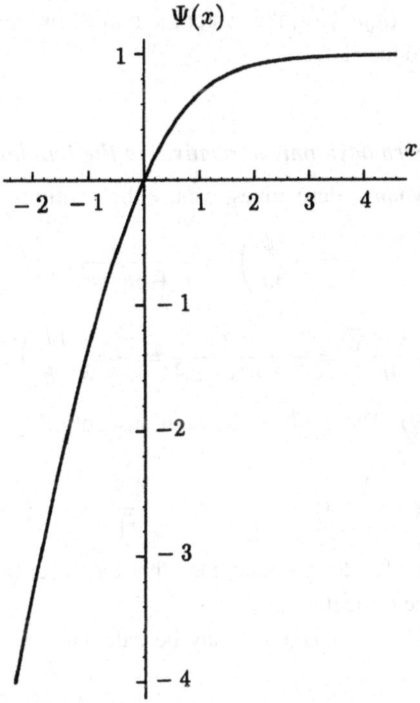

Fig. 3.3. The scaling function $\Psi(x)$ determined from (3.1.13).

for $r < 0$. As shown in Fig. 3.1, these two ordered states have the same value of $f(\psi)$. Figure 3.2 then illustrates how ψ is discontinuous between the two phases at $h = 0$ for $r < 0$. If $h \neq 0$, the degeneracy disappears and the equilibrium value ψ has the same sign as that of h. We may express it in terms of a scaling function $\Psi(x)$ as

$$\psi = \frac{h}{r}\Psi\left(\frac{r}{(u_0h^2)^{1/3}}\right). \tag{3.1.12}$$

From (3.1.9) $\Psi(x)$ satisfies

$$\Psi(x) + \frac{1}{x^3}\Psi(x)^3 = 1 \quad \text{or} \quad x = \Psi(x)/[1 - \Psi(x)]^{1/3}, \tag{3.1.13}$$

so $\Psi(x)$ behaves as (i) $\Psi(x) \cong 1$ for $x \gg 1$, (ii) $\Psi(x) \cong x$ for $|x| \ll 1$, and (iii) $\Psi(x) \cong -|x|^{3/2}$ for $x < 0$ and $|x| \gg 1$, as shown in Fig. 3.3. In case (i), where $r \gg (u_0h^2)^{1/3}$, the temperature is so high that the gaussian approximation for the Landau free energy is valid. In case (ii), where $|r| \ll (u_0h^2)^{1/3}$, the effect of the temperature deviation is negligible and

$$\psi \cong h/(u_0h^2)^{1/3} \propto |h|^{1/3}. \tag{3.1.14}$$

In case (iii), where $r \ll -(u_0 h^2)^{1/3}$, the system is almost on one side of the coexistence curve and (3.1.11) is reproduced.

3.1.3 Thermodynamic derivatives in the Landau theory

To calculate the thermodynamic derivatives, we use the relations,

$$\chi = \left(\frac{\partial \psi}{\partial h}\right)_r = \frac{1}{r + 3u_0\psi^2}, \tag{3.1.15}$$

$$\left(\frac{\partial \psi}{\partial r}\right)_h = -\frac{\psi}{r + 3u_0\psi^2} = -\psi\left(\frac{\partial \psi}{\partial h}\right)_r, \tag{3.1.16}$$

which follow from (3.1.9). The χ/T is the spin susceptibility. If $h = 0$, χ is readily calculated as

$$\chi = \frac{1}{r} \quad (r > 0), \quad \chi = \frac{1}{2|r|} \quad (r < 0). \tag{3.1.17}$$

With the variance relation (1.1.20) we recognize that the order parameter fluctuations are strongly enhanced near the critical point.

The average energy $\langle \mathcal{H} \rangle = \partial(\beta F)/\partial\beta$ may be calculated from (3.1.6). Its density is written as

$$\frac{1}{V}\langle \mathcal{H} \rangle \cong (f_c + T_c s_c) + T_c C_0 \tau - \frac{1}{2} T_c a_0 \psi^2 \tag{3.1.18}$$

where the first two terms arise from f_{reg} in (3.1.8) and the coefficient a_0 is defined in (3.1.7). Thus the energy density consists of a regular part and a term proportional to ψ^2 with a constant coefficient.[3] This will still be the case even in a more sophisticated theory in Chapter 4. Differentiation of (3.1.18) with respect to T gives the specific heat at constant H,

$$C_H = C_0 + \frac{a_0^2\psi^2}{r + 3u_0\psi^2}, \tag{3.1.19}$$

where use has been made of (3.1.16). The two relations (1.1.24) and (3.1.19) indicate that the energy fluctuations are larger in the ordered phase than in the disordered phase. While C_H has no critical divergence in the Landau theory, it is non-analytic at the critical point. In fact, for $h = 0$, it is discontinuous as

$$\Delta C_H = (C_H)_{T<T_c} - (C_H)_{T>T_c} = a_0^2/2u_0. \tag{3.1.20}$$

Next, differentiation of $\langle \mathcal{H} \rangle / V$ in (3.1.18) with respect to T at fixed ψ gives

$$C_M = C_0, \tag{3.1.21}$$

The above result also follows from (1.1.26) with the aid of (3.1.15) and (3.1.16). Therefore, C_M has no singularity for any r and h in the Landau theory.

[3] The last term ($\propto \psi^2$) in (3.1.18) is much larger than the magnetic field energy density $-H\psi$ near the critical point.

We have thus obtained singular or non-analytic behavior of the free energy F and its derivatives, starting with the analytic Landau free energy $\mathcal{H}(\psi)$ given by (3.1.8). It is important that the critical singularity has arisen from the minimization procedure of the Landau free energy with respect to ψ. In the Landau theory, the critical exponents introduced in Section 2.1 are given by

$$\gamma = 1, \quad \beta = \frac{1}{2}, \quad \alpha = 0, \quad \delta = 3. \tag{3.1.22}$$

3.1.4 Landau free energy including the energy variable

We use the notation m to denote the energy density measured from the critical value and divided by T_c. In equilibrium, (3.1.18) suggests that it is expressed in terms of ψ as

$$m = C_0\tau - \frac{1}{2}a_0\psi^2, \tag{3.1.23}$$

where $\tau = T/T_c - 1$ is the reduced temperature. In dynamics, however, the above relation holds only for quasi-static processes, because ψ and m are governed by different dynamic equations. We thus need to treat ψ and m as independent variables. The Landau free-energy density including m is of the form,

$$f(\psi, m) = f(\psi) + \frac{1}{2C_0}T_c\left[m - C_0\tau + \frac{1}{2}a_0\psi^2\right]^2, \tag{3.1.24}$$

where the first term is given by (3.1.6). Some further calculations yield

$$f(\psi, m) = f_c - T_c s_c \tau + T_c\left[\frac{1}{4}\bar{u}_0\psi^4 + \frac{a_0}{2C_0}\psi^2 m + \frac{1}{2C_0}m^2 - h\psi - \tau m\right], \tag{3.1.25}$$

where f_c and s_c are the critical values in (3.1.8), the term proportional to ψ^2 cancels to vanish on the right-hand side, and

$$\bar{u}_0 = u_0 + \frac{1}{2C_0}a_0^2. \tag{3.1.26}$$

Notice that the last two terms in the brackets of (3.1.25) linearly depend on τ and h and fulfill the requirement that their space integral coincides with $-\mathcal{M}h - \mathcal{H}_{ex}\tau/T$ which appears in the microscopic canonical distribution (1.1.9) (as can be known from (1.1.18)).

We may introduce the reduced temperature fluctuation by[4]

$$\delta\hat{\tau} = \frac{1}{T_c}\frac{\partial}{\partial m}f(\psi, m) = \frac{1}{C_0}\left[m + \frac{1}{2}a_0\psi^2\right] - \tau, \tag{3.1.27}$$

which is superimposed on the homogeneous average $\tau = T/T_c - 1$. Then, m may be removed from the Landau free-energy density in favor of $\delta\hat{\tau}$ as

$$f(\psi, m) = f(\psi) + \frac{1}{2}T_c C_0(\delta\hat{\tau})^2. \tag{3.1.28}$$

[4] The circumflex is retained here.

The above expression is consistent with the first variance relation for $\delta\tau = T_c\delta\hat{\tau}$ in (1.1.47). Neglecting the temperature fluctuation ($\delta\hat{\tau} = 0$) or, equivalently, minimizing $f(\psi, m)$ with respect to m leads to the usual free-energy density $f(\psi)$ and the equilibrium relation (3.1.23).

3.2 Tricritical behavior

As suggested by (3.1.26), when an order parameter and subsidiary variables are coupled, the coefficient u_0 of the quartic term in the Landau expansion (3.1.6) is reduced. In some cases, u_0 can be very small or even negative in a certain region of control parameters, leading to tricritical phenomena [7]–[19]. A *symmetrical tricritical point* is realized in metamagnets and ^{3}He–^{4}He mixtures, where there is no physically realizable ordering field ($h = 0$) conjugate to the order parameter and the Landau free-energy density $f(\psi)$ is a function of $|\psi|^2$. The point in the phase diagram at which $r = u_0 = 0$ is called a tricritical point. A more complicated, *unsymmetrical tricritical point* is realized in three- and four-component fluids, where the Landau free energy is not invariant with respect to $\psi \to -\psi$. Generally, at a *multicritical point* a sudden change of ordinary critical lines is encountered. We may mention *tricritical, bicritical, Lifshitz*, and *tetracritical points*, for which see Ref. [16].

3.2.1 Symmetrical tricriticality

(i) In antiferromagnets with nearest and next nearest neighbor exchange couplings (metamagnets), the order parameter ψ is the staggered magnetization. The subsidiary variables are the energy variable and the usual magnetization M. Note that these variables are eliminated in usual static theories. The control parameters are the temperature T and the magnetic field H, which are the fields conjugate to the energy density and the magnetization, respectively. A tricritical point connects a critical line $T = T_c(H)$ and a coexistence line $T = T_{cx}(H)$ separating paramagnetic and antiferromagnetic phases [15]. A representative phase diagram for $FeCl_2$ is shown in Fig. 3.4 [17]. (ii) Another notable example is ^{3}He–^{4}He mixtures at low temperatures, where the order parameter ψ is a complex number. For each pressure p a critical line of the superfluid transition $T = T_\lambda(\Delta, p)$ (the λ line) meets a first-order transition line $T = T_{cx}(\Delta, p)$ at a tricritical point $T = T_t(p)$ with an increase in the chemical potential difference $\Delta = \mu_3 - \mu_4$ or in the ^{3}He composition X, as illustrated in Fig. 3.5 [9, 19]. (iii) We will examine order–disorder phase transitions in solids to find symmetrical tricritical points in (3.3.20)–(3.3.24) below. (iv) In Subsection 10.4.7 we will investigate tricritical behavior of structural phase transition in cubic solids under uniaxial compression.

In the vicinity of a symmetrical tricritical point we need to retain the sixth-order term in the Landau expansion,

$$\frac{1}{T}[f(\psi) - f_{\text{reg}}] = \frac{1}{2}r|\psi|^2 + \frac{1}{4}u_0|\psi|^4 + \frac{1}{6}v_0|\psi|^6 + \cdots, \tag{3.2.1}$$

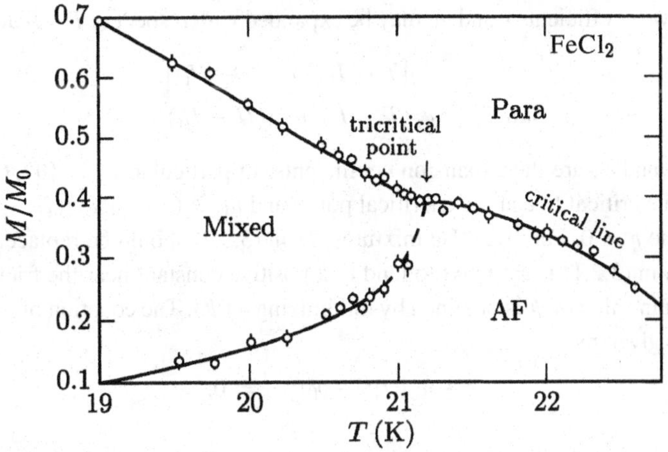

Fig. 3.4. The phase diagram of a metamagnet $FeCl_2$ [17]. M/M_0 is the reduced magnetization. There are antiferromagnetic (AF), paramagnetic (Para), and mixed states.

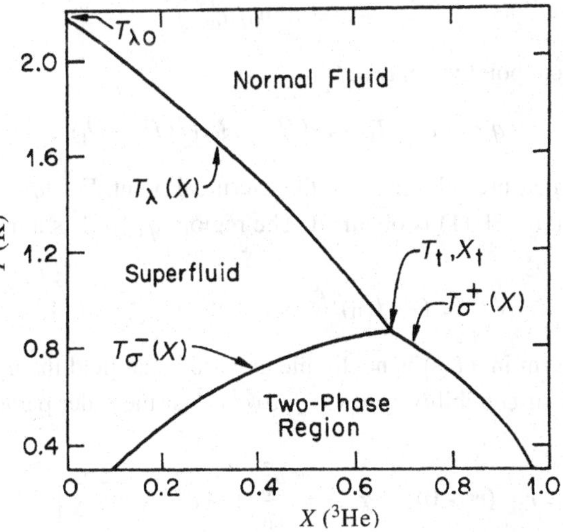

Fig. 3.5. The phase diagram of ^{3}He–^{4}He in the T–X plane at constant p. Here T_σ^+ and T_σ^- are temperatures on the coexistence curve.

where the ordering field h is assumed to be absent. The subsidiary variables, such as the magnetization in metamagnets or the ^{3}He concentration in ^{3}He–^{4}He mixtures, have been eliminated. The tricritical point in metamagnets is given by $T = T_t$ and $H = H_t$. In its vicinity, the two coefficients r and u_0 may be expanded with respect to $T-T_t$ and $H-H_t$ as

$$r \cong a_0\big[T - T_t - c_1(H - H_t)\big],$$
$$u_0 \cong c_2(T - T_t) + c_3(H - H_t), \tag{3.2.2}$$

where c_1, c_2, and c_3 are the expansion coefficients. In particular, $c_1 = [\partial T_c(H)/\partial H]_t$ is the slope of the critical line at the tricritical point, and $u_0 \cong (c_1c_2 + c_3)(H - H_t)$ close to the critical line $r = 0$. For ^{3}He–^{4}He mixtures, H in (3.2.2) should be replaced by Δ. The coefficient v_0 in (3.2.1) is assumed to tend to a positive constant near the tricritical point. The equilibrium value of ψ is obtained by minimizing $f(\psi)$. The equation of nonvanishing ψ at $h = 0$ is given by

$$r + u_0|\psi|^2 + v_0|\psi|^4 = 0. \tag{3.2.3}$$

(i) In the case $u_0 > 0$, the system is disordered ($\psi = 0$) for $r > 0$. An ordered phase appears for $r < 0$ with

$$|\psi|^2 = \frac{u_0}{2v_0}\big[(1-q)^{1/2} - 1\big], \tag{3.2.4}$$

where

$$q = 4rv_0/u_0^2. \tag{3.2.5}$$

Close to the tricritical point we may set

$$q \sim \big[T - T_t - c_1(H - H_t)\big]/(H - H_t)^2. \tag{3.2.6}$$

The inverse $1/|q|$ measures closeness to the tricritical point. For $|q| \ll 1$ the usual mean field critical behavior (3.1.11) is obtained. The region $|q| \gg 1$ is a new tricritical region, where

$$|\psi|^2 \cong (-r/v_0)^{1/2} \propto (-r)^{2\beta_t}, \quad \beta_t = 1/4, \tag{3.2.7}$$

where the quartic term in $f(\psi)$ is negligible and the mean field theory is valid for $d \geq 3$ [11]. The magnetic susceptibility $\chi = (\partial^2 f/\partial\psi^2)^{-1}$ of the order parameter[5] is calculated as

$$\chi^{-1} = r \quad (r > 0), \quad \chi^{-1} = \frac{u_0^2}{v_0}\big[1 - q - \sqrt{1-q}\big] \quad (r < 0). \tag{3.2.8}$$

If $r < 0$, $\chi^{-1} \cong 2|r|$ for $|q| \ll 1$ and $\chi^{-1} \cong 4|r|$ for $|q| \gg 1$. Thus, we have $\chi \sim |r|^{-1}$ as long as $u_0 > 0$ in the mean field theory. In Section 4.4 we shall see that the correlation length ξ is given by $(K\chi)^{1/2} \propto |r|^{-1/2}$ in the mean field theory where K is a constant. Thus we have

$$\gamma_t = 1, \quad \nu_t = 1/2, \tag{3.2.9}$$

in the tricritical region $q \gg 1$.

[5] This is the longitudinal susceptibility χ_L in ordered phases of many-component systems ($n \geq 2$), for which see Section 4.3.

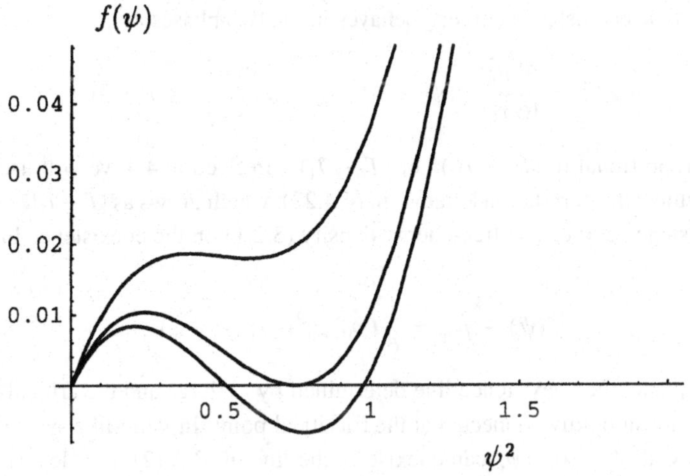

Fig. 3.6. The Landau free-energy density $f(\psi)$ near the tricritical point for (reading from top down) $r = 0.24$, $3/16$, and 0.17. Here ψ and r are scaled such that we have $u_0 = -1$ and $v_0 = 1$. There is no ordering field conjugate to ψ.

(ii) For $u_0 < 0$ we display $f(\psi)$ as a function of $|\psi|^2$ in Fig. 3.6 for three typical cases. If $q < 1$, a nonvanishing solution of (3.2.3) giving a local minimum of $f(\psi)$ is obtained as

$$|\psi|^2 = \frac{|u_0|}{2v_0}\left[\sqrt{1-q}+1\right]. \tag{3.2.10}$$

For $q < 0$ and $|q| \gg 1$, ψ becomes independent of u_0, leading to the tricritical result (3.2.7) again. The free energy at the local minimum is given by

$$f_{\min} - f_{\text{reg}} = T\frac{|u_0|^3}{24v_0}\left[\frac{3}{2}q - 1 - (1-q)^{3/2}\right], \tag{3.2.11}$$

where the right-hand side is positive for $q > 3/4$ and negative for $q < 3/4$. Thus the disordered phase is stable for $q > 3/4$ and the ordered phase is stable for $q < 3/4$. The coexistence line in the phase diagram is determined by $q = 3/4$ or

$$r = \frac{3}{16}\frac{u_0^2}{v_0}, \qquad u_0 < 0. \tag{3.2.12}$$

On the coexistence curve the absolute value of the order parameter in the ordered phase is written as

$$\psi_{\text{cx}} = (3|u_0|/4v_0)^{1/2} \propto |H - H_{\text{t}}|^{1/2} \propto |T - T_{\text{t}}|^{1/2}. \tag{3.2.13}$$

The magnetic susceptibility in equilibrium is given by

$$\chi^{-1} = r \quad (q > 3/4), \quad \chi^{-1} = \frac{u_0^2}{v_0}\left[1 - q + \sqrt{1-q}\right] \quad (q < 3/4). \tag{3.2.14}$$

Therefore, on the coexistence curve χ behaves in the two phases as

$$\chi^{-1} = \frac{3}{16}\frac{u_0^2}{v_0} \quad (\psi = 0), \quad \chi^{-1} = \frac{3}{4}\frac{u_0^2}{v_0} \quad (\psi \neq 0), \tag{3.2.15}$$

which are proportional to $(H - H_t)^2$ or $(T - T_t)^2$. In Section 4.4 we will use the above result to calculate the correlation length ξ in (4.4.22), which grows as $|T - T_t|^{-1}$ as $T \to T_t$ on the coexistence curve. The free-energy density (3.2.1) on the coexistence line is of the form,

$$f(\psi) - f_{\text{reg}} = \frac{1}{6} T v_0 |\psi|^2 (|\psi|^2 - \psi_{\text{cx}}^2)^2. \tag{3.2.16}$$

In the T–H plane, the coexistence line determined by (3.2.12) and the critical line $r = 0$ with $u_0 > 0$ are smoothly connected at the tricritical point. In Ising-like systems ($n = 1$), three phases with $\psi = 0, \pm\psi_e$ can coexist on the line of (3.2.12). In ^{3}He–^{4}He mixtures, where $n = 2$, the phase variable of the complex order parameter remains arbitrary in the ordered phase.

Nonvanishing ordering field

When the ordering field h conjugate to the order parameter ψ is nonvanishing, we should add the term $-h\psi$ on the right-hand side of (3.2.1). Here ψ is treated as a scalar variable. Then, from $\partial^2 f/\partial\psi^2 = \partial^3 f/\partial\psi^3 = 0$, we find another critical line passing through the tricritical point in the region $u_0 \leq 0$ [11], on which

$$\psi = \pm\sqrt{\frac{3}{10}\frac{|u_0|}{v_0}}, \quad h = \frac{8}{3} v_0 \psi^5, \quad r = \frac{9}{20}\frac{u_0^2}{v_0}. \tag{3.2.17}$$

We also have a coexistence surface terminating at this critical line and including the first-order phase transition line (3.2.12) for $h = 0$ in the r–u_0–h (or T–H–h) space. This field-induced critical line was observed in ferroelectric KH_2PO_4 near a tricritical point in an applied electric field [20].

3.2.2 Scaling theory around a symmetrical tricritical point

It is straightforward to develop a scaling theory near a symmetrical tricritical point [7, 9]. The singular part of the free-energy density f_{sing} as a function of r, $u_0 (\propto H - H_t$ or $T - T_t)$, and the ordering field h satisfies

$$f_{\text{sing}}(r, u_0, h) = \ell^{-\phi(2-\alpha_t)} f_{\text{sing}}(\ell^\phi r, \ell u_0, \ell^{\phi\Delta_t} h), \tag{3.2.18}$$

for any positive values of ℓ, where ϕ, α_t, and Δ_t are new exponents. By setting $\ell = |u_0|^{-1}$ we obtain

$$f_{\text{sing}}(r, u_0, h) = |u_0|^{\phi(2-\alpha_t)} F_\pm(r/|u_0|^\phi, h/|u_0|^{\phi\Delta_t}), \tag{3.2.19}$$

where $F_\pm(x, y) = f_{\rm sing}(x, \pm 1, y)$ are defined for $u_0 > 0$ and $u_0 < 0$, respectively. By differentiating $f_{\rm sing}$ with respect to h and then setting $h = 0$, we obtain

$$\begin{aligned} \psi &= |u_0|^{\phi(2-\alpha_t-\Delta_t)}\Phi_\pm(r/|u_0|^\phi), \\ \chi &= |u_0|^{\phi(2-\alpha_t-2\Delta_t)}\Xi_\pm(r/|u_0|^\phi). \end{aligned} \tag{3.2.20}$$

Here $\Phi_\pm(x) = [\partial F_\pm(x, y)/\partial y]_{y\to 0}$ and $\Xi_\pm(x) = [\partial^2 F_\pm(x, y)/\partial y^2]_{y\to 0}$. Comparing the above scaling forms and the mean field results (3.2.4)–(3.2.15) for $|q| \gtrsim 1$, we find that the scaling variable $x = r/|u_0|^\phi$ should be identified with q in (3.2.5), so that

$$\phi = 2, \quad \alpha_t = 1/2, \quad \Delta_t = 5/4. \tag{3.2.21}$$

It is well known that the mean field theory is valid for small u_0 in the region $|q| \gtrsim 1$ for $d \ge 3$, which can be concluded on the basis of the Ginzburg criterion [11]. Because the upper critical dimensionality is 3, there are logarithmic corrections in 3D [12], but they are usually negligible.

Furthermore, we may examine the singular behavior of the subsidiary variable m which is coupled to ψ (but has been eliminated in (3.2.1)). From (3.2.18) its average $\langle m\rangle$ deviates from the tricritical value m_t as

$$\langle m\rangle - m_t = |u_0|^{\phi(1-\alpha_t)}\mathcal{M}_\pm(r/|u_0|^\phi) + (\Delta m)_{\rm reg}, \tag{3.2.22}$$

where the second term is the regular part arising from $f_{\rm reg}$. The variance of m consists of the background and singular parts,

$$C = \langle m : m\rangle = C_0 + |u_0|^{-\phi\alpha_t}\mathcal{M}'_\pm(r/|u_0|^\phi), \tag{3.2.23}$$

which is either the specific heat or the usual magnetic susceptibility in metamagnets, or the concentration susceptibility in ^{3}He–^{4}He. For $|x| = |r|/|u_0|^\phi \gg 1$, C should be independent of u_0 and the tricritical specific-heat singularity follows as

$$C \propto |r|^{-\alpha_t}, \quad \alpha_t = 1/2. \tag{3.2.24}$$

On the coexistence curve (3.2.12) we have

$$\langle m\rangle - m_t \sim H - H_t \sim T - T_t, \quad C \sim |H - H_t|^{-1} \sim |T - T_t|^{-1}, \tag{3.2.25}$$

for $d \ge 3$. These results are in good agreement with experiments on metamagnets [16] and ^{3}He–^{4}He [18]. In addition, we shall see $\xi \sim |T - T_t|^{-\phi\nu_t}$ with $\phi\nu_t = 1$ on the coexistence curve in (4.4.22) below.

3.2.3 *Unsymmetrical tricriticality*

We discuss more complicated unsymmetrical tricritical points [10, 14]. For example, in three-component fluid mixtures, three-phase coexistence may be realized on a two-dimensional surface in the space of four independent field variables, and two phases become identical on a line of critical end points. Therefore, by choosing unique temperature, pressure, and two chemical potentials (or, equivalently, two mole fractions), there can be a tricritical point where all the three phases become indistinguishable and exhibit critical opalescence. More generally, in four-component fluid mixtures, we have a line of tricritical points. Around such a point, however, there is no invariance of the free energy with respect to a change of the sign, $\psi \to -\psi$, of the order parameter. Therefore, we need to add odd terms in the Landau expansion. Supposing a scalar order parameter ψ, we have [10]

$$f(\psi) = f_{\rm reg} + a_1\psi + a_2\psi^2 + a_3\psi^3 + a_4\psi^4 + a_5\psi^5 + a_6\psi^6 + \cdots, \tag{3.2.26}$$

where the coefficients a_k are functions of the field variables $T, p, \ldots$. All the subsidiary variables coupled to the order parameter have been eliminated from the minimum conditions. Here the fifth-order term vanishes if $\psi' = \psi + a_5/6a_6$ is redefined as a new order parameter, but the third-order term cannot be removed at the same time. Three-phase coexistence is realized if $f(\psi)$ is expressed as

$$f(\psi) = a_6(\psi - c_1)^2(\psi - c_2)^2(\psi - c_3)^2 + \text{const.} \tag{3.2.27}$$

In three-component fluids, we obtain lines of critical end points if two of c_1, c_2, and c_3 coincide, and a tricritical point if $c_1 = c_2 = c_3$. However, it is highly nontrivial how the critical surface and the tricritical point can be approached with changing experimental parameters.

3.3 Bragg–Williams approximation

3.3.1 *Ising systems*

We now discuss the phase transition in ferromagnetic Ising spin systems ($J > 0$) in the simplest mean field theory [2]. Let N_+ be the number of the up-spins ($s_i = 1$) and $N_- = \Omega - N_+$ the number of the down-spins ($s_i = -1$), where Ω is the total number of lattice sites. For a binary alloy forming a simple cubic lattice, N_+ and N_- are interpreted as the numbers of A and B atoms, respectively [5]. The order parameter ψ is defined by

$$\psi = (N_+ - N_-)/\Omega. \tag{3.3.1}$$

Then $N_+ = \Omega(1+\psi)/2$ and $N_- = \Omega(1-\psi)/2$. If we neglect the correlations among the spins, the probability that a neighboring pair has the same spin direction is $(N_+/\Omega)^2 + (N_-/\Omega)^2$ and the probability that a neighboring pair has different spin directions is $2N_+N_-/\Omega^2$. The exchange energy $\mathcal{H}_{\rm ex}$ between nearest neighbor pairs is approximated by the average

$$\bar{E} = -\frac{zJ}{2\Omega}\left[N_+^2 + N_-^2 - 2N_+N_-\right] = -\frac{1}{2}\Omega zJ\psi^2, \tag{3.3.2}$$

where z is the coordination number. Replacing $\mathcal{H}$ in the partition function Z in (1.1.11) by $\bar{E} - \Omega H\psi$, we obtain the approximate partition function,

$$Z = \frac{\Omega!}{N_+!N_-!}\exp\left[\Omega\beta\left(\frac{1}{2}zJ\psi^2 + H\psi\right)\right]. \tag{3.3.3}$$

By taking the logarithm of Z and using the Stirling formula $\ln M! \cong M \ln M - M$ for large $M \gg 1$, we obtain the Landau free energy $F(\psi)$ in this approximation. The free-energy density $f_{\text{site}} = F(\psi)/\Omega$ per site becomes

$$\frac{1}{T} f_{\text{site}} = \frac{1}{2}(1+\psi)\ln\frac{1+\psi}{2} + \frac{1}{2}(1-\psi)\ln\frac{1-\psi}{2} - \frac{zJ}{2T}\psi^2 - h\psi, \tag{3.3.4}$$

where $h = H/T$ and the regular part is omitted. The first two terms are the entropy contributions and are of the same form as the minus of the translational entropy $\ln(V^{N_+}/N_+!) + \ln(V^{N_-}/N_-!)$ of ideal gas mixtures [2]. (The space-filling condition $N_+ + N_- = \Omega$ is not necessarily needed for fluids, however.)

The minimization of f_{site} yields the equilibrium value:

$$\frac{1}{2}\ln[(1+\psi)/(1-\psi)] - (zJ/T)\psi = h, \tag{3.3.5}$$

which may also be transformed into

$$\psi = \tanh[h + (zJ/T)\psi]. \tag{3.3.6}$$

For this free energy the critical temperature is given by

$$T_c = zJ. \tag{3.3.7}$$

In fact, the susceptibility $\chi = (\partial\psi/\partial h)_T$ behaves as

$$\chi = T(1-\psi^2)/\big[T - zJ(1-\psi^2)\big], \tag{3.3.8}$$

which diverges as $(T - T_c)^{-1}$ at $\psi = 0$. Near the critical point, f_{site} assumes the Landau expansion form,

$$\frac{1}{T} f_{\text{site}} = \frac{1}{2}(1 - zJ/T)\psi^2 + \frac{1}{12}\psi^4 + \cdots - h\psi. \tag{3.3.9}$$

Moreover, for $T \ll T_c$ and $h = 0$, (3.3.5) is solved to give

$$\psi \cong \pm\big[1 - 2\exp(-2T_c/T)\big]. \tag{3.3.10}$$

3.3.2 Order–disorder phase transitions in bcc alloys

Let us consider an A–B binary alloy forming a body-centered-cubic (bcc) lattice such as Fe–Be and Cu–Zn [21]–[24]. The lattice may be divided into two sublattices as shown in Fig. 3.7. The concentrations of A atoms on the two sublattice sites are written as

$$c_1 = c + \frac{1}{2}\eta, \quad c_2 = c - \frac{1}{2}\eta, \tag{3.3.11}$$

where c is the concentration of A atoms averaged over the two sublattices and η is the order parameter of the order–disorder phase transition (often called the long-range order parameter). The concentrations of B atoms are $1 - c_1$ and $1 - c_2$ on the two sublattice sites. We may assume $0 < c \le 1/2$ without loss of generality; then, $|\eta| \le 2c$. The lattice

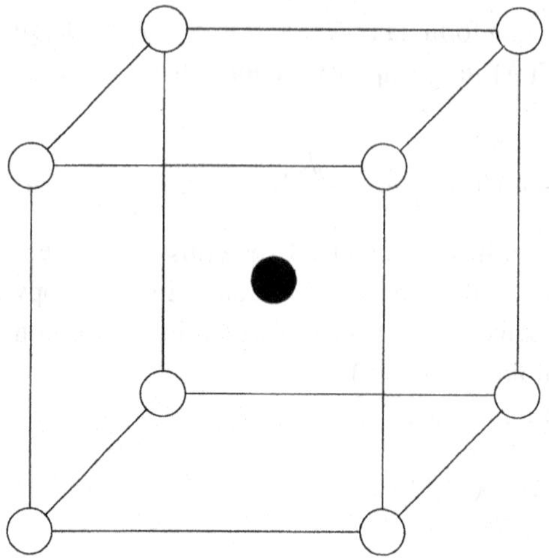

Fig. 3.7. The $L1_0$ structure on a bcc lattice.

structure in the ordered phase ($\eta \neq 0$) is called $L1_0$ or B2. The system is invariant with respect to a change of the sign of η, because the two sublattices are symmetrical, so the free energy is an even function of η. Assuming the interactions between the nearest and next nearest neighbor pairs, we obtain the free-energy density per lattice point in the form [21, 23, 24],

$$\begin{aligned} f_{\text{site}} = & \frac{T}{2}\left[\left(c+\frac{\eta}{2}\right)\ln\left(c+\frac{\eta}{2}\right)+\left(c-\frac{\eta}{2}\right)\ln\left(c-\frac{\eta}{2}\right)\right. \\ & +\left(1-c+\frac{\eta}{2}\right)\ln\left(1-c+\frac{\eta}{2}\right) \\ & \left.+\left(1-c-\frac{\eta}{2}\right)\ln\left(1-c-\frac{\eta}{2}\right)\right]-w_0c^2-w_1\eta^2, \end{aligned} \tag{3.3.12}$$

where w_0 and w_1 are combinations of the pair interaction energies. The term linear in c is not written explicitly because c is a conserved variable. Obviously, f_{site} is of the same form as (3.3.4) with $c = (1 + \psi)/2$ if there is no order ($\eta = 0$). The phase behavior is determined by the two parameters, w_0 and w_1, in a complicated manner, as illustrated in Fig. 3.8 for the case $w_1 > 0$ [23]. Generally, increasing w_0 favors phase separation, while increasing w_1 favors structural ordering. Instability curves are determined by $(\partial^2 f_{\text{site}}/\partial c^2)(\partial^2 f_{\text{site}}/\partial \eta^2) = (\partial^2 f_{\text{site}}/\partial c\partial\eta)^2$, below which homogeneously ordered or disordered states are unstable against long-wavelength perturbations of c and η. This condition is expressed as

$$\left[T-2w_0\left(c-c^2-\frac{1}{4}\eta^2\right)\right]\left[T-8w_1\left(c-c^2-\frac{1}{4}\eta^2\right)\right]=4w_0w_1(2c-1)^2\eta^2. \tag{3.3.13}$$

In the simple case of $c = 1/2$ or $\eta = 0$, the right-hand side of (3.3.13) vanishes, and we obtain two spinodal points, $T = 2w_0(c - c^2 - \eta^2/4)$ with respect to clustering and

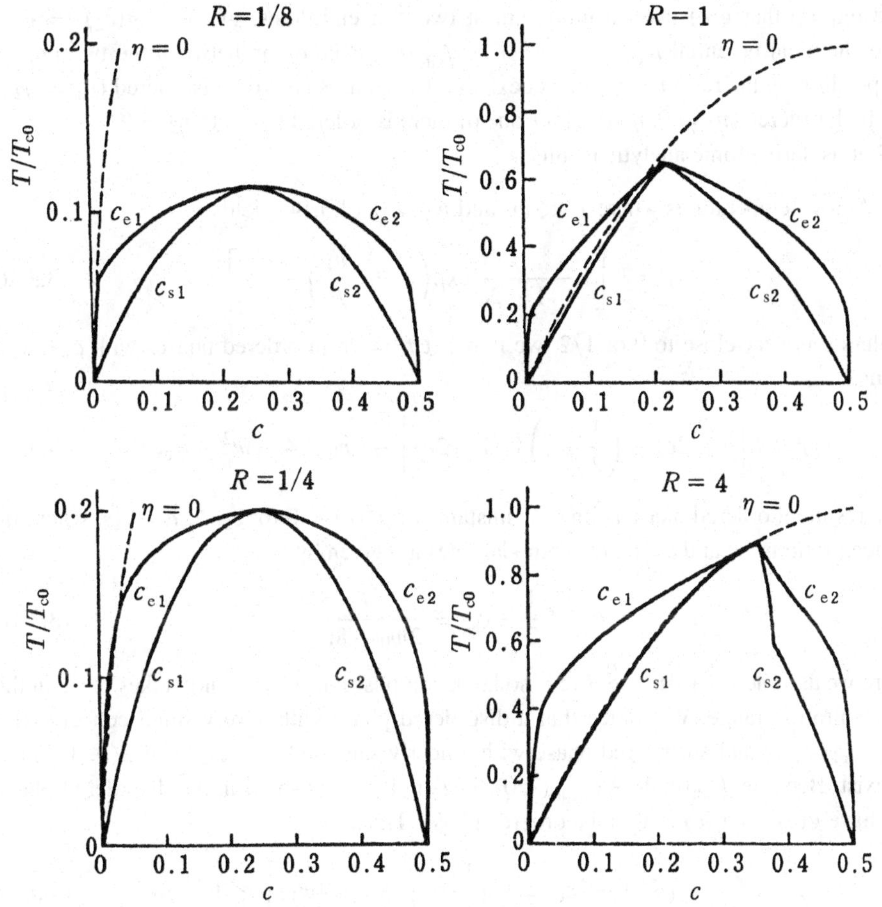

Fig. 3.8. Calculated phase diagrams of bcc alloys for $w_1 > 0$ on the basis of (3.3.12) [23]. The parameter R is defined by $w_0/w_1 = 4(R-1)/(R+1)$. The temperature is scaled by $T_{c0} = 2w_1$. The c_{e1} and c_{e2} are the solubility (coexistence) lines, and c_{s1} and c_{s2} are the spinodal lines. The instability line $T/T_{c0} = 4c(1-c)$ against ordering is also shown (broken line).

$T = 8w_1(c - c^2 - \eta^2/4)$ with respect to ordering. The equation to determine η follows from $\partial f_{\text{site}}/\partial\eta = 0$ at fixed c as

$$\ln\left[\left(c+\frac{\eta}{2}\right)\left(1-c+\frac{\eta}{2}\right)\Big/\left(c-\frac{\eta}{2}\right)\left(1-c-\frac{\eta}{2}\right)\right] = 8\frac{w_1}{T}\eta. \tag{3.3.14}$$

The solution $\eta = \eta(c)$, which gives the minimum of $f_{\text{site}}(c, \eta)$ at each c, needs to be calculated. It can be nonvanishing only for $w_1 > 0$, so we will assume $w_1 > 0$. Then $f_{\text{site}}(c, \eta(c))$ becomes a function of c only. To find two-phase coexistence we introduce

$$g(c) = f_{\text{site}}(c, \eta(c)) - \mu_{\text{cx}}c \tag{3.3.15}$$

and require that $g(c)$ takes a minimum at two concentrations, $g(c_1) = g(c_2) = g_{\min}$. The chemical potential $\mu_{cx} = f'_{\text{site}}(c_1) = f'_{\text{site}}(c_2)$ is common between the two phases. Depending on the ratio w_0/w_1, the coexisting two phases are both disordered ($\eta_1 = \eta_2 = 0$), both ordered ($\eta_1 \neq 0, \eta_2 \neq 0$), or one of them is ordered ($\eta_1 \neq 0, \eta_2 = 0$).

Let us derive some analytic results.

(i) At low temperatures where $T \ll w_1$ and $w_1 > 0$, (3.3.14) yields

$$\eta = 2c\left[1 - \frac{2}{1-2c}\exp\left(-16c\frac{w_1}{T}\right) + \cdots\right]. \tag{3.3.16}$$

Unless c is very close to 0 or 1/2, we may set $\eta = 2c$ in ordered phases with $c < 1/2$. Thus,

$$g(c) = T\left[c\ln(2c) + \left(\frac{1}{2} - c\right)\ln(1-2c)\right] - (w_0 + 4w_1)c^2 - \mu_{cx}c. \tag{3.3.17}$$

The resultant ordered phase is linearly unstable or $g''(c) < 0$ for $c_{s1} < c < c_{s2}$, where the concentrations c_{s1} and c_{s2} on the spinodal lines are given by

$$c_{s1} \cong \frac{1}{2} - c_{s2} \cong \frac{T}{2w_0 + 8w_1}. \tag{3.3.18}$$

Here we assume $w_0 + 4w_1 \gg T$. Spinodal decomposition subsequently takes place in this concentration range. We notice that a disordered phase with a very small concentration $c = c_{e1} (\ll 1)$ and an ordered phase with a nearly saturated $c = c_{e2} \cong \eta/2 (\cong 1/2)$ can coexist. Because $f_{\text{site}}(c, 0) - cf'_{\text{site}}(c, 0) \cong T\ln(1-c)$ is small in the disordered phase, we have $g(c) - cg'(c) \cong 0$ in the ordered phase. Thus,

$$c_{e1} \cong 1 - 2c_{e2} \cong \exp\left[-\frac{1}{2T}(w_0 + 4w_1)\right] \ll 1. \tag{3.3.19}$$

Disordered states in the range $c_{e1} < c < c_{s1}$ and ordered states in the range $c_{s2} < c < c_{e2}$ are metastable with respect to clustering. Nucleation of the other phase triggers phase separation, as will be discussed in Chapter 9.

(ii) If η is small under $w_1 > 0$, we expand f_{site} with respect to η as

$$\frac{1}{T}f_{\text{site}} = c\ln c + (1-c)\ln(1-c) - \frac{w_0}{T}c^2 + \frac{1}{2}r(c)\eta^2 + \frac{1}{4}\bar{u}_0\eta^4 + \frac{1}{6}v_0\eta^6 + \cdots, \tag{3.3.20}$$

where

$$r(c) = \frac{1}{4c(1-c)} - 2\frac{w_1}{T}, \quad \bar{u}_0 = \frac{1}{48}\left[\frac{1}{c^3} + \frac{1}{(1-c)^3}\right], \quad v_0 = \frac{1}{320}\left[\frac{1}{c^5} + \frac{1}{(1-c)^5}\right]. \tag{3.3.21}$$

The concentration fluctuation $\delta c = c - \bar{c}$ from the average $\bar{c} = \langle c \rangle$ plays the role of the energy variable in Ising systems. In fact, the composition dependence of $r(c)$ gives rise to a coupling term $T\gamma_0\delta c\eta^2$ in the free-energy density, as in (3.1.25), while c in $\bar{u}_0$ and v_0

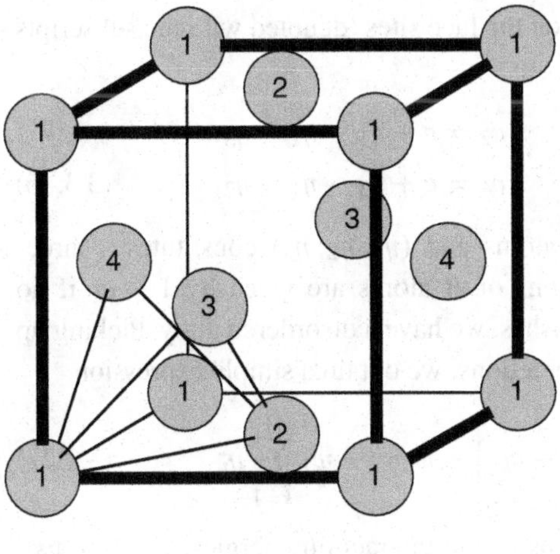

Fig. 3.9. Identification of atom sites in a fcc unit cell.

may be replaced by $\bar{c}$ for small η. Expanding $f_{\rm site}$ with respect to δc, we have

$$\gamma_0 = \frac{1}{2}r'(c) = \frac{2\bar{c}-1}{8\bar{c}^2(1-\bar{c})^2}, \quad C_0^{-1} = \frac{1}{\bar{c}(1-\bar{c})} - 2\frac{w_0}{T}. \tag{3.3.22}$$

If δc is eliminated, we obtain (3.2.1) with $u_0 = \bar{u}_0 - 2C_0\gamma_0^2$, similar in form to (3.1.26). The critical line exists in the region $c_{\rm t} < c < 1 - c_{\rm t}$, where [21]

$$c_{\rm t} = \frac{1}{2} - \sqrt{\frac{1}{12}\cdot\frac{4w_1 - w_0}{4w_1 + w_0}}, \tag{3.3.23}$$

where we assume $w_1 > |w_0|/4$. The critical temperature depends on c as

$$T_{\rm c}(c) = 8w_1 c(1-c) \tag{3.3.24}$$

and takes the highest value $2w_1$ at $c = 1/2$. There can be two symmetrical tricritical points at $c = c_{\rm t}$ and $1 - c_{\rm t}$ [21, 22]. The tricritical temperature is

$$T_{\rm t} = T_{\rm c}(c_{\rm t}) = \frac{8w_1(2w_1 + w_0)}{3(4w_1 + w_0)}. \tag{3.3.25}$$

The critical line is connected to lines of first-order phase transition in the regions, $c < c_{\rm t}$ and $c > 1 - c_{\rm t}$. However, we have only lines of first-order phase transition for $4w_1 - w_0 < 0$ and only a critical line for $2w_1 + w_0 < 0$.

3.3.3 *Order–disorder phase transitions in fcc alloys*

We next consider a binary alloy such as Al–Li or a number of Ni-based alloys having a face-centered-cubic (fcc) lattice, as in Fig. 3.9. The concentration of A atoms on the corner sites (denoted with the subscript 1) and those on the face sites (denoted with the subscripts 2, 3, 4) are expressed as [1, 21, 25]

$$c_1 = c + \eta_1 + \eta_2 + \eta_3, \qquad c_2 = c + \eta_1 - \eta_2 - \eta_3,$$
$$c_3 = c + \eta_2 - \eta_3 - \eta_1, \qquad c_4 = c + \eta_3 - \eta_1 - \eta_2, \tag{3.3.26}$$

where c is the average concentration of A atoms and (η_1, η_2, η_3) constitutes a three-component order parameter. The concentrations of B atoms are given by $1 - c_k$ if no defects are present. If the order parameter vanishes, we have a disordered alloy. Picking up the nearest and next nearest neighbor pair interactions, we obtain a simple expression,

$$f_{\text{site}} = \frac{T}{4}\sum_{k=1}^{4}[c_k \ln c_k + (1-c_k)\ln(1-c_k)] - w_0 c^2 - w_1 \sum_{k=1}^{3} \eta_k^2, \tag{3.3.27}$$

where w_0 and w_1 are appropriate combinations of the interaction energies. Because two atoms at the sites 1 and 2 (corner–face) and those at the sites 2 and 3 (face–face) are equally separated, they interact with the same potentials and the nearest neighbor interaction energy becomes proportional to $c_1(c_2 + c_3 + c_4) + (c_2c_3 + c_3c_4 + c_4c_2)$, leading to the last term of (3.3.27). The Landau expansion of f_{site} in powers of η_k becomes

$$\begin{aligned}\frac{f_{\text{site}}}{T} &= c\ln c + (1-c)\ln(1-c) - \frac{w_0}{T}c^2 \\ &+ \left[\frac{1}{2c(1-c)} - \frac{w_1}{T}\right]\sum_{k=1}^{3}\eta_k^2 + \frac{2c-1}{2c^2(1-c)^2}\eta_1\eta_2\eta_3 \\ &+ \frac{1}{12}\left[\frac{1}{c^3} + \frac{1}{(1-c)^3}\right]\left[\sum_{k=1}^{3}\eta_k^4 + 6(\eta_1^2\eta_2^2 + \eta_2^2\eta_3^2 + \eta_3^2\eta_1^2)\right] + \cdots.\end{aligned} \tag{3.3.28}$$

The free energy is isotropic up to the second-order terms. The instability curve at homogeneous c and $\eta_k = 0$ $(k = 1, 2, 3)$ is given by $T = 2w_1c(1-c)$ for $w_1 > 0$, below which small fluctuations of η_k grow. The usual spinodal is given by $T = 2w_0c(1-c)$ for $w_0 > 0$, below which disordered homogeneous solutions are unstable against fluctuations of c. More phenomenologically, we may set up the Landau expansion up the sixth-order terms from symmetry requirements of the fcc structure as [25, 26]

$$\begin{aligned}\frac{f_{\text{site}}}{T} &= \frac{f_0(c)}{T} + a_2\sum_{k=1}^{3}\eta_k^2 + \left(a_3 + a_5\sum_{k=1}^{3}\eta_k^2\right)\eta_1\eta_2\eta_3 \\ &+ \left(a_{41} + a_{62}\sum_{k=1}^{3}\eta_k^2\right)\sum_{k=1}^{3}\eta_k^4 \\ &+ a_{42}(\eta_1^2\eta_2^2 + \eta_2^2\eta_3^2 + \eta_3^2\eta_1^2) + a_{61}\sum_{k=1}^{3}\eta_k^6 + a_{63}\eta_1^2\eta_2^2\eta_3^2 + \cdots,\end{aligned} \tag{3.3.29}$$

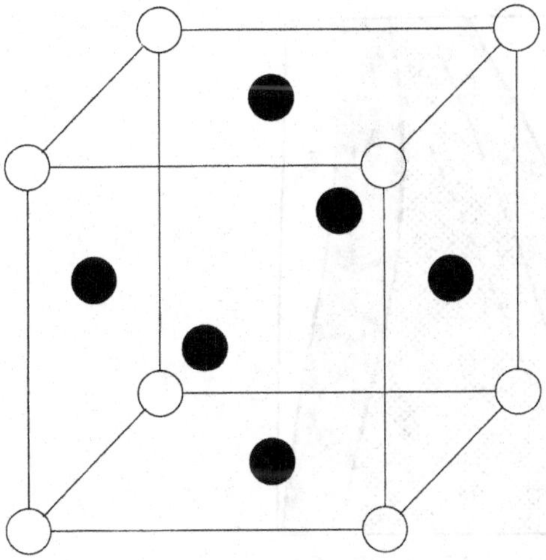

Fig. 3.10. The $L1_2$ structure of Al_3Li, Ni_3Cr, etc., on a fcc lattice (Al:•, Li:○). For Al–Li, domains of this structure appear in an Al-rich metastable, disordered phase as T is lowered or the Li concentration is increased.

where the coefficients, $a_2, a_3, \ldots$, are functions of c and T. Depending on the values of these coefficients at each c, ordered states with the form $(\eta_1, \eta_2, \eta_3) = (\pm\eta, 0, 0)$ can be stable, where we have $c_1 = c_2 = c + \eta_1$ and $c_3 = c_4 = c - \eta_1$. Then the free-energy density (3.3.27) assumes the same form as that in (3.3.12), leading to an $L1_0$ structure. Equivalently, we may set $(\eta_1, \eta_2, \eta_3) = (0 \pm \eta, 0)$ or $(0, 0, \pm\eta)$. Thus there are six variants with the $L1_0$ structure emerging in phase-ordering processes. In real fcc crystals, however, such atomic displacements in a preferred direction cause a cubic-to-tetragonal change of the lattice structure, as will be discussed in Section 10.3.

The $L1_2$ structure in Fig. 3.10 is realized for isotropic ordering $\eta_1 = \eta_2 = \eta_3 = \eta$. For a perfect $L1_2$ crystal we have $c = \eta = 1/4$. In this case, the free-energy density becomes [27, 28]

$$\begin{aligned} f_{\text{site}} &= \frac{T}{4}\big[(c+3\eta)\ln(c+3\eta) + (1-c-3\eta)\ln(1-c-3\eta) \\ &\quad + 3(c-\eta)\ln(c-\eta) + 3(1-c+\eta)\ln(1-c+\eta)\big] - w_0 c^2 - 3w_1\eta^2. \end{aligned} \tag{3.3.30}$$

Equivalently, we may set $(\eta_1, \eta_2, \eta_3) = (\eta, -\eta, -\eta)$, $(-\eta, \eta, -\eta)$, or $(-\eta, -\eta, \eta)$ from the fcc symmetry [26]. Note that (3.3.27) and (3.3.28) are invariant with respect to the change $(\eta_1, \eta_2, \eta_3) \to (-\eta_1, -\eta_2, \eta_3)$ etc. Thus there are four equivalent ordered variants. As can be seen from (3.3.27), if f_{site} is expanded in powers of η, the cubic term ($\propto \eta^3$) remains nonvanishing here, suggesting a first-order phase transition [1]. The equation to determine η follows from $\partial f_{\text{site}}/\partial\eta = 0$ as

$$\ln\big[(c+3\eta)(1-c+\eta)/(c-\eta)(1-c-3\eta)\big] = 8\frac{w_1}{T}\eta. \tag{3.3.31}$$

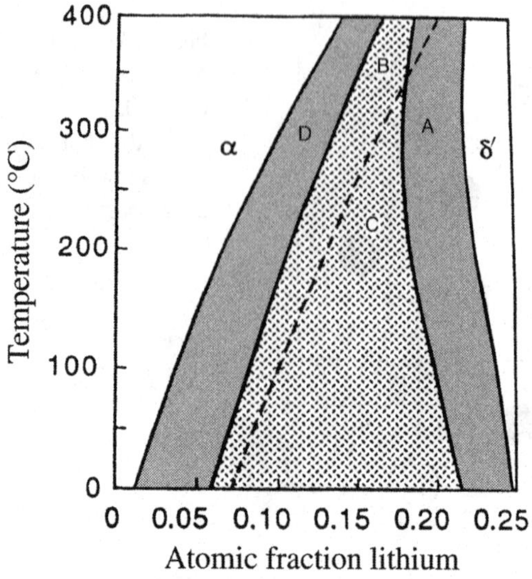

Fig. 3.11. The metastable two-phase region in Al–Li in the Bragg–Williams theory [27]. The disordered phase is stable in region α, while the δ' (Al_3Li) phase is stable in region δ'. The dashed curve represents the spinodal curve of a homogeneous disordered phase. A solution quenched into regions A and D is metastable. A solution quenched from α into region C below the dashed curve is unstable against ordering and then decomposes through a secondary spinodal. A solution quenched into region B from α is metastable with respect to ordering, but undergoes spinodal decomposition after ordering.

The instability curves are determined by

$$\left[T - \frac{w_1}{2}(3A_1 + A_2)\right]\left[T - \frac{w_0}{2}(A_1 + 3A_2)\right] = 12 w_0 w_1 \eta^2 (1 - 2c - 2\eta)^2, \qquad (3.3.32)$$

where $A_1 = (c + 3\eta)(1 - c - 3\eta)$ and $A_2 = (c - \eta)(1 - c + \eta)$. In the disordered case $\eta = 0$ we have $A_1 = A_2 = c(1 - c)$ and obtain the instability curves mentioned below (3.3.28). The resultant phase behavior is complicated and contains a rich variety of phases, depending on T, the overall composition, w_0, and w_1. Khachaturyan *et al.* [27] examined the consequences of the mean field free energy (3.3.30) setting $w_0 = -2535$ K (< 0) and $w_1 = 2030$ K (> 0) for Al–Li, as illustrated in Fig. 3.11. As in the bcc case, ordering can first take place without appreciable change of large-scale composition fluctuations and the resultant order can then induce spinodal decomposition for relatively deep quenching. In Al–Li the elastic effects to be discussed in Chapter 10 are suppressed because of very small lattice mismatch, where δ'-phase precipitates have in fact been observed to be spherical. See Ref. [29] for experiments.

As in the bcc case, we give analytic results at low temperatures. We assume $T \ll w_1$, $|w_0|/w_1 \lesssim 1$, and $c < 1/4$. From (3.3.31) we then obtain

$$\eta = c\left[1 - \frac{4}{1-4c}\exp\left(-8c\frac{w_1}{T}\right) + \cdots\right]. \tag{3.3.33}$$

Unless c is very close to 0 or $1/4$, we may set $\eta = c$ in ordered states to obtain

$$g(c) = T\left[c\ln(4c) + \left(\frac{1}{4} - c\right)\ln(1-4c)\right] - (w_0 + 3w_1)c^2 - \mu_{\text{cx}}c. \tag{3.3.34}$$

The resultant ordered phase is unstable for $c_{\text{s}1} < c < c_{\text{s}2}$, where

$$c_{\text{s}1} \cong \frac{1}{4} - c_{\text{s}2} \cong \frac{T}{2w_0 + 6w_1}, \tag{3.3.35}$$

where $w_0 + 3w_1 \gg T$ is assumed. Spinodal decomposition then takes place as indicated in Fig. 3.11. If a disordered phase with $c = c_{\text{e}1}$ and an ordered phase with $c = c_{\text{e}2} \cong \eta$ coexist, we have

$$c_{\text{e}1} \cong 1 - 4c_{\text{e}2} \cong \exp\left[-\frac{1}{4T}(w_0 + 3w_1)\right] \ll 1. \tag{3.3.36}$$

Disordered states in the range $c_{\text{e}1} < c < c_{\text{s}1}$ and ordered states in the range $c_{\text{s}2} < c < c_{\text{e}2}$ are metastable with respect to clustering. With these results, we can easily understand Fig. 3.11.

3.4 van der Waals theory

3.4.1 Thermodynamics of one-component fluids

We reconsider the van der Waals theory for one-component fluids in 3D. The pairwise potential has a hard-core volume $v_0 = \sigma^3$ and a relatively long-range attractive tail of order ϵ. See the Lenard-Jones potential given in (1.2.1) as a representative example. In calculating the partition function, we make two drastic approximations [3, 6, 30]. (i) We account for the hard-core interaction by reducing the free volume, in which each particle can move, from V to $V - Nv_0$. (ii) We estimate the number of particle pairs in contact (where $|\boldsymbol{r}_i - \boldsymbol{r}_j| \sim \sigma$) as v_0N^2/V and hence the total attractive potential energy as $-\epsilon v_0 N^2/V$. Then, the partition function for N particles in (1.2.4) is written as

$$Z_N = \frac{1}{N!\lambda_{\text{th}}^{3N}}(V - v_0N)^N \exp\left(\beta\epsilon v_0 N^2/V\right), \tag{3.4.1}$$

where $\lambda_{\text{th}} = \hbar(2\pi/m_0T)^{1/2}$ is the thermal de Broglie length (1.2.5). Therefore, the Helmholtz free energy F is given by

$$F = NT[\ln(\lambda_{\text{th}}^3 n) - 1] - NT\ln(1 - v_0n) - \epsilon v_0 nN, \tag{3.4.2}$$

where $n = N/V$ is the number density. The thermodynamic relation $p = -(\partial F/\partial V)_{TN}$ yields the van der Waals equation of state,

$$p = \frac{Tn}{1 - v_0 n} - \epsilon v_0 n^2. \tag{3.4.3}$$

From $\langle \mathcal{H} \rangle = (\partial \beta F/\partial \beta)_{VN}$, the internal energy density is written as

$$e = \frac{3}{2} nT - \epsilon v_0 n^2. \tag{3.4.4}$$

The entropy per particle $s = -(\partial F/\partial T)_{VN}/N$ is calculated as

$$s = -\ln(\lambda_{\text{th}}^3/v_0) + \ln(1/v_0 n - 1) + \frac{5}{2}. \tag{3.4.5}$$

We notice that the attractive part of the potential ($\propto \epsilon$) contributes to e and not to s, whereas the hard-core part ($\propto v_0$) contributes to s and not to e. The specific heats and the isothermal compressibility are then calculated as

$$\begin{aligned} C_V &= \frac{3}{2} n, \\ C_p &= C_V + nT/[T - T_s(n)], \\ nK_T &= (1 - v_0 n)^2/[T - T_s(n)], \end{aligned} \tag{3.4.6}$$

where $T_s(n)$ is the spinodal temperature dependent on n as

$$\begin{aligned} T_s(n) &= 2\epsilon v_0 n(1 - v_0 n)^2 \\ &= \frac{9}{4} T_c (n/n_c)(1 - n/3n_c)^2. \end{aligned} \tag{3.4.7}$$

The second line is the expression in terms of T_c and n_c, given in (3.4.16) below. In this mean field theory, K_T and C_p increase near the critical point and the spinodal curve, while C_V remains constant in a manner similar to C_M in Ising systems.

Landau free energy

The order parameter is the particle number density n measured from its critical value n_c. Its statistical distribution is given by the grand canonical ensemble. The form of the Landau free-energy density can be found from $f(n) = (F - \mu N)/V$. It is convenient to introduce the volume fraction,

$$\phi = v_0 N/V = v_0 n. \tag{3.4.8}$$

Then (3.4.2) yields a simple expression,

$$\frac{v_0}{T} f(n) = \phi \ln \phi - \phi \ln(1 - \phi) - \beta\epsilon\phi^2 - \bar{\nu}\phi, \tag{3.4.9}$$

where

$$\bar{\nu} = \mu/T - \ln(\lambda_{\text{th}}^3/v_0) = \mu/T + \frac{3}{2}\ln T + \text{const}. \tag{3.4.10}$$

From (1.2.9) the partition function for the grand canonical distribution is written as $\Xi(T,\mu) = \int_0^1 d\phi \exp[-\beta V f(n)]$. From (1.2.10) the quantity $-T\ln\Xi = -Vp$ in fluids corresponds to the Helmholtz free energy F in Ising systems. As in (3.1.5) we have $p = -f(n)$ at the minimum point $\phi = \phi(T,\mu)$ at which $\partial f/\partial\phi = 0$ and

$$\bar{\nu} = \ln[\phi/(1-\phi)] + 1/(1-\phi) - 2\beta\epsilon\phi. \tag{3.4.11}$$

We notice that $\bar{\nu}$ is removed in the combination,

$$p = n\frac{\partial}{\partial n} f(n) - f(n), \tag{3.4.12}$$

which turns out to be the van der Waals equation of state (3.4.3) in terms of T and n. Usually, $\bar{\nu}$ (or μ) is not measured and is treated as a dependent variable determined from the minimum condition as (3.4.11). In addition, in two-phase coexistence, $\bar{\nu}$ in (3.4.11) and p in (3.4.12) are common for the gas and liquid densities, $n = n_g$ and n_ℓ. At low T considerably smaller than ϵ, we find

$$v_0 n_\ell \cong 1 - T/\epsilon, \quad v_0 n_g \cong (\epsilon/T)e^{-\epsilon/T}. \tag{3.4.13}$$

As in (3.1.24) we may construct a more general Landau free-energy density $f(n,e)$ for the number and energy densities. Using $C_V = 3n/2$ we obtain

$$\begin{aligned} f(n,e) &= f(n) + \frac{1}{2TC_V}\left(e - \frac{3}{2}nT + \epsilon v_0 n^2\right)^2 \\ &= f(n) + \frac{1}{2T}C_V(\delta\hat{T})^2, \end{aligned} \tag{3.4.14}$$

where $f(n)$ is given by (3.4.9). In a manner similar to that in (3.1.27), the temperature fluctuation $\delta\hat{T}$ is defined by

$$\delta\hat{T} = T\frac{\partial}{\partial e} f(n,e) = \frac{1}{C_V}\left(e + \epsilon v_0 n^2\right) - T. \tag{3.4.15}$$

Clearly, $\partial f(n,e)/\partial e = 0$ gives (3.4.4). Obviously, $f(n,e)$ becomes consistent with (1.2.65) in the bilinear order of the deviations δn and $\delta\hat{T}$ because $\partial^2 f(n)/\partial n^2 = (\partial p/\partial n)_T/n$. The well-known formula (1.2.64) for the temperature variance can then be obtained.

Critical behavior

The usual way of finding the critical point from the van der Waals equation of state is to set $\partial p/\partial\phi = \partial^2 p/\partial\phi^2 = 0$ at the critical condition $\phi = \phi_c$ and $T = T_c$. In the Landau approach we may, equivalently, require $\partial f/\partial\phi = \partial^2 f/\partial\phi^2 = \partial^3 f/\partial\phi^3 = 0$ at fixed T and ν to obtain ϕ_c, T_c, and ν_c. Both methods lead to the critical volume fraction (or density), temperature, and pressure,[6]

$$\phi_c = v_0 n_c = \frac{1}{3}, \quad \frac{T_c}{\epsilon} = \frac{8}{27}, \quad \frac{p_c}{n_c T_c} = \frac{3}{8}. \tag{3.4.16}$$

[6] For ^{4}He, Ne, Ar, Kr, Xe, CO_2, p_c/n_cT_c is equal to 0.317, 0.305, 0.292, 0.290, 0.278, 0.287, respectively [2]. These values are systematically smaller than the van der Waals value $3/8 = 0.375$.

We expand the free energy density f in (3.4.9) in powers of the volume fraction deviation,

$$\phi_1 = \phi - \phi_c = v_0(n - n_c). \tag{3.4.17}$$

However, f is not even with respect to ϕ_1; as a result, its Taylor expansion contains a term of order ϕ_1^5 as

$$\frac{v_0}{T_c}(f - f_c) = -h_{vw}\phi_1 + \frac{1}{2}r_{vw}\phi_1^2 + \frac{1}{4}u_{vw}\phi_1^4 + w_{vw}\phi_1^5 + \frac{1}{6}v_{vw}\phi_1^6 + \cdots, \tag{3.4.18}$$

where f_c is the critical value of f, a_{vw} vanishes at the critical point, and h_{vw} vanishes along the coexistence line as

$$r_{vw} = \frac{27}{4}\left(\frac{T}{T_c} - 1\right), \quad h_{vw} = \frac{\mu - \mu_c}{T_c} - \left(\frac{\partial \mu}{\partial T}\right)_{cx}\left(\frac{T}{T_c} - 1\right). \tag{3.4.19}$$

If $\mu - \mu_c$ is removed using the pressure deviation $p - p_c$, we also have $h_{vw} = [p - p_c - (\partial p/\partial T)_{cx}(T - T_c)]/n_c T_c$, where $(\partial p/\partial T)_{cx} = (2v_0)^{-1}$ in the van der Waals model. The other coefficients are constants calculated as

$$u_{vw} = \frac{243}{16}, \quad w_{vw} = -\frac{3}{5}u_{vw}, \quad v_{vw} = \frac{27}{5}u_{vw}, \tag{3.4.20}$$

Here we may define the *true* order parameter as

$$\psi = \phi_1 + (w_{vw}/u_{vw})(\phi_1^2 - r_{vw}/u_{vw}) \quad \text{or} \quad \phi_1 = \psi - (w_{vw}/u_{vw})(\psi^2 - r_{vw}/u_{vw}), \tag{3.4.21}$$

where the terms of order ϕ_1^3 or $|r_{vw}|^{3/2}$ are neglected. If f is treated as a function of ψ, the fifth-order term vanishes in the expansion as

$$\frac{v_0}{T_c}(f - f_c) = -h_{vw}\psi + \frac{1}{2}r'_{vw}\psi^2 + \frac{1}{4}u_{vw}\psi^4 + O(\psi^6), \tag{3.4.22}$$

where $r'_{vw} = r_{vw} + 2w_{vw}h_{vw}/u_{vw}$. In the mapping relationship between fluids and spin systems in Section 2.2, ψ is the spin variable in the corresponding Ising system. From (3.4.4) the energy deviation from the critical value in fluids is written as

$$(e - e_c)/T_c = C_0(T/T_c - 1) - \frac{3}{4}v_0^{-1}\phi_1 - \frac{27}{8}v_0^{-1}\phi_1^2 + \cdots, \tag{3.4.23}$$

where $C_0 = 3n_c/2 = 1/2v_0$ is the critical value of C_V. As in (3.1.23) we may define the energy density m in the corresponding Ising system with $a_0 = 3(9/2 - w_{vw}/u_{vw})C_0$. The mapping relations (2.2.7) and (2.2.8) hold with

$$\alpha_1 = v_0^{-1}, \quad \beta_1 = \frac{2w_{vw}}{u_{vw}a_0 v_0} = -\frac{8}{51}, \quad \alpha_2 = -\frac{3}{4}v_0^{-1}, \quad \beta_2 = 1. \tag{3.4.24}$$

In this manner, if the free energy density of fluids is expanded in powers of the density deviation, the asymmetric fifth-order term arises and gives rise to the mixing term ($\beta_1 \neq 0$) in (2.2.7) and, after renormalization, the singular coexistence-curve diameter in (2.2.19). This scenario holds for general coefficients in the expansions (3.4.18) and (3.4.23). See Ref. [40] in Chapter 2.

Gradient free energy

The density $n(\mathbf{r}) = v_0^{-1}\phi(\mathbf{r})$ can be space-dependent and the particles can interact via an effective pair potential $v(r)$ extending beyond the hard-core size $\sigma = v_0^{1/3}$. These aspects can be taken into account by expressing the free energy as [6, 30]

$$F = T\int d\mathbf{r} v_0^{-1}\phi \ln[\phi/(1-\phi)] + \frac{1}{2}\int d\mathbf{r}\int d\mathbf{r}' v(|\mathbf{r}-\mathbf{r}'|)n(\mathbf{r})n(\mathbf{r}'). \tag{3.4.25}$$

The long-range part ($r \gtrsim \sigma$) of $v(r)$ can be treated separately from the short-range part ($r \lesssim \sigma$) by rewriting (3.4.25) as

$$F = \int d\mathbf{r} f(n) - \frac{1}{4}\int d\mathbf{r}\int d\mathbf{r}' v(|\mathbf{r}-\mathbf{r}'|)\left[n(\mathbf{r}) - n(\mathbf{r}')\right]^2, \tag{3.4.26}$$

where $f(n)$ is of the form of (3.4.9) (except for the term linear in n). If we set $n(\mathbf{r}) - n(\mathbf{r}') \cong (\mathbf{r}-\mathbf{r}')\cdot\nabla n$, we obtain a free energy including the gradient term,

$$F = \int d\mathbf{r}\left[f(\phi) + \frac{1}{2}C|\nabla\phi|^2\right], \tag{3.4.27}$$

where $C = -(6v_0^2)^{-1}\int d\mathbf{r} r^2 v(r)$ is assumed to be positive. Consequences of the gradient free-energy term will be discussed in the next chapter.

3.4.2 Extension to binary fluid mixtures

The van der Waals theory can be extended to mixtures of two components, 1 and 2, with N_1 and N_2 particles. Writing their hard-core volumes as v_{01} and v_{02}, we assume that the free volume is

$$V_f = V - v_{01}N_1 - v_{01}N_2, \tag{3.4.28}$$

commonly for the two species. As in (3.4.2) the Helmholtz free energy is

$$F = T\sum_\alpha N_\alpha \ln(N_\alpha \lambda_{\text{th}}^3/V_f) - TN - \sum_{\alpha\beta} w_{\alpha\beta}N_\alpha N_\beta/V, \tag{3.4.29}$$

where $N = N_1 + N_2$ and $w_{\alpha\beta}$ represent the strengths of the attractive interactions between $\alpha\beta$ pairs. The λ_{th} is assumed to be common (with the masses of the two species being the same). The van der Waals equation (3.4.3) is modified as

$$p = -\left(\frac{\partial F}{\partial V}\right)_{TN_1N_2} = TN/V_f - \sum_{\alpha\beta} w_{\alpha\beta}N_\alpha N_\beta/V^2. \tag{3.4.30}$$

The internal energy E and the total entropy S are

$$E = \frac{3}{2}NT - \sum_{\alpha\beta} w_{\alpha\beta}N_\alpha N_\beta/V, \tag{3.4.31}$$

$$S = \sum_\alpha N_\alpha \ln(V_f/N_\alpha\lambda_{\text{th}}^3) - \frac{1}{2}N. \tag{3.4.32}$$

Our model system can have a consolute critical line as well as a gas–liquid critical line in the three-dimensional space of appropriate field variables. To examine the former we assume symmetry, $v_{01} = v_{02}$ and $w_{11} = w_{22}$, between the two components, for simplicity. The free energy is then expressed as

$$F = N\{T \ln[n\lambda_{\rm th}^3/(1 - v_{01}n)] - T - w_{11}n\} + Nf_{\rm mix}, \tag{3.4.33}$$

where $n = N/V$. The first term is of the same form as the free energy for one-component fluids. The second term depends on the composition $X = N_1/N$ as

$$f_{\rm mix} = T[X \ln X + (1 - X)\ln(1 - X)] + 2n(w_{11} - w_{12})X(1 - X). \tag{3.4.34}$$

The chemical potential difference in (1.3.4) is expressed as $\Delta = (\partial f_{\rm mix}/\partial X)_T$. The concentration susceptibility in (1.3.22) becomes

$$\left(\frac{\partial X}{\partial \Delta}\right)_{pT} = X(1 - X)/[T - 4n(w_{11} - w_{12})X(1 - X)]. \tag{3.4.35}$$

Here $f_{\rm mix}$ coincides with the mixing free-energy density (3.3.4) for binary alloys if we set $X = (1 + \psi)/2$. Therefore, if $w_{11} - w_{12} > 0$, demixing can occur. The consolute critical line is characterized by the critical composition $X = 1/2$ and the critical temperature $T_{\rm c}'$ given by

$$T_{\rm c}' = n(w_{11} - w_{12}), \tag{3.4.36}$$

which depends on the number density n. As $T \to T_{\rm c}'$ at $X = 1/2$, $(\partial X/\partial \Delta)_{pT}$ diverges as $(T - T_{\rm c}')^{-1}$ from (3.4.35).[7]

3.5 Mean field theories for polymers and gels

First, we will introduce the Flory–Huggins theory for polymer solutions and polymer mixtures (blends) [31]–[33]. Second, by introduction of the classical rubber theory [31], we will discuss volume–phase transition in gels. Third, we shall see that coil–globule transition in a single chain may be understood in the same theoretical scheme as that for gels. The content here will be a basis for more advanced discussions on static critical behavior in Chapter 4, dynamics in Chapters 7–9, and nonequilibrium effects in shear flow in Chapter 11.

3.5.1 Polymer solutions

We first consider a mixture of polymer chains and low-molecular-weight particles (solvent) in 3D. The Flory–Huggins theory supposes a cubic lattice with a lattice constant a [31, 33]. The total number of lattice sites will be denoted by Ω, and then the total volume is $V = v_0\Omega$ with $v_0 = a^3$. A polymer chain consists of N beads (monomers), where N, called the polymerization index, is much larger than unity. Each lattice point is occupied by a single bead or a solvent molecule as in the Bragg–Williams approximation for A–B binary

[7] We confirm that the parameter $\epsilon_{\rm in}$ in (2.3.55) decreases as $T_{\rm c}'(\partial n/\partial p)_{TX} \ll 1$ or the degree of incompressibility increases at high densities along the consolute critical line.

alloys. Then the configuration entropy of $N_{\rm p}$ (polymer) chains and $N_{\rm s}$ (solvent) molecules is expressed as

$$\bar{S} = N_{\rm p}s_{\rm p} - N_{\rm p}\ln(N_{\rm p}/\Omega) - N_{\rm s}\ln(N_{\rm s}/\Omega). \tag{3.5.1}$$

Here $s_{\rm p}$ is the configuration entropy of a single chain calculated with one of its ends pinned at a lattice site and is a large number of order N [2, 31]. (If the conformations of each chain are those of gaussian random walks, we simply obtain $s_{\rm p} \sim N\ln z$ in terms of the coordination number z of the lattice.) As in the Bragg–Williams approximation, the two-body interaction energy is estimated as

$$\bar{E} = -\frac{z}{2\Omega}\left[\epsilon_{\rm pp}(NN_{\rm p})^2 + 2\epsilon_{\rm ps}(NN_{\rm p})N_{\rm s} + \epsilon_{\rm ss}N_{\rm s}^2\right], \tag{3.5.2}$$

where $\epsilon_{\rm pp}$, $\epsilon_{\rm ps}$, and $\epsilon_{\rm ss}$ are the attractive interaction energies between the polymer–polymer, polymer–solvent, and solvent–solvent pairs. Furthermore, it is usual to assume the space-filling condition,

$$NN_{\rm p} + N_{\rm s} = \Omega. \tag{3.5.3}$$

Namely, we do not allow the presence of vacant lattice points. Then the polymer volume fraction $\phi = NN_{\rm p}/\Omega$ is a convenient order parameter, in terms of which

$$N_{\rm p} = \Omega\phi/N, \quad N_{\rm s} = \Omega(1-\phi). \tag{3.5.4}$$

The free-energy density $f_{\rm site} = (\bar{E} - T\bar{S})/\Omega$ per lattice site is written as

$$\begin{aligned} \frac{1}{T}f_{\rm site} &= \frac{1}{N}\phi\ln\phi + (1-\phi)\ln(1-\phi) + \chi\phi(1-\phi) - \frac{\Delta}{T}\phi \\ &\cong \frac{\phi}{N}\ln\phi + \left(\frac{1}{2}-\chi\right)\phi^2 + \frac{1}{6}\phi^3 - \frac{\Delta}{T}\phi, \end{aligned} \tag{3.5.5}$$

where the second line holds for $\phi \ll 1$. The temperature-dependent coefficient,

$$\chi = \frac{z}{T}(\epsilon_{\rm pp} + \epsilon_{\rm ss} - 2\epsilon_{\rm ps}), \tag{3.5.6}$$

is called the interaction parameter (which should not be confused with the susceptibility in spin systems). The tendency for phase segregation increases with increasing χ. In the last term in (3.5.5), $\Delta = (s_{\rm p} + \ln N)/N + z(\epsilon_{\rm pp} - \epsilon_{\rm ss})/2T$ is the chemical potential difference between a bead and a solvent molecule, but it is usually omitted in the literature. The above site free-energy density reduces to that in (3.3.4) for binary alloys if we set $N = 1$ and $\phi = (1-\psi)/2$. In our system, the parameter χ is related to the temperature. From (3.5.6), the simplest dependence is $\chi = B/T$. More generally, the following form has been assumed [31]:

$$\chi = A + B/T, \tag{3.5.7}$$

where A and B are constants independent of N. The temperature at which $\chi = 1/2$ is called the theta temperature T_Θ. The second line of (3.5.5) shows that the strength of the two-body interaction is represented by [33]

$$\varepsilon = 1 - 2\chi = 2B(1/T_\Theta - 1/T). \tag{3.5.8}$$

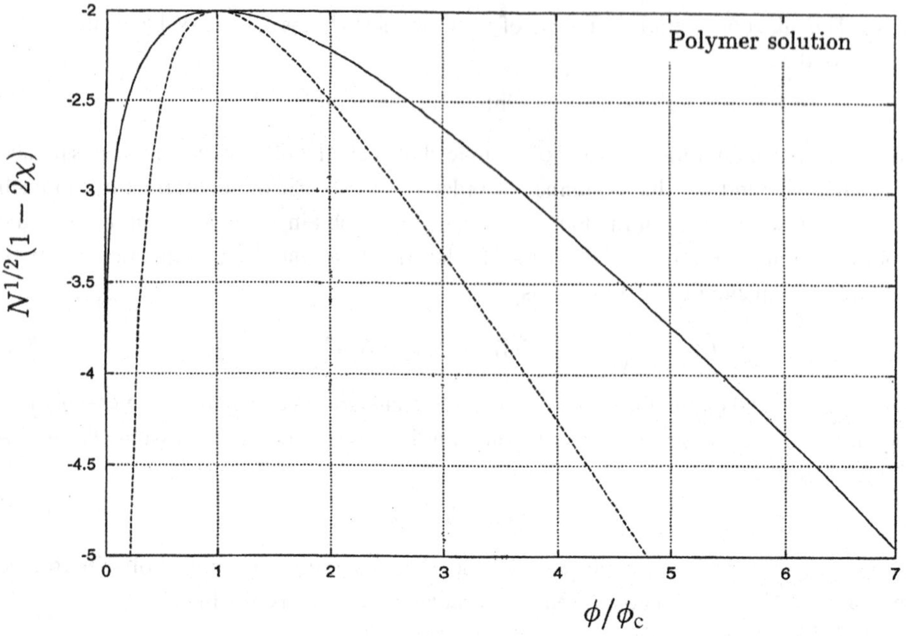

Fig. 3.12. The coexistence curve (solid line) and the spinodal curve (dashed line) for polymer solutions obtained from the second line of (3.5.5) in the plane of $N^{1/2}(1-2\chi)$ and ϕ/ϕ_c. Approximate expressions for the curves are given in (3.5.25)–(3.5.28).

We assume $T_\Theta \sim B$; then, ε decreases from of order 1 at high temperatures to negative values for $T < T_\Theta$.

The phase diagram of polymer solutions below the critical point is shown in Fig. 3.12. The critical point values of χ and ϕ are

$$\chi_c = \frac{1}{2}(1+N^{-1/2})^2 \cong \frac{1}{2} + N^{-1/2}, \quad \phi_c = N^{-1/2}. \tag{3.5.9}$$

The critical value of ε is $2N^{-1/2}$. If we assume (3.5.7), we find[8] $B(1/T_c - 1/T_\Theta) = N^{-1/2}$ and $N^{1/2}(\chi_c - \chi) = (1 - T_c/T)/(1 - T_c/T_\Theta)$. The Landau expansion near the critical point is of the form,

$$\frac{1}{T} f_{\rm site} \cong c_0 + (\chi_c - \chi)(\phi - \phi_c)^2 + \frac{1}{12} N^{1/2}(\phi - \phi_c)^4 - h_{\rm eff}(\phi - \phi_c), \tag{3.5.10}$$

where c_0 and $h_{\rm eff}$ are constants. This expansion holds for $|\phi - \phi_c| \lesssim \phi_c$ and $|\chi - \chi_c| \lesssim N^{-1/2}$. The latter condition can also be written as $|\varepsilon| \lesssim N^{-1/2}$.

Solvent quality and semidilute solutions

For $|\varepsilon| \lesssim N^{-1/2}$, the solvent will be referred to as theta solvent, where the chains assume a gaussian form with radius $R = aN^{1/2}$. The solvent quality will be said to be good for

[8] Experimentally, data of T_c have been fitted to the form $1/T_c = a_1 + a_2 M^{-1/2}$ where a_1 and a_2 are constants and M is the molecular weight [34].

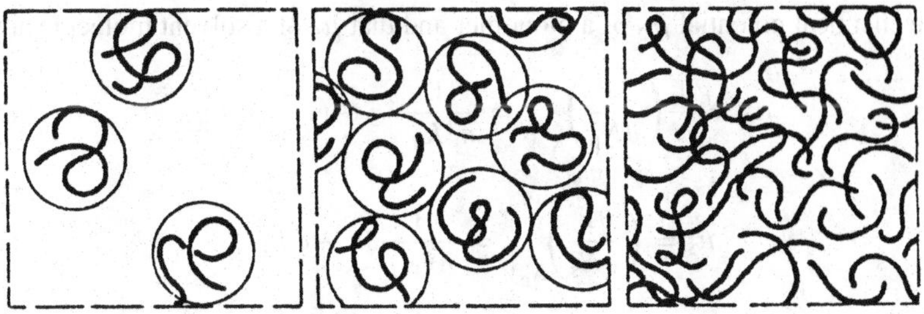

$\phi < \phi^*$ $\phi = \phi^*$ $\phi > \phi^*$

Fig. 3.13. Crossover from dilute to semidilute polymer solutions with increasing ϕ [33].

$\varepsilon \gtrsim N^{-1/2}$ and poor for $\varepsilon \lesssim -N^{-1/2}$. As ε is increased above $N^{-1/2}$, a chain becomes more expanded than in theta solvent due to the excluded volume interaction. In good solvent with $\varepsilon \sim 1$, a single chain has the Flory radius $R = aN^{3/5}$ [31, 33].[9] As illustrated in Fig. 3.13, semidilute solutions are characterized by

$$\phi^* < \phi \ll 1, \tag{3.5.11}$$

where $\phi^* = \varepsilon^{-3/5}N^{-4/5}$ in good solvent and $\phi^* = N^{-1/2}$ in theta solvent. Above the theta temperature, a semidilute polymer solution is in theta solvent for $\phi \gtrsim \varepsilon$, but in good solvent for $\phi \lesssim \varepsilon$ [33, 35]. The dynamics of a semidilute solution is severely influenced by entanglements among chains, as will be discussed in Chapter 7.

Chemical potentials

The chemical potentials of the two components can be defined unambiguously if the system has a finite but very small compressibility K_T.[10] Let the total number density $n = Nn_\text{p} + n_\text{s}$ be slightly smaller than the close-packed value v_0^{-1}, with n_p and n_s being the chain and solvent densities, respectively. This assumption means that there are a small number of vacant sites. The quantity ϕ may be re-interpreted as the composition $Nn_\text{p}/n = NN_\text{p}/(NN_\text{p}+N_\text{s})$. When a small deviation δn is created, the excess free energy of the solution is[11]

$$F = \frac{V}{2n^2K_T}(\delta n)^2 + \frac{V}{v_0} f_\text{site}(\phi). \tag{3.5.12}$$

[9] For $|\varepsilon| < 1$ let us take a region (blob) with length $\xi_\text{b} = a/|\varepsilon|$ on a single chain. The chain conformations within this region are gaussian, so the monomer number in this region is $g_\text{b} = |\varepsilon|^{-2}$. For $N > g_\text{b}$ and $\varepsilon > 0$ the blobs are under strong excluded volume interaction, leading to the Flory radius $R = \xi_\text{b}(N/g_\text{b})^{3/5} = a\varepsilon^{1/5}N^{3/5}$. We determine $\phi = \phi^*$ by $\phi R^3 = a^3 N$.

[10] Recently, highly compressible, supercritical fluids, such as CO_2, have been used as solvents.

[11] We may change the second term in (3.5.12) to $(NN_\text{p} + N_\text{s})f_\text{site}$ as another choice. Then we should add f_site to μ_s and μ_p and delete the second term in (3.5.16), but the fundamental relations, (3.5.15) and (3.5.17)–(3.5.20), remain unchanged.

Then, the chemical potential μ_p of a monomer and that μ_s of a solvent molecule are

$$\mu_p = \frac{1}{N}\left(\frac{\partial F}{\partial N_p}\right)_{N_s V} = \frac{1}{n^2 K_T}\delta n + (1-\phi) f'_{\text{site}}, \tag{3.5.13}$$

$$\mu_s = \left(\frac{\partial F}{\partial N_s}\right)_{N_p V} = \frac{1}{n^2 K_T}\delta n - \phi f'_{\text{site}}, \tag{3.5.14}$$

where $f'_{\text{site}} = \partial f_{\text{site}}/\partial\phi$. Here, μ_p and μ_s are measured from the values in pure polymer and solvent at a given pressure p_0. The chemical potential difference is simply of the form,

$$\mu_p - \mu_s = f'_{\text{site}}. \tag{3.5.15}$$

The pressure deviation $\delta p = p - p_0$ is calculated as

$$\delta p = -\left(\frac{\partial F}{\partial V}\right)_{N_p N_s} \cong \frac{1}{n K_T}\delta n - v_0{}^{-1} f_{\text{site}}, \tag{3.5.16}$$

where use has been made of $\delta V/V \cong -\delta n/n$ at constant N_p and N_s. Then we may eliminate δn in favor of δp in the chemical potentials. That is,

$$\mu_s = v_0 \delta p + f_{\text{site}} - \phi f'_{\text{site}}. \tag{3.5.17}$$

The μ_p is also expressed in terms of δp and ϕ if use is made of (3.5.15). It is now easy to check the Gibbs–Duhem relation (1.3.2) for infinitesimal changes of p, μ_p, and μ_s ,

$$dp = d(p - p_0) = N n_p d\mu_p + n_s d\mu_s, \tag{3.5.18}$$

where T and the reference pressure p_0 are fixed. This is because (3.5.13) and (3.5.14) yield $d\mu_p = v_0 dp + (1-\phi) f''_{\text{site}} d\phi$ and $d\mu_s = v_0 dp - \phi f''_{\text{site}} d\phi$ in the differential forms.

Osmotic pressure and bulk modulus

Let a polymer solution be in contact with a nearly pure solvent through a planar boundary. Such two phase coexistence can happen after phase separation far from the critical point, or when the two regions are separated by a semipermeable membrane. In such cases, the solvent chemical potential μ_s should be continuous through the two-phase boundary. The osmotic pressure $\Pi = \Pi(\phi, T)$ is defined as the pressure difference between these two regions. Here, $\mu_s \cong n^{-1}\delta p_0$ on the solvent side generally in the presence of a pressure deviation δp_0, while (3.5.17) holds on the solution side. The continuity of μ_s between the two regions gives

$$\Pi = \delta p - \delta p_0 = v_0^{-1}(\phi f'_{\text{site}} - f_{\text{site}}). \tag{3.5.19}$$

The solvent chemical potential in polymer solutions is thus expressed as

$$\mu_s = v_0(\delta p - \Pi). \tag{3.5.20}$$

The osmotic pressure is positive in the presence of a semipermeable membrane and is nearly zero on the coexistence curve far from the critical point. The isothermal osmotic bulk modulus $K_{os} = \phi(\partial \Pi / \partial \phi)_T$ is expressed as

$$K_{os} = v_0^{-1}\phi^2 f''_{site} = v_0^{-1}\phi^2 \frac{\partial}{\partial \phi}(\mu_p - \mu_s). \tag{3.5.21}$$

We may relate K_{os} to the concentration susceptibility $\chi_\phi = \langle \phi : \phi \rangle$ (= variance of the fluctuations of ϕ) as

$$\chi_\phi^{-1} = (v_0 T)^{-1} f''_{site} = T^{-1}\phi^{-2} K_{os}. \tag{3.5.22}$$

This relation is analogous to (1.3.22) for binary fluid mixtures. Note that ϕ and $\mu_p - \mu_s$ in polymer solutions correspond to X and Δ in binary fluid mixtures. The second line of (3.5.5) gives explicit expressions for Π and K_{os} for $\phi \ll 1$:

$$\Pi = T v_0^{-1}\left[\frac{1}{N\phi} + \left(\frac{1}{2} - \chi\right)\phi^2 + \frac{1}{3}\phi^3\right], \tag{3.5.23}$$

$$K_{os} = T v_0^{-1}\left[\frac{1}{N}\phi + (1 - 2\chi)\phi^2 + \phi^3\right]. \tag{3.5.24}$$

Coexistence and spinodal curves

As shown in Fig. 3.12, if ϕ is considerably larger than $\phi_c = N^{-1/2}$, the coexistence curve $\phi = \phi_{cx}$ is given by

$$\Pi \cong 0, \quad \phi_{cx} \cong 3(\chi - 1/2), \tag{3.5.25}$$

and the spinodal curve $\phi = \phi_{sp}$ by

$$K_{os} = 0, \quad \phi_{sp} \cong 2\chi - 1. \tag{3.5.26}$$

The volume fraction ϕ_{dcx} on the solvent-rich branch of the coexistence curve is obtained from $f'_{site}(\phi_{dcx}) = f'_{site}(\phi_{cx})$ and turns out to be extremely small as

$$\phi_{dcx} \sim \exp\left[-\frac{3}{8}N(2\chi - 1)^2\right], \tag{3.5.27}$$

whereas the solvent-rich branch $\phi = \phi_{dsp}$ of the spinodal curve is obtained from $K_{os} = 0$ as

$$\phi_{dsp} \cong [N(2\chi - 1)]^{-1} \cong (N\phi_{sp})^{-1}. \tag{3.5.28}$$

3.5.2 Polymer blends

The lattice theory may also be applied to mixtures of two species of polymers (polymer blends). It follows a famous expression for the free-energy density per lattice point [31, 33],

$$\frac{1}{T} f_{site} = \frac{1}{N_1}\phi \ln \phi + \frac{1}{N_2}(1 - \phi)\ln(1 - \phi) + \chi\phi(1 - \phi) - \frac{\Delta}{T}\phi, \tag{3.5.29}$$

where $\phi_1 = \phi$ and $\phi_2 = 1 - \phi$ are the volume fractions of the two components, and N_1 and N_2 are the polymerization indices of the two polymers. If we set $N_2 = 1$, the solution free energy (3.5.5) is reproduced. If both N_1 and N_2 are larger than unity, the entropic contribution, the first two terms in (3.5.29), becomes very small. This is because we are supposing chain conformations which maximize the entropy (gaussian chains). As a result, two polymers are demixed even for very small positive χ. In deriving the following calculations we may use the fact that $N_2 f_{\rm site}/T$ is of the same form as $f_{\rm site}/T$ in (3.5.5) if N and χ there are replaced by $\tilde{N} \equiv N_1/N_2 > 1$ and $\tilde{\chi} \equiv N_2\chi$, respectively.

As in polymer solutions, we may define the chemical potentials μ_1 and μ_2 per monomer of the two components. They take the same forms as (3.5.13) and (3.5.14) if the subscripts p and s are replaced by 1 and 2. As in (3.5.15) the chemical potential difference is simply of the form,

$$\mu_1 - \mu_2 = f'_{\rm site}. \tag{3.5.30}$$

The chemical potentials may be expressed in terms of δp and ϕ. As in (3.5.17) μ_2 is of the form,

$$\mu_2 = v_0\delta p + f_{\rm site} - \phi f'_{\rm site}. \tag{3.5.31}$$

The inverse susceptibility becomes

$$\begin{aligned} \chi_\phi^{-1} &= (v_0 T)^{-1} f''_{\rm site} = (v_0 T)^{-1}\frac{\partial}{\partial\phi}(\mu_1 - \mu_2) \\ &= v_0^{-1}\left[\frac{1}{N_1\phi} + \frac{1}{N_2(1-\phi)} - 2\chi\right]. \end{aligned} \tag{3.5.32}$$

The critical values of χ and ϕ are given by

$$\chi_{\rm c} = \frac{1}{2N_1N_2}(N_1^{1/2} + N_2^{1/2})^2, \quad \phi_{\rm c} = \frac{N_2^{1/2}}{N_1^{1/2} + N_2^{1/2}}. \tag{3.5.33}$$

Note that $\chi_{\rm c}$ is very small for high-molecular-weight polymers. The Landau expansion of the free-energy density near the critical point is obtained in the form,

$$\frac{1}{T} f_{\rm site} \cong c_0 + (\chi_{\rm c} - \chi)(\phi - \phi_{\rm c})^2 + \frac{1}{3}\sqrt{N_1N_2}\chi_{\rm c}^2(\phi - \phi_{\rm c})^4 - h_{\rm eff}(\phi - \phi_{\rm c}), \tag{3.5.34}$$

where c_0 is independent of ϕ and $h_{\rm eff}$ is appropriately defined. This expansion is valid for $|\tilde{\chi} - \tilde{\chi}_{\rm c}| \ll (N_2/N_1)^{1/2}$ or

$$|\chi - \chi_{\rm c}| \ll (N_1N_2)^{-1/2}, \tag{3.5.35}$$

under which the two phases have compositions close to $\phi_{\rm c}$ on the coexistence curve. However, in the region $|\chi - \chi_{\rm c}| \gtrsim (N_1N_2)^{-1/2}$, one of the two phases consists mostly of shorter chains for $N_1 > N_2$ and both phases are in strongly segregated states for $N_1 = N_2$.

Symmetric case

The coexistence and spinodal curves become simple for the symmetric case $N_1 = N_2 = N$. Phase separation occurs for $\chi > \chi_c = 2/N$ and the coexistence curve $\phi = \phi_{cx}$ is obtained from

$$\ln[\phi_{cx}/(1-\phi_{cx})] = N\chi(2\phi_{cx} - 1). \tag{3.5.36}$$

For $N\chi \gg 1$ we have strong segregation, where

$$\phi_{cx} \cong 1 - 2\exp(-N\chi) \quad \text{or} \quad 2\exp(-N\chi). \tag{3.5.37}$$

The spinodal curve $\phi = \phi_{sp}$ is explicitly calculated as

$$\phi_{sp} = \frac{1}{2}\left[1 \pm \sqrt{1 - \frac{2}{N\chi}}\right]. \tag{3.5.38}$$

3.5.3 Polymer gels

Gels are network systems composed of crosslinked polymers. They are usually in contact with solvent at zero-osmotic pressure and can swell enormously [31, 33]. It is known that gels undergo a first-order phase transition with a discontinuous change of the volume (volume-phase transition) [36]. It was predicted by Dušek and Patterson [37] and afterwards was observed by Tanaka and coworkers [38, 39] and Ilavsky [40] in ionic gels, and in non-ionic poly-N-isopropylacrylamide (NIPA) gels [41]. Here, if we consider homogeneous deformations, the polymer volume fraction ϕ and the volume V are related by

$$\phi = \phi_0 V_0/V, \tag{3.5.39}$$

where ϕ_0 and V_0 are the volume fraction and the volume at the network formation. Obviously, the total number of monomers forming the network is a constant and is written as

$$\Omega = \phi_0 V_0/v_0. \tag{3.5.40}$$

In Fig. 3.14 the chain configurations by which a gel is prepared are illustrated. The left and right diagrams show the states just before and just after network formation, respectively. The latter state will be chosen as a special reference state of a gel.

Supposing either a theta or a poor solvent, we hereafter construct the free energy F as follows.

(i) Because there is no translational entropy of the network, we may set $N = \infty$ in the first line of (3.5.5) to obtain the Flory–Huggins mixing free energy F_{mix} for a gel in the form

$$F_{mix} = v_0^{-1} VT\big[(1-\phi)\ln(1-\phi) + \chi\phi(1-\phi)\big]. \tag{3.5.41}$$

(ii) Classical rubber theory [31, 42, 43] gives the elastic free energy. For simplicity, let

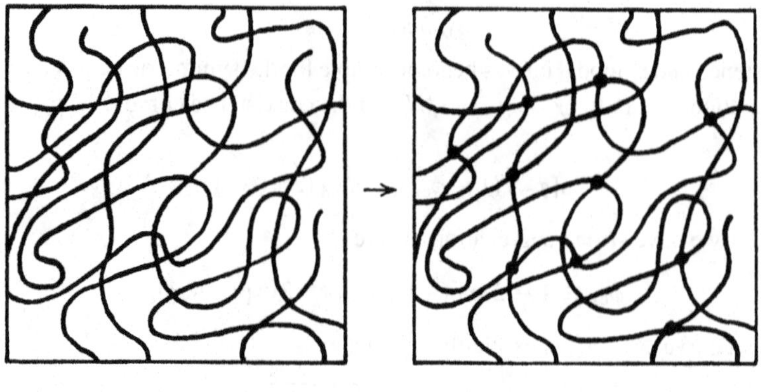

Fig. 3.14. Schematic representation of crosslinking among polymer chains [31].

a homogeneous isotropic gel with an initial cubic shape be deformed into a rectangular shape with linear dimensions along the three principal axes being elongated or compressed by α_1, α_2, and α_3. Then the volume fraction after the deformation is

$$\phi = \phi_0/(\alpha_1\alpha_2\alpha_3). \tag{3.5.42}$$

The elastic free energy needed is of the form [31],

$$F_{\text{el}} = V_0 \nu_0 T\left[\frac{1}{2}(\alpha_1^2 + \alpha_2^2 + \alpha_3^2) - B\log(\alpha_1\alpha_2\alpha_3)\right], \tag{3.5.43}$$

where ν_0 is the effective crosslink number density in the reference state and B is a coefficient. The effective polymerization index N may be defined by

$$N = \phi_0/v_0\nu_0 \quad \text{or} \quad \nu_0 = \phi_0/v_0 N. \tag{3.5.44}$$

Usually N is much larger than unity, which ensures the soft elasticity characteristic of gels. To derive the above form, we start with the equilibrium distribution of the end-to-end vector $\boldsymbol{R}$ of a single gaussian chain [33]:

$$W(\boldsymbol{R}) = (2\pi N a^2)^{-3/2}\exp(-|\boldsymbol{R}|^2/2Na^2). \tag{3.5.45}$$

We set $\boldsymbol{R} = N^{1/2}a(\alpha_1, \alpha_2, \alpha_3)$ and sum $-T\ln W$ from all the chains to obtain the term proportional to $\alpha_1^2 + \alpha_2^2 + \alpha_3^2$ in (3.5.43). The coefficient B of the logarithmic term was originally predicted to be 1 [31], but there has been some controversy and several theories predict different values of B [44, 45]. It is easy to extend the above form to more general affine deformations [46]–[48]. To this end, we represent a gel point by $\boldsymbol{x}_0 = (x_{01}, x_{02}, x_{03})$ in the reference state and by $\boldsymbol{x} = (x_1, x_2, x_3)$ after deformation using appropriate cartesian coordinates. We introduce the deformation tensor,

$$\Gamma_{ij} = \frac{\partial}{\partial x_{0j}} x_i. \tag{3.5.46}$$

The polymer volume fraction is related to the determinant of Γ_{ij} as

$$\phi = \phi_0 / \det\{\Gamma\}. \tag{3.5.47}$$

Then the elastic free energy reads

$$F_{\rm el} = V_0 \nu_0 T \left[\frac{1}{2} \sum_{ij} \Gamma_{ij}^2 + B \ln(\phi/\phi_0) \right]. \tag{3.5.48}$$

This quantity is invariant with respect to rotations, $\boldsymbol{r}_0 \to \overleftrightarrow{U}_0 \cdot \boldsymbol{r}_0$ and $\boldsymbol{r} \to \overleftrightarrow{U} \cdot \boldsymbol{r}$, for any orthogonal matrices $\overleftrightarrow{U}_0$ and $\overleftrightarrow{U}$. See Appendix 3A for a general theory of nonlinear elasticity [47].

(iii) In polymer solutions and gels it is often the case that dissociation results in charged monomers and low-molecular-weight counterions. In weakly charged gels the most important free-energy contribution arises from the translational entropy of the counterions [38],[12]

$$F_{\rm ion} = -T V_0 \nu_{\rm I} \ln(V/V_0) = T V_0 \nu_{\rm I} \ln(\phi/\phi_0), \tag{3.5.49}$$

where $\nu_{\rm I}$ is the counterion density measured in the reference state. The counterions are confined within the gel to satisfy the overall charge neutrality of the gel. The resultant osmotic pressure $T\nu_{\rm I}$ favors gel swelling at osmotic equilibria with solvent. Following Tanaka and coworkers [38, 39], we write the number of counterions per chain as

$$f = \nu_{\rm I}/\nu_0, \tag{3.5.50}$$

which should not be confused with the free-energy density.

(iv) For neutral gels [41] the presence of a first-order phase transition itself is a subtle issue. Erman and Flory [44] showed that the ϕ dependence of the interaction parameter in the expression of the osmotic pressure,

$$\chi = \chi_1 + \chi_2 \phi, \tag{3.5.51}$$

can give rise to discontinuous volume changes in neutral gels.

The total free energy F in the isotropic case is the sum of the above three contributions,

$$F = F_{\rm mix} + F_{\rm ion} + F_{\rm el} = \Omega T \left[\frac{1}{\phi} g(\phi) + \frac{1}{2N} \sum_{ij} \Gamma_{ij}^2 \right]. \tag{3.5.52}$$

Here we define a dimensionless free-energy density $g(\phi)$ by

$$g(\phi) = (1-\phi)\ln(1-\phi) - \chi_1 \phi^2 - \frac{1}{2}\chi_2 \phi^3 + \frac{f+B}{N} \phi \ln(\phi/\phi_0). \tag{3.5.53}$$

In the simplest isotropic case, we have $\partial x_i / \partial x_{0j} = \delta_{ij} (\phi_0/\phi)^{1/3}$ to obtain the usual result

[12] As will be shown in Appendix 7F, the Debye–Hückel theory yields a free-energy contribution, $(\Delta F)_{\rm DH} \propto \kappa_{\rm Db}^3$, due to the charge density fluctuations, where $\kappa_{\rm Db}^{-1}$ is the Debye screening length [1]. This theory holds in the weakly charged case, $\nu_{\rm I}\phi/\phi_0 \gg \kappa_{\rm Db}^3$, where $(\Delta F)_{\rm DH}$ is much smaller than $F_{\rm ion}$ in (3.5.49).

in the literature [31]. We may furthermore add a small shear deformation represented by $\partial x_i/\partial x_{0j} = (\phi_0/\phi)^{1/3}[\delta_{ij} + \gamma\delta_{i1}\delta_{j2}]$. Then the increase of the elastic free energy may be written as $\Delta F_{\rm el} = \frac{1}{2}V\mu\gamma^2$, where μ has the meaning of the shear modulus expressed as

$$\mu = \frac{T}{v_0 N}\phi_0^{2/3}\phi^{1/3} = \nu_0 T(\phi/\phi_0)^{1/3}. \tag{3.5.54}$$

In gels with good solvent as well as most rubber-like materials, μ is much smaller than the (osmotic) bulk modulus $K_{\rm os}$ whose explicit form will be given in (3.5.57) below. For poor solvent, however, $K_{\rm os}$ decreases and even becomes negative (in unstable states), leading to negative values of the (osmotic) Poisson ratio $(K_{\rm os} - 2\mu/3)/2(K_{\rm os} + \mu/3)$ [49, 50] in the vicinity of the transition.

Isotropically swollen gels

If a gel is swollen isotropically in a solvent, the differential form of F reads

$$dF = -S_{\rm net}dT - \Pi dV, \tag{3.5.55}$$

where $S_{\rm net}$ is the entropy supported by the network and Π is the osmotic pressure. From (3.5.52) Π and $K_{\rm os} = \phi(\partial\Pi/\partial\phi)_T$ are expressed as

$$\Pi = v_0^{-1}T\left[\phi g' - g - \frac{1}{N}\phi_0^{2/3}\phi^{1/3}\right], \tag{3.5.56}$$

$$K_{\rm os} = v_0^{-1}T\left[\phi^2 g'' - \frac{1}{3N}\phi_0^{2/3}\phi^{1/3}\right], \tag{3.5.57}$$

where $g' = \partial g/\partial\phi$ and $g'' = \partial^2 g/\partial\phi^2$. We here impose $\Pi = 0$ and $K_{\rm os} \geq 0$ and hence minimize F. For $K_{\rm os} < 0$ the gel becomes unstable against macroscopic volume changes.

If $\phi \ll 1$, we rewrite F in (3.5.52) in terms of $\Phi \equiv \phi/\phi_0$ as

$$F = \frac{\Omega}{N}T\left[\frac{\tau}{2}\Phi + \frac{w}{6}\Phi^2 + (f + B)\ln\Phi + \frac{3}{2}\Phi^{-2/3}\right]. \tag{3.5.58}$$

where $\Omega/N = V_0\nu_0$ and

$$\tau = N\phi_0(1 - 2\chi_1), \quad w = N\phi_0^2(1 - 3\chi_2). \tag{3.5.59}$$

The critical point can be sought by requiring $\partial^2 F/\partial\Phi^2 = \partial^3 F/\partial\Phi^3 = 0$. A first-order phase transition occurs for

$$(f + B)/w^{1/4} > (4/3)(5/3)^{3/4} \cong 2 \quad \text{or} \quad f > f_{\rm c} = 2w^{1/4} - B. \tag{3.5.60}$$

At the critical point we have $f = f_{\rm c}$ and $w > 0$, so that the critical values are given by

$$\Phi_{\rm c} = \phi_{\rm c}/\phi_0 = (5/3w)^{3/8}, \quad \tau_{\rm c} = N\phi_0(1 - 2\chi_{1\rm c}) = -(32/9)\Phi_{\rm c}^{-5/3}. \tag{3.5.61}$$

Even for neutral gels ($f = 0$) a first-order phase transition occurs for $w < w_{\rm c} = 0.07B^4$. In Fig. 3.15 we plot the curves of $\Pi = 0$, $K = 0$, and $K + 4\mu/3 = 0$ in the plane of the reduced temperature ($\propto \tau$) and the volume ($\propto \phi^{-1}$). In (a), (b), and (c), f is smaller

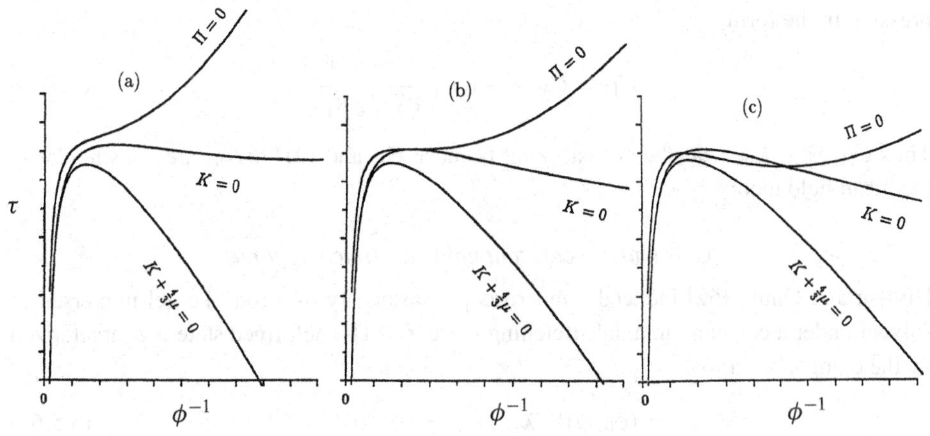

Fig. 3.15. Reduced temperature ($\propto \tau$) in (3.5.59) vs volume ($\propto \phi^{-1}$) in ionized gels, where Π, K, and μ are defined by (3.5.56), (3.5.57), and (3.5.54), respectively, and are calculated using (3.5.58) [48]. The two instability curves of $K_{\rm os} = 0$ and $K_{\rm os} + 4\mu/3$ are close near the critical volume fraction, but are much separated at large volume or swelling.

than, equal to, and larger than the critical value $f_{\rm c}$, respectively. In (c) a first-order phase transition occurs along the curve of $\Pi = 0$. Below the curves of $K = 0$ the system is unstable against macroscopic volume changes, while below the curves of $K + 4\mu/3 = 0$ spinodal decomposition occurs in the bulk region. See Chapters 7 and 8 for the dynamics of these instabilities.

In the vicinity of the critical point the Landau expansion of F becomes

$$F = \frac{\Omega}{N} T \left[-h(\Phi - \Phi_{\rm c}) + \frac{1}{2}(f_{\rm c} - f)(\Phi/\Phi_{\rm c} - 1)^2 + \frac{10}{81}(\Phi - \Phi_{\rm c})^4 + \cdots \right], \tag{3.5.62}$$

where w in (3.5.59) is treated as a constant and

$$h = \frac{1}{2}(\tau_{\rm c} - \tau) - \Phi_{\rm c}^{-1}(f - f_{\rm c}). \tag{3.5.63}$$

Here h and $f_{\rm c} - f$ play the role of a magnetic field and a reduced temperature in Ising spin systems. Thus, if f is fixed at a value unequal to $f_{\rm c}$ in experiments, the critical point can be reached just by varying the temperature.[13] However, if f is close to $f_{\rm c}$, the osmotic bulk modulus $K_{\rm os}$ becomes small around $\Phi \cong \Phi_{\rm c}$ as

$$K_{\rm os} = T v_0 \Phi_{\rm c} \left[f_{\rm c} - f + \frac{40}{27} \Phi_{\rm c}^2 (\Phi - \Phi_{\rm c})^2 \right]. \tag{3.5.64}$$

From (3.5.55) we may also calculate the specific heat of the network at zero-osmotic

[13] If the solvent is a binary mixture, we may reach a critical point by changing the composition and the temperature [51].

pressure in the form,

$$C_\Pi = C_V + +VT\left(\frac{\partial \Pi}{\partial T}\right)_\phi^2 \frac{1}{K_{os}}. \tag{3.5.65}$$

Thus $C_\Pi \sim 1/K_{os}$ near the critical point because C_V and $(\partial\Pi/\partial T)_\phi$ are nonsingular in the mean field theory.

Gels under a constant uniaxial stretching force

Hirotsu and Onuki [52] induced a macroscopic instability of a rod-like gel immersed in solvent under a constant uniaxial stretching force f_{ex}. The deformed state is characterized by the elongation ratios,

$$\alpha_\parallel = (\phi_0/\phi)^{1/3}\lambda, \quad \alpha_\perp = (\phi_0/\phi)^{1/3}\lambda^{-1/2}, \tag{3.5.66}$$

in the parallel and perpendicular directions, respectively, λ being the degree of stretching. The system volume V and the length L in the force direction are expressed in terms of $\alpha_\parallel$ and $\alpha_\perp$ as

$$V/V_0 = \phi_0/\phi = \alpha_\parallel \alpha_\perp^2, \quad L/L_0 = \alpha_\parallel, \tag{3.5.67}$$

V_0 and L_0 being the values in the relaxed, reference state. From (3.5.52) we obtain the total free energy $G = F - f_{ex}L$ in the form,

$$G = \Omega T\left[\frac{1}{\phi}g(\phi) + \frac{1}{2N}(\alpha_\parallel^2 + 2\alpha_\perp^2)\right] - f_{ex}L. \tag{3.5.68}$$

We minimize F with respect to $\alpha_\parallel$ and $\alpha_\perp$ (or ϕ and $\alpha_\parallel$ more conveniently). The first relation is obtained by differentiation with respect to $\alpha_\parallel$ with fixed ϕ:

$$f_{ex} = S_0 T \nu_0 \alpha_\parallel (1 - 1/\lambda^3) = S\mu(\lambda^2 - 1/\lambda), \tag{3.5.69}$$

where $S_0 = V_0/L_0$ and $S = S_0\alpha_\perp^2$ are the surface area of the end plate before and after the deformation, respectively, and μ is defined by (3.5.54). The above relation is well known in the classical rubber theory [31, 42, 43]. For sufficiently long experimental times, osmotic equilibration will be achieved on the side boundary, where

$$\Pi_\perp = -\left(\frac{\partial}{\partial V}F\right)_{T\alpha_\parallel} = v_0^{-1}T[\phi g' - g - \phi_0/N\alpha_\parallel] = 0. \tag{3.5.70}$$

With these relations we may examine the macroscopic phase transition. Here, for simplicity, we only calculate the (isothermal) Young's modulus,

$$E_T = \frac{L}{S}\left(\frac{\partial f_{ex}}{\partial L}\right)_T = \mu\left[\lambda^2 + \frac{2}{\lambda} - \frac{\mu}{\lambda^2(K_{os} + \mu/3)}\right], \tag{3.5.71}$$

where K_{os} is expressed as (3.5.57). The adiabatic or constant-volume Young's modulus is given by $E_V = \mu(\lambda^2 + 2/\lambda)$, which is measured before the osmotic equilibrium at the side boundary is attained. As $\lambda \to 1$, (3.5.71) becomes consistent with the well-known expression $E = 3\mu K/(K + \mu/3)$ for the Young's modulus in usual elastic theory [46]. The macroscopic instability is triggered for $E_T < 0$ or

$$K_{\rm os} + \frac{1}{3}\mu - \frac{1}{\lambda(\lambda^3+2)}\mu < 0. \tag{3.5.72}$$

This reduces to $K_{\rm os} < 0$ in the isotropic case ($\lambda = 1$).

One-dimensionally constrained gels

Dušek and Patterson [37] examined a phase transition in a constrained gel which has a fixed length in one direction and is allowed to swell in the perpendicular directions. In this case $\alpha_\parallel$ is a constant, and $\alpha_\perp$ or $\phi = (\phi_0/\alpha_\parallel)/\alpha_\perp^2$ is the order parameter. The free energy becomes

$$F = \Omega T\left[\frac{1}{\phi}g(\phi) + \frac{\phi_0}{N\alpha_\parallel}\frac{1}{\phi}\right], \tag{3.5.73}$$

where the constant term is omitted. The zero-osmotic pressure condition on the side boundary is again written as (3.5.71). The perpendicular bulk modulus reads

$$K_\perp = \phi\left(\frac{\partial}{\partial\phi}\Pi_\perp\right)_{T\alpha_\parallel} = v_0^{-1}T\phi^2 g'' = K_{\rm os} + \frac{1}{3}\mu. \tag{3.5.74}$$

A macroscopic instability thus occurs for $K_{\rm os} + \frac{1}{3}\mu < 0$. In this case the phase behavior can easily be calculated [48]. For the same $g(\phi)$ in (3.5.53), a first-order phase transition exists under the condition,

$$f + B > (9w)^{1/3}\alpha_\parallel^{-2/3}. \tag{3.5.75}$$

The critical value of $\alpha_\parallel$ is written as

$$\alpha_{\parallel c} = 3w^{1/2}/(f+B)^{3/2}. \tag{3.5.76}$$

First-order changes are favored by large $\alpha_\parallel > \alpha_{\parallel c}$, where equilibrium coexistence of shrunken and swollen phases can be realized.

Remarks

At macroscopic first-order phase transitions, gels can change their shape but still remain transparent for very small and slow temperature changes. Notice that such macroscopic changes are not possible if the gel boundary is clamped to a solid wall. However, if the temperature is changed rapidly by quenching deep into an unstable region ($K_{\rm os} + 4\mu/3 < 0$), gels become opaque, indicating the occurrence of spinodal decomposition on short spatial scales. Phase transitions in gels are thus very unlike those in simple fluids. We stress that unique aspects arise from soft elasticity or a finite, small shear modulus μ, as will be discussed in Chapter 7 in more detail.

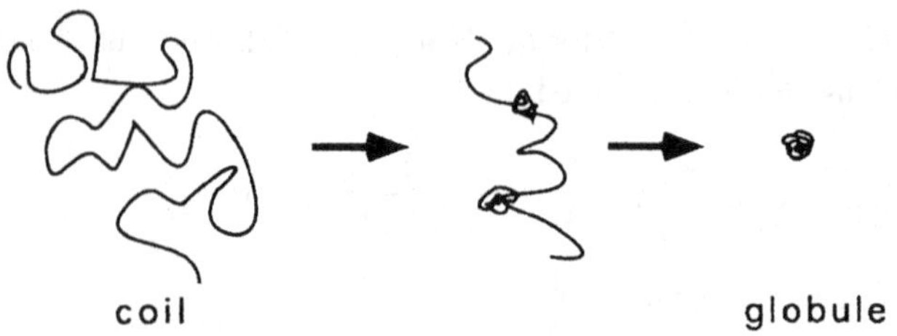

Fig. 3.16. Collapsing process of a chain [55]. Phase separation between elongated and contracted regions occurs transiently on a chain.

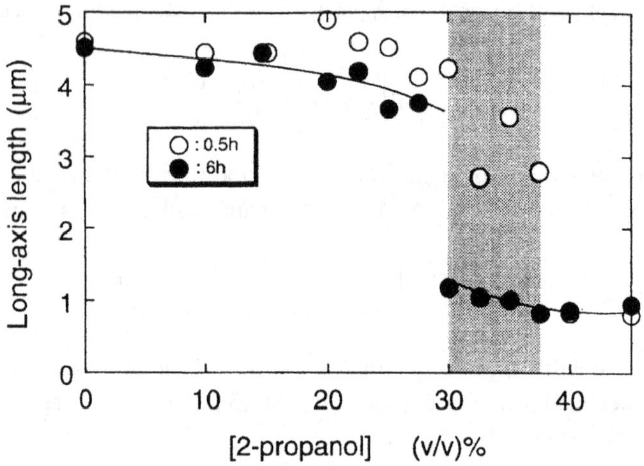

Fig. 3.17. Long-axis chain lengths of DNA with varying concentration of solvent [55]. The shaded region indicates a metastable coil with lifetime longer than 1 h. The open and closed circles are the results at 0.5 h and 6 h after sample preparation, respectively.

3.5.4 Coil–globule transition in a single chain

Much attention has been paid to the problem of coil–globule transition between elongated coils and compacted globules in a single linear chain, which is illustrated in Fig. 3.16. Theoretically, it can be either continuous or discontinuous as in gels [53]. As demonstrated in Fig. 3.17, Yoshikawa and co-workers observed a first-order phase transition of individual DNA molecules by fluorescence microscopy [54]. Let $\alpha = R_G/R_0$ be the linear expansion ratio of a chain with gyration radius R_G. The reference state corresponds to an ideal gaussian chain with radius $R_0 \propto N^{1/2}$. We set up the free energy of a single chain as

$$\frac{1}{T}F_{\rm one} = \frac{\tau}{2}\alpha^{-3} + \frac{w}{6}\alpha^{-6} + \frac{3}{2}\alpha^2 - 3B\ln\alpha + \frac{1}{T}F_{\rm ion}. \tag{3.5.77}$$

The first and second terms account for the two- and three-body interactions between the monomers, respectively. The third and fourth terms represent the elastic free energy.[14] A simple theory for the ion free energy F_{ion} is to set $F_{\text{ion}} = -3Tf \ln \alpha$ [55], as in weakly charged gels. In this case f counterions are assumed to be localized in the volume $\sim R_{\text{G}}^3$ which the chain occupies. Then, if we set $\Phi = \alpha^{-3}$ and $N = \Omega$, (3.5.77) takes the same form as the free energy (3.5.58) for gels. In theories of counterion condensation [56, 57], however, counterions are assumed to be trapped to the monomers of a chain (localized along the chain contour) and their translational entropy becomes smaller than in the weakly charged case. Furthermore, a fraction of counterions can escape from the chain [55, 57]. Our previous discussions for gels suggest that a first-order phase transition can occur for $w < w_{\text{c}}$ even without ions, but the discontinuity is much amplified in the presence of ions.

We may also examine the transition when a chain is stretched in one direction and has a fixed length. Then $\alpha_{\parallel}(> 1)$ is a constant and the relevant free energy is obtained if $3\alpha^2/2$ in (3.5.77) is replaced by $\alpha_{\perp}^2 = \phi_0/\alpha_{\parallel}\phi$ as in (3.5.73). The criterion of a first-order phase transition is again (3.5.75) for the weakly charged case. It might be satisfied even in neutral chains for sufficiently large extension $\alpha_{\parallel} \gg 1$. Furthermore, if $\alpha_{\parallel}$ is a control parameter and can be set equal to its critical value in (3.5.76), the critical point will be reached just by varying the temperature. If $\alpha_{\parallel} > \alpha_{\parallel\text{c}}$, coil and globule regions can coexist in a single chain as an equilibrium state.

Appendix 3A Finite-strain theory

Finite-strain theory is well known but is only incompletely presented in textbooks on elasticity [46]. It is a Lagrange description of finite-size deformations, where the displacement vector $\boldsymbol{u} = \boldsymbol{x} - \boldsymbol{x}_0$ is regarded as a function of the original position vector $\boldsymbol{x}_0$. We may suppose isotropic rubbers or gels as examples which can sustain large strains. Note that $\boldsymbol{x}_0$ in our notation is usually written as $\boldsymbol{x}$ in the finite-strain theory. In nonlinear elasticity, we should be careful as to whether a theory is in the Lagrange description or in the Euler description. Note that two nearby points, $\boldsymbol{x}_0$ and $\boldsymbol{x}_0 + d\boldsymbol{x}_0$, are mapped into $\boldsymbol{x}_0 + \boldsymbol{u}$ and $\boldsymbol{x}_0 + d\boldsymbol{x}_0 + \boldsymbol{u} + d\boldsymbol{u}$ after a deformation. The distances between these points is changed after the deformation according to

$$ds^2 = |d\boldsymbol{x}_0 + d\boldsymbol{u}|^2 = \sum_{ij} g_{ij} dx_{0i} dx_{0j}, \tag{3A.1}$$

where the metric tensor g_{ij} is defined in terms of the deformation tensor Γ_{ij} in (3.5.46) as

$$g_{ij} = \sum_k \Gamma_{ki}\Gamma_{kj}. \tag{3A.2}$$

In the finite-strain theory the elastic free-energy density f_{el} in the Lagrange description is assumed to be determined by the tensor g_{ij}. In the literature the nonlinear Lagrangian

[14] In the theoretical interpretation [54], another term of the form $3\alpha^{-2}/2$ was assumed in place of $-3B \ln \alpha$. This does not change the essential aspect of the transition.

strain tensor η_{ij} has been defined as

$$\eta_{ij} = g_{ij} - \delta_{ij} = \frac{\partial u_i}{\partial x_{0j}} + \frac{\partial u_j}{\partial x_{0i}} + \sum_k \frac{\partial u_k}{\partial x_{0i}} \frac{\partial u_k}{\partial x_{0j}}. \tag{3A.3}$$

If the elastic body is isotropic before deformations, it is natural to assume $f_{\rm el}$ to be a function of the following three strain invariants with respect to space rotation,

$$I_1 = g_1 + g_2 + g_3, \quad I_2 = g_1 g_2 + g_3 g_1 + g_2 g_3, \quad I_3 = g_1 g_2 g_3, \tag{3A.4}$$

where g_1, g_2, and g_3 are the eigenvalues of the tensor g_{ij} [47]. Note the relations,

$$I_3 = \det \overleftrightarrow{g} = (\det \overleftrightarrow{\Gamma})^2, \tag{3A.5}$$

$$\det\{\lambda \overleftrightarrow{I} - \overleftrightarrow{g}\} = \lambda^3 - I_1\lambda^2 + I_2\lambda - I_3. \tag{3A.6}$$

In the classical rubber theory we have $F_{\rm el} = {\rm const.} I_1 + {\rm const.} \ln I_3$ as in (3.5.48). The total elastic free energy is the space integral of its density,

$$F_{\rm el} = \int d\boldsymbol{x}_0 f_{\rm el} = \int d\boldsymbol{x} I_3^{-1/2} f_{\rm el}, \tag{3A.7}$$

where $I_3^{-1/2}$ is the jacobian $\partial \boldsymbol{x}_0/\partial \boldsymbol{x}$.

The stress tensor σ_{ij} is intrinsically a field variable defined in the deformed space or in the Euler representation. We add an infinitesimal deformation to a given deformed state as $\boldsymbol{x} \to \boldsymbol{x} + \delta \boldsymbol{u}$. Then the elastic free energy is changed as

$$\delta F_{\rm el} = \int d\boldsymbol{x} \sum_{ij} \sigma_{ij} \frac{\partial}{\partial x_j} \delta u_i, \tag{3A.8}$$

where $\boldsymbol{u}$ is regarded as a function of $\boldsymbol{x}$. Thus,

$$\left(\frac{\delta}{\delta x_i} F_{\rm el} \right)_{x_0} = -I_3^{1/2} \sum_j \frac{\partial}{\partial x_j} \sigma_{ij}. \tag{3A.9}$$

where $F_{\rm el}$ is regarded as a functional of $\boldsymbol{x} = \boldsymbol{x}(\boldsymbol{x}_0)$ and σ_{ij} as a function of $\boldsymbol{x}$. Thus the extremum condition of $F_{\rm el}$ is equivalent to the mechanical equilibrium condition. In (3A.8) we have $\partial \delta u_i/\partial x_j = \sum_\ell (\partial x_{0\ell}/\partial x_j) \delta \Gamma_{i\ell}$, so that

$$\sigma_{ij} = I_3^{-1/2} \sum_\ell \Gamma_{j\ell} \frac{\partial}{\partial \Gamma_{i\ell}} f_{\rm el}. \tag{3A.10}$$

If $f_{\rm el}$ is a function of g_{ij}, we obtain the symmetry $\sigma_{ij} = \sigma_{ji}$. Furthermore, if $f_{\rm el}$ is a function of the three rotational invariants only, it follows the Finger form of the stress tensor [58, 47],[15]

$$\sigma_{xx} = \frac{2}{\sqrt{I_3}} \left[W_{xx} C_1 + (W_{yz}^2 - W_{yy} W_{zz} + I_2) C_2 + I_3 C_3 \right], \tag{3A.11}$$

[15] For example, to derive the last term in σ_{xx}, we may use the following mathematical formula: the determinant of an arbitrary matrix $\overleftrightarrow{A} = \{A_{ij}\}$ $(1 \le i, j \le n)$ is a function of its n^2 elements. It generally holds that $\partial(\det \overleftrightarrow{A})/\partial A_{ij} = (\det \overleftrightarrow{A}) A^{ji}$ where $\{A^{ij}\}$ is the inverse matrix.

$$\sigma_{xy} = \sigma_{yx} = \frac{2}{\sqrt{I_3}}\left[W_{xy}C_1 + (W_{zz}W_{xy} - W_{yz}W_{zx})C_2\right], \tag{3A.12}$$

where $C_\alpha = \partial f_{\rm el}/\partial I_\alpha$ ($\alpha = 1, 2, 3$) and we define the symmetric tensor (the Finger tensor),

$$W_{ij} = \sum_\ell \Gamma_{i\ell}\Gamma_{j\ell}. \tag{3A.13}$$

The other stress components can be obtained by cyclic permutation of x, y, and z. Notice that W_{ij} has tensor properties with respect to rotation in the deformed space.

Representative situations are as follows. (i) For isotropic expansion $\boldsymbol{x} = \lambda \boldsymbol{x}_0$ we have $g_i = \lambda^2$ and $\sigma_{ii} = -p(\lambda)$ where

$$p(\lambda) = -2(C_1/\lambda + 2\lambda C_2 + \lambda^3 C_3). \tag{3A.14}$$

(ii) If a rod-like sample is uniaxially stretched as $x = \lambda x_0$, $y = \lambda^{-1/2} y_0$, and $z = \lambda^{-1/2} z_0$ without volume change, we have $I_1 = \lambda^2 + 2/\lambda$, $I_2 = 2\lambda + \lambda^{-2}$, and $I_3 = 1$ so that

$$\sigma_{xx} - \sigma_{yy} = 2(\lambda^2 - \lambda^{-1})(C_1 + \lambda^{-1}C_2). \tag{3A.15}$$

The total stretching force is $f_{\rm ex} = (\sigma_{xx} - \sigma_{yy})S$ where $S = S_0\lambda^{-1}$ is the surface area of the end plates after the deformation. Data of $f_{\rm ex}$ for rubbers have been fitted to this form with C_1 and C_2 being constants independent of λ, which is known as the Mooney–Rivlin form [31, 42, 43].

(iii) For shear deformation $x = x_0 + \gamma y_0$, $y = y_0$, and $z = z_0$, we have $I_1 = I_2 = \gamma^2 + 2$ and $I_3 = 1$ so that

$$\sigma_{xy} = 2\gamma(C_1 + C_2). \tag{3A.16}$$

If the displacement $\boldsymbol{u} = \boldsymbol{x} - \boldsymbol{x}_0$ is small, the usual results in isotropic linear elasticity [46] should be reproduced from (3A.11) and (3A.12). The linear stress tensor is expressed as

$$\sigma_{ij} = -p_0\delta_{ij} + K\nabla\cdot\boldsymbol{u}\delta_{ij} + \mu\left(\nabla_i u_j + \nabla_j u_i - \frac{2}{3}\nabla\cdot\boldsymbol{u}\delta_{ij}\right) + O(\boldsymbol{u}^2), \tag{3A.17}$$

where $p_0 = p(1)$ is the pressure in the undeformed state, $\nabla_i = \partial/\partial x_i$ in the Euler representation, and $K = -(\partial p(\lambda)/\partial\lambda)_{\lambda=1}$ and $\mu = 2(C_1 + C_2)_{\gamma=0}$ are the bulk and shear moduli, respectively. The elastic free energy up to the bilinear order reads

$$F_{\rm el} = \text{const.} + \int d\boldsymbol{x}\left[\left(\frac{K}{2} - \frac{\mu}{3}\right)(\nabla\cdot\boldsymbol{u})^2 + \frac{\mu}{4}\sum_{ij}(\nabla_i u_j + \nabla_j u_i)^2\right] + O(\boldsymbol{u}^3). \tag{3A.18}$$

Note that (3A.17) and (3A.18) are written in the Euler representation. There is no essential difference between the two descriptions in the lowest-order theory.

References

[1] L. D. Landau and E. M. Lifshitz, *Statistical Physics* (Pergamon, New York, 1964).

[2] R. Kubo, *Statistical Mechanics* (North-Holland, New York, 1965).

[3] L. E. Reichel, *Modern Course in Statistical Physics* (University of Texas Press, 1980).

[4] P. M. Chaikin and T. C. Lubensky, *Principles of Condensed Matter Physics* (Cambridge University Press, 1995).

[5] W. L. Bragg and E. J. Williams, *Proc. Roy. Soc. A* **145**, 699 (1934).

[6] N. G. van Kampen, *Phys. Rev. A* **135**, 362 (1964).

[7] E. K. Riedel, *Phys. Rev. Lett.* **28**, 675 (1972).

[8] E. K. Riedel and F. J. Wagner, *Phys. Rev. Lett.* **29**, 349 (1972).

[9] R. B. Griffiths, *Phys. Rev. B* **7**, 549 (1973).

[10] R. B. Griffiths, *J. Chem. Phys.* **60**, 195 (1974).

[11] R. Bausch, *Z. Physik* **254**, 81 (1972).

[12] M. J. Stephen, E. Abrahams, and J. P. Straley, *Phys. Rev. B* **12**, 256 (1975).

[13] I. D. Lawrie and S. Sarbach, in *Phase Transitions and Critical Phenomena*, Vol. 9, eds. C. Domb and J. L. Lebowitz (Academic Press, London, 1984), p. 2.

[14] C. M. Knobler and R. L. Scott, *ibid*, p. 163.

[15] K. Kincaid and E. G. D. Cohen, *Phys. Rep.* **22**, 57 (1975).

[16] *Multicritical Phenomena*, eds. R. Pynn and A. Skejeltorp (Plenum, 1984).

[17] R. J. Birgeneau, G. Shirane, M. Blume, and W. C. Koehler, *Phys. Rev. Lett.* **33**, 1098 (1974).

[18] E. K. Riedel, H. Meyer, and R. P. Behringer, *J. Low Temp. Phys.* **22**, 487 (1977).

[19] G. Ahlers, in *The Physics of Liquid and Solid Helium*, Part I, eds. K. H. Bennemann and J. B. Ketterson (Wiley, New York, 1976), p. 85.

[20] E. Courtens and R. W. Gammon, *Phys. Rev. B* **24**, 3890 (1981).

[21] A. G. Khachaturyan, *Theory of Structural Transformations in Solids* (John Wiley & Sons, New York, 1983).

[22] S. M. Allen and J. W. Cahn, *Acta. Metall.* **24**, 425 (1976).

[23] H. Ino, *Acta. Metall.* **26**, 287 (1978).

[24] H. Kubo and C. M. Wayman, *ibid.* **28**, 395 (1979).

[25] J. Braun, J. W. Cahn, G. B. McFadden, and A. A. Wheeler, *Phil. Trans. Roy. Soc. (London), A* **355**, 1787 (1997).

[26] V. Wang, D. Banerjee, C. C. Su, and A. G. Khachaturyan, *Acta Metall.* **46**, 2983 (1998).

[27] A. G. Khachaturyan, T. F. Lindsey, and J. W. Morris, Jr, *Metall. Trans. A* **19**, 249 (1988).

[28] W. A. Soffa and D. E. Laughlin, *Acta Metall.* **37**, 3019 (1989).

[29] B. J. Shaiu, H. T. Li, H. Y. Lee, and H. Chen, *Metall. Trans. A* **21**, 1133 (1989).

[30] D. ter Haar, *Lectures on Selected Topics in Statistical Mechanics* (Pergamon, Oxford, 1977).

[31] P. J. Flory, *Principles of Polymer Chemistry* (Cornell Univ. Press, Ithaca, New York, 1953).

[32] M. L. Huggins, *J. Phys. Chem.* **46**, 161 (1942).

[33] P. G. de Gennes, *Scaling Concepts in Polymer Physics* (Cornell Univ. Press, Ithaca, New York, 1980).

[34] R. Perzynski, M. Delsanti, and M. Adam, *J. Physique* **48**, 115 (1987).

[35] M. Daoud and G. Jannink, *J. Phys.* **37**, 973 (1976).

[36] T. Tanaka, *Physica A* **140**, 261 (1986); Y. Li and T. Tanaka, *Annu. Rev. Mat. Sci.* **22**, 243 (1992).

[37] K. Dušek and D. Patterson (1968) *J. Polym. Sci., Part A: Polym. Chem.* **6**, 1209 (1968).

[38] T. Tanaka, *Phys. Rev. Lett.* **40**, 820 (1978).

[39] T. Tanaka, D. Filmore, S. T. Sun, N. Izumi, G. Swislow, and A. Shah, *Phys. Rev. Lett.* **45**, 1636 (1980).

[40] M. Ilavsky, *Macromolecules* **15**, 782 (1982).

[41] Y. Hirokawa and T. Tanaka, *J. Chem. Phys.* **81**, 6379 (1984).

[42] L. R. G. Treloar, *The Physics of Rubber Elasticity*, 3rd edn (Clarendon Press, Oxford, 1975).

[43] J. E. Mark and B. Erman, *Rubberlike Elasticity* (Wiley, 1988).

[44] B. Erman and P. J. Flory, *Macromolecules* **19**, 2342 (1986).

[45] Z. S. Petrović, W. J. MacKnight, R. Koningsveld, and K. Dušek, *Macromolecules* **20**, 1088 (1987).

[46] L. D. Landau and E. M. Lifshitz *Theory of Elasticity* (Pergamon, New York, 1973).

[47] R. S. Rivlin, in *Rheology*, ed. F. Eirich (Academic, New York, 1956), Vol. 1 p. 351; *J. Polym. Sci. Symp.* **48**, 125 (1974).

[48] A. Onuki, in *Advances in Polymer Science*, Vol. 109, *Responsive Gels: Volume Transitions I*, ed. K. Dušek (Springer, Heidelberg, 1993), p. 63.

[49] S. Hirotsu, *J. Chem. Phys.* **94**, 3949 (1991).

[50] C. Li, Z. Hu, and Y. Li, *Phys. Rev. E* **48**, 603 (1993).

[51] S. Hirotsu, *J. Chem. Phys.* **88**, 427 (1988).

[52] S. Hirotsu and A. Onuki, *J. Phys. Soc. Jpn* **58**, 1508 (1989).

[53] I. M. Lifshitz, A. Yu. Grosberg, and A. R. Khoklov, *Rev. Mod. Phys.* **50**, 683 (1978).

[54] M. Ueda and K. Yoshikawa, *Phys. Rev. Lett.* **77**, 2133 (1966); K. Yoshikawa and Y. Matsuzaka, *J. Amer. Chem. Soc.* **118**, 929 (1996).

[55] E. Yu Kramarenko, A. R. Khoklov, and K. Yoshikawa, *Macromolecules* **30**, 3383 (1997).

[56] G. S. Manning, *Q. Rev. Biophys.* **2**, 179 (1978).

[57] B.-Y. Ha and A. Liu, *Phys. Rev. Lett.* **79**, 1289 (1997); *ibid.* **81**, 1011 (1998).

[58] J. Finger, *Sitzber. Akad. Wiss. Wien Mat.-naturw. Kl. Abt. IIa*, **103**, 1073 (1894).

4

Advanced theories in statics

In this chapter we will present the Ginzburg–Landau–Wilson (GLW) hamiltonian and briefly explain the renormalization group (RG) theory in the scheme of the $\epsilon = 4 - d$ expansion [1]–[12]. As unique features in this book we will introduce a subsidiary energy-like variable in addition to the order parameter, discuss GLW models appropriate for fluids, and derive a simple expression for the thermodynamic free energy consistent with the scaling theory and the two-scale-factor universality. We will try to reach the main RG results related to observable quantities in the simplest and shortest way without too much formal argument. In practice, such an approach is needed for those whose main concerns are advanced theories of dynamics. Furthermore, we will discuss inhomogeneous two-phase coexistence and the surface tension near the critical point, near the symmetrical tricritical point, and in polymer solutions and blends. In addition, we will examine vortices in systems with a complex order parameter. These topological defects are key entities in phase-ordering dynamics discussed in Chapters 8 and 9.

4.1 Ginzburg–Landau–Wilson free energy

4.1.1 Gradient free energy

When the order parameter ψ changes slowly in space, the simplest generalization of the Landau free energy is of the form,

$$\beta\mathcal{H}\{\psi\} = \int d\boldsymbol{r}\left[\frac{1}{2}r_0\psi^2 + \frac{1}{4}u_0\psi^4 - h\psi + \frac{1}{2}K|\nabla\psi|^2\right], \tag{4.1.1}$$

which is called the Ginzburg–Landau–Wilson (GLW) hamiltonian. The first three terms are of the same form as those in the Landau expansion (3.1.6). The last term in the brackets, called the gradient free energy, arises from an increase of the free energy when ψ slowly varies in space. It was first introduced by van der Waals in 1893 to describe gas–liquid interfaces (see (4.4.1) below) [13]. In their seminal theory in 1950, Ginzburg and Landau examined inhomogeneous profiles of a complex order parameter, such as the interface between normal and superconductor phases in type-I superconductors in a magnetic field [14].[1] In the same scheme, Abrikosov calculated vortex lattice structures in type-II superconductors [15] and Ginzburg and Pitaevskii calculated a vortex line in

[1] In Ginzburg and Landau's theory the free-energy density is given by $\alpha|\psi|^2 + \beta|\psi|^4 + |\hbar\nabla\psi - i(e/c)\boldsymbol{A}\psi|^2/2m + \boldsymbol{H}^2/8\pi$, where $\boldsymbol{A}$ is the vector potential and $\boldsymbol{H}$ is the magnetic field.

superfluid helium [16], while Cahn and Hilliard investigated an interface in systems with a single-component order parameter [17].

The order parameter $\psi(\boldsymbol{r})$ is a coarse-grained spin variable in Ising systems defined as follows. (i) It is natural to define it on a coarse-grained lattice with a lattice constant ℓ longer than the original lattice constant a:

$$\psi(\boldsymbol{r}) = \frac{1}{\ell^d} \sum_{i \in \text{new cell}} s_i, \tag{4.1.2}$$

where $\boldsymbol{r}$ is a representative point in each new cell, ℓ^d is the volume of a new cell, and the sum is over original lattice sites contained in each new cell. (ii) In an alternative way we may introduce an upper cut-off wave number Λ of the Fourier transform of the order parameter:

$$\psi(\boldsymbol{r}) = M + (2\pi)^{-d} \int_{k<\Lambda} d\boldsymbol{k} \psi_{\boldsymbol{k}} \exp(i\boldsymbol{k}\cdot\boldsymbol{r}), \tag{4.1.3}$$

where M is the average order parameter. These two definitions of a space-dependent $\psi(\boldsymbol{r})$ are physically equivalent, provided $2\pi/\Lambda \sim \ell$.

For n-component isotropic spin systems, where $\psi = (\psi_1, \psi_2, \ldots, \psi_n)$ and the rotational invariance in the spin space holds, we should interpret

$$\psi^2 = \sum_{j=1}^{n} \psi_j^2, \quad \psi^4 = \left(\sum_{j=1}^{n} \psi_j^2\right)^2, \quad |\nabla\psi|^2 = \sum_{j=1}^{n} |\nabla\psi_j|^2 \tag{4.1.4}$$

in (4.1.1). We will set up the GLW hamiltonians for ^{4}He and ^{3}He–^{4}He with the xy-model symmetry ($n = 2$), where there is no physically realizable ordering field ($h = 0$).

4.1.2 Gaussian approximation

We first neglect the quartic term in $\mathcal{H}$ assuming a small nonlinear coupling constant u_0 in Ising-like systems ($n = 1$) above two dimensions ($d > 2$). In disordered states with $r_0 \geq 0$ and $h = 0$, ψ obeys the gaussian distribution $\propto \exp(-\beta\mathcal{H}_0)$ with

$$\begin{aligned} \beta\mathcal{H}_0 &= \frac{1}{2}\int d\boldsymbol{r}\left[r_0\psi^2 + K|\nabla\psi|^2\right] \\ &= \frac{1}{2}\int_{\boldsymbol{k}} (r_0 + Kk^2)\psi_{\boldsymbol{k}}\psi_{-\boldsymbol{k}}. \end{aligned} \tag{4.1.5}$$

Hereafter we use the notation,

$$\int_{\boldsymbol{k}} \cdots = (2\pi)^{-d} \int d\boldsymbol{k} \cdots. \tag{4.1.6}$$

The structure factor in this approximation is given by the Ornstein–Zernike form,

$$I^0(k) = \langle|\psi_{\boldsymbol{k}}|^2\rangle_0 = \frac{1}{r_0 + Kk^2}, \tag{4.1.7}$$

where $\langle \cdots \rangle_0$ denotes the average over the gaussian distribution. The corresponding pair correlation function,

$$g_0(|\boldsymbol{r}|) = \int_{\boldsymbol{k}} \exp(i\boldsymbol{k}\cdot\boldsymbol{r}) I^0(k), \tag{4.1.8}$$

decays exponentially as $\exp(-\kappa|\boldsymbol{r}|)$ at long distances $|\boldsymbol{r}| \gtrsim \xi$, where

$$\kappa = 1/\xi = (r_0/K)^{1/2} \tag{4.1.9}$$

is the inverse correlation length. In 3D we have the famous expression

$$g_0(|\boldsymbol{r}|) = (4\pi K)^{-1} \frac{1}{|\boldsymbol{r}|} \exp(-\kappa|\boldsymbol{r}|). \tag{4.1.10}$$

Here the critical exponent ν for the correlation length is given by 1/2. In the mean field treatment of phase transitions, we use the Landau theory for the average order parameter and the gaussian approximation for the fluctuations.

4.1.3 Perturbation expansion and the critical dimension

Next we examine by perturbation calculations how the structure factor is changed in the presence of the quartic term in $\mathcal{H}$ for $n = 1$. Using the expansion,

$$\exp(-\beta\mathcal{H}) = \exp(-\beta\mathcal{H}_0)\left[1 - \beta\mathcal{H}' + \frac{1}{2}(\beta\mathcal{H}')^2 + \cdots\right] \tag{4.1.11}$$

with

$$\beta\mathcal{H}' = \int d\boldsymbol{r} \frac{1}{4} u_0 \psi^4, \tag{4.1.12}$$

we obtain

$$I(k) = I^0(k) - 3u_0[I^0(k)]^2 \int_{\boldsymbol{q}} I^0(q) + \cdots. \tag{4.1.13}$$

It is more convenient to consider the inverse,

$$1/I(k) = r_0 + Kk^2 + 3u_0 \int_{\boldsymbol{q}} I^0(q) + \cdots. \tag{4.1.14}$$

At the critical point, the susceptibility diverges, so $1/I(k) \cong Kk^2$ should tend to zero as $k \to 0$. This means that the coefficient r_0 assumes a critical value r_{0c} determined by

$$r_{0c} = -3u_0 \int_{\boldsymbol{q}} \frac{1}{Kq^2} + \cdots = -\frac{3K_d}{(d-2)} K^{-1} u_0 \Lambda^{d-2} + \cdots, \tag{4.1.15}$$

where

$$K_d = (2\pi)^{-d} 2\pi^{d/2}/\Gamma(d/2) \tag{4.1.16}$$

is the surface area of a unit sphere in d dimensions divided by $(2\pi)^d$, $\Gamma(x)$ being the Gamma function, so $K_4 = 1/8\pi^2$ and $K_3 = 1/2\pi^2$. We define r by

$$r_0 = r + r_{0c}. \tag{4.1.17}$$

Then r vanishes at the critical point, so we may assume the linear temperature dependence (3.1.7),

$$r = r_0 - r_{0c} = a_0\tau, \tag{4.1.18}$$

in terms of the reduced temperature $\tau = T/T_c - 1$. The coefficient a_0 is assumed to be positive. In the perturbation expansions it is convenient to replace the *bare* coefficient r_0 in place of the shifted coefficient r in the two-body correlation function. In this manner we can take into account the critical temperature shift due to the nonlinear fluctuation effect. This procedure of eliminating r_0 in favor of r is called *mass renormalization* (which was originally a jargon in particle physics).

With this in mind, we rewrite (4.1.14) as

$$\begin{aligned} 1/I(k) &= r + 3u_0\int_q\left[\frac{1}{r+Kq^2} - \frac{1}{Kq^2}\right] + Kk^2 + \cdots \\ &= r\big[1 - 3K_dK^{-2}u_0I_d\big] + Kk^2 + \cdots, \end{aligned} \tag{4.1.19}$$

where

$$I_d = \int_0^\Lambda dq q^{d-3}\frac{1}{\kappa^2+q^2}. \tag{4.1.20}$$

The above q integration is divergent at large q as $\Lambda \to \infty$ (ultraviolet divergence) for $d > 4$, and at small q as $\kappa \to 0$ (infrared divergence) for $d < 4$. As a result, the dominant contribution arises around the upper cut-off Λ for $d > 4$ and the lower cut-off κ for $d < 4$. In particular, if $\epsilon = 4 - d$ is small, I_d behaves as

$$I_d = \frac{1}{\epsilon}\big[\kappa^{-\epsilon} - \Lambda^{-\epsilon}\big]. \tag{4.1.21}$$

In the limit $\epsilon \to 0$ and $(\kappa/\Lambda)^\epsilon \cong 1 + \epsilon\ln(\kappa/\Lambda)$, we have logarithmic behavior, $I_d \cong I_4 = \ln(\Lambda/\kappa)$. With (4.1.19) we notice that u_0 appears in the perturbation expansion in the following dimensionless combination,

$$g = K_du_0/(K^2\Lambda^\epsilon). \tag{4.1.22}$$

For small ϵ we thus obtain

$$1/I(k) = r\left\{1 - \frac{3g}{\epsilon}\big[(\Lambda/\kappa)^\epsilon - 1\big]\right\} + Kk^2 + \cdots. \tag{4.1.23}$$

The structure of the perturbation series in powers of u_0 changes qualitatively at the marginal dimensionality $d_c = 4$. That is, if $d > 4$ or for $\epsilon < 0$, the perturbation expansion is well defined or convergent as long as $g \ll |\epsilon|$. On the contrary, if $d < 4$ or for $\epsilon > 0$, the factor $(\Lambda/\kappa)^\epsilon$ grows near the critical point and the expansion is meaningful only for

$$3g \ll \epsilon(\kappa/\Lambda)^\epsilon \quad \text{or} \quad 3K_du_0/K^2 \ll \epsilon\kappa^\epsilon. \tag{4.1.24}$$

The Ginzburg number

The condition (4.1.24) is rewritten as $|\tau| \gg Gi$ in the absence of an ordering field ($h = 0$), where

$$Gi = Ka_0^{-1}(3K_d u_0/\epsilon K^2)^{2/\epsilon} \tag{4.1.25}$$

is called the Ginzburg number expressed in terms of a_0 in (4.1.18) and the coefficients in the GLW hamiltonian in (4.1.1). Using the mean field expressions for the microscopic length $\xi_{+0} = (K/a_0)^{1/2}$ in $\xi = \xi_{+0}\tau^{-1/2}$ and the specific-heat jump ΔC_H in (3.1.20), we may also express Gi as

$$Gi = [3K_d/2\epsilon\xi_{+0}^d\Delta C_H]^{2/\epsilon}. \tag{4.1.26}$$

In particular, we write the 3D expression,

$$Gi = (3/2\pi^2)^2 u_0^2/(K^3 a_0) = (3/\pi^2)^2(\xi_{+0}^3\Delta C_H)^{-2}. \tag{4.1.27}$$

Crossover occurs around $\tau \sim Gi$ from the mean-field to asymptotic critical behavior, as has been studied theoretically with renormalization group methods [4b] [18, 19] and experimentally in various fluid systems at the critical density (or concentration) [20]–[22]. In polymer blends near the consolute critical point, Gi decreases with increasing molecular weight and can be very small [20]. In ^{3}He near the gas–liquid critical point, Gi is 2.5×10^{-3}, while in Xe it is 1.8×10^{-2} [22]. While the thermal fluctuations are asymptotically dominant in any fluids near the gas–liquid critical point, Gi becomes small in ^{3}He due to large background quantum fluctuations. We note that $\lambda_{\rm th} n_{\rm c}^{1/3}$ is equal to 1.2 for ^{3}He and to 0.048 for CO_2, where $\lambda_{\rm th}$ is the thermal de Broglie wavelength in (1.2.5) and $n_{\rm c}$ is the critical number density.

Exact relations

Because the equilibrium distribution of ψ is given by $P_{\rm eq}\{\psi\} \propto \exp(-\beta\mathcal{H})$, we notice the equilibrium relations,

$$\left\langle \frac{\delta(\beta\mathcal{H})}{\delta\psi(\boldsymbol{r})} \right\rangle = 0, \quad \left\langle \psi(\boldsymbol{r}')\frac{\delta(\beta\mathcal{H})}{\delta\psi(\boldsymbol{r})} \right\rangle = \delta(\boldsymbol{r} - \boldsymbol{r}'). \tag{4.1.28}$$

The first relation can lead to the equation of state in the form of (4.3.65) below [5], while the second one gives an equation for the pair correlation function $g(r) = \langle \delta\psi(\boldsymbol{r})\delta\psi(\boldsymbol{0})\rangle$,

$$(r_0 - K\nabla^2)g(r) + u_0\langle\psi(\boldsymbol{r})^3\delta\psi(\boldsymbol{0})\rangle = \delta(\boldsymbol{r}). \tag{4.1.29}$$

The Fourier transformation gives

$$(r_0 + Kk^2)I(k) + u_0\int d\boldsymbol{r}e^{i\boldsymbol{k}\cdot\boldsymbol{r}}\langle\psi(\boldsymbol{r})^3\delta\psi(\boldsymbol{0})\rangle = 1. \tag{4.1.30}$$

Decoupling the above four-body correlation at $\langle\psi\rangle = 0$ readily yields (4.1.14).

Fig. 4.1. The diagrammatic structure of the two-body correlation function.

4.1.4 Feynman diagram expansion

The effect of the four-body interaction $\beta\mathcal{H}'$ can be calculated systematically using well-defined Feynman diagrammatic rules. This technique is based on the fact that many-body correlations $\langle\psi\cdots\psi\rangle$ can be decoupled into sums of products of two-body correlations (because the zeroth-order distribution is gaussian). In Fig. 4.1 we display the diagrammatic structure of the two-body correlation function $I(k)$. Let the contribution from the self-energy diagrams be written as $\Sigma(r, k)$. Then we have

$$I(k) = 1/\left[r_0 - \Sigma(r, k) + Kk^2\right]. \tag{4.1.31}$$

Obviously, (4.1.30) gives the expression,

$$\Sigma(r, k) = -u_0 \int d\boldsymbol{r} e^{i\boldsymbol{k}\cdot\boldsymbol{r}} \langle\psi(\boldsymbol{r})^3 \delta\psi(\boldsymbol{0})\rangle / I(k). \tag{4.1.32}$$

The critical-point value r_{0c} is expressed generally as

$$r_{0c} = \Sigma(0, 0), \tag{4.1.33}$$

which reduces to (4.1.15) at small u_0. Elimination of the bare coefficient r_0 yields

$$I(k) = 1/\left\{r - [\Sigma(r, k) - \Sigma(0, 0)] + Kk^2\right\}. \tag{4.1.34}$$

The inverse susceptibility at $k = 0$ is expressed as

$$\chi^{-1} = r - [\Sigma(r, 0) - \Sigma(0, 0)]. \tag{4.1.35}$$

Slightly away from the critical point ($r > 0$) the self-energy part is expanded in powers of k^2 as

$$\Sigma(r, k) = \Sigma(r, 0) - (\delta K)k^2 + O(k^4). \tag{4.1.36}$$

The coefficient δK starts from the order u_0^2 because there has been no correction in the first-order calculation. By defining the renormalized coefficient K_{R} as

$$K_{\mathrm{R}} = K + \delta K, \tag{4.1.37}$$

we may express the structure factor at small k as

$$I(k) = 1/[\chi^{-1} + K_{\mathrm{R}}k^2 + O(k^4)]. \tag{4.1.38}$$

The renormalized correlation length ξ and its inverse κ are then defined as

$$\kappa = 1/\xi = (\chi K_{\mathrm{R}})^{-1/2}. \tag{4.1.39}$$

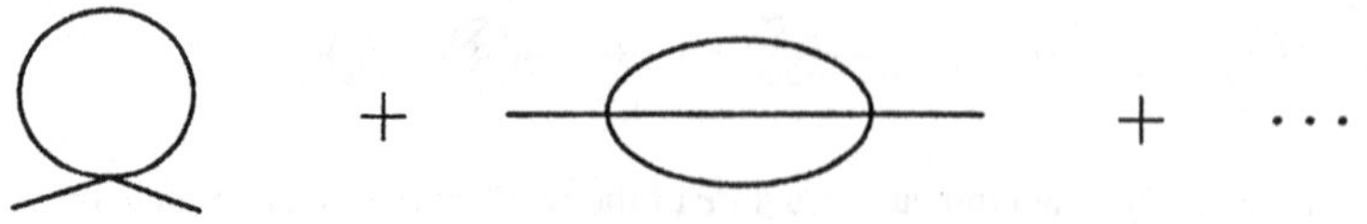

Fig. 4.2. The first- and second-order contributions to the self energy function.

On the one hand, in accord with the above expression, the scaling theory in Chapter 2 suggests the power laws, $\chi^{-1} \sim \tau^{\gamma} \sim \kappa^{2-\eta}$ and $K_R \sim \kappa^{-\eta}$. For $k \gg \kappa$, on the other hand, (2.1.9) indicates

$$1/I(k) \cong Kk^2(k/\Lambda)^{-\eta} \cong Kk^2[1 - \eta \ln(k/\Lambda)], \tag{4.1.40}$$

because η is very small. Therefore, for $\kappa \ll k \ll \Lambda$, we have

$$\Sigma(r, k) \cong \Sigma(0, 0) + \eta K k^2 \ln(k/\Lambda). \tag{4.1.41}$$

Let us calculate the second-order correction to the self-energy arising from the two-loop diagram in Fig. 4.2. For $n = 1$ it is of the form,

$$\Sigma_2(r, k) = 6u_0^2 \int_{q_1} \int_{q_2} I(q_1) I(q_2) I(|\boldsymbol{q}_1 + \boldsymbol{q}_1 - \boldsymbol{k}|). \tag{4.1.42}$$

The reader may easily derive the factor of 6 in the above expression using the decoupling procedure or the Feynman rules. See Appendix 4A for the calculation of the above double integral at $r = 0$. We shall see that $\Sigma_2(0, 0) \sim g^2 K \Lambda^2$, which contributes to r_{0c}, and[2]

$$\Sigma_2(0, k) - \Sigma_2(0, 0) \cong \frac{3}{2} g^2 K k^2 \ln(k/\Lambda). \tag{4.1.43}$$

We now compare (4.1.41) and (4.1.43) to obtain

$$\eta = \frac{3}{2} g^2 + \cdots. \tag{4.1.44}$$

Here η should be universal. Does the above relation mean that g takes a particular value? This puzzle is resolved in the renormalization group theory, which shows that g tends to a universal number g^* with decreasing Λ.

4.1.5 Inclusion of the energy density

We next introduce a subsidiary variable $m(\boldsymbol{r})$ as in Section 3.1. In Ising systems, it is the exchange-energy density measured from the critical value and divided by T_c.

[2] The calculation of $\Sigma_2(r, 0) - \Sigma_2(0, 0)$ is also straightforward, but it contains a term proportional to $[\ln(\kappa/\Lambda)]^2$ and is more complicated.

Generalization of (3.1.25) leads to the GLW hamiltonian for space-dependent $\psi(\mathbf{r})$ and $m(\mathbf{r})$ [23],

$$\begin{aligned}\beta\mathcal{H}\{\psi, m\} &= \int d\mathbf{r}\bigg[\frac{1}{2}r_{0c}\psi^2 + \frac{1}{2}K|\nabla\psi|^2 + \frac{1}{4}\bar{u}_0\psi^4 + \gamma_0\psi^2 m \\ &\quad + \frac{1}{2C_0}m^2 - h\psi - \tau m\bigg] \\ &= \beta\mathcal{H}\{\psi\} + \int d\mathbf{r}\frac{1}{2}C_0(\delta\hat{\tau})^2 - \frac{1}{2}C_0\tau^2 V. \end{aligned} \tag{4.1.45}$$

We introduce the reduced temperature fluctuation by[3]

$$\delta\hat{\tau}(\mathbf{r}) = \frac{\delta}{\delta m}\beta\mathcal{H} = \frac{1}{C_0}m + \gamma_0\psi^2 - \tau, \tag{4.1.46}$$

which obeys the gaussian distribution independent of ψ characterized by

$$\langle\delta\hat{\tau}(\mathbf{r})\delta\hat{\tau}(\mathbf{r}')\rangle = C_0^{-1}\delta(\mathbf{r}-\mathbf{r}'). \tag{4.1.47}$$

It is important that $\delta\hat{\tau}$ is statistically independent of ψ in equilibrium. The second line of (4.1.45) is written in terms of ψ and $\delta\hat{\tau}$, where the first term $\mathcal{H}\{\psi\}$ is the hamiltonian (4.1.1) for ψ only with

$$r = 2\gamma_0 C_0\tau, \quad u_0 = \bar{u}_0 - 2\gamma_0^2 C_0, \tag{4.1.48}$$

and the third term, proportional to τ^2, is the mean field contribution of the energy variable corresponding to the third term in (3.1.8). From (4.1.18) γ_0 is related to a_0 by

$$\gamma_0 = a_0/2C_0. \tag{4.1.49}$$

Note that the above definition of $\delta\hat{\tau}$ depends on the upper cut-off wave number Λ and the hydrodynamic temperature fluctuation (1.1.41) or (3.1.27) follows in the limit $\Lambda \to 0$. Also we define the magnetic field fluctuation,

$$\delta\hat{h}(\mathbf{r}) = \frac{\delta}{\delta\psi}\beta\mathcal{H} = (r_{0c} - K\nabla^2 + \bar{u}_0\psi^2 + 2\gamma_0 m)\psi - h, \tag{4.1.50}$$

whose hydrodynamic expression is (1.1.42). Then the variance relations in (1.1.43) are satisfied (if $\hat{e}$ is replaced by m).

From (4.1.46) and (4.1.47) the variance of $m(\mathbf{r})$, which is equal to the specific heat C_H at constant magnetic field, is given by

$$C_H = \langle m : m\rangle = C_0 + (\gamma_0 C_0)^2\langle\psi^2 : \psi^2\rangle, \tag{4.1.51}$$

where the second term is the singular fluctuation contribution. From (1.1.40) the specific heat C_M at constant magnetization is written as

$$C_M = C_0 + (\gamma_0 C_0)^2\big[\langle\psi^2 : \psi^2\rangle - \langle\psi : \psi^2\rangle^2/\langle\psi : \psi\rangle\big]. \tag{4.1.52}$$

[3] The circumflex is kept here because τ is used for the average reduced temperature.

For $\tau > 0$ and $h = 0$ there is no difference between C_H and C_M. For $n = 1$ the decoupling of the four-body correlation yields

$$C_H = C_0 + 2K_d\gamma_0^2 C_0^2 \int_0^\Lambda dq q^{d-1} \frac{1}{(\kappa^2+q^2)^2} + \cdots \tag{4.1.53}$$

(see Fig. 4.4). For small ϵ this becomes

$$C_H/C_0 = 1 + \frac{2v}{\epsilon}[(\Lambda/\kappa)^\epsilon - 1] + \cdots = 1 + 2v\ln(\Lambda/\kappa) + \cdots, \tag{4.1.54}$$

where v is a dimensionless coupling constant defined by

$$v = K_d\gamma_0^2 C_0/K^2\Lambda^\epsilon. \tag{4.1.55}$$

We shall see that v tends to a universal number $v^* = \alpha + O(\epsilon^2)$ with decreasing Λ, leading to the ultimate scaling behavior $C_H \propto \tau^{-\alpha}$ if the logarithmic term is exponentiated.

4.1.6 Hydrodynamic hamiltonian for $n = 1$

In (1.1.50) we introduced the hydrodynamic hamiltonian $\mathcal{H}_{\rm hyd}$ for the deviations $\delta\psi = \psi - M$ and $\delta\hat{T}/T$ for Ising-like systems, the latter being $\delta\hat{\tau}$ in the present notation. This form can be obtained after elimination of the fluctuations with sizes shorter than ξ or in the limit $\Lambda \ll \xi^{-1}$. We assume the existence of the renormalized coefficient $\gamma_{\rm R} = \lim_{\Lambda\to 0}\gamma_0$. Then the linear relation $\delta\hat{\tau} \cong \delta m/C_M + 2\gamma_{\rm R} M\delta\psi$ follows with $\delta m = m - \langle m\rangle$ from (4.1.46). Thus (1.1.50) is rewritten as

$$\frac{1}{T}\mathcal{H}_{\rm hyd} = \int d\boldsymbol{r}\left[\frac{1}{2\chi}(\delta\psi)^2 + \frac{1}{2C_M}(\delta m + 2\gamma_{\rm R} C_M M\delta\psi)^2\right] \tag{4.1.56}$$

The cross term ($\propto \delta m\delta\psi$) appears in the presence of nonvanishing average order parameter M. From $\langle\psi : m\rangle = -2\gamma_{\rm R} C_M M\chi$ we may express $\gamma_{\rm R}$ as

$$2\gamma_{\rm R} C_M = -\frac{1}{\chi M}\left(\frac{\partial M}{\partial\tau}\right)_h. \tag{4.1.57}$$

For infinitesimal h with $\tau > 0$, we have $M \cong \chi h$ and

$$2\gamma_{\rm R} C_M = \gamma(\tau\chi)^{-1} = (\gamma/\Gamma_0)\tau^{\gamma-1}. \tag{4.1.58}$$

Note that this relation is valid in general n-component systems. Here we set $\chi = \Gamma_0\tau^{-\gamma}$ for $\tau > 0$ and $\Gamma_0'|\tau|^{-\gamma}$ for $\tau < 0$ at $h = 0$. On the coexistence curve, where $M = B_0|\tau|^\beta$, we obtain

$$2\gamma_{\rm R} C_M = \beta(|\tau|\chi)^{-1} = (\beta/\Gamma_0')|\tau|^{\gamma-1}. \tag{4.1.59}$$

The coefficients on the right-hand sides of (4.1.58) and (4.1.59) are nearly the same, as can be seen from the amplitude ratio relation (4.3.83) below. From (4.1.56) we also have

$$C_H = \langle m : m\rangle = C_M + 4(\gamma_{\rm R} C_M M)^2\chi. \tag{4.1.60}$$

From (1.1.49) and (4.1.59) the universal number R_{v} on the coexistence curve is written as

$$R_{\mathrm{v}} = 4(\gamma_{\mathrm{R}} C_M M)^2 \chi / C_H = (\beta B_0)^2 / A'_0 \Gamma'_0, \tag{4.1.61}$$

where we set $C_H = A'_0 |\tau|^{-\alpha}$. This relation is consistent with (1.1.48). In addition, the coupling parameter v in (4.1.55) approaches a universal number ($\cong \alpha$) for $\Lambda \gg \kappa$, but in the region $\Lambda \ll \kappa$ it grows as

$$v = \frac{1}{4} \alpha \gamma^2 K_d R_\xi^{-d} (\xi \Lambda)^{-\epsilon}, \tag{4.1.62}$$

where $T > T_{\mathrm{c}}$ and $R_\xi (\cong (K_d/4)^{1/d})$ is defined by (2.1.45). The above relation will be used in (6.2.37) below. For many-component systems ($n \geq 2$), we will construct $\mathcal{H}_{\mathrm{hyd}}$ to account for anomalous fluctuations due to broken symmetry in Section 4.3.

4.2 Mapping onto fluids

4.2.1 One-component fluids

In one-component fluids near the gas–liquid critical point, the hamiltonian is given by (4.1.45) under the mapping relationships (2.2.2) and (2.2.3) or equivalently (2.2.7) and (2.2.8). In this scheme the temperature and pressure fluctuations may be defined by [24]

$$\delta \hat{T} = \left(\frac{\partial T}{\partial h} \right)_\tau \frac{\delta(\beta \mathcal{H})}{\delta \psi} + \left(\frac{\partial T}{\partial \tau} \right)_h \frac{\delta(\beta \mathcal{H})}{\delta m}, \tag{4.2.1}$$

$$\delta \hat{p} = \left(\frac{\partial p}{\partial h} \right)_\tau \frac{\delta(\beta \mathcal{H})}{\delta \psi} + \left(\frac{\partial p}{\partial \tau} \right)_h \frac{\delta(\beta \mathcal{H})}{\delta m}, \tag{4.2.2}$$

where $\delta \hat{\tau} = \delta(\beta \mathcal{H})/\delta \psi$ and $\delta \hat{h} = \delta(\beta \mathcal{H})/\delta m$ are the temperature and magnetic field fluctuations in the corresponding Ising system defined by (4.1.46) and (4.1.50), respectively. These expressions tend to (2.2.29) and (2.2.30) in the hydrodynamic limit. Here the first terms ($\propto \delta \hat{h}$) have variances of order r, whereas the second terms ($\propto \delta \hat{\tau}$) have those of order C_0^{-1}. Therefore, the second terms exhibit much larger fluctuations than the first terms close to the critical point.

Under the mapping relations we may regard $\mathcal{H}$ as a functional of δn and δe or that of δn and δs, where $\delta s \cong (\delta e - H_{\mathrm{c}} \delta n)/n_{\mathrm{c}} T_{\mathrm{c}}$ from (1.2.46), $H_{\mathrm{c}} = (e_{\mathrm{c}} + p_{\mathrm{c}})/n_{\mathrm{c}}$ being the enthalpy at the critical point. As the coefficients in the mapping relations (2.2.7)–(2.2.13) and the pressure expression (1.2.27), we use those at the critical point to obtain [24]

$$\delta \hat{T} = T_{\mathrm{c}} \left(\frac{\delta \mathcal{H}}{\delta e} \right)_n = n_{\mathrm{c}}^{-1} \left(\frac{\delta \mathcal{H}}{\delta s} \right)_n, \tag{4.2.3}$$

$$\delta \hat{p} = n_{\mathrm{c}} \left(\frac{\delta \mathcal{H}}{\delta n} \right)_{\mathrm{e}} + (e_{\mathrm{c}} + p_{\mathrm{c}}) \left(\frac{\delta \mathcal{H}}{\delta e} \right)_n = n_{\mathrm{c}} \left(\frac{\delta \mathcal{H}}{\delta n} \right)_{\mathrm{s}}. \tag{4.2.4}$$

The following correlation function relations are satisfied between the two sets of deviations, $\{\delta s, \delta n\}$ and $\{\delta\hat{T}, \delta\hat{p}\}$:

$$\begin{aligned}\langle\delta s(\boldsymbol{r})\delta\hat{T}(\boldsymbol{r}')\rangle &= n_c^{-1}T_c\delta(\boldsymbol{r}-\boldsymbol{r}'), \quad \langle\delta n(\boldsymbol{r})\delta\hat{T}(\boldsymbol{r}')\rangle = 0,\\ \langle\delta n(\boldsymbol{r})\delta\hat{p}(\boldsymbol{r}')\rangle &= n_c T_c\delta(\boldsymbol{r}-\boldsymbol{r}'), \quad \langle\delta s(\boldsymbol{r})\delta\hat{p}(\boldsymbol{r}')\rangle = 0,\end{aligned} \tag{4.2.5}$$

which are consistent with the thermodynamic relations (1.2.63) and (1.2.70) in the hydrodynamic limit.

4.2.2 Binary fluid mixtures

For binary fluid mixtures we introduced the third (nonsingular) variable q in addition to ψ and m in Section 2.3. The field variable ζ conjugate to q is the coordinate along the critical line. The free-energy contribution due to the fluctuation of q is simply gaussian, so the hamiltonian for the three variables is

$$\beta\mathcal{H}\{\psi, m, q\} = \beta\mathcal{H}\{\psi, m\} + \int d\boldsymbol{r}\left[\frac{q^2}{2Q_0} - \zeta q\right]. \tag{4.2.6}$$

The mapping relations are given by (2.3.9)–(2.3.11). We can see that (2.3.1)–(2.3.3) can be derived from the above hamiltonian. Also as in (2.3.33) and (2.3.34) we express the temperature and pressure variables as [25]

$$\delta\hat{T} = \left(\frac{\partial T}{\partial h}\right)_{\tau\zeta}\frac{\delta(\beta\mathcal{H})}{\delta\psi} + \left(\frac{\partial T}{\partial\tau}\right)_q\frac{\delta(\beta\mathcal{H})}{\delta m} + \left(\frac{\partial T}{\partial\zeta}\right)_{h\tau}\frac{\delta(\beta\mathcal{H})}{\delta\zeta}, \tag{4.2.7}$$

$$\delta\hat{p} = \left(\frac{\partial p}{\partial h}\right)_{\tau\zeta}\frac{\delta(\beta\mathcal{H})}{\delta\psi} + \left(\frac{\partial p}{\partial\tau}\right)_{h\zeta}\frac{\delta(\beta\mathcal{H})}{\delta m} + \left(\frac{\partial p}{\partial\zeta}\right)_{h\tau}\frac{\delta(\beta\mathcal{H})}{\delta\zeta}, \tag{4.2.8}$$

where $\delta\hat{\zeta} = \delta(\beta\mathcal{H})/\delta q = q/Q_0 - \zeta$ as in (2.3.35). The second and third terms represent weakly singular and nonsingular fluctuations. They give rise to the variance relations (2.3.36)–(2.3.38) in the hydrodynamic limit.

We regard $\mathcal{H}$ as a functional of $\{n_1, n_2, e\}$ or $\{n, X, s\}$. Similarly to (4.2.3) and (4.2.4) the temperature and pressure variables are expressed as

$$\delta\hat{T} = T_c\left(\frac{\delta\mathcal{H}}{\delta e}\right)_{n_1 n_2} = n_c^{-1}\left(\frac{\delta\mathcal{H}}{\delta s}\right)_{nX}. \tag{4.2.9}$$

$$\delta\hat{p} = \sum_{K=1,2} n_{cK}\left(\frac{\delta\mathcal{H}}{\delta n_K}\right)_e + (e_c + p_c)\left(\frac{\delta\mathcal{H}}{\delta e}\right)_{n_1 n_2} = n_c\left(\frac{\delta\mathcal{H}}{\delta n}\right)_{sX}. \tag{4.2.10}$$

As in (1.3.50) we introduce the fluctuation of the chemical potential difference by

$$\delta\hat{\Delta} = \left(\frac{\delta\mathcal{H}}{\delta n_1}\right)_{n_2 e} - \left(\frac{\delta\mathcal{H}}{\delta n_2}\right)_{n_1 e} + \Delta_c\left(\frac{\delta\mathcal{H}}{\delta e}\right)_{n_1 n_2} = n_c^{-1}\left(\frac{\delta\mathcal{H}}{\delta X}\right)_{ns}. \tag{4.2.11}$$

The two sets of deviations, $\{\delta s, \delta n, \delta X\}$ and $\{\delta\hat{T}, \delta\hat{p}, \delta\hat{\Delta}\}$, satisfy the variance relations (1.3.43), (1.3.47), and (1.3.52) as in the one-component case (4.2.5).

4.2.3 ^{4}He near the superfluid transition

For ^{4}He near the superfluid transition we may use the above hamiltonian $\mathcal{H}\{\psi, m, q\}$ under the mapping relations (2.4.8) and (2.4.9). Note that ψ is complex and h is zero in helium. Although redundant, the explicit form of the hamiltonian is

$$\begin{aligned}\beta\mathcal{H}\{\psi, m, q\} &= \int d\mathbf{r}\bigg[\frac{1}{2}r_{0c}|\psi|^2 + \frac{1}{2}|\nabla\psi|^2 + \frac{1}{4}\bar{u}_0|\psi|^4 \\ &\quad + \gamma_0|\psi|^2 m + \frac{1}{2C_0}m^2 + \frac{1}{2Q_0}q^2 - \tau m - \zeta q\bigg],\end{aligned} \tag{4.2.12}$$

where τ and ζ are defined by (2.4.5) and (2.4.23), respectively. Here m is coupled with $|\psi|^2$ and is weakly singular, whereas q is nonsingular. They are linearly related to the entropy and number density deviations, δs and δn, as (2.4.8) and (2.4.9). The coefficient K in the gradient term has been set equal to 1 because the critical exponent η is virtually zero. From (2.4.27) and (2.4.28) the temperature and pressure variables are

$$\delta\hat{T} = n^{-1}\frac{\delta\mathcal{H}}{\delta s} = (1 - \epsilon_{\text{in}}A_\lambda)^{-1}\left[\frac{\delta\mathcal{H}}{\delta m} + \epsilon_{\text{in}}\frac{\delta\mathcal{H}}{\delta q}\right], \tag{4.2.13}$$

$$\delta\hat{p} = n\frac{\delta\mathcal{H}}{\delta n} = n(1 - \epsilon_{\text{in}}A_\lambda)^{-1}\left[A_\lambda\frac{\delta\mathcal{H}}{\delta m} + \frac{\delta\mathcal{H}}{\delta q}\right]. \tag{4.2.14}$$

By setting $\delta\mathcal{H}/\delta m = \delta\mathcal{H}/\delta q = 0$ we may eliminate m and q to obtain $\mathcal{H}\{\psi\}$ in the form of (4.1.1) with (4.1.49). We can derive (4.2.5) also in this case.

4.2.4 ^{3}He–^{4}He mixtures near the λ line and the tricritical point

In ^{3}He–^{4}He mixtures near the superfluid transition the subsidiary variables are the entropy deviation per particle $m_1 = \delta s$, the number density deviation $m_2 = \delta n$, and the ^{3}He concentration deviation $m_3 = \delta X$ as in (1.3.32). From (1.3.20) the conjugate field variables are conveniently written as $h_1 = (n_c/T_c)\delta T$, $h_2 = (n_c T_c)^{-1}\delta p$, and $h_3 = (n_c/T_c)\delta\Delta$. The hamiltonian for ψ and m_j are given by [26]

$$\begin{aligned}\beta\mathcal{H}\{\psi, m_1, m_2, m_3\} &= \int d\mathbf{r}\bigg[\frac{1}{2}r_{0c}|\psi|^2 + \frac{1}{2}|\nabla\psi|^2 + \frac{1}{4}\bar{u}_0|\psi|^4 + \frac{1}{6}v_0|\psi|^6 \\ &\quad + \sum_j \gamma_{j0}|\psi|^2 m_j + \frac{1}{2}\sum_{ij} a_{ij}^{(0)} m_i m_j - \sum_j h_j m_j\bigg],\end{aligned} \tag{4.2.15}$$

where the sixth-order term $\propto |\psi|^6$ is needed near the tricritical point and $\{a_{ij}^{(0)}\}$ is a constant symmetric matrix dependent on Λ. The subsidiary fields may be eliminated by setting

$\delta(\beta\mathcal{H})/\delta m_j = 0$. After some calculations we obtain

$$\beta\mathcal{H}\{\psi\} = \int d\mathbf{r}\left[\frac{1}{2}(r_{0c}+r)|\psi|^2 + \frac{1}{2}|\nabla\psi|^2 + \frac{1}{4}u_0|\psi|^4 + \frac{1}{6}v_0|\psi|^6\right], \tag{4.2.16}$$

where

$$r = \sum_{ij}\gamma_{i0}b_{ij}^{(0)}h_j, \tag{4.2.17}$$

$$u_0 = \bar{u}_0 - 2\sum_{ij}\gamma_{i0}\gamma_{j0}b_{ij}^{(0)}, \tag{4.2.18}$$

with $\{b_{ij}^{(0)}\}$ being the inverse matrix of $\{a_{ij}^{(0)}\}$. Using the Pippard–Buckingham relations [24] we can derive

$$r = a_0\left[T - T_{\lambda 0} - \left(\frac{\partial T}{\partial p}\right)_{\lambda\Delta}(p-p_0) - \left(\frac{\partial T}{\partial \Delta}\right)_{\lambda p}(\Delta - \Delta_0)\right], \tag{4.2.19}$$

where p_0 and Δ_0 are the reference pressure and chemical potential difference, and $T_{\lambda 0} = T_\lambda(p_0, \Delta_0)$. The critical surface in the T–p–Δ space is represented by $r = 0$. The tricritical line, where $r = u_0 = 0$, is reached with Δ or an increase in the average ^{3}He concentration.

The hamiltonian (4.2.15) with three subsidiary variables is essentially the same as that in (4.2.12) with a single subsidiary variable. In fact, we may define a weakly singular variable by $m = \sum_j \gamma_{0j}m_j$ and two other nonsingular variables decoupled from $|\psi|^2$. We may also eliminate the number density variable δn or neglect the pressure fluctuations [26], retaining δs and δX. In this case it is convenient to define new variables,

$$\begin{aligned} m_1' &= \delta s + \left(\frac{\partial\Delta}{\partial T}\right)_{\lambda p}\delta X, \\ m_2' &= \frac{T}{C_\lambda}\left(\frac{\partial s}{\partial T}\right)_{\lambda p}\left[\delta X - \left(\frac{\partial X}{\partial s}\right)_{\lambda p}\delta s\right], \end{aligned} \tag{4.2.20}$$

where

$$C_\lambda = T\left(\frac{\partial s}{\partial T}\right)_{\lambda p} + T\left(\frac{\partial\Delta}{\partial T}\right)_{\lambda p}\left(\frac{\partial X}{\partial T}\right)_{\lambda p}. \tag{4.2.21}$$

Here the derivatives are performed along the λ line at fixed p, so the coefficients in the above definitions are all regular. By setting $\delta p = 0$ we then have

$$\begin{aligned} \beta\mathcal{H}\{\psi, m_1', m_2'\} = \int d\mathbf{r}\Big[&\frac{1}{2}r_{0c}'|\psi|^2 + \frac{1}{2}|\nabla\psi|^2 + \frac{1}{4}\bar{u}_0'|\psi|^4 + \frac{1}{6}v_0|\psi|^6 \\ &+ \gamma_0'|\psi|^2 m_2' + \frac{1}{2C_0'}(m_2')^2 + \frac{1}{2C_\lambda}(m_1')^2 - h_1'm_1' - h_2'm_2'\Big], \end{aligned} \tag{4.2.22}$$

where r'_{0c}, $\bar{u}'_0$, and γ'_0 are appropriately defined coefficients. The conjugate fields h'_1 and h'_2 are linear combinations of δT and $\delta\Delta$:

$$\begin{aligned} h'_1 &= \frac{n}{C_\lambda}\left(\frac{\partial s}{\partial T}\right)_{\lambda p}\left[\delta T + \left(\frac{\partial X}{\partial s}\right)_{\lambda p}\delta\Delta\right], \\ h'_2 &= -\frac{n}{T}\left(\frac{\partial \Delta}{\partial T}\right)_{\lambda p}\left[\delta T - \left(\frac{\partial T}{\partial \Delta}\right)_{\lambda p}\delta\Delta\right]. \end{aligned} \tag{4.2.23}$$

Thus h'_2 is proportional to r in (4.2.19) for $\delta p = 0$. The m'_1 is decoupled from $|\psi|^2$ and regular. From $\delta X = m'_2 + T(\partial X/\partial T)_{\lambda p} m'_1/C_\lambda$, m'_2 is the singular part of δX. The variances among m'_1 and m'_2 are written as

$$\langle m'_1 : m'_1\rangle = n^{-1}C_\lambda, \quad \langle m'_1 : m'_2\rangle = 0, \quad \langle m'_2 : m'_2\rangle = C', \tag{4.2.24}$$

where the C' is expressed in terms of the concentration susceptibility $(\partial X/\partial\Delta)_{pT}$ as

$$C' = C'_0 + (C'_0\gamma'_0)^2\langle|\psi|^2 : |\psi|^2\rangle = \frac{T}{n}\left[\left(\frac{\partial X}{\partial \Delta}\right)_{pT} - \frac{T}{C_\lambda}\left(\frac{\partial X}{\partial T}\right)^2_{\lambda p}\right]. \tag{4.2.25}$$

If the gravity effects are neglected, $(\partial X/\partial\Delta)_{pT}$ behaves logarithmically close to the λ line and as $(T_t - T)^{-1}$ on the coexistence curve near the tricritical point as derived in (3.2.24) [27].

4.2.5 Polymer solutions

We introduced the Flory–Huggins theory for polymer systems in Section 3.5, where the order parameter is the polymer volume fraction ϕ. Here we add the gradient free energy $\mathcal{H}_{\rm gra}$ using the random phase approximation [28], as summarized in Appendix 4B [29, 30]. For the Fourier components of ϕ with wave number q smaller than the inverse of the gyration radius $R_G \sim aN^{1/2}$, $\mathcal{H}_{\rm gra}$ is approximated as

$$\beta\mathcal{H}_{\rm gra} = \int d\boldsymbol{r}\frac{1}{36a\phi(1-\phi)}|\nabla\phi|^2, \tag{4.2.26}$$

where $a = v_0^{1/3}$ is the monomer size. In the reverse case $qR_G \gg 1$, however, the random phase approximation gives the structure factor $\langle|\phi_{\boldsymbol{q}}|^2\rangle = 12a\phi(1-\phi)/q^2$ [29]. This means that the factor $1/36$ in (4.2.26) should be replaced by $1/24$ at high q as

$$\beta\mathcal{H}_{\rm gra} = \int d\boldsymbol{r}\frac{1}{24a\phi(1-\phi)}|\nabla\phi|^2. \tag{4.2.27}$$

We will use (4.2.26) near the critical point and (4.2.27) in calculating the interface profile away from the critical point.

Let a polymer solution be near the critical point. From (3.5.9), (3.5.10), and (4.2.26) the coefficients in the Landau expansion are

$$r_0 = 2v_0^{-1}\left(\frac{1}{2} + N^{-1/2} - \chi\right), \quad u_0 = \frac{1}{3}v_0^{-1}N^{1/2}, \quad K = \frac{1}{18}a^{-1}N^{1/2}. \tag{4.2.28}$$

If the temperature dependence (3.5.7) is assumed, we have $a_0 = 2v_0 B/T_c$. Now we may calculate the Ginzburg number in (4.1.26) as

$$Gi \sim (N^{1/2} a_0 v_0)^{-1} \sim 1 - T_c/T_\vartheta \sim N^{-1/2}. \tag{4.2.29}$$

The asymptotic critical behavior should be observable when

$$|T/T_c - 1|/Gi \sim N^{1/2}|T/T_c - 1| \ll 1. \tag{4.2.30}$$

At the initial point of our theory, we set the upper cut-off wave number Λ equal to the inverse gyration radius $(aN^{1/2})^{-1}$; then, the initial coupling constant g in (4.1.22) is estimated as

$$g = (2^8 K_3/3)(aN^{1/2}\Lambda)^{-1} \sim 1, \tag{4.2.31}$$

indicating strong nonlinear coupling among the critical fluctuations. This means that there is no appreciable mean field critical behavior, in contrast to the polymer blend case, and simple scaling behavior is expected near the critical point. The correlation length ξ at the critical composition is scaled as

$$\xi = aN^{1/2} f_{\rm co}\big((T/T_c - 1)/Gi\big). \tag{4.2.32}$$

For $|x| \ll 1$, $f_{\rm co}(x) \sim |x|^{-\nu}$ with $\nu \cong 0.63$, so

$$\xi_{+0} \sim \xi_{-0} \sim aN^{(1-\nu)/2}. \tag{4.2.33}$$

Similarly, the volume fraction difference $\Delta\phi$ between the two coexisting phases is scaled as [31]–[34]

$$\Delta\phi = N^{-1/2} f_{\rm vo}\big((1 - T/T_c)/Gi\big), \tag{4.2.34}$$

where $f_{\rm vo}(x) \sim x^\beta$ with $\beta \cong 0.33$ for $x \ll 1$ and $f_{\rm vo}(x) \sim x^{-1}$ for $x \gg 1$. Thus, $\Delta\phi \sim N^{(\beta-1)/2}(1-T/T_c)^\beta$ near the critical point, while $\Delta\phi \sim 1-T/T_c$ is independent of N away from the critical point. Similarly, the osmotic modulus $K_{\rm os} = \phi(\partial\Pi/\partial\phi)_{\rm T}$ given in (3.5.21) behaves as the inverse susceptibility near the critical point and its asymptotic behavior is characterized by the critical exponent $\gamma \cong 1.24$ as

$$K_{\rm os} \sim T v_0^{-1} N^{(\gamma-3)/2} |T/T_c - 1|^\gamma. \tag{4.2.35}$$

4.2.6 Polymer blends

The Landau expansion holds under the condition (3.5.35) for polymer blends. From (3.5.32)–(3.5.34) we have

$$r_0 = 2v_0^{-1}(\chi_c - \chi), \quad u_0 = \frac{1}{3} v_0^{-1}(N_1 N_2)^{1/2}\chi_c^2, \quad K = 1/[18a\phi_c(1 - \phi_c)], \tag{4.2.36}$$

where ϕ_c and χ_c are given in (3.5.33). Setting $r_0 \cong a_0(T/T_c - 1)$ we obtain

$$a_0 Gi \sim (N_1^{1/2} + N_2^{1/2})^2/(N_1 N_2)^{3/2}. \tag{4.2.37}$$

$$(T/T_c - 1)/Gi \sim (N_1 N_2)^{1/2}(1 - \chi/\chi_c). \tag{4.2.38}$$

If N_1 and N_2 are both large, the asymptotic critical behavior is expected only very close to the critical point [20]. When $N_1 \geq N_2 \gg 1$, we obtain a well-defined mean field critical region given by

$$(N_1 N_2)^{-1/2}\chi_c \ll |\chi_c - \chi| \ll (N_1 N_2)^{-1/2}, \tag{4.2.39}$$

where $\chi_c \ll 1$, the lower bound arises from $T/T_c - 1 \gg Gi$, and the upper bound from (3.5.35). The correlation length at the critical composition behaves as

$$\begin{aligned} \xi/a &\sim (N_1 N_2)^{1/2}|(T/T_c - 1)/Gi|^{-1/2} \qquad (1 < |T/T_c - 1|/Gi < \chi_c^{-1}), \\ &\sim (N_1 N_2)^{1/2}|(T/T_c - 1)/Gi|^{-\nu} \qquad (|T/T_c - 1|/Gi < 1). \end{aligned} \tag{4.2.40}$$

Therefore, $N_1^{1/2} + N_2^{1/2} < \xi/a < (N_1 N_2)^{1/2}$ in the mean field critical region, while $\xi/a > (N_1 N_2)^{1/2}$ in the asymptotic critical region. These results will be used in Sections 4.4 and 9.6.

4.2.7 Gravity effects

In the presence of gravity we should include the potential energy,

$$\mathcal{H}_g = \int d\boldsymbol{r}\, \bar{g} z \delta\rho, \tag{4.2.41}$$

where $\bar{g}$ is the gravitational acceleration[4] and z is the vertical coordinate in the upward direction. From (2.2.7), (2.3.9) and (2.3.10), and (2.4.9), the mass density deviation $\delta\rho$ is expressed in terms of ψ, m, and q as

$$\begin{aligned} \delta\rho &= m_0(\alpha_1\psi + \beta_1 m) \qquad \text{(one-component fluids)} \\ &= \sum_{K=1,2} m_{0K}(\alpha_K\psi + \beta_K m + \gamma_K q) \qquad \text{(binary fluids)} \\ &= m_4(-\epsilon_{\rm in} m + q) \qquad (^4\text{He}), \end{aligned} \tag{4.2.42}$$

where m_0, m_{01}, and m_{02} are the particle masses in one- and two-component fluids and m_4 is the ^{4}He mass. The equilibrium distribution of the gross variables is

$$P_{\rm eq} \propto \exp[-\beta(\mathcal{H} + \mathcal{H}_g)]. \tag{4.2.43}$$

From (4.2.41) we notice that, on the one hand, the definitions of the temperature fluctuation $\delta\hat{T}$ in (4.2.3), (4.2.9), and (4.2.13) are unchanged even if $\mathcal{H}$ is replaced by the total hamiltonian $\mathcal{H} + \mathcal{H}_g$. On the other hand, we still define the pressure fluctuation $\delta\hat{p}$ in

[4] Here we use the notation $\bar{g}$ for the gravitational acceleration to avoid confusion with g in (4.1.22).

terms of the functional derivatives of $\mathcal{H}$ without $\mathcal{H}_g$ as in (4.2.4), (4.2.10), and (4.2.14). In equilibrium we thus obtain

$$\langle \delta \hat{T} \rangle = 0, \quad \langle \delta \hat{p} \rangle = -\rho_c \bar{g} z. \tag{4.2.44}$$

Here ρ_c should be interpreted as the λ point value ρ_λ for ^{4}He. Particular cases are as follows.

(i) For one-component fluids, we may set

$$m = C_0(\tau - \gamma_0 \psi^2 - m_0 \beta_1 \bar{g} z) \tag{4.2.45}$$

to obtain the GLW hamiltonian for ψ only in the form of (4.1.1). As a result, the parameters $r = r_0 - r_{0c} = a_0 \tau$ and h are replaced by

$$\begin{aligned} r(z) &= a_0(\tau - m_0 \beta_1 \bar{g} z), \\ h(z) &= h - m_0 \alpha_1 \bar{g} z, \end{aligned} \tag{4.2.46}$$

where β_1 is the mixing parameter discussed in Section 2.2 and is zero in the parametric model for a one-component fluid presented in (2.1.53)–(2.1.55).

(ii) For binary fluids, r and h in the GLW hamiltonian for ψ depend on z as

$$\begin{aligned} r(z) &= a_0[\tau - (m_{10}\beta_1 + m_{20}\beta_2)\bar{g} z], \\ h(z) &= h - (m_{10}\alpha_1 + m_{20}\alpha_2)\bar{g} z. \end{aligned} \tag{4.2.47}$$

It is interesting to consider a binary fluid mixture with $m_{10}\alpha_1 + m_{20}\alpha_2 \cong 0$, where the two phases after phase separation have the same mass density. Its critical behavior is influenced only through the z dependence of $r(z)$.

(iii) In ^{4}He near the superfluid transition, we have

$$m = C_0(T/T_{\lambda 0} - 1 - Gz - \gamma_0 |\psi|^2), \tag{4.2.48}$$

$$q = Q_0(\zeta - \beta m_4 \bar{g} z), \tag{4.2.49}$$

where G is defined by (2.4.32). Elimination of m and q yields the GLW hamiltonian for ψ, where the coefficient r depends on z as

$$r(z) = a_0 \tau(z) = a_0(T/T_{\lambda 0} - 1 - Gz), \tag{4.2.50}$$

in agreement with (2.4.31). The ordering field h remains zero.

Gravity-induced interface

Gravity can gives rise to coexistence of a normal fluid and a superfluid, as discussed in Section 2.4. In terms of $r(z)$ in (4.2.50) the mean field order parameter profile in ^{4}He is obtained from [35]

$$\frac{\delta}{\delta \psi^*} \beta \mathcal{H} = \left[r(z) + u_0 |\psi|^2 - \frac{d^2}{dz^2} \right] \psi = 0. \tag{4.2.51}$$

We can see that $\psi \cong 0$ in the region $r(z) > 0$ and $\psi \neq 0$ in the region $r(z) < 0$. There is no variation of the phase because no heat input is assumed. In the numerical result for $\rho_s(z) \propto |\psi(z)|^2$ in Fig. 2.17, the renormalization effect is taken into account by replacements, $r(z) \to r(z)|\xi(z)/\xi_{+0}|^{-1/2}$ and $u_0 \to u^*|\xi(z)/\xi_{+0}|^{-1}$, where u^* is a universal constant (to be discussed in Section 4.3) and $\xi(z)$ is the local correlation length. Because $\xi(z)$ should not exceed ℓ_g in (2.4.35), we have assumed the simple extrapolation form, $\xi(z) = \ell_g \tanh[\xi_{+0}|\tau(z)|^{-2/3}/\ell_g]$. With these changes the local equilibrium result (2.4.37) follows in the bulk superfluid region.

4.2.8 Electric field effects in non-ionic fluids

The electric field effects on the density fluctuations have also been discussed in the literature [36]–[53]. We apply an electric field to a non-ionic, non-polar fluid without free charges, where the static dielectric constant ε depends locally on the order parameter as

$$\varepsilon = \varepsilon_c + \varepsilon_1\psi + \frac{1}{2}\varepsilon_2\psi^2 + \cdots. \tag{4.2.52}$$

The fluid is in contact with conductors α $(= 1, 2, \ldots)$ which have surface charges Q_α and electric potentials Φ_α. The electric field is expressed as $\boldsymbol{E} = -\nabla\Phi$ in terms of the electric potential Φ, while the electric induction is given by $\boldsymbol{D} = \varepsilon\boldsymbol{E}$ with $\nabla\cdot\boldsymbol{D} = 0$. In this case we may fix either the charges Q_α or the potentials Φ_α [36]. Physically, these two boundary conditions should lead to essentially the same physical effects on the critical fluctuations [49]. Mathematically, the fixed potential condition is simpler than the fixed charge condition, so we will choose the former. That is, on the surface of the conductor α, Φ is fixed at Φ_α and the surface integral of $\boldsymbol{n}\cdot\boldsymbol{D}$ is equal to $-4\pi Q_\alpha$ and is a fluctuating quantity, where $\boldsymbol{n}$ is the normal unit vector pointed outward from the fluid region to the conductor α. The electrostatic free energy of the fluid is written as

$$\mathcal{H}_e = -\frac{1}{8\pi}\int d\boldsymbol{r}\boldsymbol{E}\cdot\boldsymbol{D} = -\frac{1}{2}\sum_\alpha Q_\alpha\Phi_\alpha, \tag{4.2.53}$$

where the space integral is within the fluid region. Then $\mathcal{H}_e$ is a functional of $\varepsilon(\boldsymbol{r})$ or $\psi(\boldsymbol{r})$ and its functional derivative is calculated as

$$\frac{\delta}{\delta\varepsilon}\mathcal{H}_e = -\frac{1}{8\pi}|\boldsymbol{E}|^2 \quad \text{or} \quad \frac{\delta}{\delta\psi}\mathcal{H}_e = -\frac{1}{8\pi}|\boldsymbol{E}|^2\left(\frac{\partial\varepsilon}{\partial\psi}\right)_T, \tag{4.2.54}$$

because $\int d\boldsymbol{r}\boldsymbol{D}\cdot\delta\boldsymbol{E} = -\int d\boldsymbol{r}\boldsymbol{D}\cdot\nabla\delta\Phi = 0$ in the fixed potential condition for small variations. In the fixed charge condition, the electrostatic energy of the fluid is the minus of (4.2.53), but its functional derivative with respect to ε is the same as (4.2.54) because $\int d\boldsymbol{r}\boldsymbol{E}\cdot\delta\boldsymbol{D} = 0$.

We assume that the average $\langle\boldsymbol{E}\rangle$ over the thermal fluctuations changes slowly in space and is nearly homogeneous on the scale of ξ. Then the electric field is written as $\boldsymbol{E} = \boldsymbol{E}_0 - \nabla\delta\Phi$, where $\boldsymbol{E}_0(\cong\langle\boldsymbol{E}\rangle)$ is the unperturbed electric field for the homogeneous dielectric constant $\bar{\varepsilon} = \langle\varepsilon\rangle$ and $\delta\Phi$ is the deviation of the electric potential induced by $\delta\varepsilon = \varepsilon - \langle\varepsilon\rangle$. From the charge-free condition $\nabla \cdot \boldsymbol{D} = 0$ inside the fluid, we have

$$\bar{\varepsilon}\nabla^2\delta\Phi = \boldsymbol{E}_0 \cdot \nabla\delta\varepsilon, \tag{4.2.55}$$

which is integrated as

$$\bar{\varepsilon}\delta\Phi(\boldsymbol{r}) = -\int d\boldsymbol{r}' G(\boldsymbol{r},\boldsymbol{r}')\boldsymbol{E}_0 \cdot \nabla'\delta\varepsilon(\boldsymbol{r}'). \tag{4.2.56}$$

The Green function $G(\boldsymbol{r},\boldsymbol{r}') = G(\boldsymbol{r}',\boldsymbol{r})$ satisfies $\nabla^2 G(\boldsymbol{r},\boldsymbol{r}') = -\delta(\boldsymbol{r}-\boldsymbol{r}')$ and vanishes as $\boldsymbol{r}$ approaches the surface of the conductors. Then $\mathcal{H}_\mathrm{e} = \mathcal{H}_\mathrm{e0} + \mathcal{H}_\mathrm{dip}$ is composed of two parts up to order $O(\delta\varepsilon^2)$ [49]. The first part is written as

$$\mathcal{H}_\mathrm{e0} = -\frac{1}{8\pi}\int d\boldsymbol{r} E_0^2\varepsilon = -\frac{1}{8\pi}\int d\boldsymbol{r} E_0^2\left(\varepsilon_\mathrm{c} + \varepsilon_1\psi + \frac{1}{2}\varepsilon_2\psi^2\right). \tag{4.2.57}$$

The second part is a long-range interaction of the form,

$$\begin{aligned}\mathcal{H}_\mathrm{dip} &= \frac{1}{8\pi}\int d\boldsymbol{r}\boldsymbol{E}_0\delta\varepsilon \cdot \nabla\delta\Phi \\ &= \frac{1}{8\pi\bar{\varepsilon}}\int d\boldsymbol{r}\int d\boldsymbol{r}'\big[\boldsymbol{E}_0 \cdot \nabla\delta\varepsilon(\boldsymbol{r})\big]G(\boldsymbol{r},\boldsymbol{r}')\big[\boldsymbol{E}_0 \cdot \nabla'\delta\varepsilon(\boldsymbol{r}')\big],\end{aligned} \tag{4.2.58}$$

which is positive-definite for the fluctuations varying along $\boldsymbol{E}_0$ and vanishes for those varying perpendicularly to $\boldsymbol{E}_0$.

Shift of the critical temperature

Let the capacitor consist of two parallel plates separated by L with the normal direction taken along the x axis. Then $\boldsymbol{E}_0$ is homogeneous between the plates. The surface charges of the lower and upper plates are Q and $-Q$, respectively, and we have $\mathcal{H}_e = -Q\Phi/2 = -QE_0L/2$, where Φ is the potential difference. First, from $\mathcal{H}_\mathrm{e0}$ in (4.2.57), we notice that there arises a small shift of the critical temperature of the form [49],

$$(\Delta\tau)_\mathrm{c} = (8\pi a_0)^{-1}\varepsilon_2 E_0^2. \tag{4.2.59}$$

Here we consider the fluctuations varying perpendicularly to $\boldsymbol{E}_0$ because $\mathcal{H}_\mathrm{dip} = 0$ for them. This result is consistent with Landau and Lifshitz's mean field calculation of the shift $(\Delta T)_\mathrm{c}$ in one-component systems [36]. However, Debye and Kleboth obtained a shift of the critical temperature in the reverse direction (= the minus of (4.2.59)) for binary fluid mixtures [37].[5] We should note that it is very difficult to detect the shift unambiguously, because a very high field is needed in one-component fluids and ohmic heating due to an electric current is usually inevitable in binary fluid mixtures [52].

[5] A recent experiment [46] on near-critical polymer solutions detected large electric field effects, apparently supporting Debye and Kleboth's prediction. An explanation of their finding is needed.

Effects on the critical fluctuations

When $\boldsymbol{E}_0$ is homogeneous, the interaction $\mathcal{H}_{\rm dip}$ in (4.2.58) becomes simple in the Fourier space as [48],

$$\mathcal{H}_{\rm dip} = \frac{1}{2} T_{\rm c} g_{\rm e} \int_{\boldsymbol{q}} \frac{1}{q^2} q_x^2 |\psi_{\boldsymbol{q}}|^2, \tag{4.2.60}$$

where q is much larger than the inverse cell width and

$$g_{\rm e} = (4\pi T_{\rm c}\varepsilon_{\rm c})^{-1} \varepsilon_1^2 E_0{}^2. \tag{4.2.61}$$

A dipolar interaction with the same form is well known in uniaxial ferromagnets [50] and ferroelectrics [51]. It can nonlinearly influence the critical fluctuations in the case $\chi^{-1} < g_{\rm e}$ or $|T/T_{\rm c} - 1| < \tau_{\rm e}$, where $\tau_{\rm e}$ is the crossover reduced temperature. Thus $\tau_{\rm e} \propto g_{\rm e}^{1/\gamma}$. If $K = \xi^2/\chi$ is regarded as a constant (or the exponent η is neglected), $\tau_{\rm e}$ is expressed as

$$\tau_{\rm e} = (\xi_{+0}^2 g_{\rm e}/K)^{1/2\nu}. \tag{4.2.62}$$

The wave number characterizing the anisotropy is thus $(g_{\rm e}/K)^{1/2} = \xi_{+0}^{-1}\tau_{\rm e}^{\nu}$. For example, $\tau_{\rm e} \sim 10^{-8}$ and $(g_{\rm e}/K)^{1/2} \sim 10^3$ cm^{-1} for typical binary fluid mixtures like aniline + cyclohexane at $E_0 = 1$ kV/cm.[6] The structure factor in the presence of $\mathcal{H}_{\rm dip}$ becomes uniaxial as

$$I(\boldsymbol{q}) = (\chi^{-1} + g_{\rm e} q_x^2/q^2 + K q^2)^{-1}. \tag{4.2.63}$$

This anisotropic form has not yet been measured by light scattering, but it can be detected, even for $\tau_{\rm e} \ll \tau$, using high-sensitivity optical techniques in electric birefringence (the Kerr effect) [43]–[45] and dichroism [47]. That is, if the local dielectric constant $\varepsilon_{\rm op}$ at optical frequencies weakly depends on ψ, we may calculate the electric field within the fluid and the fluctuation contribution to the macroscopic dielectric tensor at optical frequencies. It has slightly different xx and yy components [48]. In particular, when $k\xi \ll 1$ and $\xi(g_{\rm e}/K)^{1/2} \ll 1$, it has been predicted that

$$\Delta\varepsilon_{\rm op} \propto g_{\rm e} E_0^2 \propto E_0^2 \xi^{1-2\eta}, \tag{4.2.64}$$

to linear order in $g_{\rm e}$. Experimentally, however, $\Delta\varepsilon_{\rm op} \propto (T/T_{\rm c} - 1)^{-\Psi}$ with $\Psi \cong 0.84$ was obtained [45]. Note that $\Delta\varepsilon_{\rm op}$ generally takes a complex value, and its imaginary part is detectable in the effect of form dichroism [48]. Further discussions on this topic will be given in Section 6.1.

The macroscopic dielectric constant $\varepsilon_{\rm eff}$ is determined by $\varepsilon_{\rm eff} E_0 \Phi/L = 4\pi\langle Q\rangle/S$, where S is the surface area of the parallel plates. As $E_0 \to 0$, we have [40, 48],

$$\varepsilon_{\rm eff} = \varepsilon_{\rm c} + C_1(T/T_{\rm c} - 1)^{1-\alpha} + C_2(T/T_{\rm c} - 1) + \cdots, \tag{4.2.65}$$

where C_1 and C_2 are constants. The non-linear part $\varepsilon_{\rm eff} E_0 - \varepsilon_{\rm eff}(0)$ should behave as (4.2.64) theoretically [39, 42, 48] and [44].

[6] At $E = 1$ kV/cm we have $E^2/4\pi = 0.9$ erg/cm^3.

Critical electrostriction

Finally, we consider equilibrium in which $\langle \boldsymbol{E} \rangle \cong \boldsymbol{E}_0$ varies slowly in space. From (4.2.54) the functional derivative of the total GLW hamiltonian $\mathcal{H} + \mathcal{H}_e$ with respect to ψ should be homogeneous on average. Thus,

$$\left\langle \frac{\delta}{\delta\psi}\mathcal{H} \right\rangle - \frac{1}{8\pi}\left\langle |\boldsymbol{E}|^2 \left(\frac{\partial \varepsilon}{\partial \psi}\right)_T \right\rangle = \text{const.}, \tag{4.2.66}$$

where the first term is nearly equal to the thermodynamic chemical potential (difference) without an electric field, $\mu(T, M)$, in one-component fluids (binary mixtures) for $g_e \ll \chi^{-1}$. We are led to the well-known relation for the chemical potential in an electric field [36],

$$\mu(T, M, E_0) \cong \mu(T, M) - \frac{1}{8\pi} E_0^2 \varepsilon_1, \tag{4.2.67}$$

which is constant in space. For one-component fluids in equilibrium, an inhomogeneous average mass-density variation is induced as

$$\langle \delta\rho \rangle = K_T \frac{\langle \rho \rangle}{8\pi} E_0^2 \varepsilon_1 + \text{const.}, \tag{4.2.68}$$

where K_T is the isothermal compressibility. In binary fluid mixtures we also expect a similar relation for an inhomogeneous average composition variation, but the equilibration process is diffusive and slow. Experimentally, (4.2.68) was confirmed optically for SF_6 around a wire conductor [52], and (4.2.67) was used to determine $\mu(T, M)$ for ^{3}He in a cell within which a parallel-plate capacitor was immersed [53].

4.3 Static renormalization group theory

We have seen that the fluctuation contributions in the normal perturbation theory increase, leading to its breakdown at long wavelengths for $d < 4$, as the critical point is approached. However, a more elaborate perturbation scheme can be devised, in which the fluctuation effects are taken into account in a *step-wise manner*. That is, in the equilibrium distribution $\propto \exp(-\beta\mathcal{H})$, we take the thermal average of the fluctuations in a thin shell region in the wave-vector space,

$$\Lambda - \delta\Lambda < k < \Lambda, \tag{4.3.1}$$

with those in the long-wavelength region $k < \Lambda - \delta\Lambda$ held fixed. Let $\mathcal{H}^>$ be the part of $\mathcal{H}$ involving the fluctuations in the shell region and $\mathcal{H}^<$ be that containing only the long-wavelength fluctuations with $k < \Lambda - \delta\Lambda$. The functional integration of ψ_k in the shell region is expressed as

$$\begin{aligned} \exp(-\beta\mathcal{H}' - \beta\delta F) &= \int^> [d\psi] \exp(-\beta\mathcal{H}) \\ &= \exp(-\beta\mathcal{H}^<) \int^> [d\psi] \exp(-\beta\mathcal{H}^>). \end{aligned} \tag{4.3.2}$$

Here $\mathcal{H}' = \mathcal{H}^< + \delta\mathcal{H}$ is a new coarse-grained hamiltonian, where $\delta\mathcal{H}$ is the fluctuation contribution. The increment δF is independent of ψ, so it is the fluctuation contribution to the thermodynamic free energy and satisfies

$$\delta F = \left[\frac{\partial}{\partial \Lambda} F(\Lambda)\right]\delta\Lambda, \tag{4.3.3}$$

where $F(\Lambda) = -\int_\Lambda^\infty d\Lambda'[\partial F(\Lambda')/\partial\Lambda']$ is the contribution to the free energy from the fluctuations with wave numbers larger than Λ. Thus,

$$F = \lim_{\Lambda\to 0} F(\Lambda) = -\int_0^\infty d\Lambda\left[\frac{\partial}{\partial\Lambda}F(\Lambda)\right] \tag{4.3.4}$$

is the thermodynamic free energy. The coefficients in $\mathcal{H} = \mathcal{H}(\Lambda)$ depend on Λ and obey differential equations, called renormalization group (RG) equations [1]–[12]. It is crucial that, if $\epsilon = 4 - d$ is regarded as a small expansion parameter, the RG equations can be constructed analytically in perturbation series with respect to the coupling constants g in (4.1.22) and v in (4.1.55) which can be regarded to be of order ϵ. This ϵ expansion is easily handled, at least to first order in ϵ, and is unambiguously performed even at higher orders in ϵ in statics.[7]

We shall see that the coupling constants g and v tend to universal numbers of order ϵ as Λ is decreased sufficiently close to the critical point. The effect of the coarse-graining is then to give rise to multiplicative factors Λ^w of the coefficients in the GLW free energy. The exponent w is expanded as $w = w_1\epsilon + w_2\epsilon^2 + \cdots$ in powers of ϵ. This multiplicative effect stops when Λ is decreased down to the order of the inverse correlation length κ, giving rise to the hydrodynamic hamiltonian $\mathcal{H}_{\rm hyd}$ in (4.1.56). The coefficients thus obtained are renormalized and can be related to experimentally observable quantities. The renormalized coefficients, denoted with the subscript R, behave as

$$\begin{aligned} r_{\rm R} &\cong 2\gamma_{\rm R} C_{\rm R}\tau \sim \kappa^{2-\eta}, \\ u_{\rm R} &\sim \kappa^{\epsilon-2\eta}, \quad K_{\rm R} \sim \kappa^{-\eta}, \\ \gamma_{\rm R} &\sim \kappa^{(\epsilon+\alpha/\nu)/2-\eta}, \quad C_{\rm R} = C_H \sim \kappa^{-\alpha/\nu}. \end{aligned} \tag{4.3.5}$$

The first line holds in disordered states at $h = 0$, where $r_{\rm R}$ is equal to the inverse of the susceptibility χ. Note that the first relation can be rewritten as $\gamma_{\rm R} C_{\rm R} \sim \kappa^{(\gamma-1)/\nu}$, leading to the behavior of $\gamma_{\rm R}$ in the third line with the aid of (2.1.11) and (2.1.12). The nonlinear coupling constants u_0 and γ_0 decrease with the coarse-graining and saturate to the values given above.[8]

4.3.1 Renormalization group equations for r and g ($n = 1$)

We will set up the RG equations valid to leading order in ϵ at $h = 0$ and $r \geq 0$ in Ising-like systems. Generalization to n-component systems will be presented in Appendix 4C. To

[7] Abe developed another approach in which the inverse of the spin component number is treated as an expansion parameter (the $1/n$ expansion) [54].

[8] The strong cut-off dependence of γ_0 will turn out to be crucial in the calculation of the bulk viscosity in Section 6.2.

Fig. 4.3. The contributions to the four-body coupling u_0.

examine the behavior r and g, we may start with the GLW hamiltonian in (4.1.1) without the energy variable. We may treat the coefficient K as a constant to first order in ϵ in disordered states with $\langle\psi\rangle = 0$.[9] We thus set

$$K = 1. \tag{4.3.6}$$

Using the integration in (4.1.19), we pick up the contribution from the shell region (4.3.1) to obtain

$$\delta r = -3rg\Lambda\delta\Lambda/(r + \Lambda^2) + \cdots, \tag{4.3.7}$$

where g is defined by (4.1.22). It is convenient to introduce ℓ by

$$\Lambda = \Lambda_0 e^{-\ell}. \tag{4.3.8}$$

The parameter ℓ increases from 0 (at the starting point of the RG procedure) to ∞ (in the hydrodynamic limit). The initial wave number Λ_0 should be considerably smaller than the inverse lattice constant to assure the coarse-grained hamiltonian (4.1.1). Because $\delta\ell = \delta\Lambda/\Lambda$, the differential equation for r becomes

$$\frac{\partial}{\partial\ell}r = -3gr/(X + 1), \tag{4.3.9}$$

where

$$X = r/\Lambda^2 = re^{2\ell}/\Lambda_0^2 \tag{4.3.10}$$

is small initially ($\ell = 0$) but grows eventually ($\ell \to \infty$).

It is crucial that the coupling constant u_0 changes as Λ is decreased. Taking only the leading order correction ($\propto g^2$) in Fig. 4.3, we obtain

$$\delta u_0 = -9u_0 g\Lambda^3\delta\Lambda/(r + \Lambda^2)^2 + \cdots. \tag{4.3.11}$$

The incremental change of $g \propto u_0/\Lambda^\epsilon$ is written as

$$\delta g = \left[\epsilon g - 9g^2\Lambda^4/(r + \Lambda^2)^2 + \cdots\right]\delta\Lambda/\Lambda. \tag{4.3.12}$$

The differential renormalization group equation for g up to order $O(g^2)$ becomes

$$\frac{\partial}{\partial\ell}g = \epsilon g - 9g^2/(X + 1)^2. \tag{4.3.13}$$

[9] For $\langle\psi\rangle \neq 0$ this is not the case, as can be seen in (4.3.97) below.

Solution for $X(\ell) \ll 1$

For $r \ll \Lambda_0^2$ (or very close to the critical point) the region $0 < \ell < \ln(\Lambda_0/\sqrt{r})$ has a sizable width, in which we may set $X \ll 1$ to obtain

$$\frac{\partial}{\partial \ell} r = -3gr, \qquad \frac{\partial}{\partial \ell} g = \epsilon g - 9g^2. \tag{4.3.14}$$

To solve this equation we define

$$Q(\ell) = e^{\epsilon \ell} + g^*/g_0 - 1, \tag{4.3.15}$$

where $g_0 = g(0)$ is the initial value and

$$g^* = \frac{1}{9}\epsilon + \cdots. \tag{4.3.16}$$

is the fixed-point value. Then,

$$g(\ell) = g^* \exp(\epsilon \ell)/Q(\ell), \tag{4.3.17}$$

$$\begin{aligned} r(\ell) &= a_{00}\tau \exp\left[-3\int_0^\ell d\ell' g(\ell')\right] \\ &= a_{00}\tau\left[1 + (e^{\epsilon\ell} - 1)g_0/g^*\right]^{-1/3} \end{aligned} \tag{4.3.18}$$

where $a_{00} = a_0(0) = r(0)/\tau$ is the initial coefficient.

Mean field critical behavior

In the weak coupling case $g_0 \ll g^*$ the mean field critical behavior can be realized. This occurs if $u_0(\ell) \cong u_0(0)$ and $r(\ell) \cong r(0) = a_{00}\tau$ even at $X(\ell) = 1$ or $e^{-\ell} = r^{1/2}/\Lambda_0$. In terms of the quantities at $\ell = 0$ this condition becomes

$$e^{-\epsilon\ell} = (a_{00}\tau/\Lambda_0^2)^{\epsilon/2} \gg g_0/g^* \quad \text{or} \quad g^*(a_{00}\tau)^{\epsilon/2} \gg K_d u_0, \tag{4.3.19}$$

which is equivalent to the Ginzburg criterion (4.1.24) and is rewritten as $\tau \gg Gi$. The coefficient a_0 does not change from a_{00}.

Asymptotic critical behavior

For $\tau \ll Gi$ there appears a sizable region of ℓ in which $X(\ell) < 1$ and

$$g(\ell) \cong g^*, \quad r(\ell) = a_0(\ell)\tau \cong a_{00}(g^*/g_0)^{1/3}\tau e^{-\epsilon\ell/3}. \tag{4.3.20}$$

The fluctuations in this wave number region give rise to the asymptotic critical behavior, as given by (4.3.5). However, as ℓ is increased such that $X(\ell) > 1$, the remaining fluctuations are weakened and may be treated with a normal perturbation scheme, giving rise to corrections to the critical amplitudes and not to the exponents. We thus encounter a crossover at $\ell = \ell^*$, where $X(\ell^*) = 1$ and the lower cut-off is the inverse correlation length,

$$\kappa = \xi_{+0}^{-1}\tau^\nu = \Lambda_0 \exp(-\ell^*) \quad \text{or} \quad \exp(\ell^*) = \Lambda_0\xi_{+0}\tau^{-\nu}. \tag{4.3.21}$$

We then consider the post-crossover behavior realized in the region $\ell > \ell^*$. The dimensionless coupling parameter g starts to grow as

$$g(\ell) \cong g^* \exp[\epsilon(\ell - \ell^*)], \tag{4.3.22}$$

which means that the coefficient u_0 saturates to the renormalized value determined by

$$u_0(\ell) \to u_R \cong K_d^{-1} g^* \kappa^\epsilon. \tag{4.3.23}$$

As a result, $r(\ell)$ saturates at the inverse susceptibility as

$$r(\ell) \to r_R = a_{00}(g^*/g_0)^{1/3} \tau \exp(-\epsilon\ell^*/3). \tag{4.3.24}$$

Here we are in the hydrodynamic regime, where κ is determined by $r_R = \kappa^2$. The critical exponents ν and γ are expanded as

$$\begin{aligned} \nu &= \left(2 - \frac{1}{3}\epsilon + \cdots\right)^{-1} = \frac{1}{2} + \frac{1}{12}\epsilon + \cdots, \\ \gamma &= (2-\eta)\nu = 1 + \frac{1}{6}\epsilon + \cdots, \end{aligned} \tag{4.3.25}$$

to first order in ϵ. The microscopic length ξ_{+0} in the relation $\kappa = \xi_{+0}^{-1}\tau^\nu$ is also changed by the fluctuation effect as

$$\xi_{+0}^{-1} = \Lambda_0^{1-2\nu}[a_{00}(g^*/g_0)^{1/3}]^\nu = a_{00}^\nu (g^*/K_d u_0)^{\nu/3}, \tag{4.3.26}$$

which is different from the mean field expression $\xi_{+0} = a_0^{-1/2}$ (valid in the region $Gi < \tau < 1$ if $Gi \ll 1$). Eliminating a_{00} from r_R in favor of ξ_{+0} we have

$$a_R = r_R/\tau = \xi_{+0}^{-2}\tau^{\gamma-1}. \tag{4.3.27}$$

4.3.2 Renormalization group equation for v ($n = 1$)

We next start with the hamiltonian (4.1.45) which includes the energy density m in Ising-like systems. See Appendix 4C for the RG results in n-component systems. The two coefficients γ_0 and C_0 may be expressed in terms of v in (4.1.55) and $a_0 = 2\gamma_0 C_0$ as

$$\gamma_0 = 2K_d^{-1}\Lambda^\epsilon K^2 v/a_0, \tag{4.3.28}$$

$$C_0 = \frac{1}{4}K_d \Lambda^{-\epsilon} a_0^2/K^2 v, \tag{4.3.29}$$

where we may set $K = 1$ to first order in ϵ. We will set up the RG equation for $C_0(\ell)$, $\gamma_0(\ell)$, and $v(\ell)$ for $X(\ell) \ll 1$ and examine the asymptotic critical behavior.

From (4.1.53) or from the contribution represented by the diagrams in Fig. 4.4, we obtain the incremental increase of C_0 as

$$\delta C_0 = 2C_0 v \delta\Lambda/\Lambda + \cdots \tag{4.3.30}$$

Fig. 4.4. The contributions to the correlation function of the energy variable m or to the specific heat (4.1.51). The wavy lines on the right-hand side represent the bare two-body correlation function of m or the first term on the right-hand side of (4.1.51).

Fig. 4.5. The contributions to the three-body coupling γ_0.

leading to

$$\frac{\partial}{\partial \ell} C_0 = 2v C_0. \tag{4.3.31}$$

The diagrams in Fig. 4.5 give rise to two contributions to γ_0 in the form,

$$\delta\gamma_0 = -(3g + 2v)\gamma_0 \delta\Lambda/\Lambda + \cdots. \tag{4.3.32}$$

Its differential form is

$$\frac{\partial}{\partial \ell}\gamma_0 = -(3g + 2v)\gamma_0. \tag{4.3.33}$$

Up to second order in g and v, the RG equation for v becomes

$$\frac{\partial}{\partial \ell} v = \epsilon v - 2(3g + v)v. \tag{4.3.34}$$

Using (4.3.9), (4.3.31), and (4.3.33) we may derive the relation,

$$\frac{\partial}{\partial \ell}(2\gamma_0 C_0) = \frac{1}{\tau}\frac{\partial}{\partial \ell} r, \tag{4.3.35}$$

for $X \ll 1$. The relation $2\gamma_0 C_0 = r/\tau$ holds for $\Lambda \gg \kappa$ to all orders in ϵ in the asymptotic limit. In the limit $\Lambda \to 0$ we have $2\gamma C_R = \gamma r_R/\tau$ in (4.1.58) including the effect of the fluctuations with wave numbers on the order of κ.

Solution at criticality

From (4.3.34) v tends to a fixed-point value,

$$v^* = \frac{1}{6}\epsilon + \cdots. \tag{4.3.36}$$

However, the approach of v to v^* is much slower than that of g to g^*. In fact, if (4.3.17) is used, (4.3.34) is exactly solved to give [55][10]

$$v(\ell) = v^* \exp(\epsilon\ell)/\left[Q(\ell) + w_0 Q(\ell)^{2/3}\right], \tag{4.3.37}$$

[10] From (4.3.15) we have $Q(\ell) > g^*/g_0$, from which the denominator of (4.3.37) is positive-definite even for $w_0 < 0$.

where $Q(\ell)$ is defined by (4.3.15), $v_0 = v(0)$ is the initial value, and

$$w_0 = (g_0 v^*/v_0 g^* - 1)(g^*/g_0)^{1/3}. \tag{4.3.38}$$

Then (4.3.31) and (4.3.33) yield

$$C_0(\ell) = C_{00}(v_0/v^*)(g^*/g_0)^{2/3}\big[Q(\ell)^{1/3} + w_0\big], \tag{4.3.39}$$

$$\gamma_0(\ell) = \gamma_{00}(v^*/v_0)(g_0/g^*)^{1/3}/\big[Q(\ell)^{2/3} + w_0 Q(\ell)^{1/3}\big], \tag{4.3.40}$$

where $C_{00} = C_0(0)$ and $\gamma_{00} = \gamma_0(0)$ are the initial values. The coefficient in $C_0(\ell)$ may be rewritten in terms of ξ_{+0} as

$$C_{00}(v_0/v^*)(g^*/g_0)^{2/3} = K_d \xi_{+0}^{-d}(\xi_{+0}\Lambda_0)^{-\epsilon/3}/4v^*, \tag{4.3.41}$$

using the initial relation $C_{00}v_0 = K_d a_{00}^2/4\Lambda_0^\epsilon$ arising from (4.3.29) at $\ell = 0$ and the expression (4.3.26) for ξ_{+0}.

Furthermore, in the ℓ region where $Q(\ell) \cong \exp(\epsilon\ell)$, $v(\ell)$ and $C_0(\ell)$ behave as

$$v(\ell) = v^*/(1 + w_0 e^{-\epsilon\ell/3}), \tag{4.3.42}$$

$$C_0(\ell) = \frac{1}{4v^*} K_d \xi_{+0}^{-d}(\xi_{+0}\Lambda_0)^{-\epsilon/3}(e^{\epsilon\ell/3} + w_0). \tag{4.3.43}$$

If w_0 is not small, these quantities exhibit slow transient behavior even in the region where $g \cong g^*$.

Crossover at small, positive τ

For $\ell > \ell^*$, $C_0(\ell)$ saturates into the specific heat $C_H(= C_M)$ at zero magnetic field, which is obtained if we set $\ell = \ell^*$ in (4.3.43). We thus find the critical behavior,

$$C_H = A_0 \tau^{-\alpha} + C_{\rm B}, \tag{4.3.44}$$

with

$$\alpha = \frac{1}{6}\epsilon + \cdots. \tag{4.3.45}$$

If use is made of (4.3.21), the critical amplitude A_0 becomes

$$A_0 = \frac{1}{4v^*} K_d \xi_{+0}^{-d}. \tag{4.3.46}$$

Now the relation (2.1.45) of the two-scale-factor universality is derived in the form [56],

$$R_\xi = \lim_{\tau\to+0} \xi(\alpha\tau^2 C_H)^{1/d} = \left(\frac{1}{4}K_d\right)^{1/d}. \tag{4.3.47}$$

This result is valid only to leading order in ϵ. Nevertheless, if we set $d = 3$, we have $R_\xi = (8\pi^2)^{-1/3} \cong 0.23$ close to the reliably estimated value 0.25 (see the discussions below (2.1.45)). The background specific heat $C_{\rm B}$ is expressed as

$$C_{\rm B} = C_{00}\big(1 - g^* v_0/v^* g_0\big). \tag{4.3.48}$$

In terms of R_ξ and the Ginzburg number Gi defined by (4.1.25), C_H can also be expressed as

$$C_H = R_\xi^d \xi_{+0}^{-d} \frac{1}{\alpha}\left(\tau^{-\alpha} - Gi^{-\alpha}\right) + C_{00}. \tag{4.3.49}$$

The above form holds for $\tau \ll Gi$ in general n-component systems, as will be evident from results in Appendix 4C.

4.3.3 Perturbation theory at $g = g^*$ and $v = v^*$

Exponentiation of logarithmic terms ($n = 1$)

If $g = g^*$ from the starting point ($\ell = 0$), it does not change with the coarse-graining until the crossover at $\ell = \ell^*$ is reached. For this special choice, logarithmic terms in the usual perturbation series near four dimensions may be *exactly* exponentiated to give the correct critical behavior [5]. With this in mind, we can derive the correct asymptotic results using naive perturbation expansions. Here it is important that the upper cut-off Λ is fixed (so it may be set equal to 1 in the actual calculations). For example, (4.1.23) gives

$$\begin{aligned} \chi^{-1} &= r\left[1 - 3g\ln(\Lambda/\kappa) + O(\epsilon^2)\right] \\ &\cong r(\kappa/\Lambda)^{3g}. \end{aligned} \tag{4.3.50}$$

The exponentiation in the second line is clearly correct at $g = g^*$. Also we notice that (4.1.40) should hold just at $g = g^*$, resulting in the ϵ expansion of the critical exponent η for $n = 1$,

$$\eta \cong \frac{3}{2}(g^*)^2 = \frac{1}{54}\epsilon^2. \tag{4.3.51}$$

In the same manner, if we set $v = v^*$ from the starting point and exponentiate logarithmic terms in the simple perturbation expansions, we can obtain the correct critical behavior. As such an application, let us consider the thermodynamic free energy F for $h = 0$ and $\tau > 0$ using this strategy. Note that the gaussian integration of $\exp(-\beta\mathcal{H}_0)$ with respect the Fourier component $\psi_{\boldsymbol{k}}$ gives rise to the factor $(2\pi)^{1/2}/(r+k^2)^{1/2}$. From (4.3.2)–(4.3.4) this gives rise to

$$F/T_c V = -\frac{1}{2}C_0\tau^2 + \frac{1}{2}\int_{\boldsymbol{k}} \ln(r + k^2) + \cdots. \tag{4.3.52}$$

The first term is the mean field contribution arising from the last term in the second line of (4.1.45). The second term is the fluctuation contribution in the leading order. If use is made of (4.3.29), the singular free-energy density is of the form,[11]

$$\begin{aligned}\frac{1}{T_c} f_{\text{sing}} &\cong -\frac{1}{2} C_0 \tau^2 + \frac{1}{8} K_4 r^2 [\ln(r/\Lambda^2) + \text{const.}] \\ &\cong -\frac{1}{2} C_0 \tau^2 [1 - v^* \ln(r/\Lambda^2)]. \end{aligned} \tag{4.3.53}$$

We recognize that the fluctuation contribution is higher by ϵ than the mean field contribution and that the exponentiation at $v = v^*$ yields

$$\begin{aligned}\frac{1}{T_c} f_{\text{sing}} &\cong -\frac{1}{2} C_0 \tau^2 (\kappa/\Lambda)^{-\epsilon/3} \\ &\cong -\frac{1}{8v^*} K_d \kappa^d, \end{aligned} \tag{4.3.54}$$

where the second line follows from (4.3.29) at $g_0 = g^*$ and $v_0 = v^*$. This result is consistent with the singular specific-heat behavior in (4.3.44) and (4.3.46). The constant specific-heat contribution C_B in (4.3.48), which gives rise to the regular term $-T_c C_B \tau^2/2$ in the free-energy density, vanishes for $g/v = g^*/v^*$ and is nonexistent in the above calculation.

Higher-order perturbation calculations in n-component systems

The above exponentiation procedures can be extended to higher orders in ϵ for special values of $g = g_1\epsilon + g_2\epsilon^2 + \cdots$ [57] and $v = v_1\epsilon + v_2\epsilon^2 + \cdots$. Technically, efficient ϵ expansion calculations to higher orders can be performed without imposing a sharp cut-off in the wave number integrations but by adding the following higher-order gradient term (smooth cut-off) to the hamiltonian (4.1.1) [5, 57, 58]

$$\beta \mathcal{H}_{\text{cut-off}} = \int d\boldsymbol{r} \frac{1}{2} \Lambda^{-2} |\nabla^2 \psi|^2 = \int_{\boldsymbol{k}} \frac{1}{2} \Lambda^{-2} k^4 |\psi_{\boldsymbol{k}}|^2. \tag{4.3.55}$$

With this term the zeroth-order two-body correlation becomes $1/(r_0 + k^2 + k^4/\Lambda^2) \cong 1/(r_0 + k^2) - 1/(\Lambda^2 + k^2)$, which decays to zero rapidly for $k > \Lambda$. With this method, g^* is calculated as

$$g^* = \frac{\epsilon}{n+8} \left\{ 1 + \left[\frac{9n+42}{(n+8)^2} - \frac{1}{2} \right] \epsilon \right\} + O(\epsilon^3), \tag{4.3.56}$$

up to order ϵ^2 for n-component systems. If the sharp cut-off method is used, g^* is not given by (4.3.56), though the leading term of order ϵ is unchanged [59]. However, the observable quantities such as the critical exponents and amplitude ratios should be independent of the method used to introduce the upper cut-off. In the same manner, v^* was calculated (in the context of critical dynamics) as [23]

$$v^* = \frac{\alpha}{2n\nu} \left(1 - \frac{1}{2}\epsilon \right) + O(\epsilon^3) = g^* \left(\frac{4-n}{2n} - \frac{2}{n}\epsilon \right) + O(\epsilon^3). \tag{4.3.57}$$

[11] The second derivative of the second term of (4.3.52) with respect to r is equal to $4^{-1} K_4 \ln(\Lambda^2/r)$ at $d = 4$. Its integration gives (4.3.53). Here there arises a term linear in $r \propto \tau$, but it is incorporated into the regular part of the free energy. In fact, it gives only a constant contribution to the average energy density and no contribution to the specific heat.

We will use the expansion of v^*/g^* up to order ϵ in (4.3.77) below. The second-order corrections of the critical exponents can be known from

$$\eta = \frac{n+2}{2(n+8)^2}\epsilon^2 + O(\epsilon^3), \tag{4.3.58}$$

$$\begin{aligned}\frac{\alpha}{2\nu} &= \frac{4-n}{2(n+8)}\epsilon - \frac{(n+2)(13n+44)}{2(n+8)^3}\epsilon^2 + O(\epsilon^3) \\ &= g^*\left(\frac{4-n}{2} - \frac{n+4}{4}\epsilon\right) + O(\epsilon^3),\end{aligned} \tag{4.3.59}$$

if use is made of the exponent relations in Section 2.1.

4.3.4 Singular free energy for general h and τ in n-component systems

So far we have assumed $h = 0$ and $\tau > 0$ in Ising systems. Here we generalize our arguments for general h and τ in n-component spin systems. We perform simple perturbation calculations by setting $g = g^*$ and $v = v^*$. The upper cut-off Λ is fixed here, so it will be set equal to 1. In isotropic n-component systems the average order parameter $M_i = \langle\psi_i\rangle$ and the magnetic field h_i are vectors related by

$$M_i = \hat{h}_i M(h), \tag{4.3.60}$$

where $\hat{h}_i = h_i/h$ and $h = |\boldsymbol{h}|$. Differentiation with respect to h_j gives the spin correlation functions,

$$\langle\psi_i : \psi_j\rangle = \frac{\partial}{\partial h_j}M_i = (\delta_{ij} - \hat{h}_i\hat{h}_j)\frac{1}{h}M + \hat{h}_i\hat{h}_j\frac{\partial}{\partial h}M \tag{4.3.61}$$

where $\langle\ :\ \rangle$ is the variance defined by (1.1.35). Thus, by setting $\langle\psi_i\rangle = M\delta_{i1}$, we find general expressions for the longitudinal and transverse susceptibilities,

$$\begin{aligned}\chi_L &= \langle\psi_1 : \psi_1\rangle = \frac{\partial}{\partial h}M, \\ \chi_T &= \langle\psi_2 : \psi_2\rangle = \frac{1}{h}M.\end{aligned} \tag{4.3.62}$$

We notice that χ_T should tend to ∞ as $h \to 0$ with $\tau < 0$. In fact, we shall see that the structure factor $I_T(k)$ of ψ_2 behaves as k^{-2} at small wave number k.

It is almost trivial to set up the singular free energy including the leading order corrections in ϵ. In the mean field approximation we have the relations, $I_L(k) = \langle|\psi_1(\boldsymbol{k})|^2\rangle \cong 1/(r_L + k^2)$ and $I_T(k) = \langle|\psi_2(\boldsymbol{k})|^2\rangle \cong 1/(r_T + k^2)$, with

$$r_L = r + 3u_0M^2, \qquad r_T = r + u_0M^2. \tag{4.3.63}$$

The integrations of the Fourier components with wave vector $\boldsymbol{k}$ give the multiplicative factor $[I_L(k)I_T(k)^{(n-1)}]^{-1/2}$ to the partition function. As will be shown in Appendix 4D, we may then derive the singular free-energy density,

$$\begin{aligned}\frac{1}{T_c} f_{\text{sing}} &= \frac{1}{2} r M^2 + \frac{1}{4} u_0 M^4 - hM - \frac{1}{2} C_0 \tau^2 \\ &\quad + \frac{1}{8} K_4 \left[r_L^2 \left(\ln r_L + \frac{1}{2} \right) + (n-1) r_T^2 \left(\ln r_T + \frac{1}{2} \right) \right],\end{aligned} \tag{4.3.64}$$

where $r = a_0 \tau$ with $\tau = T/T_c - 1$. The right-hand side may be regarded as an expansion with respect to ϵ if we regard $r \sim \epsilon^0$, $M \sim \epsilon^{-1/2}$, $u_0 \sim \epsilon$, $h \sim \epsilon^{-1/2}$, $C_0 \sim \epsilon^{-1}$. Here, the logarithmic terms in the parentheses are the most important corrections, while the non-logarithmic terms proportional to r_L^2 or r_T^2 in the brackets depend on the method of introducing the upper cut-off Λ. Note that we have used the smooth cut-off introduced by $\mathcal{H}_{\text{cut-off}}$ in (4.3.55), because the ϵ expansions of g^* and v^*, (4.3.56) and (4.3.57), up to order ϵ^2 will be used in calculating the universal amplitude ratios.

Equation of state

We determine M from the minimum condition $(\partial f_{\text{sing}}/\partial M)_{h\tau} = 0$. By setting $K_4 u_0 = g$ we obtain [5, 59]

$$\begin{aligned}\frac{h}{M} &= r + u_0 M^2 + \frac{3}{2} g \left[r_L \ln(e r_L) + \frac{1}{3}(n-1) r_T \ln(e r_T) \right] \\ &= \left[r + u_0 M^2 + \frac{3}{2} g r_L \ln(e r_L) \right] \left[1 + \frac{1}{2}(n-1) g \ln(e r_T) \right].\end{aligned} \tag{4.3.65}$$

The first and second lines coincide up to first-order corrections. The second line is convenient for the case $h \to 0$ and $r < 0$. We examine some typical cases as follows.

(i) As $h \to 0$ with $\tau > 0$, h/M tends to the inverse susceptibility,

$$\chi^{-1} = \xi^{-2} = r \left[1 + \frac{n+2}{2} g \ln(er) \right] = r(er)^{(n+2)g/2}, \tag{4.3.66}$$

from the first line of (4.3.65). At $g = g^*$ this leads to the ϵ expansion,

$$\gamma = 1 + \frac{n+2}{2(n+8)} \epsilon. \tag{4.3.67}$$

(ii) We set $r = 0$ to obtain

$$h = u_0 M^3 \left[1 + \frac{n+8}{2} g \ln(e u_0 M^2) + \frac{9}{2} g \ln 3 \right] \propto M^3 (u_0 M^2)^{(n+8)g/2}, \tag{4.3.68}$$

which yields

$$\delta = 3 + (n+8) g = 3 + \epsilon. \tag{4.3.69}$$

(iii) When $h \to 0$ with $r < 0$, we have $r_L = 2|r|$ as $\epsilon \to 0$ and the second line of (4.3.63) yields

$$u_0 M^2 = |r| - \frac{3}{2} g r_L \ln(e r_L) = |r||2er|^{-3g}, \tag{4.3.70}$$

so we have

$$\beta = \frac{1}{2}(1 - 3g) = \frac{1}{2} - \frac{3}{2(n+8)}\epsilon. \tag{4.3.71}$$

Specific heat

The singular average energy density (divided by T_c) is given by $\langle m \rangle = -(\partial f_{\text{sing}}/\partial \tau)_h = -(\partial f_{\text{sing}}/\partial \tau)_M$ from $(\partial f_{\text{sing}}/\partial M)_{h\tau} = 0$ in (4.3.65). Therefore,

$$\begin{aligned} \langle m \rangle &= C_0 \tau - a_0 \left\{ \frac{1}{2} M^2 + \frac{1}{4} K_4 \big[r_L \ln(e r_L) + (n-1) r_T \ln(e r_T) \big] \right\} \\ &= \frac{C_0}{a_0} \left\{ r - \frac{2v}{g} u_0 M^2 - v \big[r_L \ln(e r_L) + (n-1) r_T \ln(e r_T) \big] \right\}, \end{aligned} \tag{4.3.72}$$

where we have set $v = K_4 a_0^2 / 4C_0$ on the second line.

(i) At $h = 0$ with $\tau > 0$, we obtain

$$\langle m \rangle = C_0 a_0^{-1} r \big[1 - nv \ln(er) \big] = C_0 a_0^{-1} r (er)^{-nv}, \tag{4.3.73}$$

which yields the specific heat at constant magnetic field,

$$C_H = \left(\frac{\partial \langle m \rangle}{\partial \tau} \right)_h = (1 - nv) C_0 (er)^{-nv}, \tag{4.3.74}$$

so

$$\alpha = nv = \frac{4-n}{2(n+8)}\epsilon. \tag{4.3.75}$$

(ii) When $h \to 0$ with $r < 0$, we use the first line of (4.3.65) to eliminate the logarithmic term $\propto r_T \ln(e r_T)$ to obtain

$$\langle m \rangle = C_0 a_0^{-1} \big[(1 + 2v/g) r + 2v r_L \ln(e r_L) \big]. \tag{4.3.76}$$

From (4.3.56) and (4.3.57) we have $v/g = (4-n)/2n - 2\epsilon/n$ up to first order in ϵ at the fixed point, so that

$$\langle m \rangle = C_0 a_0^{-1} r \left[\frac{4}{n}(1 - \epsilon) - 4v \ln(e r_L) \right] = C_0 a_0^{-1} \frac{4}{n(1+\epsilon)} r |2er|^{-nv}. \tag{4.3.77}$$

Thus,

$$C_H = C_0 \frac{4}{n(1+\epsilon)} (1 - nv) |2er|^{-nv}. \tag{4.3.78}$$

Let us write C_H at $h = 0$ as $A_0\tau^{-\alpha}$ for $\tau > 0$ and as $A'_0|\tau|^{-\alpha}$ for $\tau < 0$. Then, we derive a well-established formula for the ratio of critical amplitudes [4, 8, 12], [60]–[62],

$$A_0/A'_0 = 2^{\alpha-2}n(1+\epsilon). \tag{4.3.79}$$

The right-hand side of (4.3.79) can give a good estimate of the amplitude ratio at $\epsilon = 1$. In fact, it is about 0.5, 1, and 1.5 for $n = 1, 2,$ and 3 in 3D from experiments and reliable theories [12, 62].

4.3.5 *Results for Ising-like systems*

For $n = 1$ there is no contribution of the transverse spin fluctuations. The equation of state and the inverse susceptibility $1/\chi = (\partial h/\partial M)_\tau$ are obtained after exponentiation as

$$h/M = \left[r + u_0 M^2 (er_{\rm L})^{\epsilon/3}\right](er_{\rm L})^{\epsilon/6}, \tag{4.3.80}$$

$$\chi^{-1} = \left[r + (3+\epsilon)u_0 M^2 (er_{\rm L})^{\epsilon/3}\right](er_{\rm L})^{\epsilon/6}, \tag{4.3.81}$$

for any h and τ. The susceptibility at $h = 0$ behaves as $\chi \cong \Gamma_0\tau^{-\gamma}$ for $\tau > 0$ and as $\chi \cong \Gamma'_0|\tau|^{-\gamma}$ for $\tau < 0$ with

$$\Gamma_0/\Gamma'_0 = (2+\epsilon)2^{\epsilon/6}, \tag{4.3.82}$$

to first order in ϵ. This is consistent with a more elaborate expression,

$$\Gamma_0/\Gamma'_0 = 2^{\gamma-1}(\gamma/\beta)(1 + 0.112\epsilon^3 + \cdots), \tag{4.3.83}$$

which is valid up to order ϵ^3 [12, 60]. The above ratio is estimated to be about 4.9 for 3D Ising models.

The specific heat $C_M = (\partial\langle m\rangle/\partial\tau)_M$ at constant magnetization is readily calculated as

$$C_M \cong \left(1 - \frac{\epsilon}{6}\right)C_0(er_{\rm L})^{-\epsilon/6}, \tag{4.3.84}$$

which is valid for any h and τ. The critical amplitudes of C_M at $h = 0$, which behaves as $A_{M0}\tau^{-\alpha}$ for $\tau > 0$ and as $A'_{M0}|\tau|^{-\alpha}$ for $\tau < 0$, satisfy

$$A_{M0}/A'_{M0} = 2^\alpha \cong 1. \tag{4.3.85}$$

The correction to the above result is of order ϵ^3 [5]. The parametric model in Section 2.1 gives C_M in the simple form (2.1.51) and leads to $A_{M0}/A'_{M0} = (b^2-1)^{-\alpha}$ [24], where b is defined by (2.1.48). This agrees with (4.3.85) because $b^2 = 3/2 + O(\epsilon^2)$ [5]. On the coexistence curve we also have

$$C_M/C_H = 1 - R_{\rm v} = A'_{M0}/A'_0 = \frac{1}{4}(1+\epsilon), \tag{4.3.86}$$

where $R_{\rm v}$ is defined by (1.1.48) and behaves as $(3-\epsilon)/4$. This result also follows from $R_{\rm v} = (\partial M/\partial\tau)_h^2/C_H\chi$ using (4.3.70), (4.3.78), and (4.3.81). It is known that $R_{\rm v} \cong 0.5$ and hence $a_{\rm c}^2 = R_{\rm v}/(1-R_{\rm v}) \cong 1$ in 3D Ising systems (see footnote 2 below (2.2.37)).

The correlation length ξ can be determined from the small wave number behavior of the structure factor $I(k)$ as in the first line of (2.1.9). In Appendix 4E we will derive the following expression, valid for $k\xi \ll 1$,

$$1/I(k) = \chi^{-1} + k^2\left(1 + \frac{\epsilon}{6}u_0 M^2 \chi\right), \tag{4.3.87}$$

where χ is given by (4.3.81). Thus,

$$\xi^2 = \chi\left[1 + \frac{\epsilon}{6}u_0 M^2/(r + 3u_0 M^2)\right]. \tag{4.3.88}$$

The amplitude ratio ξ_{+0}/ξ_{-0} is equal to $2^{1/2}$ in the mean field theory, and its ϵ expansion follows from (4.3.70) and (4.3.81) as

$$\xi_{+0}/\xi_{-0} = 2^{\nu}\left(1 + \frac{5}{24}\epsilon + \cdots\right). \tag{4.3.89}$$

Its reliable estimate is 1.91 in 3D Ising models.

4.3.6 Specific heats in classical fluids

In fluids, the usually measured specific heats, C_V in one-component fluids and C_{pX} in two-component fluids, correspond to C_M in Ising systems as shown in (2.2.25) and (2.3.27) (or (2.3.63)), respectively. If we write C_V as $A(T/T_c - 1)^{-\alpha}$ on the critical isochore above T_c and as $A'(1 - T/T_c)^{-\alpha}$ on the coexistence curve below T_c, (4.3.85) yields

$$A/A' = A_{M0}/A'_{M0} \cong 1. \tag{4.3.90}$$

The same result also follows from (2.1.51) in the parametric model scheme [24]. Dahl and Moldover [63] measured C_V of ^3He in a single phase of liquid states near the coexistence curve and indeed found $A/A' \cong 1$ in agreement with (4.3.90). In other experiments on the coexistence curve, however, the specific heat has been measured in two-phase coexistence at a constant volume of the total system [64]–[69],[12] where the volume fraction of each phase adjusts to change such that the pressure and temperature stay on the coexistence curve. The critical behavior of the resultant specific heat $(C_V)_{cx}$ was first considered by Fisher [70]. In Appendix 4F we will show that it behaves as

$$(C_V)_{cx} \cong (1 + a_c^2)C_V \tag{4.3.91}$$

and asymptotically corresponds to C_H, where a_c is the universal number defined by (2.2.37). Thus, if we write $(C_V)_{cx} \cong A_{cx}(1 - T/T_c)^{-\alpha}$, we obtain

$$A_{cx}/A = A'_0/A_0 \cong 2. \tag{4.3.92}$$

[12] See Section 6.3 for discussions of C_V measurements in two-phase coexistence [65, 66]. Voronel's group [68] stirred near-critical fluids to measure C_V. See Section 11.1 for the effects of stirring.

This result has been reported widely in the literature [62, 69] as an evidence of correspondence between fluids and Ising systems, but the above delicate issue has not been recognized.

We also comment on the background specific heat in one-component fluids. In particular, on the critical isochore above T_c, C_V can be written as

$$C_V = A\big[(T/T_c - 1)^{-\alpha} + B\big]. \tag{4.3.93}$$

The constant B is about -0.5, -0.9, and 0.3 for ^{4}He [64], CO_2 [67], and SF_6 [69], respectively, whereas it is nearly zero for ^{3}He [65, 66]. If use is made of (1.2.53), the sound velocity can be written as

$$\rho c^2 = T_c \left(\frac{\partial p}{\partial T}\right)^2_{cx} A^{-1}\big[(T/T_c - 1)^{-\alpha} + B\big]^{-1}. \tag{4.3.94}$$

We have derived the above form for C_H in (4.3.44) for Ising systems (and will do so in (4C.11) for n-component systems). The C_V and C_H are related by $C_V = b_c^2 C_H$ from (2.2.25) on the critical isochore above T_c, where $\tau = b_c(T/T_c - 1)$ with $b_c = T_c(\partial\tau/\partial T)_h$ from (2.2.15) and (2.2.16) , so B is expressed as

$$B = C_B b_c^{\alpha}/A_0, \tag{4.3.95}$$

in terms of A_0 in (4.3.46) and C_B in (4.3.48). In Chapter 6 we shall see that the background specific heat C_B crucially influences the behavior of critical acoustic attenuation [71].

4.3.7 Broken symmetry for $n \geq 2$

As $h \to 0$ with $\tau < 0$ in non-Ising systems ($n \geq 2$), interesting effects arise due to the fluctuations of the transverse part $\psi_T = (\psi_2, \ldots, \psi_n)$ of the order parameter. Let $G_T(r) = \langle \psi_j(\boldsymbol{r})\psi_j(\boldsymbol{0})\rangle$ ($j = 2, \ldots, n$) be the transverse correlation function. The transverse structure factor grows at small wave numbers as

$$I_T(k) = \int d\boldsymbol{r} e^{i\boldsymbol{k}\cdot\boldsymbol{r}} G_T(r) \cong \frac{1}{h/M + K_R k^2}. \tag{4.3.96}$$

As $h \to 0$ the coefficient K_R behaves as

$$K_R = \left[1 + \frac{1}{2}g + O(\epsilon^2)\right] r_L^{-\eta\nu} \cong 1 + \frac{1}{2(n+8)}\epsilon, \tag{4.3.97}$$

as will be shown in Appendix 4E. As $h \to 0$ we have a Coulombic correlation,

$$G_T(r) \sim \frac{1}{r^{d-2}} \quad (d > 2). \tag{4.3.98}$$

It is believed that the deviation of the longitudinal part $\delta\psi_1 = \psi_1 - M$ is determined at long wavelengths by ψ_T as

$$\delta\psi_1 \cong -\frac{1}{2M}\big(|\psi_T|^2 - \langle|\psi_T|^2\rangle\big), \tag{4.3.99}$$

which follows if the amplitude deviation is neglected as $|\psi|^2 = M^2 + 2M\delta\psi_1 + \delta\psi_1^2 + |\psi_T|^2 \cong \text{const}$. For the xy model ($n = 2$) this is equivalent to introducing the phase θ,

$$\psi_1 \cong M\cos\theta \cong M\left[1 - \frac{1}{2}\theta^2 + \frac{1}{2}\langle\theta^2\rangle\right], \quad \psi_2 \cong M\sin\theta \cong M\theta. \tag{4.3.100}$$

We recognize that the transverse fluctuations are those of the angle or phase variables (for any $n \geq 2$), which exhibit slowly varying modulations without appreciable free-energy penalty. The longitudinal correlation function $G_L(r) = \langle\delta\psi_1(\boldsymbol{r})\delta\psi_1(\boldsymbol{0})\rangle$ thus behaves as [62], [72]–[74]

$$G_L(r) \cong \frac{1}{2}(n-1)M^{-2}G_T(r)^2 \sim \frac{1}{r^{2d-4}}, \tag{4.3.101}$$

which follows from (4.3.99) if ψ_T obeys the gaussian distribution at long wavelengths. In the presence of small positive h the tails of $G_T(r)$ and $G_L(r)$ are cut off at $r \sim \ell_h$, where

$$\ell_h = (h/MK_R)^{-1/2}. \tag{4.3.102}$$

Thus the longitudinal susceptibility grows for small h as [5, 7, 72, 73]

$$\chi_L \sim \int_{r<\ell_h} d\boldsymbol{r}\frac{1}{r^{2d-4}} \sim |h|^{-\epsilon/2}. \tag{4.3.103}$$

We shall see in (4.3.114) below that the longitudinal structure factor $I_L(k)$ grows as $k^{-\epsilon}$ at small k for $h = 0$.

Transverse correlation length and the superfluid density

In the literature [56, 62] a transverse correlation length ξ_T at $h = 0$ below T_c has been introduced in terms of the transverse structure factor (4.3.96) by

$$\xi_T^{d-2} = \lim_{k\to 0} I_T(k)k^2/M^2 = (K_R M^2)^{-1}. \tag{4.3.104}$$

The right-hand side is proportional to $(1-T/T_c)^{-(d-2)\nu}$ to all orders in ϵ from the exponent relation (2.1.13). Thus we may set

$$\xi_T = \xi_{T0}(1 - T/T_c)^{-\nu}. \tag{4.3.105}$$

For $n = 2$, slow modulations of the phase variable θ give rise to the following free-energy density increase,

$$\Delta f_{\text{phase}} = \frac{1}{2}T\xi_T^{-d+2}|\nabla\theta|^2. \tag{4.3.106}$$

In ^{4}He the superfluid velocity is given by $\boldsymbol{v}_s = (\hbar/m_4)\nabla\theta$ (where m_4 is ^{4}He mass) and Δf_{phase} should coincide with the kinetic energy of the superfluid component, so that the superfluid mass density turns out to be of the form,

$$\rho_s = Tm_4^2\hbar^{-2}\xi_T^{-d+2} = Tm_4^2\hbar^{-2}K_R M^2, \tag{4.3.107}$$

which leads to $\xi_{T0} \cong 3.4$ Å at SVP from data of ρ_s along the λ line [75].

As an example of the two-scale-factor universality for the case $\alpha \cong 0$, we may construct a universal number [21],

$$R_\xi^- = \xi_{T0} A'^{1/d}, \tag{4.3.108}$$

where A' is the amplitude of the logarithmic term in C_p below T_λ in (2.4.2). This universal relation is analogous to that in (2.4.4), but both ξ_{T0} and A' are measurable here. Indeed, R_ξ^- from experimental data agreed with the theoretical value $\cong 0.85$ along the λ line [75]. It is easy to derive the following ϵ expansion,

$$\begin{aligned}(\xi_{+0}/\xi_{T0})^{d-2} &= (K_R M^2)_{r<0}(\chi^{(d-2)/2})_{r>0} \\ &= 2^{-3\epsilon/(n+8)}(n+8)K_d\left[\frac{1}{\epsilon} - \frac{17n+76}{2(n+8)^2} + O(\epsilon)\right],\end{aligned} \tag{4.3.109}$$

where we can find K_R in (4.3.97), $u_0 M^2 = gM^2/K_d$ in (4.3.70) with g being expanded as (4.3.56), and χ in (4.3.66). The expansion up to $O(\epsilon)$ was performed in Ref. [58].

4.3.8 Hydrodynamic hamiltonian for $n \geq 2$

We have presented the hydrodynamic hamiltonian in (4.1.56) for the Ising case. Here it is devised in ordered states at small h for $n \geq 2$. The fluctuations with wave numbers larger than $\xi_T^{-1} \sim |\tau|^\nu$ give rise to multiplicative factors of r_L as in the disordered state. The problem is then the nonlinear interaction among the transverse fluctuations with wave numbers smaller than ξ_T^{-1}. It is convenient to introduce the following variable,

$$\varphi = \psi_1 - M + \frac{1}{2M}\left(|\psi_T|^2 - \langle|\psi_T|^2\rangle\right). \tag{4.3.110}$$

From the assumption (4.3.99), φ is decoupled from the transverse part ψ_T and should have a well-defined variance $\chi_R \propto |r|^{-\gamma}$ at long wavelengths. Setting the upper cut-off wave number at ξ_T^{-1}, we propose a hydrodynamic hamiltonian for smooth variations of the order parameter deviation $\delta\psi = (\psi_1 - M, \psi_T)$ and the energy deviation δm,

$$\frac{1}{T}\mathcal{H}_{\text{hyd}} = \frac{1}{2}\int d\boldsymbol{r}\left[\frac{1}{\chi_R}\varphi^2 + \frac{h}{M}|\psi_T|^2 + K_R|\nabla\psi_T|^2 + \frac{1}{\tilde{C}}(\delta m + A_R\varphi)^2\right], \tag{4.3.111}$$

where topological singularities are neglected.[13] The term $(h/2M)|\psi_T|^2$ arises from the magnetic field energy $-h\psi_1 = -h(\varphi - |\psi_T|^2/2M)+$const. and the cross term $A_R\tilde{C}^{-1}\delta m\varphi$ from the coupling $\gamma_0 m|\psi|^2$ in the original GLW hamiltonian. The coefficient A_R is thus related to the renormalized value of γ_0 by

$$A_R/\tilde{C} = 2M\gamma_R. \tag{4.3.112}$$

[13] This form is analogous to the Ginzburg–Landau free energy for smectic A liquid crystals [76] in terms of the layer displacement $u(z, \boldsymbol{r}_T)$, where the lateral undulation $\nabla_T u$ corresponds to ψ_T and the dilational change $\nabla_z u - |\nabla_T u|^2/2$ to φ. Note also that anomalous fluctuations of the director orientation in the nematic phase are analagous to those of the transverse components in the spin systems we are discussing in this chapter [76].

Because φ and ψ_T obey gaussian distributions independently of each other, the longitudinal susceptibility is written as

$$\begin{aligned}\chi_L &= \chi_R + (2M)^{-2}\langle|\psi_T|^2 : |\psi_T|^2\rangle \\ &= \chi_R + \frac{\pi(n-1)(2-\epsilon)}{8K_R \sin(\pi\epsilon/2)} K_d \xi_T^{2-\epsilon}\left[\left(\frac{h}{K_R M}\right)^{-\epsilon/2} - \xi_T^{\epsilon}\right],\end{aligned} \tag{4.3.113}$$

where we assume $2 < d < 4$. For $h = 0$, as in ^{4}He below T_λ, we should consider the longitudinal structure factor with nonvanishing wave number. For $k \ll \xi_T^{-1}$ it is expressed at $d = 3$ as

$$I_L(k) \cong \frac{3}{64}(n-1)K_R^{-1}\xi_T\frac{1}{k} \qquad (h = 0). \tag{4.3.114}$$

Here we should check the consistency between the result (4.3.113) derived from (4.3.111) and the ϵ expansion of the equation of state (4.3.65), from which we have

$$\chi_L^{-1} = \frac{\partial h}{\partial M} = \left[r + (3+\epsilon)u_0 M^2 (er_L)^{3g}\right](er_L)^{3g/2}\left[1 + \frac{1}{2}(n-1)g\ln(er_T)\right], \tag{4.3.115}$$

where the terms $\propto \ln(er_L)$ have been exponentiated. After some manipulations this expression is rewritten as

$$\chi_L = \chi_R + \chi_R(n-1)\frac{g}{\epsilon}\left[(r_L/r_T)^{\epsilon/2} - 1\right]. \tag{4.3.116}$$

As $h \to 0$, we have

$$\chi_R^{-1} = (2+\epsilon)|r|(2e|r|)^{(n+2)g/2} \sim |r|^{\gamma}. \tag{4.3.117}$$

If we set $n = 1$ in the above expression, it becomes of the same form as χ^{-1} in (4.3.81) for Ising systems. We can see that (4.3.113) and (4.3.116) are consistent for small ϵ if use is made of (4.3.70) and (4.3.97). The other two coefficients $\tilde{C}(\propto |\tau|^{-\alpha})$ and $A_R(\propto |\tau|^{(\gamma-\alpha)/2})$ in $\mathcal{H}_{\rm hyd}$ are determined by

$$C_H = \langle m : m\rangle = \tilde{C} + A_R^2\chi_R, \tag{4.3.118}$$

$$\frac{\partial M}{\partial \tau} = \langle m : \psi_1\rangle = -A_R\chi_R. \tag{4.3.119}$$

From $M \propto |\tau|^{\beta}$ we find

$$A_R = \beta M/|\tau|\chi_R, \quad \tilde{C} = C_H - \beta^2 M^2/|\tau|^2\chi_R. \tag{4.3.120}$$

As in the Ising case in (1.1.48), (2.2.37), and (4.3.86), we introduce the ratio,

$$R_v = A_R^2\chi_R/C_H = \beta^2 M^2/(|\tau|^2 C_H\chi_R). \tag{4.3.121}$$

Then $\tilde{C} = C_H(1 - R_v)$ holds and $R_v < 1$ is required. Its ϵ expansion is of the form,

$$R_v = \left(1 - \frac{n}{4}\right) - \frac{n}{4}\epsilon + O(\epsilon^2). \tag{4.3.122}$$

For $n = 2$, however, the sum of the first two terms in the ϵ expansion vanishes at $\epsilon = 1$.

For ^{4}He we obtain $R_v \cong 0.7(\Gamma_0/\Gamma_0')n_\lambda/C_p$ by setting $\chi = \Gamma_0\tau^{-\gamma}$ for $T > T_\lambda$ and $\chi_R = \Gamma_0'|\tau|^{-\gamma}$ for $T < T_\lambda$. If $\Gamma_0/\Gamma_0' = (2+\epsilon)2^{\gamma-1} + O(\epsilon^2)$ is not much different from the Ising value ~ 5, R_v turns out to be considerably smaller than 1 for $|\tau| \ll 1$.

Finally, we examine the singular part of the thermodynamic free energy due to the transverse fluctuations. From (4.3.111) the long-wavelength fluctuations of ψ_T give rise to the following free-energy density increase in the presence of h,

$$\begin{aligned} f(h,\tau) - f(0,\tau) &= \frac{T}{2}(n-1)K_d \int_0^{1/\xi_T} dk k^{d-1}\left[\ln(h/K_R M + k^2) - \ln(k^2)\right] \\ &\cong \text{const.}|\tau|^\beta h + \frac{\pi(n-1)K_d T\xi_T^2}{2d\sin(\pi\epsilon/2)K_R}\left[h^2 - (K_R M/\xi_T^2)^{\epsilon/2}h^{d/2}\right], \end{aligned} \tag{4.3.123}$$

where $h > 0$ is assumed. This expression is valid for $h/K_R M \ll \xi_T^{-2}$ and $\tau < 0$. The free energy thus contains the term $\propto h^{d/2}$, which gives rise to the average order parameter,

$$M = -\frac{1}{T}\left(\frac{\partial f}{\partial h}\right)_\tau = B_0|\tau|^\beta + \frac{\pi(n-1)K_d}{4\sin(\pi\epsilon/2)}(\xi_T^{d-2}/K_R)^{d/4}h^{(d-2)/2} + O(h), \tag{4.3.124}$$

and $\chi_L \propto h^{-\epsilon/2}$ in (4.3.113). We note that the singular free energy $\propto |h|^{d/2}$ is present even far below T_c for $2 < d < 4$. In addition, if the last term in (4.3.123) is expanded in powers of ϵ, we recover the term $\propto r_T^2 \ln r_T$ in the naive ϵ expansion of the singular free energy (4.3.64).

4.4 Two-phase coexistence and surface tension

In his theory of gas–liquid coexistence in 1893 [see Ref. 13], van der Waals introduced the gradient free energy and derived the equation,

$$\frac{\partial}{\partial n}f(n) = T\left[\ln\left(\frac{\phi}{1-\phi}\right) + \frac{1}{1-\phi}\right] - 2\epsilon\phi - T\bar{\nu}_\infty = K\frac{d^2}{dx^2}n, \tag{4.4.1}$$

for the number-density profile $n = n(x)$ near a gas–liquid interface, where $f(n)$ is the free-energy density in the form (3.4.9), $\phi = v_0 n(x)$ is the effective volume fraction, and $\bar{\nu}_\infty$ is a constant related to the chemical potential and the number density via (3.4.10) and (3.4.12) far from the interface. The coefficient K is assumed to be independent of n. The above equation can also be rewritten as

$$n\frac{\partial}{\partial n}f(n) - f(n) = T\frac{n}{1-\phi} - \epsilon v_0 n^2 = Kn\frac{d^2}{dx^2}n - \frac{K}{2}\left(\frac{d}{dx}n\right)^2 + p_\infty, \tag{4.4.2}$$

where p_∞ is the pressure in the bulk region. From the van der Waals equation of state (3.4.3) the above quantity may be regarded as a local pressure. If K depends on n, the right-hand sides of (4.4.1) and (4.4.2) should be appropriately changed as can be known from (4.4.16) and (4.4.17) below. Near the critical point, van der Waals found that the surface tension σ is proportional to $(T_c - T)^{3/2}$, which is the mean field result to be explained below. In 1958 Cahn and Hilliard [17] re-derived the same results in the presence of the gradient free energy. We will follow and extend their approach. However, the systems we will treat are very limited.

4.4.1 Interface profile and surface tension near the critical point

Let us note that two phases can coexist macroscopically in Ising-like systems at $h = 0$ and $\tau < 0$. We consider a planar interface whose normal direction is in the x direction. The mean field profile $\psi = \psi_{\rm int}(x)$ is calculated from

$$\frac{\delta}{\delta\psi}(\beta\mathcal{H}) = r\psi + u_0\psi^3 - K\frac{d^2}{dx^2}\psi = 0, \tag{4.4.3}$$

where use has been made of (4.1.1) and ψ depends only on x. We replace r_0 in (4.1.1) by r in (4.1.17) assuming that the fluctuations with wave numbers larger than ξ have already been coarse-grained. We multiply (4.4.3) by $d\psi/dx$ and integrate over x to obtain

$$\frac{1}{2}r\psi^2 + \frac{1}{4}u_0\psi^4 - \frac{1}{2}K\left(\frac{d}{dx}\psi\right)^2 = -\frac{|r|^2}{4u_0}. \tag{4.4.4}$$

The value on the left-hand side is determined from the boundary condition $\psi \to \pm M$ where $M = (|r|/u_0)^{1/2}$. Some manipulations yield

$$M^2(\psi^2/M^2 - 1)^2 = 4\xi^2\left(\frac{d}{dx}\psi\right)^2, \tag{4.4.5}$$

where

$$\xi = (K/2|r|)^{1/2} \tag{4.4.6}$$

is the correlation length below T_c determined from the first line of (2.1.9). We then obtain the well-known interface solution,

$$\psi_{\rm int}(x) = M\tanh(x/2\xi) = -M + 2M\big/\big[1 + \exp(-x/\xi)\big]. \tag{4.4.7}$$

The surface tension is the excess free energy stored in the interface region per unit area and is given by the following integral:

$$\begin{aligned}\sigma &= T\int_{-\infty}^{\infty} dx\left[\frac{1}{2}r\psi^2 + \frac{1}{4}u_0\psi^4 + \frac{1}{2}K\left(\frac{d}{dx}\psi\right)^2 + \frac{r^2}{4u_0}\right] \\ &= T\int_{-\infty}^{\infty} dx K\left(\frac{d}{dx}\psi\right)^2,\end{aligned} \tag{4.4.8}$$

where the free-energy density $-r^2/4u_0$ at $x = \pm\infty$ has been subtracted in the integrand on the first line and use has been made of (4.4.4) in the second line. Substitution of (4.4.6) gives[14]

$$\sigma = \frac{2}{3}TKM^2\xi^{-1} = TK^{1/2}\frac{|2r|^{3/2}}{3u_0}. \tag{4.4.9}$$

[14] Use is made of the relation $\int_0^x dx^{1/4}\cosh^{-4}x = \tanh x - \frac{1}{3}\tanh^3 x$.

Therefore, $\sigma \propto |\tau|^{3/2}$ in the mean field theory, where $\tau = T/T_c - 1$ is the reduced temperature. It is instructive to express σ in terms of the Ginzburg number Gi given in (4.1.25); in 3D, we have

$$\sigma \sim 0.1 T \xi^{-2} |\tau/Gi|^{1/2}, \tag{4.4.10}$$

which holds for $|\tau| > Gi$.

In the asymptotic critical region, $|\tau| \ll Gi$, the renormalization group theory indicates that u_0 should be replaced by the renormalized value $u_R = K_d^{-1} g^* \kappa^\epsilon$ in (4.3.23) and $|2r|^{1/2}$ by $K^{1/2}\xi^{-1}$ sufficiently close to the critical point. Then we find the scaling behavior [77, 78],

$$\sigma = A_\sigma T \xi^{-d+1}, \tag{4.4.11}$$

where

$$A_\sigma = \frac{1}{3g^*} K_d [1 + O(\epsilon)]. \tag{4.4.12}$$

The coefficient A_σ is a universal number and is known to be about 0.09 in 3D Ising-like systems [79, 80]. We note that (4.4.12) roughly gives $A_\sigma \sim 9/(3 \cdot 2\pi^2) \sim 0.1$, consistent with 0.09 mentioned above. We notice that the two limiting expressions, (4.4.9) and (4.4.11), are smoothly connected at $|\tau| \sim Gi$ from (4.4.10). The relation $\sigma \propto \xi^{-d+1}$ in the asymptotic scaling regime is analogous to that in (2.1.42) or (4.3.54).

Instability of the interface solution in many-component systems

In many-component systems ($n \geq 2$), isotropic in the spin space, two ordered states cannot be separated by a stable localized interface if these two states can be changed over only by a gradual phase variation. Let us consider a system with a complex order parameter ($n = 2$), such as ^{4}He near the superfluid transition. If we impose the boundary condition $\psi = M$ at $x = 0$ and $\psi = -M$ at $x = L$, the order parameter profile which minimizes the free energy (4.1.1) at $h = 0$ is given by

$$\psi = M \exp(i\pi x/L). \tag{4.4.13}$$

In ^{4}He this is the case in which a superfluid current is induced with the velocity $v_s = \pi\hbar/m_4 L$ in the x direction, where m_4 is the ^{4}He mass. The free-energy increase is $\rho_s v_s^2 L/2 = \pi^2 T K M^2/2L$ per unit area in the yz plane, as stated in (4.3.106) and (4.3.107). For the interface solution (4.4.7), however, it is equal to σ in (4.4.10) independent of L per unit area. To see how the localized solution becomes unstable, we superimpose a small imaginary perturbation as $\psi = M \tanh(x/2\xi) + i\delta\psi_2(x)$, where $\delta\psi_2$ is real and dependent only on x. The free energy change is written as

$$\beta\delta\mathcal{H} = \int d\mathbf{r} \delta\psi_2 \left[-\frac{K}{2}\frac{d^2}{dx^2} - \frac{|r|}{2\cosh^2(x/2\xi)} \right] \delta\psi_2. \tag{4.4.14}$$

We notice that the integrand becomes $-|r|(\delta\psi_2)^2/4 < 0$ if $\delta\psi_2 \propto 1/\cosh(x/2\xi)$. Thus amplification of this eigenmode serves to decrease the free energy.

It is worth noting that if ^{4}He is in contact with a solid surface at $x = 0$, we should impose the boundary condition $\psi = 0$ at $x = 0$. The boundary profile of ψ is given by (4.4.7) in the region $x > 0$ in the mean field theory [35].

4.4.2 Surface tension for the general free-energy density

We need the surface tension expression for the general form of the free-energy density $f(\psi)$ [17], because the Landau expansion may not be a good approximation away from criticality. Let the hamiltonian be of the form,[15]

$$\beta\mathcal{H}\{\psi\} = \int d\mathbf{r}\left[f(\psi) + \frac{1}{2}K(\psi)|\nabla\psi|^2\right], \tag{4.4.15}$$

where $f(\psi)$ has two minima at $\psi = \psi_{\rm cx}^{(1)}$ and $\psi_{\rm cx}^{(2)}$ with the same minimum value $f_{\rm min}$. The coefficient of the gradient term is allowed to depend on ψ as $K = K(\psi)$. The interface solution, which tends to $\psi_{\rm cx}^{(1)}$ as $x \to \infty$ and to $\psi_{\rm cx}^{(2)}$ as $x \to -\infty$, satisfies

$$\frac{\partial f}{\partial\psi} + \frac{1}{2}\frac{\partial K}{\partial\psi}\left(\frac{d\psi}{dx}\right)^2 - \frac{d}{dx}K\frac{d\psi}{dx} = 0. \tag{4.4.16}$$

Multiplying $d\psi/dx$ and integrating over x we find

$$f(\psi) - f_{\rm min} = \frac{1}{2}K\left(\frac{d\psi}{dx}\right)^2. \tag{4.4.17}$$

This may be integrated to give

$$x = \int_{\psi_0}^{\psi} d\psi\sqrt{\frac{K(\psi)}{2[f(\psi) - f_{\rm min}]}}, \tag{4.4.18}$$

where ψ_0 is the value of ψ at $x = 0$. At large $|x|$, we have $f(\psi) - f_{\rm min} \cong \frac{1}{2}f''(\psi_{\rm cx}^{(\alpha)})(\psi - \psi_{\rm cx}^{(\alpha)})^2$ with $\alpha = 1, 2$, so that

$$\psi(x) - \psi_{\rm cx}^{(\alpha)} \sim e^{-|x|/\xi_\alpha}, \tag{4.4.19}$$

where $\xi_\alpha = [K(\psi_{\rm cx}^{(\alpha)})/f''(\psi_{\rm cx}^{(\alpha)})]^{1/2}$ is the correlation length defined for the two phases. Note also that the structure factors in the two phases are of the Ornstein–Zernike form $\propto (k^2 + \xi_\alpha^{-2})^{-1}$. Because of (4.4.17) the surface tension is expressed as

$$\begin{aligned}\sigma &= T\int_{-\infty}^{\infty} dx K(\psi)\left(\frac{d\psi}{dx}\right)^2 \\ &= T\int_{\psi_{\rm cx}^{(2)}}^{\psi_{\rm cx}^{(1)}} d\psi\sqrt{2K(\psi)(f(\psi) - f_{\rm min})},\end{aligned} \tag{4.4.20}$$

where $\psi_{\rm cx}^{(2)} < \psi_{\rm cx}^{(1)}$ is assumed in the second line.

[15] Here Tf is the free-energy density.

4.4.3 Interface in symmetrical tricritical systems

In Section 3.2 we discussed tricritical behavior. Here let us consider the coexistence of a disordered phase and an ordered phase near a symmetrical tricritical point in general n-component systems [81, 82]. Namely, the amplitude $\psi = |\psi|$ depends on space; $\psi \to 0$ as $x \to -\infty$ and $\psi \to \psi_{\rm cx}$ as $x \to \infty$, where $\psi_{\rm cx}$ is given by (3.2.13). The free-energy density divided by T is the sum of the expression in (3.2.16) and the gradient term $K|\nabla\psi|^2/2$. Then (4.4.17) becomes

$$\frac{1}{6}v_0\psi^2(\psi^2 - \psi_{\rm cx}^2)^2 = \frac{1}{2}K\left(\frac{d}{dx}\psi\right)^2. \tag{4.4.21}$$

From (3.2.14) the correlation length in the ordered phase is obtained from $K\xi^{-2} = \chi^{-1} = 3u_0^2/4v_0$. Thus,

$$\xi = 2(Kv_0/3)^{1/2}/|u_0|. \tag{4.4.22}$$

Because $u_0 \propto T - T_{\rm t}$, we have $\xi \propto |T - T_{\rm t}|^{-1}$ for $d \geq 3$. It is easy to solve (4.4.21) in the form,

$$\psi(x) = \psi_{\rm cx}\big/\big[1 + \exp(-x/\xi)\big]^{1/2}. \tag{4.4.23}$$

The surface tension is written as

$$\sigma = TK\psi_{\rm cx}^2/8\xi = \frac{\sqrt{3}}{16}T(K/v_0)^{1/2}\xi^{-2}. \tag{4.4.24}$$

Thus, $\sigma \propto (T_{\rm t} - T)^2$ for $d \geq 3$, which was indeed confirmed for ^{3}He–^{4}He mixtures near the tricritical point [83].

4.4.4 Interface in polymer systems

We consider two-phase coexistence in polymer systems using the Flory–Huggins theory introduced in Section 3.5 and the gradient free energy (4.2.26) or (4.2.27) [85]–[88]. In all the representative cases we will study, the interface profile of the volume fraction $\phi(x)$ of the first component can be approximated by

$$\phi(x) = \phi_{\rm cx}^{(2)} + (\phi_{\rm cx}^{(1)} - \phi_{\rm cx}^{(2)})\big/\big[1 + \exp(-x/\ell)\big], \tag{4.4.25}$$

where ℓ is a suitably defined interface thickness. The solution tends to $\phi_{\rm cx}^{(1)}$ as $x \to \infty$ and $\phi_{\rm cx}^{(2)}$ as $x \to -\infty$. It obeys the differential equation,

$$(\phi_{\rm cx}^{(1)} - \phi_{\rm cx}^{(2)})\ell\frac{d\phi}{dx} = (\phi - \phi_{\rm cx}^{(2)})(\phi_{\rm cx}^{(1)} - \phi). \tag{4.4.26}$$

Semidilute polymer solutions

The phase diagram of polymer solutions is displayed in Fig. 3.12. The surface tension in polymer solutions behaves as (4.4.11) close to the critical point with ξ being scaled as (4.2.32), so it depends on N and $T - T_c$ as [32]

$$\sigma \sim T\xi^{-2} \sim Ta^{-2}N^{\nu-1}(1 - T/T_c)^{2\nu}. \tag{4.4.27}$$

Away from the critical point, a semidilute polymer solution with $\phi = \phi_{cx} > N^{-1/2}$ and a nearly pure solvent with $\phi = \phi_{dcx} \cong 0$ can coexist. The surface tension arises from a transition region of thickness $\xi \sim a/\phi_{cx}$ and is estimated as [29, 84]

$$\sigma \sim T\xi^{-2} \sim Ta^{-2}\phi_{cx}^2. \tag{4.4.28}$$

In the semidilute case, we use the second line of (3.5.5) as the free-energy density and set $K = 1/(12a\phi)$ from (4.2.27). Then (4.4.17) becomes

$$\frac{\phi}{N}\ln\phi + \left(\frac{1}{2} - \chi\right)\phi^2 + \frac{1}{6}\phi^3 - \frac{\Delta}{T}\phi - \frac{1}{T}(f_{site})_\infty = \frac{a^2}{24\phi}\left(\frac{d}{dx}\phi\right)^2, \tag{4.4.29}$$

where the two constants Δ/T and $(f_{site})_\infty/T$ are determined such that the left-hand side and its first derivative with respect to ϕ vanish as $x \to \pm\infty$. On the polymer-rich side, $\phi \to \phi_{cx} \equiv 3(\chi - 1/2)$ for $x \gg \xi$ with the correlation length,

$$\xi = \frac{1}{2}a\phi_{cx}^{-1} = \frac{1}{6}a(\chi - 1/2)^{-1}. \tag{4.4.30}$$

If ϕ_{cx} is considerably larger than ϕ_c, the volume fraction ϕ_{dcx} in the dilute region becomes very small as shown in (3.5.27). Then, we find

$$\Delta/T \cong -\frac{3}{8}(2\chi - 1)^2, \quad (f_{site})_\infty/T \cong 0. \tag{4.4.31}$$

It is obvious that the surface tension contribution arises from the spatial region where $\phi(x) \gg \phi_{dcx}$. We may then neglect the first and last terms on the left-hand side of (4.4.29) as

$$\phi^2(\phi/\phi_{cx} - 1)^2 \cong \xi^2\left(\frac{d}{dx}\phi\right)^2, \tag{4.4.32}$$

which is solved to give

$$\phi(x) = \phi_{cx}/[1 + \exp(-x/\xi)]. \tag{4.4.33}$$

Now, from (4.4.20), the surface tension is calculated as

$$\sigma = \frac{1}{24}T\phi_{cx}(a\xi)^{-1} = \frac{1}{12}Ta^{-2}\phi_{cx}^2, \tag{4.4.34}$$

in agreement with (4.4.27). Note that (4.2.27) has been used for the gradient free energy because $\xi < R_G$. Instead, if (4.2.26) had been used, σ would be multiplied by $1.5^{-1/2}$ to give $\sigma = 6^{-3/2}Ta^{-2}\phi_{cx}^2$.

Symmetric polymer blends

We first consider a symmetric polymer blend with $N_1 = N_2 = N$. In the mean field critical region $N^{-1} < \chi/\chi_c - 1 < 1$ in (4.2.39), the formula (4.4.10) gives [85]

$$\sigma = \frac{2}{3} T N^{-1/2} (\chi/\chi_c - 1)^{3/2} a^{-2}. \tag{4.4.35}$$

The right-hand side is estimated as (4.4.10) in terms of the Ginzburg number. In the asymptotic critical region $\chi/\chi_c - 1 < N^{-1}$ it is of the form,

$$\sigma \sim T\xi^{-2} \sim T N^{2\nu-2} (\chi/\chi_c - 1)^{2\nu} a^{-2}. \tag{4.4.36}$$

In the strongly segregated case $N\chi \gg 1$, ϕ_{cx} is very close to 0 or 1 as shown in (3.5.37). Using (3.5.29) for the free-energy density and (4.2.27) for the gradient free energy, we rewrite (4.4.17) as

$$\frac{1}{N}\big[\phi \ln \phi + (1-\phi)\ln(1-\phi)\big] + \chi\phi(1-\phi) - \frac{1}{T}(f_{\text{site}})_\infty = \frac{a^2}{24\phi(1-\phi)}\left(\frac{d}{dx}\phi\right)^2, \tag{4.4.37}$$

Here $(f_{\text{site}})_\infty/T$ is determined such that the right-hand side vanishes for $\phi = \phi_{cx}$, but it is estimated as $-2\exp(-N\chi)/N$ and is virtually zero. In this case we may omit the first and last terms on the left-hand side of (4.4.37) as [88]

$$\chi\phi^2(1-\phi)^2 = \frac{a^2}{24}\left(\frac{d}{dx}\phi\right)^2. \tag{4.4.38}$$

The interface profile is of the form,

$$\phi(x) = 1/\big[1 + \exp(-x/\ell)\big], \tag{4.4.39}$$

where

$$\ell = \frac{1}{\sqrt{24}}\chi^{-1/2} a \tag{4.4.40}$$

is the interface thickness. The above expression is valid in the region $|x| \lesssim N\chi\ell$, because the first term in (4.4.37) is smaller than the second in this region. If use is made of the second line of (4.4.20), σ is easily calculated as

$$\sigma = T\int_0^1 d\phi \sqrt{2K(\phi)v_0^{-1}\chi\phi(1-\phi)} = \frac{1}{\sqrt{6}} T\chi^{1/2} a^{-2}, \tag{4.4.41}$$

which agrees with the result of Helfand and Tagami [86]. If we were to use (4.2.26) as the gradient free energy, we would have $\sigma = \chi^{1/2} a^{-2}/3$ [87].

Asymmetric polymer blends

It is not difficult to examine σ in the asymmetric case $1 \ll N_2 \ll N_1$ using the results so far. We summarize its behavior:

$$\begin{aligned} \frac{a^2\sigma}{T} &\sim \chi^{1/2} && (\chi - \chi_c > 1/N_2), \\ &\sim (\chi-\chi_c)^2 N_2^{3/2} && (\sqrt{N_1/N_2} > \sqrt{N_1 N_2}(\chi-\chi_c) > 1), \\ &\sim (\chi-\chi_c)^{3/2} N_1^{-1/4} N_2^{5/4} && (1 > \sqrt{N_1 N_2}(\chi-\chi_c) > 1/N_2), \\ &\sim N_1^{\nu-1} N_2^{3\nu-1} (\chi-\chi_c)^{2\nu} && (1/N_2 > \sqrt{N_1 N_2}(\chi-\chi_c)), \end{aligned} \quad (4.4.42)$$

where $\chi_c \cong 1/2N_2$. (i) In the first line, the strong segregation limit is realized and (4.4.40) and (4.4.41) can be used. (ii) In the second line, ℓ exceeds the gyration radius $aN_2^{1/2}$ of the shorter chains. Then, the shorter chains act as solvent for the longer chains. As a result, the phase rich in the longer chains is analogous to the semidilute phase of polymer solutions. The correlation length there is $\xi = aN_2^{-1/2}/(\chi - \chi_c)$. (iii) In the third line, the mixture is in the mean field critical region, where (4.4.10) can be used. (iv) In the fourth line, $|T/T_c - 1| < Gi$ holds and asymptotic critical behavior is realized.

4.4.5 Thermal interface fluctuations

Surface undulations of a planar interface require only small free-energy cost and can be large at long wavelengths in equilibrium. We examine how $\beta\mathcal{H}$ in (4.1.1) is increased due to the deviation $\delta\psi(\boldsymbol{r}) = \psi(\boldsymbol{r}) - \psi_{\text{int}}(x)$. To the bilinear order we obtain

$$\beta\delta\mathcal{H} = \int d\boldsymbol{r}\delta\psi[-\nabla_\perp^2 + \hat{\mathcal{L}}(x)]\delta\psi, \quad (4.4.43)$$

where $\nabla_\perp^2 = \nabla^2 - \partial^2/\partial x^2$ is the laplacian operator in the yz plane. For the ψ^4 theory the operator,

$$\hat{\mathcal{L}}(x) = -K\frac{\partial^2}{\partial x^2} + |r|\left[3\tanh^2(x/2\xi) - 1\right], \quad (4.4.44)$$

is analogous to the Schrödinger operator in quantum mechanics. It is a nonnegative-definite hermitian operator, and its eigenvalues and eigenfunctions are completely known [77]. In particular, it has two discrete (or localized) eigenfunctions,

$$f_0(x) = (3/8\xi)^{1/2}\text{sech}^2(x/2\xi), \quad (4.4.45)$$

$$f_1(x) = (3/4\xi)^{1/2}\text{sech}(x/2\xi)\tanh(x/2\xi), \quad (4.4.46)$$

with the eigenvalues 0 and $3|r|/2$, respectively. The eigenfunctions with the continuous spectrum have eigenvalues larger than $2|r|$. Here notice that $f_0(x) \propto \psi'_{\text{int}}(x)$ where $\psi'_{\text{int}}(x) = d\psi_{\text{int}}(x)/dx$. In fact, differentiation of (4.4.3) with respect to x yields

$$\hat{\mathcal{L}}(x)\psi'_{\text{int}}(x) = 0. \quad (4.4.47)$$

Let the interface position be slightly displaced by $\zeta(\boldsymbol{r}_\perp)$ in the x direction, where $\zeta(\boldsymbol{r}_\perp)$ varies slowly on the surface. Then,

$$\delta\psi(\boldsymbol{r}) = \psi_{\rm int}(x-\zeta) - \psi_{\rm int}(x) \cong -\psi'_{\rm int}(x)\zeta. \tag{4.4.48}$$

Substitution of this form into (4.4.43) gives

$$\delta\mathcal{H} = \frac{1}{2}\sigma \int d\boldsymbol{r}_\perp |\nabla_\perp \zeta|^2. \tag{4.4.49}$$

Therefore, we obtain the well-known formula for the surface displacement fluctuations,

$$\int d\boldsymbol{r}_\perp \exp(i\boldsymbol{k}\cdot\boldsymbol{r}_\perp)\langle\zeta(\boldsymbol{r}_\perp)\zeta(\boldsymbol{0})\rangle = \langle|\zeta_{\boldsymbol{k}}|^2\rangle = \frac{T}{\sigma k^2}. \tag{4.4.50}$$

Here $\boldsymbol{r}_\perp = (y, z)$ is the position vector on the surface, $\boldsymbol{k}$ is the two-dimensional wave vector, and $\zeta_{\boldsymbol{k}}$ is the Fourier component.

The formula (4.4.50) has been derived near the critical point, but it holds even away from the critical point, as can be seen in the following argument. Regarding the surface as infinitesimally thin, the surface free energy is proportional to the surface area,

$$\begin{aligned} \mathcal{H} &= \sigma \int d\boldsymbol{r}_\perp \sqrt{1+|\nabla_\perp\zeta|^2} \\ &\cong \sigma \int d\boldsymbol{r}_\perp \left[1 + \frac{1}{2}|\nabla_\perp\zeta|^2\right]. \end{aligned} \tag{4.4.51}$$

The first line is obtained because the angle θ between the surface normal and the yz plane is $\cos\theta = 1/(1+|\nabla\zeta|^2)^{1/2}$ and the surface element is $d\boldsymbol{r}_\perp/\cos\theta$. The second line holds for small deformations.

In fluids, the gravity g is known to suppress the surface fluctuations with sizes longer than the so-called capillary length $a_{\rm ca}$. Let the x axis be in the reverse direction of gravity. Then the potential energy density per unit area is

$$\int_0^\zeta dx g(\Delta\rho)x = \frac{1}{2}g(\Delta\rho)\zeta^2, \tag{4.4.52}$$

where $\Delta\rho > 0$ is the mass density difference between the lower and upper phases. Thus (4.4.49) is modified as

$$\delta\mathcal{H} = \frac{1}{2}\int d\boldsymbol{r}_\perp [g(\Delta\rho)\zeta^2 + \sigma|\nabla_\perp\zeta|^2]. \tag{4.4.53}$$

The correlation length on the surface is given by the capillary length,

$$a_{\rm ca} = \sqrt{\sigma/g(\Delta\rho)}. \tag{4.4.54}$$

As is well known, this length provides the spatial scale on which the interface is deformed by gravity. It is a macroscopic length (say, 1 mm in water) far from the critical point on earth, while it decreases as $\xi^{-1+\beta/\nu}$ near the critical point but stays much longer than ξ in realistic experiments. The surface structure factor becomes

$$\langle|\zeta_{\boldsymbol{k}}|^2\rangle = \frac{T}{\sigma(a_{\rm ca}^{-2}+k^2)}. \tag{4.4.55}$$

As a result the surface position fluctuation at each point is

$$\langle\zeta(\boldsymbol{r}_\perp)^2\rangle = \frac{1}{2\pi}\int dkk\langle|\zeta_{\boldsymbol{k}}|^2\rangle = \frac{T}{2\pi\sigma}\ln(a_{\rm ca}/\xi), \tag{4.4.56}$$

where the upper limit of the k-integration is the inverse interface thickness ξ^{-1}. From (4.4.11) it follows that $\langle\zeta^2\rangle/\xi^2 \sim \ln(a_{\rm ca}/\xi)$ near the critical point.

4.4.6 Quantum interface fluctuations

The classical formulas (4.4.55)–(4.4.56) indicate that the interface fluctuations are weakened at low T. At very low temperatures the surface displacement ζ_k fluctuates quantum-mechanically. As a result, the surface structure factor $S_k = \langle|\zeta_k|^2\rangle$ is nonvanishing even for $T \to 0$. Here we assume that the low-temperature motion of ζ_k is described as a collective mode with the capillary-wave dispersion relation,

$$\omega_k = [\sigma k^3/\rho_{\rm ca}]^{1/2}, \tag{4.4.57}$$

where $\rho_{\rm ca}$ is an appropriately defined mass density and gravity is neglected. The surface tension σ is assumed to tend to a constant as $T \to 0$. We cite three observed examples.

(i) When a ^{4}He superfluid and a gas phase are separated by an interface, the capillary wave is also called *ripples* or *ripplons* [89]. In this case $\rho_{\rm ca}$ is nearly equal to the mass density of ^{4}He.

(ii) On a rough crystal–liquid surface of ^{4}He, crystallization and melting alternatively occur as the interface oscillates. This unusual oscillation is possible owing to the absence of latent heat and is called a *crystallization wave* [90, 91]. Here $\rho_{\rm ca} = (\rho_1 - \rho_2)^2/\rho_2$, where ρ_1 and ρ_2 are the mass densities of the solid and liquid phases, respectively.

(iii) Two liquid phases can coexist macroscopically in ^{3}He–^{4}He mixtures. For $T < 0.15$ K the ^{3}He-rich phase is virtually pure ^{3}He, while the ^{4}He-rich phase is a solution with the ^{3}He molar concentration X less than the upper limit X_ℓ, where $0.0637 < X_\ell < 0.094$ depending on the pressure. Here $\rho_{\rm ca}$ is nearly equal to the sum of the mass densities of the two coexisting phases as in the case of usual capillary waves on a fluid–fluid interface.

In this kind of problem, we should consider collective quantum motion, in which many particles participate. In the harmonic approximation, the hamiltonian is written in the Fourier space as [92]

$$\mathcal{H} = \frac{1}{2}\int_{\boldsymbol{k}}\left[\frac{\rho_{\rm ca}}{k}\dot{\zeta}_{\boldsymbol{k}}\dot{\zeta}_{-\boldsymbol{k}} + \sigma k^2\zeta_{\boldsymbol{k}}\zeta_{-\boldsymbol{k}}\right], \tag{4.4.58}$$

where $\dot{\zeta}_k = \partial\zeta_k/\partial t$ is the velocity of the surface displacement. Obviously, the kinetic energy is supported by incompressible flow induced around the interface. The Fourier component ζ_k is thus a harmonic oscillator with an effective mass $m_k = \rho_{ca}/k$, and its eigenfrequency is the capillary-wave frequency ω_k. The corresponding momentum p_k is defined by

$$p_k = \frac{\partial}{\partial \dot{\zeta}_{-k}} \mathcal{H} = \frac{\rho_{ca}}{k} \dot{\zeta}_k. \tag{4.4.59}$$

The equation of motion is given by

$$\frac{\partial}{\partial t} p_k = -\frac{\partial}{\partial \zeta_{-k}} \mathcal{H} = -\sigma k^2 \zeta_k. \tag{4.4.60}$$

These equations lead to the capillary-wave frequency (4.4.57). The quantization is to replace the momentum by

$$p_k = \frac{\hbar}{i} \frac{\partial}{\partial \zeta_{-k}}. \tag{4.4.61}$$

This procedure is analogous to that for phonons in low-temperature solids. In the canonical distribution the excited state with the energy $(n + \frac{1}{2})\hbar\omega_k$ of the harmonic oscillator is realized with the probability $P_n \equiv \exp(-\beta n \hbar\omega_k)[1 - \exp(-\beta\hbar\omega_k)]$, so that the equipartition of the energy between the kinetic and potential parts gives

$$\frac{1}{2} m_k \omega_k^2 \langle \zeta_k \zeta_{-k} \rangle = \frac{1}{2} \sum_{n=0}^{\infty} \left(n + \frac{1}{2}\right) \hbar\omega_k P_n. \tag{4.4.62}$$

Some further calculations yield the structure factor in the form [93],

$$S_k = \langle \zeta_k \zeta_{-k} \rangle = \frac{\hbar\omega_k}{2\sigma k^2} \coth\left(\frac{\hbar\omega_k}{2T}\right). \tag{4.4.63}$$

In the high-temperature limit $\hbar\omega_k \ll T$ the classical formula (4.4.50) is reproduced, while in the low-temperature limit $\hbar\omega_k \gg T$ we find

$$S_k = \frac{\hbar}{2\sqrt{\rho_{ca}\sigma}} \frac{1}{\sqrt{k}}, \tag{4.4.64}$$

which is the quantum fluctuation in the ground state. It is convenient to introduce a classical–quantum crossover wave number k_Q by $\hbar\omega_{k_Q} = T$. Then,

$$k_Q = (\rho_{ca} T^2/\hbar^2\sigma)^{1/3}, \tag{4.4.65}$$

which is $3 \times 10^4 T^{2/3}$ cm^{-1} for a solid–liquid interface of ^{4}He and $6 \times 10^5 T^{2/3}$ cm^{-1} for a liquid–liquid interface of ^{3}He–^{4}He with T in mK. For $k \gg k_Q$ the quantum effect is crucial and S_k is given by (4.4.63). The classical formula holds at long wavelengths $k \ll k_Q$. If $a_{ca}^{-1} \ll k_Q$ holds, the surface position fluctuation at one point, which has been measured in scattering experiments [94], is written as

$$\langle \zeta(\boldsymbol{r}_\perp)^2 \rangle \cong \frac{\hbar}{6\pi\sqrt{\rho_{ca}\sigma}} \Lambda^{3/2} + \frac{T}{2\pi\sigma} \ln(k_Q a_{ca}). \tag{4.4.66}$$

The first term represents the zero-point vibration amplitude, and Λ is the upper cut-off wave number ($\sim 10^8$ cm^{-1}) assumed to be larger than k_Q. Because the ratio of the first to second term is of order $(\Lambda/k_Q)^{3/2}$, the quantum contribution dominates over the thermal one ($\propto T$) at sufficiently low temperatures where $\Lambda \gg k_Q$.

4.5 Vortices in systems with a complex order parameter

In systems with a complex order parameter ($n = 2$) below the transition temperature ($r < 0$), a famous topological singularity is a vortex line (point) in 3D (2D) [11, 95]. In particular, in 2D xy models, vortex binding can cause the Kosterlitz and Thouless transition [11, 96]. In Section 8.10 we will examine vortex motion on the basis of the results in the present section. There can be a number of other topological defects in many-component systems ($n \geq 2$) with broken continuous symmetry for each set of n and d [11]. They play crucial roles in phase-ordering processes, as will be studied in Section 8.1.

4.5.1 Fundamental vortex solutions

Let us consider a rectilinear vortex aligned along the z direction in 3D and a vortex point in 2D. The vortex solution is written as [35, 16]

$$\psi_v(x, y) = f(\rho)e^{i\ell\varphi}, \quad (\ell = \pm 1, \ldots), \tag{4.5.1}$$

where $\rho = (x^2 + y^2)^{1/2}$ (which should not be confused with the mass density) and $\varphi = \tan^{-1}(y/x)$. The integer ℓ will be called the charge here, while it is called the winding number in the literature. From the minimum condition $\delta(\beta\mathcal{H})/\delta\psi^* = 0$ of the GLW hamiltonian (4.1.1), we obtain

$$(-\kappa^2 + u_0|\psi_v|^2)\psi_v - \left(\frac{\partial^2}{\partial x^2} + \frac{\partial^2}{\partial y^2}\right)\psi_v = 0, \tag{4.5.2}$$

where we have set $r = -\kappa^2$, $K = 1$, and $h = 0$. We notice that the amplitude $f = |\psi_v|$ is scaled as

$$f = MA_0(\kappa\rho), \tag{4.5.3}$$

where $M = \kappa/u_0^{1/2}$ is the equilibrium average order parameter. Then $A_0(s)$ satisfies

$$\left[\frac{d^2}{ds^2} + \frac{1}{s}\frac{d}{ds} - \frac{\ell^2}{s^2} + 1 - A_0^2\right]A_0 = 0. \tag{4.5.4}$$

It is easy to check the behaviors, $A_0 \sim s^{|\ell|}$ for $s \ll 1$ and $A_0 \cong 1 - \ell^2/s^2$ for $s \gg 1$. In Fig. 4.6 we plot $A_0(s)$ for $\ell = 1$ obtained numerically. As a result, the increase in the free-energy density decays as ρ^{-2} far from the vortex center, and the free-energy increase

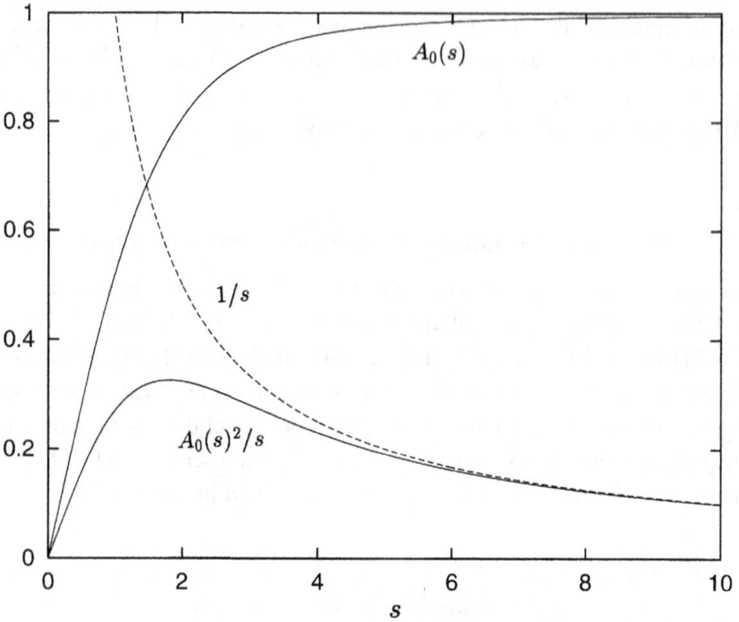

Fig. 4.6. The dimensionless amplitude $A_0(s)(\propto \rho_s^{1/2})$ and superfluid current $A_0(s)^2/s(\propto |\boldsymbol{J}_s|)$ around a vortex for $\ell = 1$, where $s = (x^2 + y^2)^{1/2}/\sqrt{2}\xi$ is the dimensionless distance from the vortex center. Here $A_0 \cong 0.58s$ for $s \ll 1$ and $A_0 \cong 1 - 1/s^2$ for $s \gg 1$. The dashed line represents $1/s(\propto |\boldsymbol{v}_s|)$.

per unit length is logarithmically dependent on the upper cut-off $R_{\max}$ as

$$\begin{aligned} E_{v\ell} &= \pi T \int_0^{R_{\max}} d\rho\rho \left[\frac{1}{2} u_0 (M^2 - f^2)^2 + \frac{\ell^2}{\rho^2} f^2 + \left(\frac{\partial f}{\partial \rho} \right)^2 \right] \\ &= \pi T M^2 \left[\ell^2 \ln \left(\frac{C_\ell R_{\max}}{\sqrt{2}\xi} \right) + O(\xi^2 / R_{\max}^2) \right], \end{aligned} \tag{4.5.5}$$

where $\xi = (\sqrt{2}\kappa)^{-1}$, and $C_1/\sqrt{2} = 1.46/\sqrt{2} \cong 1$ for $\ell = 1$ [35, 97]. If there is a single rectilinear vortex, $R_{\max}$ is of the order of the system dimension. However, if there are other vortices with opposite charges in 2D or with different directions of the tangential vector in 3D, the cut-off length becomes the characteristic distance among vortices. In 3D, the free energy for an assembly of weakly curved vortex lines with $\ell = \pm 1$ may be approximated as

$$\mathcal{H}_v^{(0)} = E_{v1} L_T = \pi T M^2 \ln(R_{\max}/\xi) L_T, \tag{4.5.6}$$

where L_T is the total length of the lines. The interaction among different line elements will be taken into account later.

In ^{4}He, the superfluid density ρ_s is expressed as (4.3.107) and the superfluid velocity $\boldsymbol{v}_s$ is equal to $(\hbar/m_4)\nabla\theta$. Around a rectilinear vortex they are of the forms,

$$\rho_s = \bar{\rho}_s A_0(\kappa\rho)^2, \tag{4.5.7}$$

$$\boldsymbol{v}_s = \frac{\hbar\ell}{m_4\rho}\boldsymbol{e}_\varphi, \tag{4.5.8}$$

where $\bar{\rho}_s = (m_4^2/\hbar^2)TM^2$ is the superfluid density far from the vortex center, m_4 being the ^{4}He mass, and $\boldsymbol{e}_\varphi = (-y/\rho, x/\rho, 0)$ is the unit vector perpendicular to $\boldsymbol{e}_\rho = (x/\rho, y/\rho, 0)$. The superfluid current is given by

$$\boldsymbol{J}_s = \rho_s\boldsymbol{v}_s = \ell\bar{\rho}_s(A_0^2/\rho)\boldsymbol{e}_\varphi, \tag{4.5.9}$$

whose profile can be seen in Fig. 4.6. The kinetic energy $E_K^{(s)}$ of the superfluid component is the space integral of $\rho_s v_s^2/2$. Around a single vortex we have $E_K^{(s)} \cong \pi M^2\ell^2 \ln(R_{max}/\xi)$, so $E_{v\ell} \cong E_K^{(s)}$ for $R_{max} \gg \xi$.

We next examine the circulation around a vortex line. From (4.5.8) we have

$$\text{rot}\,\boldsymbol{v}_s = \frac{2\pi\hbar}{m_4}\ell\delta^{(2)}(\boldsymbol{r}_\perp)\boldsymbol{e}_z, \tag{4.5.10}$$

where $\boldsymbol{r}_\perp = (x, y)$ is the 2D vector, $\delta^{(2)}$ is the 2D δ-function, and $\boldsymbol{e}_z = (0, 0, 1)$ is the unit vector along the z axis, so that[16]

$$\oint d\boldsymbol{r}\cdot\boldsymbol{v}_s = \int_0^{2\pi} d\varphi\rho\boldsymbol{e}_\varphi\cdot\boldsymbol{v}_s = \frac{2\pi\hbar}{m_4}\ell. \tag{4.5.11}$$

Vortex ring

In real 3D systems vortices appear either in the form of lines with ends attached to the boundary wall or in the form of closed rings. Figure 4.7 illustrates a vortex ring. If its radius R is much larger than the core radius ($\sim \xi$) and $\ell = 1$, the free energy needed to create such a vortex ring is expressed as [98]

$$E_{ring} = 2\pi^2 TM^2 R\left[\ln(8R/\sqrt{2}\xi) - 1.62\right]. \tag{4.5.12}$$

For vortex rings in ideal incompressible fluids we obtain nearly the same result but with 1.62 being replaced by 7/4. In Chapter 8 we shall see that generation of vortex rings leads to a decay of superfluid flow in ^{4}He.

4.5.2 Interaction between vortices

Because the phase modulation around vortices is far-reaching, the interaction between vortices is very long-ranged.

[16] In the literature $2\pi\hbar/m_4 \cong 10^{-3}$erg s/g is usually written as κ.

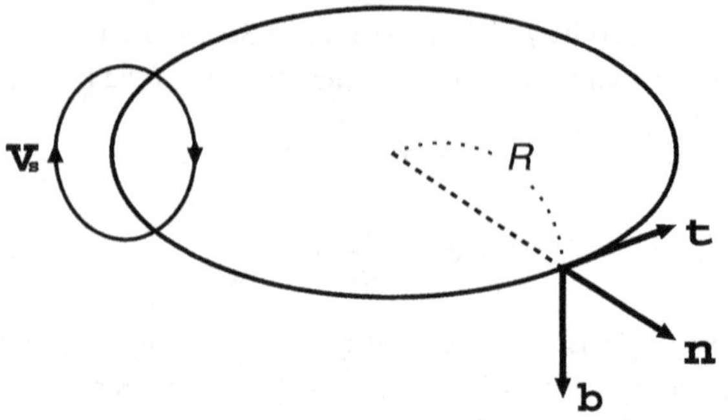

Fig. 4.7. A vortex ring with radius R in a superfluid. Here $\boldsymbol{t}$ is the tangential unit vector, $\boldsymbol{n}$ is the normal unit vector, and $\boldsymbol{b} = \boldsymbol{t} \times \boldsymbol{n}$ is orthogonal to $\boldsymbol{t}$ and $\boldsymbol{n}$.

(i) We first consider 2D xy-like systems with N_v vortices with charges ± 1 at $\boldsymbol{R}_j = (X_j, Y_j)$ $(j = 1, 2, \ldots, N_v)$. Because the phase modulation due to vortices is

$$\theta_v = \sum_j \ell_j \tan^{-1}\big[(y - Y_j)/(x - X_j)\big] \tag{4.5.13}$$

far from the vortex cores, the free energy ($\cong TM^2 \int d\boldsymbol{r} |\nabla\theta|^2/2$) may be written as [96, 11]

$$\mathcal{H}_v = -\pi T M^2 \sum_{i \neq j} \ell_i \ell_j \ln(|R_i - R_j|/\xi) + E_c N_v, \tag{4.5.14}$$

where E_c is the core (free) energy playing the role of a chemical potential of the vortices. Note that we may superimpose an arbitrary smooth, nonsingular phase modulation θ_s as $\theta = \theta_v + \theta_s$. Then the total free energy becomes the sum of $\mathcal{H}_v$ and $M^2 \int d\boldsymbol{r} |\nabla\theta_s|^2/2$ without the cross term ($\propto \theta_v\theta_s$). Because of this fact, we have neglected the smooth part in (4.5.14). Kosterlitz and Thouless [96] developed a renormalization group theory on the vortex hamiltonian (4.5.14) in 2D, in which small vortex pairs are coarse-grained in a step-wise manner giving rise to renormalization of M^2 and E_v in the long-wavelength limit.

(ii) In 3D systems, the vortices are represented by the lines $\boldsymbol{R}_j(s)$, where s is the arclength and j $(= 1, 2, \ldots)$ indicates the jth vortex. To avoid cumbersome notation we will suppress j, but the summation over different lines is implied in the following expressions. In the notation for ^{4}He the vorticity vector is defined by

$$\boldsymbol{\omega}(\boldsymbol{r}) = \frac{2\pi\hbar}{m_4} \int ds \boldsymbol{t}(s) \delta^{(3)}(\boldsymbol{r} - \boldsymbol{R}(s)), \tag{4.5.15}$$

where $\boldsymbol{t}(s) = d\boldsymbol{R}(s)/ds$ is the tangential unit vector at the point $\boldsymbol{r}(s)$.[17] Generalization of

[17] For ideal incompressible fluids, $2\pi\hbar/m_4$ should be replaced by the circulation of vortex lines.

the circulation theorem (4.5.10) yields

$$\mathrm{rot}\,\boldsymbol{v}_s = \boldsymbol{\omega}. \tag{4.5.16}$$

With the aid of the Biot–Savart law in electromagnetic theory the superfluid velocity due to vortices is obtained as

$$\begin{aligned} \boldsymbol{v}_s(\boldsymbol{r}) &= \frac{1}{4\pi}\int d\boldsymbol{r}' \frac{1}{|\boldsymbol{r}-\boldsymbol{r}'|^3}\boldsymbol{\omega}(\boldsymbol{r}') \times (\boldsymbol{r}-\boldsymbol{r}') \\ &= \frac{\hbar}{2m_4}\int ds' \frac{1}{|\boldsymbol{r}-\boldsymbol{R}(s')|^3}\boldsymbol{t}(s') \times (\boldsymbol{r}-\boldsymbol{R}(s')). \end{aligned} \tag{4.5.17}$$

The superfluid kinetic energy is written as

$$E_K^{(s)} = \frac{1}{2}\bar{\rho}_s \int d\boldsymbol{r}|\boldsymbol{v}_s|^2 = \frac{1}{8\pi}\bar{\rho}_s \int d\boldsymbol{r} \int d\boldsymbol{r}' \frac{1}{|\boldsymbol{r}-\boldsymbol{r}'|}\boldsymbol{\omega}(\boldsymbol{r}) \cdot \boldsymbol{\omega}(\boldsymbol{r}'), \tag{4.5.18}$$

where $\bar{\rho}_s = (m_4^2/\hbar^2)TM^2$ and use has been made of $\nabla \cdot \boldsymbol{\omega} = 0$. The total vortex free energy $\mathcal{H}_v$ is the sum of $E_K^{(s)}$ and the core free energy. After some calculations we obtain

$$\mathcal{H}_v = \frac{\pi\hbar^2}{2m_4^2}\bar{\rho}_s \int ds \int ds' \frac{1}{|\boldsymbol{R}(s)-\boldsymbol{R}(s')|}\boldsymbol{t}(s)\cdot\boldsymbol{t}(s') + E_c L_T, \tag{4.5.19}$$

where $L_T = \int ds$ is the total line length, and the line integrations should be performed in the regions $|\boldsymbol{R}_i(s) - \boldsymbol{R}_j(s')| > \xi_c$. The lower cut-off ξ_c is taken to be a few times larger than ξ. Then, if only a single vortex is present, we have $\mathcal{H}_v = L_T[\pi TM^2 \ln(R_{max}/\xi_c) + E_c]$. Comparing this expression with (4.5.5) we may estimate the core free energy as

$$E_c \cong \pi TM^2 \ln(\xi_c/\xi) = \frac{\pi\hbar^2}{m_4^2}\bar{\rho}_s \ln(\xi_c/\xi), \tag{4.5.20}$$

under which $\mathcal{H}_v$ becomes insensitive to the choice of ξ_c.

4.5.3 Fluid velocity at a vortex point

We are interested in the superfluid velocity at a point $\boldsymbol{R}(s)$ on a vortex line,

$$\boldsymbol{v}_{s1}(s) = \lim_{\boldsymbol{r}\to\boldsymbol{R}(s)} \boldsymbol{v}_s(\boldsymbol{r}). \tag{4.5.21}$$

The vortex moves with this velocity if there is no friction (in ideal incompressible fluids or in ^{4}He at nearly zero temperature). It is not difficult to derive the following general relation,

$$\frac{\delta}{\delta \boldsymbol{R}(s)}\mathcal{H}_v = \frac{2\pi\hbar}{m_4}\bar{\rho}_s \boldsymbol{t}(s) \times \boldsymbol{v}_{s1}(s), \tag{4.5.22}$$

which follows from (4.5.19). The derivation of this relation becomes easier if each curve is parameterized in terms of a parameter ζ such as the coordinate along an appropriate axis rather than the arclength s. Then $ds = d\zeta(\partial s/\partial\zeta)$. For example, with respect to a small deformation of a curve, $\boldsymbol{R} \to \boldsymbol{R}+\delta\boldsymbol{R}$, the line length $L = \int ds = \int d\zeta |d\boldsymbol{R}/d\zeta|$ changes as

$\delta L = \int d\zeta \boldsymbol{t} \cdot d[\delta \boldsymbol{R}]/d\zeta = -\int ds \mathcal{K} \boldsymbol{n} \cdot \delta \boldsymbol{R}$. With this relation, if $\boldsymbol{R}$ is regarded as a function of s (not ζ), we find

$$\frac{\delta}{\delta \boldsymbol{R}(s)} L = -\mathcal{K} \boldsymbol{n}. \tag{4.5.23}$$

The functional derivative of the first term (4.5.19) can also be performed similarly, though somewhat complicated, leading to (4.5.22).

Arms–Hama approximation

Although the general nonlocal form (4.5.17) for $\boldsymbol{v}_s$ looks formidable, Arms and Hama [99] noticed that most important region is the line portion close to $\boldsymbol{R}(s)$ in the s'-integration in (4.5.17). That is, in the second line of (4.5.17) we set $|\boldsymbol{R}(s) - \boldsymbol{R}(s')| \cong |s' - s|$ and $\boldsymbol{R}(s') = \boldsymbol{R}(s) + (s' - s)\boldsymbol{t}(s) + \frac{1}{2}(s' - s)^2 \mathcal{K} \boldsymbol{n}(s) + \cdots$, assuming small $s' - s$, where $\boldsymbol{n}$ is the normal unit vector and $\mathcal{K}$ is the line curvature. Then the integral is logarithmically divergent and we get the local self-induced velocity,

$$\begin{aligned} \boldsymbol{v}_{s1} &\cong \frac{\hbar}{4m_4} \int ds' \frac{1}{|s - s'|} \boldsymbol{t}(s) \times \mathcal{K}(s) \boldsymbol{n}(s) \\ &\cong \frac{\hbar}{2m_4} \ln(R_{\max}/\xi) \mathcal{K} \boldsymbol{b}, \end{aligned} \tag{4.5.24}$$

where $\boldsymbol{b} = \boldsymbol{t} \times \boldsymbol{n}$. This approximation is valid with errors of order 10%, but much simplifies the calculation of vortex motion, as will be shown in Section 8.10. Here we should note that if $\mathcal{H}_v$ in (4.5.22) is replaced by $\mathcal{H}_v^{(0)}$ in (4.5.6), we may readily reproduce (4.5.24). It is obvious that the Arms–Hama approximation is equivalent to neglecting the vortex interaction among distant line elements and setting $\mathcal{H}_v = \mathcal{H}_v^{(0)}$ in (4.5.22).

Appendix 4A Calculation of the critical exponent η

We calculate the following integral at $d = 4$,

$$\phi(k) = \int_{q_1} \int_{q_2} \frac{1}{q_1^2 q_2^2 |\boldsymbol{q}_1 + \boldsymbol{q}_2 - \boldsymbol{k}|^2}, \tag{4A.1}$$

where $q_1 < \Lambda$, $q_2 < \Lambda$, and $|\boldsymbol{q}_1 + \boldsymbol{q}_2 - \boldsymbol{k}| < \Lambda$. Using $\int_{\boldsymbol{\ell}} \exp(i\boldsymbol{\ell} \cdot \boldsymbol{m}) = \delta(\boldsymbol{m})$ for any $\boldsymbol{m}$, we rewrite this integral as

$$\begin{aligned} \phi(k) &= (2\pi)^4 \int_{q_1} \int_{q_2} \int_{q_3} \frac{\delta(\boldsymbol{q}_1 + \boldsymbol{q}_2 + \boldsymbol{q}_3 - \boldsymbol{k})}{q_1^2 q_2^2 q_3^2} \\ &= (2\pi)^4 \int_{\boldsymbol{\ell}} e^{i\boldsymbol{\ell} \cdot \boldsymbol{k}} \varphi(\ell)^3, \end{aligned} \tag{4A.2}$$

where

$$\varphi(\ell) = (2\pi)^{-4} \int_{q<\Lambda} d\boldsymbol{q} \frac{1}{q^2} \exp(-i\boldsymbol{\ell} \cdot \boldsymbol{q}) = \frac{2K_4}{\ell^2} [1 - J_0(\Lambda \ell)]. \tag{4A.3}$$

Hereafter, $J_n(z)$ $(n = 0, 1, \ldots)$ represents the Bessel function of the nth order. After the angle integration of ℓ we obtain

$$\phi(k) = 4K_4^2 \int_0^\infty d\ell \frac{J_1(k\ell)}{k\ell^4} \left[1 - J_0(\Lambda\ell)\right]^3. \tag{4A.4}$$

In particular, because $J_1(z) = \frac{1}{2}z - \frac{1}{16}z^3 + \cdots$ for $|z| \ll 1$, we have $\phi(0) = 0.214 K_4 \Lambda^2$ as $k \to 0$, and

$$\phi(0) - \phi(k) = 2K_4^2 \int_0^\infty d\ell \frac{1}{\ell^3}\left[1 - 2J_1(k\ell)/k\ell\right]\left[1 - J_0(\Lambda\ell)\right]^3. \tag{4A.5}$$

In the region $\Lambda^{-1} \ll \ell \ll k^{-1}$ the integrand of (4A.5) may be set equal to $k^2/8\ell$, so that

$$\phi(0) - \phi(k) = \frac{1}{4}K_4^2 k^2 \ln(\Lambda/k) + \cdots, \tag{4A.6}$$

which leads to (4.1.43).

Appendix 4B Random phase approximation for polymers

Let us first consider a gaussian chain with polymerization index N. Because the monomer positions $\boldsymbol{R}_i$ $(1 \le i \le N)$ on the chain satisfy $\langle |\boldsymbol{R}_i - \boldsymbol{R}_j|^2 \rangle = |i - j| a^2$, the single-chain structure factor (per volume v_0) becomes

$$I_0(q) = \frac{1}{N}\sum_{ij} \exp\left(-\frac{1}{6}a^2 q^2 |i - j|\right) = N f_{\rm D}(N a^2 q^2/6), \tag{4B.1}$$

where

$$f_{\rm D}(X) = \frac{1}{X}\left[1 - \frac{1}{X}(1 - e^{-X})\right] \tag{4B.2}$$

is called the Debye function [28]. Next we consider a mixture of two species of chains with volume fractions $\phi_1 = \phi$ and $\phi_2 = 1 - \phi$ and polymerization indices N_1 and N_2. If we set $N_2 = 1$, the results for polymer solutions are obtained. The random phase approximation gives the inverse of the structure factor in the form [29, 30],

$$I(q)^{-1} = \left[\phi_1 N_1 f_{\rm D}(N_1 a^2 q^2/6)\right]^{-1} + \left[\phi_2 N_2 f_{\rm D}(N_2 a^2 q^2/6)\right]^{-1} - 2\chi. \tag{4B.3}$$

Because $f_{\rm D}(X) \cong 1 - X/3$ for $X \ll 1$, the small-q behavior is given by

$$I(q)^{-1} \cong \frac{1}{\phi_1 N_1} + \frac{1}{\phi_2 N_2} - 2\chi + \frac{a^2}{18\phi_1\phi_2}q^2. \tag{4B.4}$$

Because $f_{\rm D}(X) \cong 2/X$ for $X \gg 1$, the large-q behavior becomes

$$I(q)^{-1} \cong \frac{a^2}{12\phi_1\phi_2}q^2 - 2\chi. \tag{4B.5}$$

These expressions yield the structure factor from the Flory–Huggins theory supplemented with the gradient term in the form of (4.2.26) at small q and (4.2.27) at large q.

Appendix 4C Renormalization group equations for n-component systems

We extend the calculations in Section 4.3 to n-component systems. It is easy to check the following RG equations at the critical point:

$$\frac{\partial}{\partial \ell} \ln r = \frac{\partial}{\partial \ell} \ln(\gamma_0 C_0) = -(n+2)g, \tag{4C.1}$$

$$\frac{\partial}{\partial \ell} g = \epsilon g - (n+8)g^2, \tag{4C.2}$$

$$\frac{\partial}{\partial \ell} C_0 = 2n v C_0, \tag{4C.3}$$

where $\Lambda = \Lambda_0 e^{-\ell}$. We may solve (4C.2) in the same form as (4.3.17), with $Q(\ell)$ being defined by (4.3.15). Then (4C.1) is integrated as

$$r(\ell) = a_{00}\tau\left[1 + (e^{\epsilon\ell} - 1)g_0/g^*\right]^{-(n+2)g^*/\epsilon}. \tag{4C.4}$$

The RG equation for v is obtained from (4C.1) and (4C.3) as

$$\frac{\partial}{\partial \ell} v = \epsilon v - 2(n+2)gv - 2nv^2. \tag{4C.5}$$

This equation is solved in the form [55],

$$v(\ell) = v^* e^{\epsilon\ell}/\left[Q(\ell) + w_0 Q(\ell)^{1-\alpha/\nu\epsilon}\right], \tag{4C.6}$$

where

$$w_0 = (g_0 v^*/g^* v_0 - 1)(g^*/g_0)^{\alpha/\nu\epsilon}. \tag{4C.7}$$

Substitution of the above result into (4C.3) gives

$$C_0(\ell) = C_{00}(v_0/v^*)(g^*/g_0)^{1-\alpha/\nu\epsilon}\left[Q(\ell)^{\alpha/\nu\epsilon} + w_0\right], \tag{4C.8}$$

where $C_{00} = C_0(\Lambda_0)$. At $\ell = 0$, the right-hand sides of (4C.6) and (4C.8) are clearly equal to v_0 and C_{00}, respectively, from $Q(0) = g^*/g_0$. If g_0 is not very small, we may set $Q(\ell) = e^{\epsilon\ell}$ to obtain

$$v(\ell) = v^*/\left[1 + w_0 e^{-\alpha\ell/\nu}\right], \tag{4C.9}$$

$$C_0(\ell) = \frac{1}{4v^*} K_d \xi_{+0}^{-d} (\xi_{+0}\Lambda_0)^{-\alpha\ell/\nu}\left[e^{\alpha\ell/\nu} + w_0\right], \tag{4C.10}$$

where C_{00} is eliminated in favor of ξ_{+0} as in (4.3.43). When τ is very small, the crossover occurs at $\Lambda = \kappa$, leading to the critical behavior,

$$C_H = A_0 \tau^{-\alpha} + C_B, \tag{4C.11}$$

with

$$A_0 = \frac{n}{4\alpha} K_d \xi_{+0}^{-d}, \quad C_B = C_{00}(1 - g^* v_0/v^* g_0). \tag{4C.12}$$

The two-scale-factor universality (2.1.45) becomes

$$R_\xi = \lim_{\tau \to 0} \xi(\alpha\tau^2 C_H)^{1/d} = \left(\frac{n}{4}K_d\right)^{1/d}. \tag{4C.13}$$

We can see that (4.3.49) holds for general n.

^{4}He near the superfluid transition

In Section 2.4 we explained critical behavior of ^{4}He near the superfluid transition. From (2.4.2) and (2.4.14) C is related to C_p as $C = C_p - A_\lambda^2 Q_0 = A\ln(\tau_0/\tau)$, where C is equal to C_H in the present notation and τ_0 is a constant. With this experimental result, let us take the limit $\alpha \to 0$ in the above RG results. Comparison of (2.4.2) and (4C.11) gives

$$A = A_0\alpha \cong \frac{1}{2}K_d\xi_{+0}^{-d}, \quad A\ln\tau_0 = B - A_\lambda^2 Q_0 = A_0 + C_{\rm B}, \tag{4C.14}$$

above T_λ. The first relation agrees with the two-scale-factor universality relation (2.4.4). We also examine the ℓ-dependence of $C_0(\ell)$ and $v(\ell)$ because such results will be needed in (6.6.71) below. For $\Lambda \gg \kappa$ we use (4C.9) and (4C.10) to obtain

$$C_0(\ell) = \nu^{-1}A\ln(\Lambda_0/\Lambda), \quad v(\ell) \cong \frac{1}{4}[\ln(\Lambda_0/\Lambda)]^{-1}, \tag{4C.15}$$

where $\Lambda_0 = \tau_0^\nu/\xi_{+0}$. For $\Lambda \ll \kappa$ the general formula (4.1.58), which is valid for any n, yields

$$v(\ell) = \gamma^2 K_d \tau^{2(\gamma-1)}/4\Gamma_0^2 C\Lambda^\epsilon \cong \frac{1.28}{4\nu\ln(\tau_0/\tau)}(\xi\Lambda)^{-\epsilon}. \tag{4C.16}$$

where we use $\Gamma_0 \cong \xi_{+0}^{-2}$ and $R_\xi = \xi_{+0}A^{1/d} \cong 0.36$ at $d = 3$.

Appendix 4D Calculation of a free-energy correction

To derive (4.3.64) we calculate the following integral at $d = 4$,

$$J(r) = \int_k \ln[(r + k^2 + \Lambda^{-2}k^4)/(k^2 + \Lambda^{-2}k^4)], \tag{4D.1}$$

where we impose a smooth cut-off using $\mathcal{H}_{\rm eff}$ in (4.3.55). The fluctuation contribution to the free-energy density is given by $[J(r_{\rm L}) + (n-1)J(r_{\rm T})]/2$. Differentiating twice with respect to r, we obtain

$$\frac{\partial^2}{\partial r^2}J(r) = -\int_k (r + k^2 + \Lambda^{-2}k^4)^{-2} = \frac{1}{2}K_4[\ln(r/\Lambda^2) + 2]. \tag{4D.2}$$

Integrating with respect to r we find two contributions,

$$J(r) = A_{\rm c}r + \frac{1}{4}K_4 r^2 \ln(e^{1/2}r/\Lambda^2), \tag{4D.3}$$

where $A_{\rm c} = \langle\psi_1^2\rangle$ at $r = 0$. The first term in (4D.3) gives rise to the contribution $A_{\rm c}nr/2 - r_{0{\rm c}}M^2/2$ to the free-energy density, where $r_{0{\rm c}} = -A_{\rm c}(n+2)u_0$ is the shift of the (scaled) critical temperature as given by (4.1.15) for $n = 1$. Here the term linear in r is regular, as stated in footnote 11 at (4.3.53), while the term proportional to r_{0c} is canceled to vanish in (4.3.64) from the mass renormalization (4.1.17). (Notice that the Landau free-energy density is written as $r_0 M^2/2 + \cdots$ with $r_0 = r + r_{0c}$.) We thus obtain the second line of (4.3.64). Instead, if a sharp cut-off at Λ is assumed, the argument of the logarithm in (4D.3) is changed to $r/(e^{1/2}\Lambda^2)$.

Appendix 4E Calculation of the structure factors

We calculate the structure factor $I(k)$ for general $M = \langle\psi\rangle$ in Ising-like systems. The correlation function on the right-hand side of (4.1.30) is rewritten as

$$\langle\psi(\mathbf{r})^3\varphi(\mathbf{0})\rangle = 3M^2\langle\varphi(\mathbf{r})\varphi(\mathbf{0})\rangle + 3M\langle\varphi(\mathbf{r})^2\varphi(\mathbf{0})\rangle + \langle\varphi(\mathbf{r})^3\varphi(\mathbf{0})\rangle, \tag{4E.1}$$

where $\varphi = \psi - M$ is the deviation. To first order in ϵ we may set

$$\langle\varphi(\mathbf{r})^3\varphi(\mathbf{0})\rangle = 3\langle\varphi(\mathbf{r})^2\rangle\langle\varphi(\mathbf{r})\varphi(\mathbf{0})\rangle, \tag{4E.2}$$

$$\langle\varphi(\mathbf{r})^2\varphi(\mathbf{0})\rangle = -6u_0 M \int d\mathbf{r}'\langle\varphi(\mathbf{r})\varphi(\mathbf{r}')\rangle^2\langle\varphi(\mathbf{r}')\varphi(\mathbf{0})\rangle. \tag{4E.3}$$

Note that the free-energy density contains the cubic term $u_0 M\varphi^3$, which leads to (4E.3). After some calculations we obtain

$$1/I(k) = r_0 + 3u_0M^2 + k^2 + 3u_0\langle\varphi^2\rangle - 18gu_0M^2J_s(k), \tag{4E.4}$$

where $r_L = r + 3u_0M^2$ and

$$\begin{aligned} J_s(k) &= K_4^{-1}\int_{\mathbf{q}} \frac{1}{(q^2 + r_L)(|\mathbf{q} - \mathbf{k}|^2 + r_L)} \\ &= -\frac{1}{2}(\ln r_L + 1) - \frac{1}{12}r_L^{-1}k^2 + \cdots. \end{aligned} \tag{4E.5}$$

The second line is the expansion valid for $k^2 \ll r_L$. Substitution of (4E.5) into (4E.4) gives (4.3.87).

Next we consider a many-component system. Let us calculate the structure factor for the transverse component ϕ_2. Analogous to (4E.4), we obtain

$$1/I_T(k) = r_0 + u_0M^2 + k^2 + u_0\big(\langle\delta\psi_1^2\rangle + (n-1)\langle\psi_2^2\rangle\big) - 4gu_0M^2J(k), \tag{4E.6}$$

where

$$\begin{aligned} J(k) &= K_4^{-1}\int_{\mathbf{q}} \frac{1}{q^2(|\mathbf{q} - \mathbf{k}|^2 + r_L)} \\ &= -\frac{1}{2}(\ln r_L + 1) - \frac{1}{4}r_L^{-1}k^2 + \cdots, \end{aligned} \tag{4E.7}$$

for small k. The second term $\propto k^2$ leads to the correction to K_R in (4.3.97).

Appendix 4F Specific heat in two-phase coexistence

We examine the specific heat when liquid and gas regions macroscopically coexist in a cell with a fixed total volume V [24] The mass densities, $\rho_\ell = m_0 n_\ell$ and $\rho_g = m_0 n_g$, and the masses, M_ℓ and M_g, are related to the volume V as

$$\frac{1}{\rho_\ell} M_\ell + \frac{1}{\rho_g} M_g = V. \tag{4F.1}$$

Here quantities with the subscript ℓ (g) are those of the liquid (gas) phase. We then change the temperature T infinitesimally to $T + \delta T$. When V is fixed, M_ℓ and M_g change as $M_\ell \to M_\ell + \delta M_\ell$ and $M_g \to M_g + \delta M_g$. Here $\delta M_\ell + \delta M_g = 0$ and

$$\delta V = \left(\frac{1}{\rho_\ell} - \frac{1}{\rho_g} \right) \delta M_\ell + M_\ell \delta \left(\frac{1}{\rho_\ell} \right) + M_g \delta \left(\frac{1}{\rho_g} \right) = 0. \tag{4F.2}$$

This mass conversion occurs at the interface and takes a long time. In the final stage, the pressure change is given by $\delta p = (\partial p/\partial T)_{\text{cx}} \delta T$, because the final state is again on the coexistence curve. We are interested in the total entropy change,

$$\delta S_{\text{total}} = (s_\ell - s_g)\delta M_\ell + M_\ell \delta s_\ell + M_g \delta s_g, \tag{4F.3}$$

where s_ℓ and s_g are the entropies per unit mass. The specific heat in two-phase coexistence per unit volume is defined by

$$V(C_V)_{\text{cx}} = T \left(\frac{\delta S_{\text{total}}}{\delta T} \right). \tag{4F.4}$$

After some calculations we obtain [24]

$$(C_V)_{\text{cx}} = \phi_\ell C_{V\ell}[1 + Z'_\ell] + \phi_g C_{Vg}[1 + Z'_g], \tag{4F.5}$$

where $\phi_\ell = M_\ell/\rho_\ell V$ and $\phi_g = M_g/\rho_g V = 1 - \phi_\ell$ are the volume fractions of the two phases, and $C_{V\ell}$ and C_{Vg} are the constant-volume specific heats per unit volume. The quantities Z'_ℓ and Z'_g are the liquid and gas values of Z' defined by [70]

$$Z' = \left(\frac{C_p}{C_V} - 1 \right) \left[\left(\frac{\partial T}{\partial p} \right)_\rho \left(\frac{\partial p}{\partial T} \right)_{\text{cx}} - 1 \right]^2 = \frac{T}{\rho^2 C_V K_T} \left(\frac{\partial \rho}{\partial T} \right)_{\text{cx}}^2, \tag{4F.6}$$

where use has been made of (1.2.53), (1.2.54), and (2.2.39). Note that (4F.5) and (4F.6) are applicable at any temperature. The positive-definiteness of $(C_V)_{\text{cx}}$ is assured by the mass conversion arising from $\delta V = 0$.

Because the thermodynamic quantities in the two phases become identical as $T \to T_c$, (1.2.54), (2.2.36), and (2.2.37) give

$$Z' \to a_c^2 = R_v/(1 - R_v), \tag{4F.7}$$

as $T \to T_c$. Thus (4F.5) becomes

$$(C_V)_{\text{cx}} \cong (1 + a_c^2) C_V = C_V/(1 - R_v), \tag{4F.8}$$

where the difference of C_V in the two phases is neglected, and R_v is the universal number in (1.1.48), which has a value close to 0.5 on the coexistence curve in 3D. Experimentally, if we apply a fixed amount of heat to a cell containing a near-critical fluid in two-phase coexistence, the fluid heat capacity will appear to be VC_V in an early stage but will be increased to $V(C_V)_{cx} = V(1 + a_c^2)C_V \cong 2VC_V$ after the mass conversion. If the cell (boundary wall + fluid) is thermally isolated from the outside after a heat input, an overshoot of the boundary temperature will occur on the timescale of the thermal diffusion, as will be illustrated in Fig. 6.14.

References

[1] K. G. Wilson and J. Kogut, *Phys. Rep.* C **12**, 76 (1974).

[2] S. Ma, *Modern Theory of Critical Phenomena* (Benjamin, New York, 1976).

[3] P. Pfeuty and G. Toulouse, *Introduction to Renormalization Group and to Critical Phenomena* (J. Wiley and Sons, New York, 1977).

[4] (a) E. Brezin, J. C. Le Guillou, and J. Zinn-Justin, *Phase Transitions and Critical Phenomena* Vol. 6, eds. C. Domb and J. L. Lebowitz (Academic, London, 1976), p. 127; (b) F. J. Wegner, *ibid.*, p. 8; (c) D. J. Wallace, *ibid.*, p. 294; (d) A. Aharony, *ibid.*, p. 358.

[5] D. J. Wallace and R. K. P. Zia, *Rep. Prog. Phys.* **41**, 1 (1978).

[6] D. J. Amit, *Field Theory, the Renormalization Group, and Critical Phenomena* (World Scientific, Singapore, 1978).

[7] G. Parisi, *Statistical Field Theory* (Addison-Wesley, 1988).

[8] C. Itzykson and J. E. Drouffe, *Statistical Field Theory* (Cambridge University Press, 1989).

[9] N. Goldenfeld, *Lectures on Phase Transitions and the Renormalization Group* (Addison-Wesley, 1992).

[10] J. J. Binney, N. J. Dowrick, A. J. Fisher, and M. E. J. Newman, *The Theory of Critical Phenomena* (Clarendon Press, Oxford, 1993).

[11] P. M. Chaikin and T. C. Lubensky, *Principles of Condensed Matter Physics* (Cambridge University Press, 1995).

[12] J. Zinn-Justin, *Quantum Field theory and Critical Phenomena* (Oxford University Press, 1996).

[13] J. S. Rowlinson, *J. Stat. Phys.* **20**, 197 (1979).

[14] V. L. Ginzburg and L. D. Landau, *Zh. Eksp. Teor. Fiz.* **20**, 1064 (1950).

[15] A. A. Abrikosov, *Zh. Eksp. Teor. Fiz.* **32**, 1442 (1957) [*Sov. Phys. JETP* **5**, 1174 (1957)].

[16] V. L. Ginzburg and L. P. Pitaevskii, *Zh. Eksp. Teor. Fiz.* **34**, 1240 (1958) [*Sov. Phys. JETP* **7**, 858 (1958)].

[17] J. W. Cahn and J. E. Hilliard, *J. Chem. Phys.* **28**, 258 (1958).

[18] M. Y. Belyakov and S. B. Kiselev, *Physica A* **190**, 75 (1992).

[19] E. Luijten and K. Binder, *Phys. Rev. E* **58**, R4060 (1998); *ibid.* **59**, 7254 (1999).

[20] D. Schwahn, K. Mortensen, and H. Yee-Madeira, *Phys. Rev. Lett.* **56**, 1544 (1987).

[21] M. A. Anisimov, A. A. Povodyrev, V. D. Kulikov, and J. V. Sengers, *Phys. Rev. Lett.* **75**, 3146 (1995).

[22] E. Luijten and H. Meyer, *Phys. Rev. E.* **62**, 3257 (2000).

[23] B. I. Halperin, P. C. Hohenberg, and S. Ma, *Phys. Rev. B* **10**, 139 (1974); *ibid.* **13**, 4119 (1976).

[24] A. Onuki, *Phys. Rev. E* **52**, 403 (1997).

[25] A. Onuki, *J. Phys. Soc. Jpn* **66**, 511 (1997).

[26] A. Onuki, *J. Low Temp. Phys.* **53**, 1 (1983).

[27] M. G. Ryschkewitsch and H. Meyer, *J. Low Temp. Phys.* **35**, 103 (1979).

[28] P. Debye, *J. Phys. Colloid Chem.* **51**, 19 (1947).

[29] P. G. de Gennes, *Scaling Concepts in Polymer Physics* (Ithaca, Cornell Univ. Press, New York, 1980).

[30] K. Binder, *J. Chem. Phys.* **79**, 6287 (1983).

[31] T. Dobashi, M. Nakata, and M. Kaneko, *J. Chem. Phys.* **72**, 6685 (1980).

[32] K. Shinozaki, T. V. Van, Y. Saito, and T. Nose, *Polymer* **23**, 728 (1982).

[33] I. Sanchez, *J. Appl. Phys.* **58**, 2871 (1985).

[34] B. Chu and Z. Wang, *Macromolecules* **21**, 2283 (1988).

[35] V. L. Ginzburg and A. A. Sobaynin, *Usp. Fiz. Nauk SSSR* **129**, 153 (1976) [*Sov. Phys. Usp.* **19**, 772 (1976)].

[36] L. D. Landau and E. M. Lifshitz, *Electrodynamics of Continuous Media* (Pergamon, Oxford, 1984), Chap. II.

[37] P. Debye and K. Kleboth, *J. Chem. Phys.* **42**, 3155 (1965).

[38] D. Bedeaux and P. Mazur, *Physica* **67**, 23 (1973).

[39] J. Goulon, J. L. Greffe, and D. W. Oxtoby, *J. Chem. Phys.* **70**, 4742 (1979).

[40] J. V. Sengers, D. Bedeaux, P. Mazu, and S. C. Greer, *Physica* **104A**, 573 (1980).

[41] D. Beaglehole, *J. Chem. Phys.* **74**, 5251 (1981).

[42] J. S. Hoye and G. Stell, *J. Chem. Phys.* **81**, 3200 (1984).

[43] W. Pyzuk, *Chem. Phys.* **50**, 281 (1980); *Europhys. Lett.* **17**, 339 (1992).

[44] S. J. Rzoska, A. D. Rzoska, M. Górny, and J. Zioło, *Phys. Rev. E* **52**, 6325 (1995).

[45] T. Bellini and V. Degiorgio, *Phys. Rev. B* **39**, 7623 (1989).

[46] D. Wirtz and G. G. Fuller, *Phys. Rev. Lett.* **71**, 2236 (1993).

[47] D. Wirtz, D. E. Werner, and G. G. Fuller, *J. Chem. Phys.* **101**, 1679 (1994).

[48] A. Onuki and M. Doi, *Europhys. Lett.* **17**, 63 (1992).

[49] A. Onuki, *Europhys. Lett.* **29**, 611 (1995).

[50] T. Garel and S. Doniach, *Phys. Rev. B* **26**, 325 (1982).

[51] A. L. Larkin and D. E. Khmel'nitskii, *Sov. Phys. JETP* **29**, 1123 (1969).

[52] G. Zimmerli, R. A. Wilkinson, R. A. Ferrell, and M. R. Moldover, *Phys. Rev. E* **59**, 5862 (1999).

[53] M. Barmartz and F. Zhong, *Proceedings of the 2000 NASA/JPL Workshop on Fundamental Physics in Microgravity*, Solvang, June 19–21, 2000.

[54] R. Abe, *Prog. Theor. Phys.* **49**, 113 (1973).

[55] D. E. Siggia and D. R. Nelson, *Phys. Rev. B* **15**, 1427 (1977).

[56] P. C. Hohenberg, A. Aharony, B. I. Halperin, and E. D. Siggia, *Phys. Rev. B* **13**, 2986 (1976).

[57] G. A. Baker, Jr, B. G. Nickel, and D. I. Meiron, *Phys. Rev. B* **17**, 1365 (1978).

[58] Y. Okabe and K. Idekura, *Prog. Theor. Phys.* **66**, 1959 (1981).

[59] E. Brezin, D. J. Wallace, and K. Wilson, *Phys. Rev. B* **7**, 232 (1973).

[60] J. F. Nicoll and P. C. Albright, *Phys. Rev. B* **31**, 4576 (1985).

[61] C. Bervillier, *Phys. Rev. B* **14**, 4964 (1976); *ibid.* **34**, 8141 (1986).

[62] V. Privman, P. C. Hohenberg, and A. Aharony, in *Phase Transitions*, Vol. 14, eds. C. Domb and J. L. Lebowitz (Academic, 1991), p. 1.

[63] D. Dahl and M. R. Moldover, *Phys. Rev. Lett.* **27**, 1421 (1971).

[64] M. R. Moldover, *Phys. Rev.* **182**, 342 (1969).

[65] G. R. Brown and H. Meyer, *Phys. Rev. A*, **6**, 364 (1972).

[66] D. Dahl and M. R. Moldover, *Phys. Rev. A* **6**, 1915 (1972).

[67] J. A. Lipa, C. Edwards, and M. J. Buckingham, *Phys. Rev. A* **15**, 778 (1977).

[68] Yu. R. Chashkin, A. V. Voronel, V. A. Smirnov, and V. G. Gobunova, *Sov. Phys. JETP* **25**, 79 (1979).

[69] A. Haupt and J. Straub, *Phys. Rev. E* **59**, 1795 (1999).

[70] M. E. Fisher, *J. Math. Phys.* **5**, 944 (1964).

[71] R. A. Ferrell and J. K. Bhattacharjee, *Phys. Lett. A* **88**, 77 (1982).

[72] E. Brezin and D. J. Wallace, *Phys. Rev. B* **7**, 1967 (1973).

[73] M. E. Fisher, M. N. Barbar, and D. Jasnow, *Phys. Rev. A* **8**, 1111 (1973).

[74] M. E. Fisher and V. Privman, *Phys. Rev. B* **32**, 447 (1985).

[75] A. Singsaas and G. Ahlers, *Phys. Rev. B* **30**, 5103 (1984).

[76] P. G. de Gennes and J. Prost, *The Physics of Liquid Crystals* (Oxford University Press, 1993).

[77] T. Ohta and K. Kawasaki, *Prog. Theor. Phys.* **58**, 467 (1977).

[78] D. Jasnow and J. Rudnick, *Phys. Rev. Lett.* **41**, 698 (1978).

[79] M. R. Moldover, *Phys. Rev. A* **31**, 1022 (1985).

[80] T. Mainzer and D. Woermann, *Physica A* **225**, 312 (1996).

[81] D. Jasnow, T. Ohta, and J. Rudnick, *Phys. Rev. B* **20**, 2774 (1979).

[82] M. san Miguel and J. D. Gunton, *Phys. Rev. B* **23**, 2317 (1981).

[83] P. Leiderer, H. Poisel, and M. Wanner, *J. Low Temp. Phys.* **28**, 167 (1977).

[84] K.-Q. Xia, C. Franck, and B. Widom, *J. Chem. Phys.* **97**, 1446 (1992).

[85] J. F. Joanny and L. Leibler, *J. Physique* **39**, 951 (1978).

[86] E. Helfand and Y. Tagami, *J. Chem. Phys.* **56**, 3592 (1971).

[87] P. G. de Gennes, *J. Chem. Phys.* **72**, 4756 (1980).

[88] K. Binder and H. L. Frisch, *Macromolecules* **17**, 2929 (1984).

[89] D. O. Edwards and W. F. Saam, in *Progress in Low Temperature Physics*, Vol. VII A, ed. D. F. Brewer (North-Holland, 1978), p. 283.

[90] A. F. Andreev and A. Ya. Parshin, *Sov. Phys. JETP* **48**, 763 (1978) [*Zh. Eksp. Teor. Fiz.* **75**, 1511 (1978)].

[91] K. O. Keshishev, A. Ya Parshin, and A. I. Shal'nikov, in *Physics Reviews*, ed. I. M. Khalatnikov (Harwood Academic, 1982) Vol. 4, p. 155.

[92] M. Uwaha, *J. Low Temp. Phys.* **77**, 165 (1989).

[93] A. Widom, *Phys. Rev. A* **1**, 216 (1970); M. W. Cole, *Phys. Rev. A* **1**, 1838 (1970).

[94] L. B. Lurio, T. A. Rabedeau, P. S. Pershan, I. F. Silvera, M. Deutsch, S. D. Kosowsky, and B. M. Ocko, *Phys. Rev. B* **48**, 9644 (1983).

[95] J. Wilks and D. S. Betts, *An Introduction to Liquid Helium* (Clarendon Press, Oxford, 1987).

[96] J. M. Kosterlitz and D. J. Thouless, *J. Phys. C* **6**, 1181 (1973).

[97] M. P. Kawatra and R. K. Pathria, *Phys. Rev.* **151**, 132 (1966).

[98] D. Amit and E. P. Gross, *Phys. Rev.* **145**, 130 (1966).

[99] R. J. Arms and F. R. Hama, *Phys. Fluids* **8**, 533 (1965).

Part two

Dynamic models and dynamics in fluids and polymers

5

Dynamic models

Slow collective motions in physical systems, particularly those near the critical point, can be best described in the framework of Langevin equations. We may set up Langevin equations when the timescales of slow and fast dynamical variables are distinctly separated. This framework originates from the classical Brownian motion and is justified microscopically via the projection operator formalism. First, in Sections 5.1–5.2, these general aspects will be discussed with a summary of the projection operator method in Appendix 5B. Second, in Section 5.3, we will examine simple Langevin equations in critical dynamics (models A, B, and C) and introduce dynamic renormalization group theory. These models have been used extensively to study fundamental problems in critical dynamics and phase ordering. Third, in Section 5.4, we will review the general linear response theory, putting emphasis on linear response to thermal disturbances.

5.1 Langevin equation for a single particle

5.1.1 Brownian motion

Most readers will be aware of the zig-zag motions of a relatively large particle, called a Brownian particle, suspended in a fluid. When its mass m_0 is much larger than those of the surrounding particles, appreciable changes of the velocity of the Brownian particle can be caused as a result of a large number of collisions with the surrounding molecules. Its velocity $u(t)$ in one direction (say, in the x direction) is governed by the Langevin equation [1]–[9],

$$\frac{\partial}{\partial t}u(t) = -\gamma u(t) + \theta(t). \tag{5.1.1}$$

If the Brownian particle is suspended in an incompressible fluid governed by the Navier–Stokes equation under the no-slip boundary condition, the relaxation rate γ may be expressed by the Stokes formula [10],

$$m_0\gamma = 6\pi\eta_0 a, \tag{5.1.2}$$

where m_0 is the mass of the Brownian particle, η_0 is the shear viscosity of the fluid, and a is the radius of the particle. The quantity $m_0\theta(t)$ is the rapidly varying (random) force arising from the numerous collisions taking place on a microscopic duration time t_{coll}. As the mathematical idealization of $t_{\text{coll}} \to 0$, its statistical properties are usually characterized by

$$\langle \theta(t) \rangle = 0, \tag{5.1.3}$$

$$\langle \theta(t_1)\theta(t_2) \rangle = 2L\delta(t_1 - t_2), \tag{5.1.4}$$

where $\langle \cdots \rangle$ is the stochastic average and the probability distribution of $\theta(t)$ is assumed to be independent of $u(t)$. The coefficient L characterizes the strength of the random force (noise) and will be related to γ in (5.1.17) below.

A precise mathematical definition of the random force can be made by specifying stochastic properties of a time integral of $\theta(t)$ [3],

$$W(t, t + \Delta t) = \int_t^{t+\Delta t} dt' \theta(t'). \tag{5.1.5}$$

Physically, the time interval Δt should be taken to be much longer than the duration time t_{coll} but much shorter than γ^{-1} [4],

$$t_{\text{coll}} \ll \Delta t \ll \gamma^{-1}. \tag{5.1.6}$$

Then $W(t, t + \Delta t)$ consists of numerous microscopic impulses, so it obeys a gaussian distribution characterized by [1]

$$\langle W(t, t + \Delta t)^2 \rangle = 2L\Delta t. \tag{5.1.7}$$

Furthermore, we assume that if two time intervals, $[t_1, t_2]$ and $[t_3, t_4]$, are disjoint ($t_1 < t_2 < t_3 < t_4$ or $t_3 < t_4 < t_1 < t_2$), the two random impulses $W(t_1, t_2)$ and $W(t_3, t_4)$ are independent of each other or have no correlation between each other. Thus,

$$\langle W(t_1, t_2)W(t_3, t_4) \rangle = 0. \tag{5.1.8}$$

This means that the random force does not remember previous random events. The stochastic process obeyed by the time-dependent variable,

$$w(t) \equiv W(0, t) = \int_0^t dt' \theta(t'), \tag{5.1.9}$$

is called the Wiener process [7], in terms of which we have $W(t, t + \Delta t) = w(t + \Delta t) - w(t)$. If random source terms in stochastic differential equations satisfy the above two properties, we will call them gaussian and markovian noises (or random forces).

Because $u(t)$ no longer has well-defined time derivatives in the limit $t_{\text{coll}} \to 0$, as can be known from (5.1.7), it is more appropriate to rewrite the original equation (5.1.1) in terms of the incremental change $\Delta u(t) \equiv u(t + \Delta t) - u(t)$ as

$$\begin{aligned} \Delta u(t) &= -\gamma \int_t^{t+\Delta t} dt' u(t') + W(t, t + \Delta t) \\ &\cong -\gamma u(t)\Delta t + W(t, t + \Delta t). \end{aligned} \tag{5.1.10}$$

[1] It would be natural to expect that a sum of many independent random variables with similar probability distributions and finite variances should obey a gaussian distribution. This asymptotic law can readily be obtained using their characteristic function expressions. A rigorous mathematical expression of this property is known as *the central limit theorem.*

In the second line use has been made of the fact that $u(t)$ is continuous with probability 1 [3].[2] The above Langevin equation may be written in the differential form as

$$du(t) = -\gamma u(t)dt + dw(t). \tag{5.1.11}$$

The stochastic differential equation in this form is sometimes called the Itô equation [7].

5.1.2 Fokker–Planck equation for the velocity

Recall that $u(t)$ is a stochastic variable obeying (5.1.1) or (5.1.10). Another equivalent description is to follow the time evolution of the probability distribution,

$$P(v,t) = \langle \delta(u(t) - v) \rangle, \tag{5.1.12}$$

which is the probability that $u(t)$ is equal to v at time t. In Appendix 5A we shall see that $P(v,t)$ obeys the Fokker–Planck equation,

$$\frac{\partial}{\partial t} P(v,t) = \mathcal{L}_{\text{FP}} P(v,t) = \frac{\partial}{\partial v}\left[\gamma v + L\frac{\partial}{\partial v}\right] P(v,t), \tag{5.1.13}$$

where $\mathcal{L}_{\text{FP}}$ is called the Fokker–Planck operator. The second-order differentiation $\partial^2/\partial v^2$ on the right-hand side arises from the random force $\theta(t)$. The conditional distribution $P(v, v_0, t)$ in which $u(0)$ at $t = 0$ is fixed at v_0 is formally written as

$$P(v, v_0, t) = \exp(t\mathcal{L}_{\text{FP}})\delta(v - v_0). \tag{5.1.14}$$

It satisfies the markovian property,

$$P(v_1, v_2, t_1 + t_2) = \int dv_3 P(v_1, v_3, t_1) P(v_3, v_2, t_2). \tag{5.1.15}$$

If the equilibrium distribution is maxwellian,

$$P_{\text{eq}}(v) = (m_0/2\pi T)^{1/2} \exp\left(-\frac{m_0}{2T} v^2\right), \tag{5.1.16}$$

it follows a fluctuation–dissipation relation,

$$\gamma = (m_0/T)L, \tag{5.1.17}$$

which relates the relaxation rate to the noise strength. The Langevin equation (5.1.1) may now be expressed in the standard form (see (5.2.1) for a general form),

$$\frac{\partial}{\partial t} u(t) = -L\frac{\partial}{\partial u}(\beta\mathcal{H}) + \theta(t), \tag{5.1.18}$$

where $\mathcal{H}$ is the *free energy* of the Brownian particle,

$$\mathcal{H}(u) = -T \log P_{\text{eq}}(u) = \frac{1}{2} m_0 u^2 + \text{const.} \tag{5.1.19}$$

[2] Note that $\Delta u(t)$ is mostly of order $(\Delta t)^{1/2}$ and $u(t)$ is not differentiable with probability 1.

We note that the average $\langle u(t)\rangle$ relaxes exponentially with the relaxation rate γ. The variance $\sigma(t) = \langle (u(t))^2\rangle - \langle u(t)\rangle^2$ obeys

$$\frac{d}{dt}\sigma(t) = \frac{d}{dt}\int dv v^2 P(v,t) + 2\gamma\langle u(t)\rangle^2 = -2\gamma\left[\sigma(t) - \frac{T}{m_0}\right]. \tag{5.1.20}$$

Thus $\sigma(t) - T/m_0$ exponentially goes to zero with the relaxation rate 2γ.

5.1.3 *Langevin equation for the position*

We now follow the space position of the Brownian particle. When there is no potential energy such as the gravity field or an electric field, the x coordinate of the Brownian particle $X(t)$ obeys

$$\frac{\partial}{\partial t}X(t) = u(t). \tag{5.1.21}$$

Because we are assuming the linear Langevin equation (5.1.1), the displacement

$$\Delta X(t) = X(t) - X(0) = \int_0^t dt' u(t') \tag{5.1.22}$$

obeys a gaussian distribution, whose variance in equilibrium is

$$\begin{aligned}\langle(\Delta X(t))^2\rangle &= 2\int_0^t dt'\,(t-t')\,\langle u(t')u(0)\rangle \\ &= 2(T/m_0\gamma)\,[t - (1-e^{-\gamma t})/\gamma].\end{aligned} \tag{5.1.23}$$

In the short- or long-time limit, the particle motion is ballistic or diffusive, respectively, as

$$\begin{aligned}\langle(\Delta X(t))^2\rangle &\cong (T/m_0)\,t^2 \qquad (t \ll \gamma^{-1}), \\ &\cong 2(T/m_0\gamma)\,t \quad (t \gg \gamma^{-1}).\end{aligned} \tag{5.1.24}$$

The diffusion constant D turns out to be given by $D = T/m_0\gamma$. If use is made of the hydrodynamic expression (5.1.2), it follows the Einstein–Stokes formula [10],[3]

$$D = T/6\pi\eta_0 a. \tag{5.1.25}$$

On timescales much longer than γ^{-1}, $u(t)$ in (5.1.21) plays the role of a gaussian and markovian random force acting on $X(t)$. To show this, we integrate (5.1.1) as

$$u(t) = u(0)\exp(-\gamma t) + \int_0^t dt' \exp[-\gamma(t-t')]\,\theta(t'). \tag{5.1.26}$$

If $t \gg \gamma^{-1}$, the first term, representing the initial memory, decays exponentially to zero and the second term becomes a stationary gaussian random variable. Neglecting the first term, we calculate the time correlation of $u(t)$ as

$$\langle u(t_1)u(t_2)\rangle = L\gamma^{-1}\exp(-\gamma|t_1 - t_2|), \tag{5.1.27}$$

[3] This formula is known to give a fair estimation of the diffusion constant of a tagged particle in a fluid even if the particle size is microscopic. However, this formula breaks down in highly supercooled liquids, as will be discussed in Section 11.4.

where t_1 and t_2 are both much longer than γ^{-1}. For $t_1 = t_2$ the equilibrium time-correlation function is obtained, while in the limit $\gamma^{-1} \to 0$ we obtain

$$\langle u(t_1)u(t_2)\rangle \cong 2L\gamma^{-2}\delta(t_1 - t_2) = \gamma^{-2}\langle\theta(t_1)\theta(t_2)\rangle. \tag{5.1.28}$$

In this mathematical idealization, $u(t)$ and $\gamma^{-1}\theta(t)$ are equivalent gaussian and markovian noises with the same variance. This formally follows from (5.1.1) if we set $\partial u/\partial t = 0$ there. However, this equivalence is not trivial, because $u(t)$ and $\gamma^{-1}\theta(t)$ are physically very different with very different timescales. The displacement $\Delta X(t)$ is then a Wiener process with its variance linearly growing as the second line of (5.1.24).

The effect of a potential $U(X)$ dependent on the particle position X can be easily incorporated in the above arguments. We change (5.1.1) as

$$\frac{\partial}{\partial t}u(t) = -\frac{1}{m_0}\frac{\partial}{\partial X}U(X) - \gamma u(t) + \theta(t). \tag{5.1.29}$$

As has been stated below (5.1.28), we are allowed to set $\partial u/\partial t = 0$ in (5.1.29) even in the presence of the potential in describing phenomena taking place on timescales much longer than γ^{-1}. It then leads to a Langevin equation for $X(t)$,

$$\frac{\partial}{\partial t}X(t) = -D\frac{\partial}{\partial X}\beta U(X) + \bar{\theta}(t), \tag{5.1.30}$$

where D is defined by (5.1.25). The noise term $\bar{\theta}(t) \equiv \gamma^{-1}\theta(t)$ satisfies the fluctuation–dissipation relation,

$$\langle\bar{\theta}(t_1)\bar{\theta}(t_2)\rangle = 2D\delta(t_1 - t_2), \tag{5.1.31}$$

which follows from (5.1.17). The Fokker–Planck equation for the probability distribution $P(x,t) \equiv \langle\delta(X(t) - x)\rangle$ is given by

$$\frac{\partial}{\partial t}P(x,t) = D\frac{\partial}{\partial x}\left[\beta\frac{\partial U(x)}{\partial x} + \frac{\partial}{\partial x}\right]P(x,t), \tag{5.1.32}$$

whose stationary solution is $P_{\text{eq}} = \text{const.}\exp[-\beta U(x)]$.

Diffusion constant in general

We may consider diffusive motion of any tagged particle, whose size may be of the same order as those of the surrounding particles, in fluids or even in solids. The simplest linear Langevin equation (5.1.1) is not applicable in many situations. Nevertheless, both in fluids and solids, the translational diffusion constant of such a tagged particle is expressed in terms of the time integration of its velocity-correlation function,

$$D = \int_0^\infty dt\langle u(t)u(0)\rangle. \tag{5.1.33}$$

The diffusion behavior $\langle(\Delta X(t))^2\rangle \cong 2Dt$ follows at sufficiently long times, as long as the above integral is convergent.

5.1.4 Compound–poissonian noise

A noise term consisting of pulse-like impacts should be treated to be poissonian rather than gaussian if even a single impact causes appreciable influence on the dynamic variable.[4] If the distribution of a dynamic variable obeys a Master equation, each sample process of the variable evolves under the influence of a compound-poissonian noise, in which the time integral of the noise term is a linear combination of independent poissonian random variables [6]–[9]. The Boltzmann equation for dilute gases may also be regarded as a Langevin equation with a compound-poissonian noise [6, 11].

As a simple example, let us consider motion of a particle caused by thermally activated jumps or hoppings in a solid or glass. The time integral of the random velocity (the particle displacement vector), $\Delta \boldsymbol{X}(t) = \int_t^{t+\Delta t} dt' \boldsymbol{u}(t')$, in a small time interval Δt consists of jumps with size $\boldsymbol{\ell}$ as

$$\Delta \boldsymbol{X}(t) = \sum_{\boldsymbol{\ell}} N(\Delta t, \boldsymbol{\ell})\boldsymbol{\ell}, \tag{5.1.34}$$

where $N(\Delta t, \boldsymbol{\ell})$ is the number of the $\boldsymbol{\ell}$-jumps obeying a poissonian distribution with average $\hat{W}(\boldsymbol{\ell})\Delta t$ independently of one another. In this case it is easy to calculate the time evolution of the van Hove time-correlation function $G(\boldsymbol{q}, t) = \langle \exp[i\boldsymbol{q} \cdot (\boldsymbol{X}(t) - \boldsymbol{X}(0)]\rangle$. To this end we note the relation,

$$\begin{aligned} G(\boldsymbol{q}, t + \Delta t) &= \langle \exp[i\boldsymbol{q} \cdot \Delta \boldsymbol{X}(t)]\rangle G(\boldsymbol{q}, t) \\ &= \prod_{\boldsymbol{\ell}} \exp\big[(e^{i\boldsymbol{q}\cdot\boldsymbol{\ell}} - 1)\hat{W}(\boldsymbol{\ell})\Delta t\big] G(\boldsymbol{q}, t). \end{aligned} \tag{5.1.35}$$

As $\Delta t \to 0$ we find

$$\frac{\partial}{\partial t} G(\boldsymbol{q}, t) = \left[\sum_{\boldsymbol{\ell}} (e^{i\boldsymbol{q}\cdot\boldsymbol{\ell}} - 1)\hat{W}(\boldsymbol{\ell})\right] G(\boldsymbol{q}, t), \tag{5.1.36}$$

which is integrated to give

$$G(\boldsymbol{q}, t) = \exp\left[\sum_{\boldsymbol{\ell}} (e^{i\boldsymbol{q}\cdot\boldsymbol{\ell}} - 1)\hat{W}(\boldsymbol{\ell})t\right]. \tag{5.1.37}$$

We notice that the tagged particle density $P(\boldsymbol{x}, t) = \langle \delta(\boldsymbol{x} - \boldsymbol{X}(t))\rangle$ is governed by the master equation [12],

$$\frac{\partial}{\partial t} P(\boldsymbol{x}, t) = \sum_{\boldsymbol{\ell}} \hat{W}(\boldsymbol{\ell})\big[P(\boldsymbol{x} - \boldsymbol{\ell}, t) - P(\boldsymbol{x}, t)\big]. \tag{5.1.38}$$

Furthermore, let the second moments $\sum_{\boldsymbol{\ell}} \hat{W}(\boldsymbol{\ell})\ell_\alpha \ell_\beta = D\delta_{\alpha\beta}$ be convergent and diagonal. Then the linear relation,

$$\langle \Delta X_\alpha(t) \Delta X_\beta(t)\rangle = 2D\delta_{\alpha\beta} t, \tag{5.1.39}$$

[4] If X is a poissonian random variable, the probability of $X = n$ $(= 0, 1, 2, \ldots)$ is given by $e^{-\langle X\rangle}\langle X\rangle^n/n!$.

holds for any t and the diffusion equation $\partial P/\partial t = D\nabla^2 P$ is obtained on long timescales in which the jump number $t\sum_{\boldsymbol{\ell}} \hat{W}(\boldsymbol{\ell})$ greatly exceeds 1 [12].

However, the particle may jump over large distances such that the second moments diverge. As such an example, let the jump distribution $P_{\text{jump}}(\boldsymbol{\ell}) = \hat{W}(\boldsymbol{\ell})/\sum_{\boldsymbol{m}} \hat{W}(\boldsymbol{m})$ obey the Lèvy distribution [8]. It has a tail at large ℓ and its characteristic function behaves as $\sum_{\boldsymbol{\ell}} e^{i\boldsymbol{q}\cdot\boldsymbol{\ell}} P_{\text{jump}}(\boldsymbol{\ell}) = \exp(-C|\boldsymbol{q}|^\sigma)$, with $\sigma \le 2$. In this case, the van Hove time-correlation function behaves as

$$G(\boldsymbol{q}, t) = \exp(-tC|\boldsymbol{q}|^\sigma), \tag{5.1.40}$$

at long times or at small $|\boldsymbol{q}|$. The displacement-distribution function obeys

$$\frac{\partial}{\partial t} P(\boldsymbol{x}, t) = -C(-\nabla^2)^{\sigma/2} P(\boldsymbol{x}, t). \tag{5.1.41}$$

We can define the fractional power of the laplacian in this manner.

5.1.5 Long-time tail

To be precise, a Brownian particle in a fluid does not obey the simple markovian Langevin equation (5.1.1) due to reaction of the flow field (backflow effect). As a result, it is known that the time-correlation function of the velocity $\phi(t) = \langle u(t)u(0)\rangle$ has a long-time tail $\sim t^{-d/2}$ [13, 14]. We assume that a Brownian particle should be convected by the fluctuating velocity field $\boldsymbol{v}(\boldsymbol{r}, t)$ at the particle position $\boldsymbol{r} = \boldsymbol{R}(t)$ on long timescales. Because the long-wavelength velocity field has long lifetimes, we have

$$\begin{aligned} \phi(t) &\cong \int d\boldsymbol{r}(4\pi Dt)^{-d/2} \exp(-r^2/4Dt)\langle v_x(\boldsymbol{r}, t)v_x(\boldsymbol{0}, 0)\rangle \\ &\sim \frac{T}{\rho}\int_{\boldsymbol{k}} \exp[-(D+\nu)k^2 t] \sim \frac{T}{\rho}[(D+\nu)t]^{-d/2}, \end{aligned} \tag{5.1.42}$$

where the diffusion of the particle is also taken into account. In the second line we have neglected the longitudinal velocity and retained the transverse velocity because the latter decays diffusively as $\exp(-\nu k^2 t)$ with $\nu = \eta_0/\rho$ at small wave numbers. In 2D, we then have $\langle(\Delta X(t))^2\rangle \sim t \ln t$ at long times. This means that the usual diffusion constant is not well defined in 2D. The other transport coefficients, such as the shear viscosity, also have logarithmic dependence on the frequency, wave number, or system size in 2D.

This flow effect can be studied analytically if the fluid particles are treated as an incompressible continuum obeying the linearized Navier–Stokes equation [10, 14]. Generally, the drag force on a sphere oscillating periodically with a small amplitude and arbitrary frequency ($\propto e^{i\omega t}$) is written as $-m_0\text{Re}[\hat{\gamma}(\omega)u]$. In 3D, under the no-slip boundary condition, the frequency-dependent friction constant $\hat{\gamma}(\omega)$ is calculated as

$$m_0\hat{\gamma}(\omega) = 6\pi\eta_0 a + \frac{2\pi}{3}\rho a^3 i\omega + 6\pi a^2(i\omega\rho\eta_0)^{1/2}. \tag{5.1.43}$$

This means that $\phi(t)$ obeys a non-markovian equation [14],

$$m_{\rm eff}\dot{\phi}(t) = -6\pi\eta_0 a\phi(t) - 6a^2(\pi\rho\eta_0)^{1/2}\int_0^t ds(t-s)^{-1/2}\dot{\phi}(s), \tag{5.1.44}$$

where $\dot{\phi}(t) = \partial\phi(t)/\partial t$ and

$$m_{\rm eff} = m_0 + \frac{2\pi}{3}\rho a^3 \tag{5.1.45}$$

is the effective mass. The Laplace transformation of $\phi(t)$ is expressed as

$$\int_0^\infty dt e^{-i\omega t}\phi(t) = \frac{T}{m_0}\frac{1}{i\omega + \hat{\gamma}(\omega)} = \frac{D}{i\Omega + 1 + 3(i\alpha\Omega)^{1/2}}, \tag{5.1.46}$$

where $D = T/6\pi\eta_0 a$, $\Omega = (m_{\rm eff}/6\pi\eta_0 a)\omega$, and

$$\alpha = 1 - m_0/m_{\rm eff} = 2\pi\rho a^3/(3m_0 + 2\pi\rho a^3). \tag{5.1.47}$$

The flow effect is thus important in liquids (where $2\pi\rho a^3/3 \sim m_0$) and small in dilute gases (where $2\pi\rho a^3/3 \ll m_0$)). The inverse Laplace transformation of (5.1.46) reproduces the long-time tail $\phi(t) \sim (T/\rho)(\nu t)^{-3/2}$, which is consistent with (5.1.42) for $D \ll \nu$. Note the relation $\lim_{t\to 0}\phi(t) = T/m_{\rm eff}$ is obtained from (5.1.46), whereas $\phi(0) = T/m_0$ is exact. This difference arises from the continuum approximation in deriving (5.1.43).

Similar long-time tails ($\propto t^{-d/2}$) can be found generally in the flux time-correlation functions in the long-wavelength limit whose time integration gives transport coefficients. They originate from nonlinear mode coupling between the hydrodynamic fluctuations.

5.2 Nonlinear Langevin equations with many variables

The theory of Brownian motion can be generalized for cases with many variables [15]–[20]. Let a set of variables $A(t) = \{A_i(t)\}$ relax slowly compared with the other degrees of freedom which constitute random forces acting on $A(t)$. They are called the gross variables [15]. The subscript i denotes the variable species and the wave vector $\boldsymbol{q}$ if $A(t)$ are fields composed of long-wavelength Fourier components ($q < \Lambda$). This framework has been widely used to study fundamental features of phase transition dynamics in various systems. Particularly for near-critical systems, the upper cut-off wave number Λ should be chosen in the region $\xi^{-1} \ll \Lambda \lesssim a^{-1}$ at the starting point of the theory, where ξ is the correlation length and a is a microscopic length such as the lattice constant. As in statics in Chapter 4, decreasing Λ is equivalent to coarse-graining of the short-wavelength fluctuations, resulting in dynamic renormalization group theory.

5.2.1 General theory

Using the projection operator method [16]–[18], which will be explained in Appendix 5B, we may construct a formal theory leading to general nonlinear Langevin equations,

$$\frac{\partial}{\partial t}A_i(t) = v_i(A) - \sum_j L_{ij}(A)F_j(A) + \theta_i(t), \tag{5.2.1}$$

in the markovian form originally presented by Green [15]. Here,

$$F_j(A) = \frac{\partial}{\partial A_j}\beta\mathcal{H}(A) \tag{5.2.2}$$

are called the *thermodynamic forces* [21]–[24]. The *potential* or hamiltonian $\mathcal{H}(A)$ is formally defined by

$$P_{\text{eq}}(a) = \left\langle \prod_j \delta(A_j - a_j) \right\rangle = \text{const.} \exp[-\beta\mathcal{H}(a)], \tag{5.2.3}$$

which is the probability of finding A at a in equilibrium (equilibrium distribution). Hereafter $\langle\cdots\rangle$ denotes the equilibrium average, and the conditional average in which A is fixed at a may be defined by

$$\langle\cdots; a\rangle = \left\langle \cdots \prod_j \delta(A_j - a_j) \right\rangle \Big/ P_{\text{eq}}(a). \tag{5.2.4}$$

with this preliminary understanding, we will now explain the physical meanings of the terms in (5.2.1).

(i) The first terms $v_i(A)$, sometimes called the streaming terms, represent the reversible, instantaneous changing rate of A_i expressed as

$$v_i(a) = \langle \dot{A}_i; a\rangle, \tag{5.2.5}$$

where $\dot{A}_i$ is the microscopic time derivative of A_i (see Appendix 5B). The linear parts of $v_i(A)$ give rise to oscillatory modes such as spin waves in magnets, sounds in fluids, or second sounds in ^{4}He [17]. The nonlinear parts of $v_i(A)$, called the mode coupling terms, lead to enhancement of the kinetic coefficients in the long-wavelength limit [25, 26]. The form of $v_i(A)$ can be determined from conservation laws or poissonian bracket relations [27], see (5B.2).

(ii) In the second terms, the kinetic coefficients $L_{ij}(A)$ can be shown to satisfy $L_{ij}(A) = \epsilon_i\epsilon_j L_{ji}(\tilde{A})$ from the formal theory in Appendix 5B, where $\tilde{A} = \{\epsilon_i A_i\}$ denotes the time-reversed gross variables (see (5.2.8) below). Here we assume that $L_{ij}(A)$ are nonvanishing only for pairs i and j with $\epsilon_i\epsilon_j = 1$ and are even functions of A_ℓ with $\epsilon_\ell = -1$. Then we may set

$$L_{ij}(A) = L_{ji}(\tilde{A}) = L_{ji}(A). \tag{5.2.6}$$

Note that L_{ij} here are bare or background coefficients, because the nonlinear terms in the dynamic equations serve to *renormalize* them into those observable in experiments [19, 25, 26].

(iii) With these assumptions we may impose the gaussian–markovian stochastic property on the last terms $\theta_i(t)$ characterized by the fluctuation–dissipation relations,[5]

$$\langle \theta_i(t)\theta_j(t'); a\rangle = 2L_{ij}(a)\delta(t-t'). \tag{5.2.7}$$

Many phenomenological dynamic equations can be treated as Langevin equations in the general form (5.2.1) if appropriate thermal noise terms satisfying (5.2.7) are added. A notable example is the usual hydrodynamic equations supplemented with random stress tensor, random energy current, and random diffusion current [10]. The symmetry of the kinetic coefficients (in the linear response regime in the original papers) is known as the Onsager reciprocal theorem, valid for various coupled transport processes [21]–[24].

Time reversal and anti-symmetric kinetic coefficients

Let A_i be changed to

$$\tilde{A}_i = \epsilon_i A_i \quad (\epsilon_i = \pm 1) \tag{5.2.8}$$

with respect to the time reversal (which is the change $(\boldsymbol{r}_j, \boldsymbol{p}_j) \to (\boldsymbol{r}_j, -\boldsymbol{p}_j)$ for classical fluids). Then $v_i(A)$ are changed to $-\epsilon_i v_i(A)$, so that the streaming terms in (5.2.1) are reversible. However, $F_i(A)$ $(= F_i(\tilde{A})$ if $\mathcal{H}(a) = \mathcal{H}(\tilde{a}))$ are changed to $\epsilon_i F_i(A)$. Thus, for $\epsilon_i\epsilon_j = 1$, the terms involving $L_{ij}(A)$ in (5.2.1) are dissipative and the kinetic coefficients are symmetric. For $\epsilon_i\epsilon_j = -1$, they are reversible and the kinetic coefficients are anti-symmetric [15, 22, 23], although this possibility is neglected in (5.2.6). The existence of reversible or anti-symmetric kinetic coefficients was first pointed out by Casimir [22]. We will encounter a situation in the critical dynamics of ^{4}He in Section 6.4, where the coarse-graining gives rise to anti-symmetric renormalized kinetic coefficients. This can happen when both reversible and dissipative nonlinear terms are present in the Langevin equations.

5.2.2 Probability distribution and Fokker–Planck equation

The Langevin equations (5.2.1) can be presented in a mathematically precise manner in the Itô scheme as

$$A_i(t+\Delta t) - A_i(t) \cong V_i(A(t))\Delta t + W_i(t, t+\Delta t), \tag{5.2.9}$$

as in (5.1.10). The last terms are gaussian random variables with variances,

$$\langle W_i(t, t+\Delta t) W_j(t, t+\Delta t); a\rangle = 2L_{ij}(a)\Delta t, \tag{5.2.10}$$

[5] This expression is misleading when L_{ij} depend on a, however. Rigorous stochastic characterization of the equations will be given in (5.2.9).

dependent on the initial $A(t) = a$.[6] Then the first term $V_i(a)$ in (5.2.9) as a function of a is given by[7]

$$V_i(a) = v_i(a) - \sum_j L_{ij}(a)F_j(a) + \sum_j \frac{\partial}{\partial a_j} L_{ij}(a). \tag{5.2.11}$$

The probability distribution $P(a, t)$ of finding $A(t)$ at $a = \{a_j\}$ is then governed by

$$\frac{\partial}{\partial t} P(a, t) = \mathcal{L}_{\text{FP}}\{a\}P(a, t) \tag{5.2.12}$$

with the Fokker–Planck operator [15, 16],

$$\mathcal{L}_{\text{FP}}\{a\} = -\sum_i \frac{\partial}{\partial a_i} v_i(a) + \sum_{i,j} \frac{\partial}{\partial a_i} L_{ij}(a) \left[\frac{\partial}{\partial a_j} + F_j(a) \right]. \tag{5.2.13}$$

This can be derived by straightforward generalization of the simplest example in Appendix 5A. The first term on the right-hand side of (5.2.13) is reversible, while the second term is dissipative. Because $\mathcal{L}_{\text{FP}}\{a\}P_{\text{eq}} = 0$, the streaming terms should satisfy

$$\sum_j v_j(a)F_j(a) = \sum_j \frac{\partial}{\partial a_j} v_j(a), \tag{5.2.14}$$

which follows from the microscopic expression (5.2.5) and is called the *potential condition*.

The statistical average of any quantity $\mathcal{Q}(A(t))$ determined by $A(t)$ at time t is expressed as

$$\langle \mathcal{Q} \rangle_t = \int da\, \mathcal{Q}(a) P(a, t), \tag{5.2.15}$$

where $\langle \cdots \rangle_t$ is the average at time t and $da = \prod_\ell da_\ell$. Its changing rate is

$$\frac{\partial}{\partial t} \langle \mathcal{Q} \rangle_t = \left\langle \sum_i v_i \frac{\partial \mathcal{Q}}{\partial A_i} + \sum_{ij} \left(\frac{\partial}{\partial A_i} - F_i \right) L_{ij} \frac{\partial \mathcal{Q}}{\partial A_j} \right\rangle_t. \tag{5.2.16}$$

For example, the equal-time variance $I_{ij}(t) = \langle A_i A_j \rangle_t$ in nonequilibrium obeys

$$\frac{\partial}{\partial t} I_{ij}(t) = \left\langle v_i A_j + L_{ij} - \sum_\ell A_j \left(L_{i\ell} F_\ell - \frac{\partial}{\partial A_\ell} L_{i\ell} \right) \right\rangle_t + (i \longleftrightarrow j), \tag{5.2.17}$$

where the last term is obtained by exchange of i and j in the first term. This is a generalization of (5.1.20) and will be used in calculating the time-dependent structure factor in (5.3.25) below.

[6] Some analytic and numerical studies were made on Langevin equations with multiplicative noise of the form $g(A(t))\theta(t)$, where $\theta(t)$ is a gaussian–markovian noise. See Ref. [28] for example.

[7] Here the last term $\sum_j \partial L_i j / \partial a_j$ arises because the Itô scheme is used [7].

5.2.3 Time-correlation functions

Let us consider the equilibrium time-correlation function between $\mathcal{Q}_1[t] = \mathcal{Q}_1(A(t))$ and $\mathcal{Q}_2[0] = \mathcal{Q}_2(A(0))$, where $\mathcal{Q}_1(A)$ and $\mathcal{Q}_2(A)$ are arbitrary functions of A. If the gross variables A are fixed at a_0 at $t = 0$, the subsequent distribution is given by $P(a, a_0, t) = \exp(\mathcal{L}_{\text{FP}}\{a\}t)\delta(a - a_0)$, in terms of which we have

$$\begin{aligned}\langle \mathcal{Q}_1[t]\mathcal{Q}_2[0]\rangle &= \int da \int da_0 \mathcal{Q}_1(a)\mathcal{Q}_2(a_0)P(a, a_0, t)P_{\text{eq}}(a_0) \\ &= \int da \mathcal{Q}_1(a)e^{\mathcal{L}_{\text{FP}}\{a\}t}\mathcal{Q}_2(a)P_{\text{eq}}(a). \end{aligned} \tag{5.2.18}$$

As will be shown in Appendix 5C, the time reversal symmetry yields

$$\langle \mathcal{Q}_1[t]\mathcal{Q}_2[0]\rangle = \langle \tilde{\mathcal{Q}}_2[t]\tilde{\mathcal{Q}}_1[0]\rangle = \int da \tilde{\mathcal{Q}}_2(a)e^{\mathcal{L}_{\text{FP}}\{a\}t}\tilde{\mathcal{Q}}_1(a)P_{\text{eq}}(a), \tag{5.2.19}$$

where $\tilde{\mathcal{Q}}_1(A) \equiv \mathcal{Q}_1(\tilde{A})$ and $\tilde{\mathcal{Q}}_2(A) \equiv \mathcal{Q}_2(\tilde{A})$ with $\tilde{A}$ being the time-reversed gross variables (5.2.8). The time-correlation functions $G_{ij}(t) = \langle A_i(t)A_j(0)\rangle$ $(t > 0)$ of the gross variables evolve as

$$\frac{\partial}{\partial t}G_{ij}(t) = \left\langle \left(v_i[t] - \sum_\ell L_{i\ell}F_\ell[t]\right) A_j(0)\right\rangle. \tag{5.2.20}$$

We differentiate the above equation with respect to t again to obtain

$$\frac{\partial^2}{\partial t^2}G_{ij}(t) = -\left\langle \left(v_i[t] - \sum_\ell L_{i\ell}F_\ell[t]\right)\left(v_j[0] + \sum_\ell L_{j\ell}F_\ell[0]\right)\right\rangle. \tag{5.2.21}$$

Here use has been made of the fact that the reversible and irreversible terms change differently with respect to the time reversal, so the latter terms appear with different signs at time t and 0.

More specifically, we consider the case in which the changing rate is divided into linear and nonlinear parts as

$$v_i(A) - \sum_\ell L_{i\ell}F_\ell(A) = -\gamma_i A_i + X_i(A), \tag{5.2.22}$$

where we assume $\langle X_i(A)A_j\rangle = 0$. As will be shown in Appendix 5C, the Laplace transformation of $G_{ij}(t)$ can be expressed as

$$\int_0^\infty dt e^{-\Omega t}G_{ij}(t) = \frac{\langle A_iA_j\rangle}{\Omega + \gamma_i} + \frac{1}{(\Omega + \gamma_i)(\Omega + \gamma_j)}\int_0^\infty dt e^{-\Omega t}\langle X_i[t]\bar{X}_j[0]\rangle, \tag{5.2.23}$$

where $X_i[t] = X_i(A(t))$ and $\bar{X}_j[0] = \epsilon_j X_j(\tilde{A})$. The above relation will be used to set up dynamic renormalization group equations for the kinetic coefficients in some dynamic models below.

5.2.4 Approach to equilibrium

If there is no externally applied perturbation such as heat flow or shear flow, the system tends to equilibrium with the distribution (5.2.3) owing to the fluctuation–dissipation theorem (5.2.7) and the potential condition (5.2.14). Let us define the total *entropy* [15] by

$$\mathcal{S}(t) = -\int \left(\prod_\ell da_\ell\right) P(a,t) \ln[P(a,t)/P_{\rm eq}(a)] = -\langle \ln[P(A,t)/P_{\rm eq}(A)]\rangle_t. \quad (5.2.24)$$

Its changing rate is nonnegative-definite as

$$\frac{\partial}{\partial t}\mathcal{S}(t) = \left\langle \sum_{i,j} L_{ij} \left[\frac{\partial}{\partial A_i}(\ln P + \beta\mathcal{H})\right]\left[\frac{\partial}{\partial A_j}(\ln P + \beta\mathcal{H})\right]\right\rangle_t \geq 0, \quad (5.2.25)$$

where the terms proportional to v_i vanish due to (5.2.14). Therefore, $\tilde{S}(t)$ monotonically decreases with time until the equilibrium $P(a,t) = P_{\rm eq}(a)$ is attained as $t \to \infty$.

In phenomenological transport equations such as the usual hydrodynamic equations, the noise terms are usually neglected and the entropy production rate is nonnegative-definite without flow from outside [23]. Note that the entropy deviation $(\Delta S)_2$ in the bilinear order in (1.2.39) or (1.2.42) corresponds to $-\beta\mathcal{H}$ in the gaussian approximation from (1.2.40). So, let us consider the changing rate of $\beta\mathcal{H}$ neglecting the noise terms:

$$\frac{\partial}{\partial t}\beta\mathcal{H}(A) = \sum_i \frac{\partial v_i}{\partial A_i} - \sum_{ij} L_{ij} F_i F_j. \quad (5.2.26)$$

The right-hand side is nonnegative-definite in the purely dissipative case $v_i = 0$ or in the divergence-free case $\sum_j \partial v_i/\partial A_i = 0$ more generally. The latter condition holds for fluid hydrodynamics and for dynamic models of phase transitions assembled in Ref. [27]. (This can be checked unambiguously in the coarse-grained lattice representation, see the next section.) In these cases $\beta\mathcal{H}$ tends to be minimized as $t \to \infty$.

5.3 Simple time-dependent Ginzburg–Landau models

First, we will construct purely dissipative Langevin equations, where a single-component order parameter $\psi(\boldsymbol{r}, t)$, called the spin variable, depends on space and time. The subscript j in the previous section is now the wave vector $\boldsymbol{k}$ with $k < \Lambda$, Λ being the upper cut-off wave number. We may equivalently suppose a coarse-grained lattice with mesh size $\ell = 2\pi/\Lambda$, as we have introduced in (4.1.2). Then $\psi(\boldsymbol{r}, t)$ may be written as $\psi_j(t)$ for $\boldsymbol{r}$ in the jth cell. The time t is explicitly written hereafter. Second, we will examine the linear dynamics. Third, we will show how the nonlinear term in the thermodynamic force serves to renormalize the kinetic coefficient near the critical point.

5.3.1 Nonconserved systems: model A

The simplest Langevin equations for $\{\psi_j(t)\}$ (j representing the lattice sites) are given by

$$\frac{\partial}{\partial t}\psi_j(t) = -L_0 \frac{\partial}{\partial \psi_j}(\beta \mathcal{H}) + \theta_j(t), \tag{5.3.1}$$

and no streaming term is assumed. The random noise terms $\theta_j(t)$ are independent of one another, gaussian, and markovian, characterized by

$$\langle \theta_j(t)\theta_\ell(t')\rangle = 2L_0\delta(t-t')\delta_{j,\ell}. \tag{5.3.2}$$

In the continuum limit, $\partial/\partial\psi_j$ is replaced by the functional derivative $\delta/\delta\psi(\boldsymbol{r})$ and the above two equations are rewritten as

$$\frac{\partial}{\partial t}\psi(\boldsymbol{r},t) = -L_0 \frac{\delta}{\delta \psi}(\beta \mathcal{H}) + \theta(\boldsymbol{r},t), \tag{5.3.3}$$

$$\langle \theta(\boldsymbol{r},t)\theta(\boldsymbol{r}',t')\rangle = 2L_0\delta(\boldsymbol{r}-\boldsymbol{r}')\delta(t-t'), \tag{5.3.4}$$

which is called model A in Ref. [27]. The GLW hamiltonian (4.1.1) yields the thermodynamic force,

$$\frac{\delta}{\delta\psi}\beta\mathcal{H} = (r + r_{0c} + u_0\psi^2 - K\nabla^2)\psi - h. \tag{5.3.5}$$

The Fokker–Planck equation for the distribution $P(\{\psi\},t)$ is

$$\frac{\partial}{\partial t}P = \int d\boldsymbol{r}\frac{\delta}{\delta\psi(\boldsymbol{r})}L_0\left[\frac{\delta}{\delta\psi(\boldsymbol{r})} + \frac{\delta}{\delta\psi(\boldsymbol{r})}(\beta\mathcal{H})\right]P. \tag{5.3.6}$$

This model describes purely dissipative dynamics of a nonconserved order parameter.

5.3.2 Conserved systems: model B

When a binary alloy consisting of A and B atoms is cooled, it phase-separates into A-rich regions and B-rich regions. If each lattice point is occupied by either an A or a B atom, the order parameter may be taken to be the concentration or density of the species A. Its local conservation law requires a continuity equation,

$$\frac{\partial}{\partial t}\psi(\boldsymbol{r},t) = -\nabla\cdot \boldsymbol{J}_\psi(\boldsymbol{r},t), \tag{5.3.7}$$

where $\boldsymbol{J}_\psi(\boldsymbol{r},t)$ represents the flux of the component A. If there is no flow from outside, the space average of the order parameter M is constant in time. Therefore, it is convenient to characterize the state of the system in terms of the reduced temperature τ and the average M, because h, representing the chemical potential difference, is not usually a controllable parameter. The simplest expression for $\boldsymbol{J}_\psi(\boldsymbol{r},t)$ is

$$\boldsymbol{J}_\psi(\boldsymbol{r},t) = -L_0\nabla\frac{\delta}{\delta\psi}(\beta\mathcal{H}) + \mathcal{G}(\boldsymbol{r},t), \tag{5.3.8}$$

where $\mathcal{G}(\boldsymbol{r}, t)$ is the random flux. Its strength is characterized by

$$\langle \mathcal{G}_j(\boldsymbol{r}, t)\mathcal{G}_k(\boldsymbol{r}', t')\rangle = 2L_0\delta_{jk}\delta(\boldsymbol{r} - \boldsymbol{r}')\delta(t - t'), \tag{5.3.9}$$

where $j, k = x, y, z$. Thus we obtain the Langevin equation (model B [27]),

$$\frac{\partial}{\partial t}\psi(\boldsymbol{r}, t) = L_0\nabla^2\frac{\delta}{\delta\psi}(\beta\mathcal{H}) + \theta(\boldsymbol{r}, t). \tag{5.3.10}$$

Here

$$\theta(\boldsymbol{r}, t) = -\nabla\cdot\mathcal{G}(\boldsymbol{r}, t) \tag{5.3.11}$$

is the noise term and its correlation is formally expressed as

$$\langle\theta(\boldsymbol{r}, t)\theta(\boldsymbol{r}', t')\rangle = -2L_0\nabla^2\delta(\boldsymbol{r} - \boldsymbol{r}')\delta(t - t'). \tag{5.3.12}$$

The corresponding Fokker–Planck equation can be obtained if we replace L_0 in (5.3.6) by $-L_0\nabla^2$.

5.3.3 Coupled systems: model C

It is usual that the order parameter ψ is coupled to a conserved variable m in the GLW hamiltonian $\mathcal{H} = \mathcal{H}\{\psi, m\}$ as (4.1.45). Then slow relaxation of m can influence the dynamics of the nonconserved ψ. The simplest dynamic equations, called model C [27], are provided by (5.3.3) for ψ and

$$\frac{\partial}{\partial t}m(\boldsymbol{r}, t) = \lambda_0\nabla^2\frac{\delta}{\delta m}\beta\mathcal{H} + \zeta(\boldsymbol{r}, t) \tag{5.3.13}$$

for m. The coefficient λ_0 is the thermal conductivity if m is the energy variable. The noise term ζ satisfies (5.3.12) with L_0 being replaced by λ_0. In this model there are no mode coupling terms, but ψ and m are coupled dissipatively because the functional derivative,

$$\frac{\delta}{\delta\psi}\beta\mathcal{H} = (r_{0c} + 2\gamma_0 m + \bar{u}_0\psi^2 - K\nabla^2)\psi - h, \tag{5.3.14}$$

contains the nonlinear term $2\gamma_0 m\psi$. Furthermore, we may use the above model to describe tricritical dynamics in metamagnets by adding the sixth-order term $v_0\psi^6$ in the free-energy density as in (3.2.1) [29].

Steady states under a temperature (chemical potential) gradient

In this coupled model we may apply a constant heat flow with a constant temperature gradient,

$$\boldsymbol{a} = \nabla\left\langle\frac{\delta}{\delta m}\mathcal{H}\right\rangle_{\mathrm{ss}}. \tag{5.3.15}$$

From (3.1.27) or (4.1.46), $\delta\mathcal{H}/\delta m$ is the temperature (or chemical potential) fluctuation. We expect the existence of a steady-state distribution $P_{\mathrm{ss}}\{\psi, m\}$, which is the solution of $\mathcal{L}_{FP}P_{ss} = 0$ under (5.3.15), $\mathcal{L}_{FP}\{\psi, m\}$ being the Fokker–Planck operator. The average $\langle\cdots\rangle_{ss}$ in (5.3.15) is taken over P_{ss}. In our system, without the mode coupling terms, P_{ss} is simply of the local equilibrium form,

$$P_{ss} = P_{local} \propto \exp(-\beta\mathcal{H}_{local}), \quad \mathcal{H}_{local} = \mathcal{H} - \int d\boldsymbol{r}(\boldsymbol{a}\cdot\boldsymbol{r})m(\boldsymbol{r}). \tag{5.3.16}$$

Then $\langle\cdots\rangle_{ss} = \int d\psi dm(\cdots)P_{local}$ and $\langle\delta\mathcal{H}_{local}/\delta m\rangle_{ss} = \langle\delta\mathcal{H}/\delta m\rangle_{ss} - \boldsymbol{a}\cdot\boldsymbol{r} = 0$, leading to (5.3.15). Because the distribution of m is gaussian for each fixed ψ, we may determine m by

$$\frac{\delta}{\delta m}\beta\mathcal{H}_{local} = C_0^{-1}m + \gamma_0\psi^2 - \tau - \beta\boldsymbol{a}\cdot\boldsymbol{r} = 0, \tag{5.3.17}$$

neglecting its fluctuations. Then P_{local} becomes a steady-state distribution for ψ only, in which the temperature coefficient r linearly depends on space as

$$r = a_0(\tau + \beta\boldsymbol{a}\cdot\boldsymbol{r}). \tag{5.3.18}$$

In this case the hamiltonian under heat flow is well defined. It is of the same form as that for ^{4}He under gravity in (4.2.50). In Chapter 6 we shall see that the steady-state distribution deviates from P_{local} in the presence of the mode coupling terms near the gas–liquid critical point, leading to critical enhancement of the thermal conductivity.

5.3.4 Mean field theory and thermodynamic stability

Models A and B

We examine the linearized dynamic equation for the deviation $\delta\psi = \psi - M$ for models A and B, where $M = \langle\psi\rangle$ is assumed to be homogeneous. The thermodynamic force (5.3.5) is given to first order in the deviation as

$$\frac{\delta}{\delta\psi}(\beta\mathcal{H}) \cong (r_{eff} - K\nabla^2)\delta\psi, \tag{5.3.19}$$

where

$$r_{eff} = r + 3u_0M^2. \tag{5.3.20}$$

The shift r_{0c} is neglected in the mean field calculation. From (5.3.3) and (5.3.10) we obtain a linearized Langevin equation,

$$\frac{\partial}{\partial t}\delta\psi = -L_0(-\nabla^2)^a(r_{eff} - K\nabla^2)\delta\psi + \theta. \tag{5.3.21}$$

Here the exponent a is 0 for the nonconserved case and 1 for the conserved case. In the Fourier space, $\psi_{\boldsymbol{k}}(t)$ are independent of one another as

$$\frac{\partial}{\partial t}\psi_{\boldsymbol{k}} = -L_0k^{2a}(r_{eff} + Kk^2)\psi_{\boldsymbol{k}} + \theta_{\boldsymbol{k}}, \tag{5.3.22}$$

with

$$\langle \theta_{\boldsymbol{k}}(t)\theta_{\boldsymbol{k}'}(t')\rangle = 2L_0 k^{2a}\delta(\boldsymbol{k}+\boldsymbol{k}')\delta(t-t'). \tag{5.3.23}$$

The decay rate is thus

$$\Gamma_k = L_0 k^{2a}(r_{\text{eff}} + Kk^2). \tag{5.3.24}$$

The system is stable in the case $r_{\text{eff}} \geq 0$ with respect to small plane-wave fluctuations. If $r_{\text{eff}} < 0$, the fluctuations grow for $k < |r_{\text{eff}}/K|^{1/2}$. We write the equation for the equal-time structure factor $I_k(t) \equiv \langle |\psi_{\boldsymbol{k}}(t)|^2\rangle$,

$$\frac{\partial}{\partial t} I_k(t) = -2\Gamma_k I_k(t) + 2L_0 k^{2a}, \tag{5.3.25}$$

which follows from (5.2.17) and is of the same form as (5.1.20). If $r_{\text{eff}} \geq 0$, $I_k(t)$ tends to the Ornstein–Zernike form,

$$I_{\text{OZ}}(k) = 1/(r_{\text{eff}} + Kk^2), \tag{5.3.26}$$

as $t \to \infty$. The spinodal point is given by $r_{\text{eff}} = 0$ or

$$a_0(T/T_{\text{c}} - 1) = -3u_0 M^2, \tag{5.3.27}$$

which forms the spinodal curve in the T–M plane placed below the mean field coexistence curve $a_0(T/T_{\text{c}} - 1) = -u_0 M^2$. In the present analysis, in which the nonlinear coupling between the fluctuations is neglected, the system is linearly unstable below the spinodal curve against long-wavelength fluctuations. Phase-ordering processes then proceed, as will be treated in Chapter 8. Between the coexistence and spinodal curves, nucleation processes are expected to take place, for which see Chapter 9.

Model C

In a disordered phase with $\langle\psi\rangle = 0$ the fluctuations of ψ and m are decoupled in the linear analysis. If the nonlinear coupling is neglected, ψ behaves as in model A and δm relaxes diffusively with the diffusion constant $D_0 = \lambda_0/C_0$. Let us then assume $M = \langle\psi\rangle \neq 0$, where $rM + u_0 M^3 = h$ with $r = a_0\tau$. The Fourier components $\psi_{\boldsymbol{k}}$ and $m_{\boldsymbol{k}}$ obey [29]

$$\frac{\partial}{\partial t}\psi_{\boldsymbol{k}} = -L_0\big[(r_1 + Kk^2)\psi_{\boldsymbol{k}} + 2\gamma_0 M m_{\boldsymbol{k}}\big] + \theta_{\boldsymbol{k}}, \tag{5.3.28}$$

$$\frac{\partial}{\partial t} m_{\boldsymbol{k}} = -\lambda_0 k^2\big[2\gamma_0 M\psi_{\boldsymbol{k}} + C_0^{-1} m_{\boldsymbol{k}}\big] + \zeta_{\boldsymbol{k}}. \tag{5.3.29}$$

Using $u_0 = \bar{u}_0 - 2\gamma_0^2 C_0$ from (4.1.48) the coefficient r_1 is written as

$$r_1 = 2\gamma_0\langle m\rangle + 3\bar{u}_0 M^2 = a_0\tau + 3u_0 M^2 + 4\gamma_0^2 C_0 M^2, \tag{5.3.30}$$

where $a_0 = 2\gamma_0 C_0$ and $r_{0\text{c}}$ is neglected. We note that the reduced temperature τ in (4.1.18) may be related to the average energy density $\langle m\rangle$ as $C_0^{-1}\langle m\rangle + \gamma_0 M^2 = \tau$ in the mean field theory. The Fourier components relax in the form of $A_1\exp(-\Omega_1 t) + A_2\exp(-\Omega_2 t)$

(if the noise terms are neglected). The linear stability ($\Omega_1, \Omega_2 \geq 0$) is assured by the nonnegativity of the following combination,

$$r_{\rm eff} = r_1 - a_0^2 M^2 = a_0\tau + 3u_0 M^2. \tag{5.3.31}$$

The spinodal is given by $r_{\rm eff} = 0$. Interestingly, as $r_{\rm eff} \to 0$ and $k \to 0$, the relaxation rates behave as

$$\Omega_1 \cong L_0 r_1, \quad \Omega_2 \cong (\lambda_0 / C_{\rm eff})k^2, \tag{5.3.32}$$

where $C_{\rm eff}$ is the specific heat at constant h written as

$$C_{\rm eff} = C_0 + a_0^2 M^2 / r_{\rm eff} = C_0 r_1 / r_{\rm eff}. \tag{5.3.33}$$

For $M \neq 0$ the diffusive mode first undergoes slowing down as $r_{\rm eff} \to 0$, where the relaxation of ψ is governed by the slow diffusive motion of m. In fact, we may set $\partial\psi/\partial t = \theta = 0$ in (5.3.28) as $k \to 0$ to obtain $\psi_{\boldsymbol k} \cong -(2\gamma_0 M/r_1)m_{\boldsymbol k}$. Substitution of this result into (5.3.29) gives the diffusion equation with the diffusion constant $\lambda_0/C_{\rm eff}$.

5.3.5 Critical dynamics in model A

We have studied the effect of the quartic term in the GLW hamiltonian in statics. However, the way it affects the purely dissipative dynamics governed by (5.3.3) is not trivial [30]–[32]. It is known that, although the dynamical effect is subtle, the kinetic coefficient L_0 is renormalized with a multiplicative factor smaller than 1 for $\epsilon = 4 - d > 0$. The upper critical dimensionality remains 4 also in dynamics. We are interested in the time-correlation function in equilibrium,

$$G(k,t) = \langle \psi_{\boldsymbol k}(t)\psi_{-\boldsymbol k}(0)\rangle. \tag{5.3.34}$$

For simplicity, we assume $h = 0$ and $\tau > 0$. Following Kawasaki [33], we rewrite (5.3.3) in the Fourier space as

$$\frac{\partial}{\partial t}\psi_{\boldsymbol k} = -\gamma_k\psi_{\boldsymbol k} - \tilde{J}_{\boldsymbol k} + \theta_{\boldsymbol k}. \tag{5.3.35}$$

Here γ_k is the linear relaxation rate defined by

$$\gamma_k = L_0/\chi_k, \tag{5.3.36}$$

where $\chi_k \equiv \langle |\psi_{\boldsymbol k}|^2\rangle$ is the static structure factor. The nonlinear part

$$\tilde{J}_{\boldsymbol k} = L_0 \frac{\partial}{\partial \psi_{-\boldsymbol k}}\beta\mathcal{H} - \gamma_k\psi_{\boldsymbol k} \tag{5.3.37}$$

is orthogonal to ψ or $\langle \tilde{J}_{\boldsymbol k}\psi_{\boldsymbol k'}\rangle = 0$ from the Fourier transformation of the second relation of (4.1.28). Therefore,

$$\frac{\partial}{\partial t}G(k,t) \to -\gamma_k\chi_k \tag{5.3.38}$$

as $t \to 0$. From (5.2.23) the Laplace transformation of $G(k, t)$ is written as

$$\frac{1}{\chi_k}\int_0^\infty dt e^{-i\omega t} G(k, t) = \frac{1}{i\omega + \gamma_k} + \frac{\gamma_k}{(i\omega + \gamma_k)^2}\phi(k, \omega), \tag{5.3.39}$$

where

$$\phi(k, \omega) = \frac{1}{L_0}\int_0^\infty dt e^{-i\omega t}\langle \tilde{J}_k(t)\tilde{J}_{-k}(0)\rangle. \tag{5.3.40}$$

Kawasaki defined the true lifetime τ_k of the fluctuations by

$$\tau_k = \frac{1}{\chi_k}\int_0^\infty dt G(k, t). \tag{5.3.41}$$

In the limit $\omega \to 0$, (5.3.39) becomes

$$\tau_k = \frac{1}{\gamma_k} + \frac{1}{\gamma_k}\phi(k, 0) = \frac{\chi_k}{L_0}\left[1 + \frac{1}{L_0}\int_0^\infty dt\langle \tilde{J}_k(t)\tilde{J}_{-k}(0)\rangle\right]. \tag{5.3.42}$$

The lifetime becomes longer than γ_k^{-1} if the nonlinearity is purely dissipative.

Dynamic renormalization group theory

Because $\tilde{J}_k(t)$ is the Fourier transformation of $u_0\psi^3$ to leading order in u_0, the function $\phi(k, \omega)$ is already of order ϵ^2 in the scheme of the ϵ expansion. As will be shown in Appendix 5D, it is given by [30]–[32]

$$\phi(k, 0) = 9\ln(4/3)g^2\ln(\Lambda/k) \tag{5.3.43}$$

for $k\xi \gg 1$ and in the limit $\omega \to 0$, Λ being the upper cut-off wavenumber. The expression for $k\xi \ll 1$ is obtained if k is replaced by $\kappa = \xi^{-1}$. The parameter g is defined by (4.1.22) and may be assumed to take the fixed-point value $g^* = \epsilon/9$ in (4.3.16). We thus find the renormalized kinetic coefficient,

$$L_R = L_0[1 - \phi(k, 0)] \cong L_0(\kappa/\Lambda)^{\bar{z}} \tag{5.3.44}$$

with

$$\bar{z} = \frac{1}{9}\ln(4/3)\epsilon^2 = 6\ln(4/3)\eta. \tag{5.3.45}$$

The dynamic exponent z is determined from $\tau_k = \chi_k/L_R \sim \xi^z$ at $k \sim \xi^{-1}$. Up to order ϵ^2 we have

$$z = 2 - \eta + \bar{z} = 2 + [6\ln(4/3) - 1]\eta, \tag{5.3.46}$$

where η is in (4.3.51). The increase of z from the mean field value 2 is of order η and is very small in 3D.

To be precise, we need to justify the exponentiation of the logarithmic term in (5.3.41) by setting up the renormalization group equation for $L_0(\Lambda)$. Obviously, the fluctuations in the shell region ($\Lambda - \delta\Lambda < q < \Lambda$) give rise to the contribution $\delta\phi = 9\ln(4/3)g^2\delta\Lambda/\Lambda$ to $\phi(k,\omega)$ at small k and ω. Because

$$\frac{1}{i\omega + \gamma_k} + \frac{\gamma_k}{(i\omega+\gamma_k)^2}\delta\phi \cong \frac{1}{i\omega + \gamma_k(1-\delta\phi)}, \tag{5.3.47}$$

we find $L_0(\Lambda - \delta\Lambda) = L_0(\Lambda)(1-\delta\phi)$, so that by setting $\Lambda = \Lambda_0 e^{-\ell}$ we obtain

$$\frac{\partial}{\partial \ell} L_0(\Lambda) = -9\ln(4/3)g^2 L_0(\Lambda), \tag{5.3.48}$$

which is integrated to give $L_0(\Lambda)\Lambda^{-\bar{z}} = L_R\kappa^{-\bar{z}}$ or (5.3.44) at $g = g^*$.

Yahata and Suzuki's calculation

Yahata and Suzuki [34] studied the kinetic Ising model [35, 36] numerically in 2D and found that the lifetime $\tau(T)$, which is τ_k in (5.3.41) in the limit $k \to 0$, behaves as

$$\tau(T) \propto (T - T_c)^{-\Delta}, \tag{5.3.49}$$

as $T \to T_c$. They obtained $\Delta \cong 2.00 \pm 0.05$, which is larger than $\gamma = (2-\eta)/\nu = 7/4$ for the 2D Ising model. If we write $1/\tau(T) = L_R/\chi$, the renormalized kinetic coefficient has a relatively large critical singularity at $d = 2$ as

$$L_R \propto \kappa^{(\Delta-\gamma)/\nu}, \tag{5.3.50}$$

with $(\Delta - \gamma)/\nu \cong 1/4 \cong \eta$. Note that η is not very small in 2D.

5.3.6 Critical dynamics in model C

Let us consider another purely dissipative dynamics, model C, governed by (5.3.3) and (5.3.13) for ψ and m in a disordered phase with $\langle\psi\rangle = 0$ [37]. In this case, $\tilde{J}_k$ in (5.3.35) contains another relevant term,

$$\tilde{J}_k = 2L_0\gamma_0 \int_q m_q \psi_{k-q} + \cdots, \tag{5.3.51}$$

which arises from $2\gamma_0 m\psi$ in (5.3.14). The fluctuation contribution to $\phi(k,\omega)$ in (5.3.40) from the shell region can be calculated using the decoupling approximation as

$$\delta\phi \cong 4L_0\gamma_0^2(K_d\Lambda^{d-3}\delta\Lambda)C_0\chi_\Lambda/(\lambda_0 C_0^{-1} + L_0), \tag{5.3.52}$$

where $\Lambda \gg \kappa$ is assumed and $\chi_\Lambda \cong 1/\Lambda^2$ is the variance at the cut-off. The RG equation for L_0 becomes

$$\frac{\partial}{\partial \ell} L_0 = -K_d(2\gamma_0 C_0 L_0)^2/[\Lambda^{\epsilon}(\lambda_0 + C_0 L_0)]. \tag{5.3.53}$$

There is no fluctuation contribution to λ_0 in the long-wavelength limit, so λ_0 is a constant. The ratio of the timescales of ψ and m is represented by

$$w = C_0 L_0/\lambda_0. \tag{5.3.54}$$

For Ising-like systems ($n = 1$), $C_0 = C_0(\Lambda)$ obeys the RG equation (4.3.31) with v being defined by (4.1.55). Its explicit form is given by (4.3.39) or (4.3.43). It is easy to rewrite (5.3.53) as

$$\frac{\partial}{\partial \ell} w = 2vw(1 - w)/(1 + w). \tag{5.3.55}$$

This equation is solved to give

$$w(\Lambda)/[1 - w(\Lambda)]^2 = A_{\text{ini}} C_0(\Lambda), \tag{5.3.56}$$

where A_{ini} is a constant determined from the initial condition at $\Lambda = \Lambda_0$. If $C_0(\Lambda)$ can grow such that $A_{\text{ini}} C_0(\Lambda) \gg 1$, then $w(\Lambda)$ approaches 1 or

$$L_R \cong \lambda_0 C_0(\kappa)^{-1} \propto \kappa^{\alpha/\nu}. \tag{5.3.57}$$

This renormalization effect can thus be sensitive to the critical behavior of the specific heat. As a result, it can be effective in 3D Ising-like systems, whereas it is expected to be negligible in 2D Ising systems (where $\alpha = 0$). For many-component systems ($n \geq 2$) the effect becomes more delicate than in single-component systems [37].

5.4 Linear response

Linear response of various physical quantities to a weak applied field represented as a small perturbation in the hamiltonian can be expressed in very compact forms in terms of the appropriate time-correlation functions [38]–[40]. Representative examples are the frequency-dependent response to a weak magnetic or electric field. Similar expressions are well known also for transport coefficients in fluids, as will be explained below. However, thermal disturbances such as spatial gradients of the velocity field and the temperature, which inevitably drive fluids away from equilibrium, cannot be expressed as perturbations in the hamiltonian. This means that nonequilibrium ensemble distributions deviate from *local equilibrium* forms for thermal disturbances.

5.4.1 Transport coefficients in fluids

Historically, transport coefficients were first systematically calculated for dilute gases on the basis of the Boltzmann equation [41, 42]. There, the one-body distribution function $f(\boldsymbol{r}, \boldsymbol{p}, t)$ in the $(\boldsymbol{r}, \boldsymbol{p})$ space is expanded around the local equilibrium maxwellian distribution in powers of gradients of the velocity field and temperature (the Enskog–Chapman expansion) [43]. Such a small deviation of the one-body distribution evolves in time with the linearized Boltzmann operator $\mathcal{L}_{\text{LB}}$ and gives rise to transport coefficients with expressions involving the inverse $\mathcal{L}_{\text{LB}}^{-1}$ from the time integration. The kinetic theory for dilute gases and the Enskog theory for non-dilute hard-sphere fluids [44] are instructive examples of nonequilibrium theories in which transport coefficients are analytically calculable.

In this subsection we will give general microscopic expressions for the shear viscosity, bulk viscosity, and thermal conductivity of fluids in terms of appropriate time-correlation functions [15]–[24], [45]–[50]. It is important that the transport coefficients naturally appear as the kinetic coefficients in the dynamics of the long-wavelength hydrodynamic variables in the scheme of linear Langevin equations [17]. As will be discussed in Appendix 5B, linear Langevin equations can be systematically derived using the linear projection operator $\mathcal{P}$ onto such gross variables. Recall that the linear projection onto the hydrodynamic variables has already been introduced at the end of Section 1.2.

Actual calculations of the transport coefficients in dense fluids can be performed via molecular dynamics simulations on the basis of the molecular expressions presented here [51, 52].

Viscosities

From (1.2.76) the random part of the stress tensor $\Pi_{\alpha\beta}(\boldsymbol{r})$ ($\alpha, \beta = x, y, z$) is given by

$$\Pi^{\mathrm{R}}_{\alpha\beta}(\boldsymbol{r}) = (1-\mathcal{P})\Pi_{\alpha\beta}(\boldsymbol{r}) = \Pi_{\alpha\beta}(\boldsymbol{r}) - (p + \delta\hat{p}(\boldsymbol{r}))\delta_{\alpha\beta}, \tag{5.4.1}$$

where p is the average thermodynamic pressure and $\delta\hat{p}(\boldsymbol{r})$ is the pressure fluctuation variable defined by (1.2.66) or (1.3.45) in the long-wavelength limit ($\Lambda \to 0$). The microscopic expression for $\Pi_{\alpha\beta}$ can be found in Appendix 5E. As (5B.9) will show, this variable evolves in time as

$$\begin{aligned} \Pi^{\mathrm{R}}_{\alpha\beta}(\boldsymbol{r}, t) &= e^{(1-\mathcal{P})i\mathcal{L}t}\Pi^{\mathrm{R}}_{\alpha\beta}(\boldsymbol{r}) \\ &\cong e^{i\mathcal{L}t}\Pi^{\mathrm{R}}_{\alpha\beta}(\boldsymbol{r}), \end{aligned} \tag{5.4.2}$$

where $i\mathcal{L}$ is the Liouville operator (5B.1) in the Γ space. In actual calculations such as in molecular dynamics [51], the modified time evolution realized by $(1-\mathcal{P})i\mathcal{L}$ is replaced by the usual newtonian time evolution realized by $i\mathcal{L}$ as in the second line of (5.4.2). This is allowable in fluids in the long-wavelength limit ($\Lambda \to 0$) of disturbances where the gross variables tend to constants of motion [20, 49]. Using the rotational invariance of the system, the frequency-dependent complex viscosities, $\eta^*(\omega)$ and $\zeta^*(\omega)$, are given by

$$\begin{aligned} &\frac{1}{T}\int_0^\infty dt \int d\boldsymbol{r} e^{-i\omega t}\langle \Pi^{\mathrm{R}}_{\alpha\beta}(\boldsymbol{r}, t)\Pi^{\mathrm{R}}_{\gamma\delta}(\boldsymbol{0}, 0)\rangle \\ &= (\delta_{\alpha\gamma}\delta_{\beta\delta} + \delta_{\alpha\delta}\delta_{\beta\gamma})\eta^*(\omega) + \delta_{\alpha\beta}\delta_{\gamma\delta}\left[\zeta^*(\omega) - \frac{2}{d}\eta^*(\omega)\right], \end{aligned} \tag{5.4.3}$$

where d is the spatial dimensionality. In particular, the following expressions are convenient:

$$\eta^*(\omega) = \frac{1}{T}\int_0^\infty dt \int d\boldsymbol{r} e^{-i\omega t}\langle \Pi_{xy}(\boldsymbol{r}, t)\Pi_{xy}(\boldsymbol{0}, 0)\rangle, \tag{5.4.4}$$

$$\zeta^*(\omega) = \frac{1}{d^2 T}\int_0^\infty dt \int d\boldsymbol{r} e^{-i\omega t}\left\langle \sum_\alpha \Pi^{\mathrm{R}}_{\alpha\alpha}(\boldsymbol{r}, t)\sum_\beta \Pi^{\mathrm{R}}_{\beta\beta}(\boldsymbol{0}, 0)\right\rangle. \tag{5.4.5}$$

In (5.4.4) we have used $\Pi^{\mathrm{R}}_{\alpha\beta} = \Pi_{\alpha\beta}$ for $\alpha \neq \beta$. Furthermore, if Π_{xy} is replaced by Π^{R}_{xx}, we have the expression for the combination $\zeta^*(\omega) + (2 - \frac{2}{d})\eta^*(\omega)$, which serves as the viscosity for one-dimensional fluid flows.

Thermal conductivity

We consider the thermal conductivity in the limit $\omega \to 0$. It is expressed in terms of the time-correlation function of the random heat current. The orthogonal part of the energy current $\boldsymbol{J}_{\mathrm{e}}(\boldsymbol{r})$ with respect to the momentum density $\boldsymbol{J}(\boldsymbol{r})$ reads

$$\boldsymbol{J}_{\mathrm{e}}^{\mathrm{R}}(\boldsymbol{r}) = (1 - \mathcal{P})\boldsymbol{J}_{\mathrm{e}}^{\mathrm{R}}(\boldsymbol{r}) = \boldsymbol{J}_{\mathrm{e}}(\boldsymbol{r}) - \frac{e+p}{\rho}\boldsymbol{J}(\boldsymbol{r}), \tag{5.4.6}$$

which may be called the *heat current.* Here e is the average energy density, p is the pressure, and ρ is the average mass density. From the microscopic expression for $\boldsymbol{J}_e$ in Appendix 5E we obtain

$$\langle J_{e\alpha} : J_\beta \rangle = \delta_{\alpha\beta} T(e + p), \tag{5.4.7}$$

where $\langle\ :\ \rangle$ denotes the correlation in the long-wavelength limit defined by (1.1.35). As in the case of the random stress in (5.4.2), the time evolution of the random heat current may be assumed to be governed by newtonian dynamics,

$$\boldsymbol{J}_{\mathrm{e}}^{\mathrm{R}}(\boldsymbol{r}, t) \cong \exp(i\mathcal{L}t)\boldsymbol{J}_{\mathrm{e}}^{\mathrm{R}}(\boldsymbol{r}). \tag{5.4.8}$$

The thermal conductivity is then expressed as

$$\lambda = \frac{1}{T^2}\int_0^\infty dt \int d\boldsymbol{r} \langle J_{\mathrm{ex}}^{\mathrm{R}}(\boldsymbol{r}, t) J_{\mathrm{ex}}^{\mathrm{R}}(\boldsymbol{0}, 0)\rangle. \tag{5.4.9}$$

Dissipative coupling in diffusion and heat conduction

In a binary fluid the momentum densities $\boldsymbol{J}_K(\boldsymbol{r})$ of the two components ($K = 1, 2$) are decomposed as

$$\boldsymbol{J}_1(\boldsymbol{r}) = \frac{\rho_1}{\rho}\boldsymbol{J}(\boldsymbol{r}) + \boldsymbol{I}^{\mathrm{R}}(\boldsymbol{r}), \quad \boldsymbol{J}_2(\boldsymbol{r}) = \frac{\rho_2}{\rho}\boldsymbol{J}(\boldsymbol{r}) - \boldsymbol{I}^{\mathrm{R}}(\boldsymbol{r}), \tag{5.4.10}$$

where ρ_K are the average mass densities with $\rho = \rho_1 + \rho_2$. The orthogonal part $\boldsymbol{I}^{\mathrm{R}}(\perp \boldsymbol{J})$ gives rise to relative motion. In addition to the thermal conductivity (5.4.9), we have additional kinetic coefficients,

$$\begin{aligned}
L_{12} &= \frac{1}{T}\int_0^\infty dt \int d\boldsymbol{r} \langle J_{\mathrm{ex}}^{\mathrm{R}}(\boldsymbol{r}, t) I_x^{\mathrm{R}}(\boldsymbol{0}, 0)\rangle, \\
L_{21} &= \frac{1}{T}\int_0^\infty dt \int d\boldsymbol{r} \langle I_x^{\mathrm{R}}(\boldsymbol{r}, t) J_{\mathrm{ex}}^{\mathrm{R}}(\boldsymbol{0}, 0)\rangle, \\
L_{22} &= \int_0^\infty dt \int d\boldsymbol{r} \langle I_x^{\mathrm{R}}(\boldsymbol{r}, t) I_x^{\mathrm{R}}(\boldsymbol{0}, 0)\rangle.
\end{aligned} \tag{5.4.11}$$

Here the relation $L_{12} = L_{21}$ follows from the microscopic time reversal invariance and is an example of the Onsager reciprocity relations. These kinetic coefficients, together with the thermal conductivity $L_{11} = \lambda$, determine diffusion and heat fluxes driven by gradients of the temperature and chemical potential difference. In Section 6.3 they will appear in coupled diffusion equations for the entropy and concentration.

In the dilute limit $\rho_2 \to 0$ of the second component, the correlations among the particles of the species 2 become negligible and

$$L_{22} \to m_{20}\rho_2 D, \tag{5.4.12}$$

where m_{20} is the particle mass of the species 2 and D is the diffusion constant of an isolated particle of the species 2 expressed in terms of the time-correlation function of the velocity, (5.1.33).

5.4.2 General linear response to thermal disturbances

Attempts have been made to seek linear response of any general dynamic variables to thermal disturbances [24, 47, 48, 53] as in the case of linear response in which the perturbation is a part of the time-dependent hamiltonian [40].[8] In this case the microscopic (Γ space) distribution $P(\Gamma)$ is expanded around a local equilibrium distribution $P_{\text{local}}(\Gamma)$ in powers of gradients of the velocity and temperature. This is analogous to the Enskog–Chapman expansion in the kinetic theory [43]. Unlike the case of perturbations which can be included in the hamiltonian [40], a set of the gross variables $\{A_j\}$ needs to be specified at the starting point of the theory (see Appendix 5B); these are long-wavelength parts of the five conserved variables in a one-component fluid. For example, let a fluid be slightly disturbed with small average velocity gradients $\partial v_\alpha(\boldsymbol{r}, t)/\partial x_\beta$ varying slowly in space and time. The nonequilibrium average $\langle \cdots \rangle_t$ at time t of any local variable $\mathcal{B}(\boldsymbol{r})$ dependent on space is written as [53]

$$\langle \mathcal{B}(\boldsymbol{r}) \rangle_t \cong \langle \mathcal{B}(\boldsymbol{r}) \rangle_\ell(t) - \frac{1}{T} \int_{-\infty}^{t} dt' \int d\boldsymbol{r}' \sum_{\alpha\beta} \langle \mathcal{B}^{\text{R}}(\boldsymbol{r}, t-t') \Pi^{\text{R}}_{\alpha\beta}(\boldsymbol{r}') \rangle \frac{\partial v_\alpha(\boldsymbol{r}', t')}{\partial x'_\beta}. \tag{5.4.13}$$

In the first term, $\langle \cdots \rangle_\ell(t)$ is the average over a local equilibrium distribution of the form,

$$P_{\text{local}}(t) \propto \exp\Big[-\beta\mathcal{H} + \sum_j A_j \Psi_j(t)\Big] \propto \Big[1 + \sum_j \delta A_j \Psi_j(t) + \cdots\Big] \exp(-\beta\mathcal{H}), \tag{5.4.14}$$

where $\delta A_j = A_j - \langle A_j \rangle$ and the coefficients $\Psi_j(t)$ are determined such that

$$\begin{aligned} \langle A_j \rangle_t &= \langle A_j \rangle_\ell(t) \\ &\cong \langle A_j \rangle + \sum_k \langle \delta A_j \delta A_k \rangle \Psi_\ell(t) \end{aligned} \tag{5.4.15}$$

hold for the gross variables. In the linear regime, the averages $\langle \cdots \rangle$ in (5.4.13)–(5.4.15) are those in equilibrium, and

$$\mathcal{B}^{\text{R}}(\boldsymbol{r}, t) \equiv e^{(1-\mathcal{P})i\mathcal{L}t}(1 - \mathcal{P})\mathcal{B}(\boldsymbol{r}) \tag{5.4.16}$$

in terms of the linear projection operator $\mathcal{P}$ onto $\{A_j\}$, so the second term in (5.4.13) identically vanishes for $\mathcal{B} = A_j$. We notice that P_{local} is analogous to the local equilibrium

[8] Formal theory in nonlinear response regimes is very complicated. Nonlinear response against shear flow in fluids has been studied via molecular dynamics simulations [52].

maxwellian distribution $f_{\rm local} = n(2\pi m_0 T)^{-d/2} \exp(-|\boldsymbol{p} - m_0 \boldsymbol{v}|^2/2m_0 T)$ in the kinetic theory of dilute gases, where the density n, the temperature T, and the velocity field $\boldsymbol{v}$ slowly depend on space and time.

Now the substitution $\mathcal{B} = \Pi_{\alpha\beta}$ gives rise to the viscosities in (5.4.3). In particular, for a simple shear flow $\partial v_\alpha(\boldsymbol{r}, t)/\partial x_\beta = \dot{\gamma}(t)\delta_{\alpha x}\delta_{\beta y}$, we obtain a well-known form,

$$\langle \Pi_{xy}(\boldsymbol{r}) \rangle_t = -\int_{-\infty}^{t} dt' G(t - t')\dot{\gamma}(t'), \tag{5.4.17}$$

where

$$G(t) = \frac{1}{T} \int d\boldsymbol{r} \langle \Pi_{xy}(\boldsymbol{r}, t)\Pi_{xy}(\boldsymbol{0}, 0) \rangle \tag{5.4.18}$$

is called the stress relaxation function. Its Laplace (one-sided Fourier) transformation is the frequency-dependent shear viscosity $\eta^*(\omega)$. In the literature the complex shear modulus is defined by

$$G^*(\omega) = i\omega \int_0^\infty dt e^{-i\omega t} G(t) = i\omega\eta^*(\omega). \tag{5.4.19}$$

The real and imaginary parts of this quantity have been measured in various materials. In many polymeric systems and supercooled liquids, $G(t)$ decays on very slow timescales and $G^*(\omega)$ exhibits singular behavior at small ω.

In the presence of a small temperature gradient, the counterpart of (5.4.13) reads

$$\langle \mathcal{B}(\boldsymbol{r}) \rangle_t \cong \langle \mathcal{B}(\boldsymbol{r}) \rangle_\ell(t) - \frac{1}{T^2} \int_{-\infty}^{t} dt' \int d\boldsymbol{r}' \sum_\alpha \langle \mathcal{B}^{\rm R}(\boldsymbol{r}, t - t') J_{e\alpha}^{\rm R}(\boldsymbol{r}') \rangle \frac{\partial T(\boldsymbol{r}', t')}{\partial x'_\alpha}. \tag{5.4.20}$$

The substitution $\mathcal{B} = \boldsymbol{J}_e$ gives rise to the thermal conductivity (5.4.9) in the steady-state limit. In a binary fluid mixture, the gradient of the chemical potential is also a thermodynamic force. In the same manner as above, we may derive the microscopic expressions (5.4.11) for $L_{12} = L_{21}$ and L_{22}. An example of deriving (5.4.20) near the gas–liquid critical point will be given in Appendix 6C in the Ginzburg–Landau scheme.

Response to sound wave

As an interesting but not well-known example, let us consider linear response to a sound wave propagating in the x direction, where $\partial v_x/\partial x \cong -(\partial \rho_1/\partial t)/\rho$ in terms of the density deviation $\rho_1(x, t)$ induced by the sound. From (5.4.13) the local equilibrium average is

$$\langle \mathcal{B}(\boldsymbol{r}) \rangle_\ell(t) \cong b + \left(\frac{\partial b}{\partial \rho}\right)_s \rho_1 + \left(\frac{\partial b}{\partial s}\right)_\rho s_1, \tag{5.4.21}$$

where b is the equilibrium average of $\mathcal{B}$. The last term, proportional to the entropy deviation s_1 is very small at long wavelengths and will be neglected here. Assuming that all the deviations depend on time as $\exp(i\omega t)$ and that the acoustic wavelength is much longer than any correlation lengths of the fluid, we obtain

$$\langle \mathcal{B}(\boldsymbol{r}) \rangle_t - b \cong \left[\rho\left(\frac{\partial b}{\partial \rho}\right)_s + i\omega \hat{K}_{\mathcal{B}}(\omega)\right]\frac{\rho_1}{\rho}, \tag{5.4.22}$$

where

$$\hat{K}_{\mathcal{B}}(\omega) = \frac{1}{T}\int_0^\infty dt \int dr' e^{-i\omega t}\langle \mathcal{B}^{\mathrm{R}}(\boldsymbol{r}, t)\Pi_{xx}^{\mathrm{R}}(\boldsymbol{r}')\rangle. \qquad (5.4.23)$$

If $\mathcal{B} = \Pi_{\alpha\alpha}$, a fundamental relation of acoustics follows,[9]

$$\langle \Pi_{\alpha\alpha}(\boldsymbol{r})\rangle_t - p \cong \left[\rho c^2 + i\omega\zeta^*(\omega) + \left(2\delta_{\alpha x} - \frac{2}{d}\right)i\omega\eta^*(\omega)\right]\frac{\rho_1}{\rho}, \qquad (5.4.24)$$

in terms of the sound velocity c and the frequency-dependent viscosities. In fluids, the normal stress difference $\langle \Pi_{xx} - \Pi_{yy}\rangle$ is equal to $2i\omega\eta^*(\omega)\rho_1/\rho$ for finite frequencies (which becomes the elastic relation $2G_0\rho_1/\rho$ if $i\omega\eta^*(\omega)$ is replaced by a shear modulus G_0). We may define the frequency-dependent adiabatic compressibility $K_s(\omega)$ by

$$K_s(\omega)^{-1} = \left[\langle \Pi_{xx}(\boldsymbol{r})\rangle_t - p\right]/(\rho_1/\rho) = \rho c^2 + i\omega\left[\zeta^*(\omega) + \left(2 - \frac{2}{d}\right)\eta^*(\omega)\right]. \qquad (5.4.25)$$

The usual adiabatic compressibility is obtained in the low-frequency limit. These relations can be used to calculate the time-dependent response of various quantities against adiabatic volume or pressure changes. Such effects become anomalously enhanced near the critical point, as will be discussed in Chapter 6.

5.4.3 Long-range correlations in steady states

When the velocity and temperature gradients tend to be stationary, (5.4.13) and (5.4.20) can be used to study steady-state fluctuations in the linear response regime. It is known that pair correlations among various quantities have a Coulombic long-range tail ($\propto 1/r$ in 3D and $\propto \ln(1/r)$ in 2D) in the steady state [53]–[58]. Its origin is the nonlinear mode coupling among the hydrodynamic fluctuations in the steady state as for the long-time tail (5.1.42) near equilibrium. For example, let us assume steady, homogeneous, incompressible velocity gradients $D_{\alpha\beta} = \partial v_\alpha/\partial x_\beta$ with $\sum_\alpha D_{\alpha\alpha} = 0$ and set $\mathcal{B} = J_\alpha(\boldsymbol{r})J_\beta(\boldsymbol{0})$, where $J_\alpha(\boldsymbol{r})$ is the momentum density. From (5.4.13) the momentum correlation in the steady state reads

$$\langle J_\alpha(\boldsymbol{r})J_\beta(\boldsymbol{0})\rangle_{\mathrm{ss}} = \rho T\delta_{\alpha\beta}\delta(\boldsymbol{r}) - \sum_{\gamma\delta} D_{\gamma\delta}G_{\alpha\beta\gamma\delta}(\boldsymbol{r}), \qquad (5.4.26)$$

where $\langle\cdots\rangle_{\mathrm{ss}}$ is the steady-state average and

$$G_{\alpha\beta\gamma\delta}(\boldsymbol{r}) = \frac{1}{T}\int_0^\infty dt \int dr' \langle J_\alpha(\boldsymbol{r}, t)J_\beta(\boldsymbol{0}, t)\Pi_{\gamma\delta}^{\mathrm{R}}(\boldsymbol{r}')\rangle. \qquad (5.4.27)$$

The second term on the right-hand side of (5.4.26) is the nonequilibrium correction. The Fourier component of $\boldsymbol{J}$ is decomposed into a longitudinal part ($\parallel \boldsymbol{k}$) and a transverse part ($\perp \boldsymbol{k}$); the former depends on time as $\exp(ickt - \frac{1}{2}\Gamma_s k^2 t)$ and the latter as $\exp(-\nu k^2 t)$ at

[9] If we retain the entropy deviation and neglect the frequency dependence of the thermal conductivity λ, we should add $i\omega\rho(1/C_V - 1/C_p)\lambda$ in the brackets of (5.4.24) for one-component fluids [10]. The acoustic dispersion relation is given by $\omega_k = ck + \frac{1}{2}i\Gamma_s k^2 + \cdots$ with the sound attenuation coefficient $\Gamma_s = (\zeta + 4\eta/3)/\rho + \lambda(1/C_V - 1/C_p)$ as $k \to 0$.

long wavelengths, where Γ_s is the sound attenuation coefficient and $\nu = \eta/\rho$ is the kinetic viscosity. Therefore, the time integration in (5.4.27) gives rise to terms proportional to $1/k^2$ in the Fourier transformation of $G_{\alpha\beta\gamma\delta}(\boldsymbol{r})$. The calculations are straightforward if use is made of the general correlation function expressions in Appendix 1A. The 3D long-range tail is of the form [55],

$$G_{\alpha\beta\gamma\delta}(\boldsymbol{r}) = \left(\frac{\rho T}{16\pi}\right)\left[\frac{1}{\nu}I^+_{\alpha\gamma}I^+_{\beta\delta} + \frac{1}{\Gamma_s}I^-_{\alpha\gamma}I^-_{\beta\delta} - \frac{1}{2}\left(\frac{1}{\nu} + \frac{1}{\Gamma_s}\right)I^-_{\alpha\beta}\hat{x}_\gamma\hat{x}_\delta\right]\frac{1}{r}, \tag{5.4.28}$$

where $I^+_{\alpha\beta} \equiv \delta_{\alpha\beta} + \hat{x}_\alpha\hat{x}_\beta$ and $I^-_{\alpha\beta} \equiv \delta_{\alpha\beta} - \hat{x}_\alpha\hat{x}_\beta$ depend on the direction $\hat{x}_\alpha \equiv x_\alpha/r$. Similar long-range correlations appear also in a temperature gradient. Originally, these steady-state long-range correlations were found via kinetic theory beyond the Boltzmann equation in the particle correlations in the $(\boldsymbol{r}, \boldsymbol{p})$ space [55]. In 2D, the steady-state pair correlations behave as $\ln(1/r)$, which indicates breakdown of the gradient expansion in the steady state.

Appendix 5A Derivation of the Fokker–Planck equation

We derive the Fokker–Planck equation from the stochastic differential equation (5.1.1). Let Δt be a time interval which satisfies (5.1.6). Then the incremental change of u is governed by (5.1.10). We introduce the characteristic function, the Fourier transformation of $P(v,t)$,

$$Q(\zeta, t) = \langle \exp[i\zeta u(t)]\rangle = \int dv P(v,t)\exp(i\zeta v). \tag{5A.1}$$

At time $t + \Delta t$, it is written as

$$\begin{aligned} Q(\zeta, t+\Delta t) &= \langle \exp[i\zeta u(t) + i\zeta\Delta u]\rangle \\ &= \langle \exp[i\zeta(1-\gamma\Delta t)u(t) - L\zeta^2\Delta t]\rangle, \end{aligned} \tag{5A.2}$$

where the random part $W(t, t+\Delta t)$ has been averaged out in the second line using the fact that it is gaussian, characterized by (5.1.7). Expanding the second line with respect to Δt, we obtain

$$\frac{\partial}{\partial t}Q(\zeta, t) = -\left(\gamma\zeta\frac{\partial}{\partial\zeta} + L\zeta^2\right)Q(\zeta, t), \tag{5A.3}$$

whose inverse Fourier transformation becomes (5.1.13).

Appendix 5B Projection operator method

The Zwanzig–Mori theory of the projection operator method [16, 17] is the statistical–mechanical basis of Langevin equations. A first idea of the method was presented by Nakajima [46]. With this scheme we may formally divide any dynamic variable into a slowly varying part and a rapidly varying part. In the following, a one-component classical fluid will be taken as a reference system. Quantum-mechanical generalization is straightforward. Any dynamic variable $\mathcal{X} = \mathcal{X}(\Gamma)$, dependent on the particle momenta and positions $\Gamma = (\boldsymbol{p}_1, \ldots \boldsymbol{p}_N, \boldsymbol{r}_1, \ldots, \boldsymbol{r}_N)$, changes in time as

$$\frac{\partial}{\partial t}\mathcal{X} = \sum_{\ell=1}^{N}\left[\frac{\partial \mathcal{H}}{\partial \boldsymbol{p}_\ell}\cdot\frac{\partial \mathcal{X}}{\partial \boldsymbol{r}_\ell} - \frac{\partial \mathcal{H}}{\partial \boldsymbol{r}_\ell}\cdot\frac{\partial \mathcal{X}}{\partial \boldsymbol{p}_\ell}\right] \equiv i\mathcal{L}\mathcal{X}, \tag{5B.1}$$

where $\mathcal{H}$ is the microscopic hamiltonian. The i is introduced to make $\mathcal{L}$ hermitian in the Γ functional space. The $i\mathcal{L}\mathcal{X}$ is expressed in terms of the Poisson bracket $\{\ ,\ \}_{\mathrm{PB}}$ as

$$i\mathcal{L}\mathcal{X} = -\{\mathcal{H}, \mathcal{X}\}_{\mathrm{PB}}. \tag{5B.2}$$

Then, by setting $\mathcal{X}(0) = \mathcal{X}$, we solve the time evolution formally as

$$\mathcal{X}(t) = e^{i\mathcal{L}t}\mathcal{X}(0). \tag{5B.3}$$

In the following we choose a set of slowly varying dynamic variables $A = \{A_j\}$. In one-component fluids they are long-wavelength Fourier components of the number, energy, and momentum densities.

Linear projection

We first define the linear projection operator $\mathcal{P}$ acting on any dynamic variable B as [17]

$$\mathcal{P}B = \sum_{jk}\langle BA_j\rangle\chi^{jk}A_k, \tag{5B.4}$$

where $\langle BA_j\rangle$ is the equilibrium equal-time correlation and χ^{jk} is the inverse matrix of $\chi_{jk} \equiv \langle A_i A_j\rangle$. Here we set $\langle A_j\rangle = 0$. The orthogonal part is written as

$$\mathcal{Q}B = B - \mathcal{P}B, \tag{5B.5}$$

where $\mathcal{Q} = 1 - \mathcal{P}$. We have $\mathcal{P}^2 = \mathcal{P}$, $\mathcal{P}\mathcal{Q} = \mathcal{P}\mathcal{Q} = 0$, and $\mathcal{Q}^2 = \mathcal{Q}$.

We next use the operator identity valid for any $i\mathcal{L}$ and $\mathcal{P}$,

$$\frac{\partial}{\partial t}e^{i\mathcal{L}t} = e^{i\mathcal{L}t}\mathcal{P}i\mathcal{L} + \int_0^t dt' e^{i\mathcal{L}(t-t')}\mathcal{P}i\mathcal{L}e^{\mathcal{Q}i\mathcal{L}t'}\mathcal{Q}i\mathcal{L} + e^{\mathcal{Q}i\mathcal{L}t}\mathcal{Q}i\mathcal{L}, \tag{5B.6}$$

from which the dynamic equation for $A_j(t) = \exp(i\mathcal{L}t)A_j(0)$ with $A_j(0) = A_j$ is written as

$$\frac{\partial}{\partial t}A_j(t) = \sum_k i\Omega_{jk}A_k(t) - \int_0^t dt'\sum_k \Gamma_{jk}(t-t')A_k(t') + F_j(t). \tag{5B.7}$$

In the first term,

$$i\Omega_{jk} = \sum_\ell \langle \dot{A}_j A_\ell\rangle\chi^{\ell k} \tag{5B.8}$$

is called the frequency matrix with $\dot{A}_j \equiv i\mathcal{L}A_j$. The last term is supposed to change relatively rapidly in time and is formally defined by

$$F_j(t) = \exp(\mathcal{Q}i\mathcal{L}t)\mathcal{Q}\dot{A}_j, \tag{5B.9}$$

and the memory kernel is expressed as

$$\Gamma_{jk}(t) = \sum_{\ell} \langle F_j(t) F_\ell(0) \rangle \chi^{\ell k}. \tag{5B.10}$$

Because $\langle F_j(t) A_k \rangle = 0$ or $\mathcal{P} F_j(t) = 0$ from the definition (5B.9), the matrix of the time-correlation functions $\Xi_{jk}(t) \equiv \langle A_j(t) A_k(0) \rangle$ satisfies

$$\frac{\partial}{\partial t} \Xi_{jk}(t) = \sum_{\ell} i\Omega_{j\ell} \Xi_{\ell k}(t) - \int_0^t ds \sum_{\ell} \Gamma_{j\ell}(t-s) \Xi_{\ell k}(s). \tag{5B.11}$$

The Fourier–Laplace transformation,

$$\hat{\Xi}_{jk}(\omega) = \int_0^\infty dt e^{-i\omega t} \Xi_{jk}(t), \tag{5B.12}$$

is the solution of the matrix equation,

$$\sum_{\ell} \left[i\omega \delta_{j\ell} - i\Omega_{j\ell} + \hat{\Gamma}_{j\ell}(\omega) \right] \hat{\Xi}_{\ell k}(\omega) = \chi_{jk}, \tag{5B.13}$$

where

$$\hat{\Gamma}_{jk}(\omega) = \int_0^\infty dt e^{-i\omega t} \Gamma_{jk}(t). \tag{5B.14}$$

When the timescales of $\Xi_{jk}(t)$ are much faster than those of $A_j(t)$, we may replace $\Gamma_{j\ell}(\omega)$ by its zero-frequency limit (markovian approximation),

$$\gamma_{jk} = \hat{\Gamma}_{jk}(0) = \int_0^\infty dt \Gamma_{jk}(t). \tag{5B.15}$$

Then $\Xi_{j\ell}(t)$ are linear combinations of $\exp(-p_k t)$ with p_k being the eigenvalues of the matrix $-i\Omega_{j\ell} + \gamma_{j\ell}$. In the linear hydrodynamic equations, γ_{jk} are proportional to the usual transport coefficients, but the frequency dependence of $\hat{\Gamma}_{jk}(\omega)$ (or the memory effect) cannot be neglected in some anomalous cases.

General symmetry relations can be derived using the invariance of the microscopic dynamics with respect to the time reversal [17, 40].[10] If A_j is changed to $\epsilon_j A_j$ ($\epsilon_j = \pm 1$), we have

$$\begin{aligned} \langle A_j(t) A_k(0) \rangle &= \epsilon_j \epsilon_k \langle A_k(t) A_j(0) \rangle, \\ \langle F_j(t) F_k(0) \rangle &= \epsilon_j \epsilon_k \langle F_k(t) F_j(0) \rangle. \end{aligned} \tag{5B.16}$$

After the Fourier–Laplace transformation we obtain

$$\begin{aligned} \hat{\Xi}_{jk}(\omega) &= \epsilon_j \epsilon_k \hat{\Xi}_{kj}(\omega), \\ \sum_{\ell} \hat{\Gamma}_{j\ell}(\omega) \chi_{\ell k} &= \epsilon_j \epsilon_k \sum_{\ell} \hat{\Gamma}_{k\ell}(\omega) \chi_{\ell j}. \end{aligned} \tag{5B.17}$$

Note that the frequency matrix $i\Omega_{j\ell}$ are nonvanishing only among the pairs A_j and A_ℓ

[10] If a static magnetic field is present, it is changed from $\boldsymbol{H}$ to $-\boldsymbol{H}$ with respect to the time reversal, so the left- and right-hand sides of (5B.16)–(5B.20) should be defined under opposite magnetic fields [23, 40].

which have the opposite signs ($\epsilon_j \epsilon_k = -1$) with respect to the time reversal. If the pairs A_j and A_k have the same sign ($\epsilon_j \epsilon_k = 1$), then the symmetric relations,

$$\sum_\ell \gamma_{j\ell} \chi_{\ell k} = \sum_\ell \gamma_{k\ell} \chi_{\ell j}, \tag{5B.18}$$

hold for the damping coefficients (5B.15). In this case $\Gamma_{jk}(t) = \Gamma_{jk}(-t)$ are even functions of t, nonvanishing at $t = 0$. If their timescales are distinctly shorter than the lifetimes of A, the markovian approximation (5B.15) can well be justified. Conversely, if A_j and A_k have opposite signs, $\Gamma_{jk}(t) = -\Gamma_{jk}(-t)$ are odd functions of t, vanishing at $t = 0$. Then, the integral (5B.15) is usually negligibly small.[11]

Nonlinear projection

To derive the nonlinear Langevin equations (5.2.1) for $A = \{A_j\}$ [16, 18], we set

$$g(A, a) \equiv \prod_j \delta(A_j - a_j). \tag{5B.19}$$

Then,

$$P_{\text{eq}}(a) = \langle g(A, a) \rangle \tag{5B.20}$$

is the equilibrium distribution of A. For any dynamic variable $B = B(\Gamma)$, its conditional average in which A is fixed at a may be defined as

$$\langle B; a \rangle = \langle \mathcal{B} g(A, a) \rangle / P_{\text{eq}}(a). \tag{5B.21}$$

Replacing a in the above expression by $A = A(\Gamma)$, we may introduce a nonlinear projection,

$$\mathcal{P}_{\text{nl}} B = \langle B; A \rangle = \int d\Gamma' P_{\text{gra}}(\Gamma') B(\Gamma') g(A(\Gamma'), A(\Gamma)) \Big/ P_{\text{eq}}(A(\Gamma)), \tag{5B.22}$$

where $P_{\text{gra}}(\Gamma)$ is the grand canonical distribution (1.2.7) for fluids.[12] Obviously, $\mathcal{P}_{\text{nl}} B$ is a functional of A, or equivalently $\mathcal{P}_{\text{nl}} B = B$ if B is a functional of A.

Mori and Fujisaka [19] noticed that the nonlinear projection $\mathcal{P}_{\text{nl}}$ onto A is the *linear* projection $\mathcal{P}$ onto $g(A, a)$. That is, by choosing $g(A, a)$ as the gross variables, we may rewrite the formal definition (5B.4) as

$$\mathcal{P} B = \prod_j \left(\int da_j \right) \langle B g(A, a) \rangle \frac{1}{P_{\text{eq}}(a)} g(A, a) = \mathcal{P}_{\text{nl}} B, \tag{5B.23}$$

where use has been made of $\langle g(A, a) g(A, a') \rangle = \delta(a - a') P_{\text{eq}}(a)$. Therefore, we will write $\mathcal{P}_{\text{nl}}$ as $\mathcal{P}$ in the following. The counterpart of (5B.7) may then be considered for $g(t, a) \equiv g(A(t), a)$ as [19]

$$\frac{\partial}{\partial t} g(t, a) = \int da' i\Omega_{aa'} g(t, a') - \int_0^t dt' \int da' \Psi_{aa'}(t - t') g(t', a') + F_a(t), \tag{5B.24}$$

[11] However, this integral can be appreciable in ^{4}He near the superfluid transition, see Chapter 6.

[12] In the original paper [16], the microcanonical distribution was used.

where

$$i\Omega_{aa'} = \langle i\mathcal{L}\delta(A-a)\cdot\delta(A-a')\rangle/P_{\text{eq}}(a') = -\sum_j \frac{\partial}{\partial a_j}[v_j(a)\delta(a-a')]. \quad (5B.25)$$

Here $v_j(a)$ is the streaming velocity defined by

$$v_j(a) = \langle \dot{A}_j\delta(A-a)\rangle/P_{\text{eq}}(a) = \langle \dot{A}_j; a\rangle. \quad (5B.26)$$

In terms of $\mathcal{Q} = 1 - \mathcal{P}$, the random force is defined by

$$F_a(t) = -\sum_j \frac{\partial}{\partial a_j}\exp(\mathcal{Q}i\mathcal{L}t)\big[(\mathcal{Q}\dot{A}_j)\delta(A-a)\big]. \quad (5B.27)$$

Assuming that A is a well-defined set of gross variables, we apply the markovian approximation. Namely,

$$\theta_j(t) \equiv \exp(\mathcal{Q}i\mathcal{L}t)\mathcal{Q}\dot{A}_j \quad (5B.28)$$

is assumed to change much more rapidly than A. Then,

$$F_a(t) \cong -\sum_j \frac{\partial}{\partial a_j}\big[\theta_j(t)\delta(A(0)-a)\big], \quad (5B.29)$$

where $A(0) = A$. The time integral of the memory kernel becomes

$$\int_0^\infty dt\Psi_{aa'}(t) \cong \sum_{jk} \frac{\partial}{\partial a_j}\frac{\partial}{\partial a'_k}\big[L_{jk}(a)P_{\text{eq}}(a)\delta(a-a')\big], \quad (5B.30)$$

with

$$L_{jk}(a) = \int_0^\infty dt\langle\theta_j(t)\theta_k(0); a\rangle. \quad (5B.31)$$

The integrand here is assumed to tend to zero rapidly while t is much shorter than the timescales of $A(t)$. The microscopic time reversal invariance leads to the symmetry relations [15, 23],

$$L_{jk}(a) = \epsilon_j\epsilon_k L_{kj}(\tilde{a}). \quad (5B.32)$$

If a steady magnetic field is present, it should also be reversed on the right-hand side [23, 40]. In (5.2.6) we have retained only the pairs j and k with $\epsilon_j\epsilon_k = 1$. See the discussion below (5B.18) to support this assumption. We now obtain the Langevin equations (5.2.1) if we multiply (5B.24) by a_j and integrate over a. Equivalently, the average of (5B.24) over a nonequilibrium ensemble gives the Fokker–Planck equation (5.2.12) with the Fokker–Planck operator (5.2.13) for the distribution $P(a,t) = \langle g(A(t), a)\rangle$.

Appendix 5C Time reversal symmetry in equilibrium time-correlation functions

First, we derive (5.2.19). To exchange Q_1 and Q_2 in (5.2.18), we rewrite it as

$$\langle \mathcal{Q}_1[t]\mathcal{Q}_2[0]\rangle = \int da\, \mathcal{Q}_2(a) \exp(\tilde{\mathcal{L}}_{\rm FP}\{a\}t)\mathcal{Q}_1(a)P_{\rm eq}(a). \tag{5C.1}$$

The operator $\tilde{\mathcal{L}}_{\rm FP}\{a\}$ is defined as

$$\tilde{\mathcal{L}}_{\rm FP}\{a\} = P_{\rm eq}\mathcal{L}_{\rm FP}\{a\}^{\dagger} P_{\rm eq}^{-1} = \sum_i \frac{\partial}{\partial a_i} v_i(a) + \sum_{i,j} \frac{\partial}{\partial a_i} L_{ij}(a)\left[\frac{\partial}{\partial a_j} + F_j(a)\right]. \tag{5C.2}$$

The superscript † denotes taking the transposed operator. Note that the first term in $\tilde{\mathcal{L}}_{\rm FP}\{a\}$ is the minus of the first term in $\mathcal{L}_{\rm FP}\{a\}$ in (5.2.13), while the second terms of the two operators coincide. After changing a to $\tilde{a}$ in the a integration of (5C.1) we find (5.2.19).

Second, we derive (5.2.23). From (5.2.20) and (5.2.22) we have

$$\left(\frac{\partial}{\partial t} + \gamma_i\right)G_{ij}(t) = \langle X_i[t]A_j(0)\rangle = \epsilon_i\epsilon_j\langle A_j(t)\bar{X}_i[0]\rangle. \tag{5C.3}$$

We again differentiate the above equation with respect to t to derive

$$\left(\frac{\partial}{\partial t} + \gamma_j\right)\left(\frac{\partial}{\partial t} + \gamma_i\right)G_{ij}(t) = \epsilon_i\epsilon_j\langle X_j[t]\bar{X}_i[0]\rangle = \langle X_i[t]\bar{X}_j[0]\rangle. \tag{5C.4}$$

The Laplace transformation of the above equation leads to (5.2.23) if use is made of the relation $(\partial/\partial t + \gamma_i)G_{ij}(t) \to 0$ as $t \to 0$.

Appendix 5D Renormalization group calculation in purely dissipative dynamics

We calculate $\phi(k, \omega)$ in (5.3.40) for $\omega = 0$ and $d = 4$ at the critical point. By decoupling the time-correlation function of the nonlinear part of $\delta(\beta\mathcal{H})/\delta\psi (\propto \psi^3)$, we obtain

$$\begin{aligned}\phi(k, 0) &= 6u_0^2(2\pi)^4 \int_{p_1}\int_{p_2}\int_{p_3} \frac{\delta(\boldsymbol{p}_1 + \boldsymbol{p}_2 + \boldsymbol{p}_3 - \boldsymbol{k})}{p_1^2p_2^2p_3^2(p_1^2 + p_2^2 + p_3^2)} \\ &= 6u_0^2(2\pi)^4 \int_{\ell}\int_0^\infty dt\, e^{i\boldsymbol{\ell}\cdot\boldsymbol{k}}\varphi(\ell, t)^3\end{aligned} \tag{5D.1}$$

In the second line we have used $\int_{\ell} \exp(i\boldsymbol{\ell}\cdot\boldsymbol{m}) = \delta(\boldsymbol{m})$ and $\int_0^\infty dt \exp(-tA) = 1/A$ with $\boldsymbol{m} = \boldsymbol{p}_1 + \boldsymbol{p}_2 + \boldsymbol{p}_3 - \boldsymbol{k}$ and $A = p_1^2 + p_2^2 + p_3^2$. Then,

$$\varphi(\ell, t) = \int_{q} \frac{\exp(i\boldsymbol{\ell}\cdot\boldsymbol{q} - tq^2)}{q^2} = \frac{2}{\pi}\int_0^\Lambda dqq \int_0^\pi d\theta \sin^2\theta \exp(iq\ell\cos\theta - tq^2), \tag{5D.2}$$

where Λ is the upper cut-off wave number. In the limit $\Lambda \to \infty$ the above integration can be performed to give

$$\varphi(\ell, t) = (2\pi\ell)^{-2}\left[1 - \exp(-\ell^2/4t)\right]. \tag{5D.3}$$

We substitute the above expression into the second line of (5D.1). The t-integration there can first be performed if use is made of $\int_0^\infty dt[1-\exp(-X/t)]^3 = 3\ln(4/3)X$, leading to

$$\phi(k,0) = \frac{9}{2}\ln(4/3)u_0^2(2\pi)^{-2}\int_{\boldsymbol{\ell}} e^{i\boldsymbol{\ell}\cdot\boldsymbol{k}}\frac{1}{\ell^4}. \tag{5D.4}$$

Here the integration at large ℓ ($\sim k^{-1}$) gives $K_4\ln(1/k)$ with $K_4 = 1/8\pi^2$, whereas it is logarithmically divergent at small ℓ. However, this divergence has arisen because we have used (5D.3). If a finite Λ is used, the divergence is removed as

$$\phi(k,0) = 9\ln(4/3)(K_4u_0)^2\ln(\Lambda/k). \tag{5D.5}$$

Appendix 5E Microscopic expressions for the stress tensor and energy current

We give microscopic expressions for the stress tensor $\overleftrightarrow{\Pi}(\boldsymbol{r},t) = \{\Pi_{\alpha\beta}(\boldsymbol{r},t)\}$ ($\alpha,\beta = x,y,z$) and the energy current $\boldsymbol{J}_e(\boldsymbol{r},t)$ in terms of the particle positions and momenta, $(\boldsymbol{r}_i,\boldsymbol{p}_i)$, ($i = 1,\ldots,N$). For simplicity, we consider one-component classical fluids interacting with a two-body potential $v(r)$. We define the stress tensor such that the momentum density

$$\boldsymbol{J}(\boldsymbol{r},t) = \sum_i \boldsymbol{p}_i\delta(\boldsymbol{r}-\boldsymbol{r}_i) \tag{5E.1}$$

exactly satisfies

$$\frac{\partial}{\partial t}\boldsymbol{J}(\boldsymbol{r},t) = -\nabla\cdot\overleftrightarrow{\Pi}(\boldsymbol{r},t). \tag{5E.2}$$

Then we have

$$\Pi_{\alpha\beta}(\boldsymbol{r},t) = \sum_i \frac{p_{i\alpha}p_{i\beta}}{m_0}\delta(\boldsymbol{r}-\boldsymbol{r}_i) - \sum_{i\neq j} v'(r_{ij})\frac{x_{ij\alpha}x_{ij\beta}}{2r_{ij}}\delta_s(\boldsymbol{r};\boldsymbol{r}_i,\boldsymbol{r}_j), \tag{5E.3}$$

where we suppress the time dependence of $(\boldsymbol{r}_i,\boldsymbol{p}_i)$. Here m_0 is the particle mass, $v'(r) = dv(r)/dr$, $r_{ij} = |\boldsymbol{r}_i - \boldsymbol{r}_j|$, and $x_{ij\alpha} = x_{i\alpha} - x_{j\alpha}$ are the cartesian components of $\boldsymbol{r}_i - \boldsymbol{r}_j$. We have introduced a symmetrized δ-function,

$$\delta_s(\boldsymbol{r};\boldsymbol{r}_i,\boldsymbol{r}_j) = \int_0^1 d\lambda\,\delta(\boldsymbol{r}-\lambda\boldsymbol{r}_i-(1-\lambda)\boldsymbol{r}_j), \tag{5E.4}$$

which is nonvanishing only on the line segment connecting $\boldsymbol{r}_i$ and $\boldsymbol{r}_j$. Its Fourier transformation is

$$\delta_s(\boldsymbol{k};\boldsymbol{r}_i,\boldsymbol{r}_j) = [\exp(-i\boldsymbol{k}\cdot\boldsymbol{r}_i) - \exp(-i\boldsymbol{k}\cdot\boldsymbol{r}_j)]/i\boldsymbol{k}\cdot(\boldsymbol{r}_j-\boldsymbol{r}_i). \tag{5E.5}$$

We may readily prove (5E.2) by using the identity,

$$(\boldsymbol{r}_i-\boldsymbol{r}_j)\cdot\nabla\delta_s(\boldsymbol{r};\boldsymbol{r}_i,\boldsymbol{r}_j) = -\delta(\boldsymbol{r}-\boldsymbol{r}_i)+\delta(\boldsymbol{r}-\boldsymbol{r}_j). \tag{5E.6}$$

We also confirm that the space integral of the stress tensor becomes (1.2.79).

Next we define the energy density $e(\boldsymbol{r}, t)$ as

$$e(\boldsymbol{r}, t) = \sum_i \frac{p_i^2}{2m_0} + \frac{1}{2}\sum_{i \neq j} v(r_{ij})\delta(\boldsymbol{r} - \boldsymbol{r}_i). \tag{5E.7}$$

The energy conservation law,

$$\frac{\partial}{\partial t} e(\boldsymbol{r}, t) = -\nabla \cdot \boldsymbol{J}_e(\boldsymbol{r}, t), \tag{5E.8}$$

is satisfied if we set

$$\begin{aligned} J_{e\alpha}(\boldsymbol{r}, t) &= \sum_i \left(\frac{p_i^2}{2m_0} + \frac{1}{2}\sum_{j \neq i} v(r_{ij}) \right) \frac{p_{i\alpha}}{m_0} \delta(\boldsymbol{r} - \boldsymbol{r}_i) \\ &\quad - \sum_{i \neq j} v'(r_{ij}) \sum_\beta \frac{x_{ij\alpha} x_{ij\beta}}{2r_{ij}} \frac{p_{i\beta}}{m_0} \delta_s(\boldsymbol{r}; \boldsymbol{r}_i, \boldsymbol{r}_j). \end{aligned} \tag{5E.9}$$

With the aid of the average pressure expression (1.2.80) supplemented with (1.2.81), this expression yields (5.4.7) in equilibrium.

References

[1] G. E. Uhlenbeck and L. S. Ornstein, *Phys. Rev.* **36**, 823 (1930); M. Chen and G. E. Uhlenbeck, *Rev. Mod. Phys.* **17**, 323 (1945).

[2] S. Chandrasekhar, *Rev. Mod. Phys.* **15**, 1 (1943).

[3] J. L. Doob, *Ann. Math.* **43**, 351 (1942).

[4] R. Kubo, in *Transport Phenomena*, Lecture Notes in Physics, Vol. 31 (Springer-Verlag, Berlin–Heidelberg–New York 1974), p. 74.

[5] R. L. Stratonovich, *Topics in the Theory of Random Noise* (Gordon and Breach, New York, 1967).

[6] M. Kac and J. Logan, in *Fluctuation Phenomena; Studies in Statistical Mechanics*, Vol. VIII, eds. E. W. Montroll and J. L. Lebowitz (North-Holland, Amsterdam, 1979), p. 1.

[7] N. G. van Kampen, *Stochastic Processes in Physics and Chemistry* (North-Holland, Amsterdam, 1992).

[8] R. Balescu, *Statistical Dynamics: Matter out of Equilibrium* (Imperial College Press, London, 1997).

[9] W. Feller, *An Introduction to Probability Theory and its Applications* (John Wiley & Sons, New York, 1957), Vol. 1.

[10] L. D. Landau and E. M. Lifshitz, *Fluid Mechanics* (Pergamon, New York, 1959).

[11] A. Onuki, *J. Stat. Phys.* **19**, 325 (1978).

[12] P. A. Egelstaff, *An Introduction to the Liquid State* (Academic, New York, 1967).

[13] B. J. Alder and T. E. Wainwright, *Phys. Rev. A* **1**, 18 (1970).

[14] R. Zwanzig and M. Bixon, *Phys. Rev. A* **2**, 2005 (1970); A. Widom, *ibid*. **3**, 1394 (1971); M. Nelkin, *Phys. Fluids* **15**, 1685 (1972).

[15] M. S. Green, *J. Chem. Phys.* **20**, 1281 (1952); **22**, 398 (1954).

[16] R. Zwanzig, *Phys. Rev.* **124**, 983 (1961)

[17] H. Mori, *Prog. Theor. Phys.* **33**, 423 (1965).

[18] K. Kawasaki, in *Critical Phenomena* (Proceedings of Enrico Fermi Summer School, Varenna, 1970), ed. M. S. Green (Academic, New York, 1971), p. 342.

[19] H. Mori and H. Fujisaka, *Prog. Theor. Phys.* **49**, 764 (1973).

[20] D. Forster, *Hydrodynamic Fluctuations, Broken Symmetry, and Correlation Functions* (Benjamin, New York, 1975).

[21] L. Onsager, *Phys. Rev.* **37**, 405 (1931); **38**, 2265 (1931).

[22] H. B. G. Casimir, *Rev. Mod. Phys.* **17**, 343 (1945).

[23] S. R. de Groot and P. Mazur, *Non-Equilibrium Thermodynamics* (Dover, 1983).

[24] D. N. Zubarev, *Non-Equilibrium Statistical Thermodynamics* (Nauka, 1971) [translation in English (Consultants Bureau, New York, 1974)].

[25] K. Kawasaki, *Phys. Rev.* **150**, 291 (1966); in *Phase Transition and Critical Phenomena.* eds. C. Domb and M. S. Green (Academic, New York, 1976), Vol. 5A, p. 165; *Ann. Phys.* (N.Y.) **61**, 1 (1970).

[26] L. P. Kadanoff and J. Swift, *Phys. Rev.* **166**, 89 (1968).

[27] P. C. Hohenberg and B. I. Halperin, *Rev. Mod. Phys.* **49**, 435 (1977).

[28] J. M. Sancho, M. San Miguel, S. L. Katz, and J. D. Gunton, *Phys. Rev. A* **26**, 1589 (1982).

[29] M. San Miguel and J. D. Gunton, *Phys. Rev. B* **23**, 2317 (1981); M. San Miguel, J. D. Gunton, G. Dee, and P. S. Sahni, *ibid.* **23**, 2334 (1981).

[30] B. I. Halperin, P. C. Hohenberg, and S. K. Ma *Phys. Rev. Lett.* **29**, 1548 (1972).

[31] Y. Kuramoto, *Prog. Theor. Phys.* **51**, 1712 (1974).

[32] H. Yahata, *Prog. Theor. Phys.* **52**, 871 (1974).

[33] K. Kawasaki, *Phys. Lett.* **54A**, 131(1975); *Physica A* **215**, 61 (1995).

[34] H. Yahata and M. Suzuki, *J. Phys. Soc. Jpn* **27**, 1421 (1969).

[35] M. Suzuki and R. Kubo, *J. Phys. Soc. Jpn* **24**, 51 (1968).

[36] K. Kawasaki, in *Phase Transition and Critical Phenomena* eds. C. Domb and M. S. Green (Academic, New York, 1972), Vol. 2, p. 443.

[37] B. I. Halperin, P. C. Hohenberg, and S. K. Ma *Phys. Rev. B* **10**, 139 (1974).

[38] R. Kubo and K. Tomita, *J. Phys. Soc. Jpn* **9**, 888 (1954).

[39] H. Nakano, *Prog. Theor. Phys.* **15**, 77 (1956).

[40] R. Kubo, *J. Phys. Soc. Jpn* **12**, 570 (1957).

[41] J. C. Maxwell, *The Scientific Papers of J. C. Maxwell* (Dover, New York, 1965).

[42] L. Boltzmann, *Lectures on Gas Theory* (S. Brush, trans., University of California Press, Berkeley, CA, 1964).

[43] S. Chapman and T. G. Cowling, *The Mathematical Theory of Non-Uniform Gases*, 3rd edn (Cambridge University Press, 1970).

[44] M. G. Velarde, in *Transport Phenomena*, Lecture Notes in Physics 31 (Springer-Verlag, Berlin–Heidelberg–New York, 1974), p. 288.

[45] R. Kubo, M. Yokota, and S. Nakajima, *J. Phys. Soc. Jpn* **12**, 1204 (1957).

[46] S. Nakajima, *Prog. Theor. Phys.* **20**, 948 (1958).

[47] J. A. McLennan, *Adv. Chem. Phys.* **5**, 261 (1963).

[48] J. M. Luttinger, *Phys. Rev. A* **135**, 1505 (1964).

[49] L. P. Kadanoff and P. Martin, *Ann. Phys.* (N.Y.) **24**, 419 (1963).

[50] B. U. Felderhof and I. Oppenheim, *Physica* **31**, 1441 (1965).

[51] J. P. Hansen and I. R. McDonald, *Theory of Simple Liquids* (Academic, London, 1986).

[52] D. Evans and G. P. Morriss, *Statistical Mechanics of Nonequilibrium Liquids* (Academic, London,1990).

[53] I. Procaccia, D. Ronis, W. A. Collins, J. Ross, and I. Oppenheim, *Phys. Rev. A* **19**, 1290 (1979); J. Machta, I. Oppenheim, and I. Procaccia, *Phys. Rev. A* **22**, 2809 (1980).

[54] G. Ludwig, *Physica* **28**, 841 (1962).

[55] A. Onuki, *J. Stat. Phys.* **18**, 475 (1978).

[56] M. H. Ernst and E. G. D. Cohen, *J. Stat. Phys.* **25**, 153 (1981).

[57] T. Kirkpatrick, E. G. D. Cohen, and J. R. Dorfman, *Phys. Rev. A* **26**, 950 (1982).

[58] A. M. S. Tremblay, M. Arai, and E. D. Siggia, *Phys. Rev. A* **23**, 1451 (1981).

6

Dynamics in fluids

In the dynamics of one- and two-component fluids near the critical point and ^{4}He and ^{3}He–^{4}He near the superfluid transition, the dynamic equations of the gross variables are nonlinear Langevin equations with reversible nonlinear mode coupling terms. These terms represent nonlinear dynamic interactions between the fluctuations, which cause critical divergence of the kinetic coefficients. We will give intuitive pictures of the physical processes leading to such enhancement of transport and review the mode coupling and dynamic renormalization group theories. New results are presented on various adiabatic processes including the piston effect and supercritical fluid hydrodynamics near the gas–liquid critical point and on nonequilibrium effects of heat flow near the superfluid transition.

6.1 Hydrodynamic interaction in near-critical fluids

In the dynamics of nearly incompressible binary fluid mixtures it is usual to take the concentration deviation δX as the order parameter ψ. In one-component fluids it is convenient to take the entropy deviation δs (per unit mass) as ψ, because δs is decoupled from the sound mode in the hydrodynamic description. In these fluids, the dynamics of the order parameter is slowed down but the kinetic coefficients are enhanced near the critical point. These features originate from random convection of the critical fluctuations by the transverse velocity field fluctuations [1]–[7].

6.1.1 Intuitive picture of random convection

The order parameter undergoes diffusive relaxation resulting from convective motion due to the velocity field fluctuations. To see this intuitively, let us examine how clusters of the critical fluctuations with linear dimension ℓ smaller than ξ are convected by the velocity field fluctuations. They are fractal objects as discussed in Chapter 2. We use the following correlation function relation for the momentum density $\boldsymbol{J} = \rho \boldsymbol{v}$,

$$\langle J_i(\boldsymbol{r}, t) J_j(\boldsymbol{r}', t)\rangle = \rho T \delta(\boldsymbol{r} - \boldsymbol{r}')\delta_{ij}, \tag{6.1.1}$$

where ρ is the mass density. This relation readily follows from (5E.1). We integrate both sides of this relation over a volume $V_\ell \sim \ell^d$ with respect to $\boldsymbol{r}$ and $\boldsymbol{r}'$ and determine the typical magnitude $v(\ell)$ of the velocity field fluctuations on the scale of ℓ as

$$v(\ell) = \left\langle \left(\frac{1}{\rho V_\ell} \int_{V_\ell} d\boldsymbol{r} \boldsymbol{J} \right)^2 \right\rangle^{1/2} \sim \left(\frac{T}{\rho \ell^d} \right)^{1/2}. \tag{6.1.2}$$

At long wavelengths we will confirm that the cluster lifetime is much longer than that of the transverse velocity fluctuations,

$$\tau_v(\ell) \sim (\rho/\eta_0)\ell^2. \tag{6.1.3}$$

For the time being, the shear viscosity η_0 is assumed to be a constant. The longitudinal part of the velocity fluctuations oscillates with much faster timescales of sound and does not affect the order parameter fluctuations. The clusters then undergo diffusive motion as a result of convection by the rapidly varying velocity field fluctuations with the diffusion constant

$$D(\ell) = v(\ell)^2\tau_v(\ell) \sim \frac{T}{\eta_0}\ell^{2-d}, \tag{6.1.4}$$

which follows from the general formula (5.1.33). From $k \sim 2\pi/\ell \gtrsim \xi^{-1}$, the resultant relaxation rate Γ_k with wave number k is of the form,

$$\Gamma_k \sim D(\ell)\ell^{-2} \sim \frac{T}{\eta_0}k^d, \tag{6.1.5}$$

which is much smaller than that of the velocity field if

$$\Gamma_k\tau_v(\ell) \sim (\rho T/\eta_0^2)k^{d-2} \ll 1. \tag{6.1.6}$$

The length $\ell^* \equiv \rho T/\eta_0^2$ in 3D is microscopic in usual binary fluid mixtures,[1] but it is much longer in polymer blends. To characterize the induced velocity field on the scale of ℓ, we introduce the Reynolds number,

$$Re(\ell) = \rho\ell v(\ell)/\eta_0 \sim (\rho T/\eta_0^2\ell^{d-2})^{1/2}. \tag{6.1.7}$$

Then $\Gamma_k\tau_v(\ell) \sim Re(\ell)^2$, and $Re(\ell) \ll 1$ holds for $\ell \gg \ell^*$.

The long-wavelength thermal fluctuations with $k \lesssim \xi^{-1}$ may be regarded to consist of clusters with sizes of order ξ. Hence the diffusion constant in the hydrodynamic regime is that of a cluster with size ξ,

$$D(\xi) \sim (T/\eta_0)\xi^{2-d}, \tag{6.1.8}$$

which is analogous to the Einstein–Stokes formula (5.1.25) for the diffusion constant of a Brownian particle. The kinetic coefficient for the order parameter relaxation in the long-wavelength limit thus grows as

$$L_R \sim D\chi \propto \xi^{4-d}, \tag{6.1.9}$$

where the susceptibility χ behaves as $\xi^{\gamma/\nu} \sim \xi^2$. Therefore, the hydrodynamic interaction is *relevant* for $d < 4$, and the critical dimensionality d_c of fluids remains 4 in dynamics as well as in statics. We also note that the reaction of ψ back on the transverse velocity $\boldsymbol{v}$ is neglected in the above picture. That is, the interaction between them is nearly *one-sided*. This is because the latter relaxes on much faster timescales, as has been confirmed in (6.1.6). However, a small reactive effect exists, leading to a nearly logarithmic dependence

[1] If we set $\rho \sim 1$ g/cm^3, $T \sim 300$ K, and $\eta_0 \sim 0.01$ poise, we find $\ell^* \sim 10^{-10}$ cm.

($\propto \ln \xi$) of the shear viscosity. The total viscosity, consisting of the background η_0 and the fluctuation contribution, will be written as η_R and called the renormalized viscosity.

It is worth noting that the dynamics of polymer solutions is also decisively governed by the hydrodynamic interaction [8, 9], as will be shown in (7.1.26)–(7.1.28) in the next chapter. Moreover, in polymer blends with large molecular weights, the hydrodynamic interaction is operative only at very long wavelengths and there are complicated dynamical crossover effects [8, 10].

6.1.2 Model H

The minimal model which describes the above dynamical behavior is given by the nonlinear Langevin equations,

$$\frac{\partial}{\partial t}\psi = -\nabla \cdot (\psi \boldsymbol{v}) + L_0 \nabla^2 \frac{\delta}{\delta \psi} \beta \mathcal{H} + \theta, \tag{6.1.10}$$

$$\rho \frac{\partial}{\partial t} \boldsymbol{v} = -\left(\psi \nabla \frac{\delta \mathcal{H}}{\delta \psi} \right)_\perp + \eta_0 \nabla^2 \boldsymbol{v} + (\boldsymbol{\zeta})_\perp, \tag{6.1.11}$$

which is called model H [11]. The equilibrium distribution for ψ and $\boldsymbol{v}$ is of the form $\exp(-\beta \tilde{\mathcal{H}})$ with

$$\tilde{\mathcal{H}}\{\psi, \boldsymbol{v}\} = \mathcal{H}\{\psi\} + \frac{1}{2} \int d\boldsymbol{r} \rho \boldsymbol{v}^2, \tag{6.1.12}$$

where the first term is the GLW hamiltonian (4.1.1). The noise terms θ and $\boldsymbol{\zeta}$ are related to the bare kinetic coefficients L_0 and η_0 by

$$\langle \theta(\boldsymbol{r}, t) \theta(\boldsymbol{r}', t') \rangle = -2L_0 \nabla^2 \delta(\boldsymbol{r} - \boldsymbol{r}') \delta(t - t'), \tag{6.1.13}$$

$$\langle \zeta_i(\boldsymbol{r}, t) \zeta_j(\boldsymbol{r}', t') \rangle = -2T \eta_0 \delta_{ij} \nabla^2 \delta(\boldsymbol{r} - \boldsymbol{r}') \delta(t - t'). \tag{6.1.14}$$

We treat the mass density ρ as a constant in (6.1.10) and (6.1.11) and neglect the longitudinal part of $\boldsymbol{v}$:

$$\nabla \cdot \boldsymbol{v} = 0. \tag{6.1.15}$$

The notation $(\cdots)_\perp$ in (6.1.11) denotes taking the transverse part of the vectors (which is perpendicular to the wave vector in the Fourier space). The first terms on the right-hand sides of (6.1.10) and (6.1.11) are the nonlinear streaming terms or the mode coupling terms. The first term in (6.1.10) is simply the convection term, while that in (6.1.11) is a nontrivial reversible force density. We confirm that both sides of the potential condition relation (5.2.14) vanish for the total hamiltonian $\tilde{\mathcal{H}}$ in (6.1.12). It is worth noting that the first term in (6.1.11) can be derived only from this requirement.[2] To examine its physical

[2] Let the reversible force density be written as $\boldsymbol{f}$ with the trivial convection term in (6.1.10) being assumed. The potential condition requires that the space integral of $(\delta\mathcal{H}/\delta\psi)\nabla \cdot (\psi \boldsymbol{v}) - \rho \boldsymbol{v} \cdot \boldsymbol{f}$ vanishes for any transverse velocity field $\boldsymbol{v}$, which determines $\boldsymbol{f}$ as given in (6.1.11).

meaning further, we set [12]

$$-\psi\nabla\frac{\delta\mathcal{H}}{\delta\psi} = -\nabla\cdot\overleftrightarrow{\Pi}. \tag{6.1.16}$$

In Appendix 6A we shall see that $\overleftrightarrow{\Pi}$ is the stress tensor induced by ψ,

$$\Pi_{ij}(\boldsymbol{r},t) = \delta\tilde{p}\delta_{ij} + TK(\nabla_i\psi)(\nabla_j\psi), \tag{6.1.17}$$

where $\nabla_i = \partial/\partial x_i$. Here $\delta\tilde{p}$ in the first diagonal part is a pressure dependent on ψ, but its form is not important under the incompressibility condition (6.1.15). The second off-diagonal part gives rise to the weak singularity of the shear viscosity in one-phase states, whereas it will lead to a large viscosity increase in phase-separating fluids, as will be discussed in Section 11.1.

6.1.3 Mode coupling theory

Relaxation of the order parameter

As has been discussed in Section 5.4, the transport coefficients of fluids in the linear response regime can be expressed as the time integral of flux time-correlation functions. Near the critical point, the nonlinear part of the reversible flux can give rise to enhancement of the transport coefficients. In the present case, the renormalized kinetic coefficient for ψ is expressed as [13, 14]

$$L_{\rm R}(k) = L_0 + \int_0^\infty dt \int d\boldsymbol{r} e^{i\boldsymbol{k}\cdot\boldsymbol{r}}\langle\psi(\boldsymbol{r},t)v_x(\boldsymbol{r},t)\psi(\boldsymbol{0},0)v_x(\boldsymbol{0},0)\rangle, \tag{6.1.18}$$

where the first term, the background kinetic coefficient, is much smaller than the second singular term near the critical point. We retain the k dependence (nonlocality) and the direction of $\boldsymbol{k}$ is taken to be along the x axis. The thermal relaxation rate is expressed as

$$\Gamma_k = L_{\rm R}(k)k^2/\chi_k = L_{\rm R}(k)k^2(1+k^2\xi^2)/\chi, \tag{6.1.19}$$

where $\chi = \lim_{k\to 0}\chi_k(\propto \xi^{\gamma/\nu})$ is the susceptibility.

In the original mode coupling theory the above four-body time-correlation function is decoupled into the product of the two-body time-correlation functions as

$$L_{\rm R}(k) = L_0 + \int_0^\infty dt \int d\boldsymbol{r} e^{i\boldsymbol{k}\cdot\boldsymbol{r}}\langle v_x(\boldsymbol{r},t)v_x(\boldsymbol{0},0)\rangle g(|\boldsymbol{r}|). \tag{6.1.20}$$

Because the timescale of $\psi(\boldsymbol{r},t)$ is much slower than that of $\boldsymbol{v}(\boldsymbol{r},t)$, $g(|\boldsymbol{r}|) = \langle\psi(\boldsymbol{r},0)\psi(\boldsymbol{0},0)\rangle$ is the static pair correlation function. In Appendix 6B the relaxation rate will be calculated for general k as

$$\Gamma_k = L_{\rm R}(k)k^2/\chi_k = L_0k^2/\chi_k + \frac{T}{6\pi\eta_{\rm R}}\xi^{-3}K_0(k\xi), \tag{6.1.21}$$

where η_R is the renormalized shear viscosity and $K_0(x)$ is called the Kawasaki function of the form [2],

$$K_0(x) = \frac{3}{4}\left[1 + x^2 + (x^3 - x^{-1})\tan^{-1} x\right]. \tag{6.1.22}$$

Here $K_0(x) \cong x^2$ for $x \ll 1$ and $(3\pi/8)x^3$ for $x \gg 1$. The ratio of the first term to the second term in the right-hand side of (6.1.21) is expressed as $(L_0/\chi)6\pi\eta_R\xi^3/T \cong A_B(T/T_c - 1)^{(1-\eta)\nu}$ for $k\xi \ll 1$ at the critical concentration, where $A_B \sim 13$ for trimethylpentane + nitroethane [15]. If this ratio is much smaller than 1, we may neglect L_0 to obtain

$$\begin{aligned} \Gamma_k &\cong (T/6\pi\eta_R)\xi^{-1}k^2 \quad (k\xi < 1), \\ &\cong (T/16\eta_R)k^3 \quad (k\xi > 1). \end{aligned} \tag{6.1.23}$$

The long-wavelength expressions read

$$D = \frac{T}{6\pi\eta_R\xi} \propto \xi^{-1}, \qquad L_R(0) = D\chi \propto \xi, \tag{6.1.24}$$

while Γ_k is nearly independent of ξ for $k\xi \gg 1$. These results agree with (6.1.5), (6.1.8), and (6.1.9) and confirm the intuitive picture of the random convection. The average lifetime of the critical fluctuations is given at $k\xi = 1$ as

$$\begin{aligned} t_\xi &= D^{-1}\xi^2 \propto \xi^3 \\ &= t_0(T/T_c - 1)^{-1.9}, \end{aligned} \tag{6.1.25}$$

where the second line holds at the critical isochore or concentration, and $t_0 = 6\pi\eta_R\xi_{+0}^3/T$ is a microscopic frequency. The dynamic exponent z in the scaling relation $t_\xi \propto \xi^z$ is equal to 3 in the mode coupling theory in 3D. The notation $\Gamma_\xi = 1/t_\xi$ will also be used. The lifetime can easily be of order 1 s in usual binary fluid mixtures close to the critical point. Figure 6.1 demonstrates remarkable agreement between the theoretical formula (6.1.21) and dynamic light scattering data [15].

Frequency-dependent shear viscosity

The shear viscosity has a weak critical singularity in one- and two-component fluids. As an example, Fig. 6.2 shows data of the shear viscosity in ^{3}He in the low-frequency limit [16]. From (5.4.4) the renormalized shear viscosity is written in terms of the off-diagonal stress time-correlation function as

$$\eta_R^*(\omega) = \eta_0 + \frac{1}{T}\int_0^\infty dt \int d\mathbf{r} e^{-i\omega t}\langle \Pi_{xy}(\mathbf{r}, t)\Pi_{xy}(\mathbf{0}, 0)\rangle, \tag{6.1.26}$$

where the first term is the background shear viscosity. The frequency dependence is retained because it can be important in experiments of oscillatory shear flow, whereas the k dependence is neglected. The xy component of the nonlinear stress tensor arises from the

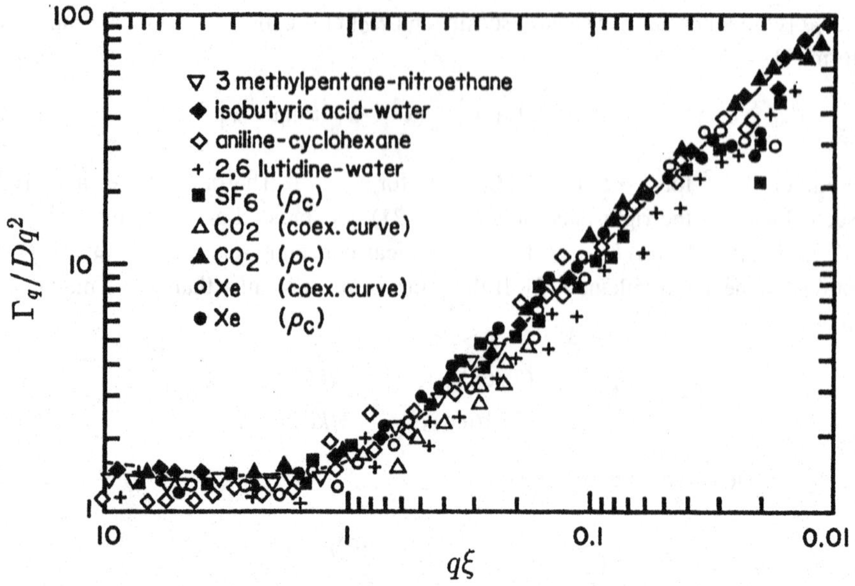

Fig. 6.1. Plot of the reduced relaxation rate Γ_q/Dq^2 as a function of $q\xi$ for various one- and two-component fluids. The solid line represents with the Kawasaki form (6.1.21) [15].

second term of (6.1.17) in the form $\Pi_{xy} = TK(\partial\psi/\partial x)(\partial\psi/\partial y)$. Again the decoupling approximation yields

$$\eta_R^*(\omega) = \eta_0 + T\int_q \frac{q_x^2 q_y^2}{(q^2+\xi^{-2})^2}\frac{1}{i\omega + 2\Gamma_q}, \tag{6.1.27}$$

where $\int_q = (2\pi)^{-d}\int d\boldsymbol{q}$. This integral is logarithmically divergent at large q for any d below 4 because $\Gamma_q \sim q^d$ at large q. Using (6.1.23) we obtain [4]–[6]

$$\begin{aligned}\frac{\eta_R^*(\omega)}{\eta_0} &\cong 1 + (8/15\pi^2)\ln(\xi/\xi_{+0}) \cong (\xi/\xi_{+0})^{\bar{x}_\eta} \qquad (\omega t_\xi \ll 1)\\ &\cong 1 - (8/45\pi^2)\ln(i\omega t_0) \cong (i\omega t_0)^{-\bar{x}_\eta/3} \qquad (\omega t_\xi \gg 1),\end{aligned} \tag{6.1.28}$$

where the upper cut-off wave number is ξ_{+0}^{-1}, and t_0 in the second line has appeared in (6.1.25). After the angle average of $\boldsymbol{q}$, $q_x^2 q_y^2$ in (6.1.27) is replaced by $q^4/15$ in 3D, yielding the small coefficients of the logarithmic terms. They may well be exponentiated with the small exponent [4],

$$\bar{x}_\eta = 8/15\pi^2 \cong 0.054. \tag{6.1.29}$$

At high frequencies $\omega t_\xi \gg 1$, the complex dynamic viscosity is independent of ξ and the ratio of the imaginary and real parts tends to a small universal number,

$$\mathrm{Im}[\eta_R^*(\omega)]/\,\mathrm{Re}[\eta_R^*(\omega)] \cong -\tan(\pi\bar{x}_\eta/6), \tag{6.1.30}$$

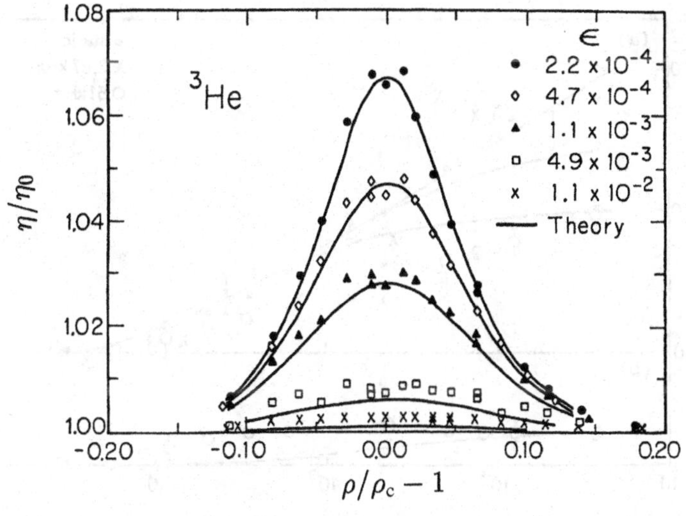

Fig. 6.2. The normalized shear viscosity η vs $\rho/\rho_c - 1$ for various reduced temperatures $\epsilon = T/T_c - 1$ in ^{3}He near the gas–liquid critical point [16].

which is equal to -0.028 from (6.1.29). We conclude that near-critical fluids are weakly viscoelastic due to the slow critical fluctuations. They can also be weakly non-newtonian in stationary shear flow, as will be discussed in Section 11.1. As shown in Fig. 6.3, the logarithmic ω dependence and the imaginary part of the shear viscosity were detected in a near-critical mixture of nitrobenzene + n-hexane [17]. Similar measurements of $\eta_R^*(\omega)$ have recently made for Xe [18].

6.1.4 Dynamic renormalization group theory

In the dynamic renormalization group (RG) theory the fluctuations in the shell region $\Lambda - \delta\Lambda < k < \Lambda$ are coarse-grained. The incremental changes δL_0 and $\delta\eta_0$ of the kinetic coefficients are readily calculated slightly below four dimensions. The correlation function expressions (6.1.18) and (6.1.26) give

$$\delta L_0 = \frac{3}{4}\frac{T}{\eta_0}\chi_\Lambda \delta V, \tag{6.1.31}$$

$$\delta\eta_0 = \frac{1}{24} T K^2 \frac{1}{\Gamma_\Lambda}\chi_\Lambda^2 \delta V, \tag{6.1.32}$$

where $\delta V = K_d \Lambda^{d-1}\delta\Lambda$ is the volume of the shell region with K_d being defined by (4.1.16), and $\chi_\Lambda = 1/K\Lambda^2$ is the structure factor. Here $\boldsymbol{v}_k$ and ψ_k at $k = \Lambda$ are assumed to decay exponentially as $\exp(-t\eta_0\Lambda^2/\rho)$ and $\exp(-tL_0K\Lambda^4)$, respectively. The factor 3/4 in (6.1.31) arises from selecting the transverse part in the velocity, while the factor 1/24

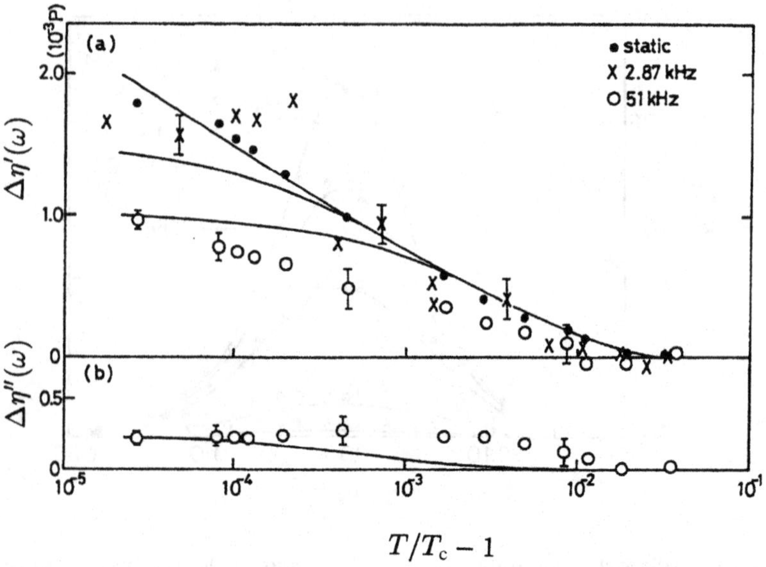

Fig. 6.3. Real and imaginary parts of the complex viscosity in nitrobenzene + n-hexane showing weak viscoelasticity of near-critical fluids. Here, $\Delta\eta'(\omega) - i\Delta\eta''(\omega) = \eta_R^*(\omega) - \eta_0$ at $\omega/2\pi = 0, 2.87$, and 51 kHz [17]. The solid lines are the theoretical results from (6.1.27).

in (6.1.32) from the angle average of $k_x^2 k_y^2$ in 4D. By setting $\Lambda = \Lambda_0 e^{-\ell}$ we find the RG equations at the critical point,

$$\frac{\partial}{\partial \ell} L_0 = \frac{3}{4} T K_d \big/ (K \eta_0 \Lambda^\epsilon) = \frac{3}{4} f L_0, \tag{6.1.33}$$

$$\frac{\partial}{\partial \ell} \eta_0 = \frac{1}{24} T K_d \big/ (K L_0 \Lambda^\epsilon) = \frac{1}{24} f \eta_0. \tag{6.1.34}$$

We notice that the following dimensionless number,

$$f = T K_d \big/ (K \eta_0 L_0 \Lambda^\epsilon), \tag{6.1.35}$$

tends to a fixed-point value f^* of order ϵ. In fact, it is governed by

$$\frac{\partial}{\partial \ell} f = \epsilon f - \frac{19}{24} f^2, \tag{6.1.36}$$

so that the fixed-point value of f is given by

$$f^* = \frac{24}{19}\epsilon + \cdots. \tag{6.1.37}$$

It is easy to solve (6.1.36) in the form,

$$f(\ell) = f_0 e^{\epsilon\ell} / [F_0(e^{\epsilon\ell} - 1) + 1] \tag{6.1.38}$$

where f_0 is the initial value of f and $F_0 = f_0/f^*$. Then (6.1.33) and (6.1.34) are solved to give

$$L_0(\Lambda) = L_0(\Lambda_0)[F_0(e^{\epsilon\ell} - 1) + 1]^{18/19}, \quad (6.1.39)$$

$$\eta_0(\Lambda) = \eta_0(\Lambda_0)[F_0(e^{\epsilon\ell} - 1) + 1]^{1/19}. \quad (6.1.40)$$

For large ℓ we have

$$L_0(\Lambda) \propto \Lambda^{-x_\lambda}, \quad \eta_0(\Lambda) \propto \Lambda^{-x_\eta}, \quad (6.1.41)$$

with

$$x_\lambda = \frac{18}{19}\epsilon + \cdots, \quad x_\eta = \frac{1}{19}\epsilon + \cdots. \quad (6.1.42)$$

In the coupled RG equations of L_0 and η_0, f is a unique expansion parameter tending to a universal number even in higher orders in ϵ. The coefficient K in the gradient free energy becomes proportional to $\Lambda^{-\eta}$ with decreasing Λ, where η is the Fisher critical exponent (not the shear viscosity). Thus, from (6.1.35) the exponent relation,

$$x_\lambda + x_\eta = \epsilon - \eta, \quad (6.1.43)$$

holds exactly or to all orders in ϵ [7, 11]. Slightly away from the critical point the above multiplicative effect stops at $\Lambda = \xi^{-1}$, yielding the renormalized kinetic coefficients,

$$L_R \sim \xi^{x_\lambda}, \quad \eta_R \sim \xi^{x_\eta}. \quad (6.1.44)$$

Because $t_\xi = \xi^2/D$ with $D = L_R(0)/\chi$, the dynamic exponent z is expressed as

$$z = 4 - \eta - x_\lambda = d + x_\eta. \quad (6.1.45)$$

The above scaling law is realized when L_R exceeds the background ($\sim L_0$ at $\Lambda \sim \xi_{+0}^{-1}$).

For classical fluids, the predictions of the mode coupling theory in 3D and those of the dynamic RG theory (even to leading order in ϵ) are in good agreement, obviously because the mode coupling between ψ and $\boldsymbol{v}$ is nearly one-sided. In particular, the exponent 0.054 in (6.1.29) from the mode coupling theory happens to be very close to that of $x_\eta = 0.053$ to first order in ϵ from the dynamic RG theory. To interpolate the two theories, Kawasaki and Gunton [19] developed the mode coupling theory slightly below four dimensions to obtain results identical to those from the dynamic RG theory (to first order in ϵ). In this way the relationship between the two theories is well understood.

6.1.5 The Stokes–Kawasaki approximation

The velocity field fluctuations relax much more rapidly than the order parameter fluctuations, so we may set $\partial\boldsymbol{v}/\partial t = 0$ in (6.1.11) as in the derivation of (5.1.30) from (5.1.29) [20]. The velocity field is determined by

$$-\eta_0\nabla^2\boldsymbol{v} = \left(-\psi\nabla\frac{\delta}{\delta\psi}\mathcal{H} + \boldsymbol{\zeta}\right)_\perp, \quad \nabla\cdot\boldsymbol{v} = 0. \quad (6.1.46)$$

In colloidal systems, the velocity field is usually determined in the same manner (the Stokes approximation). In our case it is composed of the velocity field $\boldsymbol{v}_\psi(\boldsymbol{r}, t)$ induced by ψ and the random part, $\boldsymbol{v}^{\mathrm{R}}(\boldsymbol{r}, t)$. In 3D, the above equation is solved

$$\boldsymbol{v}_\psi(\boldsymbol{r}, t) = -\int d\boldsymbol{r}' \overleftrightarrow{T}(\boldsymbol{r} - \boldsymbol{r}') \cdot \psi(\boldsymbol{r}')\nabla' \frac{\delta \mathcal{H}}{\delta \psi(\boldsymbol{r}')}, \tag{6.1.47}$$

$$\boldsymbol{v}^{\mathrm{R}}(\boldsymbol{r}, t) = \int d\boldsymbol{r}' \overleftrightarrow{T}(\boldsymbol{r} - \boldsymbol{r}') \cdot \boldsymbol{\zeta}(\boldsymbol{r}', t), \tag{6.1.48}$$

where $\overleftrightarrow{T}(\boldsymbol{r})$ is called the Oseen tensor of the form,

$$\mathcal{T}_{ij}(\boldsymbol{r}) = \frac{1}{8\pi\eta_0}\left(\frac{\delta_{ij}}{r} + \frac{x_i x_j}{r^3}\right). \tag{6.1.49}$$

We can check that the Oseen tensor becomes

$$\mathcal{T}_{ij}(\boldsymbol{k}) = \frac{1}{\eta_0}\left(\frac{\delta_{ij}}{k^2} - \frac{k_i k_j}{k^4}\right) \tag{6.1.50}$$

after the Fourier transformation. The time dependence of ψ on the right-hand side of (6.1.47) is suppressed for simplicity. The random part $\boldsymbol{v}_{\mathrm{R}}(\boldsymbol{r}, t)$ is characterized by

$$\langle v_i^{\mathrm{R}}(\boldsymbol{r}, t) v_j^{\mathrm{R}}(\boldsymbol{r}', t')\rangle = 2T\mathcal{T}_{ij}(\boldsymbol{r} - \boldsymbol{r}')\delta(t - t'). \tag{6.1.51}$$

From (6.1.10) we obtain

$$\frac{\partial}{\partial t}\psi(\boldsymbol{r}, t) = -\int d\boldsymbol{r}' L(\boldsymbol{r}, \boldsymbol{r}') \frac{\delta}{\delta\psi(\boldsymbol{r}')}\beta\mathcal{H} + \theta^{\mathrm{R}}(\boldsymbol{r}, t), \tag{6.1.52}$$

where the new kinetic coefficient,

$$L(\boldsymbol{r}, \boldsymbol{r}') = T\nabla\psi(\boldsymbol{r}) \cdot \overleftrightarrow{T}(\boldsymbol{r} - \boldsymbol{r}') \cdot \nabla'\psi(\boldsymbol{r}') - L_0\nabla^2\delta(\boldsymbol{r} - \boldsymbol{r}'), \tag{6.1.53}$$

is nonlocal and nonlinearly dependent on ψ. The random source term,

$$\theta^{\mathrm{R}}(\boldsymbol{r}, t) = -\boldsymbol{v}^{\mathrm{R}}(\boldsymbol{r}, t) \cdot \nabla\psi(\boldsymbol{r}, t) + \theta(\boldsymbol{r}, t), \tag{6.1.54}$$

satisfies the fluctuation–dissipation relation,

$$\langle \theta^{\mathrm{R}}(\boldsymbol{r}, t)\theta^{\mathrm{R}}(\boldsymbol{r}', t')\rangle = 2L(\boldsymbol{r}, \boldsymbol{r}')\delta(t - t'). \tag{6.1.55}$$

The above model was numerically solved in 3D to examine the effect of the hydrodynamic interaction in spinodal decomposition [21], as will be discussed in Section 8.5. In such applications, the fluctuations under consideration are those with spatial scales longer than ξ so that the renormalized kinetic coefficients L_{R} and η_{R} should be used in place of L_0 in (6.1.53) and η_0 in (6.1.49) with the upper cut-off wave number at ξ^{-1}.

6.1.6 Transient electric birefringence (Kerr effect)

Fluids become optically anisotropic or birefringent in the presence of an electric field, a magnetic field, a velocity gradient, or a sound wave. If constituent particles are optically anisotropic, their alignment is precisely measurable as intrinsic birefringence. As another origin, the critical fluctuations or chain molecules take anisotropic shapes, giving rise to form birefringence, even if they consist of optically isotropic particles [22, 23]. In near-critical fluids the form contribution grows near the critical point, while in polymer solutions birefringence arises from both of these two origins. Moreover, the applied field can be made to be time-dependent, and then the dynamic response can be investigated with high precision. As one example, transient electric birefringence $\Delta n(t)$ has been measured in a near-critical binary fluid mixture by applying a rectangular pulse of electric field [24]–[26]. In terms of $\Delta\varepsilon_{\rm op}$ in (4.2.64), we have

$$\Delta n(t) = n_{xx} - n_{yy} = \frac{1}{2\sqrt{\varepsilon}}\,{\rm Re}(\Delta\varepsilon_{\rm op}). \tag{6.1.56}$$

Transient birefringence was measured after an electric field was switched off at $t = 0$ [27]. If the laser wave number k in the fluid is much smaller than ξ^{-1}, the relaxation obeys

$$\frac{\Delta n(t)}{\Delta n(0)} = G(t/t_\xi) = \frac{4}{\pi}\int_0^\infty dy \frac{y^2}{(1+y^2)^2}\exp[-2K_0(y)t/t_\xi], \tag{6.1.57}$$

in the time region $t \ll 1/Dk^2$, where $K_0(x)$ is the Kawasaki function (6.1.22). The scaling function $G(x)$ behaves as

$$\begin{aligned} G(x) &\cong 1 - 2.3x^{1/3} \qquad (x \ll 1), \\ &\cong 0.2x^{-3/2} \qquad (x \gg 1). \end{aligned} \tag{6.1.58}$$

At $t \sim 1/Dk^2$, $G(t/t_\xi)$ becomes a very small number of order $0.2(k\xi)^3$. In the later time region $t \gg (Dk^2)^{-1}$, the fluctuations with wave numbers of order $(Dt)^{-1/2}$ give rise to the following signal,

$$\frac{\Delta n(t)}{\Delta n(0)} \cong 0.2(t/t_\xi)^{-3/2}(Dk^2 t)^{-1}. \tag{6.1.59}$$

In Fig. 6.4 data on butoxyetbaranol + water [25] are best-fitted to (6.1.57) for $t < t_\xi$. As another theory, Piazza *et al.* [24] derived a stretched exponential decay of $\Delta n(t)$ from a phenomenological picture based on the distribution of large clusters. Afterwards $\Delta n(t)$ was measured over three decades of t [26], but such data have not yet been compared with (6.1.59). Transient electric dichroism[3] (relaxation of Im $\Delta\varepsilon_{\rm op}$) should also be measurable for $k\xi \sim 1$.

6.2 Critical dynamics in one-component fluids

In one-component fluids near the gas–liquid transition, model H is a minimal model correctly describing critical slowing down of the entropy relaxation as observed by Rayleigh

[3] An experimental setup to measure dichroism is illustrated in Ref. [22].

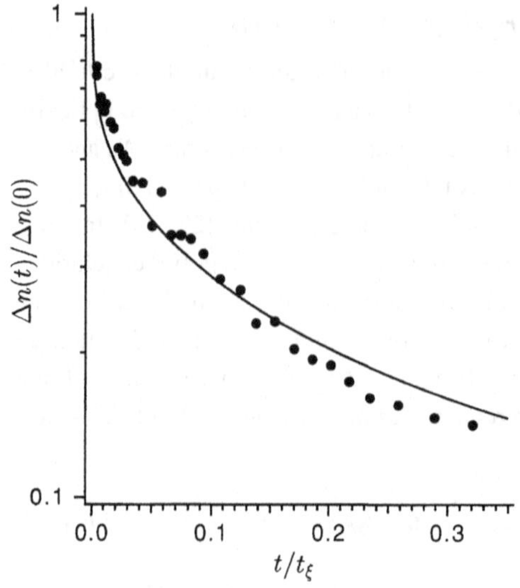

Fig. 6.4. Comparison for butoxyetbaranol + water of the theoretical decay function (solid line) defined by (6.1.57) and data (filled circles) taken from Ref. [25].

scattering, strong enhancement of the thermal conductivity, and the weak shear viscosity anomaly. However, one-component fluids are highly compressible near the critical point. As a result, a number of unique *adiabatic effects* can be predicted, in which the fluid internal state is changed by compression or expansion under constant-entropy conditions.[4] In this section we will first identify a nonlinear pressure $\hat{p}_{\rm nl}$ ($\propto \psi^2$) whose slow relaxation gives rise to a large frequency-dependent bulk viscosity $\zeta_{\rm R}^*(\omega)$. Then we may predict anomalous critical sound propagation, which has been extensively studied theoretically [28]–[38] and experimentally in one-component fluids [39]–[43] and two-component fluids [44]–[49]. Furthermore, slow relaxations can be predicted in various quantities such as the pressure, temperature, or structure factor after a macroscopic volume or pressure change. Next we will examine the effect of a thermal diffusion layer, which appears after a change in the boundary temperature and is crucial in macroscopic thermal equilibration (the piston effect) [50]–[60]. Adiabatic effects in phase separation will be discussed in Chapters 8 and 9. These effects are of fundamental importance but have not yet attracted enough attention.

6.2.1 Dynamic equations of compressible fluids

In one-component fluids near the gas–liquid critical point, the deviations of the number and energy densities, δn and δe, are linear combinations of the spin and energy densities, ψ and m, in the corresponding Ising system as given in (2.2.7) and (2.2.8), with the GLW hamiltonian $\mathcal{H} = \mathcal{H}\{\psi, m\}$ in the form of (4.1.45). The temperature and pressure fluctuations, $\delta\hat{T}$ and $\delta\hat{p}$, are given by (4.2.1)–(4.2.4) in the Ginzburg–Landau scheme.

[4] Hydrodynamically, the energy-density change $\langle\delta e\rangle$ (averaged over the thermal fluctuations) is created by the local average number-density change $\langle\delta n\rangle$ as $\langle\delta e\rangle = (e+p)\langle\delta n\rangle \cong (e_{\rm c}+p_{\rm c})\langle\delta n\rangle$.

With these preliminaries, we construct the dynamic equations for the mass and momentum densities,

$$\rho = m_0 n, \quad \boldsymbol{J} = \rho \boldsymbol{v}, \tag{6.2.1}$$

and the energy density e, where m_0 is the molecular mass and $\boldsymbol{v}$ is the velocity field. We may set $\boldsymbol{v} \cong \rho_c^{-1}\boldsymbol{J}$ hereafter, because $\boldsymbol{J}$ is already a small quantity. We set up the nonlinear Langevin equations [36],

$$\frac{\partial}{\partial t}\rho = -\nabla\cdot(\rho\boldsymbol{v}), \tag{6.2.2}$$

$$\frac{\partial}{\partial t}e = -\nabla\cdot(e\boldsymbol{v}) - p_c\nabla\cdot\boldsymbol{v} + \lambda_0 T_c\nabla^2\frac{\delta\mathcal{H}}{\delta e} + \theta, \tag{6.2.3}$$

$$\frac{\partial}{\partial t}\boldsymbol{J} = -\nabla\cdot\overleftrightarrow{\Pi} + \eta_0\nabla^2\boldsymbol{v} + \left[\zeta_0 + \left(1 - \frac{2}{d}\right)\eta_0\right]\nabla(\nabla\cdot\boldsymbol{v}) + \boldsymbol{\zeta}. \tag{6.2.4}$$

The total hamiltonian is the sum of $\mathcal{H}\{\psi, m\}$ and the kinetic energy $\frac{1}{2}\int d\boldsymbol{r}\rho_c^{-1}\boldsymbol{J}^2$.

(i) First, we explain the reversible parts. In (6.2.3) the second term on the right-hand side represents adiabatic energy changes caused by volume changes. In (6.2.4) $\overleftrightarrow{\Pi} = \{\Pi_{ij}\}$ is the reversible stress tensor arising from the fluctuations of $\delta\rho$ and δe and can be expressed in the form

$$\Pi_{ij} = (\delta\hat{p} + \delta\tilde{p})\delta_{ij} + TK(\nabla_i\psi)(\nabla_j\psi). \tag{6.2.5}$$

The first pressure term $\delta\hat{p}$ is defined by (4.2.2) or (4.2.8) and is the largest term in (6.2.5). The second pressure term $\delta\tilde{p}$ is nonlinearly dependent on $\delta\rho$ and δe and is small, so its explicit form will be given in Appendix 6A. The force density takes a rather simple form,

$$\begin{aligned}\nabla\cdot\overleftrightarrow{\Pi} &= \rho\nabla\frac{\delta\mathcal{H}}{\delta\rho} + (e + p_c)\nabla\frac{\delta\mathcal{H}}{\delta e} \\ &= \nabla\delta\hat{p} + \psi\nabla\frac{\delta\mathcal{H}}{\delta\psi} + m\nabla\frac{\delta\mathcal{H}}{\delta m},\end{aligned} \tag{6.2.6}$$

where $\mathcal{H}$ is regarded as a functional of $\delta\rho$ and δe in the first line and that of ψ and m in the second line. The nonlinear inertia part $\rho\boldsymbol{v}\boldsymbol{v}$ in the stress is neglected, because the Reynolds number is very small on relevant spatial scales in critical dynamics.

(ii) Second, we explain the dissipative parts. The λ_0, η_0, and ζ_0 are the background thermal conductivity, shear viscosity, and bulk viscosity, respectively. The random noise term $\theta(\boldsymbol{r}, t)$ in (6.2.3) satisfies (6.1.13) with L_0 being replaced by $\lambda_0 T_c^2$, while $\boldsymbol{\zeta}(\boldsymbol{r}, t)$ in (6.2.4) is characterized by

$$\langle\zeta_i(\boldsymbol{r}, t)\zeta_j(\boldsymbol{r}', t')\rangle = -2T\left[\eta_0\delta_{ij}\nabla^2 + \left(\zeta_0 + \eta_0 - \frac{2\eta_0}{d}\right)\nabla_i\nabla_j\right]\delta(t - t')\delta(\boldsymbol{r} - \boldsymbol{r}'). \tag{6.2.7}$$

Gravity effects

In gravity, (4.2.41) suggests that $\mathcal{H}$ should be replaced by

$$\mathcal{H}_{\rm T} = \mathcal{H} + \mathcal{H}_g = \mathcal{H} + \int d\boldsymbol{r} g z \delta\rho, \tag{6.2.8}$$

where g is the gravitational acceleration. The force density $-\nabla \cdot \overleftrightarrow{\Pi}$ on the fluid in (6.2.4) is increased by $-\rho g$ in the z direction by this replacement, which is nothing but the buoyancy force in gravity. In equilibrium, we obtain the pressure gradient in (4.2.44) and the gravity-induced density stratification discussed in Section 2.2. A deviation from this pressure profile induces a velocity field which drives the system towards the final equilibrium.

Slow dynamics

The full dynamic equations (6.2.2)–(6.2.4) are needed to adequately describe sound propagation. However, if we are interested in slow thermal diffusion processes, the equations may be simplified as follows. We introduce the dynamic variable representing the entropy fluctuation per particle as

$$\delta s = (n_{\rm c} T_{\rm c})^{-1}\big[\delta e - H_{\rm c}\delta n\big], \tag{6.2.9}$$

where $H_{\rm c} = (e_{\rm c} + p_{\rm c})/n_{\rm c}$ is the enthalpy per particle at the critical point. The coefficients here are those at the critical point. Then $\delta\hat{T}$ and $\delta\hat{p}$ can be expressed as (4.2.3) and (4.2.4), resulting in the correlation function relations in (4.2.5). The dynamic equation for δs is of the form

$$\frac{\partial}{\partial t}\delta s = -\nabla\cdot(\delta s \boldsymbol{v}) + (n_{\rm c} T_{\rm c})^{-1}\lambda_0 \nabla^2 \delta\hat{T} + (n_{\rm c} T_{\rm c})^{-1}\theta. \tag{6.2.10}$$

Furthermore, for slow motions (slower than the acoustic time L/c, L being the system dimension and c the sound velocity) we may assume homogeneity of the following combination,

$$p_1(t) \equiv \delta\hat{p}(\boldsymbol{r}, t) + \rho_{\rm c} g z. \tag{6.2.11}$$

Then the temperature deviation is expressed as

$$\delta\hat{T} = \left(\frac{\partial T}{\partial p}\right)_{\rm cx} p_1(t) + \alpha_s^{-1}\frac{\delta}{\delta\psi}\mathcal{H}_{\rm T}. \tag{6.2.12}$$

The time dependence of $p_1(t)$ is then determined from the macroscopic boundary condition. After such acoustic relaxation, the transverse velocity remains nonvanishing and we may set $\nabla\cdot\boldsymbol{v} = 0$.

Close to equilibrium at fixed pressure, we may set $p_1 = 0$ and hence $\delta\mathcal{H}_{\rm T}/\delta m = (\partial\tau/\partial h)_p \delta\mathcal{H}_{\rm T}/\delta\psi$ from (4.2.2), which means that the deviation of m is much smaller than that of ψ by $(C\chi)^{-1}$ and $\psi \cong n_{\rm c}\delta s/\alpha_s \cong \delta n/\alpha_1$ from (2.2.7) and (2.2.9). Thus the dynamic equations for the entropy δs and the transverse velocity $\boldsymbol{v}$ constitute model H. In accord with this conclusion, the decay rate of δs measured by dynamic light scattering

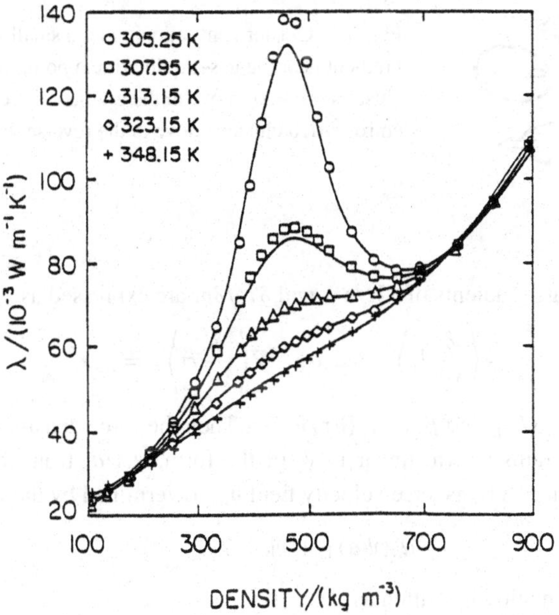

Fig. 6.5. The thermal conductivity λ vs the density for various temperatures in CO_2 near the gas–liquid critical point [28].

and the shear viscosity anomaly are well predicted by model H. In Section 6.3, however, we shall see that the time dependence of nonvanishing $p_1(t)$ is crucial in nonequilibrium thermal equilibration in a cell at fixed volume.

6.2.2 Cluster convection and enhanced heat conduction

As shown in Fig. 6.5, the thermal conductivity near the gas–liquid critical point has been observed to increase markedly near the critical point. We will examine the physical process enhancing heat conduction in more detail. In Chapter 2 the critical fluctuations were shown to emerge as large clusters. They may also be viewed as enhanced heterogeneities of the entropy (per particle) because of the linear relation $n_c \delta s \cong \alpha_s \psi$ in (2.2.9). If we apply a small temperature gradient, the clusters with $\delta s < 0$ will have a tendency to move down the gradient, whereas those with $\delta s > 0$ will tend to move in the reverse direction. This counterflow mechanism, illustrated in Fig. 6.6, should enhance heat transport.

Let a near-critical fluid be in a steady state under a small temperature gradient and a homogeneous pressure (without gravity):

$$\boldsymbol{a} \equiv \nabla \langle \delta \hat{T} \rangle_{ss}, \quad \nabla \langle \delta \hat{p} \rangle_{ss} = 0, \tag{6.2.13}$$

where $\langle \cdots \rangle_{ss}$ is the steady-state average. From (4.2.1) and (4.2.2) (or from (2.2.11) and

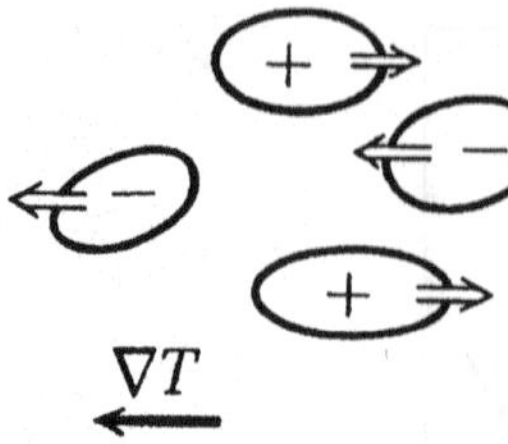

Fig. 6.6. Cluster convection under a small temperature gradient near the gas–liquid critical point. Entropy-poor clusters ($\delta s < 0$) move in the gradient direction, while entropy-rich clusters move in the reverse direction.

(2.2.12)) the average gradients of $\delta\mathcal{H}/\delta\psi$ and $\delta\mathcal{H}/\delta m$ are expressed as

$$\nabla\left\langle\frac{\delta}{\delta\psi}\mathcal{H}\right\rangle_{\rm ss} = \alpha_s \boldsymbol{a}, \quad \nabla\left\langle\frac{\delta}{\delta m}\mathcal{H}\right\rangle_{\rm ss} = \beta_{\rm s}\boldsymbol{a}, \tag{6.2.14}$$

where $\alpha_s = T_{\rm c}(\partial h/\partial T)_p$ and $\beta_{\rm s} = T_{\rm c}(\partial\tau/\partial T)_p$. Then the force density on the right-hand side of (6.2.4) contains a term linear in ψ of the form, $\alpha_s\psi\boldsymbol{a}$, from the second line of (6.2.6), which induces a transverse velocity field $\boldsymbol{v}_{\rm ind}$ determined by the force balance

$$-\alpha_s(\psi\boldsymbol{a})_\perp + \eta_{\rm R}\nabla^2\boldsymbol{v}_{\rm ind} = 0. \tag{6.2.15}$$

Its Fourier transformation is written as

$$\boldsymbol{v}_{\rm ind}(\boldsymbol{q}) = -\frac{\alpha_s}{\eta_{\rm R}q^2}[\boldsymbol{a} - (\hat{\boldsymbol{q}}\cdot\boldsymbol{a})\hat{\boldsymbol{q}}]\psi_{\boldsymbol{q}}, \tag{6.2.16}$$

where $\hat{\boldsymbol{q}} = q^{-1}\boldsymbol{q}$. See Appendix 6C to justify the above arguments.

If the fluctuations on the scale of ξ are picked up, $\boldsymbol{v}_{\rm ind}$ may be approximately expressed as

$$\boldsymbol{v}_{\rm ind} \cong \frac{-1}{6\pi\eta_{\rm R}\xi}(S_\xi\alpha_s\boldsymbol{a})_\perp, \tag{6.2.17}$$

where $S_\xi(\boldsymbol{r}) \equiv \int_{|\boldsymbol{r}'|<\xi} d\boldsymbol{r}'\psi(\boldsymbol{r}+\boldsymbol{r}')$ is the space integral of ψ around the position $\boldsymbol{r}$ over a spatial region with dimension ξ as defined in (2.1.23), at $\eta = \xi$. From (6.2.10) the heat current bilinear with respect to the gross variables is $T_{\rm c}n_{\rm c}\delta s\boldsymbol{v}$, so that the average excess heat current due to the cluster convection is

$$T_{\rm c}n_{\rm c}\langle\delta s\boldsymbol{v}_{\rm ind}\rangle_{\rm ss} = -(\Delta\lambda)\boldsymbol{a}. \tag{6.2.18}$$

Substitution of (6.2.17) gives the excess thermal conductivity,

$$\begin{aligned}\Delta\lambda &= T_{\rm c}\alpha_s^2\langle\psi S_\xi\rangle/(6\pi\eta_{\rm R}\xi)\\ &= T_{\rm c}C_p/(6\pi\eta_{\rm R}\xi),\end{aligned} \tag{6.2.19}$$

where $C_p = \alpha_s^2\chi$ is the susceptibility in the corresponding spin system from (2.2.23). The second line follows if use is made of the estimation, $\langle\psi S_\xi\rangle = \int_\xi d\boldsymbol{r}'\langle\psi(\boldsymbol{r})\psi(\boldsymbol{r}+\boldsymbol{r}')\rangle \cong \chi$. Near the critical point, $\Delta\lambda$ dominates over the background λ_0 and $\lambda_{\rm R} \cong \Delta\lambda$. For example, $\lambda_0/\Delta\lambda \cong 20(T/T_{\rm c} - 1)^\nu$ for CO_2 on the critical isochore [15]. Then thermal diffusivity $D = \lambda_{\rm R}/C_p$ is again given by the Einstein–Stokes formula in (6.1.24). In this counterflow process the clusters transfer heat from a warmer to a cooler boundary, while they have finite lifetimes of order t_ξ. Molecular dynamics simulations to confirm this picture should be informative but, to our knowledge, have not yet been performed.

6.2.3 Nonlinear pressure and temperature fluctuations

From (4.2.1) and (4.2.2) the pressure and temperature fluctuations $\delta\hat{p}$ and $\delta\hat{T}$ contain the following bilinear terms [36],

$$\hat{p}_{\rm nl} = \left(\frac{\partial p}{\partial \tau}\right)_h \gamma_0 \psi^2, \quad \hat{T}_{\rm nl} = \left(\frac{\partial T}{\partial \tau}\right)_h \gamma_0 \psi^2 = \left(\frac{\partial T}{\partial p}\right)_{\rm cx} \hat{p}_{\rm nl}, \tag{6.2.20}$$

which arise from $\delta\hat{\tau} = \delta(\beta\mathcal{H})/\delta m$. The nonlinear terms in $\delta\hat{h} = \delta(\beta\mathcal{H})/\delta\psi$ are very small. We will show that $\hat{p}_{\rm nl}$ gives rise to a strongly growing contribution to the frequency-dependent bulk viscosity. From the general formula (5.4.5) it can be written as

$$\zeta_{\rm R}^*(\omega) = \zeta_0 + \frac{1}{T}\int_0^\infty dt \int d\boldsymbol{r} e^{-i\omega t}\langle \delta\hat{p}_{\rm nl}(\boldsymbol{r}, t)\delta\hat{p}_{\rm nl}(\boldsymbol{0}, 0)\rangle, \tag{6.2.21}$$

where $\delta\hat{p}_{\rm nl} = \hat{p}_{\rm nl} - \langle\hat{p}_{\rm nl}\rangle$ is the deviation. The background bulk viscosity ζ_0 and the frequency-dependent shear viscosity $\eta_{\rm R}^*(\omega)$ are much smaller than the singular bulk viscosity near the critical point.

We consider a sound wave in which the average deviations depend on space and time as $\exp(i\omega t - ikx)$. From the general linear response formula (5.4.24) the pressure deviation $p_1 = \langle\delta\hat{p}\rangle$ in a sound is related to that of the mass density $\rho_1 = \langle\delta\rho\rangle$ as

$$p_1 = \frac{\rho_1}{K_{\rm s}(\omega)\rho} = \left[\rho c^2 + i\omega\zeta_{\rm R}^*(\omega)\right]\frac{\rho_1}{\rho}, \tag{6.2.22}$$

where $K_{\rm s}(\omega)$ is the frequency-dependent adiabatic compressibility in (5.4.25). The acoustic dispersion relation is given by

$$\omega^2/k^2 = 1/K_{\rm s}(\omega)\rho = c^2 + i\omega\zeta_{\rm R}^*(\omega)/\rho. \tag{6.2.23}$$

In usual experiments, ω is externally applied; then, the wave number $k = {\rm Re}\,k + i\,{\rm Im}\,k$ is a complex number and the frequency-dependent sound velocity and the attenuation per wavelength are expressed as

$$c(\omega) = \omega/{\rm Re}\,k, \quad \alpha_\lambda = -2\pi\,{\rm Im}\,k/{\rm Re}\,k. \tag{6.2.24}$$

In the low-frequency limit, $\omega t_\xi \ll 1$, α_λ is of the form,

$$\alpha_\lambda = \pi\omega\zeta_{\rm R}^*(0)/\rho c^2. \tag{6.2.25}$$

The contributions from the shear viscosity and the thermal conductivity in the usual hydrodynamic expression are negligible near the critical point (see footnote 9 on p. 216).

Before proceeding to the detailed calculation, we examine the magnitude of $\hat{p}_{\rm nl}$ on the critical isochore above $T_{\rm c}$. Note that the coefficient $\gamma_0 = \gamma_0(\Lambda)$ strongly depends on the upper cut-off wave number Λ, as discussed in Section 4.3. We first seek the renormalized form of $\hat{p}_{\rm nl}$ by setting $\Lambda \ll \kappa = \xi^{-1}$. We note the relations, $2\gamma_{\rm R} C_H \tau = \gamma/\chi$ from (4.1.58), $\tau^2 C_H = (T/T_{\rm c} - 1)^2 C_V$ from (2.2.27), and $\tau = (\partial\tau/\partial T)_h(T - T_{\rm c})$ from (2.2.15) and (2.2.16), where C_H and χ are the specific heat and the susceptibility in the corresponding Ising system. They readily yield [36]

$$\hat{p}_{\rm nl} = T_{\rm c} A_p \psi^2/2\chi \qquad (\Lambda \ll \kappa). \tag{6.2.26}$$

The coefficient,

$$A_p = \frac{\gamma T_{\rm c}}{(T - T_{\rm c})C_V}\left(\frac{\partial p}{\partial T}\right)_{\rm cx} = A^*(T/T_{\rm c} - 1)^{-1+\alpha}, \tag{6.2.27}$$

grows strongly as $T \to T_{\rm c}$ with A^* of order 1. The fluctuations with wave numbers smaller than κ give rise to a strongly divergent contribution to the zero-frequency bulk viscosity,

$$\zeta_{\rm R}^*(0) \sim T_{\rm c} A_p^2 K_d \kappa^d t_\xi \sim \alpha\rho c^2 t_\xi \propto \xi^{z-\alpha/\nu}, \tag{6.2.28}$$

where $K_d\kappa^d$ is the volume of the wave number space and use has been made of the thermodynamic relation (1.2.53) and the two-scale-factor universality relations (2.2.28) and (4.3.47). At the other extreme, for $\Lambda \gg \kappa$, $\hat{p}_{\rm nl}$ or γ_0 should be independent of $T - T_{\rm c}$. Assuming smooth crossover at $\Lambda = \kappa$, we have

$$\begin{aligned} \hat{p}_{\rm nl} &\cong \frac{1}{2}(\xi\Lambda)^{(\gamma+\alpha-1)/\nu} T_{\rm c} A_p \chi^{-1}\psi^2 \\ &\cong \frac{1}{2} T_{\rm c} B_p \Lambda^{(\gamma+\alpha-1)/\nu}\psi^2, \end{aligned} \tag{6.2.29}$$

where the coefficient B_p is of order $A^*\xi_{+0}^{(\gamma+\alpha-1)/\nu}/\Gamma_0$ with $\chi = \Gamma_0(T/T_{\rm c} - 1)^{-\gamma/\nu}$.

Projection operator method revisited

The general linear response theory in Section 5.4 shows that the bulk viscosity is expressed in terms of the time-correlation function of the quantity,

$$\delta P_{\rm R} \equiv \int d\boldsymbol{r}(1 - \mathcal{P})\delta\Pi_{xx}(\boldsymbol{r}), \tag{6.2.30}$$

where $\mathcal{P}$ is the hydrodynamic linear projection operator and $\delta\Pi_{xx}$ is the deviation of the xx component of the microscopic stress. In our Ginzburg–Landau theory, $\delta P_{\rm R}$ should correspond to the space integral of $\hat{p}_{\rm nl}(\boldsymbol{r})$. The original mode coupling theories [1, 31] supposed the following expansion form,

$$\delta P_{\rm R} = \int_{\boldsymbol{q}} V_q \psi_{\boldsymbol{q}}\psi_{-\boldsymbol{q}} + \cdots, \tag{6.2.31}$$

where $\psi_{\boldsymbol{q}}$ is the Fourier transformation of $\psi(\boldsymbol{r})$, and V_q is called the vertex function.

Kawasaki and Tanaka [30] confirmed microscopically that the projection of $\delta\Pi_{xx}$ onto the bilinear products of ψ is very small at long wavelengths (namely, $\langle \delta\Pi_{xx}\psi\psi\rangle \cong 0$). In fact, for the density fluctuation n_k ($\cong \alpha_1\psi_k$), (1A.12) gives

$$\langle |n_k|^2 : \Pi_{\alpha\beta}\rangle = TI(k)\delta_{\alpha\beta} - Tk_\alpha \frac{\partial}{\partial k_\beta} I(k), \tag{6.2.32}$$

$I(k)$ being the structure factor (1.2.56). However, (1.2.67) with the aid of (1.2.76) gives

$$\langle |n_k|^2 : \mathcal{P}\Pi_{\alpha\beta}\rangle = T\rho c^2 \left(\frac{\partial I(k)}{\partial p}\right)_s \delta_{\alpha\beta}. \tag{6.2.33}$$

Clearly, the latter quantity (6.2.33) is much larger than the former (6.2.32) near the critical point. We notice that this remains the case in our Ginzburg–Landau theory. That is, because $\delta\mathcal{H}/\delta m$ is statistically independent of ψ in equilibrium, the pressure fluctuation $\delta\hat{p}$ is nearly orthogonal to any powers of ψ (and hence $\langle\delta\hat{p}\psi\psi\rangle \cong 0$), whereas $(1-\mathcal{P})\delta\hat{p} \cong \hat{p}_{\rm nl}$ is bilinear in ψ.

With this finding the calculation of V_q is straightforward. Multiplying (6.2.30) by $\psi_q\psi_{-q}$ and taking the thermal average we obtain

$$2\chi_q^2 V_q \cong -\langle \psi_q\psi_{-q}\mathcal{P}\delta\Pi_{xx}\rangle = -T\rho c^2\left(\frac{\partial}{\partial p}\chi_q\right)_s, \tag{6.2.34}$$

where $\chi_q = \langle|\psi_q|^2\rangle$ and use has been made of the general thermodynamic relation (1.2.67) and (1.2.76). For $q \lesssim \kappa$ use of the Ornstein–Zernike form $\chi_q \propto 1/(\kappa^2 + q^2)$ yields a q-independent vertex function,

$$V_q \cong \frac{1}{2}T\rho c^2\left(\frac{\partial}{\partial p}\chi_q{}^{-1}\right)_s \cong \frac{\gamma T_{\rm c}}{2(T-T_{\rm c})}\rho c^2\left(\frac{\partial T}{\partial p}\right)_{\rm cx}\frac{1}{\chi}, \tag{6.2.35}$$

where the derivative at constant s has been replaced by that at constant $h = 0$. The above result turns out to coincide with our result, (6.2.26) and (6.2.27), as can be known from (1.2.53). The original Kawasaki theory [31] is based on the hydrodynamic expression (6.2.35) for the vertex function in the whole wave number region for nonlinear pressure (6.2.31). However, the vertex function strongly depends on the upper cut-off Λ as in (6.2.29) for $\Lambda > \kappa$. As a result, the Kawasaki theory is not a good approximation for high-frequency sounds.

6.2.4 The frequency-dependent bulk viscosity

In calculating the bulk viscosity in (6.2.21), we should take into account the strong Λ dependence of the coefficient $\gamma_0 = \gamma_0(\Lambda)$. As in (6.1.31) and (6.1.32) we pick up the fluctuation contribution in the shell region $\Lambda - \delta\Lambda < k < \Lambda$ and use the decoupling approximation:

$$\delta\zeta_{\rm R}^*(\omega) = \frac{2}{T_{\rm c}}\left(\frac{\partial p}{\partial \tau}\right)_h^2 \gamma_0^2 K_d\Lambda^{d-1}\delta\Lambda/[\chi_\Lambda^2(2\Gamma_\Lambda + i\omega)], \tag{6.2.36}$$

where $\chi_\Lambda = 1/K(\kappa^2 + \Lambda^2)$ and Γ_Λ is the decay rate at the cut-off. Here we have the relation $K_d\gamma_0^2 = v\Lambda^\epsilon K^2/C_0$ from (4.1.55). For $x \equiv \xi\Lambda \gg 1$, v and C_0 behave as (4.3.37) and (4.3.39). For $x \ll 1$, γ_0 and C_0 tend to the renormalized values γ_R and $C_R(= C_H$ on the critical isochore), respectively, so that v grows as $\Lambda^{-\epsilon}$. These limiting behaviors can be taken into account by the following parametrization:

$$v = v^*(1 + x^{-2})^{\epsilon/2}/(1 + Qx^{\alpha/\nu}), \tag{6.2.37}$$

$$C_0 = C_H(1 + Q)^{-1}(1 + x^2)^{-\alpha/2\nu}(1 + Qx^{\alpha/\nu}), \tag{6.2.38}$$

where $v^* = \alpha/2\nu + O(\epsilon^2)$ from (4.3.57) and

$$Q = (C_B/A_0)\tau^\alpha = B(T/T_c - 1)^\alpha \tag{6.2.39}$$

is the ratio of the background to singular parts of $C_V = b_c^2 C_H(\propto (T/T_c - 1)^{-\alpha}(1 + Q))$ in (4.3.93). The experimental values of the coefficient B were given for four fluids below (4.3.93). We further use the thermodynamic relations,

$$\frac{1}{T}\left(\frac{\partial p}{\partial \tau}\right)_h^2 C_H^{-1} = T\left(\frac{\partial p}{\partial T}\right)_{\mathrm{cx}}^2 C_V^{-1} = \rho c^2, \tag{6.2.40}$$

which follows from (2.2.26). Now integration with respect to Λ or $x = \xi\Lambda$ yields

$$\zeta_R^*(\omega) = v^*\rho c^2(1 + Q)t_\xi \int_0^\infty dx \frac{x^{3-\epsilon}}{(1 + x^2)^{(1-\alpha)/\nu}[\Gamma^*(x) + iW](1 + Qx^{\alpha/\nu})^2}, \tag{6.2.41}$$

where the decay rate Γ_q is scaled as

$$\Gamma_q = t_\xi^{-1}\Gamma^*(q\xi), \tag{6.2.42}$$

and W is a dimensionless frequency,

$$W = \omega t_\xi/2. \tag{6.2.43}$$

Here $t_\xi = t_0(T/T_c - 1)^{-z\nu}$ is the order parameter lifetime defined by (6.1.25); then, $\Gamma^*(x) \cong x^2$ for $x \ll 1$ and $\Gamma^*(x) \cong x^d$ for $x \gg 1$.

Low-frequency limit

In the low-frequency limit $\omega t_\xi \ll 1$ on the critical isochore above T_c, the bulk viscosity behaves as

$$\zeta_R^*(0) = R_B\rho c^2 t_\xi/(1 + Q) \propto \xi^{z-\alpha/\nu}/(1 + Q)^2, \tag{6.2.44}$$

including the correction Q. The coefficient R_B is a universal number of order α and its ϵ expansion is

$$R_B = \frac{1}{24}\epsilon + \cdots. \tag{6.2.45}$$

Low-frequency data of the acoustic attenuation in ^{3}He suggest $R_B \cong 0.03$ [36, 43]. Remarkably, the zero-frequency bulk viscosity $\zeta_R^*(0) \propto \xi^{2.8}$ diverges more strongly near the critical point than any other transport coefficient. For example, in ^{3}He at $T/T_c - 1 = 10^{-4}$ on the critical isochore, it is about 50 poise while the shear viscosity is 17×10^{-6} poise [43].

High-frequency limit including the background specific heat correction

In the high-frequency case $\omega t_\xi \gg 1$, the wave number region $\Lambda \gg \xi^{-1}$ or $x \gg 1$ (where $\Gamma_\Lambda \sim \omega$) is most important. Thus we may replace $1 + x^2$ and $\Gamma^*(x)$ in (6.2.41) by x^2 and x^z, respectively. Then,

$$\begin{aligned} \frac{i\omega\zeta_R^*(\omega)}{\rho c^2(1+Q)} &= 2\upsilon^* \int_1^\infty dx \frac{iW}{x^{1-\alpha/\nu}(x^z + iW)(1 + Qx^{\alpha/\nu})^2} \\ &= \int_1^\infty dx \frac{iW}{x^z + iW} \frac{\partial}{\partial x}\left(\frac{x^{\alpha/\nu}}{1 + Qx^{\alpha/\nu}}\right). \end{aligned} \tag{6.2.46}$$

We may calculate the dominant contribution for small α/ν by deforming the integration path in the complex x plane. Namely, by setting $y = (iW)^{-1/z}x$ we obtain

$$\frac{i\omega\zeta_R^*(\omega)}{\rho c^2(1+Q)} = (iW)^{\alpha/\nu z} \int_{y^*}^\infty dy \frac{1}{y^z + 1} \frac{\partial}{\partial y}\left(\frac{y^{\alpha/\nu}}{1 + Q(iWy^z)^{\alpha/\nu z}}\right), \tag{6.2.47}$$

where the lower bound $y^* = (iW)^{-1/z}$ is complex, but the integration path for $|y| > 1$ may be along the real axis in the complex y plane. For small α/ν the upper bound and $(y^z + 1)^{-1}$ may be replaced by 1, so that

$$i\omega\zeta_R^*(\omega)/\rho c^2 = (1 + Q)/\left[(iW)^{-\alpha/\nu z} + Q\right] - 1. \tag{6.2.48}$$

For $Q = 0$ or $B = 0$, which is the case for ^{3}He, the right-hand side simply becomes $(iW)^{\alpha/\nu z} - 1$. Following Ferrell and Bhattacharjee [33, 34], we may interpret the above result in terms of the frequency-dependent specific heat defined by

$$C_V^*(\omega) \cong A(i\omega t_0)^{-\alpha/\nu z} + B, \tag{6.2.49}$$

where t_0 is defined by (6.1.25). This expression simply follows if $T/T_c - 1$ in C_V is replaced by $(i\omega t_0)^{1/\nu z}$. In terms of $C_V^*(\omega)$ we have

$$\zeta_R^*(\omega) = \frac{\rho c^2}{i\omega}\left[C_V/C_V^*(\omega) - 1\right], \tag{6.2.50}$$

$$\rho\omega^2/k^2 = T\left(\frac{\partial p}{\partial T}\right)_{cx}^2 C_V^*(\omega)^{-1}. \tag{6.2.51}$$

Because $k \propto \sqrt{C_V^*(\omega)}$, (6.2.24) leads to the attenuation per wavelength [34],

$$\begin{aligned} \alpha_\lambda &= -2\pi\, \mathrm{Im}[\sqrt{C_V^*(\omega)}]/\,\mathrm{Re}[\sqrt{C_V^*(\omega)}] \\ &= 0.27/(1 + X), \end{aligned} \tag{6.2.52}$$

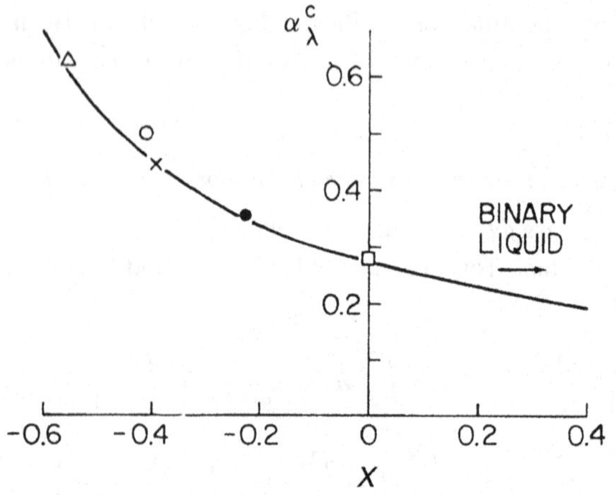

Fig. 6.7. Sound attenuation per wavelength α_λ at the critical point vs X. The parameter X defined by (6.2.53) is dependent on ω and the ratio B of the background to critical components of C_V. The solid line shows the theoretical formula (6.2.52) [33]. Experimental frequencies (in MHz) and [references]: □ ^{3}He, 0.5 and 1 [42]; • ^{4}He, 0.5 [41]; × Xe, 1 [40]; ∘ Xe, 3 [40]; △ Xe, 440 [39]. The universal relation $\alpha_\lambda \cong 0.27$ is confirmed for ^{3}He.

where we have set $i^{-\alpha/\nu z} \cong 1 - i\pi\alpha/2\nu z$ and $\pi^2\alpha/2z\nu \cong 0.27$, and the second line follows from the first line from $|\,\mathrm{Im}[C_V^*(\omega)]| \ll \mathrm{Re}[C_V^*(\omega)]$. The parameter,

$$X = QW^{\alpha/\nu z} = B(\omega t_0)^{\alpha/\nu z}, \tag{6.2.53}$$

is dependent on ω but independent of $T/T_c - 1$. If the background B is negative, it serves to increase α_λ above 0.27, as illustrated in Fig. 6.7. It is worth noting that the acoustic relation (6.2.52) is analogous to (6.1.30) for the frequency-dependent shear viscosity. The effective sound velocity is expressed as

$$c(\omega) = c(t_\xi\omega/2)^{\alpha/2\nu z}[(1+Q)/(1+X)]^{1/2} \propto \omega^{\alpha/2\nu z}(1+X)^{-1/2}, \tag{6.2.54}$$

where use has been made of $c^2 \propto \xi^{-\alpha/\nu}/(1+X)$ from (4.3.94). The dispersion relation at high frequencies is thus asymptotically independent of ξ. The correction X arising from the background specific heat is also independent of ξ and remains noticeable in real experiments except for ^{3}He on the critical isochore above T_c.

Overall behavior on the critical isochore

If we neglect the background specific heat ($B = 0$), the frequency-dependent bulk viscosity is asymptotically scaled as

$$i\omega\zeta_R^*(\omega) = \rho c^2 \mathcal{F}(W) \quad \text{or} \quad \omega^2/c^2k^2 = 1 + \mathcal{F}(W), \tag{6.2.55}$$

where W is the scaled frequency (6.2.43). We have found that $\mathcal{F}(W) \cong 2R_B iW$ for $|W| \ll 1$ and $\mathcal{F}(W) \cong (iW)^{\alpha/\nu z} - 1$ for $|W| \gg 1$. If we calculate the scaling function $\mathcal{F}(W)$ to

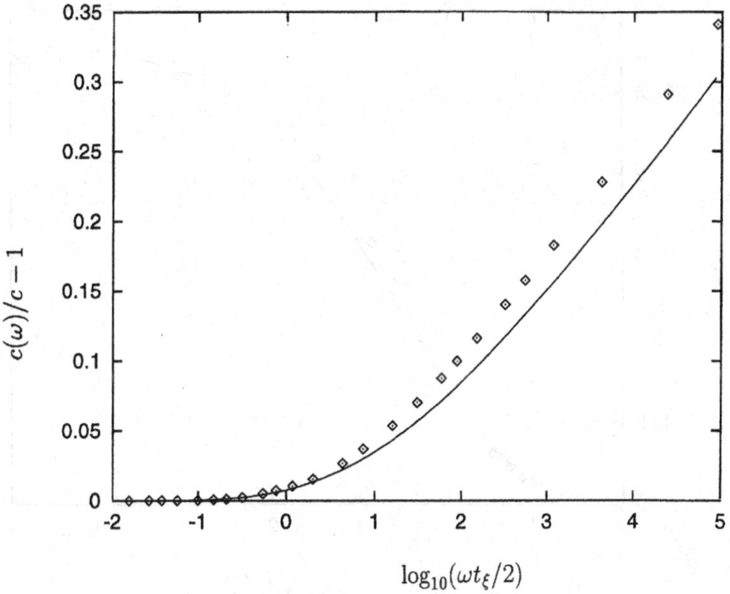

Fig. 6.8. $c(\omega)/c - 1$ vs $\omega t_\xi/2$ on the critical isochore above T_c obtained from (6.2.57) and (6.2.24) on a semi-logarithmic scale. It is compared with the data for ^{3}He of Ref. [42].

first order in ϵ, we obtain the ϵ expansion, $\mathcal{F}(W) = v^* F(W)$ with [33, 35]

$$F(W) = -1 + \frac{1}{2}\left(1 - \frac{1}{iW}\right)\ln(iW) + \frac{1}{\Delta}\left(\frac{3}{2} - \frac{1}{2iW}\right)\ln\left(\frac{1+\Delta}{1-\Delta}\right), \qquad (6.2.56)$$

where $\Delta \equiv (1 - 4iW)^{1/2}$. For $|W| \gg 1$, $F(W) \cong \frac{1}{2}\ln(iW)$ and $1 + \mathcal{F}(W) = 1 + \frac{1}{2}v^* \ln(iW) \cong (iW)^{v^*/2}$. To reproduce the Ferrell–Bhattacharjee form (6.2.49) we thus replace v^* by $2\alpha/\nu z$ (not by its ϵ expansion form) and exponentiate the logarithmic term [35] as

$$\mathcal{F}(W) = -1 + (1 + iW)^{\alpha/\nu z}\left\{1 + \frac{\alpha}{\nu z}\big[2F(W) - \ln(1 + iW)\big]\right\}. \qquad (6.2.57)$$

This form leads to the low-frequency result (6.2.44) with $R_B = \alpha/2\nu z = 0.028$ and the high-frequency result (6.2.48) with $B = 0$. Figures 6.8 and 6.9 display the sound velocity $c(\omega)$ and the attenuation per wavelength α_λ in (6.2.24) derived from the above formula with $\alpha/\nu z = 0.057$. They are compared with data of Roe and Meyer for ^{3}He on the critical isochore at 1 MHz [42]. Another theoretical formula in agreement with the data was also proposed in Ref. [38].

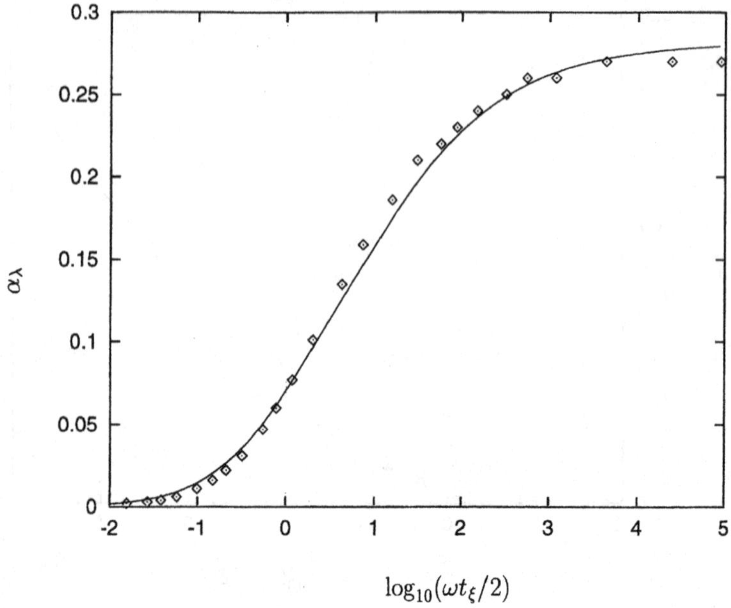

Fig. 6.9. α_λ vs $\omega t_\xi/2$ on the critical isochore above T_c obtained from (6.2.57) and (6.2.24) on a semi-logarithmic scale. It is compared with the data for ^{3}He of Ref. [42].

6.2.5 Adiabatic linear response

The time-correlation function expression (5.4.5) indicates anomalously slow relaxation of the diagonal component of the stress tensor near the critical point. The stress relaxation function is defined by

$$TG_{xx}(t) = \int d\boldsymbol{r} \langle \Pi^{\rm R}_{xx}(\boldsymbol{r},t)\Pi^{\rm R}_{xx}(\boldsymbol{0},0)\rangle \cong \int d\boldsymbol{r}\langle \delta\hat{p}_{\rm nl}(\boldsymbol{r},t)\delta\hat{p}_{\rm nl}(\boldsymbol{0},0)\rangle. \quad (6.2.58)$$

The Laplace (one-sided Fourier) transformation of $G_{xx}(t)$ is equal to the frequency-dependent bulk viscosity $\zeta^*_{\rm R}(\omega)$. If we neglect the background specific-heat correction, the following function,

$$\hat{G}(t) = \frac{1}{\rho c^2}G_{xx}(t) = \int \frac{d\omega}{2\pi i\omega}e^{i\omega t}\mathcal{F}\left(\frac{i}{2}\omega t_\xi\right), \quad (6.2.59)$$

is a universal function of t/t_ξ as displayed in Fig. 6.10 [37].

(i) For $t \lesssim t_\xi$ the high-frequency expression (6.2.48) with $B = 0$ yields the short-time behavior,

$$\hat{G}(t) \cong (t/t_\xi)^{-\alpha/\nu z} - 1. \quad (6.2.60)$$

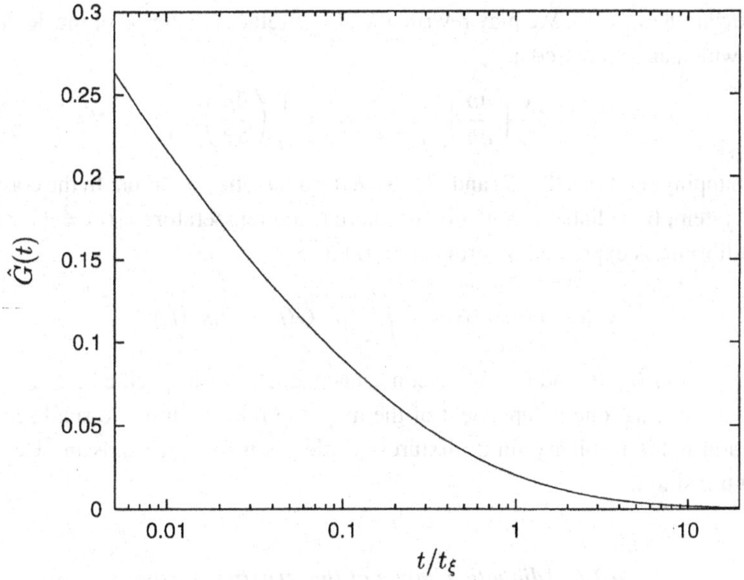

Fig. 6.10. The stress relaxation function $\hat{G}(t)$ defined by (6.2.59) near the critical point on a semi-logarithmic scale. It is calculated by the inverse Laplace transformation of (6.2.41) for $Q = 0$. It decays nearly logarithmically for $t < t_\xi$ and as $t^{-3/2}$ for $t > t_\xi$.

(ii) For $t > t_\xi$ use of (6.2.26) yields a long-time tail,

$$\hat{G}(t) \cong 2\alpha\xi^d \int_0^\kappa dq q^{d-1} \exp(-2Dq^2 t) \cong \alpha\Gamma(d/2)(2\Gamma_\xi t)^{-d/2} \propto t^{-d/2}, \tag{6.2.61}$$

where the two-scale-factor universality relations are used as in (6.2.28). This tail arises from the diffusive relaxation of the hydrodynamic fluctuations with $q < \kappa$. Note that (6.2.41) can correctly produce this tail from the small-x integration. In 3D it gives rise to a higher-order correction of order $(i\omega t_\xi)^{1/2}$ to the low-frequency bulk viscosity,

$$\zeta_{\rm R}^*(\omega) = \rho c^2 t_\xi \left[R_{\rm B} - \frac{1}{2}\pi\alpha\sqrt{i\omega t_\xi/2} + \cdots \right], \tag{6.2.62}$$

which leads to the low-frequency attenuation per wavelength,

$$\alpha_\lambda = \pi\omega t_\xi \left[R_{\rm B} - \frac{1}{4}\pi\alpha\sqrt{\omega t_\xi} + \cdots \right]. \tag{6.2.63}$$

In terms of $\hat{G}(t)$ we rewrite the acoustic relation (6.2.22) for general time-dependent pressure and density deviations, $p_1(t)$ and $\rho_1(t)$, as

$$p_1(t) = c^2 \left[\rho_1(t) + \int_{-\infty}^0 dt' \hat{G}(t-t')\dot{\rho}_1(t') \right], \tag{6.2.64}$$

where $\dot{\rho}_1(t) = \partial\rho(t)/\partial t$. The above relation holds in the adiabatic condition with vanishing entropy deviation ($s_1 = 0$). We may rewrite the above relation in terms of the deviations of ψ and m, which are expressed as

$$\psi_1 = \frac{1}{T_c}\left(\frac{\partial p}{\partial h}\right)_\tau \frac{\rho_1}{\rho}, \quad m_1 = \frac{1}{T_c}\left(\frac{\partial p}{\partial \tau}\right)_h \frac{\rho_1}{\rho}, \tag{6.2.65}$$

from the mapping relations (2.2.7) and (2.2.9) in the adiabatic condition. In the corresponding Ising system, the adiabatic deviation of the reduced temperature $\tau_1(t) = \langle\delta(\beta\mathcal{H})/\delta m\rangle$ in nonequilibrium is expressed in terms of $m_1(t)$ as

$$C_M \tau_1(t) = m_1(t) + \int_{-\infty}^{t} dt' \hat{G}(t-t')\dot{m}_1(t'), \tag{6.2.66}$$

where $\dot{m}_1(t) = \partial m(t)/\partial t$ and C_M is the constant-magnetization specific heat. The relation (6.2.66) is a universal one independent of the mapping relationship. We shall see that the same relation holds in binary fluid mixtures, while a similar one holds in ^{4}He near the superfluid transition.

6.2.6 *Adiabatic change of the structure factor*

An interesting application of the general linear response formula (5.4.21) is the adiabatic change of $\mathcal{B} \equiv \delta n(\boldsymbol{r})\delta n(\boldsymbol{0})$ in a sound, which may be calculated using (6.2.32) and (6.2.33). Its Fourier transformation with respect to $\boldsymbol{r}$ yields the structure factor $I(q,t) = I(q) + \mathrm{Re}[I_1(q,\omega)\rho_1/\rho]$, where the density deviation ρ_1 oscillates as $e^{i\omega t}$ and propagates in the x direction. Assuming an exponential relaxation of $n_{\boldsymbol{q}}(t)$ with the relaxation rate Γ_q, we have [49]

$$I_1(q,\omega) \cong \rho\left(\frac{\partial I(q)}{\partial \rho}\right)_s \frac{2\Gamma_q}{i\omega + 2\Gamma_q} + \left[I(q) - q_x\frac{\partial}{\partial q_x}I(q)\right]\frac{i\omega}{i\omega + 2\Gamma_q}. \tag{6.2.67}$$

The low-frequency limit gives the thermodynamic response $(\partial I(q)/\partial\rho)_s \rho_1$ in (1.2.67). The linear response theory holds for $|I(q,t) - I(q)| \ll I(q)$ for any q. If $\omega t_\xi \ll 1$ and $\rho = \rho_c$, this criterion becomes

$$|\rho_1|/\rho \ll |T/T_c - 1|^{1-\alpha} \quad \text{or} \quad |p_1|/p_c \ll |T/T_c - 1|, \tag{6.2.68}$$

where $p_1 = c^2\rho_1$. The nonlinear regime $|p_1|/p_c \gtrsim |T/T_c - 1|$ is then of great interest, where we expect the occurrence of periodic spinodal decomposition at low frequencies $\omega t_\xi \ll 1$, to be discussed in Section 8.8. In addition, we note that there has been no measurement of the anisotropic part ($\propto q_x\partial I(q)/\partial q_x \propto q_x^2$) in (6.2.67) in scattering or form birefringence and dichroism [49]. Similar calculations can also be made for binary fluid mixtures, and for ^{4}He near the superfluid transition.

6.3 Piston effect

One-component fluids near gas–liquid criticality are highly compressible and extremely sensitive even to a very small change of the pressure, as well as to that of the temperature.

We will also now show that thermal equilibration processes drastically depend on whether the pressure or the volume of the fluid is fixed [50]. We will show that the thermal diffusion layer near the boundary wall of the fluid container is a crucial entity in the fixed-volume condition, and that the layer acts as a *piston* causing instantaneous adiabatic changes in the interior (bulk) region. This piston is so effectively operative because of the enhanced thermal expansion that it decisively influences thermal relaxations at fixed volume. Some new predictions will be made on the resonant response of the interior temperature against oscillation of the boundary (wall) temperature.

6.3.1 Critical speeding-up at a fixed volume

Let us prepare a single-phase, near-critical fluid in a cell with a fixed volume V made of a metal with high thermal conductivity. We then change the boundary temperature by a small amount T_{1b} at $t = 0$ and keep it constant for $t > 0$. We consider only hydrodynamic variations slowly changing in space ($\gg \xi$) and time ($\gg t_\xi$) neglecting the thermal fluctuations. Pressure variations propagate very rapidly on the scale of L/c, which is the traversal time of a sound over the system length L, and can be regarded as homogeneous on much slower scales.[5] Near the boundary there arises a thermal diffusion layer with thickness,

$$\ell_D(t) = \sqrt{Dt}, \tag{6.3.1}$$

which is larger than ξ for $t > t_\xi$. Hereafter $D = \lambda/C_p$ is the thermal diffusion constant and λ is the thermal conductivity. In the isobaric condition, equilibration is achieved only for $\ell_D(t) \sim L$ or after an exceedingly long relaxation time of order $t_D \equiv L^2/D$.

The entropy variation $s_1(\boldsymbol{r}, t) \equiv \langle \delta s \rangle - (\delta s)_0$ is nonvanishing only within the layer, where $(\delta s)_0$ is the initial, homogeneous entropy deviation from the critical value. It depends on t and the distance from the boundary in an early stage in which $\ell_D(t)$ stays much shorter than the system length L. Its space integral in the cell gives the heat $Q_T(t)$ supplied through the boundary,

$$Q_T(t) = nT \int d\boldsymbol{r} s_1(\boldsymbol{r}, t) = nTV\bar{s}_1(t), \tag{6.3.2}$$

where $\bar{s}_1(t)$ is the space average of $s_1(\boldsymbol{r}, t)$. Simultaneously, a homogeneous pressure variation is produced throughout the cell,

$$p_1(t) = \left(\frac{\partial p}{\partial s}\right)_n \bar{s}_1(t) = \left(\frac{\partial p}{\partial T}\right)_n \bar{T}_1, \tag{6.3.3}$$

where use has been made of the fact that the space integral of the density deviation vanishes in the fixed-volume condition. We write the temperature variation as $T_1(\boldsymbol{r}, t)$ and its space

[5] At the end of this section we will discuss how pressure homogeneity is attained.

average as $\bar{T}_1$. Using (6.3.3) we also have

$$\begin{aligned} T_1(\mathbf{r}, t) &= \left(\frac{\partial T}{\partial s}\right)_p s_1(\mathbf{r}, t) + \left(\frac{\partial T}{\partial p}\right)_s p_1(t) \\ &= \frac{nT}{C_p} s_1(\mathbf{r}, t) + \left(\frac{nT}{C_V} - \frac{nT}{C_p}\right)\bar{s}_1(t), \end{aligned} \tag{6.3.4}$$

where use has been made of the thermodynamic identity $(\partial p/\partial T)_n(\partial T/\partial p)_s = 1 - C_V/C_p$ given in (1.2.54). The first term in (6.3.4) is localized in the thermal diffusion layer, while the second term is homogeneous and is equal to the adiabatic, interior temperature variation $T_{1\text{in}}(t)$ (the temperature deviation outside the thermal diffusion layer). In the isobaric condition, we have the first term only. Interestingly, the space average $\bar{T}_1(t) = V^{-1}\int d\mathbf{r} T_1(x, t)$ is related to $Q_T(t)$ in terms of C_V as

$$\begin{aligned} \bar{T}_1(t) &= (1 - 1/\gamma_s)^{-1} T_{1\text{in}}(t) \\ &= \frac{nT}{C_V}\bar{s}_1(t) = \frac{1}{VC_V} Q_T(t). \end{aligned} \tag{6.3.5}$$

The second line is a natural consequence at fixed volume. The specific-heat ratio,

$$\gamma_s = C_p/C_V \sim (T/T_c - 1)^{-\gamma+\alpha}, \tag{6.3.6}$$

grows strongly near the critical point, so the second homogeneous part in the second line of (6.3.4) is amplified as compared to the first localized part. Because $\bar{s}_1(t) \sim s_{1b}(t)\ell_D(t)/L$ with $s_{1b}(t)$ being the boundary value of the entropy deviation, the ratio of the second term to the first term in (6.3.4) is of order $(\gamma_s - 1)\ell_D(t)/L$ at the boundary. Thus $T_1(x, t)$ will become everywhere close to T_{1b} for $(\gamma_s - 1)\ell_D(t) \gg L$. The time t_1 of this quick temperature equilibration is determined by $(\gamma_s - 1)\ell_D(t_1) = L/2$ and is expressed as

$$t_1 = L^2/[4D(\gamma_s - 1)^2] \propto L^2\xi^{-2.7}. \tag{6.3.7}$$

As will be shown in Appendix 6D, analytic calculations of the temperature and density profiles are straightforward for the 1D geometry ($0 < x < L$). The interior temperature deviation $T_{1\text{in}}(t) = (1 - \gamma_s^{-1})\bar{T}_1(t)$ can be written in the following scaling form for $t \ll t_D = L^2/D$,

$$T_{1\text{in}}(t) = T_{1b}\big[1 - F_a(t/t_1)\big]. \tag{6.3.8}$$

The scaling function $F_a(s)$ is shown in Fig. 6.11 and is defined by

$$F_a(s) = \frac{2}{\pi}\int_0^\infty du \frac{1}{1+u^2}\exp(-su^2) = e^s\big[1 - \text{erf}(\sqrt{s})\big], \tag{6.3.9}$$

where $\text{erf}(x) = 2\pi^{-1/2}\int_0^x dz e^{-z^2}$ is the error function. Therefore, $F_a(s) = 1 - 2(s/\pi)^{1/2} + \cdots$ for $s \ll 1$ and $F_a(s) = (\pi s)^{-1/2} + \cdots$ for $s \gg 1$; hence, the short- and long-time expressions for $T_{1\text{in}}(t)$ are written as

$$\begin{aligned} T_{1\text{in}}(t)/T_{1b} &\cong 2(t/\pi t_1)^{1/2} \qquad (t \ll t_1) \\ &\cong 1 - (t_1/\pi t)^{1/2} \qquad (t_1 \ll t \ll t_D). \end{aligned} \tag{6.3.10}$$

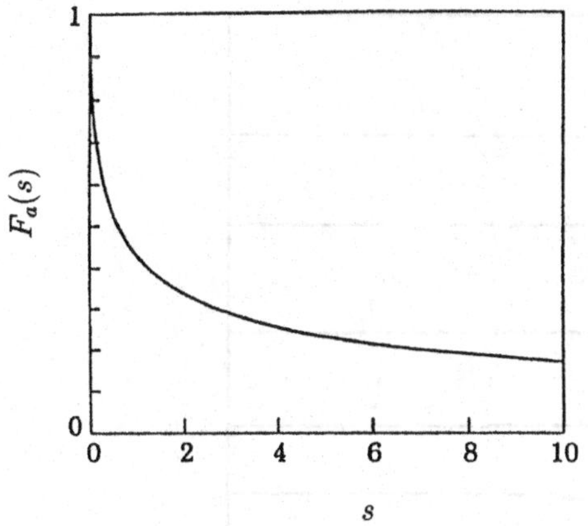

Fig. 6.11. The scaling function $F_a(s)$ defined by (6.3.9).

The pressure deviation is written as $p_1(t) = (\partial p/\partial T)_s T_{1\text{in}}(t)$. The temperature profile may be expressed in terms of a normalized temperature variation defined by

$$G(x,t) = \big[T_{1\text{b}} - T_1(x,t)\big] \big/ \big[T_{1\text{b}} - T_{1\text{in}}(t)\big], \tag{6.3.11}$$

which is zero at $x = 0$ and tends to 1 in the interior region. Some further calculations [50] yield

$$\begin{aligned} G(x,t) &\cong \text{erf}(\hat{x}) && (t \ll t_1) \\ &\cong 1 - e^{-\hat{x}^2} + \hat{x} e^{-\hat{x}^2} (t_1/t)^{1/2} && (t_1 \ll t \ll t_D), \end{aligned} \tag{6.3.12}$$

where $\hat{x} \equiv x/\sqrt{4Dt}$. Figure 6.12 displays the profile $1 - T_1(x,t)/T_{1\text{b}}$ [51]. In the final stage $t \gtrsim t_D$, a temperature variation of order $\gamma_s^{-1} T_{1\text{b}}$ relaxes exponentially. Its profile is written as

$$1 - T_1(x,t)/T_{1\text{b}} \cong \frac{2}{\gamma_s} \big[1 - \cos(2\pi x/L)\big] \exp(-4\pi^2 Dt/L^2). \tag{6.3.13}$$

It is worth noting that, if $\gamma_s \cong 1$ or with no adiabatic effect, we have the exponential relaxation $1 - T_1(x,t)/T_{1\text{b}} \sim \sin(\pi x/L) \exp(-\pi^2 Dt/L^2)$ with a relaxation time four times longer than the final relaxation time in (6.3.13).

The density variation $\rho_1(x,t)$ can also be calculated in the linear regime. In terms of $G(x,t)$ in (6.3.12) the density profile can be written as

$$\rho_1(x,t) - \rho_{1\text{in}}(t) = \left(\frac{\partial \rho}{\partial T}\right)_p \big[T_1 - T_{1\text{in}}(t)\big]\big[1 - G(x,t)\big], \tag{6.3.14}$$

where $\rho_{1\text{in}}(t) = (\partial \rho/\partial T)_s T_{1\text{in}}(t)$ is the interior density deviation. The boundary density deviation is induced at nearly constant pressure as $(\partial \rho/\partial T)_p T_{1\text{b}}$ for $t \lesssim t_1$ and slowly relaxes as $(\partial \rho/\partial T)_p T_{1\text{b}} (t_1/t)^{1/2}$ for $t_1 \lesssim t \ll t_D$. The density in the thermal diffusion layer can thus be strongly disturbed for a long time interval.

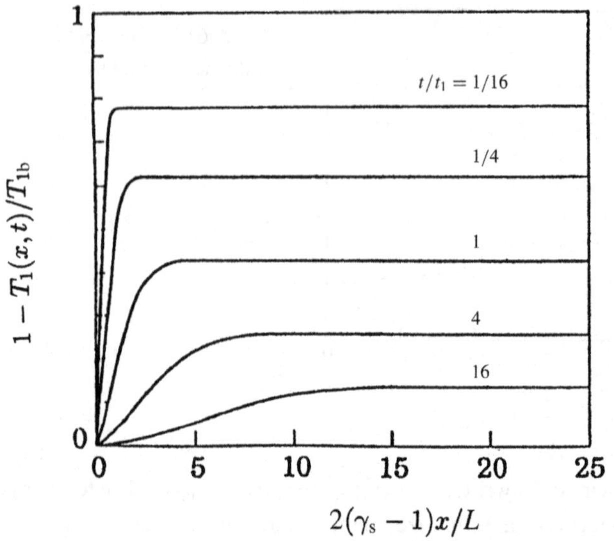

Fig. 6.12. Curves of $1 - T_1(x, t)/T_{1b}$ near the boundary. The curves are for $t/t_1 = 1/16, 1/4, 1, 4,$ and 16. The distance x from the boundary is measured in units of $L/(\gamma_s - 1)$. Note that the thickness of the thermal diffusion layer is $L/(\gamma_s - 1)$ at $t = t_1$ [51].

We recognize that the above heat transport mechanism is generally present in any compressible system. It is noticeable in gaseous systems and is exaggerated near the gas–liquid critical point. Notice that the heat transport equation $nT\partial s_1/\partial t = \lambda\nabla^2 T_1$ in one-component fluids may be rewritten as

$$\begin{aligned}\frac{\partial}{\partial t}T_1 &= \left(\frac{\partial T}{\partial p}\right)_s \frac{d}{dt}p_1 + D\nabla^2 T_1 \\ &= \left(1 - \frac{1}{\gamma_s}\right)\frac{d}{dt}\bar{T}_1 + D\nabla^2 T_1. \end{aligned} \tag{6.3.15}$$

In the first line p_1 is assumed to be homogeneous. The second line follows under the fixed-volume condition and constitutes a simple modified diffusion equation, which takes into account the adiabatic, homogeneous temperature change due to the global constraint of fixed volume. As ought to be the case, the integration of (6.3.15) over the cell gives the equation for the average temperature deviation,

$$VC_V\frac{d}{dt}\bar{T}_1 = \lambda \int da\boldsymbol{n}\cdot\nabla T_1 \tag{6.3.16}$$

The right-hand side represents the heat supply from the boundary surface, da being the surface element and $\boldsymbol{n}$ being the outward surface normal. The time integration of (6.3.16) leads to the second line of (6.3.5).

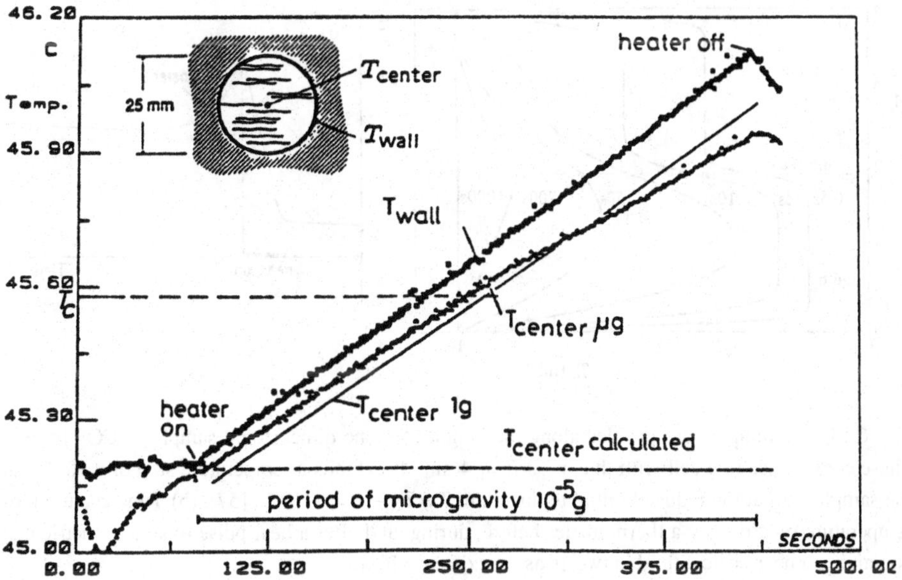

Fig. 6.13. Heating up a cell containing SF_6 at the critical density in a ballistic rocket flight [52]. The wall temperature T_{wall} was heated at a constant ramp from $T - T_c = -0.4$ K to 0.4 K within 6 min of microgravity. The temperature at the center T_{center} quickly followed T_{wall} due to the piston effect. There should have been no change of T_{center} based on the thermal diffusion mechanism only.

Experiments

Because t_1 becomes increasingly shorter as $T \to T_c$, the adiabatic heating leads to *critical speeding-up*, whereas the isobaric equilibration time $t_D \equiv L^2/D$ is usually extremely long, leading to *critical slowing-down*. For example, in CO_2 with $\Delta T = T - T_c > 0$ on the critical isochore,[6] we have $D \cong 10^{-5}(\Delta T)^{0.625}$ cm^2/s, $t_\xi \cong 2.6 \times 10^{-8}(\Delta T)^{-1.9}$ s with ΔT in units of K, so that the two equilibration times are expressed as $t_1 \sim 0.2(\Delta T)^{1.67}$ s and $t_D \sim 10^5(\Delta T)^{-0.625}$ s for $L = 1$ cm. As shown in Fig. 6.13, rapid thermal equilibration, which can now be ascribed to the piston effect, was first observed by Nitsche and Straub in their C_V measurement in a microgravity condition free from gravity-induced convection [52]. A number of experiments have subsequently followed [53]–[58]. In particular, we mention that the piston effect has also been used to induce phase separation in one-component fluids [55], as will be discussed in Section 8.6.

Piston effect in two-phase coexistence

In contrast to the critical speeding-up in one-phase states, long-duration thermal relaxations were reported even at fixed volume in the presence of an interface separating gas and liquid regions [59, 60]. This is caused by slow heat and mass transport through the interface [50].

[6] For CO_2, we use $\xi_{+0} = 1.5$ Å, $T_c = 304$ K and $\eta = 3.8 \times 10^{-4}$ poise [15].

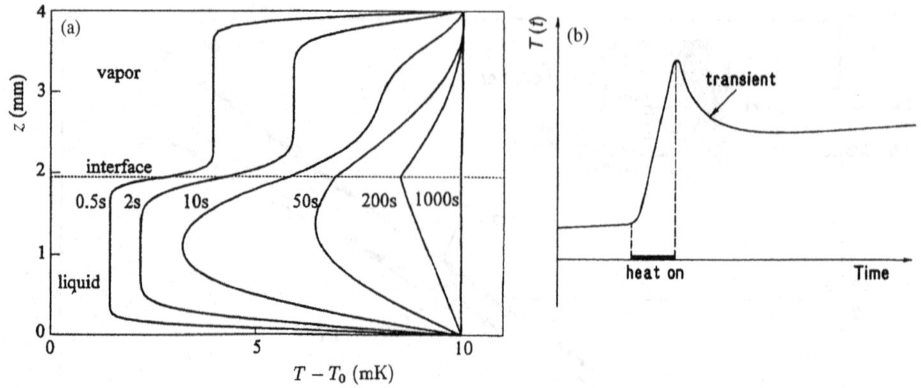

Fig. 6.14. (a) Temperature profiles along the height of a one-dimensional sample of CO_2 in two-phase coexistence in a cell with thickness $L = 4$ mm. It exhibits a temperature step of +10 mK at the sample wall at the reduced initial temperature $T_0/T_c - 1 = -10^{-2}$ [57]. (b) Time evolution of temperature recorded by a thermometer before, during, and after a heat pulse to a calorimetric cell containing a near-critical fluid in two-phase coexistence [61].

Recall that the adiabatic coefficient $(\partial T/\partial p)_s$ was calculated as (2.2.36) on the coexistence curve, which indicates that a homogeneous pressure change p_1 produces a temperature difference across the interface given by

$$(\Delta T)_{gl} = \frac{2a_c}{\sqrt{\gamma_s}} \left(\frac{\partial T}{\partial p} \right)_{cx} p_1, \tag{6.3.17}$$

where $a_c(\cong 1)$ is a universal number defined in (2.2.37). As a result, there appears a diffusive heat flux and a temperature inhomogeneity extending over $\ell_D(t)$ near the interface. For $t \ll t_D$ the interior temperature deviations far from the boundary wall and the interface are given by

$$T_{1in} \cong T_{1b}\big[1 - F_a(t/t_1)\big]\left[1 \pm \frac{a_c}{\sqrt{\gamma_s}}\right], \tag{6.3.18}$$

where the plus sign is for the gas phase and the minus for the liquid phase. Because $T_{1in}/T_{1b} - 1 \cong -(t_1/\pi t)^{1/2} \pm a_c\gamma_s^{-1/2}$ for $t_1 \ll t \ll t_D$, the main temperature inhomogeneity exists near the interface for $t \gg \gamma_s t_1 \sim t_D/\gamma_s$. Recent experiments have also confirmed slow relaxations in two-phase states [56]–[58]. Figure 6.14(a) illustrates calculated temperature profiles in two-phase coexistence [57], where the distance from the critical point is not very small and the temperature inhomogeneity is apparent. As shown schematically in Fig. 6.14(b), even if a fluid is very close to the critical point, there appears significant slow thermal relaxation on a timescale of t_D in two-phase coexistence [61]. Some discussions on this effect have already been presented in Appendix 4F.

Comments

(i) In the above example, the energy transport from the boundary to the interior takes place in the form of sound. Immediately after heating of the boundary, several traversals of sound are sufficient to heat up the interior [62]. This is analogous to the heat transport mechanism in the form of second sound in superfluid ^{4}He. Here we are assuming that the acoustic time L/c (typically 10^{-4} s for $L = 1$ cm) is much shorter than t_1. That is,

$$L/ct_1 = (\gamma_s - 1)^2 D/cL < 1. \tag{6.3.19}$$

(ii) It is worth noting that the above linear response arguments are valid only for $t_1 \gg t_\xi$ or $L \gg \xi\gamma_s$. The reverse case $L \lesssim \xi\gamma_s$ can well be realized, although the physics remains unclear. We also note that the thermal diffusion layer can easily be driven away from the linear response regime, which is suggested by the large density perturbation (6.3.14). The effect of the bulk viscosity is also neglected.

6.3.2 Relaxation after a volume change

Another impressive example of a thermal piston effect would be a simple experiment in which the volume of the cell is changed from V to $V + \delta V$ at $t = 0$ by a mechanical piston with the boundary temperature unchanged. For $t \sim L/c$ the interior temperature is adiabatically changed by $(\partial T/\partial\rho)_s(\delta\rho)_0$ with $(\delta\rho)_0 = -\rho\delta V/V$. Then there should appear a thermal diffusion layer acting as a thermal piston. The homogeneous pressure variation in this case is given by $p_1(t) = c^2(\delta\rho)_0 + (\partial p/\partial s)_n\bar{s}_1(t)$, instead of (6.3.3). The temperature deviation vanishes at the boundary and the entropy deviation is localized near the boundary. For $t \gg L/c$ the problem becomes essentially the same as the previous one by the replacement $T_1 \to T_1 + T_{1\text{b}}$ with $T_{1\text{b}} = -(\partial T/\partial\rho)_s(\delta\rho)_0$. The interior temperature variation relaxes to zero on the timescale of t_1 as

$$T_{1\text{in}}(t) = \left(\frac{\partial T}{\partial\rho}\right)_s (\delta\rho)_0 F_a(t/t_1), \tag{6.3.20}$$

where $F_a(s)$ is defined by (6.3.9). The pressure deviation $p_1(t)$ decays as $(\partial p/\partial T)_s T_{1\text{in}}(t)$ for $t \ll t_D$ and tends to the final value $(\partial p/\partial\rho)_T(\delta\rho)_0$ for $t > t_D$. The process is adiabatic for $L/c \lesssim t \ll t_1$, but becomes nearly *isothermal* for $t \gg t_1$ by the counterbalance of the effects of the mechanical and thermal pistons.

6.3.3 Rapid heat transport

We may also examine situations in which the top and bottom walls have different temperatures in the fixed-volume condition. In such cases, the heat fluxes at the two boundaries quickly become close on the timescale of t_1, whereas a relaxation time of order t_D is needed in the isobaric condition. The effective thermal conductivity in such transient states can be of order $(\gamma_s - 1)\lambda \sim (T - T_c)^{-1.77}$.

(i) For example, we switch on a heater attached to the top boundary at $x = L$ at $t = 0$ and apply a small constant heat flux Q into a fluid in a one-phase state for $t > 0$, while we keep the bottom ($x = 0$) temperature unchanged [58, 61, 63, 64]. In the time region $t \ll t_D$, we have thermal diffusion layers at the two boundaries, one being expanded and the other being contracted. After some calculations in Appendix 6D we find that the temperature deviation $T_{1\text{in}}(t)$ in the interior and that $T_{1\text{top}}(t)$ at the top are obtained as

$$T_{1\text{in}}(t) = \frac{2Q}{\sqrt{\pi}\lambda}\sqrt{Dt}\,F_b(t/4t_1), \quad T_{1\text{top}}(t) = \frac{2Q}{\sqrt{\pi}\lambda}\sqrt{Dt}\big[1 + F_b(t/4t_1)\big]. \tag{6.3.21}$$

The scaling function $F_b(s)$ is defined by

$$F_b(s) = 1 - (\pi/4s)^{1/2}[1 - F_a(s)], \tag{6.3.22}$$

which behaves as $(\pi s)^{1/2} + \cdots$ for $s \ll 1$ and as $1 - (\pi/4s)^{1/2}$ for $s \gg 1$. Note that the results in the isobaric condition are obtained if F_b in (6.3.21) is replaced by 0. In the present fixed-volume condition, $T_{1\text{in}}(t)$ tends to a half of $T_{1\text{top}}(t)$ for $t_1 \ll t \lesssim t_D$. The heat flux at the bottom is written as

$$Q_{\text{bot}}(t) = Q\big[1 - F_a(t/4t_1)\big]. \tag{6.3.23}$$

For $t_1 \ll t \ll t_D$ we have $Q_{\text{bot}}(t) \cong Q$. In this transient time region we may define the effective thermal conductivity by

$$\lambda_{\text{eff}}(t) = QL/T_{1\text{top}}(t) \cong \frac{\sqrt{\pi}}{4}(\gamma_s - 1)\lambda(t/t_1)^{-1/2}, \tag{6.3.24}$$

which changes from a value of order $(\gamma_s - 1)\lambda$ at $t \sim t_1$ to λ at $t \sim t_D$. This high rate of heat conduction realized for $\gamma_s \gg 1$ is carried by sound waves propagating through the interior region.

(ii) We may also change the top temperature by $T_{1\text{b}}$ at $t = 0$ with the bottom temperature unchanged. As will be shown in Appendix 6D, the interior temperature deviation for $t \ll t_D$ is written as

$$T_{1\text{in}}(t) = \frac{1}{2}T_{1\text{b}}\big[1 - F_a(t/t_1)\big]. \tag{6.3.25}$$

Thus $T_{1\text{in}}(t) \to T_1/2$ for $t_1 \ll t \ll t_D$. The bottom and top heat fluxes are calculated as

$$Q_{\text{bot}}(t) = \frac{\lambda T_{1\text{b}}}{L}(\gamma_s - 1)F_a(t/t_1), \quad Q_{\text{top}}(t) = \frac{\lambda T_{1\text{b}}}{\sqrt{\pi Dt}} - Q_{\text{bot}}(t). \tag{6.3.26}$$

For $t_1 \ll t \ll t_D$, both $Q_{\text{bot}}(t)$ and $Q_{\text{top}}(t)$ approach $\lambda T_{1\text{b}}/2\sqrt{\pi Dt}$. The effective thermal conductivity $\lambda_{\text{eff}}(t)$ is twice that in (6.3.24). Interestingly, we can see that $Q_{\text{bot}}(t) \to (\gamma_s - 1)\lambda T_{1\text{b}}/L$ for $t \ll t_1$ from (6.3.26), but this is valid only for $t \gtrsim L/c$ because homogeneity of the pressure has been assumed. If the fast acoustic process is accounted for, $Q_{\text{bot}}(t)$ should grow from 0 to this high value with a few traversals of sound. Finally we consider the energy $\Delta E(t) \equiv S\int_0^t dt' Q_{\text{bot}}(t')$, where S is the surface area of the parallel plates. It is the energy transferred from the top to the bottom in the time interval $[0, t]$ and is of order $\Delta E(t) \sim VC_V T_{1\mathrm{b}}(t/t_1)^{1/2}$ for $t \gg t_1$. Thus an energy of order $VC_V T_{1\mathrm{b}}$ can be transported through a macroscopic distance on the timescale of t_1.

6.3.4 Resonance induced by boundary temperature oscillation

So far we have been interested in slow motions occurring on timescales much longer than the acoustic time L/c. Here we examine sound modes in a one-dimensional geometry ($0 < x < L$) with frequency ω in the intermediate range,

$$c/L \lesssim \omega \ll t_\xi^{-1}. \tag{6.3.27}$$

The wavelength $2\pi c/\omega$ of the sound is much longer than the thickness of the thermal diffusion layer. In this case, dissipation in the thermal diffusion layer (due to the thermal conductivity) dominates over that in the interior region (due to the bulk viscosity).

We assume that all the deviations depend on time as $e^{i\omega t}$. If the bulk viscosity is neglected, the deviations ρ_1, p_1, and s_1 of the density, pressure, and entropy, respectively, satisfy

$$\omega^2 \rho_1 = -\nabla^2 p_1 = -c^2\nabla^2\left[\rho_1 - \left(\frac{\partial \rho}{\partial s}\right)_p s_1\right]. \tag{6.3.28}$$

Close to the bottom $x = 0$, we set

$$s_1 = A_0 e^{-\kappa_D x}. \tag{6.3.29}$$

Then (6.3.28) is integrated as

$$c^2\rho_1 = A\cos(kx) + B\sin(kx) - \frac{\kappa_D^2}{k^2+\kappa_D^2}\left(\frac{\partial p}{\partial s}\right)_\rho s_1, \tag{6.3.30}$$

where $k = \omega/c$ and $\kappa_D = (i\omega/D)^{1/2}$ with $\mathrm{Re}\,\kappa_D > 0$ and $k \ll |\kappa_D|$. The coefficients, A_0, A, and B, are proportional to $e^{i\omega t}$. If the boundary wall at $x = 0$ does not move, $i\omega\rho v = -\partial p_1/\partial x$ should vanish as $x \to 0$. Then B is determined as

$$B = -\frac{k}{\kappa_D}\left(\frac{\partial \rho}{\partial s}\right)_p c^2 A_0. \tag{6.3.31}$$

The pressure deviation thus becomes

$$p_1 = A\cos(kx) + \left(\frac{\partial p}{\partial s}\right)_\rho \frac{k^2}{\kappa_D^2} A_0\left[\frac{\kappa_D}{k}\sin(kx) + e^{-\kappa_D x}\right]. \tag{6.3.32}$$

If $(\gamma_s - 1)k^2 \ll |\kappa_D|^2$, the temperature deviation at $x = 0$ is written as

$$T_{1\mathrm{b}} = \left(\frac{\partial T}{\partial p}\right)_s\left[A - \left(\frac{\partial p}{\partial s}\right)_T A_0\right]. \tag{6.3.33}$$

If the boundary temperature is constant or $T_{1\text{b}} = 0$, the pressure and temperature variations in the interior region ($x \gg |\kappa_D|^{-1}$) become

$$p_1 = \left(\frac{\partial p}{\partial T}\right)_s T_{1\text{in}} = A\big[\cos(kx) - a_\text{s}\sin(kx)\big], \tag{6.3.34}$$

where

$$a_\text{s} = (\gamma_s - 1)k/\kappa_D = e^{-i\pi/4} b_\text{s}\sqrt{kL}. \tag{6.3.35}$$

The coefficient b_s sensitively depends on $T - T_\text{c}$ as

$$b_\text{s} = (\gamma_s - 1)\sqrt{D/cL} = (L/4ct_1)^{1/2}. \tag{6.3.36}$$

Typically, $b_\text{s} \sim 10^{-4}(T/T_\text{c} - 1)^{-0.87}$ for $L = 1$ cm on the critical isochore. The term proportional to $\sin(kx)$ in p_1 grows on approaching the critical point. Note that the condition $b_\text{s} < 1$ is equivalent to (6.3.19).

Eigenmodes

Now we can seek the eigenmodes of sound in a one-dimensional cell at a fixed boundary temperature. In this case ω is a complex number with $\text{Im}\,\omega > 0$. Even modes are expressed in the interior region as

$$p_1 = \left(\frac{\partial p}{\partial T}\right)_s T_{1\text{in}} \propto e^{i\omega t}\cos[k(x - L/2)]. \tag{6.3.37}$$

From (6.3.34) the dispersion relation is determined by

$$\tan(kL/2) = -a_\text{s}. \tag{6.3.38}$$

This equation is solved as $\omega = \omega_n^\text{e}$ ($n = 1, 2, \ldots$) with $\text{Im}\,\omega_n^\text{e} > 0$, where

$$\begin{aligned} \omega_n^\text{e} L/c &= 2n\pi - 2(1-i)\sqrt{n\pi}\, b_\text{s} + \cdots && (b_\text{s} \ll 1), \\ &= (2n-1)\pi + \frac{\sqrt{2}(1+i)}{\sqrt{(2n-1)\pi}} b_\text{s}^{-1} + \cdots && (b_\text{s} \gg 1). \end{aligned} \tag{6.3.39}$$

The first line holds not very close to the critical point, while the second line holds very close to the critical point. Odd modes are expressed as

$$p_1 = \left(\frac{\partial p}{\partial T}\right)_s T_{1\text{in}} \propto e^{i\omega t}\sin[k(x - L/2)], \tag{6.3.40}$$

with

$$\tan(kL/2) = a_\text{s}^{-1}. \tag{6.3.41}$$

This equation is solved as $\omega = \omega_n^\text{o}$ ($n = 1, 2, \ldots$) with

$$\begin{aligned} \omega_n^\text{o} L/c &= (2n-1)\pi - (1-i)\sqrt{2(2n-1)\pi}\, b_\text{s} + \cdots && (b_\text{s} \ll 1), \\ &= 2n\pi + \frac{1+i}{\sqrt{n\pi}} b_\text{s}^{-1} + \cdots && (b_\text{s} \gg 1). \end{aligned} \tag{6.3.42}$$

The imaginary part, $\mathrm{Im}\,\omega_n^e$ or $\mathrm{Im}\,\omega_n^o$, represents the damping rate and is much smaller than the real part for $b_s \ll 1$ and $b_s \gg 1$, but they are of the same order for $b_s \sim 1$. This damping arises from heat conduction in the thermal diffusion layer and has been assumed to be much larger than that due to the bulk viscosity ($\sim R_B\omega^2 t_\xi$ from (6.2.25) and (6.2.45)).

Resonance

It is well known that a system which supports (first) sound resonates to mechanical vibration of the boundary wall when the applied frequency matches one of the eigenfrequencies of the system. We predict similar resonance when the boundary temperature is oscillated at such high frequencies ($\sim c/L$). It is easy to expect that this effect becomes enhanced near the gas–liquid critical point with increasing $\gamma_s - 1$, because the thermal diffusion layer can effectively transform temperature variations at the boundary wall into sound in the interior region. Analogously, Peshkov realized standing second sound in superfluid ^{4}He resonantly induced by periodic temperature perturbations at a boundary plate [65].

(i) If the top and bottom temperatures are equal and depend on time as $T_0 + T_{1b}\cos(\omega t)$, the temperature in the interior region is expressed as

$$T_{1\mathrm{in}}(x,t)/T_{1b} = \mathrm{Re}\big[Z_e(\omega)e^{i\omega t}\big]\cos[k(x - L/2)], \tag{6.3.43}$$

with

$$Z_e(\omega) = 1/[\cos(kL/2) + a_s^{-1}\sin(kL/2)], \tag{6.3.44}$$

where $k = \omega/c$ and $a_s(\propto e^{-i\pi/4}\omega^{1/2})$ is defined by (6.3.35). In Fig. 6.15(a) we plot the absolute value $|Z_e(\omega)|$ as a function of $\omega L/\pi c$ for various b_s. The peaks arise from resonance with the eigenmodes given by (6.3.39). Near the peak $\omega \cong \mathrm{Re}\,\omega_n^e$, we obtain $Z_e \cong 2(-1)^n cR_n/L(\omega - \omega_e^n)$, where $R_n \cong a_s$ for $b_s \ll 1$ and $R_n \cong 1$ for $b_s \gg 1$. Thus the peak heights are much enhanced as the critical point is approached.

(ii) If the bottom temperature at $x = 0$ is kept at a constant T_0 and the top temperature at $x = L$ is oscillated as $T_0 + T_{1b}\cos(\omega t)$, we obtain the interior temperature variation in the form

$$T_{1\mathrm{in}}(x,t)/T_{1b} = \frac{1}{2}\mathrm{Re}\big[Z_e(\omega)Z_o(\omega)e^{i\omega t}[\cos(kx) - a_s\sin(kx)]\big], \tag{6.3.45}$$

where

$$Z_o(\omega) = 1/[\cos(kL/2) - a_s\sin(kL/2)]. \tag{6.3.46}$$

At the middle point $x = L/2$, only the even modes are involved in the form

$$T_{1\mathrm{in}}(L/2,t)/T_{1b} = \frac{1}{2}\mathrm{Re}\big[Z_e(\omega)e^{i\omega t}\big], \tag{6.3.47}$$

as in the previous symmetric case. At other points, we can observe resonance also with the odd modes. For example, close to the bottom, where $\sqrt{D/\omega} \ll x \ll L$, we obtain

$$\lim_{x\to 0} T_{1\mathrm{in}}(x,t)/T_{1b} = \frac{1}{2}\mathrm{Re}\big[Z_e(\omega)Z_o(\omega)e^{i\omega t}\big]. \tag{6.3.48}$$

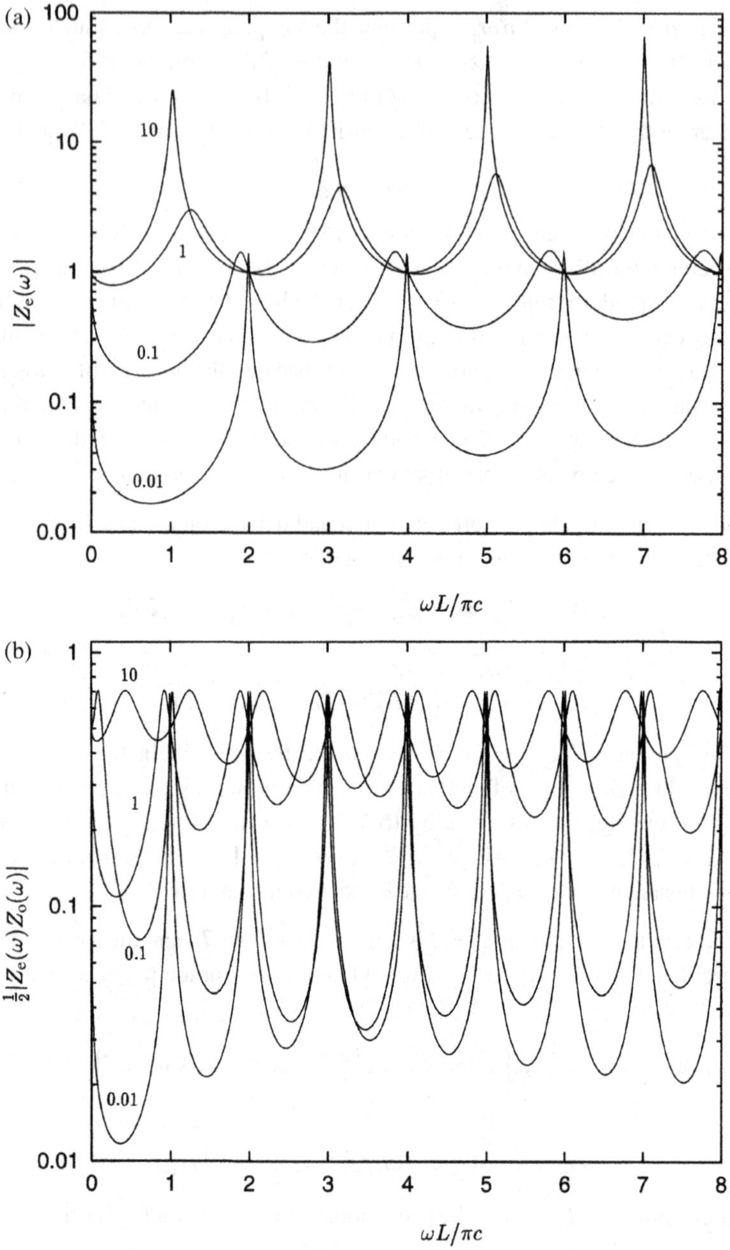

Fig. 6.15. (a) $|Z_e(\omega)|$ vs $\omega L/c$ for $b_s = 0.01, 0.1, 1, 10$ from below. This represents the maximum of $T_{1in}(L/2, t)/T_{1b}$ when the top and bottom temperatures oscillate as $T_0 + T_{1b}\cos(\omega t)$. Here b_s is defined by (6.3.36) and grows on approaching the critical point. (b) $\frac{1}{2}|Z_e(\omega)Z_o(\omega)|$ vs $\omega L/c$ for $b_s = 0.01, 0.1, 1, 10$ from below. This represents the maximum of $T_{1in}(x,t)/T_{1b}$ near the bottom, $(D/\omega)^{1/2} \ll x \ll L$, when the top temperature is oscillated with amplitude T_{1b} and the bottom temperature is fixed.

In Fig. 6.15(b) we plot the absolute value $\frac{1}{2}|Z_e(\omega)Z_o(\omega)|$, which is the maximum of T_{1in}/T_{1b} near the bottom, as a function of $\omega L/\pi c$ for various b_s. The complex response functions $Z_e(\omega)$ and $Z_o(\omega)$ have poles at $\omega = \omega_n^e$ and ω_n^o, respectively, in the upper complex ω plane from (6.6.39) and (6.6.42).

Linear response and pressure homogenization

We may now calculate the linear response to general, small time-dependent variations of the boundary temperatures using the above results. In the symmetric case, in which the top and bottom temperatures are equal and depend on time as $T_0 + T_{1b}(t)$, the temperature variation in the interior region is expressed as $T_{1in}(x,t) = \int_{-\infty}^{t} dt' \varphi_e(x, t-t')T_{1b}(t')$, where

$$\int_0^\infty dt e^{-i\omega t}\varphi_e(x,t) = Z_e(\omega)\cos[(x - L/2)\omega/c]. \tag{6.3.49}$$

For a step-wise variation, in which $T_{1b}(t)$ is equal to 0 for $t < 0$ and to a constant T_{1b} for $t > 0$, we obtain $T_{1in}(x,t)/T_{1b} = \int_0^t dt' \varphi_e(x, t')$ for $t > 0$. In this case, after a relaxation time t_{homo}, the sound-wave oscillation decays to zero, resulting in a homogeneous pressure deviation $p_1(t)$. From (6.3.39) we find

$$t_{homo} \sim L/cb_s = b_s t_1 \quad (b_s < 1), \tag{6.3.50}$$

$$\sim Lb_s/c = b_s^3 t_1 \quad (b_s > 1). \tag{6.3.51}$$

Thus $t_{homo} < t_1$ for $b_s < 1$ or under (6.3.19), but $t_{homo} > t_1$ for $b_s > 1$. For $b_s \gg 1$, however, the effect of the bulk viscosity will become important.

6.4 Supercritical fluid hydrodynamics

As the critical point is approached in supercritical fluids, the compressibility and thermal expansion grow, and hence thermal and mechanical disturbances are inseparably coupled. In such fluids the thermal diffusion constant D is small (typically less than 10^{-5} cm^2/s), while the pressure propagation is rapid. As a result, adiabatic processes are of great importance. A characteristic feature not expected in incompressible fluids is that the density heterogeneity is much more exaggerated than that of the temperature due to strong enhancement of the isobaric thermal expansion coefficient $\alpha_p = -(\partial n/\partial T)_p/n(\sim C_p/nT)$. Together with the experiments cited so far, these new features have also been revealed by simulations [62].

In the following, we will assume that a fluid is sufficiently above the critical point such that phase separation does not occur. In Sections 8.6 and 9.4, however, we will show that phase separation can easily be triggered in a thermal diffusion layer when a fluid close to the coexistence curve is heated or cooled through a boundary.

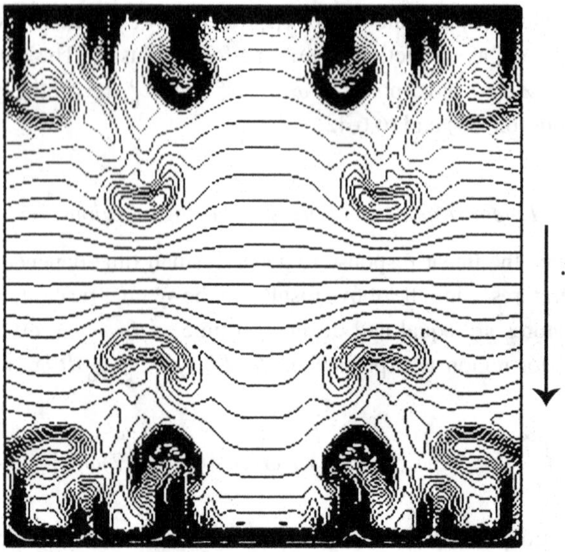

Fig. 6.16. Thermal plumes in CO_2 in a cell with 1 cm thickness in an initial stage of the Schwarzschild instability. The darkness in the figure may be interpreted to represent $|T(\mathbf{r}, t) - T_{\rm center}(t)|$ where $T_{\rm center}(t)$ is the temperature at the center. (The original figure in [67] is in color and represents $T(\mathbf{r}, t) - T_0$.) The temperature inhomogeneity is of order 0.5 mK and is much smaller than the average deviation $T - T_c = 1$ K.

6.4.1 Thermal plumes

Let us consider an expanded region with excess entropy created around a heater placed within a fluid. It will eventually rise due to gravity as a thermal *plume* [66]. If its linear dimension R is sufficiently large, such a plume has a long lifetime of order R^2/D if not deformed by the velocity field. If a plume is warmer than the ambient fluid by T_1, it has a density lower than ambient by $\rho\alpha_p T_1$. As a result, the upward velocity due to buoyancy is estimated as

$$v_{\rm plume} \sim (g\rho\alpha_p/\eta_0)T_1 R^2, \tag{6.4.1}$$

where the transverse velocity field is responsible for the drag force. The plume moves upward appreciably in the adiabatic condition if $R/v_{\rm plume} < R^2/D$ or[7]

$$T_1 R^3 > D\eta_0/(g\rho\alpha_p). \tag{6.4.2}$$

The right-hand side behaves as $(T/T_c - 1)^{\nu+\gamma}$ on the critical isochore due to the singularity of D/α_p.

As an illustration, we show a numerical simulation in 2D in Fig. 6.16 [67]. It demonstrates that thermal plumes can appear even when a flat bottom boundary is heated ho-

[7] The Reynolds number of the flow around the plume is given by $Re = \rho v_{\rm plume} R/\eta_0$. Since $Pr \gg 1$ in near-critical fluids, the condition $Re < 1$ holds near the convection onset, as will be shown in (6.4.11).

mogeneously. The fluid was initially in equilibrium at $T = T_0$ with $T_0 - T_c = 1$ K and $\rho = \rho_c$ in a cell with thickness 1 cm. At $t = 0$ the bottom boundary temperature is raised by 0.5 mK with the upper boundary temperature kept fixed. Then the thermal diffusion layer at the bottom is expanded with thickness $\sqrt{Dt} = 3 \times 10^{-3} t^{1/2}$ cm (t in s), while that at the top is contracted by the same thickness. The figure illustrates plumes at $t = 36.33$ s, where the expanded warmer plumes rise from the bottom and the contracted cooler plumes sink from the top. We can also see that the temperature differences between the plumes and the surrounding fluid become smaller far from the boundaries. In this simulation this is because a plume is adiabatically cooled (warmed) by $(\partial T/\partial p)_s \rho g (\sim 0.3$ mK/cm for $CO_2)$ per unit length as it goes upward (downward). Note that nearly the same phenomenon can be expected when the top boundary is cooled with the bottom temperature fixed.

6.4.2 Convection in supercritical fluids

Rayleigh and Schwarzschild criteria

Let a supercritical fluid column be under a temperature gradient in the downward direction or heated from below. Note that the condition of convection onset for incompressible fluids is given by $Ra > Ra_c$ (the Rayleigh criterion), where $Ra \equiv (\alpha_p \rho_c g L^3)\Delta T/\eta_0 D$ is the Rayleigh number and $Ra_c \cong 1708$ is the critical value. However, in compressible fluids another instability is well known (the Schwarzschild criterion) [68]–[70]. That is, if the cell thickness L is sufficiently large or the fluid is close to the critical point, convection sets in when the temperature gradient $|dT/dz|$ is larger than the adiabatic temperature gradient

$$a_g \equiv (\partial T/\partial p)_s \rho g. \tag{6.4.3}$$

This is the condition that the entropy per particle decreases with height as $ds/dz = (C_p/nT)[dT/dz + a_g] < 0$, under which fluid elements adiabatically convected upward are less dense than the surrounding fluid. A sufficiently large plume generated at $z = 0$ and moving upward will have a density lower than that of the surrounding fluid by

$$(\Delta\rho)_{\text{plume}} = \rho(s(0), p) - \rho(s(z), p) = -\rho\alpha_p[|dT/dz| - a_g]z. \tag{6.4.4}$$

The temperature is higher inside the plume than in the surrounding fluid by $T_1 = [|dT/dz| - a_g]z$, while there is no pressure difference. This instability is well known for large compressible fluid columns such as those in the atmosphere. Gitterman and Steinberg [69] found that the convection onset for compressible fluids is given by $Ra^{\text{corr}} > Ra_c$, where Ra^{corr} is a corrected Rayleigh number defined by

$$Ra^{\text{corr}} = Ra(1 - a_g L/\Delta T). \tag{6.4.5}$$

At the convection onset we thus have

$$(\Delta T)_{\rm onset} = a_g L + Ra_{\rm c} D\eta_0/(g\rho\alpha_p L^3). \tag{6.4.6}$$

The crossover between the two criteria is observable in near-critical fluids due to enhanced thermal expansion.[8] After its first observation in SF_6 from velocity measurements [71], it has recently been investigated with precision in ^{3}He as shown in Fig. 6.17(a) [73]. Moreover, unique transient behavior has also been observed in ^{3}He [73], which will discuss below.

Hydrodynamic equations for slow motions

In a supercritical fluid at a fixed volume in the Rayleigh–Bénard geometry, we assume that the temperature disturbance $T_1(\boldsymbol{r}, t) = T(\boldsymbol{r}, t) - T_{\rm top}$ measured from the temperature $T_{\rm top}$ at the top boundary is much smaller in magnitude than the *distance* from the critical point $T_{\rm top} - T_{\rm c}$ (written as $T - T_{\rm c}$ hereafter) and that the gravity-induced density stratification is not too severe such that the thermodynamic quantities are nearly homogeneous in the cell. To assure the latter condition we assume the temperature range (2.2.43). The bottom and top plates are made of metals with high thermal conductivity and the boundary temperature deviations are independent of the lateral coordinates. We also note that the gravity induces a pressure gradient given by $dp_{\rm eq}/dz = -\rho g \cong -\rho_{\rm c} g$ along the z axis in equilibrium. Even in nonequilibrium we may assume homogeneity of the combination $p_1(t) \equiv p(\boldsymbol{r}, t) - p_{\rm eq}(0) + \rho_{\rm c} g z$ as in (6.2.11) [50], where $p_{\rm eq}(0)$ is the pressure at $z = 0$ in equilibrium. Using the condition that the space average of the density deviation vanishes, we have $p_1(t) = (\partial p/\partial T)_n \bar{T}_1$ as in (6.3.3). The entropy $s(\boldsymbol{r}, t)$ per particle consists of the equilibrium part $s_{\rm eq}(z)$ with $ds_{\rm eq}/dz = -(\partial s/\partial p)_T \rho g = (nT)^{-1} C_p a_g$ and the nonequilibrium deviation,

$$s_1(\boldsymbol{r}, t) = (nT)^{-1} C_p \big[T_1(\boldsymbol{r}, t) - (\partial T/\partial p)_s p_1(t)\big]. \tag{6.4.7}$$

With the aid of the thermodynamic identity (1.2.54) the heat conduction equation is rewritten as

$$\left(\frac{\partial}{\partial t} + \boldsymbol{v} \cdot \nabla\right) T_1 = \left(1 - \frac{1}{\gamma_s}\right) \frac{d}{dt} \bar{T}_1 + D\nabla^2 T_1 - a_g v_z. \tag{6.4.8}$$

On the right-hand side, the first term gives rise to the piston effect, while the third term arises from $ds_{\rm eq}/dz$ and plays the role of suppressing upward (downward) motions of heated (cooled) plumes.

On long timescales ($\gg L/c$), sound waves decay to zero and the incompressibility condition $\nabla \cdot \boldsymbol{v} = 0$ becomes nearly satisfied. The timescale of the velocity field is then given by $L^2 \rho/\eta_0$ in convection. Another characteristic feature is that the Prandtl number $Pr \equiv \eta_0/\rho D$ increases in the critical region; for example, $Pr = 350$ at $T/T_{\rm c} - 1 = 10^{-3}$ in ^{3}He. This means that the timescale of the thermal diffusion is much longer than that

[8] In near-critical conditions very large Rayleigh numbers can be realized. See experiments in SF_6 [71] and in ^{4}He [72].

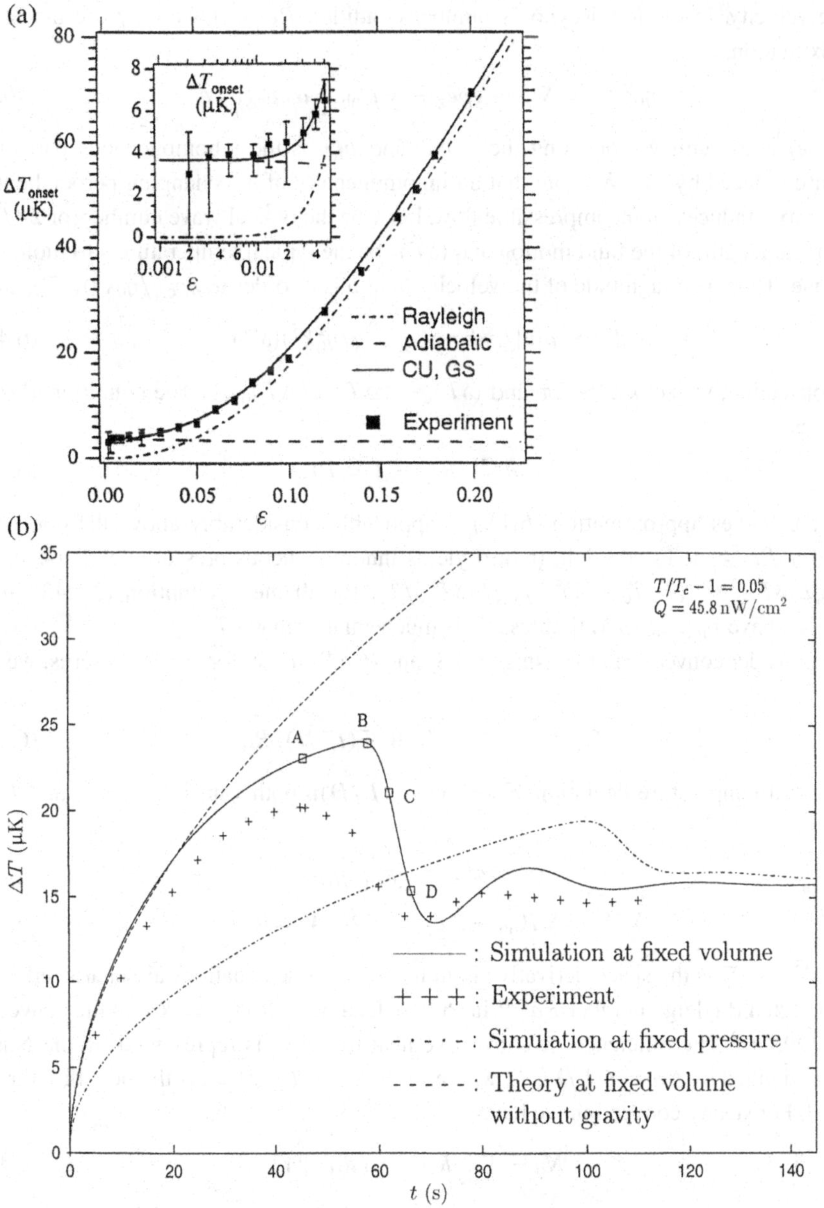

Fig. 6.17. (a) Experimental data of the temperature difference ΔT_{onset} vs $\varepsilon = T/T_c - 1$ (symbols) at convection onset measured in ^{3}He in a cell of 1 mm thickness [73]. They are compared with theory in [69] (GS) and [70] (CU). Main figure: linear plot. Inset: semi-logarithmic plot in the region where the adiabatic temperature gradient (Schwarzschild) criterion is dominant. (b) Comparison between the numerical curve (solid line) [74] and the data (+) [73] of $\Delta T(t)$ vs time at $Q = 45.8$ nW/cm^2 in the fixed-volume condition with $\gamma_s = 22.8$. The upper broken curve represents the analytic result in (6.3.21). The dot-dash curve represents the numerical curve in the fixed-pressure condition.

of the velocity. In the low-Reynolds number condition $Re < 1$ we may use the Stokes approximation,

$$\eta_0 \nabla^2 \boldsymbol{v} = \nabla p + g\rho \boldsymbol{e}_z \cong \nabla p_{\text{inh}} - \alpha_p \rho_c g \delta T \boldsymbol{e}_z, \tag{6.4.9}$$

where $\boldsymbol{e}_z$ is the unit vector along the z axis and p_{inh} is the inhomogeneous part of the pressure induced by δT. We note that an inhomogeneity of δT changing perpendicularly to the z axis induces an incompressible flow. Let k be the typical wave number (or $2\pi/k$ be the typical length) of the fluid motion and $(\delta T)_c$ be the typical temperature variation in the xy plane. Then the magnitude of the velocity field $\boldsymbol{v}$ is of order $(\alpha_p \rho_c g/\eta_0 k^2)(\delta T)_c$ and

$$Re \sim \rho |\boldsymbol{v}|/\eta_0 k \sim (\alpha_p \rho_c^2 g/\eta_0^2 k^3)(\delta T)_c. \tag{6.4.10}$$

For convection, we set $kL \sim 2\pi$ and $(\delta T)_c \sim \Delta T - (\Delta T)_{\text{onset}}$. The condition $Re < 1$ becomes

$$Ra^{\text{corr}}/Ra_c - 1 < Pr. \tag{6.4.11}$$

Thus the Stokes approximation (6.4.9) is applicable considerably above the convection onset for $Pr \gg 1$. In addition, (6.4.9) yields inhomogeneous pressure deviation $p_{\text{inh}} \sim \alpha_p \rho_c g L (\delta T)_c \sim (T/T_c - 1)^{-\gamma} \rho_c g L (\delta T)_c / T_c$. Recall the assumption (2.2.43), under which we have $p_{\text{inh}} \ll |p_1(t)|$ unless $\bar{T}_1$ is much smaller than δT.

We consider convective flow using (6.4.8) and (6.4.9). First, for steady patterns, we may set

$$T_1/\Delta T = 1 - z/L + \mathcal{F}(L^{-1}\boldsymbol{r})/Ra. \tag{6.4.12}$$

The scaled temperature deviation $\mathcal{F}$ and $\boldsymbol{V} \equiv (L/D)\boldsymbol{v}$ both vanish at $z = 0$ and L and obey

$$\boldsymbol{V} \cdot \bar{\nabla} \mathcal{F} = \bar{\nabla}^2 \mathcal{F} + Ra^{\text{corr}} V_z, \tag{6.4.13}$$

$$\bar{\nabla}^2 \boldsymbol{V} = \bar{\nabla} P_{\text{inh}} - \mathcal{F} \boldsymbol{e}_z, \qquad \bar{\nabla} \cdot \boldsymbol{V} = 0, \tag{6.4.14}$$

where $\bar{\nabla} = L\nabla$ is the space derivative in units of L. These equations are characterized by the corrected Rayleigh number Ra^{corr} in (6.4.6), leading to $Ra^{\text{corr}} = Ra_c$ at the convection onset [69, 70]. The efficiency of convective heat transport is represented by the Nusselt number defined by $Nu \equiv QL/\lambda\Delta T$, where $Q = -\lambda(\partial \delta T/\partial z)_{z=0}$ is the heat flux through the cell. For steady convection we have

$$Nu = 1 + Ra^{-1} f_\lambda (Ra^{\text{corr}}), \tag{6.4.15}$$

where $f_\lambda = -L(\partial F/\partial z)_{z=0}$ is a function of Ra^{corr}. Consistent with this result,[9] experimental curves of $Ra(Nu - 1)$ vs $Ra^{\text{corr}}/Ra_c - 1$ were fitted to a single universal curve for various densities above T_c [72] and for various $T/T_c - 1$ on the critical isochore [73].

Now we show numerical analysis of (6.4.8) and (6.4.9) [74]. We consider ^{3}He at $T/T_c - 1 = 0.05$ on the critical isochore, where $\gamma_s = 22.8$, $T\alpha_p = 26.9$, $D =$

[9] For the case of finite Pr, f_λ in (6.4.15) also depends on Pr. Its dependence should become weak once Pr considerably exceeds 1.

5.42×10^{-5} cm^2/s, and $Pr = 7.4$. The cell height is set equal to $L = 1.06$ mm, but the periodic boundary condition is imposed in the lateral direction with period $4L$. Using the experimental conditions [73], we apply a constant heat flux at the bottom for $t > 0$ with a fixed top temperature $T_{\rm top}$;[10] then, the bottom temperature $T_{\rm bot}(t)$ is a function of time. In this case we have $Ra^{\rm corr}/Ra_{\rm c} = 0.90(\Delta T/a_g L - 1)$ where $a_g L = 3.57$ μK. Figure 6.17(b) shows the numerically obtained curve of $\Delta T(t) = T_{\rm bot}(t) - T_{\rm top}$ vs time for $Q = 45.8$ nW/cm^2 (solid line). It exhibits an overshoot and a damped oscillation. In particular, the time between the maximum (point B) and the minimum (close to D) is of order $L^2/D(Ra^{\rm corr}/Ra_{\rm c} - 1)$ in accord with the experiment. This is because the arrival of thermal plumes at the top boundary causes an excess heat transfer to the boundary wall, leading to an overall temperature change, as suggested by (6.3.4). In the isobaric condition, in which the first term on the left-hand side of (6.4.8) is absent, the fluid motion approaches the final steady pattern nearly monotonically.

6.5 Critical dynamics in binary fluid mixtures

The critical dynamics of binary fluids is usually described by model H at constant pressure and in the incompressible limit. However, because the order parameter is a linear combination of the deviations of the density, entropy, and concentration, there are a number of complicated dynamical effects which are beyond the scope of model H. We will discuss (i) dissipative coupling between diffusion and heat conduction, (ii) the frequency-dependent bulk viscosity, and (iii) adiabatic relaxations.

6.5.1 Dynamic equations and renormalized kinetic coefficients

The GLW hamiltonian $\mathcal{H}\{\psi, m, q\}$ in (4.2.6) can be used for binary fluid mixtures with the linear mapping relations (2.3.12)–(2.3.14) between $\{\delta s, \delta X, \delta n\}$ and $\{\psi, m, q\}$, where q is the nonsingular variable introduced in Section 2.3 (not the wave number). All the variables are measured from their critical value at a reference critical point on the critical line. From (1.3.16) and (1.3.17) δs and δX are defined as

$$\delta s = \big[\delta e - H_1 \delta n_1 - H_2 \delta n_2\big]/n_{\rm c} T_{\rm c}, \tag{6.5.1}$$

$$\delta X = \big[(1 - X_{\rm c})\delta n_1 - X_{\rm c} \delta n_2\big]/n_{\rm c}, \tag{6.5.2}$$

where $H_i = T_{\rm c} s_{\rm c} + \mu_{i{\rm c}}$ ($i = 1, 2$). In addition to the fluctuations of the temperature and pressure introduced by (4.2.7) and (4.2.8), we introduce the fluctuating variable $\delta\hat{\Delta}$ for the chemical potential difference by (4.2.11). Regarding $\mathcal{H}$ as a functional of δs, δX, and $\delta\rho = \bar{m}_0 \delta n$ ($\bar{m}_0$ being the average mass), we may express them as

$$\delta\hat{T} = n_{\rm c}^{-1}\frac{\delta\mathcal{H}}{\delta s}, \quad \delta\hat{p} = \rho_{\rm c}\frac{\delta\mathcal{H}}{\delta\rho}, \quad \delta\hat{\Delta} = n_{\rm c}^{-1}\frac{\delta\mathcal{H}}{\delta X}. \tag{6.5.3}$$

[10] This boundary condition is usual in cryogenic heat conduction experiments. See Ref. [75] for analysis of convection onset in this case.

The dynamic equations for δs and δX are

$$\frac{\partial}{\partial t}\delta s = -\nabla\cdot(\delta s \boldsymbol{v}) + L_{0 11}\nabla^2\delta\hat{T} + L_{0 12}\nabla^2\delta\hat{\Delta} + \theta_1, \tag{6.5.4}$$

$$\frac{\partial}{\partial t}\delta X = -\nabla\cdot(\delta X \boldsymbol{v}) + L_{0 21}\nabla^2\delta\hat{T} + L_{0 22}\nabla^2\delta\hat{\Delta} + \theta_2. \tag{6.5.5}$$

The symmetric background kinetic coefficients $L_{0ij} = L_{0ji}$ are related to the noise terms as

$$\langle\theta_i(\boldsymbol{r},t)\theta_j(\boldsymbol{r}',t')\rangle = -2n_{\rm c}^{-1}T_{\rm c}L_{0ij}\nabla^2\delta(\boldsymbol{r}-\boldsymbol{r}')\delta(t-t'). \tag{6.5.6}$$

As in (6.1.18) the renormalized kinetic cofficients are expressed as the time-integrals of the correlation functions of the nonlinear fluxes $\delta s\boldsymbol{v}$ and $\delta X\boldsymbol{v}$, where we may set $\delta s \cong n_{\rm c}^{-1}\alpha_s\psi$ and $\delta X \cong n_{\rm c}^{-1}\alpha_X\psi$ in terms of the order parameter ψ. The coefficients α_s and α_X are defined in (2.3.15) and satisfy (2.3.20). It is convenient to write $\bar{\alpha}_1 \equiv \alpha_s$ and $\bar{\alpha}_2 \equiv \alpha_X$. Then (2.3.20) gives

$$\bar{\alpha}_1/\bar{\alpha}_2 = \alpha_s/\alpha_X = -(\partial\Delta/\partial T)_{p,{\rm cx}}. \tag{6.5.7}$$

Because the reversible fluxes of δs and δX are $\bar{\alpha}_1\psi\boldsymbol{v}$ and $\bar{\alpha}_2\psi\boldsymbol{v}$, respectively, we may express the k-dependent renormalized kinetic coefficients as [76]–[78]

$$L_{{\rm R}ij}(k) = L_{0ij} + (n_{\rm c}T_{\rm c})^{-1}\bar{\alpha}_i\bar{\alpha}_j L_{\rm R}(k), \tag{6.5.8}$$

where $L_{\rm R}(k)$ is the renormalized kinetic coefficient (6.1.20) for model H.

6.5.2 Diffusive relaxation and Rayleigh scattering

In the long-wavelength limit and in the isobaric condition, we write

$$\delta\hat{T} = A_{11}\delta s + A_{12}\delta X, \quad \delta\hat{\Delta} = A_{12}\delta s + A_{22}\delta X, \tag{6.5.9}$$

where A_{ij} are the thermodynamic derivatives at fixed p. Substituting these relations into (6.5.4) and (6.5.5) and using the renormalized kinetic coefficients, we obtain diffusive equations for slowly varying disturbances of δs and δX,

$$\frac{\partial}{\partial t}\delta s = \nabla^2\big[H_{\rm R11}\delta s + H_{\rm R12}\delta X\big] + \theta_{\rm R1}, \tag{6.5.10}$$

$$\frac{\partial}{\partial t}\delta X = \nabla^2\big[H_{\rm R21}\delta s + H_{\rm R22}\delta X\big] + \theta_{\rm R2}, \tag{6.5.11}$$

where $H_{{\rm R}ij} = \sum_{\ell=1}^{2} L_{{\rm R}i\ell}A_{\ell j}$. No macroscopic flow field is assumed. The noise terms $\theta_{\rm R1}$ and $\theta_{\rm R2}$ are related to $L_{{\rm R}ij}$ via the fluctuation–dissipation relations.

We next calculate the time-correlation functions around equilibrium,

$$\begin{aligned} G_{11}(k,t) &= \langle s_{\boldsymbol{k}}(t)s_{-\boldsymbol{k}}(0)\rangle, \quad G_{12}(k,t) = \langle s_{\boldsymbol{k}}(t)X_{-\boldsymbol{k}}(0)\rangle, \\ G_{22}(k,t) &= \langle X_{\boldsymbol{k}}(t)X_{-\boldsymbol{k}}(0)\rangle, \end{aligned} \tag{6.5.12}$$

which undergo double-diffusive relaxations. Their expressions in the hydrodynamic regime $k\xi \ll 1$ can be obtained from (6.5.10) and (6.5.11). They may readily be extended to the case $k\xi \gtrsim 1$ if we use the Ornstein–Zernike form for $\langle |\psi_k|^2 \rangle$ and Γ_k in (6.1.21) for the order parameter relaxation. Some further calculations yield [77]–[79]

$$G_{ij}(k,t) = \left(\bar{\alpha}_i \bar{\alpha}_j \frac{\chi}{1+k^2\xi^2} + C_{ij}^{(1)} \right) \exp(-\Gamma_k t) + C_{ij}^{(2)} \exp(-D_2 k^2 t). \tag{6.5.13}$$

Here $C_{ij}^{(1)}$ and $C_{ij}^{(2)}$ are nearly nonsingular coefficients. The two diffusion constants $D_1 = \lim_{k\to 0} \Gamma_k$ and D_2 are the eigenvalues of H_{Rij}. Here $D_1 \cong T/6\pi\eta_R\xi$ as in model H, whereas D_2 is nearly nonsingular, so $D_1 \ll D_2$ near the critical point. The slow mode may be identified with the concentration mode for nearly incompressible binary fluid mixtures and the entropy mode for nearly azeotropic binary fluid mixtures.

In particular, if the electric polarizabilities of the two components are nearly the same, as is the case in ^{3}He–^{4}He [80], dynamic light scattering detects $S_k(t) = \langle n_k(t) n_{-k}(0) \rangle$ for the density fluctuation $\delta n = (\partial n/\partial s)_{Xp}\delta s + (\partial n/\partial X)_{sp}\delta X + (\partial n/\partial p)_{sX}\delta p$. Here δs and δX give rise to Rayleigh scattering, while the pressure fluctuation leads to Brillouin scattering. From (6.5.13) we obtain

$$S_k(t) = \frac{nTK_{T\Delta}}{1+k^2\xi^2} \exp(-\Gamma_k t) + C_{\rm reg} \exp(-D_2 k^2 t) + \frac{T}{\rho c^2} \cos(ckt) \exp\left(-\frac{1}{2}\Gamma_k^s t \right), \tag{6.5.14}$$

where $K_{T\Delta} = (\partial n/\partial p)_{T\Delta} (\propto \chi)$ is the compressibility at fixed T and Δ, and $C_{\rm reg}$ is a nearly nonsingular coefficient. In the last term the sound-wave dispersion is written as $\omega_k = \pm ck + \frac{1}{2} i \Gamma_k^{\rm s}$ with k being real. Notice that the amplitude of the slowly decaying part, the first term in (6.5.13) or (6.5.14), increases markedly near the critical point, whereas the second term is insensitive to $T - T_{\rm c}$. Some experiments detected double-exponential relaxation not very close to the critical point [81, 82].

6.5.3 Heat conduction and mass diffusion

We consider nonequilibrium situations in which the deviations $T_1 = \langle \delta \hat{T} \rangle$ and $\Delta_1 = \langle \delta \hat{\Delta} \rangle$ are weakly inhomogeneous, where the averages are taken over the thermal noises. The average heat and diffusion fluxes $\boldsymbol{Q}$ and $\boldsymbol{i}$ are written in terms of the renormalized kinetic coefficients as

$$\boldsymbol{Q} = -n_{\rm c} T_{\rm c} \left[L_{\rm R11} \nabla T_1 + L_{\rm R12} \nabla \Delta_1 \right], \tag{6.5.15}$$

$$\boldsymbol{i} = -L_{\rm R21} \nabla T_1 - L_{\rm R22} \nabla \Delta_1. \tag{6.5.16}$$

In usual heat conduction experiments in a finite cell, there is no diffusive flux ($\boldsymbol{i} = \boldsymbol{0}$) in a steady state, so

$$\nabla \Delta_1 = -L_{\rm R22}^{-1} L_{\rm R21} \nabla T_1. \tag{6.5.17}$$

Then, elimination of $\nabla \Delta_1$ gives the effective thermal conductivity of the form,

$$\lambda_{\text{eff}} = -Q/\nabla T_1 = n_c T_c \big[L_{R11} - L_{R12}^2 / L_{R22} \big]. \tag{6.5.18}$$

We notice that the divergent parts in $L_{Rij} \propto \bar{\alpha}_i \bar{\alpha}_j$ are canceled in the above expression, leading to a finite thermal conductivity λ_c at the critical point. It is convenient to express λ_{eff} as

$$\frac{1}{\lambda_{\text{eff}}} = \frac{1}{\lambda_R} + \frac{1}{\lambda_c}, \tag{6.5.19}$$

where

$$\lambda_R \cong (T_c/6\pi\eta_R\xi) C_{p\Delta} + n_c T_c L_{011} \tag{6.5.20}$$

is divergent with $C_{p\Delta} = \alpha_s^2 \chi$ as in the one-component case, and

$$\lambda_c = n_c T_c \left[\left(\frac{\bar{\alpha}_1}{\bar{\alpha}_2} \right)^2 L_{022} - 2 \left(\frac{\bar{\alpha}_1}{\bar{\alpha}_2} \right) L_{012} + L_{011} \right] \tag{6.5.21}$$

is the critical-point value determined by the background kinetic coefficients L_{0ij}.

Usually, however, the chemical potential difference is not measurable and is eliminated in favor of the concentration with the aid of the linear relation $\Delta_1 = (\partial \Delta/\partial T)_{pX} T_1 + (\partial \Delta/\partial X)_{pT} X_1$, where X_1 is the average concentration deviation. We express the two fluxes in the familiar forms,

$$Q = -\lambda_{\text{eff}} \nabla T_1 - A_h \boldsymbol{i}, \tag{6.5.22}$$

$$\boldsymbol{i} = -D_T [\nabla X_1 + T^{-1} k_T \nabla T_1]. \tag{6.5.23}$$

The isothermal diffusion constant is defined by

$$D_T = L_{R22} (\partial \Delta/\partial X)_{pT}. \tag{6.5.24}$$

The cross coefficients A_h and k_T are defined by

$$A_h = nT \frac{L_{R12}}{L_{R22}} = n \left[k_T \left(\frac{\partial \Delta}{\partial X} \right)_{pT} - T \left(\frac{\partial \Delta}{\partial T} \right)_{pX} \right], \tag{6.5.25}$$

$$k_T = T \left(\frac{\partial X}{\partial \Delta} \right)_{pT} \left(\frac{L_{R12}}{L_{R22}} \right) - T \left(\frac{\partial X}{\partial T} \right)_{p\Delta}. \tag{6.5.26}$$

In particular, k_T is called the thermal diffusion ratio, because it is the ratio of the two gradients in the absence of a diffusion current. Delicate cancellation of the diverging terms also occurs in k_T. For non-azeotropic mixtures its asymptotic behavior is

$$k_T \cong D^{-1} T_c \big[L_{012} - (\alpha_s/\alpha_X) L_{022} \big], \tag{6.5.27}$$

where $D = T/6\pi\eta_R\xi$, so $k_T \propto \xi$. The predictions that $\lambda_{\text{eff}} \to \lambda_c$ and $k_T \propto \xi$ are in agreement with measurements in binary fluid mixtures near consolute critical points, such as nitrobenzene + hexane [83] and aniline + cyclohexane [84]. Figure 6.18 shows data of D and k_T for the latter system, which agree with the above results.

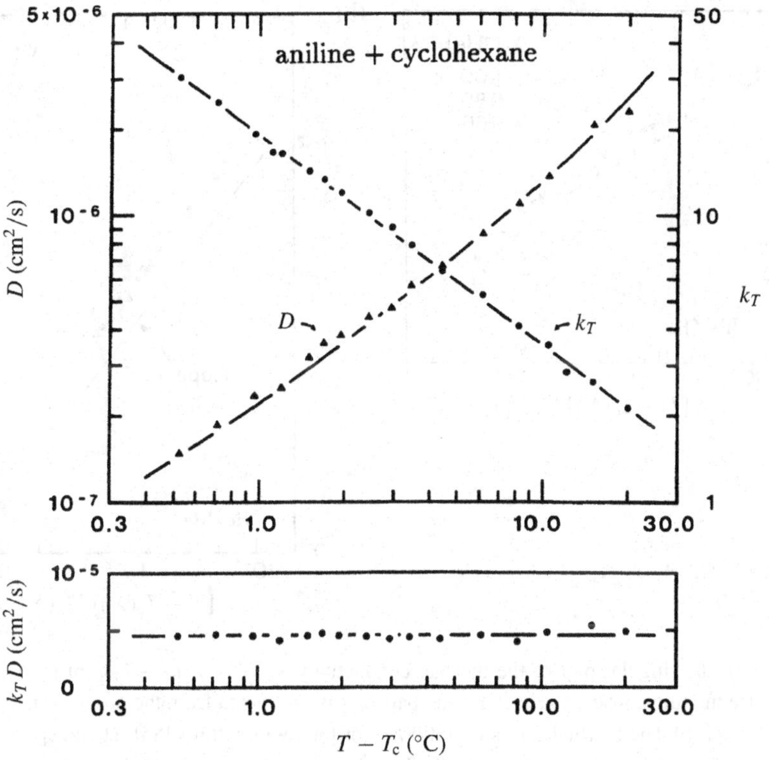

Fig. 6.18. The diffusion constant D and the thermal-diffusion ratio k_T at the critical composition in aniline + cyclohexane [84]. Here $D \propto \xi^{-1}$ was obtained from a macroscopic thermal relaxation and represents the smaller diffusion constant. In the lower panel the product $k_T D$ is shown to be independent of $T - T_c$.

Dynamic crossover in nearly azeotropic fluid mixtures

As discussed in Section 2.3, nearly azeotropic binary mixtures are characterized by small α_X, where the concentration fluctuations are weaker than those of the entropy as expressed by (2.3.50). In this case $\Delta L_{22} = L_{R22} - L_{022} \propto \alpha_X^2$ in (6.5.8) and $\lambda_c \propto \alpha_X^{-2}$ from (6.5.21). They are related by

$$\frac{\lambda_R}{\lambda_c} \cong \frac{\Delta L_{22}}{L_{022}} \sim \frac{T}{6\pi\eta_R\xi D_0}\tau_{s1}^{\gamma} \sim \tau_{s1}^{\gamma}(T/T_c - 1)^{-\nu}, \tag{6.5.28}$$

where τ_{s1} is the static crossover reduced temperature defined by (2.3.51) and

$$D_0 = L_{022} A_X / T \tag{6.5.29}$$

is the background diffusion constant (of order 10^{-5} cm²/s in ^{3}He–^{4}He). The A_X is the background part of $T(\partial X/\partial\Delta)_{Tp}$ as in (2.3.50). A dynamic crossover reduced temperature

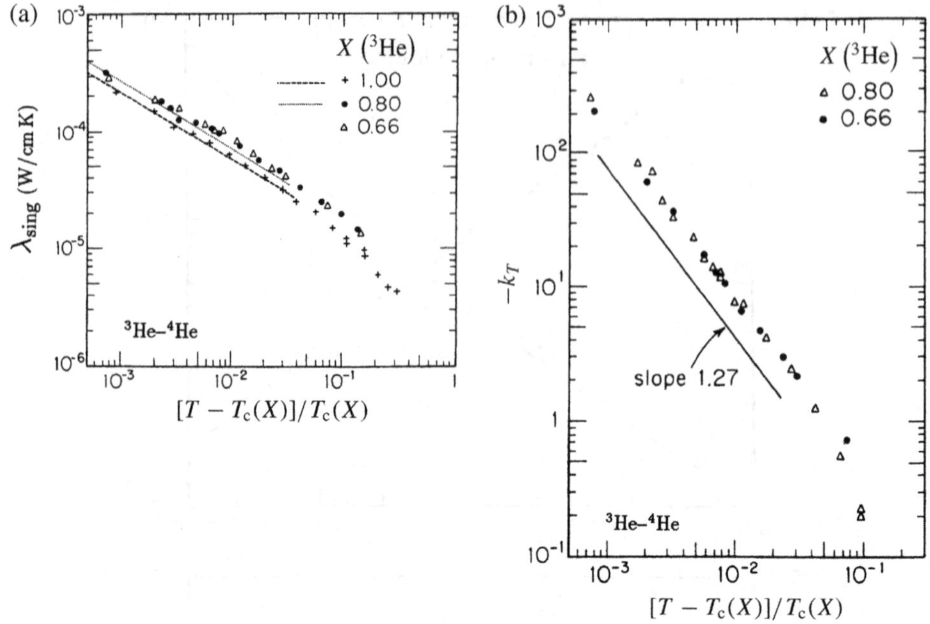

Fig. 6.19. (a) The singular part of the thermal conductivity $\lambda_{sing} = \lambda_{obs} - \lambda_{reg}$ in pure ^{3}He and two ^{3}He–^{4}He mixtures, where the background part λ_{reg} is subtracted from the observed part λ_{obs} in the steady state. (b) The thermal diffusion ratio k_T for the two mixtures [85]. Here $\lambda_{sing} \propto \xi$ and $k_T \propto \xi^2$, which is the behavior in the range $T/T_c - 1 > \tau_D$ before the dynamic crossover.

τ_D may thus be introduced by [77]

$$\tau_D^{\nu} \sim \tau_{s1}^{\gamma}. \tag{6.5.30}$$

Roughly, we have $\tau_D \sim \tau_{s1}^2$. In terms of τ_D the crossover of λ_{eff} and k_T are expressed as

$$\lambda_{eff}/\xi \propto -k_T/\xi^{\gamma/\nu} \propto \left\{1 + \left[\tau_D/(T/T_c - 1)\right]^{\nu}\right\}^{-1}. \tag{6.5.31}$$

Apparent critical divergence of the thermal conductivity has been reported in a number of fluid mixtures such as ^{3}He–^{4}He [85] and methane–ethane [86]. Very recently its saturation has also been observed in methane–ethane [87]. In Fig. 6.19 we show data of the singular part of the thermal conductivity and k_T in ^{3}He–^{4}He with the ^{3}He molar concentration at 0.80 and 0.66. The thermal conductivity behaves as in pure ^{3}He. In this temperature range the dynamic crossover was not reached, while $\tau_D \sim 10^{-4}$ for such concentrations theoretically [77].

6.5.4 The frequency-dependent bulk viscosity

It is straightforward to calculate the frequency-dependent bulk viscosity in binary fluid mixtures. From (4.2.7) and (4.2.8) the pressure and temperature fluctuations contain the following nonlinear parts [37],

$$\hat{p}_{\rm nl} = \left(\frac{\partial p}{\partial \tau}\right)_{h\zeta} \gamma_0 \psi^2, \quad \hat{T}_{\rm nl} = \left(\frac{\partial T}{\partial \tau}\right)_{h\zeta} \gamma_0 \psi^2, \tag{6.5.32}$$

which arises from the terms $\propto \delta(\beta\mathcal{H})/\delta m$. It is obvious that the bulk viscosity expression for one-component fluids remains valid if $(\partial p/\partial \tau)_h$ is replaced by $(\partial p/\partial \tau)_{h\zeta}$. The thermodynamic relations (2.3.37) and (2.3.40) give

$$\rho c^2 - \rho_{\rm c} c_{\rm c}^2 = \left(\frac{\partial p}{\partial \tau}\right)_{h\zeta}^2 \frac{1}{T_{\rm c} C_M}. \tag{6.5.33}$$

Therefore, replacing ρc^2 in the formulas for one-component fluids by $\rho c^2 - \rho_{\rm c} c_{\rm c}^2$, we obtain those for binary fluid mixtures. For simplicity, we consider a binary fluid mixture at the critical concentration and in the one-phase state. We also neglect the background specific-heat correction. From (6.2.55) we find [37, 38],

$$\zeta_{\rm R}^*(\omega) = (\rho c^2 - \rho_{\rm c} c_{\rm c}^2) \frac{1}{i\omega} \mathcal{F}(W), \tag{6.5.34}$$

where $W = \omega \tau_\xi / 2$. The scaling function $\mathcal{F}(z)$ is approximately given by (6.2.57). Thus the pressure variation p_1 and the density variation ρ_1 in a sound wave are related by

$$p_1 = \left[\rho c^2 + (\rho c^2 - \rho_{\rm c} c_{\rm c}^2)\mathcal{F}(W)\right] \frac{\rho_1}{\rho}. \tag{6.5.35}$$

We examine some representative cases.

(i) In the low-frequency limit the bulk viscosity grows as

$$\zeta_{\rm R}^*(0) = (\rho c^2 - \rho_{\rm c} c_{\rm c}^2) R_{\rm B} t_\xi \propto \xi^{z-\alpha/\nu}, \tag{6.5.36}$$

as in one-component fluids, $R_{\rm B} \cong 0.03$ being a universal number. The resultant attenuation per wavelength is of the form

$$\alpha_\lambda = \frac{\pi R_{\rm B}}{\rho c^2} (\rho c^2 - \rho_{\rm c} c_{\rm c}^2) t_\xi \omega. \tag{6.5.37}$$

(ii) From (6.2.48) and (6.2.57) the dispersion relation in the high-frequency limit $\omega t_\xi \gg 1$ becomes

$$\rho \omega^2 / k^2 = \rho_{\rm c} c_{\rm c}^2 + (\rho c^2 - \rho_{\rm c} c_{\rm c}^2)(iW)^{\alpha/\nu z}, \tag{6.5.38}$$

which is independent of ξ as ought to be the case. Let us set

$$Z(\omega) = \frac{1}{\rho_{\rm c}} (\rho c^2 - \rho_{\rm c} c_{\rm c}^2) W^{\alpha/\nu z}, \tag{6.5.39}$$

which is asymptotically independent of $T - T_{\rm c}$. Then, at the critical point we have

$$\alpha_\lambda = 0.27\left[Z(\omega)/(c_{\rm c}^2 + Z(\omega))\right]^{1/2}, \tag{6.5.40}$$

$$c(\omega) = \left[c_{\rm c}^2 + Z(\omega)\right]^{1/2}. \tag{6.5.41}$$

The above formulas reduce to those of one-component fluids at the critical point as $c_c \to 0$. As the average concentration is decreased, the critical behavior of acoustic propagation crosses over from that of binary fluid mixtures to that of one-component fluids.

(iii) In particular, $c^2 - c_c^2$ can be much smaller than c_c^2 in many nearly incompressible binary mixtures.[11] In this case, because the bulk viscosity may be treated as a perturbation in the acoustic dispersion relation, the sound attenuation per wavelength is written as

$$\begin{aligned}\alpha_\lambda &\cong \pi(\rho c^2/\rho_c c_c^2 - 1)\,\mathrm{Im}\big[\mathcal{F}(W)\big] \\ &\cong \frac{\pi\alpha}{\nu}(\rho c^2/\rho_c c_c^2 - 1)\int_0^\infty dx \frac{x^{3-\epsilon}\Gamma^*(x)W}{(1+x^2)^{(1-\alpha)/\nu}[\Gamma^*(x)^2 + W^2]},\end{aligned} \tag{6.5.42}$$

for small α/ν in our scheme. As $W \to \infty$, the high-frequency behavior is given by $\alpha_\lambda \to \alpha_{\lambda c} \equiv 0.27 Z(\omega)^{1/2}/c_c$ in accord with (6.5.40). Therefore, the ratio of α_λ to its critical value $\alpha_{\lambda c}$ becomes a universal function of W as

$$\frac{\alpha_\lambda}{\alpha_{\lambda c}} = \frac{2zW}{\pi}\int_0^\infty dx \frac{x^{3-\epsilon}\Gamma^*(x)}{(1+x^2)^{2-\epsilon/2}[\Gamma^*(x)^2 + W^2]}, \tag{6.5.43}$$

which increases from 0 to 1 with increasing W. Ferrell and Bhattacharjee proposed an approximate expression for $\alpha_\lambda/\alpha_{\lambda c}$ [33], which is obtained if the integrand in (6.5.43) is replaced by $x^3\Gamma^*(x)/(1+x^2)^2[\Gamma^*(x)^2 + W^2]$ with $\Gamma^*(x) = x^2(1+x^2)^{1/2}$. Alternatively, we may set $\epsilon = 1$ and use their $\Gamma^*(x)$ in (6.5.43). In Fig. 6.20 the two theoretical curves thus obtained are compared with some experimental data.

Frequency-dependent specific heat

As in one-component fluids, we introduce the frequency-dependent specific heat $\tilde{C}_p(\omega)$ for nearly incompressible binary mixtures where $c^2 - c_c^2 \ll c_c^2$ and $\rho \cong \rho_c$. From (6.5.35) we may express the dispersion relation as

$$\omega^2/c_c^2k^2 = 1 + \hat{g}^2\rho_c c_c^2/T_c\tilde{C}_p(\omega), \tag{6.5.44}$$

where $\tilde{C}_p(\omega)$ tends to the thermodynamic specific heat C_{pX} (per unit volume) as $\omega \to 0$ and behaves as $C_{pX}(i\omega t_\xi)^{-\alpha/\nu z}$ for $\omega t_\xi \gg 1$ as in (6.2.49). In our theory the coefficient $\hat{g}$ may be expressed in the form

$$\hat{g} = \rho_c T_c\left[\left(\frac{\partial s}{\partial p}\right)_c - \left(\frac{\partial s}{\partial X}\right)_{pT}\left(\frac{\partial X}{\partial p}\right)_c\right], \tag{6.5.45}$$

where s is the entropy per unit mass and use has been made of $(\partial T/\partial p)_{sX} = (\partial T/\partial p)_c - (\partial T/\partial s)_{pX}(\partial s/\partial p)_c - (\partial T/\partial X)_{ps}(\partial X/\partial p)_c$ with the aid of (2.3.40), (2.3.42), and (2.3.43). When the second term in the brackets of (6.5.45) is small compared to the first, we are led to Ferrell and Bhattacharjee's expression $\hat{g} = \rho_c T_c(\partial s/\partial p)_c$ [33]. Their expression was confirmed to be consistent with data of acoustic attenuation in trimethylpentane + nitroethane (where the pressure dependence of X_c is small) [47].

[11] In this case $\rho c^2/\rho_c c_c^2 - 1 \cong C_B/C_{pX}$ from (2.3.42), C_B being related to $\rho_c c_c^2$ as (2.3.40) [46].

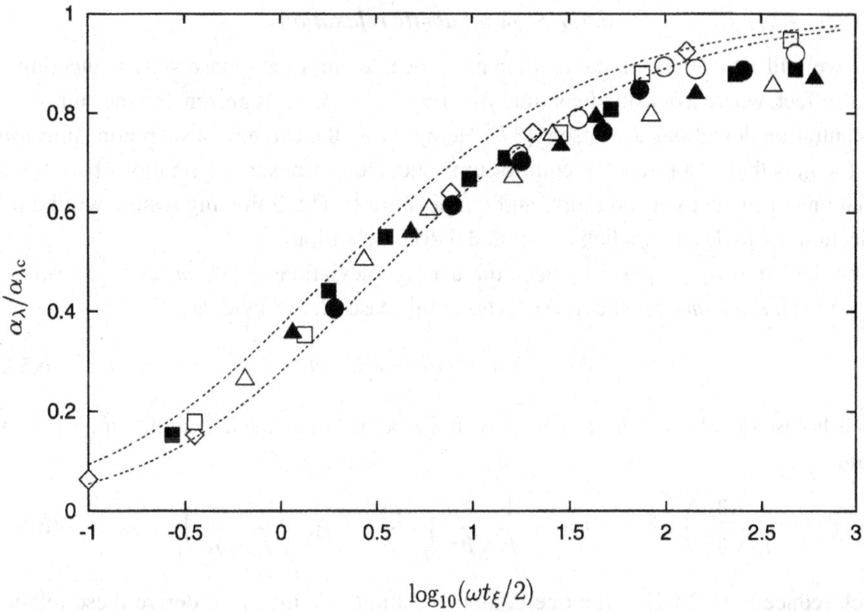

Fig. 6.20. Attenuation per wavelength relative to the critical-point value vs $W = \omega t_\xi/2$ in binary field mixtures. The lower dashed curve was given in Ref. [33], while the upper curve follows from (6.5.43) at $\epsilon = 1$. Here the relation $\rho c^2 - \rho_c c_c^2 \ll \rho_c c_c^2$ holds. Experimental frequencies (in MHz) and [references]: trimethylpentane + nitroethane, ■ 16.5, △ 27, ▲ 48, ● 80, ○ 165 [45]; trimethylpentane + nitrobenzene, ◇ 3, □ 11 [44].

6.5.5 Piston effect in binary fluid mixtures

The piston effect tends to be suppressed in binary fluids, because C_{pX} and C_{VX} play the roles of C_p and C_V in the effect and the ratio $\gamma_X = C_{pX}/C_{VX}$ grows only weakly [61]. (i) Nevertheless, in nearly azeotropic binary mixtures, the piston effect still influences thermal equilibration. In an experiment of ^{3}He–^{4}He on the critical isochore [63], thermal relaxation was measured at constant volume with a fixed bottom temperature and a fixed heat flux at the top boundary ($x = L$). If the adiabatic effect is taken into account in this geometry, the slowest relaxation rate is calculated to be $\Omega_1 \equiv \pi^2 D_1/L^2$ with $D_1 = T/6\pi\eta_R\xi$ [61], in good agreement with the experiment. At constant pressure under the same boundary conditions, however, the slowest relaxation rate is equal to $\Omega_1/4$ for $T/T_c - 1 \gg \tau_{s1}$ and Ω_1 for $T/T_c - 1 \ll \tau_{s1}$, in disagreement with the experiment, where τ_{s1} is defined by (2.3.51). (ii) In non-azeotropic fluids, the thickness of the thermal diffusion layer is given by $(D_2 t)^{1/2}$, where D_2 is the nonsingular diffusion constant appearing in (6.5.13). It can affect the interior temperature on a much longer timescale t_1' determined by

$$(D_2 t_1')^{1/2} = L/(\gamma_X - 1). \tag{6.5.46}$$

The efficiency of the piston is suppressed with a decrease in γ_X.

6.5.6 Slow adiabatic relaxations

Next we will treat nonstationary, adiabatic processes in a one-phase state neglecting the piston effect, where a density deviation, $\rho_1(t) = \langle\delta\rho\rangle - \langle\delta\rho\rangle_0$, is present but the entropy and concentration deviations are absent [37]. Here $\langle\cdots\rangle$ is the average in such nonequilibrium and $\langle\cdots\rangle_0$ is that in a reference equilibrium state. Here we examine relations between the deviations of the density, pressure, and temperature.[12] The following results will be used in Section 8.7 to describe adiabatic spinodal decomposition.

The deviation $\rho_1 = \rho_1(t)$ induces the average deviations of ψ, m, and q, written as $\psi_1 = \psi_1(t)$, $m_1 = m_1(t)$, and $q_1 = q_1(t)$. In this case (2.3.16) yields

$$h\psi_1 + \tau m_1 + \zeta q_1 = \delta p \rho_1 / T_c \rho, \tag{6.5.47}$$

which holds for arbitrary h, τ, and ζ with δp being their linear combination. Thus we obtain

$$\psi_1 = \frac{1}{T_c}\left(\frac{\partial p}{\partial h}\right)_{\tau\zeta}\frac{\rho_1}{\rho}, \quad m_1 = \frac{1}{T_c}\left(\frac{\partial p}{\partial \tau}\right)_{h\zeta}\frac{\rho_1}{\rho}, \quad q_1 = \frac{1}{T_c}\left(\frac{\partial p}{\partial \zeta}\right)_{h\tau}\frac{\rho_1}{\rho}, \tag{6.5.48}$$

which reduce to (6.2.65) in the one-component limit. We may also derive these relations by inverting the matrix relations (2.3.12)–(2.3.14). We also consider the deviations,

$$\tau_1 = \left\langle\frac{\delta(\beta\mathcal{H})}{\delta m}\right\rangle, \quad \zeta_1 = \left\langle\frac{\delta(\beta\mathcal{H})}{\delta q}\right\rangle = \frac{q_1}{Q_0}. \tag{6.5.49}$$

where $Q_0^{-1} = \rho_c c_c^2 T_c / (\partial p/\partial\zeta)_c^2$ from (2.3.40). Neglecting the average of the first terms $\propto h_1 = \langle\delta(\beta\mathcal{H})/\delta\psi\rangle \sim \psi_1/\chi$ in (4.2.7) and (4.2.8), we can express the average pressure and temperature variations, $p_1 = \langle\delta\hat{p}\rangle$ and $T_1 = \langle\delta\hat{T}\rangle$, as

$$p_1 = \left(\frac{\partial p}{\partial \tau}\right)_{h\zeta}\tau_1 + \rho_c c_c^2 \frac{\rho_1}{\rho}, \tag{6.5.50}$$

$$T_1 = \left(\frac{\partial T}{\partial \tau}\right)_{h\zeta}\tau_1 + \left(\frac{\partial T}{\partial p}\right)_c \rho_c c_c^2 \frac{\rho_1}{\rho}. \tag{6.5.51}$$

Eliminating τ_1 we may express T_1 in terms of p_1 and ρ_1. Explicitly writing the time dependence, we have a fundamental relation,

$$T_1(t) = A_c p_1(t) + (B_c - A_c)\rho_c c_c^2 \frac{\rho_1(t)}{\rho}, \tag{6.5.52}$$

where the two coefficients, A_c and B_c, are introduced in (2.3.31). Obviously, the above relation reduces to the thermodynamic relation for $(\partial T/\partial p)_{sX}$ in (2.3.43) in the quasi-static limit. It is important that (6.5.52) holds in general nonstationary adiabatic conditions.

[12] We also mention a light scattering experiment which observed relaxation of the structure factor after a pressure change in one-phase states of a critical binary fluid mixture [88].

We recall that p_1 and ρ_1 are related by (6.5.35) in a sound wave or when they depend on time as $\exp(i\omega t)$. For general time dependence they are related by

$$p_1(t) = c^2\rho_1(t) + \rho^{-1}\int_{-\infty}^{t} dt' G_{xx}(t-t')\dot{\rho}_1(t'), \tag{6.5.53}$$

where $\dot{\rho}_1(t) = \partial\rho_1(t)/\partial t$ is the time derivative. The stress relaxation function $G_{xx}(t)$ is defined by (6.2.58) also for binary fluid mixtures. If the background specific-heat correction is neglected, it is related to the universal function $\hat{G}(t)$ in (6.2.59) as

$$G_{xx}(t) = (\rho c^2 - \rho_c c_c^2)\hat{G}(t). \tag{6.5.54}$$

With (6.5.52) and (6.5.53) we may now examine temporal variations of the pressure, temperature, and density in adiabatic conditions in one-phase states. They depend only on time in the bulk fluid region far from the boundary. We also note that the relation (6.2.66) between τ_1 and m_1 holds also in binary fluid mixtures exactly in the same form, which follows from (6.5.53) and (6.5.54) with the aid of (6.5.32), (6.5.33), and (6.5.48).

Relaxation after a volume change

Let us change the volume of the cell by a small amount δV at $t = 0$ and keep it constant thereafter. The density is then changed by $\rho_1 = -\rho\delta V/V$ at $t = 0$ in a step-wise manner. From (6.5.53) the induced pressure variation is

$$p_1(t)/\rho_1 = c^2 + \rho^{-1}(\rho c^2 - \rho_c c_c^2)\hat{G}(t). \tag{6.5.55}$$

The temperature variation is

$$T_1(t)/(\rho_1/\rho) = [A_c\rho c^2 + (B_c - A_c)\rho_c c_c^2] + A_c(\rho c^2 - \rho_c c_c^2)\hat{G}(t), \tag{6.5.56}$$

where the first term is equal to $\rho c^2(\partial T/\partial\rho)_{sX}$ from (2.3.43). Thus we can directly measure the time-correlation function $G_{xx}(t)$ of the stress which relaxes as in Fig. 6.10. See Ref. [37] for the relaxation after a pressure change.

6.6 Critical dynamics near the superfluid transition

As shown in Section 2.4, ^{4}He near the superfluid transition is nearly incompressible as regards its static critical behavior. Most generally, the GLW hamiltonian is given by $\mathcal{H}\{\psi, m, q\}$ in (4.2.12), where ψ is the complex order parameter, m the weakly singular variable, and q the nonsingular variable. They are related to the density and entropy deviations, δn and δs, via the mapping relations (2.4.8) and (2.4.9). A complete set of the gross variables is then composed of ψ, δs, $\delta\rho = m_4\delta n$, and the momentum density $\boldsymbol{J}$, where m_4 is the ^{4}He mass. In the Russian literature [89]–[91], dynamic equations for ψ, $S = \rho s$(= the entropy per unit volume), ρ, and $\boldsymbol{J}$ were constructed in full nonlinear forms but without the noise terms. In the literature of critical dynamics [1, 2, 11], [92]–[95], much simpler coupled dynamic equations for ψ and δs have been used, because their mutual interaction is relevant in dynamics.

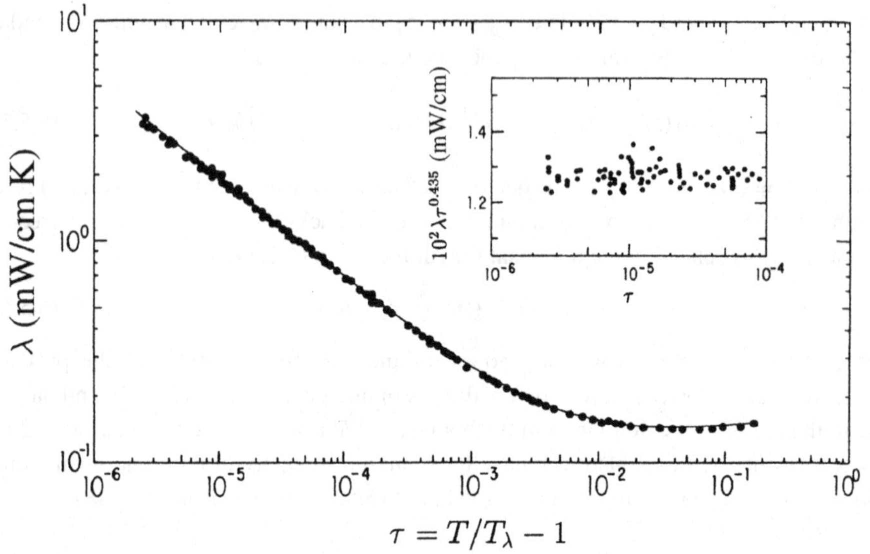

Fig. 6.21. The thermal conductivity λ vs τ for ^{4}He at SVP [98]. *Inset*: plot of $\lambda\tau^{0.435}$ vs τ showing the effective exponent for $\tau < 5 \times 10^{-3}$.

Experimentally, very precise measurements have been performed on the dynamics of ^{4}He and ^{3}He–^{4}He mixtures. We mention measurements of the thermal conductivity above T_λ [96]–[98] and those of the second-sound damping below T_λ [99, 100]. As discussed in Section 2.4, the static critical behavior in ^{4}He is characterized by the nonclassical critical exponents for $|T/T_\lambda - 1| \ll 1$, where $\gamma = 4/3$, $\nu = 2/3$, $\beta = 1/3$, $\alpha = 0$, and $\eta = 0$. However, the dynamic renormalization effect in ^{4}He turns out to be effective much closer to the transition $|T/T_\lambda - 1| \lesssim \tau_c \sim 10^{-3}$ or only on spatial scales longer than $\xi_{0+}t_c^{-\nu} \sim 10^{-6}$ cm. To demonstrate this, we show, for ^{4}He, the thermal conductivity in Fig. 6.21 and the thermal diffusivity in Fig. 6.22 [98] above T_λ.

6.6.1 Minimal model equations

Neglecting the gravity effects, we consider the coupling of ψ with the entropy deviation δs (per particle) in statics and dynamics. Here, writing

$$m = n\delta s \cong n_\lambda \delta s, \tag{6.6.1}$$

the hamiltonian will be assumed to be the form of (4.1.45) without an ordering field:

$$\beta\mathcal{H}\{\psi, m\} = \int d\mathbf{r}\left[\frac{1}{2}r_{0c}|\psi|^2 + \frac{1}{2}|\nabla\psi|^2 + \frac{1}{4}\bar{u}_0|\psi|^4 + \gamma_0|\psi|^2 m + \frac{1}{2C_0}m^2 - \tau m\right]. \tag{6.6.2}$$

The coefficient K of the gradient free energy will be set equal to 1, because its renormalized value is equal to $1 + \eta\ln(\Lambda\xi) + \cdots \cong 1$ above T_λ and to $1 + \epsilon/20 + O(\epsilon^2) \cong 1$ below T_λ

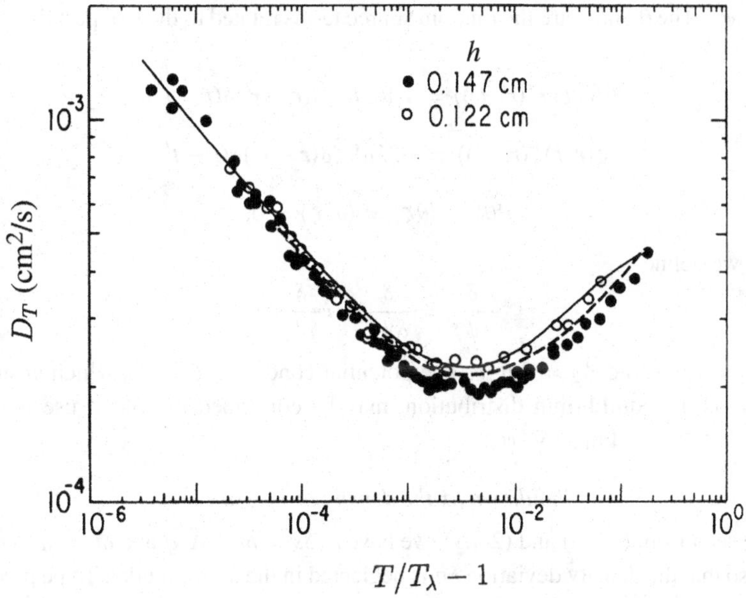

Fig. 6.22. The thermal diffusivity D_T vs $T/T_\lambda - 1$ for ^{4}He in the normal fluid phase at SVP obtained from measurements of macroscopic thermal equilibration for two cells with width $h = 0.147$ and 0.122 cm [98]. The solid curve is obtained from $D_T = \lambda/C_p$ using the thermal conductivity data. The dashed line is obtained from data of thermal equilibration times.

from (4.3.97). We will also set $\tau = T/T_\lambda - 1$, which is allowable without loss of generality. Then (4.2.13) shows that the reduced temperature fluctuation is written as

$$\delta\hat{T} = \frac{\delta}{\delta m}\mathcal{H}\{\psi, m\}. \tag{6.6.3}$$

The model F equations [11, 95] are written as

$$\frac{\partial}{\partial t}\psi = ig_0\frac{\delta(\beta\mathcal{H})}{\delta m}\psi - L_0\frac{\delta(\beta\mathcal{H})}{\delta\psi^*} + \theta, \tag{6.6.4}$$

$$\frac{\partial}{\partial t}m = g_0\,\mathrm{Im}[\psi^*\nabla^2\psi] + \lambda_0\nabla^2\frac{\delta(\beta\mathcal{H})}{\delta m} + \zeta. \tag{6.6.5}$$

The first terms ($\propto g_0$) are the reversible mode coupling terms. The L_0 is the background kinetic coefficient for the order parameter,[13] and λ_0 is the background thermal conductivity (divided by Boltzmann's constant k_B). Generally, L_0 can be complex and then the term proportional to $\mathrm{Im}\,L_0$ in (6.6.4) is reversible, because the time reversal of ψ is its complex

[13] We expect $\mathrm{Im}\,L_0 \sim \hbar/m_4$ from quantum mechanics. See the Gross–Pitaevskii equation (8.10.2).

conjugate ψ^*. The θ and ζ are the random source terms related to the real part Re L_0 and λ_0 as

$$\langle\theta(\boldsymbol{r},t)\theta^*(\boldsymbol{r}',t')\rangle = 4(\mathrm{Re}\,L_0)\delta(\boldsymbol{r}-\boldsymbol{r}')\delta(t-t'), \tag{6.6.6}$$

$$\langle\zeta(\boldsymbol{r},t)\zeta(\boldsymbol{r}',t')\rangle = -2\lambda_0\nabla^2\delta(\boldsymbol{r}-\boldsymbol{r}')\delta(t-t'). \tag{6.6.7}$$

$$\langle\theta\theta\rangle = \langle\theta\zeta\rangle = \langle\theta^*\zeta\rangle = 0. \tag{6.6.8}$$

Hereafter we define

$$\frac{\delta}{\delta\psi^*} = \frac{\delta}{\delta\psi_1} + i\frac{\delta}{\delta\psi_2}, \tag{6.6.9}$$

where $\psi_1 = \mathrm{Re}\,\psi$ and $\psi_2 = \mathrm{Im}\,\psi$. The potential condition (5.2.14), which ensures that $\exp(-\beta\mathcal{H})$ is the equilibrium distribution, may be confirmed to hold if use is made of $\mathrm{Im}[\psi^*\delta\mathcal{H}/\delta\psi^*] = -\mathrm{Im}[\psi^*\nabla^2\psi]$.

Should we fix the density or pressure?

As can be seen from (2.4.8) and (2.4.9), we have $n_\lambda\delta\hat{s} = m - A_\lambda q$ and $\delta n = q$ in the limit $\epsilon_{\mathrm{in}} \to 0$, so that the density deviation δn is neglected in the above model. To be precise, the pressure deviation $\delta\hat{p}$ should be fixed rather than δn in dynamics. As a result, the entropy relaxation rate is $\lambda_0 k^2/C_{p0}$ at wave number k in the linear approximation, where

$$C_{p0} = C_0 + A_\lambda^2 Q_0. \tag{6.6.10}$$

The constants, A_λ and Q_0, are defined by (2.4.10) and (2.4.12), respectively. Although C_{p0} will appear in the RG equation for L_0 in (6.6.62) below, the difference $A_\lambda^2 Q_0$ is relatively small compared with the logarithmic term in C_0 as shown in Section 2.4. In this sense, model F is well justified.

Two-fluid hydrodynamics

If we write $\psi = Me^{i\theta}$ with $M = |\psi|$, the phase θ in ordered states is related to the superfluid velocity in two-fluid hydrodynamics [89]–[91], [101] by

$$\boldsymbol{v}_{\mathrm{s}} = \frac{\hbar}{m_4}\nabla\theta, \tag{6.6.11}$$

where $\hbar$ is the Planck constant and m_4 is the ^{4}He particle mass. When the amplitude M is homogeneous in space, the gradient free energy in (6.6.2) should coincide with the kinetic energy of the superfluid component, so that

$$\frac{1}{2}TM^2|\nabla\theta|^2 = \frac{1}{2}\rho_{\mathrm{s}}\boldsymbol{v}_{\mathrm{s}}^2. \tag{6.6.12}$$

Therefore, the superfluid mass density is obtained in terms of M as[14]

$$\rho_{\mathrm{s}} = \frac{m_4^2 T}{\hbar^2}M^2. \tag{6.6.13}$$

[14] To be precise, ρ_{s} is defined by (4.3.107). Here we set $K_{\mathrm{R}} = 1$.

The momentum density of the superfluid component is then written as

$$\boldsymbol{J}_s = \rho_s \boldsymbol{v}_s = \frac{m_4 T}{\hbar} \mathrm{Im}[\psi^* \nabla \psi]. \tag{6.6.14}$$

The mass density and velocity of the normal fluid component are determined by

$$\rho = \rho_s + \rho_n, \tag{6.6.15}$$

$$\boldsymbol{J} = \rho_s \boldsymbol{v}_s + \rho_n \boldsymbol{v}_n, \tag{6.6.16}$$

where ρ and $\boldsymbol{J}$ are the total mass and momentum densities, respectively.

In the non-dissipative regime (at nearly zero temperatures or in the long-wavelength limit), the equation for the phase θ reads [90, 101]

$$\hbar \frac{\partial}{\partial t} \theta \cong -\delta\mu \cong s\delta\hat{T} - \frac{1}{n}\delta\hat{p}, \tag{6.6.17}$$

where $\delta\mu$ is the chemical potential deviation per particle. If the pressure fluctuation is neglected, there arises the first term of (6.6.4) with

$$g_0 = sT_\lambda/\hbar \tag{6.6.18}$$

which is 2.15×10^{11} s^{-1} at SVP. Thus the superfluid component is accelerated by the temperature gradient as

$$\frac{\partial}{\partial t} \boldsymbol{v}_s \cong \frac{s}{m_4} \nabla \delta\hat{T}. \tag{6.6.19}$$

It is also known that the entropy is supported by the normal fluid component [90, 101], so the entropy density $S = ns$ per unit volume is convected by the normal fluid velocity $\boldsymbol{v}_n$ as

$$\frac{\partial}{\partial t} S \cong -\nabla \cdot (S \boldsymbol{v}_n), \tag{6.6.20}$$

in the non-dissipative regime. Using the mass conservation equation,

$$\frac{\partial}{\partial t} \rho = -\nabla \cdot \boldsymbol{J}, \tag{6.6.21}$$

(6.6.20) is rewritten in terms of m as in (6.6.1)

$$\frac{\partial}{\partial t} m \cong -\nabla \cdot (S\boldsymbol{v}_n) + \frac{s}{m_4} \nabla \cdot \boldsymbol{J} \cong \frac{s}{m_4} \nabla \cdot \boldsymbol{J}_s, \tag{6.6.22}$$

leading to the first term on the right-hand side in (6.6.5).

The linear hydrodynamic equations below T_λ give rise to two kinds of sounds at long wavelengths ($k \ll \xi^{-1}$). That is, we linearize (6.6.17), (6.6.20), (6.6.21), and the momentum equation $\rho \partial \boldsymbol{v}/\partial t = -\nabla \delta\hat{p}$ by neglecting the dissipation. The hydrodynamic deviations are written as T_1, p_1, ρ_1, and s_1 for the temperature, pressure, density, and entropy, respectively. If they depend on space and time as $\exp(i\omega t - ikx)$, we find simple relations,

$$(\omega/k)^2 \rho_1 = p_1,$$
$$(\omega/k)^2 s_1 = (\rho_s s^2/m_4 \rho_n) T_1. \tag{6.6.23}$$

The phase velocity $u = \omega/k$ then satisfies [101]

$$u^4 - (c^2 + c_{II}^2 C_p/C_V)u^2 + c^2 c_{II}^2 = 0, \tag{6.6.24}$$

where $c = \sqrt{(\partial p/\partial \rho)_s}$ and

$$c_{II} = (\rho_s s^2 n T/\rho_n m_4 C_p)^{1/2}. \tag{6.6.25}$$

The above relations hold at any temperature below T_λ. In particular, slightly below T_λ, the first-sound mode is almost adiabatic as well as in the normal fluid because $s_1/\rho_1 \propto \rho_s$ from the second line of (6.6.23), so that the phase velocity is given by the usual expression c. However, the second-sound mode is almost isobaric, because $p_1/\rho_1 \propto \rho_s$, and its phase velocity is given by c_{II}.

6.6.2 *Intuitive pictures of enhanced heat transport above T_λ*

Random phase modulation

We will intuitively show that the mode coupling terms in (6.6.4) and (6.6.5) serve to renormalize the kinetic coefficients L_0 and λ_0 to L_R and λ_R. We consider the critical fluctuations with sizes of order ξ slightly above the transition. As in (2.1.23) we define

$$\Psi_\xi(t) = \int_\xi d\boldsymbol{r} \psi(\boldsymbol{r}, t), \tag{6.6.26}$$

where the space integral is within a region with size ξ. From (2.1.25) its amplitude variance is written in terms of the fractal dimension $D = (d + 2 - \eta)/2$ as

$$\langle |\Psi_\xi|^2 \rangle \sim \xi^{2D} \quad \text{or} \quad \langle |\xi^{-d} \Psi_\xi|^2 \rangle \sim \xi^{-2\beta/\nu}, \tag{6.6.27}$$

where $\beta/\nu = (d - 2 + \eta)/2$ from (2.1.13). We also note that $\Psi_\xi(t) = |\Psi_\xi| e^{i\theta_\xi}$ has a well-defined phase θ_ξ. From the dynamic equation (6.6.4) or (6.6.17) it is temporally modulated by the temperature fluctuations as

$$\frac{\partial}{\partial t} \Psi_\xi(t) \cong i \frac{s}{\hbar} (\delta \hat{T})_\xi(t) \Psi_\xi(t). \tag{6.6.28}$$

To pick up temperature variations on the scale of ξ we set

$$(\delta \hat{T})_\xi(t) = \xi^{-d} \int_\xi d\boldsymbol{r} \delta \hat{T}(\boldsymbol{r}, t), \tag{6.6.29}$$

which obeys the gaussian distribution with variance,

$$\langle |(\delta \hat{T})_\xi|^2 \rangle = \xi^{-2d} \int_\xi d\boldsymbol{r} \int_\xi d\boldsymbol{r}' \langle \delta \hat{T}(\boldsymbol{r}, t) \delta \hat{T}(\boldsymbol{r}', t) \rangle = \xi^{-d} T_\lambda^2 C_V^{-1}, \tag{6.6.30}$$

from (1.2.64). Hereafter we use $C_V \sim C_p$. Let us assume that the relaxation rate of the temperature fluctuations,

$$\Gamma_\lambda = \lambda_R / C_p \xi^2, \tag{6.6.31}$$

is of the same order or larger than the order parameter relaxation rate Γ_ξ. Then, a general theory of random frequency modulation [102] shows that the temporal average over the rapidly varying (random) temperature fluctuations gives rise to a damping of Ψ_ξ as

$$\langle \Psi_\xi(t) \rangle_{\text{temp}} \sim \exp(-t\Gamma_\xi)\Psi_\xi(0), \tag{6.6.32}$$

where

$$\Gamma_\xi \cong (s/\hbar)^2 \langle |(\delta \hat{T})_\xi|^2 \rangle / \Gamma_\lambda \sim g_0^2 \xi^{2-d} / \lambda_R, \tag{6.6.33}$$

with g_0 being defined by (6.6.18). The corresponding kinetic coefficient behaves as

$$L_R = \Gamma_\xi \chi \sim g_0^2 \xi^\epsilon / \lambda_R. \tag{6.6.34}$$

For more precise estimation we should set $L_R \lambda_R \cong K_d g_0^2 \xi^\epsilon$ because a dimensionless number f, to be introduced in (6.6.56) below, is of order 1 as $T \to T_\lambda$. Thus the product $L_R \lambda_R$ grows as ξ^ϵ for $\epsilon = 4 - d > 0$. We recognize that the critical dimensionality remains 4 in dynamics as well as in statics.

Enhanced heat conduction due to cluster convection

In the presence of a small average temperature gradient $\boldsymbol{a} = \langle \nabla \delta \hat{T} \rangle_{\text{ss}}$ in the disordered phase, (6.6.19) indicates that the critical fluctuations are accelerated in the gradient direction during their lifetimes. Here $\langle \cdots \rangle_{\text{ss}}$ is the steady-state average. As a result, the critical fluctuations with sizes of order ξ move with an average velocity estimated by

$$\langle \boldsymbol{v}_s \rangle_{\text{ss}} \sim s(m_4 \Gamma_\xi)^{-1} \boldsymbol{a}, \tag{6.6.35}$$

in the steady state. Then, there arises a thermal counterflow in which the fluctuations of the superfluid and background normal fluid components are convected in the opposite directions under the condition of no mass flux, $\langle \boldsymbol{J} \rangle_{\text{ss}} = \boldsymbol{0}$ or $\langle \boldsymbol{v}_n \rangle_{\text{ss}} \cong -\rho^{-1} \langle \boldsymbol{J}_s \rangle_{\text{ss}}$. The resultant average heat current is estimated as

$$\boldsymbol{Q} = \langle T S \boldsymbol{v}_n \rangle_{\text{ss}} \cong -\frac{sT}{m_4} \langle \boldsymbol{J}_s \rangle_{\text{ss}} \sim -\frac{s m_4 T^2}{\hbar^2} \langle |\xi^{-d} \Psi_\xi|^2 \rangle \langle \boldsymbol{v}_s \rangle_{\text{ss}}. \tag{6.6.36}$$

Using (6.6.35) and (6.6.36) we find the renormalized thermal conductivity,

$$\lambda_R = -Q/a \sim g_0^2 \xi^{2-d} / \Gamma_\xi, \tag{6.6.37}$$

which is equivalent to (6.6.34).

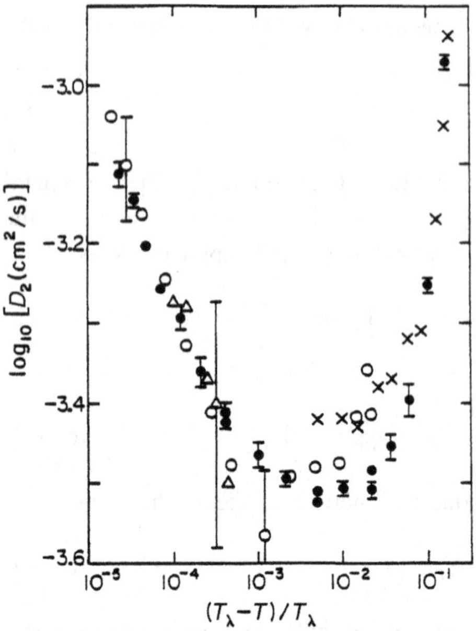

Fig. 6.23. The second-sound damping coefficient D_2 vs $1 - T/T_\lambda$ for ^{4}He in the superfluid phase at SVP [99]. Data obtained by various groups are shown.

6.6.3 Dynamic scaling behavior

Below the transition $T < T_\lambda$, the average order parameter $M = \langle \psi \rangle$ becomes nonvanishing. In this case, phase variations θ_1 and temperature variations $T_1 \cong m_1/C_p$ varying on spatial scales longer than ξ couple to form an oscillatory mode, called the second sound. If the damping is neglected, they obey

$$\frac{\partial}{\partial t}\theta_1 = \frac{g_0}{C_p} m_1, \qquad \frac{\partial}{\partial t} m_1 = g_0 |M|^2 \nabla^2 \theta_1. \tag{6.6.38}$$

The second-sound velocity is obtained as

$$c_{II} = g_0 |M| / C_p^{1/2} \propto (1 - T/T_\lambda)^{1/3}, \tag{6.6.39}$$

which is consistent with (6.6.25). The damping arises from the renormalized kinetic coefficients, $L_{\rm R}$ and $\lambda_{\rm R}$ [96, 100, 103, 104]. That is, in (6.6.38) we add $L_{\rm R}\nabla^2\theta_1$ to the right-hand side of the first equation and $(\lambda_{\rm R}/C_p)\nabla^2 m_1$ to that of the second equation to obtain the dispersion relation $\omega_k = c_{II}k - \frac{1}{2} i D_2 k^2 + \cdots$ with

$$D_2 = L_{\rm R} + \lambda_{\rm R}/C_p. \tag{6.6.40}$$

Here, however, the dissipative nonlinear coupling ($\propto \gamma_0$) is neglected. In Fig. 6.23 [99] we showed data for ^{4}He of D_2 vs $|T/T_\lambda - 1|$, which resemble those of the thermal diffusivity D_T vs $T/T_\lambda - 1$ in Fig. 6.22. For both these diffusivities, the background values are relatively large, and the crossover reduced temperatures are commonly of order 10^{-3}.

The second sound is well defined if the wave number k is much smaller than the inverse correlation length ξ^{-1}. If k become of order ξ^{-1}, the mode should becomes overdamped with a relaxation rate of order $c_{II}\xi^{-1}$. In their original work of the dynamic scaling theory, Ferrell *et al.* [92] assumed that the order parameter relaxation rate at large wave numbers with $k \gtrsim \xi^{-1}$ is indistinguishable whether $T > T_\lambda$ or $T < T_\lambda$. Then, in 3D, they predicted

$$L_R \propto \lambda_R \propto c_{II}\xi \propto |T/T_\lambda - 1|^{-1/3}. \tag{6.6.41}$$

However, precise measurements of the thermal conductivity slightly above the transition ($0 < T/T_\lambda - 1 \lesssim 10^{-3}$) exhibited a steeper power law [96]–[98],

$$\lambda_R \cong \lambda^*(T/T_\lambda - 1)^{-x_\lambda}, \quad x_\lambda \cong 0.43, \tag{6.6.42}$$

with $k_B\lambda^* \cong 125$ erg/(s cm K), as shown in Fig. 6.21. The entropy relaxation rate in (6.6.31) behaves as $\Gamma_\lambda \cong 2 \times 10^{11}(T/T_\lambda - 1)^{2\nu - x_\lambda}/(C_p/n_\lambda)$. From (6.6.33) and (6.6.34) the renormalized kinetic coefficient L_R and the order parameter relaxation rate Γ_ξ are estimated as

$$\begin{aligned} L_R &\cong A_\psi(T/T_\lambda - 1)^{-\epsilon\nu + x_\lambda}, \\ \Gamma_\xi &\cong \Omega_\psi(T/T_\lambda - 1)^{\nu z}, \end{aligned} \tag{6.6.43}$$

where $z = d - 2 + x_\lambda/\nu$ ($\cong 1.65$ in 3D). From the sentence below (6.6.34) and the value of λ^* in (6.6.42) we have $A_\psi \cong 4 \times 10^{-5}$ cm^2 s^{-1} and $\Omega_\psi \cong 2 \times 10^{11}$ s^{-1} $\sim g_0$. Thus the ratio w between the two relaxation rates behaves as

$$w = \Gamma_\xi/\Gamma_\lambda \cong (C_p/n_\lambda)(T/T_\lambda - 1)^{2x_\lambda - \nu}, \tag{6.6.44}$$

which becomes considerably smaller than 1 as $T \to T_\lambda$ because $2x_\lambda - \nu \cong 0.2$, though $C_p/n_\lambda \sim \ln(T/T_\lambda - 1)$ is larger than 1. Explanation of this apparent breakdown of the original dynamic scaling in (6.6.41) was a challenge to specialists [104]–[108].

6.6.4 Dynamic renormalization group theory

The dynamic RG equations to first order in ϵ are known to be inadequate even qualitatively (in some aspects) in ^{4}He, while a second-order theory of model F [108] was claimed to explain well the thermal conductivity data [96, 97]. In this book, we will present a derivation of the dynamic RG equations only to first order in ϵ, because they are simple and indicate the general trend of the dynamical fluctuation effects. From a fundamental statistical–mechanical point of view, model F is of great interest. This is because the simultaneous presence of the reversible and dissipative nonlinear terms in the Langevin equation (6.6.4) gives rise to a large imaginary part, Im $L_R \sim$ Re L_R, in the renormalized complex kinetic coefficient L_R. We note that Im L_R corresponds to the anti-symmetric part of the kinetic coefficients discussed in Section 5.2.

RG equations at fixed density

We calculate the incremental contributions to the kinetic coefficients from the fluctuations in the shell region $\Lambda - \delta\Lambda < q < \Lambda$ as in the case of classical fluids. To this end we rewrite (6.6.4) for the Fourier component $\psi_{\boldsymbol{k}}(t)$ as

$$\frac{\partial}{\partial t}\psi_{\boldsymbol{k}}(t) = -\Omega_k \psi_{\boldsymbol{k}}(t) + X_{\boldsymbol{k}}(t) + \theta_{\boldsymbol{k}}(t), \tag{6.6.45}$$

where

$$\Omega_k = L_0/\chi_k \cong L_0(r + k^2) \tag{6.6.46}$$

is the linear relaxation rate dependent on Λ. Then $X_{\boldsymbol{k}}$ is the nonlinear part chosen such that $\langle X_{\boldsymbol{k}}\psi_{\boldsymbol{k}}^*\rangle = 0$. Its real space representation is of the form,

$$X = (ig_0 C_0{}^{-1} - 2L_0\gamma_0)m\psi + \cdots, \tag{6.6.47}$$

where only the leading nonlinear terms are written explicitly. Now we apply the correlation function formula (5.2.23):

$$\int_0^\infty dt e^{-i\omega t}\langle\psi_{\boldsymbol{k}}(t)\psi_{-\boldsymbol{k}}^*(0)\rangle = \frac{2\chi_k}{i\omega + \Omega_k} + \frac{1}{(i\omega + \Omega_k)^2}\phi(k, \omega), \tag{6.6.48}$$

where

$$\phi(k, \omega) = \int_0^\infty e^{-i\omega t}\langle X_{\boldsymbol{k}}(t)\bar{X}_{-\boldsymbol{k}}(0)\rangle. \tag{6.6.49}$$

Here the time-reversed variable of ψ is its complex conjugate $\bar{\psi} = \psi^*$, which originates from quantum mechanics. Replacement of ψ in $X(\boldsymbol{r})$ by ψ^* yields

$$\bar{X} = (ig_0 C_0{}^{-1} - 2L_0\gamma_0)m\psi^* + \cdots. \tag{6.6.50}$$

We then calculate the contribution $\delta\phi$ from the shell region in the low-frequency and small wave number limits ($k \to 0$ and $\omega \to 0$):

$$\delta\phi \cong 2(ig_0 C_0{}^{-1} - 2L_0\gamma_0)^2 K_d \Lambda^{d-3}\delta\Lambda C_0\chi_\Lambda/(\lambda_0 C_0^{-1} + L_0). \tag{6.6.51}$$

Because of the relation,

$$\frac{2\chi_k}{i\omega + \Omega_k} + \frac{1}{(i\omega + \Omega_k)^2}\delta\phi \cong \frac{2\chi_k}{i\omega + (L_0 - \delta\phi/2)/\chi_k}, \tag{6.6.52}$$

we find $\delta L_0 = -\delta\phi/2$ and

$$\frac{\partial}{\partial \ell}L_0 = K_d(g_0 + 2i\gamma_0 C_0 L_0)^2/[\Lambda^\epsilon(\lambda_0 + C_0 L_0)], \tag{6.6.53}$$

where $\Lambda = \Lambda_0 e^{-\ell}$. The above equation reduces to (5.3.53) in the purely dissipative case ($g_0 = 0$).

The renormalized thermal conductivity can be expressed in terms of the time-correlation function of the flux $\mathcal{J}_x \equiv \mathrm{Im}[\psi^*\partial\psi/\partial x] = (\hbar/mT)J_{sx}$ as

$$\lambda_R = g_0^2\int_0^\infty dt\int d\boldsymbol{r}\langle\mathcal{J}_x(\boldsymbol{r}, t)\mathcal{J}_x(\boldsymbol{0}, 0)\rangle. \tag{6.6.54}$$

Using the decoupling approximation in the shell region, we obtain

$$\frac{\partial}{\partial \ell}\lambda_0 = K_d g_0^2/[2\Lambda^\epsilon(\mathrm{Re}\, L_0)]. \tag{6.6.55}$$

The above RG equations have been solved in terms of the following dimensionless numbers,

$$f = K_d g_0^2/[\Lambda^\epsilon(\mathrm{Re}\, L_0)\lambda_0], \tag{6.6.56}$$

$$w = C_0 L_0/\lambda_0, \tag{6.6.57}$$

where f represents the strength of the mode coupling and w the ratio of the relaxation rates of ψ and δs. In agreement with the original dynamic scaling, they tend to fixed-point values [95],

$$f^* = \frac{6}{5}\epsilon + O(\epsilon^2), \tag{6.6.58}$$

$$w^* = 0.732 + 0.480i + O(\epsilon^2), \tag{6.6.59}$$

to first order in ϵ. Here we have used the results, $\alpha/\nu = \epsilon/5$ and $v^* = \epsilon/20$, in the calculation, although α is almost zero in real 3D helium (or when the higher-order expansion terms are included).

RG equation for L_0 at fixed pressure

As we have remarked near (6.6.10), the above calculation has been obtained by neglecting the density fluctuations. We here modify (6.6.53) for the fixed pressure case. Generally, the leading nonlinear term X is of the form,

$$X = \left(\frac{i}{\hbar} s\delta\hat{T} - 2L_0\gamma_0 m\right)\psi. \tag{6.6.60}$$

At fixed pressure we have $\delta\hat{T} \cong T_\lambda C_{p0}^{-1} n_\lambda \delta s$, where C_{p0} is defined by (6.6.10). From (2.4.6), (2.4.21), and (2.4.24) we also have $\delta m = C_0\delta\hat{T}/T_\lambda$. Thus,

$$X = (ig_0 - 2\gamma_0 C_0 L_0)C_{p0}^{-1}(n_\lambda \delta s)\psi, \tag{6.6.61}$$

and these relations lead to

$$\frac{\partial}{\partial \ell}L_0 = K_d(g_0 + 2i\gamma_0 C_0 L_0)^2/[\Lambda^\epsilon(\lambda_0 + C_{p0}L_0)], \tag{6.6.62}$$

where C_0 in the denominator of (6.6.53) is replaced by C_{p0}.

More analysis of models E and F

In the case $\gamma_0 = 0$, which gives us model E [95], C_0 is a constant, the dissipative coupling vanishes, and L_0 may be treated as a real number. Then the RG equations for f and w are simplified as

$$\frac{\partial}{\partial \ell} f = \epsilon f - \left(\frac{1}{w+1} + \frac{1}{2}\right) f^2, \tag{6.6.63}$$

$$\frac{\partial}{\partial \ell} w = f w \left(\frac{1}{w+1} - \frac{1}{2}\right), \tag{6.6.64}$$

to first order in ϵ from (6.6.53) and (6.6.55). We may integrate (6.6.64) in the form,

$$w(\ell)/[1 - w(\ell)]^2 = [w_0/(1 - w_0)^2] \exp\left[\frac{1}{2}\int_0^\ell d\ell' f(\ell')\right], \tag{6.6.65}$$

where w_0 is the initial value of w at $\ell = 0$. We readily find $w \to 1$ and $f \to \epsilon$ as $\ell \to \infty$. In ^{4}He the initial or background value of f, denoted by f_0, is of order 0.02 at $\Lambda = \xi_{+0}^{-1}$ or for $T/T_\lambda - 1 = 1$, where we use $k_B\lambda_0 \sim 10^3$ erg/s K (or $\lambda_0/n_\lambda \sim \hbar/m_4$) and $L_0 \sim \hbar/m_4 = 1.6 \times 10^{-4}$ cm^2/s. Due to this weak initial coupling, the critical growth of the kinetic coefficients occurs only close to the critical point, $T/T_\lambda - 1 < \tau_c$. Because $f \cong f_0 \exp(\epsilon\ell)$ before the crossover, τ_c is determined by $\tau_c^{-\epsilon\nu} f_0 = 1$, so that $\tau_c \sim 10^{-3}$ in agreement with the thermal conductivity data. Furthermore, the calculation of model E up to second order in ϵ yielded [95], [105]–[108]

$$f^* = \epsilon - 0.16\epsilon^2 + O(\epsilon^3), \tag{6.6.66}$$

$$w^* = 1 - 1.07\epsilon + O(\epsilon^2). \tag{6.6.67}$$

The correction to w^* is rather surprising, which suggests that w^* might vanish at $d = 3$, indicating breakdown of the original dynamic scaling. It is also indicated by model F analysis up to second order [108]. We should thus treat w as a small number very close to the λ point. This is in fact consistent with (6.6.44) obtained from the thermal conductivity data.

In summary, the following unique features give rise to the observed dynamic critical behavior of ^{4}He [105]–[108]: (i) The dynamic crossover occurs at small $\tau_c \sim 10^{-3}$ because of the weak initial coupling, (ii) $w(\ell)$ decreases from $w_0 \sim 0.5$ to a small fixed-point value ($\sim 10^{-2}$), and (iii) $f(\ell)$ increases from $f_0 \sim 0.02$ to a fixed-point value of order 1 for $T/T_\lambda - 1 \ll \tau_c$.

6.6.5 The frequency-dependent bulk viscosity

We will examine the frequency-dependent bulk viscosity $\zeta_R^*(\omega)$ near the λ point [109]–[119]. In Fig. 6.24 we show data of the normalized attenuation $\alpha_\lambda/\alpha_{\lambda c}$ of first sound above and below T_λ [117], where α_λ is the attenuation per wavelength and $\alpha_{\lambda c}$ is its high-frequency λ-point limit (for which see (6.6.80) below).

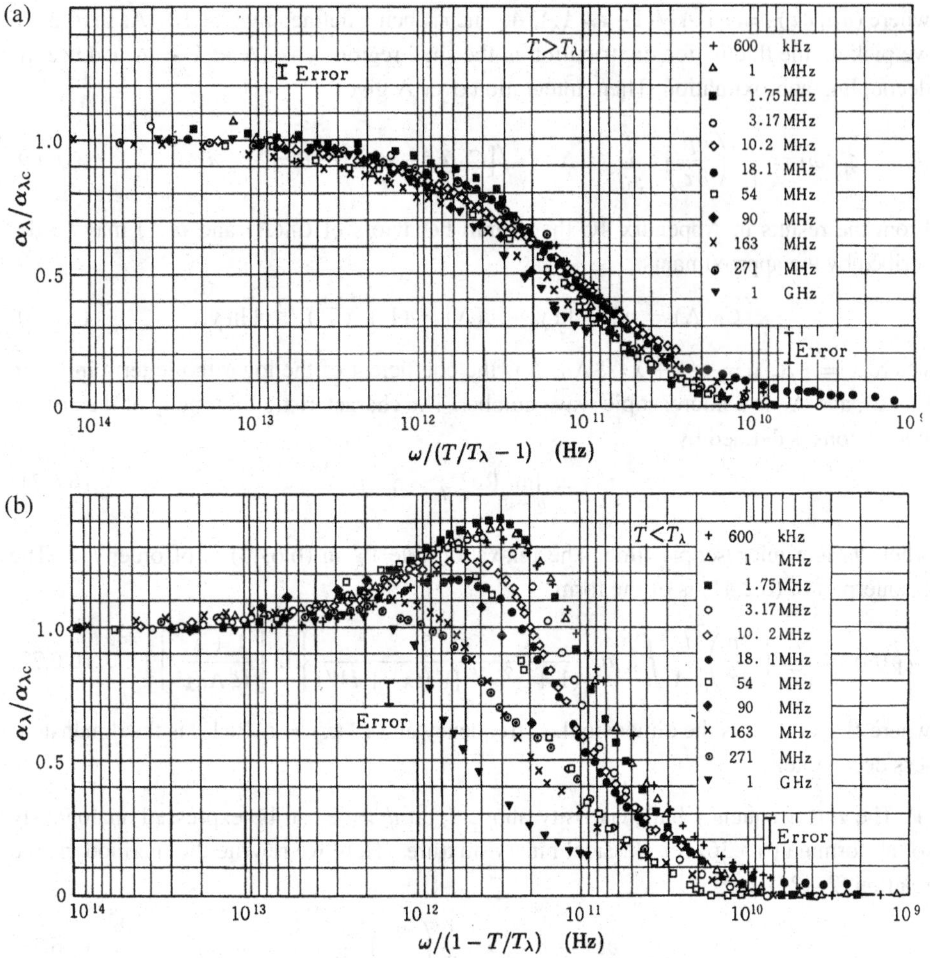

Fig. 6.24. The normalized attenuation $\alpha_\lambda/\alpha_{\lambda c}$ of first sound in ^{4}He near the λ point vs $\omega/|T/T_\lambda - 1|$ for various frequencies (a) above T_λ and (b) below T_λ [117]. The (original) dynamic scaling (with $z = 3/2$) roughly holds above T_λ. The maxima below T_λ arise from the Landau–Khalatnikov mechanism superimposed on the fluctuation mechanism.

The fluctuation mechanism above T_λ

The calculations of $\zeta_R^*(\omega)$ for $\tau = T/T_\lambda - 1 > 0$ can be performed analogously to those in classical fluids. From the expression for the fluctuating pressure $\delta\hat{p}$ in (4.2.14), we find the relevant nonlinear pressure,

$$\hat{p}_{\rm nl} = \left(\frac{\partial p}{\partial \tau}\right)_\zeta \gamma_0 |\psi|^2, \tag{6.6.68}$$

where $(\partial p/\partial\tau)_\zeta = nTA_\lambda/(1-\epsilon_{\text{in}}A_\lambda)$, A_λ and ϵ_{in} being defined by (2.4.10). As in (6.2.36) we pick up the fluctuation contribution in the shell region $\Lambda - \delta\Lambda < k < \Lambda$ and use the decoupling approximation. Then, integration over Λ gives

$$\zeta_{\text{R}}^*(\omega) = \frac{4}{T_{\text{c}}}\left(\frac{\partial p}{\partial\tau}\right)_\zeta^2 \int_0^\infty d\Lambda\Lambda^3 v/\left[C_0(\kappa^2+\Lambda^2)^2(2\,\text{Re}\,\Gamma_\Lambda + i\omega)\right]. \tag{6.6.69}$$

From the results in Appendix 4C the overall behaviors of $C_0(\Lambda)$ and $v(\Lambda)$ may be described by the approximants,

$$C_0(\Lambda) = \nu^{-1}A\ln y, \quad v(\Lambda) = (1+x^{-2})^{\epsilon/2}/4\ln y, \tag{6.6.70}$$

where $x = \xi\Lambda$, $y = \sqrt{1+x^2}/\xi\Lambda_0$, A is the coefficient of the logarithmic term in C_p in (2.4.2), and Λ_0 is a microscopic wave number. The characteristic lifetime t_ξ of the critical fluctuations is defined by

$$t_\xi^{-1} = \lim_{q\to 0}\text{Re}\,\Gamma_q = t_0^{-1}\tau^{\nu z}, \tag{6.6.71}$$

where t_0 is a microscopic time. The relaxation rate Γ_ξ in (6.6.43) is of order t_ξ^{-1}. The counterpart of (6.2.41) is of the form,

$$\zeta_{\text{R}}^*(\omega) = \frac{2}{\nu T_\lambda}\left(\frac{\partial p}{\partial\tau}\right)_\zeta^2 \frac{t_\xi}{A}\int_0^\infty dx \frac{x^{3-\epsilon}}{(1+x^2)^{d/2}[\Gamma^*(x)+iW]}\left\{\ln\left[\frac{1+x^2}{(\xi\Lambda_0)^2}\right]\right\}^{-2}, \tag{6.6.72}$$

where $W = \omega t_\xi/2$ is the dimensionless frequency and $\Gamma^*(x) = t_\xi\,\text{Re}\,\Gamma_q$ is the dimensionless decay rate.

(i) The zero-frequency bulk viscosity above T_λ at $d = 3$ can be expressed analogously to the formula (6.5.36) for classical binary mixtures. Here we rewrite the thermodynamic relation (2.4.21) as

$$\rho c^2 - \rho_\lambda c_\lambda^2 = \frac{1}{T_\lambda}\left(\frac{\partial p}{\partial\tau}\right)_\zeta^2 \frac{1}{C}, \tag{6.6.73}$$

where

$$\rho_\lambda c_\lambda^2 = T_\lambda n_\lambda^2(1-\epsilon_{\text{in}}A_\lambda)^{-2}Q_0^{-1}. \tag{6.6.74}$$

Then we obtain [93, 111],

$$\zeta_{\text{R}}^*(0) = \frac{1}{4}(\rho c^2 - \rho_\lambda c_\lambda^2)t_\xi/\ln(\tau_0/\tau). \tag{6.6.75}$$

If $\alpha > 0$, c_λ is the sound velocity at the λ point. We note that the ratio $(\rho c^2 - \rho_\lambda c_\lambda^2)/\rho_\lambda c_\lambda^2 = A_\lambda^2 Q_0/C$ is of order $0.1n_\lambda/C$ and is much smaller than 1 from (2.4.13).

(ii) At high frequencies $\omega t_\xi \gg 1$, it is convenient to introduce the frequency-dependent specific heat [112],

$$C^*(\omega) = -\frac{1}{\nu z}A\ln(i\omega t_0/2) + B - A_\lambda^2 Q_0 = -\frac{1}{\nu z}A\ln(i\omega/\omega_0), \tag{6.6.76}$$

where t_0 is defined by (6.6.72) and $A\ln(t_0\omega_0) \equiv \nu z(B - A_\lambda^2 Q_0)$. As in one-component fluids, we obtain

$$\begin{aligned}\zeta_R^*(\omega) &= (\rho c^2 - \rho_\lambda c_\lambda^2)C\left[\frac{1}{C^*(\omega)} - \frac{1}{C}\right]\frac{1}{i\omega} \\ &= (\rho c^2 - \rho_\lambda c_\lambda^2)\left[\nu z\frac{\ln(\tau_0/\tau)}{\ln(\omega_0/i\omega)} - 1\right]\frac{1}{i\omega}.\end{aligned} \tag{6.6.77}$$

The dispersion relation becomes independent of τ as

$$\rho\omega^2/k^2 = \rho_\lambda c_\lambda^2\left[1 + (A_\lambda^2 Q_0/A)\frac{\nu z}{\ln(\omega_0/i\omega)}\right]. \tag{6.6.78}$$

Here the second term in the brackets may be treated as a small perturbation because $A_\lambda^2 Q_0/A \sim 0.1$. Let us write the high-frequency limits of the attenuation per wavelength α_λ and the frequency-dependent sound velocity $c(\omega)$ as $\alpha_{\lambda c}$ and $c_c(\omega)$, respectively. Then, we have [112]

$$\alpha_{\lambda c} = \pi^2 \nu z(A_\lambda^2 Q_0/2A)\frac{1}{[\ln(\omega_0/\omega)]^2 + \pi^2/4}, \tag{6.6.79}$$

$$c_c(\omega)/c_\lambda = 1 + \nu z(A_\lambda^2 Q_0/2A)\frac{\ln(\omega_0/\omega)}{[\ln(\omega_0/\omega)]^2 + \pi^2/4}. \tag{6.6.80}$$

The above formulas are known to be in agreement with experiments [118]–[119].

The Landau–Khalatnikov mechanism below T_λ

Below T_λ we assume $M = \langle\psi_1\rangle > 0$ and $\langle\psi_2\rangle = 0$ where $\psi = \psi_1 + i\psi_2$. Then the pressure fluctuation $\hat{p}_{nl}$ in (6.6.68) contains a term *linear* in ψ_1, which leads to the Landau–Khalatnikov mechanism of sound attenuation [109, 110, 113]. As discussed in Section 4.3, the fluctuation distribution with wave numbers smaller than the inverse correlation length $\xi_T^{-1} \propto |\tau|^\nu$ is governed by the hydrodynamic hamiltonian $\mathcal{H}_{hyd}$ in (4.3.111). In the long-wavelength limit, the new pressure deviation is written as

$$(\hat{p}_{nl})_1 = 2\left(\frac{\partial p}{\partial \tau}\right)_\zeta \gamma_R M\varphi, \tag{6.6.81}$$

where φ is the deviation of $\psi_1 + \psi_2^2/2M$ as defined by (4.3.110), and $\gamma_R = \lim_{\Lambda\to 0}\gamma_0(\Lambda)$. Because φ and ψ_2^2 are orthogonal at small wave numbers, we also have

$$\lim_{q\to 0}\langle\varphi_q(t)\varphi_{-q}(0)\rangle = \lim_{q\to 0}\langle\psi_{1q}(t)\varphi_{-q}(0)\rangle = \chi_R\exp(-t/t_\xi), \tag{6.6.82}$$

where $\chi_R \propto |\tau|^{-\gamma}$ is the variance of φ appearing in $\mathcal{H}_{hyd}$ and t_ξ is defined by (6.6.71). Analogous to a classical internal relaxation mechanism [68], this order parameter relaxation gives rise to the following frequency-dependent bulk viscosity,

$$\begin{aligned}\zeta_{\mathrm{LH}}(\omega) &= \frac{4}{T_\lambda}\left(\frac{\partial p}{\partial \tau}\right)_\zeta^2 \gamma_{\mathrm{R}}^2 M^2 \chi_{\mathrm{R}} \frac{1}{i\omega + t_\xi^{-1}} \\ &\cong (\rho c^2 - \rho_\lambda c_\lambda^2)(1 - R_{\mathrm{v}})^{-2} R_{\mathrm{v}} \frac{1}{i\omega + t_\xi^{-1}},\end{aligned} \tag{6.6.83}$$

where R_{v} is the dimensionless number defined by (4.3.122) and is expected to be considerably smaller than 1. The resultant attenuation per wavelength is written as

$$(\alpha_\lambda)_{\mathrm{LH}} = \pi(A_\lambda^2 Q_0)(1 - R_{\mathrm{v}})^{-2} R_{\mathrm{v}} C^{-1} \frac{t_\xi \omega}{1 + t_\xi^2 \omega^2}. \tag{6.6.84}$$

Experimently, the attenuation below T_λ was suggested to consist of two contributions arising from (i) the fluctuation mechanism and (ii) the Landau–Khalatnikov mechanism, as can be seen in Fig. 6.24(b) [115]–[119]. The relaxation time t_ξ deduced from the data was consistent with the expectation, $t_\xi \sim \xi_{+0}|\tau|^{-\nu}/c_{II} \propto |\tau|^{-1}$ [110], which is the result of the original dynamic scaling theory [92].

6.6.6 ^{3}He–^{4}He mixtures

A detailed theory of the transport properties in superfluid mixtures was developed by Khalatnikov and Zharkov [120]. Some attempts have been made to extend the RG analysis to ^{3}He–^{4}He mixtures near the λ line and the tricritical point [121]–[123]. There, if the density or pressure deviation is neglected as in pure ^{4}He, the complex order parameter is coupled with the entropy and composition deviations in statics and dynamics. On the one hand, see (4.2.15) or (4.2.22) for the GLW hamiltonian of ^{3}He–^{4}He, where we showed that the linear combination $m_1' = \delta s + (\partial \Delta/\partial T)_{\lambda p}\delta X$ is decoupled from ψ in statics. On the other hand, in dynamics the entropy $S = ns$ per unit volume and the ^{3}He density $n_3 = nX$ are convected by the normal fluid velocity $\boldsymbol{v}_{\mathrm{n}}$ as (6.6.20) and $\partial n_3/\partial t = -\nabla \cdot (n_3 \boldsymbol{v}_{\mathrm{n}})$, respectively. We then find that the deviation,

$$c_2 = -X\delta s + s\delta X, \tag{6.6.85}$$

is not convected by $\boldsymbol{v}_{\mathrm{n}}$ and hence is decoupled from the order parameter dynamically (because $\boldsymbol{J}_{\mathrm{s}} \cong -\rho \boldsymbol{v}_{\mathrm{n}}$). As a result, c_2 relaxes diffusively with a nonsingular diffusion constant D_2 (even below T_λ).

Dilute case

In ^{3}He–^{4}He mixtures, the effective thermal conductivity λ_{eff} measured in a cell without ^{3}He flux is finite even on the λ line [120]. Slightly above the λ line, its inverse consists of two terms as [124]

$$1/\lambda_{\mathrm{eff}} = 1/\lambda_\lambda + 1/\lambda_{\mathrm{R}}, \tag{6.6.86}$$

where λ_λ is the thermal conductivity on (and slightly below) the λ line and λ_R is a singular part behaving in the same manner as the thermal conductivity in pure ^{4}He. For simplicity, we apply a small heat current in a superfluid state with small ^{3}He concentration X. In thermal counterflow with small heat flux Q, the ^{3}He concentration becomes larger at the cooler boundary, because ^{3}He molecules are convected by the normal fluid velocity $\boldsymbol{v}_n$. The steady concentration profile is determined by

$$X\boldsymbol{v}_n + D_{iso}\nabla X = \boldsymbol{0}, \tag{6.6.87}$$

where $D_{iso}(\sim 10^{-4}\ \mathrm{cm^2\ s^{-1}})$ is the diffusion constant of an isolated ^{3}He molecule in ^{4}He. Assuming that $\boldsymbol{v}_n$ is in the x direction, we obtain

$$X(x) = X(0)\exp(-v_n x/D_{iso}). \tag{6.6.88}$$

In the linear response regime we require $|v_n|h \ll D_{iso}$, where h is the cell thickness. In the superfluid phase the chemical potential μ_4 of ^{4}He is constant so that

$$s\nabla X + X\nabla\Delta = \boldsymbol{0}, \tag{6.6.89}$$

where the pressure variation is neglected and

$$\nabla\Delta \cong \left(\frac{\partial\Delta}{\partial X}\right)_{Tp}\nabla X \cong \frac{T}{X}\nabla X, \tag{6.6.90}$$

because $\Delta \cong T\ln X$ for small X. Thus,

$$\boldsymbol{v}_n \cong (sD_{iso}/XT)\nabla T. \tag{6.6.91}$$

The heat flux is equal to $Tns\boldsymbol{v}_n$, resulting in the effective thermal conductivity,

$$\lambda_{eff} = ns^2 D_{iso}/X. \tag{6.6.92}$$

This behavior was confirmed in experiments down to very small $X(< 10^{-3})$ [125], though there was disagreement in earlier measurements at such small X [126].

Crossover at $X \cong X_D$

It is interesting that m_1' and c_2 coincide when $s + X(\partial\Delta/\partial T)_{\lambda p} = 0$. This happens at an intermediate concentration X_D ($\cong 0.37$ at SVP) [122]. This means that, at $X = X_D$, $m_1' \propto c_2$ is decoupled from ψ both in statics and dynamics and the thermal fluctuations of ψ and m_2' in (4.2.20) obey the model F equations. In Section 8.10 we shall see that the Hall–Vinen mutual coefficients become divergent on the λ line at $X = 0$ and X_D. In heat-conduction problems, variations of m_1' should also be taken into account depending on the boundary condition. Let us consider a heat-conducting, steady superfluid state slightly below the λ line, neglecting gravity and assuming homogeneous pressure. The temperature gradient is given by $dT/dx = Q/\lambda_{eff}$, while the critical temperature gradient is

$$\frac{d}{dx}T_\lambda = \left(\frac{\partial T}{\partial\Delta}\right)_{\lambda p}\frac{d}{dx}\Delta = -\left(\frac{\partial T}{\partial\Delta}\right)_{\lambda p}\frac{s}{X}\frac{d}{dx}T. \tag{6.6.93}$$

The chemical potential μ_4 of ^{4}He is assumed to be homogeneous. Thus,

$$\frac{d}{dx}(T - T_\lambda) = \frac{Q}{\lambda_{\text{eff}}}\left[1 + \left(\frac{\partial T}{\partial \Delta}\right)_{\lambda p} \frac{s}{X}\right]. \tag{6.6.94}$$

If the x axis is along the temperature gradient, $d(T - T_\lambda)/dx > 0$ for $X < X_{\text{D}}$ and $d(T - T_\lambda)/dx < 0$ for $X > X_{\text{D}}$ near the λ line in the linear regime. At $X = X_{\text{D}}$, $T - T_\lambda$ is constant in the superfluid phase.

In transient cases, thermal relaxations can be very interesting. For example, if an equilibrium normal fluid state is cooled (warmed) through a boundary wall, a superfluid region will emerge and grow from that boundary for $X < X_{\text{D}}$ ($X > X_{\text{D}}$). Including gravity in the formulations will give rise to a number of intriguing, nonequilibrium effects not explored so far.

6.7 ^{4}He near the superfluid transition in heat flow

Nonlinear effects of heat flow near the superfluid transition represent one of the most dramatic heat-flow effects [127]–[137]. In addition to the well-known problem of vortex generation by heat flow, which will be discussed in Section 8.10, there is another interesting situation, in which the temperature is above T_λ at one end of the cell and below T_λ at the other end. The temperature in a superfluid should be nearly constant, whereas it has a finite gradient in a normal fluid. Then a HeI–HeII interface emerges separating the two phases, across which the temperature gradient is almost discontinuous. This interface is a very unique nonequilibrium object. It appears when ^{4}He in a normal fluid state is cooled from the boundary to below T_λ or when ^{4}He in a superfluid state is warmed from the boundary to above T_λ.

We will first clarify the condition of crossover from the linear- to nonlinear-response regime in heat flow on the basis of the scaling relations near the λ point [129]. Then we will illustrate two-phase coexistence of normal fluid and superfluid phases on the basis of numerical work. An example of self-organized states will also be given, in which the temperature gradient is equal to the transition temperature gradient in gravity.

6.7.1 Crossover between linear and nonlinear regimes

Normal fluids

On the basis of (6.6.42) we may discuss the crossover on the normal fluid side. It is convenient to introduce a characteristic reduced temperature $\bar{\tau}_Q$ and length $\bar{\xi}_Q$ by the heat conduction relation,

$$Q = (\lambda^* \bar{\tau}_Q^{-x_\lambda})(T_\lambda \bar{\tau}_Q / \bar{\xi}_Q), \tag{6.7.1}$$

where $\bar{\xi}_Q = \xi_{0+}\bar{\tau}_Q^{-\nu}$ with $\xi_{0+} = 1.4$ Å at SVP as determined below (2.4.4). Then,

$$\bar{\tau}_Q = (Q\xi_{0+}/T_\lambda\lambda^*)^{1/(1+\nu-x_\lambda)} \cong 0.48 \times 10^{-8} Q^{0.81}, \tag{6.7.2}$$

$$\bar{\xi}_Q = \xi_{0+}(Q\xi_{0+}/T_\lambda\lambda^*)^{-\nu/(1+\nu-x_\lambda)} \cong 4.9 \times 10^{-3} Q^{-0.54} \text{ cm}, \tag{6.7.3}$$

where Q is in erg/cm^2 s. The linear response to heat flow holds only for $\tau = T/T_\lambda - 1 \gg \bar{\tau}_Q$ or equivalently for $\xi \gg \bar{\xi}_Q$ in normal fluid states. In terms of $\bar{\tau}_Q$ the heat conduction equation is expressed as

$$\xi \frac{d}{dx}\tau = \tau(\bar{\tau}_Q/\tau)^{1+\nu-x_\lambda}, \tag{6.7.4}$$

which is integrated to give a temperature profile in the form,

$$\tau(x)^{1-x_\lambda} = \tau(0)^{1-x_\lambda} + (1-x_\lambda)(\bar{\tau}_Q^{1-x_\lambda}/\bar{\xi}_Q)x. \tag{6.7.5}$$

The origin $x = 0$ is taken appropriately inside the cell. The reduced temperature $\tau(0)$ at the origin is assumed to be much larger than $\bar{\tau}_Q$. For $Q > 0$ and in the warmer region $x > 0$, the system remains in the linear regime. However, in the cooler region $x < 0$, the reduced temperature can be decreased below $\bar{\tau}_Q$, where we will encounter a HeI–HeII interface.

Superfluids

In thermal counterflow, the complex order parameter ψ sinusoidally depends on x as $\exp(-ikx)$ where the wave number k is related to v_s as $v_s = \hbar k/m_4$. The heat flux Q is expressed as

$$Q = \rho s T_\lambda |v_n| \cong sT_\lambda \rho_s |v_s| = (\hbar s T_\lambda/m_4)\rho_s k, \tag{6.7.6}$$

where $\rho_s = \rho_s^*|\tau|^\nu$. Thus $k \propto Q|\tau|^{-\nu}$. As will be discussed in Section 8.10, nonlinear effects of heat flow become significant when k is increased to a value of order ξ^{-1}. We thus introduce a crossover correlation length and reduced temperature by setting

$$k = \xi_Q^{-1} = \xi_{+0}^{-1}\tau_Q^\nu, \quad \rho_s = \rho_s^*\tau_Q^\nu \tag{6.7.7}$$

in (6.7.6). At SVP we have

$$\tau_Q = (m_4\xi_{0+}/\hbar s T_\lambda \rho_s^*)^{1/2\nu} Q^{1/2\nu} \cong 0.45 \times 10^{-8} Q^{0.75}, \tag{6.7.8}$$

$$\xi_Q = (\hbar s T_\lambda \rho_s^* \xi_{0+}/m)^{1/2} Q^{-1/2} \cong 5.1 \times 10^{-3} Q^{-0.5} \text{ cm}, \tag{6.7.9}$$

with Q in erg/cm^2 s. We notice $\bar{\tau}_Q \cong \tau_Q$ comparing (6.7.2) and (6.7.8). In practice, these two reduced temperatures need not be distinguished. In superfluids the physical quantities are little affected by heat flow for $|\tau| \gg \tau_Q$, while superfluidity itself is broken for $|\tau| \lesssim \tau_Q$.

6.7.2 Renormalized mean field theory in the absence of gravity

The interface profile can be calculated approximately [129] or numerically [133] on the basis of model F for the complex order parameter ψ and the entropy deviation m. Taking a reference reduced temperature $\tilde{\tau}$, we assume that the thermal fluctuations with wave numbers smaller than the inverse of the correlation length $\tilde{\xi} = \xi_{+0}\tilde{\tau}^{-\nu}$ were coarse-grained at the starting point of the theory. The coefficients in the model are then renormalized ones proportional to some fractional power of $\tilde{\xi}$. We are interested in spatial variations varying slower than $\tilde{\xi}$. To set up the simplest theory, we first assume that $\tilde{\tau}$ is a constant independent of space. This treatment is allowable only when the amplitude of the reduced temperature stays of the order of $\tilde{\tau}$ throughout the system. In heat flow we may set $\tilde{\tau} = \tau_Q$, where τ_Q is defined by (6.7.8); then, the numerical results which follow are qualitatively valid in regions where the reduced temperature is of order τ_Q.

We make the equations dimensionless by appropriate scale changes [129]. That is, space and time are measured in units of $\tilde{\xi}$ and $\omega_{\tilde{\xi}}^{-1}$ where

$$\omega_{\tilde{\xi}} = g_0\tilde{\xi}^{-2}(u_0C_0)^{-1/2}. \tag{6.7.10}$$

We introduce a dimensionless order parameter Ψ, temperature deviation A, and entropy deviation M by

$$\Psi = (\tilde{\xi}u_0^{1/2})\psi, \quad A = \tilde{\xi}^2\tau, \quad M = (2\tilde{\xi}^2\gamma_0)m. \tag{6.7.11}$$

Then A is expressed in terms of M and Ψ as

$$A = M + \frac{1}{2}a^2|\Psi|^2, \tag{6.7.12}$$

where

$$a = 2\gamma_0(C_0/u_0)^{1/2}. \tag{6.7.13}$$

The parameter a is of order 1 as $T \to T_\lambda$ The relation (6.7.12) implies that M decreases with ordering at fixed A. The dynamic equations are written as

$$\frac{\partial}{\partial t}\Psi = ia^{-1}A\Psi - L\left[A - \nabla^2 + |\Psi|^2\right]\Psi, \tag{6.7.14}$$

$$\frac{\partial}{\partial t}M = a\nabla\cdot\boldsymbol{J}_s + \nabla\cdot\lambda\nabla A, \tag{6.7.15}$$

where L and λ are the dimensionless kinetic coefficients expected to be of order 1 and

$$\boldsymbol{J}_s = \mathrm{Im}[\Psi^*\nabla\Psi] \tag{6.7.16}$$

is the dimensionless superfluid current. The random source terms are omitted.

We are interested in steady solutions of the above equations, where we may set $\partial\Psi/\partial t = ia^{-1}A_0\Psi$ with A_0 being a constant. Then (6.7.14) becomes

$$\nabla^2\Psi = \left[A - \frac{i}{aL}(A - A_0) + |\Psi|^2\right]\Psi. \tag{6.7.17}$$

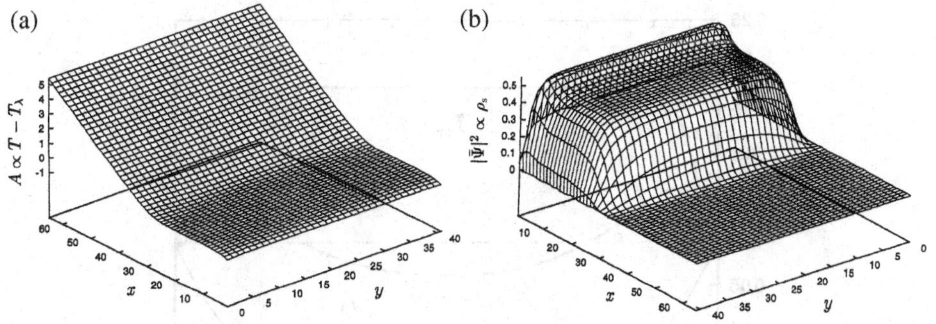

Fig. 6.25. Profiles of (a) temperature viewed from bottom and (b) superfluid density viewed from top (x being the vertical direction) in two-phase coexistence in ^{4}He in the absence of gravity. They are calculated as a steady solution of (6.7.17) and (6.7.19) in a 2D cell, $0 < x < 66$ and $0 < y < 42$. Space is measured in units of the correlation length ξ in the superfluid region.

Multiplying the above equation by Ψ^* and taking the imaginary part, we find

$$\mathrm{Im}[\Psi^* \nabla^2 \Psi] = \nabla \cdot \boldsymbol{J}_s = -\mathrm{Re}\left[\frac{1}{aL}\right](A - A_0)|\Psi|^2. \tag{6.7.18}$$

If λ is a constant, we also obtain

$$\nabla^2 A = \mathrm{Re}\left[\frac{1}{L\lambda}\right](A - A_0)|\Psi|^2. \tag{6.7.19}$$

In 1D we require the boundary conditions,

$$\begin{aligned} A &\to A_0 = -1, \quad \Psi \to (1 - K^2)^{1/2} e^{-iKx} \quad (\text{as} \quad x \to -\infty), \\ A &\to \infty, \quad\quad\quad\quad \Psi \to 0 \quad\quad\quad\quad\quad\quad\quad (\text{as} \quad x \to \infty). \end{aligned} \tag{6.7.20}$$

The coupled equations (6.7.17) and (6.7.19) are analogous to those for an interface in type-I superconductors in a magnetic field [138]. In the latter case, A is the vector potential and the right-hand side of (6.7.17) is replaced by $[-1 + A^2 + |\Psi|^2]\Psi$. The temperature difference $T - T_\lambda$, temperature gradient ∇T, and heat flow Q in the helium case correspond to the vector potential A, magnetic induction $B = \mathrm{rot}\, A$, and the externally applied magnetic field H in the superconductor case, respectively. As for the type-I superconductor case a possible analytic method is to introduce a GL parameter ($\propto [\mathrm{Re}(1/L\lambda)]^{-1/2}$) [138] and construct an approximate solution when it is small [129].

In Fig. 6.25 we show 2D steady profiles of two-phase coexistence numerically obtained at $a = L = \lambda = 1$ in the region $0 < x < h = 66$ and $0 < y < L_\perp = 42$ [133]. Here we are interested in the side-wall effect arising from the boundary condition $\Psi = 0$ at $y = 0$ and $L_\perp$. In Fig. 6.25(a) the normalized reduced temperature A, which is fixed at -1 at $x = 0$ and 5.59 at $x = h$, has an interface structure at $x \sim 30$. It turns out to be nearly

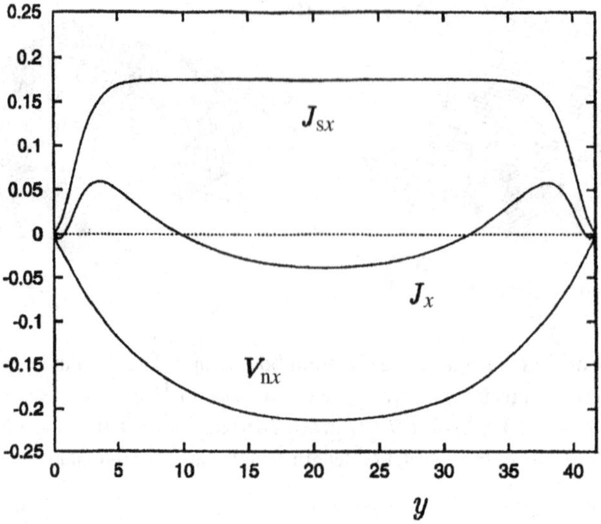

Fig. 6.26. The cross sectional curves of V_{nx}, J_{sx}, and $J_x = V_{nx} + J_{sx}$ as functions of y at $x = 21$ in the superfluid ^{4}He phase. Here $\int_0^{L_\perp} dy J_x(y) = 0$ due to mass conservation.

one-dimensional.[15] It also exhibits a drop at $x \sim 0$, corresponding to Kapitza resistance near T_λ. In Fig. 6.25(b) the scaled superfluid density $|\Psi|^2$ is displayed. It has a bump at $x \sim 5$ where conversion between a superfluid and normal fluid is taking place.

Next we calculate the scaled normal fluid velocity $\boldsymbol{V}_\mathrm{n}$ and the (total) mass current $\boldsymbol{J} = \boldsymbol{J}_\mathrm{s} + \boldsymbol{V}_\mathrm{n}$. For given $\boldsymbol{J}_\mathrm{s}$ they satisfy

$$\nabla \cdot \boldsymbol{J} = 0, \quad \nabla^2 \boldsymbol{V}_\mathrm{n} = \eta^{-1} \nabla p, \tag{6.7.21}$$

in the bulk region and vanish at the boundary walls, where η and p are the appropriately scaled viscosity and pressure, respectively. The scaled heat current is expressed as $\boldsymbol{Q} = a\boldsymbol{V}_\mathrm{n} - \lambda \nabla A$. In a superfluid region far from the top and bottom walls, $\boldsymbol{V}_\mathrm{n}$ is in the x direction and assumes a parabolic profile, $V_{nx} \propto y(L_\perp - y)$, as is well known. Figure 6.26 displays the cross sectional currents of V_{nx}, J_{sx}, and J_x at $x = 21$, where the system is in the superfluid state. Interestingly, J_x is negative in the center region $10 \lesssim y \lesssim 32$ and very close to the side walls within the distance of the correlation length ($\sim$ 1 here). The latter negative regions arise because J_{sx} and V_{nx} tend to zero quadratically and linearly, respectively, at the side walls as functions of the distance. The heat current strongly depends on the distance from the side walls (the y coordinate) in superfluids.

15 However, if a heat flow is applied in the horizontal direction in gravity, the resultant interface is curved [133].

6.7.3 Renormalized local equilibrium theory

In gravity we introduce the local reduced temperature,

$$\begin{aligned} \varepsilon &= [T - T_\lambda(p)]/T_\lambda(p) \\ &\cong (T/T_{\lambda\text{bot}} - 1) + G(x - h), \end{aligned} \tag{6.7.22}$$

where $T_{\lambda\text{bot}}$ is the λ temperature at the bottom wall ($x = h$). Here $G(\propto g)$ is defined by (2.4.33) and is of order 10^{-6} on earth, h is the cell height, and the x axis is taken downward with the origin being at the top. In heat flow and gravity, ε can be strongly inhomogeneous. We scale the reduced temperature by $\tilde{\tau} = 2.5 \times 10^{-8}$ and measure space in units of the corresponding correlation length 1.6×10^{-3} cm. Then the dimensionless gravity coefficient G (in the same notation as before) becomes $G\xi_{+0}/\tilde{\tau} = 0.04$ on earth. We propose the dynamic equations [134],

$$\frac{\partial}{\partial t}\Psi = ia^{-1}A\Psi - L\left[\varepsilon\xi^{-1/2} - \nabla^2 + \xi^{-1}|\Psi|^2\right]\Psi, \tag{6.7.23}$$

$$\frac{\partial}{\partial t}M = a\,\text{Im}[\Psi^*\nabla^2\Psi] + \nabla\cdot\lambda\nabla A, \tag{6.7.24}$$

where $A = (T - T_{\lambda\text{bot}})/T_\lambda\tilde{\tau}$ and

$$\varepsilon = A + G(x - h). \tag{6.7.25}$$

The scaled entropy deviation is expressed as

$$M = A - \frac{1}{2}a^2\xi^{-1/2}|\Psi|^2. \tag{6.7.26}$$

We define the local correlation length as

$$\xi = \ell_g \tanh(1/\ell_g|\varepsilon|^{2/3}). \tag{6.7.27}$$

The coefficients in (6.7.23) and (6.7.24) are obtained by appropriate scaling of those of model F renormalized at the local correlation length ξ. In gravity, ξ should not exceed the characteristic length $\ell_g = G^{-2/5}$(= 3.62 on earth) introduced in (2.4.36). The scaled kinetic coefficients are taken as

$$\lambda = b_\lambda\xi^{0.675}, \quad L = b_\psi\xi^{0.325}, \tag{6.7.28}$$

where b_λ and b_ψ are of order 1. The ratio $w = L/\lambda$ is considerably smaller than 1 in magnitude as $T \to T_\lambda$, as discussed below (6.6.67). Numerical results of the above model will be presented in Fig. 6.28 and discussed in Section 8.10.

6.7.4 Interface boundary condition and gravity effect

It is our main result that the reduced temperature $\tau_\infty = 1 - T_\infty/T_\lambda$ on the superfluid side is uniquely determined by the heat flow through the interface in the absence of gravity.[16]

[16] This is analogous to the equilibrium relation $T_c - T \propto H$ in type-I superconductors in two-phase coexistence [138].

It is obviously of order τ_Q in (6.7.8) or

$$\tau_\infty = R_\infty \tau_Q = A_\infty Q^{1/2\nu}. \tag{6.7.29}$$

The ratio R_∞ tends to a universal number as long as $L\lambda \sim \xi$ (insensitive to the correction to the dynamic scaling law) [130, 132]. It is related to the dimensionless wave number $K = k\xi_\infty$ with $\xi_\infty = \xi_{+0}\tau_\infty^{-\nu}$. We assume that ρ_s decreases with increasing K as $\rho_s = \rho_s^* \tau_\infty^\nu (1 - K^2)$ as in the mean field theory. Then (6.7.8) yields

$$K(1 - K^2) = (\tau_Q/\tau_\infty)^{2\nu} = R_\infty^{-2\nu}. \tag{6.7.30}$$

We require the condition $K < K_c = 1/\sqrt{3}$ for the linear stability of superfluidity.[17] Then $R_\infty \gtrsim 2$ is needed. Rough theoretical estimates [130] and numerical analysis suggest that K in two-phase coexistence is only slightly smaller than $1/\sqrt{3}$. Then $R_\infty \sim 2$ and $A_\infty \sim 10^{-8}$ with Q in cgs units in (6.7.29). In early experiments, Bhagat *et al.* observed a kink-like change of the temperature gradient at large $Q \gtrsim 10^4$ (cgs) [128]. Subsequently, Duncan *et al.* obtained $\tau_\infty \cong 10^{-8} Q^{0.81}$ for much smaller Q in the range $5 < Q < 300$ (cgs) in agreement with (6.7.29) [131].

Earlier, in (2.4.30)–(2.4.37) and in Fig. 2.17, we discussed two-phase coexistence in gravity in equilibrium. To examine competition between gravity and heat flow in the interface region, we should compare τ_Q in (6.7.8) and τ_g in (2.4.36). They are of the same order for

$$Q \sim (g/g_{\rm earth})^{2\nu/(1+\nu)} \quad ({\rm erg/cm^2\ s}), \tag{6.7.31}$$

where $g_{\rm earth}$ is the gravitational acceleration on earth. If Q is much larger than the right-hand side, gravity is negligible in the vicinity of the interface. Of course, gravity can be important on macroscopic scales (outside the interface region) even for much larger Q.

6.7.5 Balance of gravity and heat flow in normal fluid states

Intriguing nonequilibrium states are realized in the presence of both gravity and heat flow, particularly when ^{4}He is heated from above [130, 134, 136]. Hereafter we discuss one of such examples. Other examples will be presented in Section 8.10.

In a normal fluid the heat conduction equation becomes $\lambda dT/dx = -Q$ in a steady state in terms of the growing thermal conductivity λ ($= \lambda_R$). With the aid of (6.6.42) this equation is rewritten in terms of ε in (6.7.22) as

$$\frac{d}{dx}\varepsilon = G - (Q/\lambda^* T_{\lambda 0})\varepsilon^{x_\lambda}. \tag{6.7.32}$$

We notice that ε tends to a fixed-point value [130],

$$\varepsilon_c = (\lambda^* T_\lambda G/Q)^{1/x_\lambda}, \tag{6.7.33}$$

[17] Note that $K_c = 1/\sqrt{3}$ is the mean field result. More discussions on K_c can be seen below (8.10.58) and near (9.7.7).

with increasing x (in the downward direction) as $\varepsilon(x) - \varepsilon_c \propto \exp(-x/\ell_c)$, where

$$\ell_c = \varepsilon_c/(x_\lambda G). \tag{6.7.34}$$

In this case the temperature gradient due to heat flow and the critical temperature gradient due to gravity balance one another, i.e.,

$$\frac{d}{dx}T \cong \frac{d}{dx}T_\lambda(p). \tag{6.7.35}$$

The thermal conductivity spontaneously saturates into

$$\lambda \cong Q/T_\lambda G. \tag{6.7.36}$$

On earth we have

$$\varepsilon_c \cong 2\times 10^{-9}Q^{-2.2}, \quad \lambda \cong 10^6 Q, \quad \ell_c \cong 4\times 10^{-3}Q^{-2.2} \quad \text{(cgs)}. \tag{6.7.37}$$

The above results apparently suggest that ε_c can be made arbitrarily small with increasing Q in gravity, but this is not the case [134]. To show this, let us consider the steady-state correlation function $G(\boldsymbol{r}-\boldsymbol{r}') = \langle\psi(\boldsymbol{r},t)\psi^*(\boldsymbol{r}',t)\rangle$ in the mean field theory under the balance (6.7.35). Treating the kinetic coefficient L_0 as a real quantity for simplicity, we obtain

$$\left[ig_0 G(x) + 2L_0(r_0 - \nabla^2)\right]G(\boldsymbol{r}) = 2L_0\delta(\boldsymbol{r}), \tag{6.7.38}$$

where g_0 is defined by (6.6.18) and r_0 is the temperature coefficient. The Fourier transformation of $G(\boldsymbol{r})$ is expressed as[18]

$$G_k = 2L_0\int_0^\infty dt\,\exp\left[-2L_0 t\left(r_0 + k^2 + g_0 G k_x t + \frac{1}{3}g_0^2 G^2 t^2\right)\right]. \tag{6.7.39}$$

This indicates that the upper bound ξ_M of the correlation length in the x direction is determined by $g_0 G\xi_M = L_0\xi_M^{-2}$. In fact $G_k \to \xi_M^{-2}$ as $k \to 0$ and $r_0 \to 0$. Replacing L_0 by the renormalized coefficient $L_R \sim K_d g_0^2 \xi_M/\lambda$ at the cut-off ξ_M^{-1}, we obtain

$$\xi_M = \xi_{+0}(K_d g_0/\xi_{+0}^2\lambda^* G)^{\nu/(2\nu+x_\lambda)}. \tag{6.7.40}$$

The corresponding characteristic reduced temperature reads

$$\tau_M = (\xi_{+0}/\xi_M)^{1/\nu} = (\xi_{+0}^2\lambda^* G/K_d g_0)^{1/(2\nu+x_\lambda)}, \tag{6.7.41}$$

which is estimated as $\tau_M \cong 10^{-8}(g/g_{\text{earth}})^{0.56}$. The thermal conductivity arises from the steady-state average $\langle J_{sx}\rangle \propto \int_k k_x G_k$ [132, 137]. This yields scaling behavior,

$$\lambda = \lambda^*\varepsilon^{-x_\lambda} f_{\text{so}}(\varepsilon/\tau_M). \tag{6.7.42}$$

[18] The same expression can be obtained for superconductors in a dc electric field E (if G is replaced by E), where the fluctuation contribution to the electrical conductivity is suppressed by the electric field [139]. This effect is relevant for superconducting wires and films.

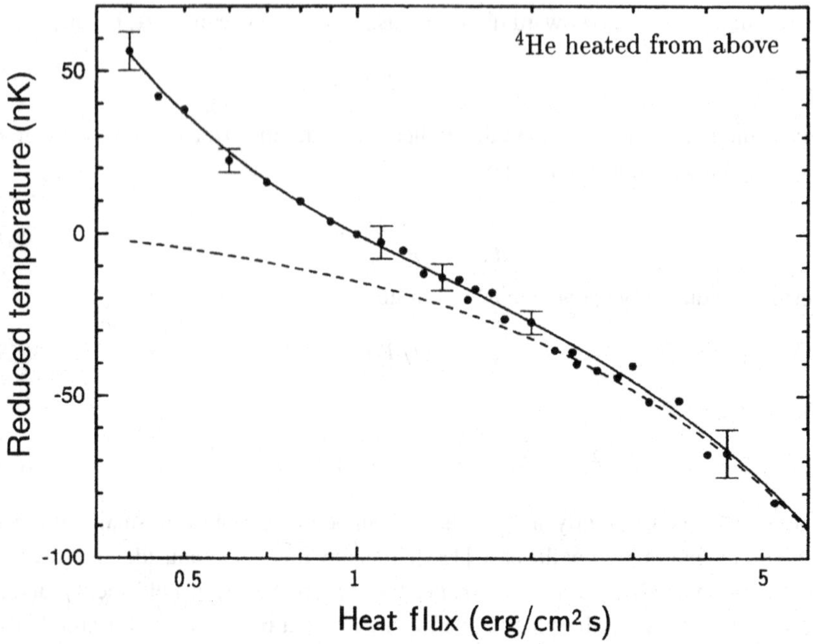

Fig. 6.27. The temperature difference $\Delta T = T - T_\lambda(p)$ in self-organized region in ^{4}He [136]. For $Q \lesssim 1$ erg/cm^2 s, the self-organized region was in a normal fluid state with $\Delta T > 0$. For larger Q, it was in a superfluid state with $\Delta T < 0$, where high-density vortices should have been produced (see Section 8.10). The dashed line is a fit to $-CQ^y$ where $y = 0.813$ and C is a constant. The solid line represents $T_\lambda \varepsilon_c - CQ^y$ with ε_c being defined by (6.7.33).

The scaling function $f_{so}(z)$ for self-organized behavior tends to 1 for $z \gg 1$ and const.z^{x_λ} for $|z| \ll 1$. Because λ cannot exceed $\lambda^* \tau_M^{-x_\lambda}$, the balance (6.7.35) can be achieved only for $\varepsilon_c \gtrsim \tau_M$ or $Q \lesssim 1$ erg/cm^2 s on earth. Thus τ_M gives the order of magnitude of the minimum reduced temperature attainable in self-organized normal fluid states.

In their experiment in the range $0.4 < Q < 65$ erg/cm^2 s, Moeur *et al.* [136] observed a self-organized region below the superfluid region. Figure 6.27 displays the measured reduced temperature ε in the self-organized region. The data can be fitted to (6.7.33) for $Q \lesssim 1$ erg/cm^2 s, but it is more surprising that the reduced temperature was negative for larger Q. In Fig. 6.28 we show numerically calculated profiles of $T/T_{\lambda\text{bot}} - 1$, ε, and ρ_s in their geometry, where we set $a = 1$, $b_\lambda = 1$, and $b_\psi = 0.2$ in (6.7.22)–(6.7.27). The case of larger heat flux will be discussed in Section 8.10.

Self-organized criticality?

We have shown that ^{4}He can spontaneously approach a homogeneous steady state, which is extremely close to the λ point, under gravity and heat flow in the same direction. Therefore, such a state has been called a *self-organized critical state* [135, 136, 140, 141].

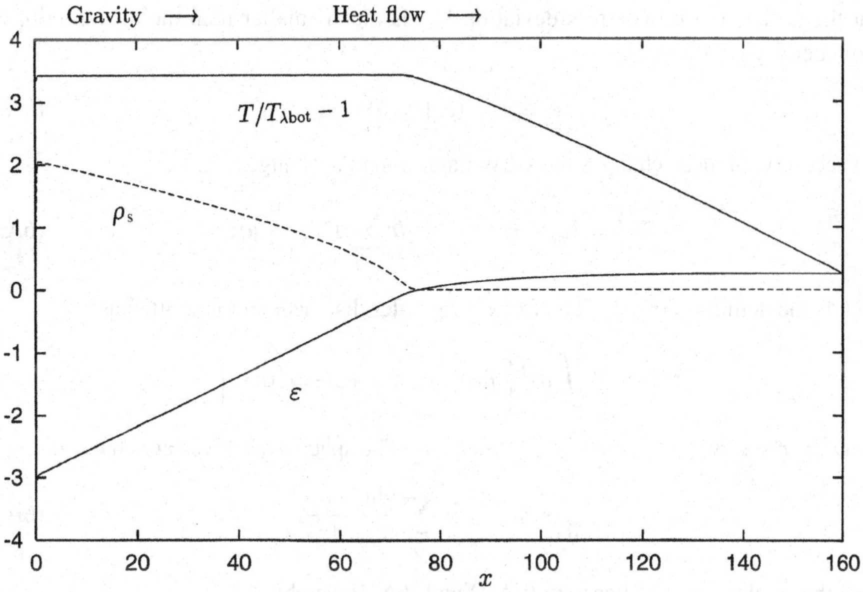

Fig. 6.28. The reduced temperatures $T/T_{\lambda\mathrm{bot}} - 1$ and ε (in units of 2.5×10^{-8}) (solid lines) and the superfluid density ρ_s (broken line) in a steady state, in ^{4}He. The lower part ($x \gtrsim 85$) is a self-organized normal fluid and the upper part is a superfluid. The curves are calculated from (6.7.22)–(6.7.27) with $Q = 0.77$ erg/cm^2 s applied from above under the earth's gravity. Space is scaled in units of 1.6×10^{-3} cm.

However, criticality is not reached in ^{4}He in this geometry owing to the lower bound ε_M in the normal fluid state. Therefore, it is simply called a *self-organized state* in this book.

Appendix 6A Derivation of the reversible stress tensor

We derive the reversible part of the stress tensor $\overleftrightarrow{\Pi} = \{\Pi_{ij}\}$ arising from the fluctuations of the scalar gross variables for near-critical binary fluid mixtures [36]. To this end we may neglect dissipation for simplicity. The reversible stress tensor is then equal to $p\overleftrightarrow{I} + \overleftrightarrow{\Pi} + \rho\boldsymbol{v}\boldsymbol{v} \cong p_c\overleftrightarrow{I} + \overleftrightarrow{\Pi}$. Adopting the Lagrange picture of fluid motion, we consider a small fluid element at position $\boldsymbol{r}$ and at time t. Due to the velocity field the element is displaced to a new position, $\boldsymbol{r}' = \boldsymbol{r} + \boldsymbol{u}$ with $\boldsymbol{u} = \boldsymbol{v}\delta t$, after a small time interval δt. From the continuity equations without diffusion, the mass densities $\rho_K = m_{0K} n_K$ ($K = 1, 2$) are changed to ρ_K' as

$$\rho_K' \cong \rho_K(1 - \nabla \cdot \boldsymbol{u}). \tag{6A.1}$$

Near the critical point the stress deviation Π_{ij} is much smaller than the deviation of the energy density δe. Therefore,

$$e' \cong e - (e + p_c)\nabla \cdot \boldsymbol{u}. \tag{6A.2}$$

In accord with these changes the GLW hamiltonian is changed as

$$\delta\mathcal{H} = \mathcal{H}' - \mathcal{H} = -\int d\boldsymbol{r} \sum_{i,j} \Pi_{ij} \frac{\partial}{\partial x_j} u_i, \tag{6A.3}$$

which is the definition of Π_{ij}. The free energy after displacement is written as

$$\mathcal{H}' = T_c \int d\boldsymbol{r}' \left[f(\rho_1', \rho_2', e') + \frac{K}{2}|\nabla'\psi'|^2 \right]. \tag{6A.4}$$

From $\boldsymbol{r}' = \boldsymbol{r} + \boldsymbol{u}$ we obtain $d\boldsymbol{r}' = d\boldsymbol{r}(1 + \nabla \cdot \boldsymbol{u})$. The space derivatives are changed as

$$\frac{\partial}{\partial x_i'} \cong \frac{\partial}{\partial x_i} - \sum_j \frac{\partial u_j}{\partial x_i} \frac{\partial}{\partial x_j}. \tag{6A.5}$$

Using these relations together with (6A.1) and (6A.2) we obtain

$$\Pi_{ij} = \left[\rho_1 \frac{\delta\mathcal{H}}{\delta\rho_1} + \rho_2 \frac{\delta\mathcal{H}}{\delta\rho_2} + (e + p_c)\frac{\delta\mathcal{H}}{\delta e} - T_c \left(f + \frac{K}{2}|\nabla\psi|^2 \right) \right] \delta_{ij} + T_c K \frac{\partial\psi}{\partial x_i}\frac{\partial\psi}{\partial x_j}, \tag{6A.6}$$

where $\mathcal{H}$ is regarded as a functional of ρ_K and e in the functional derivatives. From this expression we can confirm that the deviation Π_{ij} is very small and (6A.2) is surely a good approximation. Furthermore, under the linear relations (2.3.9)–(2.3.11) we notice the identity,

$$(\rho_1 - \rho_{1c})\frac{\delta}{\delta\rho_1} + (\rho_2 - \rho_{2c})\frac{\delta}{\delta\rho_2} + (e - e_c)\frac{\delta}{\delta e} = \psi\frac{\delta}{\delta\psi} + m\frac{\delta}{\delta m} + q\frac{\delta}{\delta q}. \tag{6A.7}$$

Thus the diagonal part of the stress tensor consists of the background p_c, $\delta\hat{p}$ defined by (4.2.8), and

$$\delta\tilde{p} = \left(\psi\frac{\delta}{\delta\psi} + m\frac{\delta}{\delta m} + q\frac{\delta}{\delta q} \right)\mathcal{H} - T_c\left(f + \frac{K}{2}|\nabla\psi|^2 \right). \tag{6A.8}$$

where $\mathcal{H}$ is regarded as a functional of ψ, m, and q. Here we set $q = 0$ in one-component fluids to obtain (6.1.17) and (6.2.5).

Appendix 6B Calculation in the mode coupling theory

We calculate (6.1.20) to reproduce Kawasaki's function $K_0(x)$ [2]. In the Fourier space the time-correlation function of the transverse velocity is written as

$$\langle v_{i\boldsymbol{q}}(t) v_{j\boldsymbol{q}'}(0) \rangle = \frac{T}{\rho}(\delta_{ij} - \hat{q}_i\hat{q}_j)(2\pi)^d \delta(\boldsymbol{q} + \boldsymbol{q}') \exp[-(\eta_R/\rho)q^2|t|], \tag{6B.1}$$

where $i, j = x, y, z$ and $\hat{\boldsymbol{q}} = q^{-1}\boldsymbol{q}$ is the direction of $\boldsymbol{q}$. Then (6.1.20) becomes

$$L_{\mathrm{R}}(k) = \frac{T\chi}{\eta_{\mathrm{R}}} \int_{\boldsymbol{q}} \frac{|\boldsymbol{q} \times \hat{\boldsymbol{k}}|^2}{|\boldsymbol{q} - \boldsymbol{k}|^4} \frac{1}{1 + \xi^2 q^2}, \tag{6B.2}$$

where $\hat{\boldsymbol{k}} = k^{-1}\boldsymbol{k}$ and $\boldsymbol{q} \times \hat{\boldsymbol{k}}$ is the vector product. Notice that the wave vector supported by the velocity field is taken to be $\boldsymbol{k} - \boldsymbol{q}$. The above integral is logarithmically divergent at $d = 4$ and is convergent for $d < 4$ at large q. Therefore, the critical dimensionality remains 4 in our dynamic problem. We perform the integration at $d = 3$. The first factor in the integrand depends on $\hat{\boldsymbol{q}}$ and its solid angle integration is performed to give

$$\int d\Omega \frac{|\boldsymbol{q} \times \hat{\boldsymbol{k}}|^2}{|\boldsymbol{q} - \boldsymbol{k}|^4} = \frac{\pi}{k^2}\left[\frac{q^2 + k^2}{2kq} \ln\left(\frac{q + k}{q - k}\right)^2 - 2\right]. \tag{6B.3}$$

By setting $z = \xi q$ and $x = \xi k$, we obtain

$$K_0(x) = \frac{3}{8\pi}(1 + x^2) \int_{-\infty}^{\infty} dz \frac{z^2}{1 + z^2}\left[\frac{x^2 + z^2}{2xz} \ln\left(\frac{z + x}{z - x}\right)^2 - 2\right]. \tag{6B.4}$$

Because the integrand goes to zero as z^{-3} at large $|z|$, we may perform the above integration by analytic continuation of the integrand to the upper complex z plane ($\mathrm{Im}\, z > 0$). We only pick up a contribution from the single pole $z = i$ using $\ln[(i + x)/(i - x)] = -2i \tan^{-1}(x)$ to obtain (6.1.22).

Appendix 6C Steady-state distribution in heat flow

In model C, where the mode coupling terms are absent, (5.3.16) shows that the steady-state distribution P_{ss} in heat flow is given by the local equilibrium distribution $P_{\mathrm{local}} \propto \exp(-\beta\mathcal{H}_{\mathrm{local}})$. In one-component fluids near the gas–liquid critical point, the Langevin equations are given by (6.2.2)–(6.2.4) with the first terms being the mode coupling terms. Then the steady distribution deviates from P_{local}. Here we calculate the deviation $\delta P_{\mathrm{ss}} = P_{\mathrm{ss}} - P_{\mathrm{local}}$, linear with respect to the temperature gradient $\boldsymbol{a}$ in (6.2.13). The second line of (1.2.47) gives

$$\mathcal{H}_{\mathrm{local}} = \mathcal{H} - \int d\boldsymbol{r} (\boldsymbol{a} \cdot \boldsymbol{r}) n_{\mathrm{c}} \delta s(\boldsymbol{r}). \tag{6C.1}$$

Using the definition of δs in (6.2.9) we find

$$\begin{aligned} \mathcal{L}_{\mathrm{FP}} P_{\mathrm{local}} &= \int d\boldsymbol{r} \beta^2 (\boldsymbol{a} \cdot \boldsymbol{r}) \left[-H_{\mathrm{c}} m_0^{-1} \nabla \cdot (\rho \boldsymbol{v}) + \nabla \cdot (e\boldsymbol{v}) + p_{\mathrm{c}} \nabla \cdot \boldsymbol{v}\right] P_{\mathrm{local}} \\ &= -\int d\boldsymbol{r} \beta (\boldsymbol{a} \cdot \boldsymbol{v}) n_{\mathrm{c}} \delta s \, P_{\mathrm{local}}. \end{aligned} \tag{6C.2}$$

Because $\mathcal{L}_{\rm FP} P_{\rm ss} = \mathcal{L}_{\rm FP}\delta P_{\rm ss} + \mathcal{L}_{\rm FP} P_{\rm local} = 0$, we obtain the deviation linear in $\boldsymbol{a}$,

$$\begin{aligned} \delta P_{\rm ss} &= \mathcal{L}_{\rm FP}^{-1}\int d\boldsymbol{r}\alpha_s\beta(\boldsymbol{a}\cdot\boldsymbol{v})\psi P_{\rm eq} \\ &= -\alpha_s\beta\int_0^\infty dt\int d\boldsymbol{r}e^{\mathcal{L}_{\rm FP}t}(\boldsymbol{a}\cdot\boldsymbol{v})\psi P_{\rm eq}, \end{aligned} \tag{6C.3}$$

where $n_c\delta s$ is replaced by $\alpha_s\psi$. The linear response of any dynamic variable $\mathcal{B}(\boldsymbol{r})$ to $\boldsymbol{a}$ in the steady state can then be written as

$$\langle\mathcal{B}(\boldsymbol{r})\rangle_{\rm ss} \cong \langle\mathcal{B}(\boldsymbol{r})\rangle_\ell - \frac{\alpha_s}{T}\int_0^\infty dt\int d\boldsymbol{r}'\langle\mathcal{B}(\boldsymbol{r},t)\psi(\boldsymbol{r}',0)\boldsymbol{v}(\boldsymbol{r}',0)\rangle\cdot\boldsymbol{a}, \tag{6C.4}$$

where the first term is the local equilibrium average and the second term is the time correlation in equilibrium defined by (5.2.18). This expression is consistent with the general linear response formula (5.4.20). Furthermore, the transverse part of $\boldsymbol{v}$ relaxes rapidly compared with ψ, so that in the Fourier space we may approximate $\delta P_{\rm ss}$ as

$$\delta P_{\rm ss} = -\frac{\alpha_s}{T}\int_q \frac{\rho}{\eta_R q^2}[\boldsymbol{a}\cdot\boldsymbol{v}_q - (\boldsymbol{a}\cdot\hat{\boldsymbol{q}})(\hat{\boldsymbol{q}}\cdot\boldsymbol{v}_q)]\psi_{-q}P_{\rm eq}, \tag{6C.5}$$

where $\hat{\boldsymbol{q}} = q^{-1}\boldsymbol{q}$. We may calculate the velocity field $\boldsymbol{v}_{\rm ind}$ induced by the fluctuations of ψ by taking the conditional average of $\boldsymbol{v}$ over $P_{\rm ss}$ with ψ held fixed. Then (6.2.16) can be obtained.

Appendix 6D Calculation of the piston effect

Here we calculate the temperature profile in the time region $t \ll t_D$ in the 1D geometry ($0 < x < L$) [50, 51]. From (6.3.15) the Laplace transformation $\mathcal{T}(x,\Omega) = \int_0^\infty dte^{-\Omega t}T_1(x,t)$ satisfies

$$\Omega[\mathcal{T} - (1-\gamma_s^{-1})\bar{\mathcal{T}}] = D\nabla^2\mathcal{T} \tag{6D.1}$$

where $\bar{\mathcal{T}} = \int_0^L dx\mathcal{T}(x,\Omega)/L$ is the space average and we have assumed $T_1(x,t) = 0$ for $t \le 0$. This equation is solved in the form,

$$\mathcal{T} = A\exp(-\kappa_D x) + B\exp[\kappa_D(x-L)] + z(A+B), \tag{6D.2}$$

where $\kappa_D = (\Omega/D)^{1/2}$ and

$$z = \frac{\gamma_s - 1}{\kappa_D L}(1-e^{-\kappa_D L}) = \frac{1}{2}(\Omega t_1)^{-1/2}(1-e^{-\kappa_D L}), \tag{6D.3}$$

where t_1 is defined by (6.3.7). If we are interested in the case $(Dt)^{1/2} \ll L$ or $t \ll t_D = L^2/D$, we may assume $\kappa_D \gg L^{-1}$. Then the first and second terms in (6D.2) are localized near $x = 0$ and L, respectively, the third term is homogeneous, and the factor $\exp(-\kappa_D L)$ in z in (6D.3) may be neglected. We confirm (6D.1) by substituting (6D.2) using $\bar{\mathcal{T}} = (\gamma_s/\kappa_D L)(A+B)$.

(i) In the first example, we have $T_1(0, t) = T_1(L, t) = T_{1b}$ for $t > 0$, so that

$$A = B = (1 + 2z)^{-1} T_{1b}/\Omega. \tag{6D.4}$$

Because the Laplace transformation of the interior temperature deviation is $z(A + B) = 2T_{1b}z/\Omega(1 + 2z) = T_{ib}[1 - 1/(1 + 2z)]/\Omega$, we obtain (6.3.9) with

$$\int_0^\infty dse^{-us} F_a(s) = (u + \sqrt{u})^{-1}. \tag{6D.5}$$

(ii) If the heat flux at the top is a constant Q for $t > 0$ and the bottom temperature is unchanged, we have

$$\lambda\kappa_D B = Q/\Omega, \quad A + z(A + B) = 0. \tag{6D.6}$$

Then the Laplace transformations of $T_{1\text{in}}(t)$, $T_{1\text{top}}(t)$, and $Q_{\text{bot}}(t)$ are $Bz/(1 + z)$, $B[1 + z/(1 + z)]$, and $Qz/\Omega(1 + z)$, respectively. The scaling function $F_b(s)$ in (6.3.21) satisfies

$$\frac{2}{\sqrt{\pi}} \int_0^\infty dse^{-su} s^{1/2} [1 - F_b(s)] = \frac{1}{u(1 + \sqrt{u})} = \frac{1}{u} - \frac{1}{u + \sqrt{u}}, \tag{6D.7}$$

which leads to (6.3.22) in terms of $F_a(s)$ in (6D.5).

(iii) If the top temperature is changed by T_{1b} at $t = 0$ with the bottom temperature unchanged, we have

$$A + z(A + B) = 0, \quad B + z(A + B) = T_{1b}/\Omega. \tag{6D.8}$$

Thus, $A = -T_{1b}z/\Omega(1 + 2z)$ and $B = T_{1b}(1 + z)/\Omega(1 + 2z)$. Then,

$$\int_0^\infty dte^{-\Omega t} T_{1\text{in}}(t) = \frac{z}{\Omega(1 + 2z)} T_{1b} = \frac{1}{2\Omega}\left[1 - \frac{1}{1 + 2z}\right] T_{1b}, \tag{6D.9}$$

which leads to (6.3.25). The Laplace transformations of $Q_{\text{top}}(t)$ and $Q_{\text{bot}}(t)$ are $\lambda B\kappa_D$ and $-\lambda A\kappa_D$, respectively, leading to (6.3.26).

References

[1] L. P. Kadanoff and J. Swift, *Phys. Rev.* **166**, 89 (1968).

[2] K. Kawasaki, in *Phase Transition and Critical Phenomena* eds. C. Domb and M. S. Green (Academic, New York, 1976), Vol. 5A, p. 165; *Ann. Phys.* (N.Y.) **61**, 1 (1970).

[3] R. Perl and R. A. Ferrell, *Phys. Rev. A* **6**, 2358 (1972).

[4] T. Ohta and K. Kawasaki, *Prog. Theor. Phys.* **55**, 1384 (1976).

[5] T. Ohta, *J. Phys. C* **10**, 791 (1977).

[6] J. Bhattacharjee and R. A. Ferrell, *Phys. Lett. A* **76**, 290 (1980).

[7] E. D. Siggia, B. I. Halperin, and P. C. Hohenberg, *Phys. Rev. B* **13**, 2110 (1976).

[8] P. G. de Gennes, *Scaling Concepts in Polymer Physics*, 2nd edn (Cornell University Press, Ithaca, 1985).

[9] M. Doi and S. F. Edwards, *The Theory of Polymer Dynamics* (Oxford University Press, 1986).

[10] P. Stepanek, T. P. Lodge, C. Kedrowski, and F. S. Bates, *J. Chem. Phys.* **94**, 8289 (1991).

[11] P. C. Hohenberg and B. I. Halperin, *Rev. Mod. Phys.* **49**, 435 (1977).

[12] B. U. Felderhof, *Physica* **48**, 541 (1970).

[13] M. S. Green, *J. Chem. Phys.* **20**, 1281 (1952); *ibid.* **22**, 398 (1954).

[14] R. Kubo, *J. Phys. Soc. Jpn* **12**, 570 (1957).

[15] H. Swinney and D. L. Henry, *Phys. Rev. A* **8**, 2586 (1973).

[16] C. Agosta, S. Wang, L. H. Cohen, and H. Meyer, *J. Low Temp. Phys.* **67**, 237 (1987).

[17] Y. Izumi, Y. Miyake, and R. Kono, *Phys. Rev. A* **23**, 272 (1981).

[18] R. F. Berg, M. R. Moldover, and G. A. Zimmerli, *Phys. Rev. E* **60**, 4079 (1999).

[19] K. Kawasaki and J. Gunton, *Phys. Rev. B* **13**, 4658 (1976).

[20] K. Kawasaki, in *Synergetics*, ed. H. Haken (Teubner, Stuttgart, 1973).

[21] T. Koga and K. Kawasaki, *Physica A* **198**, 473 (1993).

[22] A. Onuki and K. Kawasaki, *Physica A* **11**, 607 (1982).

[23] A. Onuki and M. Doi, *J. Chem. Phys.* **85** 1190 (1986).

[24] R. Piazza, T. Bellini, V. Degiorgio, R. E. Goldstein, S. Leibler, and R. Lipowsky, *Phys. Rev. B* **38**, 7223 (1988).

[25] T. Bellini and V. Degiorgio, *Phys. Rev. B* **39**, 7263 (1988).

[26] S. J. Rzoska, V. Degiorgio, T. Bellini, and R. Piazza, *Phys. Rev. E* **49**, 3093 (1994).

[27] A. Onuki and M. Doi, *Europhys. Lett.* **17**, 63 (1992).

[28] J. Luettmer-Strathmann, J. V. Sengers, and G. A. Olchowy, *J. Chem. Phys* **103**, 7482 (1995).

[29] W. Botch and M. Fixman, *J. Chem. Phys.* **42**, 196 (1965); *ibid.* **42**, 199 (1965).

[30] K. Kawasaki and M. Tanaka, *Proc. Phys. Soc.* **90**, 791 (1967).

[31] K. Kawasaki, *Phys. Rev. A* **1**, 1750 (1970); K. Kawasaki and Y. Shiwa, *Physica A* **133**, 27 (1982).

[32] L. Mistura, *J. Chem. Phys.* **57**, 2311 (1972).

[33] R. A. Ferrell and J. K. Bhattacharjee, *Phys. Lett. A* **86**, 109 (1981); *Phys. Rev. A* **24**, 164 (1981); *Phys. Rev. A* **31**, 1788 (1985).

[34] R. A. Ferrell and J. K. Bhattacharjee, *Phys. Lett A* **88**, 77 (1982).

[35] D. M. Kroll and J. M. Ruhland, *Phys. Lett. A* **80**, 45 (1980); *Phys. Rev. A* **23**, 371 (1981).

[36] A. Onuki, *Phys. Rev. E* **55**, 403 (1997).

[37] A. Onuki, *J. Phys. Soc. Jpn* **66**, 511 (1997).

[38] R. Folk and G. Moser, *Phys. Rev. E* **57**, 64 (1998); *ibid.* **57**, 705 (1998).

[39] D. S. Cannell and G. B. Benedek, *Phys. Rev. Lett.* **25**, 1157 (1970); D. Sarid and D. S. Cannell, *Phys. Rev. A* **15**, 735 (1977).

[40] D. Eden, C. W. Garland, and J. Thoen, *Phys. Rev. Lett.* **28**, 726 (1972); J. Thoen and C. W. Garland, *Phys. Rev. A* **28**, 1311 (1974).

[41] D. Roe, B. Wallace, and H. Meyer, *J. Low Temp. Phys.* **16**, 51 (1974).

[42] D. Roe and H. Meyer, *J. Low Temp. Phys.* **30**, 91 (1978).

[43] A. Kogan and H. Meyer, *J. Low Temp. Phys.* **110**, 899 (1998).

[44] Y. Harada, Y. Suzuki, and Y. Ishida, *J. Phys. Soc. Jpn* **48**, 703 (1980).

[45] D. B. Fenner, *Phys. Rev. A* **23**, 1931 (1981).

[46] H. Tanaka, Y. Wada, and H. Nakajima, *Chem. Phys.* **68**, 223 (1982); H. Tanaka and Y. Wada, *Phys. Rev. A* **32**, 512 (1985).

[47] E. A. Clerke, J. V. Sengers, R. A. Ferrell, and J. K. Bhattacharjee, *Phys. Rev. A* **27**, 2140 (1983).

[48] W. Mayer, S. Hoffmann, G. Meier, and I. Alig, *Phys. Rev. E* **55**, 3102 (1997).

[49] A. Onuki and I. Oppenheim, *Phys. Rev. E* **24**, 1520 (1981).

[50] A. Onuki and R. A. Ferrell, *Physica A* **164**, 245 (1990).

[51] A. Onuki, H. Hao, and R. A. Ferrell, *Phys. Rev. A* **41**, 2256 (1990).

[52] K. Nitsche and J. Straub, *Proc. 6th European Symp. on Material Science under Microgravity Conditions* (Bordeaux, France, 2–5 December 1986), p. 109; J. Straub and K. Nitsche, *Fluid Phase Equilibria*, **88**, 183 (1993).

[53] H. Boukari, M. E. Briggs, J. N. Shaumeyer, and R. W. Gammon, *Phys. Rev. A* **41**, 2260 (1990); *Phys. Rev. Lett.* **65**, 654 (1990).

[54] R. A. Wilkinson, G. A. Zimmerli, H. Hao, M. R. Moldover, R. F. Berg, W. L. Johnson, R. A. Ferrell, and R. W. Gammon, *Phys. Rev. E* **57**, 436 (1998).

[55] B. Zappoli, D. Bailly, Y. Garrabos, B. Le Neindre, P. Guenoun, and D. Beysens, *Phys. Rev. A* **41**, 2264 (1990); P. Guenoun, B. Khalil, D. Beysens, Y. Garrabos, F. Kammoun, B. Le Neindre, and B. Zappoli, *Phys. Rev. E* **47**, 1531 (1993); Y. Garrabos, M. Bonetti, D. Beysens, F. Perrot, T. Fröhlich, P. Carlès, and B. Zappoli, *Phys. Rev. E* **57**, 5665 (1998).

[56] H. Klein, G. Schmitz, and D. Woermann, *Phys. Rev. A* **43**, 4562 (1991).

[57] J. Straub, L. Eicher, and A. Haupt, *Phys. Rev. E* **51**, 5556 (1995); J. Straub and L. Eicher, *Phys. Rev. Lett.* **75**, 1554 (1995).

[58] F. Zhong and H. Meyer, *Phys. Rev. E* **51**, 3223 (1995); *ibid.* **53**, 5935 (1996).

[59] D. Dahl and M. R. Moldover, *Phys. Rev. A* **6**, 1915 (1972).

[60] G. R. Brown and H. Meyer, *Phys. Rev. A* **6**, 364 (1972).

[61] R. P. Behringer, A. Onuki, and H. Meyer, *J. Low Temp. Phys.* **81**, 71 (1990).

[62] B. Zappoli, S. Amiroudine, P. Carlés, and J. Ouazzani, *J. Fluid Mech.* **316**, 53 (1996); B. Zappoli and P. Carlés, *Physica D* **89**, 381 (1996).

[63] C. E. Pittman, L. H. Cohen, and H. Meyer, *J. Low Temp. Phys.* **46**, 115 (1982); L. H. Cohen, M. L. Dingus, and H. Meyer, *ibid.* **61**, 79 (1985).

[64] F. Zhong and H. Meyer, *J. Low Temp. Phys.* **114**, 231 (1999).

[65] V. P. Peshkov, *Sov. Phys. JETP* **11**, 580 (1960).

[66] H. Azuma, S. Yoshihara, M. Onishi, K. Ishii, S. Masuda, and T. Maekawa, *Int. J. Heat and Mass Transfer*, **42**, 771 (1999).

[67] S. Amiroudine, P. Bontoux, P. Larroud, B. Gilly, and B. Zappoli, *J. Fluid Mech.*, **442**, 119 (2001).

[68] L. D. Landau and E. M. Lifshitz, *Fluid Mechanics* (Pergamon, New York, 1959).

[69] M. Gitterman and V. Steinberg, *J. Appl. Math. Mech. USSR* **34**, 305 (1971); M. Gitterman, *Rev. Mod. Phys.* **50**, 85 (1978).

[70] P. Carlès and B. Ugurtas, *Physica D* **126**, 69 (1999).

[71] S. Ashkenazi and V. Steinberg, *Phys. Rev. Lett.* **85**, 3641 (1999).

[72] X. Chavanne, F. Chillà, B. Castaing, B. Hébral, B. Chabaud, and J. Chassy, *Phys. Rev. Lett.* **79**, 3648 (1997); X. Chavanne, Ph.D. thesis (Université Joseph Fourier, Grenoble, 1997, unpublished).

[73] A. B. Kogan, D. Murphy, and H. Meyer, *Phys. Rev. Lett.* **82**, 4635 (1999); A. B. Kogan and H. Meyer, *Phys. Rev. E* **63**, 056310 (2001).

[74] Y. Chiwata and A. Onuki, *Phys. Rev. Lett.* **87**, 144301 (2001); A. Furukawa and A. Onuki, *Phys. Rev. E* **66**, 016302 (2002).

[75] G. P. Metcalfe and R. P. Behringer, *J. Low Temp. Phys.* **78**, 231 (1990).

[76] L. Mistura, *Nuovo Cimento B* **12**, 35 (1972); *J. Chem. Phys.* **62**, 4572 (1975).

[77] A. Onuki, *J. Low Temp. Phys.* **61**, 101 (1985).

[78] R. Folk and G. Moser, *J. Low Temp. Phys.* **99**, 11 (1995).

[79] M. A. Anisimov, V. A. Agayan, A. A. Povodyrev, J. V. Sengers, and E. E. Gorodetskii, *Phys. Rev. E* **57**, 1946 (1998).

[80] Y. Miura, H. Meyer, and A. Ikushima, *J. Low Temp. Phys.* **55**, 247 (1984).

[81] B. J. Ackerson and J. M. J. Hanley, *J. Chem. Phys.* **73**, 3568 (1980).

[82] A. Martin, F. Ortega, and R. G. Rubio, *Phys. Rev. E* **54**, 5302 (1996).

[83] L. P. Filippov, *Int. J. Heat Transfer*, **11**, 331 (1968).

[84] M Giglio and A. Vendramini, *Phys. Rev. Lett.* **34**, 561 (1975).

[85] L. H. Cohen, M. L. Dingus, and H. Meyer, *Phys. Rev. Lett.* **50**, 1058 (1983); H. Meyer and L. H. Cohen, *Phys. Rev. A* **38**, 2081 (1988).

[86] D. G. Friend and H. M. Roder, *Phys. Rev. A* **32**, 1941 (1985).

[87] E. P. Sakonidou, H. R. van den Berg, C. A. ten Seldam, and J. V. Sengers, *J. Chem. Phys.* **109**, 717 (1998).

[88] C. M. Jefferson, R. G. Petschek, and D. S. Cannell, *Phys. Rev. Lett.* **52**, 1329 (1984).

[89] L. P. Pitaevskii, *Zh. Eksp. Teor. Fiz.* **35**, 408 (1958) [*Sov. Phys. JETP* **8**, 282 (1959)].

[90] I. M. Khalatonikov, *Introduction to the Theory of Superfluidity* (Benjamin, New York, 1965).

[91] V. L. Ginzburg and A. A. Sobaynin, *Usp. Fiz. Nauk* **120**(2), 153 (1976) [*Sov. Phys. Usp.* **19**, 773 (1976)].

[92] R. A. Ferrell, N. Meynàrd, H. Scmidt, F. Schwabl, and P. Szépfalusy, *Ann. Phys.* **47**, 565 (1968).

[93] J. Swift and L. P. Kadanoff, *Ann. Phys.* **50**, 312 (1968).

[94] P. C. Hohenberg, in *Critical Phenomena*, ed. M. S. Green, Proceedings of Enrico Fermi Summer School, Varenna, 1970 (Academic, New York, 1971), p. 285.

[95] B. I. Halperin, P. C. Hohenberg, and E. D. Siggia, *Phys. Rev. B* **13**, 1299 (1976); *ibid.*, erratum, *B* **21**, 2044 (1980).

[96] G. Ahlers, P. C. Hohenberg, and A. Kornblit, *Phys. Rev. B* **25**, 5932 (1982).

[97] W. Y. Tam and G. Ahlers, *Phys. Rev. B* **33**, 5932 (1985).

[98] M. Dingus, F. Zhong, and H. Meyer, *J. Low Temp. Phys.* **65**, 185 (1986); M. Dingus, F. Zhong, J. Tuttle, and H. Meyer, *ibid.* **65**, 213 (1986).

[99] R. Mehrotra and G. Ahlers, *Phys. Rev. B* **30**, 5116 (1984).

[100] L. S. Goldner, N. Mulders, and G. Ahlers, *J. Low Temp. Phys.* **93**, 131 (1993).

[101] S. J. Putterman, *Superfluid Hydrodynamics* (North-Holland, Amsterdam, 1974).

[102] R. Kubo, *J. Math. Phys.* **4**, 227 (1963).

[103] R. A. Ferrell and J. K. Bhattacharjee, *Phys. Rev. B* **24**, 5071(1981).

[104] V. Dohm and R. Folk, *Physica B* **109** & **110**, 1549 (1982).

[105] C. De Dominicis and L. Peliti, *Phys. Rev. Lett.* **38**, 505 (1977); *Phys. Rev. B* **18**, 353 (1978).

[106] R. A. Ferrell and J. K. Bhattacharjee, *Phys. Rev. Lett.* **42**, 1638 (1979); *Phys. Rev. B* **25**, 216 (1982); *ibid.* **28**, 120 (1983).

[107] P. C. Hohenberg, *Physica B* **109** & **110**, 1436 (1982).

[108] V. Dohm, *Phys. Rev. B* **44**, 2697(1991).

[109] L. D. Landau and I. M. Khalatnikov, *Sov. Phys. Dokl.* **96**, 469 (1951) [English translation, *Collected Papers of L. D. Landau*, ed. D. ter Haar (Gordon and Breach, New York, 1965)].

[110] V. L. Pokrovskii and I. M. Khalatnikov, *Sov. Phys. JETP. Lett.* **9**, 149 (1969); I. M. Khalatnikov, *Sov. Phys. JETP* **30**, 268 (1970).

[111] K. Kawasaki, *Phys. Lett. A* **31**, 1750 (1970).

[112] R. A. Ferrell and J. K. Bhattacharjee, *Phys. Rev. Lett.* **44**, 403 (1980).

[113] R. A. Ferrell and J. K. Bhattacharjee, *Phys. Rev. B* **23**, 2434 (1981).

[114] J. Pankert and V. Dohm, *Europhys. Lett.* **2**, 775 (1986).

[115] D. Williams and I. Rudnick, *Phys. Rev. Lett.* **25**, 276 (1970).

[116] R. Carey, Ch. Buchal, and F. Pobell, *Phys. Rev. B* **16**, 3133 (1977).

[117] K. Tozaki and A. Ikushima, *J. Low Temp. Phys.* **32**, 379 (1978).

[118] B. Lambert, R. Perzynski, and D. Salin, *Phys. Rev. B* **20**, 1025 (1979).

[119] F. Vidal, *Phys. Rev. B* **26**, 3986 (1982); in *Macroscopic Theories of Superfluids*, ed. G. Grioli (Cambridge University Press, New York, 1991), p. 95.

[120] I. M. Khalatnikov and V. N. Zharkov, *Zh. Exsp. Teor. Fiz*, **32**, 1108 (1957) [*Sov. Phys. JETP* **5**, 905 (1957)].

[121] E. D. Siggia and D. R. Nelson, *Phys. Rev. B* **15**, 1427 (1977).

[122] A. Onuki, *J. Low Temp. Phys.* **53**, 189 (1983).

[123] V. Dohm and R. Folk, *Phys. Rev. B* **28**, 1332 (1983).

[124] G. Ahlers and F. Pobell, *Phys. Rev. Lett.* **32**, 144 (1974).

[125] D. Murphy and H. Meyer, *J. Low Temp. Phys.* **107**, 175 (1997).

[126] M. Tanaka, A. Ikushima, and K. Kawasaki, *Phys. Lett. A* **61**, 119 (1977); M. Dingus, F. Zhong, and H. Meyer, *Phys. Rev. Lett.* **54**, 2347 (1985).

[127] D. Erben and F. Pobell, *Phys. Lett. A* **26**, 368 (1968); P. Leiderer and F. Pobell, *Z. Phys.* **223**, 378 (1969).

[128] M. Bhagat and R. A. Rasken, *Phys. Rev. A* **3**, 340 (1971); *ibid.* **4**, 264 (1971); *ibid.* **5**, 2297 (1972).

[129] A. Onuki, *J. Low Temp. Phys.* **50**, 433 (1983); *ibid.* **55**, 309 (1984).

[130] A. Onuki, *Jpn J. Appl. Phys.* **26**, 365 (1987) (Proc. 18th International Conference on Low Temperature Physics).

[131] R. V. Duncan, G. Ahlers, and V. Steinberg, *Phys. Rev. Lett.* **60**, 1522 (1988); G. Ahlers and R. V. Duncan in *Frontiers of Physics*, Proceedings of the Landau Memorial Conference, Tel Aviv, ed. E. Gotsman, Y. Ne'eman, and A. Voronel (Pergamon, Oxford, 1990), p. 219.

[132] R. Haussmann and V. Dohm *Phys. Rev. Lett.* **67**, 3404 (1991); *Phys. Rev.* B **46**, 6361 (1992); *Z. Phys. B* **87**, 229 (1992).

[133] A. Onuki and Y. Yamazaki, *J. Low Temp. Phys.* **103**, 131 (1996).

[134] A. Onuki, *J. Low Temp. Phys.* **104**, 133 (1996).

[135] G. Ahlers and F.-C. Liu, *J. Low Temp. Phys.* **105**, 255 (1996).

[136] W. A. Moeur, P. K. Day, F.-C. Liu, S. T. P. Boyd, M. J. Adriaans, and R. V. Duncan, *Phys. Rev. Lett.* **78**, 2421 (1997).

[137] R. Haussmann, *Phys. Rev. B* **60**, 12 349 (1999).

[138] V. L. Ginzburg and L. D. Landau, *Zh. Exsp. Teor. Fiz.* **20**, 1064 (1950).

[139] W. J. Skocpol and M. Tinkham, *Rep. Prog. Phys.* **38**, 1049 (1975).

[140] P. Bak, C. Tang, and K. Wiesenfeld, *Phys. Rev. Lett.* **59**, 381 (1987).

[141] J. Machta, D. Candela, and R. B. Hallock, *Phys. Rev. E* **47**, 4581 (1993).

7

Dynamics in polymers and gels

We will first give a theory of viscoelastic dynamics in polymeric binary systems, where a new concept of dynamic stress–diffusion coupling will be introduced in the scheme of viscoelastic two-fluid hydrodynamics. A Ginzburg–Landau theory of entangled polymer solutions will also be presented, in which chain deformations are represented by a conformation tensor. The reptation theory for entangled polymers will be summarized in Appendix 7A. We will also present a Ginzburg–Landau theory of gels to discuss dynamics and heterogeneities inherent to gels.

7.1 Viscoelastic binary mixtures

Entanglements among polymer chains impose severe topological constraints on the molecular motions. Their effects on polymer dynamics are now well described by the reptation theory in a surprisingly simple manner [1, 2]. In such systems, the stress relaxation takes place on a very long timescale τ (which should not be confused with the reduced temperature in near-critical systems). This means that a large network stress arises even for small deformations. If the timescale of the deformations is shorter than τ, the system behaves as a soft elastic body or gel. If it is longer than τ, we have a very viscous fluid.

In polymeric mixtures, it is highly nontrivial how the network stress acts on the two components and how it influences spatial inhomogeneities of the composition in various situations [3]–[5]. In this section we will introduce a mechanism of dynamical stress–diffusion coupling, which has recently begun to be recognized. In this chapter we will investigate its consequences mainly in dynamic light scattering from polymers [6]–[11]. Furthermore, we will show its relevance in viscoelastic phase separation in Chapter 8 and under shear-induced phase separation in Chapter 11. This mechanism should also be applicable to other highly viscoelastic binary mixtures such as dense colloidal suspensions [12], dense microemulsions, or fluid mixtures near the glass transition.

Before presenting the theory, we mention representative experiments. Figure 7.1 shows nonexponential relaxation of dynamic light scattering from a semidilute, aqueous borax solution of poly(vinyl alcohol) with degree of polymerization 2600 at 2 wt% polymer concentration [7]. For $q \sim 10^5$ cm^{-1}, the relaxation rate of the fast mode is diffusive as $\Gamma_f = D_m q^2$ with the mutual diffusion constant $D_m = 5.6 \times 10^{-7}$ cm^2 s^{-1}, whereas the relaxation time of the slow mode is independent of q and is about 0.3 s which is close to the stress relaxation time. As another experimental example, Fig. 7.2 displays the

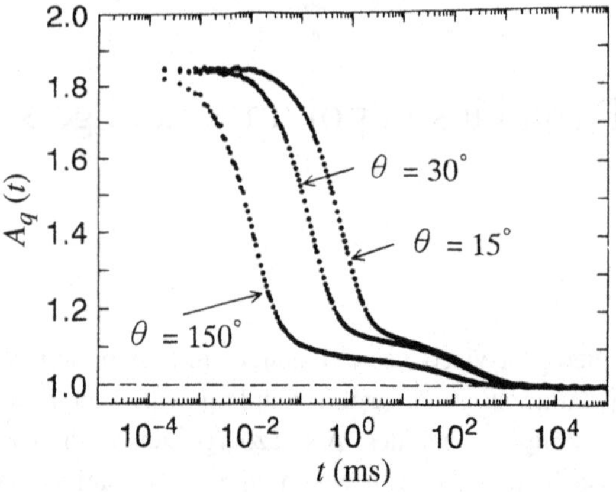

Fig. 7.1. The normalized homodyne time-correlation function $A_q(t) = 1 + \text{const.} I(q,t)^2$ at $q \sim 10^5\ \text{cm}^{-1}$ for a polymer solution [7]. Scattering angles are 15, 30, and 150 degrees as indicated. The curves exhibit the presence of two dominant decay modes with decay rates Γ_s and $\Gamma_f = D_m q^2$, in which the faster one Γ_f shifts to the left along the decay time axis with increasing θ or $q = 2q_0 \sin(\theta/2)$, while the slower one Γ_s is nearly independent of q.

two relaxation rates measured by transient light scattering from a semidilute polystyrene solution in theta solvent after cessation of shear flow [11].

7.1.1 The GLW hamiltonian and chemical potentials

Before discussing the dynamics, we give the expression for the GLW hamiltonian for polymer solutions and blends using the results in Section 3.5. We assume that the mixture is nearly incompressible and that the free-energy density is given by $v_0^{-1} f_{\text{site}}(\phi) + (2K_T)^{-1}(\delta n/n)^2$ as in (3.5.12). Furthermore, the monomers of the two components are assumed to have the same volume $v_0 = a^3$ and the same mass m_0. Then $\delta n/n = \delta\rho/\rho \cong \delta\rho/\bar{\rho}$ is very small, where $\bar{\rho}$ is the average mass density, and the mass fractions and the volume fractions coincide:

$$\rho_1/\rho = \phi_1 = \phi, \quad \rho_2/\rho = \phi_2 = 1 - \phi. \tag{7.1.1}$$

The GLW hamiltonian for the volume fraction ϕ is written as

$$\mathcal{H}\{\phi\} = \int d\boldsymbol{r} \left[v_0^{-1} f_{\text{site}}(\phi) + \frac{T}{2} C(\phi) |\nabla\phi|^2 \right], \tag{7.1.2}$$

where f_{site} is the free-energy density per site given by (3.5.5) for polymer solutions and by (3.5.29) for polymer blends. In the case of semidilute polymer solutions with $\phi \ll 1$, we treat theta or poor solvent assuming gaussian forms of chains [1]. The full hamiltonian including $\delta\rho$ and the velocities of the two components will be given in (7B.10). If variations

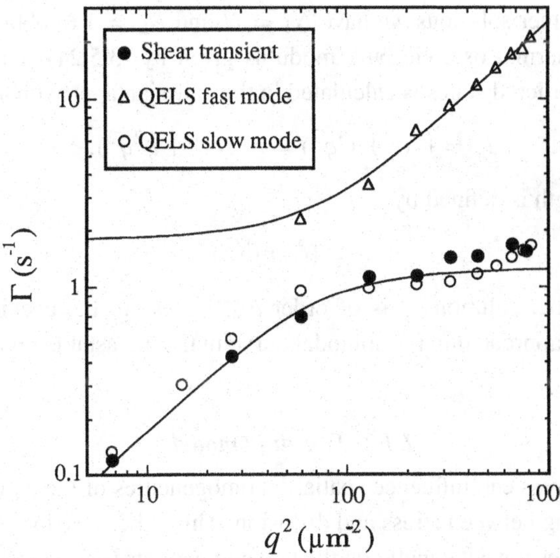

Fig. 7.2. A log–log plot of the relaxation rates Γ_s and Γ_f in a transient light scattering experiment on a semidilute solution with theta solvent after cessation of shear flow [11]. The two modes did not separate clearly in time until $q^2 \gtrsim 200\ \mu\text{m}^{-2}$. This experiment was performed in a transient situation, where the slower of the collective modes was selectively enhanced even after decay of the macroscopic flow.

of ϕ vary in space with wave numbers smaller than the inverse gyration radius R_G^{-1}, (4.2.26) suggests

$$C = 1/[18\phi(1-\phi)a]. \tag{7.1.3}$$

For the fluctuations varying on spatial scales shorter than R_G, we should replace the factor 18 in (7.1.3) by 12 from (4.2.27). In the presence of the gradient free energy, the chemical potentials of the two components (per unit mass in this chapter) are expressed as

$$\mu_1 = \frac{1}{\bar{\rho}}\left[\delta p_1 + (1-\phi)\frac{\delta\mathcal{H}}{\delta\phi}\right], \quad \mu_2 = \frac{1}{\bar{\rho}}\left[\delta p_1 - \phi\frac{\delta\mathcal{H}}{\delta\phi}\right], \tag{7.1.4}$$

where δp_1 is a pressure contribution induced by $\delta\rho$,

$$\delta p_1 = (\bar{\rho}K_T)^{-1}\delta\rho. \tag{7.1.5}$$

The chemical potential difference is then of the form,

$$\mu_1 - \mu_2 = \frac{1}{\bar{\rho}}\frac{\delta\mathcal{H}}{\delta\phi}. \tag{7.1.6}$$

These are generalized forms of (3.5.13)–(3.5.15), and (3.5.30). In this section we define

$$r = (v_0 T)^{-1} f''_{\text{site}} = v_0^{-1}\left[\frac{1}{N_1\phi} + \frac{1}{N_2(1-\phi)} - 2\chi\right]. \tag{7.1.7}$$

For semidilute polymer solutions we have $N_1 = N$ and $N_2 = 1$ to obtain $r = K_{os}/T\phi^2$, K_{os} being the isothermal osmotic bulk modulus given by (3.5.24). The structure factor $I_q = \langle|\phi_q|^2\rangle$ in disordered states is calculated in the gaussian approximation as

$$I_q = 1/(r + Cq^2) = r^{-1}/(1 + \xi^2 q^2). \tag{7.1.8}$$

The correlation length is defined by

$$\xi = (C/r)^{1/2}. \tag{7.1.9}$$

In semidilute polymer solutions ξ is of order $a\phi^{-1}$ close to the coexistence curve and grows as $K_{os}^{-1/2}$ on approaching the spinodal curve in the metastable region.

7.1.2 Two-fluid model

We show that the stress can influence spatial inhomogeneities of the composition through a dynamical coupling between stress and diffusion. This is because the stress in entangled polymer systems does not act equally on the two components (*asymmetric stress division*), and if there is an imbalance in stress, relative motion between the two components takes place. This coupling gives rise to a variety of viscoelastic effects such as nonexponential relaxation in dynamic light scattering [3]–[11], flow-induced polymer migration [13], shear-induced fluctuation enhancement, etc. (See Chapter 11 for the last topic.)

To explain the dynamical coupling we consider a two-fluid model of a very viscous two-component system [4]. The mass densities, ρ_1 and ρ_2, of the two components are convected by their velocities, $\boldsymbol{v}_1$ and $\boldsymbol{v}_2$, as

$$\frac{\partial}{\partial t}\rho_K = -\nabla\cdot(\rho_K \boldsymbol{v}_K), \quad (K = 1, 2). \tag{7.1.10}$$

The deviation of the total density obeys

$$\frac{\partial}{\partial t}\delta\rho = -\nabla\cdot(\rho\boldsymbol{v}) \cong -\bar{\rho}\nabla\cdot\boldsymbol{v}, \tag{7.1.11}$$

where $\rho = \rho_1 + \rho_2 = \bar{\rho} + \delta\rho$. The average velocity $\boldsymbol{v}$ is defined by

$$\boldsymbol{v} = \rho^{-1}(\rho_1\boldsymbol{v}_1 + \rho_2\boldsymbol{v}_2) = \phi\boldsymbol{v}_1 + (1-\phi)\boldsymbol{v}_2. \tag{7.1.12}$$

The volume fraction ϕ of the first component obeys

$$\frac{\partial}{\partial t}\phi + \boldsymbol{v}\cdot\nabla\phi = -\nabla\cdot\left[\phi(1-\phi)\boldsymbol{w}\right], \tag{7.1.13}$$

where

$$\boldsymbol{w} = \boldsymbol{v}_1 - \boldsymbol{v}_2 \tag{7.1.14}$$

is the relative velocity between the two components. The diffusion current is given by $\phi(1-\phi)\boldsymbol{w}$. The two velocities $\boldsymbol{v}_1$ and $\boldsymbol{v}_2$ are expressed as

$$\boldsymbol{v}_1 = \boldsymbol{v} + (1-\phi)\boldsymbol{w}, \quad \boldsymbol{v}_2 = \boldsymbol{v} - \phi\boldsymbol{w}, \tag{7.1.15}$$

in terms of $\boldsymbol{v}$ and $\boldsymbol{w}$. Considering only very slow motion and neglecting temperature inhomogeneities, we assume the momentum equations,

$$\rho_1 \frac{\partial}{\partial t} \boldsymbol{v}_1 = -\rho_1 \nabla \mu_1 - \zeta(\boldsymbol{v}_1 - \boldsymbol{v}_2) + \boldsymbol{F}_1, \tag{7.1.16}$$

$$\rho_2 \frac{\partial}{\partial t} \boldsymbol{v}_2 = -\rho_2 \nabla \mu_2 - \zeta(\boldsymbol{v}_2 - \boldsymbol{v}_1) + \boldsymbol{F}_2. \tag{7.1.17}$$

In the first terms, μ_1 and μ_2 are the generalized chemical potentials in (7.1.4). The second terms represent mutual friction between the two components with ζ being a friction coefficient. In the third terms, $\boldsymbol{F}_1$ and $\boldsymbol{F}_2$ are the force densities arising from the network stress $\overleftrightarrow{\sigma}$. Their sum is

$$\boldsymbol{F}_1 + \boldsymbol{F}_2 = \nabla \cdot \overleftrightarrow{\sigma}. \tag{7.1.18}$$

For polymer solutions, we need to retain the viscous stress tensor due to the background viscosity, as will be shown in (7.1.34) below.

The equation for the total momentum density $\rho \boldsymbol{v} = \rho_1 \boldsymbol{v}_1 + \rho_2 \boldsymbol{v}_2$ is the sum of (7.1.16) and (7.1.17):

$$\rho \frac{\partial}{\partial t} \boldsymbol{v} = -(\rho_1 \nabla \mu_1 + \rho_2 \nabla \mu_2) + \nabla \cdot \overleftrightarrow{\sigma}. \tag{7.1.19}$$

From (7.1.4) we derive

$$\rho_1 \nabla \mu_1 + \rho_2 \nabla \mu_2 = \nabla \delta p_1 - \frac{\delta \mathcal{H}}{\delta \phi} \nabla \phi. \tag{7.1.20}$$

The second term arises from the concentration heterogeneity, analogously to (6.1.16) derived for near-critical fluids. The relative velocity $\boldsymbol{w} = \boldsymbol{v}_1 - \boldsymbol{v}_2$ is governed by

$$\frac{\partial}{\partial t} \boldsymbol{w} = -\nabla(\mu_1 - \mu_2) - \zeta \left(\frac{1}{\rho_1} + \frac{1}{\rho_2} \right) \boldsymbol{w} + \frac{1}{\rho_1} \boldsymbol{F}_1 - \frac{1}{\rho_2} \boldsymbol{F}_2. \tag{7.1.21}$$

We are interested in slow motion with frequencies much smaller than $\omega_0 = \zeta(1/\rho_1 + 1/\rho_2)$. Then we may set $\partial \boldsymbol{w}/\partial t = \boldsymbol{0}$ and use (7.1.1) and (7.1.6) to obtain

$$\boldsymbol{w} = \frac{\phi_1 \phi_2}{\zeta} \left[-\nabla \frac{\delta \mathcal{H}}{\delta \phi} + \frac{1}{\phi_1} \boldsymbol{F}_1 - \frac{1}{\phi_2} \boldsymbol{F}_2 \right]. \tag{7.1.22}$$

The term proportional to $\nabla(\delta \mathcal{H}/\delta \phi)$ gives rise to the diffusive equation for ϕ in the usual form if substituted into (7.1.13). We may thus define the kinetic coefficient L as

$$L = \phi^2 (1 - \phi)^2 / \zeta. \tag{7.1.23}$$

In the long-wavelength limit the mutual diffusion constant is written as

$$D_{\mathrm{m}} = LTr = \zeta^{-1} T \phi^2 (1 - \phi)^2 r. \tag{7.1.24}$$

where r is given in (7.1.7). The terms proportional to $\boldsymbol{F}_K$ cancel to vanish in $\boldsymbol{w}$ when the network stress is divided between the two components *symmetrically* or *trivially* as

$$\boldsymbol{F}_K = (\rho_K/\rho) \nabla \cdot \overleftrightarrow{\sigma}, \qquad (K = 1, 2). \tag{7.1.25}$$

This will be the case if the two components are physically alike. In viscoelastic systems, however, the stress division can be *asymmetric* between the two components.

7.1.3 Dynamical coupling in semidilute polymer solutions and gels

In semidilute solutions with $\phi \ll 1$, the friction coefficient ζ is estimated as

$$\zeta = 6\pi \eta_0 \xi_b^{-2}, \tag{7.1.26}$$

where η_0 is the solvent viscosity and $\xi_b = a/\phi$ is the blob size in theta solvent ($\sim a\phi^{-3/4}$ in good solvent). Here a blob contains $g_b = (\xi_b/a)^{1/\hat{\nu}}$ monomers belonging to a single chain, where $\hat{\nu} = 1/2$ for theta solvent and $\hat{\nu} = 3/5$ for good solvent [1].[1] Then the friction coefficient on a blob is $\xi_b^3 \zeta = 6\pi \eta_0 \xi_b$ from Stokes law (5.1.2). From (7.1.24) the mutual diffusion constant D_m between polymer and solvent in the long-wavelength limit is obtained as

$$D_m = \xi_b^2 K_{os}/6\pi \eta_0. \tag{7.1.27}$$

In the semidilute region we have the Stokes formula $D_m \sim T/6\pi \eta_0 \xi_b$ above the coexistence curve, which is analogous to (6.1.24) for near-critical fluids. The characteristic time within a blob is thus written as

$$\tau_b = \xi_b^2/D_m = 6\pi \eta_0 \xi_b^3/T, \tag{7.1.28}$$

which obviously originates from the hydrodynamic interaction on the scale of ξ_b.

However, the rotational motion of chains and the diffusion rate of a tagged chain become extremely slow in semidilute solutions when the molecular weight is very large and when the polymer volume fraction ϕ exceeds the overlapping threshold $\phi^*(\sim N^{-1/2}$ for theta solvent) [1]. The entanglement number on a chain is on the order of the blob number $N/g_b = N(\xi_b/a)^{-1/\nu}$. The newtonian solution viscosity η grows as $(N/g_b)^3$ from the reptation theory in Appendix 7A and as $(N/g_b)^{3.4}$ from experiments. In terms of ϕ/ϕ^* we obtain [14, 15]

$$\eta \sim \eta_0(\phi/\phi^*)^{x_\eta}, \tag{7.1.29}$$

where the exponent x_η is large and is of order 6–7 in theta solvent. The stress relaxation time τ is estimated as

$$\tau \sim \tau_b(\phi/\phi^*)^{x_\eta}. \tag{7.1.30}$$

In the above scaling arguments the shear modulus $G = \eta/\tau$ is assumed to be proportional to $\xi_b^{-3}(\propto \phi^3$ in theta solvent). However, for theta solvent, some authors theoretically claimed that G depends on ϕ somewhat differently as $G \propto \phi^2$ [3] or $G \propto \phi^{7/3}$ [16], while $G \propto \phi^{2.25}$ in experiments [14, 15].

[1] We may require $\phi\xi_b^3 = v_0 g_b$ to obtain $\xi_b = a\phi^{\hat{\nu}/(1-3\hat{\nu})}$. We need not distinguish between the correlation length ξ and the blob size ξ_b above the coexistence curve, but they become very different near the spinodal curve.

In both polymer solutions and gels, even a small network deformation gives rise to a large stress acting *directly* on polymer chains, so that it follows a *one-sided stress division*,

$$\boldsymbol{F}_1 \cong \nabla \cdot \overleftrightarrow{\boldsymbol{\sigma}}, \qquad \boldsymbol{F}_2 \cong \boldsymbol{0}, \tag{7.1.31}$$

where the solvent viscosity is neglected. Here the subscript 1 denotes the quantities of polymer and the subscript 2 denotes those of solvent. The relative velocity $\boldsymbol{w}$ becomes

$$\boldsymbol{w} = -\frac{\phi(1-\phi)}{\zeta}\left[\nabla\frac{\delta\mathcal{H}}{\delta\phi} - \frac{1}{\phi}\nabla \cdot \overleftrightarrow{\boldsymbol{\sigma}}\right]. \tag{7.1.32}$$

The diffusive equation (7.1.13) is rewritten as

$$\frac{\partial}{\partial t}\phi + \boldsymbol{v} \cdot \nabla\phi = \nabla \cdot L\left[\nabla\frac{\delta\mathcal{H}}{\delta\phi} - \frac{1}{\phi}\nabla \cdot \overleftrightarrow{\boldsymbol{\sigma}}\right], \tag{7.1.33}$$

where L is defined by (7.1.23). We recognize that imbalance of the network stress ($\nabla \cdot \overleftrightarrow{\boldsymbol{\sigma}} \neq \boldsymbol{0}$) leads to relative motion between polymer and solvent. Originally, Tanaka *et al.* derived a linearized version of (7.1.33) for gels, where the network stress is related to the elastic displacement vector $\boldsymbol{u}$, to analyze dynamic light scattering [17]. Harden *et al.* [5] studied hydrodynamic surface modes in polymer solutions and gels on the basis of the two-fluid model. Helfand and Fredrickson [18] used the above form for sheared polymer solutions. Some authors [19, 20] tried to justify (7.1.33) using the projection operator method, where the Rouse dynamics was used, however.

From (7.1.19) and (7.1.20) the average velocity $\boldsymbol{v}$ is governed by

$$\rho\frac{\partial}{\partial t}\boldsymbol{v} = -\nabla\delta p_1 + \frac{\delta\mathcal{H}}{\delta\phi}\nabla\phi + \nabla \cdot \overleftrightarrow{\boldsymbol{\sigma}} + \eta_0\nabla^2\boldsymbol{v}. \tag{7.1.34}$$

Here we have added the last term arising from the solvent viscosity η_0 by neglecting the difference between $\boldsymbol{v}$ and the solvent velocity $\boldsymbol{v}_\mathrm{s}$ owing to $\phi \ll 1$. Furthermore, the Stokes approximation and the incompressibility assumption lead to

$$\frac{\partial}{\partial t}\boldsymbol{v} = 0, \qquad \nabla \cdot \boldsymbol{v} = 0. \tag{7.1.35}$$

On the one hand, Stokes approximation is justified if we are interested in physical processes in which the timescale is much longer than $\bar{\rho}\ell^2/\eta_0$ with ℓ being a typical spatial scale such as the domain size in spinodal decomposition. The incompressibility condition, on the other hand, automatically determines δp_1 in (7.1.34).

Constitutive equation

For small deformations the network stress can be expressed in terms of the gradient of the polymer velocity $\boldsymbol{v}_\mathrm{p} = \boldsymbol{v}_1$. That is, on spatial scales much longer than R_G, the average stress tensor in the linear response (newtonian) regime reads [2]

$$\sigma_{ij}(t) = \int_{-\infty}^{t} dt_1 G_{xy}(t-t_1)\kappa_{ij}^{(p)}(t_1), \tag{7.1.36}$$

where $\kappa_{ij}^{(p)}(t)$ is the polymer velocity gradient tensor,

$$\kappa_{ij}^{(p)} = \nabla_i v_{pj} + \nabla_j v_{pi} - \frac{2}{3}\delta_{ij}\nabla \cdot \boldsymbol{v}_p, \tag{7.1.37}$$

which is made traceless and symmetric. Hereafter $\nabla_i = \partial/\partial x_i$. The function $G_{xy}(t)$ represents relaxation of shear deformations arising from disentanglements. It relaxes from the shear modulus $G_{xy}(0) = G$ on the timescale of τ. In our theory, $\nabla \boldsymbol{v}_p$ is used in the constitutive equation rather than $\nabla \boldsymbol{v}$, which will lead to important consequences in the presence of mutual diffusion. However, there can be a diagonal stress driven by the dilation strain $\nabla \cdot \boldsymbol{v}_p$ and relaxing with disentanglements, but this effect will be neglected for simplicity.

7.1.4 Dynamical coupling in polymer blends

Next we consider stress partitioning in polymer blends. The two polymers have polymerization indices N_1 and N_2 and volume fractions $\phi_1 = \phi$ and $\phi_2 = 1 - \phi$. For simplicity, we assume that they have the same monomer size a and the same monomer number N_e between two consecutive entanglement points. Then, in the entangled case $N_1 > N_e$ and $N_2 > N_e$, the two polymers obey reptation dynamics moving in *common tubes* with diameters of order $d_t = N_e^{1/2} a$. Further discussions on entangled polymer blends will be given in Appendix 7B. We here propose an *intermediate stress division*,

$$\boldsymbol{F}_1 = \alpha_1 \nabla \cdot \overleftrightarrow{\sigma}, \quad \boldsymbol{F}_2 = \alpha_2 \nabla \cdot \overleftrightarrow{\sigma}, \tag{7.1.38}$$

where $\alpha_1 + \alpha_2 = 1$. A dynamical asymmetry parameter α may be defined by

$$\alpha = \rho\left(\frac{\alpha_1}{\rho_1} - \frac{\alpha_2}{\rho_2}\right) = \frac{\alpha_1}{\phi_1} - \frac{\alpha_2}{\phi_2}. \tag{7.1.39}$$

Then,

$$\alpha_1 = \phi_1 + \phi_1\phi_2\alpha, \quad \alpha_2 = \phi_2 - \phi_1\phi_2\alpha. \tag{7.1.40}$$

In terms of α, the relative velocity in (7.1.22) becomes

$$\boldsymbol{w} = -\frac{\phi_1\phi_2}{\zeta}\left[\nabla\frac{\delta\mathcal{H}}{\delta\phi} - \alpha\nabla \cdot \overleftrightarrow{\sigma}\right]. \tag{7.1.41}$$

Similarly to (7.1.33), the diffusive equation (7.1.13) is rewritten as

$$\frac{\partial}{\partial t}\phi + \boldsymbol{v} \cdot \nabla\phi = \nabla \cdot L\left[\nabla\frac{\delta\mathcal{H}}{\delta\phi} - \alpha\nabla \cdot \overleftrightarrow{\sigma}\right]. \tag{7.1.42}$$

In Appendix 7B we will derive the expression for α in the form,

$$\alpha = \frac{N_1\zeta_{01} - N_2\zeta_{02}}{\phi_1 N_1\zeta_{01} + \phi_2 N_2\zeta_{02}}. \tag{7.1.43}$$

Here ζ_{01} and ζ_{02} are the friction coefficients of the monomers of the two polymers and can generally be different in our theory (even if the common values of a and N_e are assumed).

In particular, the trivial stress division (7.1.25) or $\alpha = 0$ follows for $N_1\zeta_{01} = N_2\zeta_{02}$, while the one-sided division (7.1.31) or the limit of polymer solutions follows for $N_1\zeta_{01} \gg N_2\zeta_{02}$ and $N_1\zeta_{01}\phi_1 \gg N_2\zeta_{02}\phi_2$, where $\alpha = 1/\phi_1$. Furthermore, the reptation theory leads to the expression for the friction coefficient ζ in (7.1.16) and (7.1.17),

$$\frac{1}{\zeta} = \frac{N_e}{\phi_1 N_1 \zeta_{01}} + \frac{N_e}{\phi_2 N_2 \zeta_{02}}. \tag{7.1.44}$$

These expressions will be derived on the basis of a concept of a *tube* velocity $\boldsymbol{v}_t$ expressed as

$$\boldsymbol{v}_t = \alpha_1 \boldsymbol{v}_1 + \alpha_2 \boldsymbol{v}_2 = \boldsymbol{v} + \phi_1\phi_2\alpha \boldsymbol{w}, \tag{7.1.45}$$

which has the meaning of the average velocity of the entanglement structure. It is equal to the polymer velocity $\boldsymbol{v}_p$ for polymer solutions. This concept was first introduced by Brochard [21] to derive the mutual diffusion constant to be discussed below.

Constitutive equation

It is natural to claim that the network stress is determined by the gradient tensor of the tube velocity $\boldsymbol{v}_t$ [4]. In the linear response regime the network stress is then expressed as

$$\sigma_{ij}(t) = \int_{-\infty}^{t} dt_1 G_{xy}(t - t_1)\kappa_{ij}^{(t)}(t_1), \tag{7.1.46}$$

where $\kappa_{ij}^{(t)}(t)$ is the tube velocity gradient tensor written as

$$\kappa_{ij}^{(t)} = \nabla_i v_{tj} + \nabla_j v_{ti} - \frac{2}{3}\delta_{ij}\nabla\cdot\boldsymbol{v}_t. \tag{7.1.47}$$

Case in which short chains are not entangled

So far we have assumed that both N_1 and N_2 exceed N_e. The intermediate case in which $N_2 \lesssim N_e < N_1$ can also be considered. Here the shorter component acts as a solvent. Obviously, we have the one-sided stress division (7.1.31) with $\boldsymbol{v}_t \cong \boldsymbol{v}_1$, so

$$\alpha \cong 1/\phi_1, \quad \zeta \cong \phi_2\zeta_{02}, \quad L \cong \phi_1^2\phi_2/\zeta_{02}. \tag{7.1.48}$$

We may obtain these results by setting $N_2 = N_e$ in (7.1.43) and (7.1.44). It is natural that these quantities are independent of N_1, N_e, and ζ_{01}.

Symmetric case without dynamical coupling

Many theories of polymer blends have been constructed for the symmetric case, $N_1 = N_2 = N$ and $\zeta_{01} = \zeta_{02} = \zeta_0$, neglecting the stress–diffusion coupling. Then the dynamics is essentially the same as that of usual binary fluid mixtures. However, crossover effects arise from the sensitive N dependence of the static and dynamic coefficients. Here,

$$L = \phi_1\phi_2 N_e/N\zeta_0, \quad D_m = D_1(1 - 2N\phi_1\phi_2\chi), \tag{7.1.49}$$

where $D_1(\propto N_e/N^2)$ is the diffusion constant of a single chain in (7B.1). Note that the hydrodynamic diffusion constant $D_{hyd} = T/6\pi\eta\xi$ can exceed the above D_m only very close to the critical point ($\xi \gtrsim R_G N^{1.5}/N_e$ or $|1 - \chi/\chi_c| \lesssim N_e^2/N^3$) and is usually negligible [1].

7.1.5 Mutual diffusion constant in polymer blends

We examine the mutual diffusion constant D_m in (7.1.24) in polymer blends in more detail with the expression of ζ in (7.1.44). Supposing very viscous systems and neglecting the hydrodynamic diffusion constant D_{hyd}, we have [21]–[26],

$$D_m = LTr = (\phi_2 N_1 D_1 + \phi_1 N_2 D_2)\left(\frac{\phi_2}{N_1} + \frac{\phi_1}{N_2} - 2\phi_1\phi_2\chi\right), \tag{7.1.50}$$

where $D_1 \propto N_e/N_1^2\zeta_{01}$ and $D_2 \propto N_e/N_2^2\zeta_{02}$ are the single-chain diffusion constants in the reptation regime in (7B.1). If $N_1 \gg N_2$ and ϕ_1 is not small, we find

$$D_m \cong D_2\phi_1^2(1 - 2N_2\phi_2\chi). \tag{7.1.51}$$

In this case the mutual diffusion is governed by the diffusion of the shorter chains. As a result, D_m remains finite even in the gel limit $N_1 \to \infty$, as ought to be the case. However, there were some controversies before the expression (7.1.50) was established [26]. We stress that the concept of the tube velocity is essential in its derivation.

7.1.6 Relaxation of small concentration deviations

In the following theory it is important that the *network dilation rate* $\nabla \cdot \boldsymbol{v}_t$ is nonvanishing when diffusion is taking place. This is an established result for polymer gels [17], but is not trivial for other systems with transient entanglements. In the linear regime, (7.1.45) gives

$$\nabla \cdot \boldsymbol{v}_t \cong \alpha\phi_1\phi_2\nabla \cdot \boldsymbol{w} \cong -\alpha\frac{\partial}{\partial t}\delta\phi. \tag{7.1.52}$$

We assume that the system is in a one-phase state ($r > 0$) without macroscopic flow and all the deviations from equilibrium depend on space and time as $\exp(i\boldsymbol{q}\cdot\boldsymbol{r} + i\omega t)$. Then (7.1.33) may be linearized as

$$\frac{\partial}{\partial t}\delta\phi = i\omega\delta\phi = -\Gamma_q\delta\phi - L\alpha Z, \tag{7.1.53}$$

with

$$Z = \nabla\nabla : \overleftrightarrow{\sigma} = -\sum_{i,j} q_i q_j \sigma_{ij}, \tag{7.1.54}$$

where

$$\Gamma_q = LTq^2(r + Cq^2) = D_m q^2(1 + \xi^2 q^2) \tag{7.1.55}$$

is the decay rate in the absence of the dynamical coupling. In the linear regime (7.1.45)–(7.1.47) for polymer blends ((7.1.36) and (7.1.37) for polymer solutions) yield

$$Z = -\frac{4}{3}\phi_1\phi_2\alpha\eta^*(\omega)q^2(i\boldsymbol{q}\cdot\boldsymbol{w}) = \frac{4}{3}\alpha\eta^*(\omega)q^2(i\omega\delta\phi) \tag{7.1.56}$$

where use has been made of (7.1.52) and

$$\eta^*(\omega) = \frac{1}{i\omega}G^*(\omega) = \int_0^\infty dte^{-i\omega t}G_{xy}(t) \tag{7.1.57}$$

is the complex shear viscosity. We assume its behavior as

$$\begin{aligned} \eta^*(\omega) &\cong \eta \quad (\omega\tau \ll 1), \\ &\cong G/i\omega \quad (\omega\tau \gg 1), \end{aligned} \tag{7.1.58}$$

where η is the zero-shear viscosity and $G = \eta/\tau$ is the shear modulus.[2] From (7.1.54) and (7.1.57) we have

$$\left[i\omega + \Gamma_q + \left(\frac{4}{3}L\alpha^2\right)q^2 i\omega\eta^*(\omega)\right]\delta\phi = 0. \tag{7.1.59}$$

Gel-like behavior for fast motion

For $\omega\tau \gg 1$, $\delta\phi$ relaxes as in gels and the concentration decay rate is given by

$$\Gamma_{\text{gel}}(q) = \Gamma_q + \frac{4}{3}L\alpha^2 Gq^2 = L\left(Tr + \frac{4}{3}\alpha^2 G + TCq^2\right)q^2. \tag{7.1.60}$$

For small q the system relaxes diffusively with a *gel diffusion constant*,

$$\begin{aligned} D_{\text{gel}} &= L\left(Tr + \frac{4}{3}\alpha^2 G\right) \\ &= D_{\text{m}}(1 + \varepsilon_{\text{r}}{}^{-1}). \end{aligned} \tag{7.1.61}$$

On the second line we have introduced a parameter ε_{r} defined by

$$\varepsilon_{\text{r}} = Tr\Big/\left(\frac{4}{3}\alpha^2 G\right) = D_{\text{m}}\Big/\left(\frac{4}{3}L\alpha^2 G\right). \tag{7.1.62}$$

We shall see that the dynamical coupling is strong for $\varepsilon_{\text{r}} \lesssim 1$ and weak for $\varepsilon_{\text{r}} \gg 1$.

For polymer solutions, D_{m} becomes the cooperative diffusion constant (7.1.27) and $\alpha = 1/\phi$, so that

$$D_{\text{gel}} = \zeta^{-1}\left(K_{\text{os}} + \frac{4}{3}G\right), \tag{7.1.63}$$

$$\varepsilon_{\text{r}} = 3K_{\text{os}}/4G. \tag{7.1.64}$$

The above expression for D_{gel} coincides with the original one for gels [17]. It is important that $\varepsilon_{\text{r}} \sim 10^2$ for good solvent and $\varepsilon_{\text{r}} \sim 1$ for theta solvent [14]. Therefore $D_{\text{gel}} \cong D_{\text{m}}$ for

[2] The shear modulus is also written as μ for gels and solids.

good solvent, whereas $D_{\rm gel}$ can be considerably larger than $D_{\rm m}$ for theta solvent. However, in polymer blends, $D_{\rm m}$ tends to zero while $D_{\rm gel}$ remains finite near the critical point, so the strong-coupling limit $|\varepsilon_{\rm r}| \ll 1$ can be realized.

Renormalized kinetic coefficient for slow motion

For $\omega\tau \ll 1$ we may set $\eta^*(\omega) \cong \eta^*(0) = \eta$ and rewrite (7.1.59) as $[i\omega(1+\xi_{\rm ve}^2 q^2) + \Gamma_q]\delta\phi = 0$, where we define the viscoelastic length by [4]

$$\xi_{\rm ve}^2 = \frac{4}{3} L\alpha^2 \eta. \qquad (7.1.65)$$

The coupling parameter $\varepsilon_{\rm r}$ is then expressed as

$$\varepsilon_{\rm r} = D_{\rm m}\tau/\xi_{\rm ve}^2. \qquad (7.1.66)$$

The decay rate is modified as

$$\begin{aligned}\Gamma_{\rm eff}(q) &= LTq^2(r + Cq^2)/(1+\xi_{\rm ve}^2 q^2) \\ &= (L_{\rm eff}(q)/L)\Gamma_q, \end{aligned} \qquad (7.1.67)$$

with the renormalized kinetic coefficient,

$$L_{\rm eff}(q) = L/(1+\xi_{\rm ve}^2 q^2). \qquad (7.1.68)$$

An experimental result supporting the above effect is shown in Fig. 7.3 [27], which was obtained from an asymmetric polymer blend undergoing slow phase separation, as will be discussed in detail in Section 8.9. For polymer solutions we have

$$\xi_{\rm ve}^2 \sim \eta/\zeta \sim (\eta/\eta_0)\xi^2, \qquad (7.1.69)$$

where $\xi \sim a/\phi$ for theta solvent above the coexistence curve, so we confirm $\xi_{\rm ve} \gg \xi$. This length was first introduced by Brochard and de Gennes [28] for semidilute solutions with good solvent in the form $\xi_{\rm ve} = (D_m\tau)^{1/2}$. The viscoelastic length $\xi_{\rm ve}$ in polymer blends can also be much longer than ξ, as will be discussed below.

In the case $\xi_{\rm ve} \gg \xi$ the renormalized decay rate (7.1.67) behaves as

$$\begin{aligned}\Gamma_{\rm eff}(q) &\cong D_{\rm m}q^2 && (q < \xi_{\rm ve}^{-1}), \\ &\cong D_{\rm m}/\xi_{\rm ve}^2 \cong \varepsilon_{\rm r}/\tau && (\xi_{\rm ve}^{-1} < q < \xi^{-1}), \\ &\cong (\varepsilon_{\rm r}/\tau)(\xi q)^2 && (q > \xi^{-1}). \end{aligned} \qquad (7.1.70)$$

We should not forget to require $\Gamma_{\rm eff}(q)\tau < 1$ as the self-consistency condition. If $\varepsilon_{\rm r} \lesssim 1$, it is satisfied in the wide region $q\xi < \varepsilon_{\rm r}^{-1/2}$ from the third line of (7.1.70). However, if $\varepsilon_{\rm r} > 1$, it is satisfied only in the narrow region $q\xi_{\rm ve} < \varepsilon_{\rm r}^{-1/2}$ from the first line. Thus there is almost no viscoelastic renormalization effect for $\varepsilon_{\rm r} > 1$; namely, $L_{\rm eff}(q) \cong L$ for any q. The viscoelastic effect becomes important rather suddenly for $\varepsilon_{\rm r} \lesssim 1$. The simple diffusion equation cannot be used for concentration variations changing more rapidly than

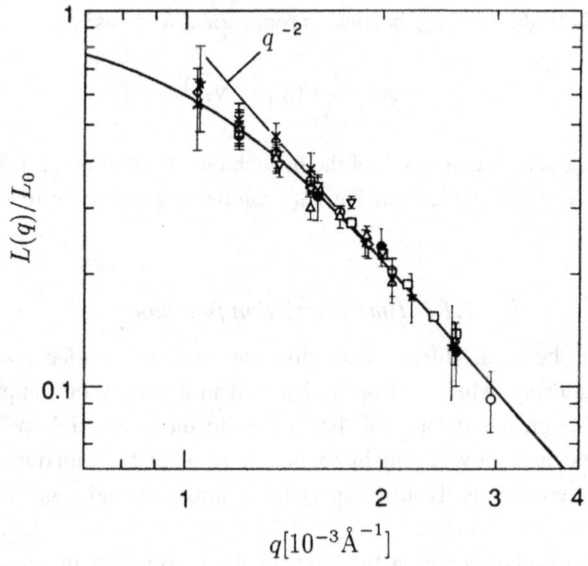

Fig. 7.3. The normalized Onsager kinetic coefficient as a function of the wave number q in an asymmetric polymer blend of PVME/d-PS observed in early-stage spinodal decomposition [27]. It may be fitted to q^{-2} for $q > R_0^{-1} = 10^{-3}$ Å^{-1}. This value of R_0 was five to seven times larger than the gyration radius R_G. In our theory it is identified with ξ_{ve} in (7.1.72).

ξ_{ve} for semidilute solutions with theta or poor solvent and entangled polymer blends with $\varepsilon_r \lesssim 1$. If the spatial scale is longer than ξ, it should be modified as

$$\left(1 - \xi_{ve}^2 \nabla^2\right) \frac{\partial}{\partial t} \delta\phi = D_m \nabla^2 \delta\phi. \tag{7.1.71}$$

Viscoelastic length for polymer blends

For polymer blends ξ_{ve} is given by

$$\xi_{ve}^2 = \frac{4}{3}(\phi_1\phi_2)^2 \left(\frac{N_1\zeta_{01} - N_2\zeta_{02}}{\phi_1 N_1 \zeta_{01} + \phi_2 N_2 \zeta_{02}}\right)^2 \left(\frac{N_e}{\phi_1 N_1 \zeta_{01}} + \frac{N_e}{\phi_2 N_2 \zeta_{02}}\right)\eta, \tag{7.1.72}$$

where use has been made of (7.1.43), (7.1.44), and (7.1.65). It is important that ξ_{ve} can be much longer than the gyration radius R_G and the correlation length ξ. To see this, let us roughly estimate it by setting $\zeta_{01} = \zeta_{02}$ in the following cases.

(i) When $N_1/N_2 - 1 \sim 1$ we find

$$\xi_{ve}^2 \sim \phi_1\phi_2 L_t^2, \tag{7.1.73}$$

where $L_t \sim N_e^{-1/2} N_1 a$ is the tube length in the reptation theory (see Appendix 7A).

(ii) In the dilute limit $\phi_2 \to 0$, ξ_{ve}^2 becomes proportional to ϕ_2 as

$$\xi_{ve}^2 \sim \phi_2 \frac{1}{N_1 N_2}(N_1 - N_2)^2 L_t^2, \tag{7.1.74}$$

where L_t is the tube length composed of the host chains. Furthermore, if $N_1 \gg N_2 \gtrsim N_e$, we obtain $\xi_{ve} \sim (N_1^3/N_2 N_e)^{1/2} \phi_2^{1/2} a$. Thus ξ_{ve} can be very long even for very small ϕ_2.

7.1.7 Time-correlation function

We now show that the stress–diffusion coupling can explain the nonexponential decay of dynamic light scattering, which has been observed in a variety of complex viscoelastic fluids [6, 7]. Although our theory will be limited to incompressible polymer solutions and blends, our mechanism will remain applicable to other fluid mixtures such as dense suspensions, microemulsions, lyotropic polymeric liquid crystals, and fluids near glass transitions.

We calculate the time-correlation function for the thermal fluctuations of the volume fraction ϕ in equilibrium in one-phase states,

$$I(q, t) = \langle \phi_{\boldsymbol{q}}(t) \phi_{\boldsymbol{q}}(0)^* \rangle, \tag{7.1.75}$$

where $\langle \cdots \rangle$ is the equilibrium average and $\phi_{\boldsymbol{q}}(t)$ is the Fourier component of $\phi(\boldsymbol{r}, t)$. The equal-time-correlation function will be assumed to be of the Ornstein–Zernike form (7.1.8) The Laplace transformation (or the one-sided Fourier transformation) with respect to time is written as

$$\widehat{I}(q, \omega) = \int_0^\infty dt e^{-i\omega t} I(q, t), \tag{7.1.76}$$

which is analytic for $\mathrm{Im}\,\omega < 0$.

In Appendix 7C, $\widehat{I}(q, \omega)$ is calculated in the following form,

$$\widehat{I}(q, \omega) = I_q \frac{1 + M^*(\omega) q^2}{i\omega[1 + M^*(\omega) q^2] + \Gamma_q}, \tag{7.1.77}$$

with

$$M^*(\omega) = \frac{4}{3} L \alpha^2 \eta^*(\omega) = \xi_{ve}^2 \eta^*(\omega)/\eta^*(0). \tag{7.1.78}$$

For polymer solutions, (7.1.77) is rewritten as

$$-i\omega \widehat{I}(q, \omega) + I_q = T\phi^2 q^2 \Big/ \left[i\omega\zeta + \left(K_{os} + \frac{4}{3} i\omega \eta^*(\omega) + \phi^2 C q^2 \right) q^2 \right]. \tag{7.1.79}$$

The above expression reduces to that for theta solvent by Brochard and de Gennes for the case of a single stress relaxation time [3]. In the real time representation, $I(q, t)$ satisfies the following non-markovian equation,

$$\dot{I}(q,t) + \Gamma_q I(q,t) + (q^2\xi_{\rm ve}^2/\eta)\int_0^t dt' G(t-t')\dot{I}(q,t') = 0, \tag{7.1.80}$$

where $\dot{I}(q,t) = \partial I(q,t)/\partial t$.

Two limiting cases are as follows.

(i) When $\Gamma_q\tau \ll 1$ holds, we may set $M^*(\omega) = \xi_{\rm ve}^2$ to obtain

$$\widehat{I}(q,\omega) \cong I_q/[i\omega + \Gamma_{\rm eff}(q)] \quad \text{or} \quad I(q,t) \cong I_q \exp[-\Gamma_{\rm eff}(q)t]. \tag{7.1.81}$$

This renormalized exponential relaxation can be observed in semidilute solutions with theta solvent and asymmetric polymer blends near the critical point.

(ii) The above formula indicates that $I(q,t)$ remains nonvanishing even for $t \gg 1/\Gamma_q$ if τ is very long. Let us assume $\tau\Gamma_q \gg 1$ and $q\xi < 1$, in which we may set $\Gamma_q \cong D_{\rm m}q^2$. In the weak-coupling case $\varepsilon_{\rm r} \gg 1$, we obtain

$$\widehat{I}(q,\omega)/I_q \cong [1 + M^*(\omega)q^2]/\Gamma_q, \tag{7.1.82}$$

for $\omega \lesssim 1/\tau$. This means that $I(q,t)$ becomes proportional to the stress relaxation function at long times as

$$I(q,t)/I_q \cong (\varepsilon_{\rm r}G)^{-1}G_{xy}(t), \tag{7.1.83}$$

for $t \gtrsim \tau$. Use has been made of the relation $M^*(\omega)/D_{\rm m} = \eta^*(\omega)T/G\varepsilon_{\rm r}$. In the strong-coupling case $\varepsilon_{\rm r} \lesssim 1$ and $\xi_{\rm ve}q \gg 1$, we may set $1 + M^*(\omega)q^2 \cong M^*(\omega)q^2$ in (7.1.77), so

$$\widehat{I}(q,\omega)/I_q \cong 1/[i\omega + \varepsilon_{\rm r}G/\eta^*(\omega)]. \tag{7.1.84}$$

Thus the decay rate is of order $\varepsilon_{\rm r}/\tau$ and is longer than $1/\tau$. In this case the relaxation is strongly governed by the viscoelastic coupling.

7.1.8 Maxwell model: single stress-relaxation time

Analytic calculations can be performed when the stress relaxes with a single relaxation time [3]. The resultant predictions are in agreement with the general trends of experiments, particularly those in Figs 7.1 and 7.2. That is, we assume that small deviations of the stress tensor are governed by

$$\frac{\partial}{\partial t}\sigma_{ij} = G\kappa_{ij}^{(\rm t)} - \frac{1}{\tau}\sigma_{ij}, \tag{7.1.85}$$

where $\kappa_{ij}^{(\rm t)}$ is defined by (7.1.47). From $\sum_{ij}\partial^2\kappa_{ij}^{(\rm t)}/\partial x_i\partial x_j = (4/3)\nabla^2\nabla\cdot\boldsymbol{v}_{\rm t}$ the quantity Z in (7.1.54) obeys

$$\frac{\partial}{\partial t}Z = -\frac{1}{\tau}Z + \frac{4}{3}G\alpha q^2\frac{\partial}{\partial t}\delta\phi. \tag{7.1.86}$$

The coupling to $\delta\phi$ arises from the network dilation relation (7.1.52). Hereafter we assume that all the deviations depend on space as $\exp(i\boldsymbol{q}\cdot\boldsymbol{r})$. From (7.1.53) $\delta\phi$ obeys

$$\frac{\partial}{\partial t}\delta\phi = -\Gamma_q\delta\phi - L\alpha Z. \tag{7.1.87}$$

Now (7.1.86) and (7.1.87) constitute a closed set of coupled equations for $\delta\phi$ and Z. If we assume $Z(0) = 0$, (7.1.87) may be integrated to give the dynamic equation of $\delta\phi(t)$ in a time-convolution form,

$$\frac{\partial}{\partial t}\delta\phi(t) = -\Gamma_q\delta\phi(t) - \frac{4}{3}\alpha^2 LGq^2\int_0^t dt'\exp[-(t-t')/\tau]\frac{\partial}{\partial t'}\delta\phi(t'). \tag{7.1.88}$$

Jäckle and co-workers [29, 30] derived a dynamic equation of the same form assuming that the chemical potential difference depends linearly on a slowly relaxing, scalar variable. Their theory was also used near to the glass transition [30]. More recently, Clarke *et al.* [31] also proposed a similar evolution equation to explain anomalous slow fluctuation growth in early-stage spinodal decomposition of a highly entangled polymer blend. General solutions of (7.1.86) and (7.1.87) are expressed as linear combinations of $\exp(-\Omega_1 t)$ and $\exp(-\Omega_2 t)$, where the two relaxation rates, Ω_1 and Ω_2, are the roots of

$$\Omega^2 - \left(\frac{1}{\tau} + \Gamma_q + \frac{4}{3}\alpha^2 GLq^2\right)\Omega + \frac{1}{\tau}\Gamma_q = 0. \tag{7.1.89}$$

In particular, $\tau\Gamma_q \ll 1$ holds at very long wavelengths, where

$$\Omega_1 \cong \Gamma_{\text{eff}}(q), \quad \Omega_2 \cong \tau^{-1}(1+\xi_{\text{ve}}^2 q^2). \tag{7.1.90}$$

We then examine the time-correlation function. The function $M^*(\omega)$ in (7.1.78) becomes

$$M^*(\omega) = \xi_{\text{ve}}^2/(1+i\omega\tau), \tag{7.1.91}$$

because the complex viscosity behaves as

$$\eta^*(\omega) = \eta/(1+i\omega\tau). \tag{7.1.92}$$

As a result $I(q,t)$ decays as

$$I(q,t)/I_q = \chi_1\exp(-\Omega_1 t) + \chi_2\exp(-\Omega_2 t), \tag{7.1.93}$$

where $\chi_1 + \chi_2 = 1$. At small q, where $\Gamma_q \ll 1/\tau$, we have the two modes given in (7.1.90). However, in dynamic light scattering experiments, the reverse condition $\Gamma_q \gg 1/\tau$ is in many cases satisfied. Here further calculations yield

$$\Omega_1 \cong \Gamma_{\text{gel}}(q), \quad \Omega_2 \cong \Gamma_q/(\tau\Gamma_q + \xi_{\text{ve}}^2 q^2) = \Gamma_q/\tau\Gamma_{\text{gel}}(q), \tag{7.1.94}$$

where $\Gamma_{\text{gel}}(q)$ is defined by (7.1.60), and

$$\chi_2 \cong \xi_{\text{ve}}^2 q^2/(\tau\Gamma_q + \xi_{\text{ve}}^2 q^2) = 1/[\varepsilon_{\text{r}}(1+\xi^2 q^2)+1]. \tag{7.1.95}$$

We furthermore assume $q \ll \xi^{-1}$ to find

$$\begin{aligned} \Omega_1 &\cong D_{\text{gel}}q^2, \quad \Omega_2 \cong \varepsilon_r/(1+\varepsilon_r)\tau, \\ \chi_2 &\cong 1/(1+\varepsilon_r), \end{aligned} \tag{7.1.96}$$

where D_{gel} is defined by (7.1.61). As $\varepsilon_r \to 0$, we have $I(q,t)/I_q \cong \exp(-\Omega_2 t)$ with Ω_2 smaller than $1/\tau$ by ε_r, which agrees with (7.1.84).

7.1.9 Viscoelastic Ginzburg–Landau theory for polymer solutions

To describe viscoelastic effects on the concentration inhomogeneities in the Ginzburg–Landau scheme, it is convenient to introduce a new dynamic variable $\overleftrightarrow{W} = \{W_{\alpha\beta}\}$, which is a symmetric tensor representing chain conformations undergoing deformations and is analogous to the Finger tensor in (3A.13). Note that ϕ changes in time more rapidly than $\overleftrightarrow{W}$ even at relatively small wave numbers for entangled systems. We need to construct a canonical form of dynamic equations or a set of Langevin equations satisfying the fluctuation–dissipation relations [32]–[34]. Such formal frameworks for viscoelastic fluids have already been presented but without discussions of phase transitions [35, 36]. In the following we consider entangled polymer solutions in the semidilute regime ($N^{-1/2} \lesssim \phi \ll 1$). We will numerically solve the resultant dynamic equations to examine viscoelastic spinodal decomposition in Section 8.9 and shear-induced phase separation in Section 11.2

For entangled polymers we may define $\overleftrightarrow{W}$ as follows. Let us consider entanglement points $\boldsymbol{R}_n$ on a chain and number them consecutively along it as $n = 1, 2, \ldots, N/N_e$, where N/N_e is the number of entanglements on a chain. Then,

$$W_{\alpha\beta} = \frac{1}{Na^2}\left\langle \sum_{n=1}^{N/N_e} (\boldsymbol{R}_{n+1} - \boldsymbol{R}_n)_\alpha (\boldsymbol{R}_{n+1} - \boldsymbol{R}_n)_\beta \right\rangle_{\text{chain}}, \tag{7.1.97}$$

where the sum is taken over entanglement points on a chain, and the average $\langle\cdots\rangle_{\text{chain}}$ is taken over all chains contained in a volume element whose linear dimension is longer than the gyration radius ($\sim N^{1/2}a$). In equilibrium we assume the gaussian distribution of $\boldsymbol{R}_{n+1} - \boldsymbol{R}_n$ to obtain $\langle W_{\alpha\beta}\rangle_{\text{eq}} = \delta_{\alpha\beta}$, where $\langle\cdots\rangle_{\text{eq}}$ is the equilibrium average. We generalize the free-energy functional in (7.1.2) to the following form

$$\mathcal{H}\{\phi, \overleftrightarrow{W}\} = \int d\boldsymbol{r}\left[v_0^{-1} f_{\text{site}}(\phi) + \frac{T}{2}C(\phi)|\nabla\phi|^2 + \frac{1}{2}G(\phi)Q(\overleftrightarrow{W})\right] \tag{7.1.98}$$

where $G(\phi)(\sim Tv_0^{-1}\phi^{\alpha_G})$ is the shear modulus. The simple scaling theory gives $\alpha_G = 3$, although $\alpha_G \cong 2.25$ in experiments [14, 15]. The $Q(\overleftrightarrow{W})$ is a nonnegative-definite function of $\overleftrightarrow{W}$. In simulations, which will be explained in Sections 8.9 and 11.2, we assume the simplest gaussian form,

$$Q(\overleftrightarrow{W}) = \frac{1}{2}\sum_{\alpha\beta}(W_{\alpha\beta} - \delta_{\alpha\beta})^2. \tag{7.1.99}$$

This form is questionable for large deformations, however. Alternatively, we may set $Q = I_1 - \ln I_3$ [33] or $Q = 3 \ln I_1 - \ln I_3$ [34], where $I_1 = \sum_\alpha W_{\alpha\alpha}$ and $I_3 = \det \overleftrightarrow{W}$. Such forms are suggested by the finite-strain theory in Appendix 3A.

Because $\overleftrightarrow{W}$ represents the network deformation, its motion is determined by the polymer velocity $\boldsymbol{v}_{\rm p}$ and its simplest dynamic equation is of the form

$$\frac{\partial}{\partial t} W_{\alpha\beta} + (\boldsymbol{v}_{\rm p} \cdot \nabla) W_{\alpha\beta} - \sum_\gamma (D_{\alpha\gamma} W_{\gamma\beta} + W_{\alpha\gamma} D_{\beta\gamma}) = -\frac{1}{\tau}(W_{\alpha\beta} - \delta_{\alpha\beta}), \tag{7.1.100}$$

where $\{D_{\alpha\beta}\}$ is the gradient tensor of the polymer velocity,

$$D_{\alpha\beta} = \frac{\partial}{\partial x_\beta} v_{{\rm p}\alpha}, \tag{7.1.101}$$

and τ on the right-hand side of (7.1.100) is the stress relaxation time which is very long in the semidilute region and behaves as (7.1.30). In the rheological literature [37, 38], the left-hand side of (7.1.100) is called the upper convective time derivative, but it is known that there is a class of time derivatives satisfying the requirement of the frame invariance. Once we have the free energy and the dynamic equation for $\overleftrightarrow{W}$, we may calculate the network stress tensor induced by $\overleftrightarrow{W}$ as

$$\begin{aligned} \overleftrightarrow{\sigma}_{\rm p} &= G \overleftrightarrow{W} \cdot \frac{\partial}{\partial \overleftrightarrow{W}} Q + \frac{1}{2} G Q \overleftrightarrow{I} \\ &= G \overleftrightarrow{W} \cdot (\overleftrightarrow{W} - \overleftrightarrow{I}) + \frac{1}{2} G Q \overleftrightarrow{I}. \end{aligned} \tag{7.1.102}$$

The first line holds for general Q and the second line for the special choice (7.1.99). (See Appendix 7D for its derivation.) With the above results and (7.1.4), the force densities $\boldsymbol{F}_{\rm p}(= \boldsymbol{F}_1)$ and $\boldsymbol{F}_{\rm s}(= \boldsymbol{F}_2)$ in the two-fluid dynamic equations (7.1.16) and (7.1.17) are obtained as

$$\boldsymbol{F}_{\rm p} = -\frac{1}{4} Q \nabla G + \nabla \cdot \overleftrightarrow{\sigma}_{\rm p}, \quad \boldsymbol{F}_{\rm s} = \eta_0 \nabla^2 \boldsymbol{v}_s \cong \eta_0 \nabla^2 \boldsymbol{v}. \tag{7.1.103}$$

The first term ($\propto \nabla G$) in $\boldsymbol{F}_{\rm p}$ arises from the concentration dependence of G (and is not included in (7.1.31)).

Noise terms

We have obtained a closed set of dynamic equations for the gross variables. We may add gaussian and markovian noise terms on the right-hand sides of the dynamic equations, (7.1.21) for $\boldsymbol{w}$, (7.1.34) for $\boldsymbol{v}$, and (7.1.100) for $\overleftrightarrow{W}$. The amplitudes of the noise terms are determined from the fluctuation–dissipation relations [32]. Then these equations are Langevin equations in the general scheme of Chapter 5.

Adiabatic approximations

When we are interested in slow motions, the relative velocity $\boldsymbol{w}$ and the average velocity $\boldsymbol{v}$ are determined in the adiabatic approximations (7.1.32) and (7.1.35). Then,

$$\boldsymbol{w} = \frac{1}{\zeta}\left[-\phi\nabla\frac{\delta\mathcal{H}}{\delta\phi} + \boldsymbol{F}_{\mathrm{p}}\right], \tag{7.1.104}$$

$$-\eta_0\nabla^2\boldsymbol{v} = \left[-\phi\nabla\frac{\delta\mathcal{H}}{\delta\phi} + \boldsymbol{F}_{\mathrm{p}}\right]_{\perp} = [\zeta\boldsymbol{w}]_{\perp}, \quad \nabla\cdot\boldsymbol{v} = 0, \tag{7.1.105}$$

where the friction coefficient $\zeta(\propto \phi^2)$ is given by (7.1.26) and $[\cdots]_{\perp}$ denotes taking the transverse part. At this stage, $\boldsymbol{w}$ and $\boldsymbol{v}$ have been expressed in terms of ϕ and $\overleftrightarrow{W}$ in the adiabatic limits, so that the independent dynamic variables are reduced from $\{\delta\rho, \phi, \boldsymbol{v}_{\mathrm{p}}, \boldsymbol{v}_{\mathrm{s}}, \overleftrightarrow{W}\}$ to $\{\phi, \overleftrightarrow{W}\}$.

Linearized equations

To linear order in the deviation $\delta W_{\alpha\beta} = W_{\alpha\beta} - \delta_{\alpha\beta}$, the network stress is expressed as

$$\sigma_{\alpha\beta} \cong G\delta W_{\alpha\beta}, \tag{7.1.106}$$

and (7.1.100) becomes

$$\frac{\partial}{\partial t}\delta W_{\alpha\beta} = D_{\alpha\beta} + D_{\beta\alpha} - \frac{1}{\tau}\delta W_{\alpha\beta}. \tag{7.1.107}$$

Thus, in the linear regime, our Ginzburg–Landau model becomes essentially the same as the Maxwell model with the dynamical stress–diffusion coupling. In the presence of weak steady shear flow $\langle\boldsymbol{v}\rangle = \dot{\gamma}y\boldsymbol{e}_x$, $\boldsymbol{e}_x$ being the unit vector in the x direction, we may use (7.1.100) to obtain $\delta W_{xy} = \dot{\gamma}\tau$. Then (7.1.106) yields the shear viscosity increase,

$$\Delta\eta = \eta - \eta_0 = G\tau, \tag{7.1.108}$$

in the linear response regime. If $\Delta\eta \gg \eta_0$, the system becomes highly viscoelastic due to deformations of $\overleftrightarrow{W}$. In the reverse case of rapid motions with characteristic frequencies much larger than τ^{-1}, we integrate (7.1.107) as

$$\delta W_{\alpha\beta} = \frac{\partial}{\partial x_\alpha}u_{\mathrm{p}\beta} + \frac{\partial}{\partial x_\beta}u_{\mathrm{p}\alpha}, \tag{7.1.109}$$

where $\boldsymbol{u}_{\mathrm{p}}(\boldsymbol{r}, t) = \int_0^t dt'\boldsymbol{v}_{\mathrm{p}}(\boldsymbol{r}, t')$ represents the displacement vector of the transient network. Then (7.1.106) assumes the form of elastic stress with G being the shear modulus.

7.2 Dynamics in gels

With the formation of a network in polymer systems, fluid (sol) states change into soft solid (gel) states. Such sol–gel phase transitions have long been studied in the literature [1]. Salient features at the transition point are singular critical behavior of the dynamic shear modulus, $G^*(\omega) \sim (i\omega)^\beta$ [39], and a power-law decay of the (homodyne) dynamic light scattering amplitude, $I_2(q,t) \sim t^{-n}$ [40]. Interplay between phase separation and gelation poses new problems in thermoreversible physical gels [41]. Field theoretical approach to vulcanization via the replica technique is also worth mentioning [42]. There have also been a number of experimental and theoretical studies on nematic elastomers [43], which exhibit unique mechanical properties due to the coupling between molecular orientation and strain. In this section, however, we will mainly treat phase transitions influenced by elasticity in chemical gels, focusing on inhomogeneous network fluctuations. To this end we will extend the mean field theory of macroscopic shape-change transitions in gels presented in Section 3.5.

Near-equilibrium dynamics of the network fluctuations has been studied by dynamic light scattering [17, 44]. Furthermore, with a lowering of the solvent quality, we encounter three kinds of instabilities occurring successively or simultaneously. (i) In isotropic gels immersed in solvent, a macroscopic instability occurs for $K_{\rm os} < 0$ against a volume change. In this process, a gel can remain transparent without small-scale phase separation if the temperature change consists of very small and slow steps. It proceeds only with absorption or desorption of solvent through the gel–solvent interface, so it is extremely time-consuming unless the gel size is very small [45]. (ii) Dynamic critical behavior detected by light scattering [44] can be expected near a bulk instability point, where $K_{\rm os} + 4\mu/3 = 0$ in isotropic gels [46, 47]. Below this point spinodal decomposition takes place, which we will investigate in Chapter 8. (iii) As the third kind of instability, a surface instability can take place on a gel–solvent interface [48]–[51]. An example is given in Fig. 7.4. It is triggered for $K_{\rm os} + \mu/3 < 0$ in isotropic gels, but it can be induced more easily on the surface of uniaxially stretched gels. Tanaka *et al.* [48] observed surface patterns consisting of numerous line segments of cusps into the gel, transiently on the surface of swelling gels and permanently on the surface of uniaxially swollen gels whose lower surface is clamped to a substrate. The physical mechanism responsible for the development of these patterns is now well understood, and they can be reproduced analytically [52, 53] and numerically [54, 55].

A variety of shape changes have also been observed in shrinking gels, which include surface bubbles [56], necklace-like bubbles, bamboos, and wrinkled tubes [57]. Representative examples are shown in Fig. 7.5. There, a sudden shrinkage produced a dense impermeable layer in the surface region, which caused an increase of the osmotic pressure inside the sample and led to internal phase separation under a fixed volume [57]. Figure 7.6 shows *stable* two-phase coexistence on a cylindrical NIPA gel in a water–methanol mixture [58], where the two phases were homogeneous and transparent and shear deformations were induced near the interfaces under the condition $\mu \ll K_{\rm os}$.

7.2.1 The GLW hamiltonian for gels

We first consider neutral gels for simplicity and will later briefly treat weakly ionized gels. A gel point is represented by $\boldsymbol{r}_0 = (x_{01}, x_{02}, x_{03})$ in the as-prepared state and by $\boldsymbol{r} = (x_1, x_2, x_3)$ after deformation. Parameterization of physical quantities in terms

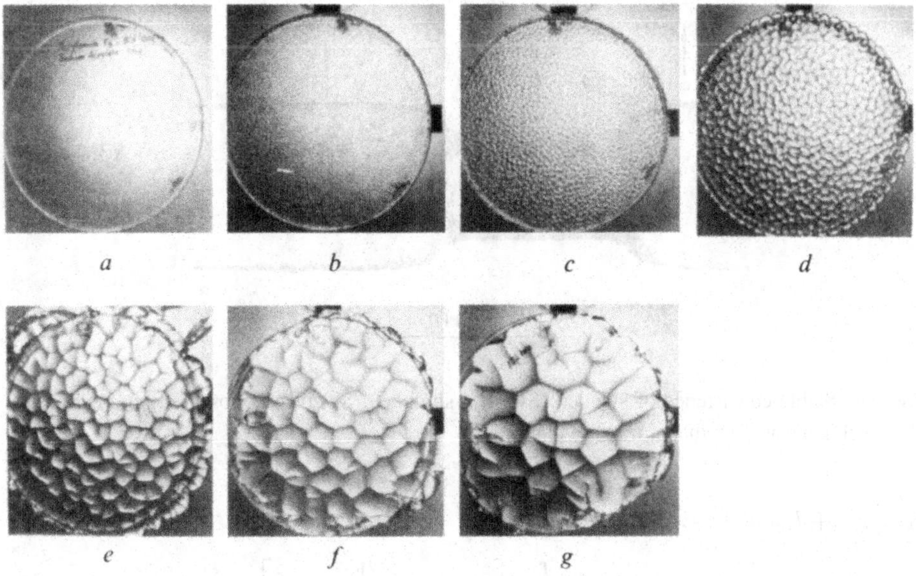

Fig. 7.4. An ionized acrylamide gel formed in a Petri dish is allowed to swell in water. An extremely fine pattern appears on the free surface of the gel, and coarsens with time (a→g) [48].

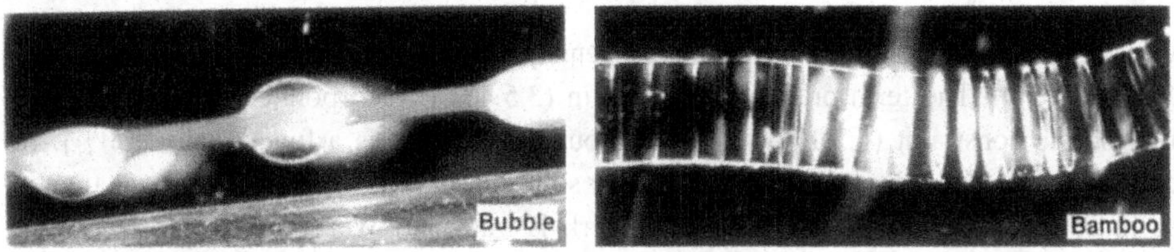

Fig. 7.5. Bubble and bamboo patterns in a shrinking cylindrical acrylamide gel immersed in an acetone–water mixture [57]. A variety of patterns emerge depending on the acetone composition and the degree of uniaxial stretching.

of $\boldsymbol{r}_0$ is called the Lagrange representation, while that in terms of $\boldsymbol{r}$ is called the Euler representation. We have already introduced the deformation tensor $\Gamma_{ij} = \partial x_i/\partial x_{0j}$ in (3.5.46) and the Finger tensor $W_{ij} = \sum_k \Gamma_{ik}\Gamma_{jk}$ in (3A.13) [46, 52, 59]. The volume fraction ϕ is expressed as (3.5.47) or

$$\phi = \phi_0/[\det\{W\}]^{1/2}. \tag{7.2.1}$$

Constructing the Ginzburg–Landau theory of gels is fairly straightforward using the free energy (3.5.52), in which the elastic free energy is included, and the gradient term for polymer solutions in (7.1.2). Using $d\boldsymbol{r}_0 = d\boldsymbol{r}\phi/\phi_0$ we thus set up the GLW hamiltonian

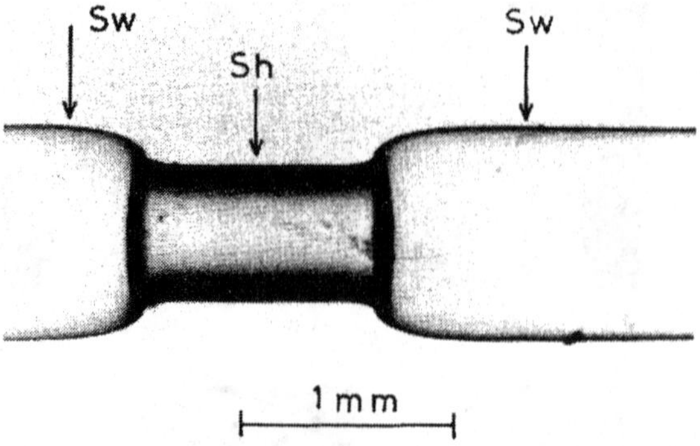

Fig. 7.6. Stable coexistence of swollen (Sw) and shrunken (Sh) phases observed on a cylindrical NIPA gel in a water–methanol mixture [58].

$\mathcal{H} = \mathcal{H}_\phi + \mathcal{H}_{\rm el}$ in the Euler representation as

$$\beta\mathcal{H}_\phi = \int d\boldsymbol{r}\left[v_0^{-1}g(\phi) + \frac{1}{2}C|\nabla\phi|^2\right], \tag{7.2.2}$$

$$\beta\mathcal{H}_{\rm el} = \int d\boldsymbol{r}\frac{\phi}{2\phi_0}v_0\sum_i W_{ii}, \tag{7.2.3}$$

where $g(\phi)$ is a dimensionless free-energy density. For neutral gels it is given by (3.5.53) (with $f = 0$). The term proportional to B in (3.5.43) is incorporated into $g(\phi)$. If we impose the constraint (7.2.1) (or (3.5.47)), our hamiltonian is a functional of $\{W_{ij}\}$. In contrast, in the viscoelastic model for polymer solutions in (7.1.98), ϕ and $\{W_{ij}\}$ are treated as independent variables because the network is transient. We note that the constraint (7.2.1) ceases to be a good approximation on spatial scales shorter than the average distance between the crosslink points.

We may calculate the stress tensor $\overleftrightarrow{\Pi} = \{\Pi_{ij}\}$ by superimposing an infinitesimal additional displacement $\delta\boldsymbol{u}$ onto $\boldsymbol{r}$ and expressing the free-energy change as (6A.6) to first order in $\delta\boldsymbol{u}$. In this calculation we use the identities, $\delta\Gamma_{ij} = \partial\delta u_i/\partial x_{0j}$ and $\partial\delta u_i/\partial x_j = \sum_k \Gamma^{kj}\delta\Gamma_{ik}$, where $\{\Gamma^{ji}\} = \{\partial x_{0j}/\partial x_i\}$ is the inverse matrix of $\{\Gamma_{ij}\}$. The incremental change of ϕ is given by

$$\delta\phi = -\phi\sum_{ij}\Gamma^{ji}\delta\Gamma_{ij} = -\phi\sum_i\frac{\partial\delta u_i}{\partial x_i}. \tag{7.2.4}$$

Note the general formula, $\partial\ln[\det\overleftrightarrow{A}]/\partial A_{ij} = A^{ji}$ for arbitrary matrix $\overleftrightarrow{A}$. Then $\overleftrightarrow{\Pi} = \overleftrightarrow{\Pi}_\phi - \overleftrightarrow{\sigma}$ consists of two terms [52]. The first term is of the same form as that for binary fluid mixtures in Chapter 6 and is determined by $\mathcal{H}_\phi$ as

$$\Pi_{\phi ij} = T\left[v_0^{-1}(\phi g' - g) - \frac{1}{2}\nabla\cdot(C\phi\nabla\phi) - \frac{1}{2}\phi C\nabla^2\phi\right]\delta_{ij} + TC\frac{\partial\phi}{\partial x_i}\frac{\partial\phi}{\partial x_j}, \tag{7.2.5}$$

where $g' = \partial g/\partial \phi$. The second term is the elastic part,

$$\sigma_{ij} = T\nu_0(\phi/\phi_0)W_{ij}. \tag{7.2.6}$$

The resultant force density acting on the network is simply of the form,

$$-\nabla\cdot\overleftrightarrow{\Pi} = -\phi\nabla\frac{\delta}{\delta\phi}\mathcal{H}_\phi + \nabla\overleftrightarrow{\sigma}. \tag{7.2.7}$$

In the Lagrange representation $\mathcal{H}$ can also be treated as a functional of $\boldsymbol{r} = \boldsymbol{r}(\boldsymbol{r}_0)$. Using (6A.6) we find

$$\left(\frac{\delta\mathcal{H}}{\delta\boldsymbol{r}}\right)_{r_0} = \frac{\phi_0}{\phi}\nabla\cdot\overleftrightarrow{\Pi}. \tag{7.2.8}$$

Thus the extremum condition $(\delta\mathcal{H}/\delta\boldsymbol{r})_{r_0} = 0$ in the Lagrange representation is equivalent to the stress balance condition in the Euler representation.

Gaussian approximation

With our model hamiltonian we examine small fluctuations around homogeneously deformed states in 3D. The deformation is represented by

$$x_i = \sum_j A_{ij}x_{0j} + u_i, \tag{7.2.9}$$

where $\boldsymbol{u}$ is a displacement vector. We are interested in the Fourier components $\boldsymbol{u}_{\boldsymbol{q}}$ with $q = |\boldsymbol{q}|$ much larger than the inverse system size. As can be known from Appendix 3A, the increase of $\mathcal{H}$ in the bilinear order is calculated as

$$\delta\mathcal{H} = \frac{1}{2}\int_{\boldsymbol{q}}\left[\left(K_{\text{os}} + \frac{\mu}{3} + TC\phi^2q^2\right)|\boldsymbol{q}\cdot\boldsymbol{u}_{\boldsymbol{q}}|^2 + \mu J(\hat{\boldsymbol{q}})q^2|\boldsymbol{u}_{\boldsymbol{q}}|^2\right], \tag{7.2.10}$$

where K_{os} and μ are defined by (3.5.57) and (3.5.54), respectively, and $J(\hat{\boldsymbol{q}})$ depends on the direction $\hat{\boldsymbol{q}} = q^{-1}\boldsymbol{q}$ of the wave vector as

$$J(\hat{\boldsymbol{q}}) = (\phi/\phi_0)^{2/3}\sum_{ijk} A_{ik}A_{jk}\hat{q}_i\hat{q}_j. \tag{7.2.11}$$

The reference volume fraction $\phi_{\text{r}} = \phi_0/\det\overleftrightarrow{A}$ is written as ϕ for simplicity. Note that the right-hand side of (7.2.10) takes the standard form of isotropic elasticity [60] in the long-wavelength limit, since $J(\hat{\boldsymbol{q}}) = 1$ in the isotropic case. The thermal structure factor in the mean field theory is written as

$$I_{\text{th}}(\boldsymbol{q}) = \langle|\phi_{\boldsymbol{q}}|^2\rangle = T\phi^2\Big/\left[K_{\text{os}} + \frac{\mu}{3} + \mu J(\hat{\boldsymbol{q}}) + TC\phi^2q^2\right], \tag{7.2.12}$$

which depends on the direction $\hat{\boldsymbol{q}}$ even in the limit $q \to 0$ in anisotropic gels.

(i) In the isotropic case, we have the usual Ornstein–Zernike form with the correlation length ξ defined by

$$\xi^{-2} = \left(K_{\text{os}} + \frac{4}{3}\mu\right)/(TC\phi^2), \tag{7.2.13}$$

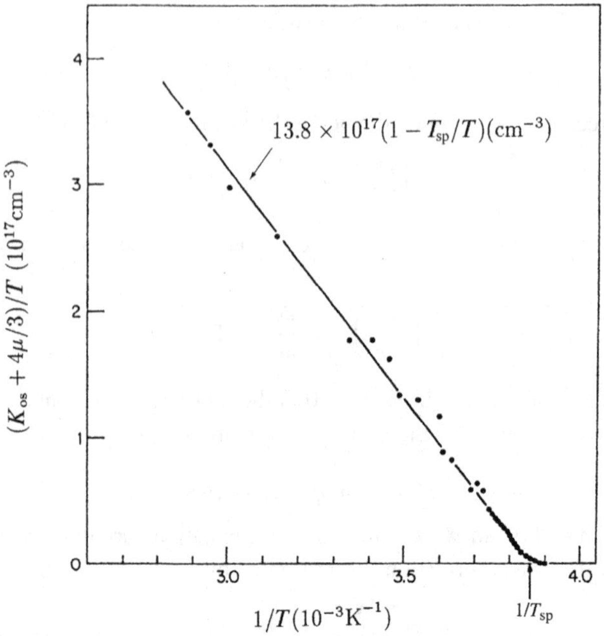

Fig. 7.7. $K_{\text{os}} + 4\mu/3$ obtained from the inverse of the scattered light intensity from a 2.5% polyacrylamide gel [44].

Thus the intensity at long wavelengths diverges as $K_{\text{os}} + 4\mu/3 \to 0$. As shown in Fig. 7.7, $K_{\text{os}} + 4\mu/3 \cong 5 \times 10^4(T/T_s - 1)$ dyn/cm^2 with the spinodal temperature $T_{\text{sp}} \cong 260$ K and $\mu \sim 10^2$ dyn/cm^2 in a 2.5% polyacrylamide gel [44].

(ii) For the uniaxial case (3.5.66), we have

$$J(\hat{\boldsymbol{q}}) = (\lambda^2 - \lambda^{-1})\hat{q}_x^2 + \lambda^{-1}. \tag{7.2.14}$$

The structure factor is then

$$I_{\text{th}}(\boldsymbol{q}) = \langle|\phi_{\boldsymbol{q}}|^2\rangle = T\phi^2 \Bigg/ \left[K_{\text{os}} + \left(\frac{1}{3} + \frac{1}{\lambda}\right)\mu + \left(\lambda^2 - \frac{1}{\lambda}\right)\mu\hat{q}_x^2 + +TC\phi^2q^2\right], \tag{7.2.15}$$

which is analogous to the structure factor of the spin fluctuations in Ising systems with dipolar interaction [61]. For $\lambda > 1$ the thermal fluctuations of the network density are weaker in the stretched direction than in the perpendicular directions even in the small-q limit. However, this is apparently in contradiction with some scattering experiments, as will be discussed in Section 7.3.

Bulk instability in uniaxial gels

With (7.2.15) we may identify the bulk spinodal point in uniaxially deformed gels. (i) For stretching $\lambda > 1$, most enhanced are the long-wavelength fluctuations varying in the plane perpendicular to the uniaxial axis ($q_x = 0$). Spinodal decomposition should take place for

$$K_{\text{os}} + \left(\frac{1}{3} + \frac{1}{\lambda}\right)\mu < 0, \tag{7.2.16}$$

where we should observe cylindrical domains elongated in the stretched direction in late stages. In agreement with this result, Horkay *et al.* [62] observed strong anisotropic domain scattering from a stretched gel in theta solvent, which we will discuss again in Section 8.9. (ii) For compression $\lambda < 1$, those varying in the uniaxial axis trigger the bulk instability for

$$K_{\text{os}} + \left(\frac{1}{3} + \lambda^2\right)\mu < 0, \tag{7.2.17}$$

where one-dimensional, lamellar-like domains should emerge.

Two-dimensionally constrained gels

A gel may change its shape only in one direction (parallel to the x axis) if its lower surface is clamped onto a plate or if it is inserted into a glass tube in a shrunken state and is swollen afterwards. Here the elongation ratio $\alpha_\perp$ in the perpendicular directions is held constant, while the elongation ratio $\alpha_\parallel$ in the x direction or $\phi = (\phi_0/\alpha_\perp^2)/\alpha_\parallel$ is the order parameter. We calculate the longitudinal osmotic modulus in the form,

$$K_\parallel = \phi\left(\frac{\partial}{\partial \phi}\Pi_\parallel\right)_{T\alpha_\perp} = K_{\text{os}} + \left(\frac{1}{3} + \lambda^2\right)\mu, \tag{7.2.18}$$

where $\Pi_\parallel = \Pi_{xx} = Tv_0^{-1}(\phi g' - g) - Tv_0(\phi/\phi_0)\alpha_\parallel^2$ from (7.2.5) and (7.2.6). (i) In the stretched case $\lambda > 1$, spinodal decomposition in the bulk region occurs under (7.2.16), while $K_\parallel > 0$ or before onset of macroscopic instability. (ii) In the compressed case $\lambda < 1$, (7.2.17) and (7.2.18) show that the two instabilities are both triggered simultaneously. In summary of this uniaxial case, we predict that spinodal decomposition in the bulk region will be observed before macroscopic shape changes both for $\lambda > 1$ and $\lambda < 1$. Another uniaxial case under a constant stretching force was considered in Section 3.5, where a macroscopic instability precedes spinodal decomposition.

7.2.2 Third-order elastic interaction: correspondence between gels and alloys

In the gaussian approximation, the longitudinal ($\parallel \boldsymbol{q}$) and transverse ($\perp \boldsymbol{q}$) components of the displacement $\boldsymbol{u}_q$ are decoupled as in (7.2.10). In Appendix 7E we shall see that they are nontrivially coupled in the third order in $\mathcal{H}$. In a gel with a clamped boundary, elimination of the transverse displacement (minimization of the free energy at fixed space-dependent volume fraction) yields

$$\mathcal{H}_{\text{el}}^{(3)} = -g_{\text{gel}} \int d\boldsymbol{r}\delta\phi \sum_{ij} \left(\nabla_i \nabla_j \frac{1}{\nabla^2}\delta\phi\right)^2, \tag{7.2.19}$$

where $1/\nabla^2$ is the inverse operator of ∇^2. This interaction is of third order in the deviation $\delta\phi = \phi - \phi_r$, where $\phi_r = \langle\phi\rangle$ is the average volume fraction. Its strength is represented by

$$g_{\rm gel} = \mu/2\phi_r^3. \tag{7.2.20}$$

This form is applicable both in 2D and 3D. It is worth noting that a third-order interaction of the same form arises in binary alloys in which the shear modulus depends on the composition as (10.1.8). From (10.1.37)–(10.1.41) in Chapter 10, we recognize that shrunken (swollen) regions in gels correspond to harder (softer) regions in binary alloys because of the minus sign in (7.2.19).[3] Although this correspondence apparently contradicts the fact that the shear modulus in gels decreases with swelling, it is supported by observations of network formation of polymer-rich regions in unstable entangled polymer solutions [63]. It will also be supported by numerical analysis of phase separation in gels in Section 8.9.

The above elastic interaction may be calculated analytically for a spheroidal domain in a metastable or unstable gel matrix. We assume that its shape is represented by $x^2/a^2 + (y^2+z^2)/b^2 = 1$ and the volume fraction changes in a step-wise manner from $\phi_{\rm in}$ inside the ellipsoid and ϕ_r outside it. The domain is shrunken for $\Delta\phi = \phi_{\rm in} - \phi_r > 0$ and swollen for $\Delta\phi = \phi_{\rm in} - \phi_r < 0$ as compared to the surrounding region. As in (10.1.62) for alloys, we obtain

$$\mathcal{H}_{\rm el}^{(3)} = -g_{\rm gel}(\Delta\phi)^3 V_{\rm e}\left[\frac{1}{3} + \frac{3}{2}\left(N_x - \frac{1}{3}\right)^2\right], \tag{7.2.21}$$

where $V_{\rm e}$ is the volume of the spheroid and N_x is the depolarization factor dependent on a/b as in Fig. 10.7. The shape factor $(N_x - 1/3)^2$ is minimum for spheres and maximum for pancake shapes for which $a \ll b$ and $N_x \cong 1$. We here present some predictions for phase-separating neutral gels. (i) Shrunken domains will eventually take compressed shapes. They will tend to touch one another to form a continuous phase even when their volume fraction is relatively small. The characteristic thickness $R_{\rm E}$ of compressed domains should be determined from a balance of the surface free energy and the third-order-interaction in the form

$$R_{\rm E} \sim \sigma/(g_{\rm gel}|\Delta\phi|^3), \tag{7.2.22}$$

where σ is the surface tension. If the volume fractions in the two phases are not close and $|\Delta\phi| \sim \phi_r$, R_e is simply expressed as

$$R_{\rm E} \sim \sigma/\mu, \tag{7.2.23}$$

which greatly exceeds the correlation length ξ for weak crosslinkage. (ii) Swollen domains will not be much deformed from sphericity particularly for small droplet volume fractions. (iii) Moreover, phase transitions in a clamped gel occur between homogeneous one-phase states and two-phase states with pinned domains. They are discontinuous or hysteretic at any network volume fraction. This means that there is no Ising-type critical point in the presence of $\mathcal{H}_{\rm el}^{(3)}$. The phase diagram of clamped neutral gels can be known from that of

[3] The minus sign arises from the constraint (7.2.1), whereas there is no such constraint between the composition and the elastic field in alloys.

alloys in Fig. 10.15, where the polymer volume fraction in gels corresponds to the volume fraction of the softer component in alloys.

7.2.3 Weakly charged gels

In Appendix 7F we will briefly explain the Debye–Hückel theory or random phase approximation (RPA) for charged polymer systems [64]–[68]. Let a small fraction $\hat{f}$ of monomers composing the chains be charged[4] and salt ions be present with a small average density $2\bar{c}_{\rm sa}$. This theory is valid when the typical electrostatic energy $e^2\bar{c}_{\rm total}^{1/3}/\epsilon_{\rm s}$ of the mobile ions is much smaller than T or equivalently $\bar{c}_{\rm total} \gg \kappa_{\rm Db}^3$ [64], where $\bar{c}_{\rm total} = 2\bar{c}_{\rm sa} + v_0^{-1}\hat{f}\phi$ is the total mobile-ion density and $\epsilon_{\rm s}$ is the dielectric constant of the solvent. In this case mobile ions are mostly moving in solvent, being not trapped by the chains, and screen the Coulomb interaction between the opposite charges on the chains. The inverse Debye screening length $\kappa_{\rm Db}$ is defined by

$$\kappa_{\rm Db}^2 = 4\pi\ell_{\rm B}\bar{c}_{\rm total}, \tag{7.2.24}$$

where

$$\ell_{\rm B} = e^2/\epsilon_{\rm s}T \tag{7.2.25}$$

is the Bjerrum length (of order 7 Å in water at 300 K). From (7F.6) the thermal structure factor $I_{\rm th}(\boldsymbol{q})$ for gels can be obtained if $1 - 2\chi$ in $K_{\rm os}$ for neutral gels is replaced by $1 - 2\chi + v_0^{-1}4\pi\ell_{\rm B}\hat{f}^2/(q^2 + \kappa_{\rm Db}^2)$, where χ is the interaction parameter assumed to be independent of ϕ. The resultant q-dependent osmotic bulk modulus becomes

$$\begin{aligned} K(q) &= K_{\rm os} + v_0^{-2}T\phi^2\frac{4\pi\ell_{\rm B}\hat{f}^2}{q^2+\kappa_{\rm Db}^2} \\ &= \frac{T\phi^2}{v_0}\left[1 - 2\chi + \phi + \frac{4\pi\ell_{\rm B}\hat{f}^2}{v_0(q^2+\kappa_{\rm Db}^2)}\right] + Tv_0\left[B\frac{\phi}{\phi_0} - \frac{1}{3}\left(\frac{\phi}{\phi_0}\right)^{1/3}\right]. \end{aligned} \tag{7.2.26}$$

The second line is the result from the Flory–Huggins theory and the classical rubber theory. As a generalization of (7.2.12), the thermal structure factor under anisotropic deformation is written as

$$I_{\rm th}(\boldsymbol{q}) = T\phi^2 \Big/ \left[K(q) + \frac{\mu}{3} + \mu J(\hat{\boldsymbol{q}}) + TC\phi^2q^2\right], \tag{7.2.27}$$

which depends on $\hat{\boldsymbol{q}}$ through $J(\hat{\boldsymbol{q}})$. For the isotropic case, where $J(\hat{\boldsymbol{q}}) = 1$, $I_{\rm th}(q)$ can have a peak at an intermediate wave number $q_{\rm m}$ determined by

$$\begin{aligned} q_{\rm m}^2 &= v_0^{-1}(4\pi\ell_{\rm B}/C)^{1/2}\hat{f} - \kappa_{\rm Db}^2 \\ &= v_0^{-1}(4\pi\ell_{\rm B}/C)^{1/2}\big[\hat{f} - (4\pi\ell_{\rm B}C)^{1/2}(2v_0\bar{c}_{\rm sa} + \phi\hat{f})\big]. \end{aligned} \tag{7.2.28}$$

[4] The number of counterions per chain was written as f in (3.5.51) and is equal to $N\hat{f}$.

The right-hand side is required to be positive here. Without salt ($\bar{c}_{\mathrm{sa}} = 0$), it is satisfied for sufficiently small ϕ because $C \propto \phi^{-1}$. On adding salt, q_{m} decreases and eventually vanishes. If the right-hand side is negative, the peak of $I_{\mathrm{th}}(\boldsymbol{q})$ is at $q = 0$ as in neutral gels. Moreover, as the solvent quality is decreased with $q_{\mathrm{m}} > 0$, we expect the occurrence of microphase separation [65, 66]. The typical size of emerging domains will be of order q_{m}^{-1} in polyelectrolyte solutions without crosslinkage. However, the network elasticity can also pin the domain growth even in neutral gels, as stated near (7.2.22). This means that we need to take into account both the charge effect and the network elasticity to determine the domain structure in polyelectrolyte gels.

7.2.4 Dynamic equations of gels

In gels swollen by solvent, network motion is highly damped by friction with the solvent, so the network velocity $\boldsymbol{v}_{\mathrm{p}} = \boldsymbol{v}_1$ with respect to the solvent velocity $\boldsymbol{v}_{\mathrm{s}} = \boldsymbol{v}_2$ is given by [17, 46, 69]

$$\boldsymbol{v}_{\mathrm{p}} - \boldsymbol{v}_{\mathrm{s}} = -\frac{1}{\zeta}\nabla\cdot\overleftrightarrow{\Pi} = -\frac{1}{\zeta}\left[\phi\nabla\frac{\delta}{\delta\phi}\mathcal{H}_{\phi} - \nabla\overleftrightarrow{\sigma}\right], \tag{7.2.29}$$

where ζ is the friction coefficient behaving as (7.1.26). In the Lagrange picture we have

$$\boldsymbol{v}_{\mathrm{p}} = \left(\frac{\partial}{\partial t}\boldsymbol{r}\right)_{\boldsymbol{r}_0} = \boldsymbol{v}_{\mathrm{s}} - \zeta^{-1}\frac{\phi}{\phi_0}\left(\frac{\delta\mathcal{H}}{\delta\boldsymbol{r}}\right)_{\boldsymbol{r}_0}. \tag{7.2.30}$$

Use has been made of (7.2.7) and (7.2.8). This equation becomes a Langevin equation if we add the noise term to the right-hand side which satisfies the fluctuation–dissipation relation. We will neglect the noise term for simplicity.

We also consider the evolution equations for the deformation tensor Γ_{ij} in (3.5.46) and the Finger tensor W_{ij} in (3A.13) in the Euler representation. The Lagrange time derivative $(\partial/\partial t)_{\boldsymbol{r}_0}$ and the Euler time derivative $\partial/\partial t \equiv (\partial/\partial t)_{\boldsymbol{r}}$ are related by $(\partial/\partial t)_{\boldsymbol{r}_0} = \partial/\partial t + \boldsymbol{v}_{\mathrm{p}}\cdot\nabla$. Further, using the relation $(\partial\Gamma_{ij}/\partial t)_{\boldsymbol{r}_0} = \partial v_{\mathrm{p}i}/\partial x_{0j} = \sum_k(\partial x_k/\partial x_{0j})(\partial v_{\mathrm{p}i}/\partial x_k)$, we obtain

$$\left(\frac{\partial}{\partial t} + \boldsymbol{v}_{\mathrm{p}}\cdot\nabla\right)\Gamma_{ij} - \sum_k D_{ik}\Gamma_{kj} = 0, \tag{7.2.31}$$

$$\left(\frac{\partial}{\partial t} + \boldsymbol{v}_{\mathrm{p}}\cdot\nabla\right)W_{ij} - \sum_k (D_{ik}W_{kj} + W_{ik}D_{jk}) = 0. \tag{7.2.32}$$

where $D_{ij} = \partial v_{\mathrm{p}i}/\partial x_j$ is the velocity gradient tensor in the deformed space. The equation for $\{W_{ij}\}$ coincides with (7.1.100) for polymer solutions in the limit $\tau \to \infty$. If $\boldsymbol{v}_{\mathrm{s}}$ is neglected, the continuity equation for ϕ follows from (7.2.4) and (7.2.29) as

$$\frac{\partial}{\partial t}\phi = -\nabla\cdot(\phi\boldsymbol{v}_{\mathrm{p}}) = \nabla\cdot L\left(\nabla\frac{\delta\mathcal{H}_{\phi}}{\delta\phi} - \frac{1}{\phi}\nabla\cdot\overleftrightarrow{\sigma}\right), \tag{7.2.33}$$

where $L = \phi^2/\zeta$ is the kinetic coefficient consistent with (7.1.23) for $\phi \ll 1$. Notice that

(7.2.33) is of the same form as that of (7.1.33) derived on the basis of the stress–diffusion coupling.

Linear dynamic equation

We consider relaxation of the displacement $\boldsymbol{u}$ in an affinely deformed state represented by (7.2.9). If $\boldsymbol{v}_s$ is neglected, from (7.2.29) the linearized dynamic equation reads [17]

$$\frac{\partial}{\partial t}\boldsymbol{u} = \zeta^{-1}\left[\left(K_{os} + \frac{1}{3}\mu - TC\phi^2\nabla^2\right)\nabla(\nabla\cdot\boldsymbol{u}) + \sum_{jk}\mu_{jk}\nabla_j\nabla_k\boldsymbol{u}\right], \tag{7.2.34}$$

where

$$\mu_{ij} = T\nu_0(\phi/\phi_0)\sum_k A_{ik}A_{jk} \tag{7.2.35}$$

is a shear modulus tensor. The deviation $\delta\phi$ then obeys

$$\frac{\partial}{\partial t}\delta\phi = \zeta^{-1}\left[\left(K_{os} + \frac{1}{3}\mu - TC\phi^2\nabla^2\right)\nabla^2 + \sum_{jk}\mu_{jk}\nabla_j\nabla_k\right]\delta\phi. \tag{7.2.36}$$

The decay rate of the Fourier component of $\delta\phi$ becomes

$$\Gamma_q = \zeta^{-1}\left[K_{os} + \frac{1}{3}\mu + \mu J(\hat{\boldsymbol{q}}) + TC\phi^2 q^2\right]q^2, \tag{7.2.37}$$

where $J(\hat{\boldsymbol{q}}) = \mu^{-1}\sum_{jk}\mu_{jk}\hat{q}_j\hat{q}_k$ equivalently with (7.2.11).

(i) In the isotropic case, the diffusion constant is given by $D_{gel} = \zeta^{-1}(K_{os} + \frac{4}{3}\mu)$ as already derived in (7.1.63). This indicates critical slowing down for $K_{os} + \frac{4}{3}\mu \to 0$ [44, 70] and spinodal decomposition for $K_{os} + \frac{4}{3}\mu < 0$. (ii) In the uniaxial case (3.5.66), the gel diffusion constant behaves as

$$D_{gel}(\hat{\boldsymbol{q}}) = \zeta^{-1}\left[K_{os} + \left(\frac{1}{3} + \frac{1}{\lambda}\right)\mu + \left(\lambda^2 - \frac{1}{\lambda}\right)\mu\hat{q}_x^2\right]. \tag{7.2.38}$$

Takebe *et al.* [71] performed dynamic light scattering from a uniaxially deformed gel with good solvent and measured anisotropy in the diffusion constant $D_{gel}(\hat{\boldsymbol{q}})$ of the density fluctuations. As displayed in Fig. 7.8, their data indicate that diffusion is faster in the stretched direction than in the perpendicular directions, in reasonable agreement with (7.2.38) (provided that the anisotropy of the friction coefficient ζ is negligible).

Two-fluid hydrodynamic equations

Following on from the work by Tanaka *et al.* [17], two-fluid hydrodynamic equations for gels in the linear regime were presented by Marqusee and Deutch [72], which contain the gel velocity $\boldsymbol{v}_1 = \partial\boldsymbol{u}/\partial t$ and the solvent velocity $\boldsymbol{v}_2$. A more general set of dynamic equations were proposed by Johnson [73], which takes into account a mass coupling effect (a mass matrix) present in the hydrodynamic theory of porous media by Biot [74]. At low frequencies these equations take essentially the same forms as (7.1.16) and (7.1.17) with $\boldsymbol{F}_1 = \nabla\cdot\overleftrightarrow{\sigma}$ and $\boldsymbol{F}_2 = \eta_0\nabla^2\boldsymbol{v}_2$ [46], where $\overleftrightarrow{\sigma}$ is the elastic stress tensor and η_0 is the

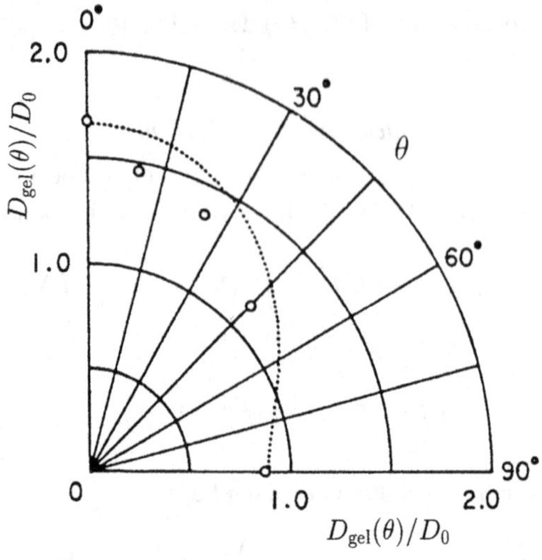

Fig. 7.8. Comparison between experimental (circles) and theoretical (dotted line) angular dependences of the relative diffusion constant $D_{gel}(\theta)/D_0$ for a swollen polyacrylamide gel stretched by $\lambda = 2$ [71]. Here $D_0 = 2.2 \times 10^{-7}\text{cm}^2/\text{s}$ is the diffusion constant for $\lambda = 1$ at the same temperature. The dotted line represents $0.78 \cos^2 \theta + 0.61$ where $\cos\theta = q_x/q$.

solvent viscosity. As an application of the two-fluid description we now examine slow transverse motion with frequencies much smaller than ζ/ρ in isotropic gels. Note that gels can support transverse sound waves with the frequency $(\mu/\rho)^{1/2}q$ at very low frequencies (or very small q), where the network and the solvent move in phase and the total mass density $\rho = \rho_p + \rho_s$ appears here.

Because the displacement $\boldsymbol{u}$ is convected by the average velocity field $\boldsymbol{v}$ (= the average of the polymer and solvent velocities as in (7.1.12)), we have

$$\frac{\partial}{\partial t}\boldsymbol{u} = \boldsymbol{v} + \zeta^{-1}\left[\left(K_{os} + \frac{1}{3}\mu\right)\nabla(\nabla \cdot \boldsymbol{u}) + \mu\nabla^2\boldsymbol{u}\right]. \tag{7.2.39}$$

As in the polymer solution case, we assume $\nabla \cdot \boldsymbol{v} = 0$. Analogously to (7.1.34), $\boldsymbol{v}$ obeys

$$\rho\frac{\partial}{\partial t}\boldsymbol{v} = \mu[\nabla^2\boldsymbol{u} - \nabla(\nabla \cdot \boldsymbol{u})] + \eta_0\nabla^2\boldsymbol{v}. \tag{7.2.40}$$

These linear dynamic equations describe two characteristic kinds of collective modes in gels; the longitudinal part of $\boldsymbol{u}$ or $\delta\phi$ obeys Tanaka's diffusion equation [17], while the transverse part and $\boldsymbol{v}$ are coupled to form slow transverse sound modes at small wave numbers. More generally, by assuming the space–time dependence as $\exp(iqx + i\omega t)$, we calculate the dispersion relation from

$$\left(i\omega + \frac{\mu}{\zeta}q^2\right)\left(i\omega + \frac{\eta_s}{\rho}q^2\right) + \frac{\mu}{\rho}q^2 = 0. \tag{7.2.41}$$

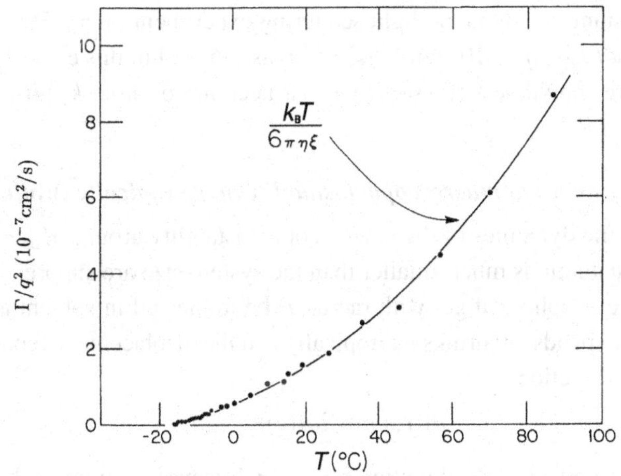

Fig. 7.9. Decay rate divided by q^2 vs T in a 2.5% polyacrylamide gel [69]. The solid line represents the Kawasaki–Stokes formula.

In the long-wavelength limit $q \to 0$, we have sound modes, $\omega \cong \pm(\mu/\rho)^{1/2}q$. For general q, we have $\omega = \omega_+$ or ω_-, where

$$i\omega_\pm = -\frac{1}{2}\left(\frac{\eta_0}{\rho} + \frac{\mu}{\zeta}\right)q^2 \pm i\left[\frac{\mu}{\rho} - \frac{1}{4}\left(\frac{\eta_0}{\rho} - \frac{\mu}{\zeta}\right)^2 q^2\right]^{1/2} q. \tag{7.2.42}$$

Because $\mu/\zeta \ll \eta_0/\rho$ in weakly crosslinked gels, the modes are oscillatory only for q smaller than

$$k_c = 2(\rho\mu)^{1/2}/\eta_0. \tag{7.2.43}$$

The corresponding crossover frequency ω_c may be introduced by

$$\omega_c = \mu/\eta_0, \tag{7.2.44}$$

which is equal to $\frac{1}{2}(\mu/\rho)^{1/2}k_c = \frac{1}{4}(\eta_0/\rho)k_c^2$ and is very small. For $q > k_c$ we have two overdamped modes. The slower mode decays with

$$i\omega_- \cong -\omega_c(1 + \eta_0\zeta^{-1}q^2) \cong -\omega_c, \tag{7.2.45}$$

where $\eta_0 q^2/\zeta \ll 1$ because $\eta_0/\zeta \sim \xi^2$ with $\xi_b \sim a/\phi$ being the blob size from (7.1.26). The faster mode is nothing but the usual shear mode, $i\omega_+ = -(\eta_0/\rho)k^2$. The k_c and ω_c are estimated to be very small, 0.5×10^3 cm^{-1} and 0.7×10^4 s^{-1}, respectively, in a 2.5% polyacrylamide gel in Ref. [44].

Hydrodynamic interaction in weakly crosslinked gels

In weakly crosslinked gels, the hydrodynamic interaction should determine the magnitude of the friction coefficient ζ as in semidilute polymer solutions or near-critical fluids. In fact, as shown in Fig. 7.9, the Kawasaki–Stokes formula $T/6\pi\eta_0\xi$ in (6.1.24) nicely explained the diffusion constant in a dynamic light scattering experiment on a 2.5% polyacrylamide gel [44]. Note that $k_c \ll q \sim 10^5\ \mathrm{cm}^{-1} \ll \xi^{-1}$ was satisfied in this experiment. However, to be precise, there should be a crossover when ξ becomes of order k_c [46].

7.2.5 Dynamics of macroscopic instability in isotropically swollen gels

We now examine the dynamics of the macroscopic instability around $K_{os} = 0$ in isotropic gels, where the fluctuations much smaller than the system size are suppressed by the finite shear modulus. Let a spherical gel with radius R be immersed in solvent at zero-osmotic pressure. The gel expands or shrinks isotropically and the displacement vector $\boldsymbol{u}$ is assumed to be in the radial direction,

$$u_i(\boldsymbol{r}, t) = \hat{x}_i u(r) e^{-\Omega t}, \tag{7.2.46}$$

where $u(r)$ is independent of the direction $\hat{\boldsymbol{r}} = r^{-1}\boldsymbol{r}$. Because we treat the linear equations, all the deviations may be assumed to depend on time as $e^{-\Omega t}$. In the dynamic equation (7.2.34) we neglect the higher-gradient term ($\propto C$). Then,

$$u(r) = Ae^{-\Omega t}\big[\sin(qr) - qr\cos(qr)\big]\frac{1}{q^2r^2}, \tag{7.2.47}$$

where A is a small amplitude and

$$q = (\Omega/D_{\mathrm{gel}})^{1/2}. \tag{7.2.48}$$

The zero osmotic pressure condition, $\hat{\boldsymbol{r}} \cdot \overleftrightarrow{\Pi} \cdot \hat{\boldsymbol{r}} = 0$ at $r = R$, becomes

$$\left(K_{os} + \frac{4}{3}\mu\right)\nabla\cdot\boldsymbol{u} - 4\mu\frac{u}{R} = 0. \tag{7.2.49}$$

By setting $Q = qR$ we readily find

$$1 + \frac{3}{4\mu}K_{os} = \frac{3}{Q^2}\left(1 - \frac{Q}{\tan Q}\right). \tag{7.2.50}$$

(i) When $|K_{os}| \ll \mu$, we have $|Q| \ll 1$. Because the right-hand side of (7.2.55) behaves as $1 + \frac{1}{15}Q^2 + \cdots$ for $|Q| \ll 1$, we obtain $Q^2 \cong 45K_{os}/4\mu$ or

$$\Omega \cong 15\zeta^{-1}R^{-2}K_{os}. \tag{7.2.51}$$

Note that the gel diffusion constant D_{gel} remains finite at $K_{os} = 0$. (ii) For $K_{os} \gg \mu$ or far above the macroscopic critical point, we have $\nabla\cdot\boldsymbol{u} \cong 0$ at $r = R$, so that $Q \cong \pi$ and [45]

$$\Omega \cong \pi^2\zeta^{-1}R^{-2}K_{os}. \tag{7.2.52}$$

In the theoretical literature, however, the distinction between the macroscopic and bulk instabilities has not been well recognized [46, 47]. Experimentally it is subtle, because considerable amounts of the critical fluctuations should be generated already at the point $K_{os} = 0$ for small μ and the observation time should be longer than the relaxation time of the macroscopic mode.

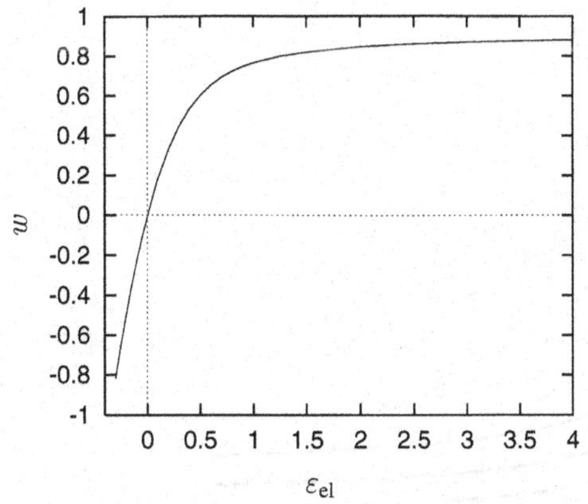

Fig. 7.10. Solution $w = w(\varepsilon_{\rm el})$ as a function of the modulus ratio $\varepsilon_{\rm el}$ in (7.2.55) obtained from the eigenvalue equation (7.2.53) for the surface mode. Here $w \to 0.9126\cdots$ as $\varepsilon_{\rm el} \to \infty$. The relaxation rate of the surface mode is expressed as (7.2.57) in isotropically swollen gels and (7.2.59) in uniaxially deformed gels.

7.2.6 Surface instability of gels

Isotropic case

We consider a slowly varying deviation $\boldsymbol{u}$ localized near a gel–solvent interface. If the higher order gradient term ($\propto C$) is neglected, the problem reduces to that of the surface (Rayleigh) sound wave on a planar stress-free surface [60]. We take the x axis in the normal direction with the gel being in the lower region $x < 0$ and the solvent being in the upper region $x > 0$. The space dependence on the plane may be assumed to be sinusoidal as e^{iqy}. Then we may set $\nabla_y = iq$ and $\nabla_z = 0$. We need to solve the following eigenvalue problem,

$$\varepsilon_{\rm el}\nabla(\nabla\cdot\boldsymbol{u}) + \nabla^2\boldsymbol{u} = -wq^2\boldsymbol{u}, \tag{7.2.53}$$

which holds in the bulk region $x < 0$. The stress-free boundary condition at $x = 0$ becomes

$$\varepsilon_{\rm el}\nabla\cdot\boldsymbol{u} + \nabla_x u_x - \nabla_y u_y = 0, \quad \nabla_x u_y + \nabla_y u_x = 0, \tag{7.2.54}$$

where

$$\varepsilon_{\rm el} = \frac{1}{\mu}K_{\rm os} + \frac{1}{3}. \tag{7.2.55}$$

As plotted in Fig. 7.10, $w = w(\varepsilon_{\rm el})$ in (7.2.53) is a function of $\varepsilon_{\rm el}$ only and is the solution of the following cubic polynomial equation [60],

$$(\varepsilon_{\rm el} + 1)(w^3 - 8w^2 + 24w - 16) = 16(w - 1). \tag{7.2.56}$$

In solids, the surface (Rayleigh) sound velocity $c_{\rm Ray}$ is related to w by $c_{\rm Ray} = (w\mu/\rho)^{1/2}$, ρ being the mass density, and the surface mode oscillates in time with the frequency $c_{\rm Ray}q$. In gels, the surface mode is overdamped as $e^{-\Omega t}$. From the linear dynamic equation (7.2.34) the relaxation rate is expressed as

$$\Omega = \zeta^{-1}\mu w(\varepsilon_{\rm el})q^2. \tag{7.2.57}$$

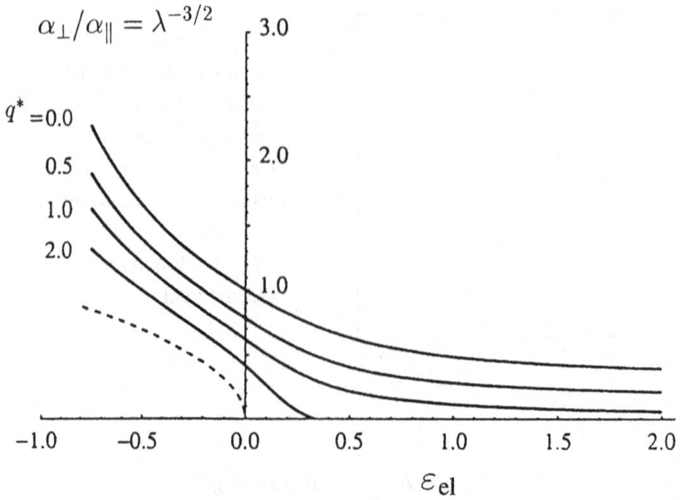

Fig. 7.11. The critical elongation ratio $\alpha_\perp/\alpha_\parallel = \lambda^{-3/2}$ vs ε_{el} for various values of the dimensionless lateral wave number $q^* = \sigma_{gs}|q|/\lambda^2\mu$ for a semi-infinite uniaxially deformed gel. They are determined by $\Omega = 0$ in (7.2.59) and represent the instability curves at wave number $(\lambda^2\mu/\sigma_{gs})q^*$ against surface undulations. The dashed curve represents the bulk spinodal line in (7.2.16) for $\lambda > 1$.

In our problem it is important that $w(\varepsilon_{el})$ tends to zero as $w \cong 2\varepsilon_{el}$ as $\varepsilon_{el} \to 0$ [52, 53], which readily follows from (7.2.56). Therefore, the surface mode is unstable for $\varepsilon_{el} < 0$. Notice also that the surface tension effect has been neglected in the above arguments. The surface tension σ_{gs} of the gel–solvent interface is expected to be of order T/ξ^2 from the scaling theory [1] and should play the role of suppressing small-scale surface disturbances. A theory accounting for its presence [53] yields

$$\Omega \cong 2\zeta^{-1}\left(K_{os} + \frac{1}{3}\mu + \frac{1}{2}\sigma_{gs}|q|\right)q^2, \tag{7.2.58}$$

for small ε_{el}. In the early stage of the instability, the unstable wave number region is bounded as $|q| < 2|K_{os} + \frac{1}{3}\mu|/\sigma_{gs}$.

Uniaxial case

We examine the surface mode on a gel deformed uniaxially as (3.5.66). (See Appendix 7G for the details of the calculation.) Generalizing (7.2.57) and taking into account the surface tension effect, we calculate the relaxation rate as [52, 53]

$$\Omega = \zeta^{-1}\mu\lambda^2\left[w(\lambda^2\varepsilon_{el}) - 1 + \frac{1}{\lambda^3} + \left(\frac{w_1\sigma_{gs}}{\lambda^2\mu}\right)|q|\right]q^2, \tag{7.2.59}$$

where w_1 is a number of order 1. In particular, for $\varepsilon_{el} \gtrsim 1$ or $K_{os} \gtrsim \mu$, the surface instability occurs for

$$\lambda^{-3} \lesssim 0.1 \quad \text{or} \quad \lambda \gtrsim 2, \tag{7.2.60}$$

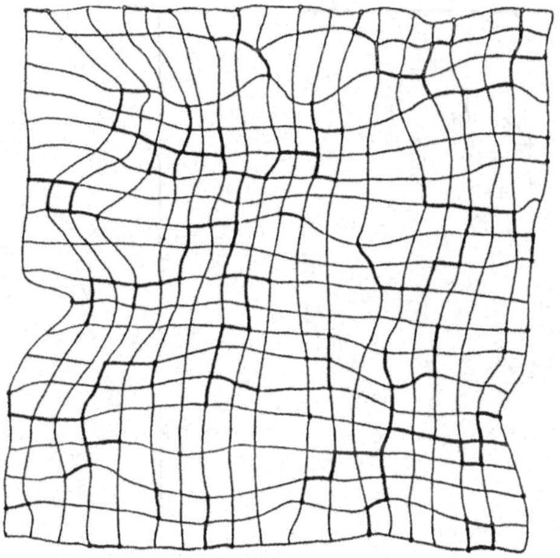

Fig. 7.12. Schematic representation of a 2D swollen network, which initially formed a periodic square lattice before swelling [81]. Here the bonds shown as thick lines cannot be elongated and represent frozen units.

because $w(\lambda^2\varepsilon_{el}) \cong 0.9$ from Fig. 7.10. In Fig. 7.11 we show the curves of $\Omega = 0$ in the plane of $\epsilon_{el} = K_{os}/\mu + 1/3$ and $\alpha_\perp/\alpha_\parallel = \lambda^{-3/2}$ for several values of $q^* \equiv \sigma_{gs}|q|/(\lambda^2\mu)$. Below these curves, surface undulations with wave numbers smaller than $(\lambda^2\mu/\sigma_{gs})q^*$ are unstable. The dashed curve is the bulk spinodal line for the case $\lambda > 1$ determined by (7.2.16).

7.3 Heterogeneities in the network structure

Heterogeneities are inherent in randomly crosslinked networks [75]. They play the role of quenched (frozen) randomness, producing quasi-static network deformations [75]–[80]. The scattering from such systems is sometimes larger than that from a semidilute solution with the same concentration at small q. As illustrated in Fig. 7.12, Bastide and Leibler [81] argued that the network heterogeneities produce quasi-static, long-range, elastic deformations $\boldsymbol{u}_R$ with swelling and the resultant concentration fluctuations ($\propto \nabla \cdot \boldsymbol{u}_R$) are the origin of the excess scattering. The scattering amplitude from heterogeneous gels then consists of the dynamic and static components; the former arises from the thermal fluctuations and decays as $\exp(-D_{gel}q^2t)$ in time, while the latter is static and does not decay in dynamic light scattering. In addition, the scattering amplitude from heterogeneous gels strongly depends on the scattering position. Interestingly, as shown in Fig. 7.13, Matsuo *et al.* [79] found that the space averages of the two components, the static $\bar{I}_S$ and the dynamic $\bar{I}_D$, grow strongly as a spinodal temperature T_{sp} is approached. Remarkably, the growth of $\bar{I}_S$ is stronger than that of $\bar{I}_D$. Furthermore, $\bar{I}_S$ depends strongly on the temperature T_{pre} at which the gel was prepared, whereas $\bar{I}_D$ is insensitive to T_{pre}. That is, $\bar{I}_S$ grows if T_{pre} is close to T_{sp}.

Recently, much attention has been paid to anomalously anisotropic quasi-static fluctuations in uniaxially stretched gels, detected by small-angle neutron scattering [75, 82, 83]. We show the isointensity curves of the scattered intensity $I(q_x, q_y)$ in Fig. 7.14(a) and

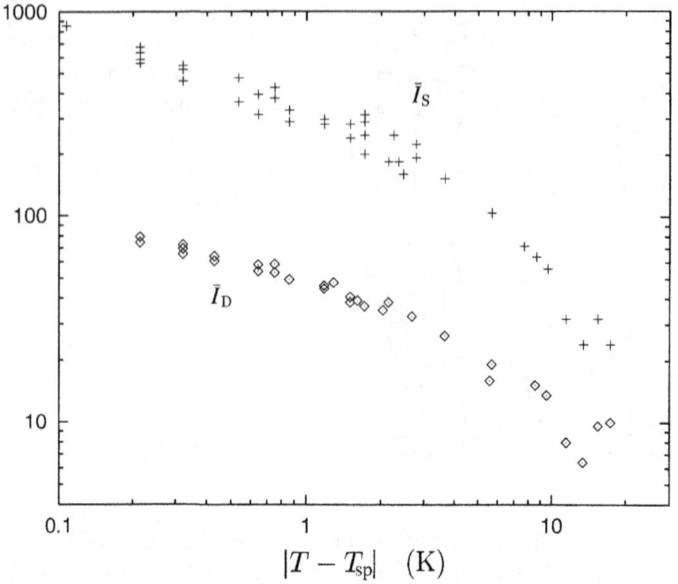

Fig. 7.13. The dynamic and static scattered light intensities, $\bar{I}_D$ and $\bar{I}_S$, respectively, from a heterogeneous NIPA gel with $T_S = 306.4$ K [79]. Both components grow strongly as the instability point is approached.

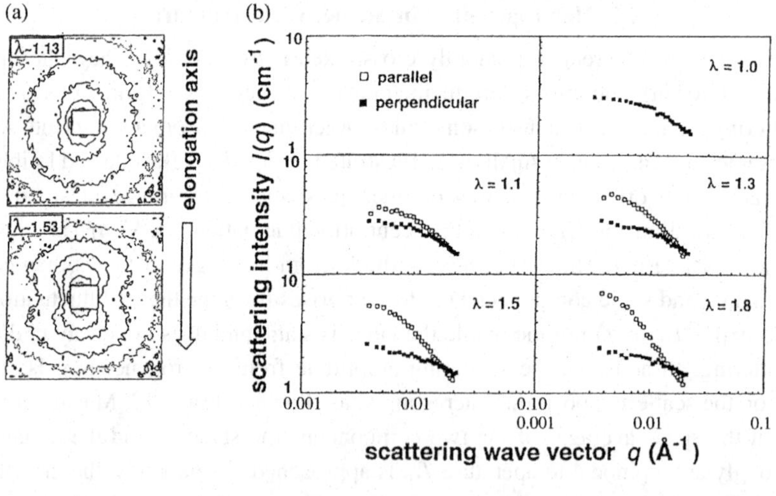

Fig. 7.14. (a) Isointensity curves of small-angle neutron scattering as a function of the elongation ratio λ in a swollen gel [82]. (b) Scattered intensities in the parallel and perpendicular directions, $I_{\|}(q)$ and $I_{\perp}(q)$, as a function of q in the same experiment.

the parallel and perpendicular components, $I_{\parallel}(q) = I(q, 0)$ and $I_{\perp}(q) = I(0, q)$, respectively, in Fig. 7.14(b) [82]. We can see that the intensity is largest at small q in the stretched direction (*abnormal butterfly pattern*). A similar trend was also found in blends of crosslinked and linear polystyrene [78]. This finding at small q apparently contradicts our theoretical intensity (7.2.15) from the thermal fluctuations, which indicates a *normal butterfly pattern*. Bastide *et al.* [84] intuitively argued that the static density fluctuations become stronger in the stretched direction than in the perpendicular directions.

We will give a simple theory of the static heterogeneities, which is a generalization of a theory by the present author [85]. We will obtain essentially the same results as those of a subsequent theory by Panyukov and Rabin [86]. However, because we will use a perturbation theory, our theory will not be applicable near the instability point where the heterogeneities are much enhanced, as in Fig. 7.13.

7.3.1 Heterogeneous crosslinkage

Let crosslinks be formed in a semidilute polymer solution at the preparation of a gel, where the polymer volume fraction is ϕ_0 on the average. The space position in the gel in the as-prepared state will be denoted by $\boldsymbol{r}_0$. Note that $\boldsymbol{r}_0$ is the original position to be shifted to $\boldsymbol{r} = \boldsymbol{r}(\boldsymbol{r}_0)$ after deformation. We then argue that there are two origins of heterogeneities in the crosslink density. (i) If the crosslinks form independently of one another, there arises an intrinsic crosslink density deviation $\nu_{\rm in}(\boldsymbol{r}_0)$ with no long-range correlation,

$$\langle \nu_{\rm in}(\boldsymbol{r}_0)\nu_{\rm in}(\boldsymbol{r}_0')\rangle = p_{\rm in}\nu_0\delta(\boldsymbol{r}_0 - \boldsymbol{r}_0'), \tag{7.3.1}$$

where ν_0 is the average crosslink density. The dimensionless coefficient $p_{\rm in}$ is expected to be of order 1. It is in fact equal to 1 if the crosslink number in a small fixed volume obeys a poissonian distribution (like the particle number in a fixed volume in a dilute gas). (ii) We note that the crosslink can be formed at points where monomers of different parts of the chains are in close contact, so that the crosslink density is proportional to the contact (or entanglement) point density $\propto \xi_{\rm pre}^{-3}$, where $\xi_{\rm pre} \propto \phi_0^{-\hat{\nu}/(3\hat{\nu}-1)}$ is the blob size or the correlation length of the semidilute solution in the as-prepared state. (See the sentences below (7.1.26) for the explanation of blobs.) Therefore, if there is a small inhomogeneous deviation $\delta\phi_0(\boldsymbol{r}_0)$ in the volume fraction at the instant of crosslink formation, it induces a crosslink density deviation $\nu_\phi(\boldsymbol{r}_0)$ proportional to the deviation of $\xi_{\rm pre}^{-3}$ as

$$\nu_\phi(\boldsymbol{r}_0) = A_0\nu_0\phi_0^{-1}\delta\phi_0(\boldsymbol{r}_0), \tag{7.3.2}$$

where the coefficient,

$$A_0 = \phi_0\frac{\partial}{\phi_0}\ln(\xi_{\rm pre}^{-3}) = 3\hat{\nu}/(3\hat{\nu} - 1), \tag{7.3.3}$$

is determined by the exponent $\hat{\nu}$, so $A_0 = 3$ in theta solvent ($\hat{\nu} = 1/2$) and $A_0 = 9/4$ in good solvent ($\hat{\nu} = 3/5$). It follows that the correlation of $\nu_\phi(\boldsymbol{r}_0)$ is proportional to that of the deviation $\delta\phi_0(\boldsymbol{r}_0)$ as

$$\langle \nu_\phi(\boldsymbol{r}_0)\nu_\phi(\boldsymbol{r}'_0)\rangle = A_0^2\nu_0^2\phi_0^{-2}\langle \delta\phi_0(\boldsymbol{r}_0)\delta\phi_0(\boldsymbol{r}'_0)\rangle. \tag{7.3.4}$$

Now the crosslink density consists of three parts as

$$\nu(\boldsymbol{r}_0) = \nu_0 + \nu_{\text{in}}(\boldsymbol{r}_0) + \nu_\phi(\boldsymbol{r}_0). \tag{7.3.5}$$

If ν_{in} and ν_ϕ are sufficiently small, they should be independent of each other. We may then write the structure factor of the deviation $\delta\nu = \nu_{\text{in}}+\nu_\phi$ immediately after the crosslink formation as

$$\int d\boldsymbol{r}_0 \langle \delta\nu(\boldsymbol{r}_0)\delta\nu(\boldsymbol{0})\rangle \exp(i\boldsymbol{q}_0\cdot\boldsymbol{r}_0) = p_{\text{in}}\nu_0 + A_0^2\nu_0^2\phi_0^{-2}I_0(\boldsymbol{q}_0), \tag{7.3.6}$$

where $I_0(\boldsymbol{q}_0)$ is the structure factor of the volume-fraction fluctuations at the instant of the gel preparation. It is important that the crosslink heterogeneity forms quenched disorder fixed to the network, so that the crosslink number $\nu(\boldsymbol{r})d\boldsymbol{r}$ in a small volume element $d\boldsymbol{r}$ in the deformed state coincides with that, $\nu(\boldsymbol{r}_0)d\boldsymbol{r}_0$, in the initial volume element $d\boldsymbol{r}_0 = d\boldsymbol{r}\phi/\phi_0$ in the original state under the mapping relation $\boldsymbol{r} = \boldsymbol{r}(\boldsymbol{r}_0)$. Therefore, if the gel is anisotropically deformed as (7.2.9), the correlations of the crosslink density deviations in the Euler representation are expressed as

$$\langle \nu_{\text{in}}(\boldsymbol{r})\nu_{\text{in}}(\boldsymbol{r}')\rangle = p_{\text{in}}\nu_0(\phi_0/\phi)\delta(\boldsymbol{r}-\boldsymbol{r}'), \tag{7.3.7}$$

$$\langle \nu_\phi(\boldsymbol{r})\nu_\phi(\boldsymbol{r}')\rangle = A_0^2\nu_0^2\phi_0^{-2}\langle \delta\phi_0(\overleftrightarrow{A}^{-1}\cdot\boldsymbol{r})\delta\phi_0(\overleftrightarrow{A}^{-1}\cdot\boldsymbol{r}')\rangle. \tag{7.3.8}$$

where we have assumed $\boldsymbol{r} = \overleftrightarrow{A}\cdot\boldsymbol{r}_0$ and $\boldsymbol{r}' = \overleftrightarrow{A}\cdot\boldsymbol{r}'_0$ neglecting the small displacement $\boldsymbol{u}$ in (7.2.9). The structure factor of $\delta\nu$ in the deformed gel is thus written as

$$\langle|\nu_{\boldsymbol{q}}|^2\rangle = \int d\boldsymbol{r}\langle \delta\nu(\boldsymbol{r})\delta\nu(\boldsymbol{0})\rangle \exp(i\boldsymbol{q}\cdot\boldsymbol{r}) = \frac{\nu_0\phi_0}{\phi}\left[p_{\text{in}} + A_0^2\frac{\nu_0}{\phi_0^2}I_0(\overleftrightarrow{A}\cdot\boldsymbol{q})\right]. \tag{7.3.9}$$

If the preparation is made in good solvent, we simply have $\langle|\nu_{\boldsymbol{q}}|^2\rangle \cong p_{\text{in}}\nu_0\phi_0/\phi$. Conversely, if the semidilute solution is close to the solution critical point at preparation [79], we have $\langle|\nu_{\boldsymbol{q}}|^2\rangle \cong 9(\nu_0^2/\phi_0\phi)I_0(\overleftrightarrow{A}\cdot\boldsymbol{q})$ by setting $A_0 = 3$ at small q.

In our theory, the elastic free energy is assumed to be given by (7.2.3) with ν_0 being replaced by $\nu_0 + \delta\nu$. Therefore, the random hamiltonian is written as

$$\mathcal{H}^{\text{R}} = \frac{1}{2}T\int d\boldsymbol{r}_0\delta\nu\sum_{ij}\Gamma_{ij}^2, \tag{7.3.10}$$

in the Lagrange description. More generally, the random hamiltonian can be of the form,

$$\mathcal{H}^{\text{R}} = \int d\boldsymbol{r}_0\sum_{ijk}\sigma_{ij}^{(\text{R})}\Gamma_{ki}\Gamma_{kj}, \tag{7.3.11}$$

where $\sigma_{ij}^{(\text{R})}$ represents a random internal stress produced at the crosslinkage [87, 88]. The deviatopic part $\sigma_{ij}^{(\text{R})} - \delta_{ij}\sum_k\sigma_{kk}^{(\text{R})}/3$ can be important in nematic networks composed of rod-like molecules [89]. There, elimination of the elastic field gives rise to quenched disorder acting on the orientational traceless tensor Q_{ij}.

7.3.2 Frozen random deformations

First we consider neutral gels. If the crosslink density is inhomogeneous as $\nu = \nu_0 + \delta\nu$, the stress tensor $\overleftrightarrow{\Pi}$ in (7.2.5) and (7.2.6) consists of that with the homogeneous part ν_0 and that proportional to the heterogeneous part $\delta\nu$. The latter can be expressed as

$$\Pi_{ij}^{\text{hetero}} = T(\phi/\phi_0)[B\delta_{ij} - W_{ij}]\delta\nu, \tag{7.3.12}$$

where B is the coefficient in the classical rubber theory in (3.5.43). The random static dilational strain $g_R = \nabla \cdot \boldsymbol{u}_R$ can be conveniently calculated from $\sum_{jk} \nabla_j \nabla_k \Pi_{jk} = 0$. To linear order in $\delta\nu$ we may use the expression (7G.1) for the deviation of the stress tensor. Together with (7.3.12), we find

$$\left[\left(K_{\text{os}} + \frac{\mu}{3} - TC\phi^2\nabla^2\right)\nabla^2 + \sum_{jk} \mu_{jk}\nabla_j\nabla_k\right] g_R = T\frac{\phi}{\phi_0}\left[B\nabla^2 - \sum_{jk} W_{jk}\nabla_j\nabla_k\right]\delta\nu. \tag{7.3.13}$$

With the aid of the definition of $J(\hat{\boldsymbol{q}})$ in (7.2.14), the Fourier transformation of (7.3.13) yields

$$g_R(\boldsymbol{q}) = \left[\frac{B\alpha^{-2} - J(\hat{\boldsymbol{q}})}{\varepsilon_{\text{el}} + J(\hat{\boldsymbol{q}}) + \hat{C}q^2}\right]\frac{1}{\nu_0}\nu_{\boldsymbol{q}}, \tag{7.3.14}$$

where $\alpha = (\phi_0/\phi)^{1/3}$, ε_{el} is defined by (7.2.55), and $\hat{C} = TC\phi^2/\mu$. The structure factor of the frozen concentration fluctuations thus becomes

$$I_R(\boldsymbol{q}) = \left[\frac{J(\hat{\boldsymbol{q}}) - B\alpha^{-2}}{\varepsilon_{\text{el}} + J(\hat{\boldsymbol{q}}) + \hat{C}q^2}\right]^2 \frac{\phi^2}{\nu_0^2}\langle|\nu_{\boldsymbol{q}}|^2\rangle, \tag{7.3.15}$$

where $\langle|\nu_{\boldsymbol{q}}|^2\rangle$ is given by (7.3.9). Although $\hat{C}$ is independent of $\hat{\boldsymbol{q}}$ from our GLW hamiltonian (7.2.2), we allow its $\hat{\boldsymbol{q}}$ dependence. It is required from the experiment [82], which showed $\hat{C}_\parallel > \hat{C}_\perp$ under uniaxial extension, $\hat{C}_\parallel$ and $\hat{C}_\perp$ being the values of $\hat{C}(\hat{\boldsymbol{q}})$ in the stretched and perpendicular directions, respectively. Notice that there should generally be higher-order gradient terms proportional to $q_i q_k u_{\boldsymbol{q}j} u_{\boldsymbol{q}\ell}^*$ in the free energy (7.2.10). For anisotropically deformed gels such terms should give rise to an angle-dependent $\hat{C}(\hat{\boldsymbol{q}})$.

The total intensity $I(\boldsymbol{q})$ is the sum of the heterogeneity contribution $I_R(\boldsymbol{q})$ and the thermal contribution $I_{\text{th}}(\boldsymbol{q})$:

$$I(\boldsymbol{q}) \cong \frac{\phi^2\alpha}{\nu_0}\left[\frac{1}{\varepsilon_{\text{el}} + J(\hat{\boldsymbol{q}}) + \hat{C}(\hat{\boldsymbol{q}})q^2} + p(\boldsymbol{q})\alpha^2\left(\frac{J(\hat{\boldsymbol{q}}) - B\alpha^{-2}}{\varepsilon_{\text{el}} + J(\hat{\boldsymbol{q}}) + \hat{C}(\hat{\boldsymbol{q}})q^2}\right)^2\right], \tag{7.3.16}$$

where

$$p(\boldsymbol{q}) = \langle|\nu_{\boldsymbol{q}}|^2\rangle\phi/\nu_0\phi_0 = p_{\text{in}} + A_0^2\nu_0\phi_0^{-2}I_0(\overleftrightarrow{A}\cdot\boldsymbol{q}). \tag{7.3.17}$$

At large q, the first term in the brackets of (7.3.16) behaves as q^{-2} and the second term as q^{-4}. Therefore, the thermal fluctuations dominate over the frozen fluctuations at large q, in agreement with the experiments. For theta solvent the above expression and that of Panykov and Rabin [86] coincide at small q ($\ll R_G^{-1}$) if comparison is made using $A_0 = 3$ in (7.3.3) and their theoretical value $p_{\rm in} = 3$.

Isotropically swollen gels

In a swollen isotropic state, where $J(\hat{\boldsymbol{q}}) = 1$ and $B\alpha^{-2} \ll 1$, the intensity takes a Debye–Bueche form [90],[5]

$$I(q) \cong \frac{\phi^2 T}{K_{\rm os} + 4\mu/3}\left[\frac{1}{1+\xi^2 q^2} + \left(\frac{\alpha^2}{\varepsilon_{\rm el}+1}\right)\frac{p(q)}{(1+\xi^2 q^2)^2}\right]. \tag{7.3.18}$$

The static contribution to the pair correlation function behaves as $\exp(-r/\xi)$ in space. In the long-wavelength limit, the ratio of the two contributions is written as

$$I_{\rm R}(0)/I_{\rm th}(0) = p(0)\alpha^2/(\varepsilon_{\rm el}+1) = p(0)(\phi_0/\phi)^{2/3}\mu/(K_{\rm os}+4\mu/3). \tag{7.3.19}$$

The excess scattering increases with increasing swelling ratio and even becomes dominant as $K_{\rm os} + 4\mu/3 \to 0$, consistent with the experimental results shown in Fig. 7.13. Here it is instructive to calculate the local variance of the static deviation $\delta\phi_{\rm R} = -\phi g_{\rm R}$ of the volume fraction. For simplicity, we set $p(\boldsymbol{q}) = p_{\rm in}$ to obtain

$$\langle \delta\phi_{\rm R}^2 \rangle = \int_{\boldsymbol{q}} I_{\rm R}(q) \sim p_{\rm in}\left[\phi_0 v_0 \nu_0/(\varepsilon_{\rm el}+1)\right]^{1/2}. \tag{7.3.20}$$

For the validity of our perturbation theory, the above variance should be much smaller than ϕ^2. However, as $K_{\rm os} + 4\mu/3 \to 0$, it grows and our theory becomes inapplicable.

Uniaxial stretching and shear deformation

For the uniaxial stretching represented by (3.5.66), $J(\hat{\boldsymbol{q}})$ is written as (7.2.14) in terms of the stretching ratio λ. We numerically calculate $I(\boldsymbol{q})$ from (7.3.16) setting $\lambda = 2$, $\varepsilon_{\rm el} = 4$, $B\alpha^{-2} = 0$, and $p(\boldsymbol{q}) = p_{\rm in}$, where the second term ($\propto I_0$) in (7.3.17) is neglected. As shown in Fig. 7.15, we obtain a normal butterfly pattern at $p_{\rm in}\alpha^2 = 0.1$ and an abnormal butterfly pattern at $p_{\rm in}\alpha^2 = 1.3$. In these two cases, $p(\boldsymbol{q})$ and $\hat{C}(\hat{\boldsymbol{q}})$ are constants independent of $\boldsymbol{q}$ for simplicity. We can see changeover from the normal to abnormal butterfly patterns for $p_{\rm in}\alpha^2 \gtrsim 1$.

As another example, let us apply a shear deformation, where $A_{ij} = \alpha(\delta_{ij} + \gamma\delta_{ix}\delta_{jy})$, γ being the shear strain. Then $\phi_0/\phi = \alpha^3$ as in the uniaxial case, and

$$J(\hat{\boldsymbol{q}}) = 1 + 2\gamma\hat{q}_x\hat{q}_y + \gamma^2\hat{q}_x^2. \tag{7.3.21}$$

We first examine the scattering in the q_x–q_y plane by setting $\hat{q}_x = \cos\theta$ and $\hat{q}_y = \sin\theta$;

[5] As an example, this form was obtained for a material with holes of varying and undetermined shapes (or in the presence of random interfaces) [90].

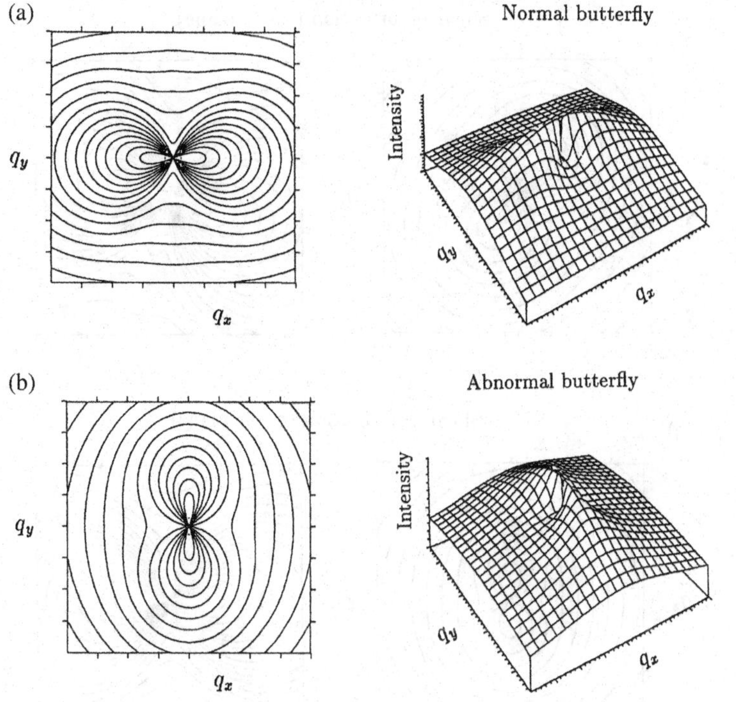

Fig. 7.15. Theoretical intensity from (7.3.16) in the uniaxial case $\lambda = 2$ in the q_x–q_y plane ($|q_x|$, $|q_y| < 2C^{-1/2}$) with $\varepsilon_{\rm el} = 4$ and $\hat{C} = 2$ [85]. (a) Normal butterfly pattern at $p_{\rm in}\alpha^2 = 0.1$. (b) Abnormal butterfly pattern at $p_{\rm in}\alpha^2 = 1.3$.

then, the maximum J_+ and minimum J_- of $J(\hat{\boldsymbol{q}})$ are given by

$$J_\pm = 1 + \frac{1}{2}\gamma^2 \pm \frac{\gamma}{2}\sqrt{4+\gamma^2}. \tag{7.3.22}$$

The maximum is attained in the most stretched direction $\theta = \theta_{\rm max} = \pi/4 - \tan^{-1}(\gamma/2)/2$, where $\theta_{\rm max}$ decreases from $\pi/4$ to 0 with increasing $\gamma(>0)$. The minimum is attained at $\theta = \theta_{\rm max} + \pi/2$. Figure 7.16(a) shows $I(\boldsymbol{q})$ for $p(\boldsymbol{q}) = p_{\rm in}$, $B\alpha^{-2} = 0$, and $\hat{C} = $ const. We can see rotation of the maximum direction from the normal to abnormal butterfly pattern in the q_x–q_y plane. In Fig. 7.16(b) we also show patterns in the q_x–q_z plane. These patterns are very analogous to those from sheared polymer solutions in theta solvent, as will be discussed in Chapter 11.

Remarks

It is surprising that even weak and short-range randomness in the crosslinkage (*elastic quenched disorder*) can produce enhanced, long-range static composition heterogeneities with large swelling or in the vicinity of the instability point. However, it remains unknown how the enhanced heterogeneities affect the phase transition and phase separation. Notice

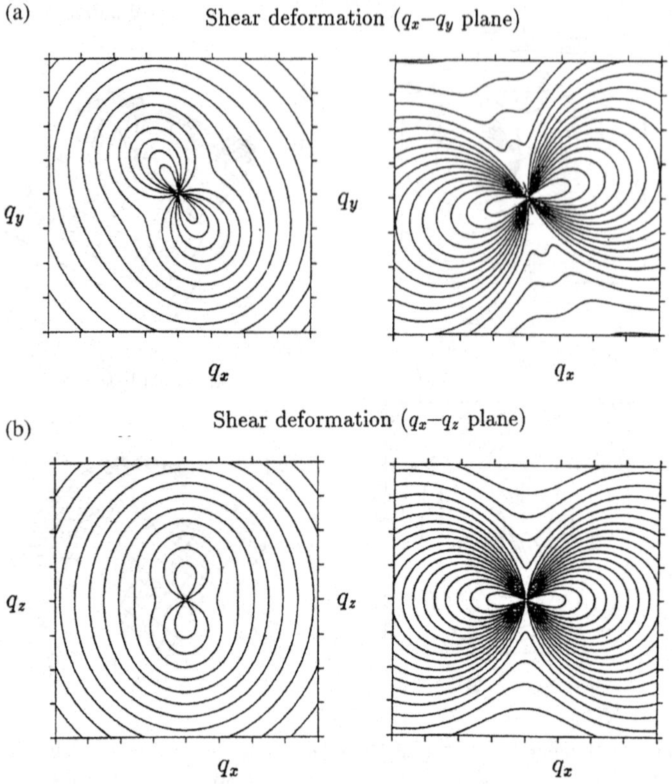

Fig. 7.16. Theoretical butterfly patterns under shear deformation at $\varepsilon_{\mathrm{el}} = 4$ [85]. (a) In the q_x–q_y plane we show the normal pattern for $p_{\mathrm{in}}\alpha^2 = 0.1$ and $\gamma = 1$ (*left*) and the abnormal one for $p_{\mathrm{in}}\alpha^2 = 1$ and $\gamma = 4$ (*right*). (b) In the q_x–q_z plane we show the normal pattern for $p_{\mathrm{in}}\alpha^2 = 0.1$ and $\gamma = 1$ (*left*) and the abnormal one for $p_{\mathrm{in}}\alpha^2 = 1$ and $\gamma = 3$ (*right*).

that the perturbation scheme used to derive (7.3.16) breaks down near the spinodal point $K_{\mathrm{os}} + 4\mu/3 \cong 0$. We can well expect that domains of a shrunken (or swollen) phase are created and pinned in regions with the crosslink density higher (or lower) than the average.

7.3.3 Heterogeneities in weakly charged gels

In anisotropically deformed, weakly charged gels, the thermal structure factor is given by (7.2.27). In the presence of crosslink heterogeneities the static density heterogeneities appear, as in neutral gels. Also in this case the charge effect can be accounted for by replacement, $K_{\mathrm{os}} \to K(q)$ in (7.2.26). As a result, we obtain the total structure factor [68],

$$I(\boldsymbol{q}) \cong \frac{\phi^2\alpha}{v_0}\left[\frac{1}{\varepsilon(q) + J(\hat{\boldsymbol{q}}) + \hat{C}(\hat{\boldsymbol{q}})q^2} + p(\boldsymbol{q})\alpha^2\left(\frac{J(\hat{\boldsymbol{q}}) - B\alpha^{-2}}{\varepsilon(q) + J(\hat{\boldsymbol{q}}) + \hat{C}(\hat{\boldsymbol{q}})q^2}\right)^2\right], \quad (7.3.23)$$

where $\varepsilon(q) = K(q)/\mu + 1/3$. This structure factor can have a peak at an intermediate wave number. For the isotropic case, its condition is given by the positivity of the right-hand side of (7.2.28). For $q \ll R_G^{-1}$ the above expression is essentially the same as a more complicated one by Rabin and Panyukov [67] and agrees with the general trend of the experiment [83].

Appendix 7A Single-chain dynamics in a polymer melt

We first consider a polymer melt composed of monodisperse long chains. An important parameter is the average monomer number N_e (~ 100) between consecutive entanglement points on a chain. For $N < N_e$, entanglements may be neglected and the single-chain motion is described by the Rouse dynamics,

$$v_0\zeta_0 \frac{\partial}{\partial t}\boldsymbol{R}_n = \frac{T}{a^2}\frac{\partial^2}{\partial n^2}\boldsymbol{R}_n + \boldsymbol{f}_n \quad (0 \le n \le N), \tag{7A.1}$$

where $v_0\zeta_0$ is the microscopic friction coefficient per monomer, $\partial \boldsymbol{R}_n/\partial n = \boldsymbol{0}$ at $n = 0$ and N, and $\boldsymbol{f}_n(t)$ is the random force characterized by

$$\langle f_{\mu n}(t) f_{\nu n'}(t')\rangle = 2T v_0\zeta_0 \delta_{\mu\nu}\delta_{nn'}\delta(t - t'). \tag{7A.2}$$

Then the slowest variation ($\boldsymbol{R}_n \propto \cos(\pi n/N)$) gives the longest relaxation time (Rouse time),

$$\tau_{\text{Rouse}} = \pi^{-2}(a^2/T)v_0\zeta_0 N^2. \tag{7A.3}$$

The diffusion constant of the mass center $\boldsymbol{R}_G = N^{-1}\sum_n \boldsymbol{R}_n$ is determined by the average random force $N^{-1}\sum_n \boldsymbol{f}_n$ and is known to be $D_{\text{Rouse}} = T/v_0\zeta_0 N$ from (7A.2) with the aid of (5.1.30) and (5.1.31).

For $N > N_e$, the single-chain motion is described by the reptation dynamics [1, 2, 91]. In the theory, each chain is regarded as a *reptile* passing through a curved tube with radius d_t and length L_t estimated as

$$d_t = N_e^{1/2}a, \quad L_t = (N/N_e)d_t = (N/N_e^{1/2})a, \tag{7A.4}$$

where a is the monomer size. The diffusion constant of a chain through a tube is inversely proportional to the polymerization index N as $D_t = T/v_0\zeta_0 N$, where $v_0\zeta_0 N$ is the friction coefficient per chain. This is of the same form as the diffusion constant of a chain in the Rouse dynamics. Thus the disentanglement (reptation) time τ in which a chain escapes from a tube is calculated as

$$\tau = L_t^2/D_t = (v_0\zeta_0 a^2/T)N^3/N_e. \tag{7A.5}$$

On this timescale, the center of mass of a chain moves a distance on the order of

$$(N/N_e)^{1/2}d_t = N^{1/2}a, \tag{7A.6}$$

so that the translational diffusion constant of a chain is

$$D = Na^2/\tau = (T/v_0\zeta_0)N_e/N^2. \tag{7A.7}$$

The effective friction constant is then equal to $T/D = v_0\zeta_0 N^2/N_e$ per chain, which is larger than that in the Rouse model by N/N_e. Next, macroscopic rheology is considered. Entangled polymers behave as gels (or soft elastic materials) with shear modulus,

$$G = T/a^3 N_e, \tag{7A.8}$$

against deformations with timescales shorter than τ. The stress relaxation time is given by τ and the zero-shear (newtonian) viscosity is estimated as

$$\eta = G\tau = (v_0\zeta_0/a)N^3/N_e^2. \tag{7A.9}$$

These expressions for D and η reduce to those of the Rouse model for $N \sim N_e$. However, a number of measurements have shown the behavior $\eta \propto N^{3.4}$ [1, 2]. The origin of the discrepancy in the exponent of N has not yet been conclusively identified.

Appendix 7B Two-fluid dynamics of polymer blends

The reptation concepts need to be applied to the dynamics of a mixture of two species of polymers, 1 and 2. The polymerization index N_e between entanglement points is assumed to be common for the two species and both species are entangled ($N_1 > N_e$ and $N_2 > N_e$). Then, each polymer undergoes reptation motion in common tubes. As in (7A.7), the diffusion constants of a single chain belonging to the two species are

$$D_K = (T/v_0\zeta_{0K})N_e/N_K^2, \quad (K = 1, 2), \tag{7B.1}$$

where ζ_{0K} are the microscopic friction coefficients of the two species. For macroscopically homogeneous deformations, the shear modulus is again given by (7A.8). If the tubes are naively assumed to be stationary during reptation motion of each chain, a simple mixing rule for the stress relaxation function is obtained as

$$G(t) = \phi_1 G_1(t) + \phi_2 G_2(t), \tag{7B.2}$$

where $G_K(t)$ are those of the pure components. However, the above form is known to only poorly explain a number of experiments presumably due to release of the tube constraints. A more successful and still simple mixing rule is known as double reptation [91]–[93], of the form

$$G(t) = \left[\phi_1\sqrt{G_1(t)} + \phi_2\sqrt{G_2(t)}\right]^2, \tag{7B.3}$$

which accounts for (i) the reptation motion of each chain and (ii) the relaxation of constraints on a given chain by reptation of its neighbors.

For not too small volume fractions, we should consider the mutual diffusion constant D_m between the two components. It may be derived from the two-fluid hydrodynamic equations. Assuming (7.1.4) we write the two-fluid dynamic equations of blends as

$$\rho_K \frac{\partial}{\partial t} \boldsymbol{v}_K = -\rho_K \nabla \mu_K - \zeta_K (\boldsymbol{v}_K - \boldsymbol{v}_t) + \boldsymbol{F}_K, \quad (K = 1, 2). \tag{7B.4}$$

The friction terms on the right-hand sides arise when the velocities are different from the velocity of the network structure or the *tube velocity* $\boldsymbol{v}_t$ [21]. Because of the force balance between the two components, we should require

$$\zeta_1(\boldsymbol{v}_1 - \boldsymbol{v}_t) + \zeta_2(\boldsymbol{v}_2 - \boldsymbol{v}_t) = 0. \tag{7B.5}$$

This equation is solved to give

$$\boldsymbol{v}_t = \frac{1}{\zeta_1 + \zeta_2}(\zeta_1 \boldsymbol{v}_1 + \zeta_2 \boldsymbol{v}_2). \tag{7B.6}$$

Because

$$\boldsymbol{v}_1 - \boldsymbol{v}_t = \frac{\zeta_2}{\zeta_1 + \zeta_2}(\boldsymbol{v}_1 - \boldsymbol{v}_2), \quad \boldsymbol{v}_2 - \boldsymbol{v}_t = \frac{\zeta_1}{\zeta_1 + \zeta_2}(\boldsymbol{v}_2 - \boldsymbol{v}_1), \tag{7B.7}$$

the friction coefficient ζ between the two components in (7.1.16) and (7.1.17) becomes

$$\zeta = \zeta_1 \zeta_2 / (\zeta_1 + \zeta_2). \tag{7B.8}$$

The friction coefficients ζ_1 and ζ_2 per unit volume should tend to be of microscopic sizes for $N_1 \sim N_2 \sim N_e$ or in the Rouse limit, and they are proportional to N_1 and N_2 in the entangled case. Thus we find

$$\zeta_1 = \phi_1 \zeta_{10} N_1 / N_e, \quad \zeta_2 = \phi_2 \zeta_{20} N_2 / N_e. \tag{7B.9}$$

which leads to the expression for ζ in (7.1.44).

The principle of positive-definiteness of the heat production rate can be conveniently used to seek fundamental dynamical relations. When the mixture is slightly displaced from equilibrium, the total free energy in our system is written as

$$\mathcal{H}_T = \mathcal{H}\{\phi\} + \int d\boldsymbol{r} \left[\frac{1}{2\bar{\rho}^2 K_T} (\delta\rho)^2 + \frac{1}{2}\rho_1 \boldsymbol{v}_1^2 + \frac{1}{2}\rho_2 \boldsymbol{v}_2^2 \right]. \tag{7B.10}$$

The first term is given by (7.1.2). Using the dynamic equations (7.1.11), (7.1.16), and (7.1.17), we obtain

$$\frac{d}{dt}\mathcal{H}_T = \int d\boldsymbol{r} [-\zeta \boldsymbol{w}^2 + \boldsymbol{v}_1 \cdot \boldsymbol{F}_1 + \boldsymbol{v}_2 \cdot \boldsymbol{F}_2]. \tag{7B.11}$$

Our simple assumption here is that the heat production rate is determined in the form

$$-\frac{d}{dt}\mathcal{H}_T = \int d\boldsymbol{r} [\zeta \boldsymbol{w}^2 + \nabla \boldsymbol{v}_t : \overleftrightarrow{\sigma}] \tag{7B.12}$$

in terms of the tube velocity $\boldsymbol{v}_t$. It then leads to the intermediate stress division (7.1.38) with

$$\alpha_1 = \zeta_1/(\zeta_1 + \zeta_2), \quad \alpha_2 = \zeta_2/(\zeta_1 + \zeta_2). \tag{7B.13}$$

We can express α as (7.1.39) and $\boldsymbol{v}_t$ as (7.1.45).

The above theory is highly phenomenological involving various assumptions. It is probably the simplest theory consistent with the existing molecular models and the experimental results. The validity of the assumptions should be critically checked in future study.

Appendix 7C Calculation of the time-correlation function

Following Ref. [3] we apply small, fictitious external fields acting on the two components in a polymer blend. The change in the free energy is

$$\delta\mathcal{H}_{\text{ext}} = \int d\boldsymbol{r}(\delta\rho_1 U_1 + \delta\rho_2 U_2), \tag{7C.1}$$

Then the forces on the two polymers are $-\rho_1 \nabla U_1$ ($K = 1, 2$), and they should be added to the right-hand sides of (7.1.16) and (7.1.17). As a result the expression for $\boldsymbol{w}$ is modified as

$$\boldsymbol{w} = (\phi_1\phi_2/\zeta)\left[-\nabla\frac{\partial\mathcal{H}}{\partial\phi} + \alpha\nabla\cdot\overleftrightarrow{\sigma} - \rho\nabla(U_1 - U_2)\right]. \tag{7C.2}$$

We hereafter assume that $U_1 - U_2 \propto \exp(i\boldsymbol{q}\cdot\boldsymbol{r} + i\omega t)$ in space and time. We modify (7.1.59) as

$$\left[i\omega + \Gamma_q + \left(\frac{4}{3}L\alpha^2\right)q^2 i\omega\eta^*(\omega)\right]\delta\phi = -\rho L q^2(U_1 - U_2). \tag{7C.3}$$

The general linear response theory [94] leads to the relation,

$$\delta\phi = -(\rho/T)\left[I_q - i\omega\widehat{I}(q,\omega)\right](U_1 - U_2), \tag{7C.4}$$

where I_q and $\widehat{I}(q,\omega)$ are given by (7.1.8) and (7.1.76), respectively. Comparison of (7C.3) and (7C.4) yields

$$-i\omega\widehat{I}(q,\omega) + I_q = TLq^2\Big/\left[i\omega + \Gamma_q + \left(\frac{4}{3}L\alpha^2\right)q^2 i\omega\eta^*(\omega)\right]. \tag{7C.5}$$

Note that both sides of (7C.5) tend to I_q as $\omega \to 0$. Some manipulations readily lead to (7.1.77).

Appendix 7D Stress tensor in polymer solutions

We derive the reversible part of the stress tensor $\overleftrightarrow{\Pi}$ arising from the deviations of ϕ and $\overleftrightarrow{W}$ neglecting dissipation for polymer solutions. We follow the method for near-critical fluids in Appendix 6A. The velocity difference $\boldsymbol{w}$ between polymer and solvent will also be neglected. We consider a small fluid element at position $\boldsymbol{r}$ and at time t. Due to the velocity field $\boldsymbol{v}$ the element is displaced to a new position, $\boldsymbol{r}' = \boldsymbol{r} + \boldsymbol{u}$ with $\boldsymbol{u} = \boldsymbol{v}\delta t$, after a small time interval δt. Then the volume element $d\boldsymbol{r}'$ and the volume fraction are changed as $d\boldsymbol{r}' = d\boldsymbol{r}(1 + \nabla \cdot \boldsymbol{u})$ and $\phi'(\boldsymbol{r}') = \phi(\boldsymbol{r})$, respectively. The change of $\overleftrightarrow{W}$ is calculated from (7.1.102) as

$$W'_{ij}(\boldsymbol{r}') = W_{ij}(\boldsymbol{r}) + \sum_k (\tilde{D}_{ik} W_{kj} + W_{ik} \tilde{D}_{jk}), \tag{7D.1}$$

where $\tilde{D}_{ij} = \partial u_i / \partial x_j$ is the strain tensor. The increment of $\mathcal{H}\{\phi, \overleftrightarrow{W}\}$ in accord with these changes is expressed as (6A.6), which yields Π_{ij}. After some calculations we obtain

$$\overleftrightarrow{\Pi} = \left[\frac{1}{\bar{\rho} K_T} \delta\rho - f + \frac{1}{2} C |\nabla\phi|^2 \right] \overleftrightarrow{I} + C(\nabla\phi)(\nabla\phi) - \overleftrightarrow{\sigma}_{\rm p}, \tag{7D.2}$$

where $\overleftrightarrow{\sigma}_{\rm p}$ is given in (7.1.102). The total stress tensor is the sum of $\overleftrightarrow{\Pi}$ and the viscous stress $-\overleftrightarrow{\sigma}_{\rm vis}$.

Appendix 7E Elimination of the transverse degrees of freedom

Here we are interested in heterogeneous fluctuations much shorter than the system size in an isotropically swollen gel with average polymer volume fraction $\phi_{\rm r}$. We may then impose the clamped boundary condition ($\boldsymbol{u} = \boldsymbol{0}$ on the boundary). Assuming no macroscopic swelling from a reference state ($\alpha = 1$), we set

$$\boldsymbol{r} = \boldsymbol{r}_0 + \boldsymbol{u} \quad \text{or} \quad \boldsymbol{r}_0 = \boldsymbol{r} - \boldsymbol{u}, \tag{7E.1}$$

where $\boldsymbol{u}$ is a small displacement vector. We use the Euler representation and treat the original gel position $\boldsymbol{r}_0$ as a function of the final position $\boldsymbol{r}$. Then the inverse matrix of Γ_{ij} in (3.5.46) is expressed as

$$\Gamma^{ij} = \nabla_j x_{0i} = \delta_{ij} - D_{ij}, \tag{7E.2}$$

where $\nabla_j = \partial/\partial x_j$ and $D_{ij} = \nabla_j u_i$. In 3D, the relative volume fraction $\Phi \equiv \phi/\phi_{\rm r}$ can be divided into four terms, each being of nth order in $\boldsymbol{u}$ ($n = 0, \ldots, 3$):

$$\Phi = \det\{\nabla_j x_{0i}\} = 1 - \nabla \cdot \boldsymbol{u} + J_2 - J_3, \tag{7E.3}$$

where

$$J_2 = \frac{1}{2} \sum_{ij} \left[D_{ii} D_{jj} - D_{ij} D_{ji} \right], \quad J_3 = \det\{D_{ij}\}. \tag{7E.4}$$

Because of the clamped boundary condition, the space averages become

$$\langle \Phi \rangle = 1, \qquad \langle \nabla \cdot \boldsymbol{u} \rangle = \langle J_2 \rangle = \langle J_3 \rangle = 0. \tag{7E.5}$$

As in (3.5.48) or (7.2.3) the elastic free energy in our theory is proportional to the first rotational invariant $I_1 = \sum_{ij} \Gamma_{ij}^2$ in (3A.4). In terms of Φ and D_{ij}, I_1 is expressed as

$$\begin{aligned}\Phi^2 I_1 &= \sum_{i\neq j}[\Gamma^{ii}\Gamma^{jj} - \Gamma^{ij}\Gamma^{ji}]^2 + \sum_{i\neq j\neq k}[\Gamma^{ij}\Gamma^{kk} - \Gamma^{ik}\Gamma^{kj}]^2 \\ &= 4\Phi - 1 + \frac{1}{2}\sum_{ij}(D_{ij} + D_{ji})^2 + X_3 + X_4. \end{aligned} \tag{7E.6}$$

In the second line the linear term $-4\nabla \cdot \boldsymbol{u}$ has been expressed in terms of Φ. Though not explicitly expressed, X_3 and X_4 are the third- and fourth-order terms. Retaining the first two terms in the above expression, we clearly obtain the usual expression for the elastic free energy (3A.18).

In order to minimize the space integral of ΦI_1 with $\Phi = \Phi(\boldsymbol{r})$ held fixed, we introduce

$$\begin{aligned} G &= \int d\boldsymbol{r}\left[\frac{1}{2}\Phi I_1 + h\left(\Phi - 1 + \nabla \cdot \boldsymbol{u} - J_2 + J_3\right)\right] \\ &= \int d\boldsymbol{r}\left[2 - \frac{1}{2\Phi}\right] + \int d\boldsymbol{r}\left[\frac{1}{4\Phi}\sum_{ij}(D_{ij} + D_{ji})^2 + h(\Phi_1 + \nabla \cdot \boldsymbol{u})\right] + \cdots, \end{aligned} \tag{7E.7}$$

where $h = h(\boldsymbol{r})$ is a space-dependent Lagrange multiplier required from the constraint (7E.3). The terms proportional to X_3, X_4, J_2, and J_3 are not written in the second line. We can show self-consistently that h is of order $\Phi_1 \equiv \Phi - 1$. To leading order in Φ_1, the minimization condition $(\delta G/\delta \boldsymbol{u})_\Phi = \boldsymbol{0}$ yields the first-order solution $\boldsymbol{u} = \boldsymbol{u}^{(1)}$ with

$$\boldsymbol{u}^{(1)} = \nabla w, \tag{7E.8}$$

where w is the solution of the Laplace equation,

$$\nabla^2 w = \Phi_1. \tag{7E.9}$$

It then follows the expansion $h = -\Phi_1 + \cdots$. Up to second order in Φ_1, we thus find $G = G_0$ with

$$G_0 = \int d\boldsymbol{r}\left[\frac{3}{2} + \frac{1}{2}\Phi_1 + \frac{1}{2}\Phi_1^2\right]. \tag{7E.10}$$

Next we consider the third-order contributions. (i) In calculating the third-order term X_3 in the second line of (7E.6), we may use (7E.8) to obtain

$$X_3 = 6J_3 + \Phi_1^3 - \Phi_1 \sum_{ij}(\nabla_i \nabla_j w)^2, \tag{7E.11}$$

where J_3 is defined by (7E.4) and its space integral vanishes as in (7E.5). (ii) From the second line of (7E.7), we find two third-order contributions; one is equal to the space integral of $-\Phi_1 \sum_{ij}(\nabla_i \nabla_j w)^2$ and the other is written as

$$G_3 = \int d\boldsymbol{r} \sum_{ij}(\nabla_i \nabla_j w)(\nabla_i u_j^{(2)} + \nabla_j u_i^{(2)}) = -2\int d\boldsymbol{r}\Phi_1 \nabla \cdot \boldsymbol{u}^{(2)}, \tag{7E.12}$$

where $\boldsymbol{u}^{(2)}$ is the displacement of order Φ_1^2. The constraint (7E.3) yields

$$\nabla \cdot \boldsymbol{u}^{(2)} = \frac{1}{2}\Phi_1^2 - \frac{1}{2}\sum_{ij}(\nabla_i \nabla_j w)^2. \tag{7E.13}$$

Up to the third order we finally have

$$G = G_0 - \frac{1}{2}\int d\boldsymbol{r}\Phi_1 \sum_{ij}(\nabla_i \nabla_j w)^2. \tag{7E.14}$$

Appendix 7F Calculation for weakly charged polymers

In weakly charged polymers in theta or poor solvents [65]–[68], we write the number density of monovalent counterions in the solvent as $n_{\rm i}$. Further assuming the presence of salt ions with unit charges $\pm e$, we set up the following free-energy functional:

$$\beta\mathcal{H}_{\rm ch} = \int d\boldsymbol{r}\big[n_{\rm i}\ln n_{\rm i} + c_{\rm i}\ln c_{\rm i} + c_{\rm sa}\ln c_{\rm sa}\big] + \frac{\ell_{\rm B}}{2}\int d\boldsymbol{r}\int d\boldsymbol{r}'\frac{n(\boldsymbol{r})n(\boldsymbol{r}')}{|\boldsymbol{r}-\boldsymbol{r}'|}, \tag{7F.1}$$

where $c_{\rm i}$ is the salt counterion density and $c_{\rm sa}$ the salt co-ion density. The last term is the electrostatic energy in a solvent with dielectric constant $\epsilon_{\rm s}$, and $\ell_{\rm B}$ is the Bjerrum length defined by (7.2.25). The charge density is written as

$$n(\boldsymbol{r}) = n_{\rm p}(\boldsymbol{r}) - n_{\rm i}(\boldsymbol{r}) + c_{\rm sa}(\boldsymbol{r}) - c_{\rm i}(\boldsymbol{r}), \tag{7F.2}$$

where

$$n_{\rm p}(\boldsymbol{r}) = v_0^{-1}\hat{f}\phi(\boldsymbol{r}) \tag{7F.3}$$

is the density of ions attached to the polymer chains. The $\hat{f}$ is the fraction of charged monomers. The space averages satisfy $\bar{n}_p = \bar{n}_i$ and $\bar{c}_s = \bar{c}_i$. In the RPA approximation we assume that the deviations, $\delta n_{\rm i} = n_{\rm i} - \langle n_{\rm i}\rangle$, $\delta c_{\rm i} = c_{\rm i} - \langle c_{\rm i}\rangle$, and $\delta n_{\rm i} = n_{\rm i} - \langle n_{\rm i}\rangle$ are small compared with the averages. Then, in the bilinear order in the deviations, the free energy is written in terms of the Fourier component as

$$\beta\delta\mathcal{H}_{\rm ch} = \frac{1}{2}\int_q \left[\frac{1}{\bar{n}_i}|n_{\rm i}(\boldsymbol{q})|^2 + \frac{1}{\bar{c}_{\rm sa}}|c_{\rm i}(\boldsymbol{q})|^2 + \frac{1}{\bar{c}_{\rm sa}}|c_{\rm sa}(\boldsymbol{q})|^2 + \frac{4\pi\ell_{\rm B}}{q^2}|n(\boldsymbol{q})|^2\right]. \tag{7F.4}$$

We define the Debye–Hückel free-energy functional $\mathcal{H}_{\rm DH}\{\phi\}$ by

$$\exp(-\beta\mathcal{H}_{\rm DH}) = \int \mathcal{D}n_{\rm i}\int \mathcal{D}c_{\rm i}\int \mathcal{D}c_{\rm sa}\exp(-\beta\mathcal{H}_{\rm ch}). \tag{7F.5}$$

In the RPA approximation the functional integrations over the thermal fluctuations of $c_{\rm i}$, $c_{\rm sa}$, and $n_{\rm i}$ may easily be performed to give

$$\beta\mathcal{H}_{\rm DH} = V\left(\bar{n}_i\ln\bar{n}_i + 2\bar{c}_{\rm sa}\ln\bar{c}_s - \frac{1}{12\pi}\kappa_{\rm Db}^3 + \cdots\right) + \int_q \frac{2\pi\ell_{\rm B}}{\kappa_{\rm Db}^2 + q^2}|n_{\rm p}(\boldsymbol{q})|^2, \tag{7F.6}$$

where $\kappa_{\rm Db}^{-1}$ is the Debye screening length defined by (7.2.24) in terms of the total mobile charge density $\bar{c}_{\rm total} = \bar{n}_i + 2\bar{c}_{\rm sa}$. For $\bar{c}_{\rm total} \gg \kappa_{\rm Db}^3$, under which the RPA approximation is valid, the dominant contribution to the thermodynamic free energy the translational entropy of the total mobile ions (counterions and salt ions), see also (3.5.49). The gaussian integrations give rise to the following contribution to the free energy:

$$\frac{1}{2}V\int_{q}\ln(1+\kappa_{\mathrm{Db}}^{2}/q^{2}) = V(\kappa_{\mathrm{Db}}^{3}/4\pi^{2})[\Lambda/\kappa_{\mathrm{Db}} - \pi/3 + \cdots]. \tag{7F.7}$$

The first term in the brackets depends on the upper cut-off wave number Λ but is simply proportional to $\bar{c}_{\mathrm{total}}$, so it can be omitted in (7F.6). The first nontrivial correction in the Helmholtz free energy is thus written as [64]

$$(\Delta F)_{\mathrm{DH}} = -\frac{1}{12}VT\kappa_{\mathrm{Db}}^{3}. \tag{7F.8}$$

Appendix 7G Surface modes of a uniaxial gel

We consider the surface mode in the uniaxial case. From (7.2.5) and (7.2.6) the deviation of the stress tensor as

$$\delta\overleftrightarrow{\Pi}_{ij} = \left[-\left(K+\frac{\mu}{3}-TC\nabla^{2}\right)\delta_{ij}+\mu_{ij}\nabla_{i}\nabla_{j}\right]\nabla\cdot\boldsymbol{u} - \sum_{k}\left[\mu_{ik}\nabla_{k}u_{j}+\mu_{jk}\nabla_{k}u_{i}\right]. \tag{7G.1}$$

Hereafter the higher-order gradient term ($\propto C$) will be neglected. From (7.2.35) we have $\mu_{ij} = \mu\delta_{ij}[(\lambda^{2}-\lambda^{-1})\delta_{ix}+\lambda^{-1}]$. Then the eigenvalue equation to be solved in the region $x < 0$ is written as

$$\varepsilon_{\mathrm{el}}\nabla(\nabla\cdot\boldsymbol{u}) + (\lambda^{2}\nabla_{x}^{2} - \lambda^{-1}q^{2})\boldsymbol{u} = -\tilde{w}q^{2}\boldsymbol{u}. \tag{7G.2}$$

Some manipulations cause the above equation to assume the same form as (7.2.53) for the isotropic case,

$$\tilde{\varepsilon}_{\mathrm{el}}\nabla(\nabla\cdot\boldsymbol{u}) + \nabla^{2}\boldsymbol{u} = -(\lambda^{-2}\tilde{w}+1-\lambda^{-3})q^{2}\boldsymbol{u}, \tag{7G.3}$$

where $\nabla^{2} = \nabla_{x}^{2} - q^{2}$ and

$$\tilde{\varepsilon}_{\mathrm{el}} = \varepsilon_{\mathrm{el}}/\lambda^{2}. \tag{7G.4}$$

In calculating the boundary condition at $x = 0$, we first note the relation, $\Pi_{ij} = (\mu_{xx} - \mu_{yy})\delta_{ij}(\delta_{jy}+\delta_{jz}) + \delta\Pi_{ij}$, which follows from $\langle\Pi_{ij}\rangle = (\phi f' - f)\delta_{ij} - \mu_{ij}$ and (7G.1). Second, the normal unit vector is written as $\boldsymbol{n} = (1, -\partial u_{x}/\partial y, 0)$. From these two relations the stress-free boundary condition $\overleftrightarrow{\Pi}\cdot\boldsymbol{n} = \boldsymbol{0}$ turns out to be of the same form as (7.2.54) with $\varepsilon_{\mathrm{el}}$ being replaced by $\tilde{\varepsilon}_{\mathrm{el}}$. Thus the results for the isotropic case can be used to give

$$\tilde{w} = \lambda^{2}[w(\tilde{\varepsilon}_{\mathrm{el}}) - 1 + \lambda^{-3}]. \tag{7G.5}$$

References

[1] P. G. de Gennes, *Scaling Concepts in Polymer Physics*, 2nd edn (Cornell University Press, Ithaca, 1985).

[2] M. Doi and S. F. Edwards, *The Theory of Polymer Dynamics* (Oxford University Press, 1986).

[3] F. Brochard and P. G. de Gennes, *Macromolecules* **10**, 1157 (1977); F. Brochard, *J. Physique I* **44**, 39 (1983).

[4] M. Doi and A. Onuki, *J. Physique II* **2**, 1631 (1992); A. Onuki, *J. Non-Cryst. Solids* **172–174**, 1151 (1994).

[5] J. L. Harden, H. Pleiner, and P. A. Pincus, *J. Chem. Phys.* **94**, 5208 (1991).

[6] E. J. Amis and C. C. Han, *Polymer* **23**, 1403 (1982); E. J. Amis, C. C. Han, and Y. Matsushita, *Polymer* **25**, 650 (1984).

[7] N. Nemoto, Y. Makita, Y. Tsunashima, and M. Kurata, *Macromolecules* **17**, 2629 (1984); N. Nemoto, A. Koike, and K. Osaki, *ibid.* **29**, 1445 (1996).

[8] M. Adam and M. Delsanti, *Macromolecules* **18**, 1760 (1985).

[9] W. Brown, T. Nicolai, S. Hvidt, and K. Heller, *Macromolecules* **23**, 357 (1990).

[10] S.-J. Chen and G. C. Berry, *Polymer* **31**, 793 (1990).

[11] P. K. Dixon, D. J. Pine, and X. L. Wu, *Phys. Rev. Lett.* **68**, 2239 (1992).

[12] H. Tanaka, *Phys. Rev. E* **59**, 6842 (1999).

[13] M. Tirrell, *Fluid Phase Equilibria* **30**, 367 (1986).

[14] M. Adam and M. Delsanti, *J. Physique I* **44**, 1185 (1983); *ibid.* **45**, 1513 (1984); M. Adam, *J. Non-Cryst. Solids*, **131–133**, 773 (1991).

[15] Y. Takahashi, Y. Isono, I. Noda, and M. Nagasawa, *Macromolecules* **18**, 1002 (1985).

[16] R. H. Colby and M. Rubinstein, *Macromolecules* **23**, 2753 (1990).

[17] T. Tanaka, L. O. Hocker, and G. B. Benedik, *J. Chem. Phys.* **59**, 5151 (1973).

[18] E. Helfand and H. Fredrickson, *Phys. Rev. Lett.* **62**, 2468 (1989).

[19] H. P. Wittmann and G. H. Fredrickson, *J. Physique II* **4**, 1791 (1994)

[20] T. Sun, A. C. Balazs, and D. Jasnow, *Phys. Rev. E* **55**, 6344 (1997).

[21] F. Brochard, in *Molecular Conformation and Dynamics of Macromolecules in Condensed Systems*, ed. M. Nagasawa (Elsevier, New York, 1988), p. 249.

[22] E. J. Kramer, P. F. Green, and C. J. Palmstrom, *Polymer* **25**, 473 (1984).

[23] H. Silescu, *Makrom. Chem. Rapid Commun.* **5**, 519 (1984).

[24] E. A. Jordan, R. C. Ball, A. M. Donald, L. J. Fetters, R. A. L. Jones, and J. Klein, *Macromolecules* **21**, 235 (1988).

[25] J. Klein, *J. Non-Cryst. Solids* **131–133**, 598 (1991).

[26] K. Binder, H.-P. Deutsch, and A. Sariban, *J. Non-Cryst. Solids* **131–133**, 635 (1991).

[27] D. Schwahn, S. Janssen, and T. Springer, *J. Chem. Phys.* **97**, 8775 (1992); G. Müller, D. Schwahn, H. Eckerlebe, J. Rieger, and T. Springer, *J. Chem. Phys.* **104**, 5826 (1996).

[28] F. Brochard and P. G. de Gennes, *Physicochem. Hydrodyn.* **4**, 313 (1983).

[29] K. Binder, H. L. Frish, and J. Jäckle, *J. Chem. Phys.* **85**, 1505 (1986).

[30] J. Jäckle and M. Pieroth, *Z. Phys. B* **72**, 25 (1988).

[31] N. Clarke, T. C. B. Mcleish, S. Pavawongsak, and J. S. Higgins, *Macromolecules* **30**, 4459 (1997).

[32] A. Onuki, *J. Phys. Soc. Jpn* **59**, 3423 (1990); *J. Phys. Condens. Matter* **9**, 6119 (1997).

[33] S. T. Milner, *Phys. Rev. E* **48**, 3674 (1993).

[34] H. Ji and E. Helfand, *Macromolecules* **28**, 3869 (1995).

[35] M. Grmela, *Phys. Lett. A* **130**, 81 (1988).

[36] A. N. Beris and B. J. Edwards, *Thermodynamics of Flowing Systems* (Oxford University Press, 1994).

[37] R. B. Bird, R. C. Armstrong, and O. Hassager, *Dynamics of Polymeric Liquids*, 2nd edn (Wiley, New York,1987), Vol. 2.

[38] R. G. Larson, *Constitutive Equations for Polymer Melts and Solutions* (Butterworths, Boston, 1986).

[39] H. H. Winter and F. Chambon, *J. Rheol.* **30**, 367 (1986); F. Chambon and H. H. Winter, *ibid.* **31**, 683 (1987).

[40] J. E. Martin and J. P. Wilcoxon, *Phys. Rev. Lett.* **61**, 373 (1988); P. Lang and W. Burchard, *Macromolecules* **24**, 814 (1991); M. Adam, M. Delsanti, J. P. Munich, and D. Durand, *Phys. Rev. Lett.* **61**, 706 (1988).

[41] F. Tanaka and W. H. Stockmayer, *Macromolecules*, **27**, 3943 (1994).

[42] P. M. Goldbert, H. Castillo, and A. Zippelius, *Adv. Phys.* **45**, 393 (1996).

[43] M. Warner and E. M. Terentjev, *Prog. Polym. Sci.* **21**, 853 (1996).

[44] T. Tanaka, S. Ishiwata, and C. Ishimoto, *Phys. Rev. Lett.* **38**, 771 (1977).

[45] T. Tanaka and D. J. Filmore, *J. Chem. Phys.* **70**, 1214 (1979).

[46] A. Onuki, *Advances in Polymer Science*, Vol. 109, *Responsible Gels: Volume Transitions I*, ed. K. Dušek (Springer, Heidelberg, 1993), p. 63.

[47] Y. Rabin and A. Onuki, *Macromolecules* **27**, 870 (1994).

[48] T. Tanaka, S. T. Sun, Y. Hirokawa, S. Katayama, J. Kucera, Y. Hirose, and T. Amiya, *Nature* (London) **325**, 796 (1987); T. Tanaka, in *Molecular Conformation and Dynamics of Macromolecules in Condensed Systems*, ed. M. Nagasawa (Elsevier, New York,1988), p. 203.

[49] T. Hayashi, H. Tanaka, T. Nishi, Y. Hirose, T. Amiya, and T. Tanaka, *J. Appl. Polymer Sci.* **44**, 195 (1989).

[50] H. Tanaka, *Phys. Rev. Lett.* **68**, 2794 (1992); H. Tanaka and T. Sigehuzi, *Phys. Rev. E* **49**, R39 (1994).

[51] Y. Li, C. Li, and Z. Hu, *J. Chem. Phys.* **100**, 4637 (1994); C. Li, Z. Hu, and Y. Li, *ibid*. **100**, 4645 (1994).

[52] A. Onuki, *Phys. Rev. A* **39**, 5932 (1989).

[53] A. Onuki and J. Harden, *Phase Transitions* (Gordon and Breach Science, S.A.), **46**, 127 (1994).

[54] K. Sekimoto and K. Kawasaki, *J. Phys. Soc. Jpn* **57**, 2594 (1988); N. Suematsu, K. Sekimoto, and K. Kawasaki, *Phys. Rev. A*, **41**, 5751 (1990).

[55] T. Hwa and M. Kardar, *Phys. Rev. Lett.* **61**, 106 (1988).

[56] E. S. Matsuo and T. Tanaka, *J. Chem. Phys.* **89**, 1695 (1988).

[57] E. S. Matsuo and T. Tanaka, *Nature* (London) **358**, 482 (1992).

[58] S. Hirotsu, *J. Chem. Phys.* **88**, 427 (1988).

[59] K. Sekimoto and K. Kawasaki, *Physica A* **154**, 384 (1989).

[60] L. D. Landau and E. M. Lifshitz, *Theory of Elasticity* (Pergamon, New York, 1973).

[61] A. Aharony, *Phys. Rev. B* **8**, 3363 (1973).

[62] F. Horkay, A. M. Hecht, and E. Geissler, *Macromolecules*, **31**, 8851 (1998).

[63] H. Tanaka, *J. Chem. Phys.* **100**, 5253 (1994).

[64] L. D. Landau and E. M. Lifshitz, *Statistical Physics* (Pergamon, New York, 1964).

[65] V. Y. Borue and I. Y. Erukhimovich, *Macromolecules* **21**, 3240 (1992).

[66] J. F. Joanney and L. Leibler, *J. Physique I* **51**, 545 (1990).

[67] Y. Rabin and S. Panyukov, *Macromolecules* **30**, 301 (1997).

[68] Y. Shiwa, *Eur. Phys. J. B* **1**, 345 (1998); Y. Shiwa and M. Shibayama, *Phys. Rev. E* **59**, 5891 (1999).

[69] T. Tanaka, *Phys. Rev. A* **17**, 763 (1978).

[70] S. Candau, J. P. Munch, and G. Hild, *J. Physique I* **41**, 1031 (1980).

[71] T. Takebe, K. Nawa, S. Suehiro, and T. Hashimoto, *J. Chem. Phys.* **91**, 4360 (1989).

[72] J. A. Marqusee and J. M. Deutch, *J. Chem. Phys.* **75**, 5239 (1981).

[73] D. L. Johnson, *J. Chem. Phys.* **77**, 1531 (1982).

[74] M. A. Biot, *J. Acoust. Soc. Am.* **28**, 168, 179 (1956).

[75] J. Bastide and S. J. Candau, in *Physical Properties in Polymer Gels*, ed. J. P. Cohen Addad (John Wiley & Sons, 1996), p. 143.

[76] R. S. Stein, *J. Polym. Sci.* **7**, 657 (1969).

[77] F. Bueche, *J. Col. Interf. Sci.* **33**, 61 (1970).

[78] B. J. Bauer, R. M. Briber, and C. C. Han, *Macromolecules* **22**, 940 (1989); R. M. Briber and B. J. Bauer, *ibid.* **24**, 1899 (1991).

[79] E. S. Matsuo, M. Orkisz, S.-T. Sun, Y. Li, and T. Tanaka, *Macromolecules*, **27**, 6791 (1994).

[80] M. Shibayama, *Macromol. Chem. Phys.* **199**, 1 (1998); F. Ikkai and M. Shibayama, *Phys. Rev. Lett.* **82**, 4946 (1999).

[81] J. Bastide and L. Leibler, *Macromolecules* **21**, 2647 (1988).

[82] E. Mendes, P. Lindner, M. Buzier, F. Boue, and J. Bastide, *Phys. Rev. Lett.* **66**, 1595 (1991); E. Mendes, R. Oeser, C. Hayes, F. Boue, and J. Bastide, *Macromolecules* **29**, 5574 (1996).

[83] M. Shibayama, K. Kawakubo, F. Ikkai, and M. Imai, *Macromolecules*, **31**, 2586 (1998).

[84] J. Bastide, L. Leibler, and J. Prost, *Macromolecules* **23**, 1821 (1990).

[85] A. Onuki, *J. Physique II* **2**, 45 (1992).

[86] S. Panyukov and Y. Rabin, *Phys. Rep.* **269**, 1 (1996).

[87] S. Alexander, *J. Physique I* **45**, 1939 (1984).

[88] L. Golubovic and T. C. Lubensky, *Phys. Rev. Lett.* **63**, 1082 (1989).

[89] N. Uchida, *Phys. Rev. E* **60**, R13 (1999).

[90] P. Debye and A. M. Bueche, *J. Appl. Phys.* **20**, 518 (1949); P. Debye, H. R. Anderson, and H. Brumberger, *ibid.* **20**, 518 (1957).

[91] M. Rubinstein, in *Theoretical Challenges in the Dynamics of Complex Fluids*, ed. T. C. B. McLeish (Kluwer Academic, Dordrecht, 1997), p. 9.

[92] J. de Cloizeaux, *Europhys. Lett.* **5**, 437 (1988).

[93] D. M. Mead, *J. Rheol.* **40**, 633 (1996).

[94] R. Kubo, *J. Phys. Soc. Jpn* **12**, 570 (1957).

Part three

Dynamics of phase changes

8

Phase ordering and defect dynamics

When an external parameter such as the temperature or the pressure is changed, physical systems in a homogeneous state often become unstable and tend to an ordered phase with broken symmetry [1]–[6]. The growth of order takes place with coarsening of domain or defect structures. Such ordering processes are observed in many systems such as spin systems, solids, and fluids. Historically, structural ordering and phase separation in alloys has been one of the central problems in metallurgy. These highly nonlinear and far-from-equilibrium processes have recently been challenging subjects in condensed matter physics. We will review various theories of phase ordering, putting emphasis on the dynamics of interfaces and vortices. As newly-explored examples we will discuss spinodal decomposition in one-component fluids near the gas–liquid critical point induced by the piston effect, that in binary fluid mixtures near the consolute critical point adiabatically induced by a pressure change, that under periodic quenching, and that in polymers and gels influenced by stress–diffusion coupling. A self-organized superfluid state will also be investigated, which is characterized by high-density vortices arising from competition between heat flow and gravity.

8.1 Phase ordering in nonconserved systems

8.1.1 Model A

We analyze the phase ordering in model A with a one-component order parameter ($n = 1$) given by (5.3.3)–(5.3.5). The temperature coefficient r is changed from a positive to a negative value at the instant of quenching $t = 0$ as

$$r = \kappa_0^2 \qquad (t < 0), \qquad r = -\kappa^2 \qquad (t > 0). \tag{8.1.1}$$

If the system is quenched from a disordered state at a relatively high temperature, we have $\kappa_0^2 \gg \kappa^2$. After the quench, ψ obeys

$$\frac{\partial}{\partial t}\psi = -L\left[-\kappa^2 - \nabla^2 + u_0\psi^2\right]\psi + Lh + \theta. \tag{8.1.2}$$

The random noise term $\theta(\boldsymbol{r}, t)$ is characterized by (5.3.4). The coefficient K of the gradient term in (5.3.5) is set equal to 1 and the shift r_{0c} in (4.1.17) is not written for simplicity. Hereafter we will examine mainly the case $h = 0$ and $\langle\psi\rangle = 0$. Effects of a small magnetic field will be examined in Section 8.2.

Our problem is highly nontrivial because of the simultaneous presence of the gradient and nonlinear terms in (8.1.2). From (5.2.17) the equal-time structure factor $I(k,t)$ evolves for $t > 0$ and $h = 0$ as

$$\frac{\partial}{\partial t} I(k,t) = 2L\{\kappa^2[1 - J(k,t)] - k^2\} I(k,t) + 2L, \tag{8.1.3}$$

where

$$J(k,t) = u_0 \int d\boldsymbol{r} e^{i\boldsymbol{k}\cdot\boldsymbol{r}} \langle \psi(\boldsymbol{r},t)^3 \psi(\boldsymbol{0},t)\rangle / \kappa^2 I(k,t) \tag{8.1.4}$$

arises from the nonlinearity. At large t we shall see that $\psi(\boldsymbol{r},t)$ locally saturates into either of $\pm\psi_{\rm eq}$ with $\psi_{\rm eq} = \kappa u_0^{-1/2}$. Then we have *domains* with a characteristic size $\ell(t)$ growing in time. The thermal correlation length in the two phases and the interface thickness are given by $\xi = 2^{-1/2}\kappa^{-1}$. If the thermal fluctuations of ψ are neglected, $\psi^2 \cong \psi_{\rm eq}^2$ holds except for the interface region. This means that $J(k,t) \to 1$ for $\ell(t) \gg \xi$ and $k \ll \ell(t)^{-1}$, but it is not obvious in what manner it tends to 1. The following theories indicate $1 - J(0,t) \propto t^{-1}$ at $h = 0$ as $t \to \infty$.

Exponential growth and the Ginzburg criterion

If u_0 is very small, the nonlinear term proportional to $u_0\psi^3$ in (8.1.2) is negligible at an early stage. As discussed in Section 5.3, the Fourier component $\psi_{\boldsymbol{k}}$ depends on time as $\exp(-\Gamma_k t)$ with

$$\Gamma_k = L(-\kappa^2 + k^2). \tag{8.1.5}$$

In the long-wavelength region $k < \kappa$, Γ_k is negative and the fluctuations grow exponentially. The maximum growth rate is attained at $k = 0$ and is written as

$$\gamma_0 = L\kappa^2. \tag{8.1.6}$$

The small-scale fluctuations with $k \gg \kappa$ decay with $\Gamma_k \cong Lk^2$ and are little affected by quenching. We integrate (8.1.3) neglecting $J(k,t)$ to obtain the structure factor $I_0(k,t) = \langle |\psi_{\boldsymbol{k}}(t)|^2\rangle$ in the linear approximation:

$$I_0(k,t) = \frac{\exp(-2\Gamma_k t)}{\kappa_0^2 + k^2} + \frac{\exp(-2\Gamma_k t) - 1}{\kappa^2 - k^2}. \tag{8.1.7}$$

At very long wavelengths ($k \ll \kappa$) and after a time of order γ_0^{-1}, the above expression is simplified as

$$I_0(k,t) \cong (\kappa_0^{-2} + \kappa^{-2}) \exp(2\gamma_0 t - 2Lk^2 t). \tag{8.1.8}$$

The exponential growth leads to eventual breakdown of the linear approximation itself. To check it, we decouple the four-body correlation function in $J(k,t)$. Then it becomes independent of k as

$$J(k,t) \cong 3u_0\kappa^{-2}\langle \psi(\boldsymbol{r},t)^2\rangle \cong 3K_d u_0 \kappa^{-2} \int_0^{\kappa} dq q^{d-1} I(q,t), \tag{8.1.9}$$

where we have picked up the growing fluctuations ($q < \kappa$) only. The K_d is the geometrical factor defined by (4.1.16). We now replace $I(q,t)$ on the right-hand side by $I_0(k,t)$ in (8.1.8). Furthermore, in the time region $t \gtrsim \gamma_0^{-1}$, we notice $\kappa \gtrsim (Lt)^{-1/2}$ and may push the upper bound of the q integration to infinity. Thus,

$$J(k,t) \sim (K_d u_0 \kappa^{-\epsilon})(\gamma_0 t)^{-d/2} e^{2\gamma_0 t}, \tag{8.1.10}$$

where $\epsilon = 4 - d$. If $K_d u_0 \kappa^{-\epsilon} \ll 1$, $J(k,t)$ remains much smaller than 1 over a sizable time region and exponential growth will be observed. Here it turns out to be small when the Ginzburg condition (4.1.24) holds and the mean field critical behavior is realized for $\epsilon > 0$.

Notice that $u_0\kappa^{-\epsilon}$ is a unique parameter characterizing the model at $h = 0$ after the model is made dimensionless. In fact, let the space, time, and ψ be measured in units of κ^{-1}, $(L\kappa^2)^{-1}$, and ψ_{eq}; then, the fluctuation–dissipation relation becomes $\langle \theta(\boldsymbol{r},t)\theta(\boldsymbol{r}',t')\rangle = 2u_0\kappa^{-\epsilon}\delta(\boldsymbol{r}-\boldsymbol{r}')\delta(t-t')$. This form indicates that the noise strength is characterized by $u_0\kappa^{-\epsilon}$. If $h \neq 0$, another relevant dimensionless parameter is the scaled magnetic field $h^* = h/\kappa^2\psi_{\text{eq}}$.

Roles of the three terms

(i) The nonlinear term ($\propto u_0\psi^3$) gives rise to saturation of ψ into ψ_{eq} or $-\psi_{\text{eq}}$. To see this, we neglect the gradient term ($\propto \nabla^2\psi$) and the random noise term, to obtain

$$\frac{\partial}{\partial t}\psi = -L(-\kappa^2 + u_0\psi^2)\psi. \tag{8.1.11}$$

Because $Y(t) \equiv \psi_{\text{eq}}^2/\psi^2$ obeys the linear equation,

$$\frac{\partial}{\partial t}Y = -2\gamma_0(Y-1), \tag{8.1.12}$$

(8.1.11) is integrated to give

$$\begin{aligned}\psi(t) &= \psi(0)\psi_e \Big/ \left\{\psi(0)^2 + [\psi_e{}^2 - \psi(0)^2]e^{-2\gamma_0 t}\right\}^{1/2} \\ &\cong \varphi(t)\Big/\left[1 + \alpha\varphi(t)^2\right]^{1/2},\end{aligned} \tag{8.1.13}$$

with

$$\alpha = 1/\psi_{\text{eq}}^2 = u_0/\kappa^2. \tag{8.1.14}$$

The second line of (8.1.13) holds for $|\psi(0)| \ll \psi_{\text{eq}}$, and

$$\varphi(t) = \psi(0)e^{\gamma_0 t}. \tag{8.1.15}$$

For $\gamma_0 t \gg 1$, $\psi(t)$ approaches ψ_{eq} or $-\psi_{\text{eq}}$ depending on the sign of the initial value $\psi(0)$. Figure 8.1 displays the behavior of $\psi(t)/\psi_{\text{eq}}$.

(ii) The role of the noise term is as follows. As (8.1.8) indicates, if $\kappa_0 \gg \kappa$, the coefficient of the exponential factor in the fluctuation intensity turns out to be κ^{-2} due to the random

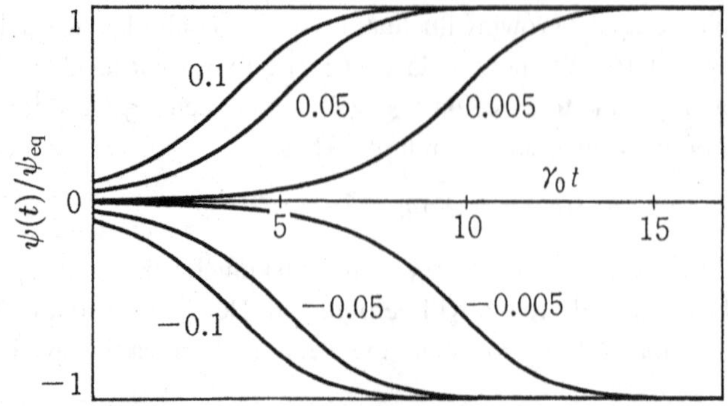

Fig. 8.1. $\psi(t)/\psi_{eq}$ as given by the first line of (8.1.13). The numbers in the figure are the initial values at $t = 0$.

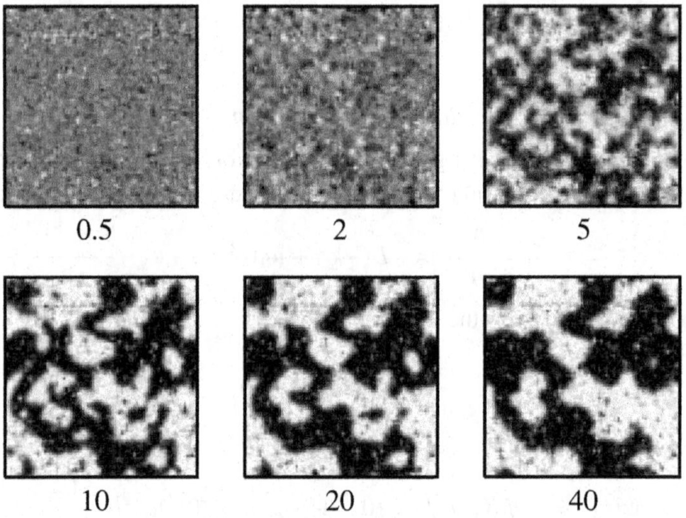

Fig. 8.2. Pattern evolution with time in model A with $\kappa = L = 1$ in the presence of the gaussian noise term θ. The dynamic equation is discretized on a 128×128 lattice with $\Delta x = 1$ and $\Delta t = 0.002$. The numbers are the times in units of γ_0^{-1} after quenching.

agitation in the time region $0 < t < \gamma_0^{-1}$. However, once the fluctuation level much exceeds the thermal order, the evolution of ψ becomes insensitive to the thermal noise. The random noise term is no longer important in the later stages of pattern evolution.

(iii) The gradient term limits the instability only in the long-wavelength region ($k < \kappa$) during the initial stage and creates the interfaces of the domains in the later stages. Coarsening is then driven in the direction of decreasing the interface area.

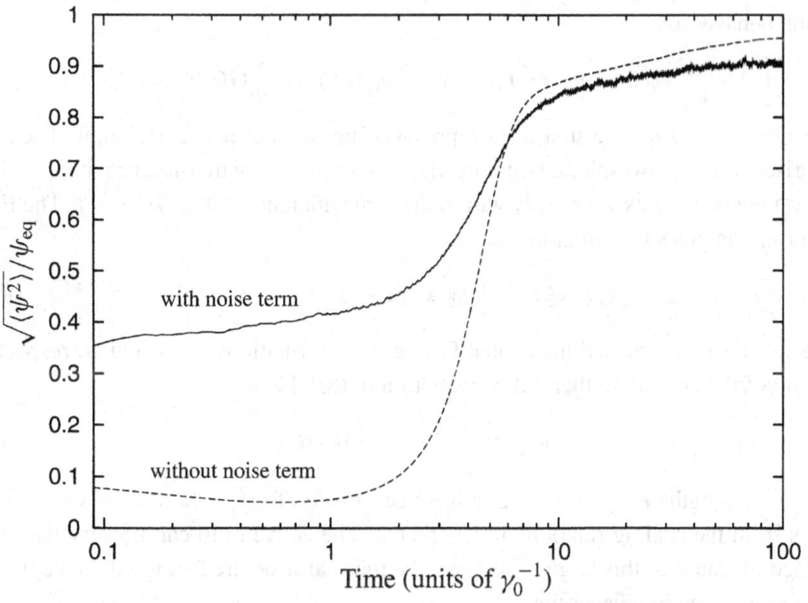

Fig. 8.3. Variance $\sqrt{\langle\psi^2\rangle}$ divided by ψ_{eq} for model A with (solid line) and without (dashed line) the noise term for $t > 0$. The initial ψ on each lattice point are commonly gaussian random numbers with variance $0.1\psi_{eq}$. Each curve is the result of a single run.

8.1.2 *Late-stage behavior and the structure factor* ($n = 1$)

Before proceeding to nonlinear theories, we show a numerical solution of (8.1.2) at $h = 0$ in the presence of the noise term θ in 2D in Fig. 8.2. We can see intricate patterns of the two-phase structure with the thermal fluctuations superimposed, which are symmetric between the two phases and are percolated throughout the system. Figure 8.3 displays the dimensionless variance $\sqrt{\langle\psi^2\rangle}/\psi_{eq}$ for the same run (solid line), together with the same quantity for another run without thermal noise. In the late stage $t \gtrsim 10\gamma_0^{-1}$, ψ saturates into either $\psi_{eq} = \kappa/u_0^{1/2}$ or $-\psi_{eq}$ except for the interfacial regions with thickness $\xi = 2^{-1/2}\kappa^{-1}$. The characteristic length of the patterns grows as

$$\ell(t) = \sqrt{Lt} \qquad \text{or} \qquad \kappa l(t) = \sqrt{\gamma_0 t}. \tag{8.1.16}$$

Note that $\ell(t)$ does not depend on the quench depth and much exceeds ξ for $t \gg \gamma_0^{-1}$. The total free energy of the system at zero magnetic field is then

$$\mathcal{H} = \sigma S(t) + \text{const.}, \tag{8.1.17}$$

where $\sigma \sim T\psi_{eq}^2/\kappa$ is the surface tension and $S(t)$ is the total surface area of order $V/\ell(t)$. The patterns are self-similar if they are scaled by $\ell(t)$. As a result, the pair correlation function behaves as

$$\langle\psi(\boldsymbol{r}_1,t)\psi(\boldsymbol{r}_2,t)\rangle \cong \kappa^{2\beta/\nu}g_{\text{th}}(\kappa r)+\psi_{\text{eq}}^2 G(r/\ell(t)), \tag{8.1.18}$$

where $r = |\boldsymbol{r}_1 - \boldsymbol{r}_2|$. The first term represents the thermal pair correlation. The second term arises from the two-phase structure and is determined by the surface pattern only. For $r \ll \ell(t)$ the two points are mostly within the same domain, so that $G(0) = 1$. The Fourier transformation gives the structure factor,

$$I(k,t) \cong \kappa^{-\gamma/\nu} f_{\text{th}}(k/\kappa)+\psi_{\text{eq}}^2 l(t)^d F(\ell(t)k), \tag{8.1.19}$$

where f_{th} and F are the d-dimensional Fourier transformations of g_{th} and G, respectively. The ratio of the domain to thermal contribution in (8.1.14) is

$$\psi_{\text{eq}}^2 \ell(t)^d/\kappa^{-\gamma/\nu} \sim [\kappa\ell(t)]^d, \tag{8.1.20}$$

at long wavelengths $k \lesssim \ell(t)^{-1}$. Use has been made of $\psi_{\text{eq}}^2 \sim \kappa^{2\beta/\nu} \sim \kappa^{d-\gamma/\nu}$ which follows from the scaling relations in Chapter 2. The above ratio can be very large in the late stage. Because of this large difference, the thermal structure factor will be neglected in much of the following discussion.

As will be shown in Appendix 8A, the domain structure factor, written as $I_{\text{dom}}(k,t)$, has the Porod tail in the region $\ell(t)^{-1} \ll k \ll \kappa$,

$$I_{\text{dom}}(k,t) = \psi_{\text{eq}}^2 \ell(t)^d F(\ell(t)k) \cong \gamma_{\text{d}}\psi_{\text{eq}}^2 \frac{A}{k^{d+1}}, \tag{8.1.21}$$

where γ_{d} is 8π in 3D and 8 in 2D, and A is the interface area (line) density for 3D (2D), so $A \sim 1/\ell(t)$. Comparing the two contributions in (8.1.18) at such large k, we find that the domain structure factor is larger than the thermal structure factor for

$$k < \kappa[\kappa\ell(t)]^{-1/(d+1)}. \tag{8.1.22}$$

The upper limit on the right-hand side decreases with time and is smaller than κ, but is much larger than $\ell(t)^{-1}$. We shall see the Porod tail obtained numerically or experimentally in Fig. 8.9 for the nonconserved case and in Figs 8.13, 8.18 and 8.24 for the conserved case.

8.1.3 The Suzuki and Kawasaki–Yalabik–Gunton theories

Nonlinear transformation leading to linear theory

Suzuki [7] presented a compact dynamic theory to describe bifurcation of a single variable ($\psi(t) \to \pm\psi_{\text{eq}}$) neglecting space dependence and taking into account the nonlinear and noise terms. Kawasaki, Yalabik, and Gunton (KYG) [8] extended Suzuki's theory to construct approximate space-dependent solutions of (8.1.2) after quenching with $h = 0$. KYG used the nonlinear transformation (8.1.13) to introduce a new field $\varphi(\boldsymbol{r},t)$,

$$\psi(\boldsymbol{r},t) = \varphi(\boldsymbol{r},t)\big/\left[1+\alpha\varphi(\boldsymbol{r},t)^2\right]^{1/2}. \tag{8.1.23}$$

The inverse relation is

$$\varphi(\boldsymbol{r},t) = \psi(\boldsymbol{r},t)/\left[1-\alpha\psi(\boldsymbol{r},t)^2\right]^{1/2}. \tag{8.1.24}$$

Then (8.1.2) at $h=\theta=0$ is rewritten as

$$\frac{\partial}{\partial t}\varphi = L\left[\kappa^2+\nabla^2\right]\varphi - \frac{3L\alpha\varphi}{1+\alpha\varphi^2}|\nabla\varphi|^2. \tag{8.1.25}$$

KYG neglected the last term of (8.1.25) to obtain

$$\frac{\partial}{\partial t}\varphi = L\left[\kappa^2+\nabla^2\right]\varphi. \tag{8.1.26}$$

This linearization approximation is valid during the very early stage of pattern evolution, but is not justified at the late stage, as discussed at the end of this section. Then, if φ is a gaussian random variable at $t=0$, it remains so at later times and the variance of its Fourier component is

$$\langle|\varphi(\boldsymbol{k},t)|^2\rangle = \chi_k \exp[2L(\kappa^2-k^2)t], \tag{8.1.27}$$

where χ_k is the initial intensity.

With this approximation a one-point distribution function is defined as

$$\begin{aligned}\rho_1(\psi,t) &= \langle\delta(\psi(\boldsymbol{r},t)-\psi)\rangle \\ &= (1-\alpha\psi^2)^{-3/2}\left\langle\delta\left(\varphi(\boldsymbol{r},t)-\frac{\psi}{\sqrt{1-\alpha\psi^2}}\right)\right\rangle.\end{aligned} \tag{8.1.28}$$

Here $\varphi(\boldsymbol{r},t)$ at each point is gaussian with the variance,

$$\begin{aligned}\beta(t) &= \langle\varphi(\boldsymbol{r},t)^2\rangle = \int_q \chi_q \exp\left[2L(\kappa^2-q^2)t\right] \\ &\cong \chi_0(8\pi Lt)^{-d/2}e^{2\gamma_0 t}.\end{aligned} \tag{8.1.29}$$

In the second line the initial correlation is assumed to be short-range and χ_q is replaced by χ_0. It then follows Suzuki's distribution function,

$$\rho_1(\psi,t) = \frac{1}{\sqrt{2\pi\beta(t)(1-\alpha\psi^2)^3}}\exp\left[-\frac{1}{2\beta(t)}\frac{\psi^2}{1-\alpha\psi^2}\right]. \tag{8.1.30}$$

This distribution is zero for $|\psi|\ge\alpha^{-1/2}=\psi_{\text{eq}}$. Its time evolution is illustrated in Fig. 8.4. At long times it has two peaks at $\psi\cong\pm\psi_{\text{eq}}$ and the peak width is determined by

$$1-\psi^2/\psi_{\text{eq}}^2 \lesssim \psi_{\text{eq}}^2/\beta(t) \sim (K_d u_0\kappa^{-\epsilon})^{-1}(\gamma_0 t)^{d/2}e^{-2\gamma_0 t}. \tag{8.1.31}$$

The right-hand side is the inverse of $J(k,t)$ in (8.1.10). For $t\gg\gamma_0^{-1}$ the peak width decreases rapidly and we may set

$$\psi(\boldsymbol{r},t)\cong\psi_{\text{eq}}\frac{\varphi(\boldsymbol{r},t)}{|\varphi(\boldsymbol{r},t)|}, \tag{8.1.32}$$

as ought to be the case except for the interface regions.

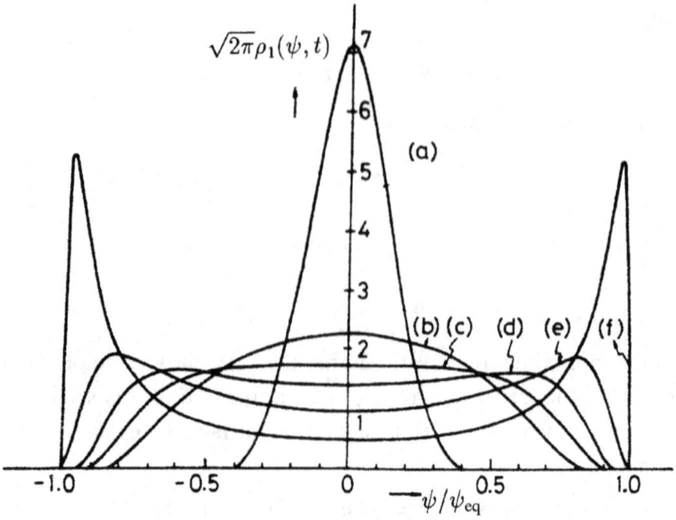

Fig. 8.4. The distribution function $\rho_1(\psi, t)$ defined by (8.1.30) for $\beta(t)/\psi_{eq}^2 = 0.02$ (a); 0.2 (b); 0.33 (c); 0.5 (d); 1 (e); and 4 (f). The function $\beta(t)/\psi_{eq}^2$ is the scaled variable τ in Suzuki's theory [7].

Pair correlation function

Next we may calculate the pair correlation function,

$$g(r, t_1, t_2) = \langle \psi(\boldsymbol{r}_1, t_1)\psi(\boldsymbol{r}_2, t_2)\rangle, \tag{8.1.33}$$

where $r = |\boldsymbol{r}_1 - \boldsymbol{r}_2|$. The two times t_1 and t_2 can be different here. Use of (8.1.23) gives

$$g(r, t_1, t_2) = \int d\varphi_1 \int d\varphi_2 \frac{\varphi_1}{(1+\alpha\varphi_1^2)^{1/2}} \frac{\varphi_2}{(1+\alpha\varphi_2^2)^{1/2}} P_0(\varphi_1, \varphi_2), \tag{8.1.34}$$

where $P_0(\varphi_1, \varphi_2)$ is the two-point distribution function of $\varphi_1 = \varphi(\boldsymbol{r}_1, t_1)$ and $\varphi_2 = \varphi(\boldsymbol{r}_2, t_2)$. It may be constructed in terms of the variances among these quantities,

$$\beta_1 = \langle \varphi(\boldsymbol{r}_1, t_1)^2\rangle, \quad \beta_2 = \langle \varphi(\boldsymbol{r}_2, t_2)^2\rangle, \quad \beta_{12} = \langle \varphi(\boldsymbol{r}_1, t_1)\varphi(\boldsymbol{r}_2, t_2)\rangle, \tag{8.1.35}$$

in the following gaussian form,

$$P_0(\varphi_1, \varphi_2) = (2\pi)^{-1} D^{-1/2} \exp\left[-\frac{1}{2D}\left(\beta_2\varphi_1^2 + \beta_1\varphi_2^2 - 2\beta_{12}\varphi_1\varphi_2\right)\right], \tag{8.1.36}$$

where $D \equiv \beta_1\beta_2 - \beta_{12}^2$. Let $t_1 \gg \gamma_0^{-1}$ and $t_2 \gg \gamma_0^{-1}$ such that (8.1.32) holds at $t = t_1$ and t_2. Then, as will be shown in Appendix 8B, we obtain

$$\begin{aligned} g(r, t_1, t_2) &= \psi_{eq}^2 \int d\varphi_1 \int d\varphi_2 \frac{\varphi_1}{|\varphi_1|} \frac{\varphi_2}{|\varphi_2|} P_0(\varphi_1, \varphi_2) \\ &= \frac{2}{\pi}\psi^2 \sin^{-1} X, \end{aligned} \tag{8.1.37}$$

where

$$X = \beta_{12}/\sqrt{\beta_1\beta_2}. \tag{8.1.38}$$

If the initial pair correlation is short-range, we may set

$$X = \left(\frac{2\sqrt{t_1t_2}}{t_1+t_2}\right)^{d/2} \exp\left[-\frac{r^2}{4L(t_1+t_2)}\right], \quad (t_1, t_2 \gg \gamma_0^{-1}). \tag{8.1.39}$$

The equal-time-correlation function ($t_1 = t_2 = t$) is written in the scaling form $\psi_{\text{eq}}^2 G_{\text{KYG}}(r/\ell(t))$ consistent with (8.1.14), where

$$G_{\text{KYG}}(x) = \frac{2}{\pi}\sin^{-1}\left[\exp\left(-\frac{1}{8}x^2\right)\right]. \tag{8.1.40}$$

At large distances $x = r/\ell(t) \gtrsim 1$ it follows the gaussian form,

$$G_{\text{KYG}}(x) \cong \frac{2}{\pi}\exp\left(-\frac{1}{8}x^2\right). \tag{8.1.41}$$

At short distances $x = r/\ell(t) \lesssim 1$ we obtain

$$G_{\text{KYG}}(x) = 1 - \frac{1}{\pi}x + O(x^3). \tag{8.1.42}$$

The term linear in $x = r/\ell(t)$ gives rise to the Porod tail (8.1.21), for which see Appendix 8A. Another interesting quantity is the equal-point correlation ($r = 0$). For $t_1 \gg t_2$, (8.1.37) and (8.1.39) yield

$$g(0, t_1, t_2) \cong \frac{2}{\pi}\psi_{\text{eq}}^2\left(\frac{4t_2}{t_1}\right)^{d/4}. \tag{8.1.43}$$

We recognize that, because the two-point correlation (8.1.38) is proportional to $t_2^{d/4}$ for $t_2 \ll t_1$, the initial variance χ_0 should come into play for $t_2 \lesssim \gamma_0^{-1}$. To check this, we set $t_2 = 0$ and $t_1 \gg \gamma_0^{-1}$ such that (8.1.32) holds at $t = t_1$. Then, the integrand of (8.1.34) may be set equal to $(\varphi_1\varphi_2/|\varphi_1|)P_0(\varphi_1, \varphi_2)$, and the double integration is performed to give

$$g(r, t_1, 0) \cong \psi_{\text{eq}}\sqrt{\frac{2\chi_0}{\pi}}(2\pi Lt_1)^{-d/4}\exp\left(-\frac{3r^2}{16Lt_1}\right), \tag{8.1.44}$$

which is indeed proportional to $\chi_0^{1/2}$.

Summary of the KYG theory

The KYG theory gives simple analytic expressions for the correlation functions via the nonlinear transformation (8.1.23). However, there is no reason to neglect the last term in (8.1.25) in the late stage. In fact, the two terms in (8.1.25) are balanced to form the interface solution, $\psi = \psi_{\text{eq}}\tanh(\kappa s/\sqrt{2})$ or $\varphi = \psi_{\text{eq}}\sinh(\kappa s/\sqrt{2})$, where $s = \boldsymbol{n}\cdot(\boldsymbol{r}-\boldsymbol{r}_a)$ is the coordinate along the normal vector $\boldsymbol{n}$ on a surface $\{\boldsymbol{r}_a\}$. It is worth noting that, if the surface is weakly curved, the above interface solution satisfies $\partial\varphi/\partial t \cong L(\nabla^2 - \partial^2/\partial s^2)\varphi$ near the interface point $\boldsymbol{r}_a$ [9]. Here, if $\nabla^2 - \partial^2/\partial s^2$ is replaced by $(1-1/d)\nabla^2$, we may reproduce a subsequent theory by Ohta, Jasnow, and Kawasaki (OJK) [10]. We shall see that the OJK theory gives a better description of the late-stage behavior than the KYG theory.

8.1.4 Periodic quench

There are some interesting nonlinear effects when the temperature coefficient r oscillates in time taking positive and negative values periodically. We start with model A with a one-component order parameter in the absence of an ordering field,

$$\frac{\partial}{\partial t}\psi = -L\left[r(t) - \nabla^2 + u_0\psi^2\right]\psi + \theta, \tag{8.1.45}$$

where $r(t)$ is a periodic function of time. For simplicity we assume a step-wise variation,

$$\begin{aligned} r(t) &= r_- \quad (0 < t - n(t_1 + t_2) < t_1) \\ &= r_+ \quad (t_1 < t - n(t_1 + t_2) < t_1 + t_2), \end{aligned} \tag{8.1.46}$$

where $n = 0, 1, 2, \ldots$ and $t_\mathrm{p} = t_1 + t_2$ is the period of the oscillation. However, our main conclusions are independent of the detailed functional form of $r(t)$. (See Ref. [11], where $r(t)$ is assumed to oscillate sinusoidally.)

We briefly summarize the scenario here. (i) If r_- and r_+ are both positive, the system is obviously in a disordered phase with vanishing order parameter $\langle\psi\rangle = 0$. (ii) If they are both negative, an ordered phase will emerge with a homogeneous average $\bar{\psi}(t) = \langle\psi(\boldsymbol{r}, t)\rangle$. If the fluctuation effect is neglected, $\bar{\psi}(t)$ obeys

$$\frac{\partial}{\partial t}\bar{\psi}(t) = -L[r(t)\bar{\psi}(t) + u_0\bar{\psi}(t)^3]. \tag{8.1.47}$$

We divide the above equation by $\bar{\psi}(t)$ and average over t in one period in a periodic state to obtain

$$\frac{1}{t_\mathrm{p}}\int_0^{t_\mathrm{p}} dt\bar{\psi}(t)^2 = \frac{1}{t_\mathrm{p}}\int_0^{t_\mathrm{p}} dt\frac{|r(t)|}{u_0}. \tag{8.1.48}$$

This is analogous to the equilibrium relation $\psi_\mathrm{eq}^2 = |r|/u_0$. (iii) However, when $r_- < 0 < r_+$, the problem is highly nontrivial. The fluctuations are much enhanced in the unstable time regions if

$$L|r_-|t_1 \gg 1. \tag{8.1.49}$$

Domains are formed during $r = r_-$, but the fluctuation level decreases exponentially during $r = r_+$ roughly by $\exp(-Lr_+t_2)$. If this factor is small enough, the phase ordering returns to its starting point and the system tends to a periodically modulated disordered state ($\langle\psi\rangle = 0$). If the domain destruction during $r = r_+$ is nearly complete, the large-scale heterogeneities remaining at $t = t_\mathrm{p}$ become weaker than the thermal level and the correlation range among the domains is cut off at

$$\ell_\mathrm{p} = (2Lt_\mathrm{p})^{1/2}. \tag{8.1.50}$$

However, with decreasing r_+t_2, the correlation range increases towards a metastability limit of the disordered phase. Eventually, domains should continue to grow over successive periods, resulting in a homogeneous, oscillating average $\bar{\psi}(t)$ of order $(|r_-|/u_0)^{1/2}$. Thus we may predict a dynamcal *first-order* phase transition with a discontinuous change in $\bar{\psi}(t)$.

Recursion relations

We outline the calculation in the case $r_- < 0 < r_+$. Supplementary discussions will be presented in Appendix 8C. With growth of the fluctuations we may set

$$\psi(\boldsymbol{r}, t) \cong b(t)^{-1/2} \frac{\varphi(\boldsymbol{r}, t)}{|\varphi(\boldsymbol{r}, t)|}. \tag{8.1.51}$$

Here $b(t) \cong u_0/|r_-|$ in the unstable time region $1/L|r_-| \lesssim t < t_1$ as in (8.1.32). In the successive time region $t_1 < t < t_1 + t_2$, (8C.2) yields

$$b(t) = u_0\left(\frac{1}{|r_-|} + \frac{1}{r_+}\right)\exp[2Lr_+(t - t_1)] - \frac{u_0}{r_+}. \tag{8.1.52}$$

Thus $\psi(\boldsymbol{r}, t)$ decays exponentially for $t - t_1 \gtrsim 1/Lr_+$. In disordered states, use of (8B.9) then gives the domain structure factor at long wavelengths,

$$I_{\rm dom}(k, t) = b(t)^{-1}\chi_k \exp(-2Lk^2 t) \Big/ \int_{\boldsymbol{q}} \chi_q \exp(-2Lq^2 t). \tag{8.1.53}$$

where $\chi_k = \langle|\psi_{\boldsymbol{k}}(0)|^2\rangle$ is the initial variance. More generally, we should allow for a nonvanishing average order parameter $\eta = \langle\psi(\boldsymbol{r}, 0)\rangle$ at $t = 0$ [11]. Under (8.1.49) we may express the next initial variance and average order parameter at $t = t_{\rm p}$, denoted by χ_k' and η', in the following recursion relations,

$$\chi_k' = (b_{\rm p}\bar{\beta})^{-1}\chi_k \exp(-\ell_{\rm p}^2 k^2 - \eta^2/\bar{\beta}) + \kappa_{\rm th}^{-2}, \tag{8.1.54}$$

$$\eta' = (2/\pi b_{\rm p})^{1/2} \int_0^{\eta/\bar{\beta}^{1/2}} dx e^{-x^2/2}, \tag{8.1.55}$$

where

$$\bar{\beta} = \int_{\boldsymbol{q}} \chi_q \exp(-\ell_{\rm p}^2 q^2). \tag{8.1.56}$$

The last term in (8.1.54) is the intensity produced by the thermal noise. From (8.1.8) we have

$$\kappa_{\rm th}^{-2} = \frac{1}{|r_-|} + \frac{1}{r_+}. \tag{8.1.57}$$

The coefficient $b_{\rm p}$ is of the form,

$$b_{\rm p} = b(t_{\rm p}) \cong u_0\kappa_{\rm th}^{-2}\exp(2Lr_+t_2). \tag{8.1.58}$$

The recursion relations (8.1.54) and (8.1.55) are independent of the functional form of $r(t)$ as long as the fluctuations are much enhanced during $r(t) < 0$. In fact, the same recursion relations were derived for a sinusoidal temperature oscillation in Ref. [11].

These equations are controlled by a unique dimensionless parameter A defined by

$$A = K_d \ell_p^{-d} \kappa_{th}^{-2} b(t_p) = K_d u_0 \ell_p^{-d} \kappa_{th}^{-4} \exp(2Lr_+ t_2), \tag{8.1.59}$$

which represents the ratio of the thermal fluctuations to the domain fluctuations at $t = t_p$ on the scale of ℓ_p. We make the above equations dimensionless by setting

$$\chi_k = \kappa_{th}^{-2} \mathcal{F}(\ell_p k), \quad \eta = (K_d/\kappa_{th}^2 \ell_p^d)^{1/2} \mathcal{G}. \tag{8.1.60}$$

The dimensionless recursion relations, $\mathcal{F} \to \mathcal{F}'$ and $\mathcal{G} \to \mathcal{G}'$, read

$$\mathcal{F}'(x) = S^{-1} \mathcal{F}(x) \exp(-x^2 - Z^2) + 1, \tag{8.1.61}$$

$$\mathcal{G}' = (2/\pi A)^{1/2} \int_0^Z dx e^{-x^2/2}, \tag{8.1.62}$$

where S and Z are determined by

$$S/A = \mathcal{G}/Z = \int_0^\infty dy y^{d-1} e^{-y^2} \mathcal{F}(y). \tag{8.1.63}$$

Periodic states are obtained by setting $\mathcal{F}'(x) = \mathcal{F}(x)$ and $\mathcal{G}' = \mathcal{G}$. In particular, disordered states ($\mathcal{G} = Z = 0$) exists only for $S \geq 1$ and

$$\mathcal{F}(x) = S/(S - e^{-x^2}), \tag{8.1.64}$$

where S and A are related by

$$A^{-1} = \int_0^\infty dy y^{d-1}/(Se^{y^2} - 1). \tag{8.1.65}$$

If S is close to 1, we have $S - 1 \sim (A - A_{c0})^2$ with $A_{c0} = 0.89$ in 3D and $S - 1 \sim \exp(-2/A)$ in 2D. The correlation length of the large-scale heterogeneities grows as $(S - 1)^{-1/2} \ell_p$ while the system stays in disordered states. Experimentally, this effect will occur as the average temperature is lowered with a fixed magnitude of the temperature oscillation.

Periodic states and first-order phase transition

We numerically examined the above recursion relations and showed that periodic states are attained after many iterations over a wide range of initial $\mathcal{F}(x)$ and $\mathcal{G}$ [11]. We plot A vs S in periodic states in Fig. 8.5(a) and A vs Z in Fig. 8.5(b) in 3D. At the point Q, where $A = 1.34$, $S = 0.80$, and $Z = 0.85$, A is locally a maximum as a function of S. The point R, where $A = 0.89$, $Z = 0$, and $S = 1$, is a metastability limit of the disordered phase, towards which the correlation length grows. At the point P, we obtain $Z = 1.82$ and $S = 0.41$, while at the point S, $Z = 0$ and $S = 1.09$.

Pasquale *et al.* [12] numerically examined periodic quench with the step-wise temperature variation (8.1.46) to confirm the first-order phase transition, but they neglected

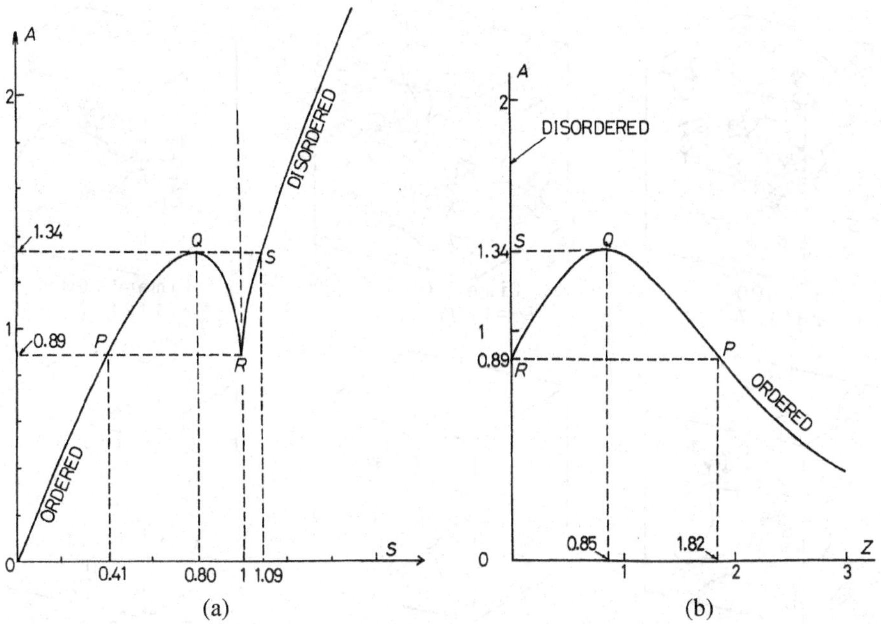

Fig. 8.5. (a) A vs S in the dimensionless recursion relations (8.1.61) and (8.1.62) in periodic states in 3D [11]. The portion of the curve with $S > 1$ corresponds to the disordered phase, while that with $S < 0.80$ to the ordered phase. The system is linearly unstable in the region QR where $0.8 < S < 1$. (b) A vs Z in periodic states in 3D. Here $Z = \mathcal{G}A/S$ vanishes in the disordered phase and is nonvanishing in the ordered phase.

the space-dependence of ψ. More numerical analysis and corresponding experiments are required.

Coarsening in many-component systems ($n \geq 2$)

When a system with continuous symmetry is quenched into an unstable temperature region, a large number of defects emerge and their number decreases as a function of time in late-stage coarsening [6], [13]–[24]. They are topologically stable singular objects for $n \leq d' = d - d_s$, where d_s is the dimension of the core structure. That is, if their positions are fixed (without pair annihilation, etc.), they cannot be eliminated by continuous deformations of the vector $\boldsymbol{\psi}$ only. Vortices are representative examples for $n = 2$ with $d_s = 1$ (line) in 3D and $d_s = 0$ (point) in 2D.

Let us consider an n-component system with the simple nonconserved dynamics,

$$\frac{\partial}{\partial t}\psi_j = -L\left[-\kappa^2 - \nabla^2 + u_0|\psi|^2\right]\psi_j + \theta_j, \tag{8.1.66}$$

where $L > 0$ and

$$\langle \theta_i(\mathbf{r}, t)\theta_j(\mathbf{r}', t')\rangle = 2L\delta_{ij}\delta(\mathbf{r} - \mathbf{r}')\delta(t - t'). \tag{8.1.67}$$

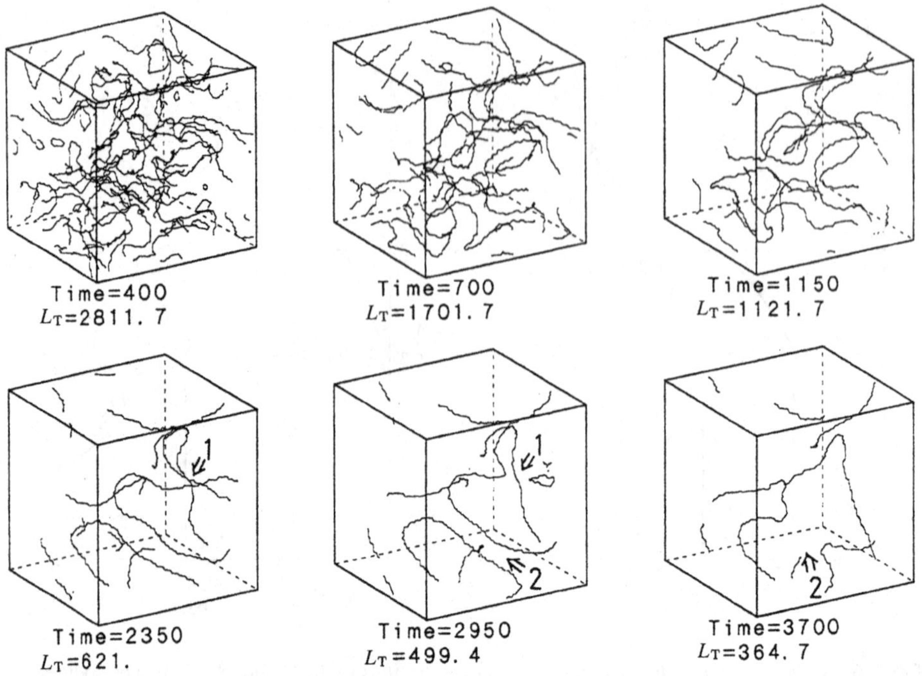

Fig. 8.6. Development with time of the configuration of vortex lines in the model (8.1.66) in the purely dissipative case $L = 1$ in a $64 \times 64 \times 64$ system under periodic boundary conditions without the noise term, with $\kappa = 1$ [16]. The times after quenching are indicated and the total line lengths $L_{\rm T}$ are given in units of the lattice spacing. All the line ends are situated at the boundary and are connected with the end at the other side. The arrows indicate reconnection of the crossing lines.

For $n = 2$ relevant singular objects are vortices, whose profile was examined in the Ginzburg–Landau theory in Section 4.5. As an example, Fig. 8.6 shows snapshots of vortex lines (with charges $\ell = \pm 1$) obtained by numerically solving (8.1.66) [16]. In 3D the typical line curvature is scaled as [14, 16, 20]

$$\mathcal{K}(t) \sim t^{-a}, \quad a \cong 0.5, \tag{8.1.68}$$

analogous to the typical surface curvature in the one-component case. It is also important that the spacing between the lines is of order $\mathcal{K}(t)^{-1}$. Then, in a volume with linear dimension $\mathcal{K}(t)^{-1}$, the lines inside are only slightly curved and their number is of order 1, so that the line length density decreases in time as

$$n_{\rm v}(t) = L_{\rm T}(t)/V \sim \mathcal{K}(t)^2 \sim t^{-2a}, \tag{8.1.69}$$

where $L_{\rm T}(t)$ is the total line length in the system with volume V. In 2D xy systems, simulations of (8.1.66) (without the noise) [13, 22] indicate that vortex pairs with opposite charges ($\ell = \pm 1$) collide and disappear, leading to a decreasing of the vortex density as $n_v(t) \sim t^{-a}$ with $a \cong 0.5$.

Generalized KYG theory

The simplest theory to investigate the coarsening in many-component systems is to generalize the nonlinear transformation (8.1.23) as [18]

$$\psi_j(\boldsymbol{r}, t) = \varphi_j(\boldsymbol{r}, t) \Big/ \left[1 + \alpha |\boldsymbol{\varphi}(\boldsymbol{r}, t)|^2\right]^{1/2} \quad (j = 1, \ldots, n). \tag{8.1.70}$$

The subsidiary vector field $\boldsymbol{\varphi} = (\varphi_1, \ldots, \varphi_n)$ obeys the gaussian distribution characterized by

$$\langle \varphi_i(\boldsymbol{k}, t)\varphi_j(-\boldsymbol{k}, t)\rangle = \chi_k \delta_{ij} \exp[2L(\kappa^2 - k^2)t]. \tag{8.1.71}$$

At long times we may set

$$\psi_j(\boldsymbol{r}, t) = \frac{1}{|\boldsymbol{\varphi}(\boldsymbol{r}, t)|}\varphi_j(\boldsymbol{r}, t), \quad (j = 1, \ldots, n), \tag{8.1.72}$$

as in (8.1.32). Then the pair correlation function $g(r, t_1, t_2) = \langle \boldsymbol{\psi}(\boldsymbol{r}_1, t_1) \cdot \boldsymbol{\psi}(\boldsymbol{r}_2, t_2)\rangle$ is expressed in the following integral form,[1]

$$g(r, t_1, t_2) = \psi_{\rm eq}^2 \frac{2\Gamma((n+1)/2)}{\sqrt{\pi}\Gamma(n/2)} \int_0^1 ds (1 - s^2)^{(n-1)/2} \frac{X}{\sqrt{1 - X^2 s^2}} \tag{8.1.73}$$

where $\Gamma(x)$ is the gamma function. The pair correlation depends on $r = |\boldsymbol{r}_1 - \boldsymbol{r}_2|$, t_1, and t_2 through the single variable X defined by (8.1.38). We notice that the one-component result (8.1.37) is recovered for $n = 1$.

The above result shows that the defect contribution to the equal-time pair correlation function can be scaled as

$$g(r, t) = \langle \boldsymbol{\psi}(\boldsymbol{r}_1, t) \cdot \boldsymbol{\psi}(\boldsymbol{r}_2, t)\rangle = \psi_{\rm eq}^2 G(r/\ell(t)), \tag{8.1.74}$$

where $r = |\boldsymbol{r}_1 - \boldsymbol{r}_2|$ and $\ell(t) = (Lt)^{1/2}$. This is consistent with the simulation result (8.1.68) for $n = 2$. It is also remarkable that the Fourier transformation of $G(r/\ell(t))$, the defect structure factor divided by $\psi_{\rm eq}^2$, behaves at large $k\ell(t) \gg 1$ as

$$\hat{I}(k) = A_{nd}\ell(t)^{-n}k^{-(d+n)}, \tag{8.1.75}$$

which is the generalization of the Porod tail (8.1.21). For $n = 2$, the short-distance behavior ($r \ll \ell(t)$) is logarithmically singular as

$$G = 1 - \left[r^2/8\ell(t)^2\right] \ln[r/\ell(t)] + \cdots. \tag{8.1.76}$$

This behavior in fact gives rise to the tail (8.1.75). For general n and d, (8.1.73) yields $A_{nd} = 2^d \pi^{d/2-1} \Gamma((n+1)/2)^2 \Gamma((n+d)/2)/\Gamma(n/2)$. Figure 8.7 numerically demonstrates the presence of the tail ($\propto k^{-5}$) in the scaled structure factor for $n = 2$ in 3D.

[1] The above integral is proportional to the hypergeometric function $F(1/2, 1/2; n/2 + 1; X^2)$ [6, 17].

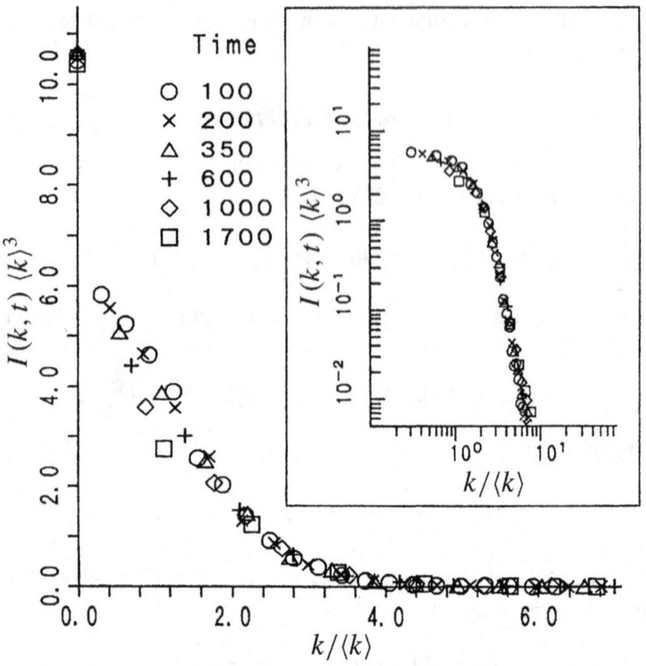

Fig. 8.7. The scaled structure factor $I(k,t)\langle k\rangle^3$ plotted on a regular scale and a logarithmic scale for $n = 2$ in 3D in the model (8.1.66) with $L = 1$ [16]. The times after quenching are indicated, and $\langle k\rangle = \sum_{k<\pi} kI(k,t)/\sum_{k<\pi} I(k,t)$, where the space is measured in units of the lattice spacing.

Generalized Porod tail

The tail ($\propto k^{-(n+d)}$) arises from the distortion of $\boldsymbol{\psi}$ around stable topological defects, which exist for $n \le d$ for point defects and for $n \le d-1$ for line defects. This will be shown in Appendix 8D for $n = 2$. In general [6, 21],

$$\hat{I}(k) = \pi^{-1}(4\pi)^{(d+n)/2}\Gamma((n+1)/2)^2\Gamma(d/2)\Gamma(n/2)^{-1}\frac{n_{\rm def}}{k^{d+n}}, \tag{8.1.77}$$

in terms of an appropriately defined defect density $n_{\rm def}$. For interfaces ($n = 1$), $n_{\rm def}$ is the surface area (line length) density A in 3D (2D). For vortices ($n = 2$), $n_{\rm def}$ is the vortex line length (number) density $n_{\rm v}$ with charge ± 1 in 3D (2D). The above formula is consistent with (8.1.21), (8D.3), and (8D.5).

Summary

The generalized KYG theory is very simple and consistent with numerical results but is not well justified. We mention an attempt to theoretically derive it [23] and a more sophisticated theory of phase ordering in many-component systems [19]. As another kind of system with a tensor order parameter, liquid crystals exhibit interesting phase-ordering processes from isotropic to nematic states [6, 16, 17]. There, the disclination line density

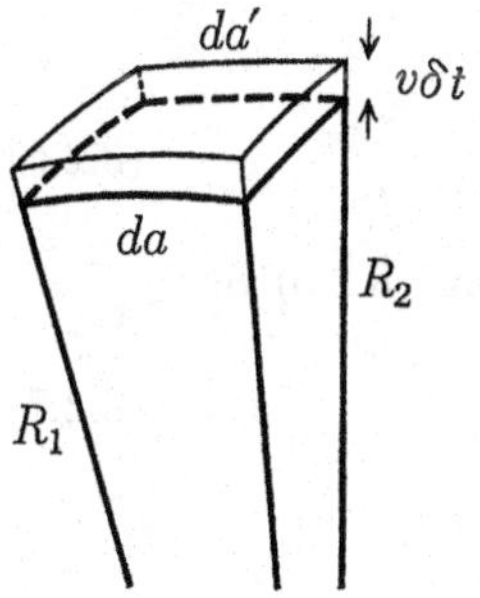

Fig. 8.8. Surface movement by $v\delta t$ in a small time interval δt, where a surface element with area da is changed to a new element with area da'. Here we can see the relation, $da'/da = (R_1 + v\delta t)(R_2 + v\delta t)/R_1 R_2 = 1 + (R_1^{-1} + R_2^{-1})v\delta t + \cdots$.

decreases in time as t^{-1} in 3D, analogous to (8.1.69) [25, 26]. The phase ordering in liquid crystals is similar to that in the xy systems, although topological singularities in nematics are more complicated.

8.2 Interface dynamics in nonconserved systems

8.2.1 The Allen–Cahn equation

In a one-component system at a late stage after quenching, phase ordering is locally completed except at the interface regions, so the problem is how to describe the interface motion in the thin limit of the interface thickness. In the nonconserved case without ordering field and thermal noise, the interface motion is governed by the Allen–Cahn equation [27],

$$v = -L\mathcal{K}, \tag{8.2.1}$$

where v is the interface velocity in the normal direction $\boldsymbol{n}$, L is the kinetic coefficient in (8.1.2), and $\mathcal{K}$ is the mean curvature multiplied by 2 or the sum of the principal curvatures,

$$\mathcal{K} = \frac{1}{R_1} + \frac{1}{R_2}. \tag{8.2.2}$$

We will call $\mathcal{K}$ simply the curvature. Then a sphere with radius R shrinks as

$$\frac{\partial}{\partial t}R = -2\frac{L}{R} \qquad \text{or} \qquad R(t)^2 = R(0)^2 - 4Lt. \tag{8.2.3}$$

In 2D, (8.2.1) remains applicable if we set $1/R_2 = 0$. From this equation we obtain the coarsening law (8.1.16) by making the following order estimations,

$$v \sim \ell(t)/t, \qquad \mathcal{K} \sim \ell(t)^{-1}. \tag{8.2.4}$$

We then show that the surface area $S(t)$ or the free energy $\mathcal{H}$ in (8.1.17) decreases monotonically in time. As shown in Fig. 8.8, if the surface is slightly moved by $\delta\zeta$ in the normal direction, a small surface element da changes to da' given by

$$da' = da(1 + \mathcal{K}\delta\zeta). \tag{8.2.5}$$

We set $\delta\zeta = v\delta t$ for a small time interval δt to obtain

$$\frac{d}{dt}S(t) = \int da\mathcal{K}v \tag{8.2.6}$$

for any v. When the Allen–Cahn dynamics (8.2.1) holds, (8.1.17) and (8.2.6) yield

$$\frac{d}{dt}\mathcal{H}(t) = \sigma\frac{d}{dt}S(t) = -L\sigma\int da\mathcal{K}^2 \le 0. \tag{8.2.7}$$

The coarsening thus proceeds, to decrease the surface energy.

8.2.2 The Ohta–Jasnow–Kawasaki theory

It is convenient to introduce a smooth subsidiary field $u(\boldsymbol{r}, t)$ to represent surfaces by $u =$ const. The differential geometry is much simplified in terms of such a field. The two-phase boundaries are represented by $u = 0$. Let all the surfaces on which $u =$ const. be governed by the Allen–Cahn equation (8.2.1) in the *whole* space. Then u obeys

$$\frac{\partial}{\partial t}u = -v|\nabla u| = L|\nabla u|\nabla\cdot\boldsymbol{n}. \tag{8.2.8}$$

From $\boldsymbol{n} = |\nabla u|^{-1}\nabla u$ the above equation is rewritten as

$$\frac{\partial}{\partial t}u = L\left[\nabla^2 - \sum_{ij} n_i n_j \frac{\partial^2}{\partial x_i \partial x_j}\right]u. \tag{8.2.9}$$

Supposing intricate surfaces, Ohta–Jasnow–Kawasaki (OJK) [10] *preaveraged* $n_i n_j$ on the right-hand side of (8.2.9) to replace it by its angle average δ_{ij}/d. The field u then obeys a diffusion equation,

$$\frac{\partial}{\partial t}u = L'\nabla^2 u \tag{8.2.10}$$

with

$$L' = (1 - 1/d)L. \tag{8.2.11}$$

Because $u = 0$ on the interfaces, ψ is expressed as

$$\psi(\boldsymbol{r}, t) \cong \psi_{\text{eq}}\frac{u(\boldsymbol{r}, t)}{|u(\boldsymbol{r}, t)|}, \tag{8.2.12}$$

on spatial scales much longer than κ^{-1}, analogous to (8.1.32). Furthermore, if the initial value $u(\boldsymbol{r}, 0)$ obeys a gaussian distribution without long-range correlation, $u(\boldsymbol{r}, t)$ remains gaussian at later times and is characterized by

$$\langle |u_{\boldsymbol{k}}(t)|^2\rangle = \chi_0 \exp(-2L'k^2 t), \tag{8.2.13}$$

where χ_0 is the initial variance assumed to be independent of k.

We notice that, if L is replaced by L', the correlation function expressions of the KYG theory in the late stage become those of the OJK theory. In other words, the OJK results

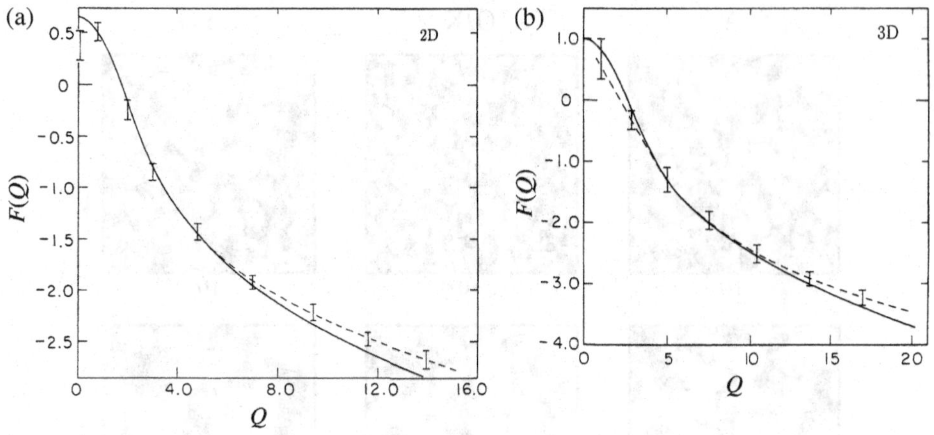

Fig. 8.9. The dimensionless structure factor $F(Q)$, the Fourier transformation of $G_{\rm OJK}(x)$, in (a) 2D and (b) 3D in the OJK theory (solid line) for the nonconserved case [10]. The broken line represents simulation results.

are obtained from the KYG results if t is replaced by $(1 - 1/d)t$. For example, the pair correlation $g(|\boldsymbol{r}_1 - \boldsymbol{r}_2|, t_1, t_2) = \langle \psi(\boldsymbol{r}_1, t_1)\psi(\boldsymbol{r}_2, t_2)\rangle$ in OJK is calculated from (8.1.37) in KYG. In particular, the equal-time correlation ($t_1 = t_2 = t$) is written in the scaling form $g(r, t, t) = \psi_{\rm eq}^2 G_{\rm OJK}(r/\ell(t))$ with

$$G_{\rm OJK}(x) = \frac{2}{\pi}\sin^{-1}\left[\exp\left(-\frac{1}{8(1-1/d)}x^2\right)\right]. \tag{8.2.14}$$

This OJK scaling function agrees excellently with simulations, as shown in Fig. 8.9. Furthermore, OJK gives the same equal-point correlation $g(0, t_1, t_2)$ as that in KYG, so (8.1.43) holds in both theories for $t_1 \gg t_2$.

Comparison of solutions of the model A, KYG, and OJK equations

It is of interest to compare actual solutions of the original model A, KYG, and OJK equations in 2D. In Fig. 8.10 we show $\psi(\boldsymbol{r}, t)$ in model A, (8.1.2), with $h = \theta = 0$ on the left, $u(\boldsymbol{r}, t) = \exp(tL\nabla^2/2)u(\boldsymbol{r}, 0)$ in OJK in the middle, and $\varphi(\boldsymbol{r}, t) = \exp(tL\nabla^2 + \gamma_0 t)\varphi(\boldsymbol{r}, 0)$ in KYG on the right in 2D. The initial values $\psi(\boldsymbol{r}, 0)$, $u(\boldsymbol{r}, 0)$, and $\varphi(\boldsymbol{r}, 0)$ on each lattice point are the same gaussian random number with variance 0.1. Here the patterns of KYG at time t and those of OJK at time $2t$ are identical. We notice the following. (i) At an early stage ($t \lesssim \gamma_0^{-1}$), the linear approximation is valid and the model A patterns coincide approximately with those of KYG. (ii) At an intermediate stage ($10\gamma_0^{-1} \lesssim t \lesssim 50\gamma_0^{-1}$), the model A patterns become very similar to those of OJK. (iii) However, the model A and OJK patterns become gradually dissimilar at a very late stage ($t \gtrsim 100\gamma_0^{-1}$). Nevertheless, the statistical properties of the patterns in these two schemes remain surprisingly close, as has already been demonstrated in Fig. 8.9. This is also demonstrated in Fig. 8.11, where

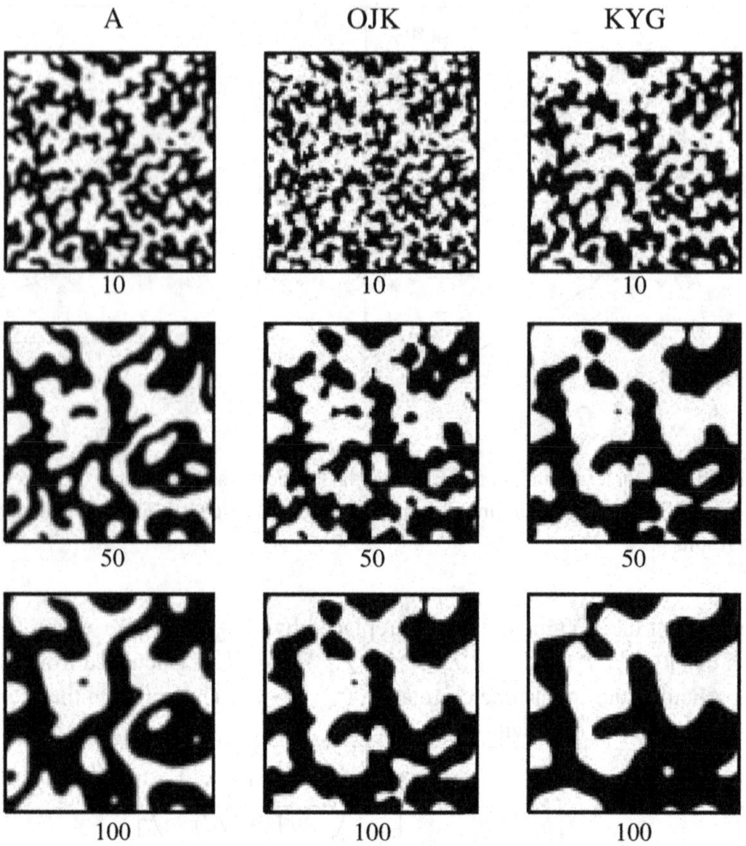

Fig. 8.10. Comparison of the time evolution of patterns in model A (*left*), OJK (*middle*), and KYG (*right*) on a 258 × 258 lattice with $\Delta x = 1$ and $\Delta t = 0.02$ without thermal noise. The numbers are the times in units of γ_0^{-1}. The patterns of model A and KYG are nearly the same at $\gamma_0 t = 10$, while those of model A and OJK are similar at $\gamma_0 t = 50$.

the perimeter density of the patterns are plotted in these three cases for the runs in Fig. 8.10.

8.2.3 Derivation of the dynamic equation for interface motion

We now derive the Allen–Cahn equation including the effects of a small magnetic field h and the random noise term θ starting with (8.1.2) [28]. We note that the average $\langle\psi\rangle$ outside the interface regions instantaneously approaches the equilibrium values,

$$\langle\psi\rangle \cong \pm\psi_{\text{eq}} + \chi h, \tag{8.2.15}$$

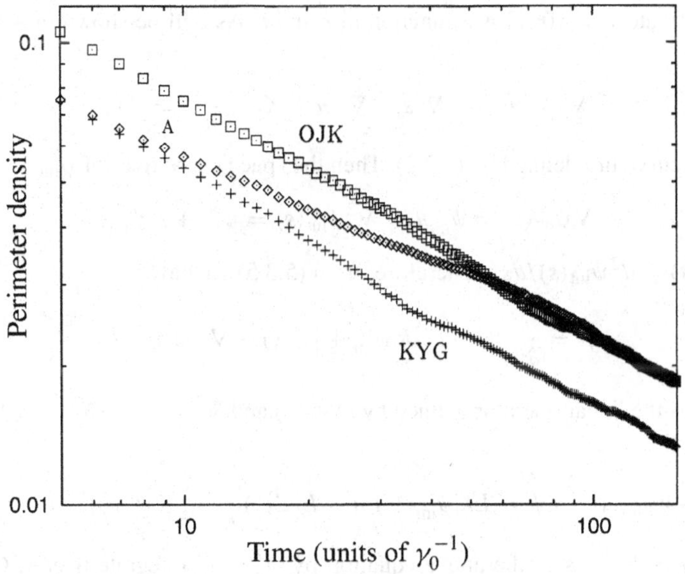

Fig. 8.11. The perimeter density for the runs in Fig. 8.10. They reveal coincidence of the curves of model A and KYG for $\gamma_0 t \lesssim 10$ and those of model A and OJK for $\gamma_0 t \gtrsim 50$. This tendency is reproducible for a sufficiently large system size.

where $\chi = (2\kappa^2)^{-1}$ is the susceptibility, $\langle\ \rangle$ being the average over the noise. This is a linear response relation valid for

$$0 \le h \ll \kappa^2 \psi_{eq}, \tag{8.2.16}$$

under which the second term in (8.2.15) is much smaller than the first. Including the interface region, we set

$$\psi(\boldsymbol{r}, t) = \psi_{int}(s) + \delta\psi(\boldsymbol{r}, t), \tag{8.2.17}$$

where s is the coordinate along the surface normal $\boldsymbol{n}$, and

$$\psi_{int}(s) = -\psi_{eq} \tanh(\kappa s/\sqrt{2}) \tag{8.2.18}$$

is the fundamental interface solution presented in Section 4.4. Therefore, we have $\langle\psi\rangle \cong -\psi_{eq} + h/2\kappa^2$ in the spatial region $s \gtrsim \kappa^{-1}$ and $\langle\psi\rangle \cong \psi_{eq} + h/2\kappa^2$ in the region $s \lesssim -\kappa^{-1}$. By suitably defining the interface position, we may assume that the deviation $\delta\psi(\boldsymbol{r}, t)$ in (8.2.17) is orthogonal to $\psi'_{int}(s) = d\psi_{int}(s)/ds$:

$$\int ds \psi'_{int}(s) \delta\psi(\boldsymbol{r}, t) = 0, \tag{8.2.19}$$

without loss of generality. This is because a small shift of the interface position by $\delta\zeta$ is equivalent to replacing $\psi_{int}(s)$ by $\psi_{int}(s - \delta\zeta) \cong \psi_{int}(s) - \delta\zeta\psi'_{int}(s)$. The s integration here is almost convergent if $|s|$ is a few times larger than κ^{-1} at the upper and lower bounds.

The coordinate $s = s(\boldsymbol{r}, t)$ is a function of $\boldsymbol{r}$ and t. As will be shown in Appendix 8E, we have

$$\nabla s = \boldsymbol{n}, \qquad \nabla^2 s = \nabla \cdot \boldsymbol{n} = \mathcal{K}, \qquad \frac{\partial}{\partial t} s = -v, \tag{8.2.20}$$

$\mathcal{K}$ being the curvature defined by (8.2.2). Then the space derivatives of $\psi_{\text{int}}(s)$ are

$$\nabla \psi_{\text{int}}(s) = \psi'_{\text{int}} \boldsymbol{n}, \quad \nabla^2 \psi_{\text{int}}(s) = \psi''_{\text{int}} + \mathcal{K} \psi'_{\text{int}}, \tag{8.2.21}$$

where $\psi''_{\text{int}}(s) = d^2 \psi_{\text{int}}(s)/ds^2$. Therefore, from (5.3.5) we have

$$\mu = \frac{\delta}{\delta \psi} \beta \mathcal{H} = -\mathcal{K} \psi'_{\text{int}} + \big[\hat{\mathcal{L}}(s) - \nabla_\perp^2\big] \delta\psi - h, \tag{8.2.22}$$

where $\hat{\mathcal{L}}(s)$ is the linear operator defined by (4.4.44) and $\nabla_\perp^2 = \nabla^2 - \partial^2/\partial s^2$. Thus (8.1.2) becomes

$$-v \psi'_{\text{int}} + \frac{\partial}{\partial t} \delta\psi = L \mathcal{K} \psi'_{\text{int}} + Lh - L\big[\hat{\mathcal{L}}(s) - \nabla_\perp^2\big] \delta\psi + \theta. \tag{8.2.23}$$

We multiply both sides of the above equation by ψ'_{int} and integrate over s. On the left-hand side, the inner product of $\partial \delta\psi/\partial t$ and ψ'_{int} is of order $O(v^2)$, because the leading contribution of order $O(v)$ vanishes from the orthogonality relation (8.2.19). We thus arrive at

$$v = -L\mathcal{K} + v_h + \theta_a, \tag{8.2.24}$$

where v_h is a constant velocity (now taken to be positive),

$$v_h = (2TL\psi_{\text{eq}}/\sigma) h, \tag{8.2.25}$$

σ being the surface tension given by (4.4.8). The θ_a is the random noise term defined at each surface point $\boldsymbol{r}_a$ as

$$\theta_a = -(T/\sigma) \int ds \psi'_{\text{int}}(s) \theta(\boldsymbol{r}, t). \tag{8.2.26}$$

From (5.3.4) its fluctuation variance is

$$\langle \theta_a(t) \theta_{a'}(t') \rangle = 2(LT/\sigma) \delta_{aa'} \delta(t - t'), \tag{8.2.27}$$

where $\delta_{aa'}$ is the δ function on the surface satisfying $\int da \delta_{aa'} = 1$.

The deviation $\delta\psi(\boldsymbol{r}, t)$ then consists of two parts as $\delta\psi(\boldsymbol{r}, t) = \psi_h(s) + \delta\psi_1(\boldsymbol{r}, t)$. The first part is of order h and is the solution of

$$\hat{\mathcal{L}}(s) \psi_h(s) = h[1 + (2\psi_{\text{eq}} T/\sigma) \psi'_{\text{int}}], \tag{8.2.28}$$

where the left-hand side is made orthogonal to ψ'_{int}. The second part is induced by the noise term θ and is determined by

$$\left[\frac{\partial}{\partial t} + L(\hat{\mathcal{L}}(s) - \nabla_\perp^2)\right] \delta\psi_1 = \theta + (T/\sigma) \psi'_{\text{int}} \theta_a, \tag{8.2.29}$$

where the left-hand side is the noise term orthogonal to $\psi'_{\rm int}$. For $|s| \gg \kappa^{-1}$, $\psi'_{\rm int}$ vanishes and $\hat{\mathcal{L}}(s)$ tends to $2\kappa^2$, so $\psi_h(s)$ tends to $h/(2\kappa^2)$ in agreement with (8.2.15) and $\delta\psi_1$ obeys the linearized Langevin equation in the bulk region,

$$\left[\frac{\partial}{\partial t} + L(2\kappa^2 - \nabla^2)\right]\delta\psi_1 = \theta. \tag{8.2.30}$$

8.2.4 Langevin equation for surfaces

At a late stage after quenching, with small h, the total free energy $\mathcal{H}$ is approximately of the form

$$\mathcal{H} = \sigma S(t) - (2T\psi_{\rm eq}h)V_+(t) + \text{const.}, \tag{8.2.31}$$

where $S(t)$ is the surface area and $V_+(t)$ is the volume of the phase with $\psi \cong \psi_{\rm eq}$. The second term is the magnetic field energy, because $2T\psi_{\rm eq}h$ is the free-energy density difference between the two phases. With respect to a small surface deformation, $\boldsymbol{r}_a$ to $\boldsymbol{r}_a + \delta\zeta\boldsymbol{n}$, the incremental change of $\mathcal{H}$ is written in the following surface integral,

$$\delta\mathcal{H} = \int da(\sigma\mathcal{K} - 2T\psi_{\rm eq}h)\delta\zeta, \tag{8.2.32}$$

with the aid of (8.2.5). The functional derivative of $\mathcal{H}$ with respect to the surface displacement ζ may thus be expressed as

$$\frac{\delta}{\delta\zeta}\mathcal{H} = \sigma\mathcal{K} - 2T\psi_{\rm eq}h \tag{8.2.33}$$

Therefore, (8.2.24) is rewritten as

$$v = -\frac{L}{\sigma}\frac{\delta}{\delta\zeta}\mathcal{H} + \theta_a. \tag{8.2.34}$$

This is a Langevin equation for the surface $\{\boldsymbol{r}_a\}$, which is a new gross variable. The fluctuation–dissipation relation between the noise term θ_a and the kinetic coefficient L/σ in (8.2.34) has been given by (8.2.27). As a generalization of (8.2.7) and also as a self-consistency relation of the model, $\mathcal{H}$ monotonically decreases in time (if the noise term is neglected) as

$$\frac{d}{dt}\mathcal{H} = \int da\left(\frac{\delta}{\delta\zeta}\mathcal{H}\right)v = -\frac{L}{\sigma}\int da\left[\sigma\mathcal{K} - (2T\psi_{\rm eq}h)\right]^2 \le 0. \tag{8.2.35}$$

Undulations of a planar interface

Because the above theory is formal, we consider a simple case of a planar interface at $h = 0$ with small disturbances superimposed. If the unperturbed interface is perpendicular to the z axis, the surface displacement $\zeta(\boldsymbol{r}_\perp, t)$ is parameterized by the two-dimensional coordinates $\boldsymbol{r}_\perp = (x, y)$. For small ζ, the normal unit vector is written as $\boldsymbol{n} = (-\partial\zeta/\partial x, -\partial\zeta/\partial y, 1)$ and the curvature is given by

$$\mathcal{K} = \nabla\cdot\boldsymbol{n} \cong -\nabla_\perp^2\zeta \tag{8.2.36}$$

where $\nabla_\perp^2 = \partial^2/\partial x^2 + \partial^2/\partial y^2$. Then ζ obeys

$$\frac{\partial}{\partial t}\zeta = L\nabla_\perp^2\zeta + \theta_\perp. \tag{8.2.37}$$

From (8.2.27) the noise term $\theta_\perp(\boldsymbol{r}_\perp, t)$ satisfies

$$\langle\theta(\boldsymbol{r}_\perp, t)\theta(\boldsymbol{r}'_\perp, t')\rangle = 2(L/\sigma)\delta(\boldsymbol{r}_\perp - \boldsymbol{r}'_\perp)\delta(t - t'), \tag{8.2.38}$$

which assures the equilibrium distribution,

$$P_{\text{eq}}(\zeta) \propto \exp\left[-\frac{\sigma}{2T}\int d\boldsymbol{r}_\perp|\nabla_\perp\zeta|^2\right], \tag{8.2.39}$$

in the gaussian approximation in accord with the results in Section 4.4. From (8.2.33) and (8.2.34) the relaxation rate of the surface displacement with wave number k is given by

$$\Gamma_k = Lk^2. \tag{8.2.40}$$

The coarsening law (8.1.16) again follows if we pick up the fluctuations with $k \sim 1/\ell(t)$ and set

$$\Gamma_k t \sim Lt/\ell(t)^2 \sim 1. \tag{8.2.41}$$

Growth of a circular or spherical domain

Let us consider a circular (in 2D) or spherical (in 3D) domain with radius R, within which $\psi \cong \psi_{\text{eq}}$ and outside of which $\psi \cong -\psi_{\text{eq}}$. In this case the free energy is a function of R:

$$\mathcal{H}(R) = S_d\left[\sigma R^{d-1} - \frac{2T\psi_{\text{eq}}}{d}hR^d\right], \tag{8.2.42}$$

where S_d is the surface area of a unit sphere in d dimensions. Using $v = \partial R/\partial t$ and $\partial\mathcal{H}(R)/\partial R = S_d R^{d-1}\delta\mathcal{H}/\delta\zeta$, we obtain a Langevin equation for R,

$$\frac{\partial}{\partial t}R(t) = -L\left[\frac{d-1}{R} - \frac{2T\psi_{\text{eq}}h}{\sigma}\right] + \theta(R, t). \tag{8.2.43}$$

The noise term $\theta(R, t)$ is the angle average of $\theta_a(t)$ in (8.2.26),

$$\theta(R, t) = S_d^{-1}R^{-d+1}\int d\Omega\, \theta_a(t), \tag{8.2.44}$$

where $d\Omega$ is the angle element. From (8.2.27) it follows that the noise amplitude relation is

$$\langle\theta(R, t)\theta(R, t'); R\rangle = 2\mathcal{L}(R)\delta(t - t'), \tag{8.2.45}$$

where $\langle\cdots; R\rangle$ is the conditional average under fixed R (see Section 5.2). The kinetic coefficient is dependent on R as

$$\mathcal{L}(R) = (L/S_d\sigma)R^{-d+1}. \tag{8.2.46}$$

It is worth noting that (8.2.43) may be expressed in the standard form of Langevin equations,

$$\frac{\partial}{\partial t}R(t) = -\mathcal{L}(R)\frac{\partial}{\partial R}\beta\mathcal{H}(R) + \theta(R,t). \tag{8.2.47}$$

We may now introduce a critical radius by

$$R_c = (d-1)(\sigma/2T\psi_{eq})h^{-1}. \tag{8.2.48}$$

If the noise term is neglected, the droplet continues to grow for $R > R_c$ and shrinks to vanish for $R < R_c$. In a weak magnetic field, which satisfies (8.2.16), we have $R_c \gg \xi$.

Diffusion of a droplet

The center of mass of a droplet undergoes diffusive motion with a radius-dependent diffusion constant. In the nonconserved case this effect is very small, but its calculation is simple and instructive. For a spherical droplet in 3D we may express $\theta_a(t) = \sum_{\ell m}\theta_{\ell m}Y_{\ell m}(\boldsymbol{n}_a)$ in terms of the spherical harmonic functions $Y_{\ell m}$. Components with $\ell = 1, m = -1, 0, 1$ arise from translational motions of the droplet. We pick them up to obtain the random velocity of the center of mass,

$$\boldsymbol{u}(t) = \frac{d}{S_d}R^{-d+1}\int da\theta_a(t)\boldsymbol{n}_a. \tag{8.2.49}$$

Use of (8.2.45) gives

$$\langle u_\alpha(t)u_\beta(t')\rangle = 2d(LT/\sigma S_d)R^{-d+1}\delta_{\alpha\beta}\delta(t-t'). \tag{8.2.50}$$

From the general relation (5.1.39) we obtain the diffusion constant in model A,

$$D_A(R) = (dLT/S_d\sigma)R^{-d+1}. \tag{8.2.51}$$

We can see that the characteristic diffusion length $[D_A(R)t]^{1/2}$ is much shorter than R on the characteristic timescale ($t \sim R^2/L$). Thus the droplet center is virtually fixed at the initial position in model A.

8.2.5 Chemical potential in the case $h = \theta = 0$

If $\theta = h = 0$, the generalized chemical potential $\mu \equiv \delta(\beta\mathcal{H})/\delta\psi$ is approximated from (8.2.22) as

$$\mu \cong -2\psi_{eq}\mathcal{K}\hat{\delta}(\boldsymbol{r}), \tag{8.2.52}$$

at a late stage. The $\hat{\delta}(\boldsymbol{r})$ is a δ-function nonvanishing only on the surface $\{\boldsymbol{r}_a\}$. For any smooth function $\varphi(\boldsymbol{r})$ we require

$$\int d\boldsymbol{r}\hat{\delta}(\boldsymbol{r})\varphi(\boldsymbol{r}) = \int da\varphi(\boldsymbol{r}_a), \tag{8.2.53}$$

da being the surface element. Then $\hat{\delta}(\boldsymbol{r})$ is well-defined mathematically in the thin-interface limit. As an illustration, Fig. 8.12 shows $\mu(\boldsymbol{r},t)$ at $\gamma_0 t = 20$.

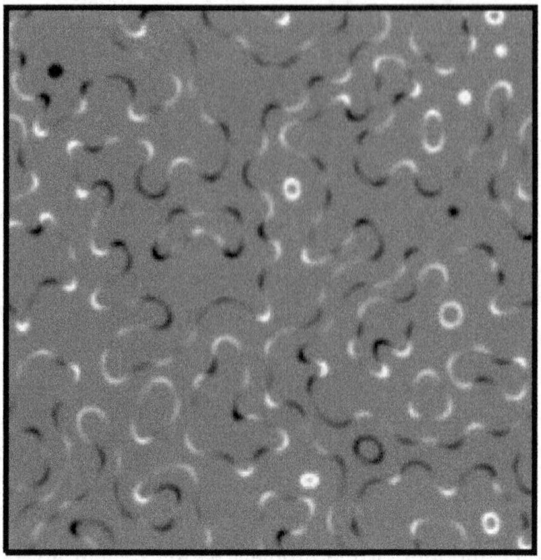

Fig. 8.12. The chemical potential $\mu(\boldsymbol{r}, t) = \delta(\beta\mathcal{H})/\delta\psi$ for model A without thermal noise at $\gamma_0 t = 20$ on a 256×256 lattice. The interfaces are located in the black regions (where $\mu > 0.2\psi_{\text{eq}}\kappa^2$) and in the white regions (where $\mu < -0.2\psi_{\text{eq}}\kappa^2$). In the gray regions μ is close to 0. The phase with $\psi \cong \psi_{\text{eq}}$ is shrinking (expanding) in the black (white) regions.

The two-point correlation function $H(r, t_1, t_2) = \langle \mu(\boldsymbol{r}_1, t_1)\mu(\boldsymbol{r}_1 + \boldsymbol{r}, t_2)\rangle$ can be expressed as

$$\begin{aligned} H(r, t_1, t_2) &= V^{-1} \int d\boldsymbol{r}_1 \int d\boldsymbol{r}_2 \mu(\boldsymbol{r}_1)\mu(\boldsymbol{r}_2)\delta(\boldsymbol{r}_1 - \boldsymbol{r}_2 - \boldsymbol{r}) \\ &= (2\psi_{\text{eq}})^2 V^{-1} \int da_1 \int da_2 \mathcal{K}_1 \mathcal{K}_2 \delta(\boldsymbol{r}_1 - \boldsymbol{r}_2 - \boldsymbol{r}), \end{aligned} \tag{8.2.54}$$

where V is the volume of the system, and $da_\alpha, \boldsymbol{r}_\alpha$, and $\mathcal{K}_\alpha$ in the second line are the surface element, position, and curvature on the surfaces at time t_α ($\alpha = 1, 2$), respectively. Because the integration of $\mathcal{K}_\alpha$ over the surface in a unit volume is of order $\ell(t_\alpha)^{-2}$, we notice the scaling relation [29],

$$H(r, t_1, t_2) = \psi_{\text{eq}}^2 \ell(t_1)^{-4} H^*(r/\ell(t_1), t_2/t_1), \tag{8.2.55}$$

where the algebraic time dependence of $\ell(t)$ ($d \ln \ell(t)/d \ln t = \text{const.}$) is assumed. From $\partial\psi/\partial t = -L\mu$ without thermal noise, the pair correlation $g(r, t) = \langle \psi(\boldsymbol{r}_1, t)\psi(\boldsymbol{r}_1, t)\rangle$ ($r = |\boldsymbol{r}_1 - \boldsymbol{r}_2|$) obeys

$$\frac{\partial}{\partial t} g(r, t) = -2L\langle \psi(\boldsymbol{r}_1, t)\mu(\boldsymbol{r}_2, t)\rangle = 2L^2 \int_0^t dt' H(r, t, t'), \tag{8.2.56}$$

where we can set $\langle \psi(\boldsymbol{r}_1, 0)\mu(\boldsymbol{r}_2, t)\rangle = 0$ in the scaling limit (or in the limit of small initial variance of ψ). If (8.2.55) is assumed, the scaling form $g(r, t) = \psi_{\text{eq}}^2 G(r/\ell(t))$ holds only for $\ell(t) \propto t^{1/2}$. By setting $\ell(t) = (Lt)^{1/2}$ as in (8.1.16), we may relate $G(x)$ and $H^*(x, s)$ as

$$-x\frac{\partial}{\partial x}G(x) = 4\int_0^1 ds H^*(x, s). \tag{8.2.57}$$

More generally, we find the two-point scaling,

$$g(r, t_1, t_2) = \psi_{\rm eq}^2 G^*(r/\ell(t_1), t_2/t_1), \tag{8.2.58}$$

with $G^*(x, 1) = G(x)$ and $G^*(x, 0) = 0$. The results of KYG and OJK clearly satisfy this scaling relation.

8.2.6 Phase ordering in small magnetic field

We next lower the temperature into the unstable region from a nearly disordered state ($\langle\psi\rangle = O(h)$ at $t = 0$) in the presence a small, positive h which satisfies (8.2.16). Here the effect of h becomes apparent after a long crossover time $t_{\rm c}$. We balance the first two terms in (8.2.24) as

$$L/\ell(t_{\rm c}) \sim v_h \sim (TL\psi_{\rm eq}/\sigma)h. \tag{8.2.59}$$

From (8.2.48) we may set $\ell(t_{\rm c}) = R_{\rm c}$ (although $R_{\rm c}$ has been defined in a different situation). Therefore,

$$t_{\rm c} = R_{\rm c}/v_h \sim \gamma_0^{-1}(\psi_{\rm eq}\kappa^2/h)^2, \tag{8.2.60}$$

where use has been made of $\sigma \sim T\psi_{\rm eq}^2\kappa$. We find $t_{\rm c} \gg \gamma_0^{-1}$ from (8.2.16). After the crossover time $t_{\rm c}$, the favored phase expands with the velocity v_h and the unfavored phase begins to disappear on the timescale of $t_{\rm c}$. The changing rate $\Gamma_{\rm c}$ of droplets with radii close to $R_{\rm c}$, which will be introduced in the next section, is of order $t_{\rm c}^{-1}$.

8.2.7 Motion of antiphase boundaries in model C

In real materials it is always the case that a nonconserved order parameter is coupled to conserved variables such as the energy or concentration. (See Section 3.4 for such examples in binary alloys.) The simplest dynamic model is model C near a critical point introduced in Section 5.3, where the free energy is given by the GLW hamiltonian $\mathcal{H}\{\psi, m\}$ in (4.1.45). A nonconserved scalar order parameter ψ obeys (5.3.3) with the kinetic coefficient L, while a conserved variable m obeys (5.3.13) with the kinetic coefficient λ. The interface profile between the two ordered phases is written as $\psi = \psi_{\rm int}(s)$ and

$$m = C_0[\tau - \gamma_0\psi_{\rm int}(s)^2], \tag{8.2.61}$$

in equilibrium, where $\psi_{\rm int}$ changes between $\pm\psi_{\rm eq}$. Here τ is the reduced temperature if m is the energy variable, while it is the chemical potential difference if m is the concentration variable. We are assuming no latent heat or no concentration difference between the two phases. This indeed happens for antiphase boundaries separating two variants of the same ordered structure in alloys [27].

Obviously, when the timescale of m is faster than that of ψ, the interface motion is nearly the same as that of model A and the domain size grows as $R(t) \sim (Lt)^{1/2}$ after quenching. This condition is given by $Dt \gg R(t)^2 \sim Lt$ or $D \gg L$, where $D = \lambda/C_0'$ is the diffusion constant of m with $C_0' = C_0(1 + 2\gamma_0^2 C_0/u_0)$ being the specific heat in the ordered phases in the mean field theory. On the other hand, in the reverse limit $D/L \to 0$, phase ordering proceeds at fixed m and we have again the growth law $R(t) \sim (Lt)^{1/2}$. In addition, if m is initially heterogeneous, it plays the role of quenched disorder in this limit. However, for slow diffusion $D \ll L$ and for strong static coupling $\gamma_0^2 C_0/u_0 \gtrsim 1$, there is some complicated transient behavior at very long times.

8.3 Spinodal decomposition in conserved systems

In conserved systems, phase-separation processes taking place in an unstable state are called spinodal decompositions [30]. Here, without flow from the boundary, the average order parameter M is fixed in time at an initial value, so it characterizes the type of quench and there can be two kinds of experiments: critical quenches are those lowering the temperature into an unstable state with $M = 0$ or through the critical point; while off-critical quenches are those with $M \neq 0$ [31]. Because M is not dimensionless, it is convenient to introduce ϕ by[2]

$$\phi = \frac{1}{2} + \frac{1}{2\psi_{\rm eq}} M. \tag{8.3.1}$$

In late stages of phase separation, the system is composed of the two phases with $\psi \cong \pm\psi_{\rm eq}$ as in the nonconserved case, and ϕ tends to the volume fraction of the phase with $\psi \cong \psi_{\rm eq}$ because $M \cong \phi\psi_{\rm eq} - (1-\phi)\psi_{\rm eq} = (2\phi - 1)\psi_{\rm eq}$. We will use ϕ rather than M to characterize the type of quench.

Experimental data on the growth of domains are usually fitted to an algebraic form, $\ell(t) \sim t^a$. Two experiments are presented here. (i) Figure 8.13 displays the scattering intensity from a phase-separating Al–Zn binary alloy, where $a \cong 0.17$ [32] (although the peak wave number $k_{\rm m}(t)$ at the largest t ($= 10^3$ min) was only one-half the initial peak value $k_{\rm m}(0)$). It could be fitted to Furukawa's phenomenological scaling function $Q^2/(2 + Q^6)$ with $Q = k/k_{\rm m}(t)$ [2]. In solids, the exponent a has often been observed to be considerably smaller than 1/3 because of elastic effects or pinning by disorder.[3] (ii) Figure 8.14 shows the scattered light intensity from a polymer blend below the spinodal temperature [33]. It grew exponentially at an early stage with a fixed peak wave number, in agreement with the linear theory presented below. At a late stage an accelerated growth rate with $a \cong 0.8$ was observed due to the hydrodynamic interaction, which we will discuss in Section 8.5.

[2] In this book we use ϕ also as the volume fraction of polymers.

[3] We will treat elastic effects on phase separation in solids in Chapter 10.

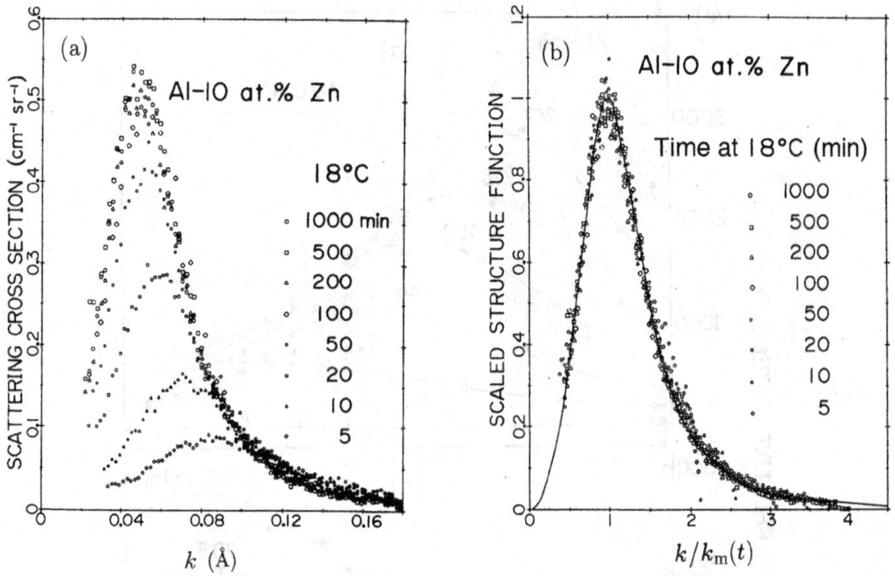

Fig. 8.13. (a) Small-angle neutron scattering intensity vs scattering wave number k for Al–10 at.% Zn polycrystals quenched from 300 °C to, and held at, 18 °C [32]. In (b) the curves are normalized and plotted vs $k_{\rm m}(t)$, the characteristic wave number.

8.3.1 Model B

We start with model B with a single conserved order parameter,

$$\frac{\partial}{\partial t}\psi = L\nabla^2[r - \nabla^2 + u_0\psi^2]\psi + \theta, \tag{8.3.2}$$

which describes the dynamics of binary alloys without elastic interactions as was discussed in Section 5.3. The temperature coefficient r is changed from a large positive value to a negative value $-\kappa^2$ at $t = 0$ as in (8.1.1). The ordering field h, if it is homogeneous in space, vanishes in the above equation. As in the nonconserved case (8.1.3) the evolution equation of the equal-time structure factor $I(k, t)$ is given by

$$\frac{\partial}{\partial t}I(k, t) = 2Lk^2\{\kappa^2[1 - J(k, t)] - k^2\}I(k, t) + 2Lk^2, \tag{8.3.3}$$

where

$$J(k, t) = u_0 \int d\boldsymbol{r}e^{i\boldsymbol{k}\cdot\boldsymbol{r}}\langle\psi(\boldsymbol{r}, t)^3\delta\psi(\boldsymbol{0}, t)\rangle/\kappa^2 I(k, t), \tag{8.3.4}$$

with $\delta\psi = \psi - M$, $M = \langle\psi\rangle$ being the space average.

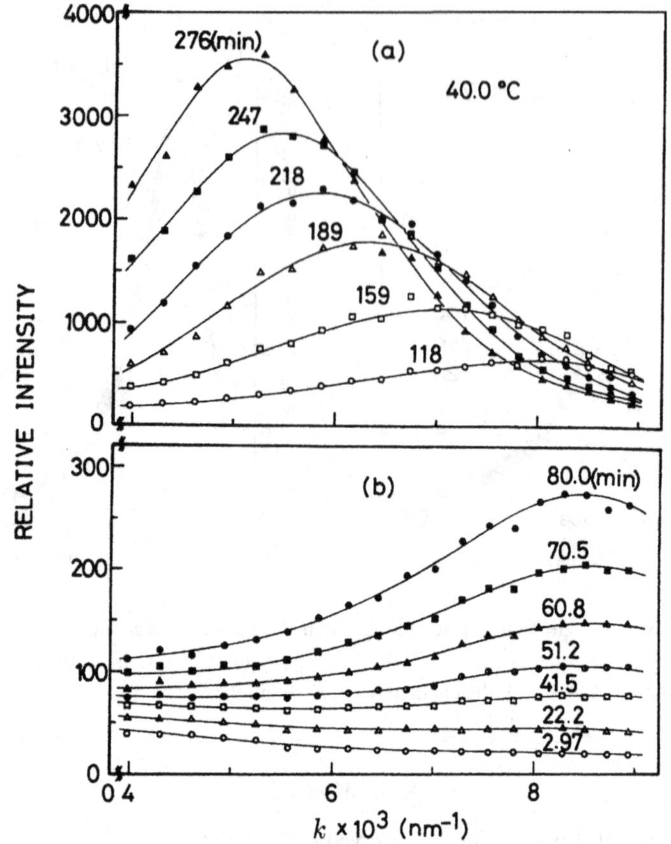

Fig. 8.14. Light scattering intensity from a polymer blend of SBR (styrene–butadiene random copolymer) (8 vol.%) + polybutadiene (30 vol.%) at an early stage ($t < 80$ min) in (b) and at a late stage ($t > 118$ min) in (a) [33]. The molecular weights are about 10^5 for the two polymers.

Linear growth

Because $J(k, t) \cong 3u_0 M^2/\kappa^2$ in the mean field theory (or for small fluctuations), spinodal decomposition occurs for

$$\kappa^2 - 3u_0 M^2 > 0, \tag{8.3.5}$$

as discussed in Section 5.3. At large t and small k, $J(k, t)$ should tend to 1 with coarsening even for $M \neq 0$. Let us suppose a critical quench ($M = 0$) and neglect $J(k, t)$ in (8.3.3). After the quench, the structure factor is again expressed as in (8.1.7) with

$$\Gamma_k = Lk^2(-\kappa^2 + k^2). \tag{8.3.6}$$

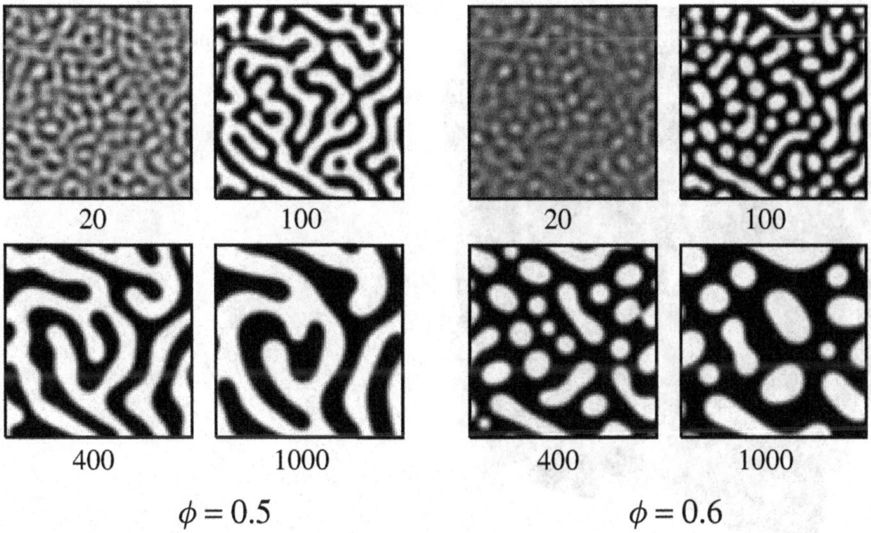

Fig. 8.15. 2D time evolution of patterns in model B after quenching at $t = 0$ without thermal noise for $\phi = 0.5$ and $\phi = 0.6$. The numbers are the times after quenching in units of $(L\kappa^4)^{-1}$.

Growth occurs for $k < \kappa$ and is maximum at an intermediate wave number $k = k_{\rm m}$ given by

$$k_m = 2^{-1/2}\kappa. \tag{8.3.7}$$

The maximum growth rate is

$$\Gamma_m = \frac{1}{4}L\kappa^4. \tag{8.3.8}$$

Near $k \sim k_{\rm m}$, the structure factor in the linear approximation grows as

$$I_0(k,t) \cong (\kappa_0^{-2} + k_{\rm m}^{-2}) \exp\left[2\Gamma_{\rm m}t - 2L(k^2 - k_{\rm m}^2)^2 t\right], \tag{8.3.9}$$

which is the counterpart of (8.1.8). The term proportional to $k_{\rm m}^{-2}$ is produced by the thermal noise term in the initial stage. We may examine the validity of the linear approximation by estimating $J(k,t)$ using the decoupling approximation as in the nonconserved case. Then,

$$J(k,t) \sim (K_d u_0 \kappa^{-\epsilon})(\Gamma_{\rm m}t)^{-1/2} \exp(2\Gamma_{\rm m}t). \tag{8.3.10}$$

Therefore, if the Ginzburg condition $K_d u_0 \kappa^{-\epsilon} \ll 1$ in (4.1.24) holds, exponential growth of the fluctuations at $k \sim k_{\rm m}$ is observable over a sizable time region. Note that model B is characterized by the two parameters, ϕ and $u_0\kappa^{-\epsilon}$, after scale changes, $\kappa \boldsymbol{r} \to \boldsymbol{r}$, $L\kappa^4 t \to t$, and $\psi/\psi_{\rm eq} \to \psi$.

Computer simulations and scaling

Figure 8.15 shows evolution patterns of model B at $\phi = 0.5$ and 0.6 in 2D [34, 35]. Figure 8.16 is a snapshot at $\phi = 0.5$ in 3D [36]. We can see bicontinuous domain structures at the

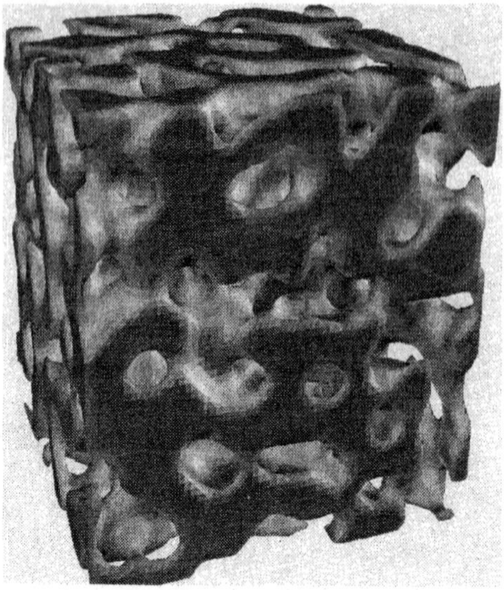

Fig. 8.16. A 3D snapshot of a two-phase structure obtained as a solution of model B [36].

critical-quench condition and droplet structures for off-critical quenches. From simulations and theories of model B, it is now established that the characteristic domain size grows as

$$\ell(t) \propto t^a \quad \text{with} \quad a = 1/3, \tag{8.3.11}$$

at long times, irrespective of the volume fraction ϕ and the space dimensionality. The equal-time pair correlation $g_{\rm dom}(r, t)$ due to the domain structure is scaled as

$$g_{\rm dom}(r, t) = \psi_{\rm eq}^2 G(r/\ell(t)). \tag{8.3.12}$$

The domain structure factor is then scaled as

$$I_{\rm dom}(k, t) = \psi_{\rm eq}^2 \ell(t)^d F(k\ell(t)), \tag{8.3.13}$$

as in the nonconserved case. To confirm the above scaling we give simulation results at $\phi = 1/2$ in 3D [36]. Namely, Fig. 8.17 shows the dimensionless pair correlation function $G(x)$ in model B, while Fig. 8.18 shows the dimensionless structure factor $F(Q)$ in model B (and model H). We recognize that the domain structure factor has a Porod tail ($\propto k^{-d-1}$) at large k as in (8.1.21). However, at small $Q = k\ell(t) \ll 1$, it goes to zero rapidly as

$$F(Q) \cong CQ^4, \tag{8.3.14}$$

from the conservation law [37]. A derivation of this small-k behavior will be given in the next section. Note that the thermal intensity $I_{\rm th}(k, t)$ tends to $\xi^2 \sim \kappa^{-2}$ for $k \ll \kappa$, so that the domain contribution is dominant for $\xi^2 \ll \psi_{\rm eq}^2 \ell(t)^d (k/\kappa)^4$ or

$$k > \ell(t)^{-1}[\xi/\ell(t)]^{d/4}. \tag{8.3.15}$$

The lower bound here is much smaller than the peak wave number $k_{\rm m}(t) \sim 2\pi/\ell(t)$. (See (8.1.22) for the upper bound, below which the Porod tail dominates over the thermal

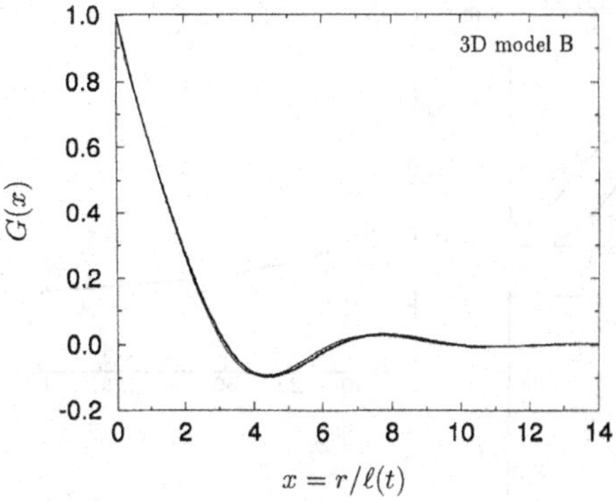

Fig. 8.17. The scaled pair correlation function $G(x)$ vs $x = r/\ell(t)$ of model B at $\phi = 0.5$ in 3D [36].

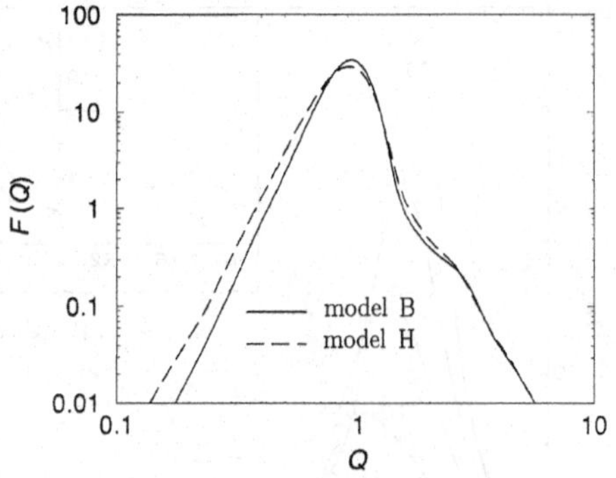

Fig. 8.18. The scaled structure factor $F(Q)$ for models B and H [36].

intensity.) The wave number region in which $I_{\text{dom}}(k,t) \gg I_{\text{th}}(k,t)$ expands with growth of $\ell(t)$.

8.3.2 The Langer–Bar-on–Miller theory

Langer, Bar-on, and Miller (LBM) [38] presented the first analytic theory for model B [4, 39, 40]. It takes into account the nonlinearity in relatively early-stage spinodal decomposition and reasonably describes the onset of coarsening. This scheme will be applied to periodic spinodal decomposition in Section 8.8. To this end, it is convenient to add a constant r_c to r as

$$r = -\kappa^2 + r_c, \tag{8.3.16}$$

where r_c is a shift of r due to the fluctuation effect in the LBM scheme. That is, we determine r_c such that the structure factor at small wave numbers grows indefinitely for $r < r_c$ and tends to a steady Ornstein–Zernike form for $r > r_c$. Then $r_c = -0.374\kappa^2$ if the upper cut-off wave number Λ is set equal to κ ($\alpha = 1$ in (8F.10)). More generally, the curve $r = r_c$ as a function of the average composition yields a spinodal curve [39, 40], but it depends on the choice of the ratio Λ/κ as an artifact of the approximation.

LBM introduced single-point and two-point distribution functions,

$$\rho_1(\psi_1, t) = \langle \delta(\psi(\mathbf{r}_1, t) - \psi_1) \rangle, \tag{8.3.17}$$

$$\rho_2(\psi_1, \psi_2, r, t) = \langle \delta(\psi(\mathbf{r}_1, t) - \psi_1)\delta(\psi(\mathbf{r}_2, t) - \psi_2) \rangle, \tag{8.3.18}$$

where ρ_2 depends on the distance $r \equiv |\mathbf{r}_1 - \mathbf{r}_2|$ and

$$\rho_1(\psi_1, t) = \int d\psi_2 \rho_2(\psi_1, \psi_2, r, t). \tag{8.3.19}$$

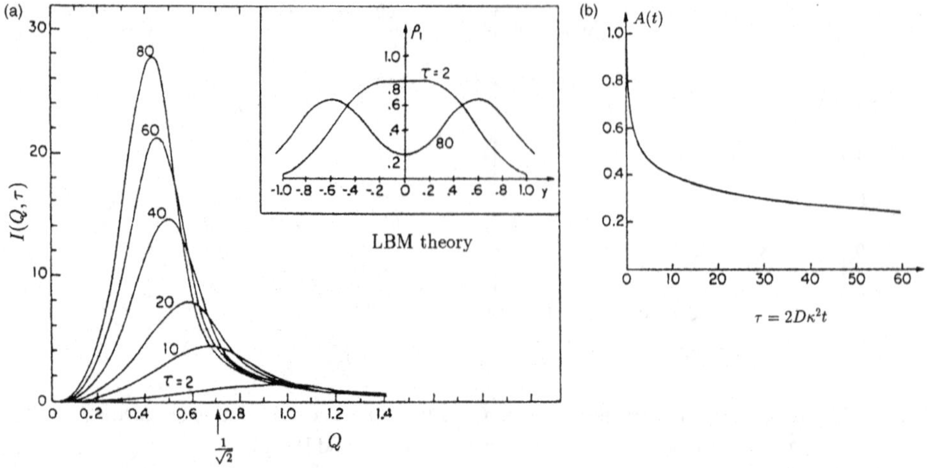

Fig. 8.19. (a) The structure factor vs $Q = k/\kappa$ for a critical quench at various $\tau = 2L\kappa^4 t$ in the LBM theory [38]. The initial peak wave number $k_m(0) = \kappa/\sqrt{2}$ is indicated below the figure. The inset shows the one-point distribution ρ_1 vs $y = \psi/\psi_{eq}$. (b) Relaxation of the parameter $A(t)$ in (8.3.22).

For any n and m, we obtain

$$\langle \delta\psi(\mathbf{r}_1, t)^n \delta\psi(\mathbf{r}_2, t)^m \rangle = \int d\psi_1 \int d\psi_2 \delta\psi_1^n \delta\psi_2^m \rho_2(\psi_1, \psi_2, r, t), \tag{8.3.20}$$

where $\delta\psi = \psi - M$. To obtain a closed set of equations for ρ_1 and $I(k, t)$, they assumed the following truncation for ρ_2,

$$\rho_2(\psi_1, \psi_2, r, t) = \rho_1(\psi_1, t)\rho_1(\psi_2, t)\left[1 + \frac{1}{\langle \delta\psi^2 \rangle^2} g(r, t)\delta\psi_1\delta\psi_2\right]. \tag{8.3.21}$$

Then $J(k, t)$ in (8.3.4) becomes independent of k. We introduce $A(t)$ by

$$A(t) = 1 - J(k, t) = 1 - r_c/\kappa^2 - u_0\langle \psi^3 \delta\psi \rangle / \kappa^2 \langle \delta\psi^2 \rangle, \tag{8.3.22}$$

where $A(t)$ is a monotonically decreasing function of t. As a result, (8.3.3) reads

$$\frac{\partial}{\partial t} I(k, t) = 2Lk^2\left[\kappa^2 A(t) - k^2\right] I(k, t) + 2Lk^2. \tag{8.3.23}$$

As will be derived in Appendix 8F, the dynamic equation for $\rho_1(\psi, t)$ is a self-consistent Fokker–Planck equation in which $I(k, t)$ is involved. Figure 8.19 is the LBM numerical result for a critical quench, where the peak wave number decreases in time with the growth exponent a about 0.2. The LBM theory can thus reproduce the initial coarsening behavior. However, it is not applicable with the formation of well-defined interfaces, because the ansatz (8.3.21) is no longer justified at such late stages.

8.4 Interface dynamics in conserved systems

At late stages after quenching, two-phase regions with $\psi \cong \pm\psi_{\rm eq}$ are distinctly separate with domain sizes much wider than the interface thickness. By analyzing the interface motion, we can explain the growth law (8.3.11). In terms of the volume fraction $q = q(t)$ of the regions with $\psi \cong \psi_{\rm eq}$, the volumes of the regions with $\psi \cong \pm\psi_{\rm eq}$ are written as

$$V_+ = Vq, \quad V_- = V(1-q), \tag{8.4.1}$$

V being the total volume. Because the interface can move only after diffusive transport of the order parameter across the interface, ψ slowly approaches the final value, $\psi_{\rm eq}$ or $-\psi_{\rm eq}$. Therefore, we define a deviation,

$$\Psi \equiv \psi - \psi_{\rm eq}\varepsilon(\boldsymbol{r}, t), \tag{8.4.2}$$

where $\varepsilon = 1$ in the regions with $\psi \cong +\psi_{\rm eq}$ and $\varepsilon = -1$ in the regions with $\psi \cong -\psi_{\rm eq}$. The space average of Ψ becomes

$$\langle\Psi\rangle = M - \psi_{\rm eq}(V_+ - V_-)/V = M - \psi_{\rm eq}(2q-1). \tag{8.4.3}$$

In the off-critical case the supersaturation $\Delta(t)$ may be introduced by

$$\Delta(t) = \langle\Psi\rangle/2\psi_{\rm eq}. \tag{8.4.4}$$

In terms of ϕ in (8.3.1) the conservation law (8.4.3) may be expressed as

$$\Delta(t) + q(t) = \phi = \text{const.} \tag{8.4.5}$$

For critical quenches ($M = 0$) we trivially have $q = 0.5$ and $\Delta = 0$, but for off-critical quenches $\Delta(t)$ is nonvanishing and slowly approaches 0.

At late stages we shall see that Ψ changes on the scale of the domain size $\ell(t)$ far from the interface regions, where we may assume $|\Psi| \ll \psi_{\rm eq}$ to obtain the diffusion equation,

$$\frac{\partial}{\partial t}\Psi = D\nabla^2\Psi, \tag{8.4.6}$$

where D is the diffusion constant in the ordered phase,

$$D = 2L\kappa^2. \tag{8.4.7}$$

8.4.1 The Gibbs–Thomson condition and the Stefan problem

As shown in Fig. 8.20, while ψ jumps by $\pm 2\psi_{\rm eq}$, the generalized chemical potential $\mu(\boldsymbol{r}, t) \equiv \delta(\beta\mathcal{H})/\delta\psi$ continuously changes at the interface even if the thin-interface limit $\ell(t)/\xi \to \infty$ is taken mathematically. Its surface value will be written as

$$\mu_a = \mu(\boldsymbol{r}_a, t), \tag{8.4.8}$$

on the surface $\{\boldsymbol{r}_a\}$. If there were discontinuities in μ across the interface, the current $-L\nabla\mu$ would change abruptly, leading to rapid temporal variations of ψ near the interface.

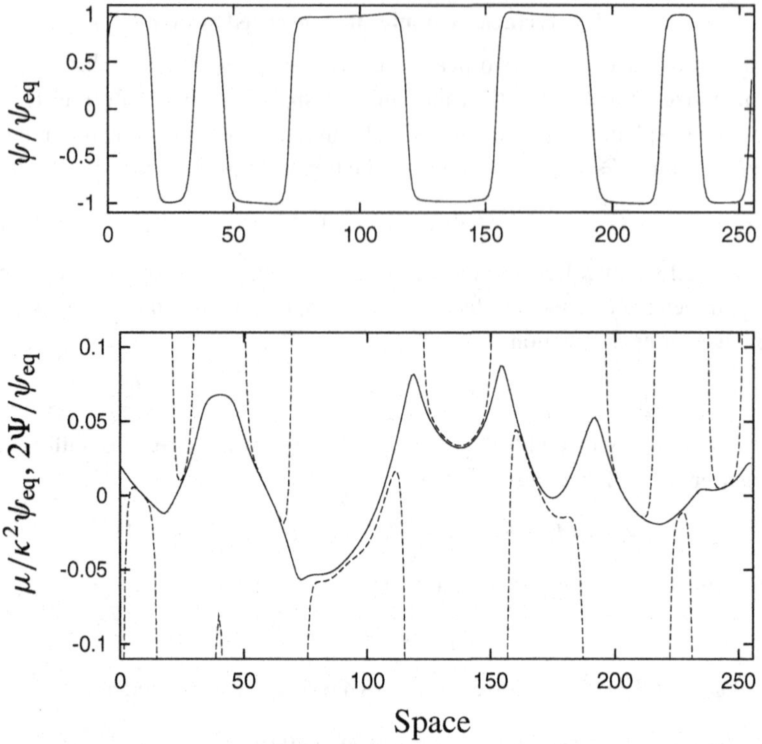

Fig. 8.20. A cross section of a 2D spinodal decomposition pattern of model B without thermal noise along the x axis in units of $(1.5^{1/2}\kappa)^{-1}$ at $t = 2200/L\kappa^4$. We can see (*upper panel*) that ψ is nearly discontinuous at the interface positions, while (*lower panel*) $\mu = \partial(\beta\mathcal{H})/\partial\psi$ (solid line) is continuous throughout the system. We also confirm that Ψ in (8.4.2) (dashed lines) nearly coincides with $\mu/2\kappa^2$ far from the interface positions.

However, as can also be seen in Fig. 8.20, the gradient $\nabla\mu$ jumps across the interface. We decompose ψ as

$$\psi(\boldsymbol{r}, t) = \psi_{\rm int}(s) + \delta\psi(\boldsymbol{r}, t) \tag{8.4.9}$$

with $\int ds\psi'_{\rm int}(s)\delta\psi = 0$, s being the coordinate along the normal $\boldsymbol{n}$, as in (8.2.17) for model A. Then $\delta\psi \to \Psi$ far from the interface. Near the interface we may rewrite (8.2.22) as

$$-\mathcal{K}_a\psi'_{\rm int} + \left[\hat{\mathcal{L}}(s) - \nabla_\perp^2\right]\delta\psi = \mu_a, \tag{8.4.10}$$

where $\hat{\mathcal{L}}(s)$ is the linear operator defined in (4.4.44). We hereafter write the curvature as $\mathcal{K}_a$ explicitly with the subscript a. Multiplication of $\psi'_{\rm int}(s) = d\psi_{\rm int}(s)/d\zeta$ and integration over s in the region $|s| \lesssim \xi$ yield

$$\mu_a = (\sigma/2T\psi_{\rm eq})\mathcal{K}_a. \tag{8.4.11}$$

This is the solvability condition of (8.4.10), which assures a unique solution for $\delta\psi$ near the interface ($|s| \lesssim \xi$). Far from the interface ($|s| \gg \xi$), on the other hand, $\delta\psi \cong \Psi$ varies slowly and its boundary value extrapolated to the interface is

$$\Psi_a = (2\kappa^2)^{-1}\mu_a = (\sigma/4T\psi_{\rm eq}\kappa^2)\mathcal{K}_a = 2\psi_{\rm eq}d_0\mathcal{K}_a, \tag{8.4.12}$$

where we define a capillary length d_0 by

$$d_0 = \sigma/(8T\psi_{\rm eq}^2\kappa^2). \tag{8.4.13}$$

Here $d_0 = \xi/6$ with $\xi = 2^{-1/2}\kappa^{-1}$ from the mean field result (4.4.9) for the surface tension. Therefore, we obtain the following order estimations,

$$\Psi_a/\psi_{\rm eq} \sim \xi\mathcal{K}_a \sim \xi/\ell(t). \tag{8.4.14}$$

The interface velocity v in the normal direction $\boldsymbol{n}$ is induced by a small discontinuity of the diffusion current across the interface. The conservation law requires

$$2\psi_e v = L[\boldsymbol{n}\cdot\nabla\mu] = D[\boldsymbol{n}\cdot\nabla\Psi], \tag{8.4.15}$$

where $[\cdots] \equiv (\cdots)_{s>0} - (\cdots)_{s<0}$ is the discontinuity across the interface. We take $\psi \cong -\psi_{\rm eq}$ in the outward region $s \gtrsim \xi$ and $\psi \cong \psi_{\rm eq}$ in the inward region $s \lesssim -\xi$. With (8.4.6), (8.4.12), and (8.4.15) we have a closed set of dynamic equations for moving interfaces. Diffusion problems with moving boundaries, which are called the Stefan problems, are nonlinear and highly nontrivial.

A circular and spherical domain

For simplicity, let us consider an isolated circular (2D) or spherical (3D) droplet with radius R, within which $\psi \cong \psi_{\rm eq}$. We assume that ψ tends to $(-1+2\Delta)\psi_{\rm eq}$ far from the interface with Δ being a small positive supersaturation. The Gibbs–Thomson condition at the interface (8.4.11) yields the boundary values,

$$\mu_a = \left(\frac{\sigma}{2T\psi_{\rm eq}}\right)\frac{d-1}{R}, \quad \Psi_a = \left(\frac{\sigma}{4T\psi_{\rm eq}\kappa^2}\right)\frac{d-1}{R}. \tag{8.4.16}$$

Within the droplet μ is fixed at μ_a and ψ is given by

$$\psi \cong \psi_{\rm eq} + \Psi_a = \psi_{\rm eq}\left[1 + 2(d-1)\frac{d_0}{R}\right]. \tag{8.4.17}$$

In Fig. 8.21 we show a growing circular domain with radius $R = 8.16\kappa^{-1}$ in a 2D simulation, where $\psi/\psi_{\rm eq}$ tends to -0.96 far from the droplet and hence $\Delta = 0.02$. The critical radius, which we will discuss below, is given by $R_c = 2.36\kappa^{-1}$. In this simulation we have $\mu = \mu_a = 0.056\kappa^2\psi_{\rm eq}$ and $\psi/\psi_{\rm eq} = 1.027$ inside the droplet. These two values are consistent with (8.4.12) and (8.4.16). In fact, the Gibbs–Thomson relation (8.4.11) gives $\mu_a = 0.058\kappa^2\psi_{\rm eq}$ if the mean field expression for σ is used. Even if we prepare a droplet within which ψ considerably deviates from $\psi_{\rm eq}$ at $t = 0$, the two relations in (8.4.16) are soon satisfied after a transient time of order R^2/D. However, violation of the

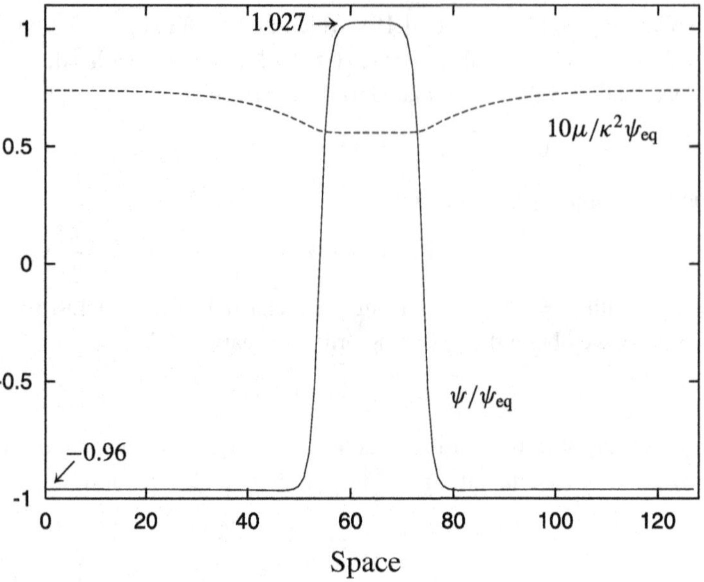

Fig. 8.21. The chemical potential μ and order parameter ψ for a growing circular solution of the model B equation without thermal noise, where $\Delta = 0.02$ and $R/R_c = 3.46$. The space is measured in units of $(1.5^{1/2}\kappa)^{-1}$. The Gibbs–Thomson relation (8.4.11) is excellently satisfied here.

Gibbs–Thomson relation becomes noticeable for large $\Delta \gtrsim 0.1$ because it holds only in the limit $\Delta \to 0$.

One-dimensional solution of the Stefan problem

The Stefan problem may be solved exactly for a one-dimensional case, where $\mathcal{K} = 0$ and an equilibrium phase with $\psi = \psi_{eq}$ expands upward into a metastable region, the interface position being at $x = x_{int}(t)$. We may envisage ice growth into metastable water from a boundary wall, where Ψ is the entropy (or temperature) deviation and $2\psi_{eq}$ in (8.4.15) corresponds to the latent heat. The boundary conditions for $\Psi = \psi + \psi_{eq}$ in the metastable region are

$$\Psi \to 0 \quad (x \to x_{int}), \quad \Psi \to 2\psi_{eq}\Delta \quad (x \to \infty). \tag{8.4.18}$$

To first order in Δ, the solution for $t > 0$ is given by

$$x_{int} = \frac{2\Delta}{\sqrt{\pi}}\sqrt{Dt}, \tag{8.4.19}$$

$$\Psi(x, t) = \frac{2\psi_{eq}\Delta}{\sqrt{\pi}} \int_0^X ds \exp\left(-\frac{1}{4}s^2\right), \tag{8.4.20}$$

where $X = (x - x_{\text{int}})/\sqrt{Dt}$. This exercise demonstrates that the interface velocity is slowed down with decreasing Δ and that the deformation of Ψ extends over the diffusion length,

$$\ell_D(t) = (Dt)^{1/2}. \tag{8.4.21}$$

In Appendix 8G, we will solve the Stefan problem for a circle in 2D and a sphere in 3D.

The Gibbs–Thomson condition in general

The boundary relation (8.4.12) is a special case of the famous Gibbs–Thomson condition. It can be derived in general statistical–mechanical contexts not necessarily close to the critical point. One notable example is crystal growth, in which the temperature at a crystal–melt interface is lowered by an amount proportional to $\mathcal{K}$ below the bulk melting temperature.

It is also straightforward to generalize (8.4.11) or (8.4.12) for the general form (4.4.15) of the free-energy density. To be specific, let us assume a generalization of model B [41],

$$\frac{\partial}{\partial t}\psi = \nabla L(\psi) \cdot \nabla \frac{\delta}{\delta \psi}\beta\mathcal{H}, \tag{8.4.22}$$

where the kinetic coefficient L may depend on ψ but the noise term is neglected. The chemical potential $\mu = \delta(\beta\mathcal{H})/\delta\psi$ is still continuous at the interface. Following the procedure which has led to (8.4.11), we obtain the surface value of μ in the form

$$\mu_a = (\sigma/T\Delta\psi)\mathcal{K}_a, \tag{8.4.23}$$

where $\Delta\psi = \psi_{\text{cx}}^{(1)} - \psi_{\text{cx}}^{(2)}$ is the difference of the order parameter values in the bulk two phases. The order parameter values extrapolated to the interface from the bulk regions are given by μ_a/χ_α ($\alpha = 1, 2$), where χ_α are the susceptibilities in the bulk. In the asymmetric case $\chi_1 \neq \chi_2$, the order parameter values Ψ_a extrapolated from the two sides are different.

8.4.2 The quasi-static approximation and scaling

We now make simple order estimations in the course of the domain growth. Let $\mathcal{K} \sim 1/\ell(t)$ and $v \sim d\ell(t)/dt \sim \ell(t)/t$ from the scaling and $\Psi_a \sim \psi_{\text{eq}}\xi/\ell(t)$ from (8.4.14). Then, (8.4.15) yields

$$\psi_{\text{eq}}\frac{d}{dt}\ell(t) \sim D\ell(t)^{-1}\Psi_a \quad \text{or} \quad \frac{d}{dt}\ell(t)^3 \sim D\xi. \tag{8.4.24}$$

Thus,

$$\ell(t) \sim (D\xi t)^{1/3} \sim (L\kappa t)^{1/3}. \tag{8.4.25}$$

If $\ell(t)$ is interpreted as the average droplet radius, the above relation also holds for off-critical quenches. The average droplet radius tends to obey (8.4.25), independently of ϕ, when $\ell_D(t)$ exceeds the distance among droplets. The coarsening in the limit of small ϕ [31] will be discussed in Chapter 8. Because $\ell(t)/\ell_D(t) = [\xi/\ell(t)]^{1/2}$, we notice that the domain size becomes shorter than the diffusion length at late stages. In such cases, we may assume the quasi-static condition,

$$\nabla^2 \Psi = 0. \tag{8.4.26}$$

To justify this equation, let us estimate the left-hand side of (8.4.6) as Ψ_a/t and the right-hand side as $D\Psi_a/\ell(t)^2$ near the interface with Ψ_a being given by (8.4.14); then, the ratio of the former to the latter is $\ell(t)^2/\ell_D(t)^2 \sim \xi/\ell(t) \ll 1$.

A spherical and circular domain

Around a spherical droplet in 3D the quasi-static condition (8.4.26) may be used to give

$$\Psi = \psi + \psi_{\text{eq}} = 2\psi_{\text{eq}}\Delta + (\Psi_a - 2\psi_{\text{eq}}\Delta)\frac{R}{r}. \tag{8.4.27}$$

As will be shown in Appendix 8G, this expression holds in the region $\xi \ll r - R \lesssim \ell_D$ only for $\Delta \ll 1$. Because the flux onto the droplet is $4\pi DR(2\psi_{\text{eq}}\Delta - \Psi_a)$, the evolution equation of R is obtained as [31]

$$\frac{\partial}{\partial t}R = D\left(\frac{\Delta}{R} - \frac{2d_0}{R^2}\right), \tag{8.4.28}$$

where the capillary length d_0 is defined by (8.4.13). For the generalized model (8.4.22), on the other hand, we should replace $2\psi_{\text{eq}}$ by $\Delta\psi$ in (8.4.27) and may suitably define Δ and d_0 as will be shown in Section 9.1. Then (8.4.28) can be used with $D = L(\psi_{\text{cx}}^{(2)})/\chi_2$ being the diffusion constant in the phase outside the droplet. In 2D, however, logarithmic corrections appear even close to the interface. As will be shown in Appendix 8G, we should replace R/r in (8.4.27) by $A\ln(r/R) + 1$ around a circular droplet with $A = 2/\ln\Delta^{-1}$ and modify the droplet evolution equation as

$$\frac{\partial}{\partial t}R = \frac{2D}{\ln(1/\Delta)}\left(\frac{\Delta}{R} - \frac{d_0}{R^2}\right), \tag{8.4.29}$$

which is valid for $\Delta \ll 1$. The critical radius in 2D and 3D is given by

$$R_{\text{c}} = \frac{d-1}{\Delta}d_0. \tag{8.4.30}$$

8.4.3 Chemical potential correlation and the Yeung relation

At a very late stage, where $\ell_D(t) \gtrsim \ell(t)$, the generalized chemical potential $\mu = \delta(\beta\mathcal{H})/\delta\psi$ varies gradually over the domain size $\ell(t)$. Conversely, it is sharply peaked in the interface regions in the nonconserved case. Let us consider the two-point correlation for the deviation $\delta\mu = \mu - \langle\mu\rangle$,

$$\begin{aligned} H(r, t_1, t_2) &= \langle\delta\mu(\mathbf{r}_1, t_1)\delta\mu(\mathbf{r}_2, t_2)\rangle \\ &= (\sigma/2T\psi_{\text{eq}})^2\ell(t_1)^{-2}H^*(r/\ell(t_1), t_2/t_1), \end{aligned} \tag{8.4.31}$$

where $r = |\boldsymbol{r}_1 - \boldsymbol{r}_2|$. The scaling relation assumed in the second line has been inferred from (8.4.11). The Fourier transformation in space yields

$$H_q(t_1, t_2) = (\sigma/2T\psi_{\rm eq})^2 \ell(t_1)^{d-2} H_Q^*(t_2/t_1), \tag{8.4.32}$$

where $Q = q\ell(t)$. Furukawa [29] examined the above correlation function in the limit $q \to 0$ numerically and found $\lim_{q\to 0} H_q(t_1, t_2)/H_q(t_1, t_1) \sim (t_2/t_1)^{0.5}$ for $t_2 < t_1$ in 2D. From $\partial\psi/\partial t = L\nabla^2\mu$ without thermal noise, the time derivative of the pair correlation function $g(r, t) = \langle\psi(\boldsymbol{r}_1, t)\psi(\boldsymbol{r}_1, t)\rangle$ (where $r = |\boldsymbol{r}_1 - \boldsymbol{r}_2|$) is written as

$$\frac{\partial}{\partial t} g(r, t) = 2L\nabla^2\langle\psi(\boldsymbol{r}_1, t)\mu(\boldsymbol{r}_2, t)\rangle = 2L^2\nabla^4 \int_0^t dt' H(r, t, t'), \tag{8.4.33}$$

where we may set $\langle\psi(\boldsymbol{r}_1, 0)\mu(\boldsymbol{r}_2, t)\rangle = 0$ in the scaling limit (or in the limit of small initial variance of ψ). If the second line of (8.4.31) is assumed, the scaling form $g(r, t) = \psi_{\rm eq}^2 G(r/\ell(t))$ holds only for $\ell(t) \propto t^{1/3}$. We define $\ell(t)$ by

$$\ell(t) = (L\sigma/2T\psi_{\rm eq}^2)^{1/3} t^{1/3}. \tag{8.4.34}$$

The Fourier transformation of (8.4.33) gives a desired relation between $F(Q)$ and $H_Q^*(s)$,

$$\left(d + Q\frac{\partial}{\partial Q}\right) F(Q) = 6Q^4 \int_0^1 ds H_Q^*(s). \tag{8.4.35}$$

Because $\lim_{Q\to 0} H_Q^*(s)$ is nonvanishing and finite [29], the above equation leads to the small-Q behavior (8.3.14) first derived by Yeung [37], which has been confirmed in simulations.

8.4.4 General solutions without thermal noise in 3D

Critical quench

Let us consider late-stage domain growth in the critical-quench ($M = 0$) case in 3D. In analogy with electrostatics, the surface boundary condition (8.4.15) may be interpreted as that of a *surface charge density* given by $\rho_a = -(2\psi_{\rm eq}/D)v_a$, where the surface velocity v in the normal direction at $\boldsymbol{r}_a$ is written as v_a. Notice that the symmetry between the two phases in the critical-quench case leads to the charge-neutrality condition, $\int da\rho_a = -(2\psi_{\rm eq}/D)\int dav_a = 0$. Then, using the 3D Green function,

$$G(\boldsymbol{r}, \boldsymbol{r}') = \frac{1}{4\pi|\boldsymbol{r} - \boldsymbol{r}'|}, \tag{8.4.36}$$

we may formally integrate (8.4.26) as

$$\Psi(\boldsymbol{r}, t) = -\frac{2\psi_{\rm eq}}{D} \int da' G(\boldsymbol{r}, \boldsymbol{r}_{a'}) v_{a'}. \tag{8.4.37}$$

The neutrality condition assures the convergence of the above surface integration at large distance, the screening length being $\ell(t)$ in (8.4.25). The Gibbs–Thomson condition (8.4.12) leads to a surface dynamic equation,

$$\int da' G(\boldsymbol{r}_a, \boldsymbol{r}_{a'}) v_{a'} = -(L\sigma/4T\psi_{\rm eq}^2)\mathcal{K}_a = -Dd_0\mathcal{K}_a, \tag{8.4.38}$$

where the length d_0 is defined by (8.4.13). Because the above equation is nonlocal, we formally define the inverse kernel $\Gamma_{aa'}$ [42] by

$$\int da'' \Gamma_{aa''} G(\boldsymbol{r}_{a''}, \boldsymbol{r}_{a'}) = \delta_{aa'}, \tag{8.4.39}$$

where $\delta_{aa'}$ is the δ-function on the surface (which satisfies $\int da' \delta_{aa'} \varphi_{a'} = \varphi_a$ for any φ_a). Then v_a is expressed as

$$v_a = -Dd_0 \int da' \Gamma_{aa'} \mathcal{K}_{a'}, \tag{8.4.40}$$

which is the counterpart of the Allen–Cahn equation (8.2.1). The nonlocality here, however, makes the problem much more complicated. The free energy in this case is equal to the surface energy as $\mathcal{H} = \sigma S(t) + \text{const}$. Its rate of change is

$$\frac{d}{dt}\mathcal{H} = -\sigma D d_0 \int da \int da' \mathcal{K}_a \Gamma_{aa'} \mathcal{K}_{a'} \le 0, \tag{8.4.41}$$

which cannot be positive because the kernel $G(\boldsymbol{r}_a, \boldsymbol{r}_{a'})$ and hence its inverse kernel $\Gamma_{aa'}$ are positive-definite. Coarsening thus occurs in order to lower the surface free energy at a late stage, where $\xi \ll \ell(t) \lesssim \ell_D(t)$.

Off-critical cases with small volume fraction

We consider the dilute case $\phi \ll 1$, in which droplets emerge in late-stage phase separation. The volume fraction $q(t)$ of the droplets slowly approaches ϕ. The free energy may be expressed in terms of the surface area $S(t)$ and the supersaturation $\Delta = \Delta(t) = \phi - q(t)$ as

$$\begin{aligned} \mathcal{H} &= \sigma S(t) + (4T\kappa^2\psi_{\rm eq}^2)\Delta^2 V \\ &= \sigma\left[S(t) + \frac{1}{2d_0}\Delta^2 V\right]. \end{aligned} \tag{8.4.42}$$

The second bulk term arises from the relation, $-\frac{1}{2}\kappa^2\psi^2 + \frac{1}{4}u_0\psi^4 \cong \kappa^2\Psi^2 \cong \kappa^2\langle\Psi\rangle^2$. Here we shift infinitesimally the surface $\boldsymbol{r}_a$ to $\boldsymbol{r}_a + \delta\zeta_a \boldsymbol{n}_a$. The subscript a is attached to all the quantities defined at $\boldsymbol{r}_a$. From the relation $\delta\Delta(t) = -\delta q(t) = -\int da\delta\zeta_a/V$, we find

$$\frac{\delta}{\delta\zeta_a}\mathcal{H} = \sigma\left(\mathcal{K}_a - \frac{\Delta}{d_0}\right). \tag{8.4.43}$$

Obviously, for a sphere with radius R, the above quantity vanishes for $R = R_{\rm c}$, $R_{\rm c}$ being the critical radius in (8.4.30).

When the diffusion length $\ell_D = (Dt)^{1/2}$ exceeds the average domain separation, we may set up the counterpart of (8.4.37) in the majority phase as

$$\Psi(\boldsymbol{r}, t) - 2\psi_{\rm eq}\Delta = -\frac{2\psi_{\rm eq}}{D}\int d\boldsymbol{r}' G(\boldsymbol{r}, \boldsymbol{r}')\big[v_{a'}\hat{\delta}(\boldsymbol{r}') + \dot{\Delta}\big], \qquad (8.4.44)$$

where $\hat{\delta}(\boldsymbol{r})$ is the surface δ-function defined by (8.2.53), and $\dot{\Delta} = \partial\Delta(t)/\partial t$. On both sides we have subtracted the space averages of Ψ and $v_a\hat{\delta}(\boldsymbol{r})$, because $\langle\Psi\rangle - 2\psi_{\rm eq}\Delta = 0$ from (8.4.4) and $\int da v_a/V + \dot{\Delta} = 0$ from the time derivative of (8.4.5). As $\boldsymbol{r} \to \boldsymbol{r}_a$, (8.4.12) holds and

$$\int d\boldsymbol{r}' G(\boldsymbol{r}_a, \boldsymbol{r}')\big[v_{a'}\hat{\delta}(\boldsymbol{r}') + \dot{\Delta}\big] = D(\Delta - d_0\mathcal{K}_a) = -\frac{L}{4\psi_{\rm eq}^2}\frac{\delta}{\delta\zeta_a}\beta\mathcal{H}, \qquad (8.4.45)$$

which is the counterpart of (8.4.38). The integration over $\boldsymbol{r}'$ should be cut off at a screening length $\ell_{\rm s}$, because domains far apart should not be correlated in their growth. More specifically, let us suppose an assembly of spheres with radii $R_i(t)$ at fixed positions $\boldsymbol{r}_i$, for which the above equation becomes [42, 43]

$$R_i\frac{\partial}{\partial t}R_i + \sum_{j\neq i,\ r_{ij}<\ell}\frac{1}{r_{ij}}R_j^2\frac{\partial}{\partial t}R_j + \frac{1}{2}\ell^2\dot{\Delta} = D\left(\Delta - \frac{2d_0}{R_i}\right), \qquad (8.4.46)$$

where $r_{ij} = |\boldsymbol{r}_i - \boldsymbol{r}_j|$ are the distances between the pairs i, j and the summation over the other spheres ($j \neq i$) is limited within a long distance cut-off ℓ. The last two terms on the left-hand side, if they are combined, should be independent of ℓ as long as $\ell > \ell_{\rm s}$. The equation without them is the starting point of the classic Lifshitz–Slyozov theory [31], which will be explained in Section 9.3. Note that the second term on the left-hand side multiplied by $-D^{-1}$ represents the fluctuation of the supersaturation seen by the sphere i and produces correlation in the droplet radii (not in the positions) in the space range $r_{ij} < \ell_{\rm s}$. It is known that this fluctuation gives rise to corrections of order $\phi^{1/2}$ to the Lifshitz–Slyozov growth law in the small-ϕ limit [44]–[46].

It is highly nontrivial how the screening length $\ell_{\rm s}$ is determined in the late stage where the diffusion length $\ell_D(t)$ exceeds the inter-domain distance $n_{\rm dom}^{-1/3} \sim \phi^{-1/3}R$ [5]. Here, $n_{\rm dom}$ is the domain density and the average radius R is determined from $4\pi R^3 n_{\rm dom}/3 = \phi$. If a sphere with radius R_0 dissolves, it results in an increase of the effective supersaturation of order $\delta\Delta \cong R_0^3/rDt_{\rm s} \sim R_0^3/r\ell_{\rm s}^2$ in its neighborhood with distance r less than $\ell_{\rm s}$, where $t_{\rm s} = \ell_{\rm s}^2/D$ is the duration time of the effect of dissolution. In this correlated region, spheres with $R > R_{\rm c}$ grow by $\delta R \sim Dt_{\rm s}\delta\Delta/R \sim R_0^3/rR$ within the time $t_{\rm s}$. Now we can determine $\ell_{\rm s}$ self-consistently by

$$(n_{\rm dom}\ell_{\rm s}^3)R^2(\delta R)_{r=\ell_{\rm s}} \sim R_0^3, \qquad (8.4.47)$$

where the left-hand side is the volume absorbed by the surrounding $n_{\rm dom}\ell_{\rm s}^3$ droplets in the correlated region. Supposing spheres with radii not much different from $R_{\rm c}(t) \sim \xi/\Delta(t)$, we set $R_0 \sim R \sim R_c$ and find

$$\ell_s \sim (n_{\rm dom} R)^{-1/2} \sim \phi^{-1/2} R, \tag{8.4.48}$$

$$t_s = \ell_s^2/D \sim t R_{c0}/R, \tag{8.4.49}$$

$$\delta R \sim \phi^{1/2} R, \tag{8.4.50}$$

where $R_{c0} \sim d_0/\phi$ is the initial critical radius, so $t_s \ll t$ and $\delta R \ll R$. The inequality, $n_{\rm dom}^{-1/3} < \ell_s < \ell_D$, follows. In the above arguments, however, we have examined the effect of a single dissolved droplet. We notice that many droplets may dissolve during the time t_s in the correlated region. Their number is estimated as

$$\delta N_{\rm dis} \sim \ell_s^3 \left| \frac{\partial n_{\rm dom}}{\partial t} \right| t_s \sim \phi^{-1/2} R_{c0}/R, \tag{8.4.51}$$

which is larger than 1 in the time region where $R/R_{c0} < \phi^{-1/2}$. The net growth of R during the time t_s is a superposition of contributions from $\delta N_{\rm dis}$ dissolved droplets:

$$(\delta R)_{\rm net} \sim \delta N_{\rm dis} \delta R \sim R_{c0}. \tag{8.4.52}$$

This increase is of order $R t_s/t$ from (8.4.49) and (8.4.50) and is consistent with the algebraic growth of R.

8.4.5 Langevin equation for surfaces

We now include the noise effect in the surface dynamic equation. It may be added to the formal solution (8.4.40) as

$$v_a = -\frac{L}{4\psi_{\rm eq}^2} \int da' \Gamma_{aa'} \frac{\partial}{\partial \zeta_{a'}} \beta \mathcal{H} + \theta_a. \tag{8.4.53}$$

where we have set $\dot{\Delta}(t) = 0$ for simplicity. The noise strength is determined from the fluctuation–dissipation relation,

$$\langle \theta_a(t) \theta_{a'}(t') \rangle = (L/2\psi_{\rm eq}^2) \Gamma_{aa'} \delta(t - t'). \tag{8.4.54}$$

This Langevin equation for the conserved case is the counterpart of (8.2.34) for the nonconserved case. A more systematic derivation of the noise term can be found in Ref. [42].

Undulations of a planar interface

We may use (8.4.53) to examine the dynamics of the surface displacement $\zeta(\boldsymbol{r}_\perp, t)$ superimposed on a planar interface at $z = 0$, where $\boldsymbol{r}_\perp = (x, y)$ is the position vector on the unperturbed surface $z = 0$. Because $\Gamma_{aa'}$ is not a usual function, it is more convenient to re-express (8.4.53) in terms of G in (8.4.36) as

$$\int d\boldsymbol{r}'_\perp G(\boldsymbol{r}_\perp, \boldsymbol{r}'_\perp) \frac{\partial}{\partial t} \zeta(\boldsymbol{r}'_\perp, t) = \frac{L\sigma}{4T\psi_{\rm eq}^2} \nabla_\perp^2 \zeta + \tilde{\theta} \tag{8.4.55}$$

with

$$\langle \tilde{\theta}(\boldsymbol{r}_\perp, t)\tilde{\theta}(\boldsymbol{r}'_\perp, t')\rangle = (L/2\psi_{\rm eq}^2)G(\boldsymbol{r}_\perp, \boldsymbol{r}'_\perp)\delta(t-t'). \tag{8.4.56}$$

The 2D Fourier transform of $G(\boldsymbol{r}_\perp, \boldsymbol{r}'_\perp) = G(|\boldsymbol{r}_\perp - \boldsymbol{r}'_\perp|)$ is

$$\int d\boldsymbol{r}_\perp \exp(i\boldsymbol{k}\cdot\boldsymbol{r}_\perp)G(|\boldsymbol{r}_\perp|) = \frac{1}{2k}, \tag{8.4.57}$$

where $k = |\boldsymbol{k}|$. Fourier transformation of the above equation yields

$$\frac{\partial}{\partial t}\zeta_k = -\frac{L\sigma}{2T\psi_{\rm eq}^2}k^3\zeta_k + \theta_k. \tag{8.4.58}$$

Here the noise term $\theta_k = 2k\tilde{\theta}_k$ satisfies

$$\langle \theta_k(t)\theta_q(t')\rangle = (L/\psi_{\rm eq}^2)k(2\pi)^2\delta^{(2)}(\boldsymbol{k}+\boldsymbol{q})\delta(t-t'), \tag{8.4.59}$$

where $\delta^{(2)}$ is the two-dimensional δ function. This assures the equilibrium distribution of the surface displacement (8.2.39). The relaxation rate of the surface undulations with wave number k is thus given by

$$\Gamma_k = (L\sigma/2T\psi_{\rm eq}^2)k^3 \sim L\kappa k^3. \tag{8.4.60}$$

As in the nonconserved case (8.2.41), the coarsening law (8.4.25) may be inferred from the above dispersion relation if we pick up the fluctuations with $k \sim 1/\ell(t)$ and set

$$\Gamma_k t \sim L\kappa t/\ell(t)^3 \sim 1. \tag{8.4.61}$$

Growth and diffusion of a spherical domain in 3D

In the dilute limit of droplets, we may consider a single droplet isolated from others. Due to its appearance, $\Delta(t)$ is slightly decreased from ϕ as $\Delta(t) = \phi - 4\pi R^3/3V$, and $\mathcal{H}$ in (8.4.42) increases by[4]

$$\mathcal{H}(R) = 4\pi\sigma\left(R^2 - \frac{\phi}{3d_0}R^3\right), \tag{8.4.62}$$

where the constant term and that of order ϕ^2 are omitted. This droplet free energy is of the same form as that in (8.2.42) if h there is replaced by $(4\psi_{\rm eq}\kappa^2)\phi$. The Langevin equation for R can also be written in the standard form (8.2.47) with the kinetic coefficient,

$$\mathcal{L}(R) = (L/16\pi\psi_{\rm eq}^2)R^{-3}. \tag{8.4.63}$$

Furthermore, we may examine the diffusion of a spherical domain by setting $\theta_a(t) = \sum_{\ell m}\theta_{\ell m}(t)Y_{\ell m}(\boldsymbol{n}_a)$ using the spherical harmonic functions and picking up the components with $\ell = 1$, $m = -1, 0, 1$. The random velocity of the center of mass is expressed as

[4] If the volume fraction of droplets $q(t) = \phi - \Delta(t)$ increases appreciably compared with ϕ, we should use $\Delta(t)$ in place of ϕ in (8.4.62), as will be discussed in Chapter 9.

(8.2.49) for the nonconserved case. The diffusion constant for model B in 3D may be readily calculated as [42]

$$D_{\rm B}(R) = \int_0^\infty dt \langle u_x(t) u_x(0) \rangle = (9L/16\pi \psi_{\rm eq}^2) R^{-3}. \tag{8.4.64}$$

This diffusion constant is again negligibly small, as in model A.

8.4.6 Interface dynamics in coupled systems

In model C, introduced in Section 5.3, the motion of a nonconserved order parameter ψ is slowed down by diffusion of a subsidiary conserved variable m [47]. This model may be used to describe order–disorder phase transitions in binary alloys near the tricritical point, as explained in Section 3.3. As another notable tricritical system, we mention ^{3}He–^{4}He mixtures [48]–[50], whose GLW hamiltonian is given in (4.2.15) or (4.2.22). In these systems, the conserved variable m can take different values, $m_{\rm e1}$ and $m_{\rm e2}$, in the two phases. Then, in late-stage phase ordering, the volume fraction of the disordered phase tends to a constant because of the conservation law, and diffusion of the conserved variable becomes the controlling factor of the coarsening. Analytic work on early-stage spinodal decomposition is straightforward but rather complicated [47, 48]. Such analysis in a simple case was given in Section 5.3. Moreover, numerical work has revealed some unique nonlinear effects [51]–[53]. In Fig. 8.22 we show typical phase-ordering patterns in model C [53]. The dynamic scaling for the structure factor was confirmed to hold both for the conserved and nonconserved variables in late-stage spinodal decomposition with the domain size growing as $t^{1/3}$ [51]. In experiments on ^{3}He–^{4}He mixtures [49, 50], the dynamic scaling behavior was indeed observed in the scattered light intensity, where the hydrodynamic interaction governs the domain growth at late stages. Furthermore, in binary alloys, phase ordering can be radically influenced by coupling to the elastic field. Such aspects will be treated in Section 10.3.

Although we will not discuss it in this book, a phase-field model similar to model C has been used to describe crystal growth in a metastable melt [54]. There, a nonconserved order parameter ψ, called a phase field, is equal to 0 in the liquid region and to 1 in the solid region, while a conserved variable m representing the entropy is related to ψ and the temperature T as $m = C_0 T_{\rm melt}^{-1}(T - T_{\rm melt} - L_{\rm lat}\psi)$. Here C_0 is the specific heat, $T_{\rm melt}$ is the melting temperature, and $L_{\rm lat}$ is the latent heat.

Model C near the tricritical point

We hereafter examine the diffusion-limited interface motion in model C, neglecting the noise terms, where a scalar nonconseved order parameter ψ and a scalar conserved variable m obey (5.3.3) and (5.3.13), respectively. The kinetic coefficients L and λ will be assumed to be constants, but our theory can readily be generalized to the case in which they differ

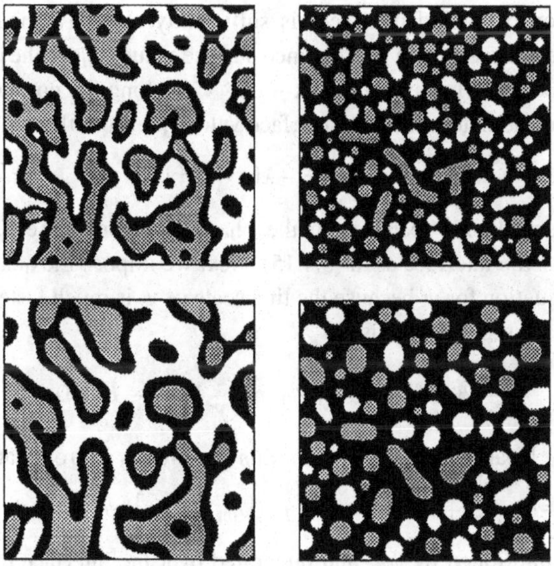

(a) (b)

Fig. 8.22. Typical time evolution of patterns in a model C system quenched into the order–disorder coexistence region [53]. As unique features, the disordered phase (black) forms a wetting layer that wraps the ordered domains of opposite sign (white or gray) in (a), and there are two kinds of domains (variants) in the ordered phase in (b). The times shown in the picture correspond to 150 and 450 from top to bottom in suitable units. The normalized concentration defined by $[2\bar{c} - (c_1 + c_2)]/(c_2 - c_1)$ is equal to $-1/3$ in (a) and $1/3$ in (b), where $\bar{c}$ is the average and c_1 and c_2 are those on the coexistence curve.

in the two phases. The GLW hamiltonian may be written as

$$\beta\mathcal{H}\{\psi, m\} = \int d\mathbf{r}\left[f(\psi) - h\psi + \frac{1}{2}|\nabla\psi|^2 + \frac{1}{2}C_0(\delta\hat{\tau})^2\right], \tag{8.4.65}$$

where a small ordering field h may be present and

$$\delta\hat{\tau} = \frac{\delta}{\delta m}\beta\mathcal{H} = C_0^{-1}m + \gamma_0\psi^2 - \tau. \tag{8.4.66}$$

The free-energy density $f(\psi)$ takes the form of (3.2.1) near a symmetrical tricritical point. The equilibrium interface profile $\psi = \psi_{\rm int}(s)$ and $m = m_{\rm int}(s)$ is obtained from minimization of $\mathcal{H}$ at $h = 0$. Then $m_{\rm int}$ is expressed as (8.2.61) in terms of $\psi_{\rm int}$, and the difference of m in the two phases is expressed as

$$\Delta m = m_{\rm e2} - m_{\rm e1} = -\gamma_0 C_0\left[(\psi_{\rm cx}^{(2)})^2 - (\psi_{\rm cx}^{(1)})^2\right], \tag{8.4.67}$$

where $\psi_{\rm cx}^{(\alpha)}$ $(\alpha = 1, 2)$ are the equilibrium values in two-phase coexistence.

When the motion of curved interfaces is sufficiently slow, $\delta\hat{\tau}$ should be continuous across the interface and assumes a well-defined surface value $\delta\hat{\tau}_a$, as the chemical potential $\mu = \delta\mathcal{H}/\delta\psi$ in model B. However, the flux $-\lambda\boldsymbol{n}\cdot\nabla\delta\hat{\tau}$ along the normal is discontinuous across the interface and determines the interface velocity v from the conservation law,

$$(\Delta m)v = -\lambda[\boldsymbol{n}\cdot\nabla\delta\hat{\tau}], \tag{8.4.68}$$

where $\boldsymbol{n}$ is the normal unit vector from the phase 1 to the phase 2, and $[\cdots]$ is the discontinuity across the interface as in (8.4.15). Next we impose the quasi-static condition on the evolution equation for ψ because the timescale of ψ is much faster than that of m at long wavelengths. This simply yields

$$\frac{\delta}{\delta\psi}\beta\mathcal{H} = f'(\psi) + a_0\delta\hat{\tau}\psi - \nabla^2\psi - h = 0, \tag{8.4.69}$$

where $a_0 = 2\gamma_0 C_0$. Near the surface point a, this equation is approximated as

$$-\mathcal{K}_a\psi'_{\text{int}} + a_0\delta\hat{\tau}_a\psi_{\text{int}} + \left[f''(\psi_{\text{int}}) - \nabla^2\right]\delta\psi - h = 0. \tag{8.4.70}$$

If this equation is multiplied by ψ'_{int} and integrated over the interface region $|s| \lesssim \xi$, the surface value of $\delta\hat{\tau}$ is determined as

$$\delta\hat{\tau}_a = -\frac{1}{\Delta m}\left[\frac{\sigma}{T}\mathcal{K}_a - (\Delta\psi)h\right], \tag{8.4.71}$$

where $\Delta\psi = \psi_{\text{cx}}^{(1)} - \psi_{\text{cx}}^{(2)}$. However, far from the interface we may neglect the gradient term ($\propto \nabla^2\psi$) in (8.4.69). Then the deviation $\delta\psi = \psi - \psi_{\text{cx}}^{(\alpha)}$ in the phase α is linearly related to h and $\delta\hat{\tau}$ as

$$\delta\psi = \chi_\alpha\left(h - \psi_{\text{cx}}^{(\alpha)}a_0\delta\hat{\tau}\right). \tag{8.4.72}$$

From (8.4.66) and (8.4.72) the deviation $\delta m = m - m_{e\alpha}$ is written as

$$\delta m = C_{0\alpha}\delta\hat{\tau} - a_0\psi_{\text{cx}}^{(\alpha)}\chi_\alpha h, \tag{8.4.73}$$

where $\chi_\alpha = 1/f''(\psi_{\text{cx}}^{(\alpha)})$ is the susceptibility of ψ and

$$C_{0\alpha} = C_0 + (a_0\psi_{\text{cx}}^{(\alpha)})^2\chi_\alpha. \tag{8.4.74}$$

If m is the entropy variable, $C_{0\alpha}$ has the meaning of the specific heat at $h = 0$. Far from the interface, $\delta\hat{\tau}$ (or δm) obeys the diffusion equation,

$$\frac{\partial}{\partial t}\delta\hat{\tau} = D_\alpha\nabla^2\delta\hat{\tau}, \tag{8.4.75}$$

where D_α is the diffusion constant in the phase α:

$$D_\alpha = \lambda/C_{0\alpha}. \tag{8.4.76}$$

This diffusion equation should be solved under the boundary condition (8.4.71) and the interface velocity is determined by (8.4.68). These equations are equivalent to those in model B except for the appearance of the ordering field h.

Droplet growth

As an application of the above relations, we consider a spherical droplet of the phase 1 growing into a metastable phase 2. The supersaturation Δ is defined by

$$\Delta = \frac{m_{e2} - m_\infty}{\Delta m} = -\frac{\delta\hat{\tau}_\infty}{C_{02}\Delta m}, \tag{8.4.77}$$

where $\delta\hat{\tau}_\infty$ and m_∞ are the values of $\delta\tau$ and m far from the droplet. Adopting the quasi-static condition $\nabla^2\delta\tau = 0$ outside the droplet, we obtain the counterpart of (8.4.28),

$$\frac{\partial}{\partial t}R = \frac{D_2}{R}\left[\Delta - d_{02}\left(\frac{2}{R} - \frac{T\Delta\psi}{\sigma}h\right)\right]. \tag{8.4.78}$$

The capillary length in the phase α is defined by

$$d_{0\alpha} = \frac{C_{0\alpha}\sigma}{(\Delta m)^2 T}. \tag{8.4.79}$$

We exchange the subscripts 1 and 2 if a droplet of the phase 2 is growing into a metastable phase 1. Near the symmetrical tricritical point, d_0 is of the order of the correlation length $\xi \propto |T - T_t|^{-1}$ from (3.2.24), (4.4.22), and (4.4.24).

Surface mode and crossover between the nonconserved and conserved cases

The above results are analogous to those in model B. However, when Δm is very small, they are valid only in the long-wavelength (or low-frequency) limit ($k \ll k_c$). For $k \gg k_c$ the conservation of m does not affect the interface motion. This crossover can be seen apparently in the surface dispersion relation. Some calculations show that the decay rate of a sinusoidal perturbation on a planar interface is written as

$$\Gamma_k = Lk^3/(k + k_c), \tag{8.4.80}$$

where k is the lateral wave number assumed to be much smaller than ξ^{-1} and

$$k_c = T(\Delta m)^2 L/2\sigma\lambda \tag{8.4.81}$$

is a crossover wave number. This decay rate coincides with the model A result (8.2.40) for $k \gg k_c$ and becomes $[2\sigma\lambda/T(\Delta m)^2]k^3$, analogous to the model B result (8.4.60), for $k \ll k_c$.

8.5 Hydrodynamic interaction in fluids

Hydrodynamic interaction plays a decisive role in the phase-separation dynamics of fluids. Representative systems are as follows. (i) Binary fluid mixtures composed of small molecules, such as isobutyric acid + water, are classic systems where light scattering experiments have been used to study asymptotic critical behavior and phase separation [55]–[60]. In such fluids, phase separation can be induced by an adiabatic pressure-quench method [56, 59] or a microwave-heating method (applicable to lutidine + water with an inverted, isobaric coexistence curve) [55]. The space and timescales of the emerging

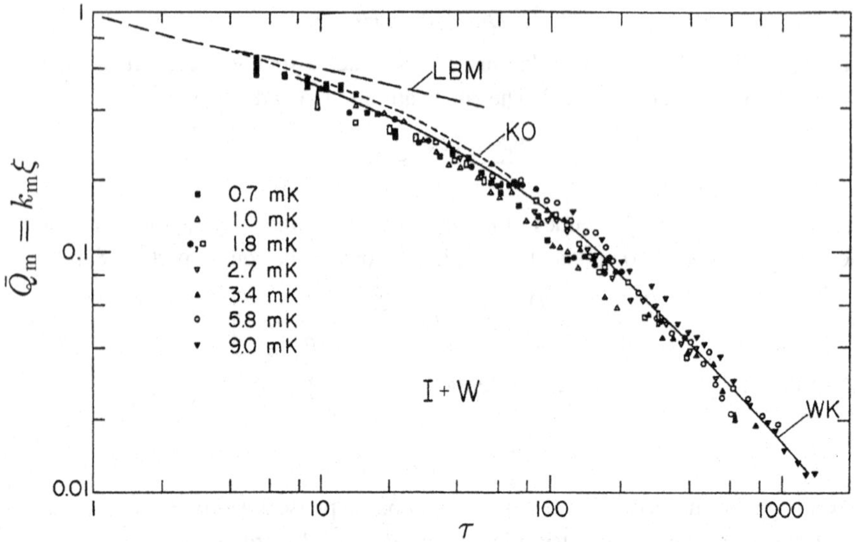

Fig. 8.23. The scaled peak wave number $\bar{Q}_m = k_m(t)\xi$ vs the scaled time $\tau = D\xi^{-2}t = (T/6\pi\eta)\xi^{-3}t$ for various quenches at the critical concentration in isobutyric acid + water (I+W) obtained by Chou and Goldburg [55], where $\xi \cong 2^{-1/2}\kappa^{-1}$. The curved solid line (WK) summarizes similar measurements on the same system by Wong and Knobler [56]. The broken and dashed lines are respectively the theoretical results of LBM [38] and of Kawasaki and Ohta (KO) [71].

concentration fluctuations can be of the order of a laser-light wavelength ($\sim 10^3$ Å) and minutes, respectively. (ii) In one-component fluids near the gas–liquid critical point, adiabatic changes occurring on the acoustic timescale can be used to induce phase separation. Here, the slow thermal diffusion controls the kinetics. As a result, the general feature of phase separation becomes very similar to that in usual binary fluid mixtures [61]. (iii) A great number of phase-separation experiments have also been performed on polymeric systems [33], [62]–[68]. In symmetric polymer blends, where constituent polymers have nearly identical molecular weights and viscoelastic properties, the hydrodynamic interaction eventually governs late-stage phase separation in the same manner as in usual binary fluid mixtures. In asymmetric polymer blends, however, viscoelastic effects unique to polymers can drastically influence phase separation, as will be discussed in Section 8.9.

We now present some representative experimental data. (i) In Fig. 8.23, the scaled peak wave number $\bar{Q}_m(\tau) = k_m(t)\xi$ is written as a function of the scaled time $\tau = (T/6\pi\eta\xi^3)t = D\xi^{-2}t$ at the critical composition in a near-critical mixture of isobutyric acid + water [55]. The growth exponent $a = -\partial \ln \bar{Q}_m/\partial \ln \tau$ is time dependent; $a \sim 0.3$ for $\bar{Q}_m \sim 0.3$, and $a \sim 1$ for $\bar{Q}_m < 0.1$. (ii) In Fig. 8.24 experimental results for the dimensionless wave number $Q_m^* = 2\pi\xi/\ell(t)$ are reported vs $\tau = D\xi^{-2}t$ for CO_2 and SF_6 in reduced gravity [61]. A decrease in the volume fraction ϕ of the gas phase to below 0.5 resulted in an interconnected morphology with $a \sim 1$ for $\phi > \phi_{hyd} \cong 0.29$ and

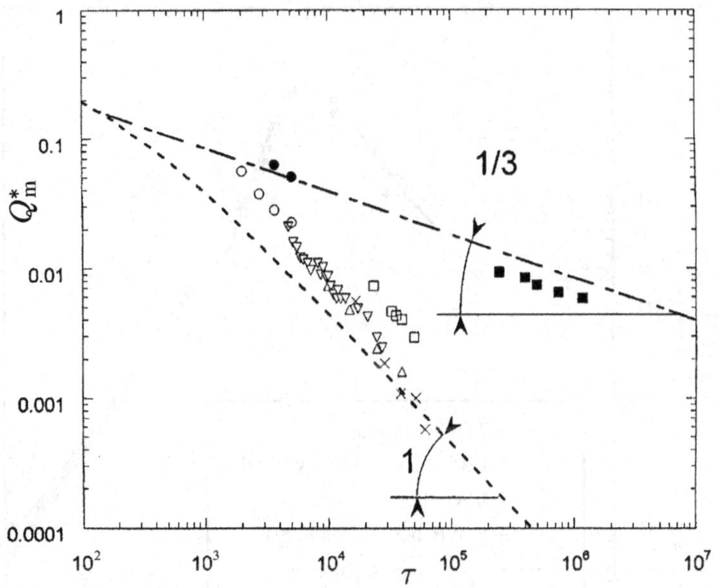

Fig. 8.24. The scaled wave number $Q^*_m = 2\pi\xi/\ell(t)$ vs $\tau = D\xi^{-2}t$ for CO_2 and SF_6 in reduced gravity [61]. The domain size $\ell(t)$ is obtained from video footage or photographs. The curves refer to an average of data obtained for binary fluid mixtures [55]–[58]. The open symbols (lower curve) correspond to interconnected-fast growth and the filled symbols (upper curve) correspond to disconnected-slow growth. The crossover between these two morphologies was found to occur when the volume fraction of the gas phase is about 0.29.

a disconnected morphology with $a \cong 1/3$ for $\phi < \phi_{\rm hyd}$. (iii) In Fig. 8.25, the scaled structure factor $F(Q)$ is written for isobutyric acid + water (I/W) [55, 56], lutidine + water (L/W) [55], and polybutadiene + polyisoprene [68]. These data demonstrate the universality of the domain morphology in fluids at late stages, in excellent agreement with 3D simulation results [36, 69, 70]. It is also worth noting that Hashimoto *et al.* took 3D images of bicontinuous domains in polymer blends using laser scanning confocal microscopy [68]. For example, from images at a very late stage of polybutadiene (50 vol%)+ polyisoprene (50 vol%), the method reproduced saddle-shaped surfaces with the statistical averages, $\langle R_1^{-1} + R_2^{-1}\rangle \cong 0$, $\langle (R_1^{-1} + R_2^{-1})^2\rangle = 8.8 \times 10^{-2}\ \mu\text{m}^2$, and $\langle (R_1 R_2)^{-1}\rangle = -6.2 \times 10^{-2}\ \mu\text{m}^2$ for the principal curvatures. These bicontinuous surfaces resemble minimal surfaces (where $R_1^{-1} + R_2^{-1} = 0$ is satisfied at each surface point), though there are considerable deviations.

8.5.1 The Kawasaki–Ohta theory

In the Stokes–Kawasaki approximation in Section 6.1, the velocity field is expressed as $\boldsymbol{v}_\psi + \boldsymbol{v}^{\rm R}$ as given by (6.1.47) and (6.1.48). Then the kinetic coefficient $L(\boldsymbol{r}, \boldsymbol{r}')$ becomes

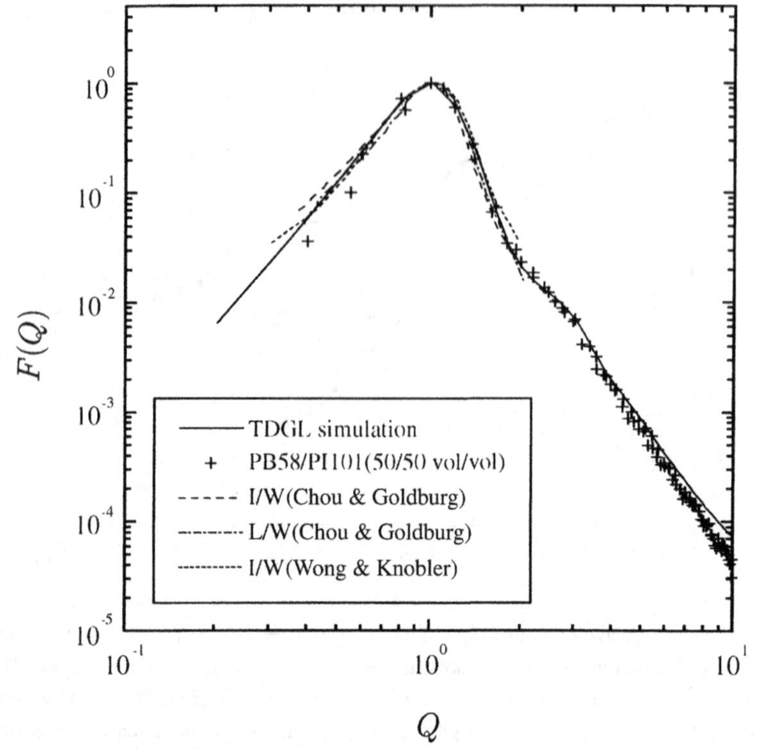

Fig. 8.25. The universal scaling function $F(Q)$ vs $Q = k/k_{\rm m}$ for binary fluids and a polymer blend at a late stage [68]. Simulation results are also shown (solid line). We can see $S(Q) \propto Q^{-4}$ for $Q \gg 1$ and $S(Q) \propto Q^4$ for $Q \ll 1$.

nonlocal and nonlinearly dependent on ψ as (6.1.53). For near-critical binary mixtures at the critical composition, Kawasaki and Ohta [71] used the LBM scheme for $J(k,t)$ as in (8.3.22) and decoupled the four-body correlation arising from the hydrodynamic interaction in the evolution equation of the intensity $I(k,t)$. The resultant equation reads

$$\begin{aligned}\frac{\partial}{\partial t}I(k,t) &= 2Lk^2\big[\kappa^2 A(t) - k^2\big]I(k,t) + 2Lk^2 \\ &\quad + 2\int_{\boldsymbol{q}} \boldsymbol{k}\cdot\overleftrightarrow{\mathcal{T}}_{\boldsymbol{k}-\boldsymbol{q}}\cdot\boldsymbol{k}\big[(q^2-k^2)I(k,t)I(q,t) - I(k,t) + I(q,t)\big],\end{aligned} \tag{8.5.1}$$

where $(\overleftrightarrow{\mathcal{T}}_{\boldsymbol{q}})_{\alpha\beta} = (T/\eta q^2)(\delta_{\alpha\beta} - q_\alpha q_\beta/q^2)$. The upper cut-off wave number of the fluctuations is taken as κ, so $L \sim T/6\pi\eta\kappa$. The viscosity η here is the renormalized one accounting for the fluctuation effect. They assumed that $A(t)$ relaxes as in the original LBM calculation in Fig. 8.19(b). As can be seen in Fig. 8.23, the last term in (8.5.1) arising from the hydrodynamic interaction considerably accelerates the coarsening. This theory turns out to agree with the experimental trend in near-critical binary mixtures in an intermediate time region, but is not applicable when the two phases are distinctly separated by sharp interfaces.

It should be noted that the size of the thermal fluctuations is very large in the asymptotic critical region of near-critical fluids. As a result, distinct domains in phase separation can be seen only when the domain size considerably exceeds ξ [72]. This is probably the reason why the Kawasaki–Ohta theory is valid for near-critical fluids over a sizable time region. In fact, in simulations without thermal noise [36, 69, 70], clear domain structures are established earlier than in the case with thermal noise, and the fast coarsening $a = 1$ soon becomes apparent for low-viscosity cases (see below).

8.5.2 Late-stage coarsening for critical quench

McMaster [63] and Siggia [73] argued that the coarsening of interconnected domain structures takes place with deformation and breakup of tube-like regions. The characteristic velocity field v_ℓ (around domains with sizes $\sim \ell$) is determined by the balance between the surface tension force density of order σ/ℓ and the viscous stress of order $6\pi\eta v_\ell/\ell$ as

$$v_\ell \sim 0.1\sigma/\eta. \tag{8.5.2}$$

The characteristic domain size $\ell(t)$ at time t may be estimated as

$$\ell(t) \sim v_\ell t \sim 0.1(\sigma/\eta)t. \tag{8.5.3}$$

In accord with this simple picture, experiments and simulations have shown that the peak wave number in the late stage is written as

$$k_{\mathrm{m}}(t) = 2\pi/\ell(t) \sim 10^2\eta/\sigma t \sim 10^2\kappa\tilde{\eta}/\tau, \tag{8.5.4}$$

where $\tau = D\xi^{-2}t$ is the dimensionless time and $\tilde{\eta}$ is a dimensionless viscosity defined by

$$\tilde{\eta} = \eta L u_0/T = \eta D/(2T\psi_{\mathrm{eq}}^2). \tag{8.5.5}$$

In particular, $\tilde{\eta}$ tends to a universal number of order 0.1 in the asymptotic critical region of near-critical fluids [42].

Tube-like regions may be regarded as aggregates of deformed spheres continuously growing into larger ones. At sufficiently high volume fractions $\phi_{\mathrm{hyd}} < \phi < 1 - \phi_{\mathrm{hyd}}$, such spheres have no time to be separated from others because new coalescence events take place before relaxation to spherical shapes. Based on this picture, Nikolaev *et al.* [74] estimated the threshold volume fraction ϕ_{hyd} to be 0.26 in agreement with the experimental value 0.29 [61]. This value is also consistent with 3D simulations of binary fluids described by the Boltzmann–Vlasov equations [75]. The interconnected patterns in fluids are thus maintained by the hydrodynamic flow produced by the surface motion. This suggests that the threshold volume fraction should be closer to 0.5 without hydrodynamics (in model B).

In the above hydrodynamic theory we have assumed low Reynolds numbers. On the scale of the domain size ℓ, the Reynolds number is estimated as

$$Re(\ell) = \rho v_\ell \ell/\eta \sim 0.1(\rho\sigma/\eta^2)\ell \sim 0.01(\rho\sigma^2/\eta^3)t. \tag{8.5.6}$$

The condition $Re(\ell) < 1$ yields $\ell < \ell_{\rm ina}$, where

$$\ell_{\rm ina} \sim 10\eta^2/\rho\sigma. \tag{8.5.7}$$

This upper bound length is very long near the critical point due to small σ and in polymer systems due to large η.

Interface dynamics

Let us use model H, as outlined in Section 6.1, to describe the interface dynamics during the late stage. A transverse velocity field can be induced around curved interfaces, where $\mu = \delta(\beta\mathcal{H})/\delta\psi$ assumes the surface value (8.4.11). From (6.1.11) the force density produced by the concentration fluctuations may be expressed near the interface as

$$-\psi\nabla\frac{\delta}{\delta\psi}\mathcal{H} \cong -\nabla\left(\psi\frac{\delta}{\delta\psi}\mathcal{H}\right) - \sigma\mathcal{K}_a\hat{\delta}(\boldsymbol{r})\boldsymbol{n}_a, \tag{8.5.8}$$

where $\hat{\delta}(\boldsymbol{r})$ is the δ-function on the surface defined by (8.2.53), and $\boldsymbol{n}_a$ is the normal unit vector. The first term on the right-hand side does not induce the transverse part of the velocity. The second term is valid on spatial scales longer than ξ and has been derived using the relations (8.2.20) and (8.4.11). The Stokes–Kawasaki approximation (setting $\partial\boldsymbol{v}/\partial t = \boldsymbol{0}$) gives the velocity field expressed in the following surface integration,

$$\boldsymbol{v}_\psi(\boldsymbol{r}, t) = -\int da' \overleftrightarrow{\mathcal{T}}(\boldsymbol{r} - \boldsymbol{r}_{a'}) \cdot \boldsymbol{n}_{a'}\sigma\mathcal{K}_{a'}, \tag{8.5.9}$$

where $\overleftrightarrow{\mathcal{T}}_{ij}(\boldsymbol{r}) = (8\pi\eta)^{-1}(\delta_{ij}r^{-1} + x_i x_j r^{-3})$ is the Oseen tensor in 3D, η being the renormalized viscosity for near-critical fluids. This velocity field is nonvanishing only when the domain shape deviates from sphericity. In fact, for a sphere placed at the origin, the second term in (8.5.8) is rewritten as $2\sigma R^{-1}\nabla\varepsilon(R - r)$, where $\varepsilon(x) = 1$ for $x > 0$ and $\varepsilon(x) = 0$ for $x < 0$, so it may be included in the pressure term. As will be shown in Appendix 8H, we can generally prove that both the velocity $\boldsymbol{v}_\psi$ and the velocity gradient $\nabla\boldsymbol{v}_\psi$ are continuous across the interface. Therefore, there is no discontinuity of the viscous stress tensor, while the pressure discontinuity is determined by the well-known Laplace law,

$$[p]_a = -\sigma\mathcal{K}_a. \tag{8.5.10}$$

For bicontinuous domain structures at very late stages, the interface velocity tends to the velocity field at the same interface position $\boldsymbol{r}_a$ [42]:

$$v_a = \boldsymbol{n}_a \cdot \boldsymbol{v}(\boldsymbol{r}_a, t) = -\int da'\left[\boldsymbol{n}_a \cdot \overleftrightarrow{\mathcal{T}}(\boldsymbol{r}_a - \boldsymbol{r}_{a'}) \cdot \boldsymbol{n}_{a'}\right]\sigma\mathcal{K}_{a'}. \tag{8.5.11}$$

In this approximation the diffusive current $-L\boldsymbol{n}_a \cdot \nabla\mu$ through the interface has been neglected. This holds for sufficiently large domain sizes, $\ell(t) \gg \tilde{\eta}^{1/2}\xi$ (for which see the comment below (8.5.13)). If we set $\mathcal{K}_a \sim \ell(t)^{-1}$, the typical magnitude of v_a becomes independent of t as $v_a \sim \sigma/\eta$. Therefore the typical domain size $\ell(t) \sim v_a t$ is known to grow as (8.5.3). We also note that, once (8.5.11) holds, the scaled structure factor $F(Q)$ should become universal, which is independent of $\tilde{\eta}$ in (8.5.5) and (probably weakly) dependent on ϕ (if ϕ is larger than $\phi_{\rm hyd}$).

Instability of a cylindrical domain

As a classic problem of hydrodynamics, it is well-known that axisymmetric perturbations superimposed on a long cylindrical domain grow to induce breakup of the cylinder into spherical droplets [76, 77]. In fact, if a long cylinder with radius a is divided into spheres with radius R, the total surface areas of the cylinder and spheres, S_0 and S, respectively, satisfy $S/S_0 = 3a/2R$ and the breakup decreases the total surface area for $R > 3a/2$.

The linear stability analysis is straightforward, in particular for the homogeneous viscosity case. Let the cylinder be along the z axis and the radius of a perturbed cylinder be written as $\tilde{a}(z) = a + \delta a + \zeta(z)$, where $\zeta(z)$ is a small perturbation with wave number k along the z axis and δa is a uniform radius change determined from the conservation of the cylinder volume, $\int dz\tilde{a}^2 = \text{const}$. Then the surface area $S = 2\pi \int dz[1 + (\partial\zeta/\partial z)^2]^{1/2}\tilde{a}$ changes by

$$\delta S = \pi \int dz\left[a\left(\frac{\partial}{\partial z}\zeta\right)^2 - \frac{1}{a}\zeta^2\right] = \frac{\pi}{a}(Q^2 - 1)\int dz\zeta^2, \tag{8.5.12}$$

which is negative for $Q = ka < 1$. For model H without thermal noise, the linear growth rate Ω stems from the concentration diffusion and the flow convection as [78]

$$\Omega = (1 - Q^2)\left(C_1\frac{L\kappa}{a^3} + C_2\frac{\sigma}{\eta a}\right), \tag{8.5.13}$$

where the dimensionless coefficients C_1 and C_2 depend on $Q = ka$ and are of order 1 for $Q \sim 1$. The first term ($\propto a^{-3}$) in the brackets is the model B result and can be important in systems with large η. The above expression indicates that the hydrodynamic interaction dominates over the diffusive processes for $a \gg \tilde{\eta}^{1/2}\xi$ with $\tilde{\eta}$ being defined by (8.5.5).

Polymer blends

In high-molecular-weight polymers the mean field critical behavior holds in statics and dynamics (except extremely close to the critical point), where u_0 remains at the mean field value in the Flory–Huggins theory and the kinetic coefficient L is determined from the Rouse or reptation theory. If the polymerization index N exceeds that $N_{\rm e}$ between entanglement points, the chains are entangled and the viscosity grows as $\eta \propto N^{z_\eta}$, where $z_\eta = 3$ follows from the reptation theory but $z_\eta = 3.4$ has been obtained experimentally, as discussed in Appendix 7A. Then we find that the dimensionless viscosity grows with increasing N as [69]

$$\tilde{\eta} \sim 10^{-2}(N/N_e)^{z_\eta - 2}. \tag{8.5.14}$$

In symmetric polymer blends exhibiting the mean field critical behavior, there will appear an intermediate time region in which the coarsening $\ell(t) \sim (D\xi t)^{1/3}$ holds. The crossover time τ^* in units of $(D\kappa^2)^{-1}$ to the appearance of hydrodynamic coarsening is then estimated as

$$\tau^* \sim (100\tilde{\eta})^{3/2}. \tag{8.5.15}$$

In entangled polymer blends, the reduced plot of $Q_m = k_m\xi$ vs $\tau = D\xi^{-2}t$ should therefore be nonuniversal; the larger N/N_e, the later is the appearance of the hydrodynamic regime. This is called the N-branching effect [79, 80].

8.5.3 Effect of the random velocity field

We may treat (8.5.11) as a Langevin equation by adding a noise term $\theta_a(t)$:

$$v_a = -\int da' \big[\boldsymbol{n}_a \cdot \overleftrightarrow{\mathcal{T}}(\boldsymbol{r}_a - \boldsymbol{r}_{a'}) \cdot \boldsymbol{n}_{a'} \big] \frac{\delta}{\delta \zeta_{a'}} \mathcal{H} + \theta_a, \tag{8.5.16}$$

with the fluctuation–dissipation relation,

$$\langle \theta_a(t)\theta_{a'}(t') \rangle = 2T \big[\boldsymbol{n}_a \cdot \overleftrightarrow{\mathcal{T}}(\boldsymbol{r}_a - \boldsymbol{r}_{a'}) \cdot \boldsymbol{n}_{a'} \big] \delta(t - t'). \tag{8.5.17}$$

The random noise term θ_a is determined by the surface value of the random velocity in (6.1.48) as

$$\theta_a(t) = \boldsymbol{n}_a \cdot \boldsymbol{v}^{\mathrm{R}}(\boldsymbol{r}_a, t). \tag{8.5.18}$$

Note that $\boldsymbol{v}^{\mathrm{R}}(\boldsymbol{r}, t)$ changes smoothly in space as can be seen from its integral form in terms of $\overleftrightarrow{\mathcal{T}}$. Then (6.1.51) yields (8.5.17). This Langevin equation can be used in the following examples.

Undulations of a planar interface

We set up the linear Langevin equation for the surface displacement $\zeta = \zeta(x, y, t)$ of a planar interface. As in (8.4.58) its Fourier component obeys

$$\frac{\partial}{\partial t}\zeta_{\boldsymbol{k}} = -\frac{\sigma}{4\eta} k \zeta_{\boldsymbol{k}} + \theta_{\boldsymbol{k}}, \tag{8.5.19}$$

where $k = |\boldsymbol{k}|$. The noise term $\theta_{\boldsymbol{k}}$ satisfies

$$\langle \theta_{\boldsymbol{k}}(t)\theta_{\boldsymbol{q}}(t') \rangle = (T/2\eta k)(2\pi)^2 \delta^{(2)}(\boldsymbol{k} + \boldsymbol{q}) \delta(t - t') \tag{8.5.20}$$

and assures the equilibrium distribution (8.2.39). The surface displacement is thus overdamped with the decay rate

$$\Gamma_k = (\sigma/4\eta)k. \tag{8.5.21}$$

Because the derivation is based on the Stokes–Kawasaki approximation, the above result is valid only when Γ_k is smaller than the viscous damping rate $\eta k^2/\rho$ or when $k \gg \rho\sigma/\eta^2$. However, the viscous damping is negligible at very small wave numbers in the region $k \ll \rho\sigma/\eta^2$, where the surface displacement oscillates as a capillary wave with the well-known dispersion relation,

$$\omega_k = (\sigma/2\rho)^{1/2} k^{3/2}. \tag{8.5.22}$$

Experimentally, the surface mode can be studied with inelastic light scattering from a surface [81]. The surface mode is well defined in the wave number region $k\xi \ll 1$ and its overall behavior can be examined with the Gibbs–Thomson relation (8.4.12), the continuity of the stress tensor, and $\partial\zeta/\partial t = v_z$ at the interface $z = \zeta \cong 0$ [81, 82]. It is worth noting that the growth law (8.5.3) can also be obtained if we set $k \sim \ell(t)^{-1}$ and $\Gamma_k \sim t^{-1}$ in (8.5.21).

For very viscous fluids with $\tilde{\eta} \gg 1$, the overdamped relaxation rate is the sum of the diffusive contribution (8.4.60) and the hydrodynamic one (8.5.21):

$$\Gamma_k = (L\sigma/2T\psi_{\rm eq}^2)k^3 + (\sigma/4\eta)k, \tag{8.5.23}$$

which is analogous to (8.5.13). The model B contribution ($\propto k^3$) arising from diffusion can be important in the intermediate wave number region $\tilde{\eta}^{-1/2}\kappa < k < \kappa$.

Diffusion of a droplet

If a spherical droplet is isolated from others and there is no average flow, the evolution equation of its radius $R(t)$ is the same as (8.4.28) for model B. However, its center of mass is convected by the random velocity field $\boldsymbol{v}^{\rm R}(t)$. As in models A and B, in model H the random velocity $\boldsymbol{u}(t)$ of the center of mass is expressed as

$$\boldsymbol{u}(t) = (3/4\pi R^2)\int da\theta_a \boldsymbol{n}_a, \tag{8.5.24}$$

where θ_a is defined by (8.5.18). The diffusion constant due to the random velocity field now reads

$$D_{\rm hyd}(R) = \int_0^\infty dt \langle u_x(t_0 + t)u_x(t_0)\rangle = \frac{T}{5\pi\eta R}. \tag{8.5.25}$$

As should be the case, this is the diffusion constant of a spherical emulsion droplet suspended in a fluid whose viscosity is the same as that inside the droplet [83].

8.5.4 Coalescence of droplets for off-critical quenches

Diffusing droplets with volumes $v = 4\pi R^3/3$ and $v' = 4\pi R'^3/3$ collide and fuse into a new droplet with volume $v + v'$. If ϕ is not too small, the characteristic droplet radius is simply determined by $D_{\rm hyd}(R)t \sim R^2$. This yields the growth law $R(t) \propto t^{1/3}$ [84] in agreement with experiments [56, 61]. Let us examine how the number density $n(v, t)$ of droplets with volume v evolves due to this mechanism [85]–[89]. The collision probability between droplets with volumes v and v' is written as $K(v, v')n(v, t)n(v', t)$ with [85, 87][5]

$$K(v, v') = 4\pi\left[D_{\rm hyd}(R) + D_{\rm hyd}(R')\right](R + R'). \tag{8.5.26}$$

Note that this collision kernel is estimated as $16\pi D_{\rm hyd}(R)R = 16T/5\eta$ for $R \sim R'$. Upon each collision, two droplets fuse into one, so the total droplet number density $n(t) = \int_0^\infty dv n(v, t)$ obeys

$$\frac{\partial}{\partial t}n(t) = -\frac{1}{2} \times 16\pi D_{\rm hyd}(R)Rn(t)^2. \tag{8.5.27}$$

which is integrated to give[6]

$$n(t)^{-1} = \frac{4\pi}{3\phi}\bar{R}(t)^3 = \frac{8T}{5\eta}(t + t_0), \tag{8.5.28}$$

where ϕ is the volume fraction and t_0 is related to the initial droplet number density by $n(0)^{-1} = (8T/5\eta)t_0$. For $t \gg t_0$ the average radius grows as

$$\bar{R}(t) = (6T\phi/5\pi\eta)^{1/3}t^{1/3}. \tag{8.5.29}$$

At very small volume fractions ($\phi \lesssim 0.03$), however, the diffusive collision of droplets becomes negligible and another mechanism of evaporation–condensation governs the coarsening, as will be discussed in Section 9.2.

It is possible to study the time evolution of the probability density $n(v, t)$ using the Smoluchowski equation [85]–[88],

$$\begin{aligned}\frac{\partial}{\partial t}n(v, t) &= -n(v, t)\int_0^\infty dv'K(v, v')n(v', t) \\ &\quad + \frac{1}{2}\int_0^v dv'K(v - v', v')n(v - v', t)n(v', t).\end{aligned} \tag{8.5.30}$$

This equation was originally constructed to describe coagulation of colloidal particles [85]. More generally, it has been used for coagulation processes in various situations if the collision kernel is appropriately redefined. Application to droplet growth in laminar and turbulent flow fields will be discussed in Section 11.1. In this evolution equation the volume fraction is fixed in time:

$$\int_0^\infty dv v n(v, t) = \phi, \tag{8.5.31}$$

which implies that the supersaturation $\Delta(t)$ is assumed to vanish. The total droplet number density $n(t)$ decreases monotonically in time as

$$\frac{\partial}{\partial t}n(t) = -\frac{1}{2}\int_0^\infty dv\int_0^\infty dv'K(v, v')n(v, t)n(v', t) < 0. \tag{8.5.32}$$

[5] The relative motion of two droplets with radii R and R' is described by the diffusion equation with $D = D_{\rm hyd}(R) + D_{\rm hyd}(R')$. One of them comes within the sphere with radius $R+R'$ enclosing the other one at a rate $4\pi D(R+R')n(v, t)$ in the quasi-static approximation (see the derivation of (8.4.28)).

[6] If the collision kernel (8.5.26) is used, the coagulation equation (8.5.30) gives $dn(t)^{-1}/dt = 1.07 \times 8T/5\eta$ numerically, which is very close to the approximate result (8.5.28) [86, 89].

We expect that $n(v,t)$ tends to the following scaling form at long times [86, 88],

$$n(v,t) = \frac{\phi}{\bar{v}(t)^2} n^*\left(\frac{v}{\bar{v}(t)}\right), \tag{8.5.33}$$

where $n^*(x)$ is a universal scaling function. This scaling holds only when $\bar{v}(t) \sim \bar{R}(t)^3 \sim t$ for the collision kernel (8.5.26).

In particular, if we set $K =$ const. as in (8.5.27), the Laplace transformation of (8.5.30) yields the equation for $f(x,t) = \int_0^\infty dv n(v,t) \exp(-xv)$,

$$\frac{\partial}{\partial t} f(x,t) = K\left[-f(0,t)f(x,t) + \frac{1}{2}f(x,t)^2\right]. \tag{8.5.34}$$

This equation is exactly solved in the form [85],

$$\frac{1}{f(x,t)} = \frac{t+t_0}{2t_0}\left[\frac{2(t+t_0)}{t_0 f(x,0)} - Kt\right]. \tag{8.5.35}$$

At long times $t \gg t_0$ the small-x ($\sim \bar{v}(t)^{-1}$) behavior is relevant, so we may set $f(x,0) = 2/Kt_0 - \phi x + O(x^2)$ to obtain $f(x,t)^{-1} = \phi^{-1}\bar{v}(t)[1+\bar{v}(t)x + O(x^2)]$. Thus we confirm the scaling (8.5.33) with simple results,

$$\bar{v}(t) = \frac{1}{2}K\phi(t+t_0), \quad n^*(y) = e^{-y}. \tag{8.5.36}$$

The first one is equivalent to (8.5.28).

8.5.5 Inertial regime and gravity effect

In the hydrodynamic regime for critical quenches, the Reynolds number on the scale of the domain size ℓ grows as (8.5.6). Eventually $Re(\ell)$ exceeds 1 for $\ell > \ell_{\rm ina}$, where $\ell_{\rm ina}$ is defined by (8.5.7). It follows a new inertial or turbulent regime, where the surface energy density σ/ℓ should be balanced with the kinetic energy density ρv_ℓ^2. By setting $\ell \sim v_\ell t$, we obtain a growth law [90, 91],

$$\ell(t) \sim (\sigma/\rho)^{1/3} t^{2/3}. \tag{8.5.37}$$

In this new regime the Reynolds number grows as

$$Re(\ell) \sim \rho\ell^2/\eta t \sim (\ell/\ell_{\rm ina})^{1/2} \propto t^{1/3}. \tag{8.5.38}$$

The growth law (8.5.37) can also be derived from the capillary-wave frequency (8.5.22) if we set $k \sim \ell^{-1}$ and $\omega_k t = 1$. This suggests that the surface deformations behave as (large-amplitude) capillary waves. A fraction of the surface free energy should then be transformed into the kinetic energy of eddies on the spatial scale of the domain size ℓ (which are the largest eddies). At high Reynolds numbers such eddies are broken into smaller ones. This cascade ends at the Kolmogorov dissipative length,

$$k_{\rm dis}^{-1} \sim \ell Re(\ell)^{-3/4} \sim \ell(\ell/\ell_{\rm ina})^{-3/8}. \tag{8.5.39}$$

Here we neglect intermittency of turbulence, for which see (11.1.70). Thus the condition $\ell \gg k_{\text{dis}}^{-1} \gg \ell_{\text{ina}}$ is needed to realize a well-defined turbulent regime. However, in near-critical fluids, $\sigma (\propto \kappa^2)$ is very small and the high Reynolds number condition is realized only at extremely late stages.

We also note that the gravity-dominated domain motion takes place for

$$\ell(t) > a_{\text{ca}} = [\sigma/g(\Delta\rho)]^{1/2}, \tag{8.5.40}$$

where a_{ca} is the capillary length in gravity introduced in (4.4.54). We have $a_{\text{ca}} \ll \ell_{\text{ina}}$ in most near-critical fluids. In the regime $a_{\text{ca}} < \ell < \ell_{\text{ina}}$, large-scale sedimentation flow accelerates the formation of macroscopic phase separation [63, 92]. In addition to experiments in space [58, 61], the gravity effect can be suppressed in a special isodensity fluid mixture (methanol + partially deuterated cyclohexane) in which the two phases have almost no mass-density difference [57]. However, even in such gravity-free experiments, there has been no indication of crossover into the inertial regime.

As will be discussed in Section 11.1, by applying laminar or turbulent shear to phase-separating fluids, we may stop spinodal decomposition at a time on the order of the inverse shear to realize dynamical steady states. For example, when two immiscible fluids with a significant surface tension are stirred (or sheared in microgravity), we will encounter a two-phase state with a high Reynolds number $Re(\ell) \gg 1$.

8.6 Spinodal decomposition and boiling in one-component fluids

In one-component fluids near the critical point, we will show that phase separation can be induced in the bulk region with the aid of the piston effect. As a new problem we will analyze boiling and condensation near a slightly heated or cooled boundary wall. However, molecular dynamics simulations of spinodal decomposition in fluids have been performed at constant temperature or energy under the periodic boundary condition [93]–[95].

8.6.1 Quench induced by the piston effect

Beysens *et al.* [61] realized spinodal decomposition in a near-critical liquid ($\rho > \rho_c$) in a cell with a fixed volume by slightly lowering the boundary temperature. Here the thermal diffusion layer is contracted and remains stable, whereas the interior region can be adiabatically expanded and cooled into a metastable or unstable state. Let T_{i} be the initial temperature and T_{bf} be the final boundary temperature. With this temperature change the pressure is decreased from the initial value p_{i} to $p(t) = p_{\text{i}} + (\partial p/\partial T)_n (T_{\text{bf}} - T_{\text{i}})[1 - F_a(t/t_1)]$, where t_1 is the quick relaxation time in (6.3.7) and $F_a(s)$ is the relaxation function in (6.3.9). For $t \gg t_1$ the pressure is decreased by $(\partial p/\partial T)_n (T_{\text{bf}} - T_{\text{i}})$. We stress that this pressure pinning is effective even during phase separation. The temperature in the interior region T_{in} is adiabatically changed from T_{i} to

$$T_{\text{in}}(t) = T_{\text{i}} + \left(\frac{\partial T}{\partial p}\right)_s [p(t) - p_i]. \tag{8.6.1}$$

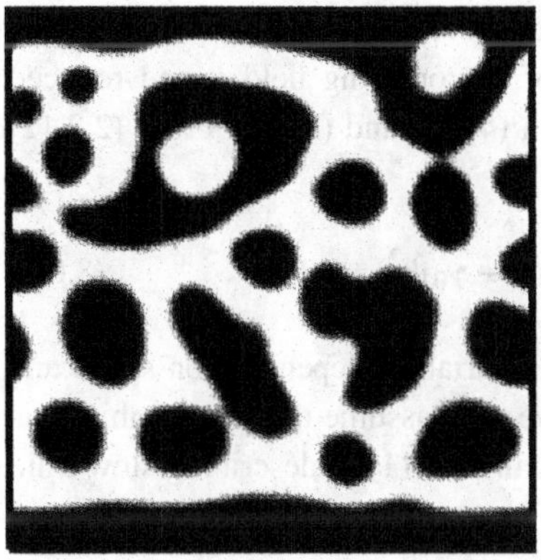

Fig. 8.26. Density deviations in spinodal decomposition in a one-component fluid in the absence of gravity at $t = 40$ after a change of the boundary temperature on a 128×128 lattice in 2D. The darkness is proportional to $\Psi(\boldsymbol{r}, t) \propto \rho(\boldsymbol{r}, t) - \rho_c$.

Because $(\partial p/\partial T)_s \cong (\partial p/\partial T)_{cx}$, the thermodynamic state in the interior region is shifted to be nearly along the coexistence line in the p–T phase diagram. The average density change $\psi_1 = \langle\psi\rangle_t - \langle\psi\rangle_0$ in the interior region is given by (6.2.65) and is much smaller than the density difference $2B(1 - T_{bf}/T_c)^\beta$ between liquid and gas, so it is negligible. The temperature coefficient $r = a_0\tau$ in (4.1.18) or (4.1.48) in the hamiltonian $\mathcal{H}$ is changed as

$$r = A_0(T_i/T_c - 1) \quad (t < 0), \quad r = -A_0(1 - T_{bf}/T_c) \quad (t \gg t_1), \tag{8.6.2}$$

in the interior region.[7]

Numerically solving dynamic equations, which will be presented below, we show a density pattern of spinodal decomposition induced by the piston effect in the gravity-free condition in Fig. 8.26. We can see the presence of wetting layers at the top and bottom. Similar two-phase patterns influenced by the boundary have been studied for model B [96] and for model H [97]. More explanations will follow.

8.6.2 Dynamic equations

The density n and entropy density s (or energy density e) in one-component fluids are related to the spin and energy variables, ψ and m, in the corresponding Ising system as in (2.2.7) and (2.2.9) (or (2.2.8)). In this section we neglect the mixing of the density and energy variables and set $\beta_1 = 0$ to analyze complex dynamics in the simplest manner, which means that the order parameter is simply the density deviation. In addition, we may set $\alpha_1 = \beta_s = 1$ without loss of generality. Then the mapping relationship near the critical point reads

$$\delta n = \psi, \quad n_c \delta s = \alpha_s \psi + m, \tag{8.6.3}$$

[7] To be precise, the coefficient A_0 depends on the reduced temperature due to the critical fluctuations.

where $\alpha_s = -(\partial p/\partial T)_{\text{cx}}/n_c$ is a negative constant. Note that the fluctuations of the pressure and temperature may be related to those of the ordering field h and reduced temperature τ in the corresponding Ising system as in (4.2.1) and (4.2.2). From (2.2.12) and (4.1.46) we may thus set[8]

$$(T - T_c)/T_c = \tau, \quad m = C_0(\tau - \gamma_0\psi^2). \tag{8.6.4}$$

In nonequilibrium, ψ and τ are fundamental dynamic variables dependent on space and time. Furthermore, in gravity along the z direction, we may assume that the combination $p_1(t) \equiv \delta p(\boldsymbol{r}, t) + \rho_c gz$ is a function of time only as in (6.2.11) in describing slow fluid motions. In a gravitational field, τ and ψ are related by

$$(n_c T_c)^{-1} p_1(t) + \alpha_s \tau = \frac{\delta}{\delta\psi}\beta\mathcal{H}_{\text{T}} = (a_0\tau - C_0\nabla^2 + u_0\psi^2)\psi + m_0 gz, \tag{8.6.5}$$

where $\mathcal{H}_{\text{T}}$ is given by (6.2.8) in gravity, $a_0 = 2\gamma_0 C_0$, C_0, and u_0 are the parameters in $\mathcal{H}$ in (4.1.45), together with (4.1.48) and (4.1.49), and m_0 is the particle mass. Notice that $p_1(t)$ is determined from the mass conservation relation $\int d\boldsymbol{r}\psi$ = const. in the fixed-volume condition. In fact, $p_1(t)$ is related to the space average of τ as $p_1(t) = T_c(\partial p/\partial T)_n\bar{\tau}$ for small disturbances in supercritical fluids, see Section 6.3.

The dynamics is governed by the heat conduction equation (6.2.10). From $n_c\delta s = \alpha_s\psi - \frac{1}{2}a_0\psi^2 + C_0\tau$ it is rewritten as

$$\begin{aligned}\left(\frac{\partial}{\partial t} + \boldsymbol{v}\cdot\nabla\right)\left(\alpha_s\psi - \frac{a_0}{2}\psi^2 + C_0\tau\right) &= \lambda_0\nabla^2\tau \\ &= \alpha_s L_0 \nabla^2(a_0\tau - C_0\nabla^2 + u_0\psi^2)\psi,\end{aligned} \tag{8.6.6}$$

where $L_0 = \lambda_0/\alpha_s^2$ is the kinetic coefficient, and the noise term is omitted. As discussed in Section 6.2, the velocity field may be assumed to be incompressible or $\nabla\cdot\boldsymbol{v} = 0$ on a long timescale. Then we may use the Stokes–Kawasaki approximation (6.1.46) to obtain

$$\eta_0\nabla^2\boldsymbol{v} = -\psi\nabla\big[(a_0\tau - C_0\nabla^2 + u_0\psi^2)\psi + m_0 gz\big] + \nabla\tilde{p}, \tag{8.6.7}$$

where $\tilde{p}$ ensures $\nabla\cdot\boldsymbol{v} = 0$. The above equation reduces to (6.4.9) if the first term on the right-hand side is set equal to $-m_0 g\psi\boldsymbol{e}_z$. Now (8.6.5)–(8.6.7) constitute a closed set of equations under the boundary conditions for τ and/or the heat flux $-\lambda_0 T_c\nabla\tau$.

To check relative magnitudes of the various terms in (8.6.6) and (8.6.7), we choose a reference reduced temperature $\tilde{\tau}$ and make the equations dimensionless by scale changes,

$$A = \tau/\tilde{\tau}, \quad \Psi = (u_0/\tilde{\tau})^{1/2}\psi, \quad \boldsymbol{V} = (\xi/D)\boldsymbol{v}, \tag{8.6.8}$$

[8] Here $\tau + \delta\hat{\tau}$ in Chapter 4 is rewritten as $\tau = \tau(\boldsymbol{r}, t)$.

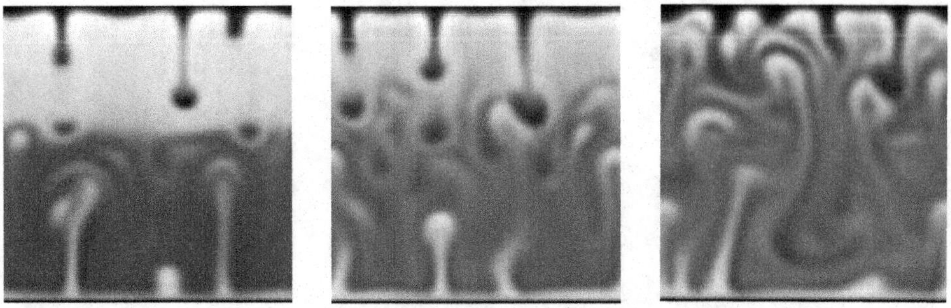

Fig. 8.27. Density deviations for $\langle\Psi\rangle = 0$ at $t = 200, 250$, and 5600 after increasing A_{bot} from -1 to 0, while A_{top} is kept at -1. Here A_{bot} and A_{top} are the boundary values of a scaled reduced temperature A.

where $D = L_0 a_0 \tilde{\tau}$. Space and time are measured in units of $\xi = \xi_{+0}\tilde{\tau}^{-\nu}$ and $t_\xi = \xi^2/D$ at $\tau = \tilde{\tau}$. We rewrite $\xi^{-1}\boldsymbol{r}$ and $t_\xi^{-1}t$ as $\boldsymbol{r}$ and t to avoid cumbersome notation. Then some calculations yield

$$A + a_c\delta_s\big[(A - \nabla^2 + \Psi^2)\Psi + Gz\big] = P(t), \tag{8.6.9}$$

$$\left(\frac{\partial}{\partial t} + \boldsymbol{V}\cdot\nabla\right)\left(\Psi + \frac{1}{2}a_c\delta_s\Psi^2 - \frac{\delta_s}{2a_c}A\right) = \nabla^2[A - \nabla^2 + \Psi^2]\Psi, \tag{8.6.10}$$

where $P(t) = p_1(t)/(|\alpha_s| n_c T_c \tilde{\tau})$ depends only on t, and a_c is the universal number close to 1 introduced in (2.2.37). The parameter δ_s is defined by

$$\delta_s = \gamma_s^{-1/2} = (C_V/C_p)^{1/2}, \tag{8.6.11}$$

where γ_s is the specific-heat ratio at $\tau = \tilde{\tau}$ on the critical isochore. Thus $\delta_s \sim \tilde{\tau}^{(\gamma-\alpha)/2} \ll 1$. The dimensionless gravitational acceleration is defined by

$$G = (m_0\xi/a_0\tilde{\tau})g \sim (\tau_g/\tilde{\tau})^{\beta\delta+\nu} \tag{8.6.12}$$

where τ_g was given by (2.2.48) with $\beta\delta + \nu \cong 2.2$. The dimensionless velocity field is determined by

$$\tilde{\eta}\nabla^2\boldsymbol{V} = -\Psi\nabla(A - \nabla^2 + \Psi^2)\Psi - G\Psi\boldsymbol{e}_z + \nabla\tilde{P}_{\text{inh}}, \tag{8.6.13}$$

where $\tilde{\eta}$ is defined by (8.5.5) and is of order 0.1 near the critical point, and $\tilde{P}_{\text{inh}}$ ensures $\nabla\cdot\boldsymbol{V} = 0$.

In the 2D simulation results in Figs 8.26–8.28, integrations are performed on a 128×128 lattice with the rigid boundaries at $z = 0$ and $z = L(= 128)$ but under the periodic boundary condition in the x (horizontal) axis. We set $a_c = 1$, $\delta_s = 0.1$ (or $\gamma_s = 100$), and $\tilde{\eta} = 0.2$. First we explain phase separation in the gravity-free case presented in Fig. 8.26. For $t < 0$, the system is in a one-phase state with $\langle\Psi\rangle = 0$ and $A = 1$. At $t = 0$,

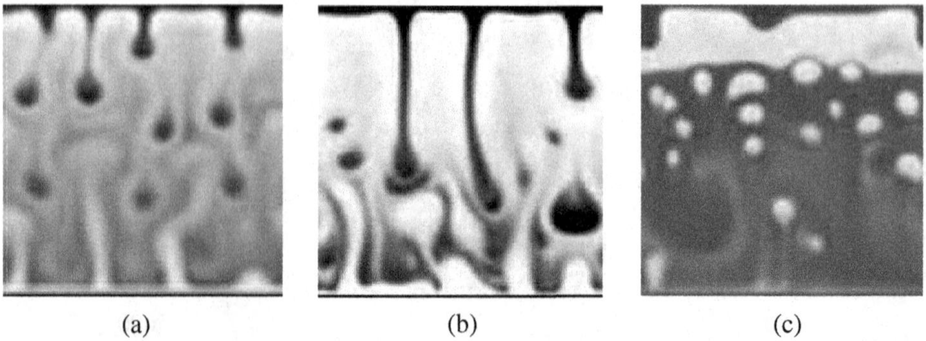

Fig. 8.28. Density deviations in dynamical steady states resulting from competition between gravity and heat flow for off-critical cases: $\langle\Psi\rangle = -0.2$ (gas-rich), $A_{\rm bot} = -1$, and $A_{\rm top} = 0$ in (a); $\langle\Psi\rangle = -0.4$ (gas-rich), $A_{\rm bot} = -2$, and $A_{\rm top} = 0$ in (b); $\langle\Psi\rangle = 0.4$ (liquid-rich), $A_{\rm bot} = -2$, and $A_{\rm top} = -1$ in (c).

the boundary values of A both at $z = 0$ and $z = L$, $A_{\rm bot}$ and $A_{\rm top}$, respectively, are decreased from 1 to -1. For $t \gtrsim t_1 = (L/\gamma_s)^2 \sim 1$ the piston effect is operative such that A approaches -1 throughout the system. The boundary values of Ψ at $z = 0$ and L are fixed at 2. This means that the boundaries are wetted by liquid in equilibrium [97].

8.6.3 Self-organized convention due to phase separation

Much more interesting are the phenomena of boiling and condensation in gravity, which occur after heating a liquid at the bottom or cooling a gas at the top. We show some numerical results by setting $G = 0.06$ and $\Psi = 1.2$ at $z = 0$ and $z = L$. (i) In Fig. 8.27 we initially prepare an equilibrium two-phase state with $\langle\Psi\rangle = 0$ and $A = -1$. At $t = 0$, $A_{\rm bot}$ is raised from -1 to 0 and is held constant thereafter, while $A_{\rm top}$ is kept at -1. For $t \gtrsim 30$, gas droplets emerge at the bottom and move upward. For $t \gtrsim 60$, liquid droplets also emerge, forming at the top and moving downward. These processes are initially *gentle*, but gradually become *violent* for $t \gtrsim 180$ with a decrease of the density difference between the upper and lower regions. For $t \gtrsim 250$, a dynamical steady state is eventually realized in the whole system with turbulent density and velocity disturbances (while the temperature disturbances are much smaller), as in the pattern at $t = 5600$. There, in the middle part of the cell, the gravity-induced density stratification is much reduced compared to that in equilibrium.[9] In these processes, heat transport is enhanced in the upward direction. The Nusselt number Nu here is about 5.5 in the final state, which is the ratio of the effective thermal conductivity in the dynamical steady state to that in the initial two-phase state (for infinitesimal heat flux). (ii) In Fig. 8.28 we show density patterns in dynamical steady states for off-critical cases. In (c) the system is relatively far from the critical point, where we can

[9] Gravity effects in stirred fluids will be discussed around (11.1.81).

see a usual picture of boiling in the lower liquid region. However, the patterns become quite unusual on approaching the critical point.

In Section 6.4 we discussed that thermal plumes move upward from the bottom when the applied temperature gradient exceeds the adiabatic temperature gradient a_g defined by (6.4.3). Similarly, gas droplets formed at the bottom move upward under the same condition. Thus, even if a fluid very close to the critical point initially consists of gas and liquid regions below T_c, convection sets in for $|dT/dz| > a_g$. In terms of the heat flux Q the condition of convection onset is also written as $Q > Q_c = \lambda a_g$, where λ is the (renormalized) thermal conductivity. A remarkable feature in the two-phase state is that heat can be transported very efficiently in the form of latent heat, where gas (liquid) droplets move upwards (downward) with positive (negative) excess entropy. As a result, in dynamical steady states with $Q > Q_c$, the temperature gradient in the middle part of the cell should be simply given by

$$(dT/dz)_{\text{middle}} \cong -a_g, \tag{8.6.14}$$

whereas the density profiles in the two-phase states are very complicated. However, the temperature gradient should become much steeper in the gas layer at the bottom and in the liquid layer at the top. Indeed (8.6.14) is excellently satisfied by the temperature profiles obtained in our simulations as long as $Q \gg Q_c$. Here the degree of phase separation is determined such that (8.6.14) is satisfied. The thickness of the boundary layers at the bottom and at the top is of the order of the capillary length a_{ca} in (4.4.54). the Nusselt number can then take a very large value of order $L/2a_{ca}$.

In summary, competition between gravity and heat flow from below produces intriguing self-organized states below T_c. Analogous self-organized heat transport is known in ^{4}He near the superfluid transition under gravity and heat flow applied from above. Note that (6.7.35) for normal fluid states and (8.10.57) for superfluid states are similar to (8.6.14).

8.7 Adiabatic spinodal decomposition

If the entropy in the ordered phase is lower than that in the disordered phase, the temperature generally rises due to internal entropy release in the course of phase ordering in the adiabatic condition. Similarly, in nearly incompressible binary mixtures, where the piston effect can be neglected, the temperature rises slowly with the progress of phase separation after a pressure quench [98]. We will develop the Ginzburg–Landau theory to account for this effect.

8.7.1 Entropy release in model C

As an illustrative example, we consider phase ordering in model C near the critical point for $h = 0$, in which a nonconserved order parameter ψ and a conserved variable m are coupled as in (5.3.3) and (5.3.13). If m is the energy density, the local reduced temperature deviation is written as

$$\delta\hat{\tau} = \frac{\delta}{\delta m}\beta\mathcal{H} = C_0^{-1}(m - m_0) + \gamma_0\psi^2. \tag{8.7.1}$$

[Note that m can be a concentration, as is usually the case in order–disorder phase transitions in solids.] Here we assume $m = m_0$ and $\psi = 0$ at the beginning of the phase ordering (neglecting the thermal fluctuations). If the system is thermally isolated or there is no flux of m from the boundary, the space average of m is fixed at the initial value m_0, so that $\int d\mathbf{r}(m - m_0) = 0$. Without noise and ordering field, we rewrite the dynamic equation for ψ as

$$\frac{\partial}{\partial t}\psi = -L[r_0 + a_0\delta\hat{\tau} - \nabla^2 + u_0\psi^2]\psi, \tag{8.7.2}$$

where $a_0 = 2\gamma_0 C_0$. The system is unstable for $r_0 < 0$ initially and the fluctuations of ψ are subsequently enhanced. However, with the development of domains, m becomes inhomogeneous around the interface as $m = -\gamma_0\psi^2 + \text{const}$. This is because $\delta\hat{\tau}$ tends to be homogeneous throughout the system at long times. Therefore, the average temperature deviation,

$$\delta\tau_1(t) = \langle\delta\hat{\tau}\rangle = \gamma_0\langle\psi^2\rangle, \tag{8.7.3}$$

starts from zero (or a small value in the presence of the thermal fluctuations) and increases with time. The effective reduced temperature deviation seen by the order parameter is given by $r(t)/a_0 = r_0/a_0 + \delta\tau_1$. As $t \to \infty$, $r(t)$ tends to r_∞ determined by

$$r_\infty = r_0 + a_0\gamma_0 \lim_{t\to\infty}\langle\psi^2\rangle = r_0 + a_0\gamma_0|r_\infty|/u_0, \tag{8.7.4}$$

where use has been made of $\psi^2 \to |r_\infty|/u_0$ as $t \to \infty$. Therefore, we find

$$r_\infty = r_0/(1 + X), \tag{8.7.5}$$

with

$$X = 2\gamma_0^2 C_0/u_0. \tag{8.7.6}$$

Here X is of order 1 in the asymptotic critical region (for which see the sentence below (8.7.15)).

8.7.2 Binary fluid mixtures after a pressure jump

In Section 6.5 we discussed adiabatic relaxations with fixed average entropy and concentration in near-critical binary fluid mixtures. To induce phase separation we change the pressure in a step-wise manner at $t = 0$ and keep it constant for $t > 0$ [56]. The average deviations ψ_1, m_1, and q_1 are related to the average density change ρ_1 as in (6.5.48). Then the average deviation $h_1 = \langle\delta(\beta\mathcal{H})/\delta\psi\rangle \sim \psi_1/\chi$ of the ordering field is negligible, leading to (6.5.50)–(6.5.52) even during phase separation. Here $\tau = (\partial\tau/\partial T)_{hp}(T - T_c)$ from (2.3.62) on the coexistence curve in the isobaric condition. As in (8.7.3) we define

$$\delta\tau_1(t) = \gamma_R(\langle\psi^2\rangle - M^2), \tag{8.7.7}$$

where the initial average order parameter $M = \langle\psi\rangle$ may be nonvanishing. We use the renormalized coefficients such as γ_R given by (4.1.59) for near-critical binary fluid mixtures in the asymptotic critical region. Then we have $\tau_1 = C_M^{-1} m_1 + \delta\tau_1$, where m_1 is related to ρ_1 as in (6.5.48) and C_M is the constant-magnetization specific heat in the corresponding Ising system. Substitution of this result into (6.5.50) and use of (6.5.33) yield

$$c^2 \rho_1 = p_1 - \left(\frac{\partial p}{\partial \tau}\right)_{h\zeta} \delta\tau_1. \tag{8.7.8}$$

Thus (6.5.52) becomes

$$T_1(t) = \left(\frac{\partial T}{\partial p}\right)_{sX} p_1 + \frac{\rho_c c_c^2}{\rho c^2}\left(\frac{\partial T}{\partial \tau}\right)_{hp} \delta\tau_1(t), \tag{8.7.9}$$

where use has been made of the second line of (2.3.46).

Let T_{ini} $(= T_i + (\partial T/\partial p)_s p_1)$ be the temperature in the initial time region where $\delta\tau_1 \cong 0$. Then the time-dependent average temperature can be expressed as

$$T(t) - T_c = (T_{ini} - T_c) + \frac{\rho_c c_c^2}{\rho c^2}\left(\frac{\partial T}{\partial \tau}\right)_{hp} \delta\tau_1(t). \tag{8.7.10}$$

As $t \to \infty$ we obtain

$$\delta\tau_1(\infty) = 4\phi(1-\phi)\gamma_R \psi_{eq}^2 = 2\beta^{-1} a_c^2 \phi(1-\phi)|\tau_f|, \tag{8.7.11}$$

similarly to (8.7.4). Here ϕ is defined by (8.3.1) and is the volume fraction at $t = \infty$, τ_f is the value of τ at $t = \infty$, and use has been made of (4.1.59). The universal number a_c defined by (2.2.37) satisfies $a_c^2 = (\beta\psi_{eq})^2/C_M\chi|\tau|$, for which see footnote 2 on p. 50. Because $\tau_f = (\partial\tau/\partial T)_{hp}(T_f - T_c)$, we obtain $T_f - T_c = (T_{ini} - T_c) + X|T_f - T_c|$, where

$$X = 2(a_c^2/\beta)\phi(1-\phi)\rho_c c_c^2/\rho c^2. \tag{8.7.12}$$

Therefore,

$$T_f - T_c = (T_{ini} - T_c)/(1+X). \tag{8.7.13}$$

The above X coincides with X in (8.7.6) in the mean field theory (where $a_c^2 = \gamma_0\psi_{eq}^2/|\tau|$) for $\phi = 1/2$ and $\rho_c c_c^2 \cong \rho c^2$. It is convenient to define

$$Z(t) = (\langle\psi^2\rangle - M^2)/[4\phi(1-\phi)\psi_{eq}^2], \tag{8.7.14}$$

where ψ_{eq} is the order parameter value in the final equilibrium state. Then $Z(t)$ grows from 0 to 1 and

$$T(t) - T_c = (T_{ini} - T_c)\left[1 - \frac{X}{1+X} Z(t)\right]. \tag{8.7.15}$$

When $\rho c^2 \cong \rho_c c_c^2$, we have $X \cong 1.5$ at the critical composition by setting $a_c = 1$ and $\beta = 1/3$. Donley and Langer [98] derived essentially equivalent results with X about 1 at $\phi = 1/2$ for 3-methylpentane + nitroethane. They calculated $Z(t)$ using the LBM scheme [38] applicable in the relatively early stage of spinodal decomposition. The resultant time evolution of the average temperature $T(t)$ is shown in Fig. 8.29. In the above theory we

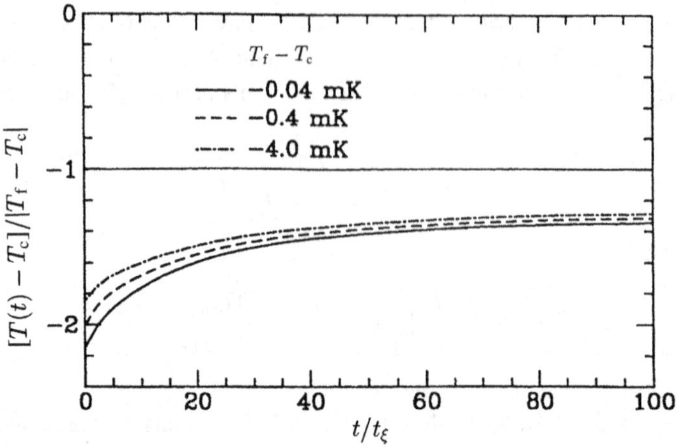

Fig. 8.29. Theoretical scaled temperature difference $[T(t) - T_c]/|T_f - T_c|$ as a function of the scaled time t/t_ξ where $t_\xi = 6\pi\eta\xi^3/T_c$ is the thermal relaxation time in the final state [98]. The horizontal line at -1 denotes the final equilibrium temperature differences.

have neglected the memory effect arising from the frequency-dependent bulk viscosity. In experiments in one-phase states [56, 99], the temperature and the scattered light intensity exhibited an overshoot as a function of time. These effects have not yet been explained.

8.8 Periodic spinodal decomposition

In Section 8.1 the effects of periodic temperature modulation were examined near the instability point in model A. Here we will consider periodic spinodal decomposition (PSD) in models B and H [40, 100] to show some new features different from those in normal spinodal decomposition (NSD). The physical processes involved are as follows. If the oscillation is sufficiently slow, domains can be formed periodically since phase separation proceeds during $T < T_c$. If the decay mechanism of domains which is effective during $T > T_c$ is strong enough, phase separation is stopped. In such a case, the system is in a one-phase state on length scales much longer than the characteristic domain size. However, if the average temperature $\bar{T}$ is lowered below a certain value T^*, domains are only partially dissipated during $T > T_c$ and continue to grow over successive periods. A salient feature is that the fluctuations at very long wavelengths are nearly constant in each period and evolve very slowly. Our theory for model B predicted $T^* < T_c$ [40, 100]. In experiments on a near-critical binary fluid mixture obeying model H [101], this dynamical phase transition was observed to be *continuous*, in contrast to the discontinuous dynamical phase transition in model A, and takes place for $\bar{T} < T^*$ with $T^* > T_c$ at critical quench. The fluctuation level in a periodically modulated one-phase state appeared to be higher than that in equilibrium at the critical point due to partial formation of domains.

8.8.1 Numerical analysis in the Langer–Bar-on–Miller scheme

Model B

We assume that the temperature coefficient $r = a_0(T - T_c)/T_c$ in the GLW hamiltonian oscillates as in the nonconserved case (8.1.46). Here we set $t_1 = t_2 = t_p/2$, where t_p is the period of the oscillation, and parameterize $r(t)$ as

$$\begin{aligned} r(t) - r_c &= r_1(\sigma - 1) \qquad \left(n < t/t_p < n + \frac{1}{2}\right) \\ &= r_1(\sigma + 1) \qquad \left(n + \frac{1}{2} < t/t_p < n + 1\right), \end{aligned} \tag{8.8.1}$$

where $n = 0, 1, 2, \ldots$, and r_c is the shift explained below (8.3.16), $r_1 = \kappa^2$ is the magnitude of the oscillation, and $r_1\sigma$ is the time average of $r(t)$. A critical quench ($M = 0$) will be assumed. We are interested in the case $|\sigma| < 1$ where the system is brought into stable and unstable temperature regions periodically. Strong fluctuation enhancement is expected when

$$\mu = Lr_1^2 t_p = L\kappa^4 t_p \tag{8.8.2}$$

is much larger than 1. In this case, if we observe only the first period, enhancement occurs in an intermediate wave number region,

$$\mu^{-1/2}\kappa < k < \kappa. \tag{8.8.3}$$

The long-wavelength fluctuations with $k < \mu^{-1/2}\kappa$ can be affected after several periods. The lower bound in (8.8.3) arises from the condition $Dk^2 t_p > 1$ with $D = L\kappa^2$.

In previous studies, we examined periodic spinodal decomposition within the LBM theory as a first nonlinear approach [40, 100]. The structure factor $I(k, t)$ then obeys (8.3.23) under the periodic temperature modulation (8.8.1). The time-dependent parameter $A(t)$ is defined by (8.3.22). One of our main results is that there is a critical value of $\sigma_c = \sigma_c(\mu)$, as shown in Fig. 8.30(a). For $\sigma > \sigma_c$ the system tends to a periodically modulated one-phase state, whereas for $\sigma < \sigma_c$ spinodal decomposition does not stop, ultimately resulting in macroscopic phase separation. The long-wavelength fluctuations with $k \ll \mu^{-1/2}\kappa$ experience only slow time evolution of $A(t)$. Hence we define the time-average of $A(t)$ in one period,

$$A_n = \frac{1}{t_p}\int_{nt_p}^{(n+1)t_p} dt A(t). \tag{8.8.4}$$

For $\sigma > \sigma_c$, a well-defined limit $A_\infty = \lim_{n\to\infty} A_n$ is attained, resulting in the limiting Ornstein–Zernike structure factor,

$$I_\infty(k, t) = \lim_{n\to\infty} I(k, nt_p + t) = 1/(\kappa^2 A_\infty + k^2), \quad (k \ll \mu^{-1/2}\kappa). \tag{8.8.5}$$

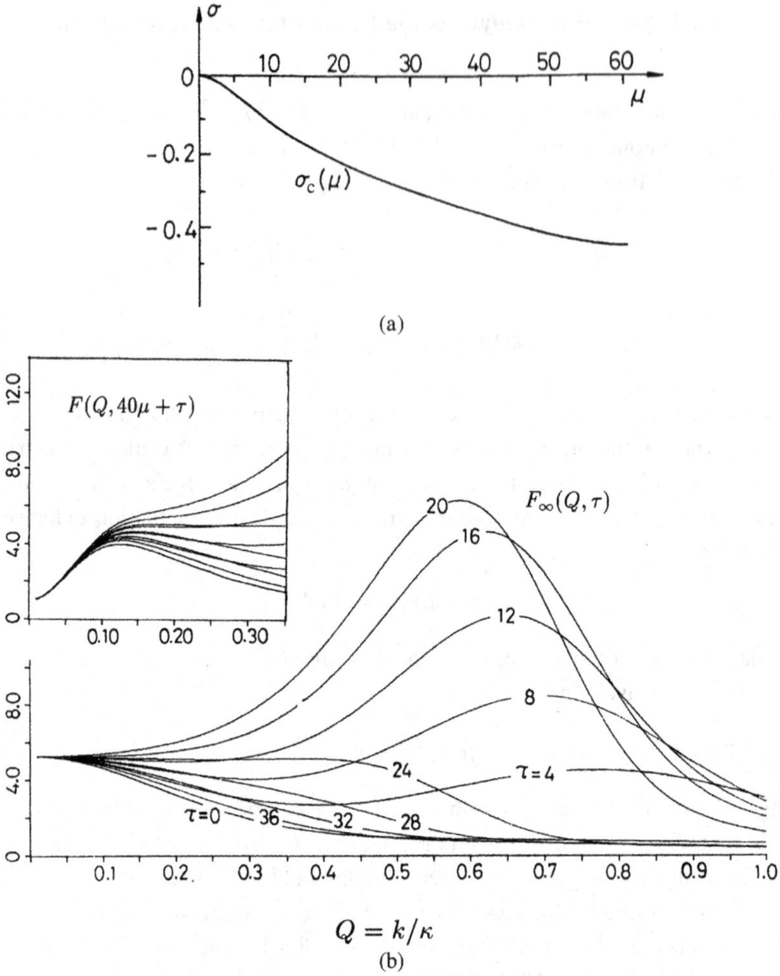

Fig. 8.30. (a) The critical value of $\sigma_c(\mu)$ as a function of μ at the critical composition obtained in the LBM scheme [100]. Phase separation occurs for $\sigma < \sigma_c(\mu)$, while the system remains in a disordered phase for $\sigma > \sigma_c(\mu)$. (b) The dimensionless structure factor $F_\infty(Q, \tau) = \lim_{n\to\infty} \kappa^2 I(k, nt_p + t)$ at various reduced times τ in one period ($0 < \tau < 40$) in a periodic disordered state with $\mu = 20$ and $\sigma = 0.174$, where $Q = k/\kappa$ and $\tau = 2L\kappa^4 t$ with $\kappa = r_1^{1/2}$ [40]. In the inset $F(Q, 40\mu + \tau) = \kappa^2 I(k, 20t_p + t)$ is shown.

We found $A_\infty \sim \sigma - \sigma_c$ for $\sigma > \sigma_c$ at each μ numerically. Thus the final correlation length grows as

$$\xi_\infty = \kappa^{-1} A_\infty^{-1/2} \propto (\sigma - \sigma_c)^{-1/2}, \tag{8.8.6}$$

as $\sigma \to \sigma_c$. In Fig. 8.30(b) we show $F_\infty(Q, \tau) = \kappa^2 I_\infty(k, t)$ in one period for $\mu = 20$ and $\sigma = 0.174$, where $Q = k/\kappa$ and $\tau = 2L\kappa^4 t$ with $\kappa = r_1^{1/2}$. For $Q \lesssim \mu^{-1/2}$ it is weakly dependent on τ and assumes the Ornstein–Zernike form (8.8.5), while for $\mu^{-1/2} \lesssim Q \lesssim 1$ it oscillates rapidly because of periodic formation and annihilation of domains. In the inset we also show $\kappa^2 I(k, 20t_p + t)$ after 20 periods. For $k/\kappa \gtrsim 0.2$ these two intensities coincide within a few percent. As a marked feature in this calculation, at finite t two peaks can emerge in the structure factor when the fluctuations in the intermediate wave numbers are enhanced.

Model H

We have also examined periodic spinodal decomposition for a critical quench on the basis of the Kawasaki–Ohta equation (8.5.1) [100]. We assume the same step-wise temperature oscillation with average $\bar{T}$ and amplitude T_1. Then we redefine μ as

$$\mu = t_p T \kappa^3 / 6\pi\eta, \tag{8.8.7}$$

where $\kappa = \xi_{+0}^{-1}(T_1/T_c)^\nu$ and η is the viscosity. In the one-phase region the intensity $I(k, t)$ again tends to the Ornstein–Zernike form, $\lim_{t\to\infty} I(k, t) = 1/(\kappa_\infty^2 + k^2)$, at long wavelengths $k \ll \mu^{-1/2}\kappa$. It is found that κ_∞^2 becomes much smaller than $A_\infty = \lim_{n\to\infty} A_n$ in model B due to the hydrodynamic interaction [40]. For example, $\kappa_\infty^2/\kappa^2 \cong 0.024$ and $A_\infty/\kappa^2 \cong 0.15$ at $\mu = 5$ and $\sigma = 0.174$. We find two peaks in $I(k, t)$ in some time regions. The hydrodynamic interaction increases the rate of the phase ordering and is crucial in PSD as well as in NSD.

8.8.2 Experiments of periodic quenches in fluids

In a PSD experiment on a binary fluid mixture of isobutyric acid + water [101], an oscillating temperature $T(t)$ was achieved by a step-wise pressure oscillation with $t_p = 1$ s. Its time-average and amplitude spanned the interval, -2.7 mK $\leq T_1\sigma = \bar{T} - T_c \leq 2$ mK and 3 mK $\leq T_1 \leq 10$ mK. Only a single peak was observed in the intensity, which diminished slowly with time. The critical value σ_c was positive and between 0.16 and 0.20. Conspicuous features are as follows. (i) In Fig. 8.31 we plot $k^2 I(k, t)$ in the one-phase region $\sigma > \sigma_c$. The limiting structure factor approaches a strongly enhanced intensity growing as

$$I(k, t) \propto 1/k^\phi, \quad (\phi \cong 2.6), \tag{8.8.8}$$

which is stronger than the equilibrium intensity at the critical point. (ii) For σ slightly smaller than σ_c, the timescale of the ring collapse (domain growth longer than the laser light wavelength) became exceedingly long. However, the peak wave number $k_m(t)$ has the same functional form as in NSD. In fact, the two sets of measurements of PSD and NSD could be mapped onto each other. To do so, Joshua *et al.* [101] rescaled k_m and t as $q_m = k_m \xi_{\text{eff}}$ and $\tau = (T/6\pi\xi_{\text{eff}}^3)t$ for PSD by introducing a new length ξ_{eff}, and thus found a mean field relation,

$$\xi_{\text{eff}} \propto (\sigma_c - \sigma)^{-\nu_{\text{eff}}} \quad (\nu_{\text{eff}} \cong 1/2). \tag{8.8.9}$$

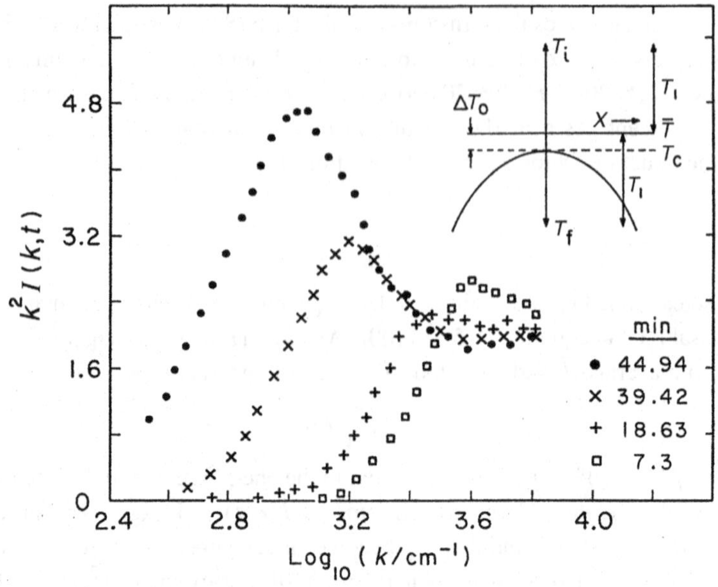

Fig. 8.31. Weighted angular distribution of scattering at various times for $\sigma > \sigma_c$ (disordered phase regime) [101]. The quench period is 1 s. The inset shows the coexistence curve in the temperature–composition plane and defines the quench parameters. This diagram corresponds to $\sigma = (\bar{T} - T_c)/T_1$.

Subsequently, Tanaka and Sigehuzi performed a PSD experiment on a polymer blend of ϵ-caprolactone (OCL) + styrene (OS) with molecular weights 2000 and 1000, respectively [102], where the timescale of phase separation was very slow even away from the critical point. The modulation could be made sinusoidal as $T(t) = \bar{T} + \Delta T \sin(2\pi t/t_p)$, where t_p and ΔT were fixed at 10 or 20 s and at 1 or 2 K, respectively. The average temperature $\bar{T}$ and the volume fraction ϕ were varied in the experiment. New findings were as follows. (i) Figure 8.32 demonstrates the presence of a two-level structure composed of an elementary structure and a large, growing superstructure at a late stage. The smaller domains are created and destroyed within each period and do not grow in time, while the larger ones grow continuously. The structure factor has two peaks, in contrast to the observation in Ref. [101]. This is probably because the unstable time interval was much longer in this experiment. Here coarsening of the larger domains occurs for $\bar{T} < T^*(\phi)$. (ii) They also examined the composition dependence of $T^*(\phi)$ to obtain a dynamic phase diagram. Most interesting is that $T^*(\phi) > T_{cx}(\phi)$ for bicontinuous domains and $T^*(\phi) < T_{cx}(\phi)$ for droplets. where $T_{cx}(\phi)$ is the equilibrium coexistence temperature.

8.9 Viscoelastic spinodal decomposition in polymers and gels

Using the reptation concepts, de Gennes [103] and Pincus [104] examined early-stage spinodal decomposition of symmetric polymer blends with equal molecular weights,

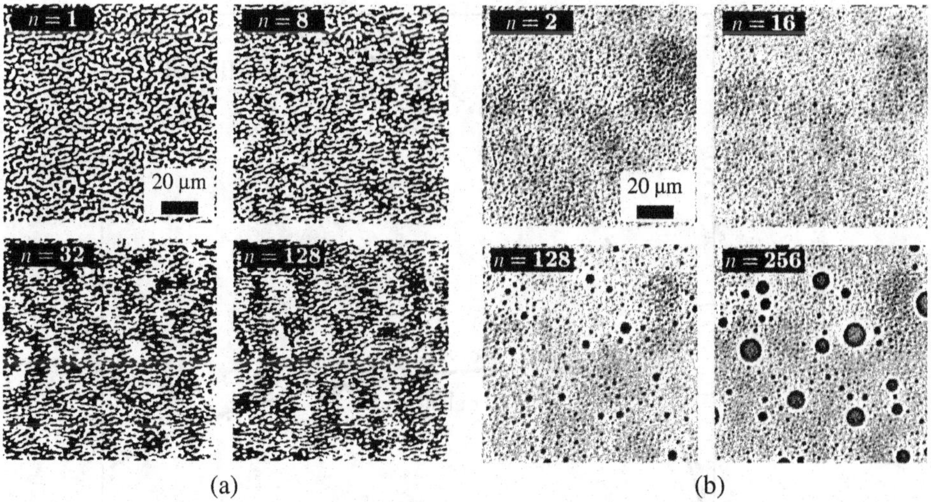

Fig. 8.32. Coarsening processes of PSD for $\bar{T} < T^*$ in a polymer blend [102]. (a) Bicontinuous in OCL/OS(35/65). The times after quenching are expressed as $t = nt_p + 9$ s with $t_p = 10$ s for various n. (b) Droplet patterns in OCL/OS(38.5/61.5) at $t = nt_p + 15$ s with $t_p = 20$ s.

$N_1 = N_2 = N$. If composition fluctuations with sizes longer than the gyration radius R_G are considered, these theories, as well as subsequent ones [105]–[109], predicted that the characteristic features of spinodal decomposition are nearly the same as those derived from simple dynamic models for usual low-molecular-weight fluids. A number of phase-separation experiments have also been performed on polymer solutions and blends [62]–[68], [110]–[118], where phase separation occurs on much slower timescales and much longer spatial scales than in usual binary fluid mixtures. In accord with the theories, if the space and time are appropriately scaled, most polymer systems studied behave like usual binary fluid mixtures. However, when the two components have distinctly different viscoelastic properties, unusual effects presumably ascribable to viscoelasticity have been detected. First, as was shown in Fig. 7.3, in early-stage spinodal decomposition of an asymmetric blend of PVME/d-PS, Schwahn *et al.* [115] found that the kinetic coefficient $L(q)$ depends on the wave number q as $L(q)/L(0) \sim q^{-2}$ even for q much smaller than the inverse of the gyration radius R_G, supporting the presence of the viscoelastic length ξ_{ve} ($\cong 7R_G$) in (7.1.68). More dramatically, Toyoda *et al.* [118] found $\xi_{ve} \sim 14R_G$ in very slow spinodal decomposition of a highly entangled 6% polystyrene in dioctyl phthalate (DOP) with molecular weight 5.5×10^6. In Fig. 8.33 their growth rate data are compared to the theory presented here. Second, as displayed in Fig. 8.34, Tanaka observed formation of sponge-like network structures composed of thin polymer-rich regions in late-stage spinodal decomposition of deeply quenched polymer solutions and asymmetric polymer blends [116, 117]. Such patterns were also reported in polymer solutions by other groups [111, 114].

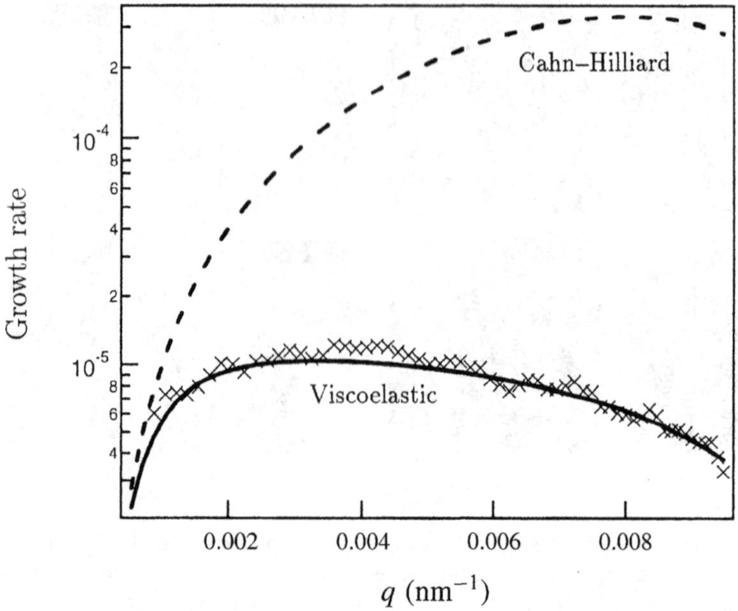

Fig. 8.33. Data for growth rate vs wavenumber (×) compared to the theoretical expression $LTq^2(|r| - Cq^2)/(1 + \xi_{ve}^2 q^2)$ in (8.9.13). The data were obtained in early-stage spinodal decomposition in a highly entangled polystyrene solution [120]. The broken line represents the usual Cahn–Hilliard form $LTq^2(|r| - Cq^2)$. We can see drastic slowing down of spinodal decomposition due to the viscoelastic effect.

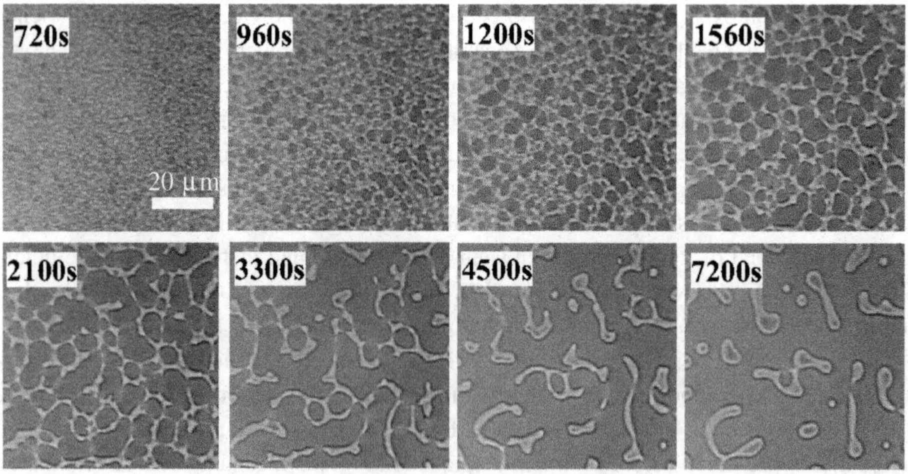

Fig. 8.34. Pattern evolution with time during phase separation of a PS/PVME mixture [117]. A network composed of more-viscous domains coarsens with time and ultimately breaks up into disconnected domains. The elapsed times after quenching are shown.

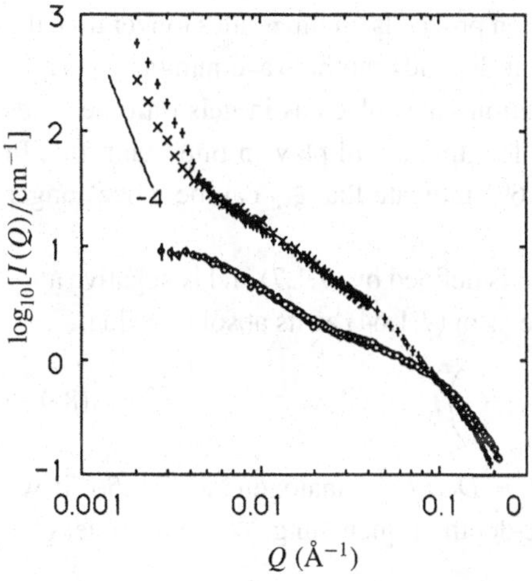

Fig. 8.35. Small-angle neutron scattering intensities from swollen gels in an isotropic state (∘), and in a uniaxially stretched state ($\times$, $I_{\parallel}(q)$; $+$, $I_{\perp}(q)$), in theta solvent at a common volume fraction [120].

Experimental reports on spinodal decomposition in gels are not abundant. However, it is often the case in experiments that, when a swollen gel is suddenly brought into an unstable temperature region, it instantly turns opaque without any appreciable volume change [119, 120]. This means that gels undergo spinodal decomposition with enhancement of small-scale fluctuations. As an example, Fig. 8.35 shows small-angle neutron scattering data from a swollen gel in theta solvent under uniaxial stretching $\lambda \sim 1.6$ [120]. Here the intensity $I_{\parallel}(q)$ in the stretched direction and that $I_{\perp}(q)$ in the perpendicular directions exhibit a Porod q^{-4} tail for $q < 0.01$ Å indicating the presence of domain structures. Because $I_{\perp}(q)/I_{\parallel}(q)$ is in excess of 2, (8A.11) suggests that the domains are elongated in the stretched directions. This behavior is consistent with the discussion below (7.2.16). Furthermore, the domain structures in gels are eventually pinned due to network elasticity. In a closely related effect, experiments have shown that the coarsening stops if crosslinks are introduced by gelation [121], chemical crosslinking reaction [122] or photo-crosslinking [123] in the course of phase separation. Theoretically, Sekimoto *et al.* demonstrated that a steady sponge-like domain structure is produced by elastic pinning [124] in a 2D microscopic network system. They also found elongation of domains under uniaxial compression. Similar results were recently reproduced from the Ginzburg–Landau model in Section 7.2 [125].

In this section we will examine early-stage viscoelastic spinodal decomposition on the basis of stress–diffusion coupling, and then we will present simulation results for polymer solutions and gels. We will defer analysis of viscoelastic nucleation to Section 9.6.

8.9.1 Early-stage viscoelastic spinodal decomposition

Using the notation of Section 7.1, we examine the initial exponential growth of the composition fluctuations in the unstable temperature region $r < 0$ in polymer solutions and blends on the basis of the Maxwell model equations (7.1.85)–(7.1.87) [126]. Our conclusions are as follows. For shallow quenching, phase separation proceeds on timescales longer than the stress-relaxation time τ and the kinetic coefficient depends on the wave number q as q^{-2} for $q\xi_{\rm ve} > 1$. For deep quenching, phase separation takes place as in gels on timescales shorter than τ. In the following, the viscoelastic length $\xi_{\rm ve}$ will play an important role. In asymmetric polymer systems, (7.1.65) and (7.1.69) indicate that $\xi_{\rm ve}$ can be much longer than the correlation length ξ.

In the present case the temperature coefficient r is defined by (7.1.7) and is negative after quenching below the spinodal curve. We redefine $\varepsilon_{\rm r}$ in (7.1.64) as its absolute value,

$$\varepsilon_{\rm r} = T|r|\Big/\left(\frac{4}{3}\alpha^2 G\right), \tag{8.9.1}$$

where G is the shear modulus. We also have $\varepsilon_{\rm r} = D_{\rm m}\tau/\xi_{\rm ve}^2$, analogous to (7.1.66) if we set $D_{\rm m} = LT|r|$. The parameter $\varepsilon_{\rm r}$ represents the depth of quenching. We measure lengths and frequencies in units of

$$\ell_{\rm e} = (3TC/4\alpha^2 G)^{1/2} = \varepsilon_{\rm r}^{1/2}\kappa^{-1}, \tag{8.9.2}$$

$$\Gamma_{\rm e} = LC\ell_{\rm e}^{-4}, \tag{8.9.3}$$

where $C = C(\phi)$ is given by (7.1.3) and $\kappa = (|r|/C)^{1/2}$ is the inverse correlation length. The parameter α, appearing in (7.1.86) and (7.1.87), represents the strength of the dynamical coupling between the composition and stress fluctuations. The stress relaxation rate $1/\tau$ and the growth rate $|\Omega_1|$ are scaled by $\Gamma_{\rm e}$ as

$$\gamma_{\rm ve} = (\Gamma_{\rm e}\tau)^{-1}, \quad R = |\Omega_1|/\Gamma_{\rm e}. \tag{8.9.4}$$

The viscoelastic length $\xi_{\rm ve}$ in (7.1.65) is related to $\gamma_{\rm ve}$ as

$$\xi_{\rm ve} = \gamma_{\rm ve}^{-1/2}\ell_{\rm e}. \tag{8.9.5}$$

We will assume

$$\gamma_{\rm ve} \ll 1 \quad \text{or} \quad \xi_{\rm ve} \gg \ell_{\rm e}, \tag{8.9.6}$$

under which the viscoelasticity can strongly affect phase separation. Because we use the Maxwell model, the equation $R = |\Omega_1|/\Gamma_{\rm e}$ follows from (7.1.90) in the form

$$R^2 + [\gamma_{\rm ve} + (1-\varepsilon_{\rm r})x + x^2]R = \gamma_{\rm ve}(\varepsilon_{\rm r} - x)x, \tag{8.9.7}$$

which depends on the wave number q through x defined by

$$x = (q\ell_{\rm e})^2. \tag{8.9.8}$$

A positive R is obtained only for $x < \varepsilon_{\rm r}$.

In semidilute polymer solutions near the coexistence curve, we estimate

$$\ell_{\rm e} \sim \xi \sim a/\phi, \quad \Gamma_{\rm e} \sim 1/\tau_{\rm b}, \quad \gamma_{\rm ve} \sim \tau_{\rm b}/\tau \sim \eta_0/\eta \ll 1, \tag{8.9.9}$$

where ξ is the thermal correlation length, a is the monomer size, ϕ is the polymer volume

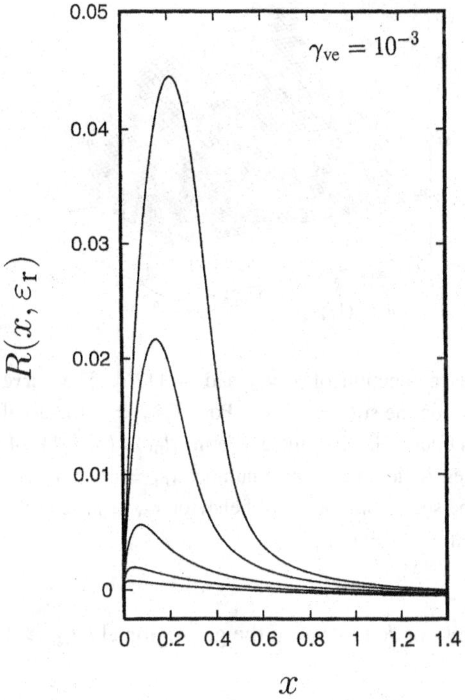

Fig. 8.36. The dimensionless growth rate $R(x, \varepsilon_r)$ vs $x = (q\ell_e)^2$ for several quench depths, $\varepsilon_r = 0.5, 0.75, 1.0, 1.25, 1.4$, from below at $\gamma_{ve} = 10^{-3}$ [126]. Here the length ℓ_e is defined by (8.9.2) and ε_r by (8.9.1), while γ_{ve} defined by (8.9.4) is the dimensionless stress-relaxation rate.

fraction, τ_b is the relaxation time within a blob defined by (7.1.28), η_0 is the solvent viscosity, and η is the solution viscosity behaving as (7.1.29). In polymer blends we consider the case in which the polymerization index N_1 of the first component is not much different from that, N_2, of the second component. Then $\xi_{ve} \sim L_t$ from (7.1.73) with L_t being the tube length in (7A.4). The reptation theory in Appendix 7A yields

$$\ell_e \sim d_t \sim N_e^{1/2} a, \quad \gamma_{ve} \sim (N_e/N_1)^2 \ll 1, \tag{8.9.10}$$

where N_e is the polymerization index between two consecutive entanglements on a chain.

Viscoelastic suppression of the growth rate

We will examine (8.9.7) and seek the maximum R_m of R attained at $x = x_m$. Then R_m and x_m are functions of ε_r and γ_{ve}. In the original units, the maximum growth rate q_m and the peak wave number Γ_m are expressed as

$$q_m = x_m^{1/2}/\ell_e = (x_m/\varepsilon_r)^{1/2}\kappa, \quad \Gamma_m = R_m \Gamma_e. \tag{8.9.11}$$

In Fig. 8.36 we plot R vs x for several ε_r at $\gamma_{ve} = 10^{-3}$. We recognize that R is much suppressed for $\varepsilon_r \lesssim 1$, compared to the usual case $R = (\varepsilon_r - x)x$ without viscoelasticity ($\gamma_{ve} = \infty$). We display R_m in Fig. 8.37(a) and x_m in Fig. 8.37(b) as functions of $1/\gamma_{ve}$ and ε_r. For $1/\gamma_{ve} \gg 1$ and $\varepsilon_r \lesssim 1$, they are much smaller than in the case $1/\gamma_{ve} \lesssim 1$. The

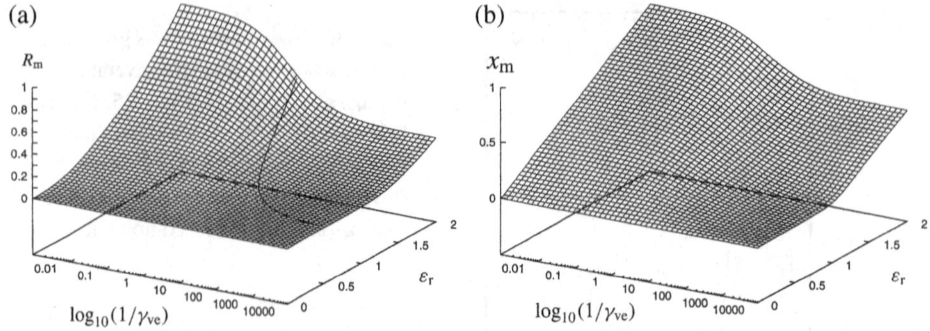

Fig. 8.37. (a) The maximum growth rate $R_{\rm m}$ as a function of $1/\gamma_{\rm ve}$ and $\varepsilon_{\rm r}$ [126]. The curve determined by $R_{\rm m} = \gamma_{\rm ve}$ (or $\Gamma_{\rm m} = 1/\tau$) is shown on the surface of $R_{\rm m}$. For $1/\gamma_{\rm ve} \ll 1$ the usual form $R_{\rm m} \cong \varepsilon_{\rm r}^2/4$ without the dynamic coupling is obtained, while the gel form, $R_{\rm m} \cong (\varepsilon_{\rm r} - 1)^2/4$, follows for $1/\gamma_{\rm ve} \gg 1$. (b) The square of the dimensionless peak wave number $x_{\rm m} = (q_{\rm m}\ell_{\rm e})^2$ as a function of $1/\gamma_{\rm ve}$ and $\varepsilon_{\rm r}$ [126]. A crossover can be seen from the usual behavior $x_{\rm m} \cong \varepsilon_{\rm r}/2$ to the gel behavior $x_{\rm m} \cong (\varepsilon_{\rm r} - 1)/2$ as $1/\gamma_{\rm ve}$ is increased.

growth rate can exceed the stress-relaxation rate or $R_{\rm m} > \gamma_{\rm ve}$ in the gel region, $1/\gamma_{\rm ve} \gg 1$ and $\varepsilon_{\rm r} \gtrsim 1$.

Shallow quenching: viscoelastic slowing-down

For shallow quenching $\varepsilon_{\rm r} \ll 1$, (8.9.7) gives

$$R \cong (\varepsilon_{\rm r} - x)x/(1 + \gamma_{\rm ve}^{-1}x). \tag{8.9.12}$$

The viscoelastic effect is to renormalize the kinetic coefficient as (7.1.67) and the growth rate in the original units reads

$$|\Omega_1| \cong LTq^2(|r| - Cq^2)/(1 + \xi_{\rm ve}^2 q^2). \tag{8.9.13}$$

The above form is in agreement with the experimental results in Figs 7.3 and 8.33 [118].

For very shallow quenching $\varepsilon_{\rm r} \ll \gamma_{\rm ve}$ the viscoelastic effect can be neglected, so that the peak position is $x_{\rm m} \cong \varepsilon_{\rm r}/2$ and the maximum of R is $R_{\rm m} \cong \varepsilon_{\rm r}^2/4$ as in the usual model B case. However, in the region $\gamma_{\rm ve} \ll \varepsilon_{\rm r} \ll 1$, the x dependence in the denominator of (8.9.12) is crucial and

$$x_{\rm m} \cong (\gamma_{\rm ve}\varepsilon_{\rm r})^{1/2}, \quad R_{\rm m} \cong \gamma_{\rm ve}\varepsilon_{\rm r}. \tag{8.9.14}$$

In the original units the peak wave number and the maximum growth rate are

$$q_{\rm m} \cong (\kappa/\xi_{\rm ve})^{1/2}, \quad \Gamma_{\rm m} \cong \varepsilon_{\rm r}/\tau. \tag{8.9.15}$$

Deep quenching: gel-like spinodal decomposition

If ε_r slightly exceeds 1 (if $\varepsilon_r - 1 \gg \gamma_{ve}^{1/3}$ more precisely), spinodal decomposition takes place as in gels in the early stage, where in accord with (7.1.60) we obtain

$$R \cong (\varepsilon_r - 1 - x)x. \tag{8.9.16}$$

Therefore,

$$x_m \cong \frac{1}{2}(\varepsilon_r - 1), \quad R_m \cong \frac{1}{4}(\varepsilon_r - 1)^2, \tag{8.9.17}$$

which are rewritten in the original units as

$$q_m \cong \frac{1}{\sqrt{2}}(\varepsilon_r - 1)^{1/2}\ell_e^{-1}, \quad \Gamma_m \cong \frac{1}{4}(\varepsilon_r - 1)^2\Gamma_e. \tag{8.9.18}$$

The growth rate in the region $\varepsilon_r - 1 < x < \varepsilon_r$ is negligibly small for $\gamma_{ve} \ll 1$. We notice that Γ_m soon exceeds $1/\tau$ for $\varepsilon_r - 1 > 2\gamma_{ve}^{1/2}$. Therefore, if τ is very long, the observed spinodal point will appear to be shifted downwards to the gel spinodal point $\varepsilon_r = 1$ (or $D_{gel} = 0$), while the true spinodal point for finite τ remains at $\varepsilon_r = 0$ (or $D_m = 0$).

8.9.2 Simulation of spinodal decomposition in polymer solutions

In Tanaka's experiments on deeply quenched semidilute polymer solutions [116], slovent-rich domains appeared at an early stage after an incubation time and grew until polymer-rich regions became thin enough to form a sponge-like network. The solvent regions were droplets enclosed by the network even if their volume fraction was considerably larger than that of the network. To explain these observations, the viscoelastic Ginzburg–Landau model of polymer solutions in (7.1.98)–(7.1.105) was numerically solved in 2D [127, 128]. We here demonstrate that a sponge-like network can appear for slow relaxation of $\overleftrightarrow{W}$ (for large τ in (7.1.100)), where the viscoelastic stress largely cancels the stress due to the surface tension and stabilizes the network structure for a long time.

We integrate (7.1.13) for ϕ and (7.1.107) for $\overleftrightarrow{W}$ on a 256×256 lattice under the periodic boundary condition, where $\boldsymbol{w}$ is given by (7.1.104) and $\boldsymbol{v}$ is calculated from (7.1.105), so the average polymer velocity $\boldsymbol{v}_p \cong \boldsymbol{v} + \boldsymbol{w}$ is expressed in terms of ϕ and $\overleftrightarrow{W}$. For $t > 0$ the system is unstable at $(1 - 2\chi)/\phi_c = 4.25$ as can be seen in Fig. 3.12, for which $\phi/\phi_c = 5.86$ in the polymer-rich phase and $\phi/\phi_c = 0.0026$ in the solvent-rich phase on the coexistence curve. Hereafter $\phi_c = N^{-1/2}$ is the critical volume fraction. In terms of the correlation length ξ and the cooperative diffusion constant D_m (see Section 7.1) in the final polymer-rich phase, space and time in Fig. 8.38 are measured in units of $\ell = 0.81\xi$, and $\tau_0 = 1.16\xi^2/D_m$, respectively. The solvent viscosity is set equal to $\eta_0 = \zeta a^2/18\phi^2$ consistent with (7.1.26). The shear modulus and the stress-relaxation time are set equal to $G = 0.2(T/v_0)\phi^3$ and $\tau = 0.1\tau_0[(\phi/\phi_c)^3 + 1]$, respectively. Because of the small coefficients (0.2 and 0.1 in G and τ), the viscoelasticity does not affect the patterns appreciably for $t \lesssim 100$ in our simulation, but it comes into play at later times within polymer-rich regions.

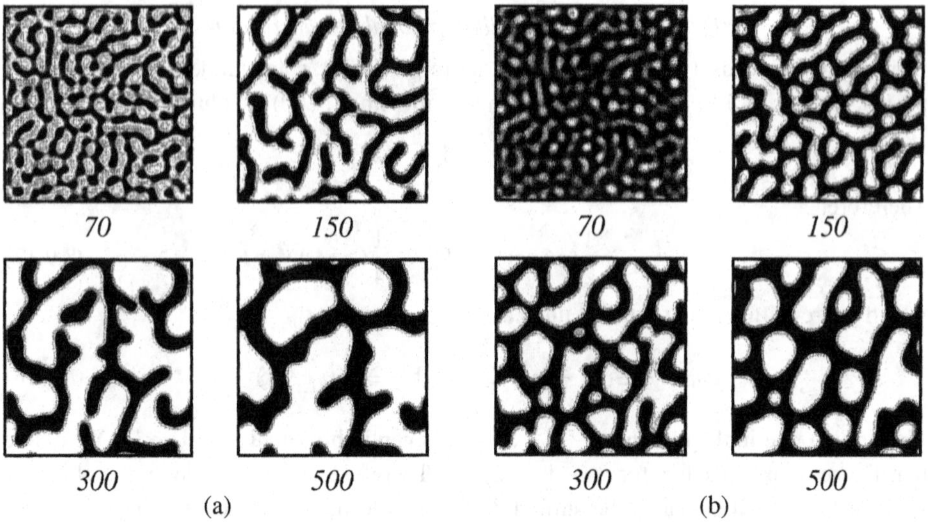

Fig. 8.38. (a) Patterns of $\phi(x, y, t)$ at $\langle \phi/\phi_c \rangle = 2$ [127]. This is the case close to the boundary between the droplet and network morphologies. (b) Patterns of $\phi(x, y, t)$ at $\langle \phi/\phi_c \rangle = 2.5$. See text for units of space and time.

In Fig. 8.38 we display patterns at $\langle \phi/\phi_c \rangle = 2$ in (a) and those at $\langle \phi/\phi_c \rangle = 2.5$ in (b), which closely resemble those of Tanaka shown in Fig. 8.34. In particular, at $\langle \phi/\phi_c \rangle = 2.5$ the interface line density $L(t)$ behaves as follows. (i) In the presence of viscoelasticity we obtain $t^{-\alpha}$ with $\alpha \sim 1/3$ for $t \gtrsim 200$. (ii) Without viscoelasticity or for the Flory–Huggins free energy with hydrodynamic interaction, the velocity field quickens the growth as $L(t) \propto t^{-2/3}$ in the region $100 \lesssim t \lesssim 400$. In this case, however, solvent droplet shapes tend to be circular for $t \gtrsim 400$ and a crossover to the droplet growth law $L(t) \propto t^{-1/3}$ appears to take place at later times. Therefore, the hydrodynamic interaction is suppressed in the presence of viscoelasticity. To support this result, we observe that the network in Fig. 8.38 does not move as a whole and $\boldsymbol{v}$ must be suppressed on longer spatial scales.

Next, we explain why polymer-rich domains do not change their elongated shapes, even after long times, in the presence of viscoelasticity [127]. In the early-stage, polymer-rich regions are elastically compressed due to desorption of solvent (as in deswelling gels). After a transient time, however, the surface tension force becomes effective at the ends of stripe-like polymer-rich regions, where the curvature is largest. If there were no viscoelasticity, circular domains would then appear. In our viscoelastic case, subsequent shape changes produce elastic expansion in the direction perpendicular to the stripe and elastic compression in the direction of the stripe. The resultant network stress largely cancels the stress originating from the surface tension (or that from $\nabla\phi$) and greatly slows down further shape changes.

8.9.3 Simulation of spinodal decomposition in gels

We next show numerical results in 2D on spinodal decomposition in gels on the basis of the Ginzburg–Landau model presented in (7.2.29)–(7.2.33) [125]. Namely, the conformation tensor $\overleftrightarrow{W}$ and the volume fraction ϕ obey (7.2.32) and (7.2.33), respectively, while the network velocity $\boldsymbol{v}$ is determined by (7.2.29) with $\zeta \propto \phi^2$. For simplicity, we assume $g(\phi) = v_0\bar{a}(-0.8\psi^2 + \psi^4)$ and C = const. with $\psi = 2\phi/\phi_0 - 1$ in the dimensionless free-energy density $g(\phi)$ in (7.2.2). The strength of crosslinkage is represented by

$$v_0^* = v_0/\bar{a} \sim v_0 T/|K_{\mathrm{os}}| \sim \mu/|K_{\mathrm{os}}|. \tag{8.9.19}$$

By measuring space and time in units of the correlation length ξ and a diffusion time (ξ^2/D_{m}), we display in Fig. 8.39 typical network domain structures in an isotropic case with $v_0^* = 0.3$ and those in a uniaxial case with $\lambda = \sqrt{2}$ and $v_0^* = 0.1$, where λ is the degree of stretching in (3.5.66). The average polymer volume fraction is $\phi_0/2$ or $\langle\psi\rangle = 0$. The domain structures for the isotropically swollen case closely resemble those observed in deeply quenched polymer solutions and asymmetric polymer blends [116]. In the uniaxially stretched case, we can see the formation of lamellar structures elongated in the stretched direction, consistent with the experimental result in Fig. 8.35. In Fig. 8.40 we plot the perimeter density $P(t)$ vs t in the isotropic case. Because $P(t)$ measures the inverse length scale of the domains, Fig. 8.40 demonstrates extreme slowing-down of the domain growth, which is consistent with the experiments [119]–[123] and the simulation [124]. Note that we are treating the case of weak network deformations without crosslink breakage and the origin of pinning is shear deformations asymmetric between the two phases.

Further remarks are as follows. (i) The patterns and pinning effect in gels are analogous to those for coherent alloys with composition-dependent elastic moduli, as can be seen in Figs 10.12 and 10.13 below. This close resemblance stems from the third-order elastic interaction (7.2.19) for gels and that in (10.1.37) for alloys, as already discussed below (7.2.19). (ii) We have neglected the effects of heterogeneities of the network structure, which was treated in Section 7.3. (iii) It is of great interest to understand how charges alter phase separation behavior when an ionized gel is quenched into an unstable region.

8.10 Vortex motion and mutual friction

Vortices in classical fluids have finite lifetimes limited by the shear viscosity. However, quantized vortices in superfluids are topological singularities, as discussed in Section 4.5, and hence are unique singular objects appearing collectively in rotating helium and in thermal counterflow [129]. Our aim here is to examine vortex motion in systems with the xy symmetry in the Ginzburg–Landau scheme. To this end, we will firstly treat a simple relaxation model and secondly review theoretical results for ^{4}He and ^{3}He–^{4}He near the superfluid transition. Defect turbulence in ^{4}He in heat flow and liquid crystalline polymers

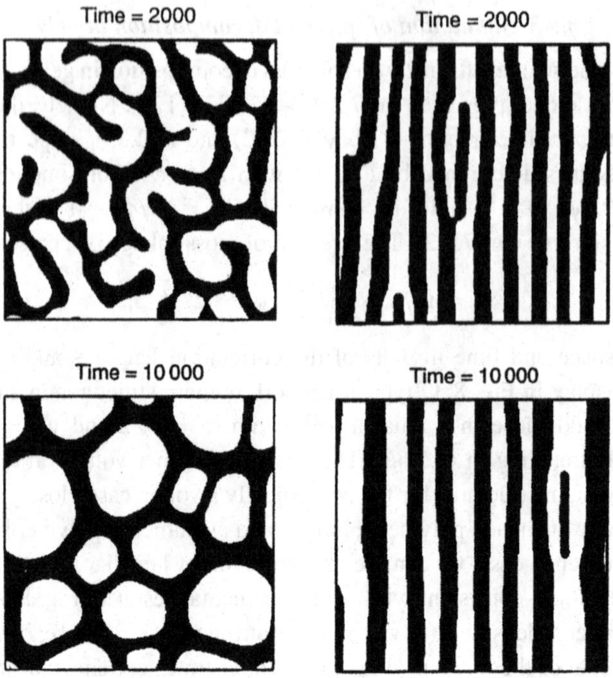

Fig. 8.39. Time evolution of domain structures for phase-separating gels [125]. The three frames on the left correspond to the isotropically swelling case with $\nu_0^* = 0.3$, and those on the right correspond to the uniaxially stretching case with $\nu_0^* = 0.1$. Polymer-rich regions are shown in black. See text for units of space and time.

in shear flow will then be briefly explained. Self-organized superfluid states with high-density vortices will be shown to be created by competition between heat flow and gravity near the superfluid transition.

Although not discussed in this book, we note that proliferation of dislocations is responsible for the plastic deformation of crystals, where the dynamics of an assembly of dislocations is strongly influenced by long-range elastic interactions on mesoscopic scales ($\sim 10^{-4}$ cm) [130].

8.10.1 Simple relaxation model

We assume that a complex order parameter $\psi = \psi_1 + i\psi_2$ obeys the simple relaxation model (8.1.66). For simplicity, we neglect the noise term and set $r = -\kappa^2$, $K = 1$, and $h = 0$, but the coefficient L is generally complex as

$$L = L_1 + iL_2, \tag{8.10.1}$$

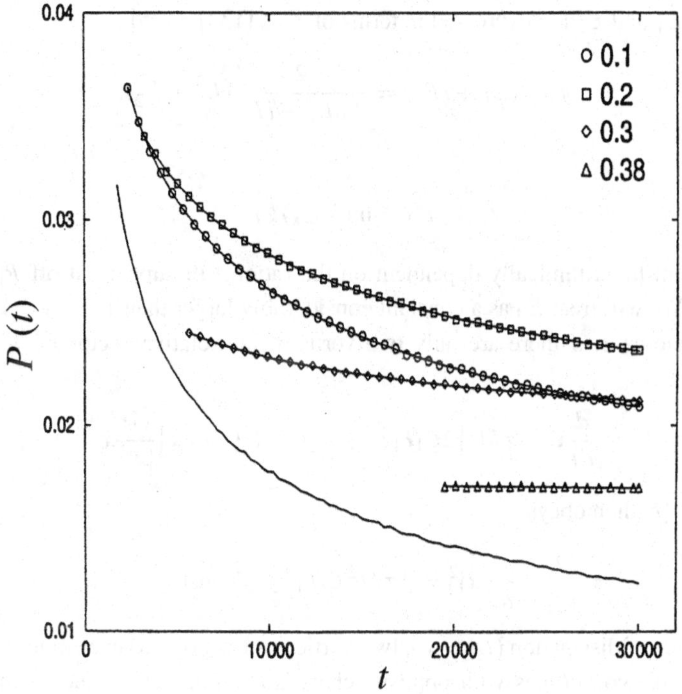

Fig. 8.40. Time dependence of the perimeter density $P(t)$ for $\nu_0^* = 0.1,\ 0.2,\ 0.3$ and 0.38 [125]. For comparison, we also plot $P(t)$ vs t (solid line) for the case without elastic effects ($\nu_0^* = 0$), which obeys $P(t) \sim t^{-1/3}$.

where $L_1 \geq 0$. If $L_1 = 0$ and $L_2 = \hbar/2m_4 > 0$, (8.1.66) reduces to the reversible Gross–Pitaevskii equation [131, 132]:

$$\frac{\partial}{\partial t}\psi = -i\frac{\hbar}{2m_4}\left[-\kappa^2 - \nabla^2 + u_0|\psi|^2\right]\psi. \tag{8.10.2}$$

2D case

We consider an assembly of vortex points, $\boldsymbol{R}_i = (X_i, Y_i)$, with charge $\ell_i = \pm 1$ in 2D. The distances between vortices are assumed to be much longer than ξ. In Appendix 8I we will derive the following vortex dynamic equation,

$$\begin{aligned}\frac{\partial}{\partial t}\boldsymbol{R}_i &= -(\mathcal{L}_1 + \ell_i\mathcal{L}_2\boldsymbol{e}_z\times)\frac{\partial}{\partial \boldsymbol{R}_i}\mathcal{H}_{\rm v}/T \\ &= \pi M^2(\mathcal{L}_1 + \ell_i\mathcal{L}_2\boldsymbol{e}_z\times)\sum_{j\neq i}\frac{\ell_i\ell_j}{R_{ij}^2}(\boldsymbol{R}_i - \boldsymbol{R}_j),\end{aligned} \tag{8.10.3}$$

where $M^2 = \kappa^2/u_0$, $\boldsymbol{e}_z$ is the unit vector along the z axis, $\boldsymbol{e}_z \times (\cdots)$ denotes taking the vector product, and $\mathcal{H}_{\rm v}$ is the vortex free energy given by (4.5.14) in 2D. The kinetic coefficients $\mathcal{L}_1$ and $\mathcal{L}_2$ are expressed in terms of L as [133]–[135]

$$\pi M^2(\mathcal{L}_1 + i\mathcal{L}_2) = \frac{2}{E_0 L_1 - iL_2}(L_1^2 + L_2^2), \tag{8.10.4}$$

where

$$E_0 \cong \ln(R_{\max}/\xi) \tag{8.10.5}$$

is a coefficient logarithmically dependent on the ratio of the upper cut-off $R_{\max}$ and the core size ξ. We will treat E_0 as a constant considerably larger than 1.

As a simple case, if there are only two vortices, the relative vector $\boldsymbol{x} = \boldsymbol{R}_1 - \boldsymbol{R}_2$ is governed by

$$\frac{d}{dt}\boldsymbol{x} = \pi M^2\big[2\mathcal{L}_1\ell_1\ell_2 + \mathcal{L}_2(\ell_1 + \ell_2)\boldsymbol{e}_z \times\big]\frac{1}{|\boldsymbol{x}|^2}\boldsymbol{x}. \tag{8.10.6}$$

The distance $|\boldsymbol{x}|$ then obeys

$$\frac{d}{dt}|\boldsymbol{x}|^2 = 4\pi M^2\mathcal{L}_1\ell_1\ell_2 = \text{const.} \tag{8.10.7}$$

In the presence of dissipation ($L_1 > 0$), two vortices attract (repel) each other for $\ell_1\ell_2 < 0$ ($\ell_1\ell_2 > 0$). If two vortices with opposite charges approach each other within the core radius ($\sim \xi$), they are annihilated.

3D case

In 3D a vortex line with unit charge is represented by $\boldsymbol{R}(s)$ where s is the arc length. The distances between different vortex line elements and the typical inverse curvature are assumed to be much longer than ξ. Similarly to (8.10.3), the vortex velocity $\boldsymbol{v}_L = \partial\boldsymbol{R}(s)/\partial t$ is written as

$$\boldsymbol{v}_L = -\big(\mathcal{L}_1 + \mathcal{L}_2\boldsymbol{t}\times\big)\frac{\delta}{\delta\boldsymbol{R}(s)}\mathcal{H}_{\mathrm{v}}/T, \tag{8.10.8}$$

where $\boldsymbol{t} = d\boldsymbol{R}(s)/ds$ is the tangential unit vector and the vortex free energy $\mathcal{H}_{\mathrm{v}}$ is given by (4.5.19). The kinetic coefficients $\mathcal{L}_1$ and $\mathcal{L}_2$ are expressed as (8.10.4) in terms of $L = L_1 + iL_2$. The above form is nonlocal and is very complicated in general.

However, if we neglect the interaction among distant vortex line elements and set $\mathcal{H}_{\mathrm{v}} = \mathcal{H}_{\mathrm{v}}^{(0)}$ in (4.5.6), we obtain a much simpler dynamic equation,

$$\boldsymbol{v}_L = \pi M^2 E_0 \mathcal{K}(\mathcal{L}_1\boldsymbol{n} + \mathcal{L}_2\boldsymbol{b}), \tag{8.10.9}$$

where $\mathcal{K}$ is the line curvature, $\boldsymbol{n}$ is the normal unit vector, and $\boldsymbol{b} = \boldsymbol{t} \times \boldsymbol{n}$. They are determined by $d\boldsymbol{t}/ds = \mathcal{K}\boldsymbol{n}$. With this local induction approximation we notice the following. (i) In the dissipationless case $L_1 = 0$ and $L_2 \neq 0$, we reproduce the Arms–Hama approximation (4.5.24). (ii) In the presence of dissipation ($L_1 > 0$), the total vortex line length decreases in time as

$$\frac{d}{dt}L_{\rm T} = -\int ds\mathcal{K}\boldsymbol{n}\cdot\boldsymbol{v}_L = -\pi M^2 E_0 \mathcal{L}_1 \int ds\mathcal{K}^2 \le 0, \quad (8.10.10)$$

which is analogous to (8.2.7). (iii) In the purely dissipative case ($L_1 > 0$ and $L_2 = 0$) (8.10.9) becomes

$$\boldsymbol{v}_L = 2L_1\mathcal{K}\boldsymbol{n}, \quad (8.10.11)$$

which resembles the Allen–Cahn equation (8.2.1) for the interface dynamic motion. The scaling behavior (8.1.68) is now derived if $\boldsymbol{v}_L$ is estimated to be of order $\mathcal{K}/t$.

Noise effect

As in the case of interface dynamics, we may regard (8.10.3) or (8.10.8) as a Langevin equation [136] by adding on its right-hand side a random noise term which is related to $\mathcal{L}_1$ via the fluctuation–dissipation relation (5.2.7). Obviously, $\mathcal{L}_1$ is a dissipative kinetic coefficient, while $\mathcal{L}_2$ is a reversible one, in the general theory of Langevin equations presented in Chapter 5. The noise effect is needed if we consider thermally activated vortices such as those in the 2D xy model near the Kosterlitz–Thouless transition [137, 138].

8.10.2 Vortex motion in a superfluid

In superfluid ^{4}He there can be different macroscopic average velocities of the normal fluid and superfluid components, $\boldsymbol{u}_{\rm n}$ and $\boldsymbol{u}_{\rm s}$, respectively. The average relative velocity will be written as

$$\boldsymbol{w} = \boldsymbol{u}_{\rm n} - \boldsymbol{u}_{\rm s}, \quad (8.10.12)$$

which is assumed to vary slowly in space as compared with the average vortex distance. Its magnitude should also be sufficiently small such that

$$|\boldsymbol{w}| \ll \hbar/m_4\xi. \quad (8.10.13)$$

Otherwise, superfluidity itself will be broken, as will be discussed in Section 9.7.

The mutual interaction between the superfluid component and the normal fluid component arises from quantized vortices. The vortex velocity $\boldsymbol{v}_L$ is determined by the local superfluid velocity $\boldsymbol{v}_{{\rm s}\ell}$ and the macroscopically averaged normal fluid velocity $\boldsymbol{u}_{\rm n}$ at the vortex point under consideration [129, 139]. First note that the lift (Magnus) force density,

$$f_{\rm M} = (2\pi\hbar/m_4)\rho_{\rm s}\boldsymbol{t}\times(\boldsymbol{v}_L - \boldsymbol{v}_{{\rm s}\ell}), \quad (8.10.14)$$

is acting on a vortex point per unit length from the superfluid component. Hereafter $\rho_{\rm s} = m_4^2\hbar^{-2}TM^2$ is the average superfluid mass density. Then $\rho_{\rm n} = \rho - \rho_{\rm s}$ is the average normal fluid mass density. Hall and Vinen [139] assumed that the drag force density is given by

$$f_{\rm D} = -\gamma_0\boldsymbol{t}\times[\boldsymbol{t}\times(\boldsymbol{u}_{\rm n} - \boldsymbol{v}_L)] + \gamma_0'\boldsymbol{t}\times(\boldsymbol{u}_{\rm n} - \boldsymbol{v}_L), \quad (8.10.15)$$

supposing collisions with normal excitations (mostly, rotons above 1 K and not close to T_λ). If we neglect the inertia of the vortex, the force balance $\boldsymbol{f}_{\rm M} + \boldsymbol{f}_{\rm D} = \boldsymbol{0}$ holds, leading to

$$\boldsymbol{v}_L = \boldsymbol{v}_{\rm s\ell} - \alpha' \boldsymbol{t} \times [\boldsymbol{t} \times (\boldsymbol{u}_{\rm n} - \boldsymbol{v}_{\rm s\ell})] + \alpha \boldsymbol{t} \times (\boldsymbol{u}_{\rm n} - \boldsymbol{v}_{\rm s\ell}). \tag{8.10.16}$$

Here the coefficients α and α' are related to γ_0 and γ_0' by

$$\alpha' + i\alpha = (\gamma_0 + i\gamma_0')/[\gamma_0 + i\gamma_0' - i(2\pi\hbar/m_4)\rho_{\rm s}]. \tag{8.10.17}$$

Hall and Vinen also introduced other mutual friction coefficients B and B' by

$$\alpha + i\alpha' = (\rho_{\rm n}/2\rho)(B + iB'). \tag{8.10.18}$$

In terms of α and α', the force densities in (8.10.14) and (8.10.15) are expressed as

$$\boldsymbol{f}_{\rm M} = -\boldsymbol{f}_{\rm D} = (2\pi\hbar/m_4)\rho_{\rm s}\{\alpha \boldsymbol{t} \times [\boldsymbol{t} \times (\boldsymbol{u}_{\rm n} - \boldsymbol{v}_{\rm s\ell})] + \alpha' \boldsymbol{t} \times (\boldsymbol{u}_{\rm n} - \boldsymbol{v}_{\rm s\ell})\}. \tag{8.10.19}$$

These coefficients can be measured by investigating second sound in rotating helium. As $T \to 0$, α and α' tend to zero or $\boldsymbol{v}_L \to \boldsymbol{v}_{\rm s\ell}$. In the temperature range $1\ {\rm K} \lesssim T \lesssim 2.1\ {\rm K}$, α is of order 1 and α' is considerably smaller than α [129]. We also note that the normal fluid (or roton) velocity $\boldsymbol{v}^{\rm R}$ near a vortex core becomes different from the average $\boldsymbol{u}_{\rm n}$ due to the viscous drag effect. The difference arises within the range of the viscous penetration length $(\eta/\rho\omega)^{1/2}$ from the core where we suppose a second sound with frequency ω. Mathieu and Simon showed that this hydrodynamic effect is the dominant mutual friction mechanism for $1.7 \lesssim T \lesssim 2.1$ K [140].

For $\boldsymbol{u}_{\rm n} = \boldsymbol{u}_{\rm s} = \boldsymbol{0}$, we have $\boldsymbol{v}_{\rm s\ell} = \boldsymbol{v}_{\rm s1}$, where $\boldsymbol{v}_{\rm s1}$ is the superfluid velocity in (4.5.24) induced at a vortex point by the local curvature. Then (8.10.16) takes the standard form (8.10.8) with

$$\mathcal{L}_1 + i\mathcal{L}_2 = \frac{m_4}{\pi\hbar\rho_{\rm s}}[\alpha + i(1-\alpha')] = \frac{2}{\gamma_0 + i\gamma_0' - i(2\pi\hbar/m_4)\rho_{\rm s}}. \tag{8.10.20}$$

In general cases with nonvanishing $\boldsymbol{u}_{\rm n}$ and $\boldsymbol{u}_{\rm s}$, we have $\boldsymbol{v}_{\rm s\ell} = \boldsymbol{v}_{\rm s1} + \boldsymbol{u}_{\rm s}$ and may rewrite (8.10.16) as

$$\boldsymbol{v}_L - \boldsymbol{u}_{\rm s} = -(\mathcal{L}_1 + \mathcal{L}_2 \boldsymbol{t}\times)\frac{\delta}{\delta \boldsymbol{R}(s)}\tilde{\mathcal{H}}_{\rm v}/T, \tag{8.10.21}$$

where we introduce a modified free energy,

$$\tilde{\mathcal{H}}_{\rm v} = \mathcal{H}_{\rm v} - \frac{\pi\hbar\rho_{\rm s}}{m_4}\int ds \boldsymbol{w} \cdot [\boldsymbol{R}(s) \times \boldsymbol{t}(s)]. \tag{8.10.22}$$

As a simple example, for a vortex ring with radius R we have

$$\tilde{\mathcal{H}}_{\rm v} = 2\rho_{\rm s}\left(\frac{\pi\hbar}{m_4}\right)^2\left[R\ln(R/\xi) + \frac{m_4}{\hbar}\boldsymbol{w}\cdot\boldsymbol{b}R^2\right], \tag{8.10.23}$$

which is analogous to the droplet free energy $\mathcal{H}(R)$ in (8.2.42) in model A in 2D.

The Arms–Hama approximation

We use the Arms–Hama approximation (4.5.24) by replacing $\mathcal{H}_v$ by $\mathcal{H}_v^{(0)}$ in (4.5.6). Then the local velocity $\boldsymbol{v}_{s\ell}$ consists of a macroscopic average $\boldsymbol{u}_s$ and the locally induced velocity $\boldsymbol{v}_{s1} = V_\ell \boldsymbol{b}$ with

$$V_\ell = \frac{\hbar}{2m_4} E_0 \mathcal{K}. \tag{8.10.24}$$

The vortex velocity (8.10.16) becomes

$$\boldsymbol{v}_L = \boldsymbol{u}_s + \boldsymbol{t} \times \left[-\alpha' \boldsymbol{t} \times \boldsymbol{w} + \alpha \boldsymbol{w}\right] + V_\ell\left[(1-\alpha')\boldsymbol{b} + \alpha \boldsymbol{n}\right]. \tag{8.10.25}$$

Using the first line of (8.10.10) we calculate the rate of change of the total line length as

$$\frac{d}{dt} L_T = \alpha \int ds \mathcal{K}(\boldsymbol{b}\cdot\boldsymbol{w} - V_\ell) - \left(\int ds \mathcal{K} \boldsymbol{n}\right)\cdot(\boldsymbol{u}_s + \alpha' \boldsymbol{w}). \tag{8.10.26}$$

If we consider only closed loops, the second term vanishes because $\int ds \mathcal{K}\boldsymbol{n} = \int d\boldsymbol{R}(s) = 0$. However, if the two ends of a line are attached to the boundary wall, the second term is in general nonvanishing.

As a simple, instructive example we apply $\boldsymbol{w}$ perpendicularly to the single vortex ring sketched in Fig. 4.7. Here $\boldsymbol{b}$ is parallel to $\boldsymbol{w}$, $\boldsymbol{n}$ points in the outward direction of the ring circle, and $\mathcal{K} = -1/R$, where R is the ring radius. Then (8.10.25) is rewritten as

$$\boldsymbol{v}_L = \boldsymbol{u}_s + (\boldsymbol{w}\cdot\boldsymbol{b})(\alpha'\boldsymbol{b} - \alpha\boldsymbol{n}) - \frac{\hbar E_0}{2m_4 R}\left[(1-\alpha')\boldsymbol{b} + \alpha\boldsymbol{n}\right]. \tag{8.10.27}$$

In the presence of dissipation ($\alpha > 0$) we notice that R changes in time as

$$\frac{\partial}{\partial t} R = \boldsymbol{n}\cdot(\boldsymbol{v}_L - \boldsymbol{u}_s) = \alpha\left(-\boldsymbol{w}\cdot\boldsymbol{b} - \frac{\hbar E_0}{2m_4 R}\right). \tag{8.10.28}$$

This is analogous to (8.2.43) for circular or spherical domain growth in the nonconserved case. In the present case, the relative velocity plays the role of a magnetic field in spin systems. Indeed, the ring can expand if $\boldsymbol{w}\cdot\boldsymbol{b} < 0$ and R is larger than the critical radius,

$$R_c = \frac{\hbar E_0}{2m_4|\boldsymbol{w}|}, \tag{8.10.29}$$

while it shrinks otherwise. Near the λ point we have $R_c \gg \xi$ from (8.10.13) and $E_0 \gtrsim 1$.

Mutual friction in ^{4}He near the superfluid transition

Near the superfluid transition, experimental data indicate divergence of $\alpha \cong B/2$ and $\alpha' \cong B'/2$ roughly as

$$\alpha \sim 1 - \alpha' \sim (1 - T/T_\lambda)^{-a_v}, \tag{8.10.30}$$

where $a_v \sim 1/3$ [141, 142]. This behavior was also derived by Pitaevskii with a simple theoretical estimate [143]. In Ref. [133] they were predicted to be on the order of the renormalized kinetic coefficient L_R in model F investigated in Section 6.5. Because we treat the fluctuations longer than ξ, L should be identified with the renormalized one, L_R, in the long-wavelength limit, which behaves as $|\tau|^{-\nu+x_\lambda}$ with x_λ being the dynamic exponent for the thermal conductivity in (6.6.42). Similarly to (8.10.4), α and α' are expressed as

$$\alpha + i(\alpha' - 1) = \frac{2m_4}{\hbar}|L_{\rm R}|^2/[({\rm Re}\, L_{\rm R})(X_1 + iY_1)], \tag{8.10.31}$$

where X_1 and Y_1 are positive, of order 1, and only weakly dependent on τ. Experimentally, α and $1-\alpha'$ are positive and grow as $T \to T_\lambda$ in agreement with the above result [129, 142]. If we adopt model F, (8.10.31) does not involve logarithmically divergent integrals (or E_0) in α and α', in contrast to (8.10.4). A deviation of the entropy variable around a moving vortex line can eliminate such divergence.

Mutual friction in ^{3}He–^{4}He mixtures near the λ line and the tricritical point

The vortex motion in ^{3}He–^{4}He mixtures is also of great interest [134], though there seems to be no experimental data so far. The behavior of α and α' sensitively depends on the average ^{3}He concentration X. This is because the linear combination c_2 of the entropy and concentration fields given in (6.6.94) relaxes diffusively with a small diffusion constant D_2 around a moving vortex line and crucially influences the vortex motion. However, as discussed near (6.6.102), the coupling between c_2 and ψ vanishes at an intermediate ^{3}He concentration $X_{\rm D} \cong 0.37$ at SVP and a mixture there behaves as pure ^{4}He [135]. Moreover, because of the slow relaxation of c_2, α and α' exhibit strong frequency dependence for $\omega \gtrsim D_1\xi^{-2}$. This effect is particularly important in the tricritical region where $D_1 \propto \xi^{-1}$.

In the low-frequency limit $\omega \ll D_1\kappa^2$ we obtain

$$\alpha + i(\alpha' - 1) = 1\bigg/\left[\left({\rm Re}\,\frac{\hbar}{2m_4 L_{\rm R}}\right)(X_{\rm A} + iY_{\rm A}) + \delta_{\rm B}\right], \tag{8.10.32}$$

where $X_{\rm A}$ and $Y_{\rm A}$ are of order 1, and ${\rm Re}(1/L) \sim |\tau|^{1/3}$ near the λ line. The quantity $\delta_{\rm B}$ is positive and depends on $|\tau|$ logarithmically, going to 0 as $\delta_{\rm B} \sim X$ for $X \ll 1$ $\delta_{\rm B} \sim (X - X_{\rm D})^2$ for $X \cong X_{\rm D}$. Thus, α and α' remain finite on the λ line. The coefficient α grows near the λ line and saturates to the following λ-line value; in particular,

$$\begin{aligned} \alpha \to \delta_{\rm B}^{-1} &\sim X^{-1} && (X \ll 1) \\ &\sim (X - X_{\rm D})^{-2} && (X \cong X_{\rm D}). \end{aligned} \tag{8.10.33}$$

However, $1 - \alpha'$ takes a maximum at $|\tau| \sim \tau_{\rm c}$ close to the λ line as

$$\begin{aligned} 1 - \alpha' &\sim {\rm Re}\, L_{\rm R} && (\tau_{\rm c} < |\tau| \ll 1), \\ &\sim \delta_{\rm B}^{-2}\, {\rm Re}(1/L_{\rm R}) && (|\tau| < \tau_{\rm c}). \end{aligned} \tag{8.10.34}$$

The crossover reduced temperature $\tau_{\rm c}$ is roughly of order $\delta_{\rm B}^3$ and is very small for $X \lesssim X_{\rm D}$, but it increases on approaching the tricritical point.

In the tricritical region, however, $\delta_{\rm B}$ grows as ξ. Therefore, $\alpha \sim \xi^{-1}$ and $1 - \alpha' \cong 0$. This behavior can be explained as follows. It is known that the fluctuations of c_2 are much enhanced with variance $\langle c_2 : c_2 \rangle \sim \xi$ as in (3.2.25). Therefore, the variation of c_2 induced around a moving vortex is very large, giving rise to a large resistance. From (8.10.32) we notice that

$$\boldsymbol{v}_L \cong \boldsymbol{u}_\mathrm{n} + (\boldsymbol{t} \cdot \boldsymbol{w})\boldsymbol{t} + \alpha \boldsymbol{t} \times (\boldsymbol{u}_\mathrm{n} - \boldsymbol{v}_{\mathrm{s}\ell}). \tag{8.10.35}$$

This behavior is in a marked contrast to that at low temperatures where $\boldsymbol{v}_L \cong \boldsymbol{v}_{\mathrm{s}\ell}$ due to small dissipation.

8.10.3 Defect turbulence in ^{4}He in heat flow

It is well known that, if $\boldsymbol{w}$ is sustained externally at a constant value, a dynamical steady state is eventually established in which a vortex tangle is generated. Vinen described the time evolution of the line density $n_\mathrm{v}(t)$ of a vortex tangle in the form [144]

$$\frac{d}{dt} n_\mathrm{v} = A_1 |\boldsymbol{w}| n_\mathrm{v}^{3/2} - A_2 n_\mathrm{v}^2. \tag{8.10.36}$$

From (8.10.28) the first term represents line stretching due to the flow with $A_1 \sim \alpha$, while the second term represents line shrinking due to the curvature effect with $A_2 \sim \alpha\hbar E_0/m_4$. The typical line curvature in the dynamical steady state is on the order of R_c and the vortex line density in the steady state is scaled as

$$(n_\mathrm{v})_\mathrm{steady} \sim R_\mathrm{c}^{-2} \sim (m_4/\hbar E_0)^2 |\boldsymbol{w}|^2. \tag{8.10.37}$$

The timescale of the tangle growth t_tan is estimated as

$$t_\mathrm{tan} \sim A_2^{-1} R_\mathrm{c}^2 \sim \frac{\hbar E_0}{\alpha m_4 |\boldsymbol{w}|^2}. \tag{8.10.38}$$

Schwarz used (8.10.25) in numerical analysis of vortex tangles in thermal counterflow [145]. He assumed that vortex lines reconnect when they encounter one another. Then reconnection gives rise to randomization of the lines as in Fig. 8.41. Subsequently, such reconnection processes were numerically studied on the basis of the dissipative dynamic equation (8.1.66) [15, 17], for which see Fig. 8.6, and the reversible Gross–Pitaevskii equation [146]. The resultant complex phenomenon has been called *vortex turbulence*, though it is very different from the usual fluid turbulence characterized by the energy cascade from large to small length scales. We also remark that fluid turbulence in superfluid ^{4}He, such as that generated by a grid, poses another fundamental problem, where we are interested in how vortices come into play in the dissipative wave number range [147, 148].

The Gorter and Mellink mutual friction force

In the presence of vortex tangles there arises mutual friction between the normal fluid and superfluid components. We take spatial averages in fluid elements with sizes much longer than the inter-vortex distance ($\sim R_\mathrm{c}$ in dynamical steady states) and assume only slow spatial variations in the averaged quantities. The average mutual force density $\boldsymbol{F}_\mathrm{sn}$ is written as

$$\boldsymbol{F}_\mathrm{sn} = \frac{1}{\Delta V} \int ds \boldsymbol{f}_\mathrm{D}, \tag{8.10.39}$$

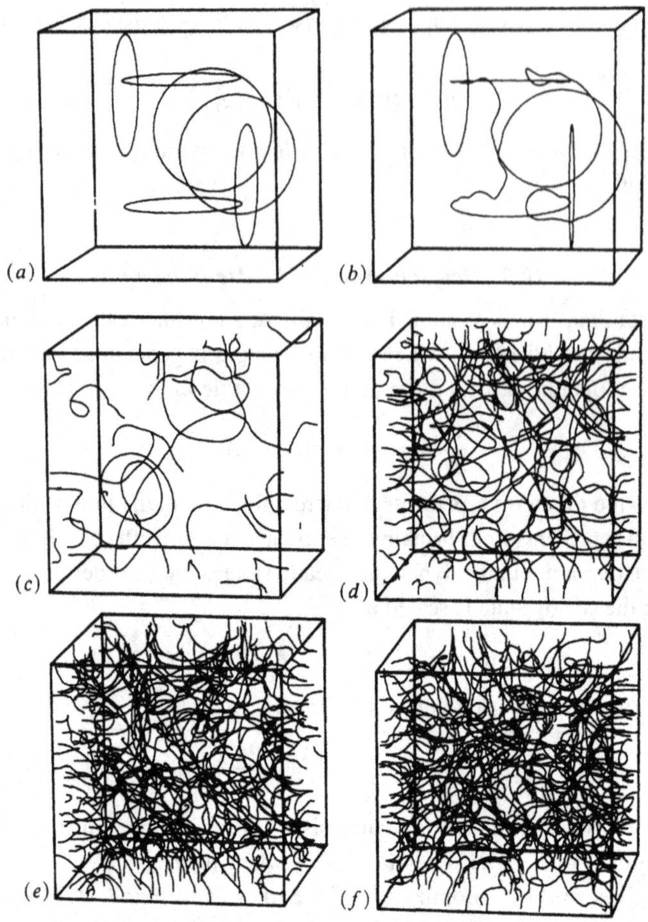

Fig. 8.41. Time evolution of a vortex tangle with $\alpha = 0.1$ starting with six vortex rings in (a) [145]. The average flow is into the front face. Reconnections of lines lead to increasingly complex patterns. A dynamical steady state is attained in (e) and (f).

where the line integral is along all the vortex lines within a fluid element with volume ΔV. From (8.10.19) we estimate its magnitude as

$$\begin{aligned} |\boldsymbol{F}_{\rm sn}| &\sim (\hbar/m_4)\rho_{\rm s}\alpha|\boldsymbol{w}|L \\ &\sim (m_4/\hbar E_0^2)\rho_{\rm s}\alpha|\boldsymbol{w}|^3. \end{aligned} \tag{8.10.40}$$

We have set $L \sim L_{\rm steady}$ in the second line. In the simplest form, the two-fluid hydrodynamic equations read

$$\rho_{\rm s}\frac{\partial}{\partial t}\boldsymbol{u}_{\rm s} = -\rho_{\rm s}\nabla\mu - \boldsymbol{F}_{\rm sn} = -\frac{\rho_{\rm s}}{\rho}\nabla p + \rho_{\rm s}s\nabla T - \boldsymbol{F}_{\rm sn}, \tag{8.10.41}$$

$$\rho_n \frac{\partial}{\partial t} \boldsymbol{u}_n = -\frac{\rho_n}{\rho} \nabla p - \rho_s s \nabla T + \boldsymbol{F}_{sn} + \eta \nabla^2 \boldsymbol{u}_n, \tag{8.10.42}$$

where $\boldsymbol{u}_n = \langle \boldsymbol{v}_n \rangle$ and $\boldsymbol{u}_s = \langle \boldsymbol{v}_s \rangle$ are the average velocities, η is the shear viscosity, and the thermal conductivity and the bulk viscosity are neglected. The μ and s are the chemical potential and entropy per unit mass, respectively. Gorter and Mellink proposed the following form [129, 149],

$$\boldsymbol{F}_{sn} = -A(T)\rho_s \rho_n |\boldsymbol{w}|^2 \boldsymbol{w}, \tag{8.10.43}$$

where $A(T)$ is a temperature-dependent coefficient of order $(m_4/\hbar E_0^2 \rho_n)\alpha$. This form is consistent with the second line of (8.10.40). In steady thermal counterflow, we obtain $\nabla p = \eta \nabla^2 \boldsymbol{u}_n$ and

$$\begin{aligned} s\nabla T &= \eta \rho^{-1} \nabla^2 \boldsymbol{u}_n + \rho_s^{-1} \boldsymbol{F}_{sn} \\ &\cong -A(T)\rho_n |\boldsymbol{w}|^2 \boldsymbol{w}. \end{aligned} \tag{8.10.44}$$

The second line holds for wide cells where the pressure gradient is small. The relation $\nabla T \propto Q^3$ has been observed in many experiments [129], where Q is the heat flux expressed as

$$\boldsymbol{Q} = T\rho s \boldsymbol{u}_n = T\rho_s s \boldsymbol{w}, \tag{8.10.45}$$

from $\rho_s \boldsymbol{u}_s + \rho_n \boldsymbol{u}_n = \boldsymbol{0}$ in 1D geometry.

Mutual friction near the superfluid transition

Near the λ point the temperature dependence of $A(T)$ is proportional to L_R ($\propto \xi/\lambda_R$) from (8.10.31). For small $|\tau| = 1 - T/T_\lambda \ll 1$, we have the behavior

$$\frac{1}{T_\lambda} \frac{d}{dx} T = -B_v |\tau|^{-m_v} Q^3. \tag{8.10.46}$$

The exponent $m_v = 4\nu - x_\lambda$ arises from $\rho_s^{-3} A(T)$. Previous experiments were fairly consistent with the above form [150, 151]. For example, Ahlers' result for $T_\lambda - T \gtrsim 10^{-4}$ K [150] was fitted to (8.10.46) with $m_v = 2.23$ and $B_v = 5 \times 10^{-29}$ in cgs units.

The assumption (8.10.13) at the starting point is equivalent to the condition $|\tau| \gg \tau_Q$, where τ_Q is the crossover reduced temperature introduced in (6.7.8). It is instructive to rewrite the above equation in terms of τ_Q in the following scaling form,

$$\frac{d}{dx} |\tau| = A_v \frac{|\tau|}{\xi} (\tau_Q / |\tau|)^{6\nu}, \tag{8.10.47}$$

where

$$A_v = B_v \xi_{+0} A_Q^{-6\nu} |\tau|^{5\nu - 1 - m_v} = (\rho/sT)(\hbar/m_4)^3 \xi_{+0}^{-2} A(T) |\tau|^{2\nu - 1}, \tag{8.10.48}$$

with $A_Q = (m_4 \xi_{0+}/\hbar s T_\lambda \rho_s^*)^{1/2\nu}$ and $\xi \equiv \xi_{+0} |\tau|^{-\nu}$. The dimensionless number A_v is theoretically of order w/E_0^2 and is expected to be much smaller than 1, where the behavior of w was discussed near (6.6.59)–(6.6.67). Ahlers' result [150] gives $A_v \sim 1.1 \times 10^{-3}$.

Hereafter we neglect the weak temperature dependence of A_v and treat it as a small number of order 10^{-3}.

Taking the origin of the x axis appropriately inside the cell, we may integrate (8.10.47) in the form

$$|\tau(x)|^{5\nu} = |\tau(0)|^{5\nu} + 5\nu A_v(\tau_Q^{6\nu}/\xi_{0+})x. \qquad (8.10.49)$$

The characteristic length over which the reduced temperature τ changes significantly due to the mutual friction is given by $\ell_{sn} \sim 10^3(|\tau|/\tau_Q)^{6\nu}\xi$. The temperature in heat-conducting superfluids may be considered to be homogeneous if the cell height h is much shorter than ℓ_{sn}. If there is a HeI–HeII interface at $x = x_{int}$, we have $\tau(x_{int}) = -\tau_\infty$ as given by (6.7.29). Then,

$$(|\tau(x)|/\tau_\infty)^{5\nu} = 1 + (5\nu R_\infty^{-6\nu} A_v)\frac{1}{\xi_Q}(x - x_{int}), \qquad (8.10.50)$$

where ξ_Q is defined by (6.7.9). Thus the height of the superfluid region where $\tau(x) \cong -\tau_\infty$ is of order 10^3–$10^4\xi_Q$ and is well defined theoretically but might be narrow experimentally.[10] For $x < x_{int}$ the system is in a normal fluid state. The vortex line density in the superfluid region is estimated as

$$n_v \sim E_0^{-2}(\tau_Q/|\tau|)^{2\nu}\xi^{-2}, \qquad (8.10.51)$$

which is much smaller than ξ^{-2} for $|\tau| \gg \tau_Q$. Finally, we compare the characteristic magnitude of the temperature gradient in normal fluid and superfluid states at the same $|\tau|$ and Q; (6.7.4) and (8.10.47) give

$$\left(\frac{d}{dx}T\right)_{\text{super}} \Big/ \left(\frac{d}{dx}T\right)_{\text{normal}} \sim 10^{-3}(|\tau|/\tau_Q)^{-13/4}, \qquad (8.10.52)$$

which is indeed very small for $|\tau| \gg \tau_Q$. The gravity effects are neglected in the above relations.

8.10.4 Self-organized states in ^{4}He heated from above

As discussed in Section 6.7, gravity and heat flow, if they are in the same direction, can compete to produce self-organized states near the superfluid transition. We introduce the local reduced temperature ε as in (6.7.22) by taking the x axis in the downward direction with the origin at the top in a cell with height h. In a superfluid state slightly below the λ line, we have $\varepsilon < 0$ and change (8.10.47) as

$$\frac{d}{dx}|\varepsilon| = -G + B_v|\varepsilon|^{-m_v}Q^3. \qquad (8.10.53)$$

We notice that there are two cases. In *regime M* the right-hand side of (8.10.53) is positive, which is realized for relatively large Q. Conversely, it is negative in *regime G*, where the gravity-induced gradient is dominant.

[10] For precise measurements of $1 - T_\infty/T_\lambda = \tau_\infty$ close to the interface, the thermometer size needs to be smaller than this length.

Regime M

As a special situation we assume that a normal fluid is in an upper region $0 < x < x_{\text{int}}$ of a cell and a superfluid is in the lower region $x_{\text{int}} < x < h$. At the interface position, where $|\varepsilon| = \tau_\infty$, the condition of regime M reads

$$\tau_g^{1+\nu} = \xi_{+0} G < A_v R_\infty^{-3+\nu} \tau_Q^{1+\nu}, \tag{8.10.54}$$

where τ_g is defined by (2.4.36). If we set $R_\infty \sim 2$, the above relation yields

$$\tau_Q \gtrsim 10^3 \tau_g \quad \text{or} \quad Q \gtrsim 10^3 (g/g_{\text{earth}})^{2\nu/(1+\nu)} \ (\text{erg/cm}^2\ \text{s}), \tag{8.10.55}$$

where g_{earth} is the gravitational acceleration on earth. In this regime the temperature profile is exemplified by curves 1 and 2 in Fig. 8.42. If the superfluid region is sufficiently wide, ε tends to a limiting value given by

$$\varepsilon_c = -(B_v Q^3/G)^{1/m_v} \sim -A_v^{1/m_v} (\tau_Q/\tau_g)^{(1+\nu)/m_v} \tau_Q. \tag{8.10.56}$$

The relaxation length is given by $|\varepsilon_c|/G$. In this self-organized state the temperature gradient due to defects becomes equal to the transition temperature gradient:

$$\left(\frac{d}{dx} T\right)_{\text{defect}} = \frac{d}{dx} T_\lambda. \tag{8.10.57}$$

The reduced temperature at the bottom ε_{bot} is larger than $10^3 \tau_g$ ($\sim 10^{-6}$ on earth) in magnitude. The vortex line density in units of ξ^{-2} is small, from (8.10.51).

Vortex turbulence in regime G

Regime G is realized under the reverse condition of (8.10.54) or (8.10.55). Further requiring (6.7.31) we have Q in the range $1 \lesssim Q/(g/g_{\text{earth}})^{2\nu/(1+\nu)} \lesssim 10^3$ (cgs) in the geometry of Fig. 8.44. The reduced temperature at the bottom satisfies $|\varepsilon_{\text{bot}}| \lesssim 10^3 \tau_g$. In this case the system approaches the λ point at constant Q with increasing distance from the interface. The dimensionless wave number $K = k\xi$ is related to $|\varepsilon|$ as

$$K(1 - K^2) = (\xi/\xi_Q)^2 = (\tau_Q/|\varepsilon|)^{2\nu}, \tag{8.10.58}$$

where τ_Q and ξ_Q are defined by (6.7.8) and (6.7.9), respectively. The value of K at $x \cong x_{\text{int}}$ is determined by (6.7.30). It increases up to a critical value K_c for $x - x_{\text{int}} \gtrsim \ell_{GQ}$ where $\ell_{GQ} = \varepsilon_\infty/G \sim 10^{-2} Q^{1/2\nu} (g_{\text{earth}}/g)$(cm) is the relaxation length. Hereafter K_c will be set equal to the mean field value $1/\sqrt{3}$, for which see (9.7.7) below. It is known that the critical fluctuations gives rise to a correction only of order 10% [153]. In the region with $K \sim 1/\sqrt{3}$, vortices should be densely generated to produce much more enhanced mutual friction than represented by the Gorter–Mellink term in (8.10.53) which holds only under the weak-flow condition (8.10.13). In the strong-flow condition $K \sim 1/\sqrt{3}$, the free energy to create a vortex line is decreased, so we propose a generalized form of (8.10.36),

$$\frac{d}{dt} n_v = A_1 |\mathbf{w}| n_v^{3/2} - A_2 (1 - 3K^2)^{\gamma_v} n_v^2, \tag{8.10.59}$$

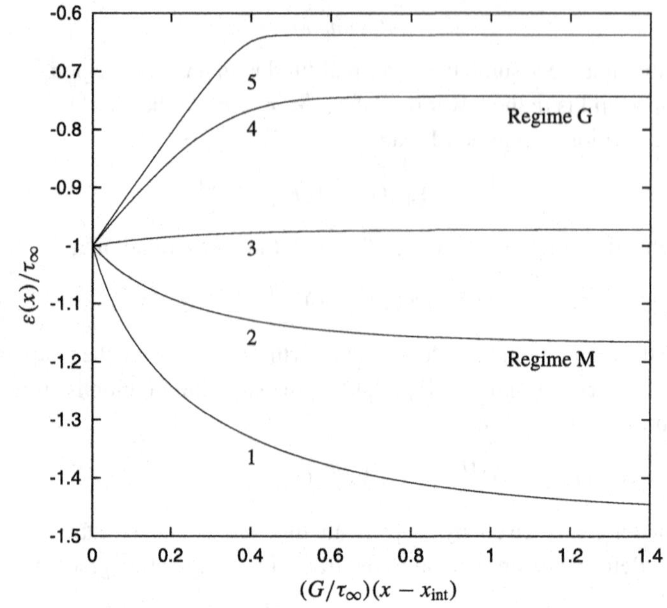

Fig. 8.42. Profiles of the local reduced temperature ε in a superfluid region on earth with $Q =$ 2160, 1240, 710, 200 and 31 erg/cm^2 s (curves 1–5) [152]. A normal fluid region is assumed to be in the region $x < x_{\rm int}$ and $\varepsilon = -\tau_\infty$ at $x = x_{\rm int}$. Regime G is realized for $Q \lesssim 10^3$ erg/cm^2 s on earth in this geometry. We can see that $\varepsilon \to \varepsilon_{\rm c}$ for $x - x_{\rm int} \gg \tau_\infty/G$, resulting in self-organized superfluid states in both regimes M and G.

where the exponent $\gamma_{\rm v}$ has not yet been calculated. In steady states, (8.10.53) is then generalized as

$$\frac{d}{dx}|\varepsilon| = -G + B_{\rm v}|\varepsilon|^{-m_{\rm v}} Q^3(1 - 3K^2)^{-2\gamma_{\rm v}}. \tag{8.10.60}$$

The last factor accounts for the growing mutual friction as $K \to 1/\sqrt{3}$. We solve the above equation at $\gamma_{\rm v} = 1$ and $R_\infty = 2.5$ to obtain the temperature curves 3–5 in Fig. 8.44. We recognize that the system tends to a self-organized superfluid state for $x - x_{\rm int} \gg \ell_{GQ}$, in which the gradient of T is equal to that of $T_\lambda(p)$ and $|\varepsilon|$ approaches a limiting reduced temperature $\varepsilon_{\rm c}$. In particular, for $Q \ll 10^3(g/g_{\rm earth})^{2\nu/(1+\nu)}$, K should be close to $1/\sqrt{3}$ and

$$\varepsilon \to \varepsilon_{\rm c} \cong -2\tau_Q. \tag{8.10.61}$$

The scaled vortex line density $n_{\rm v}\xi^2$ in (8.10.51) can be of order E_0^{-2} in regime G, while it is very small in the conventional regime M.

A self-organized superfluid region at constant ε_{bot}

We numerically solve (6.7.23)–(6.7.27) in regime G in 1D [152]. As in Fig. 6.28 we measure the space, reduced temperature, and heat flux in units of 1.6×10^{-3} cm, 2.5×10^{-8}, and 11 erg/cm^2 s, respectively. The unit of time is about 10^{-4} s. We prepare a normal fluid state heated from above and then suddenly lower the bottom reduced temperature ε_{bot} from 2 to -2 to produce an embryo of superfluid at the bottom. The heat flux at the top is fixed at 0.1. The superfluid region then grows into the upper normal fluid region. Figure 8.43 shows numerical data at $t = 45\,615$, for $\rho_s(x,t) = |\Psi(x,t)|^2$ in (a) and $A(x,t) = T(x,t)/T_{\lambda\text{bot}} - 1$ and $\varepsilon(x,t) = T/T_{\lambda\text{bot}} - 1 + G(x-h)$ in (b). We can see a number of phase slip centers [154], the one-dimensional counterpart of vortices, in the expanding superfluid region. They are rapidly varying in time and the temperature (solid line) has a gradient such that ε (dashed line) becomes flat on the average as shown in (b). In the self-organized superfluid region, the superfluid velocity $\text{Im}(\Psi^*\nabla_x\Psi)/|\Psi|^2$ is fluctuating around the critical value $1/\sqrt{3}$ and the heat flux is about 0.5 on the average. In this case the front of the superfluid region reaches the top on long timescales because a large amount of entropy is stored in the upper normal fluid region., If we increase the heat flux at the top to the value at the bottom ($Q_{\text{top}} = Q_{\text{bot}} = Q$), the interface motion can be stopped and coexistence of a normal fluid and self-organized superfluid state can be realized in a dynamical steady state. In ^{4}He in this geometry, an expanding superfluid region is in regime G only when the reduced temperature $\varepsilon_{\text{bot}}(< 0)$ at the bottom is smaller than $10^3\tau_g$ in magnitude. For deeper quenching, regime M will be realized.

A self-organized superfluid region at constant Q

As already mentioned below (6.7.42), Moeur *et al.* [155] observed a self-organized superfluid region for $Q \gtrsim 1$ erg/cm^2 s. Simulations were also performed in this geometry in 1D [156, 157]. Using the dynamic model (6.7.23) and (6.7.24) we prepare an equilibrium superfluid state for $t < 0$ and subtract a constant heat flux Q_{bot} from the bottom for $t > 0$. Then a self-organized superfluid region with defects expands upwards into the upper superfluid region without defects. If we subsequently apply the same heat flux from the top ($Q_{\text{top}} = Q_{\text{bot}} = Q$), we may realize a dynamical steady two-phase state, as shown in Fig. 8.44. In this case there is no sharp boundary between the two phases and the width of the transition region is on the order of the defect spacing. We can see continuous generation and annihilation of defects in the self-organized region. Second-sound waves are then emitted into the upper superfluid region, causing large-scale temperature perturbations.

8.10.5 Defect turbulence in liquid crystalline polymers in shear flow

It is worth noting that similar defect turbulence has been observed in nematic liquid crystalline polymers subjected to shear [158, 159], in which shear flow causes tumbling of liquid crystalline molecules. The coherence of rotating molecular alignment is broken on the spatial scale of a typical distance a_d among disclination lines. Such states in liquid

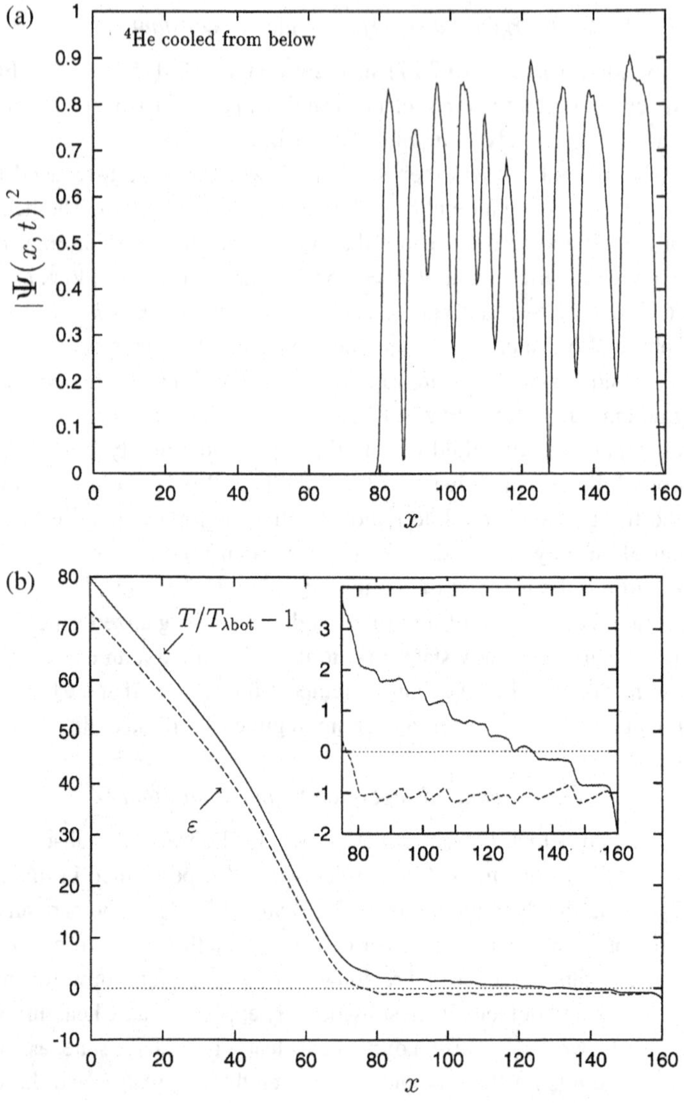

Fig. 8.43. A superfluid region created at the bottom ($x = 160$) and expanding towards the top ($x = 0$) at $t = 45\,615$ in regime G obtained by numerically solving (6.7.23) and (6.7.24) [152]. (a) The superfluid density is plotted. Because the simulation is in 1D, there are many phase slip centers in the expanding superfluid region. Space and time are scaled by 1.6×10^{-3} cm and 10^{-4} s. (b) $T/T_{\lambda\mathrm{bot}} - 1$ (solid line) and ε (dashed line) are plotted in the transient state and are expanded in the inset. They are scaled by 2.5×10^{-8}.

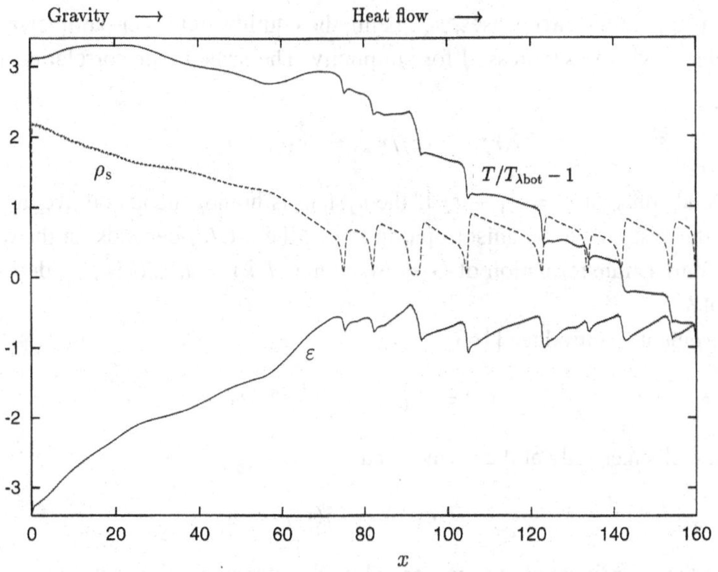

Fig. 8.44. A self-organized superfluid with defects below a superfluid without defects in regime G at $Q = 11$ erg/cm^2 s applied from above. They are obtained by numerically solving the model (6.7.23)–(6.7.28).

crystals are sometimes called polydomain states. In dynamical steady states, the viscous stress $\eta\dot{\gamma}$ is balanced with the Franck elastic energy density K/a_{d}^2, where η is the viscosity, $\dot{\gamma}$ is the shear rate taken to be positive, and K is the Franck elastic constant. Thus a_{d} is estimated [158] as

$$a_{\mathrm{d}} \sim (K/\eta\dot{\gamma})^{1/2}. \tag{8.10.62}$$

The line density $n_{\mathrm{v}} \sim a_{\mathrm{d}}^{-2}$ is then proportional to $\dot{\gamma}$ in the steady state. In transient states Larson and Doi derived the following evolution equation [159],

$$\frac{d}{dt} n_{\mathrm{v}} = B_1 \dot{\gamma} n_{\mathrm{v}} - B_2 n_{\mathrm{v}}^2, \tag{8.10.63}$$

from nematodynamic equations with B_1 and B_2 being appropriate constants. This equation is analogous to the Vinen equation (8.10.36). We notice surprising similarity between these two phenomena in which a large number of defects are generated by an externally applied flow.

Appendix 8A Generalizations and variations of the Porod law

We examine the short distance behavior ($\xi \ll r \ll \ell(t)$) or the large wave number behavior ($\xi^{-1} \gg k \gg \ell(t)^{-1}$) of the pair correlation function and the structure factor neglecting the thermal fluctuations and taking the thin interface limit [160]–[162]. Let $\epsilon(\boldsymbol{r}) = \psi(\boldsymbol{r}, t)/\psi_{\mathrm{eq}}$

take either of ± 1 in the two phases, ψ_{eq} being the equilibrium order parameter value. The time variable t will be suppressed for simplicity. The scaled pair correlation function is written as

$$G(\boldsymbol{r}) = g(r)/\psi_{\text{eq}}^2 = \langle \epsilon(\boldsymbol{r}_1)\epsilon(\boldsymbol{r}_2)\rangle, \tag{8A.1}$$

which depends only on $\boldsymbol{r} = \boldsymbol{r}_1 - \boldsymbol{r}_2$ if the system is homogeneous on average. Here we allow that the system can be anisotropic in space; then, $G(\boldsymbol{r})$ depends on the direction of $\boldsymbol{r}$ and the Fourier transformation of $G(\boldsymbol{r})$, written as $\hat{I}(\boldsymbol{k}) = I_{\text{dom}}(\boldsymbol{k})/\psi_{\text{eq}}^2$, depends on the direction of $\boldsymbol{k}$.

It is convenient to introduce [163]

$$G_1 = \nabla_1 \cdot \nabla_2 G = -\nabla_1^2 G. \tag{8A.2}$$

Notice that ϵ changes only at the surface and

$$\nabla_1 \epsilon(\boldsymbol{r}_1) = 2\delta(s_1)\boldsymbol{n}_1, \tag{8A.3}$$

where s_1 is the coordinate along the normal unit vector $\boldsymbol{n}_1$ on the surface (so $\delta(s_1)$ is the surface δ-function $\hat{\delta}(\boldsymbol{r}_1)$ in (8.2.53)). Then,

$$\begin{aligned} G_1(\boldsymbol{r}) &= \frac{1}{V}\int d\boldsymbol{r}_1 \int d\boldsymbol{r}_2 \big[\nabla_1\epsilon(\boldsymbol{r}_1)\cdot\nabla_2\epsilon(\boldsymbol{r}_2)\big]\delta^{(d)}(\boldsymbol{r}_1 - \boldsymbol{r}_2 - \boldsymbol{r}) \\ &= \frac{4}{V}\int da_1 \int da_2 (\boldsymbol{n}_1\cdot\boldsymbol{n}_2)\delta^{(d)}(\boldsymbol{r}_1 - \boldsymbol{r}_2 - \boldsymbol{r}), \end{aligned} \tag{8A.4}$$

where da_1 and da_2 are the surface elements at the surface positions $\boldsymbol{r}_1$ and $\boldsymbol{r}_2$, respectively, and the surface integrals are taken within a macroscopic volume V containing a large number of domains. Here the δ function in d dimensions is written as $\delta^{(d)}$ to avoid confusion.

If $r = |\boldsymbol{r}_1 - \boldsymbol{r}_2|$ is much smaller than the inverse curvature, the two points are mostly located on the same surface and $\boldsymbol{r}_2 - \boldsymbol{r}_1$ becomes perpendicular to $\boldsymbol{n}_1$ so that $s_1 \equiv (\boldsymbol{r}_2 - \boldsymbol{r}_1)\cdot\boldsymbol{n}_1 = -\boldsymbol{r}\cdot\boldsymbol{n}_1 = O(r^2) \cong 0$ and $\boldsymbol{n}_1 \cdot \boldsymbol{n}_2 \cong 1$. The surface integration $\int da_2 \cdots$ may then be performed to give

$$G_1(\boldsymbol{r}) \cong \frac{4}{V}\int da_1 \delta(s_1) = \frac{4A}{r}\langle \delta(\boldsymbol{n}\cdot\hat{\boldsymbol{r}})\rangle, \tag{8A.5}$$

where A is the surface area (line length) density in 3D (2D) and $\langle\cdots\rangle = \int da(\cdots)/AV$ is the average over the surface and $\hat{\boldsymbol{r}} = r^{-1}\boldsymbol{r}$ is the direction of $\boldsymbol{r}$, so it follows the short distance behavior $G_1 \propto 1/r$. We introduce the distribution function $P(\boldsymbol{n})$ for the normal unit vector $\boldsymbol{n}$ on the surface, in terms of which G_1 may also be expressed as

$$G_1(\boldsymbol{r}) \cong \frac{4A}{r}\int d\Omega P(\boldsymbol{n})\delta(\boldsymbol{n}\cdot\hat{\boldsymbol{r}}), \tag{8A.6}$$

where $d\Omega$ is the solid angle element.

Isotropic case

If the distribution of the surface normal $\boldsymbol{n}$ is isotropic (where $P = 1/4\pi$ for $d = 3$ and $P = 1/2\pi$ for $d = 2$), we obtain

$$G_1 \cong \frac{2A}{r} \quad (3D), \quad G_1 \cong \frac{4A}{\pi r} \quad (2D). \tag{8A.7}$$

In this case G behaves at short distances as

$$G = 1 - Ar + \cdots \quad (3D), \qquad G = 1 - 4\pi^{-1}Ar + \cdots \quad (2D), \tag{8A.8}$$

which follows from

$$G_1 = -\nabla^2 G = -\left[\frac{\partial^2}{\partial r^2} + (d-1)\frac{\partial}{r\partial r}\right]G \tag{8A.9}$$

for the isotropic case. This expansion form holds for thin and smooth interfaces and can also be derived from simple geometrical arguments. Namely, when r is much smaller than the typical inverse curvature, $G(r)$ can be -1 only when the two points $\boldsymbol{r}_1$ and $\boldsymbol{r}_2$ are both in the layer region where the distance to the surface is shorter than r. This probability is of order rA, giving rise to the second terms in (8A.8). The Porod tail (8.1.21) of the structure factor is now readily obtained by Fourier transformation of (8A.7).

Anisotropic case

The Fourier transformation of $G_1(\boldsymbol{r})$ is written as

$$\hat{I}_1(\boldsymbol{k}) \cong 4A \int d\Omega P(\boldsymbol{n})(2\pi)^{d-1}\delta^{(d-1)}(\boldsymbol{k}_\perp), \tag{8A.10}$$

where $\boldsymbol{k}_\perp$ is the perpendicular part of $\boldsymbol{k}$ to $\boldsymbol{n}$ and is a $d-1$ dimensional vector. Due to $\delta^{(d-1)}(\boldsymbol{k}_\perp)$ in the above integral, $\boldsymbol{n}$ must be parallel to $\boldsymbol{k}$ and $P(\boldsymbol{n})$ may be replaced by $P(\hat{\boldsymbol{k}})$, where $\hat{\boldsymbol{k}} = k^{-1}\boldsymbol{k}$ is the direction of $\boldsymbol{k}$. Then $I_1 \propto AP(\hat{\boldsymbol{k}})/k^{d-1}$. Because the Fourier transform of $G(\boldsymbol{r})$ is given by $\hat{I}(k) = \hat{I}_1(k)/k^2$, we obtain

$$\hat{I} \cong \frac{32\pi^2 A}{k^4}P(\hat{\boldsymbol{k}}) \quad (3D), \quad \hat{I} \cong \frac{16\pi A}{k^3}P(\hat{\boldsymbol{k}}) \quad (2D), \tag{8A.11}$$

which holds in the region $\ell(t)^{-1} \ll k \ll \xi^{-1}$. In the isotropic case, the above relations reduce to those known in the literature. While the Porod tail has been discussed for the isotropic case in the literature, the above formulas provide a new experimental possibility of gaining information of anisotropy of the domain structure. An example is given in Fig. 8.35, which shows the Porod tail from a uniaxially stretched gel. We note that domains in fluids are elongated in gravity and in shear flow, while domains in solids usually take anisotropic shapes due to elasticity. The Porod tail can be detected even after the spinodal ring has collapsed at very late stages.

Kirste–Porod corrections and Tomita's sum rule

In the isotropic case it is easy to calculate the second-order corrections. We consider the problem in 3D here. Expansion of G_1 in powers of r reads [164, 165]

$$G_1 = \frac{2A}{r}\left(1 - \frac{r^2}{2R_{\mathrm{m}}^2} + \cdots\right). \tag{8A.12}$$

Here $1/R_{\mathrm{m}}^2$ is written in terms of the principal curvatures $1/R_1$ and $1/R_2$ as

$$\frac{1}{R_{\mathrm{m}}^2} = \left\langle \frac{3}{8}\mathcal{K}^2 - \frac{1}{2R_1R_2}\right\rangle \tag{8A.13}$$

where $\mathcal{K} = 1/R_1 + 1/R_2$. The structure factor may then be expanded as

$$\hat{I} \cong \frac{8\pi A}{k^4}\left(1 + \frac{1}{R_{\mathrm{m}}^2 k^2} + \cdots\right). \tag{8A.14}$$

The correction term was first derived by Kirste and Porod [161]. We note that there is no constant term in G_1, which leads to Tomita's sum rule [3, 163],

$$\int_0^\infty dk[k^4\hat{I}(k) - 8\pi A] = 0. \tag{8A.15}$$

The above integral is equal to $2\pi^2 \lim_{r\to 0}[G_1(r) - 2A/r] = 0$. This sum rule has been confirmed by a simulation of spinodal decomposition without thermal noise [166].

Scattering from bilayers

Scattering experiments have been performed from fluid membranes in the so-called L_3 (sponge) phase without long-range order [168]. There, thin bilayers separate a fluid into two equivalently percolated domains and hence scattering mainly arises from surfactant molecules trapped on the surface. In this case the structure factor $\hat{I}_{\mathrm{s}}$ of the surfactant has a tail [165, 168],

$$\hat{I}_{\mathrm{s}}(\boldsymbol{k}) \cong \frac{8\pi^2 A}{k^2}P(\hat{\boldsymbol{k}}), \tag{8A.16}$$

in the region $\ell^{-1} \ll k \ll b^{-1}$, where ℓ is the typical length of the surface structure and b is the thickness of the bilayer. The above formula may be used to examine the distribution $P(\boldsymbol{n})$ of the surface normal vector $\boldsymbol{n}$, whose anisotropy may be induced by external forces.

Scattering from fractal surfaces

So far we have assumed smooth surfaces, but surfaces can be finely rugged with a surface fractal dimension D_{f}. That is, to cover such a surface with spheres of radius a, we need spheres proportional to $a^{-D_{\mathrm{f}}}$. Here $d - 1 \le D_{\mathrm{f}} < d$, and $D_{\mathrm{f}} = d - 1$ for smooth surfaces. In this case the following tail is well known [3, 169],

$$\hat{I}(k) \propto k^{-2d+D_{\mathrm{f}}}. \tag{8A.17}$$

Appendix 8B The pair correlation function in the nonconserved case

Calculation for $n = 1$

There is no essential difference in the calculation of the two-body correlation function $g(|\boldsymbol{r}_1 - \boldsymbol{r}_2|, t_1, t_2) \equiv \psi_{\rm eq}^2 G$ at a late stage between the KYG and OJK theories [8, 10]. We follow the notation in the KYG theory. The results in the OJK theory are obtained if L is replaced by L'. From (8.2.12) we express $\psi(\boldsymbol{r}, t)$ in terms of the subsidiary field $\varphi(\boldsymbol{r}, t)$ as

$$\psi(\boldsymbol{r}, t) = \psi_e \int \frac{dp}{i\pi p} \exp[ip\varphi(\boldsymbol{r}, t)], \tag{8B.1}$$

where the Cauchy principal value should be taken at $p = 0$. Then,

$$G = \int \frac{dp_1}{i\pi p_1} \int \frac{dp_2}{i\pi p_2} \langle \exp[ip_1\varphi_1 + ip_2\varphi_2] \rangle, \tag{8B.2}$$

where $\varphi_1 = \varphi(\boldsymbol{r}_1, t_1)$ and $\varphi_2 = \varphi(\boldsymbol{r}_2, t_2)$. Because φ_1 and φ_2 are gaussian, the above average can be readily performed as

$$\langle \exp(ip_1\varphi_1 + ip_2\varphi_2) \rangle = \exp\left(-\frac{1}{2}\beta_1 p_1^2 - \frac{1}{2}\beta_2 p_2^2 - \beta_{12} p_1 p_2 \right), \tag{8B.3}$$

where β_1, β_2, and β_{12} are defined by (8.1.35). From (8.1.26) and (8.1.27) we have

$$\beta_{12} = \int_{\boldsymbol{k}} \chi_k \exp\left[L(\kappa^2 - k^2)(t_1 + t_2) + i\boldsymbol{k} \cdot (\boldsymbol{r}_1 - \boldsymbol{r}_2) \right], \tag{8B.4}$$

in terms of the initial variance χ_k. By changing the integration variables to $x_1 = \beta_1^{1/2} p_1$ and $x_2 = \beta_2^{1/2} p_2$, we obtain [170]

$$G = \int \frac{dx_1}{i\pi x_1} \int \frac{dx_2}{i\pi x_2} \exp\left(-\frac{1}{2}x_1^2 - \frac{1}{2}x_2^2 - Xx_1x_2 \right), \tag{8B.5}$$

where $X = \beta_{12}/(\beta_1\beta_2)^{1/2}$. Therefore, G is a function of X only and

$$\frac{d}{dX}G = \pi^{-2} \int dx_1 \pi^{-2} \int dx_2 \exp\left(-\frac{1}{2}x_1^2 - \frac{1}{2}x_2^2 - Xx_1x_2 \right) = \frac{2}{\pi}(1 - X^2)^{-1/2}. \tag{8B.6}$$

Integration with respect to X gives (8.1.37). (i) If the lengths $\sqrt{Lt_1}$ and $\sqrt{Lt_2}$ are much longer than the initial correlation length, we may replace χ_k by its long-wavelength limit χ_0 to obtain

$$\beta_{12} = \chi_0 \left[4\pi L(t_1 + t_2) \right]^{-d/2} \exp\left[-|\boldsymbol{r}_1 - \boldsymbol{r}_2)|^2 / 4L(t_1 + t_2) \right], \tag{8B.7}$$

which leads to (8.1.39). (ii) If the initial correlation is long as in the periodic quench case, the above approximation is not valid. Focusing only on small wave number behavior, we may set

$$G \cong \frac{2}{\pi}X = \frac{2}{\pi}\beta_{12}/(\beta_1\beta_2)^{1/2}. \tag{8B.8}$$

The domain structure factor, the Fourier transformation of $\psi_{\rm eq}^2 G$ at $t_1 = t_2 = t$, now becomes

$$I_{\rm dom}(k,t) \cong \psi_{\rm eq}^2 \chi_k \exp(-2Lk^2t) \bigg/ \int_q \chi_q \exp(-2Lq^2t). \qquad (8B.9)$$

Calculation for $n \geq 2$

For many-component systems we use the identity,

$$\frac{\varphi}{|\varphi|} = A_n \int_p \frac{i\varphi}{p^{n-1}} \exp(i\boldsymbol{p}\cdot\boldsymbol{\varphi}), \qquad (8B.10)$$

where $A_n = (4\pi)^{(n-1)/2}\Gamma[(n-1)/2]$ and $\int_p = (2\pi)^{-n} \int d\boldsymbol{p}$. When (8.1.72) holds, we have

$$G = [(n-1)A_n]^2 \int_{p_1}\int_{p_2} \frac{\boldsymbol{p}_1\cdot\boldsymbol{p}_2}{(p_1p_2)^{n+1}} \exp\left(-\frac{1}{2}\beta_1 p_1^2 - \frac{1}{2}\beta_2 p_2^2 - \beta_{12}\boldsymbol{p}_1\cdot\boldsymbol{p}_2\right). \qquad (8B.11)$$

By changing the integration variables as $\boldsymbol{x}_1 = \beta_1^{1/2}\boldsymbol{p}_1$ and $\boldsymbol{x}_2 = \beta_2^{1/2}\boldsymbol{p}_2$, we notice that G depends only on X. We may perform the integrations over $x_1 = |\boldsymbol{x}_1|$ and $x_2 = |\boldsymbol{x}_2|$ in dG/dX as in (8B.6). Some calculations yield (8.1.73).

Appendix 8C The Kawasaki–Yalabik–Gunton theory applied to periodic quench

We present the calculation for periodically modulated states in the KYG scheme. More details can be found in Ref. [11]. The nonlinear transformation (8.1.23) for the normal quench case may be generalized to the periodic quench case as

$$\psi(\boldsymbol{r},t) = \varphi(\boldsymbol{r},t)\big/\left[1 + b(t)\varphi(\boldsymbol{r},t)^2\right]^{1/2}, \qquad (8C.1)$$

where

$$b(t) = 2Lu_0 \int_0^t dt' \exp\left[2L\int_{t'}^t dt'' r(t'')\right]. \qquad (8C.2)$$

Here $\varphi(\boldsymbol{r},t)$ is the solution of (8.1.45) without the nonlinear and noise terms, so it obeys

$$\frac{\partial}{\partial t}\varphi = -L[r(t) - \nabla^2]\varphi. \qquad (8C.3)$$

The space average $\bar{\varphi}(t) = \langle\varphi(\boldsymbol{r},t)\rangle$ obeys $\partial\bar{\varphi}/\partial t = -Lr(t)\bar{\varphi}$ and the deviation $\delta\varphi = \varphi(\boldsymbol{r},t) - \bar{\varphi}(t)$ is gaussian. Then (8.1.51) is justified when

$$\beta(t) = \langle\delta\varphi(\boldsymbol{r},t)^2\rangle = \int_q \chi_q \exp\left[-2Lq^2t - 2L\int_0^t dt' r(t')\right] \qquad (8C.4)$$

is much larger than $b(t)^{-1}$. In the step-wise case (8.1.46) we obtain

$$\beta(t)b(t) = u_0\left(\frac{1}{|r_-|} + \frac{1}{r_+}\right)\exp(2L|r_-|t_1)\int_q \chi_q \exp(-2Lq^2t). \qquad (8C.5)$$

in the time region $t - t_1 \gtrsim 1/Lr_+$. In a periodic disordered state, this quantity is of order $AS(S-1)^{-1}\exp(-2L\bar{r}t_p)$ at $t = t_p$, where $\bar{r}$ is the time average of $r(t)$. We are thus allowed to assume $\beta(t)b(t) \gg 1$ for $2L|\bar{r}|t_p \gg 1$ with $\bar{r} < 0$. In ordered states the average order parameter is calculated from

$$\begin{aligned}\bar{\psi}(t) &= [2\pi b(t)\beta(t)]^{-1/2}\int d\varphi \frac{\varphi}{|\varphi|}\exp\left\{-\frac{1}{2\beta(t)}[\varphi - \bar{\varphi}(t)]^2\right\} \\ &= [2/\pi b(t)]^{1/2}\int_0^{Z(t)} dx\, \exp(-x^2/2), \end{aligned} \tag{8C.6}$$

where $Z(t) = \bar{\varphi}(t)/\beta(t)^{1/2}$. At $t = t_p$ we have $\eta' = \bar{\psi}(t_p)$ to find the recursion relation (8.1.55).

Appendix 8D The structure factor tail for $n = 2$

We derive the structure factor tail at large wave number k for $n = 2$ from geometrical arguments [6, 21].

(i) In 2D, the complex order parameter $\psi = \psi_1 + i\psi_2$ close to a vortex but outside its core at the origin is expressed as

$$\psi_1/\psi_{eq} = \pm y/r, \quad \psi_2/\psi_{eq} = \pm x/r, \tag{8D.1}$$

where the charge of the vortex is assumed to be ± 1 (for which see (4.5.1) or (4.5.11)). In a system with volume V we have

$$G(r) = \psi_{eq}^{-2}\langle \boldsymbol{\psi}(\boldsymbol{r}_1)\cdot\boldsymbol{\psi}(\boldsymbol{r}_2)\rangle = \frac{n_v}{V}\int d\boldsymbol{r}_1 \int d\boldsymbol{r}_2 \frac{\boldsymbol{r}_1\cdot\boldsymbol{r}_2}{r_1 r_2}\delta^{(2)}(\boldsymbol{r}_1 - \boldsymbol{r}_2 - \boldsymbol{r}), \tag{8D.2}$$

where $\boldsymbol{r} = \boldsymbol{r}_1 - \boldsymbol{r}_2$ and n_v is the vortex number density. The Fourier transformation gives the defect structure factor (divided by ψ_{eq}^2),

$$\hat{I}(k) = n_v\left|\int d\boldsymbol{r}_1 \frac{\boldsymbol{r}_1}{r_1} e^{i\boldsymbol{k}\cdot\boldsymbol{r}_1}\right|^2 = (2\pi)^2 n_v k^{-4}. \tag{8D.3}$$

(ii) In 3D, let us consider a weakly curved vortex line with charge ± 1. We take the origin of the reference frame at a point on the line and the z axis along the tangential unit vector $\boldsymbol{t}$. Close to the line but outside the core, ψ depends on $\boldsymbol{r}_\perp = (x, y, 0)$ as in (8D.1) and is nearly independent of z. Thus,

$$G(r) = \frac{1}{V} n_v \sum_j \int d\boldsymbol{r}_1 \int d\boldsymbol{r}_2 \frac{\boldsymbol{r}_{\perp 1}\cdot\boldsymbol{r}_{\perp 2}}{r_{\perp 1} r_{\perp 2}}\delta^{(3)}(\boldsymbol{r}_1 - \boldsymbol{r}_2 - \boldsymbol{r}) \tag{8D.4}$$

The n_v is the line length density of vortices. The Fourier transformation gives

$$\begin{aligned}\hat{I}(k) &= n_v \int dz_1 (2\pi)^2 k_\perp^{-4}\exp[i\boldsymbol{t}\cdot\boldsymbol{k}(z_1 - z_2)] \\ &= n_v (2\pi)^3 k^{-5}\langle\delta(\boldsymbol{t}\cdot\hat{\boldsymbol{k}})\rangle. \end{aligned} \tag{8D.5}$$

In the first line, $\boldsymbol{k}_\perp = \boldsymbol{k} - (\boldsymbol{k}\cdot\boldsymbol{t})\boldsymbol{t}$ is the perpendicular part and becomes equal to $\boldsymbol{k}$ due to the δ-function. In the second line, $\hat{\boldsymbol{k}} = k^{-1}\boldsymbol{k}$ is the direction of the wave vector, and the average is over the direction of $\boldsymbol{t}$. If the distribution of $\boldsymbol{t}$ is isotropic, we have $\langle\delta(\boldsymbol{t}\cdot\hat{\boldsymbol{k}})\rangle = 1/2$ to obtain (8.1.75).

Appendix 8E Differential geometry

We consider the differential geometry of a smooth surface determined by $u(\boldsymbol{r}) = 0$. It is sufficient to examine the geometry in the neighborhood of a reference point $\mathbf{r}_0 = (x_0, y_0, z_0)$ on the surface, where $u(\mathbf{r})$ is expanded as

$$u(\boldsymbol{r}) = \sum_i (x_i - x_{i0})\nabla_i u + \frac{1}{2}\sum_{ij}(x_i - x_{i0})(x_j - x_{j0})\nabla_i\nabla_j u + \cdots, \tag{8E.1}$$

where $\nabla_i = \partial/\partial x_i$ and the derivatives are those at $\mathbf{r}_0$. The normal unit vector $\boldsymbol{n}$ of the surface is generally written as

$$\boldsymbol{n} = |\nabla u|^{-1}\nabla u. \tag{8E.2}$$

The normal at the point $\boldsymbol{r}_0$ will be written as $\boldsymbol{n}_0$ and the z axis will be taken along it, so $\boldsymbol{n}_0 = (0, 0, 1)$. Appropriately choosing the x and y axes at $\boldsymbol{r}_0$, we may express the distances of $\boldsymbol{r} = (x, y, z)$ to the surface as

$$s \equiv u(\boldsymbol{r})/|\nabla u| = z - z_0 + \frac{1}{2R_1}(x - x_0)^2 + \frac{1}{2R_2}(y - y_0)^2 + \cdots, \tag{8E.3}$$

where R_1 and R_2 are the principal curvatures at $\boldsymbol{r}_0$. The surface $u = 0$ is thus expressed as

$$z - z_0 + \frac{1}{2R_1}(x - x_0)^2 + \frac{1}{2R_2}(y - y_0)^2 + \cdots = 0. \tag{8E.4}$$

From (8E.2) the normal unit vector is written as

$$\boldsymbol{n} \cong \left(\frac{1}{R_1}(x - x_0), \frac{1}{R_2}(y - y_0), 1\right). \tag{8E.5}$$

For arbitrary $\boldsymbol{r} = (x, y, z)$ in the neighborhood of $\boldsymbol{r}_0$ we may check the relation,

$$s = (\boldsymbol{r} - \boldsymbol{r}_a)\cdot\boldsymbol{n} = (\boldsymbol{r} - \boldsymbol{r}_a)\cdot\boldsymbol{n}_0 = (\boldsymbol{r} - \boldsymbol{r}_0)\cdot\boldsymbol{n}, \tag{8E.6}$$

where

$$\boldsymbol{r}_a = (x, y, z_0 - (x - x_0)^2/2R_1 - (y - y_0)^2/2R_2) \tag{8E.7}$$

is the closest point on the surface $u = 0$ to $\boldsymbol{r}$. From (8E.5) the curvature relation follows as

$$\mathcal{K} = \nabla\cdot\boldsymbol{n} = \frac{1}{R_1} + \frac{1}{R_2}. \tag{8E.8}$$

The above relations together with $\partial z_0/\partial t = -v$ yield (8.2.20).

Appendix 8F Calculation in the Langer–Bar-on–Miller theory

We briefly explain the calculation scheme of the LBM theory [38]. The wave number κ in (8.3.16) and the upper cut-off wave number Λ in our notation correspond to k_c and k_{max} in LBM, respectively. In addition, $r_c = 0$ in LBM. We divide the space into cubic cells with a lattice constant a of the order of the correlation length ξ. Following LBM we relate a to Λ as $(4\pi/3)\Lambda^3 = (2\pi/a)^3$ or $a^3 = 6\pi^2/\Lambda^3$. The Langevin equation (8.3.2) is rewritten as

$$\frac{\partial}{\partial t}\psi_i = -L\sum_{j,\ell}\Delta_{ij}\big[(\Delta_{j\ell} + r\delta_{j\ell})\psi_\ell + \delta_{j\ell}u_0\psi_\ell^3\big] + \theta_i, \tag{8F.1}$$

where the matrix Δ_{ij} is the lattice representation of $-\nabla^2$. For example, in the nearest neighbor approximation we have $\Delta_{ii} = 6a^{-2}$ and $\Delta_{ij} = -a^{-2}$ for nearest neighbor pairs i and j. The noise term satisfies

$$\langle\theta_i(t)\theta_j(t')\rangle = 2La^{-3}\Delta_{ij}\delta(t-t'). \tag{8F.2}$$

We require the ansatz (8.3.21) on the lattice. Then $g_{ij}(t) = \langle\delta\psi_i(t)\delta\psi_j(t)\rangle$ obeys

$$\frac{\partial}{\partial t}g_{ij} = -L\sum_\ell[\hat{\mathcal{L}}_{i\ell}g_{\ell j} + \mathcal{L}_{j\ell}g_{\ell i}] + 2La^{-3}\Delta_{ij}, \tag{8F.3}$$

where

$$\mathcal{L}_{ij} = \sum_\ell \Delta_{i\ell}[\Delta_{\ell j} - A(t)\delta_{\ell j}], \tag{8F.4}$$

with $A(t)$ being defined by (8.3.22).

The equation for ρ_1 may be constructed from the Fokker–Planck equation for the microscopic distribution $P(\{\psi\},\ldots,t)$ for all the lattice points. It is written in the form,

$$\frac{\partial}{\partial t}\rho_1(\psi,t) = L\frac{\partial}{\partial\psi}\left[G(\psi) + \Delta a^{-3}\frac{\partial}{\partial\psi}\right]\rho_1, \tag{8F.5}$$

where $\Delta = \Delta_{ii}$, and

$$G(\psi) = W(t)\frac{\delta\psi}{\langle\delta\psi^2\rangle} + \Delta u_0\left[\psi^3 - \langle\psi^3\rangle - \langle\psi^3\delta\psi\rangle\frac{\delta\psi}{\langle\delta\psi^2\rangle}\right], \tag{8F.6}$$

with

$$W(t) = \sum_j \mathcal{L}_{ij}g_{ji} = \sum_{\ell j}\Delta_{i\ell}[\Delta_{\ell j} - \delta_{\ell j}A(t)]g_{ji}. \tag{8F.7}$$

From (8F.3) the variance $s(t) = \langle(\delta\psi)^2\rangle$ at a point obeys

$$\frac{\partial}{\partial t}s(t) = 2L\big[-W(t) + a^{-3}\Delta\big]. \tag{8F.8}$$

In the continuum limit we have

$$W(t) = (2\pi^2)^{-1}\int_0^\Lambda dk k^4\big[k^2 - \kappa^2 A(t)\big]I(k,t). \tag{8F.9}$$

Because (8F.8) should be consistent with (8.3.23), we require $(2\pi)^{-3}\int_0^\Lambda dkk^2 = \Delta/a^3$ or $\Delta = 3\Lambda^2/5$. Finally, we need to give the relation between Λ and κ. It is natural to assume

$$\Lambda = \alpha\kappa \quad \text{or} \quad a = [(6\pi^2)^{1/3}/\alpha]\kappa^{-1}, \tag{8F.10}$$

where α is a parameter of order 1. LBM set $\alpha = 1$ in their numerical calculation. Now (8.3.23) and (8F.5) constitute closed dynamic equations.

Appendix 8G The Stefan problem for a sphere and a circle

We solve the Stefan problem for a growing sphere in 3D and circle in 2D to clarify the condition under which the quasi-static approximation is valid. First, let us consider an isolated sphere in 3D whose radius at $t = 0$ is slightly larger than R_c. We assume no other droplets within the distance of the diffusion length $\ell_D(t) = (Dt)^{1/2}$ in the following. Then the initial growth rate is of order $\tau_c^{-1} = D\Delta/R_c^2 \sim D\Delta^3/d_0^2$. For $t \ll \tau_c$ we may thus solve the diffusion equation (8.4.6) at fixed R in the form,

$$\Psi(r,t) = 2\psi_{eq}\Delta + \left[(\Psi_a - 2\Delta\psi_{eq})\frac{R}{r}\right]\left[\frac{1}{\sqrt{\pi}}\int_Z^\infty ds \exp\left(-\frac{1}{4}s^2\right)\right], \tag{8G.1}$$

where Ψ_a is the boundary value given by (8.4.12) and $Z \equiv (r - R)/\sqrt{Dt}$. For $r - R \ll \ell_D(t)$ the last factor in (8G.1) may be set equal to 1, leading to the quasi-static solution (8.4.27). Second, in the time region $t \gg \tau_c$, R much exceeds R_c and the surface tension effect at the boundary condition becomes unimportant. Then we may set $\Psi(R,t) = 0$ at the interface. Expecting the growth $R(t) \propto (Dt)^{1/2}$, we set

$$R(t) = (2pDt)^{1/2}, \qquad \Psi(r,t) = 2\psi_{eq}\Delta G(r/R(t)), \tag{8G.2}$$

where p is a dimensionless number to be determined below. The scaling function $G(s)$ in the second line satisfies

$$-ps\frac{d}{ds}G(s) = \left(\frac{d}{ds} + \frac{2}{s}\right)\frac{d}{ds}G(s), \tag{8G.3}$$

which is solved to give

$$G(s) = C^{-1}\int_1^s ds_1 s_1^{-2}\exp\left(-\frac{p}{2}s_1^2\right), \tag{8G.4}$$

with

$$C = \int_1^\infty ds_1 s_1^{-2}\exp\left(-\frac{p}{2}s_1^2\right). \tag{8G.5}$$

The conservation law (8.4.15) gives

$$p = \Delta C^{-1}\exp\left(-\frac{p}{2}\right), \tag{8G.6}$$

which determines p as a function of Δ. For $\Delta \ll 1$ we have

$$p = \Delta\left[1 + \left(\frac{1}{2}\pi\Delta\right)^{1/2} + \cdots\right]. \tag{8G.7}$$

If $\Delta \ll 1$, we have $p \ll 1$, $C \cong 1$, and $G(s) \cong 1 - 1/s$ in the region $s \ll p^{-1/2}$ so that $\Psi(r,t)$ in (8G.2) approaches (8.4.27) for $p^{1/2}r/R(t) \ll 1$ or $r \ll \Delta^{-1/2}R(t) \sim \ell_D(t)$. Thus the quasi-static condition is applicable in the region $|r - R| \ll \ell_D(t)$ under $\Delta \ll 1$.

In 2D, R/r in (8.4.27) should be replaced by $A\ln(r/R) + 1$ around a circular droplet under the quasi-static condition (8.4.26). To determine the coefficient A, we consider the case $R \gg R_c$ only, neglecting the surface tension effect and assuming the scaling solution (8G.2). The 2D scaling function $G(s)$ is obtained if s_1^{-2} in (8G.4) and (8G.5) is replaced by s_1^{-1}. The equation (8G.6) holds also in 2D. For $\Delta \ll 1$ we find

$$p \cong 2\Delta/\ln p^{-1} \cong 2\Delta/\ln \Delta^{-1}. \tag{8G.8}$$

In the range $1 < r/R \lesssim p^{-1/2}$ we thus obtain $\Psi \cong 2\psi_{\rm eq}\Delta A\ln(r/R)$ with

$$A = 2/\ln p^{-1} \cong 2/\ln \Delta^{-1}. \tag{8G.9}$$

Appendix 8H The velocity and pressure close to the interface

Let a 3D incompressible fluid in a two-phase state be acted on by a force localized on a surface $\{\boldsymbol{r}_a\}$. In the Stokes approximation the velocity field $\boldsymbol{v}$ is determined by

$$\eta\nabla^2\boldsymbol{v} - \nabla p + \boldsymbol{X}_a\hat{\delta}(\boldsymbol{r}) = \boldsymbol{0}, \quad \nabla\cdot\boldsymbol{v} = 0, \tag{8H.1}$$

where $\hat{\delta}(\boldsymbol{r})$ is the surface δ-function defined by (8.2.53) and the source $\boldsymbol{X}_a$ is assumed to be smooth on the surface. We also assume that the viscosity η is homogeneous as in near-critical fluids. Then $\boldsymbol{v}$ is expressed in terms of the Oseen tensor as

$$\boldsymbol{v}(\boldsymbol{r}) = \int da' \overleftrightarrow{\boldsymbol{T}}(\boldsymbol{r} - \boldsymbol{r}_{a'})\cdot\boldsymbol{X}_{a'}. \tag{8H.2}$$

Let $\boldsymbol{r}_a$ be the closest point on the surface from $\boldsymbol{r}$ in the neighborhood of the surface. Then, we may take the local reference frame as $\boldsymbol{r} - \boldsymbol{r}_a = \zeta\boldsymbol{n}_a$ and $\boldsymbol{\rho} = \boldsymbol{r}_{a'} - \boldsymbol{r}_a$. The ζ is the coordinate along the normal $\boldsymbol{n}_a$, while $\boldsymbol{\rho}$ is perpendicular to $\boldsymbol{n}_a$. Then $da' = d\boldsymbol{\rho}$ and $|\boldsymbol{r}-\boldsymbol{r}_{a'}| = (\rho^2+\zeta^2)^{1/2}$. Using $\int_0^\infty d\rho[1-\rho/(\rho^2+\zeta^2)^{1/2}] = \int_0^\infty d\rho\rho\zeta^2/(\rho^2+\zeta^2)^{3/2} = |\zeta|$, we may perform the surface integration (that over $\boldsymbol{\rho}$) as

$$\boldsymbol{v}(\boldsymbol{r}) = \boldsymbol{v}(\boldsymbol{r}_a) - \frac{|\zeta|}{\eta}\big[\boldsymbol{X}_a - (\boldsymbol{X}_a\cdot\boldsymbol{n}_a)\boldsymbol{n}_a\big] + O(\zeta^2). \tag{8H.3}$$

In (8.5.8), $\boldsymbol{X}_a = -\sigma\mathcal{K}_a\boldsymbol{n}_a$ is parallel to the normal and the second term of (8H.3) vanishes, so that $\boldsymbol{v}(\boldsymbol{r}) - \boldsymbol{v}(\boldsymbol{r}_a)$ becomes of order ζ^2. This implies continuity of the velocity gradient tensor across the interface.[11] Next we consider the pressure p, which may be expressed as

$$p(\boldsymbol{r}) = \frac{1}{4\pi}\int da' \frac{1}{|\boldsymbol{r}-\boldsymbol{r}_{a'}|^3}(\boldsymbol{r}-\boldsymbol{r}_{a'})\cdot \boldsymbol{X}_{a'}. \tag{8H.4}$$

This quantity is generally discontinuous across the interface. As $\zeta \to 0$, the integration over $\boldsymbol{\rho}$ gives

$$p(\boldsymbol{r}) = \frac{\zeta}{2|\zeta|}\boldsymbol{n}_a \cdot \boldsymbol{X}_a. \tag{8H.5}$$

In the case $\boldsymbol{X}_a = -\sigma \mathcal{K}_a \boldsymbol{n}_a$, the above relation yields the Laplace law (8.5.10).

Appendix 8I Calculation of vortex motion

Because the calculation of vortex motion is very complicated in helium, we here present it in 2D for the simple relaxation model (8.1.66) by setting $|r| = \kappa = u_0 = 1$ and $\theta_j = 0$. Close to a vortex center $\boldsymbol{R}_i = (X_i, Y_i)$, ψ may be approximated as

$$\psi = \psi_{\rm v}(x - X_i, y - Y_i)\exp[i\boldsymbol{q}\cdot(\boldsymbol{r}-\boldsymbol{R}_i)] + \delta\psi, \tag{8I.1}$$

where $\psi_{\rm v}(x, y)$ is the fundamental vortex solution (4.5.1). The phase modulation near the vortex core is written as

$$\theta_{\rm v} = \sum_{j\neq i} \ell_j \tan^{-1}[(y - Y_j)/(x - X_j)]. \tag{8I.2}$$

The wave vector $\boldsymbol{q}$ in (8I.1) represents the gradient of $\theta_{\rm v}$ at $\boldsymbol{r} = \boldsymbol{R}_i$ due to the other vortices far away from $\boldsymbol{R}_i$, where ℓ_j is the charge of the vortex at $\boldsymbol{R}_j$. Then,

$$q_x = -\sum_{j\neq i}\frac{\ell_j}{R_{ij}^2}(Y_i - Y_j), \quad q_y = \sum_{j\neq i}\frac{\ell_j}{R_{ij}^2}(X_i - X_j), \tag{8I.3}$$

where $R_{ij} = |\boldsymbol{R}_i - \boldsymbol{R}_j|$ is the distance between the pair i and j. The deviation $\delta\psi$ is the deformation of ψ from $\psi_{\rm v}$ to be determined below.

The vortex center moves with a velocity $\boldsymbol{v}_L = (v_{Lx}, v_{Ly})$ for nonvanishing $\boldsymbol{q}$. Hereafter we take the origin of the reference frame at $\boldsymbol{R}_i$. As in the one-component case, we set $\partial\psi/\partial t = -\boldsymbol{v}_L \cdot \nabla\psi_{\rm v}$ and neglect the term of order q^2 to obtain

$$\boldsymbol{a}\cdot\nabla\psi_{\rm v} = [-1 - \nabla^2 + 2|\psi_{\rm v}|^2]\delta\psi + \psi_{\rm v}^2\delta\psi^* \tag{8I.4}$$

where the vector $\boldsymbol{a} = (a_x, a_y)$ is defined by

$$a_x = \frac{1}{L}v_{Lx} + 2iq_x, \quad a_y = \frac{1}{L}v_{Ly} + 2iq_y. \tag{8I.5}$$

If $\ell_i = 1$, the left-hand side of (8I.4) consists of two terms,

$$\boldsymbol{a}\cdot\nabla\psi_{\rm v} = a_- B_0(r)e^{2i\varphi} + a_+ C_0(r), \tag{8I.6}$$

[11] There are situations in which surfactant molecules are absorbed on a fluid interface. If they are heterogeneously distributed on it, the areal force density $\boldsymbol{X}_a$ has a lateral component and the viscous shear stress becomes discontinuous across the surface.

where $\varphi = \tan^{-1}(y/x)$ and

$$a_+ = \frac{1}{2}(a_x + ia_y), \quad a_- = \frac{1}{2}(a_x - ia_y). \tag{8I.7}$$

In terms of the amplitude $A_0(r)$ of ψ_v determined by (4.5.4), the two functions $B_0(r)$ and $C_0(r)$ in (8I.6) are expressed as

$$B_0(r) = e^{-2i\varphi}\left(\frac{\partial}{\partial x} + i\frac{\partial}{\partial y}\right)\psi_v = A_0' - \frac{1}{r}A_0, \tag{8I.8}$$

$$C_0(r) = \left(\frac{\partial}{\partial x} - i\frac{\partial}{\partial y}\right)\psi_v = A_0' + \frac{1}{r}A_0, \tag{8I.9}$$

where $A_0' = dA_0/dr$. With the form (8I.6) we notice that $\delta\psi$ may also be expressed as

$$\delta\psi = \delta B(r)e^{2i\varphi} + \delta C(r)^*, \tag{8I.10}$$

where δB and δC depend on r only. Substitution of the above form into (8I.4) yields

$$a_- B_0 = \hat{\mathcal{L}}_2 \delta B + A_0^2 \delta C, \tag{8I.11}$$

$$a_+^* C_0 = \hat{\mathcal{L}}_0 \delta C + A_0^2 \delta B, \tag{8I.12}$$

where $\hat{\mathcal{L}}_n$ $(n = 0, 2)$ are the following operators,

$$\hat{\mathcal{L}}_n = -\frac{d^2}{dr^2} - \frac{1}{r}\frac{d}{dr} + \frac{n^2}{r^2} - 1 + 2A_0^2. \tag{8I.13}$$

It is convenient to define the inner product of two functions $F(r)$ and $G(r)$, which decay sufficiently rapidly at large r, by

$$(F, G) = \int_0^\infty dr\, r F(r) G(r). \tag{8I.14}$$

Then $\hat{\mathcal{L}}_n$ are self-adjoint (or $(F, \hat{\mathcal{L}}_n G) = (\hat{\mathcal{L}}_n F, G)$). The right-hand sides of (8I.11) and (8I.12) vanish for $\delta B = B_0$ and $\delta C = C_0$; in fact, operating ∇ to (4.5.2) we have $\hat{\mathcal{L}}_2 B_0 + A_0^2 C_0 = \hat{\mathcal{L}}_0 C_0 + A_0^2 B_0 = 0$. Thus the solvability condition of (8I.11) and (8I.12) reads

$$a_-(B_0, B_0) + a_+^*(C_0, C_0) = 0, \tag{8I.15}$$

under which δB and δC are well defined. However, B_0 and C_0 decay as r^{-1} at large r, so we define

$$E_0 = \int_0^{R_{\max}} dr \left[\frac{1}{r}A_0^2 + r(A_0')^2\right] \cong \ln(R_{\max}/\xi), \tag{8I.16}$$

where $R_{\max}$ is the upper cut-off length. Then we find $(B_0, B_0) = E_0 - 1$ and $(C_0, C_0) = E_0 + 1$, because $(C_0, C_0) - (B_0, B_0) = 2$. After some calculations (8I.15) may be rewritten as

$$v_x + iv_y = \frac{2|L|^2}{E_0 L_1 - iL_2}(q_y - iq_x), \tag{8I.17}$$

which leads to (8.10.4).

References

[1] J. D. Gunton, M. San Miguel, and P. S. Sani, *Phase Transition and Critical Phenomena*, eds. C. Domb and J. L. Lebowitz (Academic, London, 1983), Vol. 8, p. 267.

[2] H. Furukawa, *Adv. Phys.* **34**, 703 (1994).

[3] H. Tomita, in *Formation, Dynamics and Statistics of Patterns*, eds. K. Kawasaki, M. Suzuki, and A. Onuki (World Scientific, Singapore, 1990), p. 113.

[4] K. Binder, in *Material Sciences and Technology*, eds. R. W. Cohen, P. Haasen, and E. J. Kramer (VCH, Weinheim, 1991), Vol. 5.

[5] J. S. Langer, in *Solids far from Equilibrium*, ed. C. Godrèche (Cambridge University Press, New York, 1992).

[6] A. J. Bray, *Adv. Phys.* **43**, 357 (1994).

[7] M. Suzuki, *Prog. Theor. Phys.* **56**, 77 (1976); *ibid.* **56**, 477 (1976); in *Advances in Chemical Physics*, ed. I. Prigogine and S. Rice (John Wiley & Sons, New York, 1981), p. 195.

[8] K. Kawasaki, M. C. Yalabik, and J. D. Gunton, *Phys. Rev. A* **17**, 455 (1978).

[9] H. Furukawa, *J. Phys. Soc. Jpn* **67**, 210 (1998).

[10] T. Ohta, D. Jasnow, and K. Kawasaki, *Phys. Rev. Lett.*, **49**, 1223 (1982).

[11] A. Onuki, *Prog. Theor. Phys.* **66**, 1230 (1981); *ibid.* **67**, 768 (1982); *ibid.* **67**, 787 (1982).

[12] F. de Pasquale, Z. Racz, M. San Miguel, and P. Tartaglia, *Phys. Rev. B* **30**, 5228 (1984).

[13] R. Loft and T. A. DeGrand, *Phys. Rev. B* **35**, 8528 (1987).

[14] H. Toyoki and K. Honda, *Prog. Theor. Phys.* **78**, 273 (1987).

[15] H. Nishimori and T. Nukii, *J. Phys. Soc. Jpn* **58**, 563 (1989).

[16] H. Toyoki, *J. Phys. Soc. Jpn* **60**, 850 (1991); *ibid.* **60**, 1153 (1991); *ibid.* **60**, 1433 (1991).

[17] H. Toyoki, in *Formation, Dynamics and Statistics of Patterns*, eds. K. Kawasaki and M. Suzuki (World Scientific, Singapore, 1993).

[18] A. J. Bray and S. Puri, *Phys. Rev. Lett.* **67**, 2670 (1991).

[19] Fong Liu and G. F. Mazenko, *Phys. Rev. B* **45**, 6989 (1992).

[20] M. Modello and N. Goldenfeld, *Phys. Rev. A* **45**, 657 (1992).

[21] A. J. Bray and K. Humayun, *Phys. Rev. E* **48**, 1609 (1993).

[22] A. D. Rutenberg and A. J. Bray, *Phys. Rev. E* **51**, R1641 (1995).

[23] K. Kawasaki, *Vistas in Astronomy* **37**, 57 (1993).

[24] M. Zapotocky and P. M. Goldbert, preprint.

[25] T. Nagaya, H. Orihara, and Y. Ishibashi, *J. Phys. Soc. Jpn* **56**, 1898 (1987); *ibid* **56**, 3086 (1987); *ibid.* **59**, 377 (1990).

[26] N. Mason, A. N. Pargellis, and B. Yurke, *Phys. Rev. Lett.* **70**, 190 (1993).

[27] S. M. Allen and J. W. Cahn, *Acta. Metall.* **27**, 1085 (1979).

[28] K. Kawasaki and T. Ohta, *Prog. Theor. Phys.* **62**, 147 (1982); R. Bausch, V. Dohm, H. K. Janssen, and K. P. Zia, *Phys. Rev. Lett.* **47**, 1837 (1981).

[29] H. Furukawa, *Phys. Rev. B* **40**, 2341 (1989); *ibid.* **42**, 6438 (1990).

[30] J. W. Cahn and J. H. Hilliard, *J. Chem. Phys.* **28**, 258 (1958); *ibid.* **31**, 688 (1959).

[31] I. M. Lifshitz and V. V. Slyozov, *J. Phys. Chem. Solids* **19**, 35 (1961).

[32] S. Komura, K. Osamura, H. Fujii, and T. Takeda, *Phys. Rev. B* **31**, 1278 (1985).

[33] T. Izumitani and T. Hashimoto, *J. Chem. Phys.* **83**, 3694 (1985); T. Izumitani, M. Takenaka, and T. Hashimoto, *ibid.* **83**, 3213 (1990).

[34] Y. Oono and S. Puri, *Phys. Rev. A* **38**, 434 (1988).

[35] T. M. Rogers, K. R. Elder, and R. C. Desai, *Phys. Rev. B* **37**, 9638 (1988); A. Chakrabarti and J. D. Gunton, *ibid.* **37**, 3798 (1988); C. Roland and M. Grant, *ibid.* **39**, 11 971 (1989).

[36] A. Shinozaki and Y. Oono, *Phys. Rev. E* **48**, 2622 (1993).

[37] C. Yeung, *Phys. Rev. Lett.* **61**, 1135 (1988).

[38] J. S. Langer, M. Bar-on, and H. D. Miller, *Phys. Rev. A* **11**, 1417 (1975).

[39] K. Binder, C. Billotet, and P. Mirold, *Z. Physik B* **30**, 183 (1978).

[40] A. Onuki, *Prog. Theor. Phys.* **66**, 1230 (1981).

[41] K. Kitahara and M. Imada, *Prog. Theor. Phys. Suppl.* **64**, 65 (1978).

[42] K. Kawasaki and T. Ohta, *Physica A* **118** A, 175 (1983).

[43] J. J. Weins and J. W. Cahn, *Materials Research* **6**, 151 (1973).

[44] J. A. Marqusee and J. Ross, *J. Chem. Phys.* **80**, 536 (1984).

[45] M. Tokuyama and K. Kawasaki, *Physica A* **123**, 386 (1984); M. Tokyama and Y. Enemoto *Phys. Rev. Lett.* **69**, 312 (1992).

[46] M. Marder, *Phys. Rev. A* **36**, 858 (1987); C. Sagui and M. Grant, *Phys. Rev. E* **59** 4175 (1999).

[47] M. San Miguel and J. D. Gunton, *Phys. Rev. B* **23**, 2317 (1981); *ibid.* **23**, 2334 (1981).

[48] P. C. Hohenberg and D. R. Nelson, *Phys. Rev. B* **20**, 2665 (1979).

[49] Th. Benda, P. Alpern, and P. Leiderer, *Phys. Rev. B* **26**, 1450 (1982).

[50] J. K. Hoffer D. N. Sinha, *Phys. Rev. A* **33**, 1918 (1986).

[51] P. S. Sahni and J. D. Gunton, *Phys. Rev. Lett.* **45**, 371 (1980); A. Chakrabarti, J. B. Collin, and J. D. Gunton, *Phys. Rev. B* **38**, 6894 (1988).

[52] T. Ohta, K. Kawasaki, A. Sato, and Y. Enomoto, *Phys. Lett. A* **126**, 93 (1987).

[53] C. Sagui, A. M. Somoza, and R. C. Desai, *Phys. Rev. E* **50**, 4865 (1994).

[54] R. Kobayashi, *Physica D* **63**, 410 (1993).

[55] Y. C. Chou and W. I. Goldburg, *Phys. Rev. A* **20**, 2105 (1979); *ibid.* **23**, 858 (1981); W. I. Goldburg, in *Scattering Techniques Applied to Supramolecular and Nonequilibrium Systems*, eds. S.-H. Chen and R. Nossal (Plenum, New York, 1981), p. 531.

[56] N.-C. Wong and C. M. Knobler, *J. Chem. Phys.* **69**, 725 (1978); *Phys. Rev. A* **24**, 3205 (1981); C. M. Knobler, in *The Fourth Mexican School on Statistical Mechanics*, eds. R. Pelalta-Fabi and C. Varea (World Scientific, Singapore, 1987), p. 1.

[57] C. Houessou, P. Guenoun, R. Gastaud, F. Perrot, and D. Beysens, *Phys. Rev. A* **32**, 1818 (1983).

[58] D. Beysens, P. Guenoun, and F. Perrot, *Phys. Rev. A* **38**, 4173 (1988).

[59] E. A. Clerke and J. V. Sengers, *Physica* **118 A**, 360 (1983).

[60] A. E. Bailey and D. Cannell, *Phys. Rev. Lett.* **70**, 2110 (1993).

[61] D. Beysens, Y. Garrabos, and C. Chabot, in *Slow Dynamics in Complex Systems*, eds. M. Tokuyama and I. Oppenheim (AIP conference proceedings 469, 1999), p. 222; F. Perrot, D. Beysens, Y. Garrabos, T. Frohlich, P. Guenon, B. Bonetti, and P. Bravais, *Phys. Rev. E* **59**, 3079 (1999).

[62] T. K. Kwei, T. Nishi, and R. F. Roberts, *Macromolecules* **7**, 667 (1974).

[63] L. P. McMaster, *Adv. Chem. Ser.* **142**, 43 (1975).

[64] H. Snyder and P. Meakin, *J. Chem. Phys.* **79**, 5588 (1983).

[65] T. Sato and C. C. Han, *J. Chem. Phys.* **88**, 2057 (1988).

[66] T. Hashimoto, *Phase Transitions* **12**, 47 (1988).

[67] F. S. Bates and P. Wiltzius, *J. Chem. Phys.* **91**, 3258 (1988).

[68] T. Hashimoto, H. Jinnai, Y. Nishikawa, T. Koga, and M. Takenaka, *Progr. Colloid Polym. Sci.* **106**, 118 (1997); H. Jinnai, T. Hashimoto, D. Lee, and S-H. Chen, *Macromolecules* **30**, 130 (1997).

[69] T. Koga and K. Kawasaki, *Physica A* **196**, 389 (1993); *ibid.* **198**, 473 (1993).

[70] T. Koga, H. Jinnai, and T. Hashimoto, *Physica A* **263**, 369 (1999).

[71] K. Kawasaki and T. Ohta, *Prog. Theor. Phys.* **59**, 348 (1978).

[72] T. Koga, private communication.

[73] E. Siggia, *Phys. Rev. A* **20**, 595 (1979).

[74] V. S. Nikolaev, D. Beysens, and P. Guenon, *Phys. Rev. Lett.* **76**, 3144 (1996).

[75] S. Bastea and J. L. Lebowitz, *Phys. Rev. Lett.* **78**, 3499 (1997).

[76] Lord Rayleigh, *Phil. Mag.* **34**, 145 (1892).

[77] S. Tomotica, *Proc. Roy. Soc.* (London) *A* **150**, 322 (1935).

[78] A. Frischknecht, *Phys. Rev. E* **58**, 3495 (1998).

[79] T. Hashimoto, M. Itakura, and N. Shimidzu, *J. Chem. Phys.* **85**, 6773 (1986).

[80] A. Onuki, *J. Chem. Phys.* **85**, 1122 (1986).

[81] M. A. Bouchiat and J. Meunier, *Phys. Rev. Lett.* **23**, 752 (1969); J. Zollweg, G. Hawkins, and G. B. Benedek, *ibid.* **27**, 1182 (1971).

[82] D. Jasnow, D. A. Nicole, and T. Ohta, *Phys. Rev. A* **23**, 3192(1981).

[83] H. Lamb Jr *Hydrodynamics* (Cambridge University Press, New York, 1975).

[84] K. Binder and D. Stauffer, *Phys. Rev. Lett.* **33**, 1006 (1974).

[85] M. Von Smoluchowski, *Z. Physik* **17**, 585 (1917); *Z. Phys. Chem.* (Leipzig) **92**, 129 (1917).

[86] S. K. Friedlander and C. S. Wang, *J. Colloid Interface Sci.* **22**, 126 (1966).

[87] S. Chandrasekhar, *Rev. Mod. Phys.* **15**, 1 (1943).

[88] H. Tomita, *Prog. Theor. Phys.* **56**, 1661 (1976).

[89] T. Ohta, *Ann. Phys.* **158**, 31 (1984).

[90] H. Furukawa, *Phys. Rev. A* **31**, 1103 (1985); *ibid.* **36**, 2288 (1987); *Physica A* **204**, 237 (1994).

[91] V. M. Kendon, J-C. Desplat, P. Bladon, and M. Cates, *Phys. Rev. Lett.* **83**, 576 (1999).

[92] C. K. Chan and W. I. Goldburg, *Phys. Rev. Lett.* **58**, 674 (1987).

[93] M. Schobinger, S. W. Koch, and F. F. Abraham, *J. Stat. Phys.* **42**, 1071 (1986).

[94] J. Farrell and O. T. Valls, *Phys. Rev. B* **42**, 2353 (1990).

[95] R. Yamamoto and K. Nakanishi, *Phys. Rev. B* **49**, 14 958 (1994); *ibid.* **51**, 2715 (1995).

[96] S. Purl and H. L. Frisch, *J. Phys. C* **9**, 2109 (1997); S. Puri and K. Binder, *Phys. Rev. E* **49**, 5359 (1994).

[97] H. Tanaka and T. Araki, *Europhys. Lett.* **51**, 154 (2000).

[98] J. P. Donley and J. S. Langer, *Phys. Rev. Lett.* **71**, 1573 (1993).

[99] C. M. Jefferson, R. G. Petschek, and D. S. Cannell. *Phys. Rev. Lett.* **52**, 1329 (1984).

[100] A Onuki, *Phys. Rev. Lett.* **48**, 753 (1982).

[101] M. Joshua, J. V. Maher, W. I. Goldburg, *Phys. Rev. Lett.* **51**, 196 (1983); M. Joshua, W. I. Goldburg, and A. Onuki, *Phys. Rev. Lett.* **85**, 1175 (1985); M. Joshua and W. I. Goldburg, *Phys. Rev. A* **31**, 3857 (1985).

[102] H. Tanaka and T. Sigehuzi, *Phys. Rev. Lett.* **75**, 874 (1995).

[103] P. G. de Gennes, *J. Chem. Phys.* **72**, 4756 (1980).

[104] P. Pincus, *J. Chem. Phys.* **75**, 1996 (1981).

[105] K. Binder, *J. Chem. Phys.* **79**, 6387 (1983).

[106] A. Onuki, *J. Chem. Phys.* **85**, 1122 (1986).

[107] K. Kawasaki and K. Sekimoto, *Macromolecules* **22**, 3063 (1989).

[108] A. Ziya Akcasu, *Macromolecules* **22**, 3682 (1989).

[109] K. Binder, *Adv. Polym. Sci.* **112**, 181 (1994).

[110] S. Nojima, K. Shiroshita, and T. Nose, *Polym. J.* **14**, 289 (1982).

[111] J. H. Aubert, *Macromolecules* **23**, 1446 (1990).

[112] J. Lal and R. Bansil, *Macromolecules* **24**, 290 (1991).

[113] N. Kuwahara and K. Kubota, *Phys. Rev. A* **45**, 7385 (1992).

[114] S. W. Song and M. Torkelson, *Macromolecules* **27**, 6390 (1994).

[115] D. Schwahn, S. Janssen, and T. Springer, *J. Chem. Phys.* **97**, 8775 (1992); G. Müller, D. Schwahn, H. Eckerlebe, J. Rieger, and T. Springer, *J. Chem. Phys.* **104**, 5826 (1996).

[116] H. Tanaka, *Phys. Rev. Lett.* **71**, 3158 (1993); *J. Chem. Phys.* **100**, 5253 (1994).

[117] H. Tanaka, *Phys. Rev. Lett.* **76**, 787 (1996).

[118] N. Toyoda, M. Takenaka, S. Saito, and T. Hashimoto, *Polymer* **42**, 9193 (2001).

[119] S. Hirotsu and A. Kaneki, in *Dynamics of Ordering Processes in Condensed Matter*, eds. S. Komura and H. Furukawa (Plenum, New York, 1988), p. 481.

[120] F. Horkay, A. M. Hecht, and E. Geissler, *Macromolecules* **31**, 8851 (1998).

[121] R. Bansil, J. Lal, and B. Carvalho, *Polymer*, **33**, 2961 (1992); D. Asnaghi, M. Giglio, A. Bossi, and P. G. Righretti, *J. Chem. Phys.* **102**, 9736 (1995).

[122] T. Hashimoto, M. Takenaka, and H. Jinnai, *Polymer Commun.* **30**, 177 (1989).

[123] A. Harada and Q. Tran-Cong, *Macromolecules* **30**, 1643 (1997).

[124] K. Sekimoto, N. Suematsu, and K. Kawasaki, *Phys. Rev. A* **39**, 4912 (1989).

[125] A. Onuki and S. Puri, *Phys. Rev. E* **59**, R1331 (1999).

[126] A. Onuki and T. Taniguchi, *J. Chem. Phys.* **106**, 5761 (1997).

[127] T. Taniguchi and A. Onuki, *Phys. Rev. Lett.* **77**, 4910 (1996).

[128] H. Tanaka and T. Araki, *Phys. Rev. Lett.* **78**, 4966 (1997); T. Araki and H. Tanaka *Macromolecules* **34**, 1953 (2001).

[129] R. J. Donnelly, *Quantized Vortices in He II* (Cambridge University Press, 1991).

[130] V. Bulatov, F. F. Abraham, L. Kubin, B. Devincre, and S. Yip, *Nature* **391**, 669 (1998).

[131] E. P. Gross, *Nuovo Cimento* **20**, 454 (1961).

[132] L. P. Pitaevskii, *Zh. Eksp. Teor. Fiz.* **40**, 646 (1961) [*Sov. Phys. JETP* **13**, 451 (1961)].

[133] A. Onuki, *J. Phys. C* **15**, L1093 (1982); *J. Low Temp. Phys.* **51**, 601 (1983).

[134] A. Onuki, *Prog. Theor. Phys.* **70**, 57 (1983).

[135] A. Onuki, *J. Low Temp. Phys.* **53**, 189 (1983).

[136] K. Kawasaki, *Physica A* **119**, 17 (1983).

[137] J. M. Kosterlitz and D. J. Thouless, *J. Phys. C* **6**, 1181 (1973).

[138] V. Ambegaokar, B. I. Halperin, D. R. Nelson, and E. D. Siggia, *Phys. Rev. B* **21**, 3204 (1980).

[139] H. E. Hall and W. F. Vinen, *Proc. Roy. Soc. (London) A* **238**, 204 (1956); W. F. Vinen, *ibid. A* **242**, 493 (1957).

[140] P. Mathieu and Y. Simon, *Phys. Rev. Lett.* **45**, 1428 (1980).

[141] P. Mathieu, A. Serra, and Y. Simon, *Phys. Rev. B* **14**, 3753 (1976).

[142] C. F. Barenghi, R. J. Donnelly, and W. F. Vinen, *J. Low Temp. Phys.* **52**, 189 (1983).

[143] L. P. Pitaevskii, *Pis'ma Zh. Eksp. Teor. Fiz.* **25**, 168 (1977) [*JETP Lett.* **25**, 154 (1977)].

[144] W. F. Vinen, *Proc. Roy. Soc. (London) A* **240**, 114 (1957); *ibid. A* **240**, 128 (1957); *ibid. A* **242**, 493 (1957);

[145] K. W. Schwarz, *Phys. Rev. B* **18**, 245 (1978); *Phys. Rev. Lett.* **49**, 283 (1982); *Phys. Rev. B* **38**, 2398 (1988).

[146] J. Koplik and H. Levine, *Phys. Rev. Lett.* **71**, 1375 (1993).

[147] S. R. Stalp, L. Skrbek, and R. J. Donnelly, *Phys. Rev. Lett.* **82**, 4831 (1999).

[148] W. F. Vinen, *Phys. Rev. B* **61**, 1410 (2000).

[149] C. J. Gorter and J. H. Mellink, *Physica* **15**, 285 (1949).

[150] G. Ahlers, *Phys. Rev. Lett.* **22**, 54 (1969).

[151] P. Leiderer and F. Pobell, *J. Low Temp. Phys.* **3**, 577 (1970).

[152] A. Onuki, *J. Low Temp. Phys.* **104**, 133 (1996).

[153] R. Haussmann and V. Dohm *Phys. Rev.* B **46**, 6361 (1992).

[154] J. S. Langer and V. Ambegaokar, *Phys. Rev.* **164**, 498 (1967).

[155] W. A. Moeur, P. K. Day, F-C Liu, S. T. P. Boyd, M. J. Adriaans, and R. V. Duncan, *Phys. Rev. Lett.* **78**, 2421 (1997).

[156] P. B. Weichman and J. Miller, *J. Low Temp. Phys.* **119**, 155 (2000).

[157] A. Onuki, *J. Low Temp. Phys.* **121**, 117 (2000).

[158] G. Marrucci and F. Greco, in *Advances in Chemical Physics*, eds. I. Prigogine and S. Rice (John Wiley & Sons, New York, 1993), p. 331.

[159] R. G. Larson and M. Doi, *J. Rheol.* **35**, 539 (1991).

[160] G. Porod, *Kolloid Z.* **124**, 82 (1951).

[161] R. Kirste and G. Porod, *Kolloid Z. Z. Polym.* **184**, 1 (1962).

[162] L. Auray and P. Auray, in *Neutron, X-ray and Light Scattering*, eds. P. Lindner and Th. Zemb (Elsevier, New York, 1991), p. 199.

[163] H. Tomita, *Prog. Theor. Phys.* **76**, 656 (1984); *ibid.* **75**, 482 (1986).

[164] M. Teubner, *J. Chem. Phys.* **92**, 4501 (1990).

[165] A. Onuki, *Phys. Rev. A* **15** (1992) R3384.

[166] A. Shinozaki and Y. Oono, *Phys. Rev. Lett.* **66**, 173 (1991).

[167] A. D. Rutenberg, *Phys. Rev. E* **54**, R2181 (1996).

[168] M. Sakouri, J. Marignan, J. Appell, and G. Porte, *J. Physique II* **1**, 1121 (1991).

[169] H. D. Bale and P. W. Schmidt, *Phys. Rev. Lett.* **53**, 596 (1984).

[170] N. F. Berk, *Phys. Rev. Lett.* **58**, 2718 (1987).

9

Nucleation

In metastable states the free energy is at a local minimum but not at the true minimum. Such states are stable for infinitesimal fluctuations. However, rare spatially localized fluctuations, called critical nuclei, can continue to grow, eventually leading to macroscopic phase separation and ordering [1]–[9]. We will discuss how critical nuclei emerge and how they grow. Then we will treat one-component fluids near the gas–liquid critical point, where bubble boiling or liquid condensation can also take place in the thermal diffusion layer as well as nucleation in the bulk region. We will also examine quantum nucleation at very low temperatures, viscoelastic nucleation in polymers, and vortex nucleation in superfluid helium.

9.1 Droplet evolution equation

9.1.1 Spherical droplets

We consider a spherical droplet emerging in a metastable state. As discussed in Chapter 8, the droplet free energy consists of the surface and bulk parts,

$$\begin{aligned}\mathcal{H}(R) &= S_d\left(\sigma R^{d-1} - \frac{1}{d}\mu_{\rm eff}R^d\right),\\ &= \mathcal{H}_{\rm c} - \frac{d-1}{2}S_d\sigma R_{\rm c}^{d-3}(\delta R)^2 + \cdots,\end{aligned} \tag{9.1.1}$$

where $S_3 = 4\pi$ and $S_2 = 2\pi$, and $\mu_{\rm eff}$ is the free-energy difference per unit volume between the metastable and stable phases. The second line is the expansion around the critical radius, $R_{\rm c} = (d-1)\sigma/\mu_{\rm eff}R_{\rm c}$, with respect to $\delta R = R - R_{\rm c}$. As shown in Fig. 9.1, $\mathcal{H}(R)$ takes a maximum at $R = R_{\rm c}$ given by

$$\mathcal{H}_{\rm c} = \mathcal{H}(R_{\rm c}) = \frac{S_d}{d}\sigma R_{\rm c}^{d-1} \sim T(R_{\rm c}/\xi)^{d-1}. \tag{9.1.2}$$

We assume $\mathcal{H}_{\rm c} \gg T$ or equivalently $R_{\rm c} \gg \xi$. We shall see that the nucleation rate I is proportional to the small factor $\exp(-\mathcal{H}_{\rm c}/T)$.

Field reversal in the nonconserved case

If the scalar order parameter ψ is nonconserved, a simple nucleation experiment is a reversal of magnetic field in ferromagnetic systems or electric field in ferroelectric systems. That is, we prepare a homogeneous state with $\psi \cong -\psi_{\rm eq}$ and apply a small positive

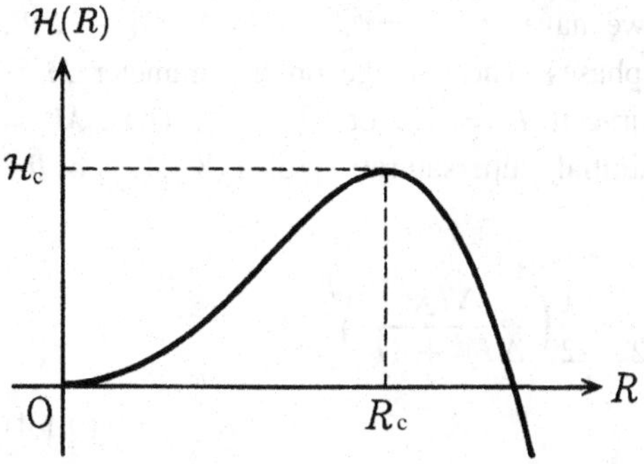

Fig. 9.1. The free energy needed to create a spherical droplet with radius R in 3D. It takes a maximum $\mathcal{H}_c$ at $R = R_c$.

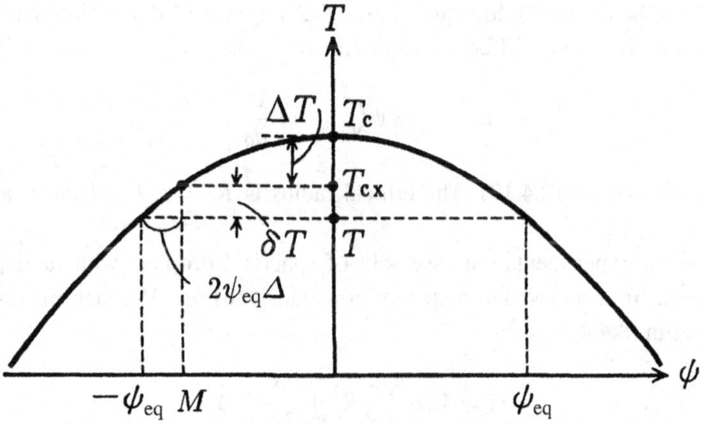

Fig. 9.2. Phase diagram of a system with a scalar conserved order parameter. The system is quenched at the average order parameter M slightly inside the coexistence curve $T = T_{cx}$. At the final temperature T, the two equilibrium phases have the average order parameters $\pm\psi_{eq}$. The supersaturation is defined by $\Delta = (M + \psi_{eq})/2\psi_{eq}$.

magnetic field h at $t = 0$. Then, as in (8.2.15), ψ instantaneously adjusts to h as $\psi \cong -\psi_{eq} + (2\kappa^2)^{-1}h$. The magnetic field is very small and satisfies (8.2.16). This initial state has a higher free energy than the true stable state with $\psi \cong \psi_{eq} + (2\kappa^2)^{-1}h$. The free-energy difference per unit volume is

$$\mu_{eff} = 2T\psi_{eq}h, \tag{9.1.3}$$

which was already used in (8.2.31). With the appearance of droplets a domain switching process slowly proceeds.

Quench experiments in the conserved case

In the conserved case (particularly in near-critical binary fluid mixtures), a metastable state is realized if the temperature difference $\delta T = T_{cx} - T$ is lowered slightly below the coexistence curve with a fixed average order parameter M, as illustrated in Fig. 9.2.

If T is held fixed near the critical point, we have $\psi = \pm\psi_{\rm eq} = \pm A_{\rm cx}(T_{\rm c} - T)^\beta$, respectively, in the two final macroscopic phases. The average order parameter M is the equilibrium value on the coexistence curve if $T = T_{\rm cx}$ or $\delta T = 0$. Thus, $M = -A_{\rm cx}(\Delta T_{\rm cx})^\beta$ with $\Delta T_{\rm cx} = T_{\rm c} - T_{\rm cx}$. The (initial) supersaturation $\varDelta$ in (8.4.4) is of the form,

$$\begin{aligned} \varDelta &= \frac{M + \psi_{\rm eq}}{2\psi_{\rm eq}} = \frac{1}{2} - \frac{1}{2}\left(\frac{\Delta T_{\rm cx}}{\Delta T_{\rm cx} + \delta T}\right)^\beta \\ &\cong \frac{\beta}{2}\delta T/\Delta T_{\rm cx}. \end{aligned} \tag{9.1.4}$$

The second line holds for shallow quenches, $\delta T/\Delta T_{\rm cx} \ll 1$. If the ψ^4 free-energy density is assumed, the free-energy difference in (9.1.3) is given by

$$\mu_{\rm eff} = 8T\psi_{\rm eq}^2\kappa^2\varDelta = \frac{\sigma}{d_0}\varDelta, \tag{9.1.5}$$

where d_0 is defined by (8.4.13). The critical radius is $R_{\rm c} = (d-1)d_0/\varDelta$ as given by (8.4.30).

In 3D quench experiments, an assembly of spherical droplets with radii $\{R_j\}$ ($j = 1, 2, \ldots$) appear in a metastable matrix after a transient time. We then express the total free energy $\mathcal{H}$ in (8.4.42) as

$$\mathcal{H} = 4\pi\sigma\sum_j R_j^2 + \frac{\sigma}{2d_0}V\varDelta^2, \tag{9.1.6}$$

with

$$\varDelta = \phi - \frac{1}{V}\sum_j \frac{4\pi}{3}R_j^3, \tag{9.1.7}$$

where V is the total volume of the system, and ϕ is defined by (8.3.1). If we select a particular droplet (say, $j = 1$), the terms related to this droplet in $\mathcal{H}$ become written as $\mathcal{H}(R_1) + (4\pi R_1^3/3)^2/V \cong \mathcal{H}(R_1)$, leading to the droplet free energy $\mathcal{H}(R)$ in (9.1.1).

Free energy of a single droplet in a finite system

While a critical droplet is unstable against its volume change, we observe a large liquid, gas, or crystal domain formed in a finite system as an *equilibrium* state in the final stage of phase separation. Indeed, in gravity-free space experiments and simulations, the final gas or liquid droplet assumes a spherical shape. Obviously, the mass conservation law brings about such a final state. To show its stability, let us consider a single spherical domain with radius R in the 3D conserved case. From (9.1.6) and (9.1.7) we obtain the free-energy

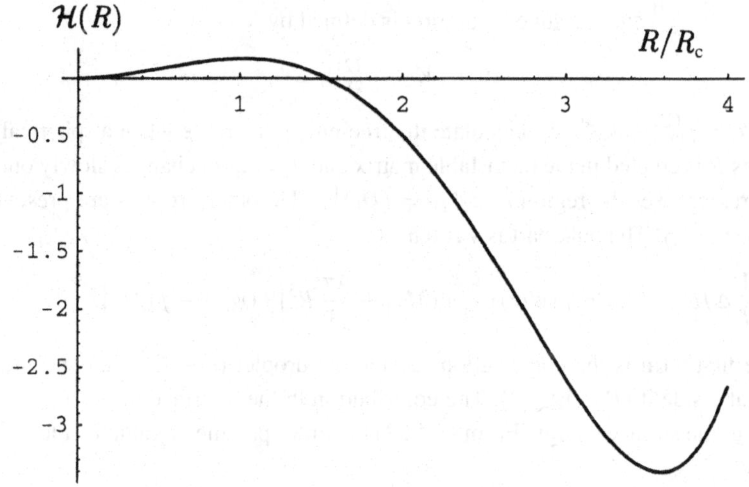

Fig. 9.3. The free energy $\mathcal{H}(R)$ needed, in units of $8\pi\sigma R_c^2$, to create a single spherical droplet in a conserved finite system. The maximum size $R_M(\propto (\phi V)^{1/3})$ is chosen to be four times larger than R_c. Then the maximum and minimum are attained at $R/R_c = 1.02$ and 3.59, respectively.

change in a finite system with volume V,

$$\begin{aligned} \mathcal{H}(R) &= 4\pi\sigma R^2 + \frac{\sigma}{2d_0}V\left(\phi - \frac{4\pi}{3V}R^3\right)^2 - \frac{\sigma}{2d_0}V\phi^2 \\ &= 8\pi\sigma R_c^2\left[\frac{x^2}{2} - \frac{x^3}{3} + \frac{1}{6}\left(\frac{R_c}{R_M}\right)^3 x^6\right], \end{aligned} \tag{9.1.8}$$

where $x = R/R_c$. The radius R cannot exceed R_M determined by $\phi V = (4\pi/3)R_M^3$. We assume that $R_M = (3V\phi/4\pi)^{1/3}$ is much larger than $R_c = 2d_0/\phi$. We notice that $\mathcal{H}(R)$ has a maximum at $R/R_c = 1 + (R_c/R_M)^3 + \cdots$ and a minimum at $R/R_c = R_M/R_c - 1/3 + \cdots$, as plotted in Fig. 9.3. The maximum corresponds to the critical droplet and the minimum to the final equilibrium droplet. Hereafter we will treat droplets with sizes much smaller than R_M and neglect the system size effect.

9.1.2 Droplets for general cases with conservation

The expressions for d_0 and μ_{eff} can be derived for more general cases. We consider a 3D system with a general free-energy density $f(\psi)$ for a scalar conserved variable ψ. The temperature T is changed from a disordered region above the coexistence curve to a metastable region slightly below it. The equilibrium values of ψ at the final temperature T are written as $\psi_{\text{cx}}^{(1)}$ and $\psi_{\text{cx}}^{(2)}$. The initial average order parameter $M = \langle\psi\rangle$ is slightly different from $\psi_{cx}^{(2)}$ and the supersaturation is defined by

$$\Delta = (M - \psi_{cx}^{(2)})/\Delta\psi, \tag{9.1.9}$$

where $\Delta\psi = \psi_{cx}^{(1)} - \psi_{cx}^{(2)}$. We calculate the free-energy increase when a spherical droplet with radius R is created in the metastable matrix and $\psi = \psi(\mathbf{r})$ changes slowly outside the interface region over the region $r < \ell_D = (Dt)^{1/2}$. No other droplets are present within the distance of ℓ_D. The bulk part is written as

$$\frac{1}{T}\Delta\mathcal{H} = \int^{>} d\mathbf{r}[f(\psi(\mathbf{r})) - f(M)] + \frac{4\pi}{3}R^3[f(\psi_{\text{in}}) - f(M)], \tag{9.1.10}$$

where the first term is the contribution outside the droplet ($r - R \gtrsim \xi$) and the second term is that inside it ($R - r \gtrsim \xi$). The contribution in the interface region ($|r - R| \lesssim \xi$) gives rise to the surface energy. From (8.4.23) the order parameter value inside the droplet is given by

$$\psi_{\text{in}} = \psi_{cx}^{(1)} + (\chi_1\sigma/T\Delta\psi)\frac{2}{R} \cong \psi_{cx}^{(1)}, \tag{9.1.11}$$

where $\chi_1 = 1/f''(\psi_{cx}^{(1)})$ is the susceptibility of the phase 1. To evaluate the first term of (9.1.10) we use the conservation of ψ,

$$\int^{>} d\mathbf{r}[\psi(\mathbf{r}) - M] + \frac{4\pi}{3}R^3[\psi_{\text{in}} - M] = 0, \tag{9.1.12}$$

and the expansion $f(\psi) - f(M) = f'(M)(\psi - M) + \frac{1}{2}\chi_2^{-1}(\psi - M)^2 + \cdots$ outside the droplet, where $\chi_2 = 1/f''(\psi_{cx}^{(2)})$ is the susceptibility in the phase 2. Then,

$$\frac{1}{T}\Delta\mathcal{H} \cong \frac{1}{2}\chi_2^{-1}\int^{>} d\mathbf{r}[\psi(\mathbf{r}) - M]^2 + \frac{4\pi}{3}R^3[f(\psi_{\text{in}}) - f(M) - f'(M)(\psi_{\text{in}} - M)]. \tag{9.1.13}$$

In the second term we use

$$f(\psi) = f_{cx} + \mu_{cx}\psi + \frac{1}{2}\chi_\alpha^{-1}(\psi - \psi_{cx}^{(\alpha)})^2 + \cdots, \tag{9.1.14}$$

where $f_{cx} = f(\psi_{cx}^{(\alpha)}) - \mu_{cx}\psi_{cx}^{(\alpha)}$ and $\mu_{cx} = f'(\psi_{cx}^{(\alpha)})$ are common in the two equilibrium phases ($\alpha = 1, 2$). This gives $f(\psi_{\text{in}}) - f(M) \cong \mu_{cx}(\psi_{\text{in}} - M)$ for $\psi_{\text{in}} \cong \psi_{cx}^{(1)}$ and $M \cong \psi_{cx}^{(2)}$. Neglecting the first term we thus obtain

$$\begin{aligned}\frac{1}{T}\Delta\mathcal{H} &\cong \frac{4\pi}{3}R^3[\mu_{cx} - f'(M)](\psi_{\text{in}} - M) \\ &\cong -\frac{4\pi}{3}R^3\chi_2^{-1}(\Delta\psi)^2\Delta. \end{aligned} \tag{9.1.15}$$

Here we need to show that the first term in (9.1.13) is really negligible. To this end we use (8.4.27) with $2\psi_{\text{eq}}$ being replaced by $\Delta\psi$. Then the first term is estimated as $2\pi\ell_D\chi_2^{-1}(\Delta\psi)^2(R\Delta - 2d_0)^2$, where ℓ_D plays the role of the large-distance cut-off. The ratio of the first to second term in (9.1.13) is very small for $R \sim R_c$, while it is of order

$\ell_D \Delta/R \sim \Delta^{1/2}$ for $R \gg R_c$ with $R \sim (\Delta Dt)^{1/2}$ being substituted. Then, the free-energy difference and the capillary length are written as

$$\mu_{\text{eff}} = T\chi_2^{-1}(\Delta\psi)^2\Delta = \frac{\sigma}{d_0}\Delta, \tag{9.1.16}$$

$$d_0 = \chi_2\sigma/T(\Delta\psi)^2. \tag{9.1.17}$$

If we set $\Delta\psi = 2\psi_{\text{eq}}$ and $\chi_2 = 1/2\kappa^2$, we reproduce (9.1.5) and (8.4.13).

Coupled systems

For model C near the tricritical point, we have already derived the interface dynamic equations (8.4.75)–(8.4.79), in which a nonconserved order parameter ψ and a conserved variable m take different values in the two phases, $\alpha = 1, 2$. The system is in a metastable state for a small ordering field h in (8.4.65) and supersaturation Δ in (8.4.77) with

$$\mu_{\text{eff}} = \frac{\sigma}{d_{02}}\Delta + T(\Delta\psi)h. \tag{9.1.18}$$

The droplet growth is slowed down by the diffusion of m in a surrounding metastable region. Nucleation in ^{3}He–^{4}He mixtures near the tricritical point is also governed by slow diffusion of the concentration with μ_{eff} being the first term in (9.1.18) as in usual fluid binary mixtures [10].

Precipitates of an ordered phase in alloys

In binary alloys treated in Section 3.3, the concentration c and the long-range order parameter η are coupled in the free-energy density $v_0^{-1} f_{\text{site}}(c, \eta)$, where v_0 is the volume of a unit cell and f_{site} is given by (3.3.12) or (3.3.28). In metallurgy, much attention has been paid to the growth of precipitates with the $\text{L}1_0$ or $\text{L}1_2$ structure in a disordered, metastable bcc or fcc matrix. In this case η is determined as a function of c because c changes slowly in time. Thus c should be identified with ψ in the relation $f(\psi) = v_0^{-1} f_{\text{site}}(c, \eta(c))$. Here $c \cong \psi_{\text{cx}}^{(1)}$ and $\eta \neq 0$ inside the droplet, while $f(\psi) = f(c, 0)$ with $c = M$ and $\eta = 0$ far from the droplet. The concentrations on the coexistence curve (which are written as $c_{\text{e}1}$ and $c_{\text{e}2}$ in Section 3.3) are here written as $\psi_{\text{cx}}^{(1)}$ and $\psi_{\text{cx}}^{(2)}$. The supersaturation Δ is then defined by (9.1.9). From (9.1.16) we have

$$\begin{aligned} \mu_{\text{eff}} &= [f'(M, 0) - \mu_{\text{cx}}](\psi_{\text{cx}}^{(1)} - M) \\ &\cong \chi_2^{-1}(\Delta\psi)^2\Delta, \end{aligned} \tag{9.1.19}$$

where $\mu_{\text{cx}} = f'(\psi_{\text{cx}}^{(1)}, \eta(\psi_{\text{cx}}^{(1)})) = f'(\psi_{\text{cx}}^{(2)}, 0)$ and $\chi_2^{-1} = f''(\psi_{\text{cx}}^{(2)}, 0)$. The second line holds for $\Delta \ll 1$ and is of the same form as (9.1.16). The capillary length d_0 is given by (9.1.17). For example, let us consider an $\text{L}1_2$ domain in Al–Li at low temperatures where $T \ll w_1 \sim -w_0(\sim 2000 \text{ K})$, $c_{\text{e}1} \cong 0$, and $c_{\text{e}2} \cong 1/4$ from (3.3.36). In terms of the average concentration M in the Al-rich matrix, we then have $\Delta\psi \cong 1/4$, $\Delta \cong 4M$, and $\chi_2^{-1} \cong (TM^{-1} + |w_0|)/v_0$, so that $\mu_{\text{eff}} = (T + M|w_0|)/4v_0$ for $M \gg c_{\text{e}1}$.

9.1.3 Droplet size distribution

In Chapter 8 we have set up the Langevin equation for a spherical droplet, which can be used when the droplet radius is much longer than ξ. Both in the nonconserved and conserved cases, it is of the standard form,

$$\begin{aligned}\frac{\partial}{\partial t}R &= -\mathcal{L}(R)\beta\mathcal{H}'(R) + \theta(R,t) \\ &= v(R) + \theta(R,t),\end{aligned} \tag{9.1.20}$$

where $\mathcal{H}'(R) = \partial\mathcal{H}(R)/\partial R$ and

$$v(R) = -\mathcal{L}(R)\beta\mathcal{H}'(R) \tag{9.1.21}$$

is the rate of change of the radius. The noise term satisfies the fluctuation–dissipation relation (8.2.45) with the kinetic coefficient $\mathcal{L}(R)$ being defined by (8.2.46) or (8.4.63). The $v(R)$ vanishes at $R = R_c$ and is of the form,

$$\begin{aligned}v(R) &= L\left(\frac{2T\psi_{eq}}{\sigma}h - \frac{d-1}{R}\right) \quad \text{(nonconserved)}, \\ &= \frac{D}{R}\left(\Delta - \frac{2d_0}{R}\right) \quad \text{(3D conserved)}.\end{aligned} \tag{9.1.22}$$

We then set up the Fokker–Planck equation for the droplet distribution $n(R,t)$ as

$$\frac{\partial}{\partial t}n = \frac{\partial}{\partial R}\mathcal{L}(R)\left[\frac{\partial}{\partial R} + \beta\mathcal{H}'(R))\right]n = \frac{\partial}{\partial R}\mathcal{L}(R)n_0\frac{\partial}{\partial R}\left(\frac{n}{n_0}\right). \tag{9.1.23}$$

We interpret $n(R,t)dR$ as the droplet number in the size interval $[R, R+dR]$ per unit volume at time t. We can see that

$$n_0(R) = n_\xi \exp[-\beta\mathcal{H}(R)] \tag{9.1.24}$$

is a steady solution of the above Fokker–Planck equation, but it grows unphysically for $R > R_c$ if $\mu_{eff} > 0$. However, on the coexistence curve ($\mu_{eff} = 0$), $n_0(R)$ has a well-defined physical meaning as the equilibrium distribution of rarely appearing droplets, where $n_0(R)$ is written as

$$n_{cx}(R) = n_\xi \exp\left(-S_d\sigma R^{d-1}/T\right). \tag{9.1.25}$$

In the asymptotic critical region, (4.4.11) gives $S_d\sigma R^{d-1}/T = S_d A_\sigma (R/\xi)^{d-1}$ with A_σ being about 0.09 in 3D. The scaling near the critical point suggests that the prefactor n_ξ is of the following order,

$$n_\xi \sim \xi^{-(d+1)}. \tag{9.1.26}$$

The number density $n_{cx}(R)$ rapidly decreases with increasing R and becomes extremely small for R several times longer than ξ. Obviously, in the nucleation problem with $\mu_{eff} > 0$ we must examine nonstationary solutions of (9.1.23) because droplets larger than R_c continue to grow. In addition, we note that the thermal noise term $\theta(R,t)$ in (9.1.20) should not be affected by a weak degree of metastability. This guarantees that the distribution of droplets of small size ($R \ll R_c$) remains almost the same as $n_{cx}(R)$ on the coexistence curve. Thus, in solving (9.1.23), we impose the following boundary condition,

$$n(R,t) \cong n_0(R) \cong n_{cx}(R) \quad \text{for} \quad \xi < R \ll R_c, \tag{9.1.27}$$

which holds at any t (> 0) after quenching.

The droplet free-energy density and irreversibility

In the conserved case, the Langevin equation (9.1.23) is supplemented with the mean-field equation for the supersaturation,

$$\Delta(t) = \phi - \int dR \frac{4\pi}{3} R^3 n(R,t), \tag{9.1.28}$$

where 3D is assumed. From (9.1.6) the droplet free-energy density f_D can be written in terms of n and Δ as

$$f_D(t) = \mathcal{H}/V = 4\pi\sigma \int dR R^2 n(R,t) + \frac{\sigma}{2d_0}\Delta(t)^2. \tag{9.1.29}$$

Because Δ depends on n, (9.1.23) becomes nonlinear with respect to n. We need to show that (9.1.23) and (9.1.28) constitute a closed set of irreversible evolution equations. The entropy of the droplet system may be defined by

$$\mathcal{S}(t) = -\int dR n \ln(n/n_\xi) - \frac{1}{T} f_D(t). \tag{9.1.30}$$

Use of (9.1.23) yields a nonnegative-definite entropy production,

$$\frac{d}{dt}\mathcal{S}(t) = \int dR \mathcal{L}(R)\left[\frac{\partial}{\partial R}\ln n + \beta\mathcal{H}'(R)\right]^2 n \geq 0. \tag{9.1.31}$$

9.1.4 Classical theory of nucleation kinetics

The kinetics of nucleation was originally formulated in a metastable fluid where the liquid and vapor number densities, n_{liq} and n_{gas}, are distinctly different [11]. In this classical theory, $n(\ell, t)$ is the number density of liquid clusters containing ℓ molecules. The cluster size changes with evaporation and condensation of molecules between the cluster and the surrounding gas phase. In terms of the frequencies of these two elementary processes, $a(\ell)$ and $b(\ell)$, the rate of change of the cluster size from $\ell - 1$ to ℓ is expressed as

$$J(\ell) = a(\ell-1)n(\ell-1,t) - b(\ell)n(\ell,t). \tag{9.1.32}$$

The rate equation for $n(\ell, t)$ is written as

$$\frac{\partial}{\partial t} n_\ell = J(\ell) - J(\ell+1). \tag{9.1.33}$$

Here we assume the detailed balance of the two processes,

$$a(\ell-1)n_0(\ell-1) = b(\ell)n_0(\ell), \tag{9.1.34}$$

where $n_0(\ell)$ is the steady distribution determined by the Boltzmann weight,

$$n_0(\ell) = n_1 \exp[-\beta\epsilon(\ell)]. \tag{9.1.35}$$

Analogous to $\mathcal{H}(R)$ in (9.1.1), $\epsilon(\ell)$ is the free energy needed to produce a cluster with size ℓ, consisting of the surface and bulk terms,

$$\epsilon(\ell) = \alpha_0(\ell - 1)^{2/3} - \delta\mu(\ell - 1). \tag{9.1.36}$$

The coefficient α_0 is proportional to the surface tension, and $\delta\mu = \mu_{\text{gas}} - \mu_{\text{liq}}$ is the chemical potential difference (per particle) between the two phases. Considering only large droplets ($\ell \gg 1$), we can take the continuum limit to obtain

$$\frac{\partial}{\partial t} n(\ell, t) = \frac{\partial}{\partial \ell} a(\ell) \left[\frac{\partial}{\partial \ell} + \frac{\partial(\beta\epsilon(\ell))}{\partial \ell} \right] n(\ell, t). \tag{9.1.37}$$

This has the same mathematical structure as (9.1.23). The kinetic coefficient $a(\ell)$ is proportional to the surface area $4\pi R^2$ for large ℓ where $4\pi R^3 n_{\text{liq}}/3 = \ell$. Thus $a(\ell)$ is of order $R^2 n_{\text{gas}} v_{\text{th}} \propto \ell^{2/3}$ where $v_{\text{th}} = (T/m_0)^{1/2}$ is the thermal velocity.

9.1.5 The Binder and Stauffer cluster dynamics

In Chapter 8 we presented the Smoluchowski equation (8.5.30) which describes coalescence of droplets due to their diffusive motions. Binder and Stauffer [6] combined the Fokker–Planck and Smoluchowski equations as

$$\begin{aligned} \frac{\partial}{\partial t} n(\ell, t) &= \frac{\partial}{\partial \ell} a(\ell) \left[\frac{\partial}{\partial \ell} + \frac{\partial(\beta\epsilon(\ell))}{\partial \ell} \right] n(\ell, t) \\ &+ \frac{1}{2} \int_{\ell_c}^{\ell - \ell_c} d\ell' K(\ell - \ell', \ell') n(\ell', t) n(\ell - \ell', t) \\ &- n(\ell, t) \int_{\ell_c}^{\infty} d\ell' K(\ell, \ell') n(\ell', t). \end{aligned} \tag{9.1.38}$$

The first term accounts for the effect of absorption and desorption of small clusters with sizes smaller than a cut-off ℓ_c. The last two terms represent the effect of coagulation of clusters with sizes ℓ and ℓ' larger than ℓ_c. Particularly near the critical point, if we consider only droplets with sizes longer than ξ, we should set $\ell_c \sim a\xi^{1/D}$, where a is a molecular size and D is the fractal dimension introduced in Chapter 2. In this case we are treating compact *droplets* below T_c (rather than fractal *clusters*) in (9.1.38). In fluids, the diffusive coagulation is described by the last two terms in (9.1.38) with $K(\ell, \ell')$ being given by (8.5.26), while the birth process and initial-stage growth are described by the first term.

9.1.6 Cluster models

Cluster theory of condensation

In a gas phase near the coexistence curve *not close* to the critical point, we may consider compact aggregates of ℓ molecules and call them *clusters*. Condensation into liquid should start with the growth of such clusters [12]. If the excluded volume interaction among the clusters is neglected, the total pressure of the system is a superposition of partial pressures from ℓ clusters,

$$p = T \sum_{\ell=1}^{\infty} n_0(\ell), \tag{9.1.39}$$

where $n_0(\ell)$ is the equilibrium number density of ℓ clusters. The total particle number density is expressed as

$$n = \sum_{\ell=1}^{\infty} \ell n_0(\ell). \tag{9.1.40}$$

From the variance relation (1.2.19) or (1.2.48) the isothermal compressibility is of the form,

$$K_T = (Tn^2)^{-1} \sum_{\ell=1}^{\infty} \ell^2 n_0(\ell). \tag{9.1.41}$$

With formation of clusters at fixed n, we can see that p decreases and K_T increases, as ought to be the case. The simplest choice of $n_0(\ell)$ is given by (9.1.35), which yields [12, 13]

$$p = n_1 T \int_0^{\infty} d\ell \, \exp\left(-\beta\alpha_0 \ell^{2/3} + \beta\delta\mu\ell\right), \tag{9.1.42}$$

where $\delta\mu = \mu_{\text{gas}} - \mu_{\text{liq}}$ and the summation $\sum_\ell$ is replaced by the integral $\int d\ell$. We regard p as a function of the gas chemical potential $\mu = \mu_{\text{gas}}$ and the temperature T. Note that the liquid chemical potential μ_{liq} is the value on the coexistence curve and is a function of T. We then differentiate p with respect to μ at fixed T to obtain (9.1.40) and (9.1.41) with the aid of (1.2.14) and (1.2.18).

Essential singularity

The right-hand side of (9.1.42) is an analytic function of $h \equiv \beta\delta\mu$ defined in the region $\text{Re}\, h \le 0$ at fixed T. However, in the region $\text{Re}\, h > 0$, the integral becomes divergent at large ℓ (or for large clusters), while all the derivatives $d^k p/dh^k$ ($k = 1, 2, \ldots$) remain finite as $h \to 0$ with $\text{Re}\, h < 0$. This implies that the thermodynamic potential p/T as a function of $\nu = \beta\mu$ has an *essential singularity* on the coexistence curve [12]–[16]. However, experimental observation of this singularity from above the coexistence curve is very difficult because there is no divergence of the thermodynamic derivatives. The above arguments can also be applied to Ising systems after magnetic field reversal, because of the correspondence between the two systems: $p \leftrightarrow -f$ and $\beta\delta\mu \leftrightarrow h$.

The Fisher model

By calculating the cluster contribution to the grand canonical partition function, Fisher proposed a more detailed form for the equilibrium cluster distribution near the coexistence curve ($T < T_c$) [13],

$$n_0(\ell) = n_1 \ell^{-(2+1/\delta)} \exp\big[-b_0(1 - T/T_c)\ell^{1/\beta\delta} + h\ell\big], \tag{9.1.43}$$

where b_0 is a positive constant and $\beta\delta\mu$ in fluids is written as h in order to apply the above expression also to Ising spin systems. Near the critical point, this form is more accurate than the classical one (9.1.24) for small clusters whose linear dimensions are shorter than ξ or

$$L_\ell \equiv a\ell^{1/D} < \xi. \tag{9.1.44}$$

Such clusters are important near the critical point and are characterized by the fractal dimension D, as discussed in Section 2.1. The algebraic power factor ($\propto \ell^{-(2+1/\delta)}$) is important close to the critical point where the region (9.1.44) of ℓ is well defined. It should be noted that the surface free-energy term in the exponent is assumed to be linear in $1 - T/T_c$. This is a natural assumption; in fact, the clusters shorter than ξ are influenced by those with lengths shorter than or comparable to L_ℓ as the renormalization group theory indicates. Hence $n_0(\ell)$ should remain analytic with respect to $1 - T/T_c$ as long as $L_\ell \ll \xi$. For $L_\ell > \xi$, however, $n_0(\ell)$ assumes the form (9.1.35) depending on fractional powers of $1 - T/T_c$. With the form (9.1.43) the critical divergence of the isothermal compressibility (or the magnetic susceptibility) can be correctly reproduced. At $h = 0$ we have

$$K_T \sim \int_1^{\ell^*} d\ell \ell^{-1/\delta} \sim \xi^{D(1-1/\delta)} \sim \xi^{\gamma/\nu}, \tag{9.1.45}$$

where the upper cut-off is $\ell^* = (1 - T/T_c)^{-\beta\delta} = (\xi/a)^D$ and use has been made of the exponent relations (2.1.7) and (2.1.28).

The Fisher cluster distribution (9.1.43) has been compared with computer simulation data of Ising systems [17]–[21], where the majority of spins are aligned in one direction (up or down) close to the coexistence curve. In the simplest definition, groups of reversed spins linked together by nearest-neighbor bonds may be called clusters. However, with this definition in 3D, *infinite clusters* percolate throughout the lattice near the critical point even in one-phase states [19]. More elaborate definitions of clusters were subsequently devised. It is known that the calculated density $n_0(\ell)$ is well characterized by the power-law factor ($\propto \ell^{-(2+1/\delta)}$) and the surface term ($\propto (1 - T/T_c)\ell^{1/\beta\delta}$) in the exponent, which are predicted by the Fisher model (9.1.43).

Binder argued that, if clusters are defined appropriately on the lattice in Ising systems, $n_0(\ell)$ should generally satisfy [20],

$$n_0(\ell) = n_1 \ell^{-(2+1/\delta)} N(L_\ell/\xi, h\ell), \tag{9.1.46}$$

in terms of a scaling function $N(x, y)$ near the critical point. Because $(1 - T/T_c)\ell^{1/\beta\delta} \sim (L_\ell/\xi)^{1/\nu}$ from (2.1.28), the Fisher form (9.1.43) is a special case of (9.1.46). Furthermore, for $L_\ell > \xi$ the clusters become compact domains or droplets with radius $R \sim L_\ell > \xi$, as discussed in Section 2.1. Then the classical droplet distribution $n_0(R)$ in (9.1.24), which is analytic in R but non-analytic in $1 - T/T_c$, becomes consistent with (9.1.46). To check it we note the relations,

$$n_1 \ell^{-(2+1/\delta)} \sim n_1 \ell^{-1} \xi^{-d}, \tag{9.1.47}$$

$$h\ell \sim h(1 - T/T_c)^\beta \ell^{d/D} \sim h\psi_{eq} R^d, \tag{9.1.48}$$

at the crossover $L_\ell/\xi = 1$. Thus $n_0(R)\partial R/\partial \ell \sim n_0(R)R/\ell$ satisfies (9.1.46).

In summary, clusters are essential entities near the coexistence curve appearing with appreciable densities. The probability that they grow into droplets with sizes several times larger than ξ is extremely small for small supersaturation, but such rare events can indeed trigger nucleation of a new phase for $h > 0$.

9.2 Birth of droplets

We consider early-stage nucleation where the droplet volume fraction is very small and interactions between droplets are nearly absent. In the conserved case the supersaturation Δ is assumed to be equal to the initial value ϕ in (8.3.1). We are interested in the time evolution of the droplet size distribution $n(R,t)$ which obeys (9.1.23) for $t > 0$ with a positive constant μ_{eff}. The initial distribution $n(R,0)$ satisfies (9.1.27) and virtually vanishes for $R > R_c$. After a long incubation time, a small number of droplets with radii larger than R_c emerge and continue to grow.

9.2.1 Evolution of droplets with $R \sim R_c$

We wish to examine how droplets with sizes close to R_c evolve. In the vicinity of R_c the Langevin equation may be linearized as

$$\frac{\partial}{\partial t}\delta R = \Gamma_c \delta R + \theta(R_c, t), \tag{9.2.1}$$

where Γ_c is the value of $\partial v(R)/\partial R$ at $R = R_c$ or

$$\Gamma_c = -\mathcal{L}_c \beta \mathcal{H}''(R_c) = (d-1) S_d \mathcal{L}_c \beta \sigma R_c^{d-3}. \tag{9.2.2}$$

Hereafter we write

$$\mathcal{L}_c = \mathcal{L}(R_c). \tag{9.2.3}$$

The growth rate Γ_c is written as

$$\begin{aligned} \Gamma_c &= (d-1) L R_c^{-2} \propto h^2 && \text{(nonconserved)} \\ &= D R_c^{-2} \Delta = \frac{1}{4} D d_0^{-2} \Delta^3 && \text{(3D conserved)}. \end{aligned} \tag{9.2.4}$$

As $h \to 0$ or $\Delta \to 0$, the timescale Γ_c^{-1} becomes very long. If the noise term were neglected, droplets with small positive (negative) δR would grow (shrink) exponentially with the growth rate Γ_c. As can be easily expected, however, the effect of the thermal noise is crucial for such near-critical droplets. To check this, we set up the corresponding linearized Fokker–Planck equation,

$$\frac{\partial}{\partial t}n = \frac{\partial}{\partial R}\left(\mathcal{L}_c\frac{\partial}{\partial R} - \Gamma_c\delta R\right)n, \tag{9.2.5}$$

where the second derivative $\partial^2/\partial R^2$ arises from the thermal noise. To represent its strength we introduce a small parameter ε by

$$\varepsilon^{-2} = \beta|\mathcal{H}_c''|R_c^2 = (d-1)S_d\beta\sigma R_c^{d-1} \sim (R_c/\xi)^{d-1}. \tag{9.2.6}$$

We have $\varepsilon \sim \xi/R_c$ in 3D. The maximum of the droplet free energy (9.1.2) reads

$$\mathcal{H}_c = \frac{1}{d(d-1)}\varepsilon^{-2}T. \tag{9.2.7}$$

We introduce a dimensionless radius deviation given by

$$x = \delta R/\varepsilon R_c \quad \text{or} \quad R = R_c(1+\varepsilon x). \tag{9.2.8}$$

Then (9.2.5) may be rewritten as

$$\frac{\partial}{\partial t}n = \Gamma_c\frac{\partial}{\partial x}\left(\frac{\partial}{\partial x} - x\right)n, \tag{9.2.9}$$

where ε is removed. Each droplet motion is sensitively affected by the thermal noise in the following narrow region,

$$|R/R_c - 1| \lesssim \varepsilon \sim (\xi/R_c)^{(d-1)/2}. \tag{9.2.10}$$

That is, if we observe a droplet in the above region, it is highly probabilistic whether it grows or shrinks. The thermal noise is also important at small $R(\ll R_c)$ where it produces the distribution in (9.1.27). In Appendix 9A general solutions of (9.2.9) will be presented.

9.2.2 Deterministic growth

For $R/R_c - 1 \gtrsim \varepsilon$, the growth rate of each droplet is nearly deterministic as

$$\frac{\partial}{\partial t}R = v(R), \tag{9.2.11}$$

and $n(R,t)$ obeys

$$\frac{\partial}{\partial t}n = -\frac{\partial}{\partial R}[v(R)n], \tag{9.2.12}$$

where the radius growth rate $v(R)$ is defined by (9.1.21). Then there is a mapping between the droplet radii at two times, R_1 slightly exceeding R_c at a time t_1 and R_2 at a later time

t_2. The droplet number conservation gives

$$n(R_2, t_2) = n(R_1, t_1)\left(\frac{\partial R_1}{\partial R_2}\right)_{t_1 t_2} = n(R_1, t_1)\frac{v(R_1)}{v(R_2)}. \quad (9.2.13)$$

Because (9.2.11) is integrated as

$$t_2 - t_1 = \int_{R_1}^{R_2} dR' \frac{1}{v(R')}, \quad (9.2.14)$$

we find $(\partial R_2/\partial R_1)_{t_1 t_2} = v(R_2)/v(R_1)$ at fixed t_1 and t_2. Furthermore, because $v(R) \cong \Gamma_c(R - R_c)$ for small $R/R_c - 1$, we rewrite (9.2.14) as

$$\Gamma_c(t_2 - t_1) = \ln\left(\frac{R_2 - R_c}{R_1 - R_c}\right) + \int_{R_1}^{R_2} dR' \left[\frac{\Gamma_c}{v(R')} - \frac{1}{R' - R_c}\right]. \quad (9.2.15)$$

For $0 < R_1/R_c - 1 \ll 1$ the lower bound R_1 of the integral in the second term may be replaced by R_c, because the integral is convergent even in the limit $R_1 \to R_c$. We obtain

$$\ln(R_1/R_c - 1) = \ln(R_2/R_c - 1) + \mathcal{G}(R_2/R_c) - \Gamma_c(t_2 - t_1), \quad (9.2.16)$$

where

$$\mathcal{G}(R/R_c) = \int_{R_c}^{R} dR' \left[\frac{\Gamma_c}{v(R')} - \frac{1}{R' - R_c}\right]. \quad (9.2.17)$$

For the models treated so far, $\mathcal{G}(R/R_c)$ turns out to be a function of $u = R/R_c$ only:

$$\begin{aligned} \mathcal{G}(u) &= u - 1 && \text{(nonconserved)} \\ &= \frac{1}{2}u^2 + u - \frac{3}{2} && \text{(3D conserved).} \end{aligned} \quad (9.2.18)$$

In Fig. 9.4 we plot $u_2 \equiv R_2/R_c - 1$ vs $u_1 \equiv R_1/R_c - 1$ for $0 < u_1 < 1$ and $\Gamma_c(t_2 - t_1) =$ 10.5, 20.5, and 40.5 in the 3D conserved case. We will use (9.2.16) in Appendix 9A.

In real nucleation experiments there is a maximum droplet radius $R_{max}(t)$ above which $n(R, t)$ is virtually zero. Because $R_{max}(t)$ obeys (9.2.11), its value at large time $t \gg \Gamma_c^{-1}$ roughly satisfies

$$\mathcal{G}(R_{max}(t)/R_c) - \Gamma_c t \cong 0. \quad (9.2.19)$$

Obviously, we have

$$\begin{aligned} R_{max}(t) &\cong R_c\Gamma_c t = v_h t && \text{(nonconserved)} \\ &\cong R_c(2\Gamma_c t)^{1/2} \sim D\Delta^{1/2} t^{1/2} && \text{(3D conserved),} \end{aligned} \quad (9.2.20)$$

where the velocity $v_h = R_c\Gamma_c$ appeared in (8.2.25). (See Fig. 9.5, p. 504, for $n(R, t)$ near $R_{max}(t)$.)

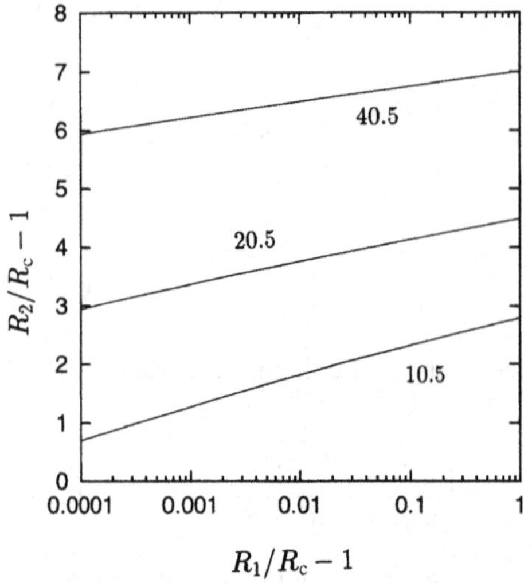

Fig. 9.4. The mapping between two radii, $R_2 > R_1$ with R_1 slightly exceeding R_c, as determined by (9.2.15) for $\Gamma_c(t_2 - t_1) = 10.5, 20.5,$ and 40.5.

9.2.3 The nucleation rate

After a transient time t_0, we observe a constant birth (nucleation) rate I of droplets with sizes larger than R_c emerging per unit volume and per unit time. From (9.2.9) $n(R, t)$ at $R \sim R_c$ is known to change on the timescale of Γ_c^{-1}. At smaller R the timescale is faster. It is then natural to estimate t_0 as [6]

$$t_0 \sim \Gamma_c^{-1}. \tag{9.2.21}$$

The meaning of the constant nucleation rate may be stated as follows. If $R - R_c \gg \varepsilon R_c$, the evolution of R is almost deterministic and

$$n(R, t)\, dR = I dt \quad \text{for} \quad dR = v(R)\, dt, \tag{9.2.22}$$

which is the number of growing droplets newly emerging in a time interval dt per unit volume and is independent of R. Thus we obtain a steady distribution in the region $R/R_c - 1 \gg \varepsilon$,

$$n_s(R) = I/v(R). \tag{9.2.23}$$

Of course, to have appreciable droplets larger than R_c in a volume V, the observation time t_{obs} must be much longer than $1/IV$. For a typical experimental volume (say, 1 cm^3), we may define a nucleation time by

$$t_N = (IV)^{-1}. \tag{9.2.24}$$

We treat the case $t_{obs} \gg t_N$ supposing slow droplet growth.[1] We also note that the droplet volume fraction $q(t)$ increases with a rate of order $R_{max}^d I$, so that

$$q(t) \sim R_{max}^d It, \tag{9.2.25}$$

[1] If the growth is rapid or ballistic with small dissipation, appearance of a single droplet with $R > R_c$ can lead to completion of macroscopic phase separation. This is the case at very low temperatures.

where the algebraic growth of $R_{\max}(t)$ is assumed. Interaction between droplets become appreciable at a particular completion time $t_{\rm co}$, as will be discussed later. In Appendix 9A we will investigate how $n(R, t)$ rapidly decays from $I/v(R)$ to 0 around $R_{\max}$. The width of this changeover region is estimated as

$$(\Delta R)_{\max} \sim v(R_{\max})/\Gamma_{\rm c}. \tag{9.2.26}$$

The steady-state distribution

Because $n_{\rm s}(R)$ is a steady solution of (9.1.23), it generally satisfies

$$\mathcal{L}(R)\left[\frac{\partial}{\partial R} + \beta\mathcal{H}'(R)\right]n_{\rm s} = -I. \tag{9.2.27}$$

Imposing the condition $n_{\rm s}(R) \to 0$ as $R \to \infty$, we integrate the above equation as

$$n_s(R) = I\int_R^\infty dR_1 \frac{1}{\mathcal{L}(R_1)} \exp\big[\beta\mathcal{H}(R_1) - \beta\mathcal{H}(R)\big]. \tag{9.2.28}$$

For $R - R_{\rm c} \gg \varepsilon R_{\rm c}$, we may replace $\beta\mathcal{H}(R_1) - \beta\mathcal{H}(R)$ and $\mathcal{L}(R_1)$ in the above integrand by $\beta(\partial\mathcal{H}/\partial R)(R_1 - R)$ and $\mathcal{L}(R)$, respectively; then, (9.2.23) is reproduced. Next we impose the boundary condition (9.1.27) at small R to obtain an equation for I,

$$n_\xi = I\int_0^\infty dR_1 \frac{1}{\mathcal{L}(R_1)} \exp\big[\beta\mathcal{H}(R_1)\big], \tag{9.2.29}$$

where n_ξ is the coefficient in (9.1.25). The integrand on the right-hand side is very sharply peaked at $R_{\rm c}$ from the second line of (9.1.1). The gaussian integration from $|R - R_{\rm c}| \lesssim \varepsilon R_{\rm c}$ yields the classical expression,

$$\begin{aligned} I &= (2\pi)^{-1/2}\mathcal{L}_{\rm c}(\beta|\mathcal{H}''|)^{1/2}n_0(R_{\rm c}) \\ &= (2\pi)^{-1/2}\Gamma_{\rm c} n_\xi \varepsilon R_{\rm c} \exp(-\beta\mathcal{H}_{\rm c}). \end{aligned} \tag{9.2.30}$$

In 3D, we have $n_\xi \varepsilon R_{\rm c} \sim \xi^{-3}$ from (9.1.26) and (9.2.6), so $I \sim \Gamma_{\rm c}\xi^{-3}\exp(-\beta\mathcal{H}_{\rm c})$. As is well known, the exponential factor $\exp(-\beta\mathcal{H}_{\rm c})$ varies over many decades even for a very small change of $\mu_{\rm eff}$ ($\propto h$ or Δ). Thus I is extremely sensitive to $\mu_{\rm eff}$, whereas it is much less sensitive to the kinetic factor $\Gamma_{\rm c}$.

Let us examine the behavior of $n_{\rm s}(R)$ close to $R_{\rm c}$. If $R < R_{\rm c}$ and $\varepsilon \ll |R/R_{\rm c} - 1| \ll 1$, we obtain

$$n_{\rm s}(R) \cong n_\xi \exp\left[-\beta\mathcal{H}_{\rm c} + \frac{1}{2}\beta|\mathcal{H}''|(\delta R)^2\right] \cong n_0(R). \tag{9.2.31}$$

However, $n_{\rm s}(R_{\rm c}) = n_0(R_{\rm c})/2$ at $R = R_{\rm c}$. Therefore, the ratio $n_{\rm s}(R)/n_0(R)$ is nearly equal to 1 in the region $1 - R/R_{\rm c} \gtrsim \varepsilon$, decreases to 1/2 at $R = R_{\rm c}$, and becomes much smaller than 1 for $R/R_{\rm c} - 1 \gtrsim \varepsilon$. The behavior in the region $|R - R_{\rm c}| \ll R_{\rm c}$ can be expressed in the following integral form,

$$n_{\rm s}(R)/n_0(R_{\rm c}) \cong (2\pi)^{-1/2}\int_0^\infty ds \exp\left(-\frac{1}{2}s^2 - xs\right), \tag{9.2.32}$$

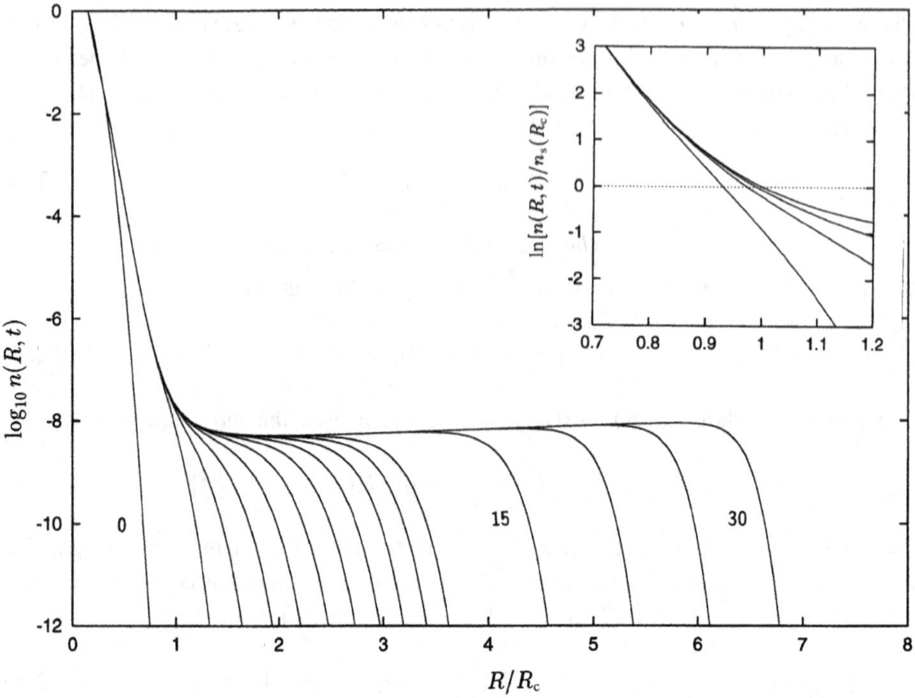

Fig. 9.5. Time evolution of the droplet size distribution $n(R, t)$ on a semi-logarithmic scale as a solution of (9.2.34) at $\varepsilon^2 = 0.0096$ in the 3D conserved case. The first 11 curves correspond to the times at $\Gamma_c t = 0, 1, \ldots,$ and 10. The last four curves are those at $t = 15, 20, 25,$ and 30. In the inset the curves of $\ln[n(R, t)/n_s(R_c)]$ are plotted at $\Gamma_c t = 1, 2, 3,$ and 15 (from below) around $R/R_c \sim 1$.

where $x = (R - R_c)/\varepsilon R_c$. This is in fact a steady solution of the linearized Fokker–Planck equation (9.2.9). It behaves as

$$
\begin{aligned}
n_s(R)/n_0(R_c) &\cong \exp\left(\frac{1}{2}x^2\right) \qquad (x \ll -1), \\
&\cong (2\pi)^{-1/2}\frac{1}{x} \qquad (x \gg 1). \qquad (9.2.33)
\end{aligned}
$$

9.2.4 Numerical analysis of the birth process

As an illustration, we show in Fig. 9.5 the time evolution of $n(R, t)$ for the 3D conserved case obtained as a solution of the Fokker–Planck equation (9.1.23),

$$
\frac{\partial}{\partial t}n = \Gamma_c \frac{\partial}{\partial r}\left[\frac{\varepsilon^2}{r^3}\frac{\partial}{\partial r} - \frac{1}{r} + \frac{1}{r^2}\right]n, \qquad (9.2.34)
$$

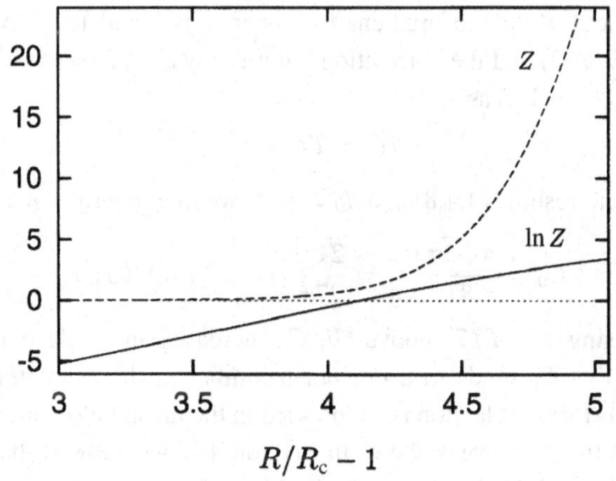

Fig. 9.6. Z and $\ln Z$ vs $R/R_c - 1$ at $\Gamma_c t = 15$ in Fig. 9.5, where we set $n(R, t) = [I/v(R)] \exp(-Z)$.

where $r = R/R_c$. From (9.2.6) we have

$$\varepsilon = (9\pi\beta\sigma R_c^2)^{-1/2} = (9\pi A_\sigma)^{-1/2}\xi/R_c, \tag{9.2.35}$$

where A_σ is the coefficient in (4.4.11). In Fig. 9.5 we choose $\varepsilon = 0.0096^{1/2} = 0.098$, though this is much larger than its typical values in real 3D nucleating systems. For this choice we have $\mathcal{H}_c/T = (6\varepsilon^2)^{-1} = 17.4$ in 3D from (9.2.7). For near-critical fluids in the asymptotic critical region, this corresponds to $R_c/\xi = 6.8$ and $\Delta = 0.3d_0/\xi \sim 0.05$ from $A_\sigma = 0.09$ and $d_0/\xi \sim 0.1$. As the initial condition of the calculation, the system is assumed to be on the coexistence curve; namely, $n(R, 0) = \text{const.} \exp(-4\pi\sigma R^2/T) \propto \exp(-r^2/2\varepsilon^2)$, which is virtually zero around $R \sim R_c$. The figure demonstrates the approach of $n(R, t)$ to a steady distribution $n_s(R)$ in the region $R < R_{\max}(t)$ on the timescale of Γ_c^{-1} in accord with (9.2.21). The upper cut-off expands in time as the second line of (9.1.22) for $\Gamma_c t \gg 1$. In addition, in Fig. 9.6 we plot $Z \equiv \ln[I/v(R)n(R, t)]$ and $\log Z$ at $\Gamma_c t = 15$ to examine the very steep decay of $n(R, t)$ around $R_{\max}$. We can see that Z roughly grows exponentially around $R_{\max}$, so that

$$n(R, t) \cong \frac{I}{v(R)} \exp\left\{-\exp\left[\frac{R - R_{\max}}{(\Delta R)_{\max}}\right]\right\}, \tag{9.2.36}$$

where the width $(\Delta R)_{\max}$ is consistent with (9.2.26).

9.2.5 The nucleation rate in the mean field critical region

The classical Landau theory of phase transition holds somewhat away from the critical point. This condition is expressed in terms of the Ginzburg number Gi in (4.1.25) in 3D as

$$1 - T/T_c > Gi = (3/2\pi^2)^2 u_0^2/a_0, \tag{9.2.37}$$

where the coefficient K of the gradient free energy is equal to 1. We may relate the coefficient a_0 in (9.2.37) and the correlation length ξ by $\xi^{-2}/2 = \kappa^2 = a_0(1 - T/T_c)$. In 3D, we rewrite $\mathcal{H}_c$ in (9.1.2) as

$$\mathcal{H}_c = TC_0/\Delta^2. \tag{9.2.38}$$

Using the mean field results, (4.4.8) and $d_0 = \xi/6$, we may rewrite C_0 as

$$C_0 = \frac{4\sqrt{2}\pi}{81}\frac{\kappa}{u_0} = \frac{2\sqrt{2}}{27\pi}\left[(1 - T/T_c)/Gi\right]^{1/2}. \tag{9.2.39}$$

Thus, with increasing $1 - T/T_c$ above Gi, C_0 increases and I decreases. In Appendix 9B, we shall see that C_0 tends to a universal number in the asymptotic critical region $T_c - T \ll Gi$. Therefore, nucleation is suppressed in the mean field critical region, because the thermal fluctuations are weak there. In Section 4.2, we showed that polymer blends with large molecular weights have very small Gi ($\propto N^{-2}$).

9.3 Growth of droplets

9.3.1 The Kolmogorov, Johnson–Mehl, and Avrami theory in nonconserved cases

In the nonconserved case (model A), the interface velocity is expressed as (8.2.24), which tends to the constant velocity in (8.2.25),

$$v_h = (2TL\psi_{eq}/\sigma)h = R_c\Gamma_c, \tag{9.3.1}$$

for $R \gg R_c$ or for $t \gg t_c \sim 1/\Gamma_c$. We assume that the nucleation rate I is very small and is independent of time. Substitution of $R_{max} = v_h t$ into (9.2.25) yields the volume fraction of the favored phase,

$$q(t) \cong \int_0^t dt\, \frac{4\pi}{3}R_{max}^3 I \cong \frac{\pi}{3}Iv_h^3 t^4 \quad \text{(3D)}, \tag{9.3.2}$$

in the early stage of nucleation where $q(t) \ll 1$ and $t \gtrsim \Gamma_c^{-1}$. The 2D version follows by replacement, $v_h^3 t^4 \to v_h^2 t^3$ in (9.3.2). The growing domains begin to touch and overlap as $q(t)$ becomes of order 1. The completion time t_{co} of this phase inversion is determined as

$$v_h^d\, t_{co}^{d+1}\, I = 1 \quad \text{or} \quad t_{co} = (v_h^d I)^{-1/(d+1)}. \tag{9.3.3}$$

Here we are assuming that t_{co} is much shorter than t_N in (9.2.24). The characteristic domain size at $t = t_{co}$ is given by

$$r_h = v_h t_{co} = (v_h/I)^{1/(d+1)}. \tag{9.3.4}$$

The finiteness of the critical radius R_c (or the surface tension effect) is important in the early stage, $t \lesssim \Gamma_c^{-1}$. Therefore, in the limit $r_h \gg R_c$ or $t_{co} \gg \Gamma_c^{-1}$, it may be neglected in the overall relaxation of $q(t)$, where Kolmogorov, Avrami, and Johnson–Mehl predicted the exponentiated form [22]–[24],

$$q(t) = 1 - \exp\left[-\frac{\pi}{3}Iv_h^d t^{d+1}\right] = 1 - \exp\left[-\frac{\pi}{3}(t/t_{co})^{d+1}\right]. \tag{9.3.5}$$

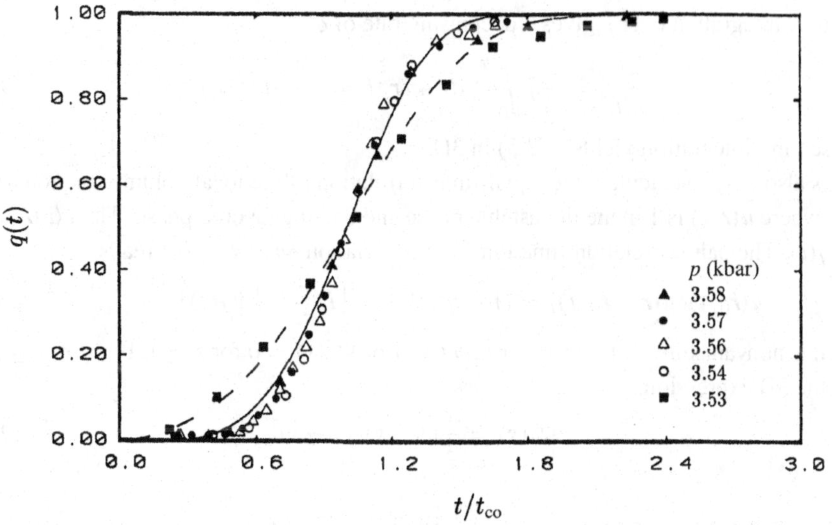

Fig. 9.7. Scaled curves of the volume fraction $q(t)$ of the CsCl (B2) structure growing from the NaCl (B1) structure in RbI after increasing the pressure above a critical value $p_c \cong 3.5$ kbar [26]. The time is measured in units of the completion time t_{co} which becomes longer with decreasing $p - p_c$. The solid and dashed curves are from (9.3.5) and (9.3.6), respectively.

This holds for 2D and 3D, with the same coefficient $\pi/3$. Subsequently, Ishibashi and Takagi phenomenologically extended the above formula taking into account finite R_c as [25]

$$q(t) = 1 - \exp\left\{-\frac{\pi}{3} I v_h^d \left[(t + t_c)^{d+1} - t_c^{d+1}\right]\right\}, \tag{9.3.6}$$

where $t_c = R_c/v_h \sim \Gamma_c^{-1}$ is defined in (8.2.60). In Fig. 9.7 we show data of time-dependent volume fractions of a new phase in a pressure-induced structural phase transition [26]. The curves 2–5 correspond to the case $r_h \gg R_c$ and nicely fall onto the (solid) theoretical curve (9.3.5). However, curve 6 corresponds to a shallow quenching and considerably deviates from (9.3.5), presumably from the effect of finite R_c. If the curve is fitted to (9.3.6), we have $R_c/r_h = t_c/t_{co} = 0.3$.

The derivation of (9.3.5) is very simple. Notice that $\bar{q}(t) = 1 - q(t)$ is the probability that no transformation has yet taken place at time t at an arbitrarily chosen point $\mathbf{r}_0$. Let the phase change be first caused in the subsequent time interval $[t, t + dt]$ by invasion of a droplet with radius in the range $[r, r + dr]$, where dt and dr are infinitesimal. The birth of such a droplet, which occurred in the time interval $[t - r/v_h, t - r/v_h + dt]$ and in the shell region $[r, r + dr]$, is a stochastic event with probability $dN = I(4\pi r^2 dr)dt$ (for $d = 3$). The inversion probability at $\mathbf{r}_0$ is given by the product $\bar{q}dN(= -d\bar{q})$, where the factor $\bar{q}$ originates from the supplemented condition that there is no transformation before t. Integration over r gives a decreasing rate of $\bar{q}$,

$$\frac{d}{dt}\bar{q} = -\bar{q}\int_0^{v_h t} dr\, 4\pi r^2 I = -\frac{4\pi}{3} v_h^3 t^4 I \bar{q}, \tag{9.3.7}$$

whose time-integration yields (9.3.5) in 3D.

It is also easy to calculate the equal-time correlation of the local volume fraction $u(\mathbf{r}, t)$ [27], where $u(\mathbf{r}, t)$ is 1 in the metastable phase and 0 in the favored phase. Then $\langle u(\mathbf{r}, t)\rangle = 1 - q(t)$. The pair correlation function for the deviation $\delta u = u - \langle u \rangle$ reads

$$\langle \delta u(\mathbf{r}_0, t)\delta u(\mathbf{r} + \mathbf{r}_0, t)\rangle = [1 - q(t)]^2 \{\exp[I v_h^d t^{d+1} \Psi_d(s)] - 1\}, \tag{9.3.8}$$

which is nonvanishing only for $s = r/2v_h t < 1$ or $\Psi_d(s) = 0$ for $s \geq 1$. For $0 \leq s \leq 1$ we give the 3D expression,

$$\Psi_3(s) = \frac{\pi}{3}(1 - s)^3(1 + s). \tag{9.3.9}$$

9.3.2 The Lifshitz–Slyozov and Wagner theory for conserved systems

The completion time

In 3D conserved systems we may define a completion time $t_{\rm co}$ such that the droplet volume fraction is some fraction of the initial supersaturation $\phi \equiv \Delta(0)$, say, 0.5. Phase separation has partially completed at $t = t_{\rm co}$. Emergence of droplets will be noticeable on the timescale of $t_{\rm co}$. From (9.2.25) and using (9.2.20), we may estimate $t_{\rm co}$ by

$$(2D\phi t_{\rm co})^{3/2} I t_{\rm co} \sim \phi. \tag{9.3.10}$$

Here we define a dimensionless nucleation rate $\tilde{I}$ by

$$I = D\xi^{-5}\tilde{I}. \tag{9.3.11}$$

Then (9.3.10) is solved to give

$$\Gamma_{\rm c} t_{\rm co} = \phi^{14/5}\tilde{I}^{-2/5}. \tag{9.3.12}$$

At $t \sim t_{\rm co}$ the maximum radius grows up to the following order,

$$\frac{R_{\rm max}(t_{\rm co})}{R_{\rm c}(0)} \sim (\Gamma_{\rm c} t_{\rm co})^{1/2} \sim \phi^{7/5}\tilde{I}^{-1/5}. \tag{9.3.13}$$

At small volume fraction we have $t_{\rm co} \gg \Gamma_{\rm c}^{-1} \sim t_0$ and

$$R_{\rm max}(t_{\rm co}) \gg R_{\rm c}(0). \tag{9.3.14}$$

Figure 9.8 shows the completion time $t_{\rm co}$ estimated for near-critical fluids [28], which increases dramatically for $\phi \lesssim 0.02$.

We also make two supplementary remarks. (i) The diffusion length at $t = t_{\rm co}$ is estimated as $\ell_D = (Dt_{\rm co})^{1/2} \sim R_{\rm max}(t_{\rm co})\phi^{-1/2}$. It becomes longer than the average inter-domain length $n_{\rm dom}^{-1/3}$, because

$$\ell_D \sim \phi^{-1/6} n_D^{-1/3}, \tag{9.3.15}$$

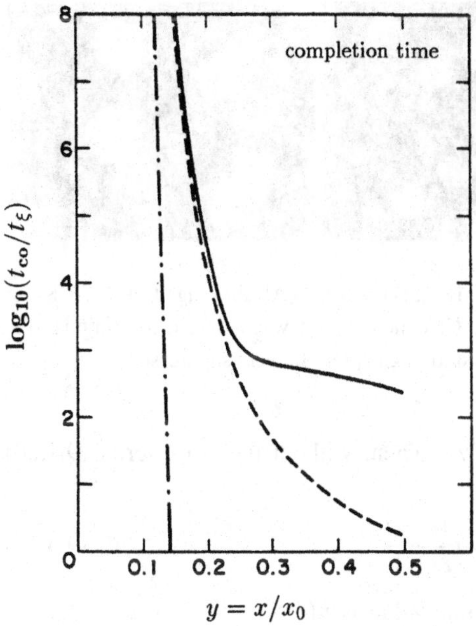

Fig. 9.8. The scaled completion time t_{co}/t_ξ vs initial relative supersaturation $y = x/x_0 \cong 6\phi$, where $t_\xi = D^{-1}\xi^2$, in a near-critical fluid [28]. The solid line is obtained from the Schwartz–Langer theory [28], and the dashed line from the Binder–Stauffer theory [6]. The dash-dot curve represents the scaled nucleation time t_N/t_ξ in (9.2.24) for V =1 cm^3.

where $n_{\rm dom} \sim \phi/R_{\max}(t_{co})^3$ is the droplet density at $t = t_{co}$. Droplet interaction may then be taken into account with the mean field constraint (9.1.28). (ii) To have a large number of droplets in the experimental cell at $t = t_{co}$, the nucleation time t_N in (9.2.24) must be much shorter than t_{co}. In 3D conserved systems this condition is realized for sufficiently large cell volume,

$$V/\xi^3 \gg \phi^{-8/15}\tilde{I}^{-3/5}. \tag{9.3.16}$$

The LSW equations

After a transient time of order t_{co}, we follow the time evolution of the droplet size distribution neglecting further emergence of new droplets and the thermal noise. Then each droplet evolves under the influence of a time-dependent supersaturation $\Delta(t)$. Here analytic theory of the droplet evolution is possible, as first presented by Lifshitz–Slyozov, and by Wagner (LSW) [29]–[31]. This theory is justified in the limit of small droplet volume faction as already discussed in Section 8.4. As an example, Fig. 9.9 shows coarsening of spherical domains with the $L1_2$ structure (illustrated in Fig. 3.10) in a Ni–18Cr–6Al metallic alloy [32], where the mean domain size distribution nicely obeyed the LSW theory with the growth law $\bar{R}(t) \propto t^{1/3}$ for the mean radius.

To make the equations simple, let us rescale the radius and time as

$$r = R/R_c(0), \quad \tau = \Gamma_c(t - t_{co}), \tag{9.3.17}$$

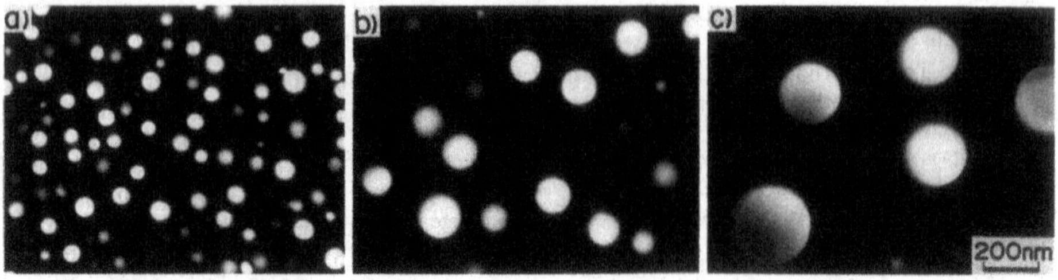

Fig. 9.9. Development of γ' precipitates for Ni–18 at.%Cr–6 at.%Al alloy aged at 1073 K for (a) 86.4 ks, (b) 691 ks, and (c) 5.2 Ms [32]. The Cr concentration was adjusted to minimize the lattice mismatch and the elastic effects, to be discussed in Chapter 10, were suppressed.

where $R_c(0) = 2d_0/\phi$ and $\Gamma_c = Dd_0^{-2}\phi^3/4$. Then, without the noise term, (9.1.20) becomes

$$\frac{d}{d\tau}r = p(\tau)\frac{1}{r} - \frac{1}{r^2}. \tag{9.3.18}$$

We write the supersaturation divided by its initial value as $p(\tau)$:

$$p(\tau) = \frac{\Delta(t)}{\Delta(0)} = \frac{R_c(0)}{R_c(t)}, \tag{9.3.19}$$

where $R_c(t)$ is the time-dependent critical radius. From (9.1.28) we express $p(\tau)$ in terms of the droplet distribution,

$$p(\tau) = 1 - \int_0^\infty dr r^3 n(r, \tau). \tag{9.3.20}$$

Here $(4\pi/3)[R_c(0)^4/\Delta(0)]n(R, t)$ is redefined as $n(r, \tau)$. The radius distribution obeys

$$\frac{\partial}{\partial\tau}n(r, \tau) = -\frac{\partial}{\partial r}\left[\frac{p(t)}{r} - \frac{1}{r^2}\right]n(r, \tau). \tag{9.3.21}$$

Now (9.3.20) and (9.3.21) constitute a closed set of evolution equations, which was examined analytically [29]–[31] and numerically [33, 34]. For finite volume fractions, however, diffusional interactions between the domains can be significant [35]–[37]. This effect was discussed in Subsection 8.4.4.

Asymptotic behavior

As the initial distribution at $\tau = 0$ (or $t = t_{co}$) we should choose the distribution behaving as (9.2.36). Then $n(r, 0)$ is broadly distributed in a wide region $r \lesssim R_{max}(t_{co})/R_c(0)$ and decays rapidly at large r as

$$\frac{d^k}{dr^k}n(r, 0) \to 0 \quad (r \to \infty) \tag{9.3.22}$$

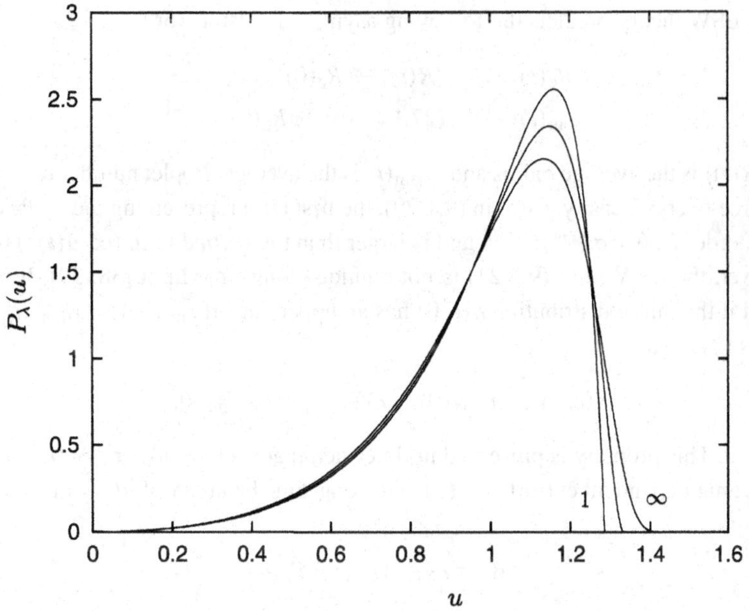

Fig. 9.10. The asymptotic scaling functions $P_\lambda(u)$ for $\lambda = 1, 2$, and ∞ reading from above at the peaks. Here $P_\infty(u)$ coincides with $P_{\rm LSW}(u)$ in (9.3.26) in the LSW theory.

for any $k = 1, 2, \ldots$. In this usual or normal case, the LSW theory holds and the long-time behavior of $p(\tau)$ and $n(r, \tau)$ are given by [29]–[31]

$$p(\tau) \to p^* \tau^{-1/3}, \tag{9.3.23}$$

$$n(r, \tau) \to A_\infty p(\tau)^4 P_{\rm LSW}(u), \tag{9.3.24}$$

where $p^* = (9/4)^{1/3}$ and

$$u = r/p(\tau) = R/R_{\rm c}(t). \tag{9.3.25}$$

As will be calculated in Appendix 9C, the scaling function $P_{\rm LSW}(u)$ is defined in a finite region $0 < u < 1.5$ and is of the form,

$$P_{\rm LSW}(u) = \frac{324u^2}{(u+3)^{7/3}(3-2u)^{11/3}} \exp\left(-\frac{u}{1.5-u}\right) \quad (0 < u < 1.5), \tag{9.3.26}$$

with the normalization conditions,

$$\int_0^{1.5} du\, u P_{\rm LSW}(u) = \int_0^{1.5} du\, P_{\rm LSW}(u) = 1. \tag{9.3.27}$$

This function is plotted in Fig. 9.10. From (9.3.18) the coefficient A_∞ is determined as

$$A_\infty = \left[\int_0^{1.5} du\, u^3 P(u)\right]^{-1} \cong 0.885. \tag{9.3.28}$$

Thus the LSW theory predicts the following asymptotic results for large $\tau \gg 1$:

$$R_c(t) \cong \langle R(t)\rangle \cong R_c(0)(4\tau/9)^{1/3}, \qquad (9.3.29)$$

$$n_{dom}(t) \cong (27A_\infty/16\pi)\phi R_c(0)^{-3}\tau^{-1}, \qquad (9.3.30)$$

where $\langle R(t)\rangle$ is the average radius and $n_{dom}(t)$ is the average droplet number density. In the droplet free-energy density $f_D(t)$ in (9.1.29), the first term representing the surface tension part is of order $(\Delta(0)^2\sigma\xi^{-1})\tau^{-1/3}$ and is larger than the second term ($\propto \Delta(t)^2$) by $\tau^{1/3}$.

However, the LSW limit (9.3.24) is not a unique long-time limit [38, 39]. It is not approached if the initial distribution $n(r, 0)$ has an upper cut-off $r_{max}(0) = R_{max}(t_{co})/R_c(0)$ and tends to zero as

$$n(r, 0) \sim (r_{max}(0) - r)^\lambda \qquad (r \to r_{max}(0)), \qquad (9.3.31)$$

with $\lambda > 0$. This property is preserved in time such that we have $n(r, \tau) \sim (r_{max}(\tau) - r)^\lambda$ around a time-dependent cut-off $r_{max}(\tau)$. The long-time behavior of $n(r, \tau)$ is then written as

$$n(r, \tau) \to A_\lambda p(\tau)^4 P_\lambda(u), \qquad (9.3.32)$$

where $P_\lambda(u)$ is a λ-dependent scaling function defined in the region $0 < u < u_1 = (3\lambda + 6)/(2\lambda + 5)$ and behaves as

$$P_\lambda(u) \cong C_\lambda(u_1 - u)^\lambda, \qquad (9.3.33)$$

as $u \to u_1$ with the normalization,

$$\int_0^{u_1} du\, u P_\lambda(u) = \int_0^{u_1} du\, P_\lambda(u) = 1. \qquad (9.3.34)$$

We will give an analytic expression for $P_\lambda(u)$ in Appendix 9C. The LSW scaling function is reproduced in the limit,

$$P_{LSW}(u) = \lim_{\lambda\to\infty} P_\lambda(u). \qquad (9.3.35)$$

The scaling relation (9.3.23) also holds with

$$p^* = \left[\frac{2\lambda + 5}{3(\lambda + 1)}\right]^{1/3} \frac{3\lambda + 6}{2\lambda + 5}. \qquad (9.3.36)$$

In Fig. 9.10 we display $P_1(u)$ and $P_2(u)$, which differ noticeably from $P_{LSW}(u)$ only in the region $1 < u < 1.5$. The coefficient C_λ in (9.3.33) is about 60 for $\lambda = 1$ and 700 for $\lambda = 2$, so the curves of finite $\lambda(\geq 1)$ are very steep at the upper cut-off u_1 and are not much different from the LSW limit ($\lambda = \infty$).

Numerical analysis of the LSW theory

Numerical integration of the LSW equations (9.3.20) and (9.3.21) has been performed by many authors. For a wide range of the initial distributions $n(r, 0)$, which decay rapidly

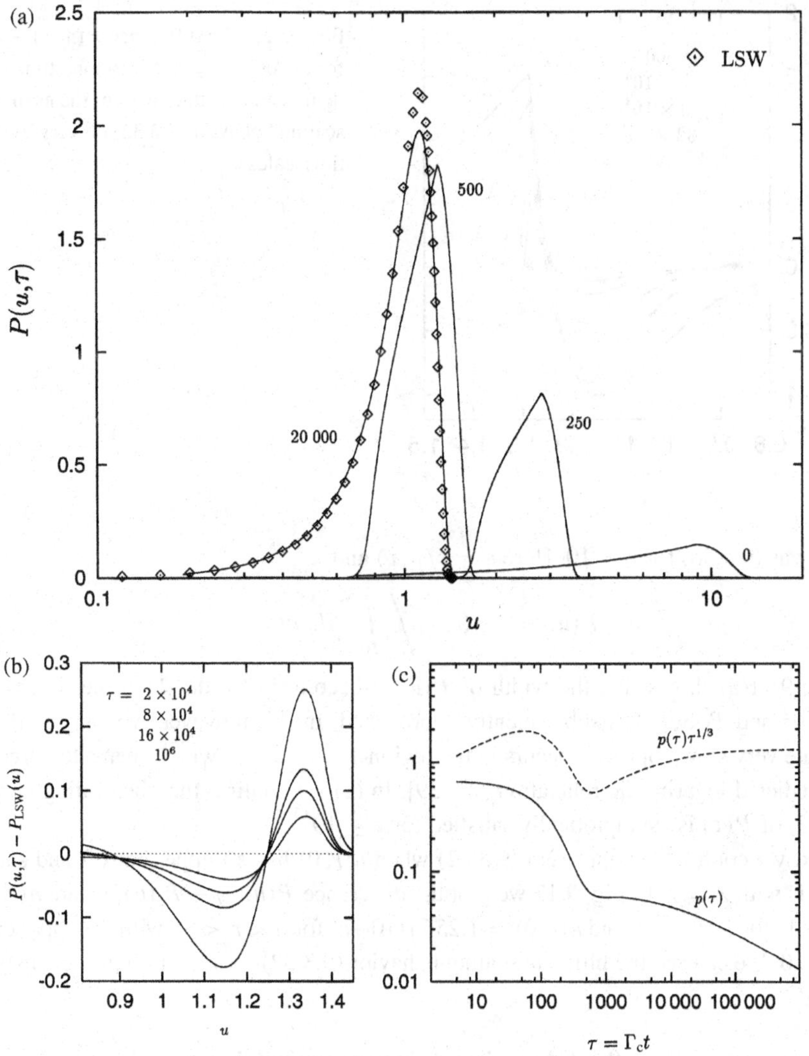

Fig. 9.11. Numerical solution of the LSW equations (9.3.20) and (9.3.21). The initial distribution at $\tau = 0$ is broadly distributed and decays rapidly for $r = R/P(0) > 14$. (a) Time evolution of the normalized distribution $P(u, \tau)$ with $u = R/R_c(\tau)$ at $\tau = 0, 250, 500$, and 2×10^4. The curves approach the LSW function $P_{LSW}(u)$ (◇) in (9.3.26). (b) The difference $P(u, \tau) - P_{LSW}(u)$ decreases to zero very slowly at very long times. (c) Time evolution of $p(\tau)$ (solid line) in (9.3.19), where $p(0) \cong 0.7$ from (9.3.20). The curve of $p(\tau)\tau^{-1/3}$ (dashed line) demonstrates the final scaling behavior (9.3.23).

and satisfy (9.3.22), the approach to the LSW limit can readily be confirmed [33, 34]. In Fig. 9.11 we show such an example, where $n(r, 0)$ is broad with $r_{max}(0) = 10$ and decays

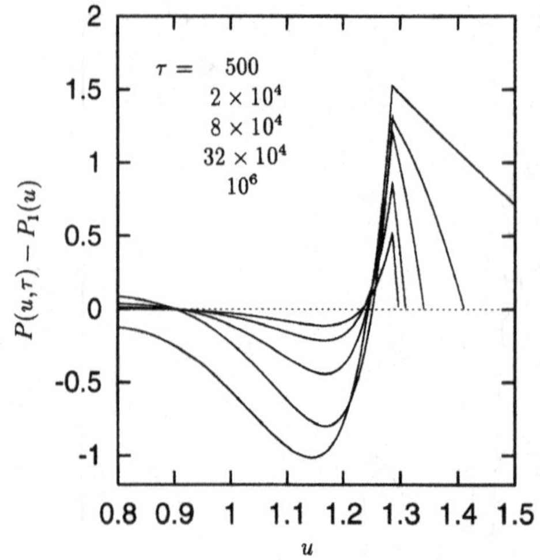

Fig. 9.12. The difference $P(u, \tau) - P_1(u)$ for the case $\lambda = 1$ at very long times. This demonstrates attainment of the asymptotic scaling behavior (9.3.32) on very long timescales.

rapidly as (9.2.36) for $r > 10$. Here $u = r/p(\tau)$ and

$$P(u, \tau) = n(r, \tau) \Big/ \int_0^\infty dr' n(r', \tau). \tag{9.3.37}$$

Figure 9.11(a) shows that the width of $P(u, \tau)$ becomes of order 1 on the timescale of $r_{\max}(0)^2$ and $P_{\rm LSW}(u)$ is subsequently approached. In (b), however, we can see that the ultimate very slow approach occurs in the region $1 < u < 1.5$, which seems to agree with the predicted logarithmic relaxation [29, 39]. In (c) we confirm that the scaling behavior (9.3.23) of $P(\tau)$ is asymptotically satisfied for $\tau \geq 10^4$.

Next we confirm the approach (9.3.32) when $n(r, 0)$ has an upper cut-off and satisfies (9.3.31) with $\lambda = 1$. In Fig. 9.12 we plot the difference $P(u, \tau) - P_1(u)$, where $n(r, 0) = C_1(r-1)$ for $1 < r < 6$ and $n(r, 0) = 1.25C_1(10-r)$ for $6 < r < 10$ with the upper cut-off being 10. We can see the ultimate scaling behavior (9.3.32) for $\lambda = 1$ unambiguously.

9.3.3 Experiments in near-critical fluids

We briefly review nucleation experiments [8, 9], [40]–[44] and theoretical interpretations [6, 28] on near-critical fluids, where the diffusion-limited coalescence discussed in Section 8.5 can be neglected at small droplet volume fraction $q(t) \lesssim 0.03$ and no elastic effects are involved. As an advantage here, if space and time are scaled by ξ and $t_\xi = D^{-1}\xi^2$, the dynamics becomes universal. In fact, the capillary length is expressed as

$$d_0 = \chi\sigma/4T\psi_{\rm eq}^2 = A_{d0}\xi, \tag{9.3.38}$$

where A_{d0} is a universal number of order 0.1, as will be shown in Appendix 9B. The growth rate of critical droplets (9.2.4) is estimated as

$$\Gamma_{\rm c} \cong 25D\xi^{-2}\phi^3 \cong 0.1(T/6\pi\eta\xi_{-0}^3)(1 - T/T_{\rm c})^{3\nu}x^3, \tag{9.3.39}$$

where we have used (6.1.24) for D and the expression for ξ in (2.1.10). x is defined in (9.3.41) below. For example, we have a very small growth rate of $\Gamma_c = 10^{-3}$ s^{-1} for isobutyric acid + water (IW) at $T/T_c - 1 = 10^{-4}$ and $x = 0.1$.

The nucleation rate

From (9.2.30) the nucleation rate in 3D behaves as

$$I \sim \xi^{-3}\Gamma_c \exp(-C_0/\phi^2), \tag{9.3.40}$$

where $\phi = \Delta(0)$ is the initial supersaturation. In near-critical fluids, the control parameter has usually been taken to be

$$x = \delta T/\Delta T_{cx} \cong (2/\beta)\phi, \tag{9.3.41}$$

where $\beta \cong 1/3$ and the second line of (9.1.4) has been used. We thus have

$$I \sim I_0(1 - T/T_c)^{6\nu} \exp\left[-(x_0/x)^2\right], \tag{9.3.42}$$

with

$$x_0 = (2/\beta)C_0^{1/2}. \tag{9.3.43}$$

As will be examined in Appendix 9B, $C_0 \cong 0.0015$ and $x_0 \cong 0.74$ are universal numbers from relations among the critical amplitudes. The exponential factor in I changes abruptly from a very small to a very large number with only a slight increase of x for $x \gg 1$. For example, if $(x_0/x)^2 = 50$, I is increased by $\exp(100\delta x/x)$ with a small increase of x to $x + \delta x$. This factor can be of order 10^3 even for $\delta x/x = 0.05$. It is also instructive to express x in terms of I and $1 - T/T_c$ as

$$x = 0.116x_0/\left[1 + 0.05\ln(1 - T/T_c) - 0.014\ln I\right]^{1/2}, \tag{9.3.44}$$

where we have used a typical value, $I_0 \sim 10^{32} \cong e^{74}$ cm^{-3} s^{-1}. Therefore, x only very weakly depends on I and $1 - T/T_c$. As a result, x remains of order 0.1 for wide ranges of experimentally accessible values of I (for instance, 10^{-2} cm^{-3} s^{-1}) and $1 - T/T_c$. Thus, if the observation time t_{obs} is sufficiently long ($> t_{co}$), we should encounter the appearance of noticeable cloudiness in the bulk fluid region at $x \sim 0.1$ and can determine a rather definite *cloud point*, $\delta T = \delta T_{BD} \sim 0.1\Delta T_{cx}$, experimentally [11].

Anomalous supercooling

Figure 9.13 shows data of cloud points measured by various groups [8, 42]. It was unexpected that the observed supercooling increased considerably at very small $|1 - T/T_c|$. However, this anomalous supercooling can now simply be ascribed to the critical slowing-down of the droplet growth. That is, the completion time t_{co} defined by (9.3.12) becomes very long near the critical point, while noticeable droplets are observable only when the observation time t_{obs} is longer than t_{co}. Then the observed cloud-point curve should be determined by equating the two times as $t_{obs} = t_{co}$ [6, 28]. The data in Fig. 9.14 are

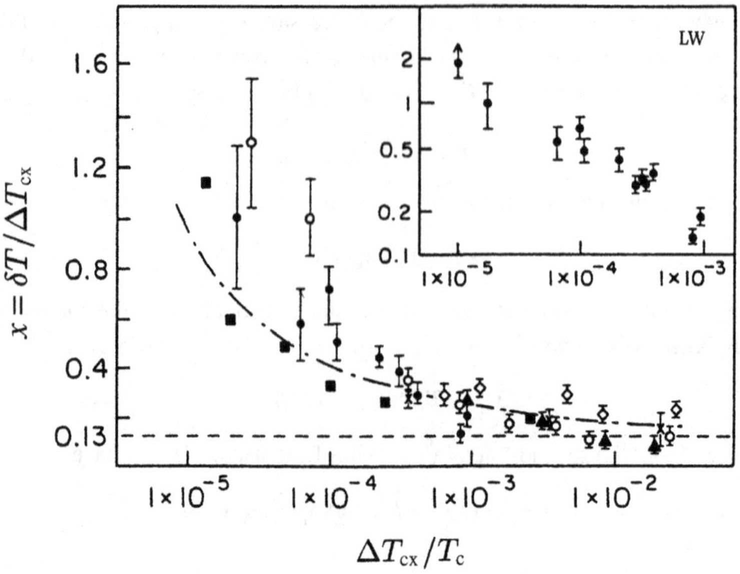

Fig. 9.13. Reduced supercooling $x = \delta T/\Delta T_{cx}$ at cloud points vs $\Delta T_{cx}/T_c$, as measured in various fluids on a log-log scale (⋄, C_7H_{14}; ▲, CO_2; ×, He^3; ○, LW; ■, IW) [8, 42]. The inset shows data for lutidine + water (LW) only. For LW, the absolute values $|\delta T/\Delta T_{ex}|$ and $|\delta T_{cx}/\Delta T_c|$ are plotted, because the coexistence curve is inverted and superheating induces metastability. The curved broken line is obtained from $t_{co} = 1$ s in the Binder–Stauffer theory [6].

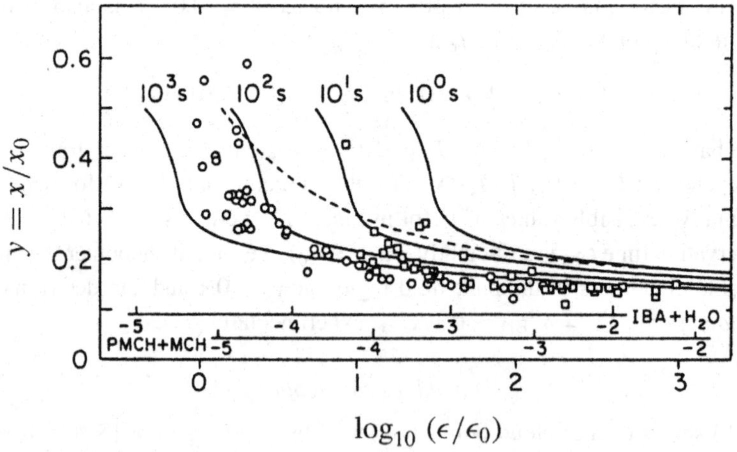

Fig. 9.14. Initial relative supersaturation x/x_0 as a function of scaled reduced temperature at four completion times t_{co} for IW (□) and PMCH + MCH (○) [43]. The solid lines are the Langer–Schwartz theoretical results [28] (which can be obtained from (9.3.12)) and the dashed one is the Binder–Stauffer prediction [6] for $t_{co} = 1$ s. Here $\epsilon = 1 - T/T_c$, while ϵ_0 is a characteristic reduced temperature dependent on fluids [28].

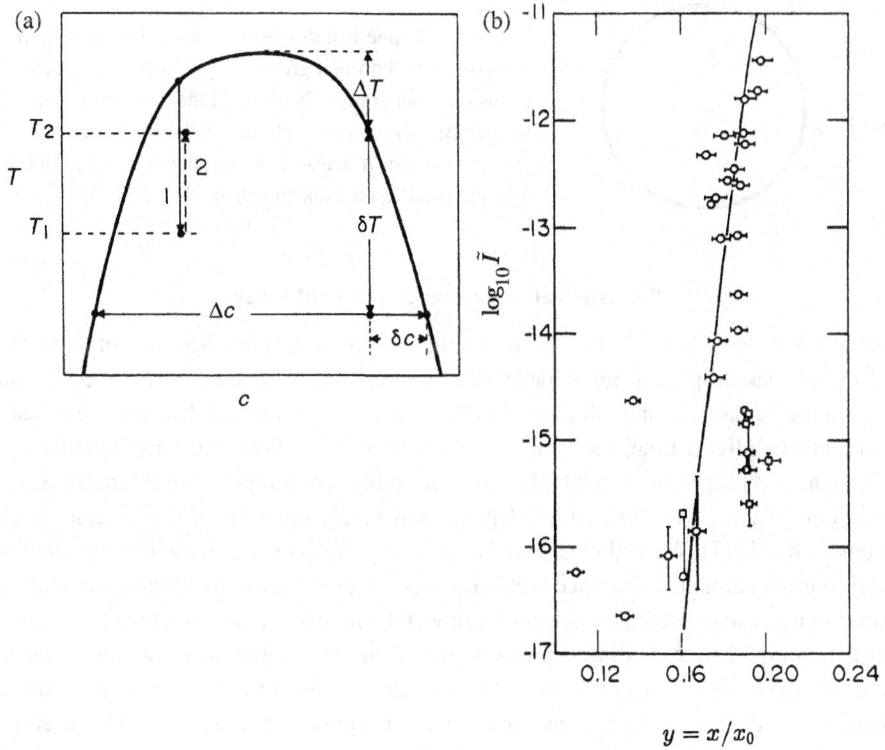

Fig. 9.15. (a) Coexistence curve in a binary fluid mixture (IW). Arrows labeled 1 and 2 show the path of the two-step quench utilized by Siebert and Knobler to determine the nucleation rate [44]. (b) Logarithm of the reduced nucleation rate vs the reduced supersaturation x/x_0 for IW.

cloud-point observations for IW and C_7H_{14}+C_7F_{14} (PMCH + MCH) in comparison with theoretical curves of t_{co} [43].

Afterwards, the validity of the classical formula (9.3.42) near the critical point was demonstrated by Siebert and Knobler with a two-step quench experiment on IW [9, 44], where both the temperature and the coexistence curve were changed adiabatically (see Subsection 6.5.6). We describe their experiment neglecting the latter change. As shown in Fig. 9.15(a), the temperature was first shifted to T_1 in the metastable region for some time t_1, where the nucleation rate $I(T_1)$ was appreciable. Then the temperature was reversely shifted to T_2, where the nucleation rate $I(T_2)$ was much smaller. Droplets were thus created at the lower temperature and their number density was $I(T_1)t_1$. In Fig. 9.15(b) the dimensionless nucleation rate $\tilde{I}$ (in units of $(T/6\pi\eta)\xi^{-6}$) is written as a very steep function of $y = x/x_0$. Its behavior is completely determined by the exponential factor $\exp(-y^{-2})$ within experimental precision.

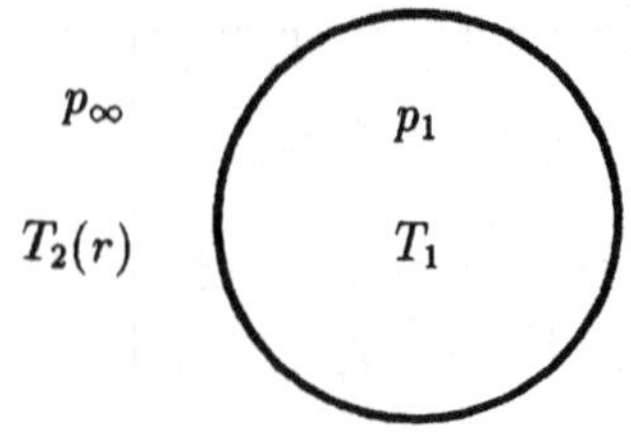

Fig. 9.16. A spherical droplet in a one-component fluid. We assume that droplet growth is governed by thermal diffusion. Then, inside the droplet, the pressure p_1 and temperature T_1 are constant; outside it, the pressure $p_\infty = p_1 + 2\sigma/R$ is also constant, but the temperature $T_2(r)$ depends on the distance from the droplet center.

9.4 Nucleation in one-component fluids

Several books have been devoted to nucleation of liquid droplets from a metastable gas and that of gas droplets from a metastable liquid [1]–[4]. In these cases, pressure and temperature variations can both control the nucleation. Furthermore, because the pressure propagates rapidly in time, the temperature changes adiabatically in most situations. To understand this aspect, we will mainly treat metastable one-component fluids near the gas–liquid critical point [45, 46], which depends sensitively on whether the pressure or the volume is fixed [47]. We will elucidate the following. (i) Let us decrease the pressure by a small constant amount with a fixed boundary temperature near the coexistence curve. If the fluid is in a gas state, isobaric nucleation can well be induced in the bulk region. However, if it is in a liquid state, boiling is easily triggered in the thermal diffusion layer near the boundary. (ii) Upon cooling of the boundary temperature under the fixed-volume condition, adiabatic nucleation can be realized in the interior region in a liquid state. However, if a gas is cooled from the boundary at a fixed volume, liquid droplets readily appear in the thermal diffusion layer, apparently suggesting no metastability in gas in agreement with previous experiments [48, 49]. (iii) If a liquid is heated at the boundary wall, boiling readily occurs both at a fixed volume and at a fixed pressure. The threshold for boiling decreases dramatically on approaching the critical point.

9.4.1 Basic nucleation formulas

Let us consider a slightly metastable, one-component fluid, in which a spherical droplet with radius R of phase 1 is growing in a metastable medium of phase 2, as illustrated in Fig. 9.16. The fluid state need not be close to the critical point. All the deviations are measured from a reference equilibrium state on the coexistence curve, whose pressure and temperature are written as p_0 and $T_0 = T_{\rm cx}(p_0)$, respectively. As is well known, the growth rate is governed by the slow thermal diffusion of latent heat absorbed or released at the interface. The pressure deviation δp_∞ outside the droplet is nearly homogeneous throughout the container of the fluid. Note that δp_∞ depends on time t under the fixed-volume condition. The pressure deviation δp_1 inside the droplet is determined by the Laplace law,

$$\delta p_1 = \delta p_\infty + \frac{2\sigma}{R}. \tag{9.4.1}$$

The temperature deviation δT_2 outside the droplet satisfies the quasi-static condition (8.4.26) near the interface, so that

$$\delta T_2(r) \cong \left[\delta T_1 - \delta T_\infty\right]\frac{R}{r} + \delta T_\infty, \tag{9.4.2}$$

where r is the distance from the droplet center, δT_∞ is the temperature deviation far from the droplet, and δT_1 is that inside the droplet. The temperature within the droplet is assumed to be homogeneous. Then, because the temperature and the chemical potential μ per particle should be continuous at the interface, we have $\delta T_2(R) = \delta T_1$ and

$$-s_2\delta T_1 + v_2\delta p_\infty = -s_1\delta T_1 + v_1\delta p_1, \tag{9.4.3}$$

using the Gibbs–Duhem relation $\delta\mu = -s\delta T + v\delta p$. Here s_α and $v_\alpha = 1/n_\alpha$ are the entropy and volume per particle, respectively, of the reference liquid or gas phase on the coexistence curve ($\alpha = 1, 2$). We may then eliminate δp_1 using (9.4.1) and express δT_1 in terms of δp_∞ as

$$\delta T_1 = \left(\frac{\partial T}{\partial p}\right)_{\text{cx}}\left[\delta p_\infty - \frac{2\sigma v_1}{R\Delta v}\right], \tag{9.4.4}$$

where $\Delta s = s_2 - s_1$ and $\Delta v = v_2 - v_1$, and use has been made of the Clausius–Clapeyron relation, $\Delta s/\Delta v = (\partial p/\partial T)_{\text{cx}}$ in (2.2.21). For $R = R_{\text{c}}$ the temperature inhomogeneity vanishes or $\delta T_1 = \delta T_\infty$, so that we obtain the well-known relation [3],

$$\delta p_\infty - \left(\frac{\partial p}{\partial T}\right)_{\text{cx}}\delta T_\infty = \frac{2\sigma}{R_{\text{c}}(v_2/v_1 - 1)}. \tag{9.4.5}$$

The free-energy difference μ_{eff} per unit volume in the droplet free energy (9.1.1) is given by

$$\mu_{\text{eff}} = n_1\left[\lim_{r\to\infty}\delta\mu_2 - \delta\mu_1\right] = (v_2/v_1 - 1)\left[\delta p_\infty - \left(\frac{\partial p}{\partial T}\right)_{\text{cx}}\delta T_\infty\right], \tag{9.4.6}$$

which is equal to $2\sigma/R_{\text{c}}$ from (9.4.5), as ought to be the case. The fluid is metastable for $\mu_{\text{eff}} > 0$, while it is stable for $\mu_{\text{eff}} \le 0$. To realize metastability in isobaric experiments with $\delta p_\infty = 0$, supercooling (superheating) is needed for a gas (liquid). From (9.1.1) the free energy to create a critical droplet is given by

$$\mathcal{H}_{\text{c}} = \frac{16\pi}{3}\sigma^3/\mu_{\text{eff}}^2 = \frac{16\pi}{3}\sigma^3\left(\frac{v_1}{\Delta v}\right)^2\Big/\left[\delta p_\infty - \left(\frac{\partial p}{\partial T}\right)_{\text{cx}}\delta T_\infty\right]^2, \tag{9.4.7}$$

slightly away from the coexistence curve.

The evolution equation for the radius R follows if we require that the heat current onto the interface $-\lambda_2(\partial\delta T_2/\partial r)$ should be balanced with the latent heat generation (or absorption) at the interface, where λ_2 is the thermal conductivity of phase 2. As in (8.4.28) for model B, energy conservation at the interface gives

$$n_1 T(\Delta s)\frac{\partial}{\partial t}R = \left(\frac{\lambda_2}{R}\right)\left[\delta T_1 - \delta T_\infty\right], \tag{9.4.8}$$

where T_0 is simply written as T on the left-hand side. (See Appendix 9D for more details.) Using (9.4.4) we may rewrite (9.4.8) in the standard form (8.4.28). We can then determine the products $D\Delta$ and Dd_0. Let the diffusion constant D be the thermal diffusivity of phase 2,

$$D = \lambda_2 / C_p^{(2)}, \tag{9.4.9}$$

where $C_p^{(2)} = n_2 T(\partial s/\partial T)_p^{(2)}$ is the constant-pressure specific heat per unit volume of the phase 2. Then the supersaturation and the capillary length are given by

$$\Delta = \frac{C_p^{(2)}}{Tn_1\Delta s}\left[\left(\frac{\partial T}{\partial p}\right)_{\rm cx}\delta p_\infty - \delta T_\infty\right], \tag{9.4.10}$$

$$d_0 = \sigma C_p^{(2)} / [T(n_1\Delta s)^2]. \tag{9.4.11}$$

Furthermore, in many experimental conditions, in which the fluid volume is changed, δp_∞ and δT_∞ are related by the adiabatic condition,

$$\delta p_\infty = \left(\frac{\partial p}{\partial T}\right)_s^{(2)} \delta T_\infty. \tag{9.4.12}$$

In this case Δ becomes

$$\Delta = \left[\frac{n_2}{n_1(\Delta s)}\left(\frac{\partial s}{\partial T}\right)_{\rm cx}^{(2)}\left(\frac{\partial T}{\partial p}\right)_{\rm cx}\right]\delta p_\infty, \tag{9.4.13}$$

where use has been made of (2.2.38). If $(\partial s/\partial T)_{\rm cx}$ in phase 2 and $\Delta s = s_2 - s_1$ have the same (opposite) sign, decompression (compression) is needed to realize metastability in phase 2.

Formulas near the gas–liquid critical point

Near the gas–liquid critical point, we rewrite Δ in (9.4.10) in terms of the universal number $a_{\rm c}$ in (2.2.40) as

$$\Delta \cong \frac{\beta}{2a_{\rm c}}\sqrt{\gamma_s}\theta_{\rm lg}\left[\left(\frac{\partial T}{\partial p}\right)_{\rm cx}\delta p_\infty - \delta T_\infty\right]\Big/(T_{\rm c} - T_{\rm cx}), \tag{9.4.14}$$

where $\gamma_s = C_p/C_V \sim (1 - T/T_{\rm c})^{\alpha-\gamma}$ is the specific-heat ratio. Hereafter $\theta_{\rm lg} = 1$ (or -1) if phase 2 is a gas (or liquid) phase. The surrounding phase 2 is metastable for $0 < \Delta \ll 0.1$ and even unstable for $\Delta \gtrsim 0.1$, while it is stable for $\Delta \le 0$. In the adiabatic condition (9.4.12) we furthermore obtain

$$\Delta \cong -\frac{\beta}{2}\delta T_\infty/(T_{\rm c} - T_{\rm cx}), \tag{9.4.15}$$

which is of the same form as (9.1.4) because δT there corresponds to $|\delta T_\infty|$. Thus cooling is needed to induce nucleation in the adabatic condition (9.4.12) both in a liquid and gas.

9.4.2 Nucleation near the critical point at fixed volume

Langer and Turski [45] presented a theory of nucleation valid at constant pressure. However, experiments on fluids near the gas–liquid critical point have been performed under the fixed-volume condition. Here we consider a near-critical fluid at a fixed total volume V_0. It was initially on the coexistence curve with $p = p_0$ and $T = T_0 = T_{\rm cx}(p_0)$ at a given pressure $p_0(\cong p_{\rm c})$. The deviation $\delta T_0 = T_0 - T_{\rm cx}(p_0)$ can be nonvanishing in experiments, but its effect is only to shift the supersaturation as will be shown in (9.4.23) below. We then slightly change the boundary temperature at $t = 0$ and fix it in later times as

$$T_{\rm b} = T_0 + T_1. \tag{9.4.16}$$

The piston effect discussed in Section 6.3 governs the subsequent relaxation process. Before the emergence of droplets the interior temperature deviation δT_∞ ($T_{1\rm in}$ in the notation of Section 6.3) relaxes to T_1 as (6.3.10) on the quick timescale of $t_1 \cong L^2/\gamma_s^2 D$ in (6.3.7). The pressure deviation δp is homogeneous and is given by $(\partial p/\partial T)_n \delta T_\infty \cong (\partial p/\partial T)_s \delta T_\infty$. Near the boundary the temperature profile is given by (6.3.11) and (6.3.12). From (9.4.14) the space-dependent supersaturation is calculated as [47]

$$\Delta(x,t) \cong \Delta_\infty \left[1 + \theta_{\rm lg} \frac{2}{a_{\rm c}} \left(\frac{\gamma_s t_1}{\pi t} \right)^{1/2} \exp\left(-\frac{x^2}{4Dt} \right) \right], \tag{9.4.17}$$

where the second term decaying as $t^{-1/2}$ in the square brackets arises from the temperature inhomogeneity in the thermal diffusion layer. In the interior region the supersaturation tends to

$$\Delta_\infty = \frac{1}{2}\beta(-T_1)\big/(T_{\rm c} - T_{\rm cx}). \tag{9.4.18}$$

However, the inhomogeneity of $\Delta(x,t)$ gives rise to important consequences in experiments, which will be discussed for the two cases $\theta_{\rm lg} = \pm 1$ separately.

We find the following.

(i) When a metastable fluid is in the liquid phase ($\theta_{\rm lg} = -1$), the supersaturation near the boundary is as shown in Fig. 9.17. If the liquid is supercooled (namely, $T_1 < 0$), $\Delta(x,t)$ becomes negative within the thermal diffusion layer in the early-stage region,

$$t < \gamma_s t_1 = t_2, \tag{9.4.19}$$

and this inhomogeneity becomes negligible for $t \gg t_2$. Fortunately in this case, controlled nucleation experiments may well be performed. That is, for $t \gg t_1$, nucleation starts from the interior liquid region initially characterized by

$$\delta T_\infty(0) \cong T_1, \quad \delta p_\infty(0) \cong \left(\frac{\partial p}{\partial T}\right)_n T_1 \cong \left(\frac{\partial p}{\partial T}\right)_s T_1. \tag{9.4.20}$$

The initial supersaturation is given by (9.4.18). In accord with this result, Moldover *et al.* were able to perform nucleation experiments at liquid densities ($n > n_{\rm c}$) in the fixed-volume condition [48, 49]. Conversely, if the liquid is heated ($T_1 > 0$) slightly above the

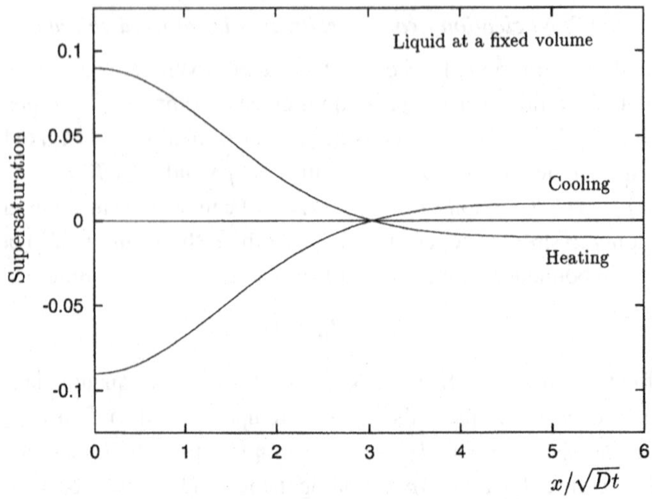

Fig. 9.17. Supersaturation in a liquid after a step-wise boundary temperature change at a fixed volume. It was initially on the coexistence curve. On cooling, the bulk region can become metastable while the thermal diffusion layer is stable. On heating, the thermal diffusion layer is easily driven into an unstable state, resulting in boiling.

coexistence curve, the thermal diffusion layer can become metastable or even unstable in the time region $t < t_2$. For $t \sim t_1$ and $x \cong 0$, $\Delta(x, t)$ attains a maximum,

$$\Delta_{\max} \sim \sqrt{\gamma_s}|T_1|/(T_c - T_{cx}). \tag{9.4.21}$$

Therefore, if $\Delta_{\max} \gtrsim 0.1$ and

$$D\xi^{-2}t_2 \sim (L/\xi)^2/\gamma_s \gg 1, \tag{9.4.22}$$

phase separation should be induced in the narrow spatial region $x \lesssim (Dt_2)^{1/2} \sim \gamma_s^{-1/2}L$ transiently in the time region $t \lesssim t_2$.

(ii) When a metastable fluid is in the gas phase ($\theta_{\text{lg}} = 1$), the supersaturation near the boundary is as shown in Fig. 9.18. Upon supercooling $\Delta(x, t)$ attains a large value within the thermal diffusion layer. Its maximum $\Delta_{\max}$ is again given by (9.4.21). This means that phase separation starts to take place within the thermal diffusion layer for $t \lesssim t_1$ except for very small $|T_1|$ ($\ll \gamma_s^{-1/2}(T_c - T_{cx})$). In realistic experimental conditions, a liquid layer will appear to wet the boundary and no appreciable metastability of the gas phase will be detected. This conclusion is consistent with the experiment by Dahl and Moldover [48], who observed no metastability in gas states ($n < n_c$) and expected preferential wetting of a liquid layer at the wall as its physical origin. In addition, upon heating, the gas phase is always stable everywhere in the cell.

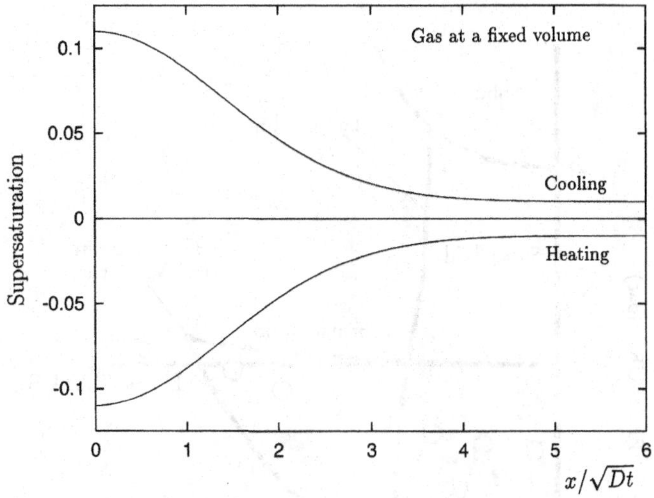

Fig. 9.18. Supersaturation in a gas after a step-wise boundary temperature change at a fixed volume. It was initially on the coexistence curve. On cooling, condensation can easily be induced in the thermal diffusion layer. On heating, the whole region remains stable.

Bubble growth in the interior liquid region

We next examine droplet growth in a bulk liquid region. In Section 6.3 we showed that the interior pressure deviation is pinned at $(\partial p/\partial T)_s T_1$ in one-phase states due to the thermal diffusion layer acting as a piston. In Appendix 9E we will show that this remains the case even in the presence of growing droplets in the interior region and that the mean field result $\Delta(t) = \Delta(0) - q(t)$ holds with

$$\Delta(0) = \frac{1}{2}\beta\left[-T_1 + \frac{\sqrt{\gamma_s}}{a_c}\theta_{lg}\delta T_0\right] \Big/ (T_c - T_{cx}), \qquad (9.4.23)$$

where $q(t)$ is the droplet volume fraction. The second term in the square brackets arises when the temperature T_0 before cooling deviates from the coexistence temperature $T_{cx} = T_{cx}(P_0)$. The LSW theory thus remains applicable without modification.[2]

9.4.3 Highly superheated and supercooled fluids

So far we have examined slightly metastable fluids particularly near the critical point. However, a large number of experiments have been performed to approach the stability limit of fluids deeply in the metastable region [3, 50]. For example, the compressibility of superheated water behaves as [51]

$$K_T \cong K_0(1 - T/T_s)^{-\gamma_c}, \qquad (9.4.24)$$

[2] An incorrect conclusion was reached in Ref. [47] in this aspect.

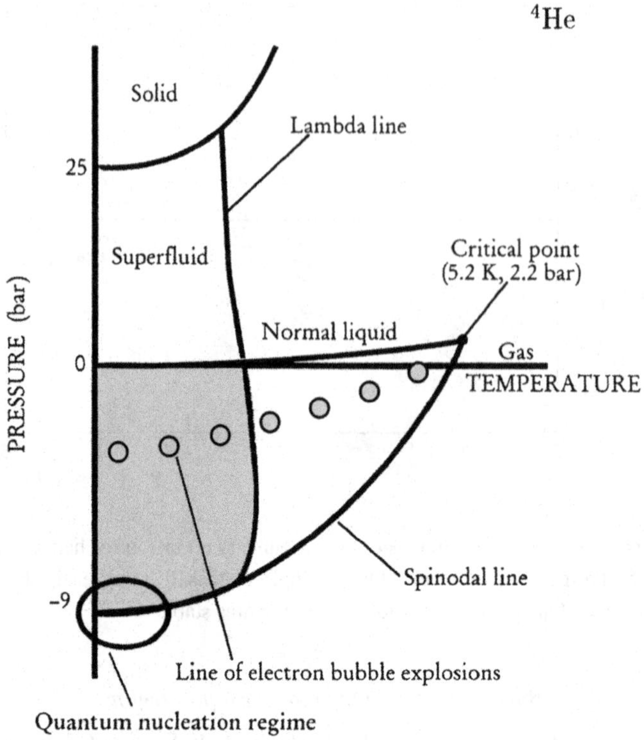

Fig. 9.19. Phase diagram of ^{4}He including the negative pressure regime [52]. The diagram is not to scale. The spinodal line indicates the pressure at which the sound velocity in the liquid becomes zero and the liquid becomes unstable against long-wavelength density fluctuations. Bubbles formed around free electrons will explode if the pressure reaches the line formed by the circles. The quantum nucleation regime is also marked (see Ref. [54]). Here 1 bar $= 10^7$ erg/cm^3.

in the range 100 °C $< T <$ 220 °C at 1 bar, where K_0 is a constant, the spinodal temperature $T_s = 315 \pm 10$ °C, and $\gamma_c \cong 1$. The heat capacity C_p grows in the same manner, while C_V does not grow. Similarly, upon supercooling, various quantities such as K_T and C_p grow as

$$X = X_0(T/T_s - 1)^{-\gamma_X} + X_b, \tag{9.4.25}$$

where X_0 and X_b are constants and γ_X is an appropriate exponent. These data indicate enhancement of those fluctuations with sizes smaller than the critical radius on approaching the metastability limit or the spinodal line $T = T_s(p)$, though it cannot be reached in practice due to the onset of nucleation.[3] In the van der Waals theory the spinodal line is given by (3.4.7), where C_p and K_T grow with $\gamma = 1$ and C_V remains nonsingular as (3.4.6).

[3] We propose experiments in shear flow where the metastability may be suppressed. See Section 11.1.

As a particularly ideal system, ^{4}He at low temperature can be supercooled considerably at negative pressures, as illustrated in Fig. 9.19 [52]. In such metastable states at $T \cong 0$, the sound velocity c decreases as [53]

$$c = c_0(p - p_s)^\nu, \tag{9.4.26}$$

where c_0 is a constant, $p_s \cong -9.5$ bar, and $\nu \cong 1/3$. Because $\rho c^2 = dp/d\rho$ as $T \to 0$, we obtain

$$p - p_s \propto (\rho - \rho_s)^\delta, \tag{9.4.27}$$

near the spinodal point, where $\delta = 1/(1 - 2\nu) \cong 3$ and $\rho_s \cong 0.095$ g cm^{-3}. Here the gas density is much lower than the liquid density and the free-energy difference μ_{eff} in the droplet free energy (9.1.1) is nearly equal to $|p|$ from (9.4.6) at negative p, so that

$$\mathcal{H}_c = 16\pi\sigma^3/3|p|^2. \tag{9.4.28}$$

At relatively high temperatures ($T \gtrsim 200$ mK), the nucleation rate is given by the classical formula $I \propto \exp(-\mathcal{H}_c/T)$ in (9.2.30). However, at lower temperatures a quantum nucleation mechanism is expected to be dominant [54, 55].

Furthermore, if electrons are injected into liquid ^{4}He, the nucleation barrier can be much reduced [56, 57]. In fact, the droplet free energy of a gas bubble around an electron is written as

$$\mathcal{H}(R) = \frac{\pi^2\hbar^2}{2m_e R^2} + 4\pi\sigma R^2 + \frac{4}{3}\pi p R^3, \tag{9.4.29}$$

where the first term arises from electron confinement, with m_e being the electron mass. Each electron expels liquid helium and forms a bubble with radius $R_{\min} = (\pi\hbar^2/8m\sigma)^{1/4} = 19$ Å even at $p = 0$. For $p < 0$ the metastable and critical radii are the solutions of $|p|/(2\sigma/R_{\min}) = x^{-1} - x^{-5} < 4/5^{5/4}$ where $x = R/R_{\min}$. The minimum becomes nonexistent for $|p| \geq (16/5)(m_e/10\pi\hbar^2)^{1/4}\sigma^{5/4} \sim 2$ bar, where all the gas droplets explode as observed [57].

9.4.4 Sound propagation in two-phase states

When systems are composed of finely divided domains, increased sound attenuation has been observed in a number of materials. Examples are polycrystals [58, 59], fluids of emulsions [60]–[63], solids undergoing martensitic transitions [64], phase-separating polymer solutions [65] and ^{3}He–^{4}He mixtures [66], and so on. Many years ago, Zener [58] and Isakovich [60] independently predicted that, when the acoustic wavelength $\lambda = 2\pi c/\omega$ is much longer than the typical domain size R, acoustic attenuation is enhanced by small-scale heat currents between adjacent crystallites or two phases. Another well-known attenuation mechanism is the scattering of sound by domains,[4] but it decreases rapidly and becomes negligible as $\lambda/R \to \infty$ [61]. Here we will examine this problem in

[4] The cross section of a droplet of radius R is of order R^6/λ^4 [61]. The attenuation per wavelength is then of order $\phi(R/\lambda)^3 \propto \phi\omega^3$ for $R \ll \lambda$ at small droplet volume fraction ϕ.

one-component fluids to predict enhanced sound attenuation in the presence of droplets at low frequencies [67]. Similar conclusions may be drawn for binary fluid mixtures [67] and ^{3}He–^{4}He mixtures [68] in two-phase states.

Temperature inhomogeneity

We assume that the acoustic frequency ω is much faster than the typical growth rate of domains but the acoustic wavelength is much longer than the typical domain size R. All the acoustic perturbations with subscript a are assumed to be very small and depend on time as $e^{i\omega t}$. Near a gas–liquid interface, the acoustic pressure perturbation δp_a may be considered to be homogeneous,[5] but the acoustic temperature deviation δT_a is inhomogeneous and is calculated from

$$i\omega\delta T_a = i\omega\left(\frac{\partial T}{\partial p}\right)_s \delta p_a + D\nabla^2\delta T_a, \tag{9.4.30}$$

which is equivalent to the first line of (6.3.15). If we require continuity of the chemical potential at the interface, we obtain

$$\delta T_a - (\partial T/\partial p)_{\rm cx}\delta p_a = 0, \tag{9.4.31}$$

at the interface. Far from the interface we also require the adiabatic condition, $\delta T_a - (\partial T/\partial p)_s\delta p_a \to 0$. We introduce a dimensionless function $\mathcal{F} = \mathcal{F}(\boldsymbol{r})$ by

$$\delta T_a - \left(\frac{\partial T}{\partial p}\right)_s \delta p_a = \left[\left(\frac{\partial T}{\partial p}\right)_{\rm cx} - \left(\frac{\partial T}{\partial p}\right)_s\right]\delta p_a \mathcal{F}. \tag{9.4.32}$$

Then $\mathcal{F}$ obeys

$$\nabla^2\mathcal{F} = (i\omega/D)\mathcal{F} = \kappa_{\rm D}^2\mathcal{F}, \tag{9.4.33}$$

where

$$\kappa_{\rm D} = \sqrt{i\omega/D}. \tag{9.4.34}$$

We may assume $\mathrm{Re}\,\kappa_{\rm D} > 0$. We solve (9.4.33) requiring that $\mathcal{F} = 1$ on the interface and $\mathcal{F} \to 0$ far from the interface. For example, near a planar interface, we have $\mathcal{F} = \exp(-\kappa_{\rm D}|x|)$, where $|x|$ is the distance from the interface. For a spherical droplet with radius R, $\mathcal{F}$ is obtained as a function of the distance r from the droplet center as

$$\begin{aligned} \mathcal{F} &= R\sinh(\kappa_{\rm D} r)/r\sinh(\kappa_{\rm D} R) && (r < R), \\ &= \frac{R}{r}\exp(\kappa_{\rm D} R - \kappa_{\rm D} r) && (r > R). \end{aligned} \tag{9.4.35}$$

Because the physical quantities are different in the gas and liquid phases, we will attach the subscripts, 1 and 2, or ℓ for liquid and g for gas, when necessary.

[5] To be precise, the droplet radius oscillates in a sound leading to a small pressure discontinuity across the interface. This effect may be neglected for large droplet sizes ($R \gg d_0$) in small-amplitude sounds [67].

Effective adiabatic compressibility

We consider a small volume element with linear dimension much shorter than the sound wavelength. But it contains many domains and its boundary moves with the fluid velocity. Its volume without sound is denoted by V and its small change due to the sound by δV_a. Then the effective adiabatic compressibility $\delta K_D(\omega)$ is defined by

$$\delta V_a/V = -K_D(\omega)\delta p_a. \tag{9.4.36}$$

Let $\bar{\rho}$ be the average mass density given by

$$\bar{\rho} = \phi_g \rho_g + \phi_\ell \rho_\ell, \tag{9.4.37}$$

where ϕ_g and $\phi_\ell = 1 - \phi_g$ are the volume fractions of the gas and liquid phases. Then, in a sound, $\bar{\rho}$ changes by $\delta\bar{\rho}_a = -\bar{\rho}\delta V_a/V$. The sound-wave dispersion relation is written as

$$\omega^2/k^2 = 1/\bar{\rho}K_D(\omega), \quad \text{or} \quad k = \omega\sqrt{\bar{\rho}K_D(\omega)}. \tag{9.4.38}$$

Hereafter we neglect the frequency-dependent bulk viscosity arising from the relaxation of the thermal fluctuations.

In each phase outside the interface regions the local volume change $\delta v_a \cong -\delta n_a/n^2$ per particle due to δT_a and δp_a is written as

$$\begin{aligned}\delta v_a &= \left(\frac{\partial v}{\partial p}\right)_s \delta p_a + \left(\frac{\partial v}{\partial T}\right)_p \left[\delta T_a - \left(\frac{\partial T}{\partial p}\right)_s \delta p_a\right] \\ &= -\frac{v}{\rho c^2}\delta p_a \left\{1 - (\gamma_s - 1)\left[\left(\frac{\partial T}{\partial p}\right)_{\text{cx}}\left(\frac{\partial p}{\partial T}\right)_s - 1\right]\mathcal{F}\right\},\end{aligned} \tag{9.4.39}$$

where use has been made of (9.4.32). In addition, in the presence of nonvanishing mass flux w through the interface, mass conversion takes place at the interface causing a volume change. The resultant total volume change consists of two parts as

$$\delta V_a = \int_V d\boldsymbol{r}\frac{\delta v_a}{v} + \frac{\Delta v}{i\omega}\int da \frac{w}{m_0}, \tag{9.4.40}$$

where $\int da$ represents the surface integrations over the interfaces contained in the volume V, $\Delta v = v_2 - v_1$ is the volume difference per particle, and m_0 is the particle mass. Using (9D.2) and the Clausius–Clapeyron relation, we may rewrite the surface integration into the bulk integration as

$$\begin{aligned}\delta V_a &= \int_V d\boldsymbol{r}\left[\frac{\delta v_a}{v} - \frac{1}{i\omega T}\left(\frac{\partial T}{\partial p}\right)_{\text{cx}} \lambda\nabla^2\delta T_a\right] \\ &= -\delta p_a \int d\boldsymbol{r}\frac{1}{\rho c^2}[1 + Z\mathcal{F}],\end{aligned} \tag{9.4.41}$$

where (1.2.53), (1.2.54) and (9.4.32) have been used in the second line, and

$$Z = (\gamma_s - 1)\left[\left(\frac{\partial T}{\partial p}\right)_{\text{cx}}\left(\frac{\partial p}{\partial T}\right)_s - 1\right]^2 = \frac{\rho c^2}{C_p}\left(\frac{\partial s}{\partial p}\right)_{\text{cx}}^2. \tag{9.4.42}$$

Separating the space into the gas and liquid regions, we may rewrite the second line of (9.4.49) as

$$K_D(\omega) = \frac{\phi_{\rm g}}{\rho_{\rm g} c_{\rm g}^2}\left[1 + Z_{\rm g}\langle\mathcal{F}\rangle_{\rm g}\right] + \frac{\phi_\ell}{\rho_\ell c_\ell^2}\left[1 + Z_\ell\langle\mathcal{F}\rangle_\ell\right], \tag{9.4.43}$$

where $Z_{\rm g}$ and Z_ℓ are the gas and liquid values of Z in (9.4.42), and $\langle\ \rangle_{\rm g}$ and $\langle\ \rangle_\ell$ are the space averages in the gas and liquid regions, respectively.

Near-critical fluids

Near the gas–liquid critical point we may set $Z = a_{\rm c}^2 \cong 1$ and obtain a simple expression,

$$K_D(\omega) = (\rho c^2)^{-1}(1 + a_{\rm c}^2\langle\mathcal{F}\rangle), \tag{9.4.44}$$

where $\langle\mathcal{F}\rangle$ is the space average of $\mathcal{F}$. (i) The dissipation occurs in the thermal diffusion layer around the interfaces in the relatively high-frequency region $D\bar{R}^{-2} \ll \omega \ll t_\xi^{-1} = D\xi^{-2}$, where $\bar{R}$ is the average droplet radius. There, we have $\langle\mathcal{F}\rangle \cong 2A/\kappa_{\rm D}$ where A is the surface area per unit volume, so that

$$K_D(\omega) \cong (\rho c^2)^{-1}\left(1 + 2a_{\rm c}^2 A\sqrt{D/i\omega}\right). \tag{9.4.45}$$

The resultant attenuation per wavelength is given by

$$\alpha_{D\lambda} \cong \pi a_{\rm c}^2 A\sqrt{2D/\omega}. \tag{9.4.46}$$

This attenuation is much larger than that due to the thermal fluctuations calculated in Section 6.2. (ii) We then consider a dilute assembly of spherical droplets with volume fraction $\phi \ll 1$. If the diffusion length $\ell_D = 1/|\kappa_{\rm D}| = \sqrt{D/\omega}$ is shorter than the screening length $\ell_{\rm s} = \phi^{-1/2}\bar{R}$ introduced in (8.4.48), $\langle\mathcal{F}\rangle$ is the sum of the space integrals of (9.4.35) for droplets in a unit volume:

$$K_D(\omega) \cong \frac{1}{\rho c^2}\left\{1 + \frac{4\pi a_{\rm c}^2}{\kappa_{\rm D}}\int dR\, n(R) R^2\left[\coth(\kappa_{\rm D} R) + 1\right]\right\}, \tag{9.4.47}$$

where $n(R)$ is the droplet size distribution per unit volume. We may reproduce (9.4.45) in the high-frequency regime. However, this expression is valid only for $\omega \gg D\ell_{\rm s}^{-2} = \phi D\bar{R}^{-2}$. For $\omega \ll D\bar{R}^{-2}$ we may devise the following approximate expression [67],

$$K_D(\omega) = \frac{1}{\rho c^2}\left[1 + a_{\rm c}^2\frac{\omega_{\rm s}}{i\omega + \omega_{\rm s}}\right], \tag{9.4.48}$$

where

$$\omega_{\rm s} = 4\pi n_{\rm dom}\bar{R}D \sim D\bar{R}^{-2}\phi. \tag{9.4.49}$$

Here $n_{\rm dom} = \int dR\, n(R)$ is the droplet number density and $n_{\rm dom}\bar{R} = \int dR\, n(R) R \sim \ell_{\rm s}^{-2}$. The resultant attenuation per wavelength becomes

$$\alpha_{D\lambda} \cong \pi a_{\rm c}^2 \frac{\omega_{\rm s}\omega}{\omega^2 + \omega_{\rm s}^2}. \tag{9.4.50}$$

This attenuation takes the maximum $\pi a_c^2/2$ at the very low frequency $\omega = \omega_s$. For $\omega \ll \omega_s$ the attenuation decreases and the sound speed tends to $c/(1+a_c^2)^{1/2}$.

Fluids far from criticality

(i) The relatively high-frequency region $\omega \gg D_g \bar{R}^{-2}$ and $D_\ell \bar{R}^{-2}$ is realized in most conditions of large droplet systems. As in (9.4.45) we have

$$K_D(\omega) = \frac{1}{\bar{\rho} c_{\rm em}^2} + \left[\frac{1}{\rho_g c_g^2} Z_g D_g^{1/2} + \frac{1}{\rho_\ell c_\ell^2} Z_\ell D_\ell^{1/2} \right] A(i\omega)^{-1/2}, \tag{9.4.51}$$

where A is the surface area density, $\bar{\rho}$ is given by (9.4.37), and $c_{\rm em}$ is the sound velocity in the Wood theory [69] determined by

$$\frac{1}{c_{\rm em}^2} = (\phi_g \rho_g + \phi_\ell \rho_\ell) \left[\frac{\phi_g}{\rho_g c_g^2} + \frac{\phi_\ell}{\rho_\ell c_\ell^2} \right]. \tag{9.4.52}$$

We may derive this sound velocity generally for composite materials in an effective medium theory neglecting heat conduction and mass conversion. In bubbly fluids, the sound velocity is known to be much decreased in the presence of a small fraction of gas bubbles. Its behavior is fairly well described by the above formula.

(ii) Although not realized in usual experiments, we may consider the very-low-frequency limit where the thermal diffusion length exceeds the inter-domain distance. In this case we have $\langle \mathcal{F} \rangle_g \cong \langle \mathcal{F} \rangle_\ell \cong 1$ and the sound velocity tends to that in Ref. [61]:

$$\frac{1}{c_L^2} = \bar{\rho} K_D(0) = (\phi_g \rho_g + \phi_\ell \rho_\ell) \left[\frac{\phi_g}{\rho_g c_g^2}(1+Z_g) + \frac{\phi_\ell}{\rho_\ell c_\ell^2}(1+Z_\ell) \right]. \tag{9.4.53}$$

As $T \to T_c$, we have $c_L \to c/(1+a_c^2)^{1/2}$.

(iii) More specifically, for liquids containing gas bubbles, we derive the counterpart of (9.4.48) valid in the low-frequency region $\omega \ll D_\ell \bar{R}^{-2}$ and $D_g \bar{R}^{-2}$:

$$K_D(\omega) = \frac{1}{\bar{\rho} c_L^2} - \left(\frac{Z_\ell}{\rho_\ell c_\ell^2} \right) \frac{i\omega}{i\omega + \omega_{s\ell}}, \tag{9.4.54}$$

where $\omega_{s\ell} = 4\pi D_\ell \int dR n(R) R \sim \phi_g D_\ell \bar{R}^{-2}$. The dissipation in the liquid mainly occurs in the liquid region. Even for small ϕ_g the attenuation per wavelength grows as ω^{-1} as ω is decreased from $D\bar{R}^{-2}$ and has a maximum at $\omega \sim \omega_s$. It goes without saying that when liquid droplets are suspended in a gas, the corresponding expression for $K_D(\omega)$ can be obtained by exchange of ℓ and g.

(iii) If the droplet size is very large or the fluid is at very low temperatures, the thermal conduction can become negligible in the droplet motion. For example, the motion of a gas bubble is governed by the Rayleigh–Plesset equation in (9.5.5) below [70]. In such cases droplets can resonate to applied sounds, leading to large oscillation of the droplet radii and enhanced acoustic attenuation [56]. This effect is not treated here.

9.5 Nucleation at very low temperatures

At very low temperatures, the thermal activation mechanism of nucleation should be replaced by a quantum mechanism. Lifshitz and Kagan constructed the first seminal theory of kinetics of first-order phase transitions at $T \cong 0$ [71]. They showed that the quantum tunneling mechanism can produce a droplet of a new phase in a metastable, ideal incompressible fluid. In ^{4}He at $T \cong 0$, consideration has been given to first-order phase transitions between solid and superfluid phases and between gas and superfluid phases [54, 55]. We also mention phase separation at nearly zero temperatures in ^{3}He–^{4}He mixtures [72]–[74]. Although a few experiments have already been performed, there still remain many unsolved problems. In this section we will briefly discuss the Lifshitz–Kagan theory of homogeneous nucleation, comparing it with the classical nucleation theory.

9.5.1 Droplet hamiltonian

The role of dissipation in the droplet motion becomes small at very low temperatures. In such cases we need to include the kinetic energy in the droplet hamiltonian as

$$\mathcal{H} = 2\pi\rho_{\text{eff}} R^3 \dot{R}^2 + 4\pi\sigma R^2 - \frac{4\pi}{3}\mu_{\text{eff}} R^3, \tag{9.5.1}$$

where $\dot{R} = \partial R/\partial t$ is the interface velocity and the first term represents the kinetic energy supported by the surrounding fluid, with ρ_{eff} being a mass density. If the surrounding fluid (phase 2) can be treated as an incompressible liquid, the fluid velocity there is written as $v(r) = A/r^2$ in the radial direction $r^{-1}\mathbf{r}$. The coefficient A is determined from the mass conservation at the interface: $\rho_2[v(R) - \dot{R}] = -\rho_1 \dot{R}$, so that

$$v(r) = (1 - \rho_1/\rho_2)\dot{R}R^2 \frac{1}{r^2} \quad (r > R), \tag{9.5.2}$$

where ρ_1 and ρ_2 are the mass densities of the inner phase 1 and the outer phase 2, respectively. Integration of $\rho_2 v(r)^2/2$ in the region $r > R$ gives the kinetic energy with

$$\rho_{\text{eff}} = (\rho_1 - \rho_2)^2/\rho_2. \tag{9.5.3}$$

In particular, $\rho_{\text{eff}} \cong \rho_2 = \rho_{\text{liq}}$ for a gas bubble. The momentum P and the mass $M(R)$ of the droplet are defined by

$$P = \left(\frac{\partial \mathcal{H}}{\partial \dot{R}}\right)_R = M(R)\dot{R}, \quad M(R) = 4\pi\rho_{\text{eff}} R^3. \tag{9.5.4}$$

In terms of R and P the kinetic energy is written as $P^2/2M(R)$. The dynamics is obtained from the canonical equations, $\dot{R} = \partial\mathcal{H}/\partial P$ and $\dot{P} = -\partial\mathcal{H}/\partial R$, leading to

$$\rho_{\text{eff}}\left(R\ddot{R} + \frac{3}{2}\dot{R}^2\right) = -\frac{2\sigma}{R} + \mu_{\text{eff}}, \tag{9.5.5}$$

where $\ddot{R} = \partial^2 R/\partial t^2$. For a gas bubble, this equation is known as the Rayleigh–Plesset equation [70], where $\mu_{\text{eff}} = p'(t) - p_\infty(t)$ is generally time dependent with $p'(t)$ and

$p_\infty(t)$ being the pressures inside and far from the droplet, respectively. Note that the timescale of the heat conduction is R^2/D, where D is the thermal diffusion constant. When $|\dot{R}| \gg D/R$, the effect of the heat conduction is negligible and the droplet expands or shrinks adiabatically.

In the metastable case $\mu_{\rm eff} > 0$ and $R \gg R_c$, the interface velocity $\dot{R}$ tends to a terminal velocity given by

$$v_\infty = (2\mu_{\rm eff}/3\rho_{\rm eff})^{1/2}, \tag{9.5.6}$$

which is slower than the sound velocity for weak metastability. For a gas bubble at a negative pressure $p = p_\infty$, we have $v_\infty = (2|p|/3\rho_{\rm liq})^{1/2}$. In particular, if the initial droplet kinetic energy is very small and the droplet is expanding, the droplet velocity is given by

$$\dot{R} = [2|U(R)|/M(R)]^{1/2} = v_\infty(1 - R_0/R)^{1/2}, \tag{9.5.7}$$

where R_0 is the radius at the turning point of the potential $U(R) = 4\pi\sigma R^2 - 4\pi\mu_{\rm eff}R^3/3$ or

$$R_0 = 3\sigma/\mu_{\rm eff} = 1.5R_c. \tag{9.5.8}$$

9.5.2 *Quantization*

Lifshitz and Kagan [71] quantized the above $\mathcal{H}$ by treating P as the following operator,

$$P = \frac{\hbar}{i}\frac{\partial}{\partial R}, \tag{9.5.9}$$

which gives the usual commutation relation $PR - RP = \hbar/i$, $\hbar$ being the Planck constant. Here the mass $M(R)$ depends on R and does not commute with P, so ambiguity arises in the form of the kinetic energy but is negligible for large R which satisfies

$$(\hbar/R)^2/2M(R) \ll 4\pi\sigma R^2. \tag{9.5.10}$$

This condition may be rewritten as $R \gg R_Q$ with

$$R_Q = \left(\hbar^2/32\pi^2\rho_{\rm eff}\sigma\right)^{1/7}, \tag{9.5.11}$$

which is of order 1 Å for ^{4}He. The quasi-classical (WKB) approximation for the wave function can be used under the condition (9.5.10) or for $R \gg R_Q$ [75]. We should also note that the Lifshitz–Kagan hamiltonian is based on the droplet picture and is meaningful only when R is longer than the interface thickness ($\sim R_Q$). In terms of R_Q, the ground-state energy of $\mathcal{H}$ on the coexistence curve ($\mu_{\rm eff} = 0$) is of the following order,

$$E_g = 4\pi\sigma R_Q^2. \tag{9.5.12}$$

In the quantum-mechanical treatment we solve the Schrödinger equation,

$$i\hbar\frac{\partial}{\partial t}\Psi = \mathcal{H}\Psi. \tag{9.5.13}$$

The wave function $\Psi(R,t)$ is localized in the region $R \sim R_Q$ at $t = 0$ but becomes nonvanishing in the region $R > R_0$ after an incubation time. The behavior of the wave function Ψ around $R \cong R_0$ is important in the quantum case, while the behavior of the droplet size distribution function near the classical critical radius R_c is important in the classical case. Assuming that Ψ is very small in the region $R > R_0$, we normalize Ψ as

$$\int_0^{R_0} dR\, |\Psi(R,t)|^2 \cong 1. \tag{9.5.14}$$

The smallness parameter is the ratio of the two lengths,

$$\Delta_Q = R_Q/R_0 = (R_Q/3\sigma)\mu_{\rm eff}, \tag{9.5.15}$$

which may be used as the quantum-mechanical supersaturation. The maximum $U_{\rm max}$ of the potential $U(R)$ is much larger than E_g in (9.5.12) from

$$U_{\rm max} \sim 4\pi\sigma R_0^2 \sim E_g/\Delta_Q^2. \tag{9.5.16}$$

9.5.3 Quantum nucleation rate

In the region $R < R_0$, Ψ is nearly independent of time and may be calculated in the WKB approximation [75]. As in the classical case, we assume that Ψ is nearly equal to the ground-state wave function for $\mu_{\rm eff} = 0$ in the region $R \lesssim R_Q$. Because E_g is small, we may set $\mathcal{H}\Psi \cong 0$ with $\Psi = \exp[iS_0/\hbar + iS_1 + O(\hbar)]$. The result up to S_1 is written as

$$\Psi = C\left[\frac{M(R)}{U(R)}\right]^{1/4} \exp\left[-\frac{1}{\hbar}\int_0^R dR'\sqrt{2M(R')U(R')}\right], \tag{9.5.17}$$

which holds for $R_Q \lesssim R < R_0$. The above result is not affected by the ambiguity in the kinetic energy arising from the R dependence of $M(R)$. The coefficient C is independent of $\mu_{\rm eff}$ and is determined from (9.5.14), so $C^2 \sim \hbar/(\rho_{\rm eff}R_Q^5)$. For $R_Q \ll R \ll R_0$, Ψ rapidly decays as

$$\Psi \sim R^{1/4}\exp\left[-\frac{2}{7}(R/R_Q)^{7/2}\right], \tag{9.5.18}$$

which is analogous to (9.1.25). As $R \to R_0$, Ψ behaves as

$$\Psi \cong C\left[\frac{M_0}{U_0'(R_0-R)}\right]^{1/4} \exp\left[-\mathcal{A} + \frac{2}{3}\left(\frac{R_0-R}{a_0}\right)^{3/2}\right], \tag{9.5.19}$$

where the coefficients near the turning point are

$$M_0 = M(R_0) = 4\pi\rho_{\rm eff}R_0^3, \quad U_0' = 4\pi\sigma R_0, \quad a_0 = (\hbar^2/2M_0U_0')^{1/3}. \tag{9.5.20}$$

The length a_0 is also expressed as $a_0 = (R_Q^7/R_0^4)^{1/3} = R_Q\Delta_Q^{4/3}$ and is very small ($\ll R_Q$), which corresponds to εR_c in (9.2.8). The $\mathcal{A}$ in (9.5.19) is the action integral,

$$\mathcal{A} = \frac{1}{\hbar}\int_0^{R_0} dR\sqrt{2M(R)U(R)}. \tag{9.5.21}$$

In the present problem the above integral is performed to give

$$\mathcal{A} = (5\pi^2/32\hbar)(2\rho_{\text{eff}}\sigma)^{1/2}R_0^{7/2} = (5\pi/128)\Delta_Q^{-7/2}. \tag{9.5.22}$$

In the region $R > R_0$, however, Ψ depends on t. It should vanish for $R > R_{\max}(t)$, where $R_{\max}(t) \cong v_\infty t$ is the upper cut-off radius after quenching the system at $t = 0$. Analogous to the classical case, the quantum-mechanical probability distribution in the region $R_0 < R < R_{\max}(t)$ behaves as

$$|\Psi(R,t)|^2 = \Gamma_Q/\dot{R}, \tag{9.5.23}$$

where $\dot{R}$ is the classical velocity in (9.5.7). Because $\int_{R_0}^{R_{\max}} dR\, \dot{R}^{-1} \cong t$, the probability that R exceeds R_0 is proportional to t as

$$\int_{R_0}^{R_{\max}} dR\, |\Psi(R)|^2 \cong \Gamma_Q t. \tag{9.5.24}$$

From (9F.6) in Appendix 9F the decay rate Γ_Q is estimated as

$$\Gamma_Q \sim \frac{1}{\hbar} E_g \exp(-2\mathcal{A}). \tag{9.5.25}$$

To calculate the nucleation rate I, Lifshitz and Kagan multiplied Γ_Q by the number density N_0 of virtual centers of precipitating droplets, which should be in the range $n_0 < N_0 < 4/3\pi R_0^3$, n_0 being the particle number density. Then,

$$I = I_0 \exp(-2\mathcal{A}), \tag{9.5.26}$$

where $I_0 \sim N_0 E_g/\hbar$ is a microscopic number. Because the droplet growth for $R > R_0$ is rapid, nucleation will be completed even with growth of a single droplet and the nucleation time is given by (9.2.24). The crossover temperature T^* from the thermal activation to quantum tunneling mechanisms may be estimated as

$$T^* \sim U_{\max}/\mathcal{A} \sim E_g \Delta_Q^{3/2}. \tag{9.5.27}$$

Experimentally, the (negative) pressure [55] or supersaturation [74] at which phase separation was observed became independent of T below a certain crossover temperature. This indicates the relevance of quantum fluctuations in nucleation. However, it is not conclusive at present whether or not the Lifshitz–Kagan theory provides the real quantum mechanism. For example, nucleation around a vortex line might be relevant.

9.6 Viscoelastic nucleation in polymers

Not enough attention has so far been paid to nucleation phenomena in polymeric systems, neither theoretically nor experimentally, whereas many experimental results have been obtained on spinodal decomposition, as described in Section 8.9. Krishnamurthy and Bansil [76] performed a light scattering experiment on a metastable polymer solution near the critical point. They found a large asymmetry between the growth of polymer-rich droplets and that of polymer-poor droplets even relatively close to the critical point. They ascribed this asymmetry to a strong concentration dependence of the viscosity and the diffusion constant. Balsara *et al.* observed early-stage nucleation in a ternary mixture of A polymers, B polymers, and A–B diblock copolymers by small-angle neutron scattering [77], where the copolymers serve to reduce the surface tension. Theoretically, nucleation in polymer blends has been treated within the traditional scheme for low-molecular-weight fluids [78], but the stress–diffusion coupling introduced in Section 7.1 has been overlooked. The aim of this section is to show that stress–diffusion coupling can drastically slow down the growth of droplets if the droplet radius is shorter than the viscoelastic length $\xi_{\rm ve}$ in (7.1.68) or (7.1.72). In our theory [79], most important will be a modified Gibbs–Thomson relation at the interface, which accounts for the relaxing network stress.

9.6.1 Supersaturation and the critical radius

Using the Flory–Huggins theory in Section 3.5 and (9.1.9)–(9.1.17), we first calculate the supersaturation $\Delta = (\phi_{\rm cx}^{(1)} - M)/\Delta\phi$, the capillary length d_0, and the critical radius $R_{\rm c} = 2d_0/\Delta$ in metastable polymer systems. Hereafter we will consider only the initial stage of nucleation and write the average volume fraction $M = \langle\phi\rangle$ simply as ϕ.

Semidilute polymer solutions

As discussed in Section 3.5, a very dilute phase with $\phi_{\rm cx}^{(2)} \cong 0$ and a semidilute phase with $\phi_{\rm cx}^{(1)} \cong 3(\chi - 1/2)$ can coexist macroscopically with almost vanishing osmotic pressure on the coexistence curve $T = T_{\rm cx}(\phi)$, where χ is the interaction parameter. If the temperature T is slightly below $T_{\rm cx}$ and the deviation $\delta T = T_{\rm cx} - T$ is increased at constant ϕ ($> \phi_{\rm c} = N^{-1/2}$), we enter into a metastable region with $\Pi \cong -K_{\rm os}\Delta < 0$. Assuming that χ depends on T as $\partial\chi/\partial T = -\chi_1$ with χ_1 being a positive constant, we obtain

$$\phi_{\rm cx}^{(1)} \cong 3\chi_1\Delta T, \quad \phi_{\rm cx}^{(1)} - \phi \cong 3\chi_1\delta T, \tag{9.6.1}$$

where $\Delta T \equiv T_{\rm c} - T_{\rm cx}$ and $\delta T \equiv T_{\rm cx} - T$ as in Section 9.1. Therefore,

$$\Delta \cong \delta T/\Delta T. \tag{9.6.2}$$

From (3.5.24) and (4.4.34) we find $K_{\rm os} \cong v_0^{-1}T\phi^3/3$ and $\sigma \cong Ta^{-2}\phi^2/12$ near the coexistence curve, where $a = v_0^{1/3}$ is the monomer size, so that

$$d_0 \cong \frac{1}{4}a\phi^{-1}, \quad R_{\rm c} \cong \frac{1}{2}a(\phi\Delta)^{-1}. \tag{9.6.3}$$

The free energy to produce a critical droplet in (9.1.2) is expressed as

$$\mathcal{H}_{\rm c} \cong \frac{\pi}{48}\Delta^{-2}T. \tag{9.6.4}$$

Polymer blends

We consider polymer blends with $N_1 \geq N_2 \gg 1$ in the mean field critical region (4.2.39). We define

$$\varepsilon_\chi = \chi/\chi_c - 1 = (N_1 N_2)^{-1/2}(1 - T/T_c)/Gi. \tag{9.6.5}$$

Some calculations yield

$$\Delta\phi \sim [\phi_c(1-\phi_c)]^{1/2}\varepsilon_\chi^{1/2}, \quad \chi_\phi^{-1} = (Tv_0)^{-1} f''_{\text{site}} \sim v_0^{-1}\chi_c\varepsilon_\chi, \tag{9.6.6}$$

where ϕ_c is given by (3.5.33). The capillary length and surface tension are estimated as

$$d_0 \sim \xi \sim a(N_1 N_2)^{1/4}\varepsilon_\chi^{-1/2}, \quad \sigma \sim Ta^{-2}(N_1 N_2)^{-1/4}\varepsilon_\chi^{3/2}, \tag{9.6.7}$$

where the behavior of σ is consistent with (4.4.42). Therefore, in accord with the general result (9.2.37)–(9.2.39), the free energy to create a critical droplet is estimated as

$$\mathcal{H}_c \sim (N_1 N_2)^{1/4}\varepsilon_\chi^{1/2}\Delta^{-2}T. \tag{9.6.8}$$

The nucleation barrier is enlarged by the factor $(N_1 N_2)^{1/4}\varepsilon_\chi^{1/2}$ in the mean field critical region, indicating suppression of the nucleation rate.

9.6.2 Viscoelastic Gibbs–Thomson relation

Polymer solutions

We set up the interfacial boundary condition around a solvent-rich spherical droplet in a semidilute polymer solution. In the presence of a nonvanishing network stress, the stress balance at the interface yields

$$\delta p - S_n + \frac{2\sigma}{R} = \delta p_0, \tag{9.6.9}$$

where

$$S_n = \boldsymbol{n} \cdot \overleftrightarrow{\sigma} \cdot \boldsymbol{n} \tag{9.6.10}$$

is the normal component of the network stress outside the droplet, δp is the pressure deviation outside the droplet, and δp_0 that inside it. We assume that S_n is nonvanishing only outside the droplet, because the network structure should be anisotropic (isotropic) outside (inside) the droplet. Here the continuity of the solvent chemical potential μ_s gives $\delta p - \delta p_0 = \Pi$, as derived in (3.5.19). It then follows a modified Gibbs–Thomson relation at the interface,

$$K_{os}(\phi/\phi_{cx}^{(1)} - 1) - S_n + \frac{2\sigma}{R} = 0, \tag{9.6.11}$$

where ϕ is the volume fraction immediately outside the droplet.

Polymer blends

We next calculate the discontinuities of the chemical potentials μ_1 and μ_2 per unit mass in (7.1.4) across the interface. To this end we assume that the two-fluid dynamic equations (7.1.16) and (7.1.17) hold even in the interface region. We divide them by $\phi_K = \rho_K/\rho$ and integrate over the region $|r - R| \lesssim \xi$. Then, using the intermediate stress division (7.1.38), we may calculate the differences $[\mu_K] \equiv (\mu_K)_+ - (\mu_K)_-$ $(K = 1, 2)$, where the subscripts, $+$ and $-$, denote the values at $r \cong R + \xi$ and $r \cong R - \xi$, respectively. Assuming (7.1.1) and using (7.1.40), we find

$$\begin{aligned} \rho[\mu_1] &= \frac{\alpha_1}{\phi_1} S_{\rm n} = \left[1 + (1 - \phi_{\rm cx}^{(1)})\alpha\right] S_{\rm n}, \\ \rho[\mu_2] &= \frac{\alpha_2}{\phi_2} S_{\rm n} = \left(1 - \phi_{\rm cx}^{(1)}\alpha\right) S_{\rm n}, \end{aligned} \tag{9.6.12}$$

where ρ is the mass density assumed to be a constant and $\phi_{\rm cx}^{(1)}$ is the volume fraction of the first component outside the droplet. Note that α_K, ϕ_K, and $S_{\rm n}$ in (9.6.12) are the values immediately outside the droplet because the network stress is assumed to vanish inside the droplet. The discontinuity of the chemical potential difference then becomes

$$\rho(\mu_1 - \mu_2) = \alpha S_{\rm n}. \tag{9.6.13}$$

Here $\rho(\mu_1 - \mu_2) = v_0^{-1} f'_{\rm site}$ holds outside the interface region from (7.1.6). Therefore, the above relation is rewritten as

$$\bar{r}_1 \Phi_1 - \bar{r}_2 \Phi_2 = \alpha S_{\rm n}. \tag{9.6.14}$$

For simplicity, we write $\bar{r}_K \equiv v_0^{-1} f''_{\rm site}$ at $\phi = \phi_{\rm cx}^{(K)}$ (equal to the values of r in (7.1.7) multiplied by T in the two phases). The deviations Φ_K are defined by $\Phi_1 \equiv \phi - \phi_{\rm cx}^{(1)}$ and $\Phi_2 \equiv \phi - \phi_{\rm cx}^{(2)}$ immediately outside and inside the droplet, respectively.

Next, from (3.5.31) we may relate the chemical potential of the second component (per unit mass) to the pressure deviation δp as $\rho\mu_2 = \delta p + (f_{\rm site} - \phi f'_{\rm site})/v_0$ outside the interface region. Expanding this expression around $\phi_{\rm cx}^{(K)}$ $(K = 1, 2)$, we find

$$\rho\mu_2 = \delta p - \phi_{\rm cx}^{(K)} \bar{r}_K \Phi_K. \tag{9.6.15}$$

Together with the second line of (9.6.12) we may express the pressure discontinuity as

$$\begin{aligned} [\delta p] &= \phi_{\rm cx}^{(1)} \bar{r}_1 \Phi_1 - \phi_{\rm cx}^{(2)} \bar{r}_2 \Phi_2 + [1 - \phi_{\rm cx}^{(1)}\alpha] S_{\rm n} \\ &= (\Delta\phi)\bar{r}_1 \Psi_1 + [1 - \alpha(\Delta\phi)] S_{\rm n}. \end{aligned} \tag{9.6.16}$$

In the second line, Φ_2 has been eliminated using (9.6.14). The stress-balance equation (9.6.9) also holds for polymer blends, leading to

$$(\Delta\phi)\bar{r}_1 \Phi_1 - \alpha(\Delta\phi) S_{\rm n} + \frac{2\sigma}{R} = 0. \tag{9.6.17}$$

If we set $\alpha = 1/\phi$ and $\Delta\phi = \phi$, the above relation reduces to (9.6.11) for polymer solutions.

9.6.3 Viscoelastic stress

Next we need to express S_n in terms of R and Δ to construct the evolution equation of R. We assume that the growth rate is much slower than the stress relaxation time τ. Then the network stress σ_{ij} may be expressed in terms of the tube velocity $\boldsymbol{v}_t$ as (7.1.46) and (7.1.47) in the linear regime. We note that the velocity fields outside the droplet can be calculated from the mass conservation relations across the interface,

$$\left[\rho_K\left(\boldsymbol{v}_K \cdot \boldsymbol{n} - \frac{\partial}{\partial t}R\right)\right] = 0 \quad (K = 1, 2), \tag{9.6.18}$$

where $\boldsymbol{n} = r^{-1}\boldsymbol{r}$ is the outward normal unit vector. Because the velocities inside the droplet vanish, these relations are rewritten as

$$\phi_{\rm cx}^{(1)}\boldsymbol{v}_1 \cdot \boldsymbol{n} = (\Delta\phi)\frac{\partial}{\partial t}R, \quad (1 - \phi_{\rm cx}^{(1)})\boldsymbol{v}_2 \cdot \boldsymbol{n} = -(\Delta\phi)\frac{\partial}{\partial t}R. \tag{9.6.19}$$

The tube velocity (7.1.45) immediately outside the droplet is given by

$$\boldsymbol{v}_t \cdot \boldsymbol{n} = (\alpha_1\boldsymbol{v}_1 + \alpha_2\boldsymbol{v}_2) \cdot \boldsymbol{n} = \alpha(\Delta\phi)\frac{\partial}{\partial t}R. \tag{9.6.20}$$

Outside the droplet we require $\nabla \cdot \boldsymbol{v}_K \cong 0$ because $\partial\rho_K/\partial t \cong 0$, so that $\boldsymbol{v}_K \propto \nabla(1/r)$ for $r > R$ or

$$\boldsymbol{v}_t(r, t) = \alpha(\Delta\phi)\left(\frac{\partial R}{\partial t}\right)\frac{R^2}{r^3}\boldsymbol{r}. \tag{9.6.21}$$

For slow motions we thus have

$$\sigma_{ij} = S_n\left(\frac{3}{2r^2}x_ix_j - \frac{1}{2}\delta_{ij}\right)\frac{R^3}{r^3}, \tag{9.6.22}$$

where the normal stress at the interface S_n is expressed as

$$S_n = -4\alpha(\Delta\phi)\eta\frac{1}{R}\left(\frac{\partial}{\partial t}R\right). \tag{9.6.23}$$

Substitution of the above result into (9.6.17) gives the desired result,

$$\frac{\Phi_1}{\Delta\phi} + \frac{4\alpha^2\eta}{\bar{r}_1 R}\left(\frac{\partial}{\partial t}R\right) + \frac{2d_0}{R} = 0, \tag{9.6.24}$$

where d_0 is the capillary length defined by (9.1.17).

9.6.4 Modified Lifshitz–Slyozov equation

The deviation $\delta\phi \equiv \phi - \phi_{\rm cx}^{(1)}$ outside the droplet obeys the diffusion equation or the modified diffusion equation (7.1.71) for slow motions. As in usual fluids, we use the quasi-static condition $\partial\delta\phi/\partial t \cong 0$ or $\nabla^2\delta\phi \cong 0$ for $r > R$ to obtain

$$\delta\phi(r, t) = \frac{R}{r}\Phi_1 - \left(1 - \frac{R}{r}\right)(\Delta\phi)\Delta. \tag{9.6.25}$$

The above solution satisfies the boundary condition at $r = R$ and tends to $-(\Delta\phi)\Delta$ for $r \gg R$. Then the evolution equation for R is given by

$$\frac{\partial}{\partial t}R = -\frac{1}{\Delta\phi}D_{\rm m}\left(\frac{\partial}{\partial r}\delta\phi\right)_{r=R} = \frac{D_{\rm m}}{R}\left[\Delta + \frac{\Phi_1}{\Delta\phi}\right], \tag{9.6.26}$$

where $D_{\rm m} = L\bar{r}_1$ is the mutual diffusion constant in phase 1 as given in (7.1.27) and (7.1.50). From (9.6.24) we arrive at

$$\frac{\partial}{\partial t}R = D_{\rm m}\left(\Delta - \frac{2d_0}{R}\right)\Big/\left(R + 3\frac{\xi_{\rm ve}^2}{R}\right), \tag{9.6.27}$$

where $\xi_{\rm ve}$ is the viscoelastic length defined by (7.1.65). This result may be interpreted as originating from renormalization of the kinetic coefficient from L to $L_{\rm eff}(R) = L/(1 + 3\xi_{\rm ve}^2/R^2)$ for droplets, which is analogous to that in (7.1.68) for plane-wave fluctuations.

Thus, the Lifshitz–Slyozov theory holds only for $R \gg \xi_{\rm ve}$, while small droplets with $R \ll \xi_{\rm ve}$ are governed by

$$\frac{\partial}{\partial t}R \cong \Gamma_{\rm c}(R - R_{\rm c}). \tag{9.6.28}$$

The growth rate is given by

$$\Gamma_{\rm c} = \frac{1}{3}D_{\rm m}\xi_{\rm ve}^{-2}\Delta = \frac{1}{3}\varepsilon_{\rm r}\tau^{-1}\Delta, \tag{9.6.29}$$

where $\varepsilon_{\rm r}$ is defined by (8.9.1). When $\varepsilon_{\rm r} \lesssim 1$, $\Gamma_{\rm c}$ can be much smaller than τ^{-1}. In particular, for polymer solutions, we have

$$\Gamma_{\rm c} = \frac{\sigma}{2\eta R_{\rm c}} = \frac{T\phi^3}{12\eta a^3}\Delta \sim \tau^{-1}\Delta. \tag{9.6.30}$$

In experiments, it is of great interest to investigate nucleation in the case $R_{\rm c} < \xi_{\rm ve}$ or $\Delta > d_0/\xi_{\rm ve}$. For polymer solutions we have $d_0/\xi_{\rm ve} \sim (\eta_0/\eta)^{1/2}$. For polymer blends with $N_1/N_2 - 1 \sim 1$, (7.1.74) gives $d_0/\xi_{\rm ve} \sim \xi/L_{\rm t} \sim (N_{\rm e}/N_1\varepsilon_\chi)^{1/2}$, where ε_χ is defined by (9.6.5). We can see that there is a crossover from the viscoelastic slowing-down into the critical slowing-down as the critical point is approached. The former is more apparent away from the critical point. In addition, we expect a considerable decrease of σ or $d_0/\xi_{\rm ve}$ in the presence of A–B diblock copolymers which come together in the interface region of A and B homopolymers [77]. Thus addition of such diblock copolymers will make the viscoelastic effect unambiguously observable.

9.7 Intrinsic critical velocity in superfluid helium

Superfluid states with a superfluid velocity $\boldsymbol{u}_{\rm s}$ are metastable in a toroidal geometry where the macroscopic normal fluid velocity $\boldsymbol{u}_{\rm n}$ vanishes. If $u_{\rm s}$ is small, it decays slowly with nucleation and growth of vortices [80]–[84].

9.7.1 Current-carrying states

We consider superfluid ^{4}He at low temperatures in a cylindrical container with cross-sectional area A_0 and length L_0. In this geometry, macroscopic superfluid currents parallel to the cylindrical axis can flow without appreciable decay within observation times if the flow velocity is below a certain critical velocity u_{sc}. Experimentally, the container can be packed with a porous substance that clamps the normal fluid component ($\boldsymbol{u}_n = \boldsymbol{0}$). In this case the complex order parameter behaves as[6]

$$\psi(x) = M \exp(ikx) \quad \text{with} \quad k = 2\pi j/L_0, \tag{9.7.1}$$

where x is the coordinate along the cylinder axis, and j is an integer ensuring the periodic boundary condition $\psi(x) = \psi(x + L_0)$. We assume $j \gg 1$ hereafter. The macroscopic superfluid velocity is given by

$$u_s = \frac{\hbar}{m_4} k. \tag{9.7.2}$$

Minimization of the GLW hamiltonian $\mathcal{H}$ in (4.1.1) yields the amplitude,

$$M = [(\kappa^2 - k^2)/u_0]^{1/2} \tag{9.7.3}$$

in the mean field theory, where $\kappa^2 = -r_0$, $K = 1$, and $h = 0$. The minimum of $\mathcal{H}$ depends on k^2 as

$$\mathcal{H}_{\min}(k) = -\frac{1}{4} V u_0 M^2 = \mathcal{H}_{\min}(0) + \frac{1}{2} V \rho_s u_s^2 + O(k^4), \tag{9.7.4}$$

where $V = A_0 L_0$ is the volume. The one-dimensional solution (9.7.1) thus represents a metastable state for small k. To examine its linear stability, we write ψ as

$$\psi = (M + w_1 + i w_2) \exp(ikx), \tag{9.7.5}$$

where w_1 and w_2 are real numbers. The GLW hamiltonian (4.1.1) is then expressed as

$$\begin{aligned} \mathcal{H} \;=\; & \mathcal{H}_{\min}(k) + \int d\boldsymbol{r} \bigg[(\kappa^2 - 3k^2) w_1^2 + \frac{1}{2} |\nabla w_1|^2 + \frac{1}{2} |\nabla w_2 + 2k w_1 \boldsymbol{e}_x|^2 \\ & + u_0 M w_1 (w_1^2 + w_2^2) + \frac{1}{4} u_0 (w_1^2 + w_2^2)^2 \bigg], \end{aligned} \tag{9.7.6}$$

where $\boldsymbol{e}_x$ is the unit vector along the cylinder axis. Therefore, current-carrying states in the form of (9.7.1) are metastable only for

$$k < \frac{1}{\sqrt{3}} \kappa, \tag{9.7.7}$$

and are linearly unstable for $k > \kappa/\sqrt{3}$ (the Eckhaus instability). This stability criterion follows in the general Ginzburg–Landau theory [82, 85]. We note that the superfluid current $J_s \propto (\kappa^2 - k^2)k$ takes a maximum at $k = \kappa/\sqrt{3}$ and the above stability criterion is equivalent to $\partial J_s / \partial k > 0$. Analogously, in superconducting wires or films, the so-called critical current has been determined by the same criterion [86].

6 We neglect variations of ψ near the boundary wall within the correlation length.

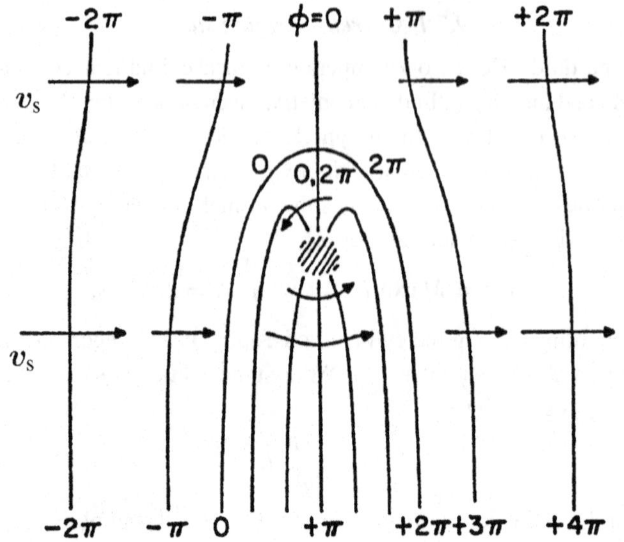

Fig. 9.20. Phase contours for a single vortex line with superimposed uniform flow normal to the line [82].

9.7.2 Nucleation of vortex rings

For $k \ll \kappa/\sqrt{3}$, the decay mechanism of the superfluid velocity has been ascribed to vortex line motion perpendicular to the flow, as illustrated in Fig. 9.20. Let a vortex ring with radius R be perpendicular to the flow with $u_s = \boldsymbol{b} \cdot \boldsymbol{u}_s > 0$ in the case $\boldsymbol{u}_n = \boldsymbol{0}$. We rewrite (8.10.28) as

$$\frac{\partial}{\partial t} R = \alpha \left(u_s - \frac{\hbar E_0}{2m_4 R} \right) = \alpha u_s \left(1 - \frac{R_c}{R} \right), \tag{9.7.8}$$

where α is the mutual friction coefficient, and $E_0 \cong \ln(R/\xi)$ is logarithmically dependent on R but will be treated as a constant considerably larger than 1. Note that u_s plays the role of a magnetic field in metastable spin systems. If R is larger than the critical radius R_c in (8.10.29), the ring will grow and eventually disappear at the boundary. In this elementary process, the phase of the complex order parameter is decreased by 2π or $j \to j - 1$ in (9.7.1). Such vortex rings can appear as rare thermal fluctuations and hence u_s decays as [82]

$$\frac{\partial}{\partial t} u_s = -\frac{2\pi\hbar}{m_4} A_0 I(u_s), \tag{9.7.9}$$

where $I(u_s)$ is the nucleation rate of vortex rings with $R > R_c$ per unit volume.

From (8.10.23) the free energy to create a vortex ring is given by

$$\tilde{\mathcal{H}}_v = \rho_s \left(\frac{\pi\hbar}{m_4} \right)^2 E_0 \big(2R - R^2/R_c \big). \tag{9.7.10}$$

The evolution equation (9.7.8) is then rewritten as

$$\frac{\partial}{\partial t}R = -\left(\frac{m_4\alpha}{4\pi\hbar\rho_s R}\right)\frac{\partial}{\partial R}\tilde{\mathcal{H}}_v. \tag{9.7.11}$$

The above equation may be treated as a Langevin equation in the standard form if we add the noise term related to the kinetic coefficient $L(R) = m_4\alpha/4\pi\hbar\rho_s T R$ via the fluctuation–dissipation relation [80]. Note that we derived $\tilde{\mathcal{H}}_v$ in (8.10.22) by analyzing the vortex motion, whereas in the literature [81] it has been derived using the relation $\epsilon = \epsilon_0 - pu_s$ between the energies, ϵ and ϵ_0, of an elementary excitation with momentum p in the moving and static reference frames, respectively. This picture is justified only without dissipation, however. For the present case of a vortex ring we have

$$\tilde{\mathcal{H}}_v = E_{\rm ring} - p_0 u_s, \tag{9.7.12}$$

where $E_{\rm ring} = E_{\rm ring}(R)$ is the vortex free energy in the static reference frame in (4.5.12) and $p_0 = p_0(R)$ is the momentum of the vortex ring. We may determine p_0 such that $\tilde{\mathcal{H}}_v$ takes a maximum at $R = R_c$, which gives[7]

$$p_0 = (2\pi^2\hbar/m_4)\rho_s R^2. \tag{9.7.13}$$

The maximum of $\tilde{\mathcal{H}}_v$ is given by

$$\mathcal{H}_{vc} = (\pi\hbar/m_4)^2\rho_s E_0 R_c = T u_0/u_s, \tag{9.7.14}$$

where u_0 is a characteristic velocity defined by

$$u_0 = (\pi E_0)^2\hbar^3\rho_s/2m_4^3 T. \tag{9.7.15}$$

Near the λ point we use the transverse correlation length $\xi_T \cong 3.4(1 - T/T_\lambda)^{-2/3}$ Å in (4.3.105). It satisfies (4.3.107), so that

$$u_0 = (\pi E_0)^2\hbar/2m_4\xi_T, \quad \mathcal{H}_{vc}/T = \pi^2 E_0 R_c/\xi_T. \tag{9.7.16}$$

Langer and Reppy [82] set

$$I(u_s) = \nu_0 \exp(-\mathcal{H}_{sc}/T) = \nu_0 \exp(-u_0/u_s), \tag{9.7.17}$$

where ν_0 is a phenomenological constant. Using (9.1.26) and the first line of (9.2.30), we estimate it near the λ point as

$$\nu_0 \sim \alpha\frac{\hbar}{m_4}R_c^{-3/2}\xi_T^{-1/2} \sim \Gamma_\xi\xi_T^{-3}(\xi_T/R_c)^{3/2}, \tag{9.7.18}$$

where Γ_ξ ($\propto \xi_T^{-z}$) is the typical order parameter relaxation rate. Its form above T_λ is given by (6.6.44).

[7] If $E_{\rm ring}$ is regarded as a hamiltonian dependent on p_0, we have a conjugate velocity $v_0 = dE_{\rm ring}/dp_0 = (\hbar E_0/2m_4)/R$. From (4.5.24) this is the velocity of a vortex ring perpendicular to the ring, in the dissipationless limit.

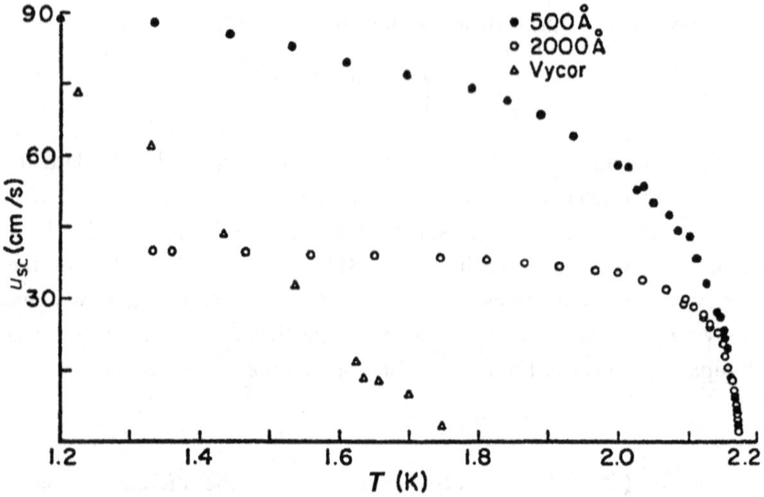

Fig. 9.21. Superfluid critical velocities v_{sc} obtained for flow through 500 Å and 2000 Å filter materials, and through Vycor glass as a function of temperature [82, 83].

9.7.3 *Critical velocity*

Experimentally, at a well-defined critical velocity $u_s = u_{sc}$, $|du_s/dt|$ takes a characteristic, observable value Φ_0, while it is not appreciable for u_s slightly below u_{sc} in realistic observation times. In Fig. 9.21 we plot the critical velocity curves, u_{sc} vs T, measured by Clow and Reppy [83]. Close to the λ point the curves for the 500 Å and 2000 Å are represented by

$$u_{sc} = 670(1 - T/T_\lambda)^{2/3} \quad \text{cm/s}. \tag{9.7.19}$$

To check the stability condition (9.7.7) we set $\kappa = \xi_{+0}^{-1}(1 - T/T_\lambda)^{2/3}$ using $\xi_{+0} = 1.4$ Å at SVP determined below (2.4.4). Then the experimental critical wave number $k_c \equiv m_4 u_{sc}/\hbar$ is written as

$$k_c = 0.062\kappa. \tag{9.7.20}$$

The coefficient is one order of magnitude smaller than $1/\sqrt{3}$ in (9.7.7). This is a natural result because (9.7.7) and (9.7.20) give the threshold of linear instability of plane-wave perturbations and nucleation of vortices, respectively.

Theoretically, the critical velocity u_{sc} is determined from (9.7.9) by [82]

$$\Phi_0 = \frac{2\pi\hbar}{m_4} A_0 I(u_{sc}) = \frac{2\pi\hbar}{m_4} A_0 \nu_0 \exp\left(-\frac{u_0}{u_{sc}}\right), \tag{9.7.21}$$

so that

$$u_{sc} = u_0/\gamma, \tag{9.7.22}$$

where

$$\gamma = \ln(2\pi\hbar A_0 v_0 / m_4 \Phi_0). \tag{9.7.23}$$

Then (9.7.9) may be expressed as

$$\frac{\partial}{\partial t} u_s = -\Phi_0 \exp\big[\gamma(1 - u_{sc}/u_s)\big]. \tag{9.7.24}$$

Note that u_{sc} depends weakly on the experimental conditions. Experimentally, if the decay of u_s is fitted to (9.7.24), γ may be determined as an adjustable parameter. In this manner Clow and Reppy obtained $\gamma \cong 46$ near the λ point. Theoretically, Langer and Reppy found $\gamma \cong 53$ and $u_{sc} = 4800(1 - T/T_\lambda)^{2/3}$ cm/s by setting $\Phi_0 = 1$ cm/s^2 and $A_0 = 10^{-8}$ cm^2 ($\sim$ the square of the pore size of the porous substance). Thus the simple homogeneous nucleation theory presented so far is not in quantitative agreement with experiment.

Appendix 9A Relaxation to the steady droplet distribution

Here we solve the (unstable) linearized Fokker–Planck equation (9.2.5) or (9.2.9) under $\varepsilon \ll 1$. The variable we use is $x = (R - R_c)/\varepsilon R_c$ and the equation is valid in the region $|x| \lesssim \varepsilon^{-1}$. Starting with an initial distribution $n_{ini}(x)$ at $t = t_0 \sim \Gamma_c^{-1}$, we calculate the subsequent solution for $t > t_0$ in the form,

$$n(x, t) = \int dx_0 \, \phi(x, x_0, t - t_0) \, n_{ini}(x_0). \tag{9A.1}$$

The $\phi(x, x_0, t - t_0)$ is the conditional probability that x is equal to the initial value x_0 at $t = t_0$:

$$\phi(x, x_0, t - t_0) \to \delta(x - x_0) \quad \text{as} \quad t \to t_0. \tag{9A.2}$$

It is calculated in the following gaussian form,

$$\phi(x, x_0, t - t_0) = \frac{1}{\sqrt{2\pi(q^2 - 1)}} \exp\left[-\frac{(x - qx_0)^2}{2(q^2 - 1)}\right], \tag{9A.3}$$

where

$$q = \exp[\Gamma_c(t - t_0)]. \tag{9A.4}$$

The initial time t_0 is chosen such that $n_{ini}(x) \cong n_s(x) \cong C_0 \exp(x^2/2)$ for $x \lesssim -1$ and $n_{ini}(x) \ll C_0$ for $x \gtrsim 1$. In the case of Fig. 9.5 we clearly have $t_0 = \Gamma_c^{-1}$.

As an illustrative example, let us assume

$$n_{ini}(x) = C_0 \exp\left(\frac{1}{2}x^2\right) \quad (x < -M), \quad n_{ini}(x) = D_0 e^{-\alpha x} \quad (x > -M), \tag{9A.5}$$

where M is of order 1. In Fig. 9.5 the above form holds with $M \sim 2$ and $\alpha \sim 1$ at

$t = \Gamma_c^{-1}$. First, the contribution from the region $x_0 < -M$ is calculated in (9A.1). By setting $y = -x_0/(q^2-1)^{1/2}$ we have

$$\begin{aligned} n^<(x,t) &= \frac{C_0}{\sqrt{2\pi}} \int_{M^*}^{\infty} dy \exp\left[-\frac{x^2}{2(q^2-1)} - \frac{qxy}{\sqrt{q^2-1}} - \frac{y^2}{2} \right] \\ &\cong \frac{C_0}{\sqrt{2\pi}} \frac{1}{x} \exp\left(-\frac{1}{2} X^2 \right), \end{aligned} \tag{9A.6}$$

where $M^* = M/(q^2-1)^{1/2}$. The second line holds for $x \gg 1$ and $q \gg 1$, and

$$X = x/q = x \exp[-\Gamma_c(t - t_0)]. \tag{9A.7}$$

Second, from the region $x_0 > -M$ and in the case $q \gg 1$, we obtain

$$\begin{aligned} n^>(x,t) &\cong \frac{D_0}{\sqrt{2\pi} q} \int_{-M}^{\infty} dx_0 \exp\left[-\frac{1}{2}(X - x_0)^2 - \alpha x_0 \right] \\ &\cong D_0 q^{-1} \exp(-\alpha X), \end{aligned} \tag{9A.8}$$

where the second line holds for $X \gtrsim 1$. Thus,

$$n(x,t) = n^>(x,t) + n^<(x,t) \cong \frac{C_0}{\sqrt{2\pi}} \frac{1}{x} F(X), \tag{9A.9}$$

where

$$F(X) = \exp\left(-\frac{1}{2} X^2 \right) + \sqrt{2\pi} \frac{D_0}{C_0} X \exp(-\alpha X). \tag{9A.10}$$

We substitute the above results into (9.2.13) replacing x and t by $x = (R_1 - R_c)/\varepsilon R_c$ and t_1. Further replacing R_2 and t_2 in (9.2.13) by R and t, we may calculate $n(R,t)$ for larger $R/R_c - 1 \gg \varepsilon$ and $t \gg t_0$ as

$$n(R,t) = \frac{I}{v(R)} F(X). \tag{9A.11}$$

Use has been made of the relation $\sqrt{2\pi} I = C_0 \Gamma_c \varepsilon R_c$ for the nucleation rate I in terms of C_0, which follows from (9.2.30) and (9.2.31). From the mapping relation (9.2.16) we have

$$X = x \exp\big[-\Gamma_c(t_1 - t_0)\big] = \frac{1}{\varepsilon}(R/R_c - 1) \exp\big[\mathcal{G}(R/R_c) - \Gamma_c(t - t_0)\big], \tag{9A.12}$$

where $\mathcal{G}(R/R_c)$ is defined by (9.2.17). We can see that X changes from a very small number ($\cong 0$) to a very large number ($\gg 1$) as R exceeds $R_{max}(t)$ in a changeover region with a width estimated as (9.2.26). From $F(0) = 1$ we also find (9.2.33).

Appendix 9B The nucleation rate near the critical point

We estimate A_{d0} in (9.3.38) and C_0 in (9.3.40) close to the critical point in 3D Ising-like systems. In terms of the universal number $A_\sigma \cong 0.09$ in (4.4.11) and the critical amplitudes, B_0 in $\psi_{eq} = B_0(1 - T/T_c)^\beta$ and Γ_0 in $\chi = \Gamma'_0(1 - T/T_c)^{-\gamma}$, we obtain

$$A_{d0} = \Gamma'_0 A_\sigma / [4B_0^2(\xi_{0-})^3] \cong 0.10 \tag{9B.1}$$

from the universal relations among critical amplitudes (see Chapter 4). This value is considerably smaller than the mean field value $1/6 = 0.145$. Then,

$$C_0 = (16\pi/3) A_\sigma A_{d0}^2 \cong 0.015, \tag{9B.2}$$

$$x_0 = (2/\beta) C_0^{1/2} \cong 0.74. \tag{9B.3}$$

Similar estimations were presented in Ref. [6], while Langer and Schwarz set $x_0 = 1.24$–1.30 in analyzing experiments [28]. There seems to be some uncertainty both in the critical amplitude ratios and in the experimental data to determine C_0 or x_0 conclusively.

Appendix 9C The asymptotic scaling functions in droplet growth

By assuming the scaling solution $n(r, \tau) \propto p(\tau)^4 P_\lambda(u)$ in (9.3.21), we obtain

$$\frac{d}{du}[\chi(u) P_\lambda(u)] = -3P_\lambda(u) \tag{9C.1}$$

with

$$\begin{aligned} \chi(u) &= u - \gamma_0\left(\frac{1}{u} - \frac{1}{u^2}\right) \\ &= \frac{1}{u^2}(u - u_1)(u - u_2)(u - u_3). \end{aligned} \tag{9C.2}$$

The parameter γ_0 is defined by

$$\gamma_0 = 3(p^*)^3, \tag{9C.3}$$

where p^* is the coefficient in (9.3.23). Let the equation $\chi(u) = 0$ have three real solutions $u = u_1, u_2$, and u_3. Then they satisfy $u_3 < 0 < u_1 \le u_2$ and may be expressed in terms of a parameter s as [39]

$$u_1 = 1 + \frac{4}{s^2 - 1}, \quad u_2 = \frac{s-1}{2} u_1, \quad u_3 = -\frac{s+1}{2} u_1, \tag{9C.4}$$

where $s \ge 3$ to guarantee $u_1 \le u_2$, and

$$\gamma_0 = -u_1 u_2 u_3 = \frac{1}{4}(s^2 - 1)^{-2}(s^2 + 3)^3. \tag{9C.5}$$

We then integrate (9C.1) as

$$\begin{aligned} P_\lambda(u) &= \frac{\text{const.}}{\chi(u)} \exp\left[-3\int_0^u du' \frac{1}{\chi(u')}\right] \\ &= D_\lambda \frac{u^2}{(u - u_3)^6}\left(\frac{u - u_3}{u_2 - u}\right)^\mu \left(\frac{u_1 - u}{u_2 - u}\right)^\lambda, \end{aligned} \tag{9C.6}$$

where D_λ is a constant and

$$\lambda = \frac{21 - s^2}{s^2 - 9}, \quad \mu = 4 - \frac{s^2 + 3}{2s(s+3)}. \tag{9C.7}$$

We notice that $P_\lambda(u)$ can have the meaning of the dimensionless distribution in the region $0 < u < u_1$ for $\lambda > 0$ or $3 \le s < \sqrt{21}$. We determine D_λ from the normalization (9.3.34). Integrating (9C.1) in the region $0 < u < u_1$ we find

$$D_\lambda = 3\gamma_0 \left(\frac{s-1}{2}\right)^{\lambda+2} \left(\frac{s+1}{s-1}\right)^{4-\mu}. \tag{9C.8}$$

Multiplying (9C.1) by u^3 and integrating in the region $0 < u < u_1$ also gives (9.3.34). The scaling function in the LWM theory can be reproduced in the limit $\lambda \to \infty$ or $s \to 3$, where $u_3 \to -3$, $u_2 - u_1 \cong (s-3)u_1/2$, $u_1 \to 3/2$, and the last factor in (9C.6) becomes

$$\left(\frac{u_1 - u}{u_2 - u}\right)^\lambda \to \exp\left(-\frac{u_1}{u_1 - u}\right), \tag{9C.9}$$

which leads to the LSW result (9.3.26). The parameter γ_0 in (9C.5) decreases from 8.64 to 6.75 as λ increases from 0 to ∞. It is known that there is no physically meaningful attractor for the case $\gamma_0 < 6.75$ [39].

Appendix 9D Moving domains in the dissipative regime

Here we consider the linear deviations in a two-phase state of a one-component fluid in the strongly dissipative regime. Let $\boldsymbol{v}_1$, $\boldsymbol{v}_2$, and $\boldsymbol{v}_{\text{int}}$ be the fluid velocities immediately inside and outside a droplet and the interface velocity, respectively. Then the mass current through the interface in the normal direction is given by

$$w = \rho_1(\boldsymbol{v}_1 - \boldsymbol{v}_{\text{int}}) \cdot \boldsymbol{n} = \rho_2(\boldsymbol{v}_2 - \boldsymbol{v}_{\text{int}}) \cdot \boldsymbol{n}, \tag{9D.1}$$

where the normal unit vector $\boldsymbol{n}$ is pointed from phase 1 to phase 2. The energy current near the interface is $(e_\alpha + p_\alpha)(\boldsymbol{v}_\alpha - \boldsymbol{v}_{\text{int}}) - \lambda_\alpha \nabla \delta T_\alpha$ ($\alpha = 1, 2$) in the reference frame moving with the interface, where e_α is the energy density, p_α is the pressure, and λ_α is the thermal conductivity in the phase α. We then use the thermodynamic identity $e + p = n(sT + \mu)$, where s and μ are the entropy and the chemical potential per particle. Because $\mu_1 = \mu_2$, the continuity of the energy current along $\boldsymbol{n}$ yields

$$T(\Delta s) m_0^{-1} w = \left[\lambda_2 \nabla \delta T_2 - \lambda_1 \nabla \delta T_1\right] \cdot \boldsymbol{n}, \tag{9D.2}$$

where (9D.1) has been used and $m_0 = \rho/n$ is the particle mass. For a spherical domain of phase 1 with radius R, we may set $\boldsymbol{v}_1 = \boldsymbol{0}$, $\boldsymbol{v}_{\text{int}} \cdot \boldsymbol{n} = \partial R/\partial t$, and $w = -\rho_1 \partial R/\partial t$. Thus,

$$T n_1 (\Delta s) \frac{\partial}{\partial t} R = -\lambda_2 \left(\frac{\partial}{\partial r} \delta T_2\right)_{r \to R}, \tag{9D.3}$$

which leads to (9.4.8).

Appendix 9E Piston effect in the presence of growing droplets

We consider a slightly metastable liquid in which gas bubbles with a small volume fraction $q(t)$ are growing in a cell with a fixed volume V_0. In the interior liquid region outside the droplets, the average pressure deviation is written as

$$\delta p_\infty(t) = A_p q(t) + \left(\frac{\partial p}{\partial s}\right)_n \frac{Q_T(t)}{nTV_0}, \tag{9E.1}$$

where the first term is the pressure increase due to the droplet formation, and the second term is that due to the heat supply $Q_T(t)$ from the boundary. (See (6.3.2) and (6.3.3) for the case without droplets.) The bubbles may be regarded as tiny pistons within the liquid. The coefficient A_p is written as

$$A_p = \left(\frac{\partial p}{\partial n}\right)_s (\Delta n) + \left(\frac{\partial p}{\partial s}\right)_n (\Delta s) = a_c \frac{n|\Delta s|T}{\sqrt{C_V C_p}} \left(\frac{\partial p}{\partial T}\right)_{cx}, \tag{9E.2}$$

where use has been made of (1.2.53), (2.2.21), and (2.2.39). The first term ($\propto \Delta n = n_\ell - n_g \cong -v^{-2}\Delta v > 0$) arises from the density difference between the two phases, and the second one ($\propto \Delta s = s_\ell - s_g < 0$) from the latent heat, where the quantities with subscript ℓ (g) are those in the liquid (gas) phase. These two terms have opposite signs and largely cancel each other. As in Appendix 6D, we perform the Laplace transformation $\int_0^\infty dte^{-\Omega t}(\cdots)$ in the case $Dt \ll L^2$ or $\Omega \gg D/L^2$, where $L \sim V_0^{1/3}$ is the system length. The condition of the constant temperature at the boundary gives

$$\int_0^\infty dte^{-\Omega t}\delta p_\infty(t) = \frac{A_p w}{1+w}\tilde{q}(\Omega) + \left(\frac{\partial p}{\partial T}\right)_s \frac{T_1}{\Omega(1+w)}, \tag{9E.3}$$

where $w = (\Omega t_1)^{1/2}$ and $\tilde{q}(\Omega) = \int_0^\infty dte^{-\Omega t} q(t)$. The interior temperature variation $\delta T_\infty(t)$ outside the droplets is given by (9.4.21) also in the present fixed-volume case. From (9.4.10) we find

$$\int_0^\infty dte^{-\Omega t}\left[\Delta(t) - \Delta(0) + q(t)\right] = -\frac{\sqrt{\Omega t_1}}{1+\sqrt{\Omega t_1}}\left[a_c^2 \tilde{q}(\Omega) + \frac{\Delta(0)}{\Omega}\right], \tag{9E.4}$$

where $\Delta(0)$ is given by (9.4.23). The inverse Laplace transformation becomes

$$\Delta(t) - \Delta(0) + q(t) = -a_c^2 \int_0^t dt' F_a(t'/t_1)\dot{q}(t-t') - F_a(t/t_1)\Delta(0), \tag{9E.5}$$

where $\dot{q}(t) = dq(t)/dt$, and $F_a(s)$ is defined by (6.3.9). The right-hand side is clearly small at long times $t \gg t_1$.

Appendix 9F Calculation of the quantum decay rate

In the vicinity of the turning point, $|R - R_0| \ll R_0$, Ψ satisfies

$$\mathcal{H}\Psi \cong \left[-\frac{\hbar^2}{2M_0}\frac{\partial^2}{\partial R^2} - U_0'(R-R_0)\right]\Psi \cong 0. \tag{9F.1}$$

The appropriate solution is uniquely expressed in terms of the Airy functions Ai(z) and Bi(z) [87] as

$$\Psi(R) = \mathcal{N}\left[\mathrm{Bi}\left(\frac{R_0 - R}{a_0}\right) + i\mathrm{Ai}\left(\frac{R_0 - R}{a_0}\right)\right], \tag{9F.2}$$

where $\mathcal{N}$ is a constant. On the left-hand side, $R_0 - R \gg a_0$, Ψ grows as

$$\Psi \cong \mathcal{N}\pi^{-1/2}\left(\frac{R_0 - R}{a_0}\right)^{-1/4} \exp\left[\frac{2}{3}\left(\frac{R_0 - R}{a_0}\right)^{3/2}\right], \tag{9F.3}$$

from the asymptotic behavior of the Airy functions. Comparing this with (9.5.19) gives

$$\mathcal{N}a_0^{1/4} \sim C(M_0/U_0')^{1/4} \exp(-\mathcal{A}). \tag{9F.4}$$

Outside the turning point, $R - R_0 \gg a_0$, Ψ behaves as

$$\Psi \cong \mathcal{N}\pi^{-1/2}\left(\frac{R - R_0}{a_0}\right)^{-1/4} \exp\left[\frac{2i}{3}\left(\frac{R - R_0}{a_0}\right)^{3/2} + \frac{\pi i}{4}\right], \tag{9F.5}$$

which yields (9.5.23) with

$$\Gamma_{\mathrm{Q}} = \mathcal{N}^2(\hbar/\pi a_0 M_0) \sim C^2 \exp(-2\mathcal{A}). \tag{9F.6}$$

References

[1] J. Frenkel, *Kinetic Theory of Liquids* (Dover, New York, 1955).

[2] F. F. Abraham, *Homogeneous Nucleation Theory* (Academic, New York, 1974).

[3] V. P. Skripov, *Metastable Liquids* (John Wiley & Sons, New York, 1974).

[4] P. G. Debenedetti, *Metastable Liquids* (Princeton University, Princeton, 1996).

[5] J. S. Langer, in *Solids far from Equilibrium*, ed. C. Godrèche (Cambridge University Press, New York, 1992), p. 297.

[6] K. Binder and D. Stauffer, *Adv. Phys.* **25**, 343 (1976).

[7] K. Binder, *Rep. Prog. Phys.* **50**, 783 (1987).

[8] W. I. Goldburg, in *Scattering Techniques Applied to Supramolecular and Nonequilibrium Systems*, eds. S.-H. Chen and R. Nossal (Plenum, New York, 1981).

[9] C. M. Knobler, in *The Fourth Mexican School on Statistical Mechanics*, eds. R. Pelalta-Fabi and C. Varea (World Scientific, Singapore, 1987).

[10] P. Alpern, Th. Benda, and P. Leiderer, *Phys. Rev. Lett.* **49**, 1267 (1982).

[11] R. Becker and W. Döring, *Ann. Phys.* (Leipzig) **24**, 719 (1935).

[12] J. E. Mayer and M. G. Mayer, *Statistical Mechanics* (John Wiley, New York, 1940).

[13] M. E. Fisher, *Physics* **3**,255 (1967); *Rep. Prog. Phys.* **30**, 615 (1967).

[14] S. Katsura, *Adv. Phys.* **12**, 391 (1963).

[15] A. F. Andreev, *Sov. Phys. JETP*, **18**, 1415 (1964) (*J. Exptl. Theor. Phys.* (U.S.S.R.) **45**, 2064 (1963)).

[16] J. S. Langer, *Ann. Phys.* **41**, 108 (1967).

[17] E. Stoll, K. Binder, and T. Schneider, *Phys. Rev. B* **6**, 2777 (1972).

[18] K. Binder and H. Müller-Krumbhaar, *Phys. Rev. B* **9**, 2328 (1974).

[19] H. Müller-Krumbhaar, *Phys. Lett. A* **48**, 459 (1974); *ibid.* **50**, 27 (1974).

[20] K. Binder, *Ann. Phys.* **98**, 390 (1976).

[21] N. Nagao, *J. Phys. A* **18**, 1019 (1985).

[22] A. N. Kolmogorov, *Bull. Acad. Sci. U.S.S.R., Phys. Ser.* **3**, 355 (1937).

[23] W. A. Johnson and P. A. Mehl, *Trans. AIMME* **135**, 177 (1939).

[24] M. Avrami, *J. Chem. Phys.* **7**, 1103 (1939).

[25] Y. Ishibashi and Y. Takagi, *J. Phys. Soc. Jpn* **31**, 506 (1971).

[26] Y. Yamada, N. Hamaya, J. D. Axe, and S. M. Shapiro, *Phys. Rev. Lett.* **53**, 1665 (1984).

[27] K. Sekimoto, *J. Phys. Soc. Jpn* **53**, 2425 (1984); *Phys. Lett. A* **105**, 390 (1984).

[28] J. S. Langer and A. J. Schwartz, *Phys. Rev. A* **21**, 948 (1980).

[29] I. M. Lifshitz and V. V. Slyozov, *J. Phys. Chem. Solids* **19**, 35 (1961).

[30] C. Wagner, *Z. Electrochem.* **65**, 581 (1961).

[31] E. M. Lifshitz and L. P. Pitaevskii, *Physical Kinetics*, Landau and Lifshitz Course of Theoretical Physics, Vol. 10 (Pergamon, 1981).

[32] T. Miyazaki and M. Doi, *Mater. Sci. Eng.* **A110**, 175 (1989).

[33] G. Venzl, *Ber. Bunsenges. Phys. Chem.* **87**, 318 (1983).

[34] M. K. Chen and P. Voorhees, *Modell. Simul. Sci. Eng.* **1**, 591 (1993).

[35] P. W. Voorhees and M. E. Glicksman, *Acta Metall.* **32**, 2013 (1984).

[36] P. W. Voorhees and R. J. Schaefer, *Acta Metall.* **35**, 327 (1987).

[37] J. H. Yao, K. R. Elder, H. Guo, and M. Grant, *Physica A* **204**, 770 (1994).

[38] L. C. Brown *Acta Metall.* **37**, 71 (1989); *Scripta Metall. Mater.* **24**, 963 (1990).

[39] B. Giron, B. Meerson, and P. V. Sasorov, *Phys. Rev. E* **58**, 4213 (1998).

[40] B. E. Sundquist and R. A. Oriani, *J. Chem. Phys.* **36**, 2604 (1962).

[41] R. B. Heady and J. W. Cahn, *J. Phys.* **58**, 896 (1973).

[42] A. J. Schwartz, S. Krishnamurthy, and W. I. Goldburg, *Phys. Rev. Lett.* **21**, 1331 (1980).

[43] R. G. Howland, N.-C. Wong, and C. M. Knobler, *J. Chem. Phys.* **73**, 522 (1980).

[44] E. D. Siebert and C. M. Knobler, *Phys. Rev. Lett.* **52**, 1133 (1984).

[45] J. S. Langer and L. A. Turski, *Phys. Rev. A* **8** (1973) 3230.

[46] L. A. Turski and J. S. Langer, *Phys. Rev. A* **22** (1980) 2189.

[47] A. Onuki, *Physica A* **234**, 189 (1996); *J. Inter. Thermophysics* 471 (1998).

[48] D. Dahl and M. R. Moldover, *Phys. Rev. Lett.* **27**, 1421 (1971).

[49] J. S. Huang, W. I. Goldburg, and M. R. Moldover, *Phys. Rev. Lett.* **34**, 639 (1975).

[50] R. J. Speedy, *J. Phys. Chem.* **86**, 982 (1982).

[51] J. Leblond and M. Hareng,*J. Physique I* **45**, 373 (1984).

[52] H. Maris and S. Balibar, *Phys. Today*, **53**, 29 (2000).

[53] H. J. Maris, *Phys. Rev. Lett.* **66**, 45 (1991).

[54] H. J. Maris, *J. Low Temp. Phys.* **98**, 403 (1995).

[55] H. Lambaré, P. Roche, S. Balibar, O. A. Andreeva, C. Guthmann, K. O. Keshishev, E. Rolley, and H. J. Maris, *Eur. Phys. J. B* **2**, 502 (1998).

[56] V. A. Akulichev, *Ultrasonics* **26**, 8 (1986).

[57] J. Classen, C.-K. Su, and H. J. Maris, *Phys. Rev. Lett.* **77**, 2006 (1996).

[58] C. Zener, *Phys. Rev.* **53**, 90 (1938); R. H. Randall, F. C. Rose, and C. Zener, *ibid.* **56**, 343 (1939); C. Zener, *Proc. Roy. Soc.* (London) **52**, 152 (1940).

[59] L. D. Landau and E. M. Lifshitz, *Theory of Elasticity* (Pergamon, 1986), Chap. 5.

[60] M. A. Isakovich, *Zh. Eksp. Teor. Fiz.* **18**, 907 (1948).

[61] L. D. Landau and E. M. Lifshitz, *Fluid Mechanics* (Pergamon, 1959), Chap. 8.

[62] S. Komura, T. Miyazawa, T. Izuyama, and Y. Fukumoto, *J. Phys. Soc. Jpn* **59**, 101 (1990).

[63] Y. Hemar, R. Hocquart, and J. F. Palierne, *Europhys. Lett.* **42**, 253 (1998).

[64] R. Wanner, H. Meyer, and R. L. Mills, *J. Low Temp. Phys.* **13**, 337 (1973); R. Banke, X. Li, and H. Meyer, *Phys. Rev. B* **37**, 7337 (1988).

[65] D. B. Fenner, *J. Chem. Phys.* **88**, 2021 (1988).

[66] J. Bodensohn and P. Leiderer, *Phys. Rev. Lett.* **65**, 1368 (1990).

[67] A. Onuki, *Phys. Rev. E* **43**, 6740 (1991).

[68] A. Onuki, *J. Phys. Soc. Jpn* **60**, 1176 (1991).

[69] A. B. Wood, *A Textbook of Sound* (Bell, London, 1941); R. J. Urick, *J. Appl. Phys.* **18**, 983 (1947); P. L. Chambré, *J. Acoust. Soc. Am.* **263**, 29 (1954).

[70] Lord Rayleigh, *Phil. Mag.* **34**, 94 (1917); D. Y. Hsieh and M. S. Plesset, *Phys. Fluids* **4**, 970 (1961); A. Crespo, *ibid.* **12**, 2274 (1969).

[71] I. M. Lifshitz and Yu. Kagan, *Zh. Eksp. Teor. Fiz.* **62** (1972) 385 [*Sov. Phys. JETP* **35** (1972) 206].

[72] I. M. Lifshitz, V. M. Polesskii, and W. A. Khokholov, *Zh. Eksp. Teor. Fiz.* **74** (1978) 268 [*Sov. Phys. JETP* **47** (1978) 137].

[73] V. A. Mikheev, E. Ya. Rudaviskii, V. K. Chagovets, and F. A. Sheshin, *Sov. J. Low Temp. Phys.* **17** (1991) 233 [*Fiz. Nizk. Temp.* **17** (1991) 444].

[74] T. Satoh, M. Morishita, M. Ogata, and S. Katoh, *Phys. Rev. Lett.* **69** (1992) 335.

[75] L. D. Landau and E. M. Lifshitz, *Quantum Mechanics* (Pergamon, 1975).

[76] K. Krishnamurthy and R. Bansil, *Phys. Rev. Lett.* **50**, 2010 (1983).

[77] N. P. Balsara, C. Lin, and B. Hammouda, *Phys. Rev. Lett.* **77**, 3874 (1996).

[78] K. Binder, *Physica A* **243**, 118 (1995).

[79] A. Onuki, *J. Physique II* **2**, 1505 (1992).

[80] S. V. Iordanskii, *Zh. Exsp. Teor. Fiz.* **48**, 708 (1965).

[81] J. S. Langer and M. E. Fisher, *Phys. Rev. Lett.* **19**, 560 (1967).

[82] J. S. Langer and J. D. Reppy, in *Progress in Low Temperature Physics* **6**, ed. C. Gorter (North-Holland, Amsterdam 1970), p. 1.

[83] J. R. Clow and J. D. Reppy, *Phys. Rev. Lett.* **67**, 29 (1967); *Phys. Rev. A* **5**, 424(1972).

[84] R. J. Donnelly, *Quantized Vortices in He II* (Cambridge University Press, 1991).

[85] L. Kramer, *Phys. Rev.* **179**, 149 (1969); H. J. Mikeska, *ibid.* **179**, 166 (1969).

[86] M. Tinkham, *Introduction to Superconductivity* (McGraw-Hill, New York, 1975).

[87] M. Abramowiz and I. A. Stegun, *Handbook of Mathematical Functions* (U.S. Government Printing Office, Washington, DC, 1968).

10

Phase transition dynamics in solids

A variety of domain structures have been observed in metals undergoing (i) phase separation, or (ii) structural phase transitions [1]–[6]. In phase separation, a difference arises in the lattice constants of the two phases (lattice misfit). At a structural phase transition, anisotropically deformed domains of a stable low-temperature phase emerge in a quenched, metastable or unstable high-temperature phase. As a consequence, elastic strains are induced which radically influence the phase transition behavior. Here we will present Ginzburg–Landau theories for phenomena (i) and (ii) under the coherent condition [1], in which the lattice planes are continuous through the interface without any coherency loss due to dislocations, as illustrated in Fig. 10.1(a). In the incoherent case, however, dislocations are accumulated at the interface regions, and the resultant elastic effects have not yet been well investigated.[1]

Among a number of important topics, here we cite examples of research on phase separation in binary alloys. In particular, experiments on Ni-base alloys are noteworthy [7]–[16]. (i) Figure 10.2 shows Ni_3Al (γ') cuboidal domains (precipitates) with the ordered $L1_2$ structure (illustrated in Fig. 3.10) in a disordered fcc Ni–Al alloy matrix [8]. Here, initially spherical domains changed their shapes into cuboids with facets in $\{100\}$ planes as they grew. (ii) As can be seen in Fig. 10.3, at a very late stage cuboids can be seen sometimes to split into two plates or eight cuboids, despite an increase in the surface energy [3], [5a], [9, 15]. (iii) Figure 10.4 shows the time evolution of Ni_4Mo domains in a Ni–16.3 at.% Mo alloy [10], where harder cuboids with a larger shear modulus C_{44} are encased in a softer matrix. With increasing aging time, the mean domain size $\bar{r}(t)$ increased but the size distribution became narrower. In Fig. 10.5 these features can be seen in the time dependence of the mean domain size $\bar{r}(t)$ in (a) and the standard deviation $\sigma(t)$ of the size distribution in (b). The coarsening virtually stopped with prolonged aging ($t \gtrsim 10^6$ s), as shown in (a). Such abnormal slowing down occurs for high solute contents under strong elastic constraints [9], [15]–[17], while the usual growth law $\bar{r}(t) \propto t^{1/3}$ has been observed for small volume fractions of precipitates and/or relatively short aging times. (iv) Application during aging of stretching or compression in the [100] direction in cubic solids is known to produce cylindrical or lamellar domains in late stages [18]–[21]. Figure 10.6 gives examples of the morphology of γ' precipitates in the absence of an applied stress and under uniaxial strain in an Ni–15at.%Al alloy [19].

[1] Grain boundaries in polycrystals are also incoherent.

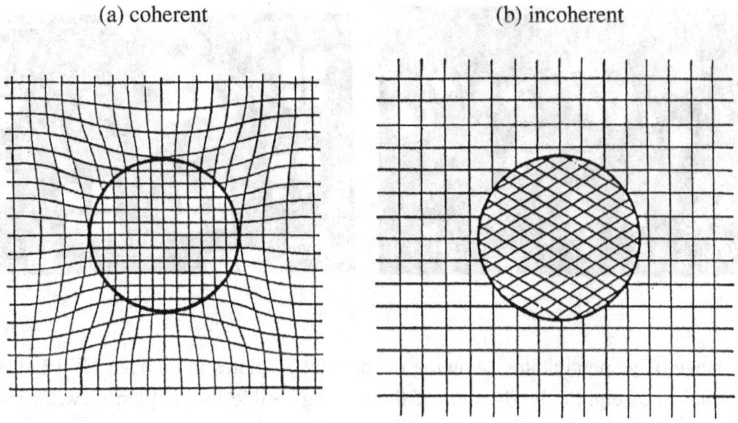

Fig. 10.1. The interface condition is coherent in (a) and incoherent in (b) in a two-phase state of a binary solid. In both cases the lattice constants of the two phases are different.

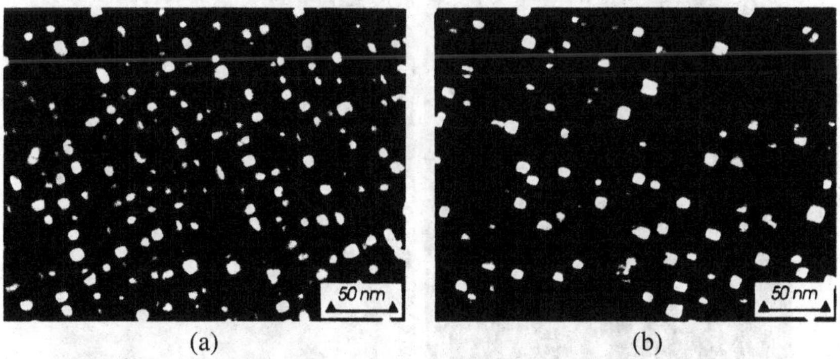

Fig. 10.2. Dark-field transmission electron micrographs of Ni–Al alloys taken using (100) γ' precipitate superlattice reflections: (a) 6.35 wt% Al aged for 92.5 h, the volume fraction of γ' being 0.13; (b) 5.78 wt% Al aged for 54 h, the volume fraction of γ' being 0.034 [8].

A short summary of theories on phase separation in binary alloys is as follows. (i) Eshelby calculated the elastic energy of ellipsoidal domains coherently embedded in a solid matrix [22]. However, an energetic theory, as such, is not suitable for describing dynamical processes in which the domain shape changes with time. (ii) Cahn presented a Ginzburg–Landau theory for the simplest case of isotropic elasticity with constant elastic moduli [23], predicting a downward shift of the coexistence curve in the temperature–concentration phase diagram after elimination of the elastic field. (iii) For cubic crystals with constant elastic moduli [24], Cahn derived a dipolar interaction, bilinear with respect to the concentration fluctuations. More refined or generalized derivations have subsequently been presented [2], [25]–[29]. This dipolar interaction is long-range and

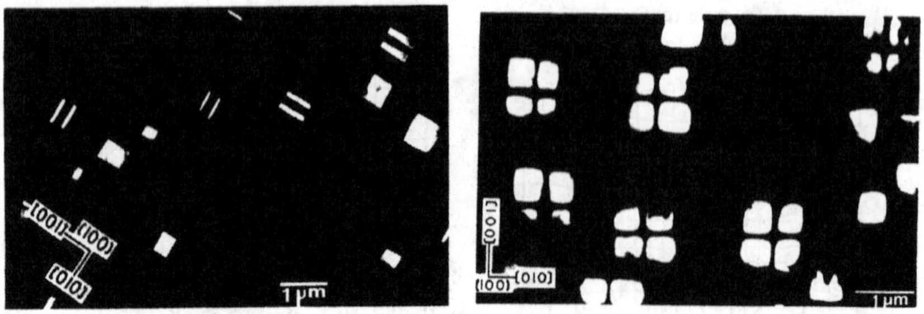

(a) (b)

Fig. 10.3. Image of γ' precipitates, showing (a) doublet γ' plates in an aged Ni–12 at.% Al alloy and (b) assemblies of eight cuboids, four of them being visible in this plan view, in an aged Ni–12 at.% Si alloy [9].

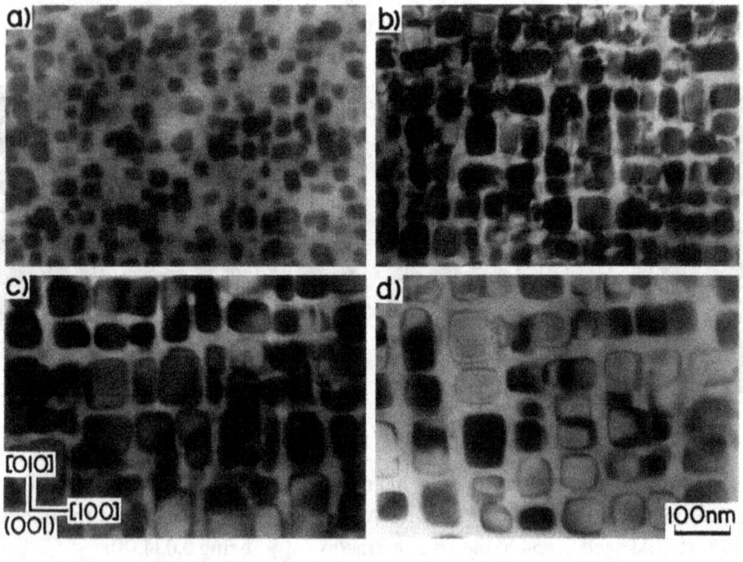

Fig. 10.4. Time evolution of Ni_4Mo precipitates in an Ni–16.3 at.%Mo alloy aged at 973 K for (a) 12.8 ks, (b) 864 ks, (c) 2.6 Ms, and (d) 5.2 Ms [10]. The domain shapes here closely resemble those in the simulation in Fig. 10.18.

angle-dependent, so it is minimized for particular shapes and configurations of precipitates. (iv) Unique effects arise when the two phases have different elastic moduli. Ardell *et al.* [7] calculated the interaction among spherical domains whose shear modulus is slightly different from that of the matrix, but this interaction loses its meaning once domains change their shape from sphericity. Johnson and Voorhees [3] and Cahn [23] predicted that a growing precipitate which is softer than the matrix should be deformed from a sphere into an ellipsoid as its radius exceeds a critical size R_E. (v) The present author extended

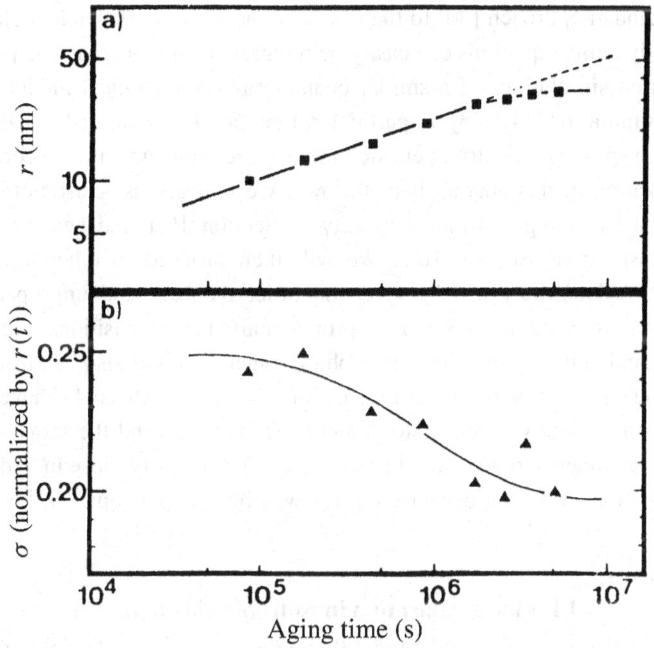

Fig. 10.5. (a) The mean particle radius $\bar{r}(t)$ and (b) the standard deviation $\sigma(t)$ vs time for Ni_4Mo particles in Ni–16.3 at.%Mo alloy aged at 973 K [10].

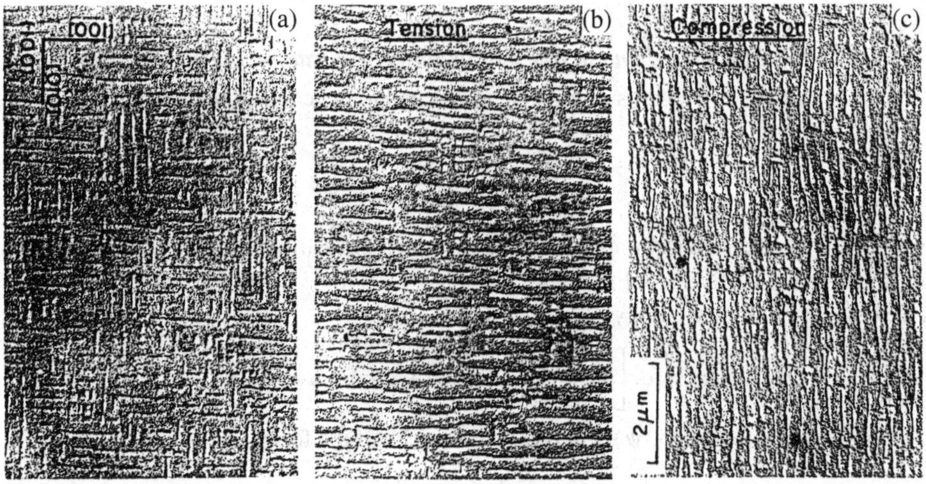

Fig. 10.6. Microphotographs of replicas taken from a (100) surface of an Ni–15 at.% Al alloy aged at 1023 K, with (a) with no external stress, (b) in tension, and (c) in compression [19]. The stress applied in (b) and (c) was 147 MPa, and there was little appreciable stress effect on the domain shapes in the early stages.

a Ginzburg–Landau approach [30] to the case of concentration-dependent elastic moduli. The resultant dynamic equations can easily be integrated using a computer [31]–[40]. (vi) We also mention simulations of a similar coarse-grained dynamical model [28, 41] and Monte Carlo simulations [42, 43]. In particular, Lee [5d, 44] examined shape changes, including domain splitting, for strong elastic inhomogeneity, using a microscopic approach.

The organization of this chapter is as follows. We will discuss elastic effects in phase separation, first assuming isotropic elasticity in Section 10.1 and next assuming cubic anisotropic elasticity in Section 10.2. We will then proceed to other topics. Reviews and some new calculations will be given on order–disorder and improper martensitic phase transitions in Section 10.3 and on proper martensitic transitions in Section 10.4. A Ginzburg–Landau theory of Jahn–Teller phase transitions will also be presented in Section 10.4. In the case of structural phase transitions experimentalists [45] have posed many problems, which are not well understood and are mostly beyond the scope of this book. We will also treat macroscopic instabilities in solids, particularly those in hydrogen–metal systems, in Section 10.5. Surface instabilities will be the last topic, to be discussed in Section 10.6.

10.1 Phase separation in isotropic elastic theory

We will describe binary alloys using model B coupled to isotropic elasticity via the Vegard law. For simplicity, we will neglect order–disorder phase transitions. Particular attention will be paid to the effect of a composition-dependent shear modulus. We will derive long-range interactions among the composition fluctuations by eliminating the elastic field.[2]

10.1.1 Ginzburg–Landau free energy for concentration and elastic field

The order parameter ψ and its average M are related to the composition c as

$$\begin{aligned} \psi &= c - c_{\rm c}, \\ M &= \langle\psi\rangle = \langle c\rangle - c_{\rm c}, \end{aligned} \tag{10.1.1}$$

where $c_{\rm c}$ is a critical concentration. The elastic field $\boldsymbol{u}$ is the displacement vector measured from an isotropic, disordered reference state at $c = c_{\rm c}$ or $\psi = 0$. In binary solids, ψ is coupled to $\boldsymbol{u}$ in the free energy $\mathcal{H}$ as

$$\mathcal{H} = \int d\boldsymbol{r}\bigg[f_0(\psi) + \frac{C}{2}|\nabla\psi|^2 + \alpha\psi\nabla\cdot\boldsymbol{u} + f_{\rm el}(\boldsymbol{u})\bigg]. \tag{10.1.2}$$

The free-energy density $f_0(\psi)$ will be assumed to be of the form,

$$f_0(\psi) = \frac{1}{2}r_0\psi^2 + \frac{1}{4}u_0\psi^4, \tag{10.1.3}$$

u_0 and C being constants. For T close to a mean field critical temperature $T_{\rm c0}$, the parameter r_0 is expressed as

$$r_0 = a_0(T - T_{\rm c0}). \tag{10.1.4}$$

[2] It is worth noting that this procedure is analogous to that of deriving the attractive interaction between electrons mediated by acoustic phonons in metals, which leads to superconductivity at low temperatures [46].

This Landau form can be accurate only for small ψ or in the weak segregation case. If we include the strong segregation case in our theory, we should use the Bragg–Williams free-energy density in Section 3.3,

$$f_0(\psi) = v_0^{-1}T\big[c \ln c + (1-c)\ln(1-c)\big] - 2v_0^{-1}T_{c0}c^2, \tag{10.1.5}$$

where v_0 is the microscopic cell volume and $c_c = 1/2$. The expansion of the above expression with respect to $\psi = c - 1/2$ yields (10.1.3) with $a_0 = 2/v_0$ and $u_0 = 4T_{c0}/3v_0$. The elastic energy f_{el} consists of the contributions from volume dilation and shear deformation:

$$f_{el}(\boldsymbol{u}) = \frac{1}{2}K|\nabla\cdot\boldsymbol{u}|^2 + \frac{1}{4}\mu\sum_{ij} e_{ij}^2, \tag{10.1.6}$$

where

$$e_{ij} = \nabla_j u_i + \nabla_i u_j - \frac{2}{d}\delta_{ij}\nabla\cdot\boldsymbol{u} \tag{10.1.7}$$

is the traceless, symmetrized strain tensor and will be called the shear strain tensor. In this section, $\nabla_i \equiv \partial/\partial x_i$, and i, j, k, ℓ stand for x, y, z in 3D (x, y in 2D). The space dimensionality d is either two or three. We assume that the bulk modulus K is a constant, but the shear modulus μ depends on ψ as

$$\mu = \mu_0 + \mu_1\psi, \tag{10.1.8}$$

where μ_0 and μ_1 are constants. Here $\mu_1 > 0$ if c is the concentration of the harder component. Moreover, $\mu_1 \ll \mu_0$ will hold if the shear moduli μ_A and μ_B of pure metals A and B are nearly the same, but $\mu_1 \sim \mu_0$ should follow in the case $|\mu_A - \mu_B| \sim \mu_A$ (or μ_B).

The elastic stress tensor $\overleftrightarrow{\sigma}_{ij}$ is calculated as follows. Against a small incremental displacement $u_i \to u_i + \delta u_i$ at fixed ψ, the change of $\mathcal{H}$ should be written as

$$\delta\mathcal{H} = \int d\boldsymbol{r}\sum_{ij}\sigma_{ij}\nabla_i\delta u_j, \tag{10.1.9}$$

leading to the expression,

$$\sigma_{ij} = (K\nabla\cdot\boldsymbol{u} + \alpha\psi)\delta_{ij} + \mu e_{ij}. \tag{10.1.10}$$

The elastic free-energy density $f_{el}(\boldsymbol{u})$ can then be expressed as

$$f_{el}(\boldsymbol{u}) = \frac{1}{2}\sum_{ij}\sigma_{ij}\nabla_j u_i - \frac{1}{2}\alpha\psi\nabla\cdot\boldsymbol{u}. \tag{10.1.11}$$

On long timescales of the concentration fluctuations, the elastic field instantaneously relaxes and adjusts to a given concentration field. This is the condition of mechanical equilibrium in the bulk region,

$$\left(\frac{\delta}{\delta u_i}\mathcal{H}\right)_{\text{bulk}} = -\sum_j \nabla_j\sigma_{ij} = 0. \tag{10.1.12}$$

The $\boldsymbol{u}$ is thus determined as a functional of ψ under each given boundary condition. In general, an average homogeneous strain can be created inside the solid as

$$\langle \nabla_j u_i \rangle = A_{ij}. \tag{10.1.13}$$

The displacement may be divided into the average and the deviation as

$$u_i = \sum_j A_{ij} x_j + \delta u_i. \tag{10.1.14}$$

Applying an external stress or clamping the boundary?

(i) A natural boundary condition will be to apply a constant external stress tensor $\overleftrightarrow{\sigma}_{\text{ex}}$ at the boundary; particularly, the stress-free boundary is given by $\overleftrightarrow{\sigma}_{\text{ex}} = 0$. For simplicity, we assume no mass exchange between the solid and the outer region, neglecting melting and crystal growth. Then we should minimize a generalized Gibbs free energy defined by [30]

$$\begin{aligned} \mathcal{H}' &= \mathcal{H} - \int da (\boldsymbol{n} \cdot \overleftrightarrow{\sigma}_{\text{ex}} \cdot \boldsymbol{u}) \\ &= \mathcal{H} - \int d\boldsymbol{r} \sum_{ij} (\overleftrightarrow{\sigma}_{\text{ex}})_{ij} \nabla_i u_j, \end{aligned} \tag{10.1.15}$$

where $\int da(\cdots)$ is the surface integral on the boundary and $\boldsymbol{n}$ is the outward normal unit vector. At fixed ψ, $\mathcal{H}'$ is minimized under the bulk condition (10.1.12) and the boundary condition,

$$\overleftrightarrow{\sigma} \cdot \boldsymbol{n} = \overleftrightarrow{\sigma}_{\text{ex}} \cdot \boldsymbol{n}. \tag{10.1.16}$$

(ii) We may alternatively clamp the solid such that $\delta \boldsymbol{u} = \boldsymbol{0}$ at the boundary. In this case $\mathcal{H}$ is minimized with respect to variations of $\boldsymbol{u}$ in the bulk region and at the boundary from (10.1.9). We note that the concentration fluctuations much shorter than macroscopic sizes are insensitive to the boundary condition. Unless we are interested in macroscopic shape changes and surface undulations, we may adopt the clamped boundary condition or even the periodic boundary condition (as in usual simulations) instead of the condition of constant applied stress. Then, using (10.1.11) and (10.1.12), we rewrite the free energy as

$$\mathcal{H} = \int d\boldsymbol{r} \left[f_0(\psi) + \frac{C}{2} |\nabla \psi|^2 + \frac{1}{2} \alpha \psi \nabla \cdot \boldsymbol{u} + \frac{1}{2} \sum_{ij} \sigma_{ij} A_{ij} \right], \tag{10.1.17}$$

under both the clamped and the periodic boundary condition. The last term in the brackets is important in the presence of elastic inhomogeneity.

Vegard law

To explain the origin of the bilinear coupling ($\propto \alpha$) in (10.1.2), let us consider an isotropic, one-phase state under the stress-free boundary condition. The average elastic deformation is isotropic as

$$\langle \nabla_i u_j \rangle = -\frac{\alpha}{dK} M \delta_{ij}, \tag{10.1.18}$$

in terms of the average order parameter M. Here the effect of the thermal expansion [47] is neglected.[3] The volume V or the lattice constant a of the system then change by

$$\delta V = -V\frac{\alpha}{K}M, \quad \delta a = -a\frac{\alpha}{dK}M. \tag{10.1.19}$$

These deviations are measured from those at $\langle c\rangle = c_c$. The lattice expansion coefficient is defined as [23]

$$\eta = \frac{d}{dc}\ln a = -\frac{\alpha}{dK}. \tag{10.1.20}$$

In real binary alloys, the lattice constant in one-phase states may be approximated as a linear function of $\langle c\rangle$ empirically in a relatively wide concentration range, which is often called the *Vegard law* [48]. In phase separation, nonvanishing α leads to a *lattice misfit* between the two phases. For precipitates in a matrix, the lattice misfit or mismatch is often defined by

$$\epsilon = (a_p - a_m)/a_m \cong \eta\Delta c, \tag{10.1.21}$$

where a_p and a_m are the lattice constants of the unconstrained (stress-free) precipitate and matrix phases, respectively, and Δc is the concentration difference between the two phases. The mismatch cannot be very large in the coherent case. As an extreme case [5a], ϵ is very small (~ 0.0008) for $L1_2$ structures in Al–Li as stated below (3.3.32). It is also known that the addition of a third component can make ϵ very small, as in the experiment in Fig. 9.9. For these cases the elastic effects become small.

10.1.2 Elimination of the elastic field for small μ_1

A general procedure of eliminating the elastic field will be given in Appendix 10A. We may calculate $\boldsymbol{u}$ by treating μ_1 as a small expansion parameter. This scheme is justified for

$$|\mu_1\psi| \ll L_0, \tag{10.1.22}$$

where L_0 is the elastic modulus for the longitudinal displacement (or sound),

$$L_0 = K + \left(2 - \frac{2}{d}\right)\mu_0. \tag{10.1.23}$$

We express $\delta\boldsymbol{u}$ in terms of $\delta\psi = \psi - M$ solving the mechanical equilibrium condition (10.1.12):

$$\alpha\nabla_i\psi + (L_0 - \mu_0)\nabla_i g + \mu_0\nabla^2 \boldsymbol{u}_i + \mu_1\sum_j \nabla_j(\psi e_{ij}) = 0, \tag{10.1.24}$$

[3] The lattice constant is dependent on the temperature as well as the composition. In this chapter we assume rapid thermal equilibration and neglect temperature inhomogeneities.

where $g \equiv \nabla \cdot \delta \boldsymbol{u}$. Taking the divergence of the above vector equation yields

$$\nabla^2[L_0 g + \alpha\psi] + \mu_1 \sum_{ij} \nabla_i \nabla_j (\psi e_{ij}) = 0. \tag{10.1.25}$$

For $\mu_1 = 0$ the zeroth-order solution is calculated as

$$\delta \boldsymbol{u}^{(0)} = -(\alpha/L_0)\nabla w, \tag{10.1.26}$$

where w is a potential determined by the Laplace equation,

$$\nabla^2 w = \delta\psi \quad \text{or} \quad w = \frac{1}{\nabla^2}\delta\psi, \tag{10.1.27}$$

where $1/\nabla^2$ is the inverse operator of ∇^2. The Fourier component of w is related to that of ψ as $w_{\boldsymbol{k}} = -k^{-2}\psi_{\boldsymbol{k}}$. The corresponding dilation strain is

$$g^{(0)} = \nabla \cdot \delta \boldsymbol{u}^{(0)} = -(\alpha/L_0)\delta\psi. \tag{10.1.28}$$

Note that L_0 is the elastic modulus for plane-wave fluctuations, whereas K is that for isotropic dilation as in (10.1.18). The zeroth-order shear strain is

$$e_{ij}^{(0)} = S_{ij} - \frac{2\alpha}{L_0}\left(\nabla_i \nabla_j - \frac{1}{d}\delta_{ij}\nabla^2\right) w, \tag{10.1.29}$$

where

$$S_{ij} = A_{ij} + A_{ji} - \frac{2}{d}\delta_{ij}\sum_{\ell} A_{\ell\ell} \tag{10.1.30}$$

is the traceless, symmetric average strain.

The first-order correction of the dilation strain is readily calculated from (10.1.25) in the form,

$$g^{(1)} = -\frac{\mu_1}{L_0}\frac{1}{\nabla^2}\sum_{ij}\nabla_i\nabla_j(\psi e_{ij}^{(0)}). \tag{10.1.31}$$

We substitute (10.1.28), (10.1.29), and (10.1.31) into (10.1.17). The free energy up to first order in μ_1 is then of the form,

$$\mathcal{H} = \int d\boldsymbol{r}\left[f(\psi) + \frac{C}{2}|\nabla\psi|^2\right] + \mathcal{H}_{\text{inh}} + \mathcal{H}_{\text{ex}}, \tag{10.1.32}$$

where the constant terms are not written explicitly. The free-energy density $f(\psi)$ includes the zeroth-order elastic contribution:

$$\begin{aligned} f(\psi) &= f_0(\psi) - (\alpha^2/2L_0)\psi^2 \\ &= \frac{1}{2}r\psi^2 + \frac{1}{4}u_0\psi^4. \end{aligned} \tag{10.1.33}$$

In the second line we have used (10.1.3). The temperature coefficient is expressed as

$$r = r_0 - \alpha^2/L_0 = a_0(T - T_c), \tag{10.1.34}$$

where $T_c = T_{c0} + \alpha^2/L_0 a_0$ is the so-called coherent critical temperature.

Cahn's theory

Cahn treated the simplest case of homogeneous moduli assuming isotropic elasticity [23]. In his theory, the displacement $\boldsymbol{u}$ is measured from the stress-free, isotropic, homogeneous state with a given concentration c, whereas our reference elastic state is that at $M = 0$ or $\bar{c} = c_c$. Thus, the dilation strain in his definition is shifted by $-(\alpha/K)\psi$ from ours. His *chemical* free-energy density $f_{\rm chem}$ is related to $f_0(\psi)$ in (10.1.2) by

$$f_{chem}(\psi) = f_0(\psi) - \frac{\alpha^2}{2K}\psi^2. \tag{10.1.35}$$

Using $f(\psi)$ in (10.1.33) we obtain

$$f(\psi) - f_{\rm chem}(\psi) = \frac{1}{2}\alpha^2\left(\frac{1}{K} - \frac{1}{L_0}\right)\psi^2, \tag{10.1.36}$$

which can also be written as $\eta^2 E(1-\nu)^{-1}\psi^2$ [23] in terms of the Young's modulus $E = 9K\mu/(3K+\mu)$ and the Poisson ratio $\nu = (3K - 2\mu)/2(3K+\mu)$ in 3D [47] with η being defined in (10.1.20). In Cahn's original theory, therefore, the elastic field only serves to shift the coexistence curve downwards by $\eta^2 E/[2(1-\nu)a_0]$ from the *chemical* coexistence curve determined by $f_{\rm chem}(\psi)$. The point $r = M = 0$ and the curve $r + u_0 M^2 = 0$ are called the coherent critical point and coexistence curve, respectively. See [49, 50] for experiments. More discussion is given for cubic solids in Subsection 10.2.1.

Elastic inhomogeneity interaction $\mathcal{H}_{\rm inh}$

The $\mathcal{H}_{\rm inh}$ in (10.1.32) arises from the elastic inhomogeneity (EI) and has a third-order dependence on ψ as

$$\mathcal{H}_{\rm inh} = g_{\rm E}\int d\boldsymbol{r}\psi\hat{Q}, \tag{10.1.37}$$

where

$$\hat{Q} = \sum_{ij}\left(\nabla_i\nabla_j w - \frac{1}{d}\delta_{ij}\nabla^2 w\right)^2 \tag{10.1.38}$$

represents the degree of anisotropic deformation with w being defined by (10.1.27). The coefficient $g_{\rm E}$ is given by

$$\begin{aligned} g_{\rm E} &= \mu_1\alpha^2/L_0^2 \\ &= 9\mu_1\eta^2/(1+4\mu_0/3K)^2, \end{aligned} \tag{10.1.39}$$

where the second line is the 3D expression. Hereafter we will neglect the higher-order interactions with respect to μ_1. Because of the relation, $\int d\boldsymbol{r}\hat{Q} = (1 - 1/d)\int d\boldsymbol{r}(\delta\psi)^2$, we may express $\mathcal{H}_{\rm inh}$ as[4]

$$\begin{aligned}\mathcal{H}_{\rm inh} &= g_{\rm E}\int d\boldsymbol{r}\left[\left(1-\frac{1}{d}\right)M(\delta\psi)^2 + \delta\psi\hat{Q}\right] \\ &= g_{\rm E}\int d\boldsymbol{r}\left[\left(1-\frac{1}{d}\right)M(\delta\psi)^2 - \frac{1}{d}(\delta\psi)^3 + \sum_{ij}\delta\psi(\nabla_i\nabla_j w)^2\right].\end{aligned} \tag{10.1.40}$$

In the first line, the first term ($\propto M(\delta\psi)^2$) in the brackets can be incorporated into the bilinear elastic term in $f(\psi)$ in (10.1.33) by replacement, $-(\alpha^2/2L_0)\psi^2 \to -(\alpha^2/2\langle L\rangle)\psi^2$, where $\langle L\rangle = L_0 + (2 - 2/d)\mu_1 M$ is the longitudinal modulus at $\psi = M$. If we assume (10.1.33) and focus our attention on the bilinear order terms, we obtain the spinodal curve of isotropic, homogeneous one-phase states in the form,

$$r + 3u_0M^2 + \left(2 - \frac{2}{d}\right)g_{\rm E}M = 0, \tag{10.1.41}$$

whose maximum point in the $r - M$ plane is given by $r = (1 - 1/d)^2 g_{\rm E}^2/3u_0$ and $M = -(1 - 1/d)g_{\rm E}/3u_0$.

Dipolar interaction $\mathcal{H}_{\rm ex}$ arising from external stress

Anisotropic deformations (10.1.13) give rise to a long-range dipolar interaction,

$$\mathcal{H}_{\rm ex} = -\frac{1}{2}g_{\rm ex}\int d\boldsymbol{r}\sum_{ij}S_{ij}(\nabla_i\psi)(\nabla_j w), \tag{10.1.42}$$

with

$$g_{\rm ex} = -2\mu_1\alpha/L_0. \tag{10.1.43}$$

In terms of the Fourier transformation $\psi_{\boldsymbol{k}}$ of $\psi(\boldsymbol{r})$ this interaction is expressed as

$$\mathcal{H}_{\rm ex} = \frac{1}{2}g_{\rm ex}\int_{\boldsymbol{k}}\sum_{ij}S_{ij}\hat{k}_i\hat{k}_j|\psi_{\boldsymbol{k}}|^2, \tag{10.1.44}$$

where $\hat{\boldsymbol{k}} = k^{-1}\boldsymbol{k}$ denotes the direction of the wave vector. For example, we apply a uniaxial deformation, for which the average strain tensor A_{ij} in (10.1.3) is expressed as $A_{xx} = \lambda_\parallel$, $A_{jj} = \lambda_\perp$ $(j \neq x)$, and $A_{ij} = 0$ $(i \neq j)$. Then this interaction becomes of the same form as the dipolar interaction in uniaxial ferromagnets [51, 52] or ferroelectrics [53],

$$\mathcal{H}_{\rm ex} = g_{\rm ex}(\lambda_\parallel - \lambda_\perp)\int_{\boldsymbol{k}}\left(\hat{k}_x^2 - \frac{1}{d}\right)|\psi_{\boldsymbol{k}}|^2. \tag{10.1.45}$$

[4] The last term in the brackets in the second line is of the same form as $\mathcal{H}_{\rm el}^{(3)}$ in (7.2.19) for gels.

Because $g_{ex} \propto \mu_1$, the concentration fluctuations can be influenced by an externally applied strain only in the presence of elastic inhomogeneity (EI). As in Fig. 10.6, several groups have performed phase separation experiments under uniaxial stress σ_A along the [100] direction in cubic solids [18]–[20]. They observed lamellar and cylindrical domain structures depending on whether the deformation is stretching or compression, respectively.

Weak and strong elastic inhomogeneity

As we approach the coherent critical point ($r = M = 0$), the third-order interaction $\mathcal{H}_{inh}$ alters the concentration fluctuations drastically (even in one-phase states). In this sense $\mathcal{H}_{inh}$ is relevant however small μ_1 is, as well as $\mathcal{H}_{ex}$. To see this at $M = 0$, we estimate the magnitude of the fluctuations of ψ as $(|r|/u_0)^{1/2}$ using the second line of (10.1.33) and compare $r\psi^2/2$ in (10.1.33) and $g_E\psi\hat{Q}$ in (10.1.37). The relative magnitude of these two terms is represented by the following dimensionless parameter,

$$g_E^* = g_E/\sqrt{|r|u_0}, \tag{10.1.46}$$

which grows as $|r|^{-1/2}$ as $r \to 0$. For $g_E^* \ll 1$ we are in the regime of weak elastic inhomogeneity (WEI), where the effects of EI can be apparent only in late-stage phase separation. For $g_E^* \gtrsim 1$, on the other hand, we are in the regime of strong elastic inhomogeneity (SEI), where even the thermal fluctuations on the scale of the correlation length are distinctly soft or hard. Using (10.1.5), (10.1.33), and (10.1.39), we rewrite the condition of SEI in terms of observable quantities as

$$v_0|\mu_1|\eta^2/T_c \gtrsim |T/T_c - 1|^{1/2}, \tag{10.1.47}$$

where v_0 is the volume of a unit cell and η is the lattice expansion coefficient in (10.1.20). Below T_c or in phase separation we also have

$$g_E^* \sim v_0\eta^2|\Delta\mu|/(T_c|\Delta c|^2), \tag{10.1.48}$$

in terms of the shear modulus difference $\Delta\mu$ and the concentration difference Δc. Alternatively, we may introduce a characteristic reduced temperature and average order parameter by

$$r_E = g_E^2/u_0, \quad M_E = g_E/u_0. \tag{10.1.49}$$

In the SEI regime we require $|r| \lesssim r_E$ and $|M| \lesssim M_E$. For $|M| \gg M_E$ (even at $\tau = 0$) the solid is in the WEI regime.

10.1.3 A nearly spherical domain

Let us suppose an isolated nearly spherical precipitate in a weakly metastable matrix in the WEI regime without externally applied stress in 3D [34], [54]–[56]. The order parameter is equal to ψ_0 within the domain and to M outside it. If we assume the free-energy density in the second line of (10.1.33), we have $\psi_0 \cong \Delta c/2$ and $c_0 \cong -\Delta c/2$ where $\Delta c = \psi_0 - M \cong 2(|r|/u_0)^{1/2}$. We are interested in how the shear modulus difference,

$$\Delta\mu = \mu_1 \Delta c, \tag{10.1.50}$$

can change the domain free energy. The typical strain around the domain is given by

$$e_0 = \alpha \Delta c / L_0. \tag{10.1.51}$$

From (10.1.39) we notice the relation,

$$g_{\mathrm{E}}(\Delta c)^3 = (\Delta\mu) e_0^2. \tag{10.1.52}$$

First, assuming that the critical domain is spherical, we introduce an effective supersaturation Δ_{eff} in the presence of EI. Because (10.1.27) is solved as $w = (\Delta c) r^2/6 + \cdots$ within a sphere, $\mathcal{H}_{\mathrm{inh}}$ in (10.1.40) is calculated as $\mathcal{H}_{\mathrm{inh}} = -(4\pi/9) R^3 g_{\mathrm{E}}(\Delta c)^3 + \cdots$. Note that $\delta\psi = 0$ outside the domain if the concentration depletion is neglected. The free-energy difference μ_{eff} in (9.1.1) thus consists of two terms as

$$\mu_{\mathrm{eff}} = 2|r|(\Delta c)^2 \Delta_{\mathrm{eff}} = 2|r|(\Delta c)^2 \Delta - \frac{1}{3} g_{\mathrm{E}} (\Delta c)^3, \tag{10.1.53}$$

where $\psi_{\mathrm{eq}}^2 \kappa^2$ in (9.1.5) is replaced by $(\Delta c)^2 |r|/4T$ in the present notation. We define the effective supersaturation [34],

$$\Delta_{\mathrm{eff}} = \Delta - \frac{1}{3} g_{\mathrm{E}}^* \theta_{\mathrm{sh}}, \tag{10.1.54}$$

including the first correction from EI. Here $\theta_{\mathrm{sh}} = 1$ for the hard domain case and $\theta_{\mathrm{sh}} = -1$ for the soft domain case, g_{E}^* being taken to be positive. Note that Δ is determined for the free-energy density (10.1.33) or for the coherent phase diagram. In the case of soft precipitates ($\theta_{\mathrm{sh}} = -1$), however, the critical domains take compressed pancake shapes if $R_{\mathrm{E}} \ll R_{\mathrm{c}}$ or $g_{\mathrm{E}}^* \gg \Delta$.

Second, we consider the free-energy increase due to deviations from sphericity ($\ell \neq 0$). We represent the domain surface by $r = R + \sum_{\ell m} \delta_{\ell m} Y_{\ell m}(\theta, \varphi)$ using the spherical harmonics $Y_{\ell m}$. Then [34, 56],

$$\Delta\mathcal{H} = \sum_{\ell=1}^{\infty} \sum_{m=\ell}^{\ell} \left[\frac{\sigma}{2}(\ell^2 + \ell - 2) + g_{\mathrm{E}}(\Delta c)^3 R \frac{\ell(\ell-1)}{2\ell+1} \right] |\delta_{\ell m}|^2. \tag{10.1.55}$$

The first term is the surface tension term, σ being the surface tension. Interestingly, the second term in (10.1.55) is negative for the softer domain case, which favors shape changes from a sphere [29]. Comparing the two terms at $\ell = 2$, we find a critical radius against shape deformations,

$$R_{\mathrm{E}} = 5\sigma / |g_{\mathrm{E}}(\Delta c)^3| = 5\sigma / (|\Delta\mu| e_0^2). \tag{10.1.56}$$

For $R > R_{\mathrm{E}}$ the spherical shape is unstable against deformations (even without external loads), leading to anisotropic shapes with larger surface areas. This instability occurs when the elastic energy ($\sim (\Delta\mu)e_0^2R^3$) from EI and the surface energy ($\sim \sigma R^2$) become of the same order. Because $\sigma \sim (\Delta c)^2|r|\xi$, we find

$$R_{\mathrm{E}}/\xi \sim 1/g_{\mathrm{E}}^* \gg 1, \tag{10.1.57}$$

where $\xi = (C/|r|)^{1/2}$ is the thermal correlation length.

10.1.4 An ellipsoidal domain

Eshelby calculated the elastic energy of an ellipsoidal domain (coherent inclusion) assuming isotropic elasticity [22]. We will reproduce his results in the WEI regime. The domain shape is represented by $\sum_{j=1}^{3} x_j^2/a_j^2 = 1$, within which $\psi = \psi_0$ and outside of which $\psi = M$. As will be shown in Appendix 10B, w takes a simple form inside the ellipsoid,

$$w = \frac{1}{2}\Delta c \sum_j N_j x_j^2 + \text{const.}, \tag{10.1.58}$$

in terms of the depolarization factors $N_j (> 0)$. They satisfy $\sum_j N_j = 1$ and are equal to $1/3$ for spheres. The zeroth-order strain in (10.1.29) is of the form

$$e_{ij}^{(0)} = S_{ij} - 2e_0\left(N_i - \frac{1}{3}\right)\delta_{ij} \tag{10.1.59}$$

within the ellipsoid. If the distance r from the center of the ellipsoid greatly exceeds the domain size (i.e., at a point some way outside of the domain), w behaves as a Coulombic potential ($\propto 1/r$) and

$$e_{ij}^{(0)} \cong S_{ij} + \frac{1}{2\pi} V_{\mathrm{e}} e_0 \nabla_i \nabla_j \frac{1}{r}, \tag{10.1.60}$$

where $V_{\mathrm{e}} = (4\pi/3)a_1a_2a_3$ is the volume of the ellipsoid. It is well known that the strain field within an ellipsoidal coherent inclusion is homogeneous for isotropic elasticity.

Under an externally applied stress, $\mathcal{H}_{\mathrm{ex}}$ in (10.1.42) may readily be calculated, because $\nabla_i\nabla_j w = \Delta c N_j \delta_{ij}$ inside the ellipsoid and $\delta\psi = \psi - \bar{\psi} = 0$ outside it. For simplicity, we assume a spheroid with $a_1 = a$ and $a_2 = a_3 = b$ with the x axis being taken along the symmetry axis. Then,

$$\mathcal{H}_{\mathrm{ex}} = -\frac{3}{2} V_{\mathrm{e}} (\Delta\mu) e_0 S_{xx}\left(N_x - \frac{1}{3}\right). \tag{10.1.61}$$

The $N_x (= N_1)$ decreases from 1 to 0 as a/b increases form 0 to ∞, as shown in Fig. 10.7. If $\alpha\mu_1 S_{xx} > 0$, $\mathcal{H}_{\mathrm{ex}}$ increases with increasing a/b. Thus, energetically favored are oblate spheroids with $a < b$ for $\alpha\mu_1 S_{xx} > 0$ and prolate spheroids with $a > b$ for $\alpha\mu_1 S_{xx} < 0$.

Next we calculate $\mathcal{H}_{\mathrm{inh}}$ produced by a spheroid. From (10.1.40) it becomes

$$\mathcal{H}_{\mathrm{inh}} = V_{\mathrm{e}} e_0^2 (\Delta\mu)\left[-\frac{1}{3} + \sum_j \left(N_j - \frac{1}{3}\right)^2\right]. \tag{10.1.62}$$

Here we have set $\psi = -\Delta c/2$ outside the domain as in (10.1.53) assuming weak

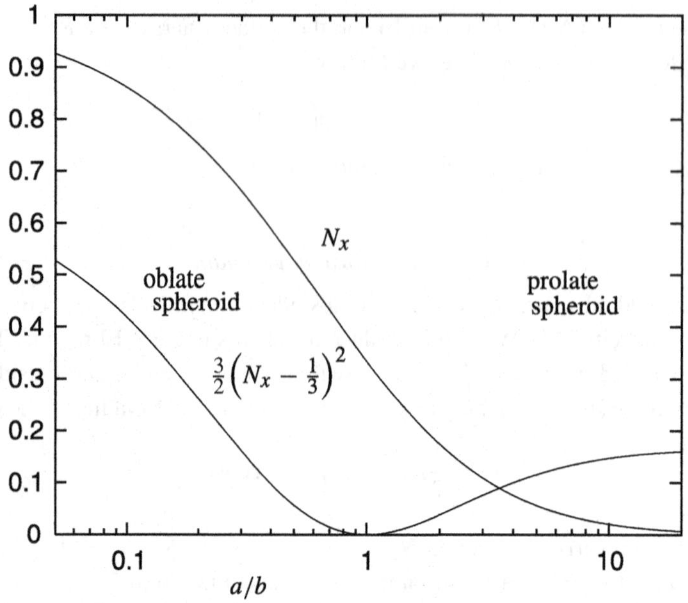

Fig. 10.7. The depolarization factor N_x vs a/b for a spheroid represented by $x^2/a^2+(y^2+z^2)/b^2 = 1$, see Appendix 10B. The function $(3/2)(N_x - 1/3)^2$ is also shown.

metastability. In Fig. 10.7 we plot $\sum_j (N_j - 1/3)^2 = (3/2)(N_x - 1/3)^2$. (i) For the soft domain case $\Delta\mu < 0$, $\mathcal{H}_{\text{inh}}$ decreases as the shape deviates from sphericity. It is minimum for compressed pancake shapes. (ii) For $\Delta\mu > 0$, the coherent inclusion is harder than the matrix and $\mathcal{H}_{\text{inh}}$ serves to stabilize a spherical shape. However, if a number of hard domains are present in a softer matrix, they interact with each other and change their shapes to minimize $\mathcal{H}_{\text{inh}}$, as will be discussed below.

10.1.5 Shape changes of hard domains

A pair of hard domains

Let two nearly spherical hard domains, A and B, be placed in a softer matrix without external stress in 3D, where $\psi = \psi_{\text{h}}$ within them and $\psi = \psi_{\text{s}}$ outside them. The shear modulus difference $\Delta\mu = \mu_1(\psi_{\text{h}} - \psi_{\text{s}})$ is positive. From $\mathcal{H}_{\text{inh}}$ in (10.1.40) the interaction energy is written as

$$\mathcal{H}_{\text{inh}} = \frac{1}{4}\Delta\mu \int_{\text{A+B}} d\boldsymbol{r} \sum_{ij} \left(e_{ij}^{\text{A}} + e_{ij}^{\text{B}}\right)^2, \tag{10.1.63}$$

where the space integral is within the domains A + B, and e_{ij}^{A} and e_{ij}^{B} are the strains (10.1.7) produced by A alone and B alone, respectively. If A and B are spheres with radii R_{A} and R_{B}, e_{ij}^{A} vanish within A and are given by the second term of (10.1.59) with $e_0 =$

$\alpha(\psi_h - \phi_s)/L_0$ outside A. We take V_A to be the volume of A, and $\boldsymbol{r}_A$ to be the center position of A. The equivalent relations hold also for B. If the distance $r_{AB} = |\boldsymbol{r}_A - \boldsymbol{r}_B|$ between the centers, $\boldsymbol{r}_A$ and $\boldsymbol{r}_B$, much exceeds R_A and R_B, the strains e_{ij}^B within A are nearly constants, given by

$$e_{ij}^B = \frac{1}{2\pi} e_0 V_B \left[\frac{3x_{ABi}x_{ABj}}{r_{AB}^5} - \frac{\delta_{ij}}{r_{AB}^3} \right], \tag{10.1.64}$$

which can be used even for non-spherical shapes. If the volume V_B of B on the right-hand side is replaced by the volume V_A of A, we obtain e_{ij}^A within B. Thus, we arrive at Eshelby's interaction between two spheres [7],

$$\mathcal{H}_{inh} \cong \frac{3}{8\pi^2}(\Delta\mu)e_0^2 V_A V_B (V_A + V_B) \frac{1}{r_{AB}^6}. \tag{10.1.65}$$

However, a crucial point is missed here [33]. That is, the assumption of spherical shapes is justified only when R_A and R_B are much smaller that R_E. If, conversely, they are much larger than R_E, the domains change their shapes such that the shear strains inside them vanish:

$$e_{ij}^A + e_{ij}^B = 0, \tag{10.1.66}$$

within A and B. Namely, vanishing of $\mathcal{H}_{inh}$ can be achieved by shape adjustment. The proof is almost obvious for $|\boldsymbol{r}_A - \boldsymbol{r}_B| \gg R_A, R_B$. Here the selected shapes after the adjustment are spheroids with the symmetry axis (the x axis) being along the relative vector $\boldsymbol{r}_{AB} = \boldsymbol{r}_A - \boldsymbol{r}_B$. From (10.1.59) $e_{ij}^A = -2e_0(N_i - 1/3)\delta_{ij}$ within A in terms of the depolarization factors N_i. Using (10.1.60) we rewrite (10.1.66) as

$$N_x - \frac{1}{3} = \frac{1}{2\pi} V_B \frac{1}{r_{AB}^3}. \tag{10.1.67}$$

For $a \cong b$ the shape of A is an oblate spheroid with

$$\frac{b}{a} - 1 \cong \frac{5}{2}\left(\frac{R_B}{r_{AB}}\right)^3. \tag{10.1.68}$$

As should be the case, A tends to a sphere for $R_B \ll r_{AB}$. The shape of B is also an oblate spheroid, for which R_B on the right-hand side is replaced by R_A. Interestingly, the above relation does not involve any material constants. We note that, on the one hand, the resultant increase of the surface free energy is of order $\Delta E_s \sim \sigma R^2(b/a - 1)^2 \sim \sigma R^2(R/r_{AB})^6$, where we assume $R_A \sim R_B \sim R$. On the other hand, the canceled elastic inhomogeneity energy is of order $\Delta E_{inh} \sim (\Delta\mu)e_0^2 R^3 (R/r_{AB})^6$ from (10.1.65). We thus estimate

$$\Delta E_{inh}/\Delta E_s \sim R/R_E. \tag{10.1.69}$$

Shape adjustment should occur for $R > R_E$, independently of the inter-domain distance, where a decrease in $\mathcal{H}_{inh}$ dominates over an increase in the surface free energy.

Many hard domains

The above arguments can be generalized to the case in which there are many hard domains in a softer matrix. Shape adjustment can cancel the elastic inhomogeneity energy when the average domain size R greatly exceeds $R_{\rm E}$. For simplicity, we assume that their separation distances greatly exceed their sizes. The condition of vanishing shear strain within a domain A is then written as

$$e_{ij}^{\rm A} + e_0 C_{ij}^{\rm A} = 0, \tag{10.1.70}$$

where

$$C_{ij}^{\rm A} = \sum_{n \neq {\rm A}} \frac{1}{2\pi} V_n \left[\frac{3x_{ni}x_{nj}}{r_n^5} - \frac{\delta_{ij}}{r_n^3} \right] \tag{10.1.71}$$

is the sum of the strain contributions from the other domains. The $\boldsymbol{r}_n$ is the relative vector between A and the nth domain and V_n is the volume of the nth domain. We then rotate the reference frame with an orthogonal matrix U_{ij} such that the matrix $C_{ij}^{\rm A}$ becomes diagonal. The condition (10.1.70) can be satisfied if N_i are determined by

$$\sum_{k\ell} U_{ki} U_{\ell j} C_{k\ell}^{\rm A} = 2\left(N_i - \frac{1}{3} \right) \delta_{ij}. \tag{10.1.72}$$

The principal axes of the ellipsoidal domain A are along the three orthogonal unit vectors $\bar{\boldsymbol{e}}_i = \sum_j U_{ij} \boldsymbol{e}_j$. In terms of the cartesian coordinates in the new axes, $\bar{x}_i = \sum_j U_{ij} x_j$, the domain boundary of A is represented by $\sum_i \bar{x}_i^2 / a_i^2 = 1$.

10.1.6 Dynamic equation

The dynamic equation for ψ is assumed to be of the diffusion type,

$$\frac{\partial}{\partial t} \psi = \frac{\lambda_0}{T} \nabla^2 \frac{\delta}{\delta \psi} \mathcal{H} + \theta, \tag{10.1.73}$$

where λ_0 is the kinetic coefficient, and $\theta(\boldsymbol{r}, t)$ is the random noise term which is negligible at late stages. Microscopically, a small number of vacancies are crucial for interdiffusion in binary alloys, because the direct exchange of A and B atoms is suppressed by a large energy barrier [57]–[59]. Here we examine the effect of the elastic inhomogeneity interaction in phase separation in the absence of an anisotropic external stress.

Shape evolution of a nearly spherical domain

We derive the evolution equation for a nearly spherical domain in the WEI regime [34, 56]. The interface is represented by $r = R + \sum_{\ell m} \delta_{\ell m} Y_{\ell m}(\theta, \varphi)$. The radius R obeys

$$\frac{\partial}{\partial t} R = \frac{D}{R} \left(\Delta_{\rm eff} - \frac{2d_0}{R} \right), \tag{10.1.74}$$

where $D = \lambda_0|r|/T$ is the diffusion constant, $d_0(\sim \xi)$ is the capillary length, and $\Delta_{\rm eff}(\ll 1)$ is the effective supersaturation defined by (10.1.54). The amplitude $\delta_{\ell m}$ obeys

$$\frac{1}{\delta_{\ell m}}\frac{\partial}{\partial t}\delta_{\ell m} = (\ell-1)\frac{1}{R}\frac{\partial R}{\partial t} - \frac{D}{R^2}(\ell^2+\ell-2)\left[(2\ell+1)\frac{d_0}{R} + \left(\frac{2\ell}{\ell+2}\right)g_{\rm E}^*\theta_{\rm sh}\right], \tag{10.1.75}$$

which reduces to the result by Mullins and Sekerka for $g_{\rm E}^* = 0$ [60].

Shape evolution of a nearly planar interface

From (10.1.75) we derive the evolution equation of disturbances on a planar interface by taking the limit, $R \to \infty$, $\ell \to \infty$ with $k = \ell/R > 0$ being fixed [60]. The deformed position is represented by $z = vt + \delta_k \cos(kx)$ where the z axis is along the normal to the interface. The region below the interface is isotropic, while the upper region is uniaxially deformed. Then (10.1.75) reduces to [34]

$$\frac{1}{\delta_k}\frac{\partial}{\partial t}\delta_k = kv - 2Dk^2(d_0k + g_{\rm E}^*\theta_{\rm sh}). \tag{10.1.76}$$

If the isotropic region is softer ($\theta_{\rm sh} = -1$), the interface is unstable even at rest ($v = 0$) at long wavelengths,

$$k < g_{\rm E}^*/d_0 \sim 1/R_{\rm E}. \tag{10.1.77}$$

The instability at $v = 0$ is of purely energetic origin in contrast to the kinetic Mullins–Sekerka instability [60]. Simulation of this instability was performed in Ref. [36].

10.1.7 Simulation with elastic inhomogeneity

We now discuss 2D simulation results of (10.1.73) on a 128×128 square lattice under the periodic boundary condition [32, 33, 36]. With appropriate scale changes of space, time, and ψ, we may set $r = -1$, $C = 1$, $u_0 = 1$, and $\lambda_0/T = 1$ in (10.1.32), (10.1.37), and (10.1.73). In the following, $g_{\rm E}$ is the dimensionless degree of elastic inhomogeneity ($= g_{\rm E}^*$). The dynamic equation without the noise term reads

$$\frac{\partial}{\partial t}\psi = \nabla^2[(-1-\nabla^2)\psi + \psi^3] + I_{\rm E}. \tag{10.1.78}$$

The second term arises from $\mathcal{H}_{\rm inh}$ in the form

$$I_{\rm E} = g_{\rm E}\nabla^2\hat{Q} + 2g_{\rm E}\sum_{ij}\nabla_i\nabla_j\psi\left[\nabla_i\nabla_j w - \frac{1}{d}\delta_{ij}\nabla^2 w\right], \tag{10.1.79}$$

where $\hat{Q}$ is defined by (10.1.38). Our system is quenched at $t = 0$ from a one-phase state. The time t after quenching will be indicated where necessary in the following figures. If $0 < g_{\rm E} \ll 1$, we have $\psi \cong -1$ in the softer regions and $\psi \cong 1$ in the harder regions. The volume fractions of the softer and harder regions are expressed as $\phi_{\rm s} = (1-M)/2$ and $\phi_{\rm h} = (1+M)/2$ in terms of the average M in the WEI regime. Hereafter, time will be measured in units of $CT/\lambda_0 r^2$.

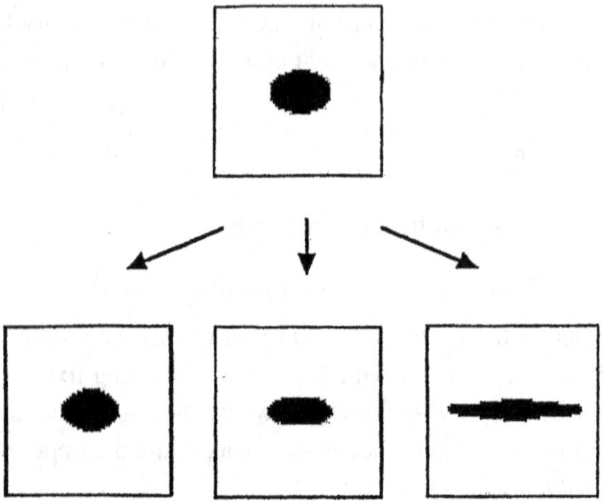

Fig. 10.8. Shape change of a single soft domain in a harder matrix assuming isotropic elasticity, where $g_E = 0.004$, 0.008, and 0.016 from the left.

Shape changes of soft domains

Figure 10.8 illustrates that a soft domain in a harder matrix undergoes a shape-change transformation for $g_E = 0.004, 0.008$, and 0.016. At $t = 0$ we prepare an ellipse slightly deformed from a circle within which $\psi = -1$ and outside of which $\psi = 0.8$. We can see that the mode $\ell = 2$ is amplified, and the domain is subsequently elongated into a slender shape for relatively large g_E. In phase separation, as soft domains are elongated, they touch and coalesce more frequently than spheres (circles in 2D), forming a percolated network even at relatively small volume fraction ϕ_s of the soft component. In Fig. 10.9 such processes are shown at $\phi_s = 0.2$ for various times after quenching with $g_E = 0.02$ and 0.07 [36]. At $g_E = 0.02$ the initial domains are close to being circles and shape deformation proceeds slowly. At $g_E = 0.07$ we find unambiguous achievement of percolation. The timescale of the network formation depends sensitively on g_E and ϕ_s. The width R_s of the elongated (black) softer regions increases with decreasing g_E, suggesting the crossover at $R \sim R_E \propto 1/g_E$. In these processes the total perimeter length first increases due to the elongation and then it begins to decrease very slowly due to coarsening.

Shape changes of hard domains

We start with two circular domains at $t = 0$ and follow their time development at $g_E = 0.05$. Figure 10.10 shows, at $t = 10^3$, the shapes of these two hard domains in a softer matrix. The distance between the domains is initially of the same order as the domain radius, but no tendency of coalescence of the domains can be seen within the computation time ($t < 10^3$). The initial values of ψ are $+1$ inside the domains and -1 outside them, with small random numbers being superimposed at $t = 0$. Then there arises no appreciable

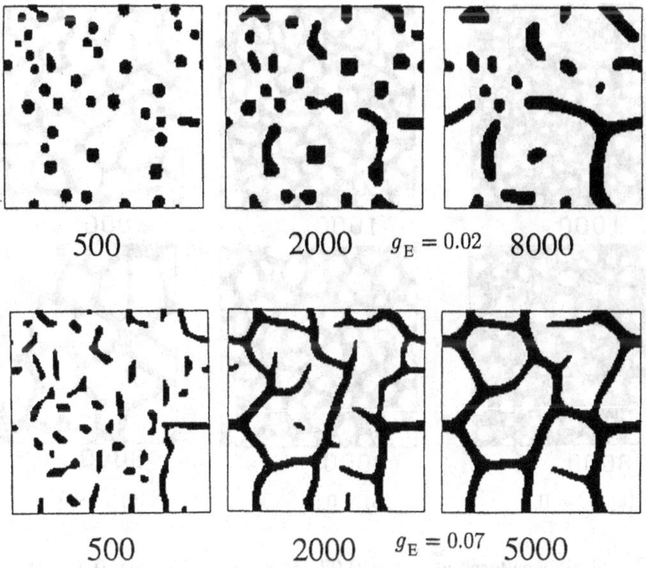

Fig. 10.9. Time evolution of softer (black) domains for $g_E = 0.02$ (*top*) and 0.07 (*bottom*) at $\phi_s = 0.2$ assuming isotropic elasticity. The numbers are the times after quenching from a disordered one-phase state [36].

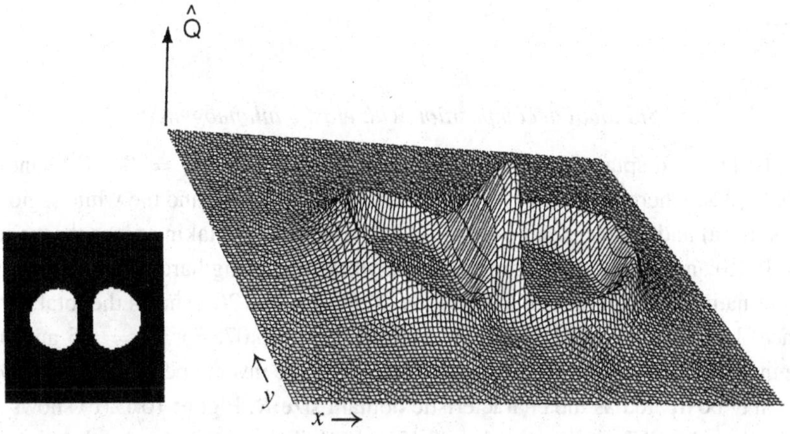

Fig. 10.10. The degree of anisotropic deformation $\hat{Q}$ defined by (10.1.38) for two hard domains at $g_E = 0.07$ in a softer matrix, at $t = 1000$ assuming isotropic elasticity. The domain profiles are also shown (*left*) [32].

change of the total area. However, if the initial mean value outside the domains is taken to be -0.8 or -0.9 (metastable values), the domains grow in time, still with no tendency to

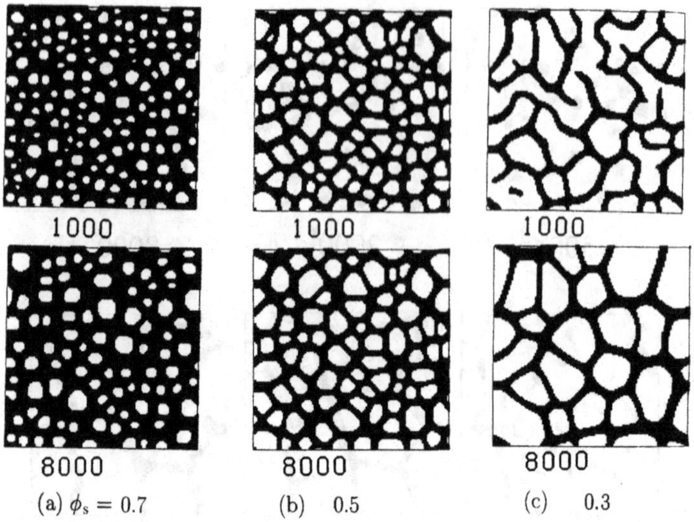

Fig. 10.11. The evolution patterns at $g_E = 0.07$ for $\phi_s = 0.7$ in (a), 0.5 in (b), and 0.3 in (c), assuming isotropic elasticity [32].

coalesce. After a transient time, the hard regions are isotropically deformed ($\hat{Q} \cong 0$), while the interfaces facing each other are flattened and the soft region between them is uniaxially deformed.

Spinodal decomposition with elastic inhomogeneity

In Fig. 10.11 we display the time evolution of domains for $\phi_s = 0.7, 0.5$, and 0.3 at $g_E = 0.07$ [32], where the black regions represent soft domains and the white regions hard domains. In (a) and (b) shape adjustment of hard domains are taking place throughout the system. In (b) and (c) the soft regions form networks enclosing hard droplet-like domains, which are natural configurations lowering $\mathcal{H}_{inh}$. Figure 10.12(a) shows the total perimeter (interface) length in the WEI regime at $g_E = 0.05$ and 0.07. For $\phi_s = 0.5$ and 0.3, the coarsening almost stops at very late stages. Note that the inverse perimeter length per unit volume may be treated as the characteristic domain size R. Figure 10.12(b) shows R thus determined in the SEI regime at $M = 0$ [5e], [38]. The inset indicates that the domain size in pinned states is inversely proportional to g_E, so $R \sim R_E$. The two-phase states here are driven into metastable states because of asymmetric shear deformations in the soft and hard regions. This picture becomes evident in Fig. 10.13, where the degree of anisotropic deformation $\hat{Q}$ is displayed in a pinned state. It exhibits *mountains* characteristic of local elastic energy barriers preventing further coarsening. In these pinned states, the surface energy ($\sim \sigma R^{d-1}$, σ being the surface tension) and the elastic inhomogeneity free energy ($\sim T g_E (\Delta\psi)^3 R^d$) per domain are of the same order. This balance leads to $R \sim R_E$.

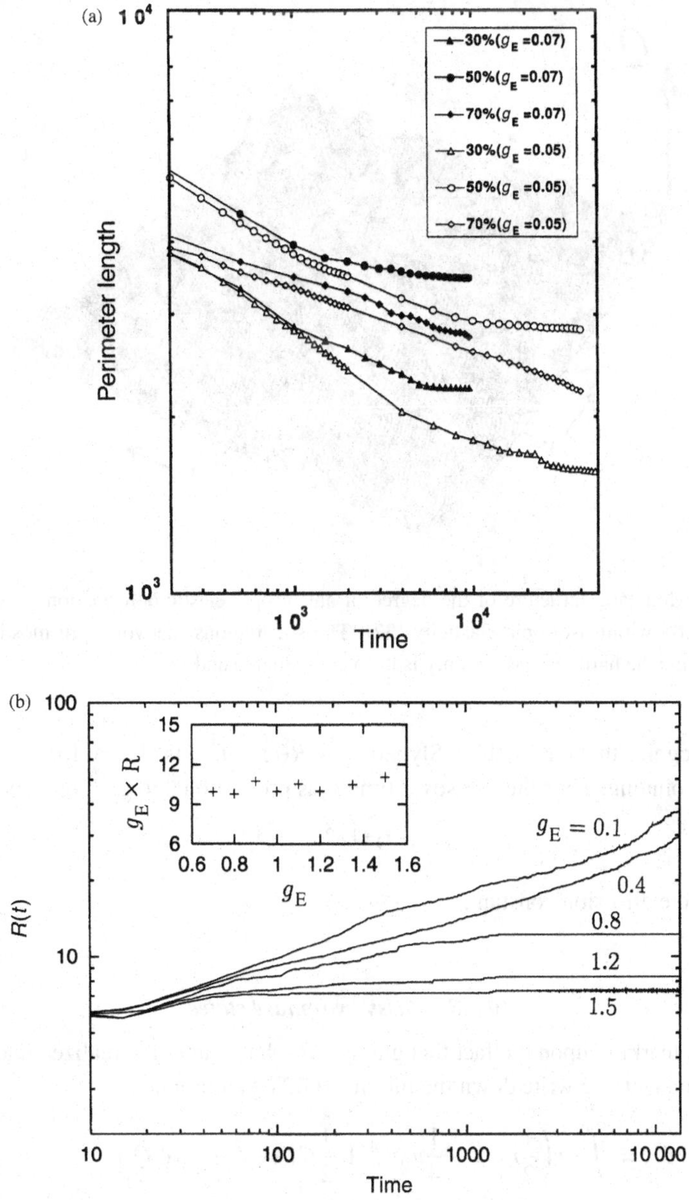

Fig. 10.12. (a) Perimeter length vs time for spinodal decomposition in the regime of weak elastic inhomogeneity [32]. Here $\phi_s = 0.3, 0.5$, and 0.7 (30, 50, and 70%) and $g_E = 0.05$ and 0.07 assuming isotropic elasticity. Pinning, evidenced by a constant length, can be seen at later stages for $\phi_s = 0.3$ and 0.5. (b) The domain size $R(t)$ obtained as the inverse of the perimeter length (in units of ξ) vs time (in units of ξ^2/D) at $M = 0$, assuming isotropic elasticity [5e]. Pinning occurs at an early stage in the case of strong elastic inhomogeneity ($g_E \gtrsim 1$). In the inset the relation $R \propto 1/g_E$ is shown to hold in pinned states.

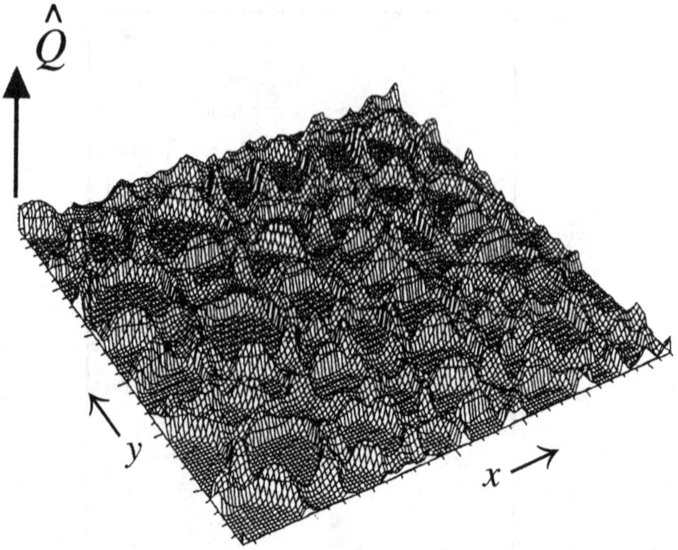

Fig. 10.13. 'Mountain' structure of the degree of anisotropic elastic deformation $\hat{Q}$ for $\phi_s = 0.5$ and $g_E = 0.05$ within isotropic elasticity [32]. The soft regions (network) are mostly uniaxially deformed, while the hard regions are only isotropically dilated and $\hat{Q} \cong 0$.

We also recognize that the Lifshitz–Slyozov law $R(t) \sim t^{1/3}$ is obeyed for $R < R_E$ before the onset of pinning. Thus the crossover time t_E is proportional to g_E^{-3}, or

$$t_E \sim D^{-1}\xi^2/(g_E^*)^3, \tag{10.1.80}$$

where D is the diffusion constant.

10.1.8 Glassy two-phase states

It is worth remarking upon the fact that glassy two-phase states are realized under EI [38]. Though redundant, we write down the minimal GLW hamiltonian,

$$\mathcal{H}_{\rm iso} = \int d\boldsymbol{r}\left(\frac{1}{2}r\psi^2 + \frac{1}{4}u_0\psi^4 + \frac{1}{2}C|\nabla\psi|^2 + g_E\psi\hat{Q}\right), \tag{10.1.81}$$

which involves the two characteristic lengths, ξ and R_E. Here we need to estimate the free-energy barrier per domain in pinned states. If $g_E^* \ll 1$, the mountain structure in Fig. 10.13 indicates

$$(\Delta\mathcal{H})_E \sim \sigma R_E^{d-1} \sim T(|r|^{2-d/2}/u_0)/(g_E^*)^{d-1}. \tag{10.1.82}$$

Thus, $(\Delta\mathcal{H})_E \gg T$, because the factor $|r|^{2-d/2}/u_0$ is large in the mean field theory (the Ginzburg criterion (4.1.24)). This means that pinning occurs even if the thermal noise

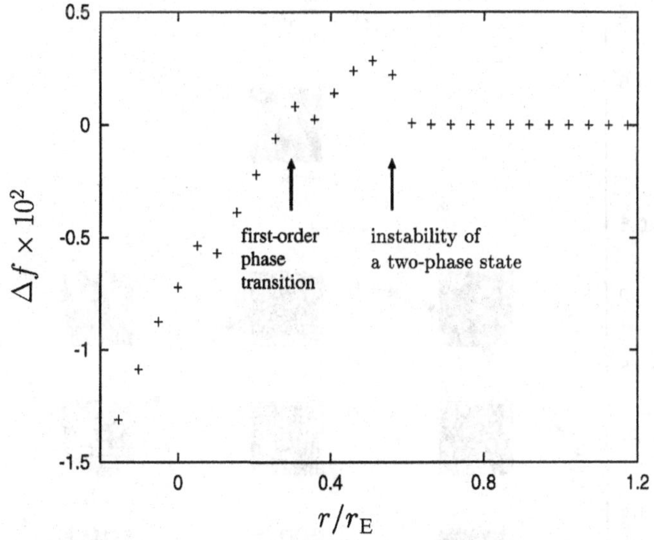

Fig. 10.14. Difference in the free-energy density between one-phase and pinned two-phase states (in units of Tg_E^4/u_0^3) as a function of r/r_E at $M/M_E = -0.21$. The difference vanishes at $r/r_E = 0.265$ on the two-phase branch, where a first-order phase transition is expected. The two-phase states are unstable for $r/r_E > 0.53$, while the one-phase states for $r/r_E < -0.89$, as predicted by (10.1.41) (not shown here).

term is added to the right-hand side of (10.1.78). Although not attained in simulations, we believe that true equilibrium two-phase states are periodic in space, in which droplet-like hard domains are enclosed by percolating soft regions.

In the SEI regime, phase transitions occur between a one-phase state and a pinned two-phase state without much growth of the domains. In Fig. 10.14 we plot the free-energy density difference $\Delta f = \mathcal{H}_{\rm iso}/V - rM^2/2 - u_0M^4/4$ relative to the value in the homogeneous phase as a function of r/r_E at $M/M_E = -0.21$ (which is close to the maximum point $M/M_E = -1/6$ of the spinodal curve (10.1.41) in 2D). The two-phase state with $\Delta f < 0$ should be stable against thermal agitations even if the thermal noise is included. Hence the point at which $\Delta f = 0$ on the two-phase branch may be treated as a first-order phase transition point. At $M/M_E = -0.21$, the value of r at the transition point thus determined is $0.26r_E$, while the values at the spinodal points are $0.53r_E$ for the two-phase states and $-0.89r_E$ for the one-phase states. As shown in Fig. 10.15, this hysteretic behavior persists at any M. Therefore, no critical point exists under EI. Around this first-order phase transition, we have $r \sim r_E$, $\psi \sim M_E$, and $\xi \sim r_E^{-1/2} \sim R_E$. Then the free-energy barrier per domain is estimated as $(\Delta\mathcal{H})_E \sim T[r_E^{2-d/2}/u_0] \gg T$ in the mean field regime. In the asymptotic critical region, if it can be reached, the barrier is weakened, suggesting the appearance of periodic two-phase states in equilibrium.

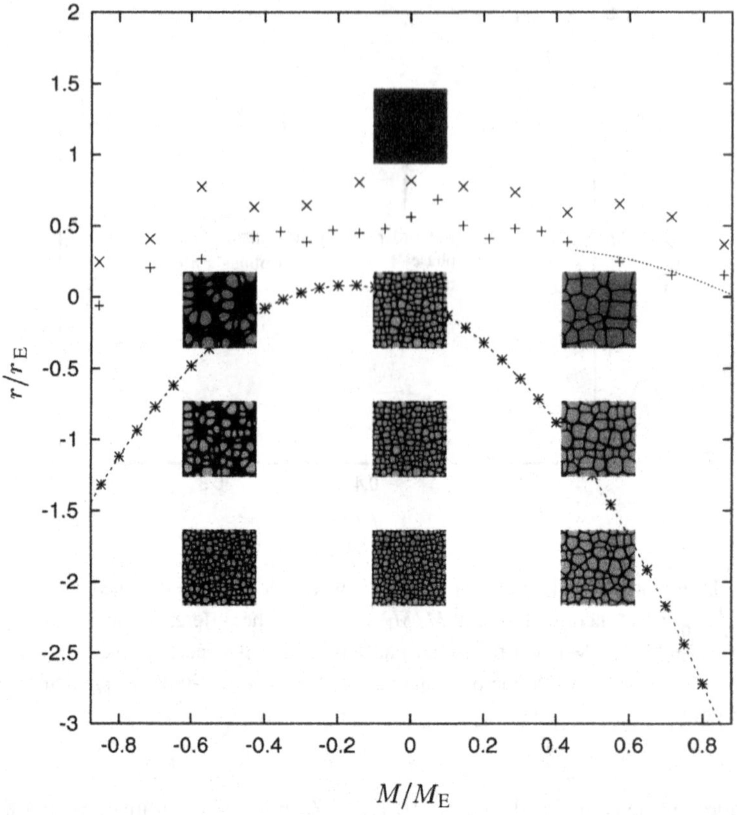

Fig. 10.15. The phase diagram in the r–M (temperature–composition) plane under elastic inhomogeneity assuming isotropic elasticity, calculated in 2D. The meanings of the data are as follows: $+$, first-order transition points; $*$, instability points of one-phase states; $\times$, instability points of pinned two-phase states. The points $*$ are on the theoretical spinodal curve (10.1.41). The dotted line is obtained from (10.1.85) at $\phi_s = 0.1$. Domain patterns in pinned states are also shown, where the soft regions are shown in black.

Figure 10.15 also shows that the soft (black) regions form a thin network at relatively small volume fractions of the soft component [38]. For such domain structures the space dependence is mostly along the interface normal $\boldsymbol{n}$ except for the junction regions. Then we may set $\nabla_i \nabla_j w \cong n_i n_j (\psi - M)$ to obtain the approximate free-energy density,

$$\frac{1}{T} f_{\text{eff}} = \frac{1}{2} r \psi^2 + \frac{1}{4} u_0 \psi^4 + \bar{g}_E \psi (\psi - M)^2, \tag{10.1.83}$$

where $\bar{g}_E = (1 - 1/d) g_E$. For this free-energy density, phase separation occurs for

$$r_{\text{eff}} = r - 4M\bar{g}_E - 3\bar{g}_E^2/u_0 < 0, \tag{10.1.84}$$

and the interface thickness is given by $\xi = |C/r_{\text{eff}}|^{1/2}$, where C is the coefficient in

(10.1.2). In the resultant two phases we have $\psi = \psi_+$ and ψ_- with $\psi_\pm = -\bar{g}_E/u_0 \pm |r_{\text{eff}}/u_0|^{1/2}$, so that

$$M = -\bar{g}_E/u_0 + |r_{\text{eff}}/u_0|^{1/2}(1 - 2\phi_s), \tag{10.1.85}$$

where ϕ_s is the volume fraction of the soft regions. The network at small ϕ_s should dissolve when the layer thickness becomes of order ξ. However, we cannot determine the mesh size ℓ_{net} of the networks in Fig. 10.15 from the quasi-1D free-energy density (10.1.83) only. In our simulation, ℓ_{net} is about ten times longer than ξ at the points of first-order phase transition at small ϕ_s in Fig. 10.15. Indeed, these points are nearly on the theoretical curve for $\phi_s = 0.1$ in (10.1.85).

10.2 Phase separation in cubic solids

For cubic solids we again suppose the GLW hamiltonian (10.1.2), where the composition and the elastic field are coupled via the Vegard law. The elastic energy density is expressed in terms of the three elastic moduli, C_{11}, C_{12}, and C_{44}, in the form [47],

$$\begin{aligned} f_{\text{el}}(\boldsymbol{u}) &= \frac{1}{2}C_{11}\sum_i(\nabla_i u_i)^2 + \frac{1}{2}C_{12}\sum_{i\neq j}(\nabla_i u_i)(\nabla_j u_j) + \frac{1}{2}C_{44}(e_{xy}^2 + e_{yz}^2 + e_{zx}^2) \\ &= \frac{1}{2}K(\nabla\cdot\mathbf{u})^2 + \frac{1}{8}(C_{11} - C_{12} - 2C_{44})\sum_i e_{ii}^2 + \frac{1}{4}C_{44}\sum_{ij} e_{ij}^2, \end{aligned} \tag{10.2.1}$$

where e_{ij} in the second line is the shear strain defined by (10.1.7). The elastic stress tensor is given by

$$\begin{aligned} \sigma_{ii} &= (C_{11} - C_{12})\nabla_i u_i + C_{12}\nabla\cdot\boldsymbol{u} + \alpha\psi, \\ \sigma_{ij} &= C_{44}e_{ij} \qquad (i \neq j). \end{aligned} \tag{10.2.2}$$

The concentration variation changes the pressure as in the isotropic case. The bulk modulus K and the shear modulus μ are given by

$$K = \frac{1}{d}C_{11} + \frac{d-1}{d}C_{12}, \quad \mu = C_{44}. \tag{10.2.3}$$

The degree of cubic anisotropy is represented by the following parameter,

$$\xi_a = (C_{11} - C_{12} - 2C_{44})/C_{44}. \tag{10.2.4}$$

Note that the transverse sound velocity is given by $c_T[100] = \sqrt{C_{44}/\rho}$ in the [100] direction and by $c_T[110] = \sqrt{(C_{11} - C_{12})/2\rho}$ in the [110] direction, where ρ is the mass density. The velocity difference in these two directions is expressed in terms of ξ_a as $c_T[110]^2 - c_T[100]^2 = \xi_a C_{44}/2\rho$. It is also well known that $(C_{11} - C_{12})/2$ and C_{44} interchange their roles for 2D deformations (homogeneous along the z axis) if the reference frame is rotated by $\pi/4$ in the xy plane. It is also well known that a cubic solid is stable under the criteria $K > 0$, $C_{44} > 0$, and $C_{11} - C_{12} > 0$ for small deformations [47]. (See Subsection 10.5.1 for more discussions.)

To examine the effects of elastic inhomogeneity (EI) we assume the linear dependence of the elastic moduli on ψ,

$$C_{ij} = C_{ij}^{(0)} + C_{ij}^{(1)}\psi. \tag{10.2.5}$$

The bulk and shear moduli are then expressed as $K = K_0 + K_1\psi$ and $\mu = \mu_0 + \mu_1\psi$, where

$$K_1 = \frac{1}{d}C_{11}^{(1)} + \frac{d-1}{d}C_{12}^{(1)}, \quad \mu_1 = C_{44}^{(1)}. \tag{10.2.6}$$

Another relevant parameter is

$$\xi_{a1} = (C_{11}^{(1)} - C_{12}^{(1)} - 2C_{44}^{(1)})/C_{44}^{(0)}. \tag{10.2.7}$$

10.2.1 Bilinear interaction for homogeneous elastic moduli

In cubic crystals, the elastic interaction among the order parameter fluctuations is already highly nontrivial even for homogeneous elastic moduli ($C_{ij} = C_{ij}^{(0)}$). It produces anisotropic domain morphologies characteristic of cubic solids. Moreover, in the absence of elastic inhomogeneity, the morphology is unaffected by an externally applied strain. As will be shown in Appendix 10A, the mechanical equilibrium condition $\nabla\cdot\overleftrightarrow{\sigma} = \mathbf{0}$ is solved to give

$$u_i = \alpha\nabla_i w_i, \tag{10.2.8}$$

in the absence of an externally applied strain. The Fourier transformation of w_j is related to that of ψ as

$$w_i(\boldsymbol{k}) = \frac{1}{[1+\varphi_0(\hat{\boldsymbol{k}})]C_{44}(k^2+\xi_a k_i^2)}\psi_{\boldsymbol{k}}, \tag{10.2.9}$$

with

$$\varphi_0(\hat{\boldsymbol{k}}) = \left(1 + \frac{C_{12}}{C_{44}}\right)\sum_j \frac{1}{1+\xi_a\hat{k}_j^2}\hat{k}_j^2. \tag{10.2.10}$$

If $\xi_a \neq 0$, $\varphi_0(\hat{\boldsymbol{k}})$ depends on the direction $\hat{\boldsymbol{k}}$ of the wave vector. The dilation strain is expressed as

$$g = \nabla\cdot\boldsymbol{u} = \alpha\sum_j \nabla_j^2 w_j. \tag{10.2.11}$$

The elastic part of $\mathcal{H}$ in (10.1.2) becomes a bilinear dipolar interaction,

$$\int d\boldsymbol{r}\frac{1}{2}\alpha\psi\nabla\cdot\boldsymbol{u} = \int_{\boldsymbol{k}}\frac{1}{2}\tau_{\rm el}(\hat{\boldsymbol{k}})|\psi_{\boldsymbol{k}}|^2, \tag{10.2.12}$$

with

$$\tau_{\rm el}(\hat{\boldsymbol{k}}) = -\frac{\alpha^2}{C_{12}+C_{44}}\left[1 - \frac{1}{1+\varphi_0(\hat{\boldsymbol{k}})}\right]. \tag{10.2.13}$$

As in (10.1.35) and (10.1.36), we are interested in the elastic contribution to the chemical free energy f_{chem} and, hence, introduce

$$B(\hat{\boldsymbol{k}}) = \tau_{\mathrm{el}}(\hat{\boldsymbol{k}}) + \frac{\alpha^2}{K}. \tag{10.2.14}$$

We then reproduce Khachaturyan's result [2]:[5]

$$B(\hat{\boldsymbol{k}}) = \frac{\alpha^2}{K} - \frac{\alpha^2(1+2\gamma_1+3\gamma_2)}{C_{11}+(C_{11}+C_{12})\gamma_1+(C_{11}+2C_{12}+C_{44})\gamma_2}, \tag{10.2.15}$$

where

$$\gamma_1 = \xi_{\mathrm{a}}(\hat{k}_x^2\hat{k}_y^2 + \hat{k}_y^2\hat{k}_z^2 + \hat{k}_z^2\hat{k}_x^2), \quad \gamma_2 = \xi_{\mathrm{a}}^2(\hat{k}_x\hat{k}_y\hat{k}_z)^2. \tag{10.2.16}$$

In Ref. [30], $\tau_{\mathrm{el}}(\hat{\boldsymbol{k}})$ was calculated for general cases with arbitrary elastic inhomogeneity and externally applied strain.

The directions of the wave vector which minimize $\tau_{\mathrm{el}}(\hat{\boldsymbol{k}})$ (and maximize $\varphi_0(\hat{\boldsymbol{k}})$) are called *elastically soft* directions. As the temperature is lowered, early-stage spinodal decomposition is triggered by those concentration fluctuations varying in the soft directions. In late-stage phase separation, the interface planes tend to be perpendicular to one of these directions. For most cubic crystals ξ_{a} is negative and the soft directions are $\langle 100\rangle$. If the solid is assumed to be inhomogeneous only in these directions, the coexistence curve is shifted downwards by [24]

$$\Delta T_{\mathrm{c}}[100] = \frac{1}{a_0}B[100] = \frac{2\alpha^2}{a_0}\cdot\frac{C_{11}-C_{12}}{(C_{11}+2C_{12})C_{11}}, \tag{10.2.17}$$

where a_0 is defined by (10.1.34). This shift was estimated to be 20 deg.K in Al–Zn [49] and 600 deg.K in Au–Ni [50].[6] On the other hand, if $\xi_{\mathrm{a}} > 0$, the softest directions are $\langle 111\rangle$.

Weak cubic elastic anisotropy

For small ξ_{a}, $\tau_{\mathrm{el}}(\hat{\boldsymbol{k}})$ is expanded as

$$\tau_{\mathrm{el}}(\hat{\mathbf{k}}) = -(\alpha^2/C_{11}) + \tau_{\mathrm{cub}}(\hat{k}_x^2\hat{k}_y^2 + \hat{k}_y^2\hat{k}_z^2 + \hat{k}_z^2\hat{k}_x^2) + O(\xi_{\mathrm{a}}^2), \tag{10.2.18}$$

where

$$\tau_{\mathrm{cub}} = -2\alpha^2 C_{44}\xi_{\mathrm{a}}/C_{11}^2. \tag{10.2.19}$$

[5] This expression agrees with the original one [24] only to first order in ξ_{a} for general $\hat{\boldsymbol{k}}$, but coincides with it exactly in the [100] direction.

[6] The latter alloy has a large lattice misfit ($|\eta| \sim 0.1$) and large elastic inhomogeneity ($C_{ij}^{(1)} \sim C_{ij}^{(0)}$), so the assumption of elastic homogeneity is inappropriate.

In this simplest form, the interaction arising from cubic elasticity is written as

$$\begin{aligned}\mathcal{H}_{\rm cub} &= \frac{1}{2}\tau_{\rm cub}\int_{\boldsymbol k}(\hat{k}_x^2\hat{k}_y^2+\hat{k}_y^2\hat{k}_z^2+\hat{k}_z^2\hat{k}_x^2)|\psi_{\boldsymbol k}|^2\\ &= \frac{1}{4}\tau_{\rm cub}\int d\boldsymbol{r}\sum_{i\neq j}(\nabla_i\nabla_j w)^2.\end{aligned}\tag{10.2.20}$$

This form has been used in computer simulations. The 2D result follows for $\hat{k}_z = 0$.

10.2.2 Third-order interactions due to elastic inhomogeneity

We calculate the third-order interaction in ψ arising from the ψ dependence of the elastic moduli in the absence of an external stress ($S_{ij} = 0$). For the elastically homogeneous case, the dilation strain is given by (10.1.28) and is written as $g^{(0)}$ here. The zeroth-order shear strain is expressed as

$$e_{ij}^{(0)} = \alpha\left[\nabla_i\nabla_j(w_i+w_j) - \frac{2}{d}\delta_{ij}\sum_{\ell}\nabla_\ell^2 w_\ell\right],\tag{10.2.21}$$

where the w_j are defined by (10.2.9). In terms of $g^{(0)}$ and $e_{ij}^{(0)}$ the elastic inhomogeneity interactions are written as

$$\mathcal{H}_{\rm inh} = \int d\boldsymbol{r}\psi\left[\frac{1}{2}K_1(g^{(0)})^2 + \frac{1}{8}\xi_{\rm a1}C_{44}\sum_j(e_{jj}^{(0)})^2 + \frac{1}{4}C_{44}^{(1)}\sum_{ij}(e_{ij}^{(0)})^2\right],\tag{10.2.22}$$

to first order in $C_{ij}^{(1)}$. However, the above expression is still very complicated, so we furthermore consider the limit $\xi_{\rm a} \to 0$. The first term ($\propto K_1$) in the brackets is then proportional to $\psi(\delta\psi)^2$ from (10.1.28) and can be incorporated into the free-energy density. In terms of w in (10.1.27) the other two terms become

$$\mathcal{H}_{\rm inh} = \frac{1}{4}g_{\rm cub}\int d\boldsymbol{r}\psi\sum_{i\neq j}(\nabla_i\nabla_j w)^2 + g_{\rm E}\int d\boldsymbol{r}\psi\sum_{ij}\left(\nabla_i\nabla_j w - \frac{\delta_{ij}}{d}\nabla^2 w\right)^2,\tag{10.2.23}$$

where

$$g_{\rm cub} = -2(\alpha/C_{11})^2\xi_{\rm a1}C_{44},\quad g_{\rm E} = \frac{1}{2}(\alpha/C_{11})^2(C_{11}^{(1)} - C_{12}^{(1)}).\tag{10.2.24}$$

The first term in (10.2.23) was first derived by Sagui *et al.* [40], while the second term has already been derived in (10.1.37) and (10.1.38) assuming isotropic elasticity.

10.2.3 Simulation with cubic anisotropy

We numerically solve the dynamic equation (10.1.73) without the noise term. For $r = -1$, $C = 1$, $u_0 = 1$, and $\lambda_0/T = 1$, it is written as

$$\frac{\partial}{\partial t}\psi = \nabla^2[(-1-\nabla^2)\psi + \psi^3] + I_{\rm cub} + I_{\rm ex} + I_E + I_{\rm E}^{\rm cub}.\tag{10.2.25}$$

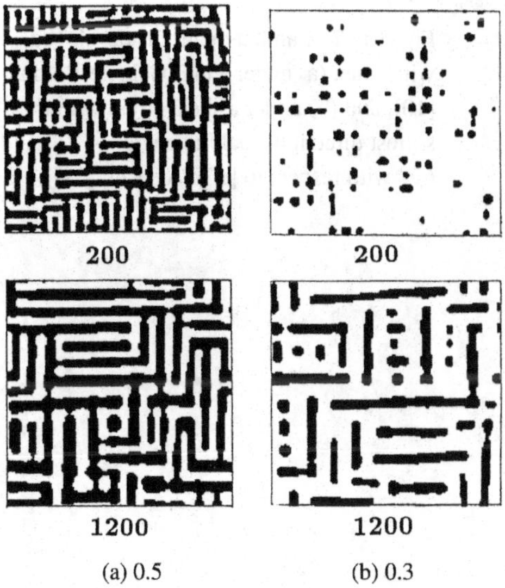

Fig. 10.16. Time evolution patterns in a cubic alloy quenched at $t = 0$ with $\tau_{\text{cub}} = 0.675$ at (a) $\phi_s = 0.5$ (*left*) and (b) $\phi_s = 0.3$ (*right*) in the absence of an external stress [31].

The cubic interaction $\mathcal{H}_{\text{cub}}$ in (10.2.20) gives rise to

$$I_{\text{cub}} = \frac{1}{2}\tau_{\text{cub}} \sum_{i \neq j} \nabla_i^2 \nabla_j^2 w, \tag{10.2.26}$$

where w is defined by (10.1.27). When anisotropic external strain is applied, $\mathcal{H}_{\text{ex}}$ in (10.1.44) follows, leading to

$$I_{\text{ex}} = -g_{\text{ex}} \sum_{ij} S_{ij} \nabla_i \nabla_j \psi. \tag{10.2.27}$$

The third-order interaction $\mathcal{H}_{\text{inh}}$ in (10.2.23) consists of two terms; as a result, one contribution in (10.2.25) is I_{E} in (10.1.79), while the other one reads

$$I_E^{\text{cub}} = \frac{1}{4} g_{\text{cub}} \nabla^2 \sum_{i \neq j} (\nabla_i \nabla_j w)^2 + \frac{1}{2} g_{\text{cub}} \sum_{i \neq j} \nabla_i \nabla_j \psi (\nabla_i \nabla_j w). \tag{10.2.28}$$

Effect of $\mathcal{H}_{\text{cub}}$ only

Taking account of $\mathcal{H}_{\text{cub}}$ or I_{cub} only, we first show that the cubic anisotropy gives rise to rectangular domains. In Fig. 10.16 we display 2D simulated domain structures at $t = 200$ and 1200 after quenching, with $\tau_{\text{cub}} = 0.675$ and $g_{\text{cub}} = g_{\text{E}} = S_{ij} = 0$ [31]. The volume fraction of one component is 0.5 in (a) and 0.3 in (b). The softest directions are [01] and [10], so domains are rectangular stripes aligned in [10] or [01] in the absence of an anisotropic external stress. The domain widths show a sharp peak at $R(t) \sim t^a$ with $a = 0.2$–0.3, while the domain lengths are broadly distributed. Other simulations including cubic anisotropy have also been performed [2, 28, 43]. In these numerical studies, no

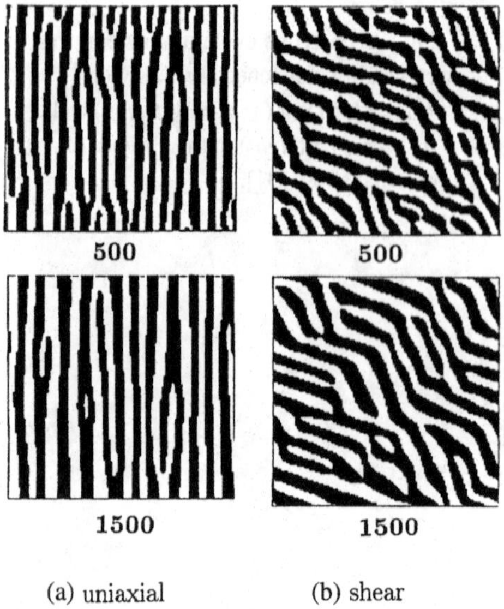

Fig. 10.17. Lamellar patterns developed with time, (a) under a uniaxial stress along [10], and (b) under a shear stress with the softest directions making angles of 21° and 69° with respect to [10] (see Ref. [31]).

pinning (freezing of coarsening) was observed and the growth law was not much different from the usual Lifshitz–Slyozov law, despite the highly anisotropic shapes of the domains.

Effect of $\mathcal{H}_{\rm cub} + \mathcal{H}_{\rm ex}$

Next we include the interaction $\mathcal{H}_{\rm ex}$ in (10.1.42) arising from an applied external stress in a cubic solid with $\tau_{\rm cub} = 0.675$, neglecting $\mathcal{H}_{\rm inh}$ (although $\mathcal{H}_{\rm ex}$ vanishes without EI). Figure 10.17(a) shows patterns under a uniaxial stress along [10] with $g_{\rm ex}S_{xx} = -g_{\rm ex}S_{yy} = 0.15$ [31]. We cite several observations of lamellar or cylindrical domain structures in cubic alloys under a uniaxial stress [18]–[20]. As another example, Fig. 10.17(b) also shows patterns under a shear stress with $g_{\rm ex}S_{xy} = -0.226$. In these cases, the domain width continues to grow with the growth exponent about 0.2, as in the case of $\mathcal{H}_{\rm cub}$ alone.

Effect of $\mathcal{H}_{\rm cub} + \mathcal{H}_{\rm inh}$

In Fig. 10.18 we display time evolution patterns and pinning of two-phase structures in the presence of $\mathcal{H}_{\rm cub}$ and $\mathcal{H}_{\rm inh}$, where we set $\tau_{\rm cub} = 0.675$ and $g_{\rm E} = 0.07$, but $g_{\rm cub} = 0$. Then the role of elastic cubic anisotropy is simply to orientate the interfaces in the preferred directions. The patterns obtained closely resemble those in Fig. 10.4 [10] observed in Ni-base fcc crystals with relatively large misfits, in which the component with smaller C_{44} forms a network [61].

Next we examine the effect of the anisotropic, third-order interaction ($\propto g_{\rm cub}$) in (10.2.26). When $\tau_{\rm cub} + g_{\rm cub}\psi$ take positive and negative values in the two phases, there arises competition in the orientation of the interfaces [40]. In Fig. 10.19 we set $g_{\rm cub} = -4$ and $g_{\rm E} = 0.5$ where $\psi < 0$ in the regions in black. Here the interfaces tend to be parallel to the x or y axis or make angles of $\pm\pi/4$ to these axes. This competition persists at any

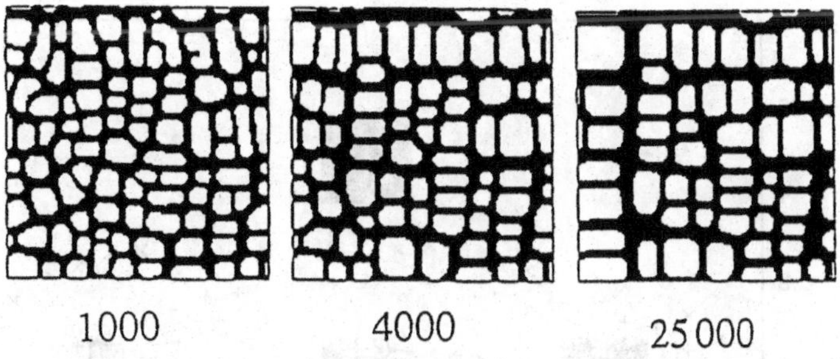

Fig. 10.18. The evolution patterns in the presence of elastic inhomogeneity in a cubic solid for $\tau_{cub} = 0.675$, $g_E = 0.07$, and $g_{cub} = 0$ at $\phi_s = 0.5$ [35]. They resemble those in Fig. 10.4.

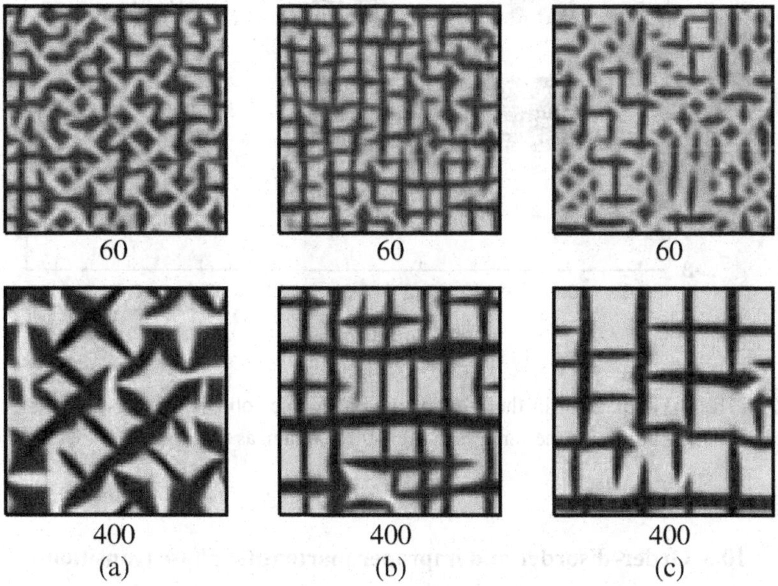

Fig. 10.19. Time evolution patterns with orientational competition for $g_E = 0.5$ and $g_{cub} = -4$. Here $\tau_{cub} = M = 0$ in (a), $\tau_{cub} = 0.1$ and $M = 0$ in (b), and $\tau_{cub} = 0$ and $M = 0.3$ in (c). We set $\psi < 0$ in the regions in black.

late stage for $\tau_{cub} = M = 0$ in (a). However, those parallel to the x or y axis gradually dominate if τ_{cub} is increased to 0.1 in (b) or if M is off-critical at 0.3 in (c).

We present the phase diagram for cubic solids in Fig. 10.20 under EI [38] (cf. Fig. 10.15 for isotropic elasticity). Here we consider the 2D case in which the total GLW hamiltonian is the sum of $\mathcal{H}_{iso}$ in (10.1.80) and $\mathcal{H}_{cub}$ in (10.2.20) with $\tau_{cub} = 0.71 r_E$, so $g_{cub} = 0$. As in the isotropic case the hysteretic behavior remains at any M, indicating no critical point.

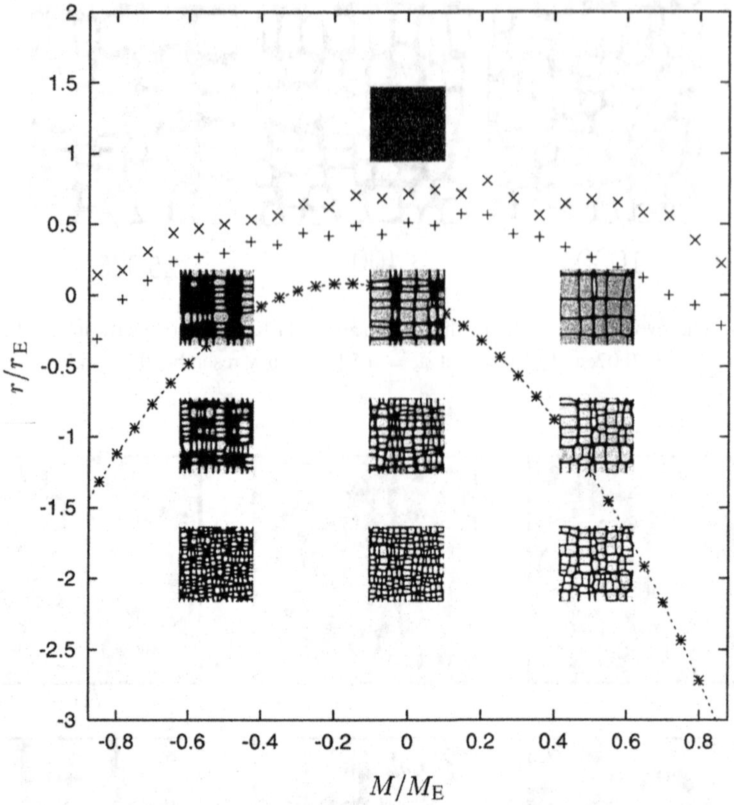

Fig. 10.20. The phase diagram in the r–M plane for a cubic solid in the presence of $\mathcal{H}_{cub}$. The meanings of the data points are the same as in Fig. 10.15. Again, as for Fig. 10.15, there is no critical point.

10.3 Order–disorder and improper martensitic phase transitions

In Section 3.3 we discussed order–disorder phase transitions due to local (optical) atomic displacements in binary alloys neglecting the elastic effects (arising from coupling to the acoustic degrees of freedom). For example, in Fe–Al alloys, a critical line separates a disordered bcc phase and an ordered bcc phase and ends at a tricritical point; below this point the transition is first order [62, 63]. Here domains in the ordered phase keep the cubic symmetry. However, $L1_0$ domains in fcc alloys are tetragonally deformed in one of the directions [100], [010], or [001], and some interesting patterns have been observed when such tetragonal precipitates are developing in a cubic matrix [64, 65]. The long-range order parameter η is a scalar quantity for bcc solids as in (3.3.11) [39] and is a vector (η_1, η_2, η_3) for fcc solids as in (3.3.26) [2], [66]–[69]. Originally, a Ginzburg–Landau model with a three-component order parameter was also presented for improper ferroelastic transitions in perovskite-structure compounds such as $SrTiO_3$ [70]. For n-component systems the chemical free-energy density $f_{\rm chem}$ may be of the form,[7]

$$f_{\rm chem} = A_1(c - c_1)^2 - A_2(c - c_2)|\eta|^2 - A_3|\eta|^4 + A_4|\eta|^6 + A_5 \sum_{\alpha<\beta} \eta_\alpha^2 \eta_\beta^2, \quad (10.3.1)$$

where $|\eta|^2 = \sum_\alpha |\eta_\alpha|^2$, c_1 and c_2 are appropriate concentrations, and A_j $(j = 1, \ldots, 4)$ are positive constants. The gradient free energy may be of the form $f_{\rm gra} = \frac{1}{2}D|\nabla c|^2 + \frac{1}{2}C\sum_\alpha |\nabla\eta_\alpha|^2$, where C and D are positive constants. The total free energy is then the sum of the chemical free energy and the elastic free energy as $\mathcal{H} = \int d\boldsymbol{r}[f_{\rm chem} + f_{\rm grad}] + \mathcal{H}_{\rm el}$, as will be discussed in Appendix 10A. For many-component systems $(n \geq 2)$, the term proportional to A_5 breaks the rotational symmetry in the vector space of $(\eta_1, \ldots, \eta_n)$ and, if $A_5 > 0$, the ordered state with $\eta_1 \neq 0$ and $\eta_j = 0$ $(j \geq 2)$ is favored over other ordered states such as the one with $\eta_1 = \cdots = \eta_n$. For $A_2 > 0$, the ordered phase has a higher value of c than in the disordered phase. There can be two kinds of elastic coupling; one is between c and the strains $\epsilon_{ij} = (\nabla_i u_j + \nabla_j u_i)/2$ as in (10.1.2), while the other involves $\eta_{\rm p}$ and is of third order as

$$\mathcal{H}_{\rm I} = -\int d\boldsymbol{r} \sum_{pij} \sigma_{ij}^0(p)\eta_p^2 \epsilon_{ij}, \quad (10.3.2)$$

where $\sigma_{ij}^0(p)$ $(p = 1, \ldots, n)$ are constant matrices. This coupling is even with respect to $\eta_{\rm p}$ from crystal symmetry. If we assume harmonic elasticity, we may eliminate the strain fields and obtain fourth-order angle-dependent interactions among $\eta_{\rm p}$ following the procedure in Appendix 10A. The simplest dynamic equations are of the forms

$$\frac{\partial}{\partial t}c = M\nabla^2 \frac{\delta}{\delta c}\mathcal{H}, \qquad \frac{\partial}{\partial t}\eta_\alpha = -L\frac{\delta}{\delta\eta_\alpha}\mathcal{H}, \quad (10.3.3)$$

where M and L are the kinetic coefficients.

At improper martensitic transitions, on the other hand, no composition field is involved, and a vector order parameter $\eta_{\rm p}$ $(p = 1, \ldots, n)$ representing optical atomic displacements is coupled to the strains as in (10.3.2) [66, 67]. The sum of the chemical free-energy density and the gradient free-energy density may be given by

$$f_{\rm chem} + f_{\rm grad} = \frac{\tau}{2}|\eta|^2 - A_3|\eta|^4 + A_4|\eta|^6 + A_5\sum_{\alpha<\beta}\eta_\alpha^2\eta_\beta^2 + \frac{C}{2}\sum_\alpha |\nabla\eta_\alpha|^2. \quad (10.3.4)$$

Interfaces between variants in this case are then under elastic constraints [71]. The dynamics of η is governed by the nonconserved (second) equation in (10.3.3). With this model, 3D simulations were performed with and without external stress [67].

In ferroelectric transitions such as those in $BaTiO_3$ [72], the order parameter is the polarization vector $\boldsymbol{P}$, and the effects of the applied electric field are of great technological importance. Here the (dipolar) electrostatic interaction and the coupling to the elastic field strongly influence the phase transition behavior [73]. These two ingredients should lead to

[7] The third-order term proportional to $\eta_1\eta_2\eta_3$ can also be present for $n = 3$ as shown in (3.3.29). Its effect in phase ordering has not yet been examined.

unique domain structures observed at long times. To study phase ordering, Nambu and Sagala performed 2D simulations [74], in which $\boldsymbol{P}$ and the strains are coupled in the form (10.3.2) but the electrostatic interaction is neglected. We predict that $\nabla \cdot \boldsymbol{P}$ should be strongly suppressed due to the electrostatic interaction, because $\rho_{\text{eff}} = -4\pi \nabla \cdot \boldsymbol{P}$ is the effective charge (as in (4.2.58)).

Anomalous elastic properties of improper and proper martensitic materials (including shape-memory effects) are of great technological importance [75]. We mention an experiment by Yamada and Uesu [76, 77] on improper martensitic $Pb_3(PO_4)_2$ with hexagonal symmetry. As an idealized condition, their system was composed of stripe domains with $\eta_2 = \pm\eta_0$ and $\eta_3 = \eta_1 = 0$ and those with $\eta_3 = \pm\eta_0$ and $\eta_2 = \eta_1 = 0$ varying along [100]. The effective shear modulus μ_{eff} was then 10^{-4}–10^{-3} of the shear modulus μ in the one-phase state. To explain such soft elasticity Yamada [77] proposed a pinning mechanism of interfaces due to defects. Ohta [78] proposed a mechanism of anomalous elasticity of twin structures, in which the domain walls are dragged by a very slowly evolving field, presumably representing a defect density.

In this section we explain some representative examples of ordering dynamics in order–disorder and improper martensitic phase transitions, though such studies are still fragmentary and insufficient.

10.3.1 Order–disorder transitions in bcc alloys with elastic inhomogeneity

Sagui *et al.* [39] examined the effect of elastic inhomogeneity in model C with scalar nonconserved and conserved variables, η and c, supposing bcc solids. If the free-energy density is even with respect to η, the simplest allowable form of the deformation stress in (10A.2) is given by

$$\sigma_{ij}^0 = -\left[\alpha_c(c - c_c) + \alpha_\eta \eta^2\right]\delta_{ij}, \tag{10.3.5}$$

where α_c and α_η are constants and c_c is a critical composition. The shear modulus depends on c and η as

$$\mu = \mu_0 + \mu_c(c - c_c) + \mu_\eta \eta^2. \tag{10.3.6}$$

Assuming isotropic elasticity, the elastic field can easily be eliminated and, to first order in the coefficients μ_c and μ_η, we obtain the elastic inhomogeneity interaction,

$$\mathcal{H}_{\text{inh}} = \frac{1}{L_0^2}\int d\boldsymbol{r}\left[\mu_c(c - c_c) + \mu_\eta \eta^2\right]\sum_{ij}\left[\nabla_i\nabla_j w - \frac{1}{d}\delta_{ij}\nabla^2 w\right]^2, \tag{10.3.7}$$

where L_0 is the longitudinal modulus in (10.1.23) and w is determined by

$$\nabla^2 w = \alpha_c(c - \langle c\rangle) + \alpha_\eta(\eta^2 - \langle\eta^2\rangle). \tag{10.3.8}$$

Recall that the morphologies from model C without the coupling to an elastic field were exemplified in Fig. 8.22. We are interested in how they are affected by $\mathcal{H}_{\text{inh}}$ using the common parameter values in f_{chem} in (10.3.1).

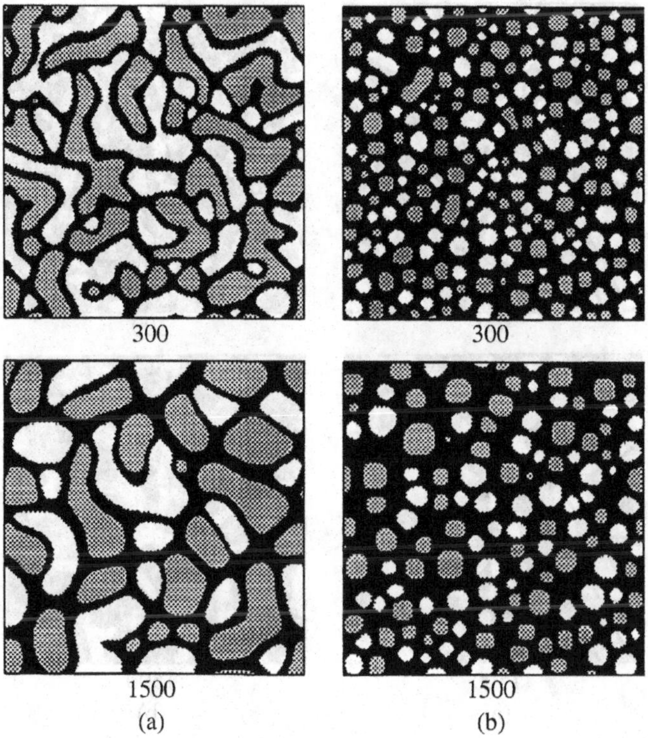

Fig. 10.21. The evolution patterns in model C when the variants of the ordered phase (white and gray) are hard [39]. The normalized concentration (see caption 8.22) is $-1/3$ in (a) and $1/3$ in (b).

In Figs 10.21 and 10.22 the normalized concentrations in (a), $-1/3$, and (b), $1/3$, are the same as those in (a) and (b), respectively, in Fig. 8.22. As in model B, we can see that the soft phase forms a percolated network at long times even if its volume fraction is relatively small. That is, in Fig. 10.21 the ordered regions (white or gray) are harder and take droplet shapes, while the disordered regions (black) are percolated. They resemble those in Fig. 10.11 (for the case without η) except that there are two variants of the ordered phase. Here we do not see antiphase boundaries (interfaces between the two variants). However, in Fig. 10.22 the disordered regions (black) are harder. In (a) the disordered regions form a wetting layer at an early stage due to the nature of the model C quench, but they tend to become droplet-like at later stages because they are hard. In (b) the soft ordered regions (white or gray) are elongated even at an early stage compared with those in Fig. 8.22 (b). Interestingly, we can see the appearance of antiphase boundaries both in (a) and (b) because the two variants touch after the shape changes of the disordered domains.

We make some remarks. (i) In the simulation, coarsening was observed to slow down considerably compared with the previous case in Fig. 8.22 without elasticity, but the reason for the asymptotic growth behavior remains unclear except for in Fig. 10.21(a) where we

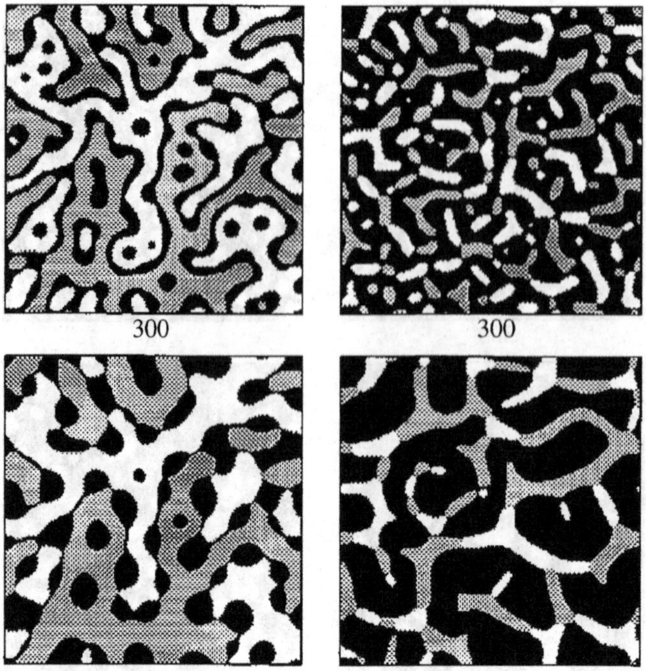

Fig. 10.22. Time evolution patterns in model C when the variants of the ordered phase (white and gray) are soft [39].

expect pinning. (ii) The anisotropy arising from cubic elasticity is neglected in the simulation, which would bring close resemblance of simulated patterns and real morphologies. (iii) Near the tricritical point, the elastic inhomogeneity interaction is marginal [38]. In fact, R_E in (10.1.56) is proportional to ξ on the line of the first-order transition phase. Thus, for sufficiently small μ_c and μ_η in (10.3.4), $R_E \gg \xi$ holds and the elastic inhomogeneity remains weak as the tricritical point is approached.

10.3.2 Chessboard-like L1$_0$ structures in fcc alloys

Bouar *et al.* [68] obtained unique chessboard patterns formed by tetragonal precipitates in a fcc matrix. The atomic configurations in a fcc cubic alloy are characterized by the concentration c and a three-component vector long-range order parameter (η_1, η_2, η_3) in (3.3.26). For simplicity, Bouar *et al.* assumed homogeneous elastic moduli and the emergence of pseudo-two-dimensional rod-like L1$_0$ microstructures with $\eta_3 = 0$ aligned along [001]. Then two pairs of variants (four variants) are considered out of three pairs of variants possible in the low-temperature phase. From the crystal symmetry, the stress-free strain in (10A.2) is diagonal and is even with respect to η_α. Its simplest form is written as

$$\epsilon^0_{ii} = \epsilon_1(\eta_1^2 + \eta_2^2) + (\epsilon_3 - \epsilon_1)(\eta_1^2\delta_{i1} + \eta_2^2\delta_{i2}) \qquad (i = 1, 2, 3), \tag{10.3.9}$$

where $\epsilon_1 = (a - a_0)/a_0$ and $\epsilon_3 = (c - a_0)/a_0$ with a_0 being the crystal lattice parameter of the cubic phase, and a and c those of the variants of the tetragonal phase. If the volume dilation is small compared with the tetragonal strain, we have $|\epsilon_3 + 2\epsilon_1| \ll |\epsilon_3|$ or $\epsilon_1 \cong -\epsilon_3/2$. Although the problem is considered in 2D, ϵ^0_{33} is nonvanishing here. In 2D, the elastic free energy $\mathcal{H}_{\text{el}}$ in (10A.5) is of the form

$$\mathcal{H}_{\text{el}} = \frac{1}{2}\int_{\boldsymbol{k}} \sum_{\alpha,\beta=1,2} B_{\alpha\beta}(\hat{\boldsymbol{k}})(\eta_\alpha^2)_{\boldsymbol{k}}(\eta_\beta^2)^*_{\boldsymbol{k}}, \tag{10.3.10}$$

where $\boldsymbol{k} = (k_1, k_2, 0)$ and $(\eta_\alpha^2)_{\boldsymbol{k}}$ are the Fourier transformations of η_α^2. Bouar *et al.* furthermore assumed isotropic elasticity on the [001] plane; then, from (10A.19) and (10A.20) we may derive

$$\mathcal{H}_{\text{el}} = \frac{\mu}{1-\nu}(\epsilon_3 - \epsilon_1)^2 \int_{\boldsymbol{k}} \left| \sum_{\alpha=1,2} (A_{\text{c}} - \hat{k}_\alpha^2)(\eta_\alpha^2)_{\boldsymbol{k}} \right|^2 + \mu\epsilon_1^2(1+\nu)\int d\boldsymbol{r}\eta^4, \tag{10.3.11}$$

where ν is the Poisson ratio in the range $-1 < \nu < 1/2$ [47] and $A_{\text{c}} = (\epsilon_3 + \nu\epsilon_1)/(\epsilon_3 - \epsilon_1)$. Orientation of domains is then selected such that the first term of (10.3.11) is minimized. We consider a tetragonal domain with $\eta_1 \neq 0$ and $\eta_2 = 0$ emerging in a cubic region, and home in on the interface between the two phases. It tends to be parallel to the habit plane whose normal $\boldsymbol{n}_0$ is in the softest direction minimizing $\mathcal{H}_{\text{el}}$. When $\epsilon_1/\epsilon_3 < 0$ and $|\epsilon_3/\epsilon_1| > \nu$, the angle θ_1 between $\boldsymbol{n}_0$ and [100] is given by

$$\cos^2\theta_1 = A_{\text{c}} = (\epsilon_3 + \nu\epsilon_1)/(\epsilon_3 - \epsilon_1). \tag{10.3.12}$$

For $\epsilon_1/\epsilon_3 < 0$ and $|\epsilon_3/\epsilon_1| \leq \nu$ we have $\theta_1 = \pi/2$. The corresponding angle θ_2 for a tetragonal domain with $\eta_2 \neq 0$ and $\eta_1 = 0$ emerging in a cubic region is given by $\theta_2 = \pi/2 - \theta_1$. In Fig. 10.23 a pattern in a simulation at $\epsilon_1/\epsilon_3 = -0.49$ and $\nu = 1/3$ is compared with an experimental image for $Co_{39.5}Pt_{60.5}$ [68]. This choice of parameters makes the angles of the chessboard pattern from the habit plane relation (10.3.11) agree with the observed ones.

10.3.3 Improper hexagonal to orthorhombic transformations

A number of unique domains have been observed in hexagonal → orthorhombic transformations [79]–[81] and hexagonal → monoclinic transformations [82]. Torres presented a general 3D form of the free-energy density for such crystal symmetry [66].

Recently, Wen *et al.* [6, 83] studied patterns emerging in improper hexagonal → orthorhombic transformations on the basis of (10.3.1) and (10.3.2). They considered the basal plane of the hexagonal lattice, illustrated in Fig. 10.24, assuming homogeneity perpendicular to the plane. Then the atomic structures of the ordered phase can be represented

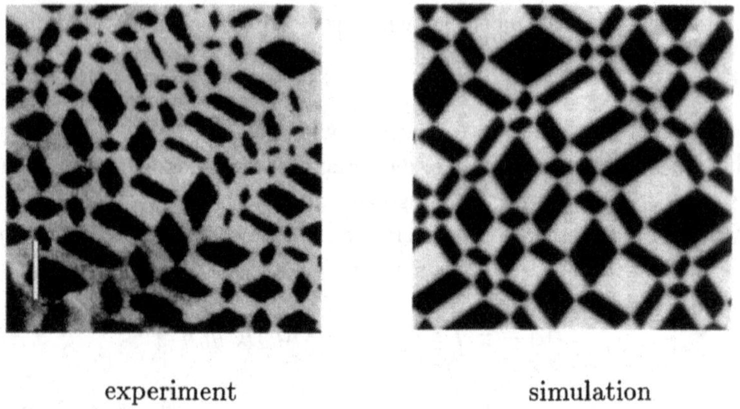

Fig. 10.23. Comparison between an experimental TEM image and a simulation result of a chessboard pattern in $Co_{39.5}Pt_{60.5}$ [68]. In the simulation, η_1 or η_2 is nonvanishing in the black regions, while $\eta_1 \cong 0$ and $\eta_2 \cong 0$ in the white regions.

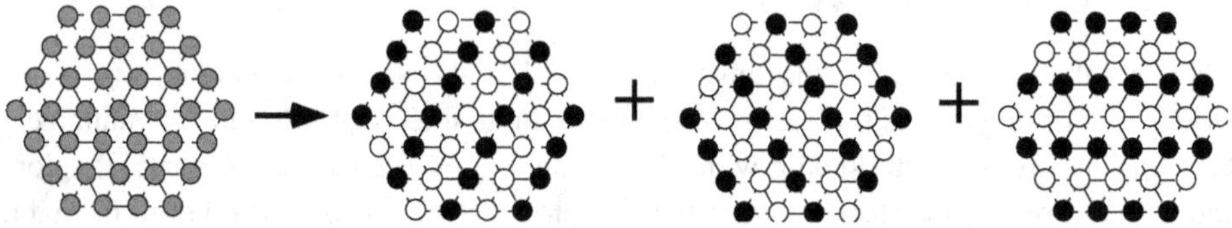

Fig. 10.24. The basal plane of the hexagonal structure (courtesy of Professor L. Q. Chen). The hexagonal disordered phase (*left*) is transformed into one of the three (pairs of) variants of the orthorhombic phase (*right*).

by a three-component long-range order parameter (η_1, η_2, η_3), giving rise to three pairs of variants (six variants) in the orthorhombic phase on the basal plane, each pair being characterized by different elastic deformations. In particular, in Ti–Al–Nb [81], where c represents the Nb concentration, the perpendicular components ϵ_{i3}^0 $(i = 1, 2, 3)$ and the dilational part of the stress-free strain are small, and the Nb concentrations in the two phases are only slightly different. On the basis of these facts, Wen *et al.* considered the problem in 2D with a concentration-independent, traceless stress-free strain,

$$\epsilon_{11}^0 = -\epsilon_{22}^0 = \frac{1}{2}\epsilon_s\left(2\eta_1^2 - \eta_2^2 - \eta_3^2\right), \quad \epsilon_{12}^0 = \epsilon_{21}^0 = \frac{\sqrt{3}}{2}\epsilon_s\left(\eta_2^2 - \eta_3^2\right), \tag{10.3.13}$$

where $\epsilon_{i3}^0 = 0$ and ϵ_s is a constant characteristic strain. This is the simplest form of the stress-free strain under the condition that the free energy is invariant with respect to rotation of the reference frame by $\theta = n\pi/3$ $(n = 1, 2, \ldots)$ in the xy plane. This can be shown as follows. From Appendix 10A, the cross term between (η_1, η_2, η_3) and $\boldsymbol{u}$ in the free-energy density is written as

$$f_{\rm I} = -\sum_{ijk\ell} \epsilon^0_{ij}\lambda_{ijk\ell}\epsilon_{ij} = -\frac{1}{2}\mu\epsilon_{\rm s}\big[(2\eta_1^2 - \eta_2^2 - \eta_3^2)e_2 + \sqrt{3}(\eta_2^2 - \eta_3^2)e_4\big], \tag{10.3.14}$$

where isotropic elasticity is assumed and

$$e_2 = \nabla_x u_x - \nabla_y u_y, \quad e_4 = \nabla_x u_y + \nabla_y u_x. \tag{10.3.15}$$

For the rotation $x' = x\cos\theta + y\sin\theta$ and $y' = -x\sin\theta + y\cos\theta$, we have

$$e_2' = e_2\cos 2\theta + e_4\sin 2\theta, \quad e_4' = -e_2\sin 2\theta + e_4\cos 2\theta. \tag{10.3.16}$$

For $\theta = \pi/3$ we set $(\eta_1', \eta_2', \eta_3') = (\eta_2, \eta_3, \eta_1)$; then, $f_{\rm I}$ is invariant in terms of the primed quantities in the rotated reference frame. We also remark that Torres' form of $f_{\rm I}$ reduces to (10.3.13) in the 2D case [66].

Because the elastic property on the basal plane of a hexagonal lattice is isotropic from the triangular symmetry, the 2D formula (10A.21) is applicable and then the elastic interaction energy becomes

$$\mathcal{H}_{\rm el} = \frac{\mu\epsilon_{\rm s}^2}{4(1-\nu)}\int_{\boldsymbol k}\big|(\hat{k}_x^2 - \hat{k}_y^2)(2\eta_1^2 - \eta_2^2 - \eta_3^2)_{\boldsymbol k} + 2\sqrt{3}\hat{k}_x\hat{k}_y(\eta_2^2 - \eta_3^2)_{\boldsymbol k}\big|^2. \tag{10.3.17}$$

With this $\mathcal{H}_{\rm el}$ Wen *et al.* numerically solved (10.3.2) with the chemical free-energy density (10.3.1). Figure 10.25 shows coarsening orthorhombic domains, where the mean concentration is 0.125 in (a) and 0.10 in (b). The resultant volume fraction of the ordered phase is about 0.69 in (a) and 0.37 in (b). We observe that the domain shapes depend sensitively on the mean concentration. The unique orientation relationship of the habit planes of the three variants is determined from minimization of (10.3.17) as a function of the angle θ of the interface normal where $\hat{k}_x = \cos\theta$ and $\hat{k}_y = \sin\theta$. Closely resembling patterns have been observed in experiments.

They furthermore examined the effect of applied strain $\epsilon^{\rm a}_{ij}$ in phase ordering. From (10A.2) and (10A.3) we notice that the applied stress $\sigma^{\rm a}_{ij} = \sum_{k\ell}\lambda_{ijk\ell}\epsilon^a_{k\ell}$ gives rise to the free-energy density of the form, $f_{\rm ex} = -\sum_{ij}\epsilon^0_{ij}\sigma^a_{ij}$. They argued that a uniaxial stress can be applied in such a direction that variants 1 and 2 are equally favored but variant 3 is unfavored. In this case $f_{\rm ex}$ is written as

$$f_{\rm ex} = -\mu\epsilon_{\rm s}\epsilon^{\rm a}(\eta_1^2 + \eta_2^2 - 2\eta_3^2), \tag{10.3.18}$$

where $\epsilon^{\rm a}$ is the strength of the applied strain and $\epsilon_{\rm s}\epsilon^{\rm a} > 0$. Obviously the applied stress is uniaxial along the direction whose angle with respect to the first principal axis (horizontal in Fig. 10.24) is $\pi/6$. With the above term included in simulations (with fixed average strain), they obtained a number of patterns in which the fraction of variant 3 diminishes with increasing $\epsilon^{\rm a}$.

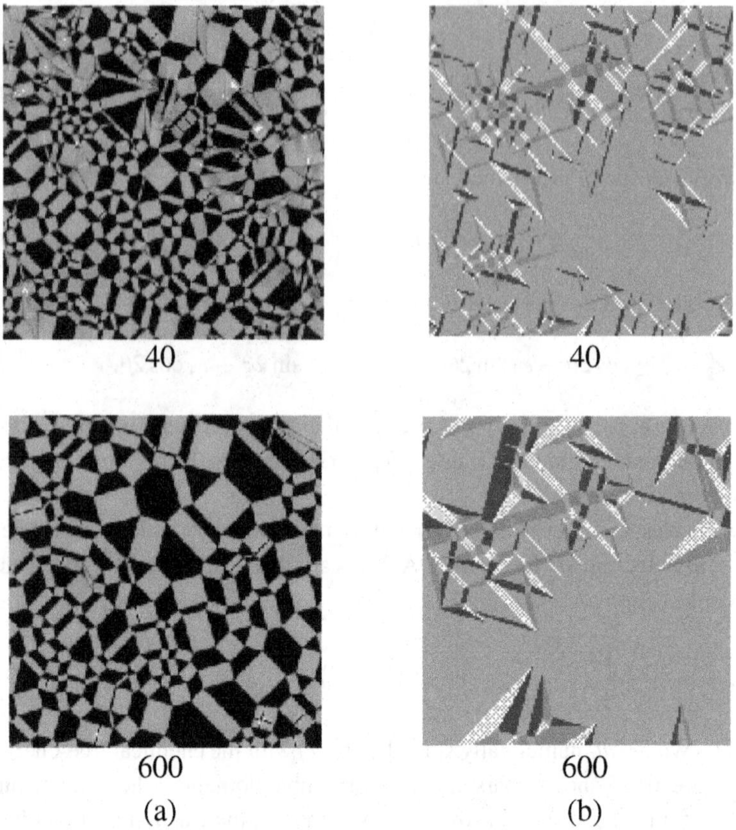

Fig. 10.25. Simulated precipitation processes of ordered orthorhombic domains from a disordered hexagonal matrix [83]. In (a) the white regions are ordered and the black regions are disordered. In (b) the shades of gray represent the values of $\eta_1^2 - \eta_2 - 2\eta_3^2$, distinguishing the three ordered variants in the majority disordered hexagonal matrix. Therefore, the four different gray levels from brightest to darkest correspond to variant 1, parent phase, variant 2, and variant 3, respectively. The volume fraction of the ordered domains is 0.69 in (a) and 0.37 in (b). The numbers are the times after quenching, scaled appropriately.

Domain pinning

Wen *et al.* also performed simulations by neglecting the concentration fluctuations or for $c = \text{const}$ [6, 83]. Then the dynamics is provided by the second equation of (10.3.3) obeyed by the three-component long-range vector order parameter (η_1, η_2, η_3). In phase ordering, the system is soon composed of three pairs of the low-temperature variants. As shown in Fig. 10.26, the late-stage pattern is characterized by fixed orientations of the interfaces between the different variants. As a result, the angles between them at the junction points are multiples of $\pi/6$. They closely resemble those observed in real alloys. Due to these

Fig. 10.26. Typical pattern of orthorhombic domains at fixed concentration in a pinned (late-stage) state on a hexagonal basal plane.

strong geometrical constraints, pinning of the domain growth is expected when the elastic energy ($\sim \mu\epsilon_s^2 R^3$) per domain exceeds the interface energy ($\sim \sigma R^2$, σ being the surface tension) [84]. Thus, the characteristic domain size R^* in pinned states is given by

$$R^* \sim \sigma/\mu\epsilon_s^2. \tag{10.3.19}$$

This relation is confirmed in Fig. 10.27, which indicates that the time to pinning becomes shorter with an increase in the characteristic elastic energy density $\mu\epsilon_s^2$.

10.4 Proper martensitic transitions

In proper martensitic materials a structural phase transition occurs without large-scale composition changes. Its representative microscopic origin is coupling between electronic orbital states and lattice distortions (called Jahn–Teller coupling [85]–[87]) For example, when one electron occupies an electronic state spanned by doubly degenerate d-orbital states at each site in the undistorted crystal structure, Kanamori introduced pseudo Pauli spin matrices $\hat{\sigma}_{nx}$ and $\hat{\sigma}_{nz}$ operating on electronic states at site n [88]. As will be derived in Appendix 10C, there arises a bilinear orbit–lattice interaction energy of the form

$$\mathcal{H}_{\rm JT} = g_{\rm K} \sum_n (\hat{\sigma}_{nz} Q_{n3} + \hat{\sigma}_{nx} Q_{n2}), \tag{10.4.1}$$

where Q_{n2} and Q_{n3} represent appropriate linear combinations of atomic displacements at site n. Their acoustic parts may be equated to e_2 and e_3 to be defined in (10.4.4) below. Then, if there is an orbital order represented by $\langle\hat{\sigma}_{nz}\rangle \propto \cos\varphi$ and $\langle\hat{\sigma}_{nz}\rangle \propto \sin\varphi$ at low

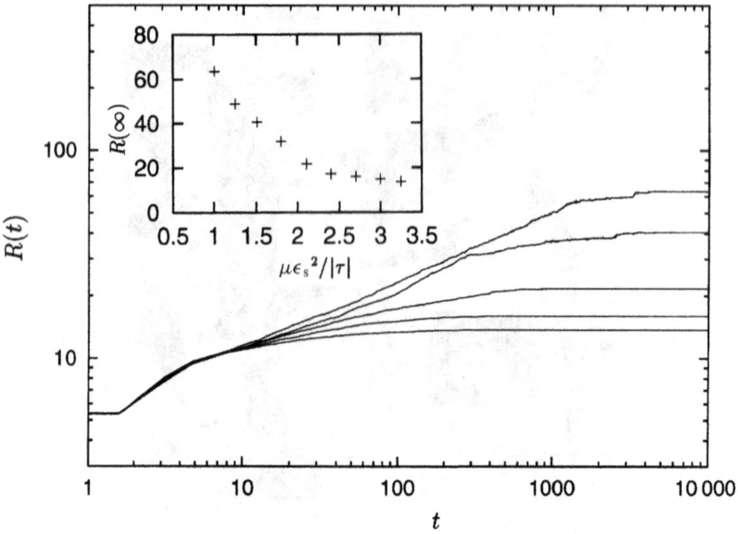

Fig. 10.27. The domain size $R(t)$ (inverse perimeter length) of orthorhombic domains vs time after quenching for $\mu\epsilon_s^2 = 1.01, 1.51, 2.11, 2.7$, and 3.25 from above [84]. In the inset the domain size in pinned states is shown. In dimensionless units we take $\tau = -1$, $A_3 = 0.5$, $A_4 = 11/6$, $A_5 = 0.5$, and $C = 1.2$ in (10.3.4) and $K/\mu = 2$ (or $\nu = 1/2$).

temperatures, a cubic to tetragonal phase transition is caused as $\langle Q_{n3}\rangle = \eta_0 \cos\varphi$ and $\langle Q_{n2}\rangle = \eta_0 \sin\varphi$. The tetragonal distortion is along the x, y, z axis for $\varphi = 2\pi/3$, $-2\pi/3$, 0, respectively. The transition becomes first order in the presence of an anharmonic energy of the form [86]

$$\mathcal{H}_3 = -A_3 \sum_n (Q_{n3}^3 - 3Q_{n3}Q_{n2}^2). \tag{10.4.2}$$

If the coefficient A_3 is small, the transition becomes weakly discontinuous.

Proper structural phase transitions are often characterized by *soft modes* [45, 90]. For example, at nearly continuous cubic to tetragonal transitions, the elastic constant against [110] sound becomes small towards the transition as

$$C' = \frac{1}{2}(C_{11} - C_{12}) = A_T(T - T_{c0}), \tag{10.4.3}$$

as a function of T at a fixed pressure p, where the solid becomes soft against tetragonal strains. As a result, the thermal fluctuations of the tetragonal strains are enhanced as $T \to T_{c0}$, as can be seen in (10.4.44) below. A representative example is given by Nb_3Sn [91, 92]. However, in KCN [93], C_{44} tends to zero and softening occurs in a two-dimensional subspace of the wave vector. (See near (10.5.1).)

10.4.1 Nonlinear elastic free energy

Following the conventional continuum theory [94]–[97], we describe proper martensitic transitions in terms of the strains, including anharmonic terms, where the microscopic *true* order parameter ($\hat{\sigma}_{nz}$ and $\hat{\sigma}_{nx}$ in the Jahn–Teller case (10.4.1)) has been eliminated. [However, it is unclear under what conditions this approach is justified.[8]]

In this book the diagonal strains are defined by

$$\begin{aligned} e_1 &= \nabla_x u_x + \nabla_y u_y + \nabla_z u_z, \\ e_2 &= \nabla_x u_x - \nabla_y u_y, \\ e_3 &= \frac{1}{\sqrt{3}}(2\nabla_z u_z - \nabla_x u_x - \nabla_y u_y). \end{aligned} \tag{10.4.4}$$

The off-diagonal components are written as

$$e_4 = \nabla_x u_y + \nabla_y u_x, \quad e_5 = \nabla_y u_z + \nabla_z u_y, \quad e_6 = \nabla_z u_x + \nabla_x u_z. \tag{10.4.5}$$

In the bilinear order the elastic energy corresponding to (10.2.1) is expressed as

$$\mathcal{H}_0 = \frac{1}{2}\int d\boldsymbol{r}\big[K e_1^2 + C'(e_2^2 + e_3^2) + C_{44}(e_4^2 + e_5^2 + e_6^2)\big]. \tag{10.4.6}$$

We hereafter consider a cubic to tetragonal phase transition, where the two tetragonal strains, e_2 and e_3 constitute a two-component order parameter. The elastic constant C' is assumed to be expressed as (10.4.3). To describe the transition we should include the following elastic energy consisting of higher-order anharmonic terms,

$$\mathcal{H}' = \int d\boldsymbol{r}\big[-\alpha e_1(e_2^2+e_3^2) - B(e_3^3 - 3e_2^2 e_3) + \frac{\bar{u}_0}{4}(e_2^2+e_3^2)^2 + \frac{v_0}{6}(e_2^2+e_3^2)^3\big], \tag{10.4.7}$$

where α is a Grüneisen constant related to the density or pressure dependence of C' as

$$\alpha = \frac{\rho}{2}\left(\frac{\partial C'}{\partial \rho}\right)_T = \frac{K}{2}\left(\frac{\partial C'}{\partial p}\right)_T. \tag{10.4.8}$$

Thus α can be known from measurements of C' for various pressures in the cubic phase (though there seem to be no available data). The third-order term proportional to B, which corresponds to $\mathcal{H}_3$ in (10.4.2), is allowable from symmetry because

$$\begin{aligned} e_3(e_3^2 - 3e_2^2) &= 12\sqrt{3}\epsilon_x^{\mathrm{D}}\epsilon_y^{\mathrm{D}}\epsilon_z^{\mathrm{D}} \\ &= \eta^3\cos(3\varphi). \end{aligned} \tag{10.4.9}$$

In the first line,

$$\epsilon_i^D = \nabla_i u_i - \frac{1}{3}e_1 \tag{10.4.10}$$

are the deviatoric diagonal strains. In the second line we have set

$$e_2 = \eta\sin\varphi, \quad e_3 = \eta\cos\varphi, \tag{10.4.11}$$

[8] In Subsection 10.4.7 we will set up another Ginzburg–Landau theory for the orbital order parameter coupled to elastic strains. It will indeed predict some effects beyond the scope of the traditional nonlinear strain theory.

where $\eta = (e_2^2 + e_3^2)^{1/2} \geq 0$, and used the relation $\cos(3\varphi) = \cos^3\varphi - 3\cos\varphi\sin^2\varphi$. Here we express ϵ_i^D in terms of η and φ as

$$\epsilon_x^{\rm D} = \frac{\eta}{\sqrt{3}}\sin\left(\varphi - \frac{\pi}{6}\right), \quad \epsilon_y^{\rm D} = -\frac{\eta}{\sqrt{3}}\sin\left(\varphi + \frac{\pi}{6}\right), \quad \epsilon_z^{\rm D} = \frac{\eta}{\sqrt{3}}\cos\varphi. \tag{10.4.12}$$

The gradient free energy can be of the form,

$$\mathcal{H}_{\rm grad} = \int d\boldsymbol{r}\left[\frac{C}{2}(|\nabla e_2|^2 + |\nabla e_3|^2) - \frac{D}{2}W\right], \tag{10.4.13}$$

with

$$\begin{aligned} W &= 4[\epsilon_y^{\rm D}(3\nabla_z^2 - \nabla^2)\epsilon_x^{\rm D} + \epsilon_z^{\rm D}(3\nabla_x^2 - \nabla^2)\epsilon_y^{\rm D} + \epsilon_x^{\rm D}(3\nabla_y^2 - \nabla^2)\epsilon_z^{\rm D}] \\ &= e_3\Delta_3 e_3 - e_2\Delta_3 e_2 - e_2\Delta_2 e_3 - e_3\Delta_2 e_2, \end{aligned} \tag{10.4.14}$$

where $\Delta_2 = \sqrt{3}(\nabla_x^2 - \nabla_y^2)$ and $\Delta_3 = 2\nabla_z^2 - \nabla_x^2 - \nabla_y^2$. Comparing (10.4.9) and (10.4.14), we notice that the gradient term $-DW/2$ in (10.4.13) is allowable as well as the third-order anharmonic term proportional to B in $\mathcal{H}'$. Because $\mathcal{H}_{\rm grad}$ should be nonnegative-definite, we require

$$C \geq |D|, \tag{10.4.15}$$

which can also be seen in (10.4.33) below. Now the total Ginzburg–Landau free energy is given by

$$\begin{aligned} \mathcal{H} &= \mathcal{H}_0 + \mathcal{H}' + \mathcal{H}_{\rm grad} \\ &= \int d\boldsymbol{r} f + \mathcal{H}_{\rm grad}. \end{aligned} \tag{10.4.16}$$

The free-energy density f (not including the gradient terms) is expressed as

$$f = \frac{K}{2}e_1^2 + \frac{C_{44}}{2}(e_4^2 + e_5^2 + e_6^2) + \left(\frac{C'}{2} - \alpha e_1\right)\eta^2 - B(e_3^3 - 3e_3e_2^2) + \frac{\bar{u}_0}{4}\eta^4 + \frac{v_0}{6}\eta^6, \tag{10.4.17}$$

where $\eta^2 = e_2^2 + e_3^2$.

Stress tensor

The stress tensor σ_{ij} can be calculated in the procedure in (10.1.9) as

$$\begin{aligned} \sigma_{xx} &= Ke_1 - \alpha\eta^2 + S_2 - \frac{1}{\sqrt{3}}S_3, \\ \sigma_{yy} &= Ke_1 - \alpha\eta^2 - S_2 - \frac{1}{\sqrt{3}}S_3, \\ \sigma_{zz} &= Ke_1 - \alpha\eta^2 + \frac{2}{\sqrt{3}}S_3, \end{aligned} \tag{10.4.18}$$

where

$$S_2 = \frac{\partial f}{\partial e_2} - C\nabla^2 e_2 + D(\Delta_3 e_2 + \Delta_2 e_3),$$
$$S_3 = \frac{\partial f}{\partial e_3} - C\nabla^2 e_3 - D(\Delta_3 e_3 - \Delta_2 e_2). \quad (10.4.19)$$

The normal stress differences are expressed as

$$\sigma_{xx} - \sigma_{yy} = 2S_2, \quad 2\sigma_{zz} - \sigma_{xx} - \sigma_{yy} = \frac{4}{\sqrt{3}} S_3. \quad (10.4.20)$$

The off-diagonal components are given by $\sigma_{ij} = C_{44}(\nabla_i u_j + \nabla_j u_i)$ $(i \neq j)$ as in (10.2.2). We readily confirm the relation,

$$\sum_j \nabla_j \sigma_{ij} = -\frac{\delta}{\delta u_i} \mathcal{H}. \quad (10.4.21)$$

Homogeneous states

In homogeneous stress-free states we have $e_4 = e_5 = e_6 = 0$, $e_1 = \alpha\eta^2/K$, and $\partial f/\partial e_2 = \partial f/\partial e_3 = 0$. Then f becomes dependent only on e_2 and e_3 or η and φ as

$$f(e_2, e_3) = \frac{1}{2}C'\eta^2 - B\eta^3 \cos 3\varphi + \frac{1}{4}u_0\eta^4 + \frac{1}{6}v_0\eta^6, \quad (10.4.22)$$

where

$$u_0 = \bar{u}_0 - 2\alpha^2/K. \quad (10.4.23)$$

We assume $B > 0$ and $u_0 > 0$ and neglect the sixth-order term ($v_0 = 0$). For $\eta > 0$, f is minimized for $\varphi = 2\pi/3, -2\pi/3$, and 0 as a function of φ, which correspond to three variants with the symmetry axis along the x, y, and z axis, respectively, from (10.4.12). In Fig. 10.28 we show a contour plot of $f(e_2, e_3)$ for $v_0 = 0$. Let $f(e_2, e_3)$ be minimized at $\eta = \eta_0$ for these values of φ. If it is nonvanishing, it satisfies

$$C' - 3B\eta_0 + u_0\eta_0^2 = 0, \quad (10.4.24)$$

which is solved to give

$$\eta_0 = (3B/2u_0)\left[1 + \sqrt{1 - 4u_0C'/9B^2}\right]. \quad (10.4.25)$$

Under the stress-free condition, the ordered phase is stable for $C' \leq 2B^2/u_0$ and metastable for $2 < C'u_0/B^2 < 9/4$. If $C' \sim B^2/u_0$, we have $\eta_0 \sim B/u_0$ and $f_{\min} \sim -B\eta_0^3$, where $f_{\min}$ is the minimum of f. The lattice constant $a_\parallel$ along the symmetry axis and that, $a_\perp$, perpendicular to it are expressed as

$$\frac{a_\parallel}{a_0} = 1 + \frac{1}{\sqrt{3}}\eta_0 - \frac{\alpha}{3K}\eta_0^2, \quad \frac{a_\perp}{a_0} = 1 - \frac{1}{2\sqrt{3}}\eta_0 - \frac{\alpha}{3K}\eta_0^2, \quad (10.4.26)$$

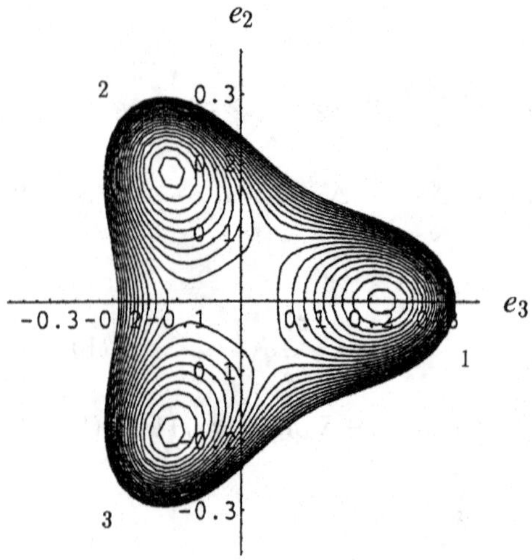

Fig. 10.28. Contour plot of the free-energy density $f(e_2, e_3)$ in (10.4.22) for $B > 0$. The variants 1, 2, and 3 correspond to the tetragonal variants stretched along the z, x, and y axis, respectively.

where the lattice constants along the three axes are set equal to $a_0(1 + \nabla_i u_i)$ with a_0 being the lattice constant in the cubic phase. We note that $a_\parallel > a_\perp$ for $B > 0$ but $a_\parallel < a_\perp$ for $B < 0$.

We may also apply a uniaxial stress $\sigma_{zz} = \sigma_a$ along the z axis under the stress-free condition in the x and y axes ($\sigma_{xx} = \sigma_{yy} = 0$). In homogeneous states we have $e_1 = (\alpha\eta_0^2 + \sigma_a/3)/K$, $\partial f \partial e_2 = 0$, and $\partial f/\partial e_3 = \sigma_a/\sqrt{3}$, so we should minimize

$$\tilde{f} = f(e_2, e_3) - \frac{1}{\sqrt{3}}\sigma_a e_3. \tag{10.4.27}$$

For small σ_a there appears a difference between the minimum free energy for $\varphi \cong 0$ and that for $\varphi \cong \pm 2\pi/3$ given by

$$\Delta f_{\min} = -\frac{1}{2}\sqrt{3}\sigma_a\eta_0 + O(\sigma_a^2). \tag{10.4.28}$$

The higher-order terms are negligible for $|\sigma_a/B\eta_0^2| \ll 1$. If σ_a satisfies this condition and has the same (opposite) sign as that of B, the variant with the symmetry axis along the z axis is energetically unfavored (favored). (See Subsection 10.4.7 for more discussions on the effect of applied stress.)

Compatibility conditions

Because the displacement vector is composed of three components, u_x, u_y, and u_z, in 3D, there are certain relations among the six strains which are satisfied identically. They are known as compatibility relations given below [99]:

$$\begin{aligned}
\nabla_x\nabla_y e_4 &= \nabla_x^2\epsilon_y+\nabla_y^2\epsilon_x, \quad 2\nabla_x\nabla_y\epsilon_z=\nabla_z(\nabla_x e_5+\nabla_y e_6-\nabla_z e_4),\\
\nabla_y\nabla_z e_5 &= \nabla_y^2\epsilon_z+\nabla_z^2\epsilon_y, \quad 2\nabla_y\nabla_z\epsilon_x=\nabla_x(\nabla_y e_6+\nabla_z e_4-\nabla_x e_5),\\
\nabla_z\nabla_x e_6 &= \nabla_z^2\epsilon_x+\nabla_x^2\epsilon_z, \quad 2\nabla_z\nabla_x\epsilon_y=\nabla_y(\nabla_z e_4+\nabla_x e_5-\nabla_y e_6),
\end{aligned} \tag{10.4.29}$$

where $\epsilon_i = \nabla_i u_i$ $(i = x, y, z)$ are the diagonal strains and can be expressed in terms of e_1, e_2, and e_3. These relations, which readily follow from the definitions of the strains, are known to be sufficient conditions for the existence of $\boldsymbol{u} = (u_x, u_y, u_z)$. That is, if they are satisfied for given strains, we may construct the corresponding $\boldsymbol{u}$ which yield these strains.

10.4.2 Interface between variants

We examine an interface between the two stress-free variants with $\varphi = 2\pi/3$ and $-2\pi/3$ [96]–[98] by assuming that all the strains depend only on

$$s = (x+y)/\sqrt{2}. \tag{10.4.30}$$

The interface normal is then in the direction [110]. We require $e_2 \to \pm\sqrt{3}\eta_0/2$ and $e_3 \to -\eta_0/2$ far from the interface $(s \to \pm\infty)$. The compatibility relations (10.4.29) indicate that ϵ_z, $e_4 - \epsilon_x - \epsilon_y$, and $e_5 - e_6$ should be constants independent of s. We thus seek the solution by setting

$$e_1 = -\sqrt{3}\left(e_3+\frac{1}{2}\eta_0\right)+\frac{\alpha}{K}\eta_0^2, \quad e_4=-\sqrt{3}\left(e_3+\frac{1}{2}\eta_0\right), \quad e_5=e_6=0. \tag{10.4.31}$$

As ought to be the case, the strain,

$$\nabla_z u_z = -\frac{1}{2\sqrt{3}}\eta_0 - \frac{1}{3K}\alpha\eta_0^2, \tag{10.4.32}$$

is constant. All the strains are now expressed in terms of e_2 and e_3. The free-energy density including the gradient terms is written as

$$\hat{f} = \frac{K}{2}\left(e_1-\frac{\alpha}{K}\eta^2\right)^2+\frac{1}{2}C_{44}e_4^2+f(e_2,e_3)+\frac{C+D}{2}\left|\frac{d}{ds}e_2\right|^2+\frac{C-D}{2}\left|\frac{d}{ds}e_3\right|^2, \tag{10.4.33}$$

where e_1 and e_4 are given by (10.4.31) and $f(e_2, e_3)$ is defined by (10.4.22). The total free energy $\mathcal{H}$ is the space integral of $\hat{f}$. From (10.4.18) and (10.4.19) the mechanical equilibrium condition $\sum_j \nabla_j\sigma_{ij} = 0$ turns out to be equivalent to the extremum condition $\delta\mathcal{H}/\delta e_2 = \delta\mathcal{H}/\delta e_3 = 0$ of $\mathcal{H}$ with respect to e_2 and e_3.

Barsch–Krumhansl solution

In $\hat{f}$ the cross terms between e_2 and e_3 are written in the form $G(e_3)e_2^2$. It then follows that $e_3 = -\eta_0/2 = \text{const.}$ if $\partial G(e_3)/\partial e_3 = 0$. This condition is rewritten as $(\sqrt{3} + 2\alpha e_3/K)\alpha + 3B + u_0 e_3 = 0$, if we assume $u_0 > 0$ and $v_0 = 0$ in f. Then,

$$C' = (9B^2/4u_0)X(2 - X), \quad \eta_0 = (3B/2u_0)X, \tag{10.4.34}$$

with $X = 4(1 + \alpha/\sqrt{3}B)/(1 + 2\alpha^2/Ku_0) > 1$. Now e_3 is constant and

$$\hat{f} = \left(\frac{u_0}{4} + \frac{\alpha^2}{2K}\right)(e_2^2 - 3\eta_0^2/4)^2 + \frac{C+D}{2}\left|\frac{d}{ds}e_2\right|^2 + \text{const.} \tag{10.4.35}$$

Thus the interface profile is simply of the form

$$e_2 = (\sqrt{3}\eta_0/2)\tanh(s/2\xi), \tag{10.4.36}$$

where

$$\xi^2 = 2(C+D)/[3\eta_0^2(u_0 + 2\alpha^2/K)]. \tag{10.4.37}$$

In their original theory Barsch and Krumhansl [96] assumed $\alpha = 0$ and $C' = -18B^2/u_0$ to obtain (10.4.36).

Small-B case

When B is small, the phase transition is weakly discontinuous and there arises a nearly critical case,

$$K + C_{44} \gg |C'| \sim B^2/u_0, \quad \eta_0 \sim B/u_0. \tag{10.4.38}$$

Notice that the first two terms in (10.4.33) give rise to the bilinear term $3(K+C_{44})(\delta e_3)^2/2$, which serves to suppress the deviation $\delta e_3 = e_3 + \eta_0/2$. Some calculations yield

$$e_3 = -\frac{1}{2}\eta_0 - \frac{\alpha}{\sqrt{3}(K + C_{44})}(\eta^2 - \eta_0^2) + O(B^3). \tag{10.4.39}$$

The free-energy density is again of the form (10.4.35) up to order B^4, leading to the profile (10.4.36). The interface thickness ξ is also given by (10.4.37), so $\xi \propto B^{-1}$.

10.4.3 Dynamic equation and linear analysis

Large-scale elastic disturbances propagate on the acoustic timescale throughout the system. We assume the dynamic equations,

$$\rho\frac{\partial^2}{\partial t^2}\boldsymbol{u} - \zeta\nabla^2\frac{\partial}{\partial t}\boldsymbol{u} = -\frac{\delta}{\delta \boldsymbol{u}}\mathcal{H} = \nabla\cdot\overleftrightarrow{\sigma}, \tag{10.4.40}$$

where ρ is the mass density and ζ is the bulk viscosity assumed to be isotropic [47]. If C' is slightly negative, the transverse acoustic sound varying along [110] (and that varying in one of the other five equivalent directions) becomes unstable.

Here we perform linear analysis assuming that all the deviations are small, depending on space and time as $\exp(i\boldsymbol{k}\cdot\boldsymbol{r}-\Omega t)$. Let the direction $\hat{\boldsymbol{k}}=k^{-1}\boldsymbol{k}$ be nearly along [110] and $\hat{\boldsymbol{k}}\cdot\boldsymbol{u}$ be very small compared to $u_x \cong -u_y$. Then the deviations are expanded in powers of $\hat{k}_x^2-1/2$, $\hat{k}_y^2-1/2$, and $\hat{k}_z^2$. It follows that $u_z \cong -(K/\mu+1)\hat{k}_z(\hat{\boldsymbol{k}}\cdot\boldsymbol{u}) \cong 0$. Thus only e_2 remains nonvanishing with the other strains being nearly zero. The relaxation rate Ω is determined by

$$\rho\Omega^2 - \zeta k^2\Omega = -k^2 C_e(\boldsymbol{k}), \tag{10.4.41}$$

where

$$C_e(\boldsymbol{k}) = C' + \mu\theta^2 + \frac{4K\mu}{K+\mu}(\varphi-\pi/4)^2 + (C+D)k^2. \tag{10.4.42}$$

Here $(\hat{k}_x,\hat{k}_y,\hat{k}_z) = (\cos\theta\cos\varphi, \cos\theta\sin\varphi, \sin\theta)$ with $\theta \cong 0$ and $\varphi \cong \pi/4$, and C and D are the coefficients in the gradient free energy (10.4.13). A growing mode with negative Ω exists for $C_e < 0$. At long wavelengths we obtain

$$\Omega \cong \pm k(|C_e(\boldsymbol{k})|/\rho)^{1/2}. \tag{10.4.43}$$

If $|C'| \ll K \sim \mu$, the instability condition $C_e < 0$ is realized only when $\hat{\boldsymbol{k}}$ slightly deviates from [110] with $|\theta|$ and $|\varphi-\pi/4|$ being smaller than $|C'/\mu|^{1/2}$. This high anisotropy of the growing fluctuations can be seen at early stages in Figs 10.33–10.35 below.

We also notice that the thermal fluctuations of $\boldsymbol{u}$ grow for small positive C' from (10A.16). In the gaussian approximation the growing part of the correlation function of $\boldsymbol{u}$ is written as

$$\langle \boldsymbol{u}_k \boldsymbol{u}_k^* \rangle \cong \frac{T}{k^2 C_e(\boldsymbol{k})} \boldsymbol{e}_{[1\bar{1}0]}\boldsymbol{e}_{[1\bar{1}0]}, \tag{10.4.44}$$

where $\boldsymbol{k}$ is assumed to be nearly along [110] and $\boldsymbol{e}_{[1\bar{1}0]} = 2^{-1/2}(1,-1,0)$.

10.4.4 Square–rectangular transition

We present nonlinear analysis of square to rectangular transformations in 2D on the basis of the following free-energy density,

$$f = \frac{K}{2}e_1^2 + \frac{\mu}{2}e_4^2 - \alpha' e_1 e_2^2 + \left(\frac{\tau}{2}e_2^2 + \frac{\bar{u}_0}{4}e_2^4 + \frac{v_0}{6}e_2^6\right) + \frac{D}{2}|\nabla e_2|^2, \tag{10.4.45}$$

where $\tau(=C')$ is a control parameter in our simulation and[9]

$$e_1 = \nabla_x u_x + \nabla_y u_y, \quad e_2 = \nabla_x u_x - \nabla_y u_y, \quad e_4 = \nabla_y u_x + \nabla_x u_y \tag{10.4.46}$$

are the dilation, tetragonal, and shear strains, respectively. This 2D free-energy density is obtained from the 3D form (10.4.17) for $\nabla_z u_z$ = const. and $e_5 = e_6 = 0$ under the

[9] In 2D the shear strain is usually written as e_3, but it is written as e_4 here to avoid confusion between the 2D and 3D cases.

condition of homogeneity along the z axis. Furthermore, the 2D compatibility relation follows from (10.4.28) in the form,

$$\nabla^2 e_1 - 2\nabla_x \nabla_y e_4 - (\nabla_x^2 - \nabla_y^2) e_2 = 0. \tag{10.4.47}$$

Experimentally, this 2D situation will be realized if a small uniaxial stress σ_a is applied such that the variant with the symmetry axis along the z axis is unfavored. If e_3 in (10.4.17) is replaced by $\langle e_3 \rangle - e_1/\sqrt{3}$ and the terms proportional to $e_1 e_2^2$ are collected, the coupling constant α' in (10.4.45) is expressed in terms of the 3D coefficients in (10.4.17) as

$$\alpha' = \alpha + \sqrt{3}B + \frac{1}{\sqrt{3}}[\bar{u}_0 + 2v_0\langle e_3 \rangle^2]\langle e_3 \rangle. \tag{10.4.48}$$

It is important that α' can be nonvanishing even for $\alpha = 0$ in solids with $B \neq 0$. With the free-energy density (10.4.45) we now rewrite (10.4.40) as

$$\begin{aligned} \rho \frac{\partial^2}{\partial t^2} u_x - \zeta \nabla^2 \frac{\partial}{\partial t} u_x &= \nabla_x [K e_1 - \alpha' e_2^2] + \mu \nabla_y e_4 + \nabla_x \mu_2, \\ \rho \frac{\partial^2}{\partial t^2} u_y - \zeta \nabla^2 \frac{\partial}{\partial t} u_y &= \nabla_y [K e_1 - \alpha' e_2^2] + \mu \nabla_x e_4 - \nabla_y \mu_2, \end{aligned} \tag{10.4.49}$$

where

$$\mu_2 = (\tau - 2\alpha' e_1 - D\nabla^2) e_2 + u_0 e_2^3 + v_0 e_2^5. \tag{10.4.50}$$

Dilation adjustment mechanism of domain pinning

First we point out an important difference in the two cases, $\alpha' = 0$ and $\alpha' \neq 0$. On the one hand, for $\alpha' = 0$, disordered regions disappear on relatively rapid timescales in phase ordering, eventually resulting in a twin structure with interfaces along [11] or [$\bar{1}$1]. This is simply because the elastic energy is minimized for twin structures, as will be discussed in Appendix 10D. On the other hand, for $\alpha' \neq 0$, the third-order coupling ($\propto e_1 e_2^2$) can give rise to nearly steady, structural intermediate states, in which domains of the tetragonal phase are coherently embedded in a cubic matrix [102]. Such elastic stabilization of domains is possible when the dilation strain e_1 is asymmetrically induced in the high- and low-temperature phases (*dilation adjustment mechanism*). Note that the effective temperature for the order parameter fluctuations is given by

$$\tau_{\text{eff}} = \tau - 2\alpha' e_1. \tag{10.4.51}$$

Domain pinning is expected if $\tau_{\text{eff}} < 0$ in ordered domains and $\tau_{\text{eff}} > 0$ in a cubic matrix. Around a tetragonal domain with $|e_2| \sim \eta_0$, the heterogeneity of e_1 is of order $\alpha' \eta_0^2/K$, resulting in a decrease of the free-energy density of order

$$\Delta f \sim -(\alpha' \eta_0^2)^2/K. \tag{10.4.52}$$

Thus the width ΔT of the temperature region in which the intermediate states are stable or metastable is obtained from

$$\Delta\tau = A_T \Delta T \sim |\Delta f|/\eta_0^2 \sim (\alpha'\eta_0)^2/K. \tag{10.4.53}$$

where A_T is defined in (10.4.3).

Simulation results

We give some numerical solutions of the dynamic equations (10.4.49) imposing the periodic boundary condition on u_x and u_y on a 128×128 lattice. This means that the space averages of e_1, e_2, and e_4 vanish in our simulations. With appropriate scale changes we set $\rho = D/2 = 1$ and assume $K = \mu$. Two representative cases are given by $u_0 = -1$ and $v_0 = 1$ in Figs 10.29, 10.31, and 10.32 (case D) and $u_0 = 1$ and $v_0 = 0.05$ in Figs 10.33 (case C). Without the cubic term ($\propto B$) in the free energy the transition remains first order in case D and becomes continuous in case C. (i) In Fig. 10.29 we start with a circular tetragonal domain with radius $R = 5$ at $t = 0$ for $\tau = -1$, $K = 5$, $\alpha' = 2$, and $\zeta = 0.5$ without noise. This seed is soon deformed into an ellipse and elastically expanded ($e_1 > 0$ since $\alpha' > 0$). Then the neighborhood outside the tips of the ellipse also become elastically expanded, which generates new elliptical domains with the opposite sign of e_2. This successive generation of domains proceeds until the whole system is covered with small-scale domains at $t \sim 180$. In this run the periodic pattern at $t = 13 \times 10^3$ lasted in the time region $10^4 \lesssim t \lesssim 4 \times 10^4$ and a large-scale twin structure was ultimately selected on the timescale of 10^5. Figure 10.30 displays slow relaxation behavior in the free-energy density. (ii) In Fig. 10.31 we confirm the rapid appearance of a twin structure at relatively deep quenching into $\tau = -4$. However, (iii) nearly steady patterns emerge for nonvanishing α' at shallow quenching in accord with (10.4.53). Examples of pinned intermediate states are given in Fig. 10.32 in case D and Fig. 10.33 in case C, where the two variants of the tetragonal phase are coherently embedded in a disordered region. In these runs the initial values of e_2 are random numbers in the range $[-0.05, 0.05]$. It is remarkable that the intermediate phase exists even for $u_0 > 0$ (case C).

The patterns indicate that the interfaces between the ordered and disordered regions are oriented in special directions. As can be known from (10D.6), if $|e_2| \sim \eta_0$ in an ordered domain, we estimate the optimal angle θ between the surface normal and the x axis as

$$\cos 2\theta \sim \alpha'\eta_0/K, \tag{10.4.54}$$

where the absolute value of the right-hand side is assumed to be smaller than 1. If it is larger than 1, we numerically find $\theta \cong 0$ and $\pi/2$ for the two variants. The above estimation is consistent with the patterns in Figs 10.32 and 10.33.

10.4.5 Structural intermediate states

In a number of metallic alloys near the martensitic phase transition, distinct domains (or *embryos*) with low-temperature tetragonal symmetry have been detected in a high-

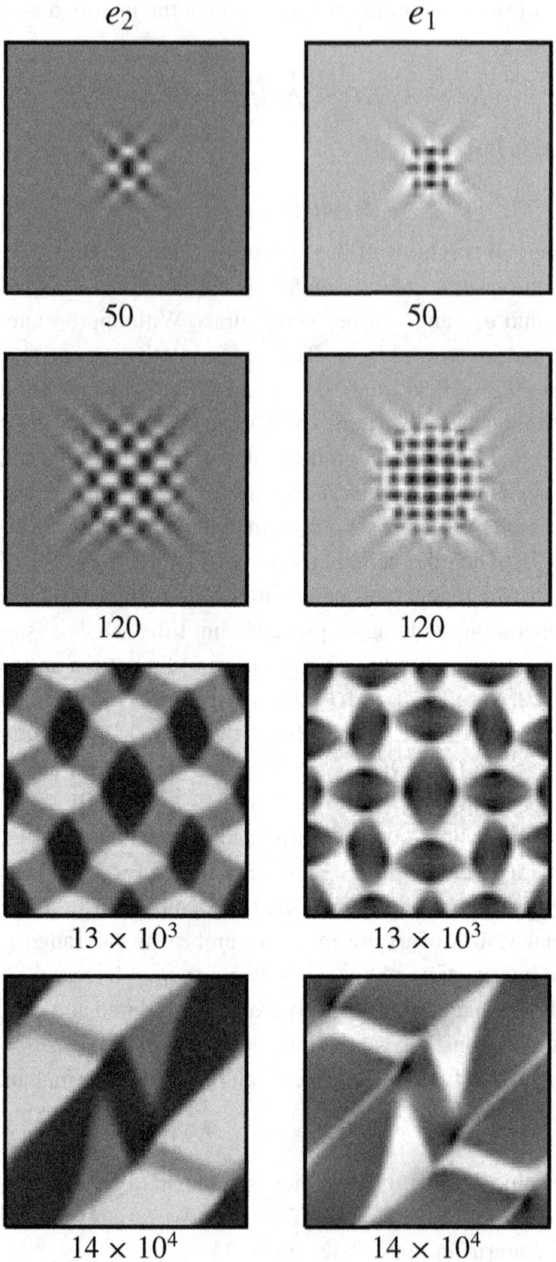

Fig. 10.29. Time evolution of e_2 (*left*) and e_1 (*right*) starting with a spherical droplet at $t = 0$. Here $\zeta = 0.5$, $\tau = -1$, $K = 5$, $\alpha' = 2$, $u_0 = -1$, and $v_0 = 1$. Elliptical domains in the tetragonal phase (in black for $e_2 \sim 1.5$ and in white for $e_2 \sim -1.5$) are then successively created. There remain disordered regions in the cubic phase with $e_2 \sim 0$ (in gray) for $t \lesssim 4 \times 10^4$. Note that e_1 is positive (in black) in the two variants of the tetragonal phase.

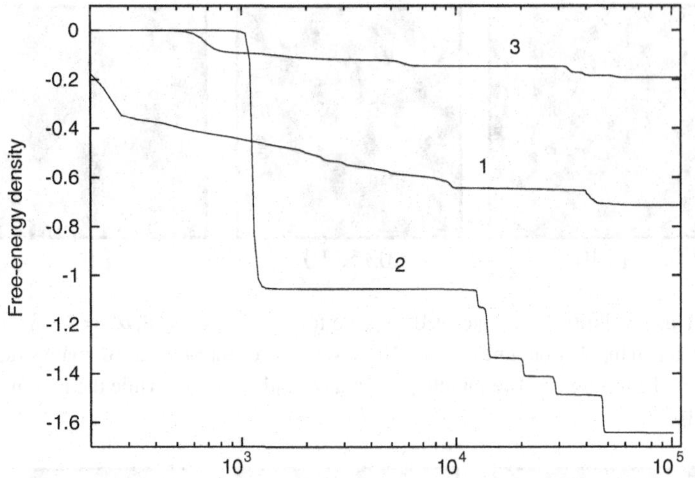

Fig. 10.30. Average free-energy density ($= \langle \mathcal{H} \rangle / V$) vs time, where the curves 1, 2, and 3 correspond to the runs in Figs 10.29, 10.32, and 10.33, respectively. Extremely slow time evolution can be seen once domains are elastically pinned. Step-like decreases represent cooperative disappearance of several domains.

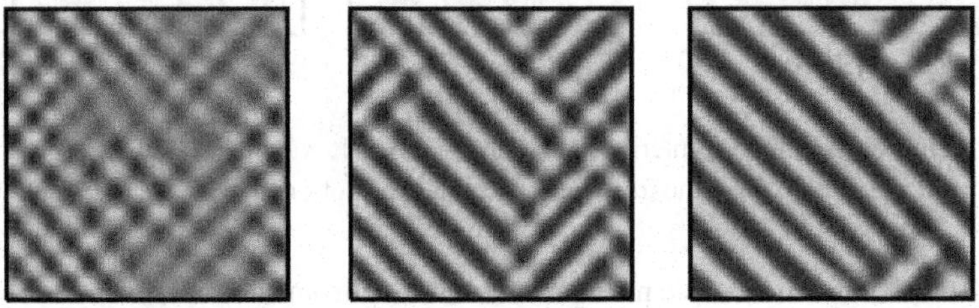

650 1×10^3 5×10^3

Fig. 10.31. Time evolution of the formation of a twin structure at relatively deep quenching at $\tau = -4$. The other parameters are $K = 8$, $\alpha' = 4$, $u_0 = -1$, $v_0 = 1$, and $\zeta = 1$.

temperature cubic phase [103]–[107]. They have been observed as *tweed patterns* by electron microscopy and an anomalous increase of sound attenuation. Presumably, they should give rise to the so-called central peak observed by neutron scattering [45]. These pretransitional (or premartensitic) effects are very unusual, but ubiquitous, and can sometimes be seen hundreds of degrees above the transition temperature at which the cubic phase disappears. Kartha *et al.* [108] have ascribed the origin of the tweed patterns to quenched disorder imposed by the compositional randomness, which is considered to strongly perturb the order parameter fluctuations. However, such impurity mechanisms can only lead to fuzzy inhomogeneous lattice distortions above the nominal transition temperature. In order to explain distinctly discernible tetragonal embryos, Fuchizaki and

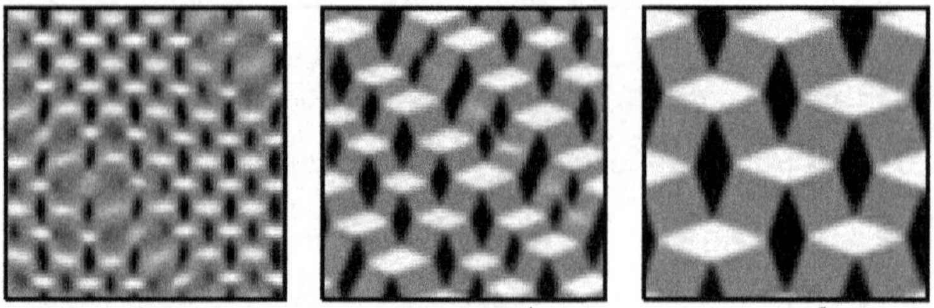

Fig. 10.32. Time evolution of an intermediate state for $\tau = -1$, $K = 8$, $\alpha' = 4$, $u_0 = -1$, $v_0 = 1$, and $\zeta = 1$. After an incubation time of order 10^3, a small-scale intermediate domain structure appears suddenly, as in the top figure. The middle pattern lasts until $t \sim 10^4$, while the bottom one is stable for $t \gtrsim 5 \times 10^4$.

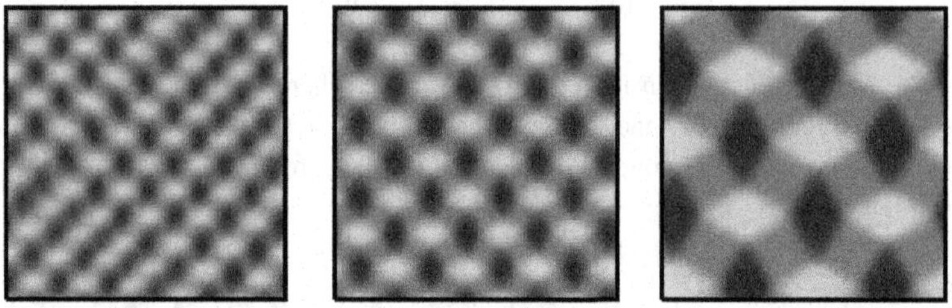

Fig. 10.33. Time evolution of an intermediate state for $\tau = -1$, $K = 5$, $\alpha' = 2$, and $\zeta = 1$. Here we set $u_0 = 1 > 0$ and $v_0 = 0.05$ in the free-energy density. The bottom pattern is stable for $t \gtrsim 5 \times 10^4$.

Yamada [109] sought an intrinsic mechanism stemming from anharmonic elasticity, though their analysis was limited to 1D. To support their claim, our 2D simulations demonstrate that third-order anharmonic elasticity can freeze tetragonal domains in a disordered matrix. As a next step, 3D simulations are strongly needed. In summary, elastic self-adjustment arising from the cubic coupling $\sim e_1 e_2^2$ can produce numerous intermediate configurations depending on τ, K, α', and the initial conditions.

We also mention recent observations of structural intermediate states in ferroelectric relaxors such as $PbMg_xNb_{1-x}O_3$ [110] and in doped manganites [111]. In two-phase coexistence near the transition, the former materials exhibit a strongly enhanced dielectric response, while the latter show a large (*colossal*) magnetoresistance. However, the importance of the elastic interactions in such phase transitions is not well recognized in the literature.

10.4.6 Proper hexagonal to orthorhombic transformations

We discussed improper hexagonal to orthorhombic transformations in Subsection 10.3.3. We here consider the proper case in 2D, where e_2 and e_4 constitute a two-component order parameter and three orthorhombic variants appear on the hexagonal basal plane. The minimal elastic free-energy density is given by [101, 112],

$$f = \frac{K}{2}e_1^2 + \frac{\tau}{2}\eta^2 - \alpha e_1\eta^2 - B(e_4^3 - 3e_4e_2^2) + \frac{\bar{u}_0}{4}\eta^4 + \frac{D}{2}(|\nabla e_2|^2 + |\nabla e_4|^2), \quad (10.4.55)$$

where $\eta^2 = e_2^2 + e_4^2$. The first two terms are those in isotropic linear elasticity in 2D, K and τ being the bulk and shear moduli. If we set $e_2 = \eta\sin\varphi$ and $e_4 = \eta\cos\varphi$, the angle φ is changed to $\varphi' = \varphi + 2\theta$ from (10.3.16) with respect to rotation, $x' = x\cos\theta + y\sin\theta$ and $x' = -x\sin\theta + y\cos\theta$. Because the cubic term in (10.4.55) is written as $-B\eta^3\cos(3\varphi)$, it is invariant with respect to the rotation by $\theta = \pi/3$ and its sign is reversed for $\theta = \pi/6$. We may thus assume $B > 0$ without loss of generality. Furthermore, if $\theta = \pi/12$, it is changed to $B\eta^3\sin(3\varphi') = -B(e_2'^3 - 3e_2'e_3'^2)$, reproducing the originally presented form [101]. In this model there are three equivalent stress-free variants in the low-temperature phase, as can be seen in Fig. 10.28. That is, variant 1 is given by $e_4 = \eta_0$ and $e_2 = 0$ ($\varphi = 0$), variant 2 by $e_2 = \sqrt{3}\eta_0/2$ and $e_4 = -\eta_0/2$ ($\varphi = 2\pi/3$), and variant 3 by $e_2 = -\sqrt{3}\eta_0/2$ and $e_4 = -\eta_0/2$ ($\varphi = -2\pi/3$), where η_0 is defined by (10.4.25) with C' being replaced by τ.

The dynamic equation (10.4.40) is rewritten as

$$\begin{aligned} \rho\frac{\partial^2}{\partial t^2}u_x - \zeta\nabla^2\frac{\partial}{\partial t}u_x &= \nabla_x[Ke_1 - \alpha\eta^2 + \mu_2] + \nabla_y\mu_4, \\ \rho\frac{\partial^2}{\partial t^2}u_y - \zeta\nabla^2\frac{\partial}{\partial t}u_y &= \nabla_y[Ke_1 - \alpha\eta^2 - \mu_2] + \nabla_x\mu_4, \end{aligned} \quad (10.4.56)$$

where

$$\begin{aligned} \mu_2 &= (\tau - 2\alpha e_1 - D\nabla^2)e_2 + 6Be_4e_2, \\ \mu_4 &= (\tau - 2\alpha e_1 - D\nabla^2)e_4 + 3B(e_2^2 - e_4^2). \end{aligned} \quad (10.4.57)$$

Linear stability analysis can readily be performed for small disturbances in a disordered homogeneous state. Assuming that all the deviations are proportional to $\exp(i\boldsymbol{k}\cdot\boldsymbol{r} - \Omega t)$, we obtain isotropic dispersion relations,

$$\rho\Omega^2 - \zeta k^2\Omega = -k^2(\tau + Dk^2) \quad \text{or} \quad -k^2(K + \tau + Dk^2). \quad (10.4.58)$$

Thus phase ordering occurs if τ is changed to a negative value, where the fluctuations start to grow isotropically in the initial stage.

Planar interfaces

We suppose a planar interface in equilibrium stress-free states. All the strains vary in one direction and are functions of

$$s = x\cos\theta + y\sin\theta, \quad (10.4.59)$$

where θ is the constant angle between the interface normal and the x axis. The 2D compatibility relation (10.4.47) gives

$$e_1(s) = e_2(s)\cos 2\theta + e_4(s)\sin 2\theta + A, \quad (10.4.60)$$

where A is a constant. For example, between the stress-free variants 2 and 3, we require $\cos 2\theta = 0$ or $\theta = \pm\pi/4$ because $e_1(\infty) = e_1(-\infty)$, $e_2(\infty) \neq e_2(-\infty)$, and $e_4(\infty) = e_4(-\infty)$. Under the stress-free condition we have $e_1 = \pm(e_4 + \eta_0/2) + \alpha\eta_0^2/K$ to obtain

$$f = \frac{K}{2}\left(e_4 + \frac{\eta_0}{2} - \frac{\alpha}{K}\delta e_2^2\right)^2 - B(e_4^3 - 3e_4e_2^2) + \frac{u_0}{4}(e_2^2 + e_4^2)^2 + \frac{D}{2}\left|\frac{d}{ds}e_2\right|^2 + \frac{D}{2}\left|\frac{d}{ds}e_4\right|^2, \tag{10.4.61}$$

where $\delta e_2^2 = e_2^2 - 3\eta_0^2/2$. As in the cubic to tetragonal case in (10.4.34)–(10.4.39), e_2 is given by (10.4.36) exactly at $\tau = -18B^2/u_0$ and approximately for small B. In the same manner, $\theta = \pi/12$ and $7\pi/12$ for the interfaces between the variants 1 and 2, and $\theta = -\pi/12$ and $5\pi/12$ for those between the variants 1 and 3. These orientation relations remain valid even for $\alpha \neq 0$ and will explain the simulation results below.

For this model we may also calculate an interface between a variant and a disordered region. Let us assume that e_4 tends to 0 as $s \to \infty$ and η_0 as $s \to -\infty$ while $e_2 = 0$ at any s. Under the stress-free condition we require $e_1 + \alpha e_4^2/K \to 0$ as $s \to \pm\infty$. In (10.4.60) we find $A = 0$ and

$$\sin 2\theta = \alpha\eta_0/K, \tag{10.4.62}$$

where we assume $|(\alpha/K)\eta_0| < 1$. Using these relations, f becomes

$$f = \frac{\tau}{2}e_4^2 - Be_4^3 + \frac{u_0}{4}e_4^4 + \frac{D}{2}\left|\frac{d}{ds}e_4\right|^2. \tag{10.4.63}$$

In particular, at $\tau = 2B^2u_0$ we obtain the equilibrium stress-free interface solution,

$$e_4 = \frac{1}{2}\eta_0[1 - \tanh(s/2\xi)], \tag{10.4.64}$$

where $\eta_0 = 2B/u_0$ and $\xi^2 = Du_0/6B^2$.

Simulation results

We present some simulation results on a 128×128 lattice with $\zeta = 1$, $K = 8$, $B = 1$, and $D = 2$. We start with a disordered state at $t = 0$ and follow subsequent structural transformations. In Fig. 10.34 the quench depth is fixed at $\tau = -1$, but α is equal to 0 in (a), 1.4 in (b), and 1.9 in (c). Notice the close resemblance between the proper case (a) and the improper case in Fig. 10.26. We also note the following. (i) Figure 10.35 shows that the domain growth is pinned at relatively early times ($t \sim 10^3$ here), analogous to Fig. 10.27. (ii) If $|\alpha|$ is larger than a critical value α_c, domains of the disordered phase do not disappear in pinned states. For the selected parameter values in Fig. 10.34, we find $\alpha_c \cong 1.2$ and the volume fraction of the disordered phase to be 0 in (a), 0.17 in (b), and 0.40 in (c). In (c) the three variants form stripes forming a network embedded in the disordered phase. (iii) The orientation relations derived above for the interfaces among the three variants are well satisfied particularly in (a), in agreement with the experiments [80, 82]. This was already reported in a previous simulation [112].

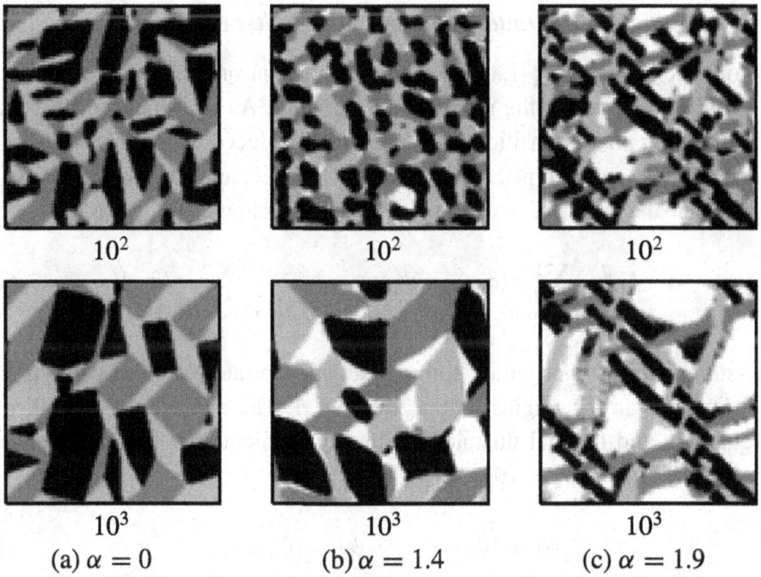

Fig. 10.34. Patterns in hexagonal to orthorhombic transformations, calculated from the 2D model (10.4.56) and (10.4.57) for $\zeta = 1$, $K = 8$, $B = 1$, and $D = 2$. We vary the coupling constant α as 0, 1.4, and 1.9. The four different gray levels from darkest to brightest correspond to variant 1 ($\varphi = 0$) (black), variant 2 ($\varphi = 2\pi/3$), variant 3 ($\varphi = -2\pi/3$), and parent phase (white), respectively. The upper figures at $t = 10^2$ represent relatively early-stage patterns emerging from the initial isotropic patterns. The lower figures correspond to nearly pinned patterns. The fraction of the parent phase increases with increasing α.

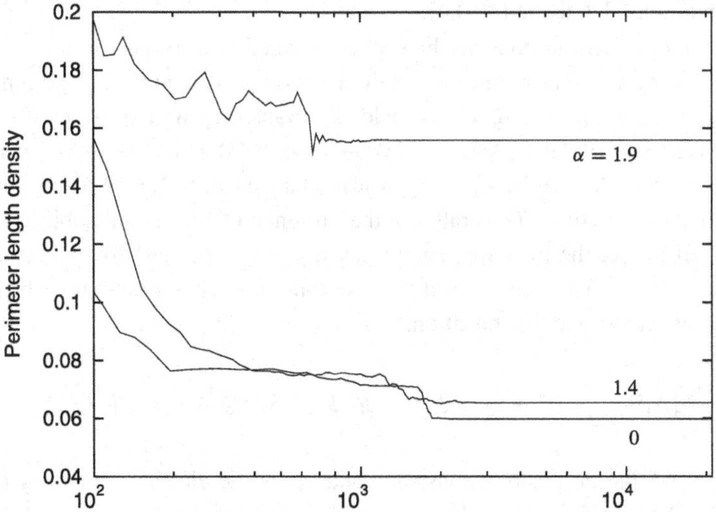

Fig. 10.35. Perimeter length (per unit volume) vs time for the runs in Fig. 10.34. We can see pinning of the domain structures for $t \gtrsim 10^3$.

10.4.7 Orbital order and Jahn–Teller coupling

We now construct a Ginzburg–Landau theory for orbital order [113], but defer analysis of phase-ordering kinetics of the model to future work. As in Appendix 10C, a cubic to tetragonal structural phase transition is assumed to be induced by the Jahn–Teller coupling (10.4.1). We suppose spinel-type crystals such as $CuFe_2O_4$ and Mn_3O_4 [88]. As in (4.1.2), we introduce a coarse-grained, two-component order parameter ψ_1, ψ_2 as

$$\psi_1(\boldsymbol{r}) = -\ell^{-d} \sum_{n \in \text{new cell}} \langle \hat{\sigma}_{nz} \rangle, \quad \psi_2(\boldsymbol{r}) = -\ell^{-d} \sum_{n \in \text{new cell}} \langle \hat{\sigma}_{nx} \rangle, \tag{10.4.65}$$

where the summation is over lattice points in an appropriately defined cell with a lattice constant ℓ longer than the original lattice constant a. The average $\langle \cdots \rangle$ is taken doubly over the quantum and thermal fluctuations. It is convenient to define the complex order parameter by $\psi = \psi_1 + i\psi_2 = |\psi| \exp(i\varphi)$ or

$$\psi_1 = |\psi| \cos\varphi, \quad \psi_2 = |\psi| \sin\varphi. \tag{10.4.66}$$

The tetragonally distorted states are given by $\varphi = 0$, $2\pi/3$, and $-2\pi/3$ for an axis of symmetry along the z, x, and y direction, respectively. The amplitude ψ is nonvanishing below the transition and increases as the temperature is lowered. Considering the average of (10.4.1), we obtain the free-energy density $f_{\rm JT}$ representing the orbit–lattice coupling:

$$f_{\rm JT} = -g_{\rm JT}(e_3\psi_1 + e_2\psi_2) = -g_{\rm JT}\,\mathrm{Re}[(e_3 + ie_2)\psi^*], \tag{10.4.67}$$

where e_2 and e_3 are defined in (10.4.4).

Here it is informative to consider how ψ is changed with respect to a rotation of the reference frame by $\pi/2$. The rotation about the x axis is equivalent to the replacement, $y \to z$, $z \to -y$, with x unchanged, which yields the transformation $\psi \to \exp(-2\pi/3)\psi^*$. The rotations about the y and z axes yield $\psi \to \exp(2\pi/3)\psi^*$ and $\psi \to \psi^*$, respectively. The complex strain defined by $e_3 + ie_2$ is also changed in the same manner and $f_{\rm JT}$ is invariant for these rotations. Generally, in the presence of the crystal cubic symmetry in the disordered phase, the total free-energy density $f = f(\psi, \boldsymbol{u})$ for ψ and the elastic strains should be invariant with respect to these rotations. Thus we propose the following Landau expansion close to the transition,

$$f = \frac{r_0}{2}|\psi|^2 + \frac{u}{4}|\psi|^4 + \frac{C}{2}|\nabla\psi|^2 - B_0(\psi_1^3 - 3\psi_1\psi_2^2) + f_{\rm JT} + f_{\rm el}, \tag{10.4.68}$$

where u, C, and B_0 are positive constants, and $f_{\rm el}$ is the elastic free-energy density of cubic solids. We assume that r_0 depends on the temperature T as $r_0 = A_0(T - T_0)$, where A_0 is a positive constant and T_0 is a constant temperature. Note that the real parts of $\psi^{3-k}(e_3 + ie_2)^k$ $(k = 0, 1, 2, 3)$ constitute four third-order invariants. More explicitly, they are written as

$$I_{30} = \psi_1^3 - 3\psi_1\psi_2^2, \qquad I_{31} = (\psi_1^2 - \psi_2^2)e_3 - 2\psi_1\psi_2 e_2,$$
$$I_{32} = \psi_1(e_3^2 - e_2^2) - 2\psi_2 e_2 e_3, \qquad I_{33} = e_3^3 - 3e_3e_2^2. \tag{10.4.69}$$

We may assume a third-order term expressed as $\sum_{j=0}^{3} B_j I_{3j}$ in the free-energy density, where B_j $(j = 0, \ldots, 4)$ are coefficients. In (10.4.68), for simplicity, we retain a third-order term proportional to I_{30} with $B_0 > 0$ and assume that f_{el} is bilinear in the strains as

$$f_{\mathrm{el}} = \frac{K}{2}e_1^2 + \frac{C_0'}{2}(e_2^2 + e_3^2) + \frac{C_{44}}{2}(e_4^2 + e_5^2 + e_6^2). \tag{10.4.70}$$

This is of the same form as the elastic free energy in (10.4.6) except that C' is replaced by a background value C_0'.

In the disordered phase, elimination of $\psi (\cong g_{\mathrm{JT}}(e_3 + ie_2)/r_0)$ yields the effective elastic moduli C' for the tetragonal strains in the form

$$C' = C_0' - g_{\mathrm{JT}}^2/r_0. \tag{10.4.71}$$

The nominal critical temperature T_{c0} in (10.4.3) is given by $T_{c0} = T_0 + g_{\mathrm{JT}}^2/A_0C_0'$. We next consider a tetragonal state stretched along the z axis in the stress-free condition, where $\psi_1 = M > 0$ and $\psi_2 = 0$. By setting $e_3 = g_{\mathrm{JT}}M/C_0'$ we obtain $r + uM^2 - 3B_0M = 0$, similar to (10.4.24), where

$$r = r_0 - g_{\mathrm{JT}}^2/C_0' = A_0(T - T_{c0}). \tag{10.4.72}$$

At $r = r_{\mathrm{tr}}$, a first-order phase transition occurs, and M and e_3 change from 0 to M_{tr} and e_{tr}, respectively, where

$$r_{\mathrm{tr}} = 2B_0^2/u, \quad M_{\mathrm{tr}} = 2B_0/u, \quad e_{\mathrm{tr}} = 2g_{\mathrm{JT}}B_0/C_0'u. \tag{10.4.73}$$

For $r < r_{\mathrm{tr}}$ the ordered phase is stable and M is expressed as

$$M = \frac{3}{4}M_{\mathrm{tr}}\left[1 + \sqrt{1 - 8r/9r_{\mathrm{tr}}}\right]. \tag{10.4.74}$$

The elastic moduli C' in (10.4.71) just *above* the transition is expressed as $C_0'w/(1 + w)$ with

$$w = 2C_0'B_0^2/g_{\mathrm{JT}}^2u. \tag{10.4.75}$$

In terms of the elastic properties, w represents weakness of the first-order phase transition.

The fluctuation effect becomes much more complicated in ordered states than in disordered states due to the two-component nature of ψ. We consider the fluctuations of ψ, e_2, and e_3. Their second-order contributions in f are written as

$$\delta f = \frac{r_{\mathrm{L}}}{2}(\delta\psi_1)^2 + \frac{r_{\mathrm{T}}}{2}\psi_2^2 + \frac{C}{2}|\nabla\psi|^2 + \frac{C_0'}{2}\left[\left(\delta e_3 - \frac{g_{\mathrm{JT}}}{C_0'}\delta\psi_1\right)^2 + \left(e_2 - \frac{g_{\mathrm{JT}}}{C_0'}\psi_2\right)^2\right] + \cdots, \tag{10.4.76}$$

where $\delta\psi_1 = \psi_1 - M$ and $\delta e_3 = e_3 - g_{JT}M/C_0'$ are the deviations. Here we do not write explicitly the second-order contributions arising from nonvanishing dilational and shear strains. As in Subsections 4.3.7 and 4.3.8, r_L and r_T are the longitudinal and transverse inverse-susceptibility, respectively. In the present case we obtain

$$r_L = 2uM^2 - 3B_0M, \quad r_T = 6B_0M. \tag{10.4.77}$$

It is worth noting that r_T is positive owing to the third-order term in f, while in Section 4.3 it was shown to vanish in (isotropic) many-component spin systems. Considering homogeneous deviations of e_3 and e_2, we obtain the effective elastic moduli, C_L' and C_T', respectively, by eliminating $\delta\psi_1$ and ψ_2. In terms of w in (10.4.75) and $m \equiv M/M_{tr}$, they are expressed as

$$\frac{C_L'}{C_0'} = \frac{w(4m^2 - 3m)}{1 + w(4m^2 - 3m)}, \quad \frac{C_T'}{C_0'} = \frac{9wm}{1 + 9wm}. \tag{10.4.78}$$

In Fig. 10.36 we display these elastic moduli near the transition for $w = 0.04$. The modulus C_L' below the transition is continuously connected to C' in (10.4.71) above the transition and is smaller (larger) than C_T' for r larger (smaller) than $-9r_{tr}$.

For inhomogeneous deviations the elastic contributions to the free energy may be calculated, using the procedure in Appendix 10A, in terms of the Fourier components $\psi_{\alpha\boldsymbol{k}}$ ($\alpha = 1, 2$) [114]. In particular, if the wave vector $\boldsymbol{k}$ is along [110], these contributions do not affect the variance of ψ_2 in the long-wavelength limit, so that $\lim_{k\to 0}\langle|\psi_{2\boldsymbol{k}}|^2\rangle = 1/r_T$. As a result, the velocity of the transverse sound propagating along [110] and polarized along $[1\bar{1}0]$ is given by $c_t[110] = (C_T'/\rho)^{1/2}$, ρ being the mass density. In passing, let us consider the limit $B_0 \to 0$ with $r(< 0)$ and g_{JT} fixed. Then we find $C_T' \to 0$ and $C_L'/C_0' \to 1/[1 + 2C_0'|r|/g_{JT}^2]$. In accord with this result, Pytte [89] found that $c_t[110]$ vanishes for all temperatures below the transition in the absence of the cubic terms.[10]

Correspondence between the two theories

We also comment on the relationship between the conventional nonlinear strain theory [94]–[97] and the present theory. We can see that only in the case $w \ll 1$ do the two theories yield essentially the same results near the transition $|r| \ll g_{JT}^2/C_0'$. In this parameter region, we have a small modulus $C'(\ll C_0')$ given in (10.4.71). From $r_0 \cong g_{JT}^2/C_0' \gg u|\psi|^2$, we here find the linear relation, $\psi \cong (g_{JT}/r_0)(e_3 + ie_2)$. Substitution of this relation into (10.4.68) gives rise to the anharmonic elastic free energy (10.4.7) with the coefficients, $B = (C_0'/g_{JT})^3 B_0$ and $u_0 = (C_0'/g_{JT})^4 u$.

Compression-induced phase transition

Furthermore, some new effects can be predicted if a uniaxial stress σ_a is applied along the z axis. As in (10.4.27) we should then minimize $\bar{f} \equiv f - \sigma_a e_3/\sqrt{3}$. By setting $e_3 =$

[10] Similar elastic softening has been observed in nematic elastomers (rubbers composed of liquid crystal molecules in the nematic phase). See Ref. [43] in Chapter 7.

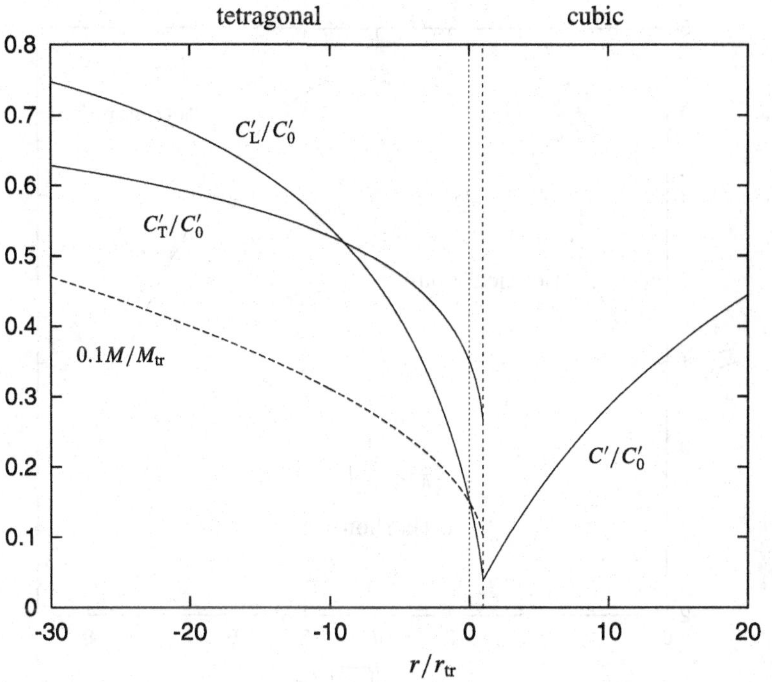

Fig. 10.36. Normalized elastic moduli for tetragonal strains vs normalized reduced temperature ($\propto T - T_{c0}$) obtained from the free-energy density (10.4.68) for $w = 0.04$. Here C'_L and C'_T are the moduli for e_3 and e_2, respectively, in the tetragonal phase stretched along the z axis. The normalized order parameter M/M_{tr} (divided by 10) is also plotted.

$(g_{JT}\psi_1 + \sigma_a/\sqrt{3})/C'_0$ we have

$$\bar{f} = \frac{r}{2}|\psi|^2 + \frac{u}{4}|\psi|^4 - B_0(\psi_1^3 - 3\psi_1\psi_2^2) - \frac{g_{JT}}{\sqrt{3}C'_0}\sigma_a\psi_1. \tag{10.4.79}$$

In the case of stretching $\sigma_a > 0$, the tetragonal variant with $\psi_1 > 0$ and $\psi_2 = 0$ is favored below the transition.[11] However, the phase behavior becomes more interesting in the case of compression σ_a, so we limit ourselves to this case. As shown in Fig. 10.37, the temperature–stress plane is divided into two regions by a line of first-order phase transition (line F) and a critical line (line C). These two lines meet at a tricritical point, where r and $|\sigma_a|$ assume the following tricritical values,

$$r_t = \frac{63}{32}r_{tr}, \quad \sigma_t = \frac{81\sqrt{3}}{64}wC'_0e_{tr}, \tag{10.4.80}$$

respectively, in terms of r_{tr} in (10.4.73) and w in (10.4.75). On line C we obtain $4r/9r_{tr} = |\sigma_a/\sigma_t|^{1/2} - |\sigma_a/8\sigma_t|$ and $M/M_{tr} = (3/8)\sigma_a/\sigma_t$. Above the lines the stable phase consists

[11] For stretching, there is still a first-order phase transition line expressed as $r/r_{tr} = 1 + (81/32)\sigma_a/\sigma_t < 3/2$ which ends at a critical point given by $r = 3r_{tr}/2$ and $\sigma_a = (16/81)\sigma_t$. Here $M = \psi_1$ is discontinuous across this line by $M_{tr}(3 - 2r/r_{tr})^{1/2}$, while there is no discontinuity for $\sigma_a > (16/81)\sigma_t$.

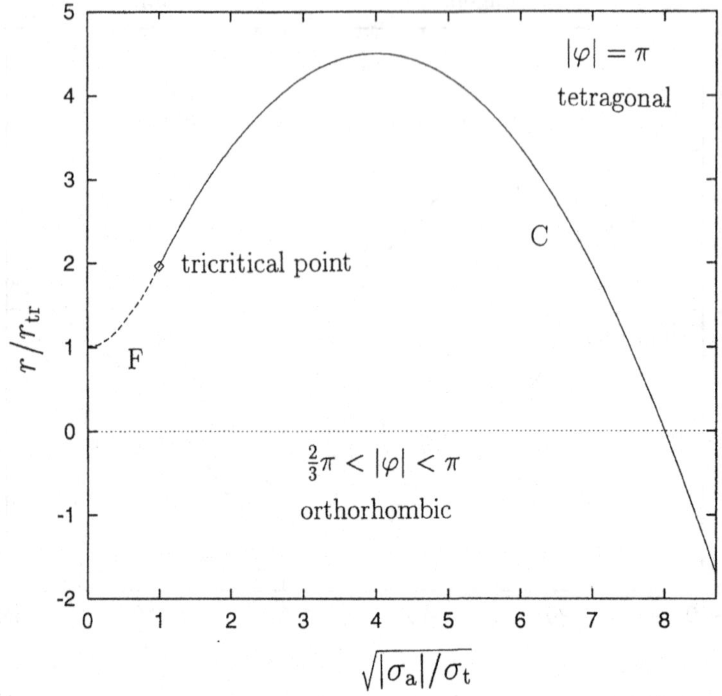

Fig. 10.37. Phase diagram of a solid under uniaxial compression ($\sigma_a < 0$) with the free-energy density (10.4.79), where Jahn–Teller coupling is responsible for the structural phase transition. The horizontal and vertical axes represent $|\sigma_a/\sigma_t|^{1/2}$ and r/r_{tr}, respectively. A line of first-order phase transition (dashed line F) starts from the point where $r = r_{tr}$ and $\sigma_a = 0$, and meets a critical line (solid line C) at a tricritical point, where $r = r_{tr}$ and $\sigma_a = -\sigma_t$. Here $\psi_1 < 0$ and $\psi_2 = 0$ above the curves, while there are two stable variants below them, as given by (10.4.81).

of a single tetragonal state expressed as $\psi_1 = -M < 0$ and $\psi_2 = 0$. Below the lines we have an orthorhombic phase with two stable variants expressed as

$$\psi_1 = -M \cos\varphi_a, \quad \psi_2 = \pm M \sin\varphi_a, \tag{10.4.81}$$

where φ_a is an angle in the range $0 < \varphi_a < \pi/3$ and satisfies $\sin^2(3/4)[1 - |\sigma_a/\sigma_t|(8M/3M_{tr})^2]$. Here $\varphi_a \to \pi/3$ as $\sigma_a \to 0$ for $r < r_{tr}$, $\varphi_a \to 0$ as line C is approached, and φ_a remains nonvanishing as line F is approached from below. The modulus C'_T for the strain e_2 tends to zero as line C is approached both from above and below.

Antiferromagnet-like order

In MnF_3 the orbital occurs alternatively in two sublattices with a symmetry axis along one of the cubic axes (say, the z axis) [88]. In such cases it is convenient to introduce the two complex order parameters, $\psi_A = \psi_{A1} + i\psi_{A2}$ and $\psi_B = \psi_{B1} + i\psi_{B2}$, for sublattices, A and

B. Then we have the antiferromagnet-like order parameter $\zeta = \psi_A - \psi_B$ in addition to the ferromagnet-like order parameter $\psi = \psi_A + \psi_B$. The orbital order in MnF_3 is represented by $\psi_A = \psi_B^* = iM\exp(i\phi)$ or by $\psi = -2M\sin\phi$ and $\zeta = 2iM\cos\phi$, where the solid is uniaxially deformed along the z axis and ϕ is the canting angle of the sublattice order. In the bilinear order, the interaction energy density between the two sublattices is of the form $f_{AB} = -r_1 M^2 \cos 2\phi$. We assume that the interaction strength r_1 is positive, because the ferromagnet-like order ($\phi = -\pi/2$) is favored for negative r_1. If the interaction between the two sublattices arises only from f_{AB}, elimination of the strains in f gives [113]

$$f = 2\left[\frac{r}{2}M^2 + \frac{u}{4}M^4 - B_0 M^3 \sin 3\phi\right] - r_1 M^2 \cos 2\phi \qquad (10.4.82)$$

where the factor 2 accounts for the presence of the two sublattices. Minimization of f with respect to M and ϕ yields a phase diagram in the r–r_1 plane. A line of first-order phase transition starts from the point $r = r_{tr}$ and $r_1 = 0$ and ends at a tricritical point where $r = r_1 = 9r_{tr}/2$. This line is expressed by $2r/r_{tr} = 1 + (2r_1/3r_{tr} + 1)^{3/2}$. For $r_1 > 9r_{tr}/2$ the phase transition becomes continuous with a critical line given by $r = r_1$. We have the disordered phase with $M = 0$ above these lines and the canted phase with $M > 0$ and $0 < \phi < \pi/6$ below them.

10.5 Macroscopic instability

10.5.1 Cowley's classification

According to Cowley [115], type-I instabilities correspond to structural phase transitions at which acoustic modes become soft in particular wave vector directions, whereas type-II are those with soft planes. At a type-0 instability, only macroscopic deformations on the sample scale ($\sim L$) become unstable without critical enhancement of small-scale fluctuations. To illustrate this classification, let us consider the sound speed c in cubic solids with $qL \gg 1$ determined by the matrix equation,

$$\rho c^2 q^2 u_i = (C_{12} + C_{44}) q_i (\boldsymbol{q} \cdot \boldsymbol{u}) + [C_{44} q^2 + 2(C' - C_{44}) q_i^2] u_i. \qquad (10.5.1)$$

If $C' = (C_{11} - C_{12})/2$ tends to zero, as in Nb_3Sn [91, 92], the transverse sound propagating along [100] and polarized along [1$\bar{1}$0] becomes soft and the instability is of type-I. If C_{44} tends to zero, as in KCN [93], softening occurs for any $\boldsymbol{q}$ on the xy plane with $\boldsymbol{u}$ being along the z axis, leading to type-II transitions. The type-0 instability occurs for negative $K = (C_{11} + 2C_{12})/3 < 0$ while $C' > 0$ and $C_{44} > 0$. As a well-known example, K approaches zero towards a transition temperature in some solids such as $Sm_{1-x}Y_xS$ or $Ce_{1-x}Th_x$ at the valence instability [116]. At such type-0 transitions, macroscopic volume changes take place on timescales of order $L|\rho/K|^{1/2}$ [117], which is obtained by replacement, $\zeta\Omega \to \rho\Omega^2$, in (7.2.56) and is much faster than in gels. In the macroscopic instability of polymer gels discussed in Section 7.2, the bulk osmotic modulus K_{os} becomes negative and the dynamics is slowed down by the network–solvent friction.

10.5.2 Hydrogen–metal systems

We will give detailed discussions on a unique example of type-0 instability in hydrogen–metal systems. Large amounts of hydrogen can be absorbed by many metals such as V, Nb, Ta, and Pd and its concentration can be of order 100 at.% (one proton per metal ion) [118]–[126]. In metals, the hydrogen molecules give their electrons to the conduction band, while the protons occupy interstitial sites. The diffusion constant of the protons strongly depends on the metal and its structure (bcc or fcc), increases with temperature, and can even be of order 10^{-4} cm^2/s at room temperature. Therefore, the protons can diffuse over macroscopic system sizes within realistic observation times. Absorption or desorption of hydrogen can also occur from the metal surface, which takes place relatively rapidly, particularly for Pd. Considerable heat is released with hydrogen absorption. Such metallic alloys can be used as efficient containers of hydrogen.

The dissolved hydrogen systems undergo phase changes involving gas (α), liquid (α'), and solid (superlattice) phases. In Fig. 10.38 the isotherms connecting the α and α' phases are shown for fcc Pd–H [119a], where η in (10.1.20) is about 0.06 and $\Delta T_c[100]$ in (10.2.17) is of order 300 deg. C [127]. The Coulomb interaction between the protons is screened by the electrons and becomes short-range, but there arises a unique elastic interaction between them because the lattice expands in the presence of proton interstitials ($\sim 10^{-24}$ cm^3 per hydrogen atom) [121]. Wagner and Horner [118b, 124] derived a Ginzburg–Landau free energy to describe the gas–liquid transition of hydrogen–metal systems. On the coarse-grained level, it turns out to be of the same form as the free energy (10.1.2) set up for usual binary metal alloys. However, a unique feature in hydrogen–metal systems is that macroscopic proton density variations can lower the free energy on experimental timescales to induce sample shape changes under the stress-free boundary condition [123]. We will show that such a macroscopic instability follows generally from the bilinear coupling ($\propto \alpha\psi\nabla\cdot\boldsymbol{u}$) in the free energy (10.1.2) or from the Vegard law. Of course, in clamped samples and on relatively deep quenching, we expect the occurrence of phase separation in which α and α' regions are separated by sharp interfaces. We mention observations of coherent plate-shaped precipitates of $PdH_{0.05}$ in an α' matrix and $PdH_{0.6}$ in an α matrix [127]. A similar changeover between macroscopic and bulk instabilities was also studied for gels in Chapter 7. We also note that various domains of ordered (superlattice) phases have also been observed [119b].

Homogeneous states in contact with a hydrogen reservoir

Let a metal be in contact with a hydrogen gas reservoir with a constant chemical potential $\mu_H = \mu_H(T, p)$. The total free energy is of the form, $\mathcal{H}_T = \mathcal{H} - \int d\boldsymbol{r}\mu_H\psi$, where $\mathcal{H}$ is given in (10.1.2) and $\psi = c_H - c_{Hc}$ represents the proton composition deviation from the critical value in the metal. For simplicity, we assume isotropic elasticity with homogeneous bulk and shear moduli, K and μ, and isotropic lattice expansion due to the proton interstitials.[12]

[12] The latter assumption is a good approximation for fcc crystals, but uniaxial deformations can be significant in bcc crystals [120].

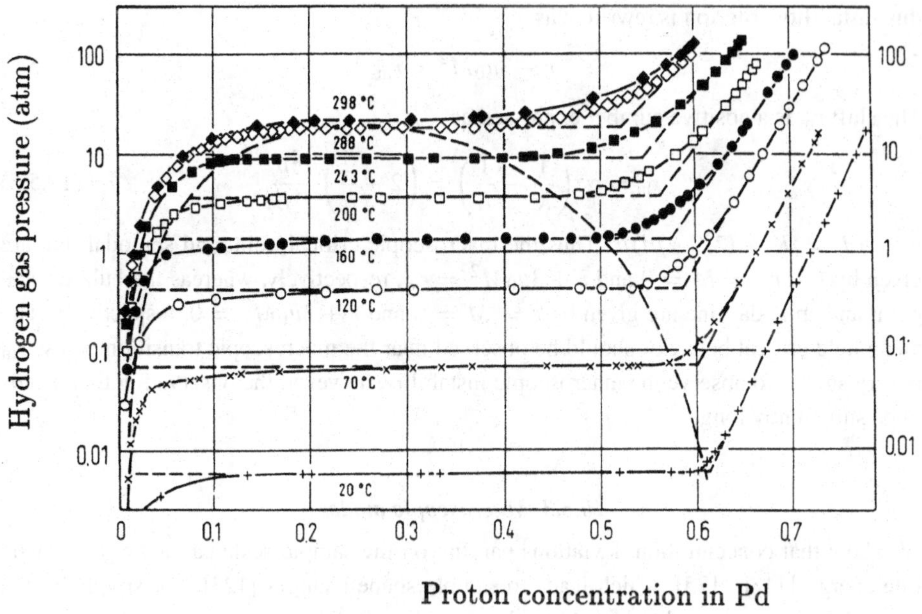

Fig. 10.38. Phase diagram of Pd–H [119a] at various temperatures showing a gas–liquid phase transition of protons in Pd. The vertical axis represents the pressure of the surrounding H gas, while the horizontal axis the relative number of protons per Pd atom.

Here we present a Landau theory neglecting spatial inhomogeneity within the solid. If the gas pressure is low, we may assume the stress-free condition $Kg + \alpha M = 0$ at the solid–gas boundary, where $M = \langle \psi \rangle$ and $g = \langle \nabla \cdot \boldsymbol{u} \rangle$. Then,

$$\begin{aligned} \mathcal{H}_{\mathrm{T}} &= V\left[\frac{r_0}{2}M^2 + \frac{u_0}{4}M^4 - \mu_{\mathrm{H}}M + \alpha g M + \frac{K}{2}g^2\right] \\ &= V\left[\frac{1}{2}(r_0 - \alpha^2/K)M^2 + \frac{u_0}{4}M^4 - \mu_{\mathrm{H}}M\right], \end{aligned} \tag{10.5.2}$$

where g has been eliminated in the second line. In chemical equilibrium with the reservoir, the equation of state is of the form,

$$(r_0 - \alpha^2/K)M + u_0 M^3 = \mu_{\mathrm{H}}. \tag{10.5.3}$$

If the hydrogen composition deviates slightly from the chemical equilibrium value by δM, the free energy increases by $\delta\mathcal{H}_{\mathrm{T}} = V[r_0 + 3u_0M^2 - \alpha^2/K](\delta M)^2/2$. Thus the system is unstable for

$$r_0 + 3u_0M^2 < \alpha^2/K, \tag{10.5.4}$$

against further absorption or desorption of hydrogen. In terms of r defined by (10.1.34), this instability criterion is rewritten as

$$r + 3u_0M^2 < r_{\rm m}, \tag{10.5.5}$$

The shift $r_{\rm m}$ is a positive number expressed as

$$r_{\rm m} = \alpha^2\left(\frac{1}{K} - \frac{1}{L}\right) = \left(2 - \frac{2}{d}\right)\frac{\alpha^2\mu}{KL}, \tag{10.5.6}$$

where $L = K + (2 - 2/d)\mu$. Thus the macroscopic critical point and spinodal line are given by $r - r_{\rm m} = M = 0$ and $r + 3u_0M^2 = r_m$, respectively, whereas the bulk critical point and spinodal line are given by $r = M = 0$ and $r + 3u_0M^2 = 0$, respectively. The mean field critical behavior should be observed near the macroscopic transition unless $r_{\rm m}$ is very small. To observe the macroscopic instability, however, the observation time needs to be sufficiently long.

10.5.3 Macroscopic modes

We show that concentration deviations varying on the sample scale can lower the elastic free energy [118b, 125], which leads to sample shape changes [123]. For simplicity the sample shape is assumed to be spherical with radius R in 3D. In the one-phase region we retain only the terms bilinear in $\delta\psi$ and neglect the gradient term in the mean field theory to obtain

$$\mathcal{H}_{\rm T} = \frac{1}{2}\int d\boldsymbol{r}\left[(r_0 + 3u_0M^2)\delta\psi^2 + \alpha\psi g\right] + \text{const}. \tag{10.5.7}$$

At $r = R$ we impose the stress-free boundary condition,

$$\sum_j \sigma_{ij}\hat{x}_j = 0. \tag{10.5.8}$$

Within the sphere $r < R$ the mechanical equilibrium condition is written as

$$\nabla_i[\alpha\psi + (L - \mu)g] + \mu\nabla^2 u_i = 0. \tag{10.5.9}$$

Taking the divergence gives

$$\nabla^2(Lg + \alpha\psi) = 0. \tag{10.5.10}$$

As will be shown in Appendix 10E, the dilation strain $g = \nabla\cdot\boldsymbol{u}$ is expressed in terms of $\psi = M + \delta\psi$ as

$$\begin{aligned} g(\boldsymbol{r}) &= -\frac{\alpha}{L}\psi(\boldsymbol{r}) - \frac{\alpha}{L}\int d\boldsymbol{r}'\mathcal{M}(\boldsymbol{r},\boldsymbol{r}')\psi(\boldsymbol{r}') \\ &= -\frac{\alpha}{K}M - \frac{\alpha}{L}\delta\psi(\boldsymbol{r}) - \frac{\alpha}{L}\int d\boldsymbol{r}'\mathcal{M}(\boldsymbol{r},\boldsymbol{r}')\delta\psi(\boldsymbol{r}'). \end{aligned} \tag{10.5.11}$$

The first two terms in the second line correspond to (10.1.18) and (10.1.28), respectively, and the last term arises from the macroscopic modes. From (10.5.10) the kernel satisfies

$$\nabla^2\mathcal{M}(\boldsymbol{r},\boldsymbol{r}') = 0. \tag{10.5.12}$$

It is written in terms of the eigenfunctions as

$$\mathcal{M}(\boldsymbol{r}, \boldsymbol{r}') = \sum_{\ell m} \mathcal{M}_\ell \chi_{\ell m}(\boldsymbol{r}) \chi_{\ell m}(\boldsymbol{r}')^*, \tag{10.5.13}$$

with

$$\chi_{\ell m}(\boldsymbol{r}) = \left(\frac{2\ell+3}{R^3}\right)^{1/2} \left(\frac{r}{R}\right)^\ell Y_{\ell m}(\theta, \varphi), \tag{10.5.14}$$

where $Y_{\ell m}(\theta, \varphi)$ are the spherical harmonics with $\ell = 0, 1, 2, \ldots$ and $-\ell \le m \le \ell$. The eigenvalues are given by

$$\mathcal{M}_\ell = \frac{\mu(\ell+1)(\ell+2)}{L(\ell^2 + 2\ell + 3/2) - \mu(\ell+1)(\ell+2)}. \tag{10.5.15}$$

The eigenfunctions satisfy $\nabla^2 \chi_{\ell m}(\boldsymbol{r}) = 0$ and are orthogonal and normalized as

$$\int d\boldsymbol{r} \chi_{\ell m}(\boldsymbol{r}) \chi_{\ell' m'}(\boldsymbol{r})^* = \delta_{\ell\ell'} \delta_{mm'}, \tag{10.5.16}$$

but they do not form a complete set. The first two eigenvalues are

$$\mathcal{M}_0 = \mathcal{M}_1 = 4\mu/3K. \tag{10.5.17}$$

For $\ell > 1$, $\mathcal{M}_\ell$ is a decreasing function of ℓ and, as $\ell \to \infty$, it tends to

$$\mathcal{M}_\infty = \mu/(L-\mu) = \mu/(K + \mu/3). \tag{10.5.18}$$

Substitution of (10.5.11) into (10.5.7) gives

$$\mathcal{H}_{\mathrm{T}} = \frac{1}{2} \int d\boldsymbol{r} (r + 3u_0 M^2) \delta\psi(\boldsymbol{r})^2 - \frac{\alpha^2}{2L} \int d\boldsymbol{r} \int d\boldsymbol{r}' \delta\psi(\boldsymbol{r}) \mathcal{M}(\boldsymbol{r}, \boldsymbol{r}') \delta\psi(\boldsymbol{r}'), \tag{10.5.19}$$

where the constant terms are not written explicitly. We decompose $\delta\psi(\boldsymbol{r})$ as

$$\delta\psi(\boldsymbol{r}) = \delta\psi(\boldsymbol{r})_\perp + \sum_{\ell m} \Psi_{\ell m} \chi_{\ell m}(\boldsymbol{r}), \tag{10.5.20}$$

where $\delta\psi_\perp$ is orthogonal to $\chi_{\ell m}$. Then we obtain

$$\mathcal{H}_{\mathrm{T}} = \frac{1}{2} \int d\boldsymbol{r} (r + 3u_0 M^2) \delta\psi(\boldsymbol{r})_\perp^2 + \frac{1}{2} \sum_{\ell m} \left(r + 3u_0 M^2 - \frac{\alpha^2}{L} \mathcal{M}_\ell \right) |\Psi_{\ell m}|^2. \tag{10.5.21}$$

The modes characterized by ℓ become unstable for

$$r + 3u_0 M^2 < \frac{\alpha^2}{L} \mathcal{M}_\ell, \tag{10.5.22}$$

which becomes (10.5.5) for $\ell = 0$ in 3D.

10.5.4 Dynamics at macroscopic instability

We examine linear dynamics of the macroscopic modes in the one-phase region when the sample shape is spherical [118b, 125, 126]. From (10.1.73) and (10.5.7) $\delta\psi$ obeys the diffusion equation,

$$\frac{\partial}{\partial t}\delta\psi = \frac{\lambda_0}{T}\nabla^2\frac{\delta}{\delta\psi}\mathcal{H} \cong D\nabla^2\delta\psi, \tag{10.5.23}$$

where the long-range part ($\propto \mathcal{M}$) vanishes from (10.5.11) and (10.5.12) and $D = \lambda_0(r + 3u_0M^2)/T$ is the diffusion constant. Let the deviation $\delta\psi$ be of the form

$$\delta\psi(\mathbf{r}, t) = e^{-\Omega_\ell t}Y_{\ell m}(\theta, \varphi)\Psi(r). \tag{10.5.24}$$

If this form is substituted in the diffusion equation, we find $\Psi(r) \propto j_\ell(qr)$, where $j_\ell(z) = (2/\pi z)^{1/2}J_{\ell+1/2}(z)$ is the spherical Bessel function of order ℓ and

$$q = (\Omega_\ell/D)^{1/2}. \tag{10.5.25}$$

The dilation deviation is calculated from (10.5.11) in the form

$$\delta g(\mathbf{r}, t) = -(\alpha/L)e^{-\Omega_\ell t}Y_{\ell m}(\theta, \varphi)\mathcal{G}(r), \tag{10.5.26}$$

where

$$\mathcal{G}(r) = \Psi(r) + \mathcal{M}_\ell(2\ell+3)\frac{r^\ell}{R^{2\ell+3}}\int_0^r dr_1 r_1^{2+\ell}\Psi(r_1). \tag{10.5.27}$$

Solids under the chemical equilibrium

For hydrogen–metal systems we impose the chemical equilibrium condition

$$(r_0 + 3u_0M^2)\delta\psi + \alpha\delta g = 0 \tag{10.5.28}$$

as the boundary condition at $r = R$. Using (10.5.25)–(10.5.27) we obtain

$$r + 3u_0M^2 = \frac{\alpha^2}{L}\mathcal{M}_\ell F_{1\ell}(qR), \tag{10.5.29}$$

where

$$\begin{aligned} F_{1\ell}(z) &= (2\ell+3)\int_0^1 dx x^{2+\ell} j_\ell(xz)/j_\ell(z) \\ &= 1 + \frac{1}{(2\ell+3)(2\ell+5)}z^2 + O(z^4). \end{aligned} \tag{10.5.30}$$

Near the instability point, Ω_ℓ is small and $|qR| \ll 1$, so

$$\Omega_\ell \cong (2\ell+3)(2\ell+5)DR^{-2}\big[(r + 3u_0M^2)L/(\alpha^2\mathcal{M}_\ell) - 1\big]. \tag{10.5.31}$$

Particularly for the isotropic mode $\ell = 0$, we have

$$\Omega_0 \cong 15DR^{-2}[(r + 3u_0M^2)/r_{\rm m} - 1], \tag{10.5.32}$$

where r_{m} is defined by (10.5.6). For $\ell = 0$, Ω_0 tends to zero as the macroscopic spinodal line is approached, indicating a slowing-down of volume changes. If the temperature is lowered slightly below this line but above the bulk spinodal line, additional hydrogen will be absorbed (desorbed) if the initial average M is positive (negative). This transition proceeds monotonically towards the final equilibrium state where the equation of state (10.5.3) is satisfied. The proton density heterogeneity involved remains on the system size scale, so no two-phase coexistence can be expected and the spinodal line keeps its mean field character. We stress that a mass flux through the surface is needed to induce the instability of the uniform dilation mode $\ell = 0$.

Solids under no mass exchange

We consider the macroscopic modes in usual binary metal alloys, neglecting mass exchange at the boundary. This condition is written as $\boldsymbol{n} \cdot \nabla \delta \mathcal{H}/\delta\psi = 0$ at the boundary, where $\boldsymbol{n}$ is the surface normal. For a spherical sample we obtain

$$\frac{\partial}{\partial r}(r_0 \delta\psi + \alpha\delta g) = 0 \tag{10.5.33}$$

at $r = R$. For the isotropic case $\ell = 0$, no effect from the long-range part appears, resulting in the usual boundary condition,

$$j_0'(qR) = 0 \quad \text{or} \quad \tan(qR) = qR, \tag{10.5.34}$$

where $j_\ell'(z) = dj_\ell(z)/dz$. This isotropic mode slows down near the bulk instability where $D \to 0$. For $\ell \neq 0$, however, the macroscopic modes can become unstable before the bulk instability. Some calculations yield

$$r + 3u_0 M^2 = \frac{\alpha^2}{L} \mathcal{M}_\ell F_{2\ell}(qR), \tag{10.5.35}$$

analogous to (10.5.29), where

$$\begin{aligned} F_{2\ell}(z) &= \ell(2\ell+3)\int_0^1 dx x^{2+\ell} j_\ell(xz)/z j_\ell'(z) \\ &= 1 + \frac{2\ell^2 + 5\ell + 4}{2(2\ell+3)(2\ell+5)} z^2 + O(z^4). \end{aligned} \tag{10.5.36}$$

Near the instability point with $\ell \neq 0$, the counterpart of (10.5.31) reads

$$\Omega_\ell \cong \frac{2(2\ell+3)(2\ell+5)}{2\ell^2+5\ell+4} D R^{-2} \big[(r + 3u_0 M^2) L / (\alpha^2 \mathcal{M}_\ell) - 1 \big]. \tag{10.5.37}$$

In real binary metal alloys, the resultant timescales are exceedingly long for macroscopic samples and the effects of crystal anisotropy should also be taken into account.

Gorsky effect

We may generally assume (10.5.19) for the free energy in one-phase states for any sample shape, although the kernel $\mathcal{M}(\boldsymbol{r}, \boldsymbol{r}')$ is difficult to calculate except for spheres. We recognize that the elastic field $\boldsymbol{u}$ and the stress tensor are determined by the overall distribution of the concentration. This is the origin of anelastic relaxation called the Gorsky effect [128], as observed in hydrogen–metal systems. That is, if external forces are applied at $t = 0$, an elastic strain is instantaneously induced. If the dilation applied is inhomogeneous, the protons start to diffuse from a locally compressed to an expanded region to achieve homogeneity of the chemical potential ($= \delta\mathcal{H}/\delta\psi$). As a result, a slowly relaxing additional strain $\epsilon_{\rm ad}(t)$ appears [118c, 121].

10.6 Surface instability

10.6.1 Surface modes

The modes with $\ell \gg 1$ in (10.5.14) represent deformations localized near the surface, because $\chi_{\ell m}(\boldsymbol{r})$ is appreciable only for $R - r \lesssim R/\ell$ from $(r/R)^\ell = (1 + (r - R)/R)^\ell \cong \exp((r - R)\ell/R)$. The wave number of the surface corrugations is given by ℓ/R. We shall see below that a surface instability is triggered for

$$r + 3u_0 M^2 < r_{\rm s}, \tag{10.6.1}$$

where

$$r_{\rm s} = \alpha^2 \mathcal{M}_\infty / L = \frac{3}{4}[K/(K + \mu/3)] r_{\rm m}, \tag{10.6.2}$$

in 3D. It follows the relation $0 < r_{\rm s} < r_{\rm m}$, where $r_{\rm m}$ is defined by (10.5.6).

Let us now examine the surface modes localized near a planar interface at $z = 0$, where a binary solid is placed in the lower region ($-\infty < z < 0$) and a gas in the upper region ($0 < z < \infty$). We treat the problem assuming isotropic elasticity and neglecting crystal growth and melting. The elastic field is induced to satisfy the mechanical equilibrium condition against the concentration deviation. As will be shown in Appendix 10F, the Fourier transformations of $\delta\psi$ and δg in the yz plane are related by

$$g_{\boldsymbol{k}}(z) = -\frac{\alpha}{L}\left[\psi_{\boldsymbol{k}}(z) + \frac{\mu}{L - \mu} e^{kz} \Phi_{\boldsymbol{k}}\right], \tag{10.6.3}$$

where $\boldsymbol{k}$ is the wave vector in the xy plane, $k = |\boldsymbol{k}|$, and

$$\Phi_{\boldsymbol{k}} = \int_{-\infty}^{0} dz' e^{kz'} \psi_{\boldsymbol{k}}(z'). \tag{10.6.4}$$

The resultant free energy is written in the form

$$\mathcal{H} = \frac{1}{2}\int d\boldsymbol{r}(r + 3u_0 M^2)\delta\psi(\boldsymbol{r})^2 + \int_{\boldsymbol{k}} (-r_{\rm s} k + \tilde{\sigma} k^2)|\Phi_{\boldsymbol{k}}|^2, \tag{10.6.5}$$

where

$$\tilde{\sigma} = \frac{1}{2}\sigma[\alpha/(L-\mu)]^2. \tag{10.6.6}$$

The term proportional to the surface tension σ arises from the surface displacement calculated in (10F.7). The normalized eigenfunction is given by $\chi_k(z) = \sqrt{2k}e^{kz}$ defined in the region $z < 0$. If we set $\psi_{\boldsymbol{k}}(z) = \Psi_{\boldsymbol{k}}\chi_k(z)$, we obtain $\Psi_{\boldsymbol{k}} = \sqrt{2k}\Phi_{\boldsymbol{k}}$ and

$$\mathcal{H} = \frac{1}{2}\int_{\boldsymbol{k}} (r + 3u_0M^2 - r_\mathrm{s} + \tilde{\sigma}k)|\Psi_{\boldsymbol{k}}|^2, \tag{10.6.7}$$

in agreement with the criterion (10.6.1). When $r + 3u_0M^2 < r_\mathrm{s}$, the surface undulations grow with the characteristic wave number given by

$$k_\mathrm{m} = [r_\mathrm{s} - (r + 3u_0M^2)]/\tilde{\sigma}. \tag{10.6.8}$$

10.6.2 Dynamics at surface instability

We examine linear dynamics of sinusoidal disturbances proportional to $e^{ikx-\Omega_k t}$ in the long-wavelength limit ($k \ll k_\mathrm{m}$). The diffusion equation in the region $z < 0$ reads

$$-\Omega_k\psi_{\boldsymbol{k}} = D(\nabla_z^2 - k^2)\psi_{\boldsymbol{k}}, \tag{10.6.9}$$

and is integrated to give

$$\psi_{\boldsymbol{k}}(z,t) = \Psi e^{qz-\Omega_k t}, \tag{10.6.10}$$

where q with $\mathrm{Re}q > 0$ is determined by

$$q^2 = k^2 - \Omega_k/D. \tag{10.6.11}$$

From (10.6.3) the dilation strain at $z = 0$ is written as

$$g_{\boldsymbol{k}}(0) = -\frac{\alpha}{L}\exp(i\boldsymbol{k}\cdot\boldsymbol{r}_\perp)e^{-\Omega_k t}\left[1 + \frac{2\mu k}{(L-\mu)(k+q)}\right]\Psi, \tag{10.6.12}$$

where Ψ is the amplitude in (10.6.10).

(i) In hydrogen–metal systems we require the chemical equilibrium condition at $z = 0$, which results in

$$r + 3u_0M^2 = \frac{2k}{k+q}r_\mathrm{s}. \tag{10.6.13}$$

Near the instability we find

$$\Omega_k \cong 4Dk^2(r + 3u_0M^2 - r_\mathrm{s})/r_\mathrm{s}. \tag{10.6.14}$$

(ii) This surface instability still exists even if we assume the condition of no mass flux at the interface, $\partial(r_0\delta\psi + \alpha\delta g)/\partial z = 0$ at $z = 0$. Some calculations yield

$$r + 3u_0M^2 = \frac{2k^2}{q(k+q)}r_\mathrm{s}. \tag{10.6.15}$$

Near the instability this equation is solved to give

$$\Omega_k \cong \frac{4}{3} D k^2 (r + 3u_0 M^2 - r_s)/r_s. \tag{10.6.16}$$

10.6.3 Surface instability in growing films

So far we have neglected crystal growth and melting. However, a variety of patterns have been observed in growing thin films, where elastic effects arise from a lattice misfit with the substrate and the deposition rate is a new control parameter of the growth [129]. When the film consists of a one-component metal, surface patterns between the film and the surrounding vapor or melt are of primary concern [130]. When the film is composed of an alloy, there can also be phase separation within the film influenced by elasticity and coupled with the surface undulations [131]. Although these problems are beyond the scope of this book, we here briefly discuss the Asalo–Tiller–Grinfeld instability [132, 133] for uniaxially deformed films composed of a one-component metal at zero deposition rate. This instability was observed on a superfluid–crystal interface in ^{4}He [134] and on an interface of a polymer melt and its crystal [135].

At long wavelengths we may adopt the hydrodynamic approach. Let us write the stress tensor in the solid as $p_s \delta_{ij} - \sigma_{sij}$, where the first term represents a pressure dependent on the solid mass density ρ_s, and the second term arises from the anisotropic deformations and is proportional to the shear modulus. The mechanical equilibrium condition at the interface gives

$$p_f + \sigma \mathcal{K} = p_s - \boldsymbol{n} \cdot \overleftrightarrow{\sigma}_s \cdot \boldsymbol{n}, \tag{10.6.17}$$

where p_f is the fluid pressure, σ is the surface tension, $\mathcal{K}$ is the curvature, and $\boldsymbol{n}$ is the unit normal at the surface. The chemical equilibrium at the interface yields [136, 137]

$$\rho_s \mu_f = p_s - \boldsymbol{n} \cdot \overleftrightarrow{\sigma}_s \cdot \boldsymbol{n} + f_s = p_f + \sigma \mathcal{K} + f_s, \tag{10.6.18}$$

where μ_f is the chemical potential of the fluid and f_s the free-energy density of the solid including the elastic energy. In particular, in ^{4}He at low temperatures, p_f and μ_f may be treated as constants and their deviations from those in equilibrium two-phase coexistence are related by $\delta\mu_f = \delta p_f/\rho_f$ from the Gibbs–Duhem relation. Then (10.6.18) gives

$$\sigma \mathcal{K} + (f_s - f_s^{(0)}) = (\rho_s/\rho_f - 1)\delta p_f, \tag{10.6.19}$$

where $f_s^{(0)}$ is the value of f_s in unstrained solids in equilibrium two-phase coexistence and the right-hand side is an externally controllable parameter. We notice that the chemical equilibrium condition still holds for positive $\mathcal{K}$ and decreasing f_s. Such deformations can release a fraction of the stored elastic energy, thereby overcoming the surface energy increase and thus leading to a surface instability. The characteristic wave number of the growing undulations k_m is determined by balance of the two terms on the right-hand side of (10.6.19) as $k_m \sim (f_s - f_s^{(0)})/\sigma \sim \mu\epsilon_a^2/\sigma$, where $\epsilon_a = \epsilon_{zz} - \epsilon_{xx}$ is the applied anisotropic strain. See (10.6.22) below for k_m from the linear theory.

We are then interested in small undulations upon a planar surface of a uniaxially deformed solid at a fixed $p_{\rm f}$. If the characteristic wavelength of the undulations in the lateral directions is much shorter than the sample thickness, we may assume that the solid occupies the semi-infinite region $-\infty < z < \zeta(x, y)$. The surface position ζ changes as a result of crystal growth or melting. The calculations are then similar to those in Appendix 10F assuming isotropic elasticity [138]. We obtain an increase of the total free energy of the system bilinear with respect to ζ,

$$\Delta\mathcal{H} = \int_{k}\left(-J\sigma_{\rm a}^2 k + \frac{1}{2}\sigma k^2\right)|\zeta_k|^2, \tag{10.6.20}$$

with

$$J = (K + 4\mu/3)/[4\mu(K + \mu/3)] = (1 - \nu^2)/E, \tag{10.6.21}$$

where $\sigma_{\rm a} = \sigma_{zz} - \sigma_{xx} = 2\mu\epsilon_{\rm a}$ is the applied uniaxial stress, E is the Young's modulus, and ν is the Poisson ratio. The characteristic wave number is given by

$$k_{\rm m} = 2J\sigma_{\rm a}^2/\sigma. \tag{10.6.22}$$

The result (10.6.22) is applicable only when the sample thickness H is much greater than $k_{\rm m}^{-1}$. In experiments on ^{4}He, Torii and Balibar [134] applied a very small anisotropic strain ($\epsilon_{\rm a} \sim 10^{-5}$) and observed macroscopic patterns with wavelength about 7 mm, where the gravity contribution in (4.4.53) should also be taken into account. In epitaxial films, $k_{\rm m}^{-1}$ can be microscopic with much larger $\epsilon_{\rm a}$. Grinfeld [133] extended the above result to the case of finite H on a rigid substrate and found a critical thickness $H_{\rm c}$, above which a film becomes unstable against undulations and below which it can adjust coherently to the substrate. Some simulations [139]–[141] have been performed to investigate the nonlinear pattern formation to find growing grooves which serve to release the stored elastic energy. In particular, Müller and Grant [140] set up a free-energy density $f(\phi, \boldsymbol{u})$ for a phase field ϕ and an elastic field $\boldsymbol{u}$, in which the shear modulus $\mu(\phi)$ is zero in liquid ($\phi = 0$) and positive in solid ($\phi = 1$). Assuming that $\mu(\phi)$ is much smaller than the bulk modulus K, they eliminated $\boldsymbol{u}$ in terms of ϕ to obtain an elastic inhomogeneity interaction similar to that in (10.1.37) and solved the resultant dynamic equation of ϕ. Figure 10.39 shows a typical 3D pattern from their simulation.

Appendix 10A Elimination of the elastic field

We eliminate the elastic field $\boldsymbol{u}$ in general anisotropic elasticity characterized by the fourth-rank elastic constant tensor $\lambda_{ijm\ell}(= \lambda_{jim\ell} = \lambda_{ij\ell m} = \lambda_{m\ell ij})$ [47]. If the elastic constants are homogeneous, this procedure is almost trivial [2, 5c], readily leading to the final result in (10A.15) below. The case with inhomogeneous elastic constants is more complicated and was analyzed in Ref. [30].

Defining the elastic strain by

$$\epsilon_{ij} = \frac{1}{2}(\nabla_i u_j + \nabla_j u_i), \tag{10A.1}$$

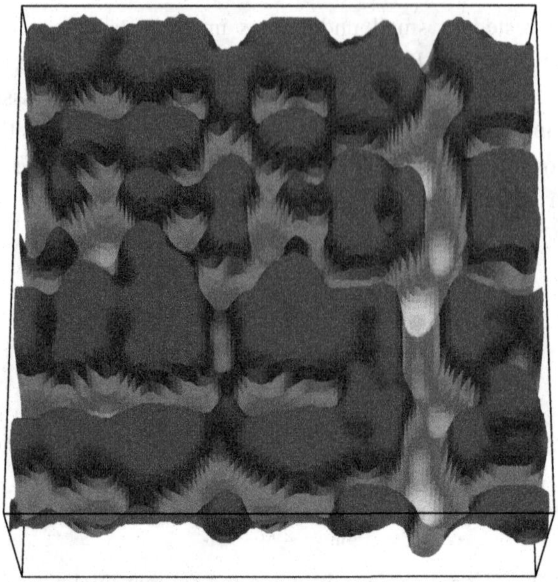

Fig. 10.39. Simulated surface pattern on a uniaxially strained solid growing into a melt [140].

we assume the total free energy in the form

$$\mathcal{H}\{\psi, \boldsymbol{u}\} = \mathcal{H}\{\psi\}_0 + \int d\boldsymbol{r}\left[-\sum_{ij} \sigma_{ij}^0 \epsilon_{ij} + \frac{1}{2} \sum_{ijm\ell} \lambda_{ijm\ell} \epsilon_{ij} \epsilon_{m\ell}\right]. \tag{10A.2}$$

The first term on the right-hand side is a functional of ψ independent of $\boldsymbol{u}$. We may generally treat ψ as a set of the important gross variables such as the concentration c and the long-range order parameter η. The tensor σ_{ij}^0 depends on ψ and can be arbitrary. It is convenient to express it as

$$\sigma_{ij}^0 = \sum_{m\ell} \lambda_{ijm\ell} \epsilon_{m\ell}^0. \tag{10A.3}$$

In the literature ϵ_{ij}^0 is called the *stress-free strain, transformation strain, intrinsic strain*, or *spontaneous deformation*. The elastic stress tensor is then written as

$$\sigma_{ij} = \sum_{m\ell} \lambda_{ijm\ell} \epsilon_{m\ell} - \sigma_{ij}^0 = \sum_{m\ell} \lambda_{ijm\ell} [\epsilon_{m\ell} - \epsilon_{m\ell}^0]. \tag{10A.4}$$

The elastic free energy is usually defined in the form,

$$\mathcal{H}_{\rm el} = \frac{1}{2} \int d\boldsymbol{r} \sum_{ij} [\epsilon_{ij} - \epsilon_{ij}^0] \sigma_{ij} = \frac{1}{2} \int d\boldsymbol{r} \sum_{ijm\ell} \lambda_{ijm\ell} [\epsilon_{ij} - \epsilon_{ij}^0][\epsilon_{m\ell} - \epsilon_{m\ell}^0], \tag{10A.5}$$

which is nonnegative-definite in stable states. Then the total free energy is expressed as

$$\mathcal{H}\{\psi, \boldsymbol{u}\} = \mathcal{H}\{\psi\}_{\rm c} + \mathcal{H}_{\rm el}. \tag{10A.6}$$

The first term may be called the *chemical* free energy and is of the form,

$$\mathcal{H}\{\psi\}_{\rm c} = \mathcal{H}\{\psi\}_0 - \frac{1}{2}\int d\boldsymbol{r} \sum_{ij} \sigma_{ij}^0 \epsilon_{ij}^0. \tag{10A.7}$$

If we assume the Vegard law and adopt the coupling in (10.1.2), we have

$$\sigma_{ij}^0 = -\alpha\psi\delta_{ij}, \qquad \epsilon_{ij}^0 = -\frac{\alpha}{dK}\psi\delta_{ij}, \tag{10A.8}$$

where ψ is the composition deviation and K is the bulk modulus. Let the free-energy density in $\mathcal{H}\{\psi\}_{\rm c}$ be $f_{\rm chem}$; then, (10A.7) indicates that it is related to f_0 in $\mathcal{H}\{\psi\}_0$ as in (10.1.35). In order–disorder phase transitions, σ_{ij}^0 depends on η as in (10.3.2) for fcc alloys and as in (10.3.5) for bcc alloys, for example.

For simplicity, we assume homogeneous $\lambda_{ijm\ell}$ (independent of ψ). In the presence of general homogeneous strain $\langle\epsilon_{ij}\rangle$, $\mathcal{H}$ is expressed as

$$\mathcal{H}\{\psi, \boldsymbol{u}\} = \mathcal{H}\{\psi\}_0 - \int d\boldsymbol{r} \sum_{ij} \langle\epsilon_{ij}\rangle\sigma_{ij}^0 + \frac{1}{2}V \sum_{ijm\ell} \lambda_{ijm\ell}\langle\epsilon_{ij}\rangle\langle\epsilon_{m\ell}\rangle + \delta\mathcal{H}. \tag{10A.9}$$

The second term can be important when σ_{ij}^0 contains terms proportional to η_α^2 as in (10.3.9) or (10.3.14). The third term is simply a constant (if the solid shape is fixed during phase separation). The last term $\delta\mathcal{H}$ is obtained if ϵ_{ij} in the second term of (10A.2) is replaced by the deviation $\delta\epsilon_{ij} = \epsilon_{ij} - \langle\epsilon_{ij}\rangle$. If $\langle\epsilon_{ij}\rangle = 0$, we simply have $\mathcal{H}_{\rm el} = \delta\mathcal{H} + \frac{1}{2}\int d\boldsymbol{r} \sum_{ij} \sigma_{ij}^0\epsilon_{ij}^0$. To express $\delta\mathcal{H}$ in terms of ψ, we impose the mechanical equilibrium condition:

$$\sum_j \nabla_j \sigma_{ij} = \sum_{m\ell} \lambda_{ijm\ell}\nabla_j\nabla_\ell u_m - f_i = 0, \tag{10A.10}$$

where

$$f_i = \sum_j \nabla_j \sigma_{ij}^0 \tag{10A.11}$$

is the force density created by the order parameter fluctuations. In terms of the Fourier components $u_{i\boldsymbol{k}}$ and $f_{i\boldsymbol{k}}$, the above equation is expressed as

$$k^2 \sum_j \Omega^{ij}(\hat{\boldsymbol{k}}) u_{j\boldsymbol{k}} = -f_{i\boldsymbol{k}}, \tag{10A.12}$$

where

$$\Omega^{ij}(\hat{\boldsymbol{k}}) = \sum_{m\ell} \lambda_{imj\ell}\hat{k}_m\hat{k}_\ell. \tag{10A.13}$$

The vector $\hat{\boldsymbol{k}} = k^{-1}\boldsymbol{k}$ denotes the direction of the wave vector. Let $\Omega_{ij}(\hat{\boldsymbol{k}})$ be the inverse matrix of $\Omega^{ij}(\hat{\boldsymbol{k}})$. Then (10A.11) is solved in the form

$$u_{ik} = -\frac{1}{k^2}\sum_j \Omega_{ij}(\hat{\boldsymbol{k}}) f_{jk}. \tag{10A.14}$$

Substitution of the above result into (10A.2) yields the desired result,

$$\delta\mathcal{H} = -\frac{1}{2}\int_{\boldsymbol{k}} \frac{1}{k^2}\sum_{ij} \Omega_{ij}(\hat{\boldsymbol{k}}) f_{ik}(f_{jk})^* = -\frac{1}{2}\int_{\boldsymbol{k}} \sum_{ijm\ell} \Omega_{ij}(\hat{\boldsymbol{k}})\hat{k}_m\hat{k}_\ell\sigma^0_{imk}(\sigma^0_{j\ell k})^*. \tag{10A.15}$$

Note that the eigenvalues of $\Omega^{ij}(\hat{\boldsymbol{k}})$ are $\rho c_\alpha(\hat{\boldsymbol{k}})^2$ ($\alpha = 1, 2, 3$), where $c_\alpha(\hat{\boldsymbol{k}})$ are the sound velocities, as can be seen in (10.5.1) for cubic solids. It is also worth noting that the correlation functions of the thermal fluctuations of the elastic field $\boldsymbol{u}$ can be expressed in terms of Ω_{ij} as

$$\langle u_{ik}u^*_{jk}\rangle = \frac{T}{k^2}\Omega_{ij}(\hat{\boldsymbol{k}}) + \frac{T}{k^4}\sum_{\ell m} \Omega_{i\ell}(\hat{\boldsymbol{k}})\Omega_{jm}(\hat{\boldsymbol{k}})\langle f_{\ell k} f^*_{mk}\rangle, \tag{10A.16}$$

where the vector $\boldsymbol{f}$ is defined by (10A.11). Thus the fluctuations of $\boldsymbol{u}$ are enhanced with softening of sound modes.

In cubic solids we have

$$\Omega^{ij}(\hat{\boldsymbol{k}}) = C_{44}(1 + \xi_a\hat{k}_i^2)\delta_{ij} + (C_{12} + C_{44})\hat{k}_i\hat{k}_j. \tag{10A.17}$$

The inverse is written in terms of $\varphi_0(\hat{\boldsymbol{k}})$ in (10.2.10) as

$$\Omega_{ij}(\hat{\boldsymbol{k}}) = \frac{1}{C_{44}(1 + \xi_a\hat{k}_i^2)}\left[\delta_{ij} - \left(1 + \frac{C_{12}}{C_{44}}\right)\frac{\hat{k}_i\hat{k}_j}{[1 + \varphi_0(\hat{\boldsymbol{k}})](1 + \xi_a\hat{k}_j^2)}\right]. \tag{10A.18}$$

The calculation of $\mathcal{H}_{\rm el}$ is very complicated except for the scalar case (10A.8) treated in Section 10.2.

In the isotropic case $\xi_a = 0$ we have

$$\Omega_{ij}(\hat{\boldsymbol{k}}) = \frac{1}{\mu}\left[\delta_{ij} - \frac{1}{2(1-\nu)}\hat{k}_i\hat{k}_j\right], \tag{10A.19}$$

where $\mu = C_{44}$ and $\nu = C_{12}/(C_{11}+C_{12})$ (the Poisson ratio). This expression is applicable in 2D and 3D with this definition of ν. To study salient features arising from the anisotropy of ϵ^0_{ij}, particularly in simulations, use has been made of the expression (10A.19) even for cubic solids. (i) For example, if the transformation strain is orthorhombic or tetragonal, we have diagonal transformation strain and stress tensors, $\epsilon^0_{ij} = \delta_{ij}\epsilon^0_{ii}$ and $\sigma^0_{ij} = \delta_{ij}\sigma^0_{ii}$. It is easy to derive

$$\delta\mathcal{H} = \frac{1}{4\mu}\int_{\boldsymbol{k}}\left[\frac{1}{1-\nu}\left|\sum_j \hat{k}_j^2\sigma^0_{jjk}\right|^2 - 2\sum_j \hat{k}_j^2|\sigma^0_{jjk}|^2\right]. \tag{10A.20}$$

This expression leads to (10.3.11) for the case (10.3.9). (ii) In particular, in 2D we set $\epsilon^0_{i3} = 0$ $(i = 1, 2, 3)$ to obtain a simple expression,

$$\mathcal{H}_{\rm el} = \frac{\mu}{4(1-\nu)} \int_k \left| \epsilon^0_{11k} + \epsilon^0_{22k} - (\hat{k}_1^2 - \hat{k}_2^2)(\epsilon^0_{11k} - \epsilon^0_{22k}) - 4\hat{k}_1\hat{k}_2\epsilon^0_{12k} \right|^2, \tag{10A.21}$$

where $\hat{k}_1 = \hat{k}_x$ and $\hat{k}_2 = \hat{k}_y$. This expression leads to (10.3.17) for the case (10.3.13).

Appendix 10B Elastic deformation around an ellipsoidal domain

We calculate elastic deformations around an ellipsoidal inclusion assuming isotropic elasticity. We define the ellipsoidal coordinate $\xi = \xi(x, y, z)$ as the solution of the equation [142]

$$\frac{1}{a_1^2 + \xi} x^2 + \frac{1}{a_2^2 + \xi} y^2 + \frac{1}{a_3^2 + \xi} z^2 = 1, \tag{10B.1}$$

with $\xi > -a_1^2, -a_2^2, -a_3^2$. The ellipsoidal surface is represented by $\xi = 0$, while $\xi > 0$ outside it and $\xi < 0$ inside it. We also write $x_1 = x, x_2 = y$, and $x_3 = z$. For large $r^2 = x^2 + y^2 + z^2$ we obtain $\xi \cong r^2$. We use the following relations,

$$\nabla_i \xi = \frac{2x_i}{a_i^2 + \xi} \Big/ \left[\sum_j \frac{1}{(a_j^2 + \xi)^2} x_j^2 \right]. \tag{10B.2}$$

As $\xi \to 0$, the gradient vector $\nabla\xi$ tends to $|\nabla\xi|\boldsymbol{n}$ where $\boldsymbol{n} = (n_1, n_2, n_3)$ is the normal unit vector at the surface, $n_i = x_i/a_i^2 \left[\sum_j x_j^2/a_j^4 \right]^{1/2}$. We introduce the depolarization factors N_i by

$$N_i = \frac{1}{2} a_1 a_2 a_3 \int_0^\infty \frac{ds}{(s + a_i^2)R(s)}, \tag{10B.3}$$

where

$$R(s) = \sqrt{(s + a_1^2)(s + a_2^2)(s + a_3^2)}. \tag{10B.4}$$

From $\partial \ln R(s)/\partial s = \frac{1}{2} \sum_i 1/(s + a_i^2)$, we can easily find

$$N_1 + N_2 + N_3 = 1. \tag{10B.5}$$

For a spheroid ($a_1 = a$ and $a_2 = a_3 = b$) we have $N_1 = b^2(g(e) - 1)/(a^2 - b^2)$, where $e = |1 - b^2/a^2|^{1/2}$ is the eccentricity, $g(e) = \ln[(1 + e)/(1 - e)]/2e$ for $a > b$, and $g(e) = e^{-1} \tan^{-1} e$ for $a < b$ [142]. In Fig. 10.7 we show $N_1 (= N_x)$.

We next introduce the following vector,

$$\begin{aligned} D_i &= \frac{1}{2}(a_1 a_2 a_3) x_i \int_\xi^\infty \frac{ds}{(s + a_i^2)R(s)} && (\xi > 0), \\ &= N_i x_i && (\xi < 0). \end{aligned} \tag{10B.6}$$

From the definition of N_i the continuity of D_i holds across the ellipsoidal surface $\xi = 0$. It is known that $D_i - N_i x_i$ is proportional to the dipolar field around an ellipsoidal conductor in an electric field in the x_i direction [142].[13] The gradient ∇D_i has a discontinuity across the interface proportional to $\boldsymbol{n}$ as

$$[\nabla D_i] = -a_i^{-2} x_i \nabla \xi = -n_i \boldsymbol{n}. \tag{10B.7}$$

Furthermore, using (10B.2), we confirm $\nabla_j D_i = \nabla_i D_j$. Therefore, D_i turns out to be expressed as $D_i = \nabla_i W$. Then $\nabla^2 W = 1$ inside the ellipsoid and $\nabla^2 W = 0$ outside it. The field w in (10.1.27) may be written as $w = (\Delta\psi)W$, resulting in (10.1.58).

Appendix 10C Analysis of the Jahn–Teller coupling

We derive the Jahn–Teller interaction $\mathcal{H}_{\rm JT}$ in (10.4.1) for doubly degenerate d-orbital states around Cu^{2+} or Mn^{3+} [88], whose wave functions are represented as linear combinations of the wave functions proportional to $2z^2 - x^2 - y^2$ and $x^2 - y^2$. The index n denoting the lattice site will be dropped. The orbital states proportional to $2x^2 - y^2 - z^2$, $2y^2 - z^2 - x^2$, and $2z^2 - x^2 - y^2$ are written as $|x^2\rangle$, $|y^2\rangle$, and $|z^2\rangle$, respectively, while those with wave functions proportional to $x^2 - y^2$, $y^2 - z^2$, and $z^2 - x^2$, are written as $|x^2 - y^2\rangle$, $|y^2 - z^2\rangle$, and $|z^2 - x^2\rangle$, respectively. As orthogonal, complete bases, we define

$$|1\rangle = |x^2 - y^2\rangle = \frac{1}{\sqrt{3}}(|x^2\rangle - |y^2\rangle), \quad |2\rangle = |z^2\rangle. \tag{10C.1}$$

Because the electronic orbit is elongated in the x, y, and z axes in the states $|x^2\rangle$, $|y^2\rangle$, and $|z^2\rangle$, respectively, the orbit–lattice (Jahn–Teller) coupling energy at each site in cubic solids should be of the form,

$$H_{\rm JT} = -\bar{g}_{\rm K}\left[Q_x|x^2\rangle\langle x^2| + Q_y|y^2\rangle\langle y^2| + Q_z|z^2\rangle\langle z^2| - \frac{1}{2}(Q_x + Q_y + Q_z)\right], \tag{10C.2}$$

in the bra-ket representation of quantum mechanics, where $\bar{g}_{\rm K}$ is a positive coupling constant and Q_i ($i = x, y, z$) represent the atomic displacements whose acoustic parts are $\nabla_i u_i$. We then use $|x^2\rangle = 2^{-1}(\sqrt{3}|1\rangle - |2\rangle)$ and $|y^2\rangle = -2^{-1}(\sqrt{3}|1\rangle + |2\rangle)$ to express $H_{\rm JT}$ in the form of (10.4.1) with

$$Q_2 = Q_x - Q_y, \quad Q_3 = \frac{1}{\sqrt{3}}(2Q_z - Q_x - Q_y), \tag{10C.3}$$

and $g_{\rm K} = (\sqrt{3}/4)\bar{g}_{\rm K}$. From (10.4.4) the acoustic part of Q_2 and Q_3 are e_2 and e_3, respectively. The pseudo-Pauli matrices are defined at each lattice site and are expressed as

$$\hat{\sigma}_z = |1\rangle\langle 1| - |2\rangle\langle 2|, \quad \hat{\sigma}_x = |1\rangle\langle 2| + |2\rangle\langle 1|. \tag{10C.4}$$

Thus $\mathcal{H}_{\rm JT}$ in (10.4.1) is the sum of the Jahn–Teller contributions from all the lattice sites. If one electron is in a d-orbital state at a lattice site, its state is expressed as a linear combination of the two bases $|1\rangle$ and $|2\rangle$ in the form $c_1|1\rangle + c_2|2\rangle$, where c_1 and c_2 are

[13] Let us assume the Laplace equation $\nabla^2[x_i F(\xi)] = 0$ outside the ellipsoid $\xi > 0$; then, $F(\xi)$ satisfies $d^2F/d\xi^2 + dF/d\xi \cdot d[\ln R(\xi)(\xi + a_i^2)]/d\xi = 0$, leading to either of $F =$ const. or the first line of (10B.6).

complex coefficients with $|c_1|^2 + |c_2|^2 = 1$. For example, if the orbital state is purely $|x^2\rangle$, we have $c_1 = \sqrt{3}/2$ and $c_2 = -1/2$. If Q_2 and Q_3 are treated as constants, the eigenvalues of $H_{\rm JT}$ in (10C.2) are calculated as $\pm g_{\rm K}(Q_2^2 + Q_3^2)^{1/2}$, and the eigenstate corresponding to the lower eigenvalue is given by $\sin\theta_{23}|1\rangle - \cos\theta_{23}|2\rangle$ where $\tan(2\theta_{23}) = Q_2/Q_3$.

Appendix 10D Nonlocal interaction in 2D elastic theory

For the 2D free-energy density (10.4.45) we may easily express e_1 and e_4 in terms of the order parameter $\psi = e_2$ such that they minimize $\mathcal{H} = \int d\mathbf{r} f$ at fixed ψ. Under this constraint of fixed ψ we consider the part of $\mathcal{H}$ which involves e_1 and e_4,

$$\Delta\mathcal{H} = \int d\mathbf{r}\left[\frac{K}{2}e_1^2 + \frac{\mu}{2}e_4^2 - \alpha' e_1\psi^2 + \lambda(\nabla_x u_x - \nabla_y u_y - \psi)\right], \tag{10D.1}$$

where λ is a space-dependent Lagrange multiplier. From $\delta\Delta\mathcal{H}/\delta u_x = \delta\Delta\mathcal{H}/\delta u_y = 0$ we obtain

$$\begin{aligned} K\nabla_x e_1 + \mu\nabla_y e_4 &= -\nabla_x\lambda + \alpha'\nabla_x\psi^2, \\ K\nabla_y e_1 + \mu\nabla_x e_4 &= \nabla_y\lambda + \alpha'\nabla_x\psi^2. \end{aligned} \tag{10D.2}$$

Some calculations yield

$$e_1 = \frac{\alpha'}{K}\psi^2 + \mathcal{L}^{-1}\nabla^2 W, \quad e_4 = -\frac{2\mu}{K}\mathcal{L}^{-1}\nabla_x\nabla_y W, \tag{10D.3}$$

where

$$W = (\nabla_x^2 - \nabla_y^2)\psi - \frac{\alpha'}{K}\nabla^2\psi^2, \tag{10D.4}$$

and $\mathcal{L}^{-1}$ is the inverse operator of

$$\mathcal{L} = \nabla^4 + \frac{4}{\mu}K\nabla_x^2\nabla_y^2. \tag{10D.5}$$

We also obtain $\lambda = -\mathcal{L}^{-1}(\nabla_x^2 - \nabla_y^2)W$. The $\Delta\mathcal{H}$ is expressed in terms of ψ as

$$\Delta\mathcal{H} = \int d\mathbf{r}\left(\frac{1}{2}KW\mathcal{L}^{-1}W - \frac{1}{2K}\alpha'^2\psi^4\right). \tag{10D.6}$$

If $\alpha' = 0$, we simply obtain $W = (\nabla_x^2 - \nabla_y^2)\psi$ [108], so that $\Delta\mathcal{H}$ is lowered when the space variations are along [11] or [1$\bar{1}$], leading to the formation of twin structures. If $\alpha' \neq 0$, we are led to the estimation (10.4.54) for the interface orientation in intermediate states. We also note that Kartha *et al.* [108] added a term $\lambda' g$ to the free-energy density for the case $\alpha' = 0$, where $g \equiv \nabla^2 e_1 - 2\nabla_x\nabla_y e_4 - (\nabla_x^2 - \nabla_y^2)e_2$. From the elastic compatibility relation (10.4.47) we identically have $g = 0$ in 2D. Then they minimized the free energy by taking the functional derivatives with respect to e_1 and e_4 and treating λ' as a space-dependent Lagrange multiplier at fixed e_2. There is no essential difference between their method and the one presented above.

Appendix 10E Macroscopic modes of a sphere

We calculate the macroscopic modes for the free energy (10.1.2) under the stress-free boundary condition in the absence of crystal growth and melting. For simplicity, we assume that the crystal shape is a sphere with radius R and the elastic moduli are homogeneous.

Isotropic case ($\ell = 0$)

First we assume that $\psi = \psi(r)$ is independent of the direction $\hat{\boldsymbol{r}}$. Then the elastic field is isotropic as

$$u_i(\boldsymbol{r}) = u(r)\hat{x}_i, \tag{10E.1}$$

where $u(r)$ depends only on $r = (x^2 + y^2 + z^2)^{1/2}$. The dilation strain is written as

$$g = \nabla \cdot \boldsymbol{u} = u' + 2\frac{u}{r}, \tag{10E.2}$$

where $u' = \partial u/\partial r$. From the mechanical equilibrium condition (10.5.10) we obtain

$$Lg + \alpha\psi = A = \text{const.} \tag{10E.3}$$

The stress tensor (10.1.10) is expressed $\sigma_{ij} = A\delta_{ij} - 2\mu(\delta_{ij} - \hat{x}_i\hat{x}_j)g + 2\mu(\delta_{ij} - 3\hat{x}_i\hat{x}_j)u/r$. The displacement at $r = R$ is equal to $u_R = (A/4\mu)R$ from the stress-free boundary condition (10.5.8). The space integral of g is the volume change $\delta V = \int d\boldsymbol{r} g = 4\pi R^2 u_R$, which follows from (10E.2). The space integration of (10E.3) gives

$$A = -\frac{4\mu}{3K}\alpha M, \quad u_R = -\frac{1}{3K}R\alpha M, \tag{10E.4}$$

in terms of $M = \langle\psi\rangle$. From (10E.2) we find

$$u(r) = -\frac{\alpha}{Lr^2}\int_0^r d\rho\rho^2\left[\psi(\rho) + \frac{4\mu}{3K}M\right]. \tag{10E.5}$$

Representation of a vector in spherical coordinates

As a preparation for general anisotropic cases, we introduce a general representation in which an arbitrary vector variable, written as $\boldsymbol{u}$, is expressed in terms of three scalar functions h, Q, and S as

$$\boldsymbol{u}(\boldsymbol{r}) = \nabla h + Q\boldsymbol{r} + (\boldsymbol{r} \times \nabla)S. \tag{10E.6}$$

As a simplifying result, we will find $S = 0$ in our present problem. The dilation strain becomes

$$g = \nabla^2 h + 3Q + rQ', \tag{10E.7}$$

where $Q' = \partial Q/\partial r$. Along the three orthogonal unit vectors,

$$\begin{aligned} \boldsymbol{e}_1 &= \hat{\boldsymbol{r}} = (\sin\theta\cos\varphi, \sin\theta\sin\varphi, \cos\theta), \\ \boldsymbol{e}_2 &= \frac{\partial}{\partial\theta}\boldsymbol{e}_1 = (\cos\theta\cos\varphi, \cos\theta\sin\varphi, -\sin\theta), \\ \boldsymbol{e}_3 &= \boldsymbol{e}_1 \times \boldsymbol{e}_2 = (-\sin\varphi, \cos\varphi, 0), \end{aligned} \tag{10E.8}$$

the vector $\boldsymbol{u}$ has three components expressed as

$$\begin{aligned} u_1 &= \boldsymbol{e}_1 \cdot \boldsymbol{u} = h' + Qr, \\ u_2 &= \boldsymbol{e}_2 \cdot \boldsymbol{u} = \frac{1}{r}\nabla_\theta h - \nabla_\varphi S, \\ u_3 &= \boldsymbol{e}_3 \cdot \boldsymbol{u} = \frac{1}{r}\nabla_\varphi h + \nabla_\theta S, \end{aligned} \tag{10E.9}$$

where $h' = \partial h/\partial r$, $\nabla_\theta = \partial/\partial\theta$, and $\nabla_\varphi = (\sin\theta)^{-1}\partial/\partial\varphi$. Therefore, h and S can be expressed in terms of u_2 and u_3 as

$$\begin{aligned} -\hat{\Lambda} h &= \frac{r}{\sin\theta}\nabla_\theta(\sin\theta)u_2 + r\nabla_\varphi u_3, \\ -\hat{\Lambda} S &= -\nabla_\varphi u_2 + \frac{1}{\sin\theta}\nabla_\theta(\sin\theta)u_3, \end{aligned} \tag{10E.10}$$

where

$$\hat{\Lambda} = \frac{1}{\sin\theta}\frac{\partial}{\partial\theta}\sin\theta\frac{\partial}{\partial\theta} + \frac{1}{\sin\theta^2}\frac{\partial^2}{\partial\varphi^2} \tag{10E.11}$$

is the angle part of the laplacian operator $\nabla^2 = \partial^2/\partial r^2 + (2/r)\partial/\partial r + \hat{\Lambda}/r^2$.

To impose the boundary condition at $r = R$, we need to calculate the following components of the shear strain e_{ij} in (10.1.7):

$$\begin{aligned} \boldsymbol{e}_1 \cdot \overleftrightarrow{e} \cdot \boldsymbol{e}_1 &= 2u_1' - \frac{2}{3}g = \frac{4}{3}g - \frac{4}{r}h' - \frac{2}{r^2}\hat{\Lambda}h - 4Q, \\ \boldsymbol{e}_2 \cdot \overleftrightarrow{e} \cdot \boldsymbol{e}_1 &= \nabla_\theta\left(\frac{2h'}{r} - \frac{2h}{r^2} + Q\right) + \nabla_\varphi\left(S' - \frac{S}{r}\right), \\ \boldsymbol{e}_3 \cdot \overleftrightarrow{e} \cdot \boldsymbol{e}_1 &= \nabla_\varphi\left(\frac{2h'}{r} - \frac{2h}{r^2} + Q\right) - \nabla_\theta\left(S' - \frac{S}{r}\right), \end{aligned} \tag{10E.12}$$

where $S' = \partial S/\partial r$. An advantage of our representation is that different ℓ and m are not mixed in the bulk relations ($r < R$) and the boundary conditions ($r = R$) if they are expressed in terms of h, Q, and S. We may thus assume that h, Q, and S commonly depend on the angles θ and φ as $Y_{\ell m}(\theta,\varphi)$, whereas u_2 and u_3 are not proportional to $Y_{\ell m}(\theta,\varphi)$.

Anisotropic case ($\ell > 0$)

We now use the above representation by assuming h, Q, $S \propto Y_{\ell m}(\theta,\varphi)$, and

$$\psi(\boldsymbol{r}) = \psi_{\ell m}(r)Y_{\ell m}(\theta,\varphi). \tag{10E.13}$$

We may then replace $\hat{\Lambda}$ by $-\ell(\ell+1)$. We take the inner products between the bulk vector relation (10.5.9) and $\boldsymbol{e}_i$. Those with $i = 2, 3$ simply give

$$\nabla^2 Q = 0, \quad S = 0. \tag{10E.14}$$

Therefore, we may set

$$Q = Q_{\ell m} r^\ell Y_{\ell m}(\theta, \varphi), \tag{10E.15}$$

where $Q_{\ell m}$ is a constant. Taking the product with $\boldsymbol{e}_1$ gives

$$Lg + \alpha\psi = \mu(Q + rQ') = \mu(\ell + 1)Q. \tag{10E.16}$$

From (10E.7) the equation for h is obtained in the form,

$$\nabla^2 h = -\frac{\alpha}{L}\psi + \left[(\ell+1)\frac{\mu}{L} - (\ell+3)\right]Q. \tag{10E.17}$$

Because $h(\boldsymbol{r}) = h_{\ell m}(r)Y_{\ell m}(\theta, \varphi)$ should be finite at $r = 0$, we may integrate the above equation as

$$\begin{aligned} h_{\ell m}(r) &= \left[(\ell+1)\frac{\mu}{L} - (\ell+3)\right]\frac{Q_{\ell m} r^{\ell+2}}{4\ell+6} \\ &+ \frac{\alpha}{(2\ell+1)L}\left[\frac{\Psi^<_{\ell m}(r)}{r^{\ell+1}} + r^\ell \Psi^>_{\ell m}(r)\right] + H_{\ell m} r^\ell, \end{aligned} \tag{10E.18}$$

where $H_{\ell m}$ is a constant and

$$\Psi^<_{\ell m}(r) = \int_0^r d\rho\, \psi_{\ell m}(\rho)\rho^{\ell+2}, \quad \Psi^>_{\ell m}(r) = \int_r^R d\rho\, \psi_{\ell m}(\rho)\frac{1}{\rho^{\ell-1}}. \tag{10E.19}$$

From the strain relations (10E.12) the stress-free boundary condition yields two relations at $r = R$,

$$\frac{2h'}{r} - \frac{2h}{r^2} + Q = 0, \tag{10E.20}$$

$$Lg + \alpha\psi - \mu\left(\frac{4}{r}h' + \frac{2}{r^2}\hat{\Lambda}h + 4Q\right) = 0. \tag{10E.21}$$

We notice that the combination $X_{\ell m}(r) \equiv rh'_{\ell m} - \ell h_{\ell m}$ does not involve the last term ($\propto H_{\ell m}$) in (10E.18) and satisfies a simple boundary condition,

$$(\ell+2)X_{\ell m}(R) = -\frac{3}{2}R^2 Q_{\ell m}, \tag{10E.22}$$

which readily yields

$$\left\{\left[(\ell+1)\frac{\mu}{L} - (\ell+3)\right]\frac{1}{2\ell+3} + \frac{3}{2\ell+4}\right\}Q_{\ell m} = \frac{\alpha}{L}\int_0^R d\rho\, \psi_{\ell m}(\rho)\rho^{\ell+2}. \tag{10E.23}$$

From (10E.16) we may now express g in terms of ψ. Some manipulations yield (10.5.11)–(10.5.16).

Appendix 10F Surface modes on a planar surface

Supposing a semi-infinite elastic system, we examine small surface undulations under the mechanical equilibrium condition in the absence of crystal growth and melting. We take the z axis in the normal direction and the x axis on the horizontal plane. All the deviations are proportional to e^{ikx} and independent of y. They decay as e^{kz} far below the interface ($z \to -\infty$), where k is assumed to be positive for simplicity. This semi-infinite approximation is allowable if the wavelength $2\pi/k$ is much shorter than the thickness of the solid. From (10.5.10) we obtain

$$L\delta g + \alpha\delta\psi = Ae^{kz+ikx}, \tag{10F.1}$$

where A is a constant. Then (10.5.9) becomes

$$\mu\nabla^2 u_i + \nabla_i[(\alpha\mu/L)\psi + (L-\mu)Ae^{kz}] = 0, \tag{10F.2}$$

and the displacement is of the form,

$$u_x = \frac{i\alpha}{2L}e^{ikx}\left[G(z) + (\beta_x + \gamma z)e^{kz}\right], \tag{10F.3}$$

$$u_z = \frac{\alpha}{2L}e^{ikx}\left[\frac{1}{k}\frac{\partial}{\partial z}G(z) + (\beta_z + \gamma z)e^{kz}\right], \tag{10F.4}$$

where $\gamma = -[L(L-\mu)/\alpha\mu]A$ and

$$G(z) = \int_{-\infty}^{0} dz' e^{-k|z-z'|}\delta\psi(z'). \tag{10F.5}$$

We determine the three coefficients, β_x, β_z, and A from (10F.1) and the stress-free boundary condition $\sigma_{zz} = \sigma_{xz} = 0$ at $z = 0$. Some calculations yield

$$\beta_x = -\beta_z = \frac{L+\mu}{L-\mu}G(0), \quad \gamma = 2kG(0), \tag{10F.6}$$

where $G(0)$ coincides with Φ_k in (10.6.4). The surface displacement is given by

$$u_z(0) = -\frac{\alpha}{L-\mu}e^{ikx}G(0). \tag{10F.7}$$

References

[1] J. W. Cahn, in *Critical Phenomena in Alloys, Magnets, and Superconductors*, eds. R. I. Jafee *et al.* (McGraw-Hill, New York, 1971), p. 41.

[2] A. G. Khachaturyan, *Theory of Structural Transformations in Solids* (John Wiley & Sons, New York, 1983).

[3] W. C. Johnson and P. W. Voorhees, *Solid State Phenomena* **23**, 87 (1992).

[4] P. Fratzl, O. Penrose, and J. L. Lebowitz, *J. Stat. Phys.* **95**, 1429 (1999).

[5] (a) A. J. Ardell and D. M. Kim, in *Phase Transformations and Evolution in Materials*, eds. P. E. A. Turch and A. Gonis, TMS (TMS SIAM, 2000), p. 309; (b) A. G. Khachaturyan, *ibid.*, p. 309; (c) L. Q. Chen, *ibid.* p. 209; (d) K. Thornton, N. Akaiwa, and P. W. Voorhees, *ibid.*, p. 73; (d) J. K. Lee, *ibid.*, p. 237; (e) A. Onuki and A. Furukawa, *ibid.*, p. 221.

[6] *Proceedings of the International Conference on Solid–Solid Phase Transformations '99* (JIMIC-3), eds. M. Koiwa, K. Otsuka, and T. Miyazaki (The Japan Institute of Metals, 1999).

[7] A. J. Adrell, R. B. Nicholson, and J. D. Eshelby, *Acta Metall.* **14**, 1295 (1966).

[8] A. Maheshwari and A. J. Ardell, *Phys. Rev. Lett.* **70**, 2305 (1993).

[9] T. Miyazaki, H. Imamura, and T. Kozakai, *Mater. Sci. Eng.* **54**, 9 (1982); M. Doi, T. Miyazaki, and T. Wakatsuki, *Mater. Sci. Eng.* **67**, 247 (1984).

[10] T. Miyazaki and M. Doi, *Mater. Sci. Eng. A* **110**, 175 (1989); T. Miyazaki, M. Doi, and T. Kozakai, *Solid State Phenomena* **384**, 227 (1988).

[11] R. A. Ricks, A. J. Porter, and R. C. Ecob, *Acta Metall.* **31**, 43 (1983).

[12] J. G. Conley, M. E. Fine, and J. R. Weertman, *Acta Metall.* **37**, 1251 (1989).

[13] W. Hein, *Acta Metall.* **37**, 2145 (1989).

[14] M. Fährmann, P. Fratzl, O. Paris, E. Fährmann, and W. C. Johnson, *Acta. Metall.* **43**, 1007 (1995).

[15] H. A. Calderon, G. Kostorz, Y. Y. Qu, H. J. Dorantes, J. J. Cruz, and J. G. Cabanas-Moreno, *Mater. Sci. Eng. A* **238**, 13 (1997).

[16] A. Ges, O. Rornaro, and H. Palacio, *J. Mater. Sci.* **32**, 3687 (1997).

[17] R. W. Carpenter, *Acta Metall.* **15**, 1567 (1967).

[18] J. K. Tien and S. M. Copley, *Metall. Trans.* **2**, 215, 543 (1971).

[19] J. Miyazaki, K. Nakamura, and H. Mori, *J. Mat. Sci.* **14**, 1827 (1979).

[20] M. V. Nathal and L. J. Ebert, *Scripta Metall.* **17**, 1151 (1983); R. A. Mackay and L. J. Ebert, *Scripta Metall.* **17**, 1217 (1983).

[21] O. Paris, M. Fährmann, E. Fährmann, T. M. Pollock, and P. Fratzl, *Acta Metall.* **45**, 1085 (1997).

[22] J. D. Eshelby, *Proc. Roy. Soc. (London) A* **241**, 376 (1957).

[23] J. W. Cahn, *Acta Metall.* **9**, 795 (1961).

[24] J. W. Cahn, *Acta Metall.* **10**, 179 (1962).

[25] A. G. Khachaturyan and G. A. Shatalov, *Soviet Physics – Solid State* **11**, 118 (1969).

[26] J. E. Hilliard, in *Phase Transformations* (American Society for Metals, Cleveland, 1970), p. 497.

[27] H. Yamauchi and V. de Fontaine, *Acta Metall.* **27**, 763 (1979).

[28] Y. Wang, L. Q. Chen, and A. G. Khachaturyan, *Scripta Metall. Mater.* **25**, 1387 (1991); *Acta Metall.* **41**, 279 (1993).

[29] W. C. Johnson and J. W. Cahn, *Acta Metall.* **32**, 1925 (1984).

[30] A. Onuki, *J. Phys. Soc. Jpn* **58**, 3065, 3069 (1989).

[31] H. Nishimori and A. Onuki, *Phys. Rev. B* **42**, 980 (1990).

[32] A. Onuki and H. Nishimori, *Phys. Rev. B* **43**, 13 649 (1991).

[33] A. Onuki and H. Nishimori, *J. Phys. Soc. Jpn* **60**, 1 (1991).

[34] A. Onuki, *J. Phys. Soc. Jpn* **60**, 345 (1991).

[35] H. Nishimori and A. Onuki, *J. Phys. Soc. Jpn.*, **60**, 1208 (1991).

[36] H. Nishimori and A. Onuki, *Phys. Lett. A* **162**, 323 (1992).

[37] A. Onuki, in *Mathematics of Microstructure Evolution*, eds. L. Q. Chen *et al.* (TMS SIAM, 1996), p. 87.

[38] A. Onuki and A. Furukawa, *Phys. Rev. Lett.* **86**, 452 (2001).

[39] C. Sagui, A. M. Somoza, and R. C. Desai, *Phys. Rev. E* **50**, 4865 (1994).

[40] C. Sagui, D. Orlikowski, A. M. Somoza, and C. Roland, *Phys. Rev. E* **58**, R4092 (1998); D. Orlikowski, C. Sagui, A. M. Somoza, and C. Roland, *Phys. Rev. B* **59**, 8646 (1999); *ibid.* **62**, 3160 (2000).

[41] T. Koyama, T. Miyazaki, and A. Mebed, *Metall. Mater. Trans.* A **26**, 2617 (1995); T. Miyazaki and T. Koyama, in *Mathematics of Microstructure Evolution*, eds. L. Q. Chen *et al.* (TMS SIAM, 1996), p. 111.

[42] J. Gayda and D. J. Srolovitz, *Acta Metall.* **37**, 641 (1989).

[43] C. A. Laberge, P. Fratzl, and J. L. Lebowitz, *Phys. Rev. Lett.* **75**, 4448 (1995); *Acta Metall.* **45**, 3949 (1998).

[44] J. K. Lee, *Scripta Metall. Mater.* **32**, 559 (1995).

[45] *Structural Phase Transitions I*, eds. K. A. Müller and H. Thomas (Springer, 1981); R. A. Cowley, *Adv. Phys.* **29**, 1 (1980); A. D. Bruce, *ibid.* **29**, 111 (1980).

[46] J. Bardeen, L. N. Cooper, and J. R. Schriffer, *Phys. Rev.* **108**, 1175 (1957).

[47] L. D. Landau and E. M. Lifshitz, *Theory of Elasticity* (Pergamon, New York, 1973).

[48] L. Vegard, *Z. Phys.* **5**, 17 (1921); *Z. Kristallogr.* **67**, 239 (1928).

[49] K. B. Rundman and J. E. Hilliard, *Acta. Metall.* **15**, 1025 (1967).

[50] B. Golding and S. C. Moss, *Acta Metall.* **15**, 1239 (1967).

[51] A. Aharony, in *Phase Transitions in Critical Phenomena* (Academic, New York, 1976), Vol. 6, p. 358.

[52] T. Garel and S. Doniach, *Phys. Rev. B* **26**, 325 (1982).

[53] A. L. Larkin and D. E. Khmel'nitskii, *Sov. Phys. JETP* **29**, 1123 (1969).

[54] A. Pineaw, *Acta Metall.* **24** 559 (1976).

[55] W. C. Johnson, M. B. Berkenpas, and D. E. Laughlin, *Acta Metall.* **36**, 3149 (1988).

[56] P. H. Leo and R. F. Sekerka, *Acta Metall.* **37**, 3139 (1989).

[57] K. Yadim and K. Binder, *Acta Metall.* **39**, 707 (1991).

[58] P. Frazl and O. Penrose, *Phys. Rev. B*, **50**, 3447 (1994).

[59] S. Puri, *Phys. Rev. E*, **55**, 1752 (1997).

[60] W. W. Mullins and R. F. Sekerka, *J. Appl. Phys.* **34**, 323 (1963); *ibid.* **35**, 444 (1964).

[61] T. Miyazaki, private communication.

[62] S. M. Allen and J. W. Cahn, *Acta Metall.*, **23**, 1017 (1975); *ibid.* **24**, 425 (1976).

[63] K. Oki, H. Sagane, and T. Eguchi, *J. Physique I* **38**, C7–414 (1977).

[64] C. Leroux, A. Loiseau, D. Broddin, and G. van Tendeloo, *Phil. Mag. B* **64**, 57 (1991).

[65] K-I. Udoh, A. M. El Araby, Y. Tanaka, K. Hisatsune, K. Yasuda, G. van Tendeloo, and J. Van Landuyt, *Mater. Sci. Eng. A* **203**, 154 (1995).

[66] J. Torres, *phys. stat. sol.* **71**, 141 (1975).

[67] Y. Wang and A. G. Khachaturyan, *Acta Metall.* **45**, 759 (1997); M. Artemev, Y. Wang, and A. G. Khachaturyan, *ibid.* **48**, 2503 (2000).

[68] Y. Le Bouar, A. Loiseal, and A. G. Khachaturyan, *Acta Metall.* **46**, 2777 (1998).

[69] Y. Yamazaki, *J. Phys. Soc. Jpn* **66**, 2628 (1997).

[70] H. Thomas and K. A. Müller, *Phys. Rev. Lett.* **21**, 546 (1968); J. C. Slonczewski and H. Thomas, *Phys. Rev. B* **1**, 3599 (1970).

[71] W. Cao and G. R. Barsch, *Phys. Rev. B* **41**, 4334 (1990); W. Cao, G. R. Barsch, and J. A. Krumhansl, *ibid.* **42**, 6396 (1990),

[72] F. Jona and G. Shirane, *Ferroelectric Crystals* (Dover, New York, 1962).

[73] W. Kinase and H. Takahashi, *J. Phys. Soc. Jpn* **12**, 464 (1957).

[74] S. Nambu and S. A. Sagala, *Phys. Rev. B* **50**, 5838 (1994).

[75] K. Otsuka and K. Shimizu, *Int. Met. Rev.* **31**, 93 (1986).

[76] Y. Yamada and Y. Uesu, *Solid State Commun.* **81**, 777 (1992).

[77] Y. Yamada, *Phys. Rev. B* **46**, 5906 (1992).

[78] T. Ohta, *J. Phys. Soc. Jpn* **68**, 2310 (1999).

[79] R. Sinclair and J. Dutkiewicz, *Acta Metall.* **25**, 235 (1977).

[80] Y. Kitano, K. Kifune, and Y. Kimura, *J. Physique I* **49**, C5–201 (1988).

[81] K. Muraleedharan, D. Banaerjee, S. Banerjee, and S. Lele, *Phil. Mag. A* **71**, 1011 (1995).

[82] C. Manolikas and S. Amelinckx, *phys. stat. sol.* **60**, 607 (1980); *ibid.* **61**, 179 (1980).

[83] Y. H. Wen, Y. Wang, and L. Q. Chen, *Phil. Mag. A* **80**, 1967 (2000); Y. H. Wen, Y. Wang, L. A. Bendersky, and L. Q. Chen, *Acta Metall.* **48**, 4125 (2000).

[84] A. Furukawa and A. Onuki, to be published.

[85] H. A. Jahn and E. Teller, *Proc. Roy. Soc. (London) A* **161**, 220 (1937).

[86] U. Opik and M. H. L. Pryce, *Proc. Roy. Soc. (London) A* **238**, 425 (1957).

[87] G. A. Gehring and K. A. Gehring, *Rep. Prog. Phys.* **38**, 1 (1975).

[88] J. Kanamori, *J. Appl. Phys.* **31**, 14S (1961); M. Kataoka and J. Kanamori, *J. Phys. Soc. Jpn* **32**, 113 (1972).

[89] E. Pytte, *Phys. Rev. B* **3**, 3503 (1971).

[90] F. Schwabl and U. C. Täuber, *Phil. Trans. Roy. Soc. (London) A* **354**, 2847 (1996).

[91] G. Shirane and J. D. Axe, *Phys. Rev. Lett.* **27**, 1803 (1971).

[92] W. Rehwald, M. Rayl, R. W. Cohen, and G. D. Cody, *Phys. Rev. B* **6**, 363 (1972).

[93] S. Haussühl, *Solid State Commun.* **13**, 147 (1973); K. Knorr, A. Loidl, and J. K. Kjems, *Phys. Rev. Lett.* **55**, 2445 (1985).

[94] P. W. Anderson and E. I. Blount, *Phys. Rev. Lett.* **14**, 217 (1965).

[95] M. W. Finnis and V. Heine, *J. Phys. F: Metal Phys.* **4**, 960 (1974).

[96] G. R. Barsch and J. Krumhansl, *Phys. Rev. Lett.* **53**, 1069 (1984).

[97] A. E. Jacobs, *Phys. Rev. B* **46**, 8080 (1992); *ibid.* **61**, 6587 (2000).

[98] S. H. Curunoe and A. E. Jacobs, *Phys. Rev. B* **62**, R11 925 (2000).

[99] E. A. H. Love, *A Treatise on the Mathematical Theory of Elasticity* (Dover, New York, 1944), p. 49.

[100] G. S. Bale and R. J. Gooding, *Phys. Rev. Lett.* **67**, 3412 (1991).

[101] A. C. E. Reid and R. J. Gooding, *Physica A* **239**, 10 (1997).

[102] A. Onuki, *J. Phys. Soc. Jpn* **68**, 5 (1999).

[103] L. R. Tanner, *Phil. Mag.* **14**, 111 (1966).

[104] I. M. Robertson and C. M. Wayman, *Phil. Mag. A* **48**, 421 (1983); *ibid.* **48**, 443 (1983); *ibid.* **48**, 629 (1983).

[105] R. Oshima, M. Sugiyama, and F. E. Fujita, *Metall. Trans. A* **19**, 803 (1988).

[106] S. Muto, R. Oshima, and F. Fujita, *Acta Metall. Mater.* **4**, 685 (1990).

[107] H. Seto, Y. Noda, and Y. Yamada, *J. Phys. Soc. Jpn* **59**, 965 (1990); *ibid.* **59**, 978 (1990).

[108] S. Kartha, J. Krumhansl, J. Sethna, and L. K. Wickham, *Phys. Rev. B* **52**, 803 (1995).

[109] K. Fuchizaki and Y. Yamada, *Phys. Rev. B* **40**, 4740 (1989).

[110] M. Yoshida, S. Mori, N. Yamamoto, Y. Uesu, and J. M. Kiat, *J. Korean Phys. Soc.* **32**, S993 (1998).

[111] A. Machida, Y. Morimoto, E. Nishibori, M. Takata, M. Sakata, K. Ohoyama, S. Mori, N. Yamamoto, and A. Nakamura, *Phys. Rev. B* **62**, 3883 (2000).

[112] S. H. Curunoe and A. E. Jacobs, *Phys. Rev. B* **63**, 094110 (2000).

[113] A. Onuki, *J. Phys. Soc. Jpn.* **70**, 3479 (2001).

[114] M. Kataoka and Y. Endoh, *J. Phys. Soc. Jpn.* **48**, 912 (1980).

[115] R. A. Cowley, *Phys. Rev. B* **13**, 4877 (1976).

[116] J. M. Lawrence, M. C. Croft, and R. D. Parkes, *Phys. Rev. Lett.* **35**, 289 (1975).

[117] A. Onuki, *Phys. Rev. B* **39**, 12 308 (1989).

[118] (a) J. Völkl and G. Alefeld, in *Hydrogen in Metals I*, eds. G. Alefeld and J. Völkl (Springer-Verlag, Heidelberg, 1978), p. 32; (b) H. Wagner, *ibid.*, p. 5; (c) H. Peisl, *ibid.*, p. 53.

[119] (a) E. Wicke and H. Brodowsky, *Hydrogen in Metals II*, eds. G. Alefeld and J. Völkl (Springer-Verlag, Heidelberg, 1978), p. 73; (b) T. Schober and H. Wenzl, *ibid.*, p. 11.

[120] Y. Fukai, *The Metal–Hydrogen System* (Springer-Verlag, Heidelberg, 1993).

[121] G. Shaumann, J. Völkl, and G. Alefeld, *Phys. Rev. Lett.* **21**, 891 (1968); G. Alefeld, *phys. stat. sol.* **32**, 67 (1969).

[122] J. Tretkowski, J. Völkl, and G. Alefeld, *Z. Physik B* **28**, 259 (1977).

[123] H. Zabel and H. Peisl, *Phys. Rev. Lett.* **42**, 511 (1979).

[124] H. Wagner and H. Horner, *Adv. Phys.* **23**, 587 (1974).

[125] R. Bausch, H. Horner, and H. Wagner, *J. Phys. C* **8**, 2559 (1975).

[126] H.-K. Janssen, *Z. Physik B* **23**, 245 (1976).

[127] E. Ho, H. A. Goldberg, G. C. Weatherly, and F. D. Manchester, *Acta Metall.* **27**, 841 (1979).

[128] W. S. Gorsky, *Phys. Z. Sowjun.* **8**, 457 (1935).

[129] A. Pimpinelli and J. Villan, *Physics of Crystal Growth* (Cambridge University Press, 1988).

[130] E. Guyer and P. W. Voorhees, *Phys. Rev. Lett.* **74**, 4031 (1995).

[131] F. Léonard and R. C. Desai, *Phys. Rev. B* **56**, 4955 (1997); *ibid.* **57**, 4805 (1998).

[132] R. Asaro and W. Tiller, *Metall. Trans.* **3**, 1789 (1972).

[133] M. Grinfeld, *Sov. Phys. Dokl.* **31**, 831 (1986); *Europhys. Lett.* **22**, 723 (1993).

[134] R. H. Torii abd S. Balibar, *J. Low Temp. Phys.* **89**, 391 (1992).

[135] J. Berréhar, C. Caroli, C. Lapersonne-Meyer, and M. Scott, *Phys. Rev. B* **46**, 13 487 (1992).

[136] F. C. Larché and J. W. Cahn, *Acta Metall.* **21**, 1051 (1973); *ibid.* **26**, 1579 (1978).

[137] M. Uwaha and G. Baym, *Phys. Rev. B* **26**, 4928 (1982); P. Nozières and D. E. Wolf, *Z. Phys. B* **70**, 399 (1988).

[138] P. Nozières, *J. Physique II* **3**, 681 (1993).

[139] W. H. Yang and D. J. Srolovitz, *Phys. Rev. Lett.* **71**, 1593 (1993).

[140] J. Müller and M. Grant, *Phys. Rev. Lett.* **82**, 1736 (1999).

[141] K. Kassener and C. Misbah, *Europhys. Lett.* **46**, 217 (1999).

[142] L. D. Landau and E. M. Lifshitz, *Electrodynamics of Continuous Media* (Pergamon, New York, 1984), Vol. 8.

11

Phase transitions of fluids in shear flow

In recent years, much attention has been focused on the nonlinear effects of shear flow in which a certain internal structure of fluids is strongly affected by flow field [1]–[3]. As shown in Fig. 11.1, the simplest flow profile is $\dot{\gamma} y \boldsymbol{e}_x$ (simple shear flow), where the flow direction is taken to be along the x axis, $\boldsymbol{e}_x$ being the unit vector along the x axis, and the mean velocity varies in the y or shear gradient direction, while the z direction is called the vorticity direction. Effects of elongational flow have also been studied for polymeric systems [4]. Such nonlinear, nonequilibrium effects have been known for some time in polymer science with no satisfactory explanations [4]–[10], and are now becoming notable topics in the study of (i) fluids near the critical point (near-critical fluids) and (ii) various complex fluids such as polymers, liquid crystals, colloidal systems, and amphiphilic systems. This trend has developed out of the foundation of a deeper understanding of dynamic critical phenomena, kinetics of first-order phase transitions, and polymer physics. Experimentally, the investigation has been accelerated through the recent application of scattering techniques to nonequilibrium phenomena under shear. As will be shown in Appendix 11A, the equal-time-correlation functions of scalar variables satisfy the translational invariance in a flow field with a homogeneous average velocity gradient, giving rise to the proportionality of the light and small-angle neutron scattering intensity and the structure factor. Other optical effects such as birefringence and dichroism have also provided sensitive techniques with which to detect spatial anisotropy of concentration fluctuations and molecular alignment. The information gained by these means can then be combined with rheological data of the shear stress and normal stress differences, which in many cases exhibit unusual behavior in nonlinear response regimes of shear. Though the study of complex fluids under shear has often been conducted with the goal of producing engineering oriented results, it is now developing into a new interdisciplinary field embracing engineering and physics. Here rheology and phase transitions are closely and uniquely related.

We will treat near-critical fluids under shear in Section 11.1 and shear-induced phase separation in polymer solutions in Section 11.2, where we will also discuss analogous effects in other fluids, as much as possible. In Section 11.3 we briefly mention, for various complex fluids, a number of shear flow problems which are still being studied and are not yet well understood. Finally, the subject of Section 11.4 will be supercooled liquid dynamics with (and without) shear on the basis of recent simulations.

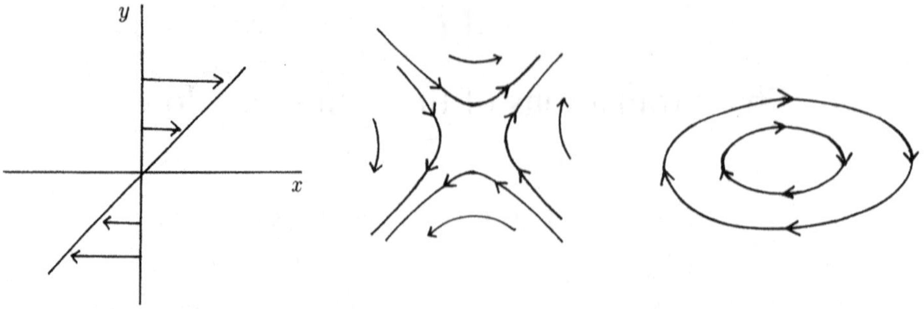

(a) Shear flow (b) Elongational flow (c) Rotational flow

Fig. 11.1. If the average flow is expressed by $\langle v_x \rangle = (S + A)y$, $\langle v_y \rangle = (S - A)x$ and $\langle v_z \rangle = 0$, we have a simple shear flow for $S = A = \dot{\gamma}/2$ in (a), an elongational flow for $S > A > 0$ characterized by hyperbolic stream lines in (b), and a rotational flow for $A > S > 0$ characterized by elliptic stream lines in (c). Suppression of the concentrated fluctuations is strong in (b) and weak in (c), while it is intermediate in (a) [12].

11.1 Near-critical fluids in shear

We consider nearly incompressible fluid binary mixtures near the consolute critical point under shear flow [11]–[17]. (See Ref. [12] for the other types of flow shown in Fig. 11.1.) The concentration fluctuations are greatly deformed as they are convected by a spatially varying velocity field. The deformation time is given by the inverse shear $1/\dot{\gamma}$, so the deformation is strong or nonlinear when the so-called Deborach number De, defined by

$$De = \dot{\gamma} t_\xi \tag{11.1.1}$$

exceeds 1 (the strong shear case), where t_ξ is the characteristic lifetime of the critical fluctuations given in (6.1.25). The dynamic equation (6.1.10) for ψ is rewritten as

$$\frac{\partial}{\partial t}\psi = -\dot{\gamma} y \frac{\partial \psi}{\partial x} - \nabla \cdot (\psi \boldsymbol{v}) + L_0 \nabla^2 \frac{\delta}{\delta \psi} \beta \mathcal{H} + \theta, \tag{11.1.2}$$

where $\mathcal{H}$ is the GLW hamiltonian given by (4.1.1). The velocity field is divided into the average $\dot{\gamma} y \boldsymbol{e}_x$ and the deviation $\boldsymbol{v}$, where $\boldsymbol{v}$ obeys (6.1.11) with $\nabla \cdot \boldsymbol{v} = 0$. We will investigate the effects of the first term on the right-hand side of (11.1.2) in various situations.

11.1.1 Strong shear regime in one-phase region

Deformations by shear are weak for $\dot{\gamma} t_\xi < 1$ and strong for $\dot{\gamma} \tau_\xi > 1$. It is convenient to introduce a characteristic wave number k_c by $\Gamma_{k_c} = \dot{\gamma}$. The decay rate Γ_k in (6.1.21) in the mode coupling theory yields

$$k_c = (6\pi \eta / T)^{1/3} \dot{\gamma}^{1/3}, \tag{11.1.3}$$

in strong shear. The viscosity will be written as η. Then, by setting $k_c\xi = k_c\xi_0\tau_s^{-\nu} = 1$, we may introduce a crossover reduced temperature τ_s in shear flow by

$$\tau_s = (6\pi\eta\xi_0^3/T)^{1/3\nu}\dot{\gamma}^{1/3\nu} \propto \dot{\gamma}^{0.54}. \tag{11.1.4}$$

Slightly different and essentially the same definitions of k_c and τ_s follow if use is made of the expressions in the dynamic renormalization group theory. The critical fluctuations are strongly deformed by shear in the long-wavelength region $q < k_c$. For example, $k_c = 2.3 \times 10^{-4}\dot{\gamma}^{1/3}$ cm^{-1} and $\tau_s = 10^{-5}\dot{\gamma}^{0.54}$ with $\dot{\gamma}$ in s^{-1} in isobutyric acid + water.

Mean field structure factor

Let us then calculate the steady-state structure factor for positive temperature coefficient r_0 at the critical composition. We start with the mean field approximation or linearizing of the dynamic equation (11.1.2). Its Fourier transformation yields

$$\frac{\partial}{\partial t}\psi_{\boldsymbol{q}} = \dot{\gamma}q_x\frac{\partial}{\partial q_y}\psi_{\boldsymbol{q}} - L_0q^2(r_0+q^2)\psi_{\boldsymbol{q}} + \theta_{\boldsymbol{q}}. \tag{11.1.5}$$

The fluctuations are simultaneously convected by shear and thermally dissipated with the decay rate (in the mean field theory) given by

$$\Gamma(q) = L_0q^2(r_0+q^2). \tag{11.1.6}$$

The steady-state structure factor $I(\boldsymbol{q})$ satisfies

$$\left(2\Gamma(q) - \dot{\gamma}q_x\frac{\partial}{\partial q_y}\right)I(\boldsymbol{q}) = 2L_0q^2. \tag{11.1.7}$$

The right-hand side arises from the thermal noise term $\theta_{\boldsymbol{q}}(t)$, giving rise to the Ornstein–Zernike form $I_{\mathrm{eq}}(q) = 1/(r_0+q^2)$ without shear. The simplest way to examine the shear effect is to expand $I(\boldsymbol{q})$ in powers of $\dot{\gamma}$ as

$$I(\boldsymbol{q}) = I_{\mathrm{eq}}(q)\left[1 - 2q_xq_y\dot{\gamma}/I_{\mathrm{eq}}(q)\Gamma(q) + \cdots\right]. \tag{11.1.8}$$

In this linear regime, the intensity increases most in the directions in which $q_x = -q_y$ and $q_z = 0$. Clearly, this expansion is valid in the (mean field) weak shear condition $\dot{\gamma} < L_0r_0^2$.

Generally, taking into account the convection due to shear, we may solve (11.1.7) in the following integral form,[1]

$$\begin{aligned} I(\boldsymbol{q}) &= \int_0^\infty dt\,\exp\left[-2t\Gamma(|\boldsymbol{q}|) + t\dot{\gamma}q_x\frac{\partial}{\partial q_y}\right]2L_0q^2 \\ &= \int_0^\infty dt\,\exp\left[-2\int_0^t dt_1\Gamma(|\boldsymbol{q}(t_1)|)\right]2L_0\boldsymbol{q}(t)^2, \end{aligned} \tag{11.1.9}$$

[1] This follows from the mathematical identity, $\exp[\lambda U(x) + \lambda\partial/\partial x] = \exp[\int_0^\lambda d\lambda' U(x+\lambda')]\exp[\lambda\partial/\partial x]$, where $U(x)$ is an arbitrary function of x.

in terms of a deformed wave vector defined by[2]

$$\boldsymbol{q}(t) = \boldsymbol{q} + \dot{\gamma} t q_x \boldsymbol{e}_y. \tag{11.1.10}$$

We find $I(\boldsymbol{q}) \to 1/r_0$ as $q \to 0$ and $I(\boldsymbol{q}) \cong 1/(r_0 + q^2)$ for $q \gg k_c$ even in shear flow. The effect of shear is significant in the region $r_0 \cong 0$ and $q \ll k_c$, where we have $\Gamma(|\boldsymbol{q}(t_1)|) \cong L_0(\dot{\gamma} t_1 q_x)^4$ and

$$I(\boldsymbol{q}) = \int_0^\infty dt \exp\left[-\frac{2}{5} L_0(\dot{\gamma} q_x)^4 t^5\right] 2L_0(\dot{\gamma} t q_x)^2 \sim k_c^{-8/5} |q_x|^{-2/5}, \tag{11.1.11}$$

where k_c is determined by $L_0 k_c^4 = \dot{\gamma}$ in the mean field theory. These limiting cases can be interpolated by the following approximate expression,

$$I(\boldsymbol{q}) \cong 1/(r_0 + c k_c^{8/5} |q_x|^{2/5} + q^2), \tag{11.1.12}$$

where $c \cong 0.76$. Unless q_x is very small, (11.1.11) holds for $\boldsymbol{q}$ much smaller than k_c in strong shear (which is $r_0 < k_c^2$ in the mean field theory).

Renormalization effects

We thus find that $I(\boldsymbol{q})$ is suppressed below the equilibrium level. As a result, the critical dimensionality d_c is lowered from 4 to 2.4 in strong shear. To see this, let us consider the renormalized kinetic coefficient L_R in strong shear for general spatial dimensionality d. From (6.1.18), L_R in the long-wavelength limit is written as

$$\begin{aligned} L_R &\cong \frac{(d-1)T}{d\eta} \int_{\boldsymbol{q}} \frac{1}{q^2} I(\boldsymbol{q}) \\ &\cong C_1 \frac{T}{\eta k_c^{4-d}} + C_2 \frac{T}{\eta} \int_{\boldsymbol{q}}^{<} \frac{1}{q^2 (r_0 + |q_x|^{2/5})}, \end{aligned} \tag{11.1.13}$$

where the first term in the second line is the contribution from the wave number region $q > k_c$ and the second term that from the region $q < k_c$, with C_1 and C_2 being dimensionless constants. Notice that the second term grows as $r_0 \to 0$ for $d < 2.4$ and converges to a value of order $T/\eta k_c^{4-d}$ even at $r_0 = 0$ for $d > 2.4$. This means $d_c = 2.4$. In 3D, therefore, the lower cut-off wave number of the singular fluctuation contributions becomes k_c in strong shear, whereas it is $\xi^{-1} = \xi_{+0}^{-1}(T/T_c - 1)^\nu$ in the linear response regime. Thus the renormalized kinetic coefficient behaves in 3D as

$$L_R = T/6\pi\eta k_c \propto \dot{\gamma}^{-1/3}. \tag{11.1.14}$$

The shear viscosity η is treated as a constant here, because its singularity is weak. We shall see its weak non-newtonian behavior later. The structure factor after the renormalization is roughly of the form,

$$I(\boldsymbol{q}) \cong 1/(r_R + c k_c^{8/5} |q_x|^{2/5} + q^2), \tag{11.1.15}$$

[2] This wave number is equal to $\tilde{q}(-t)$ introduced below (11A.7).

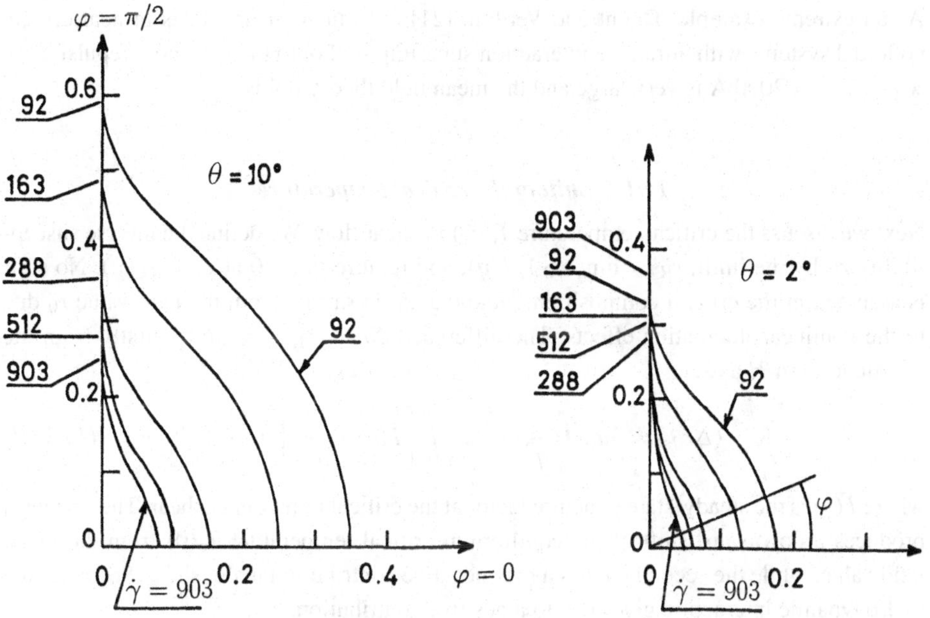

Fig. 11.2. Reduced scattering intensity $I(\boldsymbol{q})/I_{\text{eq}}(q)$ for aniline + cyclohexane as a function of $\varphi = \tan^{-1}(q_x/q_y)$ in the polar coordinate at $T - T_{\text{c}} = 1.5$ mK [15]. The horizontal axis ($\varphi = 0$) is parallel to the flow ($\parallel x$), while the vertical axis ($\varphi = \pi/2$) is in the velocity gradient direction. Results for two scattering angles $\theta = 2°$ ($q = 5200\ \text{cm}^{-1}$) and $\theta = 10°$ ($q = 26\,000\ \text{cm}^{-1}$) are shown. Shear rates $\dot{\gamma}$ are in units of s^{-1}.

with

$$r_{\text{R}} = \xi_{+0}^{-2}\tau_{\text{s}}^{2\nu-1}[T - T_{\text{c}}(\dot{\gamma})]/T_{\text{c}}, \tag{11.1.16}$$

where k_{c} and τ_{s} are defined by (11.1.3) and (11.1.4), respectively, and $T_{\text{c}}(\dot{\gamma})$ is the critical temperature in shear to be discussed below. Obviously, if τ_{s} is replaced by $(T - T_{\text{c}})/T_{\text{c}}$ and the $\dot{\gamma}$ dependence of $T_{\text{c}}(\dot{\gamma})$ is neglected, the equilibrium result $r_{\text{R}} \cong \xi^{-2}$ is reproduced. At small r_{R} ($\ll k_{\text{c}}^2$), $I(\boldsymbol{q}) \propto |q_x|^{-2/5}$ in most directions of $\boldsymbol{q}$ for $q < k_{\text{c}}$. This means substantial suppression of the fluctuations below the equilibrium critical intensity $I_{\text{eq}}(q) = 1/(\xi^{-2} + q^2)$.

In Fig. 11.2 we show $I(\boldsymbol{q})/I_{\text{eq}}(q)$ as measured by Beysens' group in a critical mixture of aniline + cyclohexane which agrees with (11.1.15) [15]. A small-angle neutron scattering experiment was also performed in a low-molecular-weight polymer blend under shear [18]. Form birefringence and dichroism have also been used to detect anisotropy of concentration fluctuations under shear in agreement with theory [19, 20].

In usual (low-molecular-weight) near-critical fluids, the static and dynamic renormalization effects are crucial, leading to multiplicative fractional powers of τ_{s} as in (11.1.14) and (11.1.16). There are also systems in which the renormalization effects are negligible.

As an extreme example, Dhont and Verduin [21] examined shear effects in near-critical colloidal systems with attractive interaction superimposed onto the hard-core repulsion, in which $\xi_{+0} = 2000$ Å is very large and the mean field theory holds.

11.1.2 Shift of the critical temperature

Next we discuss the critical temperature $T_c(\dot\gamma)$ in shear flow. We define the inverse susceptibility r_R by the limit, $r_R = \lim_{q\to 0} 1/I(\boldsymbol{q})$, and require $r_R = 0$ at $T = T_c(\dot\gamma)$. No shift is assumed in the critical composition. Note that r_R is shifted from the bare value r_0 due to the nonlinear fluctuation effects. The difference $\Delta r = r_R - r_0$ arises firstly from the quartic term in $\mathcal{H}$ as

$$(\Delta r)_{\rm st} = 3u_0 \int_q \frac{1}{q^2} + 3u_0 \int_q \left[I(\boldsymbol{q}) - \frac{1}{q^2} \right] + \cdots, \tag{11.1.17}$$

where $I(\boldsymbol{q})$ is the steady-state structure factor at the critical point under shear. The first term produces a downward shift of the equilibrium critical temperature $T_c(0)$ from the mean field value, while the second term is a new negative contribution under shear. Secondly the hydrodynamic interaction gives rise to a positive contribution,

$$(\Delta r)_{\rm hyd} = \left(1 - \frac{1}{d}\right) \frac{T}{\eta_0 L_0} \int_q \frac{1}{q^2} [1 - q^2 I(\boldsymbol{q})] + \cdots, \tag{11.1.18}$$

which vanishes in equilibrium as ought to be the case. We obtain the above result if we start with the Kawasaki equation (6.1.52) and construct the equation for $I(\boldsymbol{q})$. Note that the ratio of the second term in (11.1.17) and the first term in (11.1.18) is written as $-3[d/(d-1)]g/f$ in terms of g in (4.1.22) and f in (6.1.35), so it tends to a universal number ($= -19/54 + O(\epsilon)$) in the asymptotic critical region.

The ϵ expansion of the shift in near-critical fluids

We may calculate the shift using the ϵ expansion in low-molecular-weight near-critical fluids. The fluctuation effects are strong in such fluids and hence, if the upper cut-off wave number Λ becomes much smaller than the microscopic wave number ξ_{+0}^{-1}, the dimensionless coefficients g and f approach the universal fixed-point values g^* in (4.3.16) and f^* in (6.1.37), respectively. Because the dominant contributions in the last two integrals of (11.1.17) and (11.1.18) arise from $q \sim k_c$, we may set $\Lambda = k_c$ to obtain

$$(\Delta r)_{\rm st} = (\Delta r)_{\rm eq} - 0.044\epsilon k_c^2, \tag{11.1.19}$$

$$(\Delta r)_{\rm hyd} = 0.127\epsilon k_c^2, \tag{11.1.20}$$

where the first term on the right-hand side of (11.1.19) represents the shift in equilibrium. Summing the two contributions proportional to k_c^2, we find a downward shift,

$$T_c(\dot\gamma) - T_c(0) = (0.044 - 0.127)\epsilon\tau_s T_c = -0.083\epsilon\tau_s T_c, \tag{11.1.21}$$

where τ_s is defined by (11.1.4). In 3D we thus expect $T_c(\dot{\gamma}) - T_c(0) \sim -0.1\tau_s T_c$. It is important that the hydrodynamic interaction does not affect the equilibrium properties, but gives rise to the downward shift (11.1.20) in strong shear. Beysens *et al.* [16] detected a downward shift from the turbidity and the structure factor with $\boldsymbol{q}$ perpendicular to flow. It was proportional to $\dot{\gamma}^{0.53}$ but four times smaller than the result (11.1.21) at $\epsilon = 1$ in a few critical binary fluid mixtures, so that this aspect remains unsettled. It is difficult to determine a small shift definitely in usual binary fluid mixtures, because scattering is suppressed even at $T = T_c(\dot{\gamma})$ as in (11.1.15) and domains do not grow indefinitely below $T_c(\dot{\gamma})$ as will be explained in Subsection 11.1.4.

Polymer A + polymer B in common solvent

Hashimoto *et al.* observed a large downward shift and notable shear-induced mixing in ternary mixtures of polystyrene (PS) and polybutadiene (PB) in a common solvent of dioctylphthalate (DOP) [22]–[29]. In their system the polymer volume fraction $\phi = \phi_{PS} + \phi_{PB}$ is of the order of the overlapping value and the fluid may be treated as a binary fluid mixture of weakly interacting PS-rich blobs and PB-rich blobs with $\xi_{+0} \sim 50$ Å [29]. Thus the space and timescales are much more enlarged than in usual binary fluid mixtures; for example, $t_\xi \sim 1$ s even for $|T - T_c| \sim 1$ deg. K. In the temperature region investigated, the static properties are described by the mean field theory, but the hydrodynamic interaction is operative [23]. As a result, the crossover reduced temperature τ_s from weak to strong shear is three or four orders of magnitude larger than in usual binary fluid mixtures. They obtained a downward shift given by $T_c(\dot{\gamma}) - T_c(0) \sim -A_c\tau_s T_c$ with $\tau_s \propto \dot{\gamma}^{0.5}$ and $A_c \cong 0.06$ using the following two methods [23]. First, they could express the scattered intensity above $T_c(\dot{\gamma})$ perpendicular to flow ($q_x = 0$) as

$$1/I(q) \cong \xi_{+0}^{-2}[T - T_c(0)]/T_c + A_c k_c^2 + q^2, \tag{11.1.22}$$

where $k_c = \xi_{+0}^{-1}\tau_s^{1/2}$. Second, if shear was increased from a two-phase state at fixed T below $T_c(0)$, scattering gradually decreased and shear-induced homogenization eventually took place at the critical condition $T_c(\dot{\gamma}) = T$. Subsequently, Yu *et al.* [30] used fluorescence and phase-contrast microscopy on a similar ternary mixture of PS + PB in DOP and reported that the shift tends to saturate at very high shear.

11.1.3 Transition temperature shift in diblock copolymers

By slightly changing the calculations presented so far, we may readily examine the shear effect on A–B diblock copolymers, where each chain is composed of A and B blocks [31]–[36]. In such systems the equilibrium structure factor in the disordered phase has a maximum at an intermediate wave number k_0 and is expressed in the region $q \sim k_0$ as

$$I_{eq}(q) \cong 1/[r + (q - k_0)^2], \tag{11.1.23}$$

where $r \propto T - T_c$ with T_c being a nominal transition temperature from a disordered to an ordered phase. The volume fraction deviation $\psi = \phi_A - \langle\phi_A\rangle$ of type-A blocks is assumed to obey the dynamic equation (11.1.2) with a Ginzburg–Landau free energy $\mathcal{H}$ expanded up to $O(\psi^4)$. For simplicity, we treat the problem in the disordered phase in the symmetric case where A and B blocks have the same lengths. Then the free-energy density is even with respect to $\psi = \phi_A - 1/2$, and lamellar domain structures emerge at low temperatures.

In mean field calculations of shear effects [32, 34], the steady-state intensity $I(\boldsymbol{q})$ is expressed in the integral form (11.1.9) with $\Gamma(q) = L_0 q^2[r + (q - k_0)^2]$. Then we can see that the linear response regime is given by $\dot{\gamma} < \Gamma_c(r/k_0^2)^{3/2}$ where $\Gamma_c = L_0 k_0^4$ is the noncritical relaxation rate. Only in this regime may $I(\boldsymbol{q})$ be expanded in powers of $\dot{\gamma}$. In the region $\Gamma_c(r/k_0^2)^{3/2} < \dot{\gamma} < \Gamma_c$, nonlinear deformations occur on the fluctuations with $q \cong k_0$ as

$$I(\boldsymbol{q}) \cong 1/[r + (q - k_0)^2 + c_1 k_0^2 |\dot{\mu}\hat{q}_x\hat{q}_y|^{2/3} + c_2 k_0^2 |\dot{\mu}\hat{q}_x|^{4/5}], \tag{11.1.24}$$

where $\dot{\mu} = \dot{\gamma}/\Gamma_c$, $\hat{\boldsymbol{q}} = k_0^{-1}\boldsymbol{q}$, and c_1 and c_2 are positive numbers of order 1. As in (11.1.17) the shift of the temperature coefficient is written as

$$r - r_0 \cong 3u_0 \int_{\boldsymbol{q}} I(\boldsymbol{q}), \tag{11.1.25}$$

where $r_0(\propto T - T_{c0})$ is the bare coefficient and u_0 is the coefficient of the quartic term in $\mathcal{H}$ assumed to be small ($\ll k_0$). The hydrodynamic interaction is not relevant for the fluctuations with $q \sim k_0$ [34].[3] In equilibrium, the fluctuation contribution grows as $r - r_0 \cong (3/2\pi)u_0 k_0^2 r^{-1/2}$ at small r because of the singular integral $\int dq[r + (q - k_0)^2]^{-1} \sim r^{-1/2}$. Brazovskii [35] concluded that a first-order phase transition into a lamellar phase should take place at

$$r_0 = r_c \cong -(3u_0 k_0^2/2\pi)^{2/3} \qquad (\dot{\gamma} = 0). \tag{11.1.26}$$

In shear flow, Cates and Milner [36] predicted that the first-order phase transition curve is shifted upwards as

$$r_c(\dot{\gamma}) - r_c \cong (\dot{\gamma}/\dot{\gamma}^*)^2 |r_c| \qquad (\dot{\gamma} \ll \dot{\gamma}^*), \tag{11.1.27}$$

where $\dot{\gamma}^* = \Gamma_c u_0/k_0$. This is because the fluctuations with $q \cong k_0$ are suppressed below the equilibrium level as in (11.1.24). In addition, the spinodal curve (metastability limit of the disordered phase) is given by

$$r_0 = r_s(\dot{\gamma}) \cong (\dot{\gamma}/\dot{\gamma}^*)^{-1/3} r_c \qquad (\dot{\gamma} \ll \Gamma_c), \tag{11.1.28}$$

which tends to $-\infty$ as $\dot{\gamma} \to 0$. These results were in qualitative agreement with a subsequent small-angle neutron scattering experiment [37].

11.1.4 Spinodal decomposition in shear

More dramatic are the effects of shear in the unstable temperature and composition region. Beysens and Perrot performed a spinodal decomposition experiment in a near-critical

[3] In (11.1.18) the integrand is replaced by $q^{-2}[1 - I(q)/I_0(q)]$ for diblock copolymers. Then the integrand is nonsingular at $q = k_0$ and $(\Delta r)_{\rm hyd}$ becomes negligible.

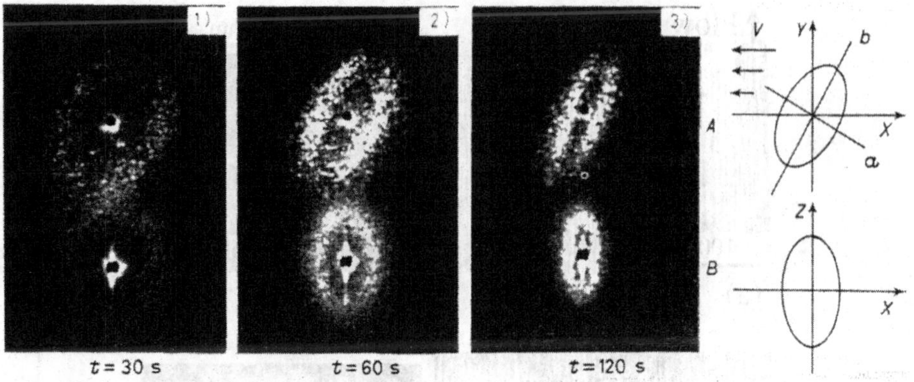

Fig. 11.3. Time evolution of light scattering patterns from a phase-separating near-critical binary fluid mixture at the critical composition [17]. Here $\dot{\gamma} = 0.035 \text{ s}^{-1}$ and $T_c - T \sim 1$ mK, so $\dot{\gamma} t_\xi \sim 0.01$. The upper patterns (*A*) are those in the q_x–q_y plane, while the lower ones (B) are those in the q_x–q_z plane.

binary fluid mixture below T_c by periodically tilting a quartz pipe container [38]. Such a periodic shear was found to prevent domain growth, resulting in a permanent spinodal ring of the scattered light. For steady shear, domains are elongated in the flow direction as $\xi \dot{\gamma} t$ in an initial stage [17, 39], but are eventually broken by shear. In Fig. 11.3 we show light scattering patterns from a phase-separating fluid in shear, which are characterized by strong anisotropy (streak patterns) even in weak shear $\dot{\gamma} t_\xi \ll 1$ below T_c [17, 40].

Computer simulations with various methods, though in 2D, have also shown strong deformations of bicontinuous domain structures just after quenching [41]–[48]. Experimentally, it has also been observed that spinodal decomposition is stopped in steady shear at a particular stage [22, 23], giving rise to dynamical stationary states. Such states can be realized by a balance between the thermodynamic instability and flow-induced deformation. In these two-phase states we may neglect the gravity effect when the domain size R is very small compared with the so-called capillary length a_g in (4.4.54). The Reynolds number Re of a domain is given by $Re = \rho \dot{\gamma} R^2 / \eta$ and is very small near the critical point. However, we may well encounter the opposite limit $Re \gg 1$ far from the critical point, where the inertia effect is crucial [13].

Unfortunately, detailed information cannot be gained from scattering alone, so some theoretical speculations were made on the domain morphology giving rise to streak patterns [49]. Hashimoto *et al.* [50] have taken optical microscope images from a DOP solution to investigate the ultimate bicontinuous morphology in shear, as shown in Fig. 11.4. They have found that domains are elongated into extremely long cylinders in steady states except when they are under extremely weak shear. For $\dot{\gamma} t_\xi < 1$ such string-like domains still contain a number of random irregularities undergoing frequent breakup, interconnection, and branching, while the overall structure is kept stationary. For $\dot{\gamma} t_\xi > 1$ the continuity of

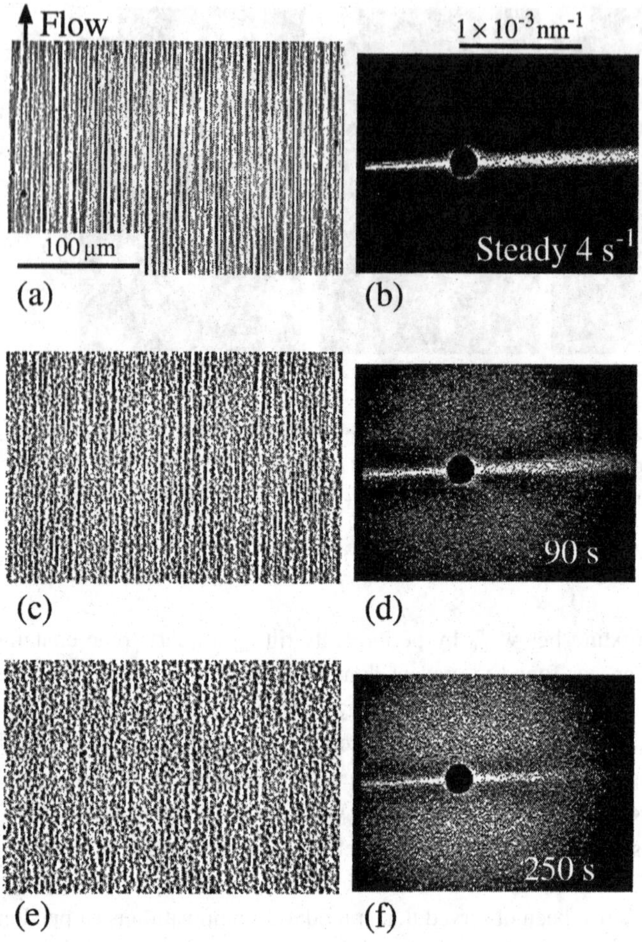

Fig. 11.4. Optical microscopic images (a, c, e) and corresponding light scattering patterns (b,d,f) for a PS/PB(80:20)/DOP 3.3 wt% solution at $T_c - T = 10$ K (taken by Hashimoto's group [50]). Here (a) and (b) were obtained under steady shear at 4 s^{-1}, while (c) to (f) were obtained at 90 s and 250 s after cessation of shear. We can see breakup of cylindrical domains into droplets, which occurs on a timescale of $\eta\xi_\perp/\sigma$, where σ is the surface tension and $\xi_\perp$ is the cylinder diameter.

the strings increases and even extends macroscopically in the flow direction (string phase). The scattering intensity perpendicular to the flow is proportional to the squared lorentzian form $1/[1 + (q\xi_\perp)^2]^2$ due to cylindrical domains, where $\xi_\perp$ represents the diameter of the cylinders and decreases with shear as

$$\xi_\perp \cong [2\pi/q_m(0)](\dot{\gamma}t_\xi)^{-\alpha}, \tag{11.1.29}$$

where $q_{\mathrm{m}}(0)$ is the peak wave number in spinodal decomposition without shear and $\alpha = 1/4$–$1/3$. Thus we have $\xi_\perp \sim 2\pi/k_{\mathrm{c}}$ where k_{c} is determined from $\Gamma_{k_{\mathrm{c}}} = \dot{\gamma}$ in strong shear. For very large shear $\dot{\gamma} \gtrsim 10^2/t_\xi$, the diameter ultimately becomes of the order of the interface thickness and the contrast between the two phases vanishes, resulting in shear-induced homogenization (at $T = T_{\mathrm{c}}(\dot{\gamma})$ if at the critical composition). Afterwards, Hobbie *et al.* [51] studied the dynamics of formation of the string phase in a DOP solution after application of shear. Note that the streak scattering patterns in DOP solutions closely resemble those in usual binary fluid mixtures, so strong elongation of domains should also occur in usual binary fluid mixtures [17, 40]. We should also mention that optical microscope images of string-like domains have been reported for polymer blends [52, 53].

Note that cylindrical domains are unstable in the absence of shear against surface undulations, resulting in the breakup of cylinders into droplets (the Tomotika instability [54]), as discussed in Section 8.5.1. Frischknecht [55] examined the linear stability of cylindrical domains in the presence of shear and showed that shear can suppress growth of surface undulations under the condition $R \gtrsim \sigma/\eta\dot{\gamma}$ ($\sim \xi/\dot{\gamma}t_\xi$ for near-critical fluids). We note that the surface tension is extremely small ($\lesssim 10^{-4}$ cgs) in Hashimoto's case, as in near-critical fluids. Figure 11.4 is a dramatic example of the Tomotika instability observed after cessation of shear. This capillary-driven instability is, in essence, the coarsening mechanism of late-stage spinodal decomposition at the critical composition, as discussed in Section 8.5. Rheologically, there should be no appreciable increase $\Delta\eta$ of the macroscopic viscosity in the string phase because the surfaces do not resist flow.

We may also consider spinodal decomposition under oscillating shear $\dot{\gamma}(t) = \dot{\gamma}_0 \cos(\omega t)$ [56], where we may predict a new bifurcation effect under periodic shear. That is, if the maximum shear strain $\gamma = \dot{\gamma}_0/\omega$ is larger than a critical value γ_{c}, the shear distortion is effective enough and the domain growth can be halted, resulting in a periodic two-phase state. If $\gamma < \gamma_{\mathrm{c}}$, the shear cannot stop the growth, leading to macroscopic phase separation. A similar bifurcation was found in periodic spinodal decomposition in Section 8.8.

11.1.5 Nucleation in shear

Droplet breakup in shear

We slightly lower the temperature T below the coexistence temperature T_{cx} by $\delta T = T_{\mathrm{cx}} - T$ at an off-critical composition. The initial supersaturation Δ is much smaller than 1 and is related to δT and $\Delta T = T_{\mathrm{c}} - T_{\mathrm{cx}}$ by $\Delta \cong (\delta T/\Delta T)/6$ near criticality as in (9.1.4) for $\beta = 1/3$. Appreciable droplets of the new phase can appear only when the critical droplets are not torn by shear. This indicates that the critical radius of nucleation $R_{\mathrm{c}} \sim \xi/\Delta$ must satisfy

$$R_{\mathrm{c}} < R^*, \tag{11.1.30}$$

where

$$R^* \cong C_{\mathrm{b}}\sigma/\eta\dot{\gamma} \tag{11.1.31}$$

is the Taylor breakup size in shear flow [57]–[59]. The coefficient C_b is of order 3 for near-critical fluids. It is known that the droplet shape at the breakup condition deviates from a sphere and may be approximated as a spheroid with the ratio $R_\parallel / R_\perp$ between the longer and shorter radii depending on the viscosity ratio η_1/η_2 between the viscosities inside and outside the droplet. Then there follows a necessary condition of observing noticeable droplets [13, 49, 60],

$$\dot{\gamma} t_\xi < \Delta \ll 1. \tag{11.1.32}$$

This gives an upper limit of shear, $\dot{\gamma}^* \sim \Delta / t_\xi$, at each δT or a lower limit of the quench depth,

$$\delta T^* \sim \dot{\gamma} t_\xi(\Delta T) \propto \dot{\gamma}(\Delta T)^{1-3\nu}, \tag{11.1.33}$$

at each $\dot{\gamma}$ in order to have droplets. This simple criterion has been confirmed in binary mixtures under gentle stirring [61, 62] and uniform shear [63]. Very sensitive dependence of the droplet density with $R > R_c$ on $\dot{\gamma}$ around $\dot{\gamma}^*$ was observed, for example, by dynamic light scattering after cessation of shear [63]. This suggests that the droplets become monodisperse in shear flow.

The key quantity in the initial stage of nucleation is the nucleation rate I in (9.3.42). It is known that I can be of order 1 when δT is equal to the classical Becker–Döring limit $\delta T_{\rm BD}$ ($\cong 0.13\Delta T$ from Fig. 9.13) as discussed in Section 9.3.3. We note $\delta T^* < \delta T_{\rm BD}$ for very weak shear which satisfies (11.1.32). If this inequality holds, droplets will emerge at $\delta T = \delta T_{BD}$ on increasing δT from zero, but droplets will disappear at $\delta T = \delta T^*$ on decreasing δT from a state in which droplets preexist. This hysteretic behavior was observed by Min and Goldburg [63] as shown in Fig. 11.5.

Spinodal in flow field?

We raise a fundamental question as to the existence of metastability itself in relatively large shear for which (11.1.32) is not satisfied. Namely, if δT is increased in such shear, droplet formation will be suppressed, because localized droplets larger than R^* cannot be stable. In particular, if $\dot{\gamma} t_\xi \sim 1$, R_c becomes of order ξ and the suppression is complete in the sense that phase separation can be triggered only by instability of plane-wave fluctuations. This suggests that a spinodal point becomes well defined in such shear as the onset point of phase separation. Recall that the spinodal point for the off-critical case obtained in the mean field theory has no definite theoretical meaning in quiescent fluids.

To investigate this effect, we suggest that an experiment be undertaken to measure the light scattering intensity from off-critical binary fluid mixtures under weak shear below the equilibrium coexistence curve. If droplet formation can be suppressed, we expect growth of the intensity as $\lim_{q\to 0} I(q) \sim (T - T_s)^{-\gamma}$ on approaching a spinodal temperature T_s. We also mention experiments by a Uzbekistan group [64]. They detected a peak in the specific heat C_{VX} far below the coexistence curve in gently stirred off-critical binary mixtures of methanol + heptane. They claimed that a spinodal point can be reached in the presence

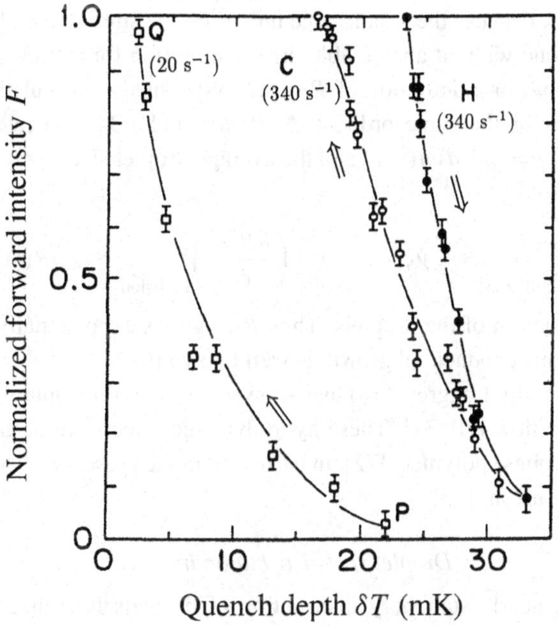

Fig. 11.5. The normalized forward intensity F (the transmittency of light) from an off-critical binary fluid mixture as a function of the quench depth δT [63]. The curves H, C and PQ correspond to $\dot{\gamma} = 340$, 340, and 20 s^{-1}, respectively. These shear rates are much smaller than $1/t_\xi = 1.3 \times 10^4$ s^{-1}. The lines are a viewing guide. On the branch PQ the experiment was started at the point P in an opaque state and was ended at the point Q where droplets disappeared due to the breakup mechanism. The branch H was started at the point $F \cong 1$ where the nucleation rate is appreciable. The branch C was ended at the point $F \cong 1$ due to the breakup mechanism.

of stirring. More experiments, including light scattering, on stirred off-critical fluids in the metastable temperature region would be very informative.

Flow-induced coagulation

Another important mechanism is coagulation of droplets induced by shear [65, 66]. As discussed in Section 8.5, such coalescence becomes important in late-stage droplet growth, where the droplet volume fraction saturates to the initial supersaturation ϕ. It is known that in flow, both laminar and turbulent, a droplet collides with others on the timescale of order $1/\phi\dot{\gamma}$ (the mean free time) where ϕ is the droplet volume fraction. In a flow field we may set up the Smoluchowski equation (8.5.30) with the collision kernel estimated as

$$K(v, v') \sim \dot{\gamma}(R + R')^3 \sim \dot{\gamma}(v + v'), \qquad (11.1.34)$$

where $R \sim v^{1/3}$ and $R' \sim v'^{1/3}$ are the radii of the colliding droplets [65]. However, the above estimation (11.1.34) is valid only when the sizes of the colliding droplets are of the same order. It is known that flow-induced collisions rarely occur between droplets with very different sizes, because the smaller one moves on the stream line of the velocity field around the larger one without appreciable diffusive motion for a Peclet number $Pe \gg 1$ (see (11.1.37) below for a definition of Pe) [67, 68]. Thus, if coagulation occurs among droplets with sizes of the same order, (8.5.30) and (11.1.34) indicate that the droplet number density $n(t) = \int_0^\infty dv n(v,t)$ and the average droplet size $R(t) = [3\phi/4\pi n(t)]^{1/3}$ obey [66, 69]

$$\left(\frac{\partial}{\partial t}n(t)\right)_{\text{collision}} \sim -\dot{\gamma}\phi n(t), \qquad \left(\frac{\partial R(t)}{\partial t}\right)_{\text{collision}} \sim \dot{\gamma}\phi R(t). \tag{11.1.35}$$

ϕ is the volume fraction of the droplets. Thus $R(t)$ grows exponentially. For aggregating colloidal systems this exponential growth is well known [69].

Simulations of colloid aggregates have shown deformation, rupture, and coagulation of clusters in shear flow [70, 71]. These hydrodynamic effects are of great technological importance in two-phase polymers [72], in particular in the presence of copolymers (which lower the surface tension) [73].

Droplet size distribution in shear

Under (11.1.32) a nearly stationary distribution of droplets is realized after a long relaxation time. As stated above, Min and Goldburg [63] found the results indicating a monodisperse distribution of droplets peaked at $R \cong R^*$ and, once such a distribution is established, further time development of the droplet distribution becomes extremely slow. Though such a state is nearly stationary, there is still a diffusive current onto each droplet from the surrounding metastable region. It will grow above R_c and break into smaller droplets, which will then start to grow again or dissolve into the metastable region depending on whether their radii are larger or smaller than R_c. Each droplet will also collide with another one on the timescale of $1/\dot{\gamma}\phi$. The evolution of the droplet size distribution is therefore very complex and the observed quasi-stationarity is produced by a delicate balance among these processes. Alternatively, we may also start with an opaque state at a sufficiently large δT characterized by a small shear-dependent supersaturation $\Delta(\dot{\gamma})$. Then, by gradually decreasing δT at fixed $\dot{\gamma}$, a nearly stationary state will be obtained, which corresponds to the branch C in Fig. 11.5. Interestingly, it has been found to be more opaque and has a larger droplet volume fraction (or a smaller supersaturation) than in the reverse case of increasing δT from zero.

Figure 11.6 show multiple peaks in the scattered light intensity characteristic of very monodisperse droplets in the q_x–q_z plane taken by Hashimoto *et al.* [26, 27]. They also observed similar hysteresis by increasing or decreasing $\dot{\gamma}$ over a wide range with δT fixed. First, they increased $\dot{\gamma}$ from an opaque state with droplets to reach a transparent state without droplets at $T_{spi}(0) - T \propto \dot{\gamma}$, where $T_{spi}(0)$ is the cloud-point temperature at zero shear. We believe this disappearance of droplets to have been caused by the Taylor breakup mechanism (though the difference of $T_{spi}(0)$ and the temperature T_{cx} on the coexistence curve was not clarified in their work). Second, they decreased $\dot{\gamma}$ from a one-phase state homogenized by large shear to reach a spinodal-like point at which $T_{spi}(0) - T \propto \dot{\gamma}^{1/2}$

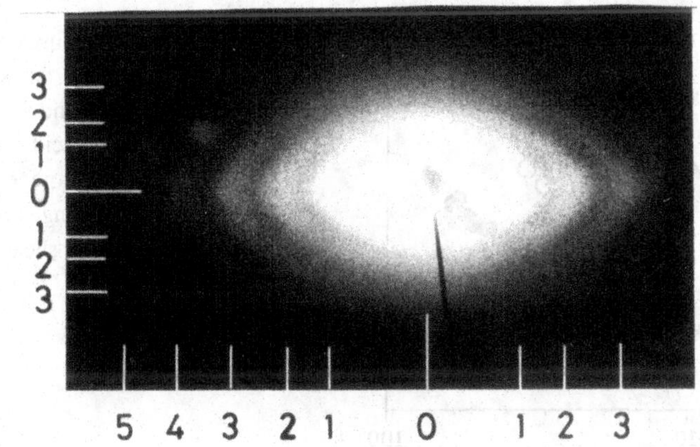

Fig. 11.6. Multiple peaks in the scattered light intensity from monodisperse droplets in the q_x–q_z plane in an off-critical PS/PB/DOP solution at $\dot{\gamma} = 0.33\ \mathrm{s}^{-1}$ (taken by Hashimoto's group [26]–[28]). Here the lines with numbers n (= 0, 1, ...) indicate the positions of the nth peak.

and below which droplets appear. However, they found that quasi-steady states reached in the decreasing branch are still slowly evolving towards the steady states reached in the increasing branch on timescales of several hours. The experiments by Hashimoto *et al.* and those by Min and Goldburg are consistent with each other.

Acceleration of droplet growth in shear

To analyze their experimental findings Baumberger *et al.* [74] argued that growth of an isolated droplet in a metastable fluid can be considerably accelerated even in very weak shear by an advection mechanism. If the growth is slow, the composition ψ outside the droplet is determined by a quasi-static condition,

$$\boldsymbol{u} \cdot \nabla\psi + D\nabla^2\psi = 0, \tag{11.1.36}$$

where $\boldsymbol{u}$ is the average flow tending to a simple shear flow far from the droplet. If we assume that ψ changes on the scale of the droplet radius, the relative importance of the two terms in (11.1.36) is given by the Peclet number,

$$Pe = \dot{\gamma} R^2/D = \dot{\gamma} t_\xi (R/\xi)^2. \tag{11.1.37}$$

We have $Pe > \dot{\gamma} t_\xi/\Delta^2$ for $R > R_c$ and $Pe \sim 1/\dot{\gamma} t_\xi$ at the breakup size $R \sim R^*$. Thus $Pe \gg 1$ can hold in a wide time interval even under (11.1.32). The deviation from the spherical shape is small for $R \ll \sigma/\eta\dot{\gamma}$ or for $R \ll R^*$. For $Pe \gg 1$ it is important that the concentration gradient is localized in a thin layer with a thickness $\ell_{\dot{\gamma}}$ given by

$$\ell_{\dot{\gamma}} = (D/\dot{\gamma})^{1/2} = R/\sqrt{Pe} \tag{11.1.38}$$

around the droplet. This relation follows from a balance between the two terms in (11.1.33). As a result, the diffusion current onto the droplet from the metastable fluid is increased by

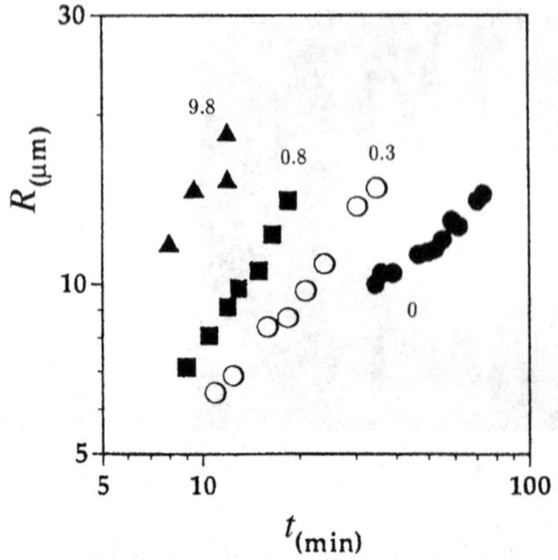

Fig. 11.7. Droplet radius R as a function of time for $\delta T = 8$ mK and several shear rates $\dot{\gamma} = 0, 0.3, 0.8$, and 9.8 in an off-critical isobutyric acid + water [74]. The effective growth exponent $\partial \ln R/\partial \ln t$ increases with increasing shear from the usual value 1/3 at zero shear.

$R/\ell_{\dot{\gamma}} \sim Pe^{1/2}$ as compared to the case $Pe \ll 1$ [75, 76], so that the usual Lifshitz–Slyozov equation is modified as

$$\frac{\partial}{\partial t} R \sim \frac{D}{R}\left(\Delta - \frac{2D_0}{R}\right)\sqrt{Pe} \sim \sqrt{\dot{\gamma} D}\left(\Delta - \frac{2d_0}{R}\right), \tag{11.1.39}$$

where d_0 is the capillary length ($\sim \xi$) in (8.4.13) or (9.1.17). Thus, as shown in Fig. 11.7, the timescale of the initial stage can be considerably accelerated by the convection effect. As the supersaturation around the droplets decreases, however, the probability of droplet encounters will become the dominant mechanism of the droplet growth. Interestingly, all the data in Fig. 11.7 obey $Pe \sim (\phi \dot{\gamma} t)^b$ with $b \sim 4/3$. Note also that the critical radius $R_c = 2d_0/\Delta$ is unchanged by very weak shear and there seems to be no drastic change in the nucleation rate.

The above mechanism is important in systems with a small diffusion constant such as polymer blends. As a similar effect we note that, if surfactant molecules are added to an oil–water two-phase system, they can be advected onto the oil–water interfaces efficiently in shear flow, leading to shear-induced emulsification. Systematic experimentation in these cases should be interesting.

11.1.6 Rheology in near-critical fluids

From (6.1.17) the fluctuations of the order parameter ψ give rise to the following additional shear stress,

$$\begin{aligned} \dot{\gamma}\Delta\eta &= -T\langle(\nabla_x\psi)(\nabla_y\psi)\rangle \\ &= -T\int_q q_x q_y I(\boldsymbol{q}), \end{aligned} \tag{11.1.40}$$

where $\nabla_i = \partial/\partial x_i$ and $\Delta\eta$ is the fluctuation contribution. Other important quantities are the normal stress differences,

$$\begin{aligned} N_1 &= \sigma_{xx} - \sigma_{yy} = T\langle(\nabla_x\psi)^2 - (\nabla_y\psi)^2\rangle, \\ N_2 &= \sigma_{yy} - \sigma_{zz} = T\langle(\nabla_y\psi)^2 - (\nabla_z\psi)^2\rangle. \end{aligned} \tag{11.1.41}$$

These quantities can also be expressed in terms of the structure factor as in the second line of (11.1.40).

Strong shear regime in one-phase states

In the one-phase region, the above quantities may be expressed as integrals in the wave vector space using the structure factor $I(\boldsymbol{q})$. We find that $\Delta\eta$ is nearly logarithmic as $\ln(\xi/\xi_{+0})$ in weak shear and as $\ln(1/k_c\xi_{+0})$ in strong shear. This crossover was first predicted by Oxtoby [77]. If use is made of the ϵ expansion in strong shear, the steady-state viscosity is of the form [78],

$$\eta = \eta_0 + \Delta\eta \propto (k_c\xi_{+0})^{-x_\eta} \propto \dot{\gamma}^{-x_\eta/d}, \tag{11.1.42}$$

where $x_\eta = \epsilon/19 + \cdots$ is the small dynamic exponent in (6.1.42). This shear-rate dependence was measured by Hamano *et al.* [79]. In weak shear the normal stress differences are proportional to $\dot{\gamma}^2$ and are very small. In strong shear, the right-hand sides of (11.1.40) and (11.1.41) after the wave vector integrations are of order Tk_c^d, so that

$$N_1 = 0.046\epsilon\eta\dot{\gamma}, \qquad N_2 = -0.032\epsilon\eta\dot{\gamma}, \tag{11.1.43}$$

to first order in ϵ [78]. Note that N_1 and N_2 are even functions of $\dot{\gamma}$, while the shear stress σ_{xy} is odd. If we allow the case $\dot{\gamma} < 0$, we should use $|\dot{\gamma}|$ in (11.1.42) and (11.1.43).

Weak shear regime in two-phase states

When a near-critical fluid is undergoing phase separation, larger stress contributions arise from interface deformations because $(\nabla\phi)(\nabla\phi)$ behaves like a δ function near the interface multiplied by the tensor $\boldsymbol{nn}$, where $\boldsymbol{n} = (n_x, n_y, n_z)$ is the normal unit vector. In weak shear, the interfaces are sharp and (11.1.40) yields a well-known expression [80]–[82],

$$(\Delta\eta)_{\text{sur}} = -\frac{1}{\dot{\gamma}}\sigma \int da\, n_x n_y, \tag{11.1.44}$$

where σ is the surface tension, da is the surface element, and the surface integral is within a unit volume containing many domains. This surface contribution is the sole change of the macroscopic viscosity in newtonian two-phase fluids with the same viscosity. Similarly,

$$\begin{aligned} (N_1)_{\text{sur}} &= \sigma \int da\, (n_x^2 - n_y^2), \\ (N_2)_{\text{sur}} &= \sigma \int da\, (n_y^2 - n_z^2). \end{aligned} \tag{11.1.45}$$

If we suppose an assembly of largely deformed droplets near the breakup condition $R \sim R^*$ in (11.1.31) with volume fraction ϕ, we estimate $\langle -n_x n_y \rangle \sim \langle n_x^2 - n_y^2 \rangle \sim 1$ to obtain [83]

$$(\Delta\eta)_{\mathrm{sur}} \sim \phi\sigma/\dot{\gamma}R \sim \phi\eta, \tag{11.1.46}$$

$$(N_1)_{\mathrm{sur}} \sim (N_2)_{\mathrm{sur}} \sim \eta\dot{\gamma}\phi, \tag{11.1.47}$$

where the surface area density is of order ϕ/R. Because (11.1.46) is independent of shear, it is analogous to well-known expressions for the macroscopic viscosity of suspensions or emulsions in the zero-shear limit [81]. However, in our case droplets are largely deformed, so the rheology is strongly nonlinear. The behavior of N_1 and N_2 is marked because they are nearly zero in one-phase states and jump to large values after quenching.

Doi and Ohta set up dynamic equations for the interfacial stress tensor $\sigma \int da n_i n_j$ and the surface area density [84]. In steady states, their equations reproduced (11.1.46) and (11.1.47). Furthermore, they can reasonably describe transient stress relaxation after a step increase in the shear rate. Simulations of simple fluids in 2D also showed a considerable increase of the viscosity in spinodal decomposition under shear [41]–[47]. Such studies in complex fluids with internal structures should be of great importance.

We mention related experiments. (i) Krall *et al.* [56] measured the viscosity increase $\Delta\eta(t)$ in a near-critical binary fluid mixture of isobutyric acid + water using a viscometer in which shear was oscillated and damped in time. After a pressure quench at $t = 0$, $\Delta\eta(t)$ increased on the timescale of t_ξ, in accord with (11.1.46). While it tended to a constant for the droplet case, it slowly decayed to zero at the critical (bicontinuous) case after a long time ($\sim$ 20 s). In Fig. 11.8 we show their data of the viscosity increase and the shear modulus ($= \mathrm{Re}\, G^*(\omega)$) at a critical quench. Hamano *et al.* [79] subsequently observed the same decay of $\Delta\eta(t)$ in steady shear in a rotational viscometer. Because a sharp streak scattering pattern emerges with $\Delta\eta(t) \to 0$, we may conclude that a string phase (see Fig. 11.4) was realized in their critical-quench cases. For such highly elongated domains the interfaces are mostly parallel to the flow and $n_x \cong 0$ in (11.1.44), leading to $\Delta\eta \cong 0$. Notice that N_1 is still given by (11.1.47) even in the string phase. (ii) The rheology of phase-separating polymer blends has also been studied [85]–[88] particularly when the two phases are newtonian and have almost the same viscosity. The observed $\Delta\eta$ and N_1 were in excellent agreement with the scaling relations (11.1.46) and (11.1.47). Figure 11.9 shows data of N_1 in a blend of PS + poly(vinyl methyl ethyl)) (PVME) with a molecular weight of order 10^5 in two-phase states [86], in which the viscosities of the two components were of the same order. The linear behavior $N_1 \propto \dot{\gamma}$ was seen even at low shear rates where shear-thinning of the viscosity was still weak.

11.1.7 Rheology in two-phase binary fluid mixtures with viscosity difference

Let us consider phase-separating newtonian binary fluid mixtures in which the two phases have different viscosities η_1 and η_2 [89]. Batchelor [75] derived a formal expression

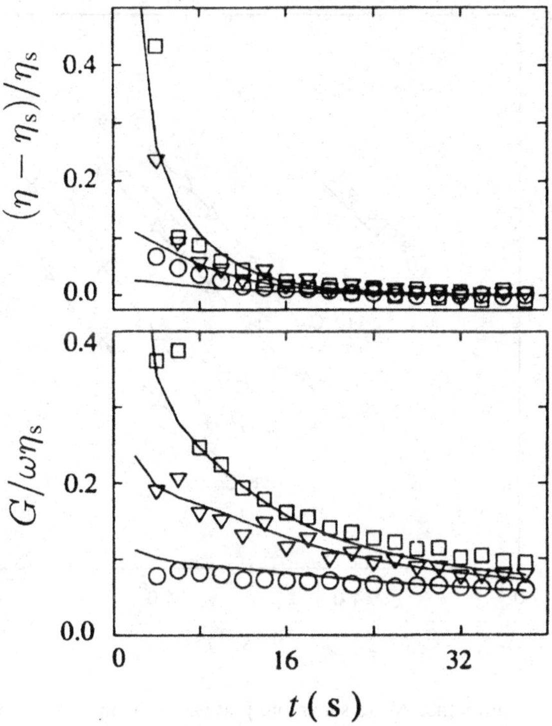

Fig. 11.8. Reduced viscosity increase $(\eta - \eta_s)/\eta_s$ and the elastic shear modulus $G/\omega\eta_s$ in a critical binary mixture of isobutyric acid + water [56], where η_s is the viscosity in one-phase states at the same temperature. The symbols ○, ▽, and □ correspond to quenches of depths of 11, 19, and 64 mK, respectively. The smooth curves are those from the Doi–Ohta constitutive equations [84].

for the average stress tensor in two-phase states in the low-Reynolds number limit. In incompressible flow $\langle \boldsymbol{v} \rangle = \boldsymbol{u}$ with velocity gradient $D_{ij} = \partial u_i/\partial x_j$, its space average is written in the following surface integral form [81],

$$\begin{aligned} \langle \sigma_{ij} \rangle \quad &= \quad -p\delta_{ij} + (\phi_1\eta_1 + \phi_2\eta_2)(D_{ij} + D_{ji}) \\ &\quad + (\eta_1 - \eta_2) \int da(v_i' n_j + v_j' n_i) - \sigma \int da n_i n_j, \end{aligned} \tag{11.1.48}$$

where p is a pressure, da is the surface element, $\boldsymbol{n}$ is the normal unit vector at the interface from phase 1 to phase 2, $\boldsymbol{v}'$ in the third term is the velocity deviation $\boldsymbol{v} - \boldsymbol{u}$ immediately inside the droplets, and ϕ_1 and $\phi_2 = 1 - \phi_1$ are the volume fractions of the two phases. The surface integral is performed over the surfaces within a unit volume. The last term arises from the surface tension force and remains nonvanishing even for $\eta_1 = \eta_2$, leading to (11.1.44) and (11.1.45). Its contribution to the shear viscosity becomes negligible for high elongation of the domains, such as in the string phase. However, in a transient process

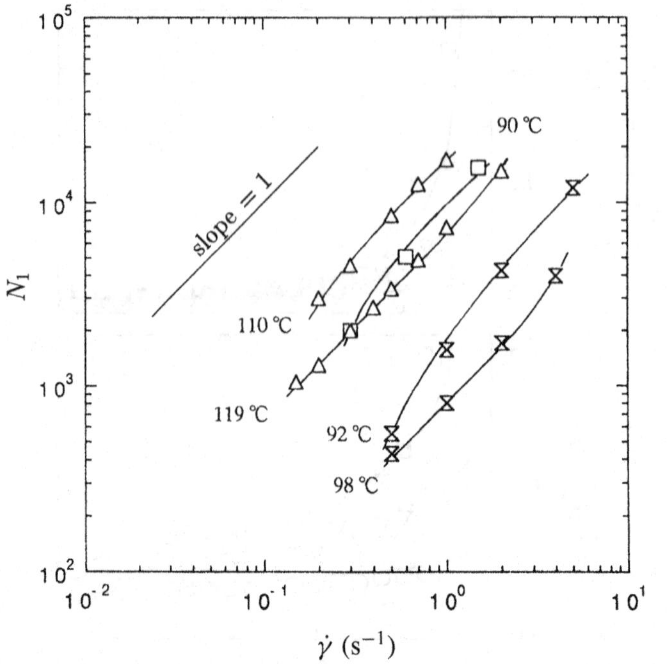

Fig. 11.9. Normal stress difference N_1 vs shear rate $\dot{\gamma}$ at various temperatures in a polymer blend in two-phase states [86]. These data demonstrate the relation $N_1 \propto \dot{\gamma}$ in (11.1.47).

or in the droplet case, we have $\langle -n_x n_y \rangle \sim 1$ and

$$(\Delta\eta)_{\text{sur}} \sim \sigma A/\dot{\gamma}, \tag{11.1.49}$$

where A is the surface area density.

For simplicity, we consider nearly steady two-phase states under shear in the case $\eta_1 \gg \eta_2$. Then phase 2 is compressed into layers with thickness R_2, and the distance between two neighboring domains of phase 1 with size R_1 is equal to R_2. The two lengths R_1 and R_2 are related to the volume fractions as

$$AR_1 \sim \phi_1, \quad AR_2 \sim \phi_2. \tag{11.1.50}$$

The typical velocity gradients $\dot{\gamma}_1$ and $\dot{\gamma}_2$ in the two phases satisfy

$$\eta_1\dot{\gamma}_1 \sim \eta_2\dot{\gamma}_2 \sim \langle \sigma_{xy} \rangle. \tag{11.1.51}$$

The macroscopic shear rate $\dot{\gamma}$ is given by

$$\dot{\gamma} \sim (R_1 + R_2)^{-1}(R_1\dot{\gamma}_1 + R_2\dot{\gamma}_2) \sim \phi_1\dot{\gamma}_1 + \phi_2\dot{\gamma}_2. \tag{11.1.52}$$

These relations yield the effective viscosity,

$$\eta_{\text{eff}} = \langle \sigma_{xy} \rangle/\dot{\gamma} \sim \left(\phi_1/\eta_1 + \phi_2/\eta_2\right)^{-1}. \tag{11.1.53}$$

When $\eta_1 \gg \eta_2$, this relation means that even a small fraction of the second phase can drastically reduce η_{eff} from η_1 to η_2/ϕ_2 for $\phi_2 \gtrsim \phi_1\eta_2/\eta_1$. Obviously, the second phase acts as a lubricant. Furthermore, the typical velocities in the two phases are estimated as

$$v_1 = \dot\gamma_1 R_1 \sim \frac{\phi_1}{\eta_1}\frac{\langle\sigma_{xy}\rangle}{A}, \quad v_2 = \dot\gamma_2 R_2 \sim \frac{\phi_2}{\eta_2}\frac{\langle\sigma_{xy}\rangle}{A}. \tag{11.1.54}$$

We assume that the typical velocity of the droplet phase is smaller than that in the continuous phase. When the two phases are both percolated, we require $v_1 \sim v_2$ to obtain the condition of bicontinuity,

$$\phi_1/\eta_1 \sim \phi_2/\eta_2. \tag{11.1.55}$$

This relation has been known as an empirical law for polymer mixtures in the engineering literature [90].

In particular, we consider three cases in more detail. (i) When $\phi_1/\eta_1 < \phi_2/\eta_2$ and the more viscous phase 1 forms a droplet phase, the velocity gradient is mainly supported by the less viscous phase 2 and $\dot\gamma \sim \phi_2\dot\gamma_2$ even for $\phi_2 \ll 1$. Because the two mechanisms of aggregation and breakup should balance in the steady state, typical droplets will be close to the breakup condition. They are only slightly deformed from a sphere and the stress due to the surface tension and that due to the viscosity are of the same order. Therefore, we have $\langle\sigma_{xy}\rangle \sim \sigma/R_1 \sim \dot\gamma_2\eta_2$ and

$$R_1 \sim \sigma/(\eta_{\text{eff}}\dot\gamma) \sim \sigma\phi_2/(\eta_2\dot\gamma), \tag{11.1.56}$$

which decreases down to $\sigma/\eta_1\dot\gamma$ with increasing ϕ_1 at fixed $\dot\gamma$. To the normal stress differences the last two terms in (11.1.48) both give rise to contributions of the same order,

$$N_1 \sim N_2 \sim \sigma\phi_1/R_1 \sim (\phi_1\eta_2/\phi_2)\dot\gamma, \tag{11.1.57}$$

which increases up to order $\eta_1\dot\gamma$ at $\phi_2 \sim \eta_2/\eta_1$. (ii) When ϕ_2 is very small, phase 2 forms a droplet phase. In the case $\eta_2 \ll \eta_1$ an isolated droplet of phase 2 is elongated into a slender shape prior to breakup [58, 59]. The ratio of the longest radius $R_\parallel$ and the shortest radius $R_\perp$ deviates from 1 appreciably for $R_\parallel \sim \sigma/\eta_1\dot\gamma_1$ and is of order $(\eta_1\dot\gamma_1/\sigma)^{3/4}V^{1/4}$ in the steady state, where $V(\sim R_\parallel R_\perp^2)$ is the droplet volume. Here $\dot\gamma_1$ may be set equal to $\dot\gamma$. At the breakup we have

$$R_\parallel \sim \sigma/(\eta_2\dot\gamma), \qquad R_\perp \sim (\eta_2/\eta_1)^{1/2}R_\parallel. \tag{11.1.58}$$

Here $\eta_{\text{eff}} \sim \eta_1$, consistent with (11.1.52). The behavior of N_1 and N_2 is complicated. Let us consider N_1. The contribution from the last term in (11.1.48) is of order σA with $A \sim \phi_2/R_\perp$, which is $(\eta_1\eta_2)^{1/2}\dot\gamma\phi_2$ at the breakup. However, the third term in (11.1.48) is of order $\eta_1 A v_2 \cos\theta$, where $v_2 \sim R_\perp\dot\gamma_2 \sim R_\perp\eta_1\dot\gamma/\eta_2$ is the typical velocity within the slender droplet and θ is the angle between the normal $\boldsymbol{n}$ and the x axis (which is parallel to the flow). We may set $\cos\theta \sim R_\perp/R_\parallel$. Thus,

$$N_1 \sim (\eta_1^{3/2}/\eta_2^{1/2})\phi_2\dot\gamma, \tag{11.1.59}$$

which is larger than the surface tension contribution by η_1/η_2. The above relation is still a conjecture because the velocity field around a slender droplet is very complicated and more systematic analysis is needed. We also notice that N_1 seems to be discontinuous where the slender droplets become percolated near $\phi_2 \sim \eta_2/\eta_1$.

11.1.8 Rheology in diblock copolymers

In diblock copolymers the fluctuations with $q \sim k_0$ can be strongly enhanced as (11.1.23) at small r. Note that the transition becomes first order due to the fluctuation effect as stated near (11.1.26). By calculating the complex shear modulus $G^*(\omega)$ in the disordered phase, Fredrickson and Larson found that these fluctuations give rise to anomalous rheological properties [33]. In particular, they predicted that the fluctuation contribution $\Delta\eta$ in the zero-frequency shear viscosity grows as $r^{-3/2}$. In the linear regime, a subsequent mode coupling theory [34] yielded

$$G^*(\omega)/i\omega = \eta_0 + A_c r^{-3/2}(1+\sqrt{1+i\Omega})^{-2}, \tag{11.1.60}$$

where η_0 is the background viscosity, A_c is a constant, and $\Omega = (k_0^2/8\Gamma_c)\omega/r$, with Γ_c being the noncritical relaxation rate introduced above (11.1.24). With increasing shear, however, the fluctuation contribution $\Delta\eta$ in the steady state decreases as

$$\Delta\eta \propto \dot{\gamma}^{-1} \tag{11.1.61}$$

in the region $(r/k_0^2)^{3/2} < \dot{\gamma}/\Gamma_c < 1$. The form birefringence was also predicted to grow towards the transition [34, 91].

If the temperature is cooled below the transition, lamellar ordered grains appear and evolve slowly [92] (and their orientation may be achieved by application of shear or an electric field). In such locally ordered states, Rosendale and Bates [93] found anomalous low-frequency behavior,

$$G^*(\omega) \sim (i\omega)^{1/2}, \tag{11.1.62}$$

as shown in Fig. 11.10. To explain their finding, Kawasaki and Onuki examined the dynamics of mesoscopic phases with locally lamellar morphology with disorder [94]. Because lamellar systems behave like solids in the direction normal to the lamellae, a large stress arises as

$$\sigma_{ij} = Bn_i n_j(\boldsymbol{n}\cdot\nabla)u, \tag{11.1.63}$$

for variations of the lamellar spacing. Here B is a compression elastic constant and $u(\boldsymbol{r},t)$ is the local displacement field of lamellae. The local normal unit vector $\boldsymbol{n}(\boldsymbol{r})$ is assumed to be stationary and varies randomly in space on the scale of the defect distance $\ell_{\rm def}$ much longer than the lamellar spacing λ. The proposed mechanism is associated with overdamped collective modes with wave vector $\boldsymbol{k} = \boldsymbol{k}_\perp + k_\parallel \boldsymbol{n}$ in the region represented by $|k_\parallel| \ll |\boldsymbol{k}_\perp| \ll \lambda^{-1}$ and $k \gg (\rho\omega/\eta_0)^{1/2}$. The decay rate of $u_{\boldsymbol{k}}$ is then given by

$$\Gamma(\boldsymbol{k}) \cong (B/\eta_0)(\cos^2\theta + \lambda^2 k^2), \tag{11.1.64}$$

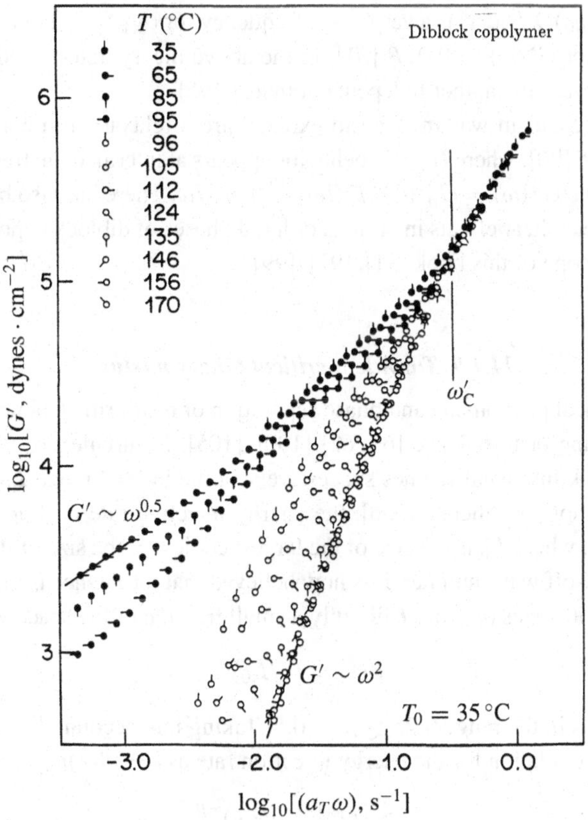

Fig. 11.10. Dynamic shear modulus $G'(\omega) = \text{Re}\, G^*(\omega)$ as a function of reduced frequency for a symmetric diblock copolymer near the microphase separation point [93]. The temperature-dependent parameter a_T is chosen such that $a_T = 1$ at $T = T_0$ and the curves at various T coincide for $\omega > \omega'_c$. Filled and open symbols correspond to the ordered and disordered states, respectively.

with $\cos\theta = k_\parallel/k$ being small. A mode coupling expression for the frequency-dependent viscosity is written as

$$\eta^*(\omega) = \eta_0 + B \int_{\boldsymbol{k}} \varphi(k) \frac{1}{i\omega + \Gamma(\boldsymbol{k})} \cdot \frac{\cos^2\theta}{\cos^2\theta + \lambda^2 k^2}. \tag{11.1.65}$$

The $\varphi(k)$ is defined by

$$\varphi(k) = \int d\boldsymbol{r} e^{-i\boldsymbol{k}\cdot\boldsymbol{r}} \langle n_x(\boldsymbol{r}) n_y(\boldsymbol{r}) n_x(\boldsymbol{0}) n_y(\boldsymbol{0}) \rangle, \tag{11.1.66}$$

where the average is taken over the random distribution of $\boldsymbol{n}$. Notice that $\varphi(k)$ behaves as $(2\pi)^d \delta(\boldsymbol{k})/12$ in the limit $\ell_{\text{def}} \to \infty$. Then the angle integration over $\theta (\cong \pi/2)$ yields

$$G^*(\omega) = i\omega\eta_0 + \frac{\pi}{24}(B\eta_0 i\omega)^{1/2} \quad (\omega > \omega_\ell) \tag{11.1.67}$$

where $\omega_\ell = (B/\eta_0)(\lambda/\ell_{\rm def})^2$ is a very small frequency for $\ell_{\rm def} \gg \lambda$. For $\omega < \omega_\ell$ we obtain solid-like behavior $G^*(\omega) \sim 0.01B$ [31]. In the above theory defect motion is neglected, whereas it is relevant in another independent theory [95].

The above mechanism was invoked to explain stress relaxation in concentrated emulsions by Liu *et al.* [96], where $(i\omega)^{1/2}$ behavior appears at intermediate frequencies in their empirical formula, $G^*(\omega) = G_{\rm p} + A(\phi)(i\omega)^{1/2} + \eta_\infty i\omega$. There has also been a number of observed nonlinear shear effects in various ordered phases of diblock copolymers, but they are beyond the scope of this book [31], [97]–[99].

11.1.9 Turbulent critical binary mixtures

We examine critical phenomena and phase separation of near-critical binary fluid mixtures in vigorous stirring or turbulence [61, 62] [100]–[105]. In turbulence, eddies with linear dimension ℓ break into smaller ones successively in the inertial range $L_0 > \ell > k_{\rm d}^{-1}$. In the original Kolmogorov theory [106], the energy injection rate $\bar{\epsilon} \sim u_\ell^3/\ell$ is a constant independent of ℓ, where L_0 is the size of the largest eddies ($\sim$ the size of the stirrer) and $k_{\rm d}$ is the viscous cut-off wave number. It is now believed that turbulence is intermittent [107]. That is, eddies with sizes of order ℓ fill only a small fraction of the space which is of order

$$\beta(\ell) = (\ell/L_0)^\mu. \tag{11.1.68}$$

The exponent μ is in the range $0.25 \lesssim \mu \lesssim 0.5$. Taking into account the intermittency, we should modify the relation for the energy injection rate as $\bar{\epsilon} \sim \beta(\ell)u_\ell^3/\ell$, which yields

$$u_\ell \sim (\bar{\epsilon}\ell)^{1/3}(\ell/L_0)^{-\mu/3}. \tag{11.1.69}$$

In the dissipative range $\ell < k_{\rm d}^{-1}$ the velocity fluctuations are dissipated by the shear viscosity η. Thus $u_\ell/\ell \sim (\eta/\rho)\ell^{-2}$ at $\ell \sim k_{\rm d}^{-1}$. Using the definition of the Reynolds number $Re = L_0u_0/(\eta/\rho)$, we may express $k_{\rm d}$ as

$$k_{\rm d} = L_0^{-1}Re^{3/(4-\mu)}. \tag{11.1.70}$$

The maximum shear rate $\dot{\gamma}_{\rm dis}$ in turbulence is given by

$$\dot{\gamma}_{\rm dis} = (\eta/\rho)k_{\rm d}^2 \sim (\eta/\rho L_0^2)Re^{6/(4-\mu)}. \tag{11.1.71}$$

To make rough estimates we set $Re \sim 10^4$, $L_0 \sim 1$ cm, $\eta/\rho \sim 10^{-2}$ cm^2 s^{-1}, and $\mu = 0$ to obtain $k_{\rm d} \sim 10^3$ cm^{-1} and $\dot{\gamma}_{\rm dis} \sim 10^4$ s^{-1}.

Droplet sizes in turbulence

For simplicity, we assume droplets with sizes R ($\gg \xi$) with sharp interfaces in a two-phase state below $T_{\rm c}$. Near the critical point, the droplets are broken into smaller sizes in the dissipative range ($< k_{\rm d}^{-1}$):

$$R \sim \sigma/\eta\dot{\gamma}_{\rm dis} \sim (\rho\sigma/\eta^2)k_{\rm d}^{-2}, \tag{11.1.72}$$

where the shear stress $\eta\dot{\gamma}_{\rm dis}$ is balanced with the capillary force density ($\sim \sigma/R$) as for laminar shear [13]. The condition $R < k_{\rm d}^{-1}$ is equivalent to

$$\sigma \sim T/\xi^2 < (\eta^2/\rho)k_{\rm d}. \tag{11.1.73}$$

However, away from the critical point, the surface tension increases and the reverse of (11.1.73) holds. Then R is in the inertial range ($> k_{\rm d}^{-1}$) and is determined by a balance between the typical pressure variation ($\sim \rho u_R^2$) over the distance R and the capillary force density ($\sim \sigma/R$) in the form,

$$R \sim k_{\rm d}^{-1}(\rho\sigma/\eta^2 k_{\rm d})^{3/(5-2\mu)} > k_{\rm d}^{-1}. \tag{11.1.74}$$

This expression (with $\mu = 0$) was originally derived by Kolmogorov [13, 66, 108].

Critical fluctuations and spinodal decomposition in turbulence

The concentration fluctuations in near-critical fluids have sizes much shorter than the size of the smallest eddies ($\sim 1/k_{\rm d}$) and are most effectively strained by the smallest eddies. These eddies turn over on the timescale of $1/\dot{\gamma}_{\rm dis}$, during which the concentration fluctuations are acted on by the eddies. The concentration fluctuations encounter them intermittently, and the mean free time is determined by

$$1/t_{\rm mf} = \beta(k_{\rm d})\dot{\gamma}_{\rm dis} \sim (\eta/\rho)L_0^{-2}Re^{(6-3\mu)/(4-\mu)}. \tag{11.1.75}$$

This time $t_{\rm mf}$ should be compared with the thermal relaxation time t_ξ. In a one-phase state, the critical fluctuations are not much affected in the weak shear regime $t_\xi < t_{\rm mf}$, while they are strongly suppressed in the wave number region $k < k_{\rm c}$ in the strong shear regime $t_\xi > t_{\rm mf}$. As in the laminar shear case, the characteristic wave number $k_{\rm c}$ is defined by

$$k_{\rm c} = (6\pi\eta/Tt_{\rm mf})^{1/3} \sim (6\pi\eta^2/\rho TL_0^2)^{1/3}Re^{(2-\mu)/(4-\mu)}. \tag{11.1.76}$$

The crossover reduced temperature $\tau_{\rm s}$ in shear flow is defined by $\tau_{\rm s} = (\xi_{+0}k_{\rm c})^{1/\nu}$. For example, if $1/t_{\rm mf} \sim 10^4$ s^{-1}, we have $\tau_{\rm s} \sim 10^{-3}$ for isobutyric acid + water. As shown in Fig. 11.11, Chan *et al.* [101] observed that there is no sharp phase transition in turbulence. As T is slightly lowered below $T_{\rm c}$, the scattered light intensity increases gradually but dramatically. As shown in Fig. 11.12, the steady-state intensity I_k has a peak at $k = 0$, even below $T_{\rm c}$. Moreover, they did not detect the Porod tail in the intensity for $T/T_{\rm c} - 1 \sim -10^{-4}$ mK and $Re \sim 10^4$, where the two-phase boundaries should have been blurred by the turbulent shear.

We will now present numerical examples on turbulent critical binary mixtures [104]. If the sizes of the concentration fluctuations are in the dissipative range, the velocity field $\boldsymbol{v}$ may be expanded as

$$\boldsymbol{v}(\boldsymbol{r}, t) = \boldsymbol{v}(\boldsymbol{r}_0, t) + \sum_j D_{ij}(t)(x_j - x_{j0}) + \cdots, \tag{11.1.77}$$

where $\boldsymbol{r}_0$ represents an appropriate reference point. To examine the fluctuations of a passive scalar c convected by $\boldsymbol{u}$ as $\partial c/\partial t = -\boldsymbol{u}\cdot\nabla c + D\nabla^2 c$, Batchelor [109] made an analysis

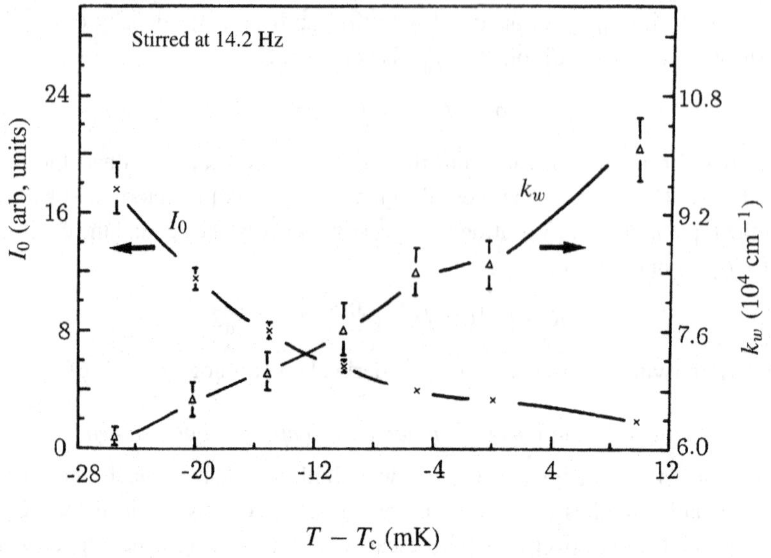

Fig. 11.11. The scattered light intensity $I_0 = \lim_{k\to 0} I_k$ in the long-wavelength limit and the wave number k_w vs $T - T_c$ in dynamical steady states of a critical binary fluid mixture of isobutyric acid + water stirred at 14.2 Hz [101], which corresponds to $Re = 10^4$. Here k_w is determined by $I_{k_w} = I_0/2$.

assuming that $\{D_{ij}(t)\}$ is nearly stationary, while Kraichnan [110] investigated the reverse case in which $\{D_{ij}(t)\}$ changes rapidly as a white noise. The latter will be the case in near-critical fluids. Then the intensity $I(k,t)$ obeys

$$\frac{\partial}{\partial t}I(k,t) = B\left(k^2\frac{\partial^2}{\partial k^2} + 4k\frac{\partial}{\partial k}\right)I(k,t) + X(k,t), \tag{11.1.78}$$

where

$$B = \frac{1}{15}\sum_{ij}\int_{-\infty}^{t} ds\langle[\nabla_j u_i(\boldsymbol{r},t)][\nabla_j u_i(\boldsymbol{r},s)]\rangle. \tag{11.1.79}$$

Here B is estimated as $(\bar{\epsilon}/15\nu)t_{\rm rel} \sim \beta(k_{\rm d})\dot{\gamma}_{\rm dis}^{-2}t_{\rm rel}$ where $t_{\rm rel}$ is the relaxation time of the time-correlation function in (11.1.77). If we set $t_{\rm rel} = \dot{\gamma}_{\rm dis}^{-1}$, we find $B \sim 1/t_{\rm mf}$. In the passive scalar case, Kraichnan set $X(k,t) = -2Dk^2I(k,t)$ [110]. For near-critical fluids below T_c we may set $X(k,t)$ equal to the right-hand side of the Kawasaki–Ohta equation (8.5.1) [104]. Then, as $t \to \infty$, $I(k,t)$ tends to a steady intensity I_k scaled as

$$I_k = k_{\rm w}^{-3}I^*(k/k_{\rm w}, B^*), \tag{11.1.80}$$

where $k_{\rm w}$ is determined by $I_{k_{\rm w}} = I_0/2$, and $B^* = Bt_\xi$ is the dimensionless turbulent shear rate. We found that the curve of $k_{\rm w}\xi$ vs $1/B^*$ is analogous to the curve of $k_{\rm m}(t)\xi$ vs t/t_ξ in normal spinodal decomposition in Fig. 8.22. This suggests that spinodal decomposition

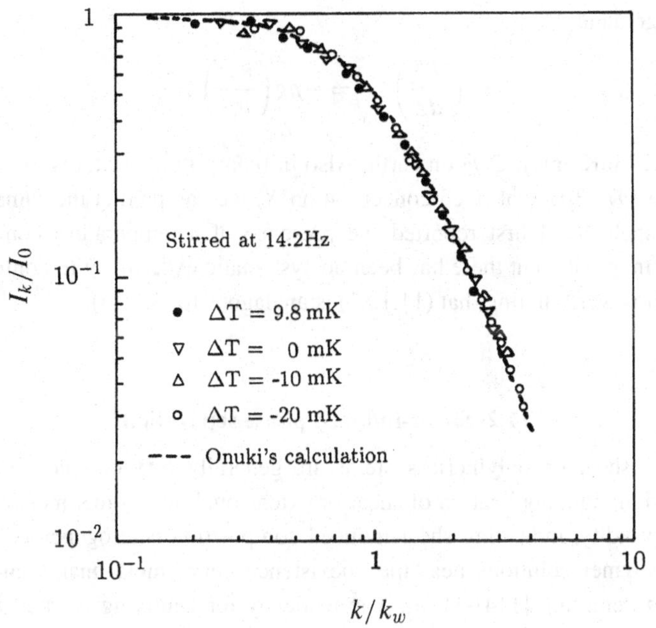

Fig. 11.12. The scaled intensity I_k/I_0 vs $k/k_{\rm w}$ in a stirred critical binary mixture on a log–log scale [101]. The dashed curve is obtained as the steady solution of (11.1.78) at $B = 0.1/t_\xi$, with $X(k, t)$ being the right-hand side of the Kawasaki–Ohta equation (8.5.1).

is stopped at a time of order $1/B(\gg t_\xi)$. These results are in good agreement with data of the scattered light intensity, as demonstrated in Fig. 11.12.

11.1.10 Gravity effect in stirred fluids

Fluids can be mixed efficiently even by gentle stirring, so it is often used in experiments and in everyday life. Chashkin *et al.* [111] measured the specific heat C_V under gentle stirring in one-component fluids at the critical density near the gas–liquid critical point. As shown in Fig. 2.4, they could observe a sharp peak of C_V even very close to the critical point ($|T/T_c - 1| \gtrsim 10^{-4}$); this would have been masked by the gravity effect in a quiescent fluid. (See also Ref. [68] cited in Chapter 4.) To support their finding, we may argue [112] that the density stratification in gravity is much reduced from $-\rho g(\partial\rho/\partial p)_T$ to $-\rho g(\partial\rho/\partial p)_s$ under stirring. The ratio of these two quantities is $(\partial\rho/\partial p)_T/(\partial\rho/\partial p)_s = C_p/C_V$ and is very large near the gas–liquid critical point. This means that the entropy per unit mass tends to be homogenized ($s(p, T) \cong$ const.) in stirred fluids despite the presence of a pressure gradient. This is because of the instantaneous pressure equilibration and the slow thermal diffusion, as discussed in Section 6.4. As a result, there should arise a small vertical temperature gradient,

$$\left(\frac{dT}{dz}\right)_{\text{stirr}} = -\rho g \left(\frac{\partial T}{\partial p}\right)_s, \tag{11.1.81}$$

which is -0.27 mK/cm in CO_2 on earth. Also in binary fluid mixtures, if use is made of the derivative $(\partial T/\partial p)_{sX}$ at fixed concentration X, we can predict the same temperature gradient. Cannell [113] first reported the presence of a temperature nonuniformity in stirred fluids in gravity, but there has been no systematic experiment to confirm the above predictions. It is worth noting that (11.1.81) is analagous to (8.6.14).

11.2 Shear-induced phase separation

The effects of shear on polymeric systems are generally very complex [9]. As well as shear-induced mixing, application of shear or extensional flow sometimes induces a large increase in turbidity, indicating shear-induced composition heterogeneities or demixing. Semidilute polymer solutions near the coexistence curve most unambiguously exhibit shear-induced demixing [114]–[116]. The tendency for demixing is intensified with increasing molecular weight $M(\gtrsim 2 \times 10^6)$ and the polymer volume fraction above the overlapping value, as can be seen in Fig. 11.13 [117]. The fluctuation enhancement becomes more remarkable at non-newtonian shear, where shear-thinning behavior is significant. Large stress fluctuations have also been reported upon demixing by shear [6, 7, 118], suggesting formation of gel-like aggregates under shear.

Recently, a number of scattering experiments have detected shear-induced demixing in high-molecular-weight PS in DOP [119]–[128]. In addition, van Egmond and co-workers used form birefringence and dichroism [129, 130]. As representative examples we show scattering patterns in the q_x–q_y plane in Fig. 11.14 [119] and in the q_x–q_z plane in Fig. 11.15 [120], and data of form dichroism in Fig. 11.16 [129]. Also elongational (extensional) flow was applied to PS in DOP [131], where fluctuation enhancements were even more dramatic, giving rise to fourfold symmetry in the scattered intensity with intensity maxima on the axes at 45° to the principal axes. Rheological behavior of polymer solutions at phase separation was also studied. In particular, semidilute PS/DOP solutions display shear-thickening at high shear [120, 129] and a second overshoot in the shear stress after application of shear [123, 132]. The latter is caused by the onset of large concentration enhancement, as will be demonstrated by simulations to follow.

Intensive theoretical efforts have been made to understand this complex problem. We mention three main theoretical ingredients being established. They are (i) the dynamical coupling mechanism first applied to sheared polymer solutions by Helfand and Fredrickson [133, 136, 137], (ii) the viscoelastic Ginzburg–Landau scheme [134, 135, 138, 139] with a conformation tensor as a new independent dynamical variable, and (iii) computer simulations [140, 141] which give insights of the behavior of strongly fluctuating polymer solutions under shear. The first two ingredients were discussed in Chapter 7. However, a

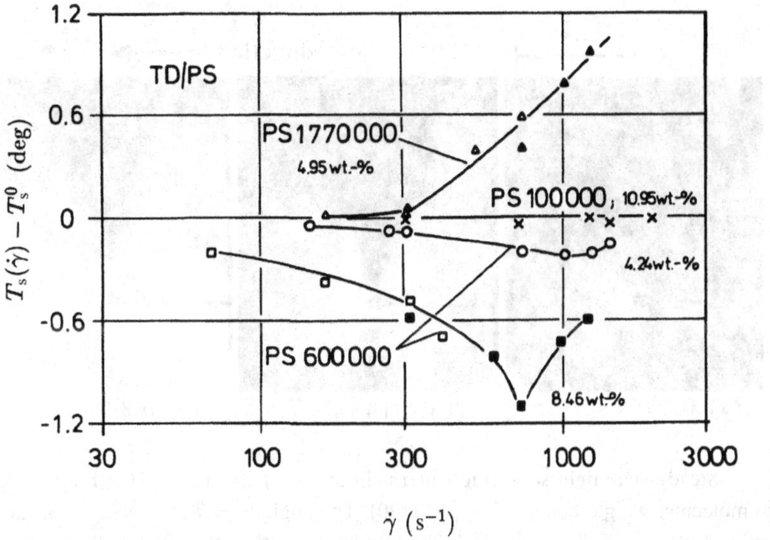

Fig. 11.13. Difference $T_s(\dot{\gamma}) - T_s^0$ vs $\dot{\gamma}$ in PS + trans-decalin (TD), where $T_s(\dot{\gamma})$ is the demixing temperature in shear and T_s^0 is that for the solution at rest. The molecular weights and concentrations are indicated [117]. For positive (negative) values of the difference, shear-induced demixing (homogenization) occurs.

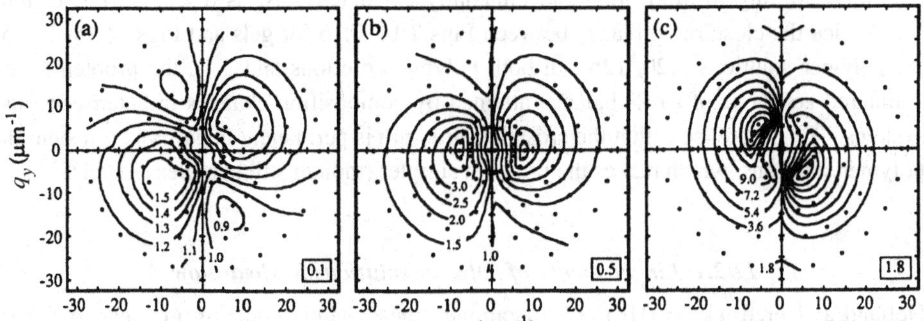

Fig. 11.14. Contour plots of steady-state light scattering amplitudes in the q_x–q_y plane from 4% PS in DOP at 15 °C for (a) $\dot{\gamma} = 0.4\ s^{-1}$, (b) $\dot{\gamma} = 1.2\ s^{-1}$, and (c) $\dot{\gamma} = 10\ s^{-1}$ [119]. The molecular weight is 1.8×10^6 and $\tau = 0.6$ s. In the newtonian regime (a) we can see an abnormal butterfly pattern aligned in the direction of $q_x \cong q_y$. With increasing shear in the region $\dot{\gamma}\tau \gtrsim 1$, the fluctuations are gradually rotated due to convective motion. Numbers to the lower right indicate contour increments.

number of puzzles remain unexplored in spinodal decomposition and nucleation with (and even without) shear, close to and below the coexistence curve.

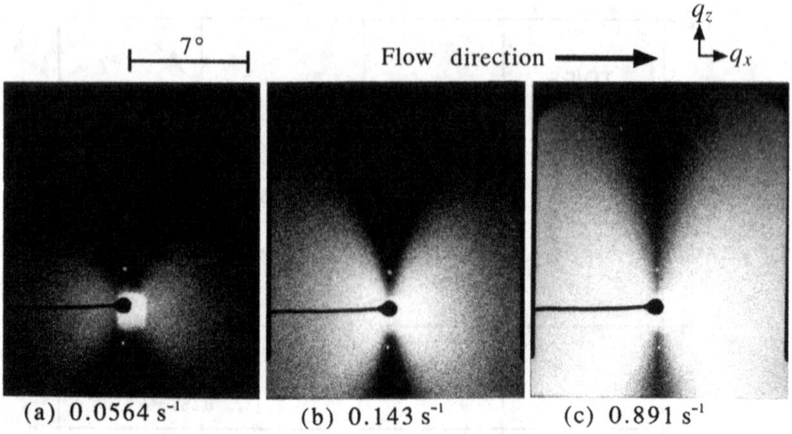

Fig. 11.15. Steady-state light scattering patterns in the q_x–q_z plane from 6% PS in DOP at 27 °C with the molecular weight being 5.5×10^6 [120]. The angle $\theta = 7°$ refers to the scattering angle and corresponds to $q = 1.8 \times 10^4$ cm^{-1}. The number beneath each pattern indicates the shear rate. The patterns have a strong intensity along the flow direction and a *dark streak* along the vorticity direction.

We also remark that similar scattering patterns have been observed by small-angle neutron scattering from swollen and uniaxially expanded gels, as discussed in Section 7.2. Notice the close resemblance between Figs 7.14–7.16 for gels and Figs 11.14–11.15 for polymer solutions [120, 126]. In both polymer solutions and gels, the problems encountered are those of stress balance attained by composition changes in heterogeneous systems. The difference is that the crosslink structure is permanent in gels and transient in polymer solutions, which makes the problem simpler (though still complex) in gels.

11.2.1 Linear theory of polymer solutions in shear flow

Helfand and Fredrickson (HF) [133] examined the dynamic coupling in shear to linear order in the concentration fluctuations by assuming that the stress fluctuations instantaneously follow the concentration fluctuations. Their theory most simply illuminates the mechanism of shear-induced fluctuation enhancement, but it is applicable only at very long wavelengths. Here we present a more general linear theory which is valid in a wider wave vector region and is still analytically tractable. We first consider the newtonian regime [10, 142],

$$\dot{\gamma}\tau \lesssim 1, \tag{11.2.1}$$

where τ is the stress relaxation time behaving as (7.1.30). The fluctuation enhancement is rather mild in the newtonian regime, but it can be drastic in the non-newtonian regime $\dot{\gamma}\tau \gtrsim 1$. Interestingly, such effects become apparent even when $\dot{\gamma}$ is still much smaller

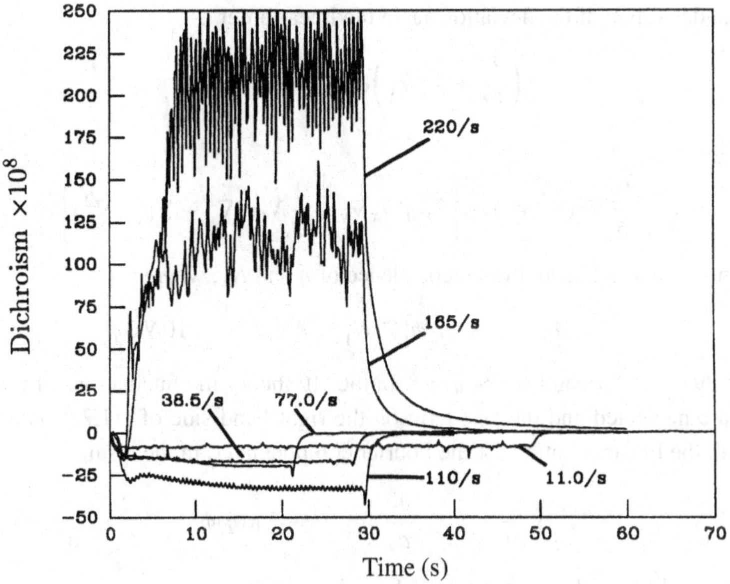

Fig. 11.16. Dichroism vs time from PS in DOP at $T = 20$ °C with the laser beam along the shear gradient (y) axis [129]. Shear was applied at $t = 0$ and stopped at $t = 30$ s. At high shear rates the dichroism changes from negative to positive values because of orientation of the concentration fluctuations along the flow (x) direction.

than the inverse of the diffusion time $t_\xi = \xi^2/D_{\rm m}$ in contrast to the case of near-critical fluids. The correlation length ξ and the mutual diffusion constant $D_{\rm m}$ are defined by (7.1.9) and (7.1.27), respectively. We have $t_\xi \ll \tau$ except very close to the critical point and thus assume $\dot{\gamma} t_\xi \ll 1$ in the semidilute concentration region with theta solvent. In this subsection the temperature region is assumed to be above the coexistence curve, where phase separation does not occur in the absence of shear.

We linearize (7.1.33) with respect to the deviation $\delta\phi$ around a homogeneous state under shear flow. When τ does not exceed the timescale of the deformations under consideration and the inverse shear rate $1/\dot{\gamma}$, we express the network stress $\overleftrightarrow{\sigma}$ as

$$\sigma_{ij} \cong \eta \left(\nabla_j v_{pi} + \nabla_i v_{pj} - \frac{2}{3}\delta_{ij} \nabla \cdot \boldsymbol{v}_p \right) + \frac{1}{3} N_1 \delta_{ij} (2\delta_{ix} - \delta_{iy} - \delta_{iz}), \tag{11.2.2}$$

where η is taken to be the newtonian shear viscosity in the regime $\dot{\gamma}\tau < 1$ dependent on ϕ as in (7.1.29) and $N_1 = \sigma_{xx} - \sigma_{yy} \sim \eta\tau\dot{\gamma}^2$ is the first normal stress difference. We neglect the second normal stress difference $N_2 = \sigma_{yy} - \sigma_{zz}$ and assume $\sum_i \sigma_{ii} = 0$. The key relation arising from using the polymer velocity $\boldsymbol{v}_p$ in (11.2.2) is that $\nabla \cdot \boldsymbol{v}_p$ is related to the time derivative of the deviation $\delta\phi$ in the linear order as

$$\left(\frac{\partial}{\partial t} + \dot{\gamma} y \nabla_x\right)\delta\phi \cong -\phi\nabla\cdot\boldsymbol{v}_p. \tag{11.2.3}$$

Then we find

$$\nabla\cdot\nabla\cdot\overleftrightarrow{\boldsymbol{\sigma}}_p \cong \frac{4\eta}{3}\nabla^2(\nabla\cdot\boldsymbol{v}_p) + \left[2\dot{\gamma}\eta'\nabla_x\nabla_y + \frac{1}{3}N_1'(2\nabla_x^2 - \nabla_y^2 - \nabla_z^2)\right]\delta\phi. \tag{11.2.4}$$

The second term arises from the ϕ dependence of η and N_1, where

$$\eta' = \partial\eta/\partial\phi \sim 6\eta/\phi, \quad N_1' = \partial N_1/\partial\phi \sim 10N_1/\phi \tag{11.2.5}$$

from (7.1.29), (7.1.30), and $N_1 \sim \eta\tau\dot{\gamma}^2$. In the HF theory the fluctuations of the velocity gradient are neglected and the first term on the right-hand side of (11.2.4) is absent. We then obtain the linear equation for the Fourier component $\phi_{\boldsymbol{q}}$ in the form,

$$\left(\frac{\partial}{\partial t} - \dot{\gamma} q_x \frac{\partial}{\partial q_y}\right)\phi_{\boldsymbol{q}} = -\Gamma_{\rm eff}(\boldsymbol{q})\phi_{\boldsymbol{q}}, \tag{11.2.6}$$

where the (modified) relaxation rate is defined by

$$\Gamma_{\rm eff}(\boldsymbol{q}) = \frac{L}{1+\xi_{\rm ve}^2 q^2}\left[q^2(r + Cq^2) - \frac{2\eta'}{\phi}\dot{\gamma}q_x q_y - \frac{N_1'}{3\phi}(2q_x^2 - q_y^2 - q_z^2)\right]. \tag{11.2.7}$$

Here the kinetic coefficient is modified as $L_{\rm eff}(q) = L/[1+\xi_{\rm ve}^2 q^2]$ as in (7.1.68) due to the first term on the right-hand side of (11.2.4), where the viscoelastic length $\xi_{\rm ve}$ defined by (7.1.65) is much longer than ξ as estimated in (7.1.69). The explicit form of the coefficient $r = K_{\rm os}/\phi^2$ is given in (7.1.7), where $K_{\rm os}$ is the osmotic bulk modulus. The mutual diffusion constant $D_{\rm m}$ is related to the kinetic coefficient L as $D_{\rm m} = Lr \sim T/6\pi\eta_0\xi$.

To calculate the steady-state structure factor we add a random source term $\theta_{{\rm R}\boldsymbol{q}}$ on the right-hand side of (11.2.6), where

$$\langle\theta_{{\rm R}\boldsymbol{q}}(t)\theta_{{\rm R}\boldsymbol{q}'}(t')\rangle = 2(2\pi)^d\delta(\boldsymbol{q}+\boldsymbol{q}')L_{\rm eff}(q)q^2\delta(t-t'). \tag{11.2.8}$$

This form assures the Ornstein–Zernike structure factor $I_0(q) = 1/(r + Cq^2)$ in equilibrium. The expression for the steady structure factor $I(\boldsymbol{q})$ is obtained from (11.1.9) if $\Gamma(q)$ and L_0 are replaced by $\Gamma_{\rm eff}(\boldsymbol{q})$ and $L_{\rm eff}(q)$, respectively, where $\boldsymbol{q}(t)$ depends on t as (11.1.10). If we expand $I(\boldsymbol{q})$ in powers of $\dot{\gamma}$ as in (11.1.8), we obtain for $q\xi \ll 1$

$$I(\boldsymbol{q})/I_0(q) = 1 + 2q_x q_y\left[\eta' L_{\rm eff}(q)/\phi - \xi^2\right]\dot{\gamma}/\Gamma_{\rm eff}(\boldsymbol{q}) + \cdots. \tag{11.2.9}$$

Comparing this with (11.1.8), we notice a surprising result even in the linear order with respect to $\dot{\gamma}$. That is, the correction due to the viscoelasticity is much larger than and has a sign opposite to that due to the convection, in accord with a light scattering experiment by Wu *et al.* [119] at small shear $\dot{\gamma}\tau < 1$. The ratio of these contributions is about $-6(\xi_{\rm ve}/\xi)^2$ for $q\xi_{\rm ve} < 1$ from (7.1.65) and (11.2.5). This suggests that the concentration fluctuations tend to be aligned in the directions opposite to those for near-critical fluids.

Similar abnormal alignment perpendicular to the stretched direction has been observed in heterogeneous gels, treated in Chapter 7.

From (11.2.7), $\Gamma_{\text{eff}}(\boldsymbol{q})$ can be negative even for positive r, indicating growth of the fluctuations even above the spinodal curve. In terms of $K_{\text{os}} = \phi^2 r$ this condition becomes

$$K_{\text{os}} < \phi\eta'\dot{\gamma} \sim 6\eta\dot{\gamma} \quad \text{or} \quad K_{\text{os}} < 2\phi N_1'/3 \sim 7N_1, \tag{11.2.10}$$

where the first relation is obtained for $q_x = q_y$ and the second for $q_x \neq 0$ and $q_y = q_z = 0$. In particular, in the newtonian limit $\dot{\gamma}\tau \ll 1$ we may neglect the normal stress effect and the critical shear rate $\dot{\gamma}_{\text{c}}$ is given by

$$\dot{\gamma}_{\text{c}} = \phi r/\eta' \cong K_{\text{os}}/6\eta. \tag{11.2.11}$$

Then some calculations show that the maximum of $-\Gamma_{\text{eff}}(\boldsymbol{q})$ is attained at $q_x = q_y = \pm q_{\text{m}}$ and $q_z = 0$ with

$$q_{\text{m}} \sim (\dot{\gamma} - \dot{\gamma}_{\text{c}})^{1/4} D_{\text{m}}^{-1/4} \xi^{-1/2}, \tag{11.2.12}$$

which increases from 0 and becomes of order $(\xi\xi_{\text{ve}})^{-1/2}$ for $\dot{\gamma} \sim \tau^{-1}$. The maximum growth rate is given by $\Gamma_{\text{m}} = D_{\text{m}}\xi^2 q_{\text{m}}^4$ and becomes of order K_{os}/η for $\dot{\gamma} \sim \tau^{-1}$. Thus the above estimates are self-consistent in theta and poor solvent where $G = \eta/\tau \gtrsim K_{\text{os}}$. However, note that the growth of the fluctuations is transient because $\Gamma_{\text{eff}}(\boldsymbol{q})$ is negative only in a limited wave vector region and the convection brings the wave vector outside this unstable region. We should thus regard the above results to be very approximate.

The above linear theory can explain the experiment by Wu *et al.* at small shear [119], but cannot adequately explain those by Hashimoto's group. In particular, on PS/DOP solutions with $M \sim 5.5 \times 10^6$ [128], Saito *et al.* took data of critical shear rates, $\dot{\gamma}_{\text{c}x}$ and $\dot{\gamma}_{\text{c}z}$, above which the scattering amplitudes in the x and z directions grow abruptly above the thermal level with increasing shear. They found $\dot{\gamma}_{\text{c}x} \sim \tau^{-1}$ consistent with (11.2.11), but $\dot{\gamma}_{\text{c}z}$ was systematically larger than $\dot{\gamma}_{\text{c}x}$ by a factor of 3. As a result, the ratios of the stress values at these two critical shear rates were $(\sigma_{xy})_{\text{c}z}/(\sigma_{xy})_{\text{c}x} \sim 1.6$ and $(N_1)_{\text{c}z}/(N_1)_{\text{c}x} \sim 3.4$. Thus, at $\dot{\gamma} = \dot{\gamma}_{\text{c}z}$, the system was in the non-newtonian regime and the linear theory is inapplicable. More seriously, the diffusion time $1/D_{\text{m}}q^2$ was shorter than $\tau(\sim 50$ s) for most values of q observed. When $Dq^2\tau \gg 1$, sheared polymer solutions should behave like gels under shear strain and another theory is needed. Notice the close resemblance between Fig. 7.16 for gels and Fig. 11.15.

11.2.2 Normal stress effect

We next examine how the normal stress causes diffusion perpendicular to the flow direction [134, 135]. We will first treat polymer solutions, but the following theory is also applicable to polymer blends [143] and dense colloidal suspensions. A similar theory was also developed to discuss flow instability of layered structures aligned in the flow direction [144].

Because the convection makes the mathematics very complex, we assume that all the deviations are small and vary only in the y (velocity gradient) direction as e^{iqy}. To linear order in the composition deviation $\delta\phi$, (7.1.33) becomes

$$\frac{\partial}{\partial t}\delta\phi = -\frac{Lq^2}{1+q^2\xi_{\rm ve}^2}\left(r\delta\phi - \frac{1}{\phi}\delta\sigma_{yy}\right), \tag{11.2.13}$$

where $r = K_{\rm os}/\phi^2$ and σ_{yy} is the yy component of the polymer stress on the order of the first normal stress difference N_1. The deviations are related as

$$\delta\sigma_{yy} = -A_n\delta N_1, \tag{11.2.14}$$

where A_n is of order 1 (equal to 1/3 if the diagonal part $\sum_j \sigma_{jj}/3$ and the second normal stress difference N_2 are neglected). For slow motions we may assume the mechanical equilibrium condition along the x direction,

$$\rho\frac{\partial}{\partial t}v_x = \nabla_y\sigma_{xy} \cong 0. \tag{11.2.15}$$

Hence the shear stress σ_{xy} should be constant in the y direction and the deviation of the shear rate is expressed as

$$\delta\dot\gamma = \left(\frac{\partial\dot\gamma}{\partial\phi}\right)_{\sigma_{xy}}\delta\phi = -\left[\left(\frac{\partial\sigma_{xy}}{\partial\phi}\right)_{\dot\gamma}\Big/\left(\frac{\partial\sigma_{xy}}{\partial\dot\gamma}\right)_{\phi}\right]\delta\phi. \tag{11.2.16}$$

Now the relaxation rate in (11.2.13) is written as $q^2D_y/(1+q^2\xi_{\rm ve}^2)$, where the diffusion constant in the y direction is of the form,

$$D_y = L\left[r + \frac{A_n}{\phi}\left(\frac{\partial N_1}{\partial\phi}\right)_{\sigma_{xy}}\right]. \tag{11.2.17}$$

Next we consider the concentration fluctuations varying in the z direction as e^{iqz}. They induce no velocity gradients varying in the y direction in the linear order. Thus the relaxation rate in the z direction is written as $q^2D_z/(1+q^2\xi_{\rm ve}^2)$ with

$$D_z = L\left[r + \frac{A_n}{\phi}\left(\frac{\partial N_1}{\partial\phi}\right)_{\dot\gamma}\right]. \tag{11.2.18}$$

These diffusion constants may also be expressed as $D_j = (L/\phi)(\partial\Pi^*/\partial\phi)$ $(j = y, z)$, where Π^* is a generalized osmotic pressure defined by

$$\Pi^* = \Pi(\phi) - \sigma_{yy}. \tag{11.2.19}$$

The above Π^* is analogous to the right-hand side of (9.6.9) or (9.6.11) (if the inverse curvature R^{-1} there is set equal to zero).[4]

[4] Here we propose the following experiment. Let us apply a shear flow to a two-phase state with a planar interface parallel to the flow; then, owing to the normal stress effect, the semidilute region will expand and the polymer volume fraction will decrease by $(\dot\gamma\tau)^2\phi \sim [\sigma_{xy}/G]^2\phi$ for $\dot\gamma\tau \lesssim 1$.

(i) In the newtonian regime $\dot\gamma\tau \lesssim 1$, we may set [10, 142]

$$\begin{aligned}\sigma_{xy} &= \eta\dot\gamma = G\tau\dot\gamma,\\ N_1 &= A_1 G(\tau\dot\gamma)^2 = A_1\sigma_{xy}^2/G,\end{aligned} \tag{11.2.20}$$

where A_1 is a constant of order 1. Here $G \propto \phi^p$ with $p = 2$–3 and $\eta \propto \phi^{x_\eta}$ with $x_\eta \sim 6$ for semidilute solutions with theta solvent from (7.1.29) and (7.1.30), so

$$\begin{aligned}D_y &= L[r - pA_n N_1/\phi^2],\\ D_z &= L[r + (2x_\eta - p)A_n N_1/\phi^2].\end{aligned} \tag{11.2.21}$$

Thus the fluctuations varying in the y axis become linearly unstable for

$$K_{\text{os}} = \phi^2 r > pA_n N_1 \sim N_1. \tag{11.2.22}$$

This means that shear-induced demixing occurs in the y direction with increasing shear even for $r > 0$ or in one-phase states [134]. However, (11.2.10) suggests that the fluctuations with $q_x \sim q_y$ and those varying along the x axis should have already been enhanced at this instability point.

(ii) In the non-newtonian regime $\tau\dot\gamma > 1$, it is known that $N_1 > \sigma_{xy}$ and $N_1 \sim G(\dot\gamma\tau)^\beta$ with β smaller than 1 [10, 142]. We conjecture that the composition fluctuations should grow when the typical value of N_1 exceeds K_{os} or

$$G(\dot\gamma\tau)^\beta \gtrsim K_{\text{os}} \quad \text{or} \quad (\dot\gamma\tau)^\beta \gtrsim (T - T_{\text{s}})/(T_{\text{cx}} - T_{\text{s}}), \tag{11.2.23}$$

where T_{s} is the spinodal temperature and T_{cx} is the coexistence temperature (see Section 3.5). Thus fluctuation enhancement readily occurs in highly entangled polymer solutions as the temperature approaches the coexistence curve where $K_{\text{os}} \sim G$.

Polymer blends

In polymer blends, shear-induced mixing and demixing can both occur in the same polymer mixture depending on the composition, temperature, and the shear rate [145]–[148]. To predict the shear effect we should know a number of complex factors such as the strength of the hydrodynamic interaction, the degree of viscoelasticity, or the asymmetry between the two components. For entangled polymer blends, if we assume spatial variations along the y or z direction, (7.1.42) is linearized as

$$\frac{\partial}{\partial t}\delta\phi = -\frac{Lq^2}{1 + q^2\xi_{\text{ve}}^2}(r\delta\phi - \alpha\delta\sigma_{jj}), \tag{11.2.24}$$

where r is given by (7.1.7), α is defined by (7.1.43), and $j = y$ or z. In their theory Clarke and McLeish [143] set $\delta\sigma_{jj} = -\delta N_1/3$ to obtain

$$D_j = L\left(r + \frac{\alpha}{3}\frac{\partial N_1}{\partial\phi}\right), \tag{11.2.25}$$

where the derivative is at fixed σ_{xy} for $j = y$ and at fixed $\dot\gamma$ for $j = z$. In calculating N_1 they used the double reptation model in Appendix 7B and obtained shear-dependent spinodal curves for the fluctuations varying perpendicularly to the flow direction. In a scattering experiment seen along the z direction ($\boldsymbol{q} \parallel$ the z axis) on a blend of PS + PVME by Gerard *et al.* [148], shear-induced demixing exhibited features remarkably similar to normal spinodal decomposition in quiescent states. These include an initial increase of scattered intensity with time and a maximal growth rate at $q = q_{\rm m}$. They analyzed their data using the diffusion constant along the z direction of the form $D_z = a_1(T - T_{\rm s}) + a_z\dot{\gamma}^2$, where $a_z < 0$ for shear-induced demixing. Theoretically [143], a_z can be both positive and negative in polymer blends, while it is positive in polymer solutions as given in (11.2.21).

Slipping layer formation in colloidal suspensions

For some time, considerable attention has been paid to the migration or diffusion of polymers [149, 150] or colloidal particles [151]–[155] in the velocity gradient direction. Simulations have also been performed on this effect for colloids including the hydrodynamic interaction [156]. In particular, to describe plug-flow formation in concentrated colloidal suspensions flowing through a capillary, Nozières and Quemada [155] proposed a diffusion equation for the colloid volume fraction ϕ varying in the y direction:

$$\frac{\partial}{\partial t}\phi = \nabla_y\left[a\phi\left(\nabla_y\mu + \frac{1}{2}b\nabla_y\dot{\gamma}^2\right)\right], \tag{11.2.26}$$

where $\mu = \mu(\phi)$ is the chemical potential of colloids dependent on ϕ. The coefficients a and b may depend on ϕ. This equation is analogous to (11.2.13); the term proportional to $\dot{\gamma}^2$, which was called a *lift force*, corresponds to that proportional to $\delta\sigma_{yy}/\phi$ in (11.2.13). For slow motions, the mechanical equilibrium condition $\sigma_{xy} = \eta\dot{\gamma} =$ const. is satisfied. The viscosity grows sharply towards a close packing volume fraction $\phi_{\rm m}$ as

$$\eta = \eta_0(1 - \phi/\phi_{\rm m})^{-x_\eta}, \tag{11.2.27}$$

where $x_\eta \sim 2$. The diffusion constant for infinitesimal deviations around a homogeneous steady state is then written as

$$D_y = a\phi\left(\frac{\partial\mu}{\partial\phi} - b\dot{\gamma}^2\frac{\partial}{\partial\phi}\ln\eta\right), \tag{11.2.28}$$

where the second term is negative as in (11.2.21). Hence, with increasing $\dot{\gamma}$, D_y becomes negative, leading to the formation of a slipping layer containing a small volume fraction of colloids near the boundary. In such a phase-separated state Nozières and Quemada introduced a modified chemical potential,

$$\mu^* = \mu(\phi) + \frac{1}{2}b\dot{\gamma}^2 = \mu(\phi) + \frac{1}{2}b\sigma_{xy}^2/\eta^2, \tag{11.2.29}$$

where b was treated as a constant. They could then write a schematic phase diagram at fixed σ_{xy}. Furthermore, they assumed homogeneity of $\mu^* - C\nabla_y^2\phi$ in space including the interface region, which is analogous to the interface equation (4.4.1) for the gas–liquid phase transition. In order to obtain a more plausible phase diagram, we here assume $\mu(\phi) = v_0^{-1}T\{\ln[\phi/(1 - \phi/\phi_{\rm m})] + 1/(1 - \phi/\phi_{\rm m})\}$, which is the result in the van der

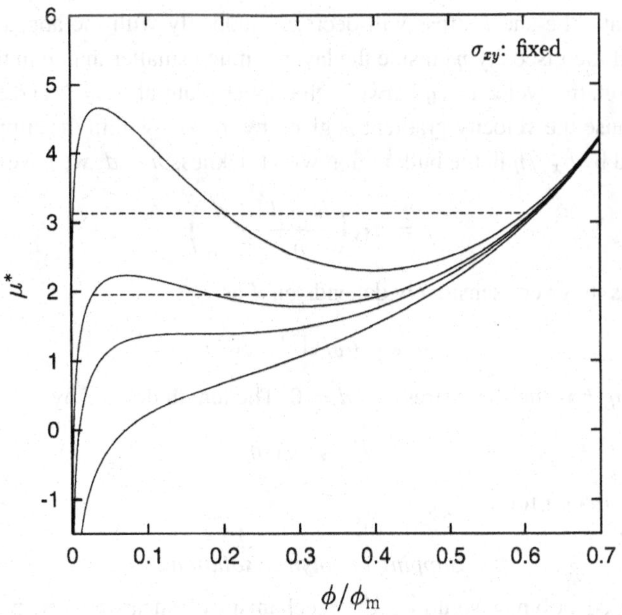

Fig. 11.17. The effective chemical potential μ^* in (11.2.30) (in units of $v_0^{-1}T$) vs ϕ/ϕ_m, where $A = 2$, A_c, 5, and 8 reading from below. The portion of the curves with $\partial\mu^*/d\phi < 0$ is produced by the lift force, and D_y in (11.2.28) vanishes at the spinodal points where $\partial\mu^*/\partial\phi = 0$. The dashed lines are obtained from the Maxwell construction.

Waals theory in Section 3.4 with v_0 being the volume of a colloidal particle. Together with the assumption $b = $ const. (which is problematic, however), we obtain

$$\mu^* = v_0^{-1}T\left[\ln\left(\frac{\Phi}{1-\Phi}\right) + \frac{1}{1-\Phi} + A(1-\Phi)^4\right], \tag{11.2.30}$$

where $\Phi = \phi/\phi_m$, $A = v_0 b\sigma_{xy}^2/2T\eta_0^2$, and use is made of (11.2.27). In Fig. 11.17 we plot μ^* vs Φ for various A. Two-phase coexistence is achieved for $A > A_c = (3/2)(6/5)^5 \cong 3.73$.

In colloidal suspensions, however, $\dot\gamma^{-1}$ becomes the only timescale for $Pe \gg 1$ [156] where

$$Pe = (6\pi\eta_0 a^3/T)\dot\gamma, \tag{11.2.31}$$

is the Peclet number, a being the particle radius. As a result the normal stress difference N_1 should behave as

$$N_1 \sim \eta_0\dot\gamma \quad (Pe \gg 1). \tag{11.2.32}$$

In this case we should have $\dot\gamma$ instead of $\dot\gamma^2$ in (11.2.26) and (11.2.28), but this does not change the results qualitatively.

In experiments, the shear stress will decrease suddenly with the appearance of a thin slipping layer if the viscosity η_0 inside the layer is much smaller than η in the bulk region. Let us fix the relative velocity v_0 between the upper plate at $y = h$ and the lower plate at $y = 0$. Because the velocity gradient is given by σ_{xy}/η_0 within the slipping layer with thickness d and by σ_{xy}/η in the bulk region with thickness $h - d$, we have

$$v_0 = \sigma_{xy}\left(\frac{h-d}{\eta} + \frac{d}{\eta_0}\right). \qquad (11.2.33)$$

The shear stress thus very sensitively depends on d as

$$\sigma_{xy} = \sigma_0/(1 + d/d^*), \qquad (11.2.34)$$

where $\sigma_0 = v_0\eta/h$ is the shear stress for $d = 0$. The length defined by

$$d^* = h\eta_0/\eta \qquad (11.2.35)$$

is much shorter than h for $\eta_0 \ll \eta$.

Slipping in polymer solutions

Also in entangled polymer solutions the mechanism of shear-induced phase separation might be relevant to slippage, though the above simple theory for colloids will be inadequate. In an experiment on entangled PS in *good* solvent [157], marked composition inhomogeneities varying on the wall plane appeared close to the wall and traveled into the bulk with the occurrence of slippage. This phenomenon was found to be strongly influenced by the interaction between the polymer and the surface.

11.2.3 Thermodynamic theory on sheared polymer solutions

Rangel-Nafaile *et al.* [116] developed a thermodynamic theory of shear-induced phase separation. They assumed that the total free-energy density consists of the Flory–Huggins free-energy density and a stored elastic energy density f_{el} on the order of N_1. Such a form of the free energy was suggested by Marrucci's work [158] on the dumbbell model. Then a spinodal curve was determined by $K_{os} + \phi^2\partial^2 f_{el}/\partial\phi^2 = 0$, where the derivative with respect to ϕ was performed with the shear stress held fixed. However, the second derivative of f_{el} is positive, leading to a downward shift of the spinodal if ϕ is much larger than a critical entanglement volume fraction ϕ^*. Conversely, if the problem is treated dynamically, the shift due to the stress–diffusion coupling is definitely upward. Nevertheless, the absolute value of the shift is determined by $|K_{os}| \sim N_1$ from the thermodynamic assumptions in accord with (11.2.10) or (11.2.22). We believe that it is appropriate to introduce the concept of the stored free energy or the elastic free energy to describe viscoelastic fluids. In the thermodynamic theories [116, 159, 160], however, the usual scheme of thermodynamics is assumed and space-dependent fluctuations are not adequately taken into account. Jou and co-workers [160] stressed that thermodynamic arguments, if improved, can be useful in understanding shear effects in polymers.

11.2.4 Simulation of shear-induced phase separation: elastic turbulence

We need a numerical approach to understand the nonlinear regime of shear-induced phase separation. To this end, the viscoelastic Ginzburg–Landau model in (7.1.98)–(7.1.105) was solved in the presence of shear flow in 2D [14, 140] using a numerical scheme [161] similar to that used by Lee and Edwards for nonequilibrium molecular dynamics (MD) simulations [162]. A simpler approach based on smoothed particle hydrodynamics also produced similar results [141]. We integrate (7.1.33) for ϕ and (7.1.100) for the conformation tensor $\overleftrightarrow{W}$ on a 128×128 square lattice, where the relative velocity $\boldsymbol{w}$ and the average velocity $\boldsymbol{v}$ are given by (7.1.104) and (7.1.105), respectively. A shear flow $\langle v_x \rangle = \dot{\gamma} y$ is applied at $t = 0$. We shall see that the small-scale fluctuations emerging due to the viscoelastic instability grow in magnitude and spatial size but are eventually broken by the flow. Phase separation is then only partially completed, resulting in a chaotic dynamical steady state with large fluctuations in the composition and stress. Here we will present simulated physical images, which can be obtained only through numerical work at present, mentioning related experiments.

Shear-induced composition fluctuations above the coexistence curve

We first assume that our system is above the coexistence curve as [14]

$$\langle \Phi \rangle = 2, \quad T = T_c \quad \text{or} \quad \chi - \frac{1}{2} = N^{-1/2}, \tag{11.2.36}$$

where N is the polymerization index. Hereafter we set

$$\Phi = \phi/\phi_c, \tag{11.2.37}$$

where $\phi_c = N^{-1/2}$. The coefficient of the gradient free energy is written as $C = (T/v_0)C_0/\phi$ with $C_0 = a^2/18$ from (4.2.26). The thermal correlation length and the mutual diffusion constant in the equilibrium state determined by (11.2.36) are written as $\xi = (NC_0/5)^{1/2}$($\sim$ the gyration radius) and D_m, respectively. We measure space and time in units of $\ell = (5/3)^{1/2}\xi$ and $\tau_0 = (25/6)\xi^2/D_m$. The shear modulus and the stress relaxation time are set equal to

$$G = T v_0^{-1} \phi^3, \quad \tau = 0.3\tau_0(\Phi^4 + 1). \tag{11.2.38}$$

The solvent viscosity is taken to be $\eta_0 = (T/v_0)\phi_c^3\tau_0$, which is equivalent to assuming the friction coefficient as $\zeta = \eta_0\phi^2/C_0$. Then the newtonian solution viscosity and the relaxation time are written as

$$\eta/\eta_0 = 0.1\Phi^3(\Phi^4 + 1), \quad \tau/\tau_0 = 0.3(\Phi^4 + 1). \tag{11.2.39}$$

We have $\eta/\eta_0 = 13.6$ and $\tau/\tau_0 = 5.1$ in equilibrium determined by (11.2.36), but G, τ, and η are fluctuating quantities in nonequilibrium. In our case G is considerably larger and τ is much smaller than those in the real experiments. We also add random source terms in the dynamic equations; as a result, the equilibrium distribution is expressed as

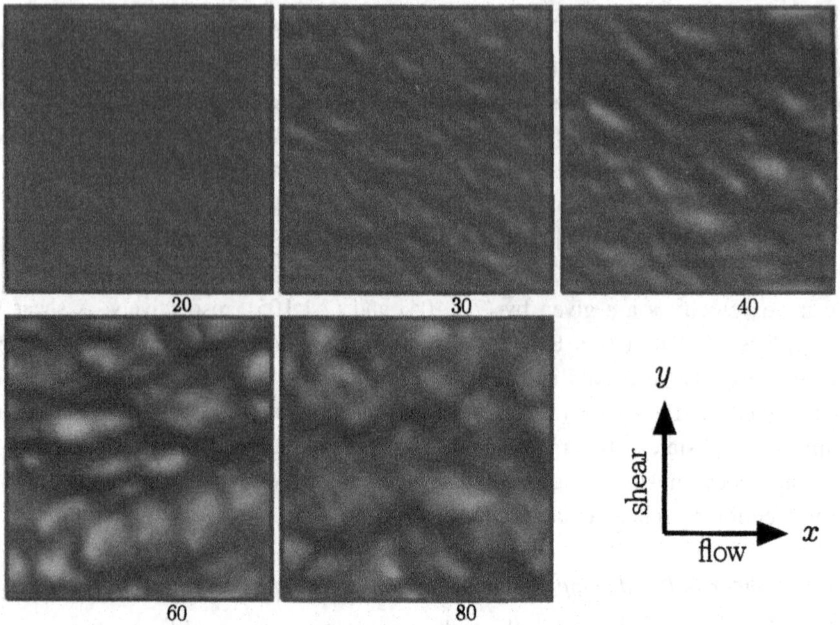

Fig. 11.18. Time evolution of $\Phi(x, y, t) = \phi(x, y, t)/\phi_c$ after application of shear $\dot{\gamma}\tau = 0.25$ at $t = 0$ [140]. The numbers below the figures are the times measured in units of $\tau_0 = 2.5\ell^2/D_m$. The space region in our simulations is given by $0 < x, y < 128$, where the space coordinates are measured in units of $\ell = (5/3)^{1/2}\xi$, ξ being the correlation length in equilibrium. The shading represents $[\Phi(x, y, t) - \Phi_{min}]/(\Phi_{max} - \Phi_{min})$ with $\Phi_{max} \cong 3.6$ and $\Phi_{min} \cong 0.38$ being the maximum and the minimum of $\Phi(x, y, t)$ at these times.

$\exp(-\tilde{\mathcal{H}}/\epsilon^2)$, where $\tilde{\mathcal{H}}$ is a dimensionless Ginzburg–Landau free energy with the space unit being ℓ. In this work we set $\epsilon = 0.1$ and the variance

$$\mathcal{V} = \sqrt{\langle(\Phi - \langle\Phi\rangle)^2\rangle} \tag{11.2.40}$$

taken over all the lattice points turns out to be 0.038 in thermal equilibrium.

We display snapshots of $\Phi(x, y, t)$ at various times in Figs 11.18 and a 3D graphical representation in Fig. 11.19, respectively, after application of shear $\dot{\gamma}\tau = 0.25$ at $t = 0$. At an early stage ($t \lesssim 40$) we can see growth of the fluctuations with wave vectors with $q_x \cong q_y$ in agreement with the linear theory. At a later time the polymer-rich regions are elongated into long stripes forming a transient network and are continuously deformed by hydrodynamic convection on the timescale of $1/\dot{\gamma}\tau_0 (= 20)$. We also notice that $\Phi(x, y, t)$ varies irregularly even on the mesh size scale ℓ in regions with $\Phi(x, y, t) \gtrsim 2$, whereas it varies smoothly in space in regions where $\Phi(x, y, t)$ is considerably below 2. This is obviously because viscoelasticity is weakened in the latter regions. The structure factor $I(q_x, q_y, t)$ is much enhanced at small q, but is fluctuating in time in our calculation

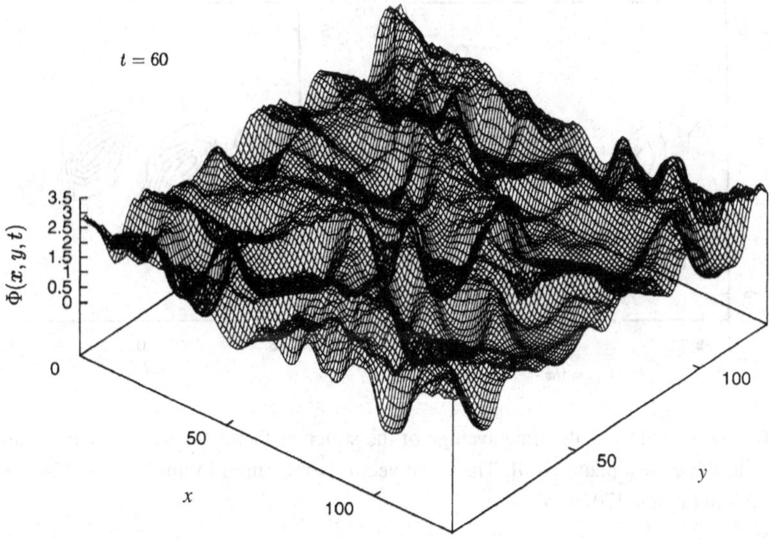

Fig. 11.19. 3D graphical representation of $\Phi(x, y, t)$ for $\dot{\gamma}\tau = 0.25$ at $t = 60$ under the same conditions as in Fig. 11.18, showing turbulent enhancement of the concentration fluctuations comparable to those in spinodal decomposition [140].

because of the small system size. In Fig. 11.20 its time average taken over the interval $150 < t < 1000$ is shown for $\dot{\gamma}\tau = 0.1$ in (a) and for $\dot{\gamma}\tau = 0.25$ in (b). We can see two peaks in the q_x–q_y plane in the steady state in accord with the scattering experiment [119]. At smaller shear they are located at $q_x \cong q_y$, while they approach the q_x axis as shear is increased. In Fig. 11.21 we show the variance $\mathcal{V}$ defined by (11.2.40), which increases from the equilibrium value 0.038 and fluctuates around 0.5 in the dynamical steady state.

Stress fluctuations

In Fig. 11.21 we show the space averages of the shear stress and normal stress difference,

$$\begin{aligned} \bar{\sigma}_{xy} &= \langle \sigma_{xy} \rangle - \langle C(\nabla_x\phi)(\nabla_y\phi) \rangle, \\ \bar{N}_1 &= \langle \sigma_{xx} - \sigma_{yy} \rangle + \langle C[(\nabla_y\phi)^2 - (\nabla_x\phi)^2] \rangle, \end{aligned} \tag{11.2.41}$$

where the tensor σ_{ij} is treated as a fluctuating quantity defined by (7.1.102). At high shears the stress components due to viscoelasticity ($\propto \langle \sigma_{ij} \rangle$) are much larger than those from the gradient free energy ($\propto \langle C\nabla\phi\nabla\phi \rangle$), while the latter ones are dominant singular contributions in low-molecular-weight fluids. The average shear stress first grows linearly in time up to the order of $\eta\dot{\gamma}$ at $t \sim \tau$, but it begins to decrease with growth of the shear-induced fluctuations. The normal stress difference grows as t^2 initially. After the transient stage they both exhibit chaotic fluctuations. Figure 11.22 displays $\bar{N}_1$ (divided by $\eta_0/3\tau_0$) for $0 < t < 800$ at $\dot{\gamma}\tau = 0.05, 0.1$, and 0.25. At the largest shear $\dot{\gamma}\tau = 0.25$, the network composed of elongated polymer-rich regions is often extended throughout the

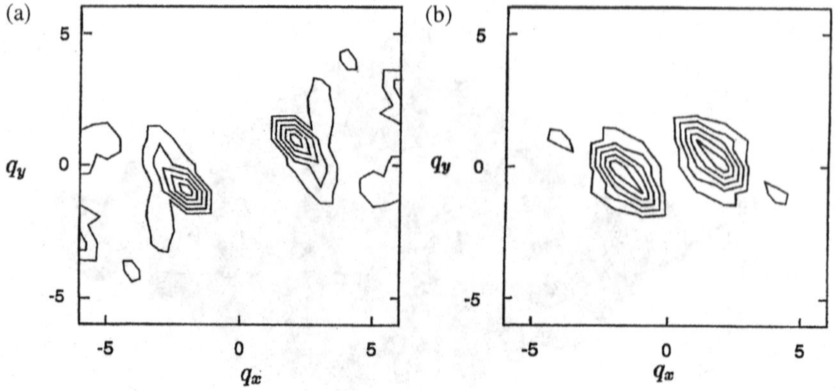

Fig. 11.20. Contour plots of the time average of the structure factor for $\dot{\gamma}\tau = 0.1$ in (a) and $\dot{\gamma}\tau = 0.25$ in (b) in the q_x–q_y plane [140]. The wave vector is measured in units of $2\pi/128\ell$. The peak height is 15.7 in (a) and 470 in (b).

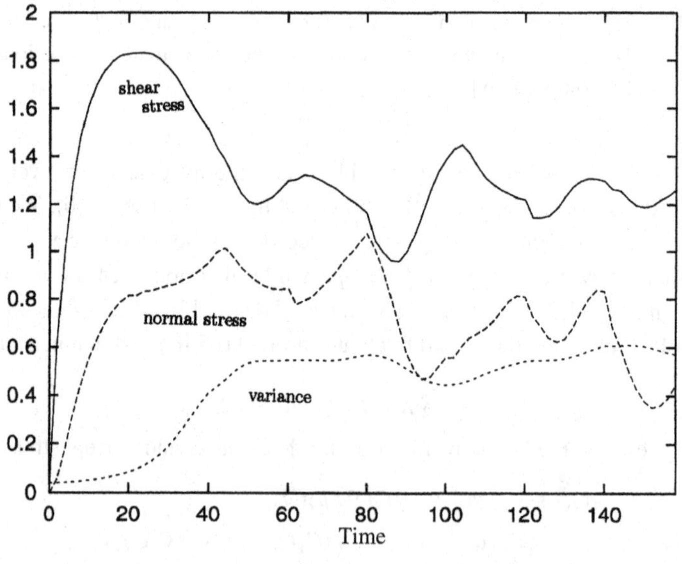

Fig. 11.21. Time evolution of the variance $\mathcal{V}$ defined by (11.2.40) (dotted line), the average shear stress $\bar{\sigma}_{xy}$ (solid line), and the average normal stress difference $\bar{N}_1$ (broken line) for $\dot{\gamma}\tau = 0.25$ [140]. The stress components and the time are scaled appropriately.

system but is subsequently disconnected. Because the stress is mostly supported by such a network, this process produces abnormal fluctuations of the stress. Interestingly, in many cases the normal stress difference takes a maximum (or minimum) when the shear stress takes a minimum (or maximum) [14].

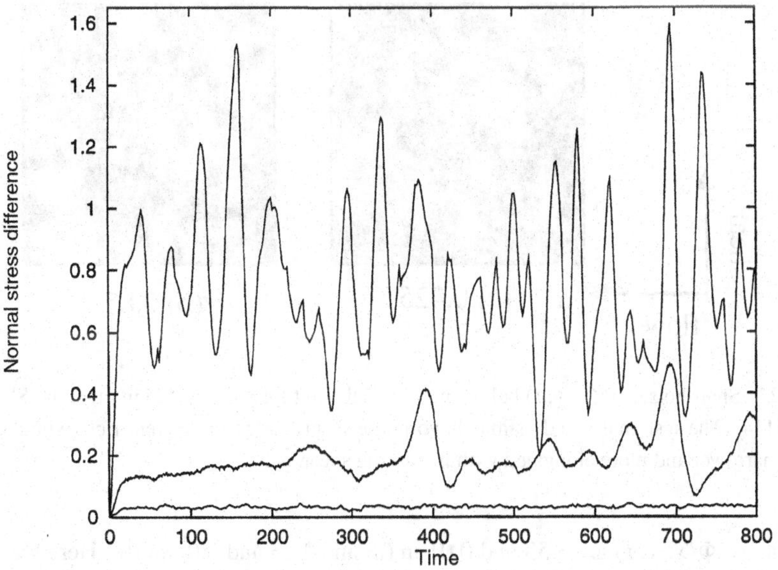

Fig. 11.22. Chaotic time evolution of the average normal stress difference as a function of time for $\dot{\gamma}\tau = 0.05$, 0.1, and 0.25 reading from below [140]. The average shear stress also exhibits similar behavior.

In experiments, the stress components are measured as the force density acting on a surface with a macroscopic linear dimension h. If h is much longer than the characteristic size of the network structure, the observed stress components will exhibit only small temporal fluctuations. More than four decades ago Lodge [6] reported abnormal temporal fluctuations of the normal stress difference at a hole of 1 mm diameter for polymer solutions contained in a cone-plate apparatus. He ascribed its origin to growth of inhomogeneities or gel-like particles. Peterlin and Turner [7] suggested temporary network formation in sheared polymer solutions to explain their finding of a maximum in the shear stress after application of shear. In subsequent measurements [118, 123, 132], σ_{xy} and N_1 have exhibited a peak at a relatively short time (first overshoot) arising from transient stretching of polymer chains and a second peak (second overshoot) arising from shear-induced phase separation. In our dynamic model we are neglecting the former relaxation process, so our first overshoots correspond to the observed second overshoots. It would be informative if further rheological experiments could be performed at various temperatures including the case below the spinodal point or in small spatial regions, as in Lodge's case [6].

Strongly deformed composition fluctuations below the coexistence curve

We also simulated a quench of the system at $N^{1/2}(2\chi - 1) = 3$, which is below the classical spinodal value $N^{1/2}(2\chi - 1) \cong 2.5$, with the same volume fraction $\langle\Phi\rangle = 2$ [14]. Figure 11.23 shows snapshots for $\dot{\gamma}\tau = 0.425$ in (a) and 0.85 in (b) at $t = 200$. The maximum and

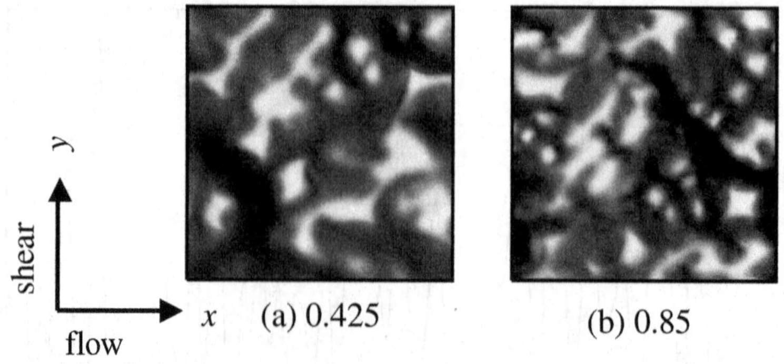

Fig. 11.23. Snapshots of $\Phi(x, y, t)$ below the spinodal point for $\dot{\gamma}\tau = 0.425$ in (a) and 0.85 in (b) at $t = 200$ [14]. The system is in a dynamical two-phase steady state. The solvent-rich (white) domains become narrower and more elongated with increasing shear.

minimum of $\Phi(x, y, t)$ are 3.53 and 0.01 in (a) and 3.23 and 0.01 in (b). Here we can see formation of sharp interface structures and continuity of the polymer-rich (dark) regions. In the droplet-like solvent-rich regions Φ becomes very small, whereas in the continuous polymer-rich regions it increases slowly in time because deswelling of solvent is taking place there, as in gels. At relatively large shear, the system tends to a two-phase dynamical steady state, where the solvent-rich regions are narrow and compressed.

However, for very small shear and deep quenching, we found that the system is ultimately divided into two regions, one mostly with solvent and the other being polymer-rich. In transient time regions in such cases, solvent-rich regions are very easily deformed by shear into extended shapes and the shear stress decreases abruptly once such solvent-rich regions are percolated throughout the system. (A gas droplet in a newtonian viscous liquid can be elongated into a slender shape in shear flow [57]–[59].) Here thin solvent-rich regions should act as a lubricant serving to diminish the measured viscosity. This picture was originally presented by Wolf and Sezen [115] to interpret their finding of a viscosity decrease which signals the onset of phase separation at small shear in semidilute solutions.

11.3 Complex fluids at phase transitions in shear flow

There are a large number of intriguing examples of nonlinear shear effects in complex fluids undergoing some kind of phase transition [1]–[10]. We mention them here without detailed discussions. (i) In colloidal systems, even when a relatively weak experimentally producible shear is applied, the structure of the phase can be changed drastically. In particular, shear-induced melting of crystal structures has been studied by scattering experiments [163]–[165]. Some theoretical approaches have also been presented [166]. At a gas–liquid critical point in colloidal systems, the critical fluctuations are extremely sensitive to shear [21]. The viscosity was reported to increase strongly in such colloidal systems [167] and

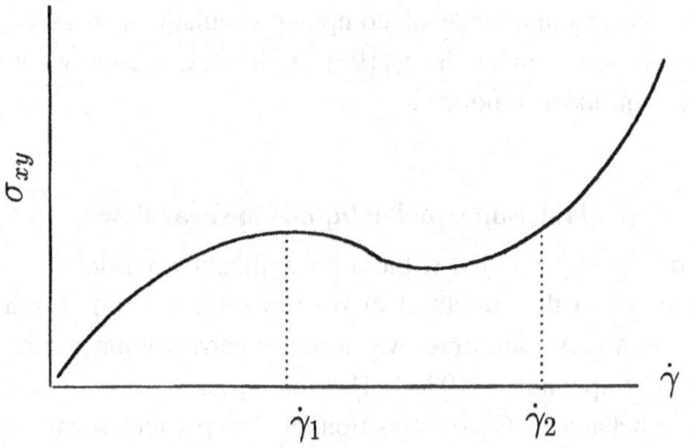

Fig. 11.24. A theoretical schematic diagram of shear stress vs shear rate in the steady state for an entangled micellar system [173]. With increasing shear a shear-banding instability occurs at $\dot{\gamma} = \dot{\gamma}_1$.

in dense microemulsions near the percolation threshold [168]. (ii) Phase transitions in fluids with complex internal structure and long-range order are very sensitive to shear. Examples are various mesoscopic phases of liquid crystals [169]–[172], amphiphilic systems [173]–[176], and block copolymers [97]–[99]. It is obvious that structures such as lamellae or cylinders are easily aligned by relatively weak shear. Even their structures and phase behavior can be altered by shear near the transition point. For example, shear can induce transitions between phases of lamellae and monodisperse multilamellar vesicles [174] and between isotropic and nematic phases, giving rise to two-phase coexistence in inhomogeneous flow [175]. The latter phenomenon can be understood from a steady-state stress–strain curve of the type shown in Fig. 11.24 [173]. Also spectacular is the shear-thickening behavior in worm-like micelles induced by shear-induced structures [177, 178]. (iii) We also mention electro-rheological and ferromagnetic fluids, in which string-like structures of colloidal particles are formed due to dipolar interaction under an electric or a magnetic field. They exhibit unique rheology and phase behavior in shear flow [179, 180]. (iv) Less studied in physics, but important in polymer science are crystallization [181, 182] and gelation [4], [183]–[186] of polymers in a flow field. In particular, molecular theory of thermoreversible gels in shear flow is worth mentioning [185]. In aqueous surfactant solutions, marked increases of the viscosity and N_1 were observed, which were interpreted as arising from shear-induced aggregate formation or gelation [177]. In aqueous agarose solutions, huge viscosity enhancement was also observed, in which gelation was probably induced upon phase separation [187]. (v) We also mention boundary effects such as slipping between a viscoelastic fluid and a solid boundary [188, 189]. Furthermore, application of shear has become possible on molecular systems inserted between two solid plates of spacing on the order of 10 Å. In such confined systems, measurements of the shear stress give information on shear-induced melting of a solid phase and nonlinear rheology of a fluid phase [190, 191].

Finally, we stress the importance of computer simulations in understanding various complex problems of fluids under shear [192]. In the next section we will discuss a new examples of the use of this technique.

11.4 Supercooled liquids in shear flow

When fluids are deeply supercooled without crystallization, particle motions are severely restricted or jammed and the structural or α relaxation time τ_α increases dramatically from a very short to a very long time over a rather narrow temperature range ($T \sim T_g$, the glass transition temperature) [193, 194]. Since $\eta \propto \tau_\alpha$, a high value of τ_α leads to highly viscoelastic behavior.[5] Glass transitions are of particular importance in polymer science [195, 196]. Recently, much attention has been paid to the mode coupling theory of glass transitions [197, 198], which is the first analytic scheme to describe the onset of slow structural relaxations at temperatures considerably above T_g. For a long time, however, it has been expected [199]–[201] that rearrangements of particle configurations in glassy materials should be cooperative, involving many molecules, owing to configuration restrictions. In other words, such events occur only in the form of *clusters* whose sizes increase at low temperatures. In normal liquid states, on the contrary, rearrangements are frequent and uncorrelated among one another in space and time. Such an idea was first put forward by Adam and Gibbs [199], who invented a frequently used jargon, *cooperatively rearranging regions*. A number of molecular dynamics (MD) simulations have detected mobile clusters or strings in coexistence with immobile regions in supercooled model binary fluid mixtures using various visualization methods [202]–[207]. We shall see that such heterogeneities are analogous to the critical fluctuations in Ising systems.

Most previous papers on glass transitions are concerned with near-equilibrium properties such as relaxations of the density time-correlation functions or dielectric response. From our point of view, these quantities are too restricted or indirect, and there remains a rich group of unexplored problems in far-from-equilibrium states. Here we shall see that shear is a relevant perturbation, drastically changing the glassy dynamics when $\dot{\gamma}$ exceeds τ_α^{-1} [205, 208]. In this sense, applied shear is analagous to a magnetic field in Ising systems. In near-critical fluids and various complex fluids, nonlinear shear regimes emerge when $\dot{\gamma}$ exceeds some underlying relaxation rate. However, uniquely in supercooled liquids, even very small shear can greatly accelerate the *microscopic* rearrangement processes. Similar effects are usually expected in systems composed of very large elements such as colloidal suspensions.

As shown in Fig. 11.25, Simmons *et al.* [209] observed shear-thinning behavior roughly represented by

$$\eta(\dot{\gamma}) = \sigma_{xy}/\dot{\gamma} \cong \eta(0)/(1 + \dot{\gamma}\tau_\eta), \tag{11.4.1}$$

in steady states in the range $6 \times 10^{13} > \eta(0) > 7 \times 10^5$ poise in soda–lime–silica glass.

[5] In experiments, the glass transition temperature T_g is determined such that the (zero-shear) viscosity η becomes 10^{13} poise at $T = T_g$ [194]. In the simulations cited here, η (or τ_α) is *only*, at most, 10^4 times larger than that far above the glass transition.

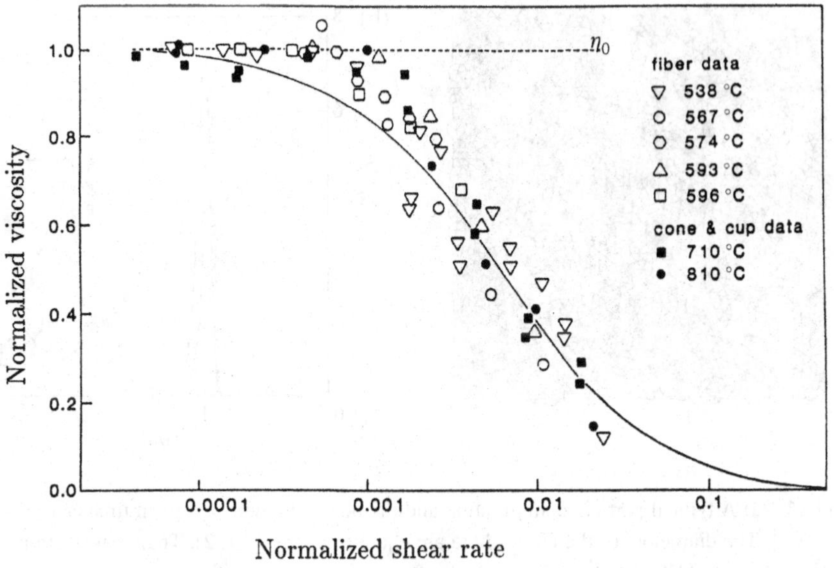

Fig. 11.25. Normalized viscosity $\eta(\dot{\gamma})/\eta(0)$ vs normalized shear rate $\dot{\gamma}\tau_0$ measured in viscous flow in soda–lime–silica glass [209]. Here τ_0 is equal to $\eta(0)/G = \tau_\eta \sigma_{\text{lim}}/G \sim 10^{-2}\tau_\eta$ in terms of the shear modulus G and the limiting shear stress σ_{lim}. The solid curve represents (11.4.1).

The characteristic time τ_η is expected to be of order τ_α. Remarkably, σ_{xy} tends to a limiting shear stress, $\sigma_{\text{lim}} = \eta(0)/\tau_\eta$, of order $10^{-2}G$, G being the shear modulus. After application of shear, they also observed an overshoot of the shear stress before approach to a steady state. As a closely related problem, understanding of the mechanical properties of amorphous metals has been of great technological importance [210]. They are usually ductile in spite of their high strength. At low temperatures $T \lesssim 0.6 \sim 0.7T_{\text{g}}$, localized bands, where zonal slip occurs, have been observed above a yield stress. Their thickness grows from microscopic to mesoscopic lengths. At relatively high temperatures $T \gtrsim 0.6 \sim 0.7T_{\text{g}}$, however, shear deformations are induced *homogeneously* (on macroscopic scales) throughout samples, giving rise to viscous flow with strong shear-thinning behavior. In particular, in a model amorphous metal in 3D, Maeda and Takeuchi [202] followed atomic motions after application of a small shear strain to observe heterogeneities among poorly and closely packed regions (on microscopic scales), which are essentially the same entities that we will discuss.

Another interesting issue is as follows. Several experiments have revealed that the translational diffusion constant D of a tagged particle in a fragile glassy matrix becomes increasingly larger than the Einstein–Stokes value $D_{\text{ES}} = T/6\pi\eta a$ with lowering T, where a is the radius of the particle [211, 212]. At sufficiently low temperatures power law behavior is observed,

$$D \propto \eta^{-\nu}, \tag{11.4.2}$$

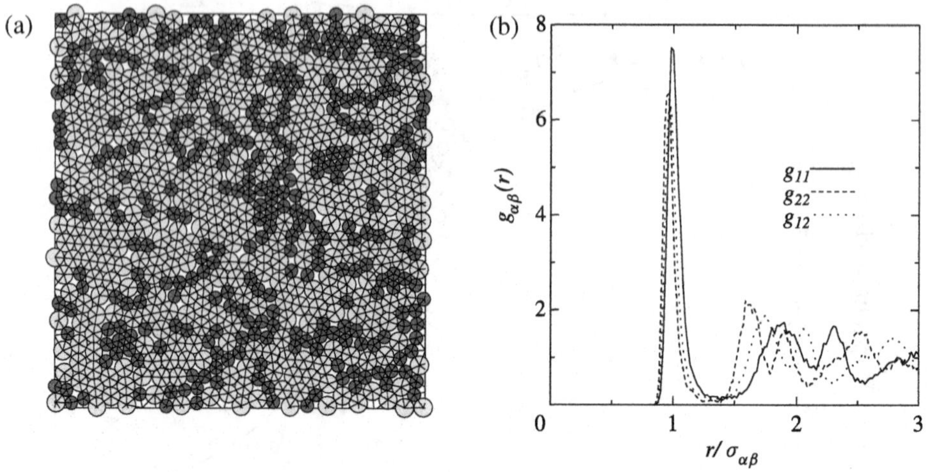

Fig. 11.26. (a) A typical particle configuration and the bonds defined at a given time at $T = 0.337$ in 2D [205]. The diameters of the circles here are equal to σ_α ($\alpha = 1, 2$). The areal fraction of the soft-core regions is 93%. (b) The pair correlation functions $g_{\alpha\beta}(r)$ in quiescent states as functions of $r/\sigma_{\alpha\beta}$ at $T = 0.337$ in 2D [205].

where $\nu \cong 0.75$. Thus $D/D_{\rm ES}$ increases from of order 1 up to order $10^2 \sim 10^3$ in supercooling experiments. The same tendency has been confirmed by MD simulations [213]–[215]. Its origin is now ascribed to the coexistence of relatively active and inactive regions within which the diffusion constant varies significantly.

11.4.1 Model system and glassy slowing-down

We will discuss dynamic heterogeneity detected in simulations of a model binary fluid mixture consisting of $N_1 = N_2 = 5000$ particles and interacting via the soft-core potential [216],

$$v_{\alpha\beta}(r) = \epsilon(\sigma_{\alpha\beta}/r)^{12}, \qquad \sigma_{\alpha\beta} = \frac{1}{2}(\sigma_\alpha + \sigma_\beta), \tag{11.4.3}$$

where r is the distance between two particles and $\alpha, \beta = 1, 2$. Space and time are measured in units of σ_1 and $\tau_0 = (m_1\sigma_1^2/\epsilon)^{1/2}$, where m_1 is the mass of the species 1. The temperature T will be measured in units of ϵ, so it will be a dimensionless number. The size ratio σ_2/σ_1 is chosen to prevent crystallization at low T (which is 1.4 in 2D and 1.2 in 3D in the following). The pressure p and the number density $n(\sim\sigma_1^{-d})$ need to be high to realize jammed particle configurations. We apply shear in nonequilibrium MD simulations imposing the Lee–Edward boundary condition [162]. For our model, no essential differences have been found between 2D and 3D (except for a difference in the dynamic exponent z in (11.4.7) below). Binary fluid mixtures interacting via the Lenard-Jones potential have also been used to study glassy dynamics [207].

Because of the convenience of visualization in 2D, we first present a snapshot of particles at $T = 0.337$ in 2D in Fig. 11.26(a), which gives an intuitive picture of the particle configurations. We can see that each particle is touching mostly six particles and infrequently five particles at distances close to $\sigma_{\alpha\beta}$. Similar jammed particle configurations can also be found in 3D, where the coordination number of other particles around each particle is about 12. Then it is natural that the pair correlation functions $g_{\alpha\beta}(r)$ have a very sharp peak at $r \cong \sigma_{\alpha\beta}$, as displayed in Fig. 11.26(b) for 2D.

11.4.2 Bond breakage and dynamic heterogeneity

Owing to the sharpness of the first peak of the pair correlation functions $g_{ab}(r)$, we can unambiguously define *bonds* between particle pairs at distances close to the first peak position [205]. That is, the particle pair i and j is bonded if $r_{ij}(t_0) = |\boldsymbol{r}_i(t_0) - \mathbf{r}_j(t_0)| \le \ell_{1ab}$ where $i \in a$ and $j \in b$. After a lapse of time Δt, the bond is broken if $r_{ij}(t_0 + \Delta t) > \ell_{2ab}$. Here ℓ_{1ab} is longer than the first peak position of $g_{ab}(r)$, and $\ell_{2ab}(\ge \ell_{1ab})$ is shorter than the second peak position. The number of the unbroken bonds may be fitted to $\exp[-(\Delta t/\tau_\mathrm{b})^c]$ as a function of the time interval Δt with $c \lesssim 1$ ($c \sim 0.6$ at $T = 0.234$ in 3D). Thus we determine the bond breakage time τ_b both in quiescent and sheared conditions. It may be fitted to a simple formula,

$$1/\tau_\mathrm{b}(\dot{\gamma}) \cong 1/\tau_\mathrm{b}(0) + A_\mathrm{b}\dot{\gamma}, \tag{11.4.4}$$

where A_b is a constant of order 1. In the strong shear condition $\dot{\gamma}\tau_\mathrm{b}(0) > 1$, we have $\tau_\mathrm{b}(\dot{\gamma}) \sim \dot{\gamma}^{-1}$. This means that jump motions are induced by applied shear on the timescale of $\dot{\gamma}^{-1}$.

Following the bond breakage process, we can visualize the kinetic heterogeneity without ambiguity and quantitatively characterize the heterogeneous patterns. In Fig. 11.27 we show spatial distributions of broken bonds in a time interval of $[t_0, t_0 + 0.05\tau_\mathrm{b}]$ in 2D, where about 5% of the initial bonds defined at $t = t_0$ have been broken. The dots are the center positions $\mathbf{R}_{ij} = \frac{1}{2}[\boldsymbol{r}_i(t_0) + \boldsymbol{r}_j(t_0)]$ of the broken pairs at the initial time t_0. The broken bonds are seen to form *clusters* of varying size. While the heterogeneity is weak for a liquid case (a) at $T = 2.54$ and $\dot{\gamma} = 0$, it is marked in a glassy case (b) at $T = 0.337$ and $\dot{\gamma} = 0$. The bond breakage time τ_b is 17 in (a) and 5×10^4 in (b). In (c) we set $\dot{\gamma} = 0.25 \times 10^{-2}$ and $T = 0.337$ with $\tau_\mathrm{b} = 32 \sim 1/\dot{\gamma}$. The heterogeneity becomes much suppressed by shear, while its spatial anisotropy remains small. Notice that even in normal liquids bond breakage events frequently occur in the form of strings involving a few to several particles, obviously because of the high density of our system. In glassy states such strings become longer and aggregate, forming large-scale clusters.

We next define the structure factor of the broken bonds as

$$S_b(q) = \left\langle \left| \sum_{\text{broken bonds}} \exp(i\mathbf{q}\cdot\boldsymbol{R}_{ij}) \right|^2 \right\rangle, \tag{11.4.5}$$

where $\boldsymbol{R}_{ij} = \frac{1}{2}[\boldsymbol{r}_i(t_0) + \boldsymbol{r}_j(t_0)]$. The summation is over the broken pairs in a time interval

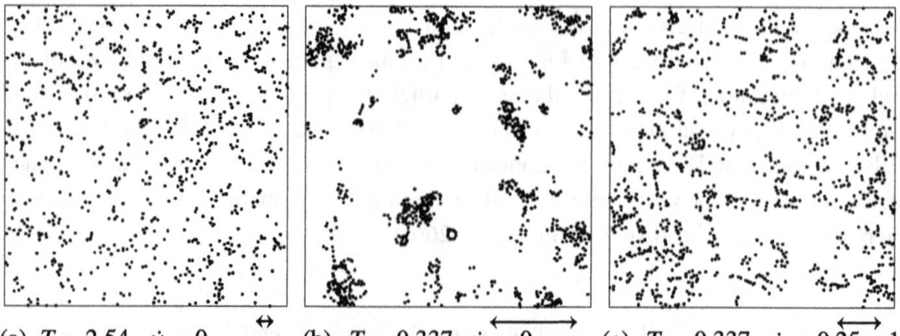

Fig. 11.27. Snapshots of the broken bonds in 2D [205]. Here $T = 2.54$ with weak heterogeneity in (a), and $T = 0.337$ with enhanced heterogeneity in (b) in the absence of shear. In (c), where $\dot{\gamma} = 2.5 \times 10^{-2}$ and $T = 0.337$, the heterogeneity is much suppressed. The flow is in the upward direction and the velocity gradient is in the horizontal direction from left to right. The arrows indicate the correlation length ξ obtained from (11.4.6).

$[t_0, t_0 + \Delta t]$. Then $S_b(q)$ can be fitted to the Ornstein–Zernike form,

$$S_b(q) = S_b(0)/(1 + \xi^2 q^2), \tag{11.4.6}$$

both in 2D and 3D, as shown in Fig. 11.28 for 3D where $\Delta t = 0.05\tau_{\rm b}$. The correlation length ξ is determined from this expression. We can also see that $S_b(0) \sim \xi^2$ leading to weak temperature dependence of $S_b(q)$ at large q. The clusters of the broken bonds are analogous to the critical fluctuations in Ising systems. As in critical dynamics, we have furthermore confirmed a dynamical scaling relation,

$$\tau_{\rm b} \sim \xi^z, \tag{11.4.7}$$

where $z = 4$ in 2D and $z = 2$ in 3D. This relation holds even in strong shear $\dot{\gamma}\tau_b(0) \gg 1$, where $\xi \sim \dot{\gamma}^{-1/z}$. At present, we cannot explain the origin of these simple numbers for z. We can only argue that z should be larger in 2D than in 3D because of stronger configurational restrictions in 2D. Because $\dot{\gamma}$ suppresses the heterogeneity, it is analogous to a magnetic field h in Ising systems.

11.4.3 The α relaxation time

In the literature it is usual to follow the motion of tagged particles. The self-part of the density time-correlation function is defined by

$$F_{\rm s}(q, t) = \frac{1}{N_1}\left\langle \sum_{j=1}^{N_1} \exp[i\boldsymbol{q} \cdot \Delta \boldsymbol{r}_j(t)] \right\rangle, \tag{11.4.8}$$

where $\Delta \boldsymbol{r}_j(t) = \boldsymbol{r}_j(t) - \boldsymbol{r}_j(0)$, and the summation is taken over all the particles of the species 1. As shown in Fig. 11.29(a) in 3D, this function has a plateau at low temperatures,

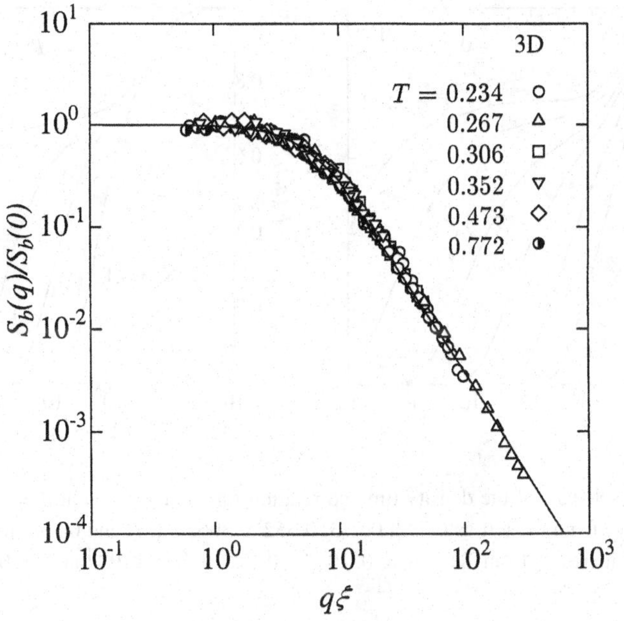

Fig. 11.28. $S_b(q)/S_b(0)$ vs $q\xi$ on logarithmic scales for various T and $\dot{\gamma}$ in 3D [205]. The solid line is the Ornstein–Zernike form $1/(1+x^2)$ with $x = q\xi$.

during which the particle is trapped in a *cage* formed by the surrounding particles. After a long time the cage eventually breaks, resulting in diffusion with a very small diffusion constant D. We may define the α relaxation time such that

$$F_s(q, \tau_\alpha) = e^{-1} \tag{11.4.9}$$

holds at $q = 2\pi$. Thus τ_α represents the cage breakage time on the microscopic spatial scale ($\sim \sigma_1$). Figure 11.29(a) shows that τ_α grows strongly at low T.

We also generalize the density time-correlation function (11.4.8) in the presence of shear flow by introducing a new displacement vector of the jth particle as

$$\Delta \boldsymbol{r}_j(t) = \boldsymbol{r}_j(t) - \dot{\gamma} \int_0^t dt' y_j(t') \boldsymbol{e}_x - \boldsymbol{r}_j(0), \tag{11.4.10}$$

where $\boldsymbol{e}_x$ is the unit vector in the flow direction. In this displacement, the contribution from convective transport by the average flow has been subtracted. Then, $F_s(q, t)$ only slightly depends on the angle of the wave vector $\boldsymbol{q}$ for $\dot{\gamma} \ll 1$ in our model. In Fig. 11.29(b) we shown its relaxation at $q = 2\pi$ and $T = 0.267$ for various $\dot{\gamma}$. Comparing the two figures with and without shear, we recognize that applying shear is equivalent to raising the temperature.

We can thus determine the α relaxation time also in shear. We found that the bond

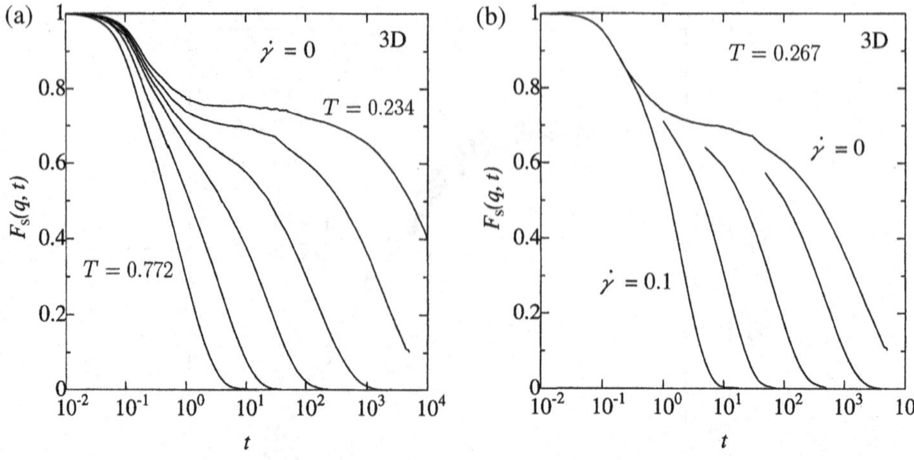

Fig. 11.29. The self-part of the density time-correlation function $F_S(q, t)$ at $q = 2\pi$ in 3D [205]. In (a) T decreases from the left as 0.772, 0.473, 0.352, 0.306, 0.267, and 0.234 in quiescent states ($\dot\gamma = 0$). In (b) $\dot\gamma$ increases from the right as 0, 10^{-4}, 10^{-3}, 10^{-2}, and 10^{-1} at $T = 0.267$. Increasing $\dot\gamma$ is equivalent to raising T.

breakage time and the α relaxation time are simply related in 3D by

$$\tau_\alpha \cong 0.1\tau_b, \tag{11.4.11}$$

which holds at any T and $\dot\gamma$ in any supercooled state in our simulations. In this section, however, we use the notation τ_α for the usual α relaxation time in quiescent states ($\dot\gamma = 0$). Remarkably, the particle motion out of the cage takes place on the timescale of $\dot\gamma^{-1}$ in the case $\dot\gamma\tau_\alpha > 1$. We propose that dielectric relaxation measurements be carried out on glass-forming fluids under shear, where $\tau_\alpha(\dot\gamma)$ should be observed.

11.4.4 Heterogeneity in diffusion

As $q \to 0$, $F_s(q, t)$ decays diffusively as

$$F_s(q, t) \cong \exp(-2Dq^2t) \quad (q \ll 1). \tag{11.4.12}$$

In shear flow, Dq^2 in the above expression should be replaced by $\sum_{\mu\nu} D_{\mu\nu} q_\mu q_\nu$, where

$$\langle[\Delta \boldsymbol{r}_j(t)]_\mu[\Delta \boldsymbol{r}_j(t)]_\nu\rangle = 2D_{\mu\nu}t \quad (\mu, \nu = x, y, z) \tag{11.4.13}$$

at long times ($t \gg \tau_\alpha$). However, we confirmed for our 3D model fluid that the tensor $D_{\mu\nu}$ is nearly diagonal as $D\delta_{\mu\nu}$ for $\dot\gamma \ll \tau_0^{-1}$ $(= 1)$ in supercooled states.

In Fig. 11.30 we show 3D simulation results of the diffusion constant of a tagged particle of the species 1 [215]. The data can be fitted to $D \propto \eta^{-0.75}$ at low T in agreement with the experiments [211, 212]. However, the zero-shear viscosity η is proportional to τ_α as

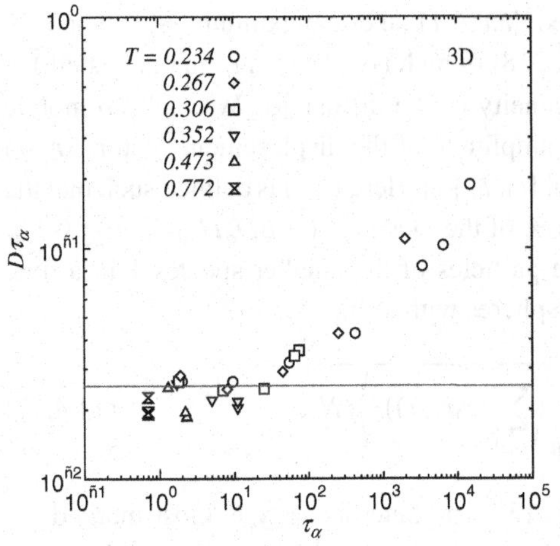

Fig. 11.30. $D\tau_\alpha$ vs τ_α in a model 3D fluid binary mixture [215]. The solid horizontal line represents the Stokes–Einstein value $D_{ES}\tau_\alpha = (2\pi)^{-2}$.

$\eta \sim T\tau_\alpha$. Both τ_α and D can be obtained from $F_s(q,t)$ in (11.4.8); τ_α from the relaxation behavior at $q = 2\pi$ as in (11.4.9) and D from that in the region $q \ll 1$ as in (11.4.12).

To understand the different dependences of D and η on τ_α, let us consider the van Hove correlation function $G_s(r,t)$, whose 3D Fourier transformation is equal to $F_s(q,t)$ in (11.4.8). It is the probability that a tagged particle moves over a distance r in time interval t, so it is nonnegative-definite and normalized as $4\pi\int_0^\infty drr^2 G_s(r) = 1$. The mean-square displacement is related to D as

$$\begin{aligned} \langle|\Delta \boldsymbol{r}(t)|^2\rangle &= 4\pi\int_0^\infty drr^4 G_s(r,t) \\ &\cong 6Dt. \end{aligned} \tag{11.4.14}$$

The second line holds for $t \gtrsim 0.1\tau_\alpha$ in our system, while the first line $\cong (3T/m)t^2$ for $t \lesssim 0.1\tau_\alpha$. Numerically, however, $G_s(r,t)$ deviates considerably from the asymptotic gaussian form, $(4\pi Dt)^{-3/2}\exp(-r^2/4Dt)$, even when the second line of (11.4.14) holds. We found that the scaled function $\sqrt{6\pi Dt}4\pi r^2 G_s(r,t)$ has a large r-tail which can be scaled in terms of $r/t^{1/2}$ for $t \lesssim 3\tau_\alpha$ and gives a dominant contribution to D [215]. Because this tail vanishes for $t \gtrsim 3\tau_\alpha$, $3\tau_\alpha$ is the lifetime of the heterogeneity in our system. We may thus conjecture that $G_s(r,t)$ is expressed in terms of the local diffusion constant $D(\boldsymbol{x},t)$ as

$$G_s(r,t) = \left\langle [4\pi D(\boldsymbol{x},t)t]^{-3/2}\exp\left[-r^2/4D(\boldsymbol{x},t)t\right]\right\rangle, \tag{11.4.15}$$

where $\boldsymbol{x}$ denotes the space position and the average is taken over space. Here the space variation of $D(\boldsymbol{x},t)$ is significant for $t \lesssim 3\tau_\alpha$, but its average is fixed as $\langle D(\boldsymbol{x},t)\rangle \cong D$ for $t \gtrsim 0.1\tau_\alpha$. To the mean-square displacement in (11.4.14) the contributions from regions with large $D(\boldsymbol{x},t)$ are expected to be dominant. From (11.4.15) the so-called non-gaussian parameter defined by $\mathcal{A}(t) = 3\langle|\Delta\boldsymbol{r}(t)|^4\rangle/\langle|\Delta\boldsymbol{r}(t)|^2\rangle^2 - 1$ is written as

$$\mathcal{A}(t) = \langle D(\boldsymbol{x},t)^2\rangle/\langle D(\boldsymbol{x},t)\rangle^2 - 1. \tag{11.4.16}$$

In accord with this result, it has been expected that $\mathcal{A}(t)$ takes a maximum (~ 3) when the heterogeneity structure is most marked [217, 218] (which is at $t \sim 0.1\tau_\alpha$ in our system).

In the following we visualize the heterogeneity of the diffusivity. We pick up mobile particles of each species a (1 or 2) with the amplitude of the displacement vector $\Delta \boldsymbol{r}_j(t)$ exceeding a lower limit $\ell_c(t)$ in a time interval $[t_0, t_0+t]$. Here $\ell_c(t)$ is defined such that the sum of $[\Delta \boldsymbol{r}_j(t)]^2$ of the mobile particles is 66% of the total sum ($\cong 6D_a t N_a$ for $t \gtrsim 0.1\tau_\alpha$ with $a = 1, 2$). In Fig. 11.31(a) the mobile particles of the smaller species 1 in a time interval of $[t_0, t_0 + 0.125\tau_\alpha]$ are depicted as spheres with radii,

$$a_j(t) = |\Delta \boldsymbol{r}_j(t)| \Big/ \sqrt{\Big\langle \sum_{\ell \in 1} (\Delta \boldsymbol{r}_\ell(t))^2 \Big\rangle / N_1}, \tag{11.4.17}$$

located at $\boldsymbol{R}_j(t) = \frac{1}{2}[\boldsymbol{r}_j(t_0) + \boldsymbol{r}_j(t_0 + t)]$ [215]. The heterogeneity is most marked for that time interval at which the so-called non-gaussian parameter is maximum. Next we represent the displacement vectors of the mobile particles of both the species 1 and 2 by cones with the base center and the tip being the initial and final positions, respectively. We then group the mobile particles into clusters with particle number $n = 1, 2, \ldots$, where the mobile particles $i \in a$, $j \in b$ belong to the same cluster if either of $|\boldsymbol{r}_i(t_0) - \boldsymbol{r}_j(t_0 + t)|$ or $|\boldsymbol{r}_i(t_0 + t) - \boldsymbol{r}_j(t_0)|$ is shorter than $0.3(\sigma_a + \sigma_b)$. In Fig. 11.31(b) we pick up those belonging to the clusters with $n \geq 5$ [206]. They are 5% of the total particle number N, but they contribute 40% to the sum $\langle \sum_\ell [\Delta \boldsymbol{r}_\ell(t)]^2 \rangle$ of all the particles. The mobile particles thus form chains, as also reported by Donati and coworkers [207]. Moreover, these chains aggregate to form large-scale heterogeneities on the scale of ξ. Note also that the above visualization method sensitively depends on the time interval t. Indeed, the diffusion process becomes homogeneous if t is longer than the lifetime of the heterogeneity structure ($\sim 3\tau_\alpha$).

Shear-induced diffusion in supercooled liquids and dense suspensions

It is remarkable that the relation $D \propto \eta^{-0.75}$ at low T holds even under strong shear. Thus,

$$D \propto \dot{\gamma}^{0.75} \quad (\dot{\gamma}\tau_\alpha \gg 1), \tag{11.4.18}$$

in the simulations. We mention similar observations in concentrated suspensions under shear. When the Peclet number Pe in (11.1.37) is much larger than 1 [219], the motion of the colloidal particles is predominantly caused by shear-induced changes of the particle configurations. The self-diffusion constant in the shear gradient direction D_y and that in the vorticity direction D_z both behave as

$$D_j \cong \hat{D}_j(\phi) a^2 \dot{\gamma} \quad (j = y, z), \tag{11.4.19}$$

where $\hat{D}_j(\phi)$ is a dimensionless number dependent on the colloid volume fraction ϕ and is of order 0.1 at $\phi \sim 0.4$.

(a)

(b)

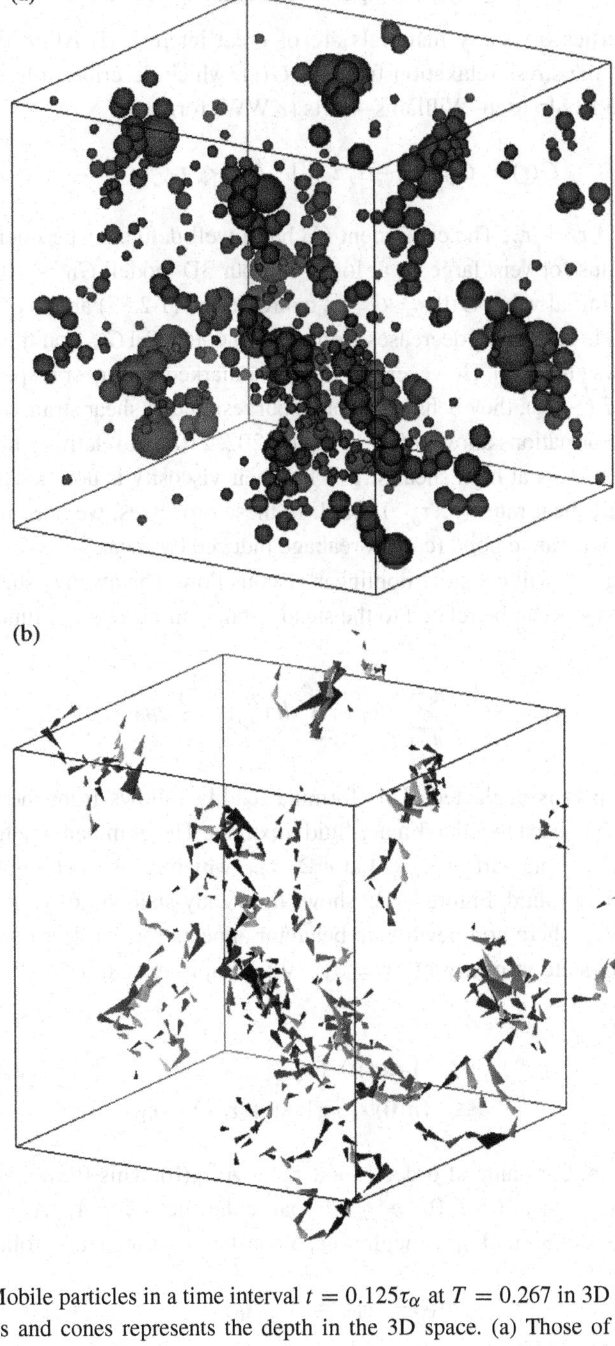

Fig. 11.31. Mobile particles in a time interval $t = 0.125\tau_\alpha$ at $T = 0.267$ in 3D [206]. The darkness of the spheres and cones represents the depth in the 3D space. (a) Those of the smaller species 1 represented by spheres with radii $a_j(t)$ in (11.4.17). (b) Those belonging to clusters with sizes $n \geq 5$.

11.4.5 Rheology in a supercooled binary fluid mixture

Mechanical properties of glassy materials are of great interest. (i) After a microscopic transient time $t_{\rm tra}$, the stress relaxation function $G(t)$, which describes a linear response, can be fitted to the Kohlrausch–Williams–Watts (KWW) form,

$$G(t) = G_0 \exp[-(t/\tau_{\rm s})^{\beta}] \quad (t_{\rm tra} \ll t \lesssim \tau_{\rm s}), \tag{11.4.20}$$

where $\beta \sim 0.5$ and $\tau_{\rm s} \sim \tau_\alpha$. The coefficient G_0 has a well-defined experimental meaning as the shear modulus for very large τ_α at low T. In our 3D model, $G_0 \sim 10$ at $T \sim 0.2$. Note that the true initial value $G(0) = G_\infty$ is expressed as (1.2.85) and is of order 10^2 at low T in our model. Thus $G(t)$ decreases from G_∞ to $G_0(\sim 0.1G_\infty)$ on the timescale of $t_{\rm tra}$ and then follows (11.4.20). However, (ii) there is a marked nonlinear response in glassy materials. At low $T(\lesssim T_{\rm g})$, they behave as solids but respond to shear strain nonlinearly or undergo plastic deformations above a few % strain [202, 210]. At relatively high $T(\gtrsim T_{\rm g})$, they can be made to flow at high shear stress, but their viscosity is non-newtonian except for extremely small shear rates ($< \tau_\alpha^{-1}$) [209]. In these processes, we need to understand the dynamics of cooperative bond (cage) breakage induced by shear.

In the following, we will consider nonlinear viscous flow. The average shear stress σ_{xy} in sheared steady states can be related to the steady-state pair correlation functions $g_{\alpha\beta}(\boldsymbol{r})$ as

$$\sigma_{xy} = -\frac{1}{2} \sum_{\alpha,\beta=1,2} n_\alpha n_\beta \int d\boldsymbol{r} v'_{\alpha\beta}(r) \frac{xy}{r} g_{\alpha\beta}(\boldsymbol{r}), \tag{11.4.21}$$

where the kinetic part is neglected. This formula readily follows from the microscopic expression (5E.3) if it is extended to binary fluid mixtures. The dominant contribution here arises from the anisotropic part of $g_{\alpha\beta}(r)$ at $r \cong \sigma_{\alpha\beta}$, which is, at most, only a few % of the isotropic part in our fluid. Figure 11.32 shows the steady-state viscosity $\eta(\dot{\gamma}) = \sigma_{xy}/\dot{\gamma}$ in our system in 3D, where non-newtonian behavior appears for $\dot{\gamma}$ larger than $\tau_{\rm b}(0)^{-1} \sim 0.1\tau_\alpha^{-1}$. The steady-state viscosity $\eta(\dot{\gamma}) = \sigma_{xy}/\dot{\gamma}$ is simply related to the bond breakage time in (11.4.4) as

$$\begin{aligned} \eta(\dot{\gamma}) &\cong A_\eta \tau_{\rm b}(\dot{\gamma}) + \eta_{\rm B} \\ &\cong [\eta(0) - \eta_{\rm B}]/(1 + \tau_\eta \dot{\gamma}) + \eta_{\rm B}, \end{aligned} \tag{11.4.22}$$

where A_η and $\eta_{\rm B}$ are constants of order 1, and $\tau_\eta = A_{\rm b}\tau_{\rm b}(0)$. This form agrees with the experimental result (11.4.1) for $\eta(0) \gg \eta_{\rm B}$. In particular, $\eta(\dot{\gamma}) \cong (A_\eta/A_b)/\dot{\gamma} + \eta_{\rm B}$, for $\dot{\gamma}\tau_{\rm b}(0) \gg 1$. If the background $\eta_{\rm B}$ is negligible, a constant limiting stress follows as

$$\sigma_{xy} \cong \sigma_{\rm lim} = A_\eta/A_b, \tag{11.4.23}$$

which holds for $1/\tau_b(0) \ll \dot{\gamma} \ll \sigma_{\rm min}/\eta_{\rm B} \sim 0.1/\tau_0$. Here $\sigma_{\rm lim}$ is of order 0.5 and is considerably smaller than the shear modulus G_0.

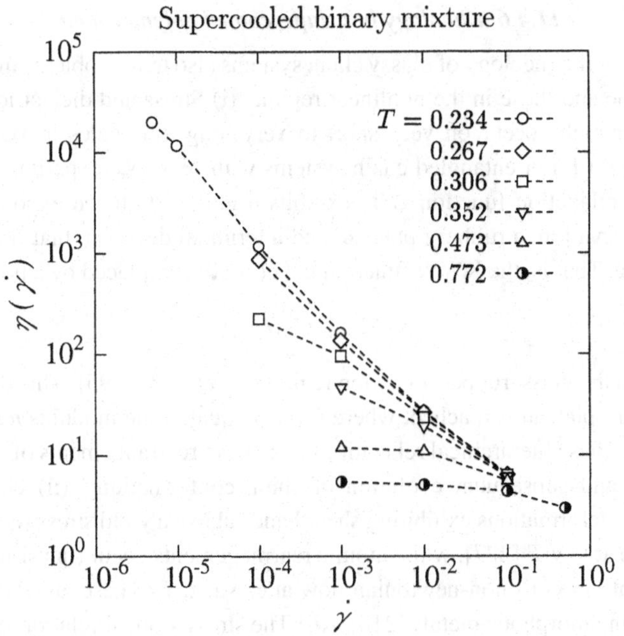

Fig. 11.32. The steady-state viscosity $\eta(\dot\gamma)$ in units of $\epsilon\tau_0/\sigma_1^3$ vs the shear rate $\dot\gamma$ in units of $1/\tau_0$ at various T in a model 3D binary fluid mixture [208]. The data tend to become independent of T at high shear.

The physical mechanism of this strong shear-thinning behavior is as follows. Upon each bond breakage induced by shear, the particles involved release a potential energy of order ϵ. This is then changed into energies of random motions supported by the surrounding particles. The heat production rate is estimated as

$$\dot{Q} \sim n\epsilon/\tau_b(\dot\gamma) \sim n\epsilon\dot\gamma, \tag{11.4.24}$$

where n is the number density. Because $\dot{Q}$ is related to the viscosity by $\dot{Q} = \sigma_{xy}\dot\gamma$, we obtain $\sigma_{xy} \sim n\epsilon$ in high shear.

Jamming rheology and plastic flow

Similar *jamming rheology* has begun to be recognized in granular materials and foams composed of large elements [220]–[224]. A simple phenomenological theory taking into account disorder and metastability was also presented [225]. Shear-thinning behavior and heterogeneities in configuration rearrangements are universally observed experimentally and numerically from microscopic to macroscopic constrained systems. See an assembly of related papers [226]. Plastic flow in solids has long been studied and is an example of jamming rheology for amorphous solids [210]. In crystals, slips due to dislocation motions are relevant causing plastic flow.

11.4.6 Rheology in a supercooled polymer melt

Interpretations of the rheology of glassy chain systems also treat problems in both the linear response regime and those in the nonlinear regime. (i) Stress and dielectric relaxations of glassy polymer melts occur on very short to very long timescales in very complicated manners [195, 196]. For entangled chain systems with $N > N_e$, experiments have shown that the stress relaxation function $G(t)$ exhibits a glassy stretched-exponential decay, a glass–rubber transition, a rubbery plateau, and a terminal decay, in that order, over many decades of time. That is, the KWW function in (11.4.20) is replaced by a power-law decay,

$$G(t) \cong e^{-1} G_0 (t/\tau_s)^{-\nu}, \tag{11.4.25}$$

with $\nu \sim 0.5$ in the glass–rubber transition region $t \gtrsim \tau_s \sim \tau_\alpha$ [195]. This decay continues until the rubbery plateau is reached, where $G(t)$ is equal to the modulus nT/N_e of entangled polymers. These hierarchical relaxations arise from rearrangements of jammed atomic configurations and subsequent evolution of chain conformations. (ii) Glassy polymers undergo plastic deformations exhibiting shear bands above a yield stress (corresponding to a few % strain) at low T [227], while atomic rearrangements occur (quasi)homogeneously leading to highly viscous non-newtonian flow at elevated T. These effects are commonly observed also in amorphous metals [210]. (iii) The stress–optical relation (proportionality between birefringence and stress) has been used in experiments on polymers both in the linear and nonlinear regimes. However, it is violated as T is approached T_g [228, 229], obviously owing to enhancement of the glassy part of the stress. Note that the stress in polymers consists of the glassy and entropic parts; the former is usually negligible (on not very fast timescales) as compared to the latter far above T_g, but becomes important near and below T_g.

In the following we will present simulation results on a model melt composed of short chains with polymerization index $N = 10$, obeying the Rouse dynamics in quiescent states [230]. The monomers interact via a Lenard-Jones potential (characterized by ϵ and σ as in (1.2.1)) and consecutive beads interact via a nonlinear spring potential [231]. The temperature T, the time, and the viscosity are scaled in units of ϵ, $\tau_0 = (m\sigma^2/\epsilon)^{1/2}$, and $\epsilon\tau_0/\sigma^3$. For such N the longest relaxation time of the chains is the Rouse time,

$$\tau_R \sim N^2 \tau_\alpha. \tag{11.4.26}$$

The α relaxation time τ_α characterizes the decay of the correlation function $F_s(q, t)$ at $q \sim 2\pi/\sigma$ as in (11.4.9) and represents the timescale of monomeric structural relaxation. Figure 11.33 shows $G(t)$ in our system containing 100 chains. At $T = 0.2$ it behaves as

$$G(t) = G_0 \exp[-(t/\tau_s)^\beta] + G_R(t) \quad (t \gg t_{tra}), \tag{11.4.27}$$

where $t_{tra} \sim 1$ and $\tau_s \sim \tau_\alpha \sim 10^2$. The first term is of the same form as (11.4.25), while $G_R(t)$ is the Rouse relaxation function decaying as $nTN^{-1}\exp(-t/\tau_R)$ for $t \gtrsim \tau_R$. On relatively short timescales ($< \tau_R$), the first term is important and the stress–optical relation, valid at high T, is violated. At $T = 0.2$ we can see distinct differences in the following

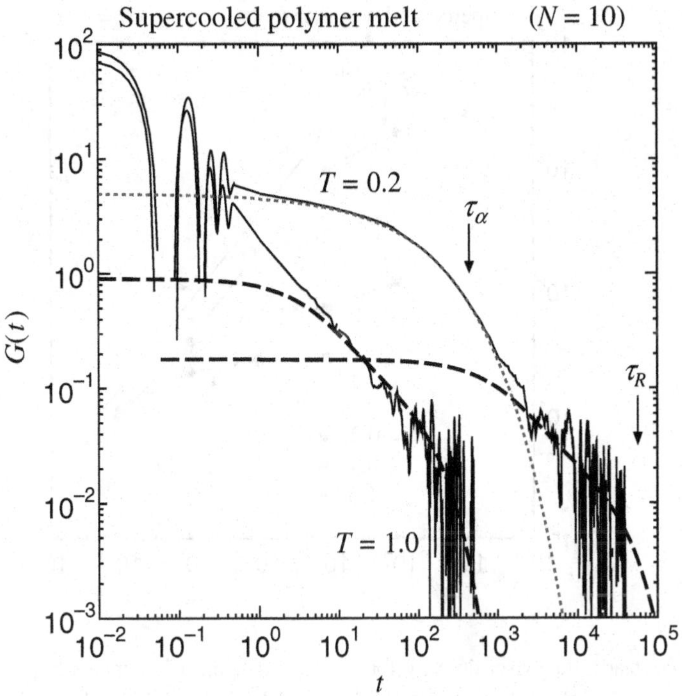

Fig. 11.33. The stress relaxation function $G(t)$ (thin solid lines) at $T = 0.2$ in a supercooled state and $T = 1$ in a normal liquid state in a model polymer melt [206]. It may be fitted to the stretched-exponential form (dotted line) at relatively short times and tends to the Rouse relaxation function $G_R(t)$ (bold dashed lines) at long times. The noisy behavior of the curves of $G(t)$ at long times $(t \geq \tau_R)$ arises from the thermal fluctuations of the stress in a finite system [230].

moduli: $G_\infty = G(0) \sim 10^2$, $G_0 \sim 5$, $G_R(0) = nT \sim 0.2$, and $G_R(\tau_R) \sim nTN^{-1} \sim 0.02$.

In shear flow, polymer chains are significantly elongated when $\dot{\gamma}$ becomes of order τ_R^{-1} for $N < N_e$. (This criterion becomes $\dot{\gamma} \geq \tau_{rep}^{-1}$ for $N > N_e$, where τ_{rep} is the disentanglement time estimated as (7A.5) in the reptation theory.) Such shear rates are extremely small in supercooled states. Marked shear-thinning behavior then takes place for larger shear rates. In Fig. 11.34 we display the steady-state viscosity $\eta(\dot{\gamma})$ in our model system [206]. The horizontal arrows indicate the linear viscosity $\eta_R = \int_0^\infty dt G_R(t)$ $(\propto N^{-1}\tau_R)$ from the Rouse model, and the vertical arrows indicate the points at which $\dot{\gamma} = \tau_R^{-1}$. In particular, the curve of $T = 0.2$ may be fitted to $\eta \propto \dot{\gamma}^{-\nu}$ with $\nu \sim 0.7$ for $\dot{\gamma}\tau_R \gtrsim 1$. The shape changes of chains occurring for $\dot{\gamma} \gtrsim \tau_R^{-1}$ should be observable by scattering experiments. It would be interesting to know how the monomeric relaxation time τ_α is affected by shear, particularly for very long chain systems. Thus dielectric relaxation measurements in shear [229] seem to be informative.

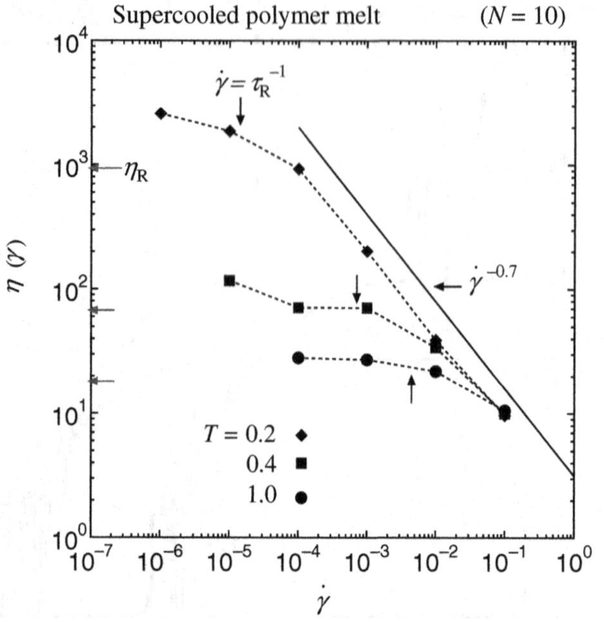

Fig. 11.34. The steady-state viscosity vs $\dot{\gamma}$ for $T = 0.2$, 0.4, and 1 in a model polymer melt [206]. A line of slope -0.7 is also drawn as a guide. The model exhibits marked shear-thinning behavior for $\dot{\gamma}\tau_R \gtrsim 1$ and becomes independent of T for very high shear rates.

Appendix 11.A Correlation functions in velocity gradient

We consider time-correlation functions in steady states under flow with a homogeneous velocity gradient

$$\boldsymbol{u}(\boldsymbol{r}) = \boldsymbol{u}_0 + \overleftrightarrow{D} \cdot \boldsymbol{r}, \tag{11A.1}$$

where $\boldsymbol{u}_0$ is a constant and $\overleftrightarrow{D}$ is the velocity gradient tensor, assumed to be constant. In this case the time-correlation function of any scalar variable $\psi(\boldsymbol{r}, t)$ satisfies [12]

$$\langle \psi(\boldsymbol{r}, t)\psi(\boldsymbol{r}', t')\rangle = \langle \psi(\boldsymbol{r} - e^{\overleftrightarrow{D}(t-t')} \cdot \boldsymbol{r}', t - t')\psi(\boldsymbol{0}, 0)\rangle. \tag{11A.2}$$

The equal-time-correlation function ($t = t'$) depends only on the relative position $\boldsymbol{r} - \boldsymbol{r}'$. Its Fourier transformation yields the steady-state structure factor

$$I(\boldsymbol{q}) = \int d\boldsymbol{r} \exp[i\boldsymbol{q} \cdot (\boldsymbol{r} - \boldsymbol{r}')]\langle \psi(\boldsymbol{r}, t)\psi(\boldsymbol{r}', t)\rangle, \tag{11A.3}$$

which is observable by scattering experiments.

In particular, in shear flow the derivation of the above relation is obvious. We note that a shift of the origin of the reference frame by a in the y axis is equivalent to a Galilean transformation to a new reference frame moving with a velocity $-a\dot{\gamma}\boldsymbol{e}_x$. This implies that, in homogeneous stationary states, the time-correlation function of any density variable $\psi(\boldsymbol{r}, t)$ may be written as

$$\langle \psi(\boldsymbol{r}, t)\psi(\boldsymbol{r}', t')\rangle = \langle \psi(\boldsymbol{r} - \boldsymbol{r}' - \dot{\gamma}(t - t')y'\boldsymbol{e}_x, t - t')\psi(\boldsymbol{0}, 0)\rangle. \tag{11A.4}$$

It is instructive to rewrite (11A.4) in terms of the Fourier components,

$$\langle \psi_{\boldsymbol{q}}(t)\psi_{\boldsymbol{k}}(t')\rangle = (2\pi)^d \delta(\boldsymbol{q} + \boldsymbol{k} + q_x\dot{\gamma}(t - t')\boldsymbol{e}_y)I(\boldsymbol{q}, t - t'), \tag{11A.5}$$

where

$$I(\boldsymbol{q}, t) = \int d\boldsymbol{r} \exp(i\boldsymbol{q}\cdot\boldsymbol{r})\langle \psi(\boldsymbol{r}, t)\psi(\boldsymbol{0}, 0)\rangle. \tag{11A.6}$$

The first factor in (11A.5) is the delta function in d dimensions. To understand its origin we note that a plane-wave concentration fluctuation ($\propto \exp(i\boldsymbol{q}\cdot\boldsymbol{r})$ at $t = 0$) with a small amplitude changes in time into a plane wave with a time dependent wave vector given by

$$\tilde{\boldsymbol{q}}(t) = \boldsymbol{q} - \dot{\gamma} t q_x \boldsymbol{e}_y, \tag{11A.7}$$

if nonlinear couplings among the fluctuations are neglected. Then (11A.4) is nonvanishing only for $\tilde{\boldsymbol{q}}(-t + t') = -\boldsymbol{k}$ on the average over the fluctuations, yielding the above delta function. It would be informative to measure the time-dependence of $I(\boldsymbol{q}, t)$ in (11A.6), but dynamic light scattering in shear flow has not yet been successful for binary fluid mixtures. This is probably because the Doppler shift of scattered light depends on the y coordinate of the scattering position and the observed signal strongly depends on the thickness of the scattering region in the y direction [232, 233]. Recently, however, dynamic light scattering experiments were performed on lyotropic lamellar phases of brine and surfactant in shear flow [234].

References

[1] *Physics of Complex and Supermolecular Fluids*, eds. S. A. Safran and N. A. Clark (Wiley, New York, 1987).

[2] *Dynamics and Patterns in Complex Fluids*, eds. A. Onuki and K. Kawasaki (Springer, Berlin, Heidelberg, 1990).

[3] *Complex Fluids*, eds. E. B. Sirota, D. Weitz, T. Witten, and J. Israelachvili (Materials Research Soc., Pittsburgh, 1992).

[4] A. Keller and J. A. Odell, *Colloid Polym. Sci.*, **263**, 181 (1985).

[5] A. Silberberg and W. Kuhn, *J. Polym. Sci.* **13**, 21 (1954).

[6] A. S. Lodge, *Polymer* **2**, 195 (1961).

[7] A. Peterlin and D. T. Turner, *J. Polym. Sci., Polym. Lett.* **3**, 517 (1965).

[8] L. A. Utracki, *Polymer Alloys and Blends. Thermodynamics and Rheology* (Hanser Publishers, Munich, 1990).

[9] R. G. Larson, *Rheol. Acta* **31**, 497 (1992).

[10] R. G. Larson, *The Structure and Rheology of Complex Fluids* (Oxford University Press, 1999).

[11] A. Onuki and K. Kawasaki, *Ann. Phys.* (NY) **121**, 456 (1979); A. Onuki, K. Yamazaki, and K. Kawasaki, *Ann. Phys.* (NY) **131**, 217 (1981).

[12] A. Onuki and K. Kawasaki, *Prog. Theor. Phys.* **63**, 122 (1980); *Supplement of Prog. Theor. Phys.* **69**, 146 (1980).

[13] A. Onuki, *Int. J. Thermophys.* **10**, 293 (1989).

[14] A. Onuki, *J. Phys. C* **9**, 6119 (1997).

[15] D. Beysens, M. Gbadamassi, and L. Boyer, *Phys. Rev. Lett.* **43**, 1253 (1979); D. Beysens and M. Gbadamassi, *Phys. Rev. A* **22**, 2250 (1980).

[16] D. Beysens, M. Gbadamassi, and B. Moncef-Bouanz, *Phys. Rev. A* **28**, 2491 (1983).

[17] F. Perrot, C. K. Chan, and D. Beysens, *Europhys. Lett.* **65** (1989); C. K. Chan, F. Perrot, and D. Beysens, *Phys. Rev. A* **43**, 1826 (1991); T. Baumberger, F. Perrot, and D. Beysens, *Physica A* **174**, 31 (1991).

[18] E. K. Hobbie, D. W. Hair, A. I. Nakatani, and C. C. Han, *Phys. Rev. Lett.* **69**, 1951 (1992).

[19] A. Onuki and K. Kawasaki, *Physica A* **11**, 607 (1982); A. Onuki and M. Doi, *J. Chem. Phys.* **85**, 1190 (1986).

[20] Y. C. Chou and W. I. Goldburg, *Phys. Rev. Lett.* **47**, 1155 (1981); D. Beysens and M. Gbadamassi, *Phys. Rev. Lett.* **47**, 846 (1981); D. Beysens, R. Gastand, and F. Decrupppe, *Phys. Rev. A* **30**, 1145 (1984); T. A. Lenstra and J. K. Dhont, *Phys. Rev. E* **63**, 061401 (2001).

[21] J. K. G. Dhont and H. Verduin, *J. Chem. Phys.* **101**, 6193 (1994); *Physica A* **235**, 87 (1997).

[22] T. Hashimoto, T. Takebe, and S. Suehiro, *J. Chem. Phys.* **88**, 5874 (1988).

[23] T. Takebe, R. Sawaoka, and T. Hashimoto, *J. Chem. Phys.* **91**, 4369 (1989).

[24] T. Takebe, K. Fujioka, R. Sawaoka, and T. Hashimoto, *J. Chem. Phys.* **93**, 5271 (1990).

[25] K. Fujioka, T. Takebe, and T. Hashimoto, *J. Chem. Phys.* **96**, 717 (1993).

[26] T. Hashimoto, T. Takebe, and K. Asakawa, *Physica A* **194**, 338 (1993).

[27] K. Asakawa and T. Hashimoto, J. Chem. Phys. **105**, 5216 (1996).

[28] T. Hashimoto, K. Matsuzaka, and K. Fujioka, *J. Chem. Phys.* **108**, 6963 (1998).

[29] A. Onuki and T. Hashimoto, *Macromolecules* **22**, 879 (1989).

[30] J.-W. Yu, J. F. Douglas, E. K. Hobbie, S. Kim, and C. C. Han, *Phys. Rev. Lett.* **78**, 2664 (1997).

[31] G. H. Fredrickson and F. S. Bates, *Ann. Rev. Mater. Sci.* **26**, 501 (1996).

[32] G. H. Fredrickson, *J. Chem. Phys.* **85**, 5306 (1986).

[33] G. H. Fredrickson and R. G. Larson, *J. Chem. Phys.* **86**, 1553 (1987).

[34] A. Onuki, *J. Chem. Phys.* **87**, 3692 (1987).

[35] S. Brazovskii, *Zh. Eksp. Teor. Fiz.* **68**, 175 (1975) [*Sov. Phys. JETP* **41**, 85 (1975)].

[36] M. E. Cates and S. T. Milner, *Phys. Rev. Lett.* **62**, 1856 (1989).

[37] K. A. Koppi, M. Tirrell, and F. S. Bates, *Phys. Rev. Lett.* **70**, 1449 (1993).

[38] D. Beysens and F. Perrot, *J. Physique* **45**, L-31 (1994).

[39] T. Imaeda, A. Onuki, and K. Kawasaki, *Prog. Theor. Phys.* **71**, 16 (1984).

[40] F. Fukuhara, K. Hamano, N. Kuwahara, J. V. Sengers, and A. H. Krall, *Phys. Lett. A* **176**, 344 (1993).

[41] C. K. Chan and L. Lin, *Europhys. Lett.* **11**, 13 (1990).

[42] T. Ohta, H. Nozaki, and M. Doi, *J. Chem. Phys.* **93**, 2664 (1990).

[43] D. H. Rothman, *Europhys. Lett.* **14**, 337 (1991).

[44] B. D. Butler, H. J. M. Hanley, D. Hansen, and D. J. Evans, *Phys. Rev. B* **53**, 2450 (1996).

[45] T. Okuzono, *Phys. Rev. E* **56**, 4416 (1997).

[46] P. Padilla and S. Toxvared, *J. Chem. Phys.* **106**, 2342 (1997).

[47] R. Yamamoto and X. C. Zeng, *Phys. Rev. E* **59**, 3223 (1999).

[48] A. J. Wagner and J. M. Yeomans, *Phys. Rev. E* **59**, 4366 (1999); L. Berthier, *ibid.* **63**, 051503 (2001).

[49] A. Onuki, *Phys. Rev. A* **34**, 3528 (1986).

[50] T. Hashimoto, K. Matsuzaka, E. Moses, and A. Onuki, *Phys. Rev. Lett.* **74**, 126 (1995).

[51] E. K. Hobbie, S. Kim, and C. C. Han, *Phys. Rev. E* **54**, R5909 (1996).

[52] M. A. van Dijk, M. B. Eleveld, and A. van Veelen, *Macromolecules* **25**, 2274 (1992); Z. J. Chen, M. T. Shaw, and R. A. Weiss, *Macromolecules* **28**, 2274 (1995).

[53] K. B. Migler, *Phys. Rev. Lett.* **86**, 1023 (2000); H. S. Jeon and E. K. Hobbie, *Phys. Rev. E* **63**, 061403 (2001).

[54] S. Tomotika, *Proc. Roy. Soc.* (London) *A* **150**, 322 (1932).

[55] A. Frischknecht, *Phys. Rev. E* **58**, 3495 (1998).

[56] A. H. Krall, J. V. Sengers, and K. Hamano, *Phys. Rev. E* **48**, 357 (1993).

[57] G. I. Taylor, *Proc. Roy. Soc.* (London) *A* **146**, 501 (1934).

[58] J. M. Rallison, *Ann. Rev. Fluid Mech.* **16**, 46 (1984).

[59] E. J. Hinch and A. Acrivos, *J. Fluid Mech.* **98**, 305 (1980); A. Acrivos, in *Physicochemical Hydrodynamics, Interfacial Phenomena*, ed. M. G. Velarde (Plenum, New York, 1988), p. 1.

[60] A. Onuki and S. Takesue, *Phys. Lett. A* **114**, 133 (1986).

[61] K. Y. Min, J. Stavans, R. Piazza, and W. I. Goldburg, *Phys. Rev. Lett.* **63**, 1070 (1989).

[62] N. Eswar, *Phys. Rev. Lett.* **68**, 186 (1992).

[63] K. Y. Min and W. I. Goldburg, *Phys. Rev. Lett.* **70**, 469 (1993); *ibid.* **71**, 569 (1993); *Physica A* **204**, 246 (1994).

[64] O. M. Atabaev, Sh. O. Tursunov, P. A. Tadzhibaev *et al.*, *Dokl. Akad. Nauk* (SSSR) **315**, 889 (1990); P. K. Khabibullaev, M. Sh. Butabaev, Yu V. Pakharukov, and A. A. Saidov, *Dokl. Akad. Nauk* (SSSR) **320**, 1372 (1991); [*Sov. Phys. Dokl.*] **36**, 712 (1991).

[65] P. G. Saffman and J. S. Turner, *J. Fluid Mech.* **1**, 16 (1956).

[66] V. G. Levich, *Physicochemical Hydrodynamics* (Prentice Hall, Englewood Cliffs, NJ, 1962).

[67] P. M. Alder, *J. Colloid Interface Sci.* **83**, 106 (1981).

[68] H. Wang, A. Z. Zinchenko, and R. H. Davis, *J. Fluid Mech.* **265**, 161 (1994).

[69] D. J. Swift and S. K. Friedlander, *J. Colloid Sci.* **19**, 621 (1964).

[70] M. Doi and D. Chen, *J. Chem. Phys.* **90**, 5271 (1989).

[71] A. H. West, J. R. Melrose, and R. C. Ball, *Phys. Rev. E* **49**, 4237 (1994).

[72] C. M. Roland and G. G. A. Böhm, *J. Polym. Sci. Polym. Phys. Edn.* **22**, 79 (1984).

[73] S. T. Milner and H. Xi, *J. Rheol.* **40**, 663 (1996).

[74] T. Baumberger, F. Perro, and D. Beysens, *Phys. Rev. A* **46**, 7636 (1992).

[75] G. K. Batchelor, *J. Fluid Mech.* **95**, 369 (1979).

[76] V. N. Kurdyumov and A. D. Polyamin, *Fluid Dyn.* (USSR) **24**, 611 (1990).

[77] D. W. Oxtoby, *J. Chem. Phys.* **62**, 1463 (1975).

[78] A. Onuki and K. Kawasaki, *Phys. Lett. A* **75**, 485 (1980).

[79] K. Hamano, S. Yamashita, and J. V. Sengers, *Phys. Rev. Lett.* **68**, 3578 (1992).

[80] C. E. Rosenkilde, *J. Math. Phys.* **8**, 84 (1967).

[81] G. K. Batchelor, *J. Fluid Mech.* **410**, 545 (1970).

[82] M. Doi, in *Physics of Complex and Supermolecular Fluids*, eds. S. A. Safran and N. A. Clark (Wiley, New York, 1987), p. 611.

[83] A. Onuki, Phys. Rev. A **35**, 5149 (1987).

[84] M. Doi and T. Ohta, J. Chem. Phys. **95**, 1242 (1991).

[85] K. Hamano, T. Ishi, M. Ozawa, J. V. Sengers, and A. H. Krall, *Phys. Rev. E* **51**, 1254 (1995).

[86] Y. Takahashi, N. Kawashima, I. Noda, and M. Doi, *J. Rheol.* **38**, 699 (1994); Y. Takahashi, H. Suzuki, Y. Nakagawa, and I. Noda, *Macromolecules* **27**, 6476 (1994).

[87] I. Vinckier, P. Moldenauers, and J. Mewis, *J. Rheol.* **40**, 613 (1996).

[88] J. Läuger, C. Laubner, and W. Gronski, *Phys. Rev. Lett.* **75**, 3576 (1995).

[89] A. Onuki, *Europhys. Lett.* **28**, 175 (1994).

[90] G. M. Jordhano, J. A. Mason, and L. H. Sperling, *Polym. Eng. Sci.* **26**, 517 (1986); I. S. Miles and A. Zurek, *Polym. Eng. Sci.* **28**, 796 (1988); H. S. Jeon and E. K. Hobbie, *Phys. Rev. E* **64**, 049901 (2001).

[91] G. H. Fredrickson, *Macromolecules* **20**, 3017 (1987).

[92] H. J. Dai, N. P. Balsara, B. A. Garetz, and M. C. Newstein, *Phys. Rev. Lett.* **77**, 3677 (1996).

[93] J. H. Rosendale and F. S. Bates, *Macromolecules* **23**, 2329 (1990).

[94] K. Kawasaki and A. Onuki, *Phys. Rev. A* **42**, 3664 (1990).

[95] M. Rubinstein and S. P. Obukov, *Macromolecules* **26**, 1740 (1993).

[96] A. Liu, S. Ramaswamy, T. G. Mason, H. Gang, and D. A. Weitz, *Phys. Rev. Lett.* **76**, 3017 (1996).

[97] H. H. Winter, D. B. Scott, W. Gronski, S. Okamoto, and T. Hashimoto, *Macromolecules* **26**, 7236 (1993); K. I. Winley, S. S. Patel, L. G. Larson, and H. Watanabe, *ibid.* **26**, 2542 (1993); *ibid.* **26**, 4373 (1993); G. A. McConnell, M. Y. Lin, and A. P. Gast, *ibid.* **28**, 6754 (1995); V. K. Gupta, R. Krishnamoorti, J. A. Kornfield, and S. D. Smith, *ibid.* **29**, 1359 (1995); Y. Zhang and U. Wiesner, *J. Chem. Phys.* **103**, 4784 (1995).

[98] M. Doi, J. L. Harden, and T. Ohta, *Macromolecules* **26**, 4935 (1993).

[99] A. V. Zvelindovsky, G. J. A. Sevink, B. A. C. van Vlimmeren, N. M. Maurits, and J. G. E. M. Fraaije, *Phys. Rev. E* **57**, R4879 (1998).

[100] D. J. Pine, N. Eswar, J. V. Maher, and W. I. Goldburg, *Phys. Rev. A* **29**, 308 (1984).

[101] C. K. Chan, W. I. Goldburg, and J. V. Maher, *Phys. Rev. A* **35**, 1756 (1987).

[102] P. Tong, W. I. Goldburg, J. Stavans, and A. Onuki, *Phys. Rev. Lett.* **62**, 2472 (1989).

[103] R. Ruiz and D. R. Nelson, *Phys. Rev. A* **23**, 3224 (1981).

[104] A. Onuki, *Phys. Lett. A* **101**, 286 (1984).

[105] A. M. Lacasta, J. M. Sancho, and S. Saguès, *Phys. Rev. Lett.* **75**, 1791 (1995); L. L. Berthier, J. L. Barrat, and J. Kurchan, *ibid.* **86**, 2014 (2001).

[106] A. N. Kolmogorov, *Dokl. Akad. Nauk.* (SSSR) **30**, 301 (1941); *ibid.* **31**, 538 (1941); *ibid.* **32**, 16 (1941).

[107] V. Fritch, P. L. Sullem, and M. A. Nelkin, *J. Fluid Mech.* **87**, 719 (1978).

[108] A. N. Kolmogorov, *Dokl. Akad. Nauk.* (SSSR) **66**, 825 (1949).

[109] K. Batchelor, *J. Fluid Mech.* **5**, 113 (1959).

[110] R. H. Kraichnan, *Phys. Fluids* **11**, 113 (1968); *J. Fluid Mech.* **64**, 737 (1968).

[111] Yu. R. Chashkin, A. V. Voronel, V. A. Smirnov, and V. G. Gobunova, *Sov. Phys. JETP* **25**, 79 (1979).

[112] A. Onuki, *Prog. Theor. Phys. Suppl.* **99**, 382 (1990).

[113] D. S. Cannell, *Phys. Rev. A* **12**, 225 (1975).

[114] G. Ver Strate and W. Philippoff, *J. Polym. Sci. Polym. Lett.* **12**, 267 (1974).

[115] B. A. Wolf and M. C. Sezen, *Macromolecules* **10**, 1010 (1977).

[116] C. Rangel-Nafaile, A. B. Metzner, K. F. Wissburn, *Macromolecules* **17**, 1187 (1984).

[117] H. Krämer and B. A. Wolf, *Macromol. Chem. Rapid Commun.* **6**, 21 (1985).

[118] Z. Laufer, H. L. Jalink, and A. J. Staverman, *J. Polym. Sci., Polym. Chem.* Edn. **11**, 3005 (1973).

[119] X. L. Wu, D. J. Pine, and P. K. Dixon, *Phys. Rev. Lett.* **68**, 2408 (1991).

[120] T. Hashimoto and K. Fujioka, *J. Phys. Soc. Jpn* **60**, 356 (1991); T. Hashimoto and T. Kume, *J. Phys. Soc. Jpn* **61**, 1839 (1992).

[121] E. Moses, T. Kume, and T. Hashimoto, *Phys. Rev. Lett.* **72**, 2037 (1994).

[122] H. Murase, T. Kume, T. Hashimoto, Y. Ohta, and T. Mizukami, *Macromolecules* **28**, 7724 (1995).

[123] T. Kume, T. Hattori, and T. Hashimoto, *Macromolecules* **30**, 427 (1997).

[124] J. van Egmond, D. E. Werner, and G. G. Fuller, *J. Chem. Phys.* **96**, 7742 (1992); J. van Egmond and G. Fuller, *Macromolecules* **26**, 7182 (1993).

[125] A. I Nakatani, J. F. Douglas, Y.-B. Ban, and C. C. Han, *J. Chem. Phys.* **100**, 3224 (1994).

[126] F. Boue and P. Lindner, *Europhys. Lett.* **25**, 421 (1994); I. Morfin, P. Linder, and F. Boue, *Macromolecules* **32**, 7208 (1999).

[127] K. Migler, C. Liu, and D. J. Pine, *Macromolecules* **29**, 1422 (1996).

[128] S. Saito, T. Hashimoto, I. Morfin, P. Linder, and F. Boue, *Macromolecules* **35**, 445 (2002).

[129] H. Yanase, P. Moldenauers, J. Mewis, V. Avetz, J. W. van Egmond, and G. G. Fuller, *Rheol. Acta* **30**, 89 (1991).

[130] J. van Egmond, *Macromolecules* **30**, 8057 (1997).

[131] J. W. van Egmond and G. G. Fuller, *Macromolecules* **26**,7182 (1993).

[132] J. J. Magda, C.-S. Lee, S. J. Muller, and R. G. Larson, *Macromolecules* **26**, 1696 (1993).

[133] E. Helfand and H. Fredrickson, *Phys. Rev. Lett.* **62**, 2468 (1989).

[134] A. Onuki, *Phys. Rev. Lett.* **62**, 2472 (1989).

[135] A. Onuki, *J. Phys. Soc. Jpn* **59**, 3423 (1990); **59**, 3427 (1990).

[136] M. Doi, in *Dynamics and Patterns in Complex Fluids*, eds. A. Onuki and K. Kawasaki (Springer, 1990), p. 100.

[137] M. Doi and A. Onuki, *J. Physique II* **2**, 1631 (1992).

[138] S. T. Milner, *Phys. Rev. E* **48**, 3874 (1993).

[139] H. Ji and E. Helfand, *Macromolecules* **28**, 3869 (1995).

[140] A. Onuki, R. Yamamoto, and T. Taniguchi, *J. Physique II* **7**, 295 (1997); *Progress in Colloid & Polymer Science* **106**, 150 (1997).

[141] T. Okuzono, *Modern Phys. Lett.* **11**, 379 (1997).

[142] R. B. Bird, R. C. Armstrong, and O. Hassager, *Dynamics of Polymeric Liquids*, 2nd edn, Vol. 2 (Wiley, New York, 1987).

[143] N. Clarke and T. C. B. McLeish, *Phys. Rev. E* **57**, R3731 (1998).

[144] V. Schmitt, C. M. Marques, and F. Lequeux, *Phys. Rev. E* **52**, 4009 (1995).

[145] J. D. Katsaros, M. F. Malone, and H. H. Winter, *Polym. Eng. Sci.* **29**, 1434 (1989); S. Mani, M. F. Malone, and H. H. Winter, *Macromolecules* **25**, 5671 (1992).

[146] R.-J. Wu, M. T. Shaw, and R. A. Weiss, *J. Rheol.* **36**, 1605 (1992).

[147] Z. J. Chen, R.-J. Wu, M. T. Shaw, R. A. Weiss, M. L. Fernandez, and J. S. Higgins, *Polym. Eng. Sci.* **35**, 92 (1995); M. L. Fernandez, J. S. Higgins, R. Horst, and B. A. Wolf, *Polymer* **36**, 149 (1995).

[148] H. Gerard, J. S. Higgins, and N. Clarke, *Macromolecules* **32**, 5411 (1999).

[149] R. H. Shafer, N. Laiken, B. H. Zimm, *Biophysical Chemistry* **2**, 180 (1974); K. A. Dill, and B. H. Zimm, *Nucleic Acids Research* **7**, 735 (1979).

[150] M. Tirrell and M. F. Malone, *J. Polym. Sci. Polym. Phys. Edn.* **15**, 1569 (1977).

[151] G. Sergé and A. Silberberg, *J. Fluid Mech.* **14**, 115 (1962).

[152] A. Karnis, H. Goldsmith, and S. Mason, *J. Colloid Interface Sci.* **22**, 41 (1966).

[153] P. G. de Gennes, *J. Physique I* **40**, 783 (1979).

[154] C. J. Hoh, P. Hookham, and L. G. Leal, *J. Fluid Mech.* **266**, 1 (1994).

[155] P. Nozières and D. Quemada, *Europhys. Lett.* **2**, 129 (1986).

[156] P. R. Nott and J. F. Brady, *J. Fluid Mech.* **275**, 157 (1994).

[157] V. Mhetar and L. A. Archer, *Macromolecules* **31**, 6639 (1998).

[158] G. Marrucci, *Trans. Soc. Rheol.* **16**, 321 (1972).

[159] R. Horst and B. A. Wolf, *Macromolecules* **26**, 5676 (1993).

[160] J. Casas-Vazquez, M. Criado-Sancho, and D. Jou, *Europhys. Lett.* **23**, 469 (1993); D. Jou, J. Casas-Vazquez, and M. Criado-Sancho, *Adv. Polym. Sci.* **120**, 207 (1995).

[161] A. Onuki, *J. Phys. Soc. Jpn* **66**, 1836 (1997).

[162] D. J. Evans and G. P. Morriss, *Statistical Mechanics of Nonequilibrium Liquids* (Academic, New York, 1990).

[163] N. A. Clark and B. J. Ackerson, *Phys. Rev. Lett.* **44**, 1005 (1980); B. J. Ackerson and N. A. Clark, *Phys. Rev. A* **30**, 906 (1984); L. B. Chen, C. F. Zukoski, B. J. Ackerson, H. J. M. Hanley, G. C. Straty, J. Barker, and C. J. Glinka, *Phys. Rev. Lett.* **69**, 688 (1992).

[164] W. D. Dozier and P. M. Chaikin, *J. Physique* **43**, 843 (1982).

[165] M. J. Stevens and M. O. Robbins, *Phys. Rev. E* **48**, 3778 (1993); H. Komatsugawa and S. Nosé, *Phys. Rev. E* **51**, 5944 (1995).

[166] S. Ramaswarmy and S. R. Renn, *Phys. Rev. Lett.* **56**, 945 (1986); R. Lahiri and S. Ramaswarmy, *ibid.* **73**, 1043 (1994); B. Baguchi and D. Thirumalai, *Phys. Rev. A* **37**, 2530 (1988); D. Ronis, *Phys. Rev. A* **29**, 1453 (1984); J. F. Scharwl and S. Hess, *Phys. Rev. A* **33**, 4277 (1986).

[167] I. Bondár and J. K. G. Dhont, *Phys. Rev. Lett.* **77**, 5304 (1996).

[168] C. Cametti, P. Codastefano, G. D'Arrigo, P. Tartaglia, J. Rouch, and S. H. Chen, *Phys. Rev. A* **42**, 3421 (1990).

[169] P. G. de Gennes, *Mol. Cryst. Liq. Cryst.* **34**, 91 (1976).

[170] R. F. Bruinsma and C. R. Safinia, *Phys. Rev. A* **43**, 5377 (1991); R. F. Bruinsma and Y. Rabin, *Phys. Rev. A* **45**, 994 (1992).

[171] P. D. Olmsted and P. M. Goldbart, *Phys. Rev. A* **46**, 4966 (1992).

[172] J. P. Decruppe, E. Capplelaere, and R. Cressely, *J. Physique II* **7**, 257 (1997).

[173] N. A. Spenley, M. E. Cates, and T. C. B. McLeish, *Phys. Rev. Lett.* **71**, 939 (1993); M. E. Cates, *J. Phys.: Condens. Matter* **8**, 9167 (1996).

[174] O. Diat, D. Roux, and F. Nallet, *J. Physique II* **3**, 1427 (1993); P. Sierro and D. Roux, *Phys. Rev. Lett.* **78**, 1496 (1997).

[175] H. Rehage and H. Hoffmann, *Mol. Phys.* **74**, 933 (1988); J. F. Berret, R. C. Roux, G. Porte, and P. Linder, *Europhys. Lett.* **25**, 521 (1994); G. Porte, J. F. Berret, and J. L. Harden, *J. Physique II* **7**, 459 (1997); J. F. Berret, G. Porte, and J. P. Decruppe, *Phys. Rev. E* **55**, 1668 (1997).

[176] J. Yamamoto and H. Tanaka, *Phys. Rev. Lett.* **74**, 932 (1996).

[177] H. Rehage, I. Wunderlich, and H. Hoffmann, *Prog. Colloid Polymer Sci.* **72**, 51 (1986).

[178] C. Liu and D. J. Pine, *Phys. Rev. Lett.* **77**, 2121 (1996); Y. T. Hu, P. Boltenhagen, and D. J. Pine, *J. Rheol.* **42**, 1185 (1998).

[179] T. C. Halsey and W. Toor, *Phys. Rev. Lett.* **65**, 2820 (1990).

[180] J. E. Martin and J. Odinek, *Phys. Rev. Lett.* **75**, 2827 (1995).

[181] R. B. Williamson and W. F. Busse, *J. Appl. Phys.* **38**, 4187 (1967); A. J. McHugh and E. H. Forrest, *J. Macromol. Sci. Phys.* **11**, 219 (1975).

[182] M. Okamoto, H. Kubo, and T. Kotaka, *Polymer* **39**, 3135 (1998); H. Kubo, M. Okamoto, and T. Kotaka, *ibid.* **39**, 4827 (1998).

[183] M. E. Cates and M. S. Turner, *Europhys. Lett.* **11**, 681 (1990).

[184] R. Bruinsma, W. M. Gelbert, and Ben-Shaul, *J. Chem. Phys.* **96**, 7710 (1992).

[185] F. Tanaka and S. F. Edwards, *Macromolecules* **25**, 1516 (1992); *J. Non-Newtonian Fluid Mech.* **43**, 247, 273, 289 (1992).

[186] P. G. Khalatur, A. R. Khokhlov, and D. A. Mologin, *J. Chem. Phys.* **109**, 9602 (1998); *ibid.* **109**, 9614 (1998).

[187] A. Emanuele and M. B. Palma-Vittorelli, *Phys. Rev. Lett.* **69**, 81 (1992).

[188] F. Brochard and P. G. de Gennes, *Langmuir* **8**, 3033 (1992).

[189] K. B. Migler, H. Hervet, and L. Léger, *Phys. Rev. Lett.* **70**, 219 (1992).

[190] J. N. Israelachvili and P. M. McGuiggan, *Science* **241**, 795 (1988).

[191] H.-W. Hu, G. A. Carson, and S. Granick, *Phys. Rev. Lett.* **66**, 2758 (1991).

[192] S. Hess, C. Aust, L. Bennet, M. Kröger, C. P. Borgmeyer, and T. Weider, *Physica A*, **240** 126 (1997).

[193] J. Jäckle, *Rep. Prog. Phys.* **49**, 171 (1986).

[194] M. D. Ediger, C. A. Angell, and S. R. Nagel, *J. Phys. Chem.* **100**, 13 200 (1996).

[195] S. Matsuoka, *Relaxation Phenomena in Polymers* (Oxford University Press, New York, 1992).

[196] G. R. Stroble, *The Physics of Polymers* (Springer, Heidelberg, 1996).

[197] U. Bengtzelius, W. Götze, and A. Sjölander, *J. Phys. C* **17**, 5915 (1984).

[198] E. Leutheusser, *Phys. Rev. A* **29**, 2765 (1984).

[199] G. Adam and J. H. Gibbs, *J. Chem. Phys.* **43**, 139 (1965).

[200] M. H. Cohen and G. S. Grest, *Phys. Rev.* B **20**, 1077 (1979).

[201] F. H. Stillinger, *J. Chem. Phys.* **89**, 6461 (1988).

[202] K. Maeda and S. Takeuchi, *Phil. Mag. A* **44**, 643 (1981).

[203] T. Muranaka and Y. Hiwatari, *Phys. Rev. E* **51**, R2735 (1995); *Suppl. Prog. Theor. Phys.* **126**, 403 (1997).

[204] M. M. Hurley and P. Harrowell, *Phys. Rev. E* **52**, 1694 (1995); D. N. Perera and P. Harrowell, *Phys. Rev. E* **54**, 1652 (1996).

[205] R. Yamamoto and A. Onuki, *J. Phys. Soc. Jpn* **66** 2545 (1997); *Phys. Rev. E* **58**, 3515 (1988).

[206] R. Yamamoto and A. Onuki, *J. Phys. C* **29**, 6323 (2000).

[207] W. Kob, C. Donati, S. J. Plimton, P. H. Poole, and S. C. Glotzer, *Phys. Rev. Lett.* **79**, 2827 (1997); C. Donati, J. F. Douglas, W. Kob, S. J. Plimton, P. H. Poole, and S. C. Glotzer, *Phys. Rev. Lett.* **80**, 2338 (1988).

[208] R. Yamamoto and A. Onuki, *Europhys. Lett.* **40**, 61 (1997).

[209] J. H. Simmons, R. K. Mohr, and C. J. Montrose, *J. Appl. Phys.* **53**, 4075 (1982); J. H. Simmons, R. Ochoa, K. D. Simmons, and J. J. Mills, *J. Non-Cryst. Solids* **105**, 313 (1988).

[210] H. S. Chen and M. Goldstein, *J. Appl. Phys.* **43**, 1642 (1971); F. Spaepen, *Acta Metall.* **25**, 407 (1977); A. S. Argon, *Acta Metall.* **27**, 47 (1979); A. S. Argon, in *Material Sciences and Technology*, edited by R. W. Cohen, P. Haasen and E. J. Kramer (VCH, Weinheim, 1993), Vol. 6.

[211] I. Chang, F. Fujara, B. Geil, G. Heuberger, T. Mangel, and H. Silescu, *J. Non-Cryst. Solids* **172–174**, 248 (1994).

[212] M. T. Cicerone and M. D. Ediger, *J. Chem. Phys.* **104**, 7210 (1996).

[213] D. Thirumalai and R. D. Mountain, *Phys. Rev. E* **47**, 479 (1993).

[214] D. Perera and P. Harrowell, *Phys. Rev. Lett.* **81**, 120 (1988).

[215] R. Yamamoto and A. Onuki, *Phys. Rev. Lett.* **81**, 4915 (1988).

[216] B. Bernu, Y. Hiwatari, and J. P. Hansen, *J. Phys. C* **18**, L371 (1985); B. Bernu, J. P. Hansen, Y. Hiwatari, and G. Pastore, *Phys. Rev. A* **36**, 4891 (1987).

[217] R. Zorn, *Phys. Rev. B* **55**, 6249 (1987).

[218] T. Kanaya, I. Tsukushi, and K. Kaji, *Suppl. Prog. Theor. Phys.* **126**, 133 (1997).

[219] D. Leighton and A. Acrivos, *J. Fluid Mech.* **109**, 177 (1987); V. Breedveld, D. van den Ende, A. Tripathi, and A. Acrivos, *J. Fluid Mech.* **375**, 297 (1998).

[220] B. Miller, C. O'Hern, and R. P. Behringer, *Phys. Rev. Lett.* **77**, 3110 (1996).

[221] T. Okuzono and K. Kawasaki, *Phys. Rev. E*, **51**, 1246 (1995).

[222] D. J. Durian, *Phys. Rev. E* **55**, 1739 (1997); S. A. Langer and A. J. Liu, *J. Phys. Chem. B* **101**, 8667 (1997).

[223] P. Hébrand, F. Lequeux, J. P. Munch, and D. J. Pine, *Phys. Rev. Lett.* **78**, 4657 (1997).

[224] R. S. Farr, J. R. Melrose, and R. C. Ball, *Phys. Rev. E* **55**, 7203 (1997).

[225] P. Sollich, *Phys. Rev. E* **58**, 738 (1998).

[226] A. J. Liu and S. R. Nagel (eds.), *Jamming and Rheology* (Taylor & Francis, London and New York, 2001).

[227] P. H. Mott, A. S. Aragon, and U. W. Suter, *Phil. Mag. A* **67** (1993) 931; A. S. Aragon, V. V. Bulatov, P. H. Mott, and U. W. Suter, *J. Rheol.* **39** (1995) 377.

[228] R. Muller and J. J. Pesce, *Polymer* **35**, 734 (1994); M. Kröger, C. Luap, and R. Muller, *Macromolecules* **30**, 526 (1997).

[229] T. Inoue, D. S. Ryu, and K. Osaki, *Macromolecules* **31**, 6977 (1998); M. Matsuyama, H. Watanabe, T. Inoue, K. Osaki, and M.-L. Yao, *ibid.* **31**, 7973 (1998).

[230] R. Yamamoto and A. Onuki, *J. Chem. Phys.* **117**, 2359 (2002).

[231] K. Binder, J. Baschnagel, C. Bennemann, and W. Paul, *J. Phys. C* **11**, A47–A55 (1999).

[232] A. Onuki and K. Kawasaki, *Phys. Lett. A* **72**, 233 (1979).

[233] B. J. Ackerson and N. A. Clark, *J. Physique I* **42**, 929 (1981).

[234] A. Al Kahwaji, O. Greffier, A. Leon, J. Rouch, and H. Kellay, *Phys. Rev. E* **63**, 041502 (2001).

Index